THE BIOLOGICAL SCIENCES

THE BIOLOGICAL SCIENCES

Alvin Silverstein
The City University of New York

Rinehart Press / Holt · Rinehart · Winston

SAN FRANCISCO

For Professor Mordecai Gabriel, who inspired me to the study of life,
and for my wife, Virginia, who makes life worth living.

Library of Congress Cataloging in Publication Data

Silverstein, Alvin.
The biological sciences.

Includes bibliographies.
1. Biology. I. Title. [DNLM: 1. Biology.
QH308.7 S587b 1974]
QH308.2.S54 574 73–18474
ISBN 0-03-003341-1

5643 Paradise Drive
Corte Madera, Calif. 94925

A division of Holt, Rinehart and Winston, Inc.

PRINTED IN THE UNITED STATES OF AMERICA

4 5 6 7 032 9 8 7 6 5 4 3 2

ILLUSTRATION CREDITS

Chapter 1 1-1 (a), (d), (e), (f), (g) E. Baitel, Omikron; (b) A. Lopel, Omikron; (c) NASA. 1-3 Omikron. 1-5 USDA. 1-6 (a) Louvre, Paris; (b) German Information Center; (c) Museum S. Marco, Florence.

Chapter 2 2-1 (a) Brenda Gaber, Omikron; (b) U.S. Atomic Energy Commission; (c) Pfizer Inc.; (d) Gabriele Wunderlich; (e) A. Lopez, Omikron. 2-3 (a), (f) U.S. Department of the Interior, National Park Service; (b), (c), (d), (e), (g) E. Baitel, Omikron.

Chapter 3 3-2 U.S. Navy 3-5 National Audubon Society. 3-7, 3-8 From *Cell Structure and Function*, 2/E, by Ariel G. Loewy and Philip Siekevitz. Copyright © 1969 by Holt, Rinehart and Winston. Reprinted by permission of Holt, Rinehart and Winston. 3-9 John Kendrew, Medical Research Council. 3-10 Bruce Merrifield, Rockefeller University. 3-12 Abbott Laboratories.

Chapter 4 4-1 (a) Pfizer Inc.; (b), (c), (d) Carl Zeiss; (e) Fisher Scientific Co.; (f) F. W. Hoffman, USAEC. 4-5 Metler Instrument Corp. 4-6 Fisher Scientific Co. 4-9 Abbott Laboratories. 4-11 Lawrence Radiation Laboratory, Univ. of California, Berkeley. 4-12 USAEC. 4-13 The Granger Collection. 4-15 (a), (b) Philip L. Feinberg, Omikron. 4-16 Pfizer Inc. 4-17 Hitachi Ltd. 4-19 (a), (b) L. M. Beidler. 4-22 Bell Telephone Laboratories.

Chapter 5 5-1, 5-2 Eric V. Gravé. 5-3 (a) Loewy/Siekevitz; (b) G. E. Palade, Rockefeller University. 5-4 (a) Loewy/Siekevitz; (b) M. C. Ledbetter, Brookhaven National Laboratory. 5-5 Dr. J. F. Hoffman, Yale University. 5-7 D. Soifer, Omikron. 5-13 From *Cells and Organelles* by Alex Novikoff and Eric Holtzman. Copyright © 1970 by Holt, Rinehart and Winston. Reprinted by permission of Holt, Rinehart and Winston. 5-16 (a) Cynthia Jones and Dr. George B. Chapman, Georgetown University; (b) Sandra F. Lane and Dr. George B. Chapman, Georgetown University. 5-17 Department of Zoology, University of Michigan. 5-18 Willowbrook State School. 5-19, 5-20 Eric V. Gravé. 5-22 University of Connecticut. 5-23 Arthur Ambler, National Audubon Society. 5-24 Dr. Robert L. Barchi and Dr. George B. Chapman, Georgetown University. 5-26 (a) Dr. George B. Chapman, Georgetown University; (b) Edward L. Anderson and Dr. George B. Chapman, Georgetown University. 5-27 (a), (b) Loewy/Siekevitz. 5-29 Sister Anne Susalla and Dr. George B. Chapman, Georgetown University. 5-27 (a) Philip Feinberg, Omikron; (c) Eric V. Gravé. 5-28 Carolina Biological Supply Co. 5-29 Sister Anne Susalla and Dr. George B. Chapman, Georgetown University. 5-30 Eric V. Gravé.

Chapter 6 6-1 (a), (d), (e) Philip L. Feinberg, Omikron; (b), (c) Hitachi, Ltd; (f) Eric V. Gravé. 6-3 (a) Eric V. Gravé; (b), (c) Dr. George B. Chapman, Georgetown University. 6-4, 6-5, 6-7 (a) Eric V. Gravé; (b) Dr. L. Shettles, The American Museum of Natural History. 6-8 Eric V. Gravé. 6-9 Richard F. Trump. 6-10 Philip L. Feinberg, Omikron. 6-11 Jack Dermid. 6-12 Philip L. Feinberg, Omikron.

(Continued on p. 715)

Preface

For two decades following World War II, science and technology occupied a position of the highest regard in our society. The technological achievements of World War II had demonstrated the potential of science for solving many of the world's problems. Radar, penicillin, and spectacular developments in electronics quickly found applications in the civilian arena after the war was over. The launching of the Soviet *Sputnik* in 1957 resulted in a frenzied effort by the United States to "catch up" with the USSR not only in space, but in other scientific endeavors. New science courses were instituted at all levels of education, and students flocked by the millions into the fields of science and technology.

In the late sixties, the fortunes of science and technology turned dramatically downward. The public became increasingly aware of such problems as pollution, overcrowding, and shortages of critical resources. Predictions of doom for the earth and its peoples became commonplace. Science and technology in general, and scientists in particular, were singled out as the culprits. During this critical time, basic research was shunted aside and new orders of priority were established. Only those areas of science and technology that could solve our immediate problems were considered worthwhile. Indeed, this basic attitude prevails today.

And yet, it is basic research that ultimately holds the hope of solving man's most urgent problems.

The public and government officials do not appreciate the value of basic research and the huge impact it can have upon our world. The main cause of this lack of understanding is ignorance. The people who must make crucial decisions affecting our world are not sufficiently aware of the basic principles of science, and hence their decisions often are not intelligent ones.

The purpose of this book is to help to close that gap of ignorance to some small degree. One of the most critical areas in our society today is the problems in the realm of the biological sciences, such as pollution, ecological imbalances, limitations of living resources, and overpopulation. An understanding of the principles of life helps to put such problems in proper perspective. This book presents such basic principles and indicates their applications to the problems of the present day.

The first chapter presents an introduction to the nature of science, its potentials and limitations.

Chapters 2 and 3 provide a fundamental background of principles of the physical world and the chemical basis of life, needed to understand and appreciate the exciting advances in the most actively researched fields of life science today.

Further background is provided by Chapter 4, a discussion of the tools and techniques biologists use in their work.

In Chapters 5 and 6, the cell is described as the basic unit of life, and insights are given into the organization of higher organisms.

The causes of the present crisis on "Spaceship Earth"—the forces that threaten the delicate balance of living nature—cannot be properly understood without some appreciation of the diversity of life forms and the characteristics, adaptations, and special needs of different types of organisms. This is the subject of Chapters 7 through 9.

Chapters 10 through 12 emphasize the human systems. A knowledge of how our bodies work is essential for an understanding of the effects that various factors of the modern environment, from air pollution to food additives, can have on us.

Chapter 13, discussing recent findings in the fields of behavior and psychology, provides keys to understanding the "whys" of human and animal actions.

Modern research into the molecular basis of life is providing man with the potential for changing his own heredity and perhaps shaping the future development of his species. Chapters 14 and 15 provide a firm grounding in basic genetic principles, both on the macro-level and on the molecular level; the story of today's most active field of biological research, genetics, is continued in Chapter 22.

Reproduction and the principles and patterns of development of living organisms, from conception to death, are the subjects of Chapters 16 and 17.

Chapters 18 and 19 discuss the major theories concerning the origin and development of the world we live on and the living organisms that share our planet with us.

We live in the "atomic age." The discovery of atomic energy has provided man with the potentials for a life of unprecedented ease and plenty, or the annihilation of all life on earth. With the current energy crisis, it is particularly important that citizens and their governments make intelligent decisions on the development of nuclear power installations. Such intelligent decisions can be made only with a knowledge of the nature of atomic energy and radioactivity and their effects on living organisms, as provided by Chapter 20.

Utilizing the basic principles conveyed in earlier chapters, Chapters 21 and 22 provide a sampling of the findings and speculations in some of the most exciting frontier areas of modern biology.

One of the areas of greatest personal concern to humans is health and disease. Chapter 23 presents important background on the nature of disease and modern avenues to combatting it.

Is man the master of his planet, or a custodian entrusted with ensuring the welfare of all the species of earth's living organisms? Either way, his activities are daily affecting the living world—both intentionally and unintentionally. Only with an understanding of the principles of ecology—the interrelationships of living organisms with one another and with their physical environment—can we learn what we are doing to our world and to ourselves—and what we *should* be doing. Chapters 24 through 26 discuss this most crucial area of modern concern and the special responsibilities of the scientifically informed citizen.

Many things go into the writing of a textbook—a lifetime of study and teaching, the interest and encouragement of friends, relatives, and colleagues, and the competent aid of skilled professionals in a variety of areas. I would like to take this opportunity to thank J. Clovesko, Wilbur A. Gilbert, Laszlo Hanzely, Richard F. Heller, Richard C. Saunders, and R. Talbert for their thoughtful comments on the manuscript. Warm thanks are also due to my editors, Richard P. Thiel and Gregory Hubit, for their unflagging enthusiasm and help; to the copy editor, William S. Green, for his painstaking work; to the artist, Richard Kassouf, to photo researchers, Peggy Simsarian and Gabriele Wunderlich, and page designer and layout editor, Sally Thompson, for their heroic efforts to make this book a special achievement; and to Richard Heller for compiling the glossary. And, finally, a loving tribute to my children, Robert, Glenn, Carrie, Sharon, Laura, and Kevin, for their patience, and to my wife, Virginia, without whose encouragement and aid this book would never have been written.

A.S.

Table of Contents

CHAPTER 1 **The Nature of Science** **1**

Science in Our Lives 1
The Accelerating Rate of Scientific Progress 2
The Nature of Science 4
Scientific Method 5
The Limitations of Science 8
The Divisions of Science and Their Interdependence 10

PART I **THE WORLD ABOUT US**

CHAPTER 2 **Matter, Energy, and Space** **15**

The Nature of Matter and Energy 15
Mixtures 19
The Kinetic Theory of Matter 21
The Structure of the Atom 24
Chemical Compounds 27
Chemical Reactions 32

CHAPTER 3 **The Chemistry of Life** **34**

Introduction 34
Inorganic Constituents 35
Organic Constituents 36
Miscellaneous Constituents 46
Metabolism 47
Enzymes 48

CHAPTER 4 **Tools and Techniques of Scientific Research** 53
Introduction 53
Basic Tools 54
Chemical Analysis 57
Chromatography and Electrophoresis 58
Centrifugation 61
Radiation Sources and Radioisotopes 62
Microscopy 65
X-Ray Crystallography 70
Computers 71

PART II **THE LIVING WORLD**

CHAPTER 5 **The Cell** 74
Introduction 74
The Cell Theory 75
An Overview of the Cell 76
Membranes of the Cell and the Transport of Matter 79
Plasmolysis and Turgor 85
The Cytoplasm and the Colloidal Properties of Cells 87
The Nucleus and Cell Division 88
The Ribosomes and Protein Synthesis 97
Manufacture and Storage: The Plastids and the Golgi Complex 98
Energy for the Cell's Activities: Mitochondria and Lysosomes 99
Organelles for Locomotion: Cilia and Flagella 101
Other Cellular Organelles 101
Extracellular Substances 104

CHAPTER 6 **From Cell to Organism** 107
The Organization of Cells 107
Animal Tissues 108
Plant Tissues 114
Organs and Organ Systems 117
The Organism 118
Communities 118

CHAPTER 7 **The Diversity of Life** 122
Introduction 122
Systems of Classification 123
Fundamental Characteristics of Life 130

Criteria for the Classification of Living Things 137

CHAPTER 8 **Panorama of the Protist and Plant Kingdoms** 141
The Borderline of Life 141
Bacteria: Phylum Schizomycophyta 145
Protozoa: Phylum Protozoa 146
Algae: Phyla Cyanophyta, Chlorophyta, Euglenophyta, Chrysophyta, Pyrrophyta, Phaeophyta, and Rhodophyta 151
True Fungi: Phylum Eumycophyta 153
Slime Molds: Phylum Myxomycophyta 154
The Plant Kingdom 155
Mosses and Liverworts: Phylum Bryophyta 156
Vascular Plants: Phylum Tracheophyta 157

CHAPTER 9 **Panorama of the Animal Kingdom** 173
The Animal Kingdom 173
Sponges: Phylum Porifera 174
Coelenterates: Phylum Cnidaria 176
Flatworms: Phylum Platyhelminthes 179
Roundworms: Phylum Nematoda 182
Segmented Worms: Phylum Annelida 184
Mollusks: Phylum Mollusca 186
Echinoderms: Phylum Echinodermata 191
Arthropods: Phylum Arthropoda 193
Chordates: Phylum Chordata 200

PART III **BASIC PROCESSES OF LIFE**

CHAPTER 10 **Matter and Energy for Life: Input and Output** 218
Foods for Life 218
Modes of Nutrition in Living Organisms 219
Photosynthesis 221
Nutrition 225
Digestion 238
The Human Digestive System 244
Excretion 257

CHAPTER 11 **Matter and Energy for Life: Circulation and Utilization** 267
Introduction 267
Circulatory Systems 268
The Human Circulatory System 271

The Blood 275
Accessory Organs of the Circulatory System 280
The Lymphatic System 281
Defenses Against Disease 281
Blood Types and Blood Transfusions 284
Respiration 287
The Human Respiratory System 290
Cellular Respiration: The Energy Cycles 294

CHAPTER 12 **Interaction with the Environment** 297
Introduction 297
Sense Organs 298
The Nervous System 308
The Autonomic Nervous System 320
The Endocrine System 322
Movement and Locomotion 330
Responsiveness in Plants 338

CHAPTER 13 **Behavior** 347
Introduction 347
Unlearned Behavior; Instincts 350
Learning 355
Biological Clocks and Rhythms 367
Territorial Behavior: Migration and Homing 373
Imprinting and Affection 377
Courtship and Family Life 379
Social Behavior 382
Communication and Language 384

PART IV FROM EGG TO OLD AGE

CHAPTER 14 **Heredity** 390
Introduction 390
Gregor Mendel, the Father of Modern Genetics 391
Chromosomes and Genes 395
Meiosis 396
Some Basic Concepts of Genetics 403
Some Basic Techniques of Genetics on the Macro-Scale 409
Sex Determination and Sex Linkage 414

CHAPTER 15 **The Molecular Basis of Genetics** 420
Redefining the Gene 420
Mutations 424

Experiments in Microbial Genetics 429
DNA—The Master Chemical 432

CHAPTER 16 **Reproduction and Embryology** 439
Introduction 439
Asexual Reproduction 441
Sexual Reproduction 443
The Human Reproductive System 447
Copulation and Fertilization 451
Early Embryonic Development 453
Pregnancy and the Placenta 456
Birth and Lactation 462
Reproduction in Higher Plants 463

CHAPTER 17 **Growth and Senescence** 472
Introduction 472
Growth 472
Metamorphosis 477
Repair and Regeneration 481
Senescence 484
Death 487

PART V THE UNIVERSE

CHAPTER 18 **The Birth of the Universe and the Orgin of Life** 492
The Origin of the Universe 492
The Origin of the Solar System 494
The Origin of Life on Earth 495

CHAPTER 19 **The Evolution of Life on Earth** 500
Introduction 500
Darwin: The Theory of Natural Selection 501
The Evidence for Evolution 506
Modern Concepts of Evolution 518
The Story of Life on Earth 527
The Precambrian Eras (Ca. 4.5 Billion to 600 Million Years Ago) 528
The Paleozoic Era (Ca. 600 to 230 Million Years Ago) 529
The Mesozoic Era (Ca. 230 to 63 Million Years Ago) 534
The Cenozoic Era (Ca. 63 Million Years Ago to the Present) 537
The Evolution of Man 539

PART VI OUR MODERN AGE

CHAPTER 20 **Atomic Energy and Radioactivity** 548
Introduction 548
Nuclear Fission 549
Nuclear Fusion 550
Radioactive Fallout 552
Effects of Radiations on Living Organisms 553
Peaceful Uses of Atomic Energy 559
Nuclear Power: Energy for the Future 561

CHAPTER 21 **The Interdisciplinary Revolution** 565
Introduction 565
Bionics 566
New Tools for Biology 573
The Computer Revolution 579

CHAPTER 22 **Frontiers of Biology** 589
Introduction 589
A Germfree World 589
Genetic Engineering and the Creation of Life 593
Oceanography 597
The Exploration of Space 601
Exobiology: The Search for Life on Other Worlds 603

CHAPTER 23 **Frontiers of Medicine** 608
Microbes and Disease 608
Bacterial Diseases 609
Virus Diseases 612
Fungus Diseases 615
Other Parasitic Diseases 616
Treatment of Infectious Diseases 618
Cancer 622
Heart Disease 626
Organ Transplants—A Promising Dead End? 628
When Does a Man Cease to Be a Man? 630
Medical Electronics 631

PART VII MAN IN HIS WORLD

CHAPTER 24 Ecology: The Interrelatedness of Life on Earth **637**

Introduction 637
The Physical Environment 637
Biomes 643
Cyclic Utilization of Matter and Energy 648
The Biotic Community 655

CHAPTER 25 Man Changes His World **669**

Introduction 669
Demography: The Study of Populations 670
Food for a Hungry World 673
Maintaining the Balance of Nature: Conservation 674
Man Tampers with His World: Pollution 678
Man Tampers with Himself: Food Additives 685
Drugs in the Modern World 687

CHAPTER 26 Science and Society **694**

Science and the Citizen 694
Scientific Attitudes and Everyday Life 695
The Government and Research 696
Priorities of Research 696
The Future 698

Glossary **701**

Suggestions for Further Reading **712**

Illustration Credits (Cont.) **715**

Index **717**

(a)

(b)

(e)

(d)

(f)

(g)

CHAPTER 1

The Nature of Science

Science in Our Lives

The impact of science and technology upon our lives is profound. In the urban world of today, every waking moment, virtually every activity, involves an interaction with or a utilization of some product of science and technology. As the day begins, the raucous and discordant sounds of city traffic permeate the room, ceasing now and then to give way, not to silence, but to the hum of the electric clock, the whine of the refrigerator, the blare of the radio of an inconsiderate neighbor. From the electric toothbrush with the latest fluoridated toothpaste, to the chemically impregnated wrinkleproof, wash 'n' wear garments, to the vitamin-fortified breakfast, to the deafening roar and violent swayings of the subways, to the clockwork splendor of the modern office with its background of tapping typewriters, wet and dry duplicators and clicking computers, all interconnected through an intricate network of telephones and intercoms—the world of today would bewilder past generations.

The advance of science and technology has also had a profound impact upon the rural environment. The use of fertilizers and pesticides, for example, has led to land productivity undreamed of a few generations ago. But along with the bumper crops, there have been wider effects in the world of nature, and not all of them positive. Some pesticides have killed off helpful insects, and birds, and even poisoned the fish in nearby streams.

As populations have multiplied and industry, aided by science and technology, has expanded, man has been increasingly tampering with the delicate balance of nature in even more pervasive ways. He has dammed rivers and drained swamps. He has heedlessly dumped his waste products into rivers, lakes, and oceans, or belched them out into the air from smokestacks. Radioactive debris from nuclear explosions and from peaceful applications of atomic energy has been added indiscriminately to our environment. Meanwhile we are consuming a fantastic and ever-growing array of chemicals, from drugs to food additives, many of which have no parallel in nature. All these new influences are having far-reaching effects upon the world of nature, and upon man himself, the ramifications of which we are only beginning to comprehend.

Figure 1–1 Today's technological world.

(a), (d), (e), (f), (g), E. Baitel, Omikron; (b), A. Lopez, Omikron

Meanwhile, numerous advances in biology and medicine have literally saved the lives of millions of humans. Antibiotics, new techniques in heart surgery, transplants of organs, the use of radioisotopes over a wide gamut of diagnosis, treatment, and research, and the growing use of computers and other electronic instrumentation in biology and medicine have resulted in a profound revolution in medical practice and given new hope for the ailing and dying. More babies are being born and living to maturity, leading healthier and more productive lives. This is not an unmixed blessing: the decreased death rate has led to increased pressures of population throughout the world, with concomitant shortages of food and living space.

Science and technology themselves may hold answers to the problems that they have helped to create, but whether man will take advantage of these answers is another matter.

The progressive conquest of space may not seem to have any immediate practical impact upon the "man in the street." Yet the horizons opened up by this great adventure may have an ultimate impact upon man fully as powerful as the voyages of discovery of Marco Polo, Columbus, and Magellan.

The Accelerating Rate of Scientific Progress

We are living in an exciting age. Hardly a week goes by without the announcement of some important new achievement in science and technology: a rocket landing on Venus shedding new light on the mysteries of the solar system, a new vaccine to fight some dread disease, a satellite launching providing a quantum jump in improved communications, or the elucidation of basic chemical reactions in the cell providing added glimpses into the secrets of life.

But this age of simmering excitement is one of quite recent vintage. For millions of years man has groped and bumbled about in a darkness of ignorance and superstition. In some distant age billions of years ago, in the warm, soupy seas, an anomaly appeared: a chemical capable of reproducing itself. From this beginning, through a long process of evolution encompassing millions and billions of years, forms of life arose which eventually spread throughout the earth: through its waters, on its lands, in its air.

At some point—who can say when?—a creature appeared, capable of passing on the information and experiences it had acquired in its lifetime to its offspring, not by inheritance, but by a process of teaching and learning. This new ability doubtless proved to be a great advantage in the constant struggle for survival. The development of greater and greater brain capacity permitted more efficient exploitation of this ability. But even with a superior brain, there is a limit to the amount of information that can be passed on from one individual to another. Another quantum jump was achieved with the formation of the human community, in which specialization of individual members permitted a far larger store of information to be passed on to succeeding generations. Progress in the acquisition of human knowledge took another great step forward when man, in his newly civilized state, discovered the advantages of recorded knowledge.

Although the rate of acquisition of this knowledge ebbed and flowed for thousands of years, it took a sudden and startling upturn in the fifteenth and sixteenth centuries. This sudden upturn coincided with the Renaissance, the great reawakening and rechanneling of human thought and activity. The rate continued to accelerate and is still doing so (Fig. 1–2). For instance, a survey conducted by the U.S. Air Force in the early 1960s indicated that the amount of knowledge acquired in science and technology during the 250-year period from 1500 to 1750 was equivalent to that acquired throughout all the previous history of mankind. This knowledge doubled again in the 150-year span from 1750 to 1900. In the 50 years from 1900 to 1950, the amount of knowledge acquired in the fields of science and technology once more doubled. Astounding though it may seem, according to this

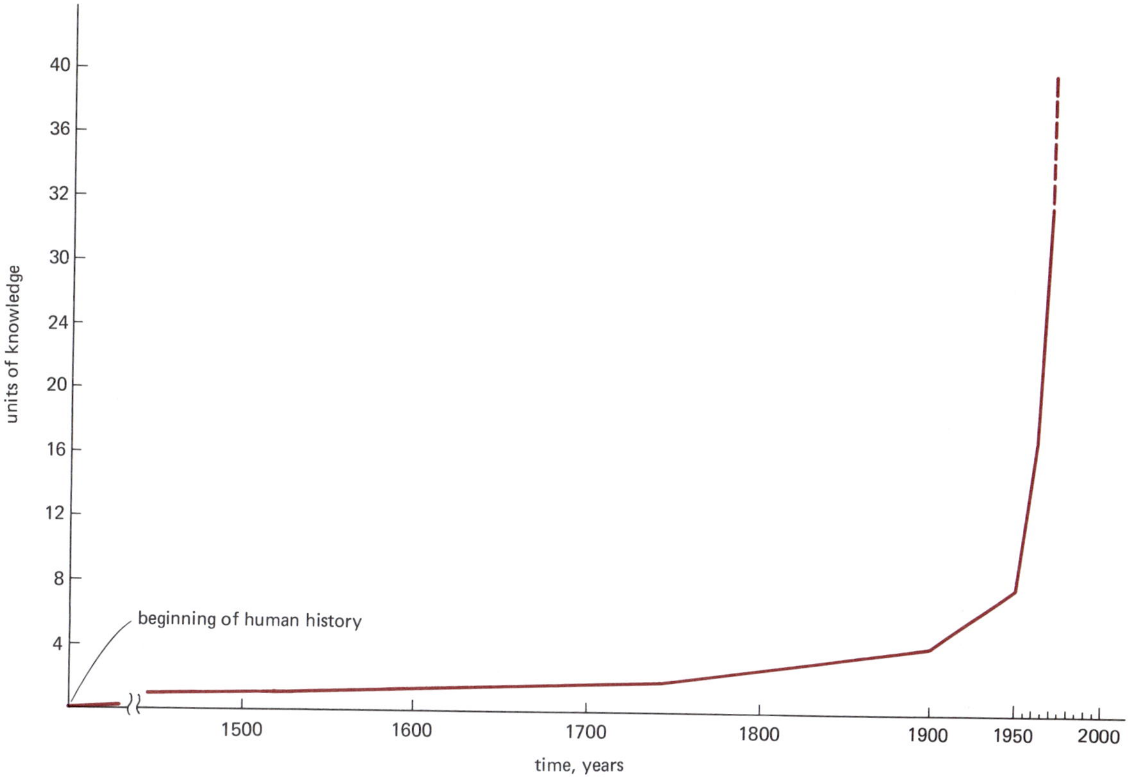

Figure 1–2 The acquisition of scientific knowledge is currently in a "log phase" of accelerating increase.

survey, the amount of scientific and technological knowledge acquired by man in the single *decade* from 1950 to 1960 equaled all the knowledge amassed by man in the preceding million years!

Furthermore, it was estimated that the knowledge accumulated in the seven-year period from 1960 to 1967 would once more represent a doubling. Some corroborative evidence has been contributed by *Chemical Abstracts,* a journal surveying virtually all of the major chemical and biochemical literature of our day. The editors of this journal have stated that in recent years, the chemical and biochemical literature has been accumulating at a rate equivalent to a doubling approximately every eight years.

If one were to assume that this rate of accumulation of knowledge—i.e., doubling every seven to eight years—will continue, then the amount of knowledge amassed in the next few generations would be incredibly huge.* Let us take the year 1900 as the year for comparison, and the amount of knowledge acquired by that date as the reference unit. According to the above figures, the amount of knowledge amassed by 1967 would be equivalent to eight units. A little over fifty years later (A.D. 2020), the world would have accumulated a thousand times as much knowledge as was available in 1900, and before the end of the twenty-first century the people of Earth would have acquired the incredible sum of a million units of knowledge. This would be equivalent to an addition to our knowledge of the staggering sum of 8 billion new words each minute, or enough to fill the equivalent of the *Encyclopaedia Brittanica* three hundred thousand times *every day.*

*For a number of reasons, we cannot expect this rate to continue indefinitely.

The Nature of Science

Curiosity has been a driving force in man's unceasing search into the unknown. Through the ages men have felt that there must be some unifying patterns in the seemingly random and discordant universe about them. Over the years men have sought in numerous ways to uncover these patterns. At the root of these quests have been the questions "why?" and "how?". Often the "why" has led to explorations in philosophical and spiritual directions. Frequently the "answers" provided by such approaches, although they sometimes prove deeply satisfying to the immediate personal needs of the individual, have essentially served to block any further inquiry into the "how." The belief in spirits by primal societies and in a supreme being by others has provided explanations that removed the need for and, indeed, often prohibited further investigation into the unknown. The philosophical approach of the ancient Greeks, with its emphasis upon mental methods of inquiry, was just as effective in preventing the actual testing of the ingenious explanations they devised.

But with the Renaissance the ascendancy of the "how" was assured. The atmosphere of the time seethed with intellectual ferment. Men became less satisfied with blind acceptance of tradition with its abundance of ready-made answers. Instead, they sought the truth actively, employing methods which heralded the beginnings of a new age in human thought, the age of science.

We are still in this age of science—indeed, more so than ever before. With today's headlong plunge into the unknown at an ever-accelerating rate, science and technology are providing man with tools and techniques for gaining a better understanding of himself and his environment and for ensuring his survival and comfort. (Of course, some of these very same tools may be employed for his ultimate destruction.)

Today, more than ever before, science and technology are becoming the mainstays of our society. These two interdependent branches of human endeavor are often confused in the mind of the layman. The newspapers hail a new rocket launching or an improved version of a vaccine as a "great scientific feat." Yet these are actually *technological* developments, whose accomplishment ultimately rests upon basic scientific research, often performed decades before. Hence, we see a distinction between science and technology. The one, **science,** stems from basic human curiosity, the desire to discover the "how" of things, often with no practical purpose in mind, simply from a compelling need to *know.* The other, **technology,** feeds upon the fundamental knowledge garnered by science, turning this knowledge to practical ends. But this is not a one-way process. Often a technological development (for instance, the electron microscope) opens the way for the discovery of new fundamental knowledge. **Basic research** is the term applied to the activity of the scientist in his explorations into the unknown, seeking knowledge purely for its own sake without necessarily thinking of potential practical applications. **Applied research** describes the activity of the technologist in applying basic knowledge to practical ends.

These are the extremes. In practice, one rarely encounters pure basic research. In our pragmatically oriented world of today, a basic researcher often has to seek out potential practical applications for his project in order to justify it to those—industry, governmental agencies, or private foundations—who can provide the necessary funds for its implementation. For example, a researcher studying cell division in a rare marine organism might "justify" his project by contending that such research may help elucidate the mechanism by which cancer cells reproduce. This "subterfuge" is not necessarily as devious as it might seem at first, for numerous important discoveries in science have been the result of serendipity—scientists seeking the answer to a particular problem have frequently quite accidentally made findings of far-reaching impact, even of a nature entirely unrelated to their original pursuits.

In 1928 Alexander Fleming, a British microbiologist, accidentally discovered the first of the "miracle drugs." One day, as he entered

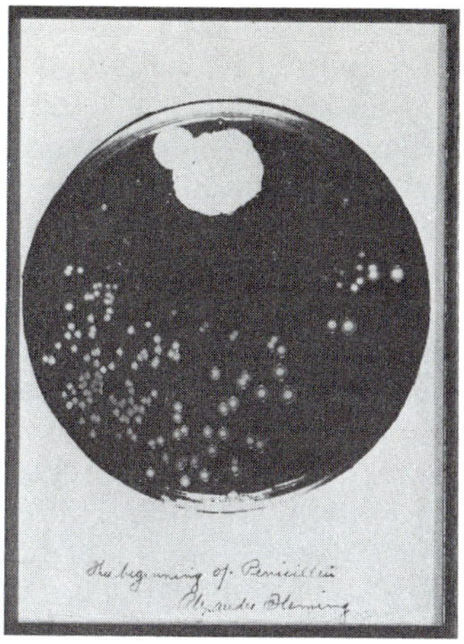

Figure 1–3 This culture dish held the key to a great discovery.

his laboratory, he noticed that one of his bacterial cultures was contaminated with a blue green mold. Instead of discarding the agar plate in which the culture had grown, as is commonly done with "spoiled" specimens, Fleming examined this one more closely. He noticed that a circular region around the mold was entirely free of the bacterial growth. He prepared extracts from the mold, and even in great dilutions this material, which he named penicillin, exhibited antibacterial properties. Although Fleming published his findings in a medical journal the following year, the world did not recognize the importance of his discovery at that time. It was not until a concentrated effort was launched to find an effective way of treating the wounded in the Second World War that the true potential of this new antibiotic was realized and fully developed.

This episode also illustrates the fact that the significance of a basic discovery may lie dormant for years, decades, or even generations, until it is suddenly utilized as a foundation for numerous practical applications. Another classic instance of this phenomenon, on another level, was the general public's unawareness of the significance of Einstein's special theory of relativity (1905), which contained his now-famous equation expressing the interconvertibility of mass and energy; $E = mc^2$. It was not until nearly forty years later that the full import of this equation was demonstrated to the world by one of its practical applications, the explosion of the atom bomb.

Scientific Method

There is no one scientific method. The problems encountered by scientists in the various scientific disciplines are so varied in scope that numerous avenues of approach are adopted in attempts to find solutions to them. Nonetheless, certain basic principles may be cited that are common to the vast majority of these approaches.

First, one must recognize that a problem exists whose solution is desired. This factor may appear too obvious to warrant discussion. Yet many problems that now loom large in our society at one time were unrecognized as such for years or even centuries. The cacophony that accompanies modern city life had, until quite recently, been accepted as a necessary concomitant of modern civilization. Now, however, it is becoming increasingly evident that such "ordinary noise" can have far-reaching psychological (and even physiological) effects upon man. And now the whole area of "noise pollution" is recognized as a very real problem, indeed.

Once a problem is recognized, a valid scientific investigation of it cannot be begun until this problem is properly defined. Phrasing the problem in a meaningful way—"asking the right question"—often determines the ultimate chances for the success or failure of a prospective line of inquiry. In considering the factors contributing to city noise, for example, one might be tempted to ask, "Are cars, trucks, and busses really necessary?" (Indeed, this question has already been raised, and traffic has been severely restricted in the business areas of some American communities. For a time, a similar measure was taken in the heart of Rome, but the clamor of irate shopkeepers soon led to restoration of the vexing and earsplitting traffic jams.) In general, especially in large cities, this proves to be a rather impractical question. More meaningful questions, then, might be: "Can some other means of transportation be substituted?" or "Can some measures be taken to reduce the noise level generated by these vehicles?"

After the problem has been defined properly, the scientist analyzes the mass of accumulated observations or **data**, trying to determine which are pertinent to the problem and to organize and arrange them in such a manner that a pattern or some unifying order will emerge. On the basis of this sorting and collating process, the scientist formulates a tentative explanation or statement of apparent interrelationships among the observed phenomena. This statement is called a **hypothesis.**

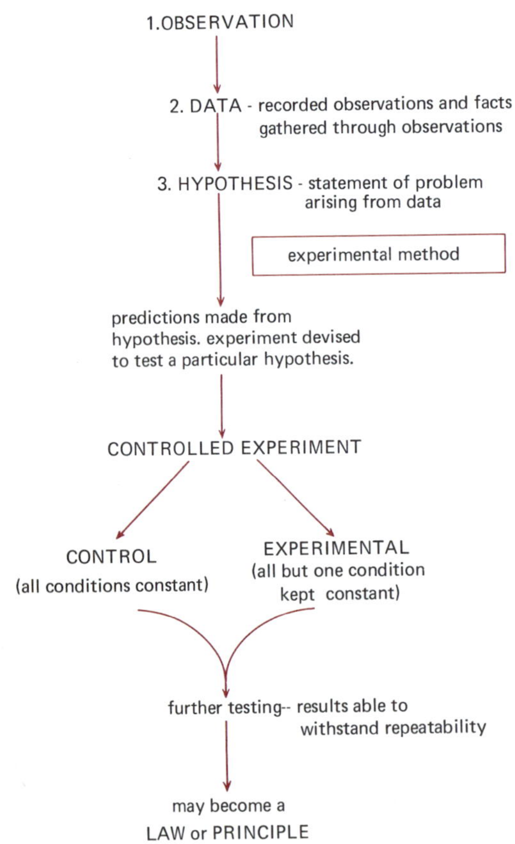

Figure 1–4 Steps in the establishment of scientific principles.

Its formulation is a highly individual process, in which the scientist uses his background of experience, his skill and intelligence, and often a sudden flash of insight to fit the pieces of the puzzle together, fill in the gaps, and come up with an explanation that covers all the facts. If several different hypotheses could be used to explain all the data, he will choose the simplest. (The principle of choosing the simplest of the available hypotheses that will satisfactorily explain all the data is known as **Occam's razor,** after the fourteenth-century philosopher who first formulated the concept.)

When the scientist has a satisfactory hypothesis consistent with all the known data on the problem, he does not stop here. His hypothesis is only a working explanation that must be tested over and over again. New experiments may yield data that fit perfectly into the hypothesis or data that refute part or all of it.

A good hypothesis is one that can be used as a basis for further generalization. Thus, it is used to predict the characteristics of related phenomena that have not yet been observed. These predictions are tested by new experiments, yielding new data. Indeed, the validity of a hypothesis is often measured by its usefulness in correctly predicting the characteristics of new phenomena and the new interrelationships among phenomena already observed.

A good hypothesis should not be static. After it is tested, if necessary it must be modified to bring it into alignment with any new facts uncovered as a result of the testing process. Often a hypothesis must be revised many times before a satisfactory formulation is achieved. Frequently a hypothesis must be completely abandoned when new facts are uncovered which are inconsistent with any possible variation of it. Since scientists are human beings, parting with a cherished hypothesis may be an extremely painful process. One of the most important characteristics of science is that the researcher must be prepared at all times to set aside his personal prejudices to accept the facts, wherever they may lead.

Some hypotheses must remain (for the moment, at least) beyond the scope of scientific investigation. The statement, "Man possesses an immortal soul," does not readily lend itself to scientific investigation. However, other hypotheses that in the past have appeared to be unverifiable have yielded to testing with improved techniques of measurement. For example, for a time the existence of a cell membrane was doubted by some scientists. Although it seemed obvious that there should be *some* covering surrounding the cell, no such structure was visible under the most powerful optical microscope. It was only after the development of the electron microscope that the cell membrane could at last be seen.

Several important features generally characterize well-planned and properly conducted scientific experiments. The first of these is strict **control of variables.** For a valid conclusion to be reached, only one characteristic or condition should be varied at a time, and the effects of its variation should be compared with a similar experiment or test in which it is kept constant. The group of specimens or series of tests in which conditions are varied in some specific manner is generally referred to as the **experimental group,** while the "normal" (unvaried) group against which it is matched is termed the **control group.**

The reason for the need to vary only one specific condition at a time is obvious. Suppose, for example, that two colorless liquids were mixed in a beaker at room temperature and nothing happened. Next suppose that the mixture was simultaneously exposed to a beam of ultraviolet light and heated, whereupon shiny yellow crystals formed and settled to the bottom of the beaker. It would be impossible to tell, on the basis of this single experiment, whether the reaction was stimulated by the exposure to ultraviolet light, by the application of heat, or by an interaction of the two influences. Additional experiments in which the mixture was only heated or only exposed to ultraviolet light would be needed to resolve the question.

Generally, living organisms exhibit far more aspects of individual variation than does nonliving matter. Consequently, controlled experiments are usually more difficult to conduct with biological systems. However, techniques have been worked out to conduct experiments yielding useful data in this field also.

For instance, to test the effect of an injection of insulin on the blood sugar level of rats, the following experiment might be set up. Six dozen highly inbred female rats are divided into two equal groups, with care taken to equalize them with respect to body weight and age. Group I, the controls, are injected with 1cc of physiological saline (a solution which resembles the blood in terms of salt content and is known not to influence blood sugar level). Group II, the experimentals, are injected with 1 cc of a specified concentration of insulin solution. At definite intervals, blood samples are taken from each group of rats, and the blood sugar levels are determined.

The above experiment illustrates a number of factors vital to scientific validity. As can be noted, such potentially influential variables as sex, age, weight, and heredity, have been kept constant. In addition, any physiological complications resulting from the very act of injecting the animals have been taken into account by subjecting the controls to the same stress. The wide scope of possible variation among individuals has been allowed for by the utilization of large numbers (i.e., thirty-six in each group). Samples of this size permit a statistical treatment of the results, by which the variations *within* the groups offset each

Figure 1–5 A white rat weighs in at the USDA Food and Nutrition Laboratories. Large numbers of animals are commonly used in biological experiments to establish the repeatability of the results. Both experimental and control groups are observed.

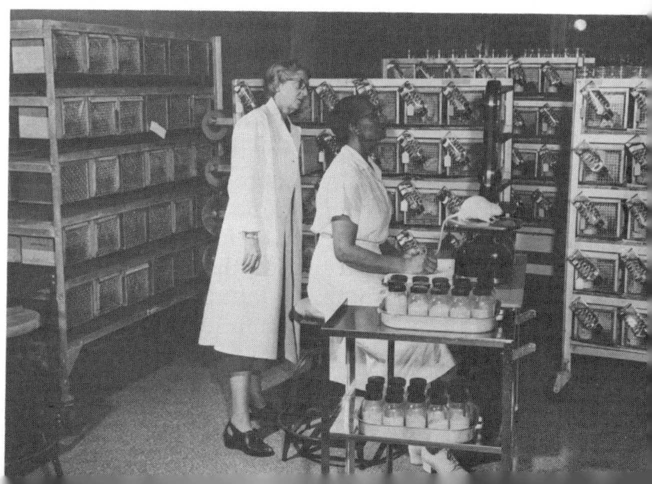

other, permitting one to discern differences *between* the groups.

To be valid, any experiment must be **repeatable.** That is, Professor Ivan Ivanovich, in Novosibirsk, USSR, should be able to perform the same laboratory experiment and obtain essentially the same results as Dr. Frank Adams, in Madison, Wisconsin. Hence, in the experiment described above, such factors as the concentration of insulin, the room temperature, the humidity, lighting conditions, etc., should be reported. Even the source of the insulin should be noted (for slight impurities might lead to significant differences in the final results of the experiment).

If a series of well-planned and -implemented experiments corroborates the working hypothesis, which in turn is found to be valid as a basis for further predictions, this hypothesis then becomes known as a **theory.** Although the term *theory* is often used rather loosely, actually only the most comprehensive and insightful hypotheses are corroborated to the point of being accepted as theories. And even theories, by their very nature, can never be proved absolutely.

A term that is sometimes used synonymously with *theory* is **natural law.** Such a law represents a statement of a universally observed phenomenon, just as a theory does. However, a *law* generally refers to a more limited concept than the broader *theory*. Thus, we speak of the "laws of planetary motion," the "laws of thermodynamics," and the "Mendelian laws of heredity," but the "theory of relativity," the "cell theory," and the "theory of evolution."

To speak of natural laws as statements of absolute truth is to misjudge the meaning of the scientific approach. Science is in essence a pragmatic discipline—i.e., it is a description of "that which works." Science does not profess to describe the absolute truth. Hence, we cannot really speak of "facts," for this term seems to imply a description of the absolute. Rather, we should speak of evidence or data disclosed by scientific methods, which indicate *probabilities* of an *approximation* to the truth. In common speech, the term *fact* is often used as a synonym for datum. Hence, we may speak of the "fact" that an increase of insulin in the blood will lead to a drop in the blood sugar level. But we must bear in mind that some of the "facts" of today may be disproved tomorrow, as the march of scientific experimentation continues to expand, modify, and discard hypotheses, theories, and even natural laws.

The Limitations of Science

Today the term *scientific progress* has a magical ring to it. Witnessing a flood of discoveries following one another in rapid succession, the layman has begun to believe that the potential of science is unlimited, that science can ultimately solve all the problems afflicting man. The elegant techniques of scientific research have been utilized to unlock one secret after another, from the age of the earth to the mysteries of the atom, until even some scientists themselves have begun to believe in its infallibility.

The majority of scientists, however, have always maintained that there are definite limits to the probings of science, that some areas must forever remain outside its boundaries. For instance, the concept of "beauty" is so subjective as to make any *controlled* investigations of it virtually impossible. One might study an oil painting in terms of the chemical composition of the pigments, the wavelengths and light intensities reflected by various regions of the picture, the degree of symmetry of design, and other technical features. But no value judgment can be made based on such an analysis, for the concept of beauty is ultimately a matter of individual taste, which itself is rather fragile and ephemeral. For instance, a work of art may have been highly lauded in the nineteenth century, only to be relegated to oblivion in the twentieth century. Even more familiar is the story of the artist who, scorned by his contemporaries, lives and dies in poverty and misery, only to be resurrected by some future generation as a neglected genius. Even within the same era, esthetic values may vary considerably: some of the music of the Far East sounds strange and discordant to Western ears. Very likely,

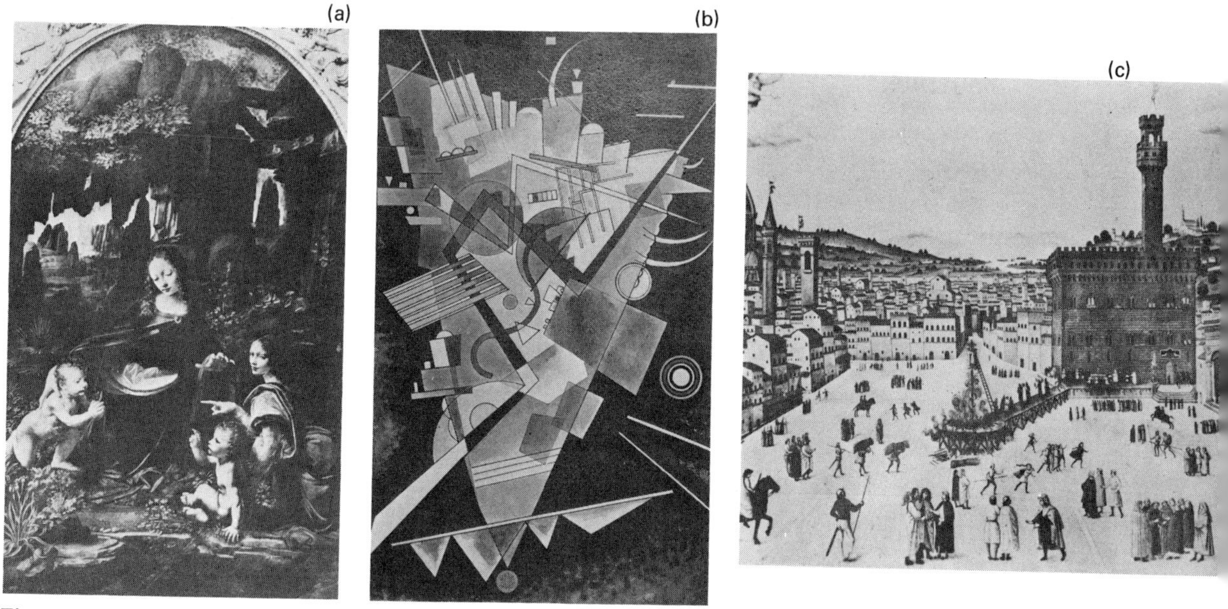

Figure 1–6 Outside the realm of science—esthetics: (a) *Madonna of the Rocks*, by Leonardo da Vinci; (b) *Weighted Suspension*, by Wassily Kandinsky; (c) *Death of Savonarola*, by an unknown fifteenth-century artist.

this feeling is shared by some Orientals with regard to Western music.

Such highly subjective areas as esthetic taste, religious faith, and such concepts as liberty, justice, and mercy, must, for the present, remain beyond the limits of science. This statement must be qualified, however, for areas that in the past appeared to be beyond the limits of scientific investigation have subsequently become fruitful areas of research as more sophisticated techniques have been developed. Less than a hundred fifty years ago, the mainstream of scientific thought maintained that the chemicals of life, such as carbohydrates, fats, and proteins, could be produced only by living things and could never be synthesized in the laboratory. This belief was shattered in 1828 by the experiments of a German chemist, Friedrich Wöhler, in which he synthesized a common metabolic waste product, urea. Now carbohydrates, fats, and other biochemicals are routinely synthesized in the laboratory. And some scientists are even boldly speculating on the possibility of synthesizing life itself.

Although some of the apparent limitations upon the probings of science may eventually yield to new methods of inquiry, there seem to be some definite limitations inherent within the very framework of the physical universe. In 1927, the German physicist Werner Heisenberg propounded what has come to be called the **Heisenberg uncertainty principle.** This principle states that it is not possible to measure simultaneously both the position and the velocity of an electron. This has been interpreted by scientists to mean that we can never be completely aware of the exact status of the subatomic world. Further interpretation has led to the thesis that one cannot unerringly predict future events, since in order to do so one would have to be aware of all aspects of present conditions.

A still more sobering limitation on the potentialities of scientific probings was revealed by the work of Kurt Gödel, a Czech mathematician. In 1932 he propounded his **incompleteness theorem.** Although the intricacies of his reasoning can be fully appreciated only by mathematicians and logicians, briefly, he stated that any mathematical system must of necessity be inherently incomplete, since another, entirely separate system must be enlisted to demonstrate its consistency. Furthermore, any system of reasoning derived from such a mathematical system must also be inherently incomplete. Since all of the physical sciences (and even huge areas of biology) are based on a mathematical foundation, Gödel's theorem seems to imply that the validity of any theory or law whose ultimate foundation rests upon mathematics can never be ultimately proven.

The Divisions of Science and Their Interdependence

Nature does not divide her domain into separate compartments, neatly labeled and carefully sealed to prevent contamination. In essence, there is *one* Science, the study of reality (or what man *believes* to be reality). But man, in his need to categorize and classify, has devised an intricate scheme of divisions and subdivisions of scientific exploration.

Science may be divided into two broad categories: the **social sciences** and the **natural sciences.** Examples of the former group include such areas of endeavor as psychology, sociology, political science, and economics. Some scientists maintain that these are not really sciences in the same sense as the natural sciences, since human relationships present so many uncontrollable variables that rigidly controlled experiments are difficult if not impossible to implement, and measurement of these variables is at best ambiguous. Even in

Figure 1–7 The divisions of science.

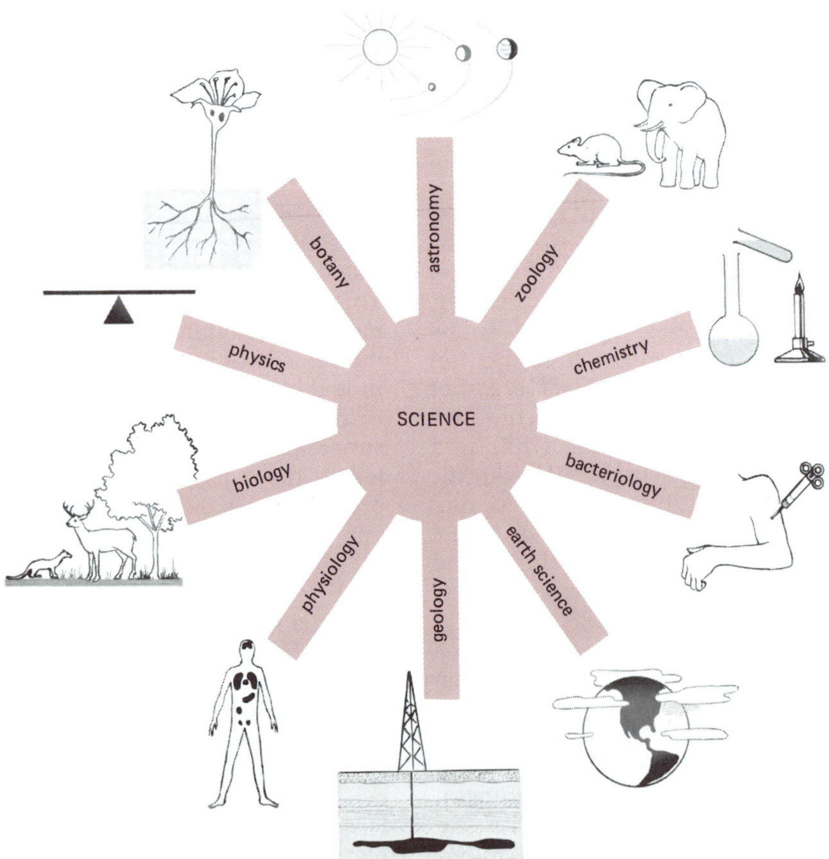

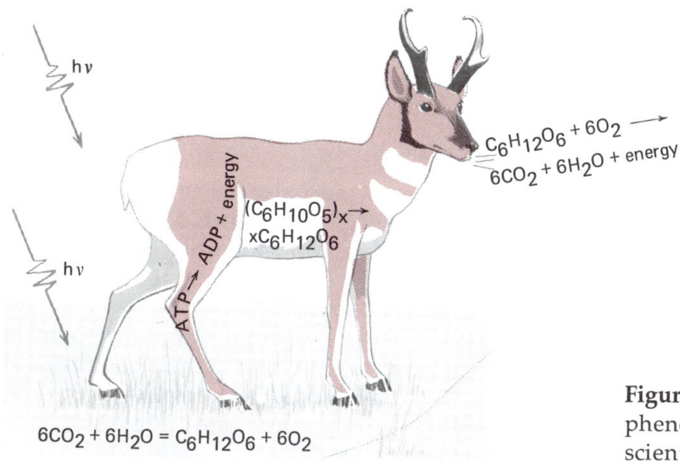

Figure 1–8 Life includes phenomena of interest to scientists in many fields.

such a highly quantitative discipline as economics, there are often so many uncontrollable factors that it is a common phenomenon to have many economists propounding highly contradictory theories, all purporting to describe the same set of economic conditions and accompanied by an equally divergent series of forecasts. On the other hand psychology is rapidly becoming more of a disciplined science, especially with regard to well-controlled experiments with animals.

The natural sciences, on the other hand, are frequently both highly disciplined and extremely quantified. This is particularly true of what are known as the **physical sciences.** Among this group are **physics,** the study of matter and energy; **astronomy,** the study of the universe; and **chemistry,** the study of chemical interactions of matter. **Geology,** the study of the earth, and **meteorology,** the study of weather, are also examples of physical sciences. The **life sciences,** including **biology,** the study of living organisms, and **medicine,** the study and treatment of disease, represent the other major branch of the natural sciences.

As scientific research has yielded an ever-growing wealth of information, there have been two important trends in the sciences. First, there has been an increasing degree of specialization. For example, physics has been broken down into such areas of study as mechanics, electricity, and nuclear physics. Biology splintered first into botany, the study of plants, and zoology, the study of animals, and then into narrower subdivisions such as ichthyology (the study of fish), bacteriology (the study of bacteria), physiology (the study of bodily functions of living organisms), and genetics (the study of hereditary mechanisms).

The increasing specialization in the sciences, however, has been paralleled by a second important trend—an increasing recognition of the interdependence of the sciences. New fields of study have grown up at the boundaries between sciences, such as physical chemistry, astrophysics, geophysics, and geochemistry. Though the life sciences were for a long time purely descriptive, recent years have seen an increasing quantification of biology and an influx of scientists and technologists from the physical sciences. Now the newer interdisciplinary sciences, such as biochemistry, biophysics, and biomedical engineering, are active fields of research.

The interdependence of the sciences might be illustrated by the following example.

The slender shadow of an antelope glides through the tall grass, pausing now and then as its graceful head bends to nibble delicately at some succulent morsel. Suddenly the sharp echo of a snapped twig rivets the animal to the spot; then with a sudden spring it is bounding away through the grass. Such a scene would fascinate a watching naturalist, whose primary aim is to observe animals and plants in their natural settings and to note their responses to various stimuli.

A biochemist, on the other hand, might see the *same* scene in the following aspects:

(1) $6CO_2 + 6H_2O \xrightarrow{h\nu} C_6H_{12}O_6 + 6O_2$

(2) $(C_6H_{10}O_5)_x \xrightarrow[H_2O]{\text{cellulase}} xC_6H_{12}O_6$

(3) ATP → ADP + energy

(4) $C_6H_{12}O_6 + 6O_2 \rightarrow 6CO_2 + 6H_2O$ + energy

Equation (1) represents the photosynthetic reactions occurring in the grass, which, with the aid of sunlight ($h\nu$), converts atmospheric carbon dioxide and water into organic nutrients ($C_6H_{12}O_6$). Equation (2) describes one of the processes occurring in the digestive tract of the antelope. Equation (3) symbolizes the rapid production of energy, which enabled the frightened antelope to bound away. And equation (4) symbolizes the respiration of the antelope as it pants frantically.

A physicist would also see some interesting features in the scene. The velocity of the running antelope and its acceleration as it speeds away can be symbolized by the equations:

(5) $v = \Delta s/\Delta t$

(6) $a = \Delta v/\Delta t$

The physicist would also be interested in the spectrum of sunlight shining down upon the grass, the wavelengths absorbed by the grass, and the value of the light quantum ($h\nu$) utilized in photosynthesis.

A specialist in electronics might be helpful in designing instruments to measure and record the sound of the snapping twig, the light falling on the grass, and the minute currents of electricity pulsing through the body of the antelope as it responds to the changing conditions of its environment.

The different sciences, working together, can thus contribute to a fuller understanding of the world about us.

QUESTIONS FOR REVIEW AND THOUGHT

1. Has the idea of "progress" always been an inherent part of man's philosophy and outlook? Explain.
2. Is counting the number of publications produced in a given period necessarily an accurate indicator of the amount of new information amassed by a group or society? What role, if any, should *quality* play? And how can quality of achievement be evaluated?
3. Can one expect the accelerating rate of scientific progress to continue unabated indefinitely? Explain.
4. Can science or the fruits of science be considered either "good" or "bad"? Explain.
5. List a dozen items commonly seen in our present world the very existence of which would have been considered impossible, or at best highly unlikely, one hundred years ago.
6. Give examples of questions or problems the solutions to which lie outside the domain of scientific investigation.
7. Distinguish between science and technology, basic research and applied research.
8. Can man hope to know all that there is to be known? Explain.
9. Distinguish between the social sciences and the natural sciences. Can the social sciences ever become "exact sciences"? If so, in what areas?
10. What characteristics of the modern world are responsible for the development of two apparently contradictory trends in science: increasing specialization and the interdisciplinary emphasis?

PART I

THE WORLD ABOUT US

(a)

(b)

(c)

(d)

(e)

CHAPTER 2

Matter, Energy, and Space

The Nature of Matter and Energy

All that we can perceive, either through our own senses or with the aid of instruments, is composed of matter or represents some form of energy. A stone, a building, a child, the moon, and the sun are examples of **matter.** Each has a definite amount of substance, or mass, and occupies space.

Matter can exist in a variety of states. The book you are reading now and the chair on which you are sitting are forms of **solid** matter. They have a fixed shape and a volume that would remain essentially constant even if great pressure were applied. Water, milk, alcohol, gasoline, and blood normally occur in the **liquid** state. Although their volume remains relatively constant over wide variations of temperature and pressure, they do not have a fixed shape. Instead, a liquid flows to assume the shape of its container.

It is easy to associate solids such as wood and steel and liquids such as water and gasoline with the concept of matter. In the case of **gases,** however, the connection may not seem so obvious. Take the most common gas we know, for example—the air around us. We cannot see it or taste it or smell it. (That is true of pure air, at least—a citydweller might have a somewhat different working definition of "air.") It seems to have no substance at all. Yet under certain conditions it becomes obvious that air does have substance after all. If you pump air into a rubber tire, the walls of the tire swell and round out until the tire is firm enough to support the weight of an automobile. If there is a leak in the tire, and air escapes from it, the walls collapse and you have a "flat." A gas, like a solid or a liquid, has a definite amount of substance and occupies space. But it has neither a fixed volume nor a fixed shape. A gas tends to expand until it assumes the shape of its container. But if pressure is applied (and/or the temperature is lowered), gases can be greatly compressed. Gases at room temperature, for instance, may be compressed to about a thousandth of their original volume.

Figure 2–1 The states of matter.

(a), Brenda Graber, Omikron; (e), A. Lopez, Omikron

We tend to think of water as a liquid, iron as a solid, and air as a gas. These are the states that these substances assume under the familiar condition of "room temperature." Yet if the temperature of a bowl of water is lowered, it turns into a solid block of ice; if it is heated, it gradually turns into a gas and disappears into the air. Enough heat can also melt solid iron so that it flows like water and can be poured into molds. If air is cooled, and pressure is applied to it, this gas will be transformed first into a liquid and then into an icelike solid.

The various states of matter thus form a continuous spectrum through which substances can pass depending on their temperature and the pressure applied to them. With the addition of heat, a solid can **melt** into a liquid and ultimately **evaporate** into a gas. A gas that is cooled can be **condensed** into a liquid and, as the temperature is further lowered (or the pressure raised), **frozen** into a solid. The specific temperatures and pressures at which changes of state occur vary among substances, and they are constant for any particular substance. The freezing and boiling points of water (32°F and 212°F) have been used as the basis for another temperature scale, the **centigrade** scale (Fig. 2–2). The freezing point of water is taken as 0° in the centigrade scale, while its boiling point is 100°C; the entire temperature range in between is divided evenly into a hundred degrees, each of which is a bit less than twice as large as a Fahrenheit degree.

Matter is presumably distributed throughout the universe. Furthermore, observations of distant stars and galaxies reveal that the same basic components of matter are found in virtually every known part of the universe.

For thousands of years man has speculated on the nature of matter. Long before recorded history, he certainly observed that some solids are hard while others are soft, that some can readily be bent and molded into a variety of shapes, while others break into fragments when they are hit with a hard object. Some solids melt when they are heated, turning into flowing liquids, while others char or burn. Liquids, too, differ widely. Water flows freely,

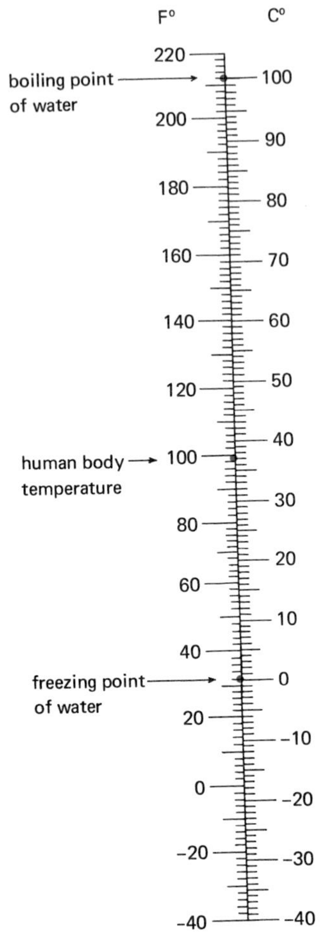

Figure 2–2 Comparison of the Fahrenheit and centigrade temperature scales.

while molasses oozes sluggishly; some liquids freeze in the winter cold, while others do not. Man gradually learned that substances can be changed by various influences, such as heating or mixing with other substances.

This growing knowledge about the nature of matter is reflected in the history of mankind. The Bronze Age signaled an advance in technology: the discovery that bronze, a combination of copper and tin, makes stronger weapons than copper alone. Empires toppled and new ones arose as a result of this technological advance, and then the process was repeated in the Iron Age, when the secret of extracting metallic iron from its ores was learned. Another revolution resulted from the discovery that two naturally occurring substances, sulfur and saltpeter, can be mixed with charcoal to produce gunpowder, which, when heated, yields an explosive release of energy.

Gradually a growing body of knowledge about the nature of matter and the changes it undergoes was amassed. At times there were rapid advances, while at other times there were periods of stagnation spanning centuries. Some pursuits led up blind alleys, such as the ceaseless quest of the medieval alchemists for the legendary "philosopher's stone" that would transmute base metals such as lead into precious gold. But even the blind alleys often yielded valuable treasures of empirical knowledge.

The present **atomic theory of matter,** although anticipated by the Greek philosopher Democritus about twenty-four hundred years ago, is less than two centuries old. This theory, in its present formulation, visualizes matter as composed of various kinds of building blocks or **elements.** More than a hundred different elements have been found in nature or made in the laboratory. Each element is composed of minute individual particles called **atoms.** Each atom of a particular element has properties similar to those of any other atom of that element and differs from the atoms of other elements in internal structure and usually in mass (quantity of matter).

Atoms often combine with other atoms to form aggregates known as **molecules.** Such combinations form an almost infinite variety of molecules, but each type of molecule is characterized by a definite assortment and arrangement of atoms. Each unique grouping of atoms of two or more different elements (i.e., molecule) represents a specific substance called a **compound.**

Matter in the universe is continually undergoing physical changes (such as changes of state) and chemical changes (changes in the groupings of atoms in molecules). All of these changes and interactions involve an input or outflow of **energy.** Energy, though quite familiar to all of us in its practical manifestations, is a difficult concept to visualize. Unlike matter, energy does not have substance or take up space. It may be defined as a capacity to do work.

Some energy is associated with motion. This type of energy is referred to as **kinetic energy.** For example, a hammer swung through the air possesses kinetic energy. If it is brought down forcibly onto the head of a nail, its motion is stopped and its energy is partly converted to heat (another form of energy) and partly used to do work, driving the nail into a board or wall.

Another basic form of energy is stored or latent energy **(potential energy),** which under the proper conditions can be converted to kinetic energy. A book held up in the air has potential energy owing to its position: if it is let go, it will fall to the floor, and its potential energy will be transformed into kinetic energy. The energy stored in chemical bonds can be used to power flashlights and to "run" the thousands of intricate chemical reactions in our bodies that permit us to move and speak and think. Heat, light, and electricity are other familiar forms of energy in our world.

Under normal conditions, both matter and energy may be converted from one of their forms to another. In the process of such conversions, nothing is lost—that is, the sums of both matter and energy remain constant. When a log is burned in a fireplace, for example, part of it seems to disappear, leaving only a small residue of ashes. But in the burning, part of the substances of wood are converted to gases—oxides of carbon and water. If these were collected and weighed along with the ashes, the total would exactly equal the original weight of the log. In the energy equation, the stored energy in the wood chemicals is released as an equivalent amount of energy in the form of heat and light. Thus, both matter and energy are conserved.

Under special conditions, such as those which are believed to prevail in the sun and which have been reproduced on a small scale here on earth in thermonuclear reactions, matter can be converted to energy. The con-

(a)

(b)

(d)

(c)

(e)

(f)

(g)

version occurs according to a very precise quantitative relationship, rendered by Einstein's famous formula, $E = mc^2$, where E represents energy, m represents mass, and c the velocity of light. The destruction wrought by atomic bombs on earth and inherent in the more modern varieties of fusion weapons is puny in comparison with the mass-to-energy conversion that goes on continuously in the sun. It has been computed that the sun is converting approximately 300 million tons of matter into energy every minute, generating about 400 sextillion* kilowatts of power. This fantastic conversion has been going on for more than 5 billion years and presumably will continue for at least another 5 billion.

Mixtures

Nothing on earth is absolutely pure. All natural materials are blends. For example, an iron ore deposit containing more than 60% iron is considered "rich." (Metallic meteors, the purest natural sources of iron known, may consist of as much as 90% iron.) The iron and steel industry converts iron ores into finished products containing a much higher percentage of iron—perhaps 99.5%. But even this is far from pure. Possibly the purest substance that has been produced in the world is 99.99999999% pure germanium. But even this fantastically pure material contains an enormous number of impurity atoms. Ten grams of this ultrapure germanium (about half the size of a lump of sugar) would still contain about 10 trillion atoms of impurities.

For practical purposes, if a substance is about 99.9% pure, it is often considered a "pure" substance. But even by this arbitrary yardstick pure substances are rare in nature. Instead, combinations of substances are found together in blends known as **mixtures.** Various types of mixtures are known: gases with gases, liquids with liquids, solids with solids, and various combinations of two or more of these states.

*Written out, this number is:
400,000,000,000,000,000,000,000

Figure 2–3 Forms of energy.

(b), (c), (d), (e), (g), (e), Baitel, Omikron

Sea water is an interesting example of an extremely complex mixture. The basic medium, a liquid, is permeated with dissolved gases (oxygen, nitrogen, and carbon dioxide), dissolved solids (sodium chloride and numerous other substances), and various colloidal particles (minute particles in stable suspension). Where the ocean bottom is stirred up, larger particles of sand and debris may also be suspended in sea water (although they may settle out in time).

Other common mixtures are carbonated beverages (gases and solids in a liquid) and milk (liquids and solids in a liquid). The very air we breathe is a mixture of gases (oxygen, nitrogen, carbon dioxide, water vapor, and various minor constituents), together with small amounts of suspended solid particles—dust motes.

Generally, the most interesting and important from the chemical and biological standpoints are mixtures in which the major component is a liquid. Depending on the size of the particles, such a mixture may be either a suspension, a colloidal dispersion, or a solution (Fig. 2–4).

If the size of the particles mixed with the liquid component significantly exceeds 100 $m\mu$ (one hundred millimicrons—about 0.000004 inch, just barely perceptible under an ordinary optical microscope), they will remain suspended in the liquid for various periods of time, depending on the particle size. For example, if sand is shaken up in a liquid, most of the larger grains will settle out immediately and the smaller grains more slowly. Such a mixture is known as a **suspension.** The physical properties of the liquid (e.g., its freezing and boiling points) are not significantly changed by the presence of suspended material (in this case, sand particles).

When particles in the range from 1 $m\mu$ to about 100 $m\mu$ (invisible under an optical microscope but visible under an electron microscope) are suspended in a liquid medium, the mixture takes on unique characteristics. Such mixtures, known as **colloidal dis-**

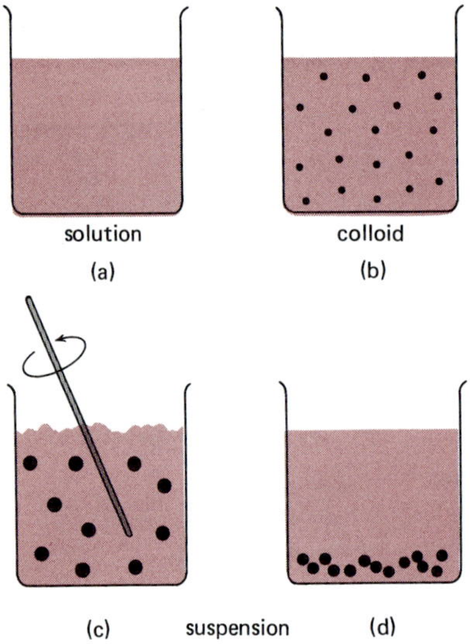

Figure 2–4 Three types of mixtures: solution, colloid, and suspension. The large particles of a suspension will settle out on standing. The intermediate-sized particles of a colloid will remain dispersed under normal conditions. The tiny solute particles of a solution will remain dispersed in the solvent indefinitely.

persions, play important roles in the functioning of living systems. The particles are kept in a state of uniform dispersion by the motions of the molecules of the liquid in which they are suspended.

Homogenized milk is a mixture which is both a colloidal dispersion and an **emulsion** (a suspension of tiny droplets of one liquid in another). The fat in milk is normally present in relatively large globules and tends to rise to the top, separating from the water phase of the suspension. Homogenization breaks down the fat globules to colloidal-size particles, which can remain suspended indefinitely. Casein, a protein naturally found in milk, helps to prevent the minute fat droplets from coalescing back into larger globules, and thus acts as an emulsifying agent.

When the size of suspended particles in a mixture falls significantly below 1 mμ, entirely different phenomena will be observed. In many cases these particles will **dissolve**—that is, their *molecules* or *ions* (charged particles) will become uniformly dispersed throughout the medium. This type of homogeneous mixture is known as a **solution,** and the dissolved molecules or ions, which may be of a solid, liquid, or gas, will remain uniformly dispersed indefinitely (assuming that the prevailing conditions, such as concentration, temperature, etc., remain constant). The dissolved particles in this mixture constitute the **solute,** while the liquid medium in which they are dissolved is called the **solvent.** Hence, if salt is dissolved in water, the salt is the solute and the water is the solvent. Dissolved oxygen in the ponds and oceans of the world provides the inhabitants of these aqueous (water) media with virtually their only supply of vital gas.

Some substances, such as alcohol and water, can dissolve in each other in unlimited amounts. Others show only limited ability to dissolve in various solvents. Large quantities of sodium chloride (salt) and sucrose (sugar), but only minute quantities of such compounds as barium sulfate and camphor, will dissolve in water. Fats, such as the butterfat in milk, are generally not soluble in water but dissolve readily in organic (carbon-containing) solvents such as chloroform. In general, the solubility of a solid in a liquid increases when the temperature is raised, while gases are more soluble at low temperatures. If an unopened bottle of a carbonated beverage is heated, it may explode!

The physical (and often the chemical and electrical) properties of a solvent are significantly altered by the presence of a solute, and the change is usually directly proportional to the concentration of the solute. Dissolved substances, for example, raise the boiling point of the solvent and lower its freezing point. The reason for sprinkling rock salt on icy sidewalks is that the salt lowers the freezing point of water and so causes the ice to melt faster.

The molecules of some substances break down into two or more electrically charged particles **(ions)** when they dissolve in a solvent. In such cases the solution as a whole remains electrically neutral, even though the individual particles are charged. Salts, such as sodium chloride, and various other compounds dissociate in this manner. The molecules of other substances, such as sugars and alcohols, are unchanged when they dissolve.

Water is often referred to as the "universal solvent." This is hardly true, but water does possess the ability to dissolve a great number and variety of solutes including both organic and inorganic substances. Water is a particularly effective solvent for substances that dissociate into ions. The water molecule is shaped something like an Australian boomerang (Fig. 2–5), with the two hydrogen atoms on the ends and the oxygen atom in the middle. The oxygen end has a slight negative electric charge, while the hydrogens are slightly positive. (Such molecules, in which there is a partial separation of electric charges, are referred to as **polar** molecules. Water molecules line up around charged ions in such a way that opposite charges are balanced. In so doing, they force the ions apart and insulate them with layers of water molecules, thus keeping them in solution. This process is known as **solvation. Nonpolar** substances, such as vegetable oil and gasoline, do not possess solvating ability, and most salts do not dissolve in them.

An interesting application of the solvating ability of water is the use of soaps. These are generally made up of large molecules, one end of which is of the ionic type, readily soluble in water, while the other end is fatlike, soluble in oils and other organic substances. The fatlike end of the soap molecule dissolves in oily dirt, while the ionic end dissolves in water and carries away both the fatty end of the molecule and the oily dirt along with it. The warmer the water, the greater the solubility of the soap molecules in water, and the better the cleaning process works.

Figure 2–5 The polarization of the water molecule (a) permits it to solvate ionic substances (b,c). The charged ions, surrounded by shells of water molecules, are shielded from the electrical effects of ions of opposite charge and thus kept in a state of dispersion in solution.

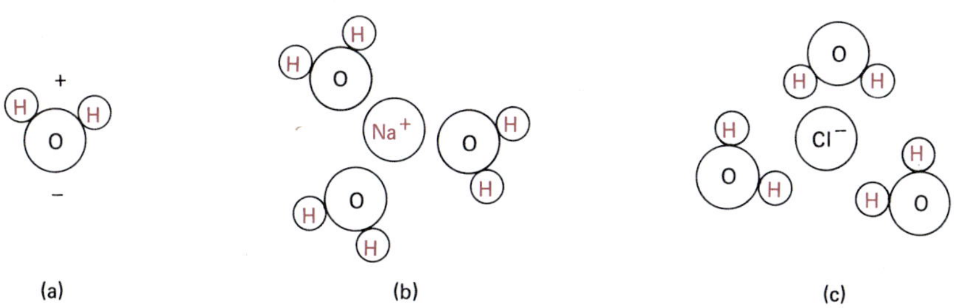

The Kinetic Theory of Matter

If a bottle of ionone (a synthetic chemical with the odor of violets) were spilled over home plate in the huge domed stadium in Houston, Texas, the fans in the farthest reaches of the stands would eventually detect the odor of violets. Particles of this volatile chemical would ultimately have permeated the entire Astrodome. Yet the stadium is not open to winds that could blow the ionone about. Apparently some sort of motion must be inherent in the molecules of the substance themselves.

More evidence to support such an idea can be obtained by watching dust motes dancing in a ray of sunlight. Or if a drop of india ink is suspended in water, a microscope will reveal the tiny dye particles jiggling about erratically. (This phenomenon is referred to as **Brownian movement,** after the scientist who first observed it.)

The **kinetic theory of matter** was postulated to explain such puzzling phenomena, and it has since been used to predict and explain many aspects of the nature and interactions of matter and energy. According to this theory, the molecules that make up any substance are in constant random motion. The speeds at which these molecules move may be quite high—of the same order as the speed of a rifle bullet. Yet the distance that a particular particle travels along any one path may be quite minute.

In solids, the molecules are packed so closely that forces of attraction and repulsion among them keep the individual molecules from wandering far from their equilibrium positions. The molecules of a solid might be visualized as a collection of solid bodies, suspended on a three-dimensional network of springs, which bounce and jiggle constantly about their places in the network.

In liquids, the molecules are not so closely packed and are freer to move. But frequent collisions with other moving molecules and with the walls of the container tend to impede the progress of any particular molecule.

In gases, the molecules are so far apart that electromagnetic forces of attraction and repulsion are of relatively minor consequence. For example, the gas molecules moving about in a cubic centimeter of air might be compared to some fifty basketballs bouncing about in a gymnasium 50 feet by 100 feet by 15 feet high. The basketballs are scarcely crowded, and yet even here collisions play a major role in impeding the progress of any particular basketball (or gas molecule) from its starting point. For example, in hydrogen gas at 0°C, a single hydrogen molecule will be involved in more than 10^{10} (i.e., 10,000,000,000) collisions each *second.* With each collision, the moving molecule is deflected randomly according to the combination of its velocity and that of the particle with which it collided. As a result, it traces an extremely erratic zigzag path. In a second, such a hydrogen molecule might cover a total distance of a mile, and yet at the end of this time it probably will still be rather close to where it started (Fig. 2–6).

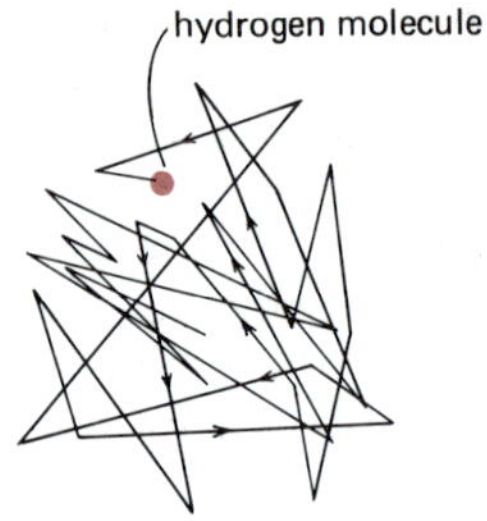

Figure 2–6 Under the influence of collisions with other molecules, a moving molecule traces an erratic, zigzag path.

The random movements of molecules sometimes result in an apparent paradox: a *directed* movement of a substance from a region of high concentration to a region of low concentration. For example, if you place a drop of ink in a glass of water, you can see the color gradually spreading throughout the water until the entire contents of the glass is a uniform color. This seeming paradox is easily resolved. Picture two rooms, *A* and *B* (Fig. 2–7).

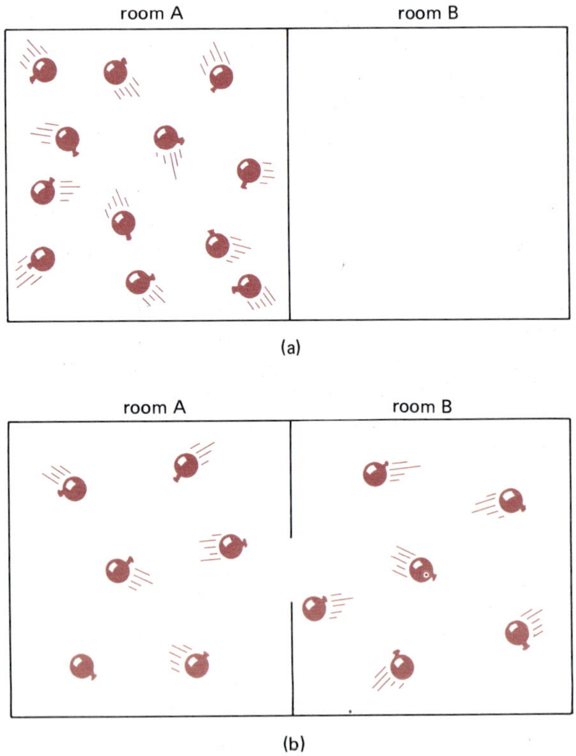

Figure 2–7 The bouncing balloon analogy of diffusion. When the connecting door is open, balloons randomly stray in and out of the formerly empty room *B,* until at the equilibrium state (b) the concentrations of balloons in the two rooms are equal.

Room *A* is filled with bouncing balloons, while room *B* is empty. If the connecting door between them is opened, some of the balloons from room *A* will randomly bounce into room *B*. As time goes by, some of the balloons in room *B* will randomly bounce back into room *A*. But in general there will be a *net* flow of balloons into room *B*, simply because, with more balloons to start with in room *A*, there are more opportunities for *A*-to-*B* transfers than the reverse. The directed movement will continue until there is approximately the same number of balloons in room *B* as in room *A*—an **equilibrium** has been reached. This process of directed movement of molecules from a region of high concentration to a region of low concentration, due to the random kinetic motions of the molecules, is called **diffusion.**

Diffusion is a widespread phenomenon in the nonliving world, occurring in gases, liquids, and even to some degree in solids. It also plays a basic role in biological processes. A particularly important aspect is the diffusion of molecules through the membranes of living systems. The thin membranes that line the digestive tract, the lungs, and the blood vessels, and the membranes that cover the basic units of life—cells—are selectively permeable, permitting only certain types of substances to pass through (permeate) them. In the simplified example discussed above, the partition between rooms *A* and *B* would be analogous to such a **semipermeable membrane** if there were several different sizes of bouncing balloons, not all of which were small enough to fit through the open door. The smaller balloons would ultimately reach an equilibrium, with equal concentrations in the two rooms; but the large ones would all remain in room *A*. The diffusion of a solvent through a semipermeable membrane is referred to as **osmosis,** and it is the basis of many of the most fundamental functions and adaptations of living organisms (Fig. 2–8).

In some cases, molecules of the solute may also pass through a semipermeable membrane. This variety of diffusion is called **dialysis.** Its principle is utilized in modern kidney machines, which remove dangerous wastes from the blood of patients whose own kidneys have been damaged by disease.

The molecules that pass through a semipermeable membrane by diffusion are in general small molecules, such as water and sugar molecules. But under certain conditions living cells can also bring about a transfer of large molecules through such membranes, even though theoretically they would seem to be too large to pass. Moreover, living systems also have means of transporting molecules against the **concentration gradient** (i.e., from regions of lower concentration to regions of higher concentration), opposite to the normal direction of diffusion. These processes of active transport require an input of energy, which is supplied by chemical reactions inside the living cells.

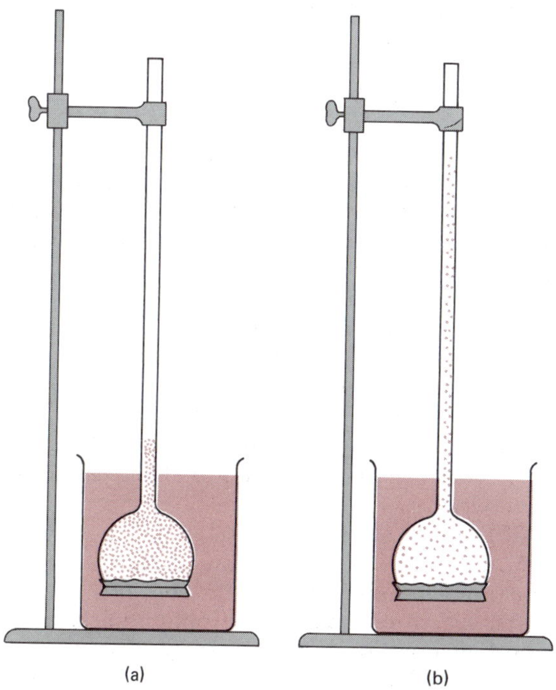

Figure 2–8 Osmosis. The thistle tube initially contained a concentrated sugar solution (a). Water diffused through the semipermeable membrane into the tube (b), lowering the concentration of the solution and increasing the volume of liquid in the tube (b). The membrane does not permit passage of sugar molecules out into the beaker.

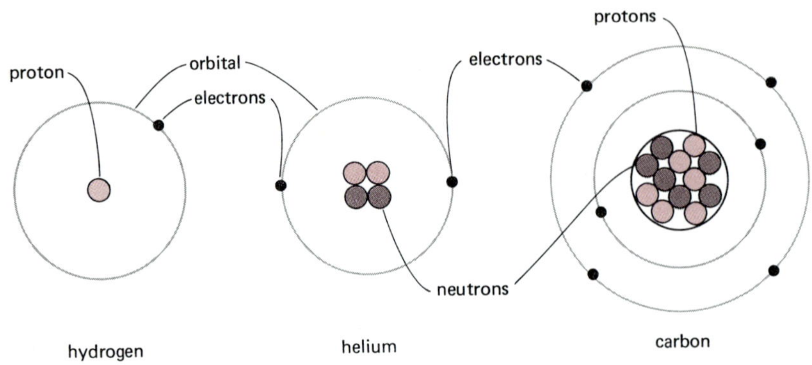

Figure 2–9 A simplified view of atomic structure. The electrons move in orbitals around a central nucleus.

The Structure of the Atom

The word *atom* is derived from the Greek *atomos*, "indivisible." Although today man is capable of "splitting the atom," the term atom is still quite valid. For the atom is the smallest unit of an element which still retains the chemical properties of that element. When an atom is "split," as in atomic fission reactions in nuclear reactors, the resultant products no longer bear any resemblance to the original atom.

In a sense, we might liken the structure of an atom to that of an onion. If an onion is cut in half, an arrangement of concentric circles around a central core is observed on the cut surface. In an atom, the central core or **nucleus** is comprised of positively charged particles called **protons** and neutral particles termed **neutrons** (Fig. 2–9). **Electrons** (negatively charged particles) may be visualized as moving within concentric spherical orbits, known as shells, arranged around the nucleus.

But the onion analogy is valid only up to a point. In an onion, the concentric rings are filled with matter. In an atom, the electron shells* consist mainly of empty space! The nucleus occupies a small region in the center of the atom. The electrons are tiny bits of matter that move about within the regions circumscribed by the electron shells. Each electron has a mass only 1/1836 as great as that of a proton or 1/1837 as great as that of a neutron. Thus, only a minute fraction of the atom is actually occupied by particles of matter. The atom, on its micro-scale, is thus nearly as vast and uncluttered as our universe, with light-years between neighboring stars and millions of light-years between galaxies.

The Bohr atom should be viewed as an oversimplification in another important respect. According to the modern views, electrons are not simply particles moving about a central nucleus in circular planetlike orbits. Each electron moves in a specified path called an **orbital** within its respective shell. If somehow an image of a moving electron could be photographed using a time exposure, a blurred "electron cloud" would appear on the photograph. Depending on the particular electron, this cloud would have a spherical, dumbbell, or even more complex shape (Fig. 2–10). The densest portions of the cloud would represent the regions in which the probability of finding the moving electron at any particular moment is the greatest. (Recall that the Heisenberg uncertainty principle states that the velocity of an electron and its position cannot be both determined precisely at the same time—the closer one comes to determining one quantity, the greater the uncertainty about the other.) Depending on their distance from the nucleus, the electron shells can hold a maximum of 2, 8, 18, 32, etc., electrons (Fig. 2–11). Each electron orbital holds a certain amount of energy (in general, the farther from the nucleus, the higher the energy). Sometimes electrons may move from one orbital to another, absorbing or emitting energy.

Atoms, by definition, are electrically neutral. This means that the total positive charge

*In Fig. 2–9 the electron shells are shown as circles, something like the orbits of the planets about our sun. This is merely a graphical simplification in which the atom is depicted in two dimensions. This kind of depiction is commonly referred to as the **Bohr atom,** after its originator, Danish physicist Niels Bohr.

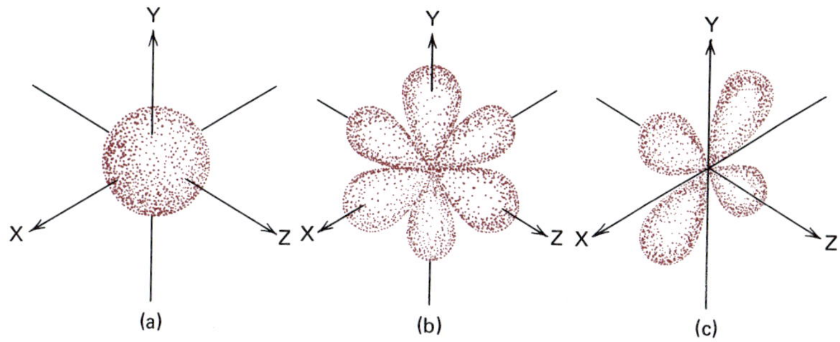

Figure 2–10 Three types of electron orbitals: (a) *s* orbital; (b) *p* orbitals; (c) *d* orbital.

of the protons in the nucleus must be exactly equal to the total negative charge of the electrons moving in the electron shells about it. Although the proton has a mass 1836 times as great as that of an electron, the electrical charge of a proton is exactly equal (and opposite) to the charge of an electron. Thus, any atom has exactly as many electrons in its shells as it has protons in its nucleus.

It has been found that the atoms of different elements differ from one another in the number of protons (or electrons) that they contain. For example, a hydrogen atom has a single proton in its nucleus and a single electron orbiting about it. A carbon atom, on the other hand, has 6 protons and 6 electrons; a nitrogen atom has 7 protons and 7 electrons; and so forth. The elements known to man can be arranged in a continuous series, running from hydrogen up to element number 105, according to the number of protons (or electrons) that they contain (Fig. 2–12). The *number of protons (or electrons)* of an element is referred to as its **atomic number.**

In addition to the charged protons and electrons, atoms may contain various numbers of electrically neutral neutrons. Atoms of the same element may contain different numbers of neutrons; these variants are called **isotopes** of the element. For example, there are three isotopes of hydrogen. The ordinary variety contains 1 proton, 1 electron, and no neutrons at all. Each atom in the isotope of hydrogen called deuterium contains a single neutron in addition to the proton and the electron. The third isotope, tritium, contains 2 neutrons per atom, together with 1 proton and 1 electron.

While the isotopes of hydrogen have special names, the various isotopes of the other elements are denoted merely by attaching a numerical suffix to the name of the element, indicating the sum of the protons and neutrons in its nucleus. The most common variety of oxygen, for example, is oxygen-16 (8 protons, 8 electrons, and 8 neutrons). The less common isotopes with 9 and 10 neutrons are called oxygen-17 and oxygen-18, respectively. Note that each of the three isotopes of oxygen has 8 protons and 8 electrons; it is precisely this fact that permits us to denote them as varieties of oxygen, rather than some other element.

In addition to its atomic number, each element has a characteristic **atomic weight.** This is the *sum of the protons and neutrons* in the nucleus of an "average" atom of the element, the average being weighted to take into account the percentage of each of the naturally occurring isotopes. Thus, hydrogen has an atomic weight of slightly over 1 (there are only

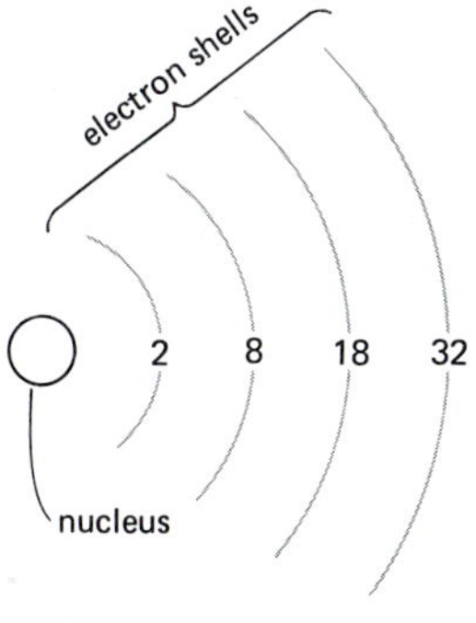

Figure 2–11 Each electron shell can accommodate a specific quota of electrons. The numbers given are maximum figures: one or more orbitals in a particular shell may be partly or entirely vacant, depending on the type of atom or ion and its energy content.

Elements essential for living organisms (shaded)

Key: 6 (atomic number) / C (symbol for element (here, Carbon)) / 12.011 (atomic weight)

IA	IIA	IIIB	IVB	VB	VIB	VIIB	VIII	VIII	VIII	IB	IIB	IIIA	IVA	VA	VIA	VIIA	0
1 H 1.00797																	2 He 4.0026
3 Li 6.939	4 Be 9.012											5 B 10.811	6 C 12.011	7 N 14.007	8 O 15.9994	9 F 18.998	10 Ne 20.183
11 Na 22.990	12 Mg 24.312											13 Al 26.98	14 Si 28.086	15 P 30.97	16 S 32.064	17 Cl 35.453	18 Ar 39.95
19 K 39.102	20 Ca 40.08	21 Sc 44.96	22 Ti 47.90	23 V 50.94	24 Cr 52.00	25 Mn 54.94	26 Fe 55.85	27 Co 58.93	28 Ni 58.71	29 Cu 63.54	30 Zn 65.37	31 Ga 69.72	32 Ge 72.59	33 As 74.92	34 Se 78.96	35 Br 79.91	36 Kr 83.80
37 Rb 85.47	38 Sr 87.62	39 Y 88.91	40 Zr 91.22	41 Nb 92.91	42 Mo 95.94	43 Tc 99	44 Ru 101.07	45 Rh 102.91	46 Pd 106.4	47 Ag 107.87	48 Cd 112.40	49 In 114.82	50 Sn 118.69	51 Sb 121.75	52 Te 127.60	53 I 126.90	54 Xe 131.30
55 Cs 132.90	56 Ba 137.34	57–71 La series*	72 Hf 178.49	73 Ta 180.95	74 W 183.85	75 Re 186.2	76 Os 190.2	77 Ir 192.2	78 Pt 195.1	79 Au 196.97	80 Hg 200.59	81 Tl 204.37	82 Pb 207.19	83 Bi 208.98	84 Po 210	85 At 210	86 Rn 222
87 Fr 223	88 Ra 226	89– Ac series†															
*Lanthanide series		57 La 138.91	58 Ce 140.12	59 Pr 140.91	60 Nd 144.24	61 Pm 147	62 Sm 150.35	63 Eu 151.96	64 Gd 157.25	65 Tb 158.92	66 Dy 162.50	67 Ho 164.93	68 Er 167.26	69 Tm 168.93	70 Yb 173.04	71 Lu 174.97	
†Actinide series		89 Ac 227	90 Th 232.04	91 Pa 231	92 U 238.03	93 Np 237	94 Pu 239	95 Am 241	96 Cm 242	97 Bk 249	98 Cf 252	99 Es 254	100 Fm 253	101 Md	102 No	103 Lw	

Figure 2–12 The periodic table. The elements are arranged horizontally according to the number of electrons in the outermost shell. Because these outer **valence electrons** determine many of the chemical properties of elements, the elements in each period (vertical column) tend to be chemically similar to one another.

small amounts of naturally occurring deuterium and tritium); carbon has an atomic weight of 12; oxygen has an atomic weight of 16; and chlorine has an atomic weight of about 35.5. Note that even though some of the elements have atomic weights that are not whole numbers, any particular atom will have a whole number of protons and a whole number of neutrons.

When atoms combine to form molecules, the resultant compound also has a characteristic weight, which is simply the sum of the atomic weights of its constituent atoms. This weight is called the **molecular weight** of the compound. The molecular weight of water (H_2O), for example, is approximately 18 (the sum of two hydrogens, about 1 each, and one oxygen, with an atomic weight of 16). Some of the complicated chemicals of life, the proteins and nucleic acids, have molecular weights in the thousands or even millions.

Ordinary hydrogen, oxygen-16, and the other common naturally occurring isotopes are stable—they do not change spontaneously to any other form, nor can they readily be changed. But some isotopes, such as tritium, carbon-16, and strontium-90 are unstable. Their nuclei can break down, emitting various subatomic particles and transforming the isotope into another one—possibly one of a different element altogether. Three main types of **radiations** may be emitted by such unstable isotopes: **alpha rays** (positively charged particles consisting of two protons and two neutrons); **beta rays** (electrons or particles of the same mass but opposite charge); and **gamma rays** (electromagnetic radiation with deep penetrating powers).

The particular series of disintegrations that a radioisotope (radioactive isotope) undergoes is specific for it alone; carbon-16, for example, emits a beta ray (or particle) and decays to nitrogen-16, which in turn loses a beta particle and is transformed into the stable form, oxygen-16. The pattern followed by the decay of all radioactive substances, however, is universal. In every case, half the mass of the radioisotope is converted to some other substance(s) within a specific period of time. This time is called the **half-life;** it is constant for any given isotope, regardless of its origin or amount, but each radioactive substance possesses its own characteristic half-life. Indeed, radioisotopes are commonly identified according to measurements of their half-lives. The half-life of carbon-16 is only $^3/_4$ of a second; that of uranium-235 is 713 million years.

The materials of the earth's crust contain a certain amount of naturally occurring radioactive materials. Radioisotopes are also formed through the action of cosmic and solar radiations on the matter of our atmosphere, land, and waters. But in the time since man achieved the capability of bringing about nuclear fission reactions, in which the nuclei of heavy atoms such as uranium isotopes are split into the nuclei of lighter substances, the level of radioactivity on earth has risen sharply. After explosions of nuclear weapons there are local and widespread rises in the radioactivity levels not only in the atmosphere, but also in the soil, plant cover, and animal products such as milk. This radioactive "fallout" poses a severe menace to the life of our planet. On the other hand, radioisotopes have also been put to many valuable uses in a variety of fields, from medicine and agriculture to industrial manufacturing.

Chemical Compounds

The salt you sprinkle on your food to improve its taste is made up of a viciously corrosive, poisonous gas (chlorine) and a soft, shiny metal (sodium) so reactive that it explodes on contact with water. Yet you need not regard your saltshaker with apprehension. For when atoms of sodium and chlorine combine, they form a stable salt, sodium chloride, which is totally unlike either of its constituent elements. It is a solid that crystallizes in the form of shiny white cubes, meekly dissolves in water, does nothing more drastic than melt when it is heated, is nonpoisonous when consumed in reasonable amounts, and indeed is absolutely essential to life.

Sodium chloride typifies a common class of inorganic compounds, the **salts.** These are

Figure 2–13 The radioactivity of earth's environment comes from varied sources.

electrically neutral combinations of a metal and a nonmetal. Many of them are readily soluble in water, and in it they can dissociate into positively and negatively charged ions. (When sodium chloride dissociates, the sodium ions have a positive and the chloride ions a negative electrical charge.)

Another common class of compounds is the **acids.** Typically, acids are sour to the taste, release hydrogen ions (hydrogen atoms that have lost an electron and thus acquired a positive charge) when they are dissolved in water, and turn litmus paper (paper impregnated with a colored compound sensitive to the presence of hydrogen ions) from blue to red. A common inorganic acid is hydrogen chloride, denoted as HCl in the chemist's shorthand.* This acid is normally a constituent of the gastric juice in the stomach that helps to digest food.

*The study of chemical reactions would be hopelessly cumbersome if full formal names had to be used for each compound. Instead, each of the elements is given a brief letter symbol, generally taken from its name. Hydrogen becomes H, oxygen O, carbon C, chlorine Cl, sodium Na (from the Latin name for the element, *natrium*), and so forth.

The **bases** are another major class of compounds. Inorganic bases are commonly bitter tasting, turn litmus paper from red to blue, and have a displaceable hydroxyl group (OH). Dissolved in water, such a base will dissociate, liberating one or more hydroxyl ions; for example:

$NaOH$	$\rightarrow$	Na^+	+	OH^-
sodium hydroxide		sodium ion		hydroxyl ion

Bases and acids can react with one another, to form a salt and water:

$$NaOH + HCl \rightarrow NaCl + H_2O$$

Interactions involving acids, bases, and salts form one of the most important and extensive classes of reactions in all of chemistry. In addition, the presence of hydrogen or hydroxyl ions and their relative concentrations are factors that often have a major effect on the reactions of other substances. Hydrogen-ion concentration is a key factor, in particular, in the reactions that take place in living organisms. For example, there is a certain "normal" concentration of hydrogen ions in the blood and tissues of man and other animals. Should this concentration decrease significantly, severe convulsions would set in, leading to death. On the other hand, a significant increase in the hydrogen-ion concentration in the blood may result in coma and, if prolonged, in death.

In view of the importance of the hydrogen-ion concentration in a variety of chemical situations, a convenient scale has been devised for expressing this quantity. It is called the **pH scale,** and it permits comparisons of hydrogen-ion concentrations in solutions within the range of greatest practical interest.

The pH of a solution is defined as the negative logarithm of the hydrogen-ion concentration in that solution—i.e., $pH = -\log [H^+]$, or $pH = \log 1/[H^+]$. It has been found experimentally that one liter (a little over a quart) of pure water contains 0.0000001 gram of hydrogen ions. The pH of pure water thus works out to 7. If more hydrogen ions are added, the solution becomes acid and the pH *decreases* (because the pH expresses an inverse relationship to the hydrogen-ion concentration). If hydroxyl ions are added, part of them react with the hydrogen ions present to form water, and the solution becomes basic. Then the pH increases. Human blood has a normal pH of 7.3 to 7.4, which is slightly on the alkaline (basic) side of the neutral value of 7.

Human blood shows great resistance to changes in its hydrogen-ion concentration. It is able to maintain such constancy of its pH because of the action of a special chemical system, called a **buffer system** (Fig. 2–14). A buffer system is able to adjust for an influx of either hydrogen or hydroxyl ions through an appropriate chemical reaction. In blood, for example, dissolved carbon dioxide (CO_2) reacts with water to form a weak acid, carbonic acid (H_2CO_3). This acid can dissociate to form bicarbonate ions (HCO_3^-). Both H_2CO_3 and HCO_3^- are present in the blood, and together they can take care of either hydrogen or hydroxyl ions:

$$H_2CO_3 + OH^- \rightarrow H_2O + HCO_3^-$$
$$H^+ + HCO_3^- \rightarrow H_2CO_3$$

If excess carbonic acid accumulates, the body has another mechanism to cope with that problem. The excess carbonic acid dissociates into carbon dioxide and water. The blood carries the excess carbon dioxide to the lungs, whence it is eliminated from the body by exhalation.

Most of the chemical compounds mentioned thus far belong to the class of **ionic compounds.** Their elements are held together by the electrical forces of an **ionic bond,** formed by a transfer of electrons from one atom to another. A sodium atom, for example, has a single electron in its outermost shell, which it can readily give to certain other atoms. After the sodium atom has given away its electron, it becomes a positively charged ion. The chlorine atom, on the other hand, has an outermost electron shell with room for an additional electron. It readily accepts electrons, to become an electrically negative chloride ion. Sodium and chlorine react explosively when they are brought together; sodium atoms give electrons to chlorine atoms, and then the ions are held together by the mutual attraction of their opposite charges (Fig. 2–15). Their combination, a salt, is extremely stable and shows no tendency for further chemical reactions under normal conditions.

Figure 2–14 The buffer system in the blood compensates for additions of hydrogen or hydroxyl ions. Excess OH^- reacts with H_2CO_3 to form HCO_3^-; excess H^+ reacts with HCO_3^- to form H_2CO_3. Through the dynamic equilibrium depicted a constant pH is maintained.

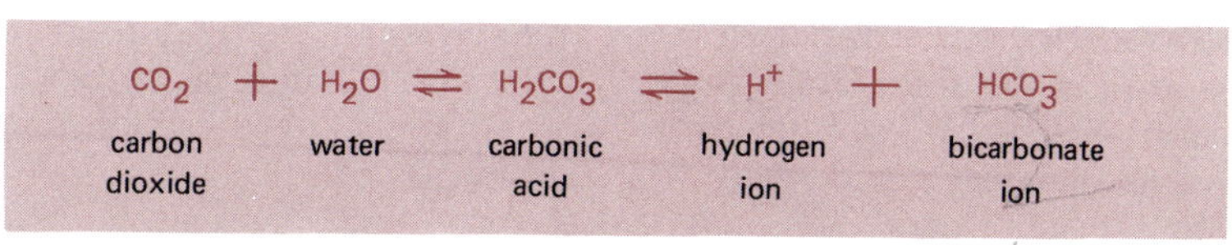

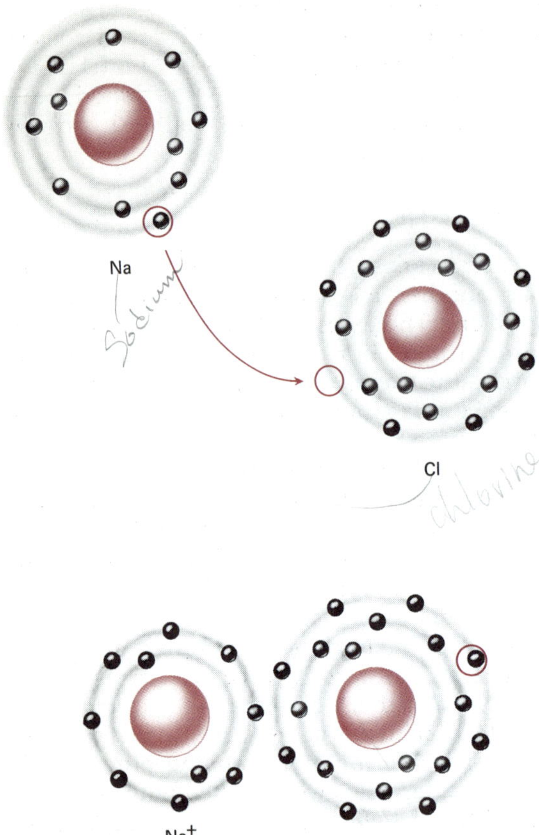

Figure 2–15 The ionic bond. A sodium atom donates a valence electron to a chlorine atom to form the ionic compound NaCl. The transferred electron travels exclusively around the chlorine nucleus and is indistinguishable from the other electrons in its outermost shell.

There is another major type of chemical bond, the **covalent bond.** This type is formed by a mutual sharing of electrons between two atoms. Neither atom completely donates or accepts an electron, and thus neither acquires a full electrical charge. The shared electrons travel in precisely defined paths about the two atoms, referred to as molecular orbitals (Fig. 2–16).

Energy may be absorbed or liberated when atoms combine to form molecules. The energy stored in chemical bonds can be readily converted to other forms of energy and made to

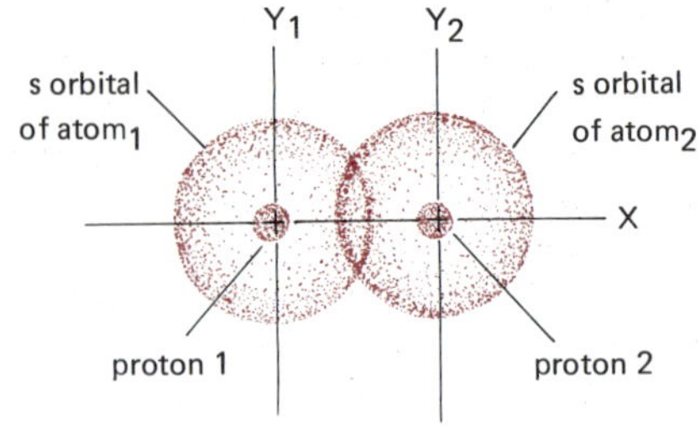

Figure 2–16 A convalent bond: the hydrogen molecule. The two electrons (one originally derived from each atom) are held in common and both travel around each nucleus in a molecular orbital.

do useful work. Flashlight batteries are a common example of stored chemical energy. In biological systems, energy is commonly stored in such molecules as sugars, starches, and ATP (adenosine triphosphate).

The carbon atom, with four electrons in its outermost electron shell, has a strong tendency to share electrons, rather than giving or receiving them. The compound carbon dioxide is a covalent compound, formed by sharing of electrons between a carbon atom and two oxygen atoms (Fig. 2–17).

Carbon atoms have such a strong tendency to form bonds by sharing electrons that they can link up with other carbon atoms to form chains and rings. Combinations of these carbon chains and rings with certain other elements, especially hydrogen, oxygen, and nitrogen, produce an incredible variety of different compounds. These carbon-containing compounds are so unique and numerous that scientists class them in a separate group, **organic** compounds (as opposed to **inorganic** compounds, which do not contain carbon). The number of all inorganic compounds is in the thousands, yet it is estimated that some 6 *million* organic compounds are already known to science, and more are being discovered or synthesized in laboratories each day.

The reason these compounds are termed *organic* is that many of them were first isolated from the tissues of body products of living organisms. These carbon compounds are the chemicals of life.

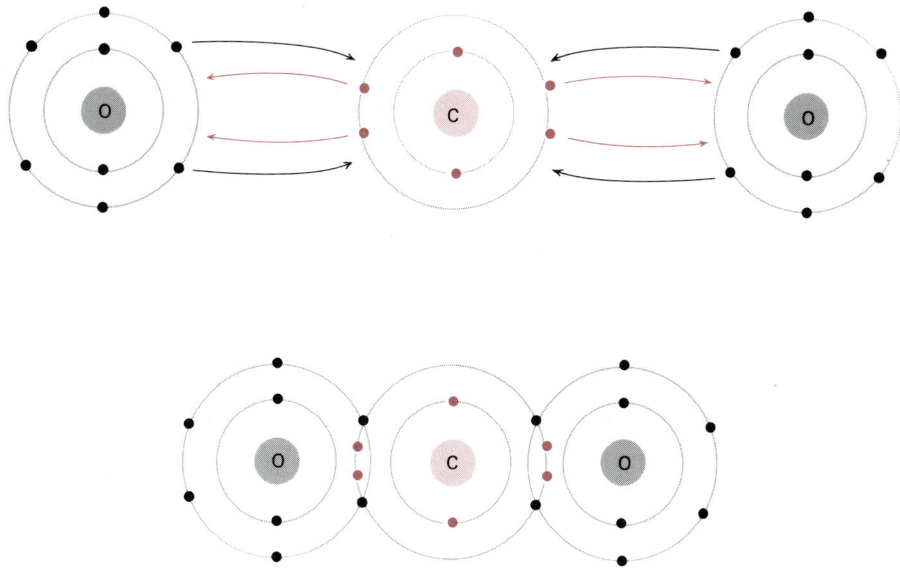

(a)

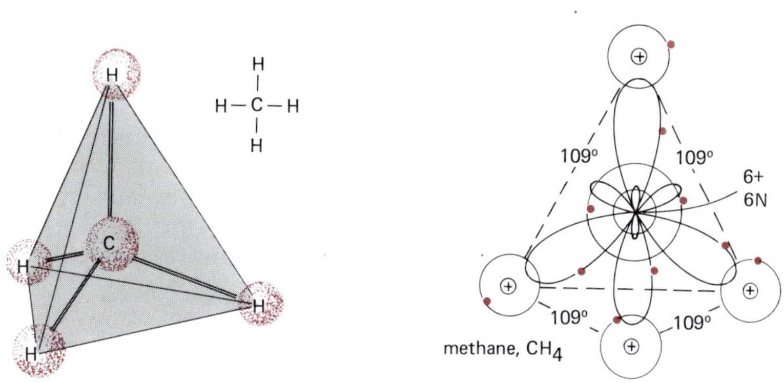

(b)

Figure 2–17 Carbon atoms form covalent bonds. (a) In carbon dioxide, carbon shares two of its valence electrons with each oxygen atom, while the oxygen atoms, in turn, each share two electrons with carbon. Thus, each atom effectively has a completed shell of eight electrons. (b) Carbon's unique chemical properties are determined to a large extent by its ability to combine its outermost electron orbitals into special hybrid orbitals. Instead of the normal one *s* orbital and three *p* orbitals, the four valence electrons of carbon may be found in four identical sp^3 orbitals, directed toward the vertices of a tetrahedron. Each of these may form a covalent bond with another atom. (In the compound methane depicted, carbon is thus bonded to four hydrogen atoms.) sp^2 and sp combinations of three and two electron orbitals are also possible. With these variations, carbon is capable of forming a virtually infinite variety of rings, chains, and branched structures—the chemicals of life.

Chemical Reactions

Picture two fruit flies, fluttering about in a bottle with a layer of mashed bananas on the bottom. Will they meet? And if they meet, will they perhaps mate and give rise to a new generation of fruit flies?

The chemical compounds discussed in this chapter are formed by chemical reactions, in which molecules and atoms join or split apart, exchange portions, or are changed in some other manner. Like two fruit flies with the potential for mating, two molecules or atoms must be brought together before they can react. Like mating fruit flies, molecules and atoms can react only if they are suitably shaped and have sufficient energy. But unlike fruit flies, whose union produces fruit fly eggs but leaves the mating partners essentially unchanged, chemical reactants can give rise to a variety of products and often are themselves unrecognizably changed in the process. The nature of the products of chemical reactions is strictly determined by chemical laws arising from the nature of the reactants—the numbers and arrangements of their electrons, the three-dimensional geometry of the atoms, and energy relationships within and among the reacting molecules.

The ceaseless random kinetic motions of atoms and molecules provide a means of bringing them together in collisions that might lead to reactions. Movements of atoms and thus opportunities for such fruitful collisions are limited in closely packed solids. Gases, on the other hand, are often so diffuse that far-ranging atoms or molecules may meet only rarely. (Two fruit flies in a large bottle might flutter about for some time before meeting. If fifty fruit flies were placed in the same bottle, encounters among them would be much more frequent.) In liquids, such as the fluid media of living organisms, opportunities for collisions leading to chemical reactions are maximized. When a solid surface is present in contact with the liquid—particularly if this solid is capable of holding solute molecules by physical or chemical forces—it may enhance the potentialities for reaction. (In the fruit fly analogy, the mashed bananas at the bottom of the bottle might attract both flies and thus help to bring them together.) This is the basic principle of industrial **catalysts,** which are used to help various chemical substances react more rapidly and effectively while themselves remaining unchanged. It is also a key factor in the chemical reactions that take place in living organisms: such reactions frequently take place not in the actual fluid but at the surface of solid membranes, bathed by the fluid medium.

Chemical reactions are influenced by a variety of factors, including the nature of the reactants (two male fruit flies could not mate!), their concentration, and the temperature of the medium. The higher the temperature, the more rapid the kinetic motions of molecules and atoms. These particles not only have greater opportunities for collision when they are moving more rapidly, but their greater kinetic energy provides the energy needed for forming and breaking chemical bonds. In general, raising the temperature increases the rate of chemical reactions, up to a certain optimum point, beyond which the reactants themselves are altered or destroyed. Reactions in chemical laboratories often take place at temperatures of hundreds or even thousands of degrees; reactions in living organisms usually proceed optimally at about 37°C. Much beyond that, the intricate organic catalysts called **enzymes** are inactivated and then destroyed.

The enormous variety of chemical reactions that take place in the nonliving world and among living organisms can be classified in a few basic types:

$A + B \rightarrow AB$	**(combination or addition reactions)**
$AB \rightarrow A + B$	**(decomposition reactions)**
$A + BC \rightarrow AC + B$	**(single displacement reactions)**
$AB + CD \rightarrow AD + CB$	**(double displacement reactions)**
$A \rightarrow A'$	**(isomerization reactions)**

The combination of sodium and chlorine to form the salt sodium chloride is a combination reaction ($2Na + Cl_2 \rightarrow 2NaCl$). If an electric current is passed through water, water molecules decompose to their constituents, hydrogen and oxygen ($2H_2O \rightarrow 2H_2 + O_2$). If a bit of sodium metal is dropped into water, it will react explosively in a single displacement reaction ($2Na + 2HOH \rightarrow 2NaOH + H_2$). In **hydrolysis** reactions, common both in the nonliving and in the living world, water enters into a double displacement reaction with another molecule ($AB + HOH \rightarrow AOH + HB$). In isomerization reactions, the atoms of a molecule are rearranged into a new grouping, often with quite different physical and chemical properties. The sugar molecule, glucose ($C_6H_{12}O_6$), for example, can be converted to nearly a dozen different **isomers,** each containing exactly the same numbers of carbon, hydrogen, and oxygen atoms in slightly different arrangements. (One of the isomers is not even a sugar.)

In many of these reactions, there is an exchange of electrons between the reacting molecules or atoms. The reactant that loses an electron (or electrons) is said to be **oxidized,** while the reactant that gains an electron (or electrons) is said to be **reduced.** Such reactions are referred to as **oxidation-reduction reactions.** Oxygen is a frequent participant in oxidation-reduction reactions. (It oxidizes the other reactant and is itself reduced.) The rusting of iron is an oxidation-reduction reaction in which iron oxide is formed from the combination of iron and oxygen. The rancidity of oils is caused by oxidation-reduction reactions in which oxygen is added to unsaturated bonds between carbon atoms (C=C). Often hydrogen is added to the unsaturated bonds of vegetable oils to protect them from oxidation. Such hydrogenated oils can be kept longer without spoiling.

The countless molecules in living organisms are engaged in an endless flow of chemical reactions, combining, recombining, and interacting in an astonishing variety of ways. These reactions are basic to our every action and even thought. In recent years, scientists have begun to gain a growing knowledge of the interactions among the chemicals of life.

QUESTIONS FOR REVIEW AND THOUGHT

1. Distinguish between matter and energy.
2. Give examples of interconversions among the various states of matter. Among different forms of energy. Between matter and energy.
3. Distinguish between the terms *atom* and *element.*
4. Is it probable that other planets or other stars contain elements that do not exist on earth?
5. If people, rocks, and other common forms of matter are really more than 99.99999% empty space, then why aren't they invisible?
6. How do isotopes of an element differ from one another? How does one determine that they are all forms of the same element?
7. Explain why the half-life of a radioactive isotope is an important factor in determining its relative danger to human health.
8. Reconcile the phenomena of random motions of molecules and their apparent directed movement in diffusion down a concentration gradient (i.e., from regions of high concentration to regions of low concentration).
9. What is the difference between a solution and a suspension?
10. If time is a factor in waiting for a pot of water to come to a boil, would it be more efficient to add salt to the water before turning on the gas or to wait until the water is already boiling? Explain.

CHAPTER 3

The Chemistry of Life

Introduction

In September 1965 Charles C. Price, then president of the American Chemical Society, proposed a national goal: the synthesis of life in the laboratory. Calling for a coordinated national program of basic research on the cell and its components, Dr. Price stated:

> We have been making fantastic strides in uncovering the basic chemistry of the life process and the structure of many of the key components of living systems.
>
> It seems to me we may be no further today from at least partial synthesis of living systems than we were in the 1920's from the release of nuclear energy—or in the 1940's from a man-in-space. The political, social, biological, and economic consequences of such a breakthrough would dwarf those of either atomic energy or the space program. Success could lead to modified plants and algae for synthesis of food, fibers, and antibiotics, to improved growth or properties of plants and animals, or even to improved characteristics for man himself.

According to some estimates, the human body may contain over a hundred thousand different chemical compounds. The exact chemical structures of the vast majority of these compounds is still unknown. But every day new discoveries are bringing us closer to the time when the intricate processes of the body will be clarified. This knowledge may some day be used to alter these processes, hopefully for the betterment of mankind.

In this chapter we shall discuss the chemical components of **protoplasm,** the complex mixture which comprises all living cells. Protoplasm consists of a multitude of substances including water, salts, carbohydrates, lipids, steroids, proteins, nucleic acids, hormones, and vitamins. Although the proportions and the exact nature of these chemical components vary among cells from different organisms (and even among cells from the same organism), they are essential components of all protoplasm.

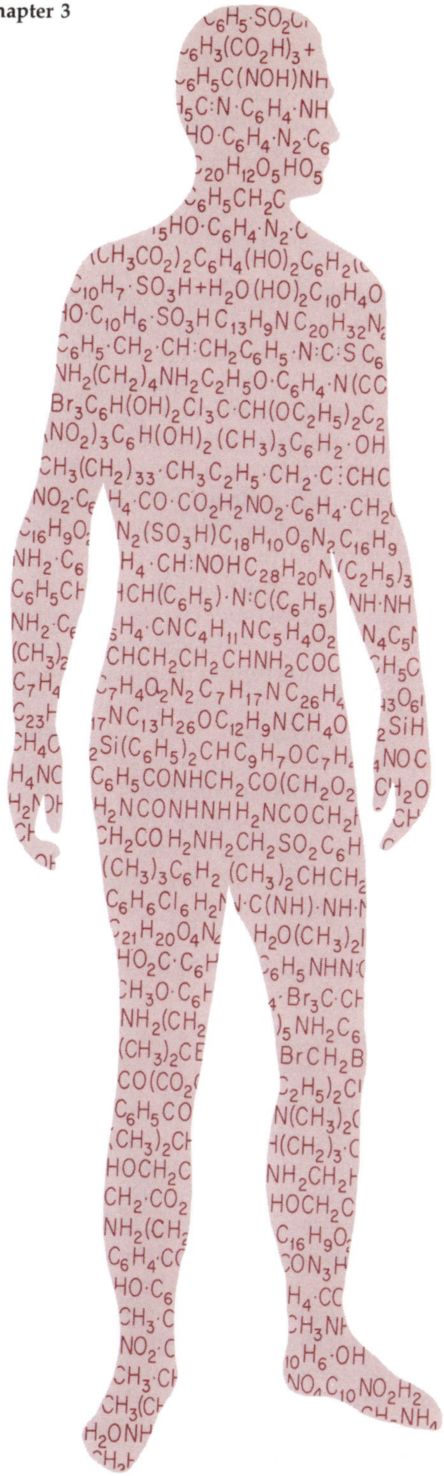

Figure 3–1

Inorganic Constituents

Although the vast majority of the chemical constituents of protoplasm contain carbon, a number of inorganic substances are also of vital importance. No living organism could exist without water, salts, acids, bases, and gases.

WATER

Tissues in the human body vary greatly in water content, from 5% in teeth to more than 90% in blood plasma. The human body as a whole consists of about 70% by weight of water; the bodies of some organisms contain an even larger proportion of water.

Water serves many vital functions in living organisms. It takes part in numerous reactions that occur in protoplasm, either as a reactant in or product of metabolic activities. Perhaps even more important is the function of water as a solvent for other chemical substances. Most substances will not react unless they are in a liquid medium. Some will react only in an ionic state, a state that may be brought about by the solvating properties of water.

The fluidity of water, along with its great solvent powers, permits it (as a major constituent of blood and lymph) to act as a vehicle for the transportation of food, gases, waste products, and metabolites throughout the body.

Water also absorbs much of the heat produced as a by-product of cell metabolism. It thus effectively protects the delicate tissues from being damaged while itself undergoing no drastic changes in temperature. Water is characterized by an unusually high **heat capacity,** defined as the amount of heat required to raise a substance's temperature by 1°C. Water also participates in another major cooling mechanism of the body: sweating. Excess body heat is used to change water from the liquid to the vapor state; when it evaporates from the sweat pores, the heat is carried away.

Water also serves as a lubricant in various organisms and even as a supportive medium (in the human eye and in the roots, stems, and leaves of nonwoody plants).

SALTS, ACIDS, AND BASES

In the cells of man and the other vertebrates (animals with backbones), there is a remarkable similarity in the types of salts present and their relative distribution. Some scientists have speculated that the salt content of the ancient seas, in which the first life forms presumably appeared, closely resembled the salt content of present-day vertebrates' blood. Despite the fact that today's sea water contains about 4% salt, whereas the blood of vertebrates contains less than 1%, there are still marked similarities. But there are a few significant differences: the relative potassium-ion concentration in sea water is lower than that found in vertebrate blood, and sea water contains a greater relative concentration of chloride and magnesium ions.

Despite their relatively low concentrations in living organisms, salts play a vital role in many of their normal metabolic activities. Iron in the hemoglobin molecule makes possible the transport of oxygen in red blood cells; muscle contraction and nerve impulse transmission could not occur without a flow of calcium, sodium, and potassium ions; many enzyme systems are activated by the presence of tiny amounts of magnesium, zinc, and other metal ions. Calcium salts largely make up the shells of snails, clams, and other mollusks; human bones are also composed mainly of calcium salts.

Figure 3–2 A skin diver gains first-hand experience in the properties of gases and liquids.

Small amounts of acids and bases are also found in living organisms—for example, in the digestive juices of the stomach.

GASES

With skindiving growing in popularity, more and more people must learn to beware of the "bends." If a diver rises to the surface too rapidly, there is a sudden decrease in pressure within the body. Since the solubility of a gas in a liquid usually decreases with decreasing pressure, the nitrogen normally dissolved in the blood becomes less soluble as the pressure on the diver's body drops. As a result, air bubbles are formed which may block small veins and arteries and lead to a painful death.

Nitrogen gas is dissolved in essentially all protoplasm. In most species it is inert, but some soil bacteria can bind atmospheric nitrogen into compounds which plants and animals can utilize.

Of more importance to most organisms is the oxygen dissolved in protoplasm, essential in the energy-releasing reactions of the cell. Carbon dioxide is a by-product of these reactions, and most of it is eventually eliminated from the cell and the organism. Green plants utilize carbon dioxide in a process called photosynthesis, converting it to organic materials and forming oxygen as a by-product.

Organic Constituents

About 1½ million different species of plants and animals are known to man. Some scientists suggest that each of these species has at least one chemical compound which is unique to that species. And yet, even 1½ million compounds is only a small fraction of the total number of compounds that can conceivably exist. Most of these are organic, since the peculiar chemical nature of the carbon atom allows for an almost infinite number of combinations with itself and other elements.

Carbohydrates, lipids, proteins, and nucleic acids are major groups of organic compounds that are found in virtually all living cells.

CARBOHYDRATES

A **carbohydrate** may be defined as a substance possessing carbon, hydrogen, and oxygen in the ratio $C_yH_{2x}O_x$. That is, a carbohydrate molecule has twice as many hydrogen atoms as oxygen atoms. Frequently the number of carbon atoms equals the number of oxygen atoms, but, as the formula shows, this is not a necessary condition.

The group of carbohydrates includes the simple sugars or **monosaccharides,** such as **glucose** or blood sugar ($C_6H_{12}O_6$); **oligosaccharides** (more complex sugars), in which two or several simple sugar molecules are chemically linked together (e.g., as in **sucrose,** table sugar); and **polysaccharides,** consisting of many chemically linked sugar units (hundreds or even thousands). Common examples of polysaccharides are **starch, glycogen** (animal starch), and **cellulose.**

Sugars are the basic food materials in living organisms. A mere 5 grams (about a hundredth of a pound) of glucose in a person's bloodstream provides immediate nourishment to the trillions of cells in his body. If this amount should fall by little more than a single gram, convulsions, coma, and even death may follow. Another simple sugar, **ribose** (with five carbon atoms in its molecules), forms an integral part of the RNA (ribonucleic acid) molecule, a key component of cell functioning.

Both plants and animals link sugar molecules together into more readily stored forms, starch and glycogen. Plant starches—those of wheat, rice, corn, and potatoes, for example—are mixtures of the polysaccharide **amyloses** and **amylopectins** (Fig. 3–3). Amyloses contain between three hundred and two thousand glucose units linked together in a single long chain. In amylopectins, up to five thousand glucose units are arranged in a branching network.

Animals cannot utilize the large plant starch molecules as such. They must first be broken down into glucose units by special systems of enzymes. Part of the glucose is used in energy-liberating reactions in the body. The rest is transported by the blood to the liver, where it is converted to the animal starch, glycogen. When the blood sugar level drops, some of the stored glycogen is broken down into glucose again and used for the body's energy needs.

The most abundant organic compound on the earth is the carbohydrate **cellulose,** the major structural substance in plant cells. Its molecules resemble amylose in that they are constructed of a single long chain of glucose units. But the starch-digesting enzyme systems of most animals cannot break down cellulose. A cow would starve to death in a meadow full of grass if it were not for the presence of special cellulose-digesting bacteria in her rumen (first stomach). Termites, too, are able to feed on the cellulose of wood because their intestines contain colonies of protozoa whose enzymes can digest this carbohydrate.

It is estimated that the second most abundant organic chemical is another polysaccharide, **chitin.** This material is found in the integument (tough outer covering) of insects, spiders, crabs, and lobsters.

LIPIDS

The rounded contours of the human body are due in part to deposits of adipose ("fatty") tissue beneath the skin. This adipose tissue is filled with globules of those fat and fatlike substances that are grouped under a broad category known as **lipids.** In addition to fats, the fatty acids, phospholipids, steroids, fat-soluble vitamins (A, D, E, and K), and many other substances are classified as lipids.

Chemically, fats belong to the group of esters. They are formed in the chemical combination of the alcohol glycerol (this alcohol has three separate —OH groups) and a fatty acid (*R*—COOH, where *R* is a long hydrocarbon chain) (Fig. 3–4).

Today physicians and nutritionists are hotly debating the role of saturated fats in heart disease and the advisability of replacing the saturated fats in our diet by unsaturated fats. Unsaturated fats contain double bonds (chemical bonds in which carbon atoms share two pairs of electrons instead of one pair),

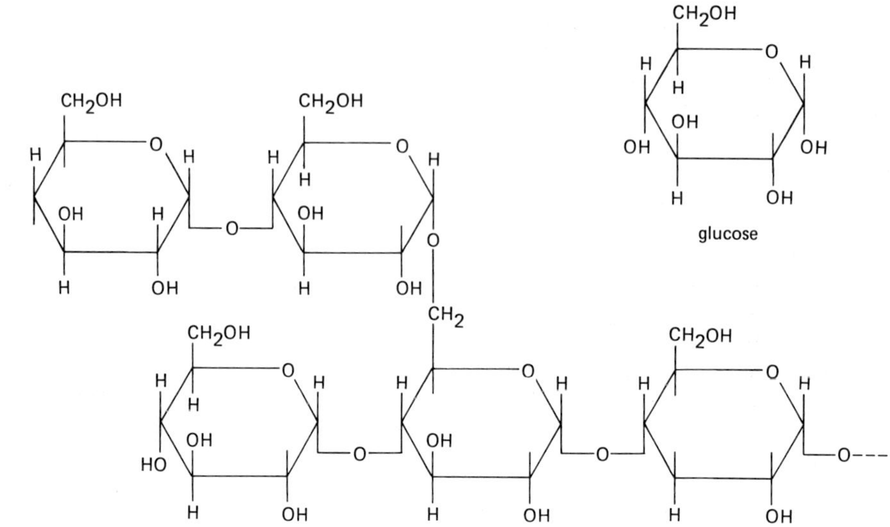

glucose

amylopectin

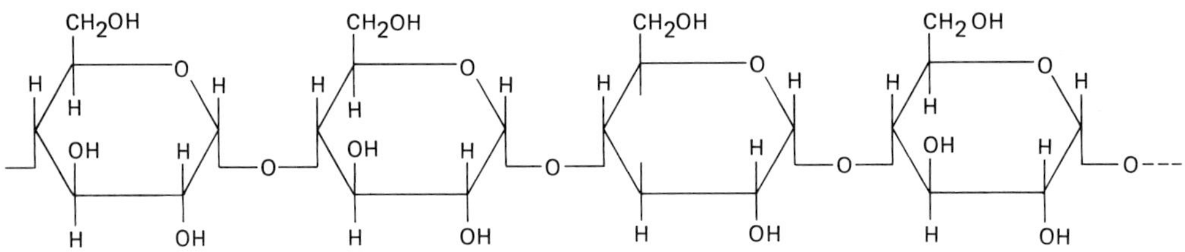

amylose

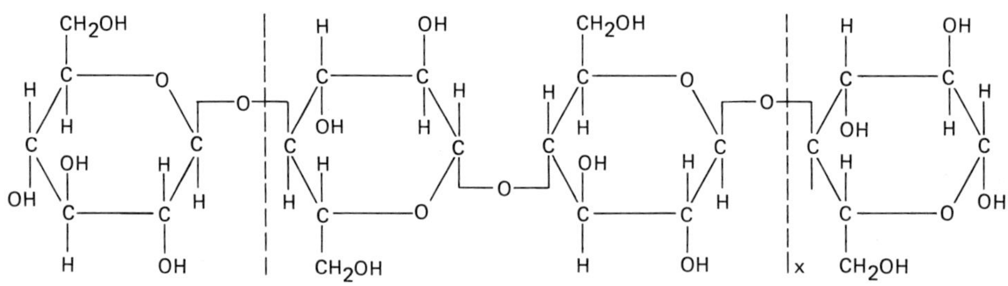

cellulose

Figure 3–3 Both starch and cellulose are polymers consisting of long chains of glucose units, linked together. The constituents of starch, amylose and amylopectin differ in the arrangement of the basic glucose units, in straight or branched chains.

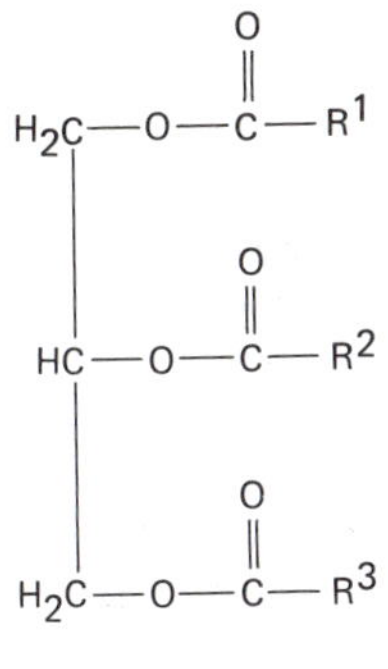

fat

Figure 3–4 Chemical structure of a fat. R^1, R^2, and R^3 represent long-chain hydrocarbons. They may differ, or two or more of them may be the same. (Tristearin, a fat found in beef tallow, contains three stearic acid residues combined with glycerin.)

which are extremely reactive chemically. They are readily oxidized, combining with the oxygen of the air and causing the fat to "spoil" or turn rancid. Thus, commercial processing methods often add hydrogen to the double bonds of natural unsaturated oils such as soybean oil, converting them to saturated forms. This treatment greatly increases their shelf life, but may reduce their nutritive value if the saturated-fat theories of heart disease are correct.

One of the most important of the numerous functions of fats is to supply a reserve of energy for the organism. One gram of fat supplies more than twice as many calories as roughly the same amount of carbohydrate or protein and thus provides a more economical means for energy storage. Carbohydrates such as glycogen can be converted to glucose much more rapidly and readily, however, and so better serve the immediate energy needs of the organism. Fats are used more for long-term energy storage, and thus are particularly useful for hibernating animals. Woodchucks and other hibernators store up fat in their bodies during the summer, when food is abundant, and then live on their stored fat during the winter. Fats are also a major form of stored energy in plant seeds.

Deposits of fat beneath the skin also serve as excellent insulators against both cold weather and sudden shocks or blows that might otherwise injure vital internal organs. Among humans, women have a more concentrated distribution of fat than men (particularly in the hips and breasts). The blubber of a whale is sometimes more than a foot thick and may weigh many tons.

Fatty substances are also important constituents of the cell membranes and of the myelin sheaths that encase many of the nerves of the body. Other fatty substances are found in muscle and brain tissue.

PROTEINS

Proteins are found in every cell, in every living organism. They serve almost every conceivable type of function, from support (in cell membranes, skin, and bones) to coordination (as hormones and basic units of neuron structure), sight (as the basic structural unit of

Figure 3–5 Old-time whalers "cutting the blanket" peel off huge strips of blubber and hoist them aboard the ship.

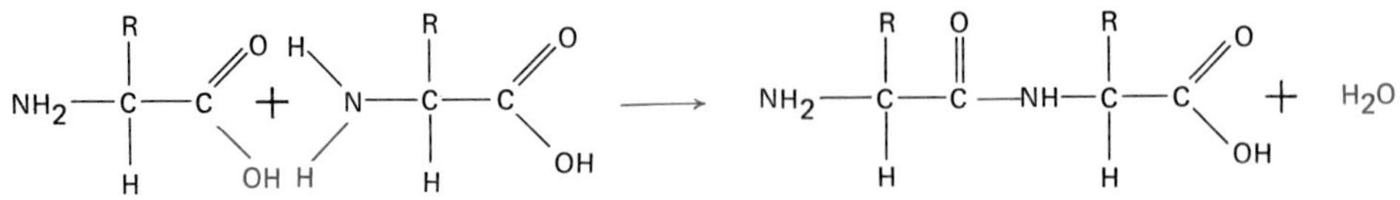

Figure 3–6 Amino acids have two reactive groups, the carboxyl group (—COOH) and the amino group ($—NH_2$). A reaction of the carboxyl group of one amino acid molecule with the amino group of another can link them together into a peptide. Proteins are long chains of amino acid units linked together by peptide bonds.

the crystalline lens), locomotion (in muscle contraction), and oxygen and carbon dioxide transport (as hemoglobin).

One of the most important and far-reaching protein functions is that of acting as **catalysts** —chemicals that help other chemicals to react. Special proteins known as **enzymes** act as catalysts in controlling the rates of thousands of chemical reactions which occur in living systems. The number of enzymes in the human body is estimated to be in the tens of thousands. If just a few key enzymes ceased to function, the result could be devastating. Cyanide, for example, stops the functioning of a key enzyme, cytochrome oxidase, and results in instant death.

Like polysaccharides, proteins are **polymers**—large molecules comprised of long chains of smaller units (Fig. 3–6). In the case of proteins, these units are **amino acids** (NH_2 —*R*—COOH). The size of some proteins is staggering. Hemocyanins, oxygen-carrying pigments found in some invertebrates (animals without backbones), have molecular weights measured in the millions. Hemoglobin, the red pigment found in human red blood cells, is a relatively small protein, with a molecular weight of only about 67,000. (Compare these figures with the molecular weight of water: 18.)

If all the millions of different proteins found in nature were placed in one huge pot and carefully heated in a dilute acid solution, they would all be broken down to their constituent amino acids. But from all these millions of different proteins, significant amounts of only about twenty amino acids would be recovered. In other words, these twenty amino acids form the basic building blocks of essentially all of the natural proteins on earth.

How could millions of different proteins be formed from only twenty basic units? The answer lies in the specific orders in which the amino acid units are arranged in the protein chain. The situation is quite analogous to the twenty-six-letter English alphabet, which can be used to spell out a virtually infinite number of words.

It has been calculated that for a rather small protein with a molecular weight of 34,000 (about half the size of hemoglobin), containing 288 amino acids of twelve different types, there are about 10^{300} possible variants.* If all these variants existed and we were able to collect just one molecule of each of them, their total mass would come to about 10^{280} grams. (By comparison, the entire mass of the sun, the earth, and the rest of the planets in our solar system is only about 10^{33} grams! In fact, the mass of the entire universe is estimated to be only 10^{55} grams!)

In a protein, individual amino acids are joined to one another by means of chemical bonds known as **peptide bonds** $\left(-\overset{\overset{\displaystyle O}{\|}}{C}-\overset{\overset{\displaystyle H}{|}}{N}-\right)$. Depending on its length, a chain of amino acids linked together in this way is referred to as a peptide, a polypeptide, or a protein. Just when a polypeptide becomes a protein is rather nebulous. Insulin, with its fifty-one amino acids, represents just about the lower limit of a protein; by some definitions, its molecular weight of about 5700 would make it too small to qualify.

The nature of proteins and the properties they manifest are determined by three types of structure. The nature of the constituent amino acids and their sequence along the protein chain (or chains) constitute the protein's **primary structure.**

X-ray diffraction studies have established

*When written out, 10^{300} is a one followed by three hundred zeroes.

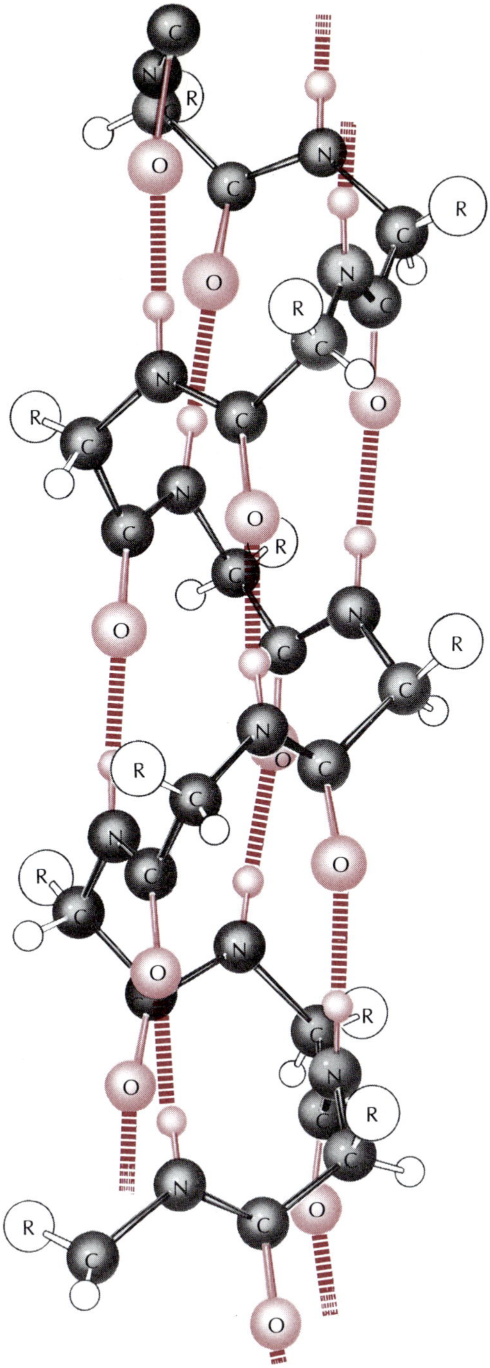

Figure 3–7 The α - helix of proteins. Hydrogen bonds link >N—H groups to neighboring >C=O groups to maintain the coiled structure. The hydrogen bonds are not as strong as the peptide bonds that hold together the amino acids in sequence along the chain.

that the long chains of amino acids in a protein are often coiled into a **helix** or spiral (Fig. 3–7). This helical arrangement is referred to as the **secondary structure** of the protein.

In addition, many proteins are folded upon themselves in some unique manner. This is the **tertiary structure,** one of the most crucial aspects of the protein. This folding gives the protein its unique three-dimensional configuration, which doubtless is responsible for most of the widely divergent characteristics of proteins as a group. These unique configurations often give rise to "active sites" which, in enzymes, are the areas most responsible for the specific activity of the enzyme.

Green plants, as well as some microorganisms, can manufacture all of the twenty naturally occurring amino acids. But in higher animals a number of these amino acids cannot be synthesized by the body, or they are synthesized in inadequate amounts. These are called *essential* amino acids. This group varies somewhat from species to species. In man, eight of the twenty amino acids are essential and must be consumed in food.

With the incredible complexity of proteins, it came as a great surprise to many scientists when, in 1954, Frederick Sanger announced his discovery of the complete amino sequence of the protein insulin (Fig. 3–8). (Insulin is a hormone secreted by the pancreas and plays a role in carbohydrate metabolism. A deficiency of this protein may lead to a disease called diabetes mellitus.) This feat, which eventually won Sanger a Nobel Prize, required more than ten years of intensive work.

The "protein barrier" was pierced. In the years that followed, scientists in other laboratories gradually delineated the amino acid sequences of other, larger proteins, including the complete sequence of an antibody (a disease-fighting protein of the blood).

A more recent important milestone in the study of protein structure came with the announcement by John C. Kendrew of the complete amino acid sequence *and* the three-dimensional configuration of myoglobin (153 amino acids, molecular weight ca. 17,000) (Fig. 3–9). Previously only the amino acid sequence of proteins (i.e., their primary structure) had been elucidated. Now, for the first

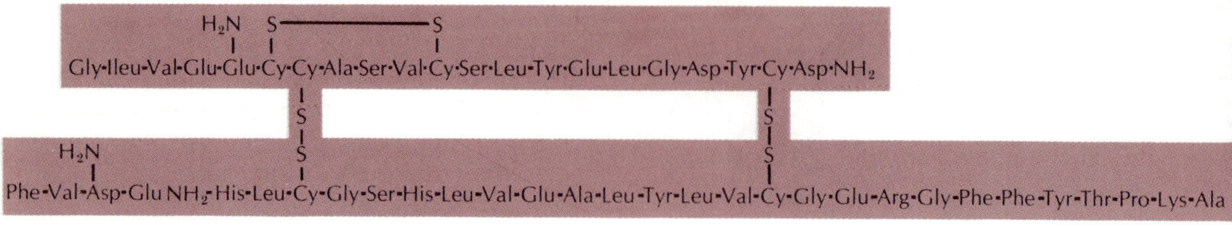

Figure 3–8 The amino acid sequence of the protein insulin. The two chains are held together by disulfide (—S—S—) bonds. (The sequence is printed in linear fashion for clarity; the actual molecule is curled and folded into an intricate and characteristic three-dimensional structure.)

time, the entire structural picture of a protein had been determined. Together with M. F. Perutz (who determined the three-dimensional structure of hemoglobin), Kendrew was awarded the Nobel Prize in chemistry in 1962. In the years that followed, the techniques of these researchers and others were used to elucidate the three-dimensional structure of a number of other proteins.

Another milestone in protein chemistry was the complete synthesis of the insulin molecule by a team of chemists on mainland China. Other syntheses followed, but if we consider (as some scientists do) the lower limit of protein size to be a molecular weight of 10,000, the first complete protein synthesis was not accomplished until 1969. In that year, two separate teams of American researchers, using two different methods, announced the synthesis of the 124–amino-acid enzyme ribonuclease. (This enzyme catalyzes the breakdown of the nucleic acid RNA.) One team at Rockefeller University, headed by Robert B. Merrifield, used an automated process in which the first amino acid of the chain was attached to a tiny bead of polystyrene plastic and the following amino acids were added one by one. The other team, working in the research laboratories of Merck, Sharp, and Dohme, synthesized fragments of the RNase molecule, containing from six to seventeen amino acids each, and then put them together. In each case, forming of the proper three-dimensional structure proved surprisingly simple: when the synthetic RNase was placed in solution, it promptly folded itself up into the appropriate globular form and exhibited the characteristic activity of natural RNase.

The encouraging progress in recent years in elucidating the structure of proteins and in synthesizing them in the laboratory offers great promise for the ultimate unraveling of the secrets of life.

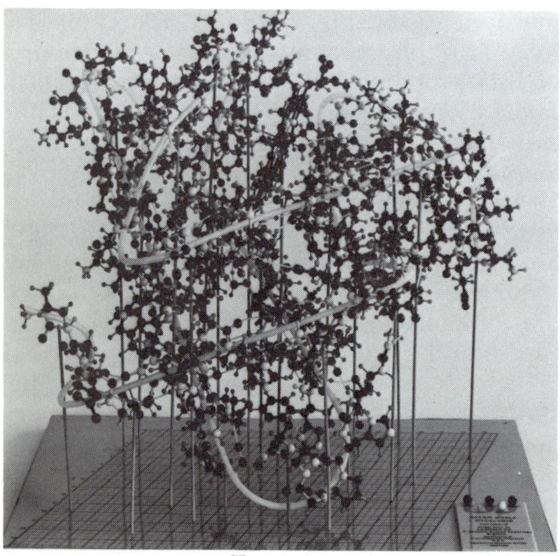

Figure 3–9 A model of the three-dimensional structure of myoglobin.

NUCLEIC ACIDS

Within living things there are "magic chemicals" potentially capable of making any living thing in this world: a dazzling variety of flowers, towering trees of huge dimensions, microscopic bacteria, and man himself. These chemicals are thus indirectly responsible for the manufacture of the thousands of ships that ply the seas, the rockets that circle the

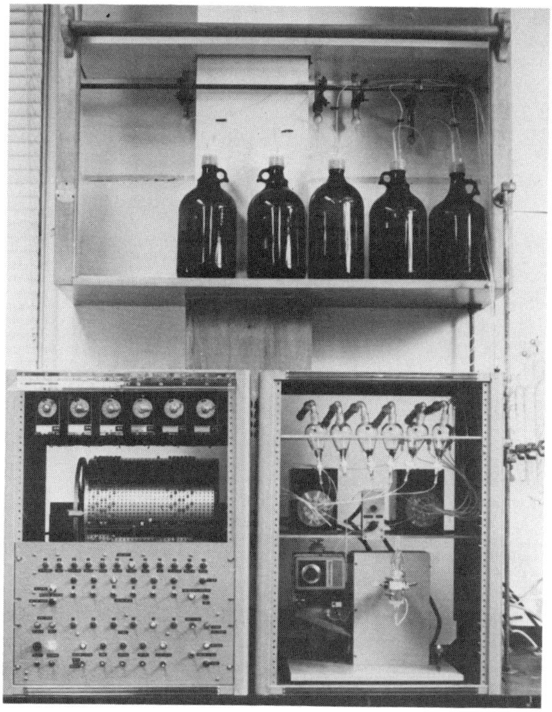

Figure 3–10 The Merrifield group used this setup in the synthesis of ribonuclease A.

earth, and the never-ending stream of new gadgets and devices that pours out of the laboratories and factories of this world. These "magic chemicals" are **nucleic acids.**

Nucleic acids are the essence of the genes, the units of hereditary information which determine and shape our lives. They are at the same time immensely complex and beautifully simple structures—complex because of their huge size, simple insofar as the fantastic variety of life forms on earth is due essentially to variations in a nucleic acid "alphabet" of only four "letters." The information of heredity is stored in the **DNA** molecule. (The molecular weight of this gigantic molecule may run into the billions!) In addition, the cells of organisms contain several varieties of a second type of nucleic acid known as **RNA,** the molecular weight of which ranges from the thousands to the tens of millions.

The current view of the structure of the nucleic acids was first formulated in 1953 by James Watson and Francis Crick. Their proposal, based on indirect experimental data, created an instant sensation in the scientific community. Their model was at first viewed with skepticism, but has since been found to check beautifully with all the experimental findings (including recent high-resolution electron micrographs of the DNA molecule itself) and has been of great value in gaining an understanding of the workings of nucleic acid molecules. Watson and Crick shared the 1962 Nobel Prize in physiology and medicine with Maurice Wilkins for their landmark achievement. The events leading up to the formulation of the Watson-Crick model have been vividly described by Dr. Watson in his informal autobiography, *The Double Helix.*

In several ways, nucleic acids resemble proteins. Both belong to the chemical class of polymers, compounds consisting of long chains of repetitive units. In the case of nucleic acids, each unit is called a **nucleotide** and consists of a nitrogen-containing organic base (either a **purine** or a **pyrimidine**), a five-carbon sugar **(pentose),** and a **phosphate** group (Fig. 3–11). Nucleic acid molecules may contain hundreds, thousands, or even millions of nucleotides, and hence they are called polynucleotides.

A typical nucleic acid contains a backbone of alternating sugar and phosphate groups (Fig. 3–11). The purine and pyrimidine bases are attached to the pentose sugar. According to the Watson-Crick model, the nucleic acid chains, like proteins, are coiled into a spiral-shaped helix.

The full names of RNA and DNA, the two basic types of nucleic acids, are descriptive of the contents of these compounds. The first, ribose nucleic acid (now more commonly referred to as **ribonucleic acid**) contains the pentose **ribose,** while **deoxyribonucleic acid** contains the pentose **deoxyribose.**

There are two other important structural differences between DNA and RNA. One difference lies in the purine and pyrimidine bases they contain. Both DNA and RNA contain the two purines **adenine** (A) and **guanine** (G). But of the pyrimidines, whereas both DNA and RNA contain **cytosine** (C), only DNA contains **thymine** (T) and only RNA contains **uracil** (U). It is the nature of the bases

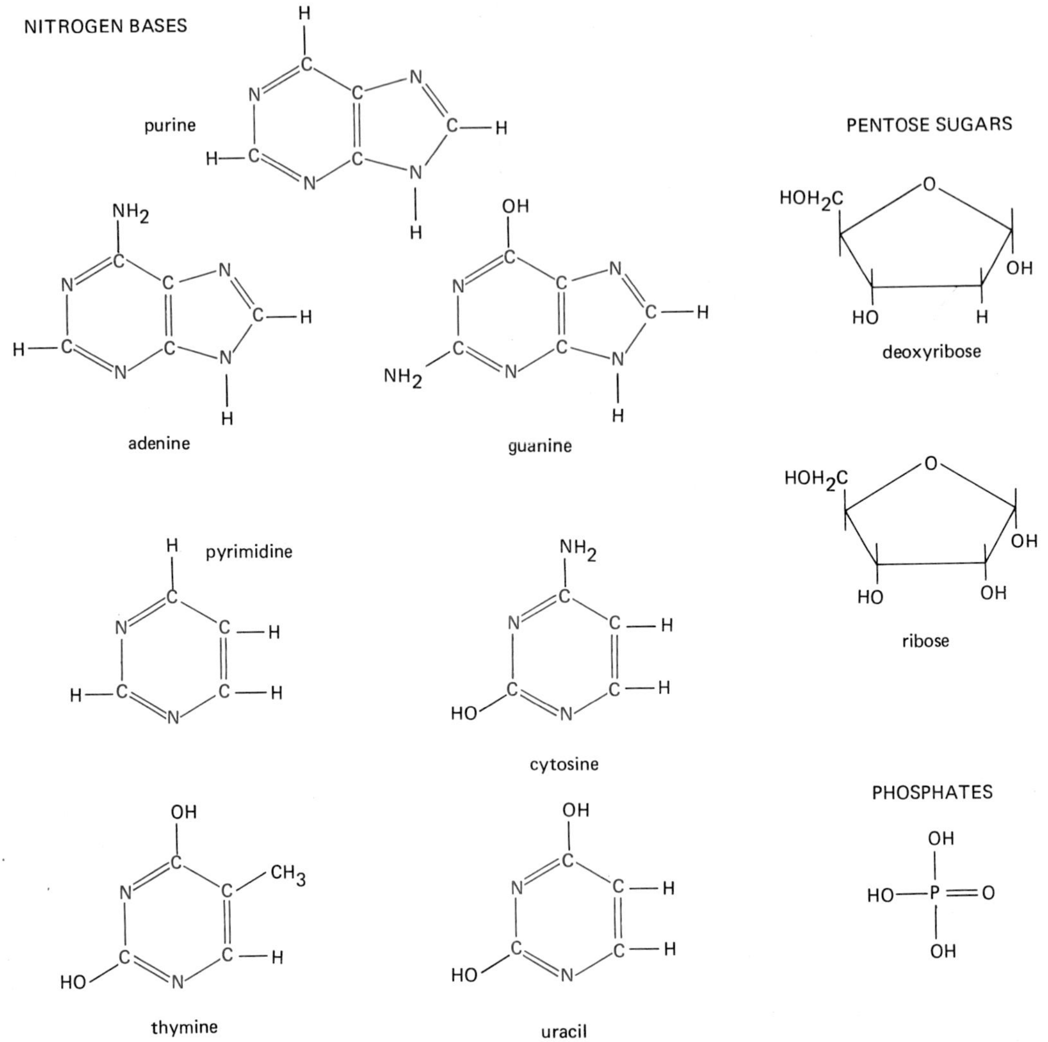

Figure 3–11 The constituents of nucleic acids: nitrogen bases (purines and pyrimidines), pentose sugars (deoxyribose and ribose), and the phosphate radical.

and their arrangement in the molecule that give the different forms of nucleic acids their unique properties. Sequences comprised of just four kinds of bases spell out the difference between a whale and a mouse, a tree and a man.

The other important structural difference between DNA and RNA lies in the nature of the coiled chains. The "double helix" that provided the title for Watson's book is the DNA molecule. It consists of two chains, twisted upon one another to form a structure that resembles a spiral staircase (Fig. 3–12). The two chains of the double helix are held together by chemical bonds (hydrogen bonds) between pyrimidines of one strand and purines of the other (and vice versa). This bonding is quite specific: a guanine is always

Figure 3–12 The Watson-Crick model of the DNA double helix. In the schematic drawing below, the base pairs are indicated by crossbars. The blowup above shows the pentose-phosphate skeletons of the two helices and the linkages between base pairs. The dotted lines denote hydrogen bonds. (The molecular diagram depicts the molecule unwound and flattened to show the structural details.) The inset photo shows a three-dimensional model of a portion of the DNA double helix.

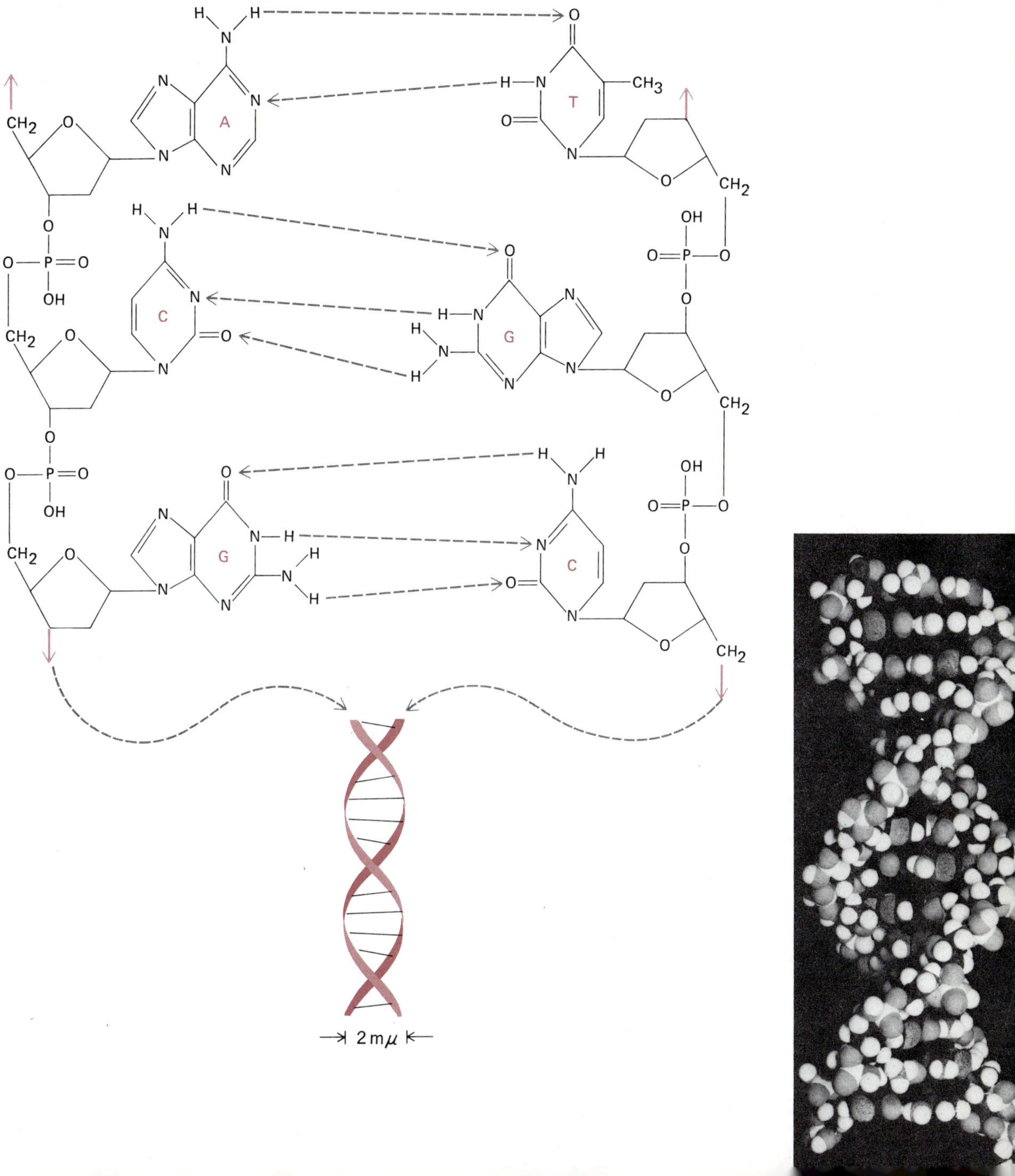

linked with a cytosine (G—C) and a thymine with an adenine (T—A). The strands may thus be looked upon as complementary. For example, a sequence ATTGCTG in one strand of DNA would be accompanied by TAACGAC in the opposite strand.

The purines and pyrimidines of RNA are capable of forming similar linkages between chains, but usually they do not. Instead, RNA occurs in the form of a single chain, a single helix similar to the helix of protein structure.

DNA is the ultimate source of control in the cell and, hence, in the organism. As might be guessed from its name, it is found in the **nucleus** of the cell. Here, as an integral part of the chromosomes, it is associated with special types of proteins in compounds called **nucleoproteins.**

DNA may be likened to a foreman directing the operation of a large shop, such as a bakery, in which a number of different activities are taking place. Each individual worker performing his own unique task might be compared to RNA and its functions in the cell. The final products completed by the individual workers (bread, rolls, pastry, etc.) would then be analogous to the proteins produced by these RNA molecules.

There are several types of RNA in a cell. **Messenger RNA,** which appears to be manufactured directly by the DNA molecules in the nucleus, passes out into the cytoplasm, carrying portions of the DNA "master plan." Another type of highly mobile RNA originating in the nucleus and passing out into the cytoplasm is **transfer RNA (tRNA).** There are a number of types of the latter, all much smaller than the messenger RNA molecules. Each type can combine with a specific amino acid. Messenger RNA, in conjunction with transfer RNA molecules, functions in the synthesis of the cell proteins, from amino acid constituents, in special cytoplasmic structures called **ribosomes.** These structures contain still a third type of RNA, **ribosomal RNA.**

For a time biochemists were beset by feelings of frustration, similar to those that plagued the protein chemists before the insulin breakthrough, over the prospects of determining the detailed structure of RNA. Then, in 1965, after about nine years of intensive research, Robert W. Holley announced the complete nucleotide sequence (seventy-seven nucleotides) of a tRNA (specific for the amino acid alanine). Intensive efforts were then launched to determine the three-dimensional structures of several forms of transfer RNA. In 1973 an MIT team announced an L-shaped double helix structure for a yeast tRNA, determined by X-ray crystallography.

Miscellaneous Constituents

No complete account of the chemical substances found in the body could be given in this text, if for no other reason than that we do not now know the exact chemical makeup of the body. We have already briefly discussed some of the major groups of chemicals found in living systems: inorganic salts, carbohydrates, proteins, nucleic acids, and so on. Here we shall briefly mention some substances that do not fall readily into any of these categories.

Hormones are a group of substances that are produced by specialized cells, secreted into the bloodstream, and subsequently carried to regions of the body where they exert specific and often profound effects. Some hormones are proteins; others, such as the sex hormones, belong to a special class of lipids called steroids (Fig. 3–13).

Vitamins play a vital role in a number of enzyme systems. They belong to a variety of chemical classes; some, such as ascorbic acid (vitamin C), are relatively simple organic mol-

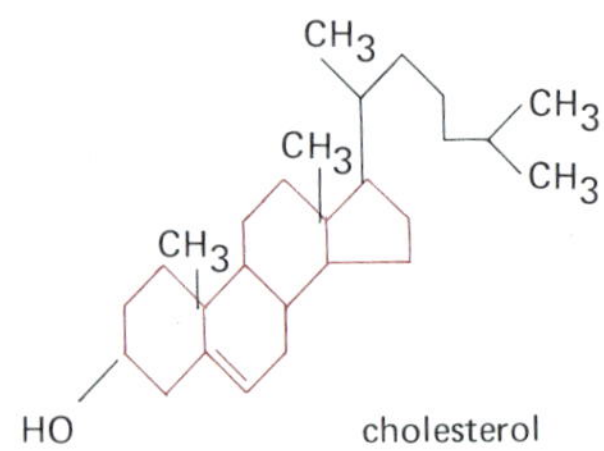

Figure 3–13 A typical steroid: cholesterol. Other steroids share the same basic condensed ring structure (color) and differ in substituents.

ecules. Although green plants can synthesize all of the vitamins, most animals and many protists can synthesize only a limited number of vitamins and must obtain the others from an external source, such as food.

Two of the most important chemicals in living systems, hemoglobin and chlorophyll, are strikingly similar chemically. Each substance contains an organic structure, called a **porphyrin** group, linked with a metal. In the case of chlorophyll, the metal is magnesium; for hemoglobin it is iron. Similar metal-porphyrin groups are found in other biologically important substances, including enzymes involved in the reactions of cell respiration.

Thus is completed our thumbnail sketch of the major groups of chemicals that constitute virtually every living thing on earth. We can look forward to the day when we may have the opportunity of examining material from other worlds for possible evidence of alien life. Whether these life forms will share the same basic chemical constituents is a question that only time can resolve.

Metabolism

The living cell is a seething cauldron of chemical interactions in which a fantastic variety of chemicals are organized into complex and highly specific patterns of interaction coordinated in a delicate balance.

The sum of all the chemical processes that go on within the living organism is known as **metabolism.** Many of these reactions result in a breakdown or degradation of complex compounds into simpler components, with a release of stored chemical energy. Such reactions are collectively termed **catabolism.** Other reactions that occur within the organism lead to a buildup of complex compounds from simpler substances and frequently require energy to drive them. These reactions are collectively called **anabolism.** It is safe to say that within every living organism, both catabolic and anabolic processes take place simultaneously and constantly. In fact, the components of the living organism are in a continuous state of flux, though the metabolic rates of the various tissues vary greatly: that of the brain, for example, is extremely high, whereas that of the bone is relatively low.

The driving force for anabolic and even for many catabolic reactions has its origin in the solar energy captured by photosynthetic plants and protists (Fig. 3–14). This energy is stored in various organic materials in plants, which may then be consumed by animals. As these organic compounds of plant origin are digested by the animals, much of the energy that they contain is liberated and stored in various organic materials in plants, compounds, subsequently to be used as the driving force for other metabolic reactions.

Various modern analytical laboratory techniques, including the use of radioactive tracers and chromatography, have greatly facilitated (and indeed in many instances have made possible) the isolation of various intermediate products, thereby elucidating the course of various complex metabolic processes. These studies have revealed a striking similarity in many basic metabolic processes in virtually *all* cells—animal, plant, and protist. Of course, there are significant differences

Figure 3–14 The flow of energy on planet earth.

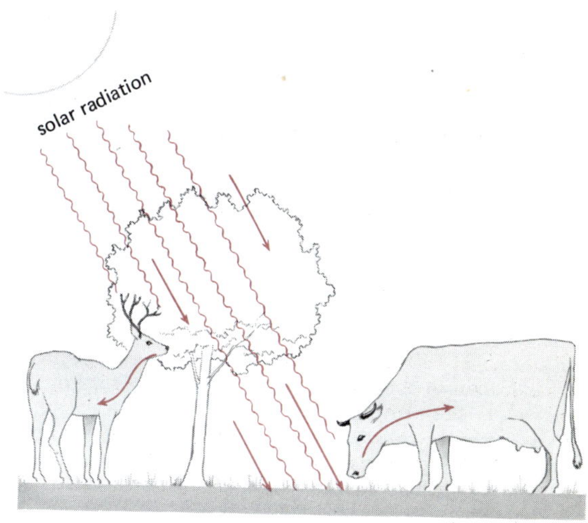

in the metabolism of different groups of organisms, even among different though closely related species (and to a lesser extent even among different individuals of the same species). But the basic similarities of metabolism are so striking that studies of such distantly related organisms as the bread mold *Neurospora crassa* and the intestinal bacillus *Escherichia coli* have cast much light on the metabolism of man himself.

It has been estimated that literally tens of thousands of chemical reactions take place within the organism. These reactions follow the usual rules of organic chemistry. But the organic reactions that take place in living things are distinguished by several main factors. With few exceptions, they take place within a rather narrow temperature range of 0 to 40°C and at normal pressures; often they involve extremely minute amounts of reactants, so small as to be below the sensitivity of the usual chemical tests; and the reactions that take place in living things generally are not discrete entities, but are linked into chains of reactions, each of which depends upon products and/or energy supplied by previous reactions in the chain, supplies starting materials for subsequent reactions, and may be influenced by their products.*

There is a final distinguishing feature: most of the metabolic reactions could not possibly proceed under the conditions prevailing in living organisms if it were not for the action of highly specialized organic catalysts known as enzymes.

*Scientists refer to this interlinking of different chemical reactions as a **feedback mechanism.** There are two main kinds of feedback: positive and negative (Fig. 3–15). In **negative feedback,** the accumulation of the product of a reaction (or series of reactions) causes the original reaction to slow down and ultimately stop. This sort of self-regulating system is the basis for the operation of many of the interrelated hormone systems of the body. In **positive feedback,** on the other hand, an increase in the concentration of a reaction product signals an increase in the output of the original reaction, in turn causing the concentration of the product to increase even further. If unchecked, the positive-feedback mechanism leads to a runaway process called a **vicious cycle.** Some diseases and other pathological conditions are examples of runaway positive feedback, which may ultimately lead to death of the organism if the cycle is not broken.

Enzymes

A solution of pure hydrogen peroxide, if permitted to remain in the dark, will very slowly decompose, forming water and gaseous oxygen. If a pinch of iron is added, the rate will pick up somewhat. An industrial catalyst such as finely divided platinum will speed up the decomposition even more. But a tiny quantity of the enzyme catalase will result in such a rapid reaction that a froth of bubbles of escaping oxygen will quickly form. This reaction proceeds at such a furious rate because a single molecule of catalase can catalyze the decomposition of as many as 5 million molecules of hydrogen peroxide each minute at the temperature of ice water (0°).

Virtually every movement we make, every thought that passes through our minds—indeed, essentially every process, conscious or unconscious, that goes on within our bodies—is controlled or mediated by **enzymes.** Without these ubiquitous catalysts, life as we know it could not exist.

Yet despite their critical importance to living systems, scientists have barely scratched the surface in their efforts to learn about the ultimate structures and modes of activity of enzymes.

All the enzymes that have thus far been isolated have been found to be proteins or proteins in a complex with some smaller organic compound (frequently a vitamin) and/or a metal ion. The protein portion of an enzyme complex is referred to as the **apoenzyme,** and the smaller components are commonly termed **cofactors.** Neither component separately exhibits any significant enzyme activity.

The substance on which an enzyme acts is known as its **substrate.** Most enzymes are exceedingly specific—that is, they will act only on a single substrate or group of closely related substrates; in fact, the names of enzymes are generally formed by combining the name of the substrate on which they act and the suffix *-ase*. For example, sucr*ase* is an enzyme that breaks down the sugar sucrose. (An enzyme may also be named for the kind of reaction it catalyzes: a dehydrogen*ase*, for example, is active in dehydrogenation reactions, in which hydrogen is removed.)

Enzymes are fantastically efficient catalysts —far more efficient than any catalysts that have been synthesized in the laboratory or in chemical industry. But like other catalysts, the enzymes themselves are unchanged by the reaction they influence—they are reusable.

Like other chemical reactions, enzyme-catalyzed reactions in living organisms are modified by a variety of factors. An increase in the temperature, for example, tends to speed up the reaction. (But only to a certain point. If the medium gets too hot, the enzyme will be **denatured**—its structure will be irreversibly changed—and it will lose its activity.) The concentrations of the enzyme and its cofactors, and of the substrate and product(s), will also have an effect, as will the pH and the presence of possible inhibitors of the reaction.

How do enzymes work so efficiently and so selectively? The theory most widely accepted by biologists today had its beginnings well before James Sumner succeeded in purifying and crystallizing an enzyme (urease) for the first time, in 1926, and long before the chemical nature of enzymes was even known. Back in 1913, the German scientist Leonor Michaelis, studying with his associate M. L. Menten the rates of enzyme-catalyzed reactions, proposed that the enzyme combines temporarily with the substrate to form an enzyme-substrate complex, which then decomposes to yield the products and regenerate the free enzyme. Michaelis and Menten derived an equation relating the rate of an enzyme-catalyzed reaction to the concentration of the substrate, and as more and more enzyme reactions were studied their equation was found to describe these reactions quite well. Meanwhile, more was learned about the nature of enzymes themselves, and the modern **"lock-and-key theory"** was gradually evolved.

Although an individual enzyme molecule is microscopic, in comparison with the typical substrate molecule it is gigantic. Catalase, for example, has a molecular weight of about 250,000, whereas its substrate, hydrogen peroxide, has a molecular weight of only 34. The protein framework of the enzyme molecule is generally coiled, twisted, and folded into a highly intricate and characteristic three-dimensional structure. The surface geometry of the molecule, far from being smooth and featureless, is covered with numerous odd-shaped bumps and hollows, knobs, hooks, and so on. The variety of the surface is not just geometrical; chemical functional groups may be found in the surface, providing unique and specific reaction sites.

Each reaction site, with its unique geometry and chemical functional groups, may be exactly complementary to a particular substrate molecule, which fits into it like a key into a lock (Fig. 3–15). If an enzyme happens to collide with a substrate that fits its reaction site, the two can combine to form an enzyme-substrate complex. However, it is believed that the fit in such a complex is not quite perfect and that there is consequently a slight strain on the bonds of the substrate. The bonds are thus weakened at the points of strain and break much more readily than in the absence of the enzyme. When the weakened bonds of the substrate break and the substrate is subsequently released from the complex, the products of the catalytic reaction are formed, and the enzyme is free to react with another substrate molecule. A single substrate molecule may be transformed into a product or products by its reaction with an enzyme. Or two substrate molecules may combine with adjacent reaction sites of the enzyme and thus in juxtaposition react much more readily than could occur in the cell medium, where random collisions between them, in just the right orientation, would be rare events.

The lock-and-key theory thus neatly explains the unique features of enzyme-catalyzed reactions: the amazing specificity of enzymes, their reusability, the fact that they catalyze reactions in either direction, and the tremendous increase in reaction rate that they produce, often making possible reactions that could not proceed at all without them or could theoretically proceed only at temperatures so great as to destroy the reactants.

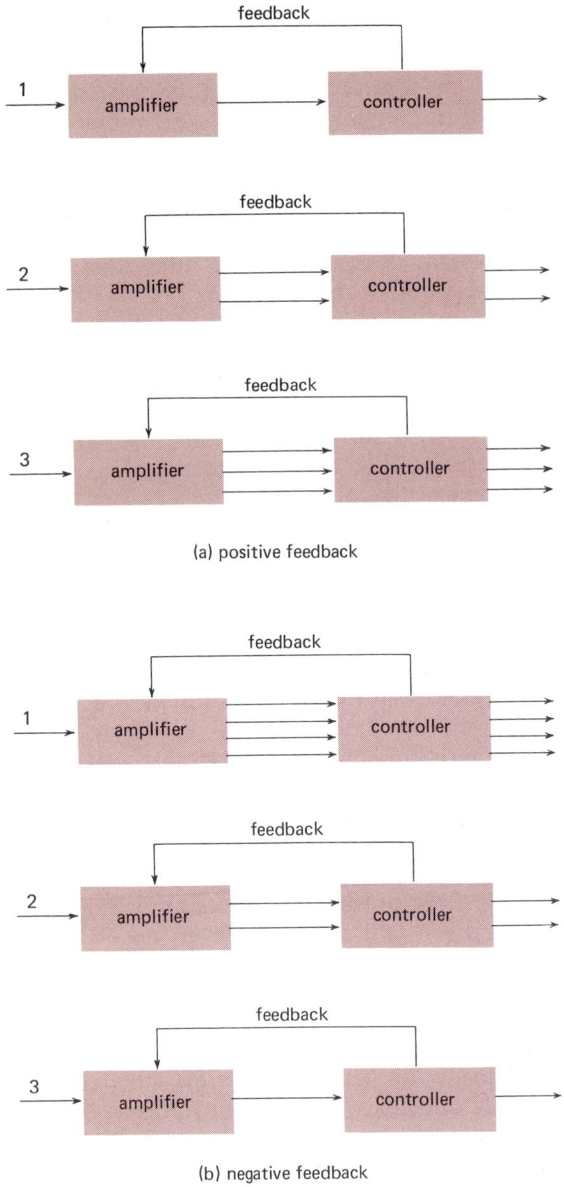

Figure 3–15 Processes in living organisms are controlled by positive and negative feedback interrelationships. In positive feedback (a) the reaction product enhances the reaction and thus stimulates further accumulation of the product. In negative feedback (b) the reaction product has an inhibiting effect on the reaction; as it is accumulated, the reaction tends to come to a halt, thus imposing a self-limiting regulation.

Figure 3–16 (*below*) The lock and key model of enzyme action. The molecular configuration of the substrate exactly fits a corresponding active site on the enzyme molecule. The enzyme and substrate combine reversibly, and the reactivity of the substrate is enhanced (or in the case of two substrates, they are brought into a suitable mutual orientation for reaction; note that the enzyme-catalyzed reaction depicted can proceed in both directions).

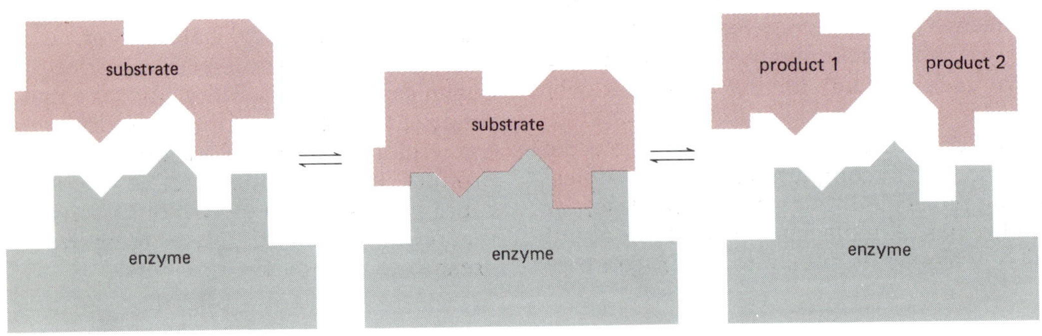

Studies of enzyme action have usually concentrated on a specific reaction. But it should be remembered that in the seething potpourri of the protoplasm, enzyme-catalyzed reactions are frequently linked into reaction chains, with energy-yielding reactions supplying the driving force for energy-requiring reactions and the products of one reaction governing the rate of another through feedback mechanisms. This great complexity and interrelationship, often a trial to the research biochemist, is a vital factor in the exceedingly fine control exerted by the cell over its continuous, many-faceted metabolic "juggling act."

QUESTIONS FOR REVIEW AND THOUGHT

1. Discuss the merits and shortcomings of Dr. Price's proposal for a "crash program" for the synthesis of life.
2. Discuss the various functions of water in the human body.
3. Differentiate among the following: carbohydrate, monosaccharide, disaccharide, oligosaccharide, polysaccharide. Give examples.
4. Give the functions of some lipids found in living systems.
5. In what ways does the structure of proteins resemble that of polysaccharides?
6. Why are nucleic acids considered by some to be the most important compounds in living organisms?
7. Briefly describe the DNA molecules as envisioned by Watson and Crick. Compare with protein structure.
8. How does the DNA molecule differ from the RNA molecule?
9. How does an enzyme differ from an inorganic catalyst?
10. Describe the workings of positive- and negative-feedback mechanisms.

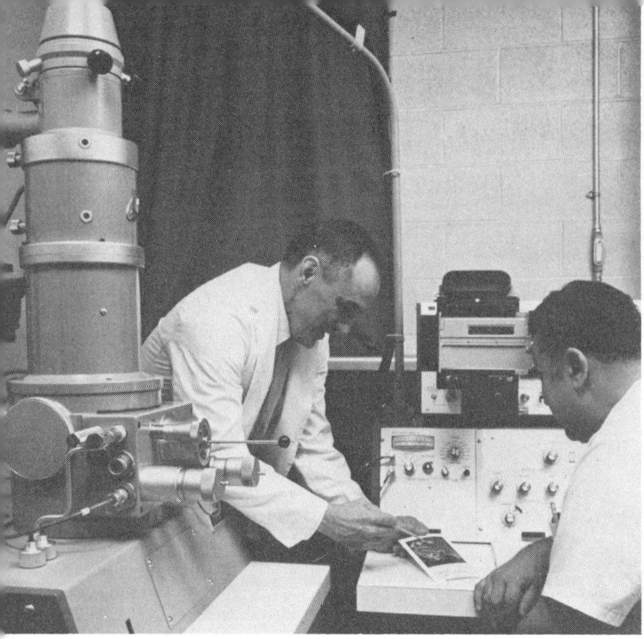

(a)

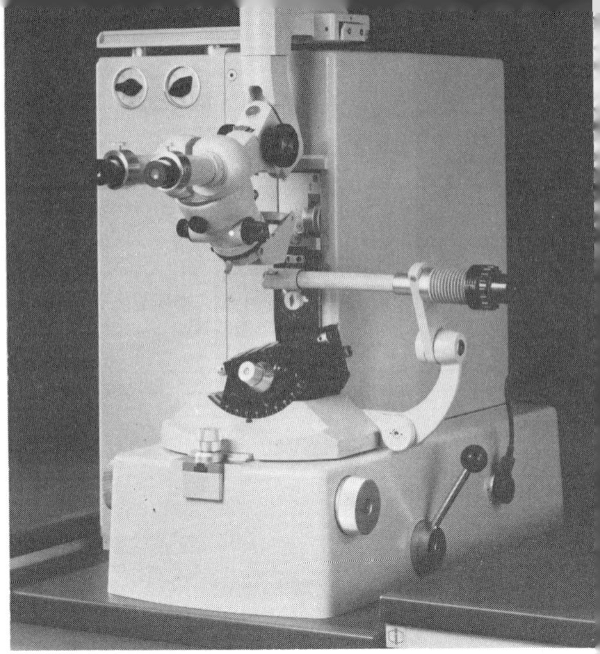

(b)

(c)

CARL ZEISS
ZEISS

(d)

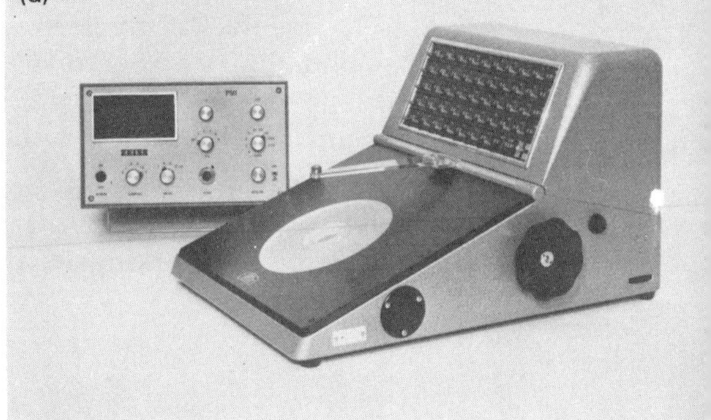

(e)

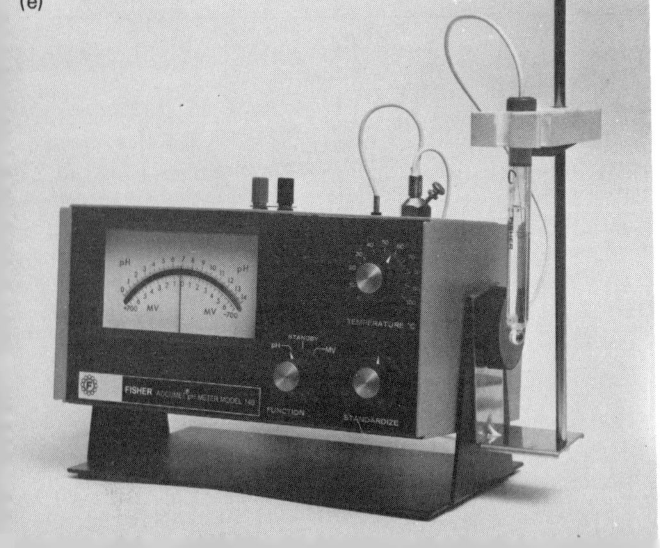
pH
MV
TEMPERATURE °C
STANDBY
FISHER
FUNCTION
STANDARDIZE

(f)

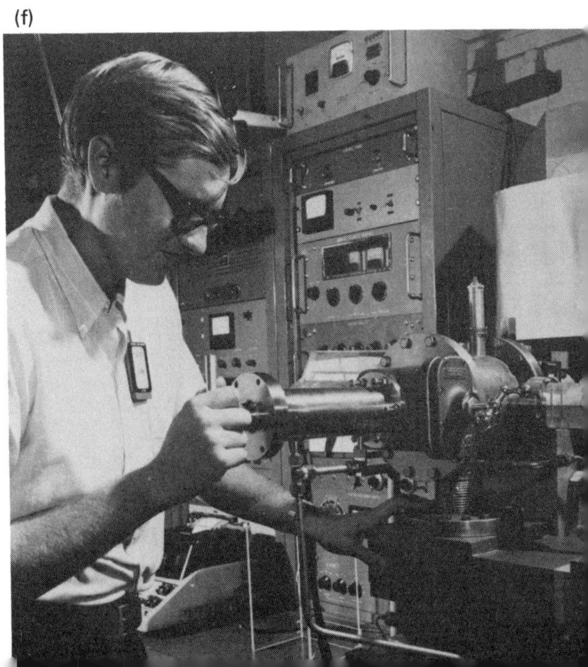

CHAPTER 4

Tools and Techniques of Scientific Research

Introduction

The accomplishments of science in our society are diverse, complex, and sometimes spectacular. For the most part these accomplishments have been a direct result of the creative activity of the human mind. And yet, in many instances, even the greatest of human geniuses would have failed in bringing forth these new discoveries were it not for tools—extensions of the human mind. Perhaps instrumentation has had the greatest impact. For without microscopes to peer into the unseen world of the ultrasmall, man would doubtless be ignorant of what it holds—perhaps even of its very existence. Without telescopes to peer out at the unseen macrocosm beyond, we would doubtless be unaware of the immensity of the universe and the incredible number of stars that populate it. Without various analytical tools, we would be ignorant of the chemical composition of our world and our own bodies and of the physical principles of their interactions. But with these analytical tools, man may now rationally hope to unravel the chemical and physical mysteries of his world and of his body. The rewards already abound in the fruits of modern technology, and even greater rewards are promised for the future.

In the following pages we shall briefly discuss some of the tools used by scientists in their study of the living world. Some of these tools, such as the optical microscope, have been used by man for centuries. Others, such as chromatography and electrophoresis, have been concertedly used for less than three decades. Indeed, each year old techniques and instruments are modified and wholly new techniques and instruments are devised.

Although space hardly permits us to run through the entire gamut of tools and techniques used by biologists, we can nevertheless discuss briefly a few of the more noteworthy ones. In Chapter 21 we shall discuss some additional tools and techniques, ones that are part of a new era in science: the explosive growth of new technologies.

Figure 4–1 Some tools of the modern researcher: (a) scanning electron microscope; (b) a microtome for cutting thin sections; (c) a revolver microprojector and a cytoscope with projection screen; (d) a particle size analyzer; (e) a pH meter; (f) a mass spectrometer.

Basic Tools

The United States Congress is currently in the process of working out a program for changing over from our present ("English") system of weights and measures, which includes pounds, inches, and quarts, to the metric system, which includes grams, centimeters, and liters (see Table 4–1).

There are numerous reasons for considering such a change. It is believed that our present system is costing American businessmen literally billions of dollars each year in lost sales and conversion expenses. Nearly every other country in the world has adopted the metric system as its national system of weights and measures, and the metric system is also the international system of weights and measures used throughout the scientific community.

The reason for preferring the metric system of weights and measures is that conversions from one unit to another can be performed more conveniently in the metric system—multiples and fractions of the standard measures may be obtained merely by moving the decimal point.

On the macro-scale, measurements of length are made in either system with a very simple device: a rod or tape that is marked off in subdivisions of some standard unit of measurement. Thus, the foot ruler and yardstick are marked off in inches and fractions of inches (usually down to eighths or sixteenths of an inch), while the meter rule and small metric measuring sticks are marked in centimeters and millimeters (hundredths and thousandths of a meter, respectively).

On the micro-scale, such as in measurements of living cells and their constituents, measurements of length are usually made in conjunction with a microscope, which magnifies both the object to be measured and the measuring scale to a convenient size for view-

TABLE 4-1 Some common metric units of measurement and their (metric and English) equivalents

Length

1 angstrom (Å) = 10^{-10} meter (m) = 3.94×10^{-9} inch

1 micron (μ) = 10^{-6} m = 3.94×10^{-5} inch

1 millimeter (mm) = 0.001 m = 0.0394 inch

1 centimeter (cm) = 0.01 m = 0.394 inch; 1 inch = 2.54 cm

1 meter (m) = 100 cm = 39.37 inches = 3.28 feet = 1.09 yards

1 kilometer (km) = 1000 m = 3280.84 feet = 0.62 miles; 1 mile = 1.61 km

Volume

1 cubic centimeter (cc *or* cm^3) = 0.999972 milliliter (ml) = 0.061 cubic inch = 0.0338 fluid ounce

1 milliliter (ml) = 0.001 liter = 0.061 cubic inch = 0.0338 fluid ounce; 1 fluid ounce = 29.57 ml

1 liter = 1000 ml = 33.81 fluid ounces = 1.0567 quarts; 1 quart = 0.946 liter

Weight

1 milligram (mg) = 0.001 gram = 0.0000353 ounce

1 gram (g) = 0.0353 ounce; 1 ounce = 28.35 g

1 kilogram (kg) = 1000 g = 35.27 ounces = 2.205 pounds

Temperature

$°C = \frac{5}{9}(°F - 32)$ (centigrade)

$°F = \frac{9}{5}(°C) + 32$ (Fahrenheit)

$°K = °C + 273.15$ (Kelvin or absolute)

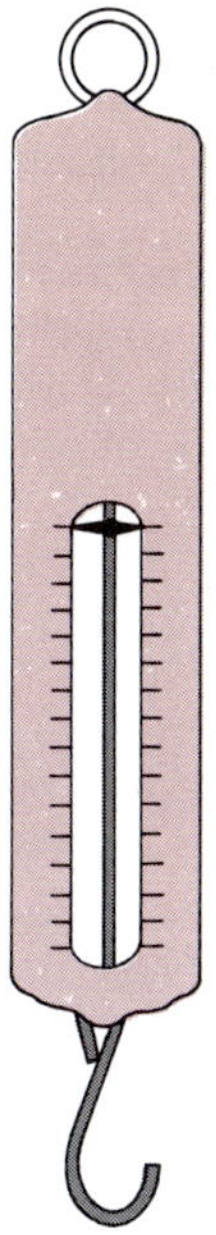

Figure 4–2 A spring balance: the pointer moves a distance proportional to the weight of the object.

ing. A preparation may be placed on a **stage micrometer,** consisting of a standard-sized microscope slide, a small area of which is ruled in intervals of 0.01 mm. Special attachments that can be incorporated into the microscope itself, in place of or supplementary to the lenses, can be used for further refinement of micromeasurements.

The measurement of mass is another area in which standard laboratory practice requires a variety of instruments of differing precision for use in different ranges of magnitude. Gross measurements can be made on a simple **spring balance** (Fig. 4–2). Strictly speaking, the spring balance measures weight rather than mass: the object to be weighed is suspended from the hook at the bottom, and its weight (the action of the force of gravitational attraction on its mass) tends to stretch the spring downward. The distance through which the spring is stretched is proportional to the force applied (the weight of the object). Thus, the scale can be graduated directly in weight units.

Rough weighings in the laboratory may be performed on a **platform balance** (Fig. 4–3a). This may consist of two platforms mounted on a lever beam. The object to be weighed is placed on one platform and counterbalanced by calibrated weights placed on the other platform. A small pointer indicates when the two platforms are balanced. In the single-pan platform balance, the object on the single platform is counterbalanced by sliding weights along a series of three beams until the downward-acting forces on the two sides of the lever are exactly balanced (Fig. 4–3b).

Finer instruments are needed for analytical work, for which greater accuracy is required.

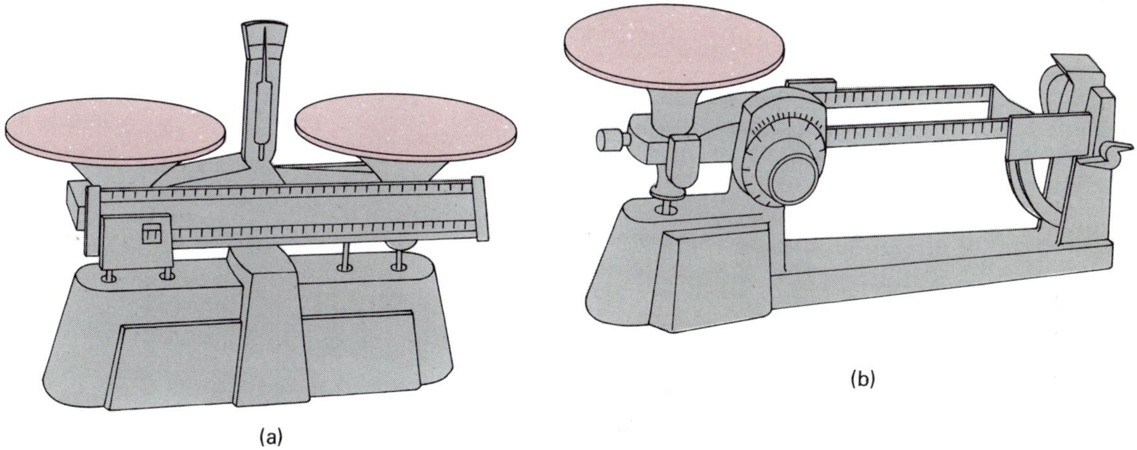

Figure 4–3 (a) A double-pan platform balance; (b) A single-pan balance with digital readout.

A typical laboratory **analytical balance** can take a total capacity of about 200 grams and has a sensitivity of 0.01 milligram. (The **sensitivity** of a balance is generally defined as the smallest weight that will produce an appreciable displacement in the position of the pointer.) The laboratory analytical balance also works on the lever principle (Fig. 4–4). The entire balance is enclosed in a glass-windowed case to minimize perturbations by air currents, gross temperature and humidity differences, etc.

The analytical balance of this type is very similar to the models used for the past several centuries, and they are fairly effective in weighing relatively small samples with good accuracy. Indeed, microbalances of this design can be used to measure samples in the milligram range, with a sensitivity of a microgram or so. But the design does have a number of inherent disadvantages. The counter-balancing weights must be carefully placed on the pan with forceps, and the rider has an annoying tendency to fall off the beam. The final value is obtained by laboriously adding up the various weights used, and this may be a source of errors.

A newer type of balance that eliminates these shortcomings has gained wide acceptance in modern laboratories. This is the **single-pan balance** with a window that shows a digital readout of the final value. Instead of establishing the zero point with no load on the pan, as in the traditional analytical balance, in the single-pan type the zero point is established under a maximum load. The sample is placed on the pan, and then, by simply turning knobs, internal counterweights are successively removed until the zero point is reestablished. The counterweights removed correspond to the weight of the sample on the pan, and their sum is automatically shown in the digitial readout (Fig. 4–5).

Both semimicro- and microbalances in the single-pan design are available. They have total capacities of 160 and 20 grams and permit the determination of weights to the nearest 0.01 mg and 0.001 mg, respectively. The microbalances, particularly, are extremely sensitive to all sorts of influences. Not only is the balance enclosed in a glass-windowed case, but the researcher must even wear a cotton glove in placing specimens on the pan, so that the heat from his hand will not disturb the zero reading of the balance. These balances are usually placed in a special room, separate from the general laboratory area; the balance room may even have its own air conditioning system, providing constant temperature, humidity, and electrostatic filtering to remove dust particles.

With modern advances in the study of biological constituents and the synthesis of new compounds and isotopes which may be available only in microgram quantities, even the microbalance sometimes is not sensitive

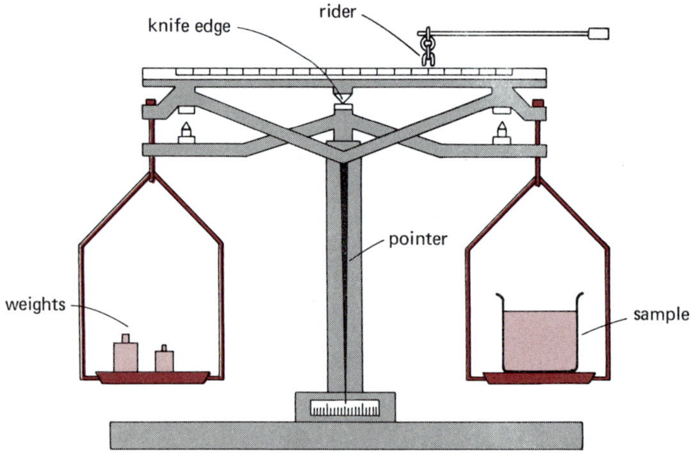

Figure 4–4 Working parts of a typical laboratory analytical balance.

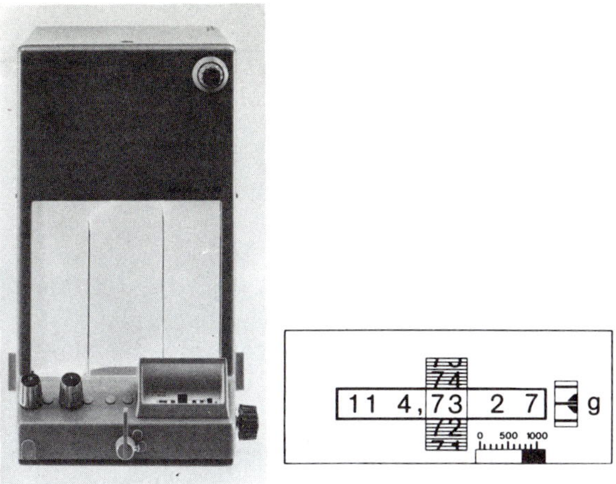

Figure 4–5 A single-pan analytical balance. Note the sliding doors on the sides, which keep out air currents and help to maintain constant conditions inside the weighing chamber. The inset shows the reading 114.7327g as it would actually appear on the balance.

enough. **Ultramicrobalances** have been devised for such purposes, utilizing a coiled fiber of fused quartz instead of a beam. When a minute sample is placed at the end of the fiber, it produces a twisting tendency or torque. The application of a counterbalancing torque to the fiber restores it to its original position, and from the amount of torque applied the weight of the sample can be calculated. Ultramicrobalances of this kind have a sensitivity of 0.01 microgram (i.e., 10^{-8} gram). This is an incredibly small figure—the dried ink on the period at the end of this sentence weighs about 2 μg, far above the level of sensitivity of the ultramicrobalance. And yet even such an ultrasensitive balance is not fine enough to weigh a human red blood cell (about 0.0001 μg), much less a bacterium (about 0.000001 μg).

Chemical Analysis

Over the past centuries man has evolved numerous techniques for detecting and identifying various chemical substances, including the chemicals of life. Today we have a rather sophisticated array of such techniques of **chemical analysis** that are frequently used in conjunction with instrumental techniques. (The latter will be discussed in more detail in the following sections.)

In any analysis there are certain basic steps which must be followed. First, after deciding what a particular substance is to be tested for, one must obtain an appropriate representative sample of that substance. The sample must then be brought into a form suitable for the proposed analysis. For example, most chemical reactions take place only in a liquid medium. Therefore, in chemical analyses of samples, whether biological specimens or inorganic materials, **extracts** (solutions of substances obtained by treating the sample with a suitable solvent) or **homogenates** (dispersions of a pulverized sample in a fluid medium) are frequently used. Finally, a suitable analytical technique must be chosen and the appropriate procedure carried out.

For instance, in determining the nitrogen content in dandelion leaves, a weighed sample of leaves must first be boiled with concentrated sulfuric acid, which converts the nitrogen present to ammonium sulfate. This is then converted to free ammonia by the addition of excess sodium hydroxide. The ammonia is distilled over into a measured volume of carefully standardized acid, and the amount of excess acid remaining is determined by adding a solution of a standard base

Figure 4–6 Setup for the Kjeldahl nitrogen determination.

drop by drop until the endpoint of the reaction is reached. The amount of nitrogen in the original sample is then calculated from the measured amounts of reagents used, with the aid of suitable conversion equations.

This method of nitrogen determination, called the Kjeldahl method, was originally performed manually; as may well be imagined, it was laborious and time-consuming. With the addition of modern instrumental techniques, such analyses are largely or wholly automated, and results are obtained far more rapidly and accurately. Various modifications have brought such analyses as the Kjeldahl method down to the micro-range, permitting the determination of substances in samples weighing a few milligrams or even less.

Methods of chemical analysis, increasingly in conjunction with instrumental techniques, are widely used in biology and medicine. All the basic elements and various trace elements in tissues and cell fractions can be determined by chemical-instrumental methods of analysis, and specific simple and complex organic compounds, such as proteins (total protein and various specific proteins), DNA and RNA, and vitamins, can also be detected and assayed.

An important technique of chemical analysis as applied to biology is that of tissue staining. Preparations of tissues and cells for microscopic examination are treated with specific chemical reagents that produce a characteristic color with specific components of the cell. For instance, the Feulgen reagent lends a purple color to the DNA in a cell, but not to its RNA or proteins. The field of histochemistry—the study of the chemical composition of tissues using staining techniques—has blossomed greatly in recent years.

Chromatography and Electrophoresis

One of the greatest discoveries ever made by man went unrecognized and lay dormant for a generation. In 1906 a Russian botanist, M. S. Tswett, passed a petroleum ether extract of plant pigments through a column of calcium carbonate packed into a narrow glass tube and obtained a series of layers of different-colored pigments distributed down the column. When he subsequently passed a stream of pure solvent through the column, each colored zone was dissolved and emerged from the tube in sequence and could be collected separately. As a result Tswett discovered that the plant pigments actually consist of two different green pigments (chlorophylls a and b) and several yellow pigments (the carotenoids).

Tswett's account of his discovery, published in a prominent German journal, was virtually ignored. In the 1930s, following the publication of another landmark paper on the use of **column chromatography** by R. Kuhn and E. Lederer, this technique of chromatographic analysis was finally recognized as a useful and ultimately indispensable separating tool. As the result of still more research in the 1940s and early 1950s by the English researchers A. J. P. Martin and R. L. M. Synge and their coworkers in the development of **partition chromatography** and an important variation, **paper chromatography,** the techniques of chromatography became more and more indispensable in chemistry, biology, and medicine.

Chromatographic techniques permit the separation of numerous, sometimes very closely related, components of a mixture. Before the advent of chromatography such separations were often literally impossible. Now mixtures of amino acids, nucleotides, various lipid fractions, and a host of others are separated routinely in the laboratory.

An indication of the extreme importance of these techniques was the awarding of the 1953 Nobel Prize in chemistry to Martin and Synge. Subsequently, other important variations of chromatography were developed, including **gas chromatography** and **thin-layer chromatography.** In addition, related techniques, including **ion-exchange chromatography,** the use of **molecular sieves,** and **electrophoresis,** are now in wide use.

The basic principle of column chromatography is relatively simple (Fig. 4–7). A mixture of substances is separated by distribution between two phases. One phase is the solid column packing (or a liquid tightly bound to it), the **stationary phase.** The other is a **mobile phase,** a liquid solvent. The mixture to be separated is applied to the column, and its components become bound to the stationary phase. In the case of adsorption chromatography, in which the column is packed with some inert material such as charcoal, silica gel, or alumina, the components of the mixture are bound to the surface of the solid particles by physical forces of adsorption. Similarly, in partition chromatography there is a physical affinity between the stationary phase and the components of the mixture. In ion-exchange chromatography, the stationary phase is composed of an ion-exchange resin, a solid substance which possesses charged (ionic) functional groups. These groups react with the corresponding ionic groups of the components of the mixture, which are therefore held to the stationary phase by chemical forces.

Figure 4–7 A column chromatography setup. The sample to be separated is applied to the top of a column of packed adsorbent, then a series of solvents is passed through the column. The components of the sample are distributed down the column in characteristic bands, each with its own relative affinities for the solvents and column packing, and ultimately emerge in successive fractions of the eluting liquid.

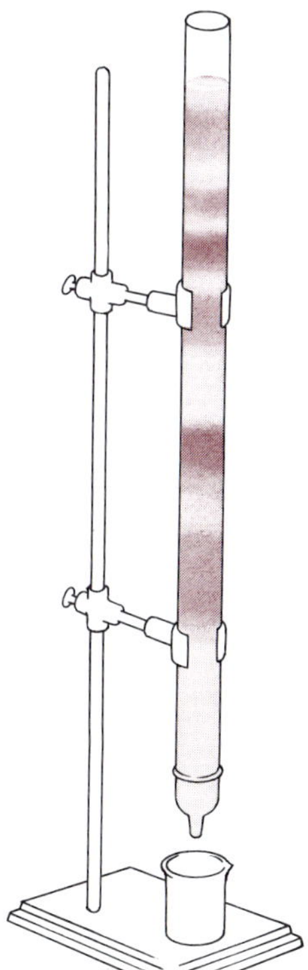

The essence of the separation is that the components of the mixture are not all held identically by the stationary phase. Each has its own characteristic affinity for the stationary phase and is held with its own particular strength. Thus, when a solvent (mobile phase) is passed through the column, the more loosely held molecules or ions pass down the column much more rapidly than those held more tightly, and are removed from the column **(eluted)** by the early portions of solvent that emerge. The affinity of the substances for the eluting solvent also plays an important role: the water-soluble substances of the cell, for example, would tend to be washed from a column much more rapidly by water or some other polar solvent than by ether or carbon tetrachloride. Oil- or lipid-soluble substances would show the opposite behavior. The rate of elution of each particular substance from the column is therefore determined by the interaction of its affinities for the stationary and mobile phases.

The original work on chromatography was performed on mixtures of colored substances, which formed clearly distinguishable colored bands as they passed down the column. This is how the term *chromatography* ("color mapping") came to receive its name. But the chromatographic technique is by no means limited to colored substances. Colorless substances may sometimes be detected by characteristic fluorescence under ultraviolet light. Or the eluent (the emerging solvent with dissolved components of the mixture) may be collected in small fractions, and each fraction tested by chemical or physical procedures to determine the presence and quantity of each component.

But the various methods of column chromatography have certain disadvantages. One important disadvantage is that they are not well suited to the separation of small samples, often a prime requisite in biological work. Because of the large volumes of solvent that must be used, the components of a sample weighing only a few milligrams would be hopelessly difficult to detect in the fractions of eluent obtained.

A major breakthrough in this area was the development of paper chromatography. In this variation of partition chromatography, the stationary phase is not a column of adsorbent or ion-exchange resin, but rather a liquid phase adsorbed on a sheet of filter paper. A tiny sample of the mixture to be separated, dissolved in a solvent, is applied, a drop at a time, at a set point on the paper and allowed to dry. The paper is then set up in a chromatography chamber (Fig. 4–8a), and a solvent or mixture of solvents is then allowed to flow past the drop and along the paper, either upward by capillary action **(ascending paper chromatography)** or downward by the combined forces of gravity and capillary action **(descending paper chromatography).** The various components of the mixture are carried along by the solvent, each at its own particular rate determined by its relative affinities for the stationary and mobile phases. When the solvent front has reached a predetermined point (often the end of the sheet), the **chromatogram** is dried, and the locations of the spots of the components are determined. If these substances are colored or fluoresce under ultraviolet light, this is a simple matter. If not, the entire chromatogram is sprayed with a suitable chemical reagent, which reacts specifically with certain classes of compounds to give a colored product; the latter then shows up as a spot on the chromatogram. For example, the reagent ninhydrin is generally used as the developer in paper chromatographic separations of amino acids. Each amino acid reacts with ninhydrin to give a purple-colored spot that moves a characteristic distance along the paper relative to the solvent front. The ratio of the distance of a spot from the start to the distance of the solvent front from the starting line is called the R_f (Fig. 4–8b).

Sometimes, if the separation obtained is not good enough, the chromatogram may be turned sidewise and run again, either with the same solvent system or with a different one. This variation is known as **two-dimensional paper chromatography.**

Figure 4–8 Paper chromatography. Shown at the left are schemes of apparatus for descending and ascending paper chromatography. The chromatogram at the right was obtained by passing two different pairs of solvents through the paper sheet successively at right angles.

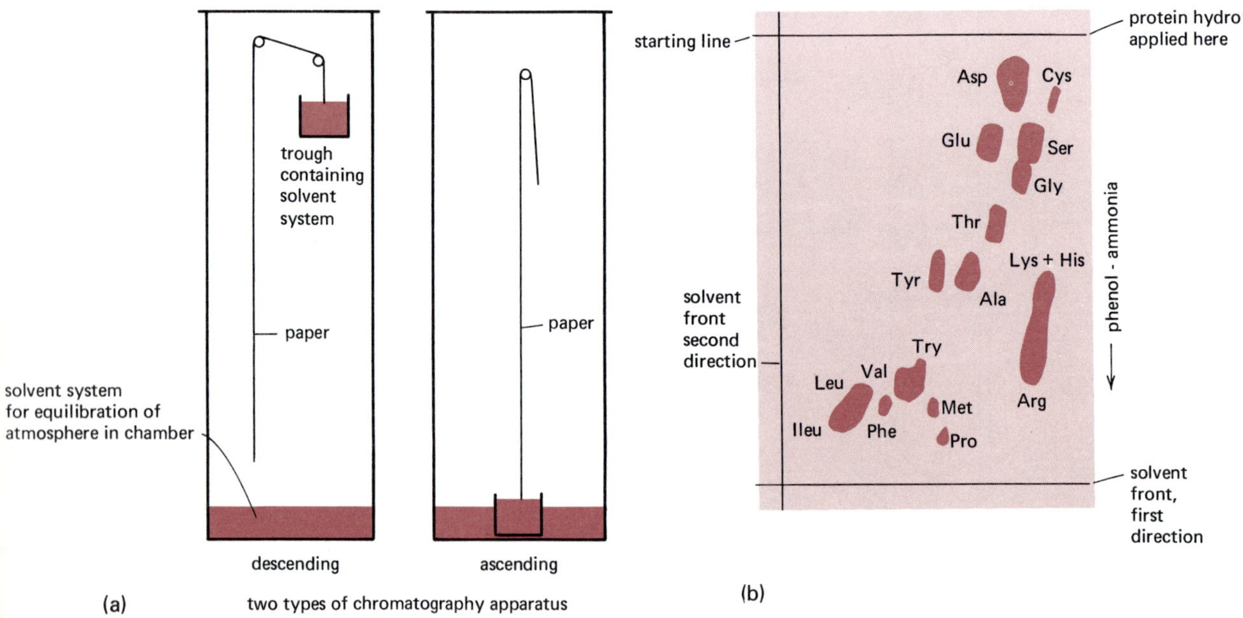

One of the newest modifications of the chromatographic technique, **thin-layer chromatography,** provides a great increase in the resolution of separations of mixtures, as well as a great increase in speed over paper chromatography. (A thin-layer separation may be completed in a matter of minutes, while a typical paper chromatographic separation often takes more than twenty-four hours.) In this method, the separation is run similarly to paper chromatography, but on a thin layer of some adsorbent, such as silica gel, alumina, or cellulose powder, uniformly spread over a plate of glass. Like paper chromatography, thin-layer chromatography can be used for the separation of mixtures or for the detection and quantitative determination of specific components in extremely small samples.

Another variation of chromatography of growing importance is **gas chromatography.** In this method, the stationary phase is a solid adsorbent, such as activated charcoal or silica gel, or an inert material, such as firebrick coated with some nonvolatile liquid such as a polyester or silicone resin. The sample, which may be in the microgram range, is volatilized and is swept along by a stream of a **carrier gas** such as hydrogen, helium, argon, or nitrogen. Each component of the mixture is carried along at a different rate, depending on the strength with which it is retained by the column packing. A thermal conductivity cell is usually used to detect the presence of a substance emerging from the column. Different substances are identified according to their characteristic times of emergence. Gas chromatography is finding increasing use, not only for the analysis of mixtures, but also for obtaining fractions of extremely pure substances to be used in the production of various compounds.

Very similar to methods of chromatography is the technique of **electrophoresis.** The important difference is that in electrophoresis the components of the mixture migrate under the influence of an applied electric field. Electrophoresis is generally conducted on paper or on starch gel (Fig. 4–9). When an electric field is turned on, each electrolyte (electrically charged particle) migrates toward the electrode of the opposite charge. The mobility of the electrolyte is determined by the characteristics of the molecule. This technique has been useful in separating charged high-molecular-weight compounds, such as blood serum proteins, and can be modified for both analytical and preparative applications.

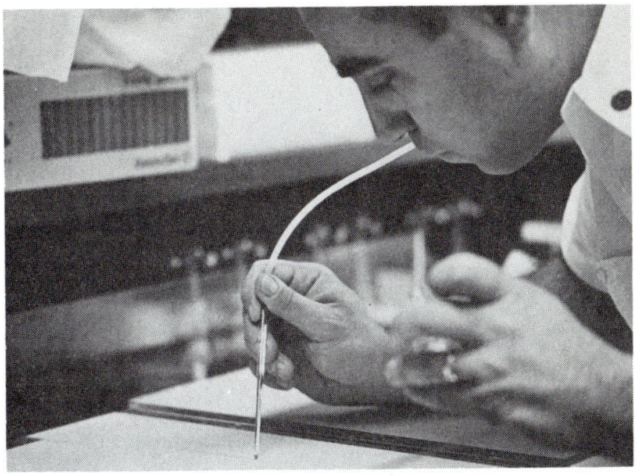

Figure 4–9 Using electrophoresis in molecular biology studies.

Centrifugation

Part of the training that astronauts must undergo in preparing for space travel includes spending a certain amount of time in a huge centrifuge, which whirls them around at dizzying speeds. As the number of revolutions per minute increases (i.e., as the speed increases), the astronauts feel heavier and heavier, until they are subjected to the high gravity loads characteristic of acceleration in space vehicles.

This example illustrates a basic characteristic of centrifugation: there is a direct relationship between the speed of rotation (number of revolutions per minute—rpm) and the centrifugal force, which is exerted in an outward direction perpendicular to the direction of the motion (Fig. 4–10). Moreover, the greater the distance of the revolving body from the central axis, the greater the force.

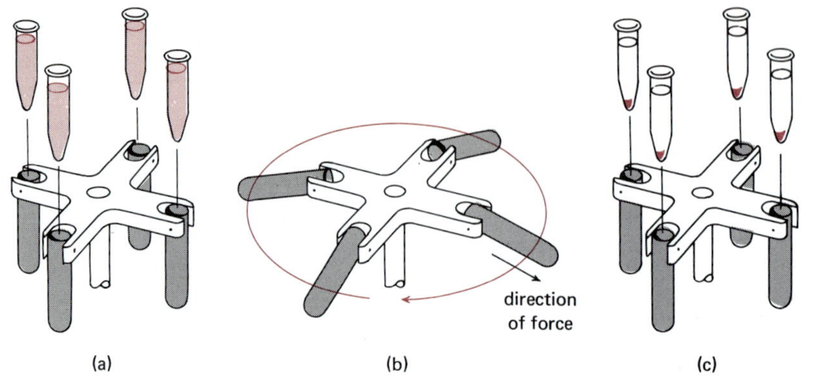

Figure 4–10 When the centrifuge rotates, a centrifugal force is created, effectively concentrating the heaviest components in the centrifuge tubes to the bottom of the tubes (i.e., the points at the greatest distance from the axis of the rotating centrifuge).

In a chemical centrifuge, these two principles are effectively used to separate mixtures of substances with different densities. In a simple laboratory model, this separation is of a gross nature: the denser solids are **sedimented** to the bottom of the centrifuge tube (which is the part farthest from the central axis; see Fig. 4–10), and the liquid, called the **supernatant,** remains in the upper portion of the tube.

The faster the rate of rotation of the centrifuge head, the more distinct the separation. A simple, relatively inexpensive laboratory centrifuge generally spins at speeds of several thousand rpm (with a maximum of about 5000). High-speed centrifuges, with maximum speeds of about 20,000 rpm, can give a much finer degree of separation, differentiating among solid components in a particular homogenate. And ultracentrifuges, with speeds up to about 75,000 rpm, can give a still finer separation.

Centrifugation is extremely useful in chemical preparation and analysis for separating a precipitate that is slow to settle out under the influence of gravity. This process has also been of inestimable value in studying the components of living cells. In separating the components of a homogenate of a biological specimen with an ultracentrifuge, one generally starts by spinning at a fairly slow speed (e.g., about 5000 rpm) to remove the gross cellular fragments. Then by spinning at successively higher speeds, one can sediment progressively lighter fractions.

An extremely useful modification of centrifugation has been the **density gradient** technique. A concentrated solution of cesium chloride (CsCl) is centrifuged in an ultracentrifuge at high speeds until an equilibrium is reached among the opposing processes of sedimentation and diffusion. At equilibrium, there is a stable concentration gradient of the CsCl along the centrifuge tube. If the solution also contains a small amount of a high-molecular-weight substance, such as DNA, when equilibrium is reached the DNA molecules will be concentrated in bands at the zones of the centrifuge tube at which their density and the density of the medium are exactly equal. The resolving power of this method is so great that DNA containing the normal isotope of nitrogen (N^{14}) can be separated from DNA containing the heavy isotope of nitrogen (N^{15}).

Radiation Sources and Radioisotopes

If a small drop of water were to be added to a large beaker of water, it would be virtually impossible to trace the subsequent fate of the water molecules comprising the original drop.

If the water in the beaker were boiled, how could we tell whether any of the added molecules were being volatilized? If some of the water were poured down the drain, how could we know how many of the added molecules went down the drain and how many remained in the beaker? In fact, how could we even know whether, after the initial addition of the drop, the molecules were distributed uniformly through the beaker, or all sank to the bottom, or all floated on the top, or perhaps all stayed on one side?

What if a dye were added to the original water? Then the path of the colored dye could be followed. But we would now be following the dye molecules, not the water molecules of our original sample. Perhaps the original water molecules could be treated with some substance that would react chemically with them and thus render them different enough from normal molecules so that they could be distinguished from the larger sample of water into which they were poured. But such a chemical reaction would yield a product that was no longer water, and the behavior of the new molecules might not be typical of water molecules in general. The results obtained by watching them might therefore be quite misleading if we were trying to draw conclusions about the behavior of water molecules. It seems that what is needed is some sort of "marker" that will make the water molecules different enough to be distinguished from the other water molecules, yet similar enough to the usual variety so that their behavior will still be "normal" for water molecules.

Scientists in a variety of fields, from the study of the structure of metals to the study of living organisms, are faced with problems of distinguishing and following a few individual entities among a multitude of similar ones. The naturalist out in the field solves his problem by capturing a representative sample of organisms, banding or tagging them, and then releasing them again. (Even here he runs some risk of disturbing their normal behavior.) But how do you "tag" an atom or molecule that cannot even be seen under the most powerful electron microscope? Problems exist even for structures readily visible under the optical microscope. For instance, how does one distinguish one red blood cell from another—they all seem to look alike! With the use of radioactive isotopes as **tracers,** a seemingly impossible problem becomes quite manageable.

For example, it was long known that red blood cells are continually being produced in the bone marrow and are continually being destroyed primarily in the liver and spleen. How, then, to determine the average life span of the red blood cell, since one red blood cell looks pretty much like any other? The problem was eventually solved by administering a radioisotope that was incorporated into the hemoglobin of the red blood cells being manufactured. As the red cells produced at this time were gradually destroyed, the radioactivity of the mass of red cells dropped. The average radioactivity of the red blood cells was followed over the succeeding days and the average life span determined by appropriate mathematical calculations. Such radioisotope tracer studies were conducted with various isotopes, including radioactive iron (Fe^{59}). The various determinations yielded an average of about 120 days.

Even more often, biologists and chemists desire to know the fate of specific molecules in various systems. For example, using radioactive tracers it has been found that iron tends to concentrate in the hemoglobin, phosphorus in the bones, and iodine in the thyroid gland. Indeed, the latter property of radioiodine has been used extensively for both diagnostic and therapeutic purposes in diseases of the thyroid gland.

In addition, radioactive tracers can reveal much valuable information about the metabolic reactions that take place within the organism. The sequence of products of photosynthesis has been worked out largely through the use of radioactive carbon (C^{14}): the plant is placed in a chamber whose atmosphere contains radioactive carbon dioxide. As the radioactive $C^{14}O_2$ is taken in by the leaves, it is incorporated into various intermediate products, which can be determined by ascertaining (for instance, by chromatographic techniques) the substances containing the bulk of the radioactivity at any particular time after the exposure was begun.

Radioactive isotopes of hydrogen (H^3: tritium), carbon (C^{14}), phosphorus (P^{32}), sulfur (S^{35}), and iodine (I^{131}) are commonly used as radioactive tracers in biological studies.* Radioactive sulfur and carbon are commonly used to label amino acids, which are then incorporated into proteins. Tritium and phosphorus-32 are frequently used to label nucleotides, which are then incorporated into nucleic acids.

The radioactivity of substances into which radioisotopes have been incorporated can be determined with a radiation counter—such as a scintillation counter, a proportional counter,

*Certain nonradioactive isotopes, such as heavy hydrogen (H^2: deuterium), nitrogen (N^{15}), and oxygen (O^{18}), are also frequently used as tracers. Their presence is determined with a mass spectrometer (an instrument for detecting extremely small differences in mass) on the basis of their heavier-than-normal mass. For instance, determinations of the red blood cell life span by the administration of the amino acid glycine labeled with heavy nitrogen have been conducted similar to those described above with radioactive iron, and have yielded a value in fairly good agreement, 125 days.

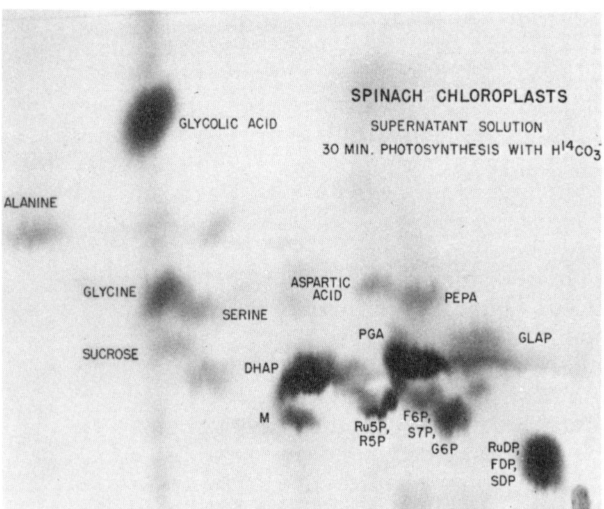

Figure 4–11 Dark spots on the autoradiogram correspond to spots of radioactive products of photosynthesis on a chromatogram.

Figure 4–12 The Bevatron at Lawrence Radiation Laboratory, Berkeley.

or a Geiger counter—which counts the decay products of the radioactive isotope. Another ingenious technique, **autoradiography,** utilizes the property of radiations that first led to their discovery: their ability to blacken a photographic emulsion. An organism or section of tissue that has incorporated a radioactive isotope is placed in contact with an unexposed photographic plate. When the plate is developed after a sufficient exposure time, the pattern of absorption of radioactivity will be traced out on the autoradiogram. This technique can also be applied to chromatograms of various samples of specimens containing radioisotopes (Fig. 4–11).

The effects of radiations on both inorganic materials and living systems are active areas of research. The radioactive isotope of cobalt, Co^{60}, an emitter of beta and gamma radiations, is frequently used as a radiation source in biological studies. Another radiation source that has been in regular use since the turn of the century is the X-ray machine. The physical characteristics of gamma rays and X rays are quite similar, their major difference being their origin: gamma rays result from the nuclear disintegration of various radioactive isotopes, whereas X rays are produced by bombarding a stationary target in an evacuated tube with high-speed electrons. Nuclear reactors and particle accelerators are also used as sources of radiations.

Additional uses of radiation sources in biology and medicine include the use of X rays to produce photographs and fluoroscope images of internal structures. X rays and other radiations such as those emitted by cobalt-60 are used in medicine to destroy cancerous tumor cells. In genetic studies X rays are used to induce changes in individual organisms that are passed on to their offspring, some of which may be valuable for selective breeding. Radiations are also used to induce sterility in various insects as a form of biological pest control.

Geology, paleontology (the study of geological history as revealed by fossils), and archaeology (the study of material remains of past cultures) have benefited enormously from the use of radioisotopes. By analyzing an object for residual carbon-14, which is known to have a half-life of 5700 years, scientists have been able to determine the age of ancient weapons, tools, and bones, in turn making it possible to date the rock strata in which such specimens are found. For older samples, isotopes with longer half-lives can be utilized as dating devices. For instance, rubidium-87, with a half-life of 47 billion years, has been used to date rocks some 3.5 billion years old. The uniform rate of decay of radioactive elements has thus provided a convenient "clock" for clarifying the history of our planet and its life forms.

Microscopy

A simple microscope, in the form of a single magnifying lens, has been used by man for thousands of years to make the world of the small more easily discernible. In fact, nearly two millennia ago the Roman philosopher Seneca used water-filled glass globules for looking at small objects; by the time of Leonardo da Vinci, nearly fifteen hundred years later, the use of glass lenses was well established.

Indeed, it was a lensmaker who made the first really significant advance toward the development of the modern **compound microscope.** This first microscope was devised by Zaccharias Janssen in the Netherlands around 1590 and gave only a ninefold magnification.

A quantum jump in the perfection of the microscope lens was taken during the next century by another Dutchman, Anton van Leeuwenhoek. Although he used a single lens, rather than the compound arrangement introduced by Janssen, Leeuwenhoek's lenses were so exquisitely ground that they were able to magnify objects clearly more than two hundred times. Through these lenses, some of them barely larger than the head of a pin, Leeuwenhoek gazed in fascination at myriad forms of protozoa swarming in a drop of water from a roadside ditch. He examined red blood corpuscles from his own blood, the fine structure of muscle, skin, hair, and ivory. He watched the circulation of blood in the tail of the tadpole, and discovered tiny parasites living on parasitic fleas. Leeuwenhoek even saw bacteria in scrapings from his teeth. But it was to be more than a century before the microscope was perfected to the point where others could see bacteria.

Granger collection

Figure 4–13 Anton van Leeuwenhoek fascinated seventeenth-century scientists with his descriptions of microorganisms seen through the lenses he ground as a hobby.

Gradually the resolving power of the compound microscope was improved, from the crude model—maximum magnification, 42× —with which Robert Hooke viewed the cellular structure of cork in 1665 to the modern optical microscope, the finest models of which may reach a magnification in the area of 2000×.

The basic construction of the modern compound microscope is clear from Fig. 4–14.

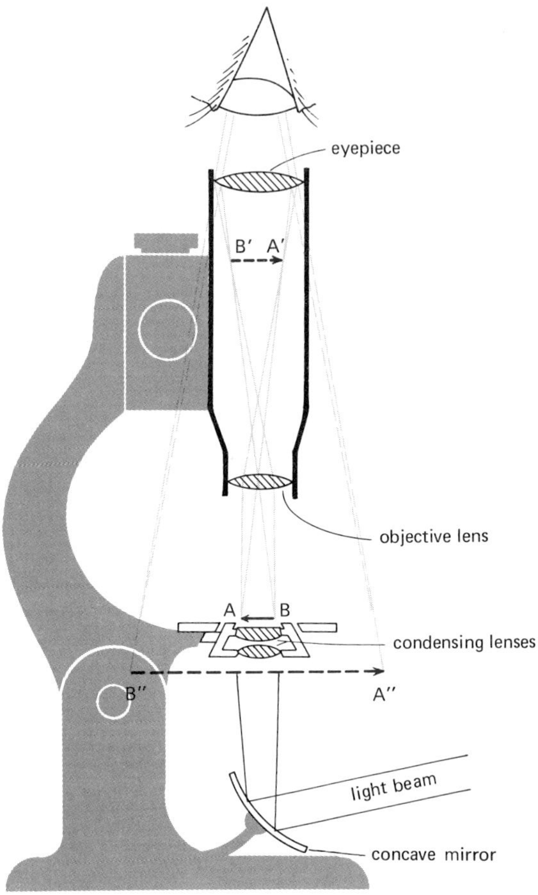

Figure 4–14 The compound microscope. The combination of lenses magnifies the field *A-B* to an apparent size *B″-A″*. Note that the image is inverted.

Light, coming from an illuminator below the stage of the microscope, is focused by a condenser, and thence passes through an aperture in the stage. Continuing its upward path it passes *through* the specimen, through the objective lens, up the body tube, through the eyepiece, and finally to the eye of the observer. As can be seen from the diagram, the image seen by the observer actually is an inverted and reversed image of the actual specimen. In other words, a protozoan that seemed to be swimming diagonally toward the upper right of the microscope field would in reality be swimming toward the lower left in its drop of water on the microscope slide.

Because light must travel *through* a specimen, the latter must be extremely thin. Special instruments, known as **microtomes,** are used to cut extremely thin sections of plant and animal tissue, down to a mere 50–100 Å thick. (This is roughly equivalent to a few score atoms thick.)

In addition, since the components of living systems are usually colorless and often of similar densities (with resultant very low visual contrast), it is usually necessary to **stain** them with appropriate dyes. There are certain disadvantages to staining techniques: they usually kill the tissue or organism being studied, and they may introduce extraneous changes (artifacts) into the structures that obscure the true nature of the material under study.

Over the years, the optical microscope has proven to be a fundamental tool in the study of life. With its maximum magnification in the area of 2000×, it has allowed biologists to see not only individual cells but also such cell structures as the nucleus, mitochondria, and various plastids and, in addition, even smaller structures such as chromosomes.

But with the rapid advances of science in the twentieth century and the focusing of the biologist's attention upon smaller and smaller structures, down to the ultrasubcellular and even molecular level, an urgent need grew for even greater magnifications. To some extent this was satisfied by the development of new modifications of the basic principle of the optical microscope.

In the **ultramicroscope,** for example, particles too tiny to be seen under the ordinary optical microscope are made visible as bright specks by illumination from a strong beam of light directed parallel to the surface of the stage, at right angles to the direction of vision through the microscope. This is in contrast to the operation of the optical microscope, in which light shines up through the specimen from below the stage.

Another modification permitting greater magnification is **dark-field microscopy.** The

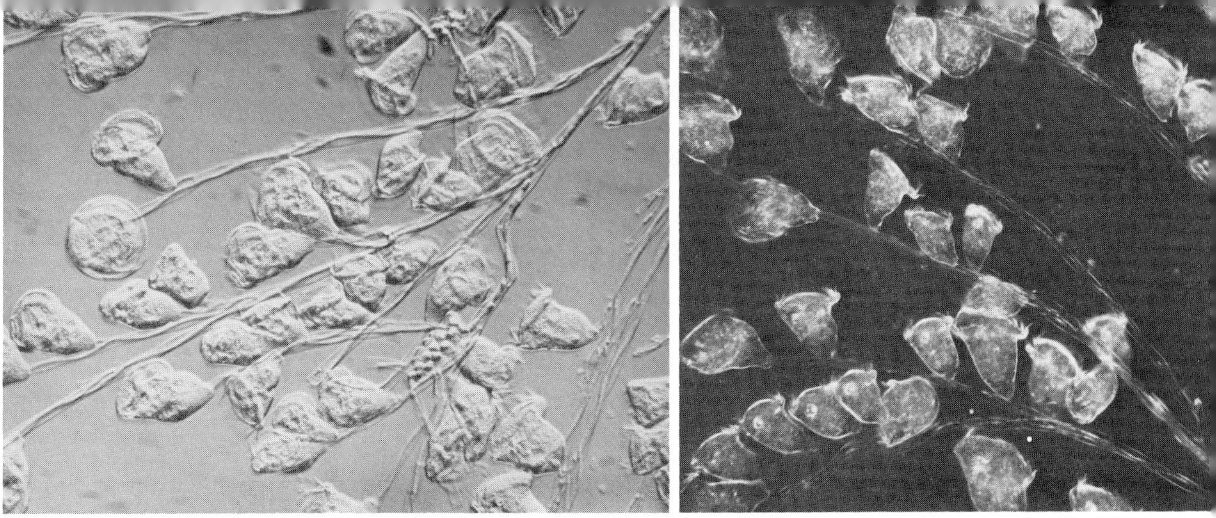

(a) Philip L. Feinberg, Omikron (b) Philip L. Feinberg, Omikron

Figure 4–15 Two views of *Carchesium*, a colonial protozoan: (a) bright field with oblique illumination (the organisms are too transparent to record well in ordinary bright field) and (b) dark field illumination. Both 141×.

dark-field condenser prevents any direct light from entering the objectives (lenses). A cone of oblique light rays is focused on the specimen, which appears as a very bright body against a dark background (Fig. 4–15).

The **phase-contrast method** of microscopy, proposed by Fritz Zernike of the Netherlands in the early 1930s, depends upon a combining of light waves and the detection of subtle differences in refractive indices through the use of special objectives and other special equipment. Greater visual contrast is achieved by phase-contrast microscopy, which permits the direct viewing of unstained, living materials such as yeasts, bacteria, and fungi.

Figure 4–16 An electron microscope in action.

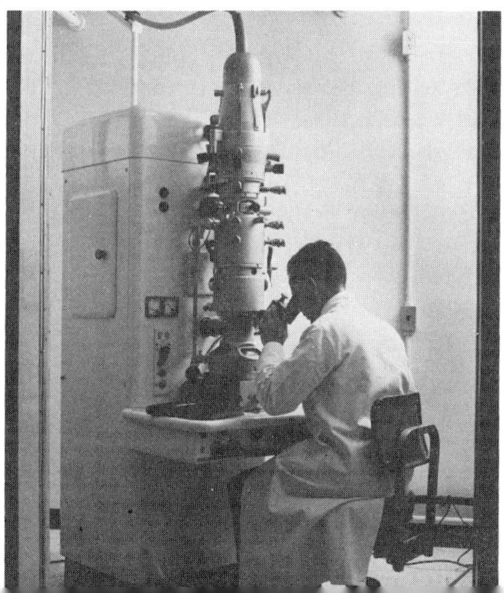

Ultraviolet microscopy utilizes shorter-wavelength ultraviolet light instead of normal visible light to obtain higher magnification. The use of this "invisible" light requires photographing the specimens (since the magnified image cannot be seen) and the use of ultraviolet-transmitting quartz lenses in place of the normal glass lenses.

Another modification of the basic microscope design that has greatly aided in biological research is the **stereoscopic microscope,** which is fundamentally a combination of two microscopes, one for each eye, producing a stereoscopic view. The image seen through such a microscope is three-dimensional, with much greater depth and a larger field than that seen through the monocular (single-eyepiece) compound microscope. In addition, the image is not inverted. All these factors make the stereoscopic microscope extremely valuable for dissection and micromanipulation.

Of all the developments in microscopy, perhaps the most exciting and potentially revolutionary has been the development of the **electron microscope.** Since the first electron microscope was developed in 1932 by Knoll and Ruska, there has been an accelerating rate of development of this powerful tool. Today electron microscopists have achieved a resolving power* that has already permitted them to view individual atoms!

Basically the same principle that operates in the optical microscope operates in the electron microscope: the highest resolution obtainable is about half the wavelength of the electromagnetic radiation being used. Since the lower limit of the visible range is in the region of 4000 Å, we would expect that the theoretical limit of the resolving power of the optical microscope would be of the order of 2000 Å. This is indeed the case. However, the wavelength of electrons is thousands of times shorter than this (and indeed is inversely proportional to the energy of the electrons—and thus to the accelerating voltage applied to them). Hence the theoretical limit of resolving power of the electron microscope is much less than 1 Å. Progress has been so dramatic that each year electron microscopists have been able to achieve greater and greater resolution. In 1969 a Japanese model (HU-11E) that achieved a voltage of 1.1 million volts actually demonstrated a resolution of about 1 Å (Fig. 4–17). However, at present these ultrahigh resolutions have been achieved for viewing inorganic substances only. Biological specimens present numerous special problems, including a greater tendency for contamination of the specimen and difficulty in obtaining ultrathin sections. As a result, progress in obtaining ultrahigh resolving powers for biological specimens has not been nearly as rapid as it has been for inorganic specimens. Nonetheless, with the heightened interest and research effort in the field of electron microscopy, we can look forward to dramatic progress in the immediate future.

The electron microscope, as we have said, works on the same basic principle as the optical microscope. However, instead of light, a beam of electrons is used as the illuminating source. And instead of finely ground glass

*The resolving power of an optical instrument refers to its ability to distinguish between two points set a specified distance apart. For instance, if a particular microscope has a resolving power of, say, 2 Å, it is capable of distinguishing two points as separate entities if they are 2 or more angstroms apart.

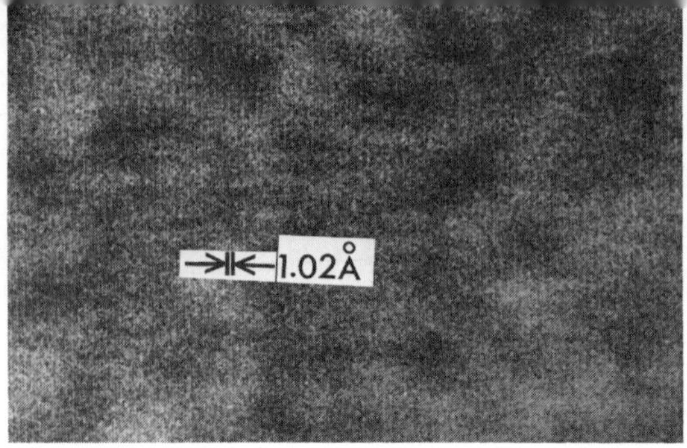

Figure 4–17 A high-resolution electron micrograph of an evaporated gold film. Although a 1 Å resolution has been achieved on inorganic materials such as this, technical difficulties of various kinds have so far limited the best resolutions on organic materials to several angstroms. 1,600,000×.

lenses, magnetic "lenses" are used to focus the electrons (Fig. 4–18).

Two promising avenues for improvement of the electron microscope are now receiving a great deal of attention. First, dramatic increases in the voltage applied to the electrons permit the use of thicker specimens. In some laboratories scientists have observed living organisms, such as bacteria, in special chambers under relatively natural conditions, permitting them to follow the life cycle of these organisms. (With lower-voltage electron microscopes, specimens can be observed only under high vacuum, but the walls of any chamber housing living specimens would absorb the slow-moving electrons. This limits the use of low-voltage electron microscopy to dead, dehydrated specimens.) Dupouy and Perrier in Toulouse, France, are currently working with the most powerful electron microscope in the world, a 3-million-volt instrument.

Another promising approach is the use of superconducting lenses operating at cryogenic (ultralow) temperatures. This approach, pioneered by H. Fernandez-Moran at the University of Chicago, permits exceptional stability and high resolution, in addition to significantly reducing contamination and specimen damage. Indeed, this method is quite compatible with the ultrahigh-voltage approach described above. One of the future developments is expected to be a combination of these two approaches.

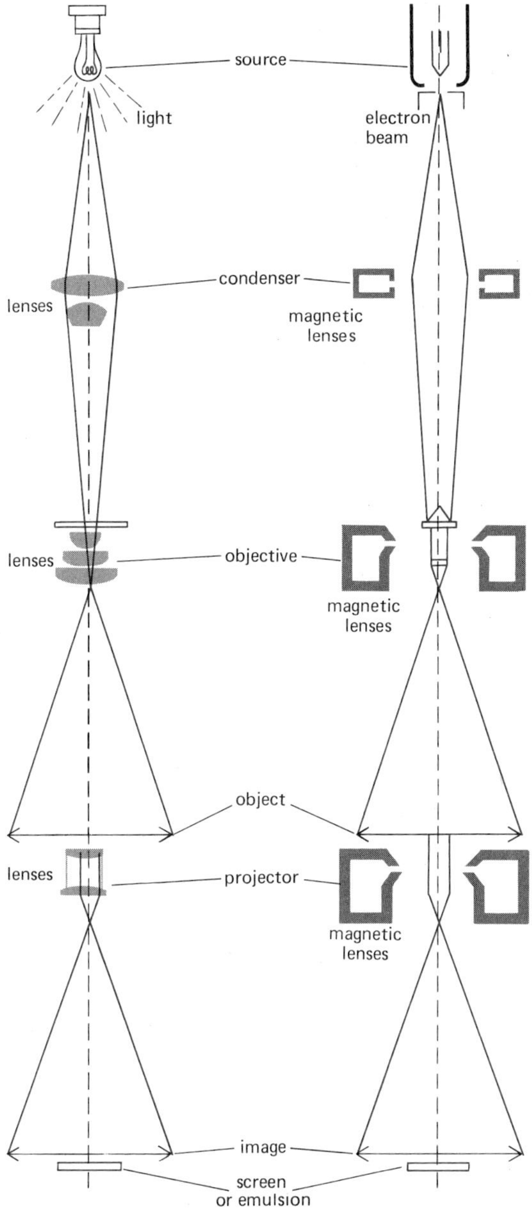

Figure 4–18 Comparison of the light microscope and the electron microscope. In the light microscope, visible light is focused by ground glass lenses and the image is viewed with the eye or recorded on a photographic emulsion. In an electron microscope, a beam of electrons is focused with magnetic lenses, and the image is registered on a fluorescent screen or recorded on a photographic emulsion.

An exciting new field of electron microscopy is the development of a **scanning electron microscope.** This type of microscope provides several advantages over the conventional electron microscope. It gives a three-dimensional view of specimens, providing up to 300 times greater depth of field (impression of distance) than the usual electron microscope (Fig. 4–19). Indeed, whole cells can be observed without the necessity of preparing ultrathin sections. The use of the scanning electron microscope is expanding so meteorically that it is rapidly becoming standard equipment in many laboratories throughout the world.

Why all the keen interest in electron microscopy? The prospects are breathtaking. At present the ultramicroscopic structure of the body—its structure at the molecular level—is a vast mystery to the biologist. It is like trying to discern the faces of a huge crowd in a dark room, with but the faintest glow to guide one. But the development of an electron microscope capable of resolving down to 1 Å with living specimens would be like turning on a floodlamp. The world of life on the molecular level—which is indeed the very essence of life—would be illuminated for man to see.

Figure 4–19 The scanning electron microscope can produce vividly three-dimensional pictures of the micro-world: (a) pine pollen grains, 500×; (b) a single storage cell of a potato, with starch granules, 275×.

(a)

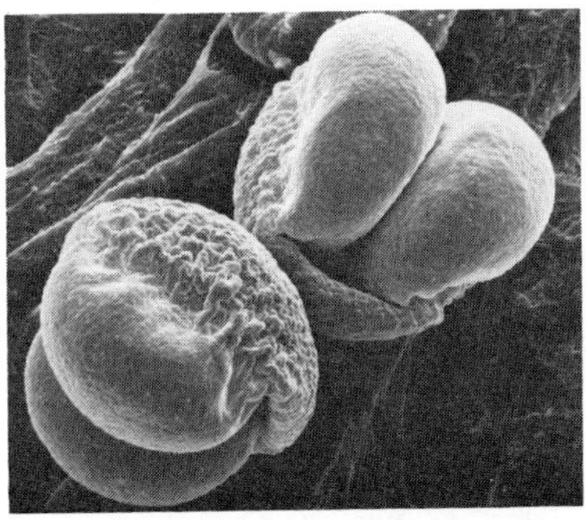

(b)

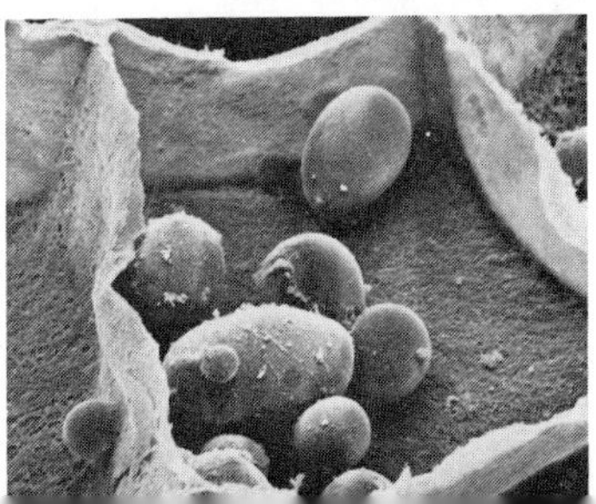

X-ray Crystallography

The knowledge of the three-dimensional structures of the molecules of life—proteins and nucleic acids—will provide biologists with powerful tools for solving many of the mysteries of life. X-ray crystallography provides a potent means of determining the three-dimensional structures of these organic molecules.

If a beam of light passes through a series of closely spaced pinhole openings or thin slits, the resultant light waves interfere with one another to produce a characteristic **diffraction pattern** (Fig. 4–20). Back in 1912 the German physicist Max von Laue proposed that crystals, which consist of highly regular ordered arrangements of atoms or ions, might be used as natural diffraction gratings for the study of another type of electromagnetic radiation, X rays. The implementation of Laue's suggestion was a brilliant success and has been used not only to provide information about the nature of X rays, but also for the study of the structure of crystals.

Figure 4–21 shows two arrangements of apparatus for **X-ray crystallography,** the study of crystal structure by X-ray diffraction methods. In the Laue method (Fig. 4–21a), a narrow beam of X rays is passed through a small, thin crystal, and the characteristic **Laue diffraction pattern** is recorded on a photographic plate. In the **Bragg spectrometer** (Fig. 4–21b), the beam of X rays strikes a large (perhaps 1- or 2-cm) single crystal, and X rays are scattered from it. The receiving device, a photographic plate or perhaps a Geiger counter or scintillation detector, is set up so that it receives only the scattered X rays coming from the face of the crystal. The crystal is then rotated, and additional recordings are made of the diffraction pattern in the new position. Gradually a picture is built up of the arrangement in space of the crystal lattice, which may be visualized as a minute, three-dimensional diffraction grating. Procedures have also been developed for X-ray diffraction study of powders of crystalline materials, which are made up of enormous numbers of microcrystals. The powder method has greatly expanded the applicability of X-ray crystallography, since it is extremely difficult and frequently impossible to grow single crystals large enough to be used.

Figure 4-20 Light rays shining through a diffraction grating interfere to produce a characteristic diffraction pattern.

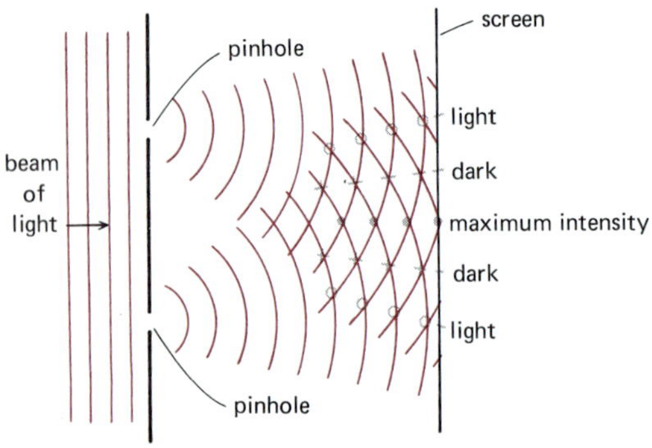

The earliest work in X-ray crystallography was done with simple inorganic crystals. The use of this technique has expanded dramatically in recent years, however, and has been extended to the study of complex organic and bioorganic chemicals. The Allied war effort during the Second World War was helped greatly by the development of penicillin in a massive crash program which included the delineation of its three-dimensional structure through the use of X-ray crystallography. This success brought the process to the attention of industry, and the following years saw the determination of the structures of a variety of drugs and other substances, such as vitamin B_{12} and saccharin. More and more complicated substances yielded to the probings of X-ray crystallographers, and now the three-dimensional configurations of a number of proteins have been determined through the use of this invaluable tool.

An enormous amount of work goes into the X-ray diffraction study of a complex bioorganic molecule. The more measurements taken, the higher the resolution of the picture obtained. Since the study of a protein molecule usually involves X-ray diffraction measurements of a number of heavy-metal–containing derivatives, more than a hundred thousand separate measurements may be required for the really fine resolution that

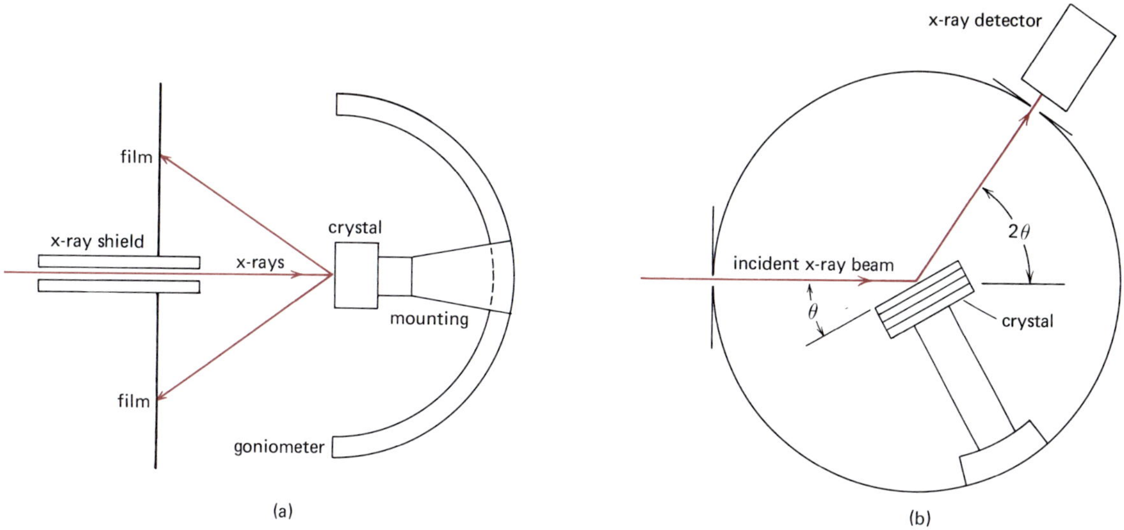

Figure 4–21 Diagrams of Laue spectrometer (a) and Bragg spectrometer (b). In the variation of the Laue camera depicted here, the X rays are reflected back by the atoms of the crystal rather than transmitted through the crystal. In both methods, the positions of the atoms in the crystal lattice can be deduced by computations of the diffraction patterns based on the laws of physics.

yields the image of a complete three-dimensional structure. Considering the fact that detailed mathematical computations are required for the analysis of each measurement, the X-ray diffraction study of a large molecule is a herculean task that may take years to accomplish.

An exciting new advance in X-ray crystallography techniques has been achieved by linking up the X-ray diffractometer with a computer, which not only performs the tedious computations but also directs the measurement taking itself, automatically rotating the crystal according to its set program. The use of a computer in conjunction with an X-ray diffractometer can cut out more than 95 percent of the heretofore tedious work involved in such studies. For example, crystallographers at the University of Indiana working on the structure of the protein lysozyme were able to perform only about eighty measurements per week using manual equipment. By linking up a diffractometer with a computer, they were able to perform a thousand measurements *per day*. (The complete three-dimensional structure of lysozyme derived from the white of chicken eggs was eventually announced in 1965 by another group of researchers, working at the Royal Institution in London and also using automated X-ray crystallography.)

Computers

The development of electronic computers, machines able to perform varied computations automatically with high accuracy and extraordinary speed, has placed in man's hands an unparalleled array of new tools. Each year these electronic devices grow in sophistication and potential. The concerted use of computers is still in its infancy, and already they have gone beyond the stage of simple replacement of less efficient devices to the performance of tasks never before possible.

The use of computers in scientific research is fast-growing. Mention has already been made of the revolution that their use has brought about in X-ray crystallography. Computers are now used not only to perform the cumbersome calculations involved and to control the program of rotation of the crystal and obtaining of photographs, but also to produce electron-density maps and various other displays that help to interpret the data ob-

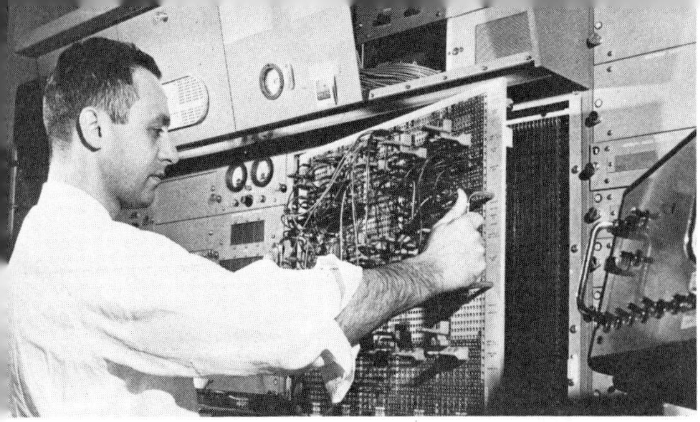

Figure 4-22. Inserting a plug-in program board into an analog computer.

tained. Chemical analyzers, spectrographs, gas chromatographs, and other instruments are now automated with the aid of computers, thus permitting rapid routine analyses of a variety of substances. Indeed, there is a growing trend of providing modern instruments with computer facilities so that they can be linked up into diverse automated systems, adapted to meet the needs of the particular laboratory.

Computers are used in basic research to simulate various chemical and physical processes and thus solve problems, in areas as diverse as biochemistry and hydraulic engineering, without actually conducting the experiments in the laboratory. The storage of data in computer memory banks is already providing researchers with a ready reserve of relevant knowledge, and this use should expand dramatically in the coming years. In medicine, computer-based diagnostic aids are being devised, and computer-controlled monitoring systems now keep a continuous watch over various indices of the condition of seriously ill patients.

Computers have also been programmed to simulate various human mental processes. It is hoped that computer study of such activities as playing checkers and chess, solving problems, reading, recognizing patterns, writing poetry, and scoring music will bring new insight into the workings of the human mind.

Computers have become such an important force in modern life that they will be discussed in greater detail in Chapter 21.

QUESTIONS FOR REVIEW AND THOUGHT

1. What would be the advantages and disadvantages of an immediate changeover to the metric system in the United States?
2. Give examples of instruments used to measure length and mass. Which are particularly applicable to biological studies?
3. How can chemical reactions provide information about living organisms?
4. The introduction of paper chromatography created a revolution in biochemistry. What special advantages does it have over other techniques of separation and analysis?
5. Which substances are sedimented to the bottom of a centrifuge tube, the heavier or the lighter? Why?
6. Give examples of the use of radioisotopes as tools in various sciences. What special precautions must be taken in dealing with radioactivity? Why?
7. Was the microscope really necessary in order to establish that a world of microscopic life exists? Explain.
8. Why would the future development of an electron microscope capable of observing organic materials at resolutions in the neighborhood of 1 Å be a discovery of the first magnitude?
9. In what ways do modern crystallographic techniques fill in gaps that other analytical techniques fail to fill or do so inadequately?
10. Discuss some areas in which computers are aiding biological research.

PART II

THE LIVING WORLD

CHAPTER **5**

The Cell

Introduction

"Why can't scientists create life?" This question is often asked by the layman. It is argued that since we know what is in the cell, why then don't we just add the proper ingredients and cook awhile? This line of thinking is misguided on several accounts. First, biologists are far from knowing even a small fraction of the exact nature of the constituents of cells. There may be as many as a hundred thousand different proteins in the human body, yet we know the exact three-dimensional configuration of perhaps only a dozen of these! This is not to speak of the nucleic acids and other complex constituents of the **cell,** the fundamental structural unit of living things. The cell is undoubtedly the most complex "building block" in the universe. It is very likely far more than the mere sum of its parts. Simply combining the proper number of carbon atoms, oxygen atoms, etc., and stirring over medium-low heat will obviously not result in a living cell. To build an automobile, we would not dream of just throwing together some raw metal, rubber, glass, and plastics and expect a car suddenly to form. Even randomly putting together the finished parts (fenders, wheels, hood, engine, etc.) would be little better. An automobile is a highly complex and integrated device requiring special methods of construction and coordination.

Despite its microscopic size, a living cell is far more complex than an automobile or any other man-made device. And the thousands of interrelated processes that go on within living cells require a delicate precision of coordination and balance far beyond anything man has achieved in his finest mechanisms.

Some cells are capable of existing and functioning independently. Each of the thousands of kinds of single-celled organisms that inhabit the ponds, rivers, lakes, and oceans of the world is capable of growing and reproducing itself, responding to its environment, adapting, feeding or manufacturing its own food—all the activities that characterize living beings.

But far more numerous than the single-celled organisms are the millions of different kinds of multicellular organisms. In these animals and plants, thousands, millions, or even trillions of cells are associated in an integrated whole, each not only carrying out its own life processes but also cooperating with other cells to keep the organism alive. In multicellular organisms, the cells often are highly specialized, much like the factory workers or professionals in our own society. Each specialized cell depends upon others to provide it with the various means of sustenance and to protect it from the ever-hostile environment.

Using another analogy, each cell of a multicellular organism may be likened to a highly integrated and well-organized factory located in the midst of a complex of other factories in related, but nonetheless different, lines of endeavor. A gland cell, for example, may receive oxygen and nutrients from the blood cells and plasma, which in turn take away the gland cell's waste products. It may be stimulated by special message-carrying nerve cells or by chemicals produced by other cells in the complex of the organism, and in turn may produce its own hormones—special chemicals that influence the functioning of other types of cells in other parts of the body.

The Cell Theory

Today even elementary school children are aware of the fact that virtually all living things are made up of microscopic cells. But only 150 years ago this was still a point of controversy among scientists. When Robert Hooke first applied the term *cell* to organic matter back in 1665 (after seeing the cell walls of once-living cork tissue and noting their resemblance to the cells in the honeycomb of a beehive), he and the other scientists of his day never realized the widespread implications of this discovery. For to Hooke the "cell" signified an empty cavity, not its living contents.

At about the same time (1674), the Dutch lensmaker Leeuwenhoek reported seeing "animalcules" or single-celled "animals" through his carefully ground lenses. Later Leeuwenhoek reported seeing what was undoubtedly bacteria. But still scientists did not realize the wider applications of these findings. The concept of cells as basic living units of virtually all organisms had yet to evolve.

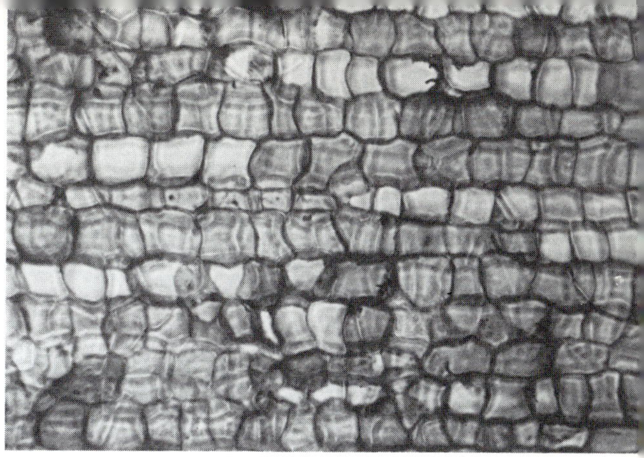

Figure 5–1 Cork, magnified 160×.

In time, the quality of microscopes improved. Thousands of individual observations of cells were made in the eighteenth century. Mirbel in France made numerous observations of plant structures, and in 1802 he concluded that plants are made up of cells. The "cell theory" continued to evolve as a number of scientists, including Lorenz Oken in Germany (1805) and Jean Baptiste Lamarck (1809) and René Dutrochet (1824) in France, postulated that cells are a common factor in plants and animals.

Thus, in the early 1800s the concepts of the cell theory were beginning to gain currency in scientific circles. Two German biologists, Mathias Schleiden (a botanist) and Theodor Schwann (a zoologist), wove the various loose ends into a coherent cell theory. In 1838 and 1839 these two scientists independently published works indicating the fundamental nature of the cell as a basic constituent of all living organisms. They emphasized the cell as the fundamental unit, both structural and functional, in the organization and development of life, and their landmark works captured the attention and support of the scientists of their day. Thus, though Schleiden and Schwann did not originate the theory that they were so instrumental in popularizing, they are now generally recognized as the fathers of the cell theory.

With the attention of the scientific community focused upon the cell, its structure and contents became the objects of intensive scrutiny. In 1831 the English scientist Robert Brown described and named the *nucleus* of the cell. At the end of the decade (1839) Johannes Purkinje in Bohemia proposed using the term *protoplasm* to describe the material within the cell.

Another advance after the formulation of the cell theory itself came about twenty years later, in 1858, when the German physician Rudolf Virchow stated an important principle: all cells arise from preexisting cells. This principle implies a continuity of life; although new materials are taken in from the environment and various products of cell activity are given off, life itself goes on through a continuous succession of cellular generations. The continuity of life extends even through successive generations of entire organisms, since each new organism arises from a single cell or pair of cells (sperm and egg) from its parent or parents. Thus, Virchow's postulate implies an unbroken chain of life, stretching back from the cells of present-day organisms through billions of years of evolution to the single-celled ancestors of the varied life forms of today.

The major generalizations of the cell theory, formulated in the mid-nineteenth century, served as a basis and impetus for further studies of the cell. In the past hundred years or so, enormous strides have been made in staining and other techniques of microanalysis, climaxing in the use of the ultracentrifuge and the ultrahigh resolutions obtained with the electron microscope. As a result, our picture of the cell has been greatly clarified. Many mysteries remain, to be sure, but these are slowly giving way under the worldwide and persistent efforts of cytologists (cell biologists) and microscopists.

An Overview of the Cell

The concept of the cell has come a long way since Hooke first used the word to describe the lifeless remnants of the honeycomblike cell walls of cork. When scientists learned that cells are actually the living units of nearly all the organisms on earth, they thought it natural to assume that each cell contains a unique living substance, which came to be called the **protoplasm.** This term, and the concept that it represented, were for some time quite useful, insofar as they emphasized the fundamental similarities of living things.

But refined techniques and technology led to the knowledge that protoplasm is not really a uniform "living substance," or even a simple mixture of various substances. It is rather a highly intricate and well-integrated complex of specific structures, each making its own contribution to the functions of life. As a result, the use of the term *protoplasm* has tended to become somewhat more limited. Scientists frequently prefer to use somewhat more precise derivatives of the word, such as the **protoplast,** which is sometimes used to refer to the living cell as an entity, and **cytoplasm,** referring to a definite region within the cell.

It is tempting, and yet a great oversimplification, to speak of the structure of "the" cell—as though there were some typical cell that embodies all the major characteristics of cells in general. This is not at all the case. Cells are fantastically diversified. First of all, they vary widely in size and shape (Fig. 5–2). For example, the cells of tiny microorganisms called mycoplasmas are about 0.1 μ in diameter (a tenth of a micron: *ca.* $^{4}/_{1,000,000}$ of an inch), while those of many protozoa reach 200–300 μ (actually visible to the naked eye as tiny specks), and some of the nerve cells of elephants and whales, while of microscopic thickness, may be stretched out for yards.

Even within the same organism there are great differences in the gross structure of cells. In man, for example, the nerve cells are usually long and threadlike, some cells in the skin may be shaped like flat paving stones, red blood cells are like little round doughnuts without the hole completely through the middle, and white blood cells are amorphous blobs that can change their shape readily. Within the cell there are also great variations, particularly in multicellular organisms, in

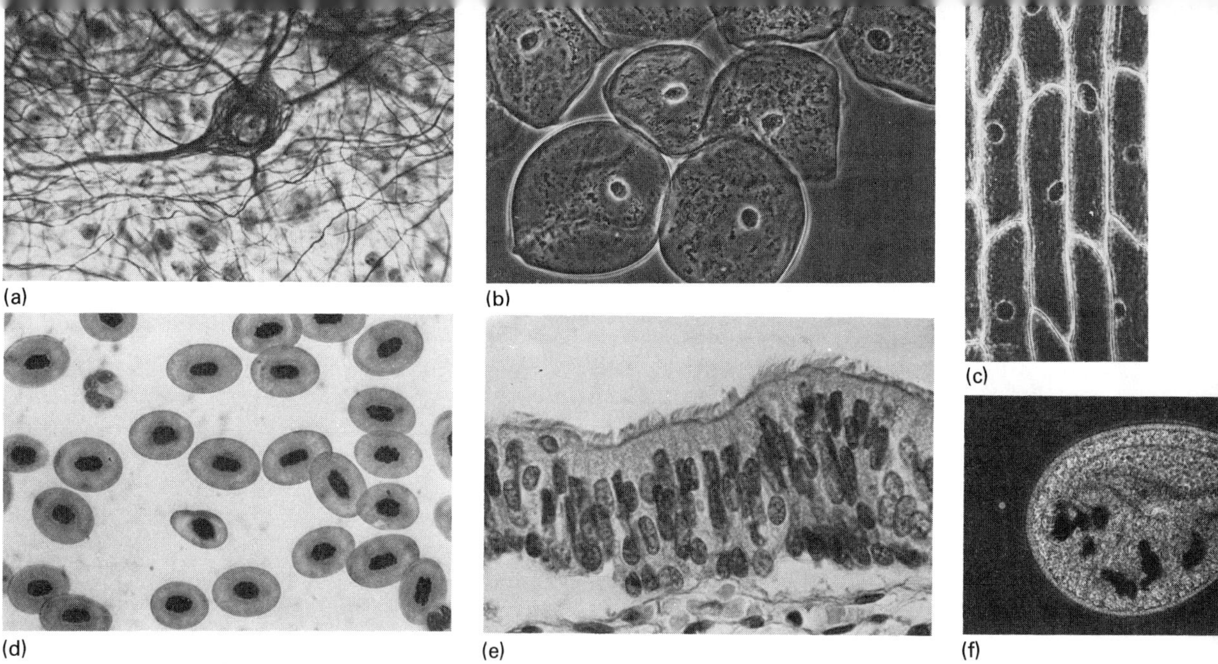

Figure 5–2 The diversity of cells: (a) a neuron in a mouse spinal cord ganglion, 389×; (b) human cheek cells; (c) cells in an onion bulb, 70×; (d) frog blood cells, 389×; (e) cells in a section of human trachea, 389×; (f) *Bursaria truncatella*, a single-celled organism.

which groups of cells are highly specialized for one particular function or complex of functions. Even in single-celled organisms, there may be highly specialized structures modified for particular functions such as locomotion or food-getting.

Despite their great diversity, cells in general have many basic features in common. When viewed under a microscope, a cell is seen to be in one of two fundamental states: **dividing** or **nondividing.** During cell division, or **mitosis,** the structures of the cell undergo a regular cycle of changes, culminating in the formation of two smaller duplicates of the original cell. The nondividing state is sometimes referred to as the "resting" state. This term is to some degree misleading, for in this state the cell is not at rest. There is a seething whirl of activity—synthesis and breakdown of chemical compounds, transport of matter within the cell and between the cell and its surroundings.

The nondividing cell is a three-dimensional structure, with a definite shape and a highly complex and integrated substructure (Fig. 5–3). It is characteristically enclosed within a thin envelope of protoplasm, the **cell membrane** or **plasma membrane,** which serves to keep the contents of the cell together. This membrane is semipermeable: it permits the free passage of certain substances into or out of the cell, while retaining or barring the passage of others. The cell membrane thus also regulates the inflow and outflow of materials—the exchange of matter between the cell and its surroundings. The semifluid material within the cell membrane is not a mere watery dispersion, but rather a highly organized and structured complex of chemical substances, some circulating freely and others aggregated into larger units with distinctive structures and specialized functions. Within the cell is suspended a smaller, roughly spherical body called the **nucleus.** The protoplasm outside the nucleus is commonly referred to as the **cytoplasm.**

The structural subunits within a cell are sometimes referred to as **organelles.** Although the basic functions of many of these structures may remain relatively constant from cell to cell, their appearances and positions within the cell often vary considerably.

In addition to the nucleus and cell membrane, several other organelles are found almost universally in cells. Within the nucleus are the **chromosomes,** nucleoprotein struc-

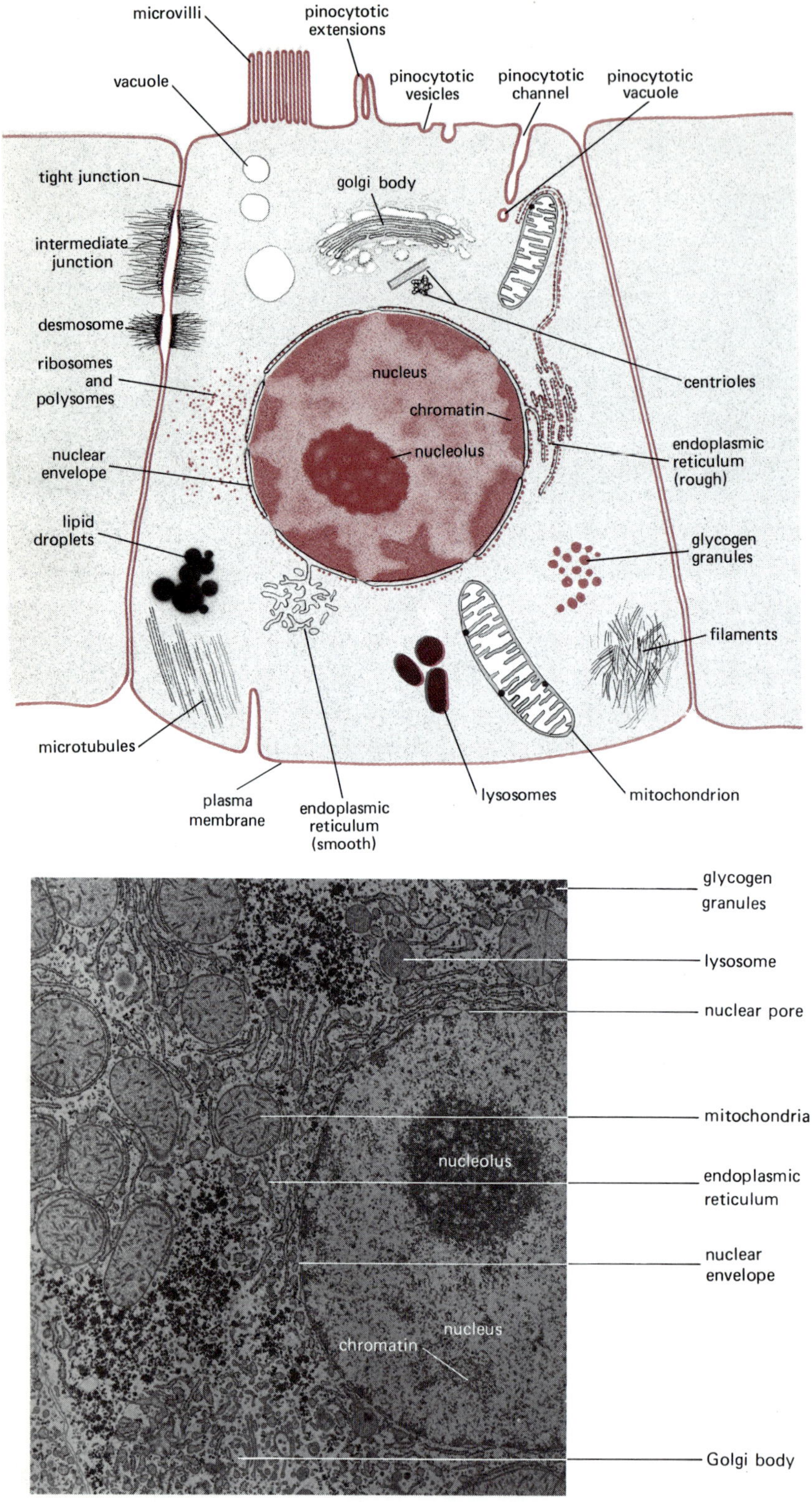

Figure 5–3 (a) Drawing of a generalized animal cell. (b) Electron micrograph of a cell from the liver of a newborn rat (9000×).

tures that carry the basic hereditary information of the individual organism. Extending out from the **nuclear membrane** or **envelope,** which encases the nucleus, is an elaborate complex of membranes, called the **endoplasmic reticulum.** This is a saclike system, which to some degree organizes and compartmentalizes the cytoplasm and links the nuclear membrane with the outer plasma membrane of the cell. Associated with the endoplasmic reticulum are specialized structures called **ribosomes,** which are intimately involved in protein synthesis in the cell. Other membranous structures of the cytoplasm, called **mitochondria,** are the centers of cell respiration, where fuel substances are broken down into smaller fragments and where the energy released is stored in compact chemical packets —ATP molecules. The **lysosomes** are organelles that digest materials within the cell.

In addition to the organelles that are almost universally found in cells of all kinds, there are various specialized organelles found in certain types of cells but not in others. Plant cells (Fig. 5–4) generally contain structures called **plastids,** which function in the manufacture and storage of food substances. Many cells, particularly those of single-celled organisms, contain **vacuoles,** saclike structures that are used for the temporary storage of materials or for the elimination of excess water and waste products from the cell.

Just as it is often tempting to speak of the structure of the "typical" cell, it is also sometimes convenient to speak of "animal cells" and "plant cells" as though they were quite separate and distinctive entities. This, too, is a great oversimplification, since plant and animal cells have far more points of similarity than of difference. A few distinctions are nevertheless valid in most cases. For example, plant cells in general are encased in a thick, rigid outer **cell wall,** made up chiefly of the nonliving material **cellulose.** Animal cells, on the other hand, do not have such a cell wall; they are enclosed only within the thin, flexible, semipermeable cell membrane. (Plant cells have a similar cell membrane, but it lies just inside the cell wall.) Another common distinction is the fact that many (but not all) plant cells contain the pigment **chlorophyll,** which gives them their characteristic green color; no animal cells contain this pigment.

The citation of distinctions between plant and animal cells runs into an important difficulty, in that there are large groups of organisms that do not fit clearly into either of the two categories. Most bacteria, for example, do not possess chlorophyll, and though they possess cell walls they are not of the same cellulose nature as is found in the higher plant forms. Many protozoa, single-celled "animals," are ensheathed in a protective flexible covering known as a **pellicle.** Like the cell walls of plant cells, the pellicle provides some support and maintains the shape of the animal, but it is not as rigid and permits ephemeral distortions that may be necessary in the normal functioning of the protozoan. Euglena and other similar single-celled organisms normally possess chlorophyll and manufacture their own food, as typical plant cells do—but they move about actively with the aid of a lashing whip-like structure called a flagellum and have no cell wall.

After many years of wrestling with such perplexing "in-between" creatures, taxonomists (specialists in the classification of organisms) are tending more and more to consider the living world to be divided into *three* major kingdoms: animals, plants, and **protists.** In the last category are classified such controversial organisms as bacteria, protozoa, algae, fungi, and slime molds.

Membranes of the Cell and the Transport of Matter

The human body has more than 100 trillion "mouths" to feed and attend to, each with separate and urgent needs. These living cells continuously require nutrients, react sensitively to varied stimulants and stimuli, and rid themselves of a continual buildup of waste products. In humans, the blood plays a vital role in bringing oxygen, nutrients, and "information" (in the form of hormones and other chemical messengers) to the cells. But in

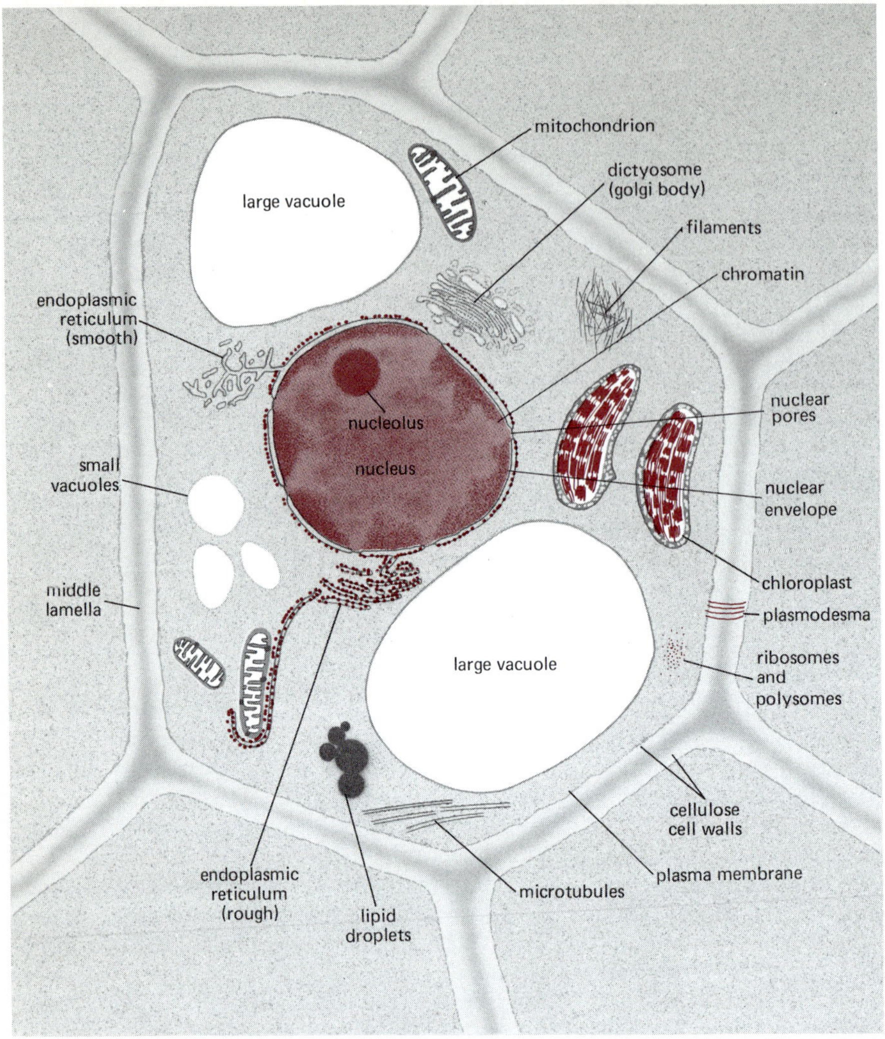

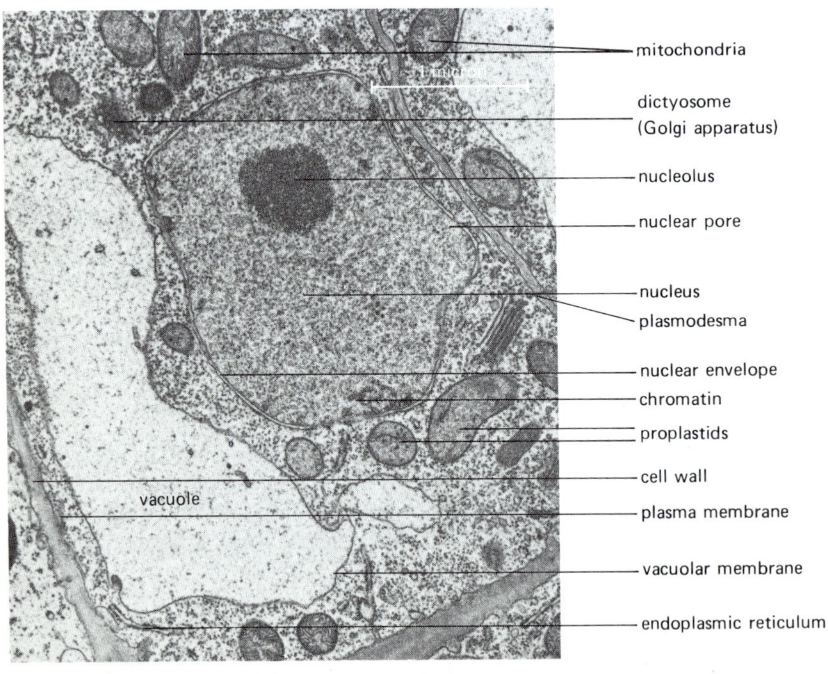

Figure 5–4 (a) Drawing of a generalized plant cell. Note the characteristic cellulose cell wall and large vacuoles. (b) Electron micrograph of a cell from the root tip of the plant *Arabidopsis thaliana* (21,000×). This cell is about 2 microns in diameter.

order to be effective, these materials must penetrate a barrier, the plasma membrane.

Every living cell is ensheathed in an invisible envelope. In animal cells, this envelope, the **plasma membrane,** is generally the outermost covering of the cell. In plant cells the plasma membrane lies just inside the rigid outer covering, the cell wall.

Before the advent of the electron microscope, scientists had postulated the existence of a cell membrane enveloping the living cell and had even formulated ideas on the composition of this membrane from studies of the behavior of the cell surface. But they could not be entirely sure of its existence, because the plasma membrane is invisible even under the best light microscopes. Electron micrographs revealed that this membrane does indeed exist and explained earlier failures to see it: plasma membranes vary in thickness from about 50 to 200 Å, well below the resolving power of ordinary light microscopes (Fig. 5–5).

By combining the information regarding cell properties from electron micrographs with that gathered by other means, scientists have developed a coherent representation of the molecular structure of the plasma membrane. J. D. Robertson has reported that a similar molecular arrangement is observed at the surfaces of many kinds of animal cells, plant cells, and bacteria. He postulates that such a **unit membrane** is "a biological constant common to the surfaces of all kinds of cells." In Robertson's interpretation, the unit membrane is a sandwichlike layered structure consisting of three basic regions (Fig. 5–6): a thin, dense outer layer (ca. 20 Å thick) consisting of protein molecules; a central region made up of a double layer of phospholipids (ca. 35 Å); and an inner proteinaceous region (ca. 20 Å).

The most recent studies seem to indicate that although this view is basically correct, the layer arrangement is not continuous. In some places, the protein component may penetrate the entire membrane. Some researchers even visualize the proteins of the membrane as floating in a fluid layer of lipid matter.

The membrane system of the cell does not stop at its surface. Extending into the cytoplasm is an elaborate system of membranes,

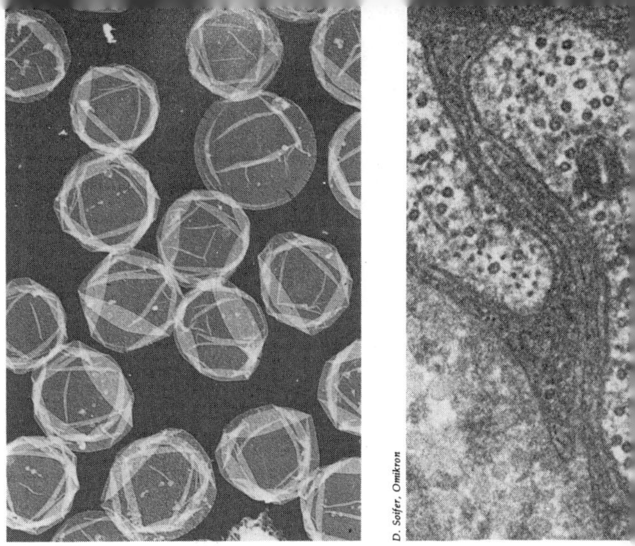

Figure 5–5 (a) When red blood cells are placed in distilled water, water diffuses into them by osmosis until they burst, releasing their contents. The "red cell ghosts" that remain are the empty cell membranes. 1920×. (b) Electron microscopy reveals a distinct unit membrane structure, shown here in a transverse section of unmyelinated axons in a cat sciatic nerve, 50,000×.

Figure 5–6 Schematic drawing of the unit membrane. Fibrous proteins are believed to make up the outer and inner layers of the "sandwich," while the middle layer consists of lipid molecules, arranged at right angles to the surface.

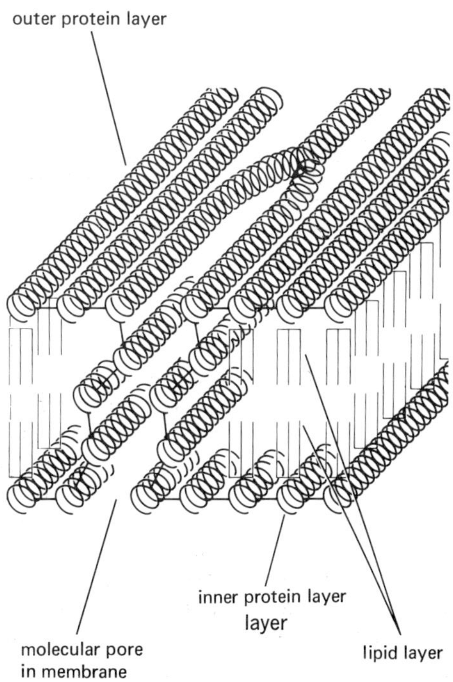

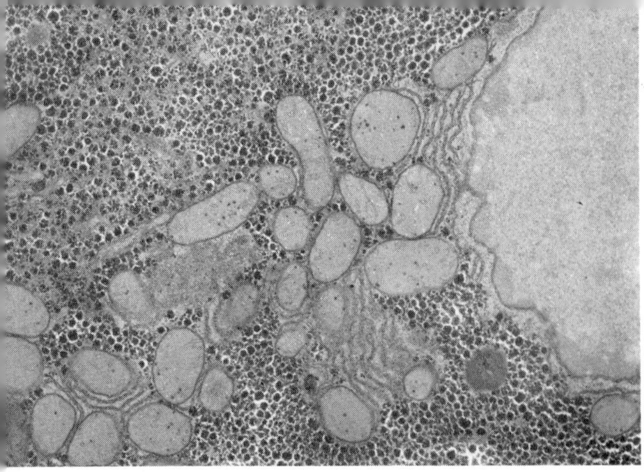

D. Soifer, Omikron

Figure 5–7 This portion of a frog liver cell shows an abundance of smooth ER among glycogen granules (dark spots) and sheets of rough ER associated with mitochondria. (9,315×)

the **endoplasmic reticulum (ER).** Electron micrographs reveal that the endoplasmic reticulum forms a system of interconnecting canals enclosing a number of vacuoles or spaces (Fig. 5–7). The structure of this membrane system varies greatly among different kinds of cells. The membranes of the ER may be tightly packed within the cytoplasm, or they may be loosely distributed throughout the cell.

Two types of endoplasmic reticulum have been observed: rough and smooth. The **rough ER** is especially abundant in cells in which protein synthesis is particularly active. Ribosomes—organelles concerned with protein synthesis—are often found associated with the rough ER, thus permitting a compartmentalization of protein-synthesizing activities in the cell. The **smooth ER,** on the other hand, does not have associated ribosomes and occurs most extensively in cells active in the synthesis of lipids and steroids. Compact arrangements of smooth ER are also found as a structural component in certain specialized cells. For example, an intricate layered network of smooth ER membranes containing light-sensitive pigments is found in the light-receptor cells of the eyes of many different organisms.

In general, the membranous network of the ER provides a vastly increased surface area within the cytoplasm upon which varied chemical reactions may take place. Since enzymes are believed to be associated with the membranes of the cell, local concentrations of specific enzymes may permit a localization and compartmentalization of specific reactions within the cell. In addition, the ER provides pathways for the transport of materials, and its compartmentalization may permit the collection and temporary storage of various chemical products.

Electron micrographs have revealed that the endoplasmic reticulum is continuous not only with the plasma membrane at the outer surface of the cell, but also with the **nuclear membrane** or **envelope** within the cell. The latter membrane possesses the same unit-membrane structure as the other membranes of the cell, except that it is only double-layered. (It is the outer layer that is continuous with the endoplasmic reticulum.) The nuclear membrane possesses another peculiarity, in that it has a series of pores that may permit the passage of large molecules.

In order for a cell to survive, there must be an interchange of materials with the environment and among the various parts of the cell itself. It is the system of membranes that we have just described that regulates these flows of materials. For the plasma membrane, ER, and nuclear membrane are permeable to many types of molecules, permitting their free passage through the membrane barrier. However, this is a selective permeability: though certain substances pass freely through the membrane, others are retained and normally do not pass through at all. Hence, the membranes of the cell are referred to as **semipermeable** or **differentially permeable.**

Through studies with various nonliving models, such as cellophane, and through studies of the properties of cells themselves, scientists have learned much about the principles underlying the transport of materials through semipermeable membranes.

This transport of materials occurs largely through diffusion, as was discussed in Chapter 2. The constant random motion of molecules results in a tendency for the concentrations of substances on each side of the semipermeable membrane to be equalized.

When solvent molecules move through the membrane, the process is known as **osmosis;** the transport of solute molecules through a membrane is referred to as **dialysis.**

A semipermeable membrane has been likened to a tiny sieve (Fig. 5–8): in general, smaller molecules move through the membrane with greater ease than do larger molecules. But when the membranes of actual cells are considered, numerous exceptions to this general rule are found. Many larger molecules of certain quite definite types are able to penetrate cell membranes with relative ease. Moreover, the transport of materials often occurs *against* the concentration gradient, the antithesis of diffusion. Hence, there must be some other, perhaps even more important, mechanism (or mechanisms) of the passage of materials through the semipermeable membranes of the living cell.

It has been shown repeatedly that there is a direct relationship between the fat solubility of a substance and the ease with which it penetrates the plasma membrane. This fact seems to imply that the fat-soluble substance must somehow be aided in its passage through the plasma membrane by the high lipid content of the membrane.

Certain molecules and ions can pass into a cell quite readily against a concentration gradient. Marine animals called tunicates, for example, are capable of accumulating the metal vanadium to such an extent that the concentration of this element within the animals' bodies may be 2 million times as great as that in the surrounding sea water. Or to take another example, the functioning of human nerves is aided by a "sodium-potassium pump," in which potassium ions pass into the nerve cell and sodium ions pass out, each against the concentration gradient for the particular ion.

The movement of molecules and ions in a direction opposite to that of the concentration gradient is known as **active transport.** This is an energy-requiring process, the needs of which are supplied by energy-generating chemical reactions of the cell. The exact mechanism involved in active transport is still undetermined. One of the most widely held hypotheses is that some sort of "carrier" molecules form complexes with ions or other substances and transport them through the membrane (Fig. 5–9).

Large molecules and fluids may on occasion be taken into the cell by an infolding of the plasma membrane extending over the particle until it is entirely enclosed. Then a saclike cavity is formed, probably by a pinching-off of the plasma membrane. In some way the particle is then released into the cell. When solid particles are taken into the cell in this way, the process is referred to as **phagocytosis,** which literally means "cell eating." The taking of liquids into the cell is called **pinocytosis,** or "cell drinking." White blood cells capture bacteria in the human bloodstream by a process of "phagocytosis," and this is also the method used by the microscopic pond dweller, amoeba, to capture its prey (Fig. 5–10).

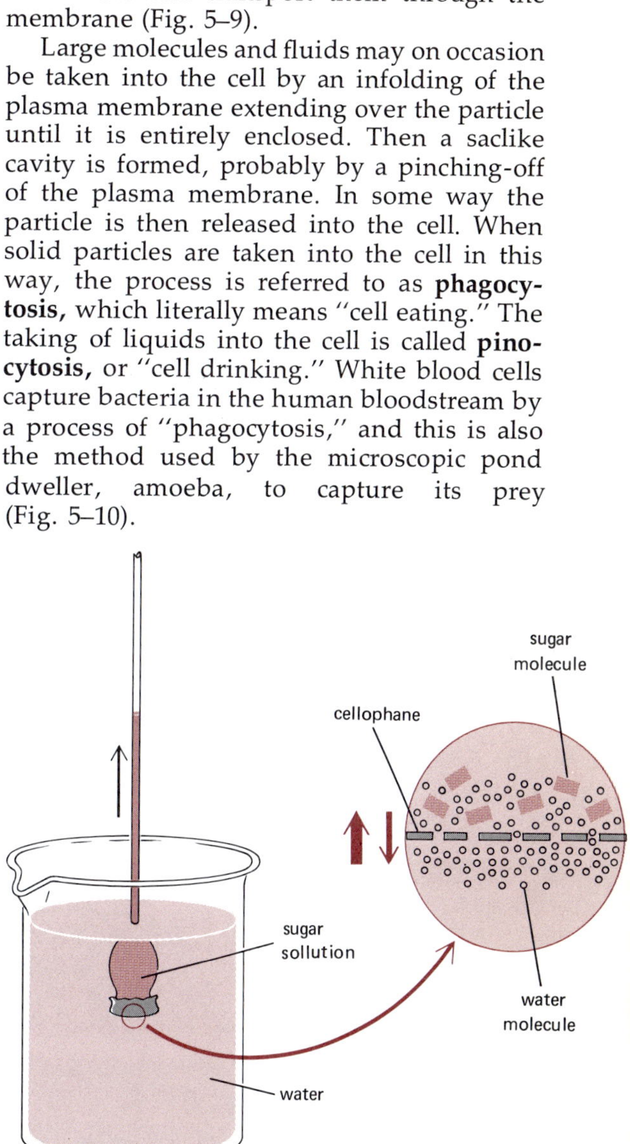

Figure 5–8 How a semipermeable membrane works in osmosis. The tiny water molecules diffuse easily through the pores of the membrane, while the sugar molecules are too large to pass through. As the arrows indicate, there is a net transfer of water molecules into the thistle tube (transport along the concentration gradient).

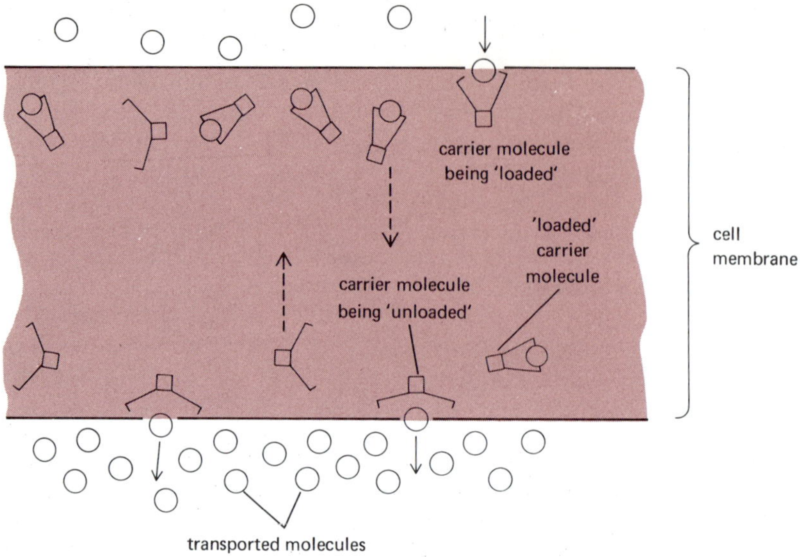

Figure 5–9 Active transport across the cell membrane. A currently accepted hypothesis explains transport against the concentration gradient by the functioning of fat-soluble carrier molecules, which combine with molecules or ions at the cell surface; the loaded carrier molecules diffuse through the membrane from a region of high concentration at the outer surface to a region of lower concentration at the inner surface, then discharge their load into the cell and diffuse back to the outer surface where they can be reloaded. Both the "loading" and the "unloading" reactions are catalyzed by enzymes and consume energy.

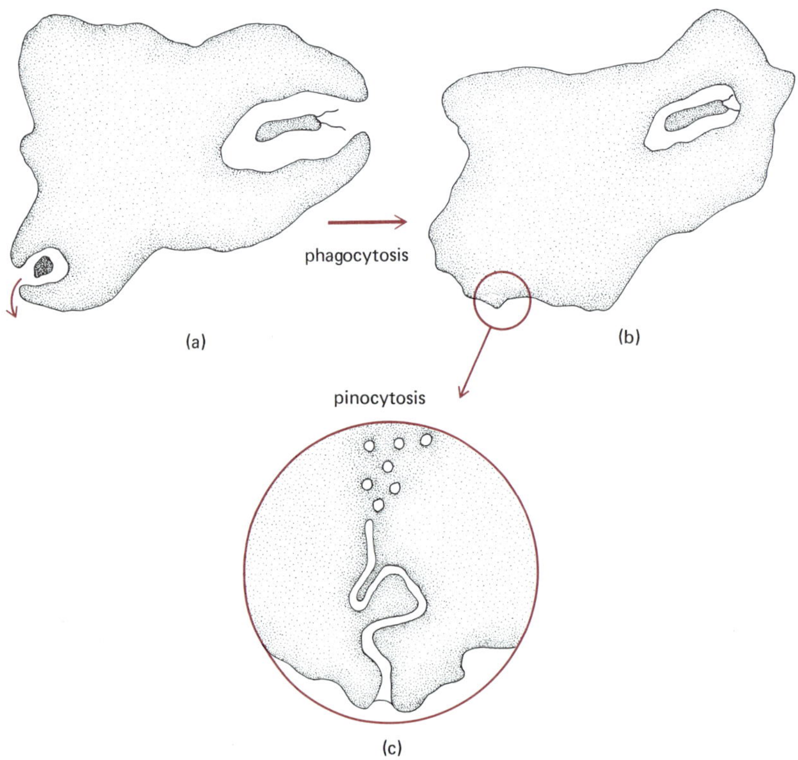

Plasmolysis and Turgor

If human red blood cells are placed in a glass of ordinary drinking water, they will quickly begin to change shape. Their normal biconcave forms will swell until they resemble tiny balloons. Suddenly, one by one, these red blood cells will explode, spewing their cellular contents throughout the watery medium. The reason for this is that when the cells are first placed in water the higher concentration of water molecules outside the cells results in an influx of these water molecules into the cells. As the number of water molecules within the cell continues to increase, their very presence causes an increase in internal pressure. Although the plasma membrane of the red blood cells is relatively elastic, there is a limit beyond which they cannot stretch—and so they burst.

A solution which contains less dissolved material than a particular cell (i.e., has a greater water concentration than the cell) is said to be **hypotonic** to that cell. Under such circumstances, water will tend to diffuse into the cell (osmosis). As a result, the cell becomes firm or **turgid** (Fig. 5–11a). In animal cells, as we have seen, the cell may be destroyed if the process continues unchecked.

In plant cells, on the other hand, the presence of the strong cell wall prevents the cell from bursting. At some point, the pressure within the cell balances the tendency for osmosis, and an equilibrium is reached, even though the concentrations within and outside the cell are not equal. The increased pressure due to the influx of water—known as **turgor pressure**—provides support to the plant by giving the individual cells a firm and solid consistency. The loss of this water would result in the wilting of the plant.

If the external medium of the cell contains a higher concentration of solute than the cell, it is said to be **hypertonic** to that cell. In this case, water will tend to diffuse out of the cell, and the cell first will become flaccid and then, as dehydration proceeds, a condition known as **plasmolysis** sets in. A **plasmolyzed** animal cell shrivels up to a small fraction of its normal size (Fig. 5–11b). In the case of plant cells, the cell wall maintains the general outline of the cell, and it is superficially unchanged. But a closer examination of a section under a micro-

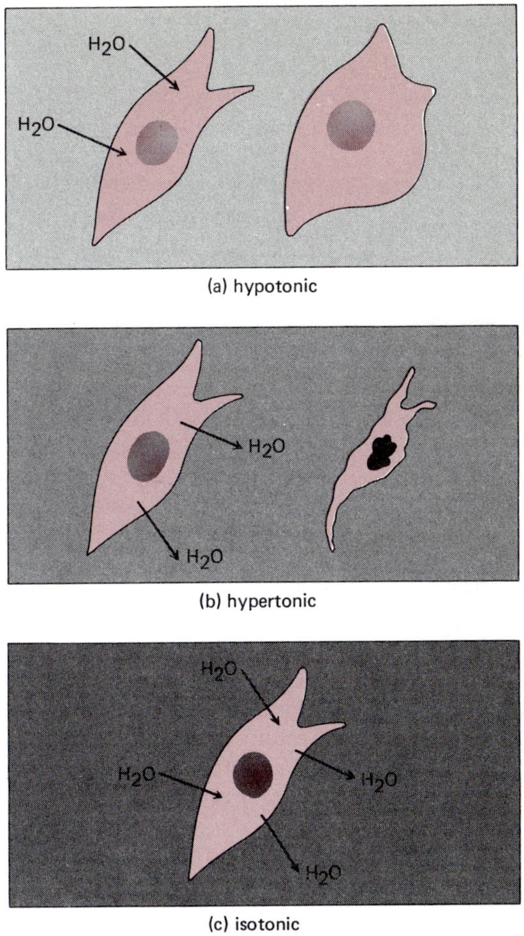

Figure 5–11 In a hypotonic medium (a), water diffuses into the cell; it swells and may ultimately burst. In a hypertonic medium (b), a cell tends to lose water by osmosis and shrivels. In an isotonic medium (c), the net inflow of water is equal to the net outflow, and the cell is at equilibrium with its surroundings.

Figure 5–10 (*opposite*) Phagocytosis and pinocytosis in an amoeba. In (a), pseudopods are surrounding a small protist. In (b), the protist has been completely enclosed in a vacuole; (c) shows a pinocytotic channel, terminating in several small vesicles, permitting the intake of liquids and small adsorbed solid particles.

scope reveals that the cellular material is concentrated near the center of the cavity enclosed by the cell wall.

When a cell is placed in a medium with an osmotic pressure equivalent to that found within the cell, the medium is said to be **isotonic** with respect to the cell. That is, so long as these conditions exist, there will be no *net* inflow or outflow of water (Fig. 5–11c). A 0.9% NaCl solution, for example, is isotonic to human cells. Such an isotonic solution, sometimes referred to as **physiological saline,** will leave red blood cells suspended in it essentially unchanged, in contrast to distilled water, which will cause them to burst, or a 2% salt solution, which will cause the same cells to shrivel.

Organisms that live in a watery environment spend their lives in a constant struggle to maintain isotonic equilibrium. Fresh-water dwellers are immersed in a medium that is generally hypotonic to their cells. In fresh-water plants, the water that diffuses into the cells sets up turgor pressure, and the cells are kept from bursting by their strong cell walls. But in fresh-water animals and protozoa, the cells are surrounded only by a thin plasma membrane, which will burst if the turgor exceeds a certain limit. Such organisms cope with the constant influx of water by pumping out the excess with mechanisms that vary with the type of organism, although all require energy generated within the cell or organism (Fig. 5–12a,b).

Quite different problems are presented to the organisms that live in seas and oceans. Sea water has a salt content of about 3%, which is approximately isotonic to the cytoplasm of marine plants and invertebrates. But this salt concentration is hypertonic for the bony fishes, which thus must cope with a continual loss of water from their cells by osmosis. To prevent plasmolysis, such fishes must have a continual intake of water to replace that loss. But their only source of water is the hypertonic sea. These fishes have solved the problem by special (energy-requiring) desalting mechanisms, which remove salt from the sea water that they drink and excrete it back into the environment through the gills (Fig. 5–12c).

Figure 5–12 Water-dwelling protozoa and animals must wage a constant battle to maintain equilibrium with an external medium that may be hypotonic or hypertonic to their cells. The amoeba (a) uses a contractile vacuole to pump out excess water that diffuses in from the hypotonic medium. The fresh-water fish (b) uses more complicated mechanisms for the same purpose, while the salt-water fish (c) must excrete the excess salt it takes in from the hypertonic sea water.

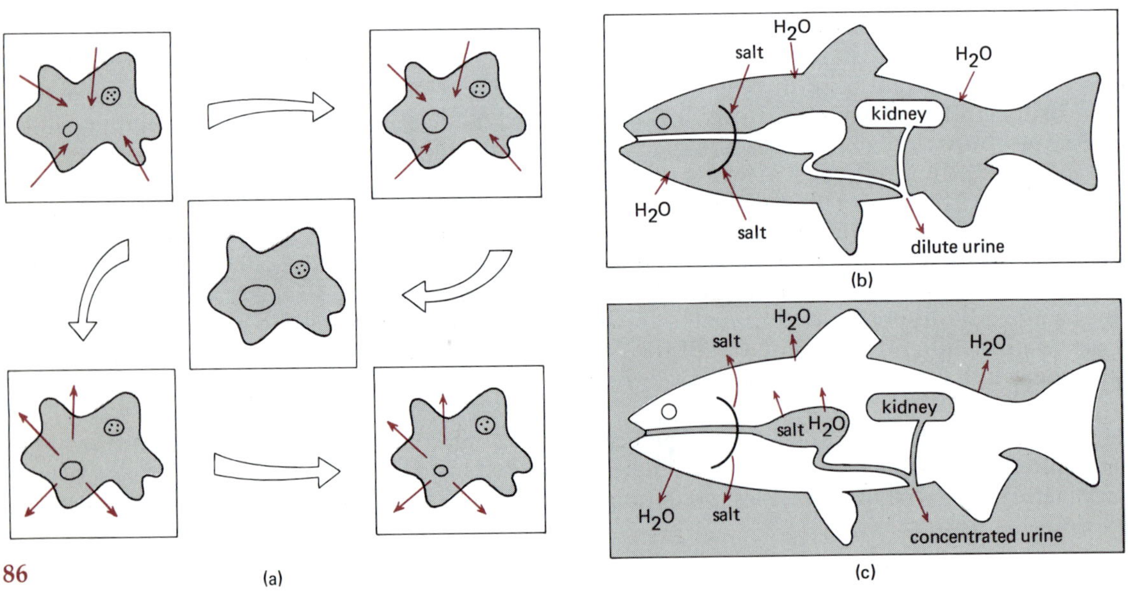

The Cytoplasm and the Colloidal Properties of Cells

By far the largest volume of the cell is taken up by the **cytoplasm,** the region outside the nucleus. Furthermore, most of the organelles of the cell are situated in this region. The cytoplasmic fluid, shorn of its organelles, is referred to as the **ground substance.** In the outer regions of most cells, this ground substance is relatively clear and rigid or viscous, and is often called the **ectoplasm.** The more watery inner region, known as the **endoplasm,** is permeated with granular material and contains such cytoplasmic structures as the ribosomes, mitochondria, lysosomes, Golgi bodies, and others.

The ground substance of the cytoplasm is composed of a complex **colloidal dispersion:** within the watery medium are suspended numerous protein molecules, oil droplets, and granules, all particles of such a size that they will neither dissolve nor settle out upon standing.

Colloidal dispersions possess a number of unique properties, an understanding of which can provide an insight into some of the characteristics of living cells. For example, colloidal dispersions can alter their viscosity rather suddenly. A sudden increase in the viscosity of a colloidal dispersion is known as **gelation,** and the resultant viscous state is termed a **gel.** The reverse process—that is, a sudden decrease in the viscosity of a colloidal dispersion—is referred to as **solation,** and the resultant more liquid state is called a **sol.** It is believed that the unique ability of colloids to change rapidly from sol to gel and back again is utilized by the amoeba in its characteristic amoeboid motion.

The oozing of an amoeba across the visual field of a microscope is a fascinating sight. A relatively featureless blob with no apparent means of locomotion begins to swell in an unexpected direction, the tiny projection **(pseudopod)** steadily growing until the entire mass of the cell has been transferred to the new position. It is believed that this movement occurs through reversible sol $\longleftrightarrow$ gel interconversions: at the site where a pseudopod is forming, the highly viscous (gelated) cytoplasm liquefies and flows out, stretching the elastic plasma membrane. The advancing plasmasol thickens again, forming plasmagel, while more plasmasol flows forward into the pseudopod formed. According to a recent hypothesis, contraction of the plasmagel helps to pull the plasmasol forward (Fig. 5–13).

Interconversions between sol and gel phases are also responsible for other cytoplasmic phenomena: the protoplasmic streaming, known as **cyclosis,** which is readily observed in numerous living cells and produces a sort of intracellular circulation; the contraction of muscles; and the sudden impermeability of a newly fertilized egg to the penetration of further sperm.

Another property of colloids, one whose importance would be difficult to overempha-

Figure 5–13 (a) The movement of an amoeba involves a succession of sol-gel transformations. (b) Cyclosis, the rapid streaming of cytoplasm around a central vacuole in a plant cell, also involves sol-gel interactions.

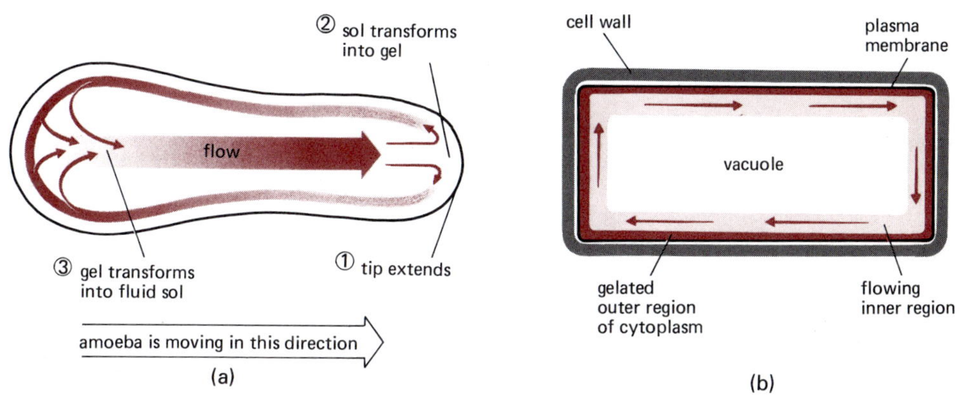

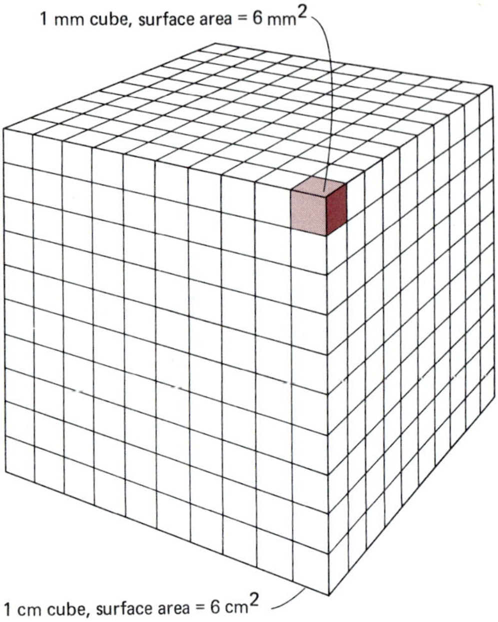

Figure 5–14 Reduction of particle size results in an increase in surface area. The 1 cm cube pictured has a surface area of 6 cm^2. If it were cut apart into one thousand 1mm cubes, its surface area would be increased to 6000 mm^2 or 60 cm^2.

size, is the huge increase in intracellular surface area that is afforded by these particles. To illustrate this point, let us consider the surface area on a 1-cm cube. Since each of the six sides measures 1 centimeter × 1 centimeter, the total surface area comes to 6 cm^2. But if we divided up this 1-cm cube into tiny cubes of the dimensions found among some of the smaller colloidal particles, let us say, into 1-mμ cubes, the total surface area would be about 60,000,000 cm^2—about the size of one-and-a-half football fields! Since numerous chemical reactions occur optimally at interfaces—in this case, on the surface of colloidal particles—the fantastic amount of surface area presented by these particles greatly accelerates such reactions.

The Nucleus and Cell Division

If the nucleus of an amoeba is removed by careful micromanipulation (Fig. 5–15), and the denucleated protozoan is then cultured in a normal manner, it at first seems to have been unaffected by the delicate surgery. It moves about in a seemingly normal manner. But then its movements slow down, and it cannot digest its food. Soon it becomes noticeable that growth has ceased, and in a few days the amoeba dies without having reproduced in the intervening time (as is otherwise normal). If, on the other hand, an amoeba is simply pierced with a microdissector and the contents of the cell are stirred around, but the nucleus is not removed, the protozoan, after recovering from the operation, continues all its normal activities, feeding, growing, and ultimately dividing to produce two daughter cells that live out their life cycle in turn.

By experiments such as these it has been clearly demonstrated that the nucleus is the organelle responsible for growth and reproduction in the cell.

There are some cells, such as those of bacteria and the primitive blue-green algae, which seem to lack a definitive nucleus. Closer studies have revealed "nuclear regions" and "nucleoid bodies" in such cells (Fig. 5–16). Although these regions are not bounded by a nuclear membrane, as is the nucleus in higher organisms, they do possess the properties characteristic of nuclei in general.

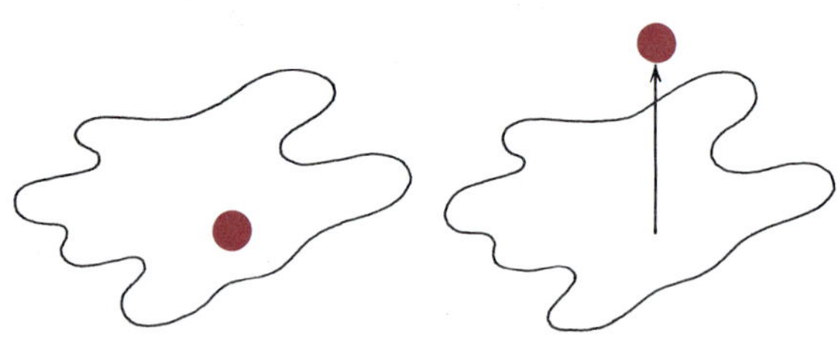

Figure 5–15 The nucleus of an amoeba can be removed by careful microdissection without killing the organism. The anuclear amoeba at first appears normal, but cannot function normally and ultimately dies without reproducing.

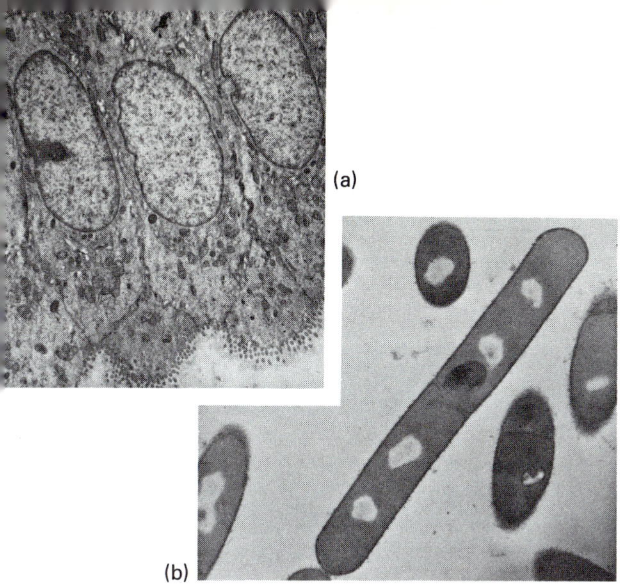

Figure 5–16 (a) Typical animal cells from a cat gallbladder show well defined nuclei, bounded by a nuclear membrane (2442×); (b) an ultrathin section of *Bacillus subtilis,* stained to show the nuclear region (12,500×).

Mature mammalian red blood cells do not possess a nucleus. (The immature red blood cell is nucleated, but its nucleus degenerates after a certain stage of development has been reached.) It is interesting to note that the red blood cell, unlike most other cells, is incapable of self-duplication and eventually dies (in man, in about 120 days).

CHROMOSOMES

In a normal "resting" cell, the nucleus is filled with a diffuse network of thin fibers called the **chromatin network,** so called because when the cell is subjected to a chemical reaction called the Feulgen reaction, the chromatin (from the Greek *chroma,* "color") is stained a deep purple and is seen under a microscope to resemble a fishnet. During cell division, the long fibers of the chromatin network untangle and become tightly coiled into discrete rod-shaped bodies. This is the form in which the **chromosomes** are generally visualized.

The chromosomes are composed of DNA, low-molecular-weight proteins called **histones,** more complex proteins, and RNA. The DNA and histones are chemically associated in complexes called **nucleoproteins.** The interrelationship of DNA and histones and the manner in which the other constituents fit into the structure and functions of the chromosomes are phenomena not yet completely understood. But it is known that incorporated into the structure of the DNA is the "master plan" of the cell, the genetic information that determines the development and functioning of the cell—indeed, of the entire organism.* This information is normally perfectly duplicated and passed on to each of the daughter cells formed when a cell divides.

The DNA molecules of the chromosomes are extensively coiled and twisted upon themselves. It has been estimated, for example, that the DNA in some species of bacteria is about five hundred times longer than the bacterium itself. In some regions this coiling and folding is apparently more pronounced than in others, giving rise to areas of varying visual density. When the chromosomes are stained in certain phases of cell division, these areas show up under the microscope as characteristic beadlike arrangements, knobs and constrictions, and other distinctive formations (Fig. 5–17).

*Every cell of every type in a given organism contains the genetic information relevant to the structure and functions of the *entire* organism. The genes that do not relate to the needs of a particular cell are "turned off." How this regulation is accomplished and how genes can be "turned on" at will is an active area of research today.

Figure 5–17 A stained chromosome preparation, viewed under high magnification, reveals a characteristic pattern of bands, knobs, and constrictions.

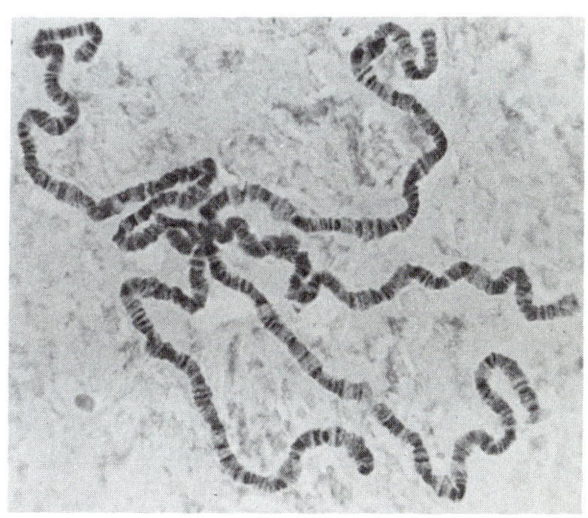

Chromosomes vary greatly in size, not only among different species (from 0.1 μ in some fungi to more than 30 μ in some plants), but even within the same cell (the longest chromosome in a cell may be as much as fifty times longer than the shortest). This is not to suggest any sort of impermanence in the size and shape of a particular chromosome. On the contrary, the stability of chromosomes in maintaining their integrity is one of the most important characteristics of these structures. Even the knobs and peculiar bends of a chromosome remain as permanent features, faithfully repeated in every cell of the organism, and can be used for reliable identification of particular chromosomes.

Each species seems to have a highly specific, characteristic set of chromosomes, found in all members of the species. The number of chromosomes in the set, while constant for a given species, varies widely among the plant, animal, and protist kingdoms. The range of variation extends from two chromosomes in the cells of certain types of roundworms to about sixteen hundred in some radiolarians (single-celled marine protozoa). Despite this wide range, most species exhibit a chromosome number between ten and fifty. In man we find forty-six chromosomes. The chromosome numbers of some representative species of plants, animals, and protists are given in Table 5-1.

TABLE 5-1 Chromosome number of representative species

PLANTS	
Garden pea	14
Red clover	14
Onion	16
Corn	20
Tomato	24
Lily	24
Durum wheat	28
ANIMALS	
Roundworm (some species)	2
Fruit fly	8
Fox	34
Mouse	40
Rat	42
Rhesus monkey	42
Rabbit	44
Man	46
Cebus monkey	54
Silkworm	56
Cattle	60
Dog	78
PROTISTS	
Radiolarian (some species)	1600

We can make a number of deductions from this table. First, the number of chromosomes in a given species has little relation to the degree of complexity of its structure and is quite independent of the size of the organism. For instance, a chihuahua has the same number of chromosomes in each of its cells as is found in a Great Dane's cells (78), and each dog possesses more than man (46). The tiny single-celled radiolarian has a larger number of chromosomes than any of the animal or plant species.

Note also that each of the species listed has an *even* number of chromosomes. This is a reflection of the process of sexual reproduction. It has been found that the male and female sex cells that unite to produce a **zygote** (the fertilized egg that marks the beginning of a new individual) each contain only *half* the usual number of chromosomes. (The zygote resulting from their fusion thus again has the full set characteristic of the species.) Hence, each parent, through its sex cells, contributes only half the genetic information to the new individual. To ensure that each new offspring will have a complete set of genetic information, it seems logical to assume that the full set of chromosomes is a set of *pairs* of chromosomes. Each member of each chromosome pair would thus contain information determining the same kind of traits, and would be quite similar to the other member. Microscopic studies reveal that this is indeed the case. When the individual chromosomes of a particular set are carefully cut out of a photograph of a dividing cell, appropriately stained and highly magnified, and are sorted according to size, shape, and distinctive markings, a set of similar chromosome pairs is obtained

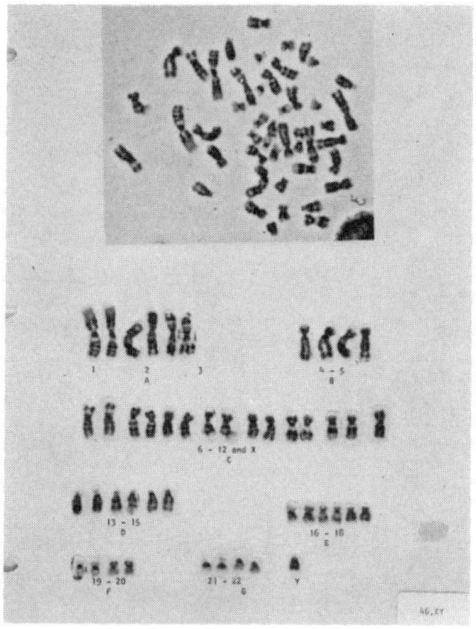

Figure 5–18 A male human chromosome set, as it appears on a metaphase plate, and with the individual chromosomes cut out and sorted. Note the single X chromosome (at the right end of the second line) and the Y chromosome. A female would have two X chromosomes and no Y chromosomes.

(Fig. 5–18). The members of each pair generally are closely matched in size, shape, and general appearance.* The paired chromosomes are known as **homologous chromosomes,** and should not be confused with the **chromatids** (identical chromosomes, one derived from the other).

As we have already indicated, the "resting" or nondividing cell is not really at rest at all: it is a bustling whirl of metabolic activities. And even before there are any visible signs that a cell is going to divide, the DNA of the chromatin network is reduplicated—a perfect copy is made of each DNA molecule. As the visible manifestations of cell division begin, and the chromatin strands coil into the thickened, discrete bodies called chromosomes, careful observation reveals that the chromosomes have doubled: each now consists of a pair of exact duplicate chromosomes called **sister chromatids.** These "identical twin" chromatids are held together at a constricted region referred to as the **centromere.** The position that the centromere occupies on the chromosome determines the basic shape of the chromosome as seen in micrographs taken during the process of cell division. For instance, if the centromere is centrally located, the chromosome will assume the shape of a "V." If the centromere occupies a position between the center and one end, the chromosome will appear somewhat like the letter "J." A chromosome with a terminal centromere (or with one very close to the end) will appear rod-shaped. The position of the centromere provides an additional means of identifying particular chromosomes in the characteristic set of a given organism or species.

The field of **genetics,** the study of the transmission of hereditary information from parent(s) to offspring and how it controls the development, structure, and function of cells and organisms, has burgeoned in recent decades. With advances in microscopy and staining techniques, and with the introduction of the use of radiations and other agents that can (sometimes selectively) change portions of the chromosomes themselves and thus indirectly modify the traits of the cell or organism, the study of genetics at the molecular level has become feasible. The concept of the **gene,** the basic unit of heredity, has undergone a long evolution. At first each gene was thought to be a portion of a chromosome that controlled a specific trait, such as hair color, eye color, or the number of limbs or digits. Because microscopic studies revealed that the chromosomes have a beadlike structure during certain phases of cell division, it was at first assumed that each "bead" on the chromosome chain represents a gene. Finer studies and observations have revealed that the picture is far more complex: there is no such simple one-to-one correspondence between genes and "beads," either in the internal structure of the chromosome as revealed by micrographs or regarding the control of specific traits; rather, many traits have been found to be influenced by several genes, and

*There is usually one exception to the rule that the chromosomes of each pair are closely matched in general appearance. When the chromosomes of a set are matched up, there may be two chromosomes "left over"; they do not match either each other or the other chromosomes of the set. These are the sex chromosomes.

specific sites on chromosomes have sometimes been found to influence more than one trait.

Though there is not yet complete agreement on exactly how the "genes" correspond to the DNA structure, much has been learned about which chromosomes and even which regions of the individual chromosomes control the inheritance of certain traits. As we shall see in Chapter 15, chromosome "maps" have been drawn up for many species, including man, and more and more details on these maps are being filled in with each passing year.

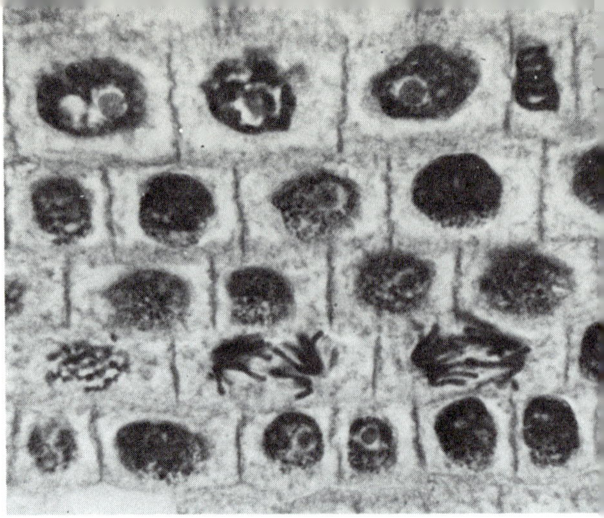

Figure 5–19 Section of a hyacinth root, showing cells in various phases of the mitotic cycle, 700×.

THE NUCLEOLUS

In addition to the chromosomes, the nucleus contains another important organelle, the **nucleolus.** This is a highly compact and darkly staining spherical body, which is generally in contact with certain specific chromosomes. The nucleolus is particularly rich in RNA and protein, although it lacks DNA. In certain species, the nucleus may contain several nucleoli (the number is constant for a given species).

The high RNA content of the nucleolus and the fact that particularly large nucleoli are characteristically found in rapidly growing or protein-secreting cells are indications that the nucleolus is closely linked with protein synthesis within the cell. It is now believed that the region of contact between the nucleolus and its associated chromosomes, the **nucleolar organizer,** is a site of large-scale production of ribosomal RNA, which is stored in the nucleolus and then sent out into the cytoplasm, where it plays a key role in protein synthesis. The nucleolus also plays an important role in cell division.

THE REPLICATION OF CELLS: MITOSIS

In much of the foregoing discussion, we have referred to structures and functions of nondividing cells. But we have also made frequent reference to the process of cell division. For basic to the concept that "all cells arise from preexisting cells" is the fact that most cells do divide, producing two smaller but otherwise faithful copies of the original single cell. That this process of cell division, or **mitosis,** does occur is evidenced by countless microscopic observations of cell preparations. In a specimen taken from almost any tissue of a multicellular organism, for example, some cells will be in the nondividing state at the moment when the slide was prepared, and others will be in various phases of a continuous process that ultimately results in the production of two cells from one (Fig. 5–19). Generally the most actively growing tissues contain the greatest proportion of dividing cells.

The mitotic process involves events occurring both in the nucleus and in the cytoplasm. Sometimes it is convenient to distinguish between these events. Thus, the division of the nucleus is referred to as **karyokinesis. Cytokinesis,** on the other hand, is the division of the cytoplasm.

Using special techniques, cytologists (specialists in cell study) have been able to synchronize various cultures—that is, to grow cultures in which all the cells begin to divide at the same time and pass through each successive stage of division simultaneously. In this and other ways, such as through examination of micrographs of normal, randomly dividing cultures and specimens, cytologists have learned a great deal about the sequence

But the significance of mitosis in the kingdoms of life goes far beyond merely replacing damaged, diseased, or worn-out cells. Essentially all new life is brought into this world by processes involving mitosis. A single-celled organism may reproduce simply by dividing in two. A multicellular organism begins life as a single cell, which must divide many times before becoming a recognizable member of the species.

Virtually all growth, whether it be the addition of another foot or so to the height of a tree or the addition of another inch or so to a human child, and virtually all healing, accompanied by the proliferation of new cells, involve the mitotic process. Mitosis is intimately related to the mysterious and miraculous process of regeneration whereby, for instance, a salamander that has lost a limb may grow an entirely new one and a decapitated flatworm can even grow a new head!

A key factor in the impact of mitosis on the living world is the fact that cell division results not only in the production of two cells where before there was only one, but also in the production of two cells, which, barring occasional slight errors, contain the *same* genetic information as their parent cell. Mitosis thus not only ensures the continuance of life itself but also ensures that the forms of life will be consistent with those that went before.

The Ribosomes and Protein Synthesis

When the contents of a cell are spun in a centrifuge at ultrahigh speeds, various fractions may be obtained. One of these fractions, known as the **microsomal fraction,** consists of fragments of the endoplasmic reticulum and other cellular constituents. The **microsomes** are not, as was first thought, organelles of the intact cell, but merely represent breakdown products of the endoplasmic reticulum and associated structures.

Finer methods of fractionation have isolated these associated structures; one major type is the **ribosomes.** These tiny (0.02–0.2 μ) spherical bodies contain relatively large amounts of RNA and are sites of protein synthesis in the cell. It has recently been discovered that protein synthesis involves the association of a molecule of messenger RNA, sent out from the nucleus, with a cluster or group of ribosomes known as a **polyribosome** or **polysome** (Fig. 5–22).

In the intact cell, many of the ribosomes appear to be physically attached to the endoplasmic reticulum; others float freely in the cytoplasm. It has been found that cells involved in active protein synthesis for "export" to other cells of the organism, such as cells that secrete digestive enzymes or hormones, are particularly rich in endoplasmic reticulum and associated ribosomes. But in cells that synthesize proteins mainly for internal consumption—such as immature red blood cells, which are engaged in active synthesis of hemoglobin—there is relatively little of the rough endoplasmic reticulum, and the polysomes lie free in the cytoplasm. It is believed that the endoplasmic reticulum and its associated structures are involved in "repackaging" of the proteins synthesized in the ribosomes for eventual transport out of the cell.

Figure 5–22 Polysome isolated from a chick embryo.

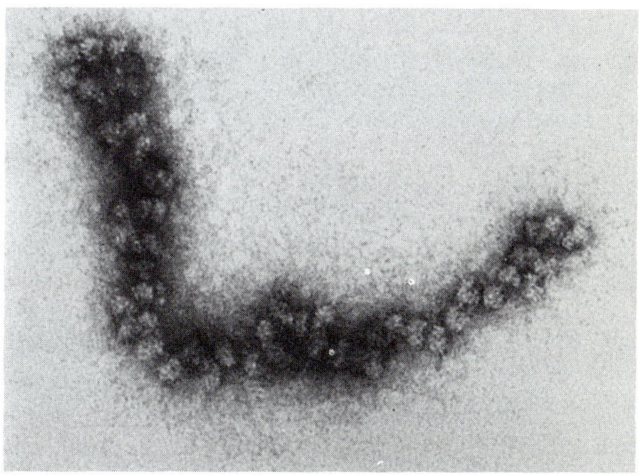

Figure 5–23 As tomatoes ripen, the plastids in the fruits are successively colorless leucoplasts, green-pigmented chloroplasts, and red-pigmented chromoplasts.

Manufacture and Storage: The Plastids and the Golgi Complex

When a tomato plant reaches a certain stage of development, clusters of yellow flowers appear. In time these are replaced by tiny button-sized straw-colored fruits. As the days pass, these fruits grow larger and take on a pale green color. After a while, the green color of the tomatoes gradually gives way to a yellowish, then orange, and at last to a bright scarlet red. These succulent fruits* are now ripe and ready for the dinner table.

The change in color described above is due to changes in pigments within organelles located in the cells of the tomato plant. These organelles are known as **plastids** and are found throughout the plant kingdom (as well as in a number of photosynthetic protists).

*Botanically, tomatoes are considered fruits, but in 1893 the U.S. Supreme Court officially designated them vegetables.

The young tomato fruits contain plastids essentially devoid of pigments. These colorless plastids are known as **leucoplasts.** As the fruit develops, deposits of starch are laid down in some of the leucoplasts. Other leucoplasts in different parts of the plant, particularly the roots, also serve as storage deposits for starch. But as the tomatoes grow, some leucoplasts are transformed into **chloroplasts,** plastids bearing the light-sensitive pigment **chlorophyll.** This ubiquitous pigment, which is responsible for the characteristic green color of higher plants, plays an important role in the capture of part of the energy of sunlight and its storage in the form of chemical bonds in sugars and starches. (This energy can be utilized for the immediate needs of the plant or stored as starch in leucoplasts.) As the tomatoes ripen, there is a further change in the plastids, resulting in the final change in the color of the fruits. The chlorophyll breaks down, and the chloroplasts are transformed into red-pigmented **chromoplasts.** (A chromoplast may be defined as a plastid containing a pigment; hence, a chloroplast is a special type of chromoplast.)

There are many types of chromoplasts in nature, including the brown-pigmented plastids of brown algae and diatoms and the red-pigmented plastids of red algae. But in the plant kingdom, the most abundant and important chromoplasts are the green chloroplasts.

Another organelle important in the synthetic activities of the cell is the **Golgi complex,** consisting of *Golgi bodies.* These structures have been a storm center of controversy in scientific circles ever since their discovery in 1898 by Camillo Golgi, an Italian physician. Because Golgi bodies are difficult to see in the living cell, many scientists at first insisted that the structures pointed out by Golgi were not organelles of the living cell, but merely artifacts, introduced into the cell as a result of the silver-staining technique originated by Golgi. It was not until the use of the electron microscope in the 1940s and 1950s that this dispute was resolved and Golgi bodies were generally acknowledged to be true organelles.

There is a great variability in the appearance of the Golgi complex in different species and even among different cells in the same organism. This diversity of forms has contributed to some confusion of terminology among cytologists, who have given various names to this structure, including Golgi bodies, Golgi complex, Golgi apparatus, Golgi substance, dictyosomes, lipochondria, and idiosomes. In any case, three basic components are usually observed in the Golgi complex: flattened sacs which appear as parallel membranes; clusters of small, tightly packed sacs or vesicles; and large, clear vacuoles (Fig. 5–24).

Though many of its aspects are still uncertain, a picture of the functions of the Golgi complex is gradually emerging. The fact that Golgi bodies are particularly well developed in secretory cells, such as the cells of the pancreas, salivary glands, and endocrine glands, indicates that the Golgi complex plays an important role in processes of secretion. Evidence has been amassed indicating that the

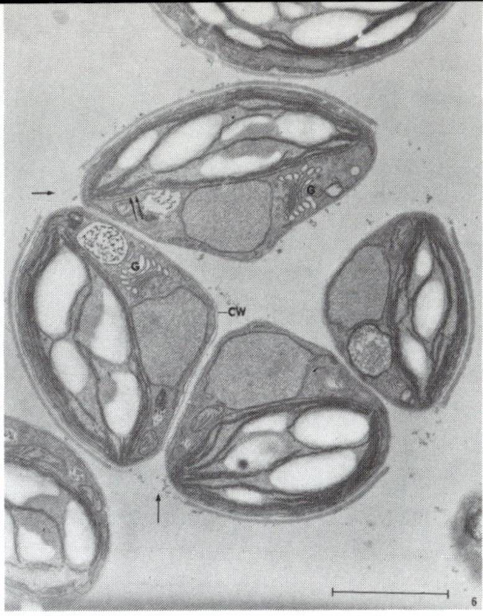

Figure 5–24 Sections of *Chlorella* cells show the details of the Golgi complex (G) and chloroplasts.

vesicles and vacuoles of the Golgi complex provide a temporary storage area and repackaging center for substances such as proteins and possibly lipids synthesized in other cellular organelles, especially the rough endoplasmic reticulum. Within the Golgi complex, water is removed from the products of synthesis, which are then concentrated and packaged in membrane-enclosed granules. It is believed that some of the membranous canals of the Golgi complex are connected to the plasma membrane and that the products of secretion pass out of the cell through these canals.

The Golgi complex may also have a synthetic function. Recent studies indicate that the Golgi bodies secrete carbohydrates. Thus, the Golgi complex of plant cells may be responsible for secreting the cellulose of the cell walls.

Energy for the Cell's Activities: Mitochondria and Lysosomes

Nearly all the activities of the cell and of the organism require energy. This energy is obtained by a breakdown of the energy-rich chemical bonds of foodstuffs, organic chemicals such as carbohydrates, fats, and proteins, which are either synthesized in the cell itself or in other cells of the organism or are taken in through a process of eating.

The breakdown of the complex organic compounds of food into their ultimate products, water and carbon dioxide, with a controlled release of the energy they contain for the uses of the cell, is a long and intricate process involving many successive steps. For the immediate energy needs of the cell, it is more convenient to have a reserve of intermediate energy storehouses that can be quickly broken down to supply small packets of energy as they are required for particular metabolic reactions. In living cells, the chemical compound ATP (adenosine triphosphate), containing energy-rich phosphate bonds (Fig. 5–25), acts as such a readily available source of energy.

The vital processes of cellular respiration, in which organic chemicals are successively oxidized and the energy released stored in ATP molecules, take place in organelles of the cell called the **mitochondria.**

Under the optical microscope the mitochondria may be seen as rodlike or filamentous structures (Fig. 5–26a). Under the electron microscope the mitochondrion is shown to be surrounded by a double membrane. The inner membrane extends into the cavity of the mitochondrion, forming inner folds known as **cristae** and thus providing a greatly increased surface area (Fig. 5–26b).

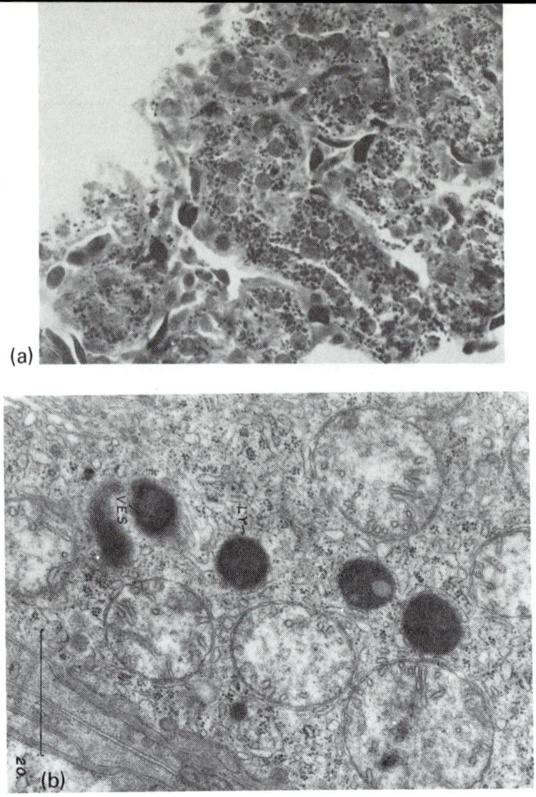

Figure 5–26 (a) Light micrograph of liver cells, 366×. Mitochondria are visible but not in fine detail. (b) An electron micrograph of the rat adrenal cortex (16,100×) reveals fine details in the structure of the mitochondrion.

Figure 5–25 ATP, the energy currency of the cell. Note the similarity to the nucleotide units of the nucleic acids. Energy is stored in the phosphate bonds, which can be broken one by one.

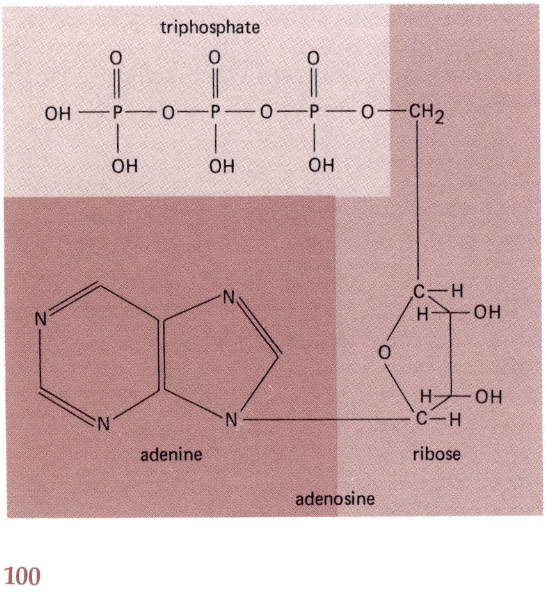

A number of vital enzymes are intimately associated with the membranes of the mitochondrion, and others are found within the fluid-filled cavity. These enzymes are responsible for carrying out a series of chemical reactions ultimately leading to the formation of ATP. This chemical "energy packet" is intimately involved in many of the most important reactions of the body, such as muscle contraction, nerve impulse transmission, and protein manufacture. Hence, it is not surprising that poisons such as cyanide that disrupt the processes of cellular respiration in the mitochondria and thus cut off the supply of ATP cause instant death. Because of its relationship to ATP, the mitochondrion is often referred to as the "powerhouse" of the cell.

The organic materials that are subjected to the step-by-step oxidation reactions of cellular respiration in the mitochondria are not the original carbohydrates, fats, and proteins synthesized by or taken into the cell. These materials must first be broken down into smaller fragments—pyruvic acid, fatty acids, and amino acids—which can pass through the mitochondrial membrane. This breakdown occurs by a process of intracellular digestion accomplished by enzymes associated with another type of cellular organelle, the **lysosomes.**

Lysosomes have been observed primarily in animal cells and protists, though some published reports have noted their presence in plant cells as well. In contrast to other organelles, the lysosome is encased in only a single membrane, and it has no internal cristae such as are possessed by the mitochondria.

The most striking characteristic of the lysosomes is their high content of **hydrolytic enzymes.** These enzymes catalyze the breakdown of complex organic substances through reactions involving the addition of water; various types are capable of digesting nucleic acids, proteins, and some carbohydrates. The activities of the lysosomes are involved in many special processes that go on in higher organisms under specific conditions: in addition to aiding in the digestion of materials normally stored within cells, the enzymes of the lysosomes take part in the phagocytic activity of white blood cells (wherein bacteria and other foreign particles are engulfed and broken down), the degeneration of the tadpole's tail during metamorphosis, and the digestion of bone by certain bone-digesting cells (a curious process that paradoxically aids in growth, by enlarging the cavities within the bones and thus making more room for bone marrow).

The lysosomes even contain enzymes capable of digesting the very substance of the cell. Indeed, when a cell dies, the hydrolytic enzymes of the lysosomes are usually released from these organelles, causing self-digestion or **autolysis** of the cell. Thus, in its lysosomes, each of the cells within us houses the means of its own destruction.

Organelles for Locomotion: Cilia and Flagella

As a paramecium glides gracefully through the water, hundreds of tiny hairlike projections known as **cilia** beat in a finely synchronized, flowing movement (Fig. 5–27a). Many other protozoa also are equipped with cilia, in numbers ranging from three hundred up to the staggering total of fourteen thousand in a single unicellular (single-celled) organism. It is these cilia, working in unison, which are responsible for the fact that ciliates (protozoa possessing cilia) are among the fastest moving of the protists. Some of these organisms reach speeds in excess of 2 mm per second. Although this speed may not at first appear to be very fast, if body size and speed were extrapolated to the human scale, these creatures would swim ten times faster than the best human swimming champion.

Cilia are by no means restricted to protists. Many cells of higher animals are equipped with these structures. For instance, the cells lining human respiratory passages are supplied with cilia. Their rhythmical beating propels dust and other foreign materials to a position where they can be readily expelled from the body by a cough or sneeze.

Many protists contain only one or two elongated cilia termed **flagella** (Fig. 5–27b). These organisms (such as euglena, many bacteria, and sperm cells of numerous plants and animals) are propelled through liquid media by the whiplike motion of their flagella.

Other Cellular Organelles

Beneath the lens of a microscope an amoeba—a mere speck to the naked eye—is resolved into an amorphous slithering mass. A rhythmically pulsating body within the amoeba continually returns water to the environment. This is the **contractile vacuole,** and it is vital to the amoeba's existence. A minute amount of cyanide introduced in the immediate vicinity of this amoeba will result in a cessation of

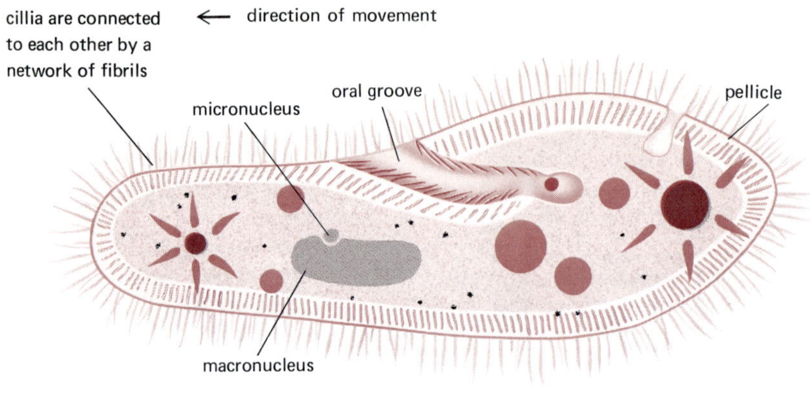

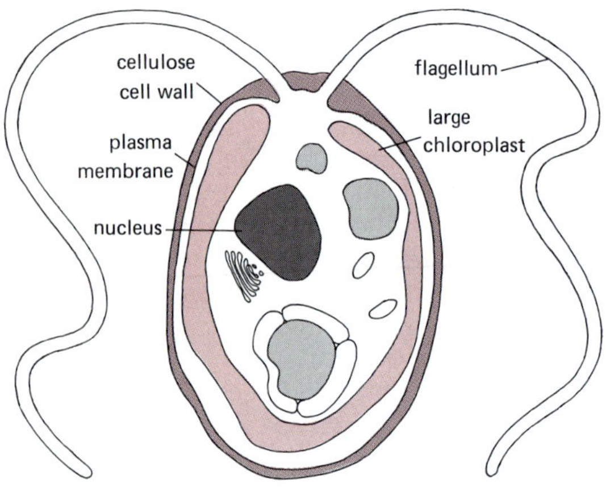

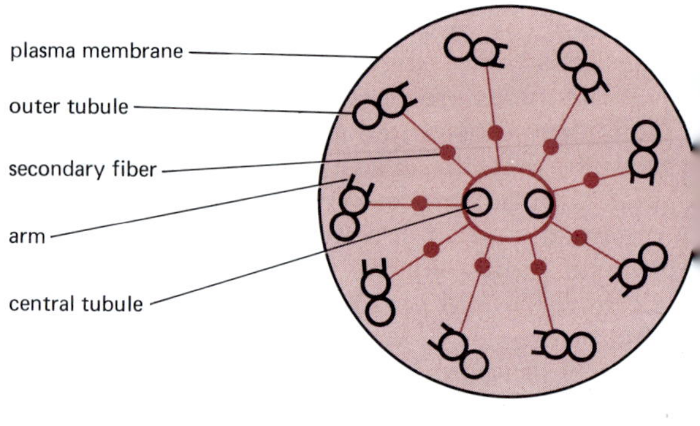

Figure 5–27 The protozoan *Paramecium* (a) is covered with about 2500 cilia, tiny hairlike structures that beat rhythmically to propel the organism through the water. The alga *Chlamydomonas* (b) is equipped with a pair of flagella, which lash back and forth in a whiplike motion. (c) Cross section of a cilium or flagellum.

these pulsations. Quickly the amoeba swells as water pours into the animal. With a sudden burst the inner contents are strewn haphazardly.

Contractile vacuoles are found in numerous fresh-water protozoa. As in the amoeba, their primary function is to maintain an equilibrium with the hypotonic environment, pumping out the excess water that diffuses in through the semipermeable membrane enclosing the cell.

Other vacuoles often found in protozoa serve the function of **food vacuoles.** Ingested food particles are enclosed in membranous sacs, derived from the plasma membrane. As the food vacuoles float about passively in the

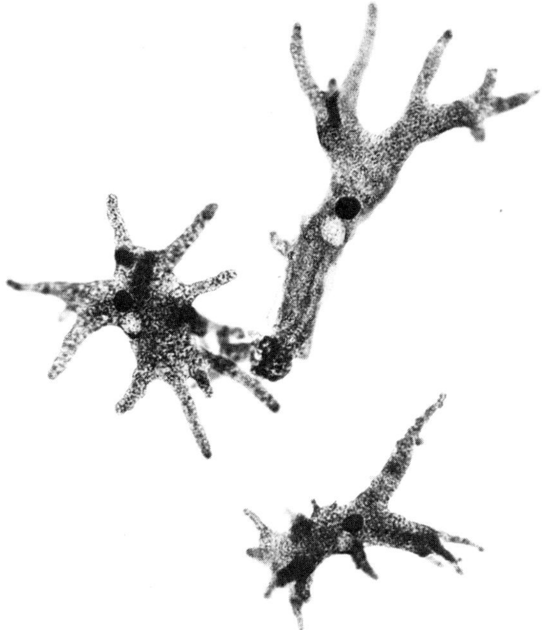

Figure 5–28 *Amoeba proteus.* Note the vacuoles and various inclusions.

cytoplasm, in the form of isolated globules, digestive enzymes break down their contents. As digestion progresses, the smaller fragments of usable food materials pass through the vacuolar membrane into the cytoplasm, while wastes accumulate in the vacuole. The latter, now a **waste vacuole,** serves as an organ of excretion. It may drift through the cytoplasm to the edge of the cell, where it merges with the plasma membrane and opens, discharging its contents to the external medium. Or the waste materials may be excreted through special fixed pores in the outer surface of the cell.

Vacuoles, or fluid-filled "sacs," are also commonly found in the cells of plants. These vacuoles vary greatly in size, but in many cases they occupy a considerable proportion of the cell volume (Fig. 5–29). This is true, for example, of the cells of the water plant elodea. The fluid found within these vacuoles is known as **cell sap.** Salts, sugars, pigments and other materials are dissolved in this fluid. (The red, blue, or purple coloration of flowers, leaves, and stems is often due to pigments located in the cell sap.) Enclosing the vacuole is a membrane similar in composition and structure to the other membranes of the cell.

Vacuoles serve several important functions in plant cells. The presence of a vacuole within the cell forces the cytoplasm toward the outer margins of the cell and thus facilitates an exchange of gases with the surrounding medium. Vacuoles also help to maintain the rigidity of the cell structure. The cell sap is generally hypertonic to the environment, and water tends to diffuse into the vacuole by osmosis, thus setting up turgor pressure. Plant vacuoles may also serve a sort of excretory function, acting as a repository for wastes or toxic by-products of the plant's activities. These may remain within the cell, safely sealed away in vacuoles where they cannot harm the living contents of the cell. Spinach leaves, for example, often contain large deposits of oxalic acid crystals. These are eventually gotten rid of by the plant when the old leaves die and drop off.

In addition to substances deposited in vacuoles, other deposits of various types are also frequently scattered through the cytoplasm. Oil droplets of varying size are found in both plant and animal cells. These droplets, along with pigment and secretion granules, crystals of inorganic and organic salts, resins, and miscellaneous other products, are known as **cytoplasmic inclusions.** For the most part the cytoplasmic inclusions are randomly distributed throughout the cytoplasm and are products of the metabolic functions of the cell.

Figure 5–29 As this electron micrograph of *Kalanchoe tubiflora* (4675×) illustrates, vacuoles frequently take up a large fraction of the plant cell contents.

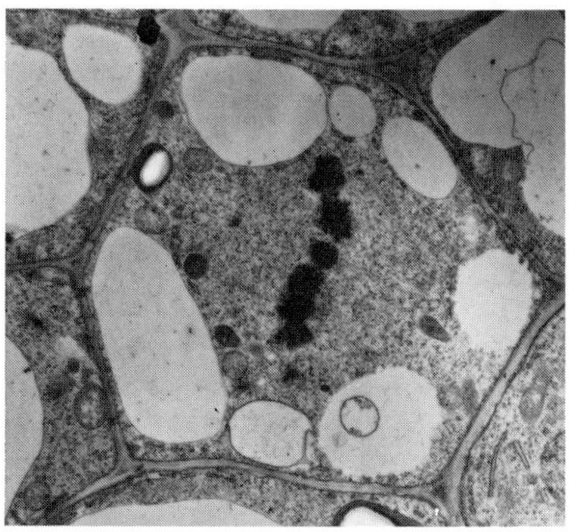

Extracellular Substances

In animal cells, the living plasma membrane is usually the outer boundary of the cell. In plant cells, as we have pointed out, the plasma membrane is characteristically surrounded by a tougher, more rigid boundary, the **cell wall.** The cell walls of plants and many protists are made up of deposits of nonliving substances secreted by the cytoplasm. The major constituent is the high-polymer carbohydrate cellulose. Other components of the cell wall include lignin and pectin (complex organic substances which act as binding agents), salts, and waxy waterproofing substances such as suberin.

Though the cell wall may grow to rather substantial dimensions, it is quite porous, permitting a wide variety of substances to pass through without much difficulty. Unlike the cell membrane, it exhibits little, if any, selective permeability.

The main function of the cell wall is support. The relative rigidity of the walls of the individual cells, together with other mechanisms such as turgor pressure, helps to keep plants firm. So effective is the support provided by the cell wall that even a loss of substantial amounts of water will not ordinarily shrivel plant cells. Cell walls tend to retain their shape even after the plant has died, a property that man has taken advantage of in his use of plant products such as wood and natural fibers (Fig. 5–30).

Some protists, such as algae, fungi, and bacteria, also have cell walls. But the structure of the cell walls in these organisms is not always the same as that in plants. In bacteria, for example, the cell wall does not contain cellulose, but rather is a complex structure composed of proteins and carbohydrates. In many bacteria, a polysaccharide layer (known as a **capsule**) surrounds the cell wall.

Although animal cells in general do not possess cell walls, numerous animal cells do secrete a **matrix** material which, as in bone cells, often forms a permanent surrounding medium. The matrix generally contains organic substances that serve to connect and support the cells and, in certain cases, deposits of inorganic salts, such as the calcium salts in bone.

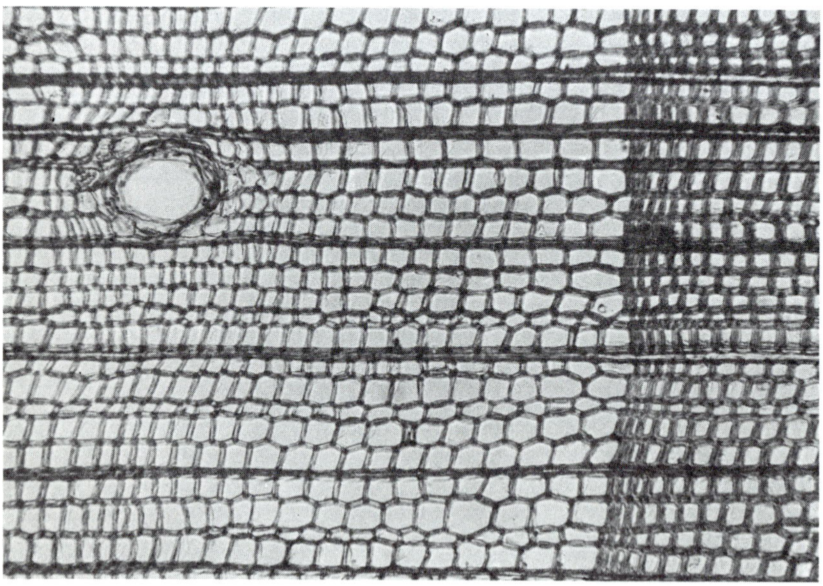

Figure 5–30 Magnified cross section through pine wood. The cells have died, but their reinforced cell walls remain.

QUESTIONS FOR REVIEW AND THOUGHT

1. Trace the history of the development of the cell theory. What recent developments have given the study of the cell an added impetus?
2. Explain the following paradox: some cells are several feet long and yet are considered to be microscopic structures.
3. Draw a diagram of a typical cell and label important structures.
4. What are the distinguishing characteristics of animal cells? Of plant cells? Can all cells be classified as either plant or animal?
5. What are the functions of the plasma membrane?
6. Contrast scientists' conceptions of the cell before and after the advent of the electron microscope.
7. Describe the phenomena of plasmolysis and turgor.
8. What unique properties of colloidal dispersions contribute to the characteristics of the cytoplasm?
9. Why is the nucleus looked upon as the "hub" of the cell?
10. Name the stages of mitosis and describe the sequence of events that occur during cell division.

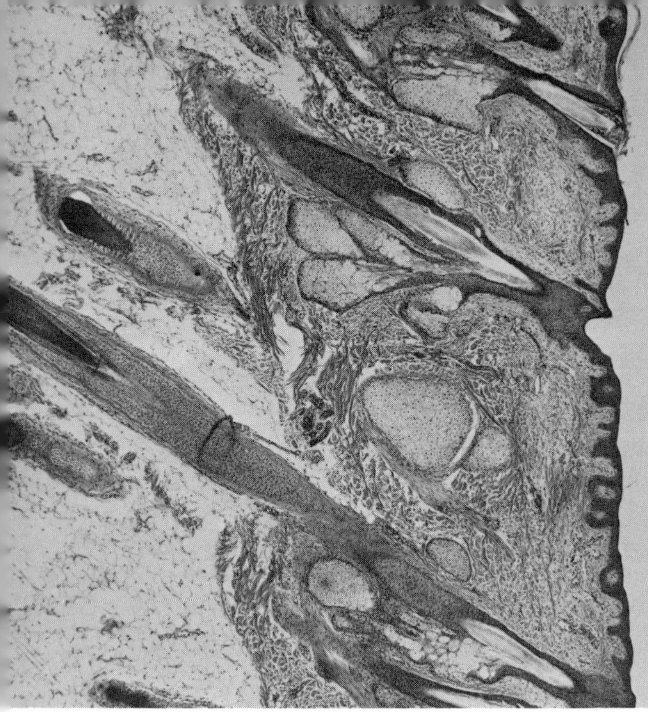

(a)

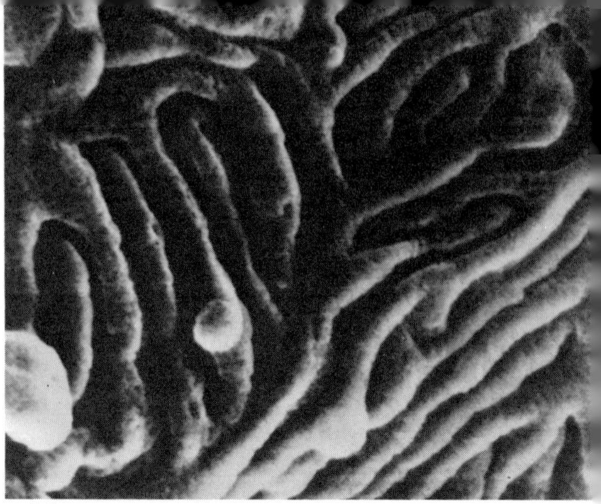

(b)

(c)

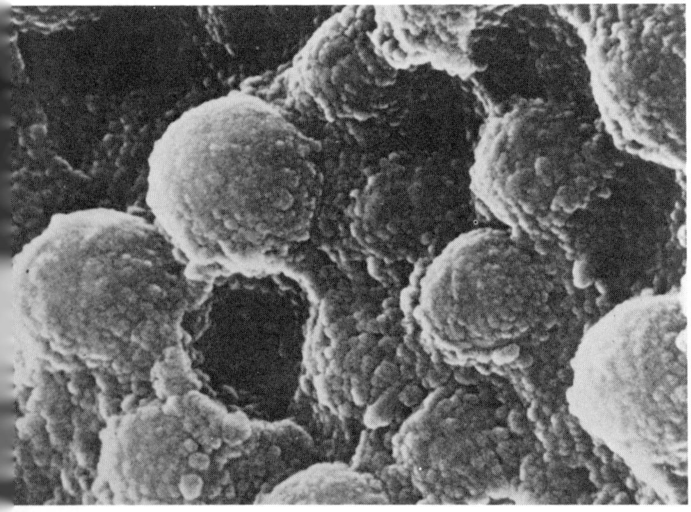

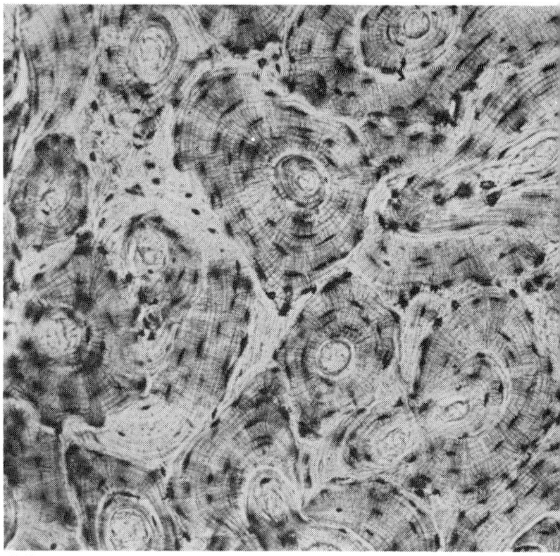

(d)

(e)

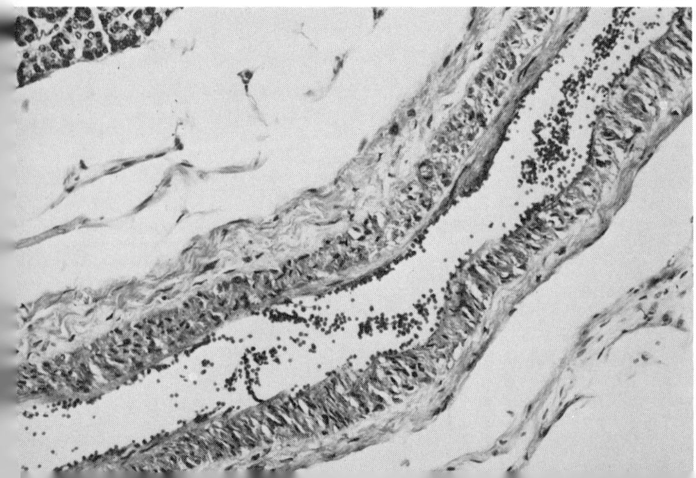

(f)

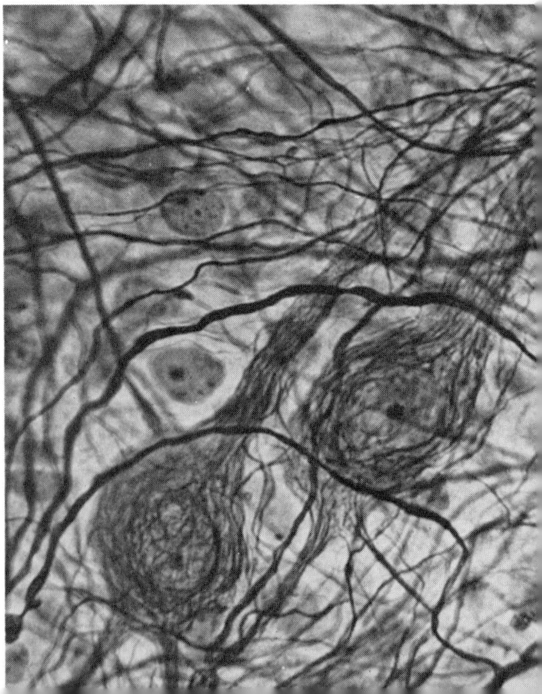

CHAPTER 6

From Cell to Organism

The Organization of Cells

The living cell may be small, but it is far from simple. Indeed, a single cell is a highly intricate aggregate of specialized organelles and is capable of performing a great variety of tasks, both for its own well-being and for the overall functioning of enormous numbers of other cells if it is part of a multicellular organism.

Although for the sake of simplification we have sometimes spoken of "the cell," "the animal cell," or "the plant cell," this was done with the understanding that real cells, even from different parts of the same organism, may differ widely in their basic structure and major functions. Indeed, this diversity is so marked that cells from the same organism frequently differ more from one another in gross anatomy than do cells that perform similar functions in widely divergent species. For example, the circular, wafer-shaped, red-pigmented, nonnucleated human red blood cell bears little resemblance to the elongated, threadlike, white nerve cells ramifying throughout the human body. On the other hand, vertebrate nerve cells are basically the same in gross anatomical structure, whether in fish or in humans.

Just as there is great diversity of structure among cells, there is also a wide variation in activities. Groups of cells that cooperate to perform a uniform function constitute a **tissue.** The cells of a tissue frequently are not contiguous; instead, they are separated from one another by an intercellular fluid (known as the **matrix**), which is secreted by the cells themselves. The amount of matrix varies considerably depending on the type of tissue. In connective tissue (e.g., bone or cartilage tissue), the matrix constitutes the preponderance of the tissue mass, and the cells are frequently widely scattered. In epithelial tissue (e.g., skin), on the other hand, there is a minimum of matrix, as the cells are closely packed.

Figure 6–1 The diversity of tissues: (a) human scalp tissue, 29×; (b) kidney of a cat, 26,640×; (c) pancreas of a dog, 31,500×; (d) section of human bone, 86×; (e) longitudinal section of a small human artery with blood corpuscles, 158×; (f) nerve tissue in a mouse spinal cord ganglion, 646×.

(a), (d), (e), Philip L. Feinberg, Omikron

There is a wide variety of different types of tissue within a single multicellular organism. Frequently several different types of tissue are associated in performing a single function or several related functions. Such an association of tissues is called an **organ.** For instance, the heart is an organ which consists of outer and inner coverings of epithelial tissue and a solid core of muscle tissue, interspersed with blood, nerve, and connective tissues (Fig. 6–2). Often, structurally dissimilar organs may be associated in the performance of closely related functions, thus forming **organ systems.** The heart, along with the blood vessels and such accessory organs as the spleen and bone marrow, constitute the **circulatory system.**

Even an organ system is too specialized for independent existence. The **organism** is composed of a number of organ systems. Its very existence and its ability to adapt to the varying demands of its environment are dependent upon the cooperative interplay of its organ systems.

Throughout the world numerous examples may be seen of organisms living together in large numbers, in a cooperative social structure known as a **community.** Such diverse creatures as bees, ants, and humans form highly structured communities.

Animal Tissues

The little girl lies motionless on the table. A sudden hush falls over the operating room as the surgeon makes a neat incision in her tiny chest, revealing a sac through which the child's beating heart can be seen. For a long moment all that can be heard is the hissing of the anesthetic and the gurgling of the blood as it pulses through the life-sustaining heart-lung machine. Carefully and deftly the surgeon works in the now-quiescent heart, until at last his work is done. Sutures are applied, quickly the chest is closed, and the tiny heart beats once more—better than before.

In the scene described above, the surgeon and his associates encounter and must deal with nearly all of the basic types of tissue found within the body. The initial incision slices first through an outer, protective covering of skin **(epithelial tissue),** and thence through underlying layers of **connective tissue.** The blood that wells out from the incision and the blood rushing through the heart-lung machine represent a portion of the **vascular** or **blood tissue.** The beating heart represents one of the major types of **muscle tissue.** The child felt no pain because the anesthetic administered effectively prevented the impulses

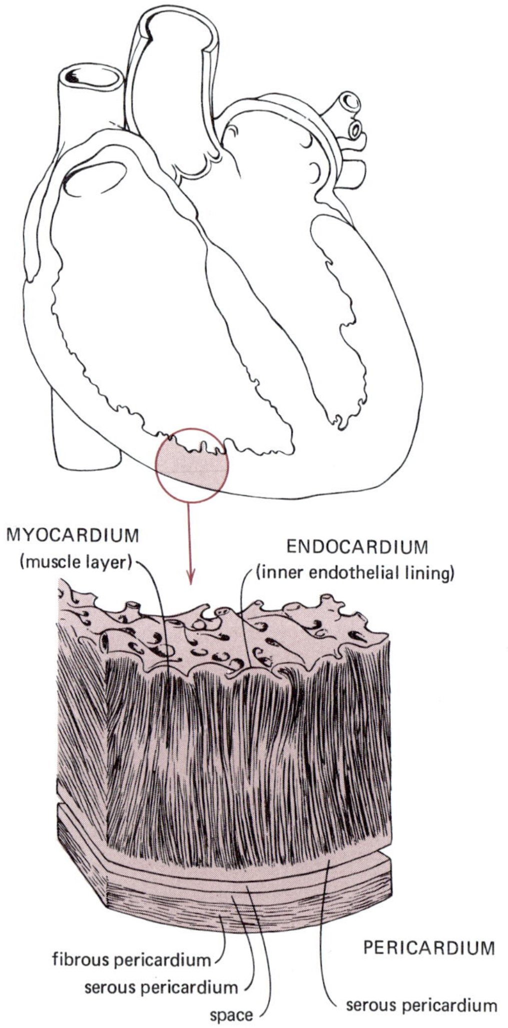

Figure 6–2 The heart consists of a number of kinds of tissues associated in performing the function of pumping blood through the system of blood vessels.

transmitted through the nerves **(nervous tissue)** from being meaningfully interpreted by the brain.

A sixth type of animal tissue, which in this child is not yet fully developed, is **reproductive tissue.**

EPITHELIAL TISSUE

A newborn mouse bears little resemblance to the sleek, furry creature that is the nemesis of the fainthearted housewife. Coming upon a litter of newborn mice, one might think at first glance that one had stumbled upon a nest of earthworms. The reddish hue of their bodies is an indication of the thinness of the bare and delicate epithelial covering which ensheaths them. As the tiny mice grow, this covering thickens, the redness lightens, and the characteristic fur makes its appearance.

Epithelial tissue is essentially covering tissue. It serves as a protective covering for the body and its organs and also lines the internal cavities of these organs and the various body cavities. Epithelial tissues are comprised of closely packed cells, which are frequently arranged in broad, flat sheets. The numerous tasks performed by epithelial tissues may be grouped into several main categories: protection, absorption, secretion, excretion, sensation, and temperature regulation. Some epithelial tissue may perform several of these functions simultaneously. In addition to its protective role, the skin, for instance, excretes waste materials (in the form of sweat), houses various types of sensory organs, and takes part in temperature regulation.

Epithelial tissues are composed of various types of cells, which are named according to their shapes (Fig. 6–3). The outer surface of the body is covered with layers of **squamous epithelium,** composed of flat, closely packed cells somewhat resembling pavement stones (the word *squamous* comes from a Latin root meaning "scaly"). Squamous epithelium also lines the body cavities such as the mouth cavity.

Cubelike epithelial cells line the kidney tubules and the ducts of many glands and are appropriately called **cuboidal epithelium.** The stomach, the small and large intestines, and the ducts of many glands are lined with elongated, column-shaped cells known as **columnar epithelium.** The free surface of some columnar cells, such as those lining the respiratory passages, the uterus, and the oviducts, may be equipped with cilia.

Any of the three basic structural types of epithelial cells may be found either in a single layer or in a multilayered arrangement, the latter being typified by the skin.

CONNECTIVE TISSUE

Connective tissue is the most widespread tissue in the body. If suddenly a man were to be miraculously stripped of all tissues other than connective tissue, he would stand before us eerily similar to his former self. His skeletal framework, enfleshed with masses of tissue and permeated with a network of flaccid blood-laden vessels, would give a basically humanoid form. If, on the other hand, a man were stripped of all of his connective tissue, leaving intact all the remaining tissues, an

Figure 6–3 (a) Squamous epithelial cells from the human cheek; (b) cuboidal epithelium from the human kidney; (c) columnar epithelium from the dog stomach.

(a) (b) (c)

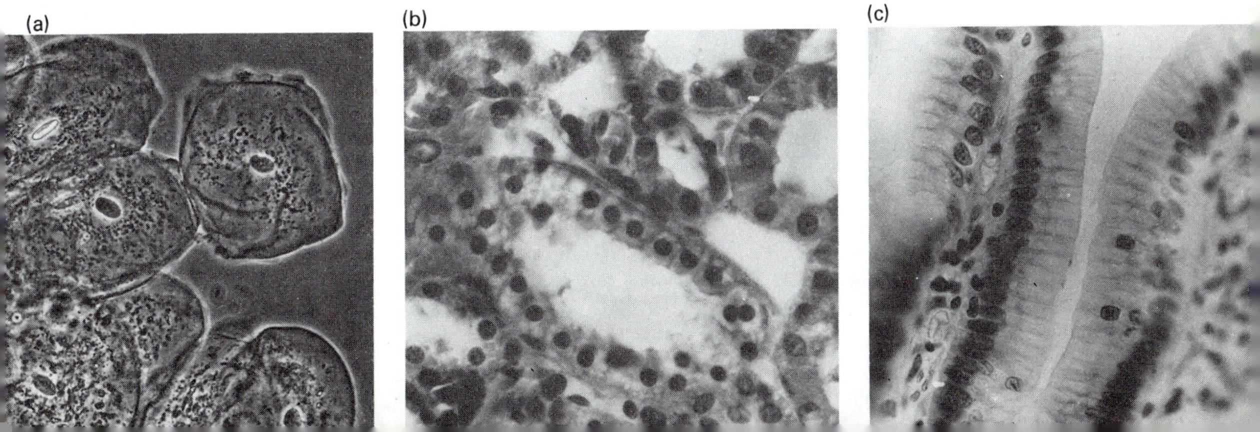

entirely different picture would be seen. For a second, a wraith of his former self would linger. A half-empty bag of skin sagging with masses of unsupported muscle tissue, permeated with an intricate network of nerve tissue, would gently collapse into a shapeless heap.

Like all other types of tissues in the body, connective tissue contains cells. But the distinguishing feature of connective tissue is that the cells here represent only a relatively minor fraction of the mass of the tissue. The remaining portion is made up of the intercellular material, or matrix (Fig. 6–4). In addition, many types of connective tissue are interwoven with proteinaceous fibers.

The connective tissue cells themselves provide the source for the matrix. In some cases, such as the bone, this material is secreted as a liquid, only subsequently to be transformed into a solid matrix.

Depending on the varying nature of the matrix, connective tissue may be pliable or rigid, soft or hard, distensible or inelastic. And hence this tissue is well suited for its diverse functions: (1) support (bones and cartilage); (2) connecting two distinct organs (tendons and ligaments); (3) filling cavities and spaces between and within organs (loose and dense connective tissues); (4) protection of delicate internal organs and structures (loose and dense connective tissue and adipose tissue); and (5) fat storage (adipose tissue).

Scars resulting from severe wounds are evidence of an important property of connective tissue. Whereas the cells of muscle and nerve tissue cease to divide after a time, connective tissue cells retain their powers of multiplication and replace the dead tissue.

Superficially the **bones** of the body may appear inanimate. On the contrary, bones are so sensitive to alterations in their normal mechanical functions that disuse of them will lead to atrophy, while excessive use may lead to an increase in the bone mass or hypertrophy. Due to the nature of its matrix, bone tissue is the most rigid of all of the connective tissues. This matrix consists of proteinaceous fibers (chiefly **collagen**) and inorganic material in the form of deposits of calcium phosphate (Ca_3PO_4) and calcium carbonate ($CaCO_3$). Bone tissue contains far less water than most other cells.

Cartilage tissue bears some resemblance to bone tissue. As in the bone, the tissue is reinforced by numerous collagen fibers. But the matrix of cartilage tissue consists of a protein-polysaccharide mixture known as chondrin, which, in contrast to the calcified cementlike rigidity of bone matrix, is firm but quite flexible.

Cartilage is found in the tip of the nose, the external portions of the ear, and the articular surfaces of bones (i.e., at their juncture with other bones). The skeleton of the early human fetus is essentially all cartilage, most of which is gradually replaced by bone tissue.

Ligaments and **tendons** bind various body structures together. For instance, ligaments bind bones to one another, while tendons connect muscles to other muscles or to bones. Both tissues contain collagen fibers. These fibers are arranged in a more parallel orientation in tendons, giving these structures a flexible but inelastic nature. Ligaments, on the

Figure 6–4 Human connective tissues: (a) bone cross section, 120×; (b) cartilage from trachea, 200×; (c) adipose tissue, 160×.

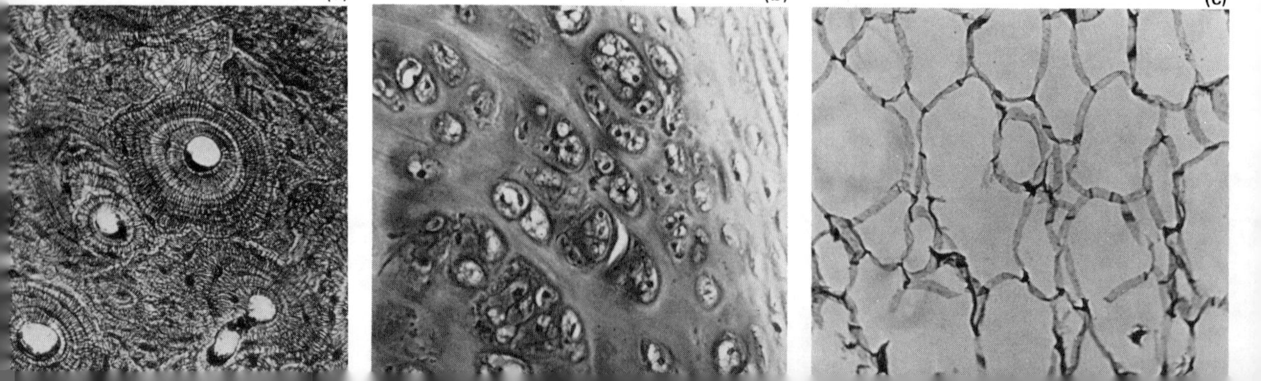

other hand, contain the protein elastin, which lends them elasticity.

Large regions of the body are packed with what may be termed **loose connective tissue.** This tissue encircles and supports the other tissues of the body. It serves as packing material by filling in the spaces *between* organs and other tissues and *within* organs and other tissues. Because other tissues are embedded in loose connective tissue, the nutrient materials passing from the blood to the various tissues that lie within connective tissue must first pass through the connective tissue. The reverse process is true of any waste and secretory products emerging from these embedded tissues.

Still another type of connective tissue, sometimes referred to as **dense connective tissue,** is found in the dermis (lower layer of the skin) and in parts of the intestinal and urinary tracts.

Adipose or **fat tissue** may be thought of as a modified type of loose connective tissue in which the constituent cells are filled with fat, leaving only a tiny space for the nuclear and cytoplasmic materials. Adipose tissue is found throughout the body, but it is especially concentrated beneath the skin and in the abdominal cavity.

MUSCLE TISSUE

Nearly 50% of the human body, by weight, consists of muscle tissue, and the forces exerted by these muscles are phenomenal. For instance, the human calf muscle can withstand forces exceeding a ton. During a convulsive fit, the uncoordinated contracture of the muscles may exert forces so great as to rip tendons and snap bones.

Under more normal circumstances the contractions of muscles are utilized in movement, while muscle tissue also provides support for various body structures. In general three types of muscle tissue are recognized: skeletal, smooth, and cardiac.

Each of these types of muscle tissue is comprised of the same fundamental units. Within the cells of each are found long fibers, which under the proper conditions are capable of contracting.

Groups of **skeletal muscle** cells are bound together by thicker, tougher membranes to form bundles, which in turn are bound together to form larger and larger bundles until the familiar whole muscle is enveloped in a connective tissue sheath. As the name implies, skeletal muscles are associated with the skeleton and are responsible for virtually all of the voluntary movements of vertebrates and many lower forms of animals.

In man, more than 80% of the muscle tissue is skeletal muscle tissue. This tissue is sometimes referred to as striated muscle tissue. It derives this name from the fact that the individual cells appear to have a series of parallel darkened bands or striations running transversely through their entire length (Fig. 6–5a). Each of these striated cells or fibers contains numerous nuclei, which are distributed at random along its surface just beneath the muscle cell membrane.

Smooth muscle tissue, in contrast to skeletal muscle tissue, is not under voluntary control. Its contractions may be stimulated by nervous or hormonal influences. This tissue is found within the walls of blood vessels and of the intestines and other internal organs of the body. The spindle-shaped smooth muscle cells have but a single nucleus (Fig. 6–5b). Unlike the skeletal muscle tissue, the smooth muscle exhibits no striations.

Cardiac muscle tissue, as the name implies, is found in the heart. It is characteristically capable of self-initiated contraction. That is, unlike other muscle tissues, cardiac muscle tissue will contract without the aid of external stimuli. This is not to imply that it is insensitive to external stimuli. On the contrary, mechanisms exist within the body which ensure a fine regulation of cardiac or heart contraction.

Cardiac muscle tissue resembles skeletal muscle tissue in that the individual cells of both varieties are striated and multinucleated. However, unlike the skeletal cells, the nuclei in the cardiac cells are located deep

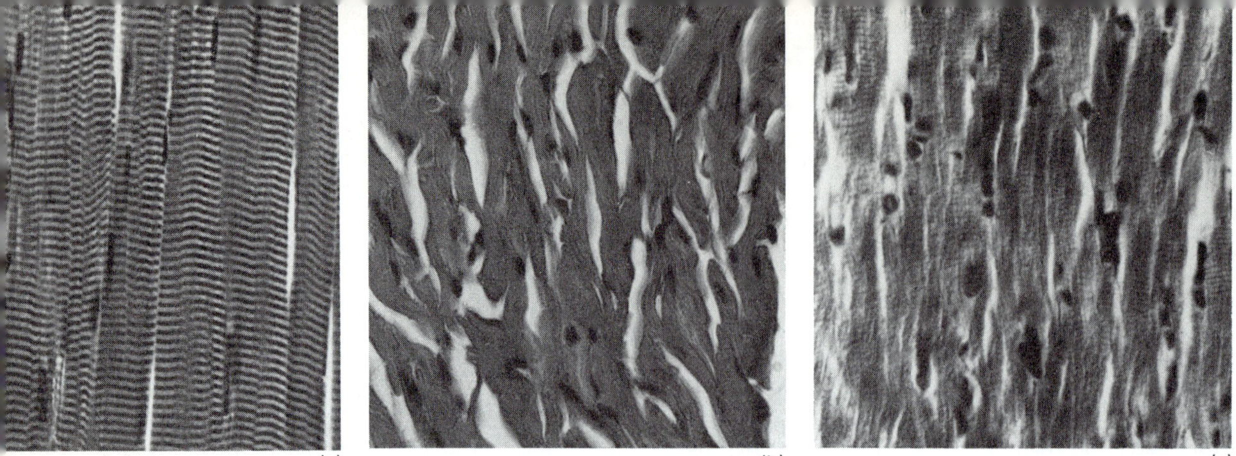

(a) (b) (c)

Figure 6–5 Human muscle tissue: (a) skeletal, 350×; (b) smooth, 350×; (c) cardiac, 350×.

within the boundaries of the cell. A second important difference in gross anatomy between these two types of cells is that whereas skeletal muscle cells retain their individuality, the cardiac cells appear to merge, forming a complicated network (Fig. 6–5c). The advantage in this type of arrangement is that a single stimulus may cause all of the cardiac muscle cells in a given area to contract simultaneously, thus giving rise to a single powerful contraction.

The relative rates of contraction of these three types of muscle tissue vary considerably. In general, the rate of contraction of skeletal muscle is fastest, while that of the smooth muscle is slowest of the three. Though the smooth muscle may be rather slow to contract, it can maintain its state of contraction for longer periods than skeletal muscle. In addition, it is less subject to the effects of fatigue than is the skeletal muscle.

NERVOUS TISSUE

It is through the nervous system that we are made aware of the world without and within us. Each moment the brain is bombarded with countless thousands of bits of information. Each moment the brain sifts through this information, retaining the pertinent and screening out that which is not. We are made aware of the world without and the world within through an intricate network of sensory organs and nerve fibers which interlaces our bodies.

The basic unit of the nervous system is the individual nerve cell or **neuron,** billions of which extend to almost every imaginable portion of the body. Although the bodies of animals contain various types of neurons in sizes ranging from several thousandths of an inch to several feet, the same basic units exist in each nerve cell. Each neuron consists of a cell body and one or more threadlike extensions (Fig. 6–6). The cell body houses the nucleus and such structures as Golgi bodies, mitochondria, and RNA-laden Nissl bodies (whose exact function is unknown). Each neuron is provided with an extension, or "process," which carries impulses away from the cell body. This process is known as an **axon.** Axons generally terminate in branchlike networks commonly referred to as **terminal branches.** Most neurons have several processes that carry impulses *to* the cell body. These processes are called **dendrites.**

Neurons which conduct impulses from some sensory organ to a second neuron are called **sensory neurons.** Neurons which innervate some muscle or gland are referred to as **motor neurons.** Often these two types are joined by still a third neuron, a **connector neuron.** There is no actual physical contact between neurons, however. Instead, a **synapse,** or a region of nonprotoplasmic intercellular material, separates the two neurons. Nerve impulses are carried across synapses from one neuron to another by chemical messengers.

REPRODUCTIVE TISSUES

The potentialities of the reproductive processes are enormous. For instance, within the

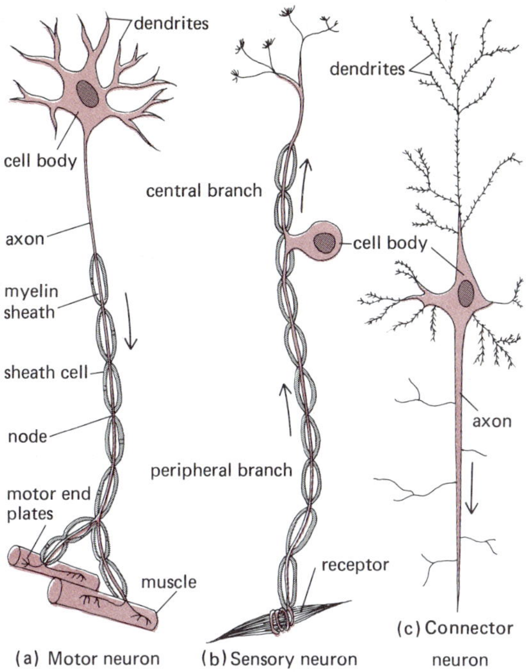

Figure 6–6 Structures of human nerve cells. (a) Motor neuron. Stimuli are received by the dendrites, which conduct the nerve impulse to the cell body and then through the long axon to an effector such as a muscle. (b) Sensory neuron. A single branch of the nerve fiber leads from a sensory receptor to the cell body. From there the nerve impulse is conducted along another branch to synapses in the spinal cord or brain. (c) Connector or interneuron. Nerve impulses are conducted along the dendrites to the cell body and then along the axon. Note the absence of a myelin sheath on the axon, which may be much shorter than the axon of a motor neuron.

space of a year, under ideal conditions, a female mouse might be the progenitress of well over a hundred thousand descendants. Even more startling is the potential of a pregnant fruit fly. It has been estimated that in little more than a year such a fly and its offspring could *theoretically* give rise to more than 10^{58} fruit flies. This fantastic brood would together weigh more than the entire universe!

In man and in higher animals and plants, there are two types of reproductive or sex

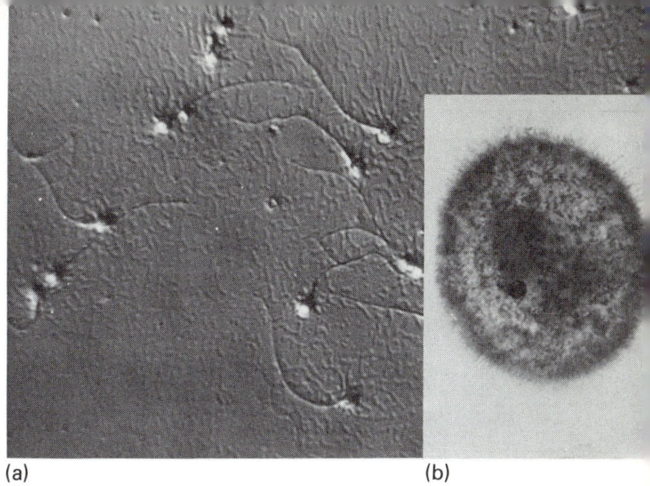

Figure 6–7 The human gametes: (a) sperm, 580× and (b) an ovum surrounded by sperm.

cells, called **gametes** (Fig. 6–7). The male gamete or **sperm** usually is quite small and frequently is motile, equipped with a flagellum. The female gamete or **ovum** is generally much larger and contains varying amounts of stored food materials or **yolk,** depending on the species.

The tiny sperm and the ovum or egg each contain half of the normal complement of chromosomes for the species. When a sperm and an egg join, they form a new cell, a **zygote.** This cell contains the complete set of chromosomes for the species and divides again and again to form a new organism.

In the higher vertebrates, sperm cells are produced by the hundreds of millions in organs known as testes. In the female, ovaries produce large numbers of immature forms of eggs. It has been estimated that the ovaries of a young woman contain hundreds of thousands of immature eggs. Only a few hundred of these will ever mature into ova during her lifetime, and only a small numer of these will be fertilized (join with a sperm to produce a zygote).

BLOOD TISSUE

The term *tissue* may at first call to mind a solid mass of some particular type of cells held together by a *firm* matrix. And yet blood appears to be a watery fluid. Nonetheless, blood meets the criteria normally specified for a tissue: it is comprised of cells which cooperate to perform a mutual function. In fact, cells constitute as much as 45% of the total volume of whole blood. The fact that blood cells are produced by connective tissue has led some biologists to classify blood as a connective tissue.

Whole blood may be divided into two fractions: a liquid fraction, the **blood plasma,** and a solid fraction, the **formed elements.** Thousands of different compounds are dissolved in the blood plasma. The great majority of these substances are proteins, performing a wide range of functions, from protecting the body against invading microbes to participating in the clotting of the blood. In addition to proteins, such diverse substances as dissolved gases, salts, steroids, carbohydrates, and lipids are to be found in the blood plasma.

The formed elements are comprised of the **red blood cells, white blood cells,** and **blood platelets** (Fig. 6–8). The mature red blood cells are anucleated, disc-shaped, and resemble tiny doughnuts without the hole going completely through the middle. These red blood cells are indispensable in transporting oxygen throughout the body and also play an important role in carbon dioxide transport.

Several different types of white blood cells are found. These cells are colorless, amorphous, and nucleated. A strange fact about some of these cells is that they are not restricted to the bloodstream. Often, white cells will squeeze through the inner lining of the blood vessels and invade the tissues, where they seek out and devour foreign intruders. Thus a major function of white blood cells is the protection of the body against invading bacteria and viruses, both in the blood and in the tissues.

Figure 6–8 Human blood cells, showing red cells, a white cell, and platelets, 1220×.

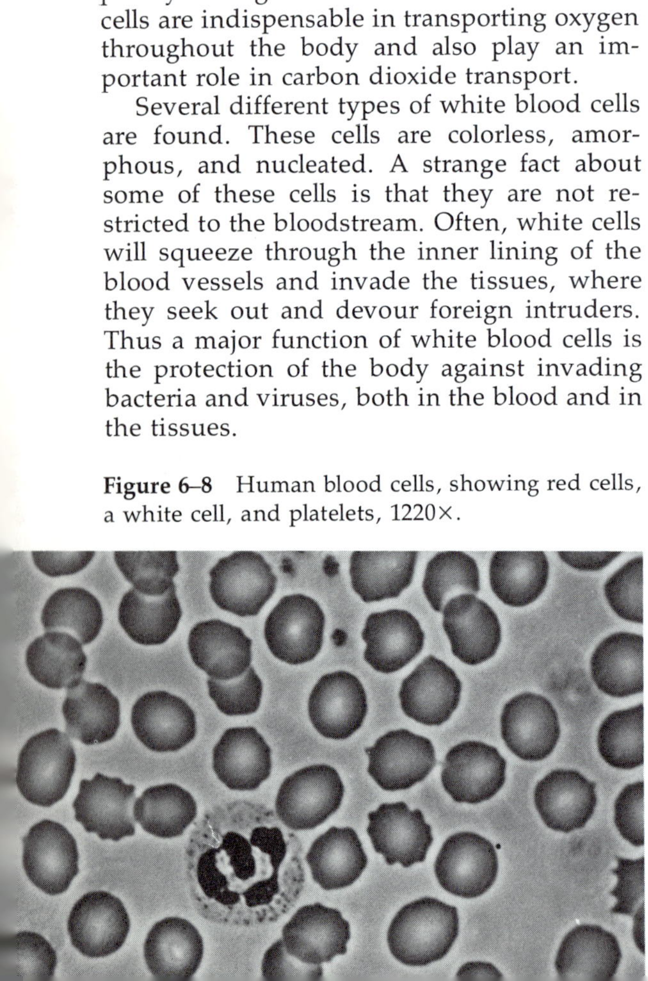

Blood platelets play a vital role in the mechanism of blood clotting. Ironically, it is when they are destroyed that they become of value to the body. For the shattering of their delicate membranes when they brush against the jagged walls of a ruptured blood vessel triggers the release of a substance into the blood that initiates a complicated series of chemical reactions ultimately resulting in the formation of a mesh of fibers and blood cells known as a clot.

Plant Tissues

Like animals, plants are organisms that function through the coordinated activities of organ systems, organs, and tissues. Although there are a number of possible systems of classification of plant tissues, in this text they will be grouped in the following categories: undifferentiated (meristematic); protective (epidermis and cork); food production and storage (parenchyma); supporting (collenchyma and sclerenchyma); and conducting (xylem and phloem).

MERISTEMATIC TISSUE

Popular among amateur gardeners is the unusual plant *Kalanchoe pinnata*, the "leaf of life" plant. Drop a single leaf on the soil, water, and wait. Within a few days tiny green rosettes appear at the notches on the margin of the leaf (Fig. 6-9). Day by day these tiny plantlets grow, sending down roots into the soil and rising on slender green stems. Snip them away from the mother leaf, replant them, and they are ready for life on their own.

In the notches along the edge of the *Kalanchoe* leaf is a special type of undifferentiated tissue known as **meristematic tissue.** This tissue may be likened to the embryonic tissue of plants and animals, in that it is capable of giving rise to literally all the varied types of tissue in the plant.

Basically two types of meristematic tissues are found in plants: apical meristem and lateral meristem. **Apical meristem** is normally

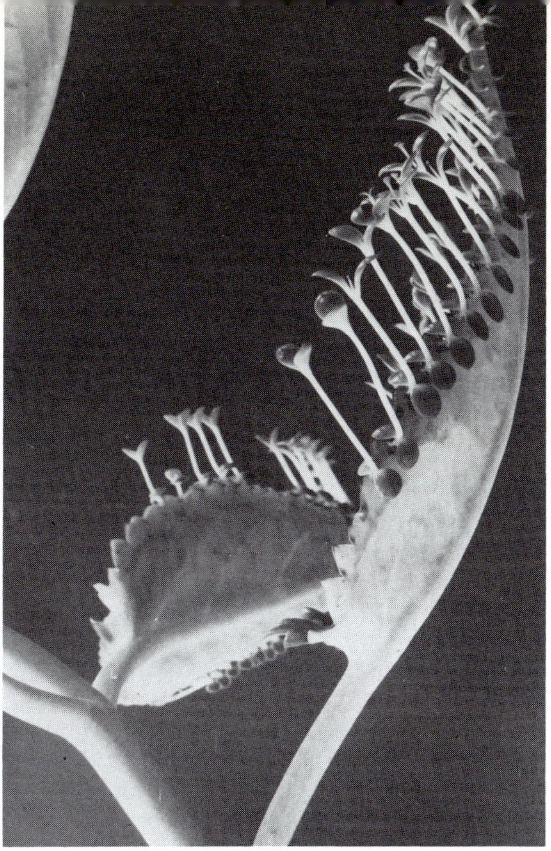

Figure 6–9 Tiny plantlets grow from meristematic tissue on the margin of the *Kalanchoe* leaf.

found at the tips of roots and stems. The second type of meristem, often referred to as **cambium,** is located along the sides of the stem and roots. The cambium is responsible for the increase in the girth of the plants; one type gives rise to the bark of trees and shrubs.

Thus, a plant grows in basically two ways: the mitotic activity of its apical meristem results in an increase in height and an extension of its root system, while the activity of its cambium results in an increase in width. Together, these two types of meristem give rise to virtually all other types of plant tissue; protective, storage, supporting, conductive, and even more meristematic tissue.

PROTECTIVE TISSUE: EPIDERMIS AND CORK

Like the animal, the plant may fall prey to infectious agents and disease. Like the animal, the plant is sensitive to excessive heat and excessive cold. Like the animal, the plant is susceptible to the damaging effects of radiations. And so it is not surprising that the plant, like the animal, is protected by layers of tissue shielding it from its potentially hostile environment.

This **protective tissue** may be found virtually everywhere in the plant kingdom as an outer epidermis and, in most trees and shrubs, in a deeper, much thicker layer known as **cork.**

The epidermal cells usually form a layer of tissue only one cell thick, covering the surface of leaves and the younger parts of the stems and roots (Fig. 6–10). (In the older parts of the stems and roots the epidermis is usually worn away or sloughed off.) The walls of the cells are generally rather thick. These cells often secrete a waxy substance, **cutin,** which

Figure 6–10 (a) Section of a carnation leaf, 58×, showing stomata (two open and two closed on the underside of the leaf and one closed on the upper surface); (b) epidermis of dayflower leaf, 120×, showing the surface appearance of the stomata.

(a) Philip L. Feinberg, Omikron

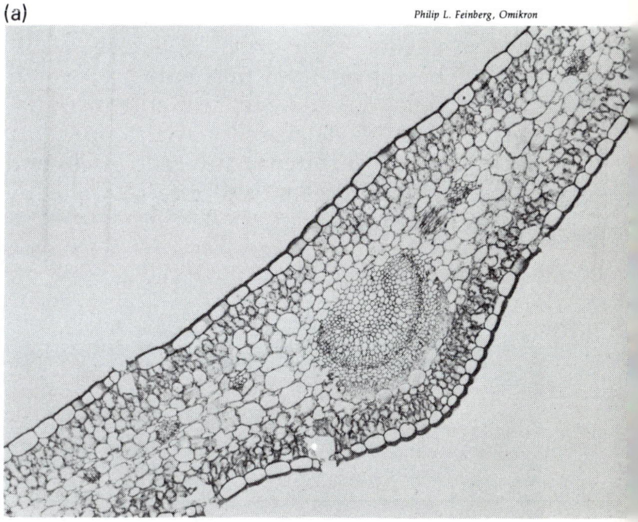

(b)

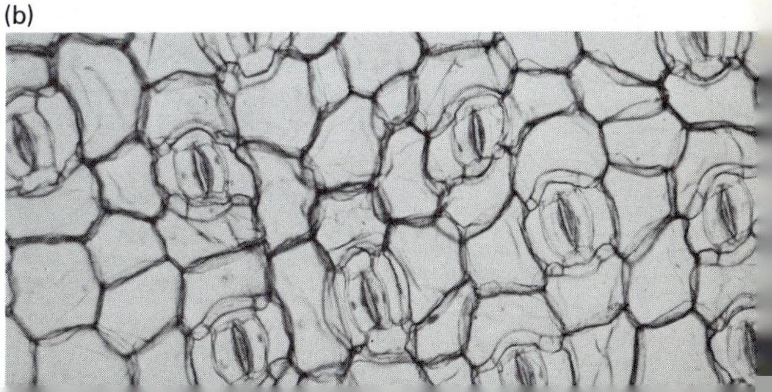

forms a waterproof film on the surface of the leaves, stems, and roots. This covering aids the plant in the retention of water and prevents dehydration. In combination with the relatively thick walls, it also affords some degree of protection from mechanical injury and prevents the penetration of potentially harmful microorganisms into the plant.

Some specialized epidermal cells called **guard cells** are found in the leaves, usually concentrated on the underside of the leaf (Fig. 6–10). These bean-shaped cells, occurring in pairs, contain the green pigment chlorophyll. They regulate the inflow and outflow of gases through tiny openings or pores known as **stomata.**

Tiny projections, called **root hairs,** grow out from the epidermal cells in the roots (Fig. 6–11). It is through these root hairs that most of the water and nutrients for the plant are absorbed. Without the enormous increase in surface area provided by these root hairs, the plant could scarcely fulfill the requirements of all of its metabolizing cells.

Cork forms the outer bark of trees and shrubs. As the cambium matures into cork cells, the latter rapidly secrete a waterproof substance into their thick cellulose walls. Being thus cut off from nutrients and water, the cork cells soon die. Hence mature cork tissue consists of dead cells, which form an impervious layer (often many cells thick) and thus effectively prevent the loss of water by the inner tissues.

Figure 6–11 A young radish seedling. Note the many delicate root hairs.

FOOD-PRODUCING AND STORAGE TISSUE: PARENCHYMA

Among the most important of the activities in the plant are the production and storage of organic food materials. So dependent is man upon this production that his very existence would come to an end if plants were to cease their activities.

Parenchyma is a widely distributed tissue in plants. It produces organic food materials by way of photosynthesis and stores these materials. The leaves (and often the stems) of many plants contain numerous parenchyma cells, which possess chlorophyll-containing chloroplasts. It is within these green structures that the process of photosynthesis takes place. The roots, and often the stems and tubers, of plants contain numerous parenchyma cells, which in turn contain starch-storing leucoplasts. Leucoplasts are found in great abundance in many common food items such as potatoes, bananas, beans, rice, radishes, and turnips. If a leucoplast is exposed to sunlight, it may be converted into a chloroplast, with the development of chlorophyll. In some plants these chloroplasts are ultimately converted into chromoplasts as the plant, or its fruit, takes on a characteristic hue (e.g., the unripe green tomato turning to a luscious red).

Parenchyma tissue may also serve as packing tissue, enveloping and protecting other plant tissues.

SUPPORTING TISSUES: COLLENCHYMA AND SCLERENCHYMA

As the howling wind swirls through the open meadow, the grasses and wild flowers are seen to bend and sway in billows as the gusts sweep by. That few stems snap in the windy onslaught is a tribute to the resilience and strength of their supporting tissues.

Two types of supporting tissues are commonly found in plants. **Collenchyma,** which is seen in stems, the midribs or veins and petioles or stalks of leaves, and some flowering parts of the plants, represents the more pliable of the two.

Philip L. Feinberg, Omikron

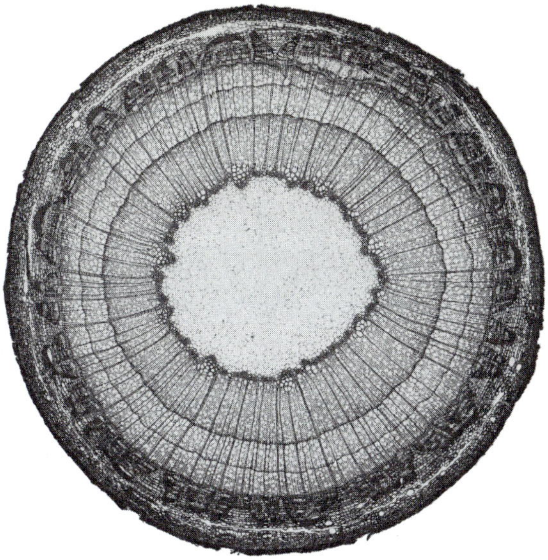

Figure 6–12 Cross section of a tulip tree stem, 14×.

Sclerenchyma, the second type of supporting tissue, is composed of cells which are more uniformly thick-walled than collenchyma cells (Fig. 6–12). The fibers of linen (from flax) and those of hemp are two examples of sclerenchyma. Other types of sclerenchyma are found in walnut shells and the gritty masses in the skin and pulp of pears and other fruits. Sclerenchyma cells also are an important constituent of the bark of trees.

CONDUCTING TISSUES: XYLEM AND PHLOEM

If a large tree is felled and its cut surface is examined carefully, an interesting record of the past can be deciphered. From numerous concentric rings, each ring corresponding to a year's growth, can be inferred not only the age of the tree, but also some measure of the degree of rainfall in the region for each year. The tissue examined is xylem tissue. At one time this tissue was alive, carrying water and salts from the soil.

Xylem and phloem represent the major conducting tissues found in the plant kingdom. **Xylem** conducts water and dissolved inorganic salts up from the roots to the various regions of the plant, while **phloem** carries dissolved organic nutrients up and down the plant, either for storage in the roots or for distribution to metabolizing cells of the organism.

Xylem tissue is, in reality, a complex of several different types of cells, including parenchyma, sclerenchyma, and specialized tubular cells with thickened, reinforced walls.

Like xylem tissue, phloem tissue is a complex, formed of parenchyma, sclerenchyma, specialized conducting cells, and companion cells, which appear to influence the metabolic activities of the conducting tubes. Support for this hypothesis lies in the fact that the cytoplasm of the nucleated companion cells is joined to the sieve tube via tiny pores. The walls of both the sieve tubes and the companion cells are relatively thin. Phloem sclerenchyma provides mechanical support while phloem parenchyma not only provides storage facilities but also takes part in the lateral transport of food materials and water (as **phloem rays**).

Organs and Organ Systems

In the evolution of higher organisms various parts of the organism have developed an advanced degree of specialization of function. These areas have assumed key responsibilities, and upon their optimum functioning the very existence of the organism depends. Hence, a division of labor may be said to have been established.

An **organ** may be defined as a group of tissues associated in the performance of similar or closely related functions. However, in practice this definition is not closely adhered to. For instance, the liver, which is considered to be an "organ," performs scores of different tasks, many of which are not directly related. An organ usually contains many diverse types of tissues. For instance, within the liver may be found blood vessels filled with blood, nerves, muscle tissue, connective tissue, and epithelial tissue, each performing in such a manner as to ensure the proper functioning of the liver. A similar picture is seen in the heart, the kidneys, the lungs, etc.

Each organ, in turn, functions as a member of a larger integrated group comprising several other organs, which functions as an **organ system.** For instance, such widely separated organs as the heart, the spleen, the bone marrow, and the liver team up with the blood vessels (and blood) to form the **circulatory system.** Other systems found in man and the other vertebrates (as well as in most species of the "lower" animals) include the **nervous system, endocrine system, respiratory system, digestive system, excretory system, integumentary system, skeletal system, muscular system, reproductive system,** and **sense organs.**

In the plant kingdom there is a general lack of agreement among plant morphologists concerning what is and what is not an organ. The typical plant may be considered to consist of two basic organ systems: the **vegetative** and the **reproductive** (Fig. 6–13). The former may be subdivided into the following organs: the **roots, stems,** and **leaves.** The reproductive organs include the parts of **cones** and **flowers. Fruits** and **seeds** may also be considered as plant reproductive organs.

In both plants and animals the individual organ systems cooperate in an integrated manner to promote the optimum functioning of the living organism.

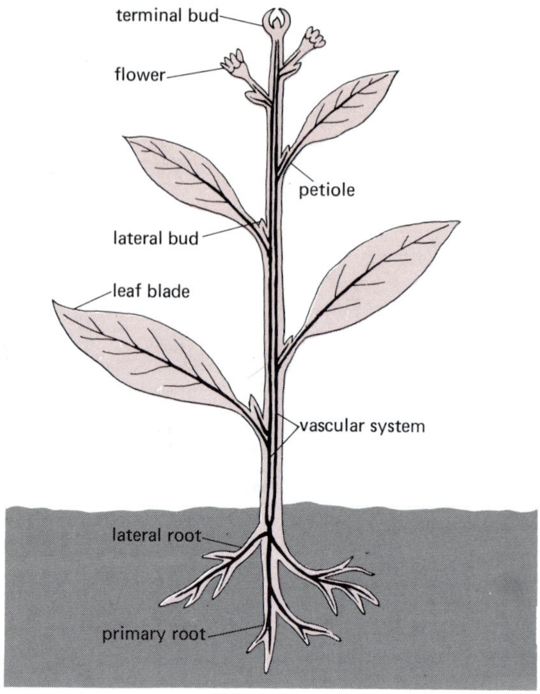

Figure 6–13 Structure of a typical higher plant. The organs—roots, stems, leaves, and flower—are linked together by the vascular system.

The Organism

The study of organisms is the essence of biology. For any part of an organism, be it cell, tissue, organ, or even organ system, by definition is not truly "alive." The organism is the smallest living unit that is capable of independent existence.

There are more than 1½ million different kinds of organisms on the earth. They vary from single cells to multicellular organisms composed of intricate complexes of trillions of cells or more and integrated into interrelated systems of organs.

But despite the great diversity of life, the organisms of our world have a surprising number of basic characteristics in common. Together with gross structural differences, there are also patterns of similarity, such as the general layout of the skeleton and internal organs among many quite diverse groups. Even more striking are the similarities that exist on the microscopic and molecular levels. Not only do the cells of varied organisms contain similar organelles performing similar functions, but similar enzyme systems and modes of transmitting biological information are also to be found throughout the plant, animal, and protist kingdoms.

Communities

Nowhere in the world do we find a living organism isolated from the rest. Isolation in the purest sense does not exist on this planet; virtually all species are interdependent, in some way, with others. In a quiet sunlit

Figure 6–14 Any living community is populated by a variety of organisms, linked together directly and indirectly in numerous interrelationships.

meadow, where a cow is grazing on the grass, we see a whole chain of interrelationships among living organisms. The cow is nourished by the grass, whose growth is in turn abetted by the action of burrowing worms, which loosen the soil and bring nutrients down from the surface. The overhanging trees provide shade and a rookery for the twittering birds, which feed mainly on insects which, if allowed to multiply unchecked, could do serious damage to the grass, trees, and other plant life of the meadow. The root systems of the trees prevent rapid runoff from a recent rainfall, thereby encouraging the growth of the nearby grass. Chemicals from the cow droppings and bird droppings find their way into the soil, thereby returning nourishment which eventually leads to a more vigorous growth of grass. This **biotic community,** or local region of animal and plant life, is repeated in varying forms throughout the world.

But the term *community* has still another meaning. Frequently, it is used as a synonym for the term *society.* A **society** may be defined as an interdependent group of members of the same species. Thus in addition to a society of men, we may have a society of bees (in a beehive), a society of ants (in an anthill), or a society of termites (perhaps under one's back porch). In each of these there is a specialization and division of labor among the individuals. (To some extent this is similar to the division of labor among the parts of the organism.) In ant species, for instance, the division of labor is reflected in differences in morphological development (Fig. 6–15). In addition to the queen ant, various types of worker ants develop. In many ant species, special soldier ants perform defensive (and sometimes offensive) duties. For instance, in some ant species, the soldier ant plugs up the doorway to the anthill after a worker has passed through, opening it again for the next worker. The door is barred, however, to all other species, as well as to members of another colony of the same species!

With the advance of technology, the individual members of human societies are becoming more and more specialized. This

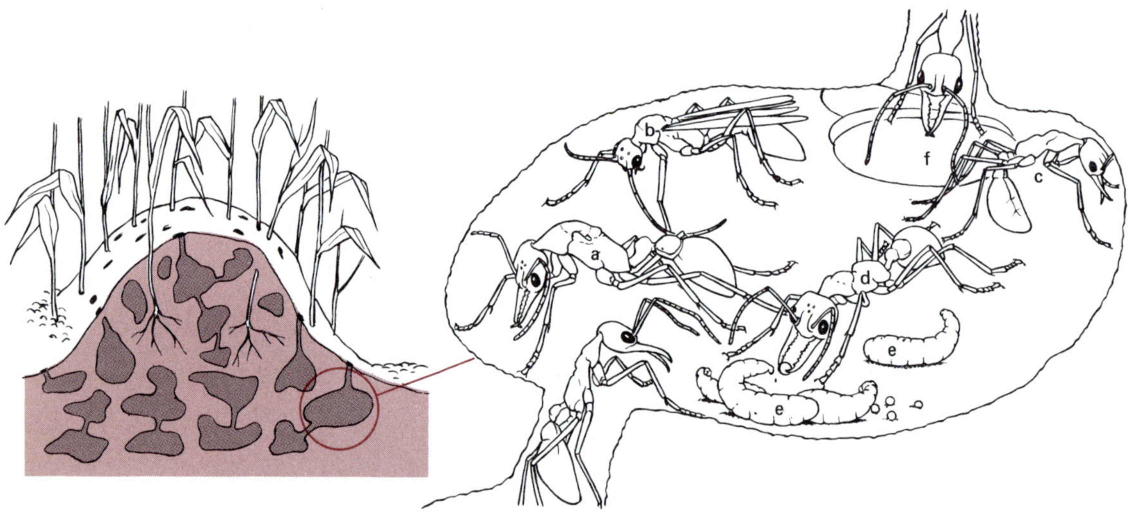

Figure 6–15 An anthill. Worker ants have hollowed out the chambers, digging away a grain of soil at a time. The ants shown in the magnified view are of the species *Myrmecia gulosa*. At left in the chamber are the queen (a) and a male (b). The worker at right (c) is laying a trophic egg, which will be used to feed the larvae (e), which hatch from eggs laid by the queen. Another worker (d) is offering a trophic egg to a larva. The cocoons (f) at upper right contain pupae undergoing the change from larvae to adults.

has progressed to such a degree that cries of alarm have been raised in an attempt to halt its inexorable advance. But specialization appears to be one of the necessary prices which must be paid for the fruits of a highly complex and technologically oriented society.

QUESTIONS FOR REVIEW AND THOUGHT

1. Are the cells of a particular tissue all of the same type?
2. Is the matrix of the tissue a living substance?
3. What elements of structure and function do the skin and the lining of the intestines have in common?
4. List the major types of connective tissue and discuss their similarities and differences.
5. How do skeletal muscle tissues differ from smooth muscle tissues?
6. Draw and label the basic structure of nervous tissue.
7. Should blood be classified as a tissue? Discuss.
8. Describe the major types of plant tissues. Could a plant lacking one or more of these tissue types survive?
9. Distinguish among: tissue, organ, organ system, and organism.
10. A beehive is sometimes referred to as a single organism. Does this metaphor have any validity?

CHAPTER 7

The Diversity of Life

Introduction

Picture a world without names. How would we identify our friends? civic leaders? How would we go about buying or ordering food or clothing? getting from one place to another? In short, our society—our very lives—are inextricably involved with the practice of giving objects and people "tags" (i.e., names) so that they may be referred to quickly and accurately.

Now picture a world in which each person used different names to refer to the same objects. Here too, we would have chaos. No one would be quite sure what anyone else was referring to. To some extent this situation exists today. Some linguists estimate that there are nearly five thousand distinct languages in the world today (not including dialects). Each object, each animal, each concept has its own particular designation in each of these languages. For example, "man's best friend" is a *dog* in English, a *chien* in French, a *Hund* in German, a *cane* in Italian, a *sobaka* in Russian, and a *perro* in Spanish. This diversity of names has been a definite hindrance to the progress of science, which requires an exact designation for each entity with which it deals.

This problem has been recognized to some degree by the fact that the U.S. and Soviet governments heavily subsidize programs of translations of scientific studies. But even these programs cover only a very small fraction of the world's scientific literature. If a single, uniform worldwide language existed, the "language barrier" would no longer be a bar to free interchange of scientific ideas among nations.

To some extent there is just such a worldwide language in existence today. The binomial classification system used in biology assigns a unique designation to each type of living organism known to man. This system is utilized throughout the world. Thus, "man's best friend" is recognized by scientists everywhere as *Canis familiaris.* Another example of a "universal language" is the system of chemical nomenclature used by scientists the world over. From Irkutsk, Siberia, to Sioux City, Iowa, the chemical symbol for sodium is "Na," and the formula for aspirin is $CH_3CO_2C_6H_4CO_2H$.

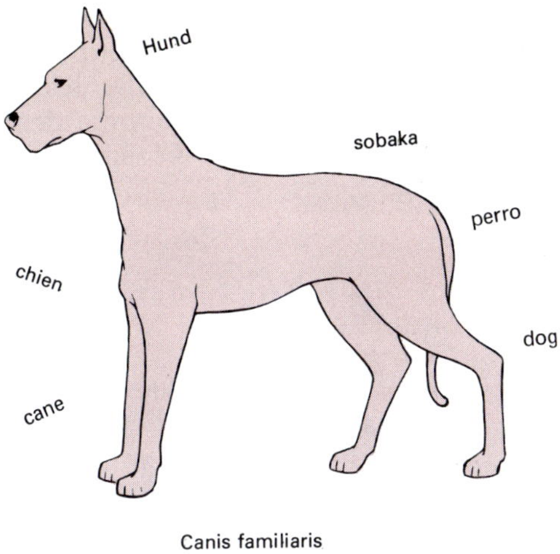

Figure 7–1 "Man's best friend" is known by many names. Scientists need standardized names that will be recognizable to their colleagues of all nationalities.

However, a uniform scientific language is not enough. Effective utilization of the ever-expanding volume of scientific information requires that it be meaningfully organized into some system from which needed information can be easily retrieved.

Picture a telephone book in which the names of subscribers are printed in the order in which they have paid their bills. Such a telephone book might be of some interest to a credit investigator, but for a person desiring to call a John Smith such a telephone book would be virtually useless; the names would be a hopeless jumble. For example, there are nearly a million entries in the Manhattan *Telephone Directory*, with more than eighteen hundred pages of names. Without some means of meaningful organization (in this case, alphabetization), the search for a particular name would be hopeless.

A similar problem exists in the biotic world. More than $1\frac{1}{2}$ million animals and plants have already been discovered and named by biologists. An enormous amount of labor goes into compiling descriptions of these organisms, their characteristics and habits. Yet if such information is not readily retrievable, it is virtually worthless. Furthermore, if a scientist wanted to verify whether or not he had indeed discovered a new species, he would require a quick and accurate means of comparing the various features of his prospective new species with those already discovered.

Man has recognized the need to classify living organisms for thousands of years, and many attempts have been made to do so. In general, we may divide classification systems into two rough categories: the "artificial" and the "natural" systems. In reality, all of these systems are, at least to some degree, artificial. But the natural systems profess to take into account the evolutionary aspects of species development: species which are closely related in their evolutionary development are classified closer together than those which bear a more distant evolutionary relationship. For example, the horse and the zebra would be included in a much smaller subdivision of such a classification system—i.e., they would be grouped closer together—than the horse and, let us say, the pig.

The artificial systems of classification, on the other hand, are essentially based upon certain arbitrary standards of classification. In these systems the evolutionary relationships are often ignored. Thus, the dolphin and the fish might be classed together because both live in the water, even though they are structurally and physiologically quite different.

The science which deals with the classification of living things is called **taxonomy.** The basic aims of this science are the classification of organisms according to similarities and differences in structure, function, development, and evolution.

Systems of Classification

Taxonomy is probably as old as man himself. In prehistoric times, when man first became aware of his surroundings, when he first noticed the diverse forms of life, he must have observed certain similarities and differences:

some creatures flew; others swam in the waters, but died when removed from those waters. Still others spent their lives on the land. Surely man must have realized that plants "lived" too; for as the days went by, they grew larger in size, and in times of drought they withered and died—just as his neighbors and children did. Mentally (and perhaps verbally) he must have grouped these creatures into definite categories. Under such a primitive system, birds, bats, and butterflies were probably grouped together.

We know much more about how man has looked upon the world of nature since the advent of written records. In the fourth century B.C., Aristotle and his student Theophrastus classified the animals as land, water, and air dwellers. Plants were grouped into herbs, shrubs, and trees by these two scientists from ancient Greece. In the typical man-oriented psychology of a later period, Saint Augustine (fourth century A.D.) classified animals as useful, harmful, or superfluous to man.

These simple systems of classifications may have served their purpose in the pre-Christian era, but with the advent of the Crusades and the expansion of trade, the people of the world were exposed to a greater variety of flora and fauna. Furthermore, with the steady rise of a middle and upper class, more leisure time was reflected, in part, in more time devoted to exploration of the secrets of nature. With the European discovery of the Western Hemisphere and the onset of the Renaissance, these explorations of nature took on an added impetus.

With the scientific advances in the understanding of the anatomical and physiological similarities and differences among living things, systems of classification took on a greater degree of sophistication and utility. When the principles of evolution became accepted in the late nineteenth and early twentieth centuries, and the details of this important phenomenon began to accumulate, these systems reached a still higher degree of sophistication and utility. Recently, biochemical evidence has been used with ever-increasing frequency in taxonomy. With our growing ability to decipher the amino acid sequence of proteins and the structural configuration of other complex molecules, the science of taxonomy should take on a degree of objectivity never thought possible just a few decades ago.

LINNAEUS'S BINOMIAL SYSTEM OF CLASSIFICATION

Having written more than 180 works concerning the classification of plants and animals, the Swedish physician and biologist Carolus Linnaeus (1707–1778) has well earned the title "father of taxonomy." Although Linnaeus was unaware of the evolutionary aspects of development, his classification of animals and plants according to structural similarities came surprisingly close to the modern system of "natural classification."

The name of the system used by Linnaeus, the **binomial** (Latin, "double name") **system,** is derived from the fact that each animal or plant in his system could be designated by two names: a **generic name** and a **specific name** (usually in Latin, at one time the standard language of scholars). For example, in this system man is called *Homo sapiens,* "*Homo*" being the generic **(genus)** name and "*sapiens*" the specific **(species)** name. Similarly, *Lumbricus terrestris* designates an earthworm. (This system may be likened to our own system of naming persons. For example, in the name John Smith, the family name, Smith, would correspond to the generic name, and the given name, John, would correspond to the specific name.) In addition to classifying each organism in terms of its genus and, within that genus, its species, Linnaeus included two larger groups: **order** and **class** (the latter representing his largest group). In the years following the death of Linnaeus, numerous additional discoveries of animals and plants have necessitated the incorporation of still other categories. First, two groups were added, one **(family)** between genus and order, and the second **(phylum**—plural, phyla) a more comprehensive category than class. Then the system was expanded still further,

Figure 7–2 Systems of classification attempt to find orderly patterns in the diversity of life.

until today we recognize the following categories (from largest to smallest): kingdom, phylum, subphylum, class, subclass, order, family, subfamily, genus, subgenus, species, subspecies. A classification of man according to this system would be: kingdom Animalia, phylum Chordata, subphylum Vertebrata, class Mammalia, subclass Eutheria, order Primates, family Hominidae, genus *Homo*, species *sapiens*. In classifying the corn plant: kingdom Plantae, phylum Tracheophyta, subphylum Pteropsida, class Angiospermae, subclass Monocotyledonae, order Gramminales, family Graminaceae, genus *Zea*, species *mays*.

Much confusion reigns in the field of taxonomy as to the exact number of species of animals and plants. There are several reasons for this. First, there is such a vast number of different species that it is often difficult to establish positively whether a supposed new species is in reality new. This difficulty is compounded by the fact that the same animal may exhibit slightly different characteristics depending on environmental factors. These slight variations may be taken by some over-eager investigator to be evidence of a new species. According to common agreement, the person who first finds and studies a particular species has the privilege of naming that particular species. In fact, even under similar conditions, there is often enough variation within the same species to create some confusion among taxonomists. Hence, it is not surprising to find that estimates of the total number of plant species range from two hundred fifty thousand to over four hundred thousand and those of the number of animal species from about 1 million to about 1.3 million.

THE CONCEPT OF SPECIES

Each form of life, each type of living creature on this earth is set apart from every other by certain unique qualities. Though they may graze together, or share the same water, each is separated from the rest of the world by a barrier—a genetic barrier.

Each type, or **species,** of animal or plant may interbreed successfully only with members of its own species. Some exceptions to this are to be found. For instance, a horse and donkey are considered to be separate species. And yet a viable offspring, a mule, results from a mating between these two animals (Fig. 7–3). The mule, however, is always found to be sterile. So, we might modify our definition of a species by stating that two animals (or plants) are of different species if no viable and fertile offspring results from their mating. But even here, we run into difficulty, for many organisms reproduce asexually. That is, a single organism gives rise to offspring.

The problem is compounded by the fact that occasionally within the same species certain physical obstacles may prevent normal matings. For example, in the case of dogs (all

Figure 7–3 The interspecies mating of a mare and a male donkey produces a mule, which is sterile.

dogs are of the same species, but different "breed") the size differential might present insurmountable difficulties. Breeding of a male chihuahua with a female great Dane would be practicable only through artificial insemination; but in the reverse insemination, normal development of the fetuses would probably not occur.

Speciation, the formation of new species, is a vital and integral part of the process of evolution. Closely related species are believed to have common ancestors and hence are closely related genetically, structurally, and physiologically.

THE KINGDOMS OF LIFE

Looking at a tree, one tends automatically to think of this organism as "a plant." A cow grazing beneath this tree would be readily classified as "an animal." These two groups represent the two great kingdoms of living organisms under which Linnaeus classified the varied forms of life with which he was familiar. For nearly two hundred years after his death the use of these kingdoms for classifying the organisms of the earth dominated the field of taxonomy. But as man's knowledge of the world he lived in grew and his appreciation of the evolutionary relationships among organisms increased, these two categories seemed to become increasingly inadequate to encompass the varied forms which were discovered and investigated.

For instance, it was generally agreed that a plant should possess the following characteristics: it should (1) contain a cell wall (in addition to a plasma membrane, of course); (2) be relatively nonmotile; (3) possess the green pigment chlorophyll, and hence manufacture its own food via photosynthesis; and (4) in many cases possess indeterminate growth capabilities (i.e., grow indefinitely). (Animals were expected to possess characteristics antithetic to these.) But it was found that there are numerous organisms that do not fulfill all these criteria. For example, one of the most "obvious" differences between plants and animals is the fact that the former are usually sessile, while the latter, as a rule, are highly motile. Yet corals, most sponges, and many other organisms universally recognized as animals spend their days moored to solid objects, ingesting their food as it floats or swims by in the water.

The criterion of indeterminate growth likewise is not an absolute criterion of a plant by any means. It is well known that many trees continue to grow, year after year, until some blight or accident or the depredations of man bring about their death. But some types of animals also seem to exhibit this type of indeterminate growth. It is thought that some fish continue to grow throughout their lifetimes, as may the giant Galápagos tortoises, which can exceed five hundred pounds in weight.

Among the single-celled organisms, firmly distinguishing between the seeming opposites of "plant" and "animal" sometimes becomes an insuperable problem. *Euglena,* for example, is a microscopic one-celled organism which normally contains chlorophyll and hence can manufacture food via photosynthesis. However, the euglena does not possess a cell wall, is highly motile, and reaches a definite size beyond which no further growth takes place. (At this point, the organism undergoes binary fission, i.e., splits in two.) Botanists have classified this organism in the kingdom of plants and, as you might expect, many zoologists have placed the euglena among the protozoa, i.e., considered it an animal. Today there are a number of systems in use which classify the euglena neither as an animal nor as a plant, but in a third kingdom, known as **Protista,** or protists. One such system places all single-celled organisms, including protozoa, algae, bacteria, and viruses, under this grouping. In addition, fungi and slime molds are also grouped under this category by some taxonomists.

The pull of tradition is strong. The simple-seeming "either-or" classification into plant and animal kingdoms, familiar to all from early childhood, has many attractions. But the tide of taxonomic thought appears to be turning toward a three-kingdom approach, in some ways more intellectually satisfying,

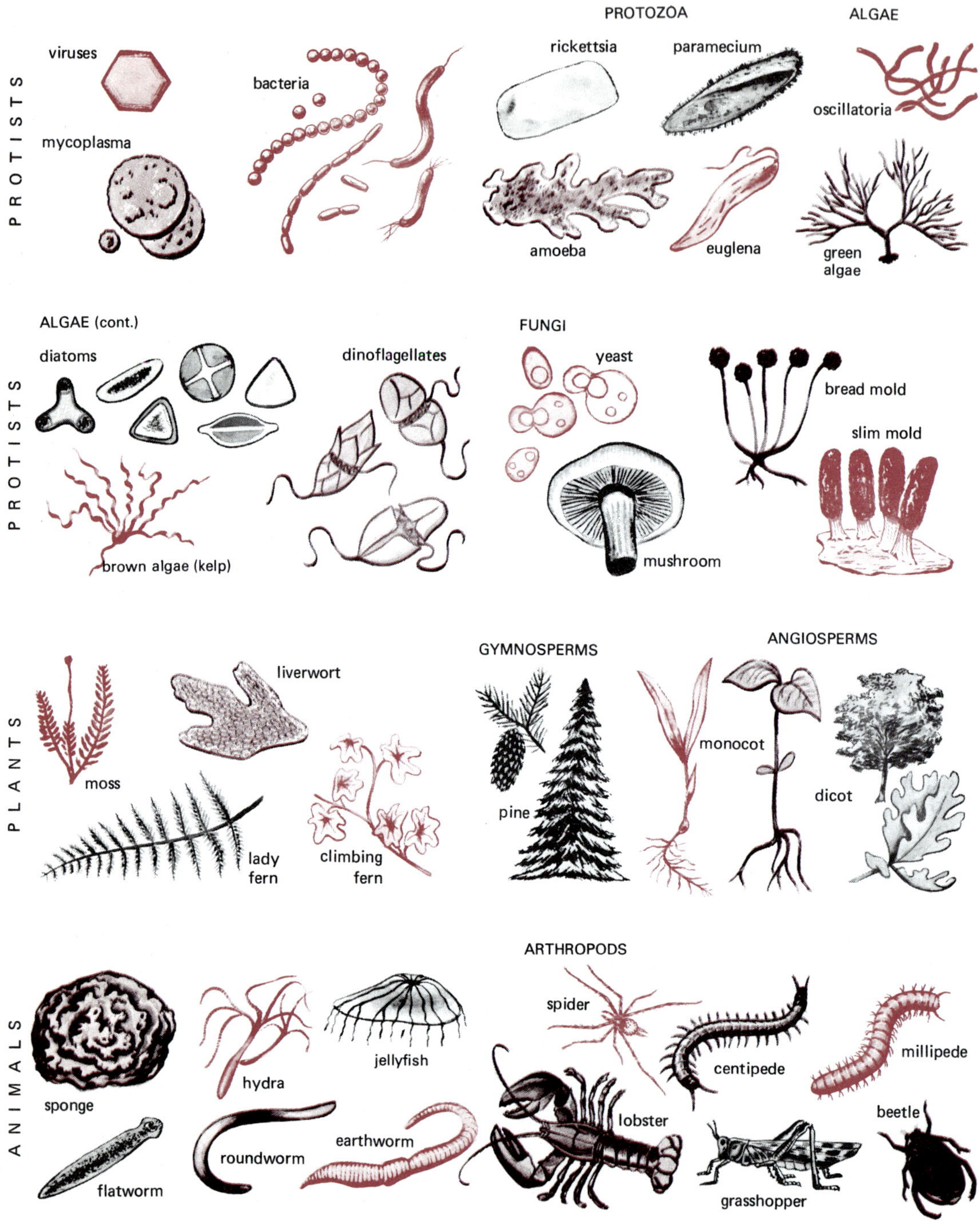

Figure 7–4 The kingdoms of life.

ANIMALS

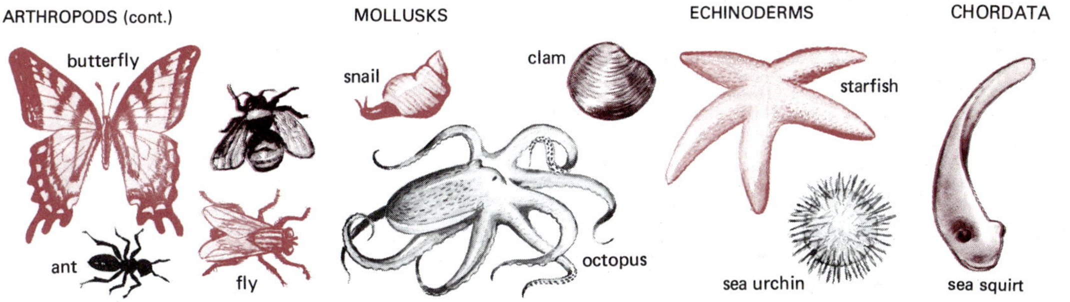

ANIMALS

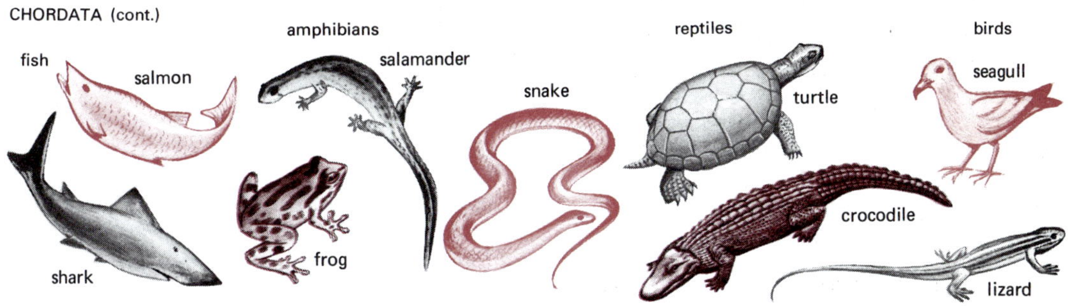

ANIMALS

ANIMALS

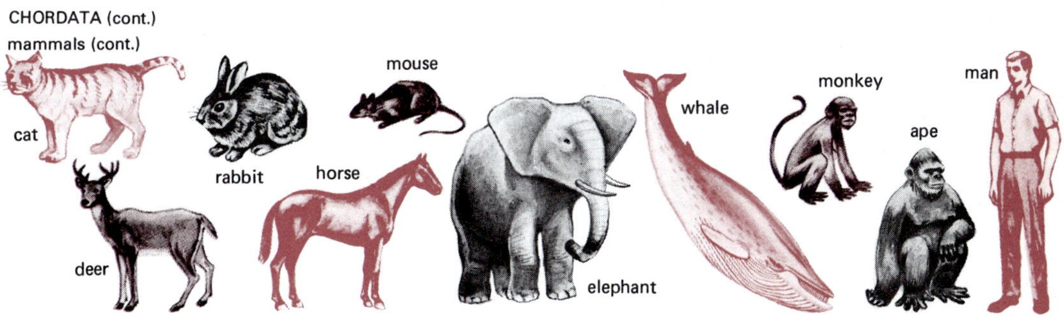

which recognizes the existence not only of clear-cut "animals" and "plants," but also of a vast area of "in-between" organisms. These organisms, many just single cells, seem to offer hints to the early evolutionary history of life on earth, to the time before the development of the varied organisms of the world had diverged to produce true "animals" and "plants."

It is this three-kingdom system of classification that will be used in this text. A summary of the major phyla (and some of their classes) that constitute the three kingdoms of life—protist, plant, and animal—is given in Fig. 7–4.

Fundamental Characteristics of Life

Before embarking upon a panoramic survey of the kingdoms of life by covering the salient characteristics of the members of the major taxonomic groups and their interrelationships, it might be well to pause a moment for a consideration of a more fundamental point: Just what is life?

This seemingly simple question has puzzled great minds for millennia. Empirical criteria have been devised which (nearly always) distinguish between the living and the nonliving. But still the essence of life eludes us.

For many years a school of thought known as **vitalism** held sway. It was believed that living things are characterized by a "vital force," which sets them apart from the nonliving world. This force could not be readily described purely in terms of the physical and chemical laws applicable to the nonliving world. An extension of the vitalist philosophy maintained that organic substances, being the substances of life, are qualitatively different from inorganic chemicals and can be produced only by living organisms.

Vitalism sustained a severe blow in 1828, when Wohler synthesized the organic compound urea. Numerous subsequent experiments weakened the position of the vitalists still further. As the barriers of ignorance were pushed back, as new insights into the secrets of the cell and organism were gained, a newer school of thought, **mechanism,** gained ascendancy.

The mechanists believe that life, as well as the secrets of the entire universe, are ultimately explainable in terms of physical and chemical principles. Of course, we are still ignorant of many aspects of these basic principles. But mechanism holds that if all these factors were known, all the observable phenomena in the universe would be found to be governed by physical and chemical laws.

We are still faced with our initial question: What is life? Turning from the philosophical aspects of the problem, let us discuss some of the basic characteristics that we normally associate with life, and hence the properties and activities of living organisms. These include metabolism, growth, reproduction, mutation, irritability, movement, and adaptation. Many of these characteristics, or semblances of them, are observed in the inanimate world as well. But only living things possess the entire complex of characteristics.

METABOLISM

The living organism has frequently been likened to a chemical factory. For instance, in the petroleum industry, such raw materials as crude oil fractions are catalytically decomposed into smaller units. These smaller units are separated into still smaller fractions. Some of these fractions may be burned as fuel to provide energy to run the factory. Other petroleum by-products may be utilized within the same factory complex for the synthesis of new chemical substances, ultimately yielding such useful products as fibers, plastics, and drugs. Waste products, resulting from the combustion of the fuels and arising in the various manufacturing processes, are often disposed of into the atmosphere and into nearby streams.

In the living organism, ingested food is catalytically decomposed into smaller units.

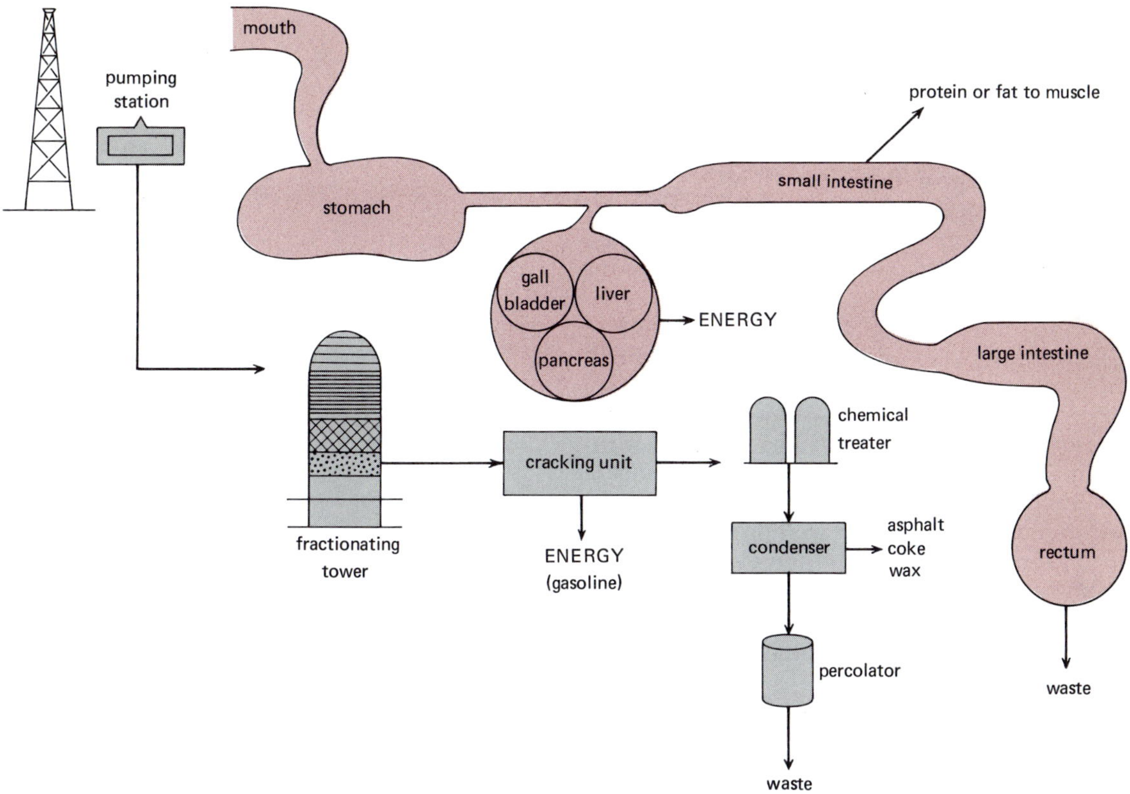

Figure 7–5 A living organism can be compared to a complex chemical factory.

Some of these units are oxidized as fuel, thereby yielding the energy needed to sustain the organism. (In plants, the initial organic materials are synthesized from simple raw materials with the aid of sunlight, rather than ingested as such.) Other units are incorporated into the substance of the organism by processes of synthesis and assimilation. Invariably waste products are formed and are eliminated by means peculiar to the particular organism.

The sum of all the chemical processes that take place in an organism is known as **metabolism.** Essentially two basic types of chemical processes may be said to occur in an organism: **catabolism** (i.e., destructive processes) and **anabolism** (constructive processes). The digestive processes, as a group, are considered catabolic, while such processes of assimilation as the building of new tissues are classified as anabolic.

Just how does the metabolism of a living organism differ from the chemical reactions and energy interconversions that occur in a factory complex such as that outlined above? A main distinction lies in the incredible efficiency of metabolic processes. In the compact system of the living cell—a structure so small that, with a few rare exceptions, it cannot even be seen without a microscope—numerous highly specific reactions proceed under conditions of normal pressure and moderate temperatures. Many of these reactions are beyond the scope of our present technological capability to reproduce, while others can be accomplished in manmade systems only in enormous installations and under great extremes of temperature and pressure.

GROWTH

Growth, one might argue, is as much a part of the nonliving world as it is of the living world. The snowball thundering down the hill, increasing in size with every revolution, is certainly an example of growth. Similarly, a blue crystal of copper sulfate in a saturated copper sulfate solution will continue to "grow" indefinitely (as long as the solution remains saturated).

But it may be argued that these forms of "growth" are not identical with that growth observed in living organisms. Indeed, there are definite distinctions. One of the most important of these is that growth in the inanimate world usually occurs by addition of material to the outer surface of the object—in other words, growth by **accretion** (Fig. 7–6a). The growth experienced by living organisms, from the amoeba to man, occurs by entirely different processes. Material *enters* the individual cell in a simple form and is organized so as to become an integral part of that cell. This type of growth is referred to as growth by **intussusception** (Fig. 7–6b).

There are certain fascinating aspects of growth in the living world. Most animals cease to grow after reaching a definite size. At this point a dynamic equilibrium is maintained. For although new cells are constantly being formed throughout the organism, the rate of replacement equals the death rate of the cells. (During active growth, the rate of new cell formation exceeds the death rate.) In many plants, however, there is a continuous process of growth. This phenomenon is seen in the case of many trees. For instance, some of the giant redwood trees on the Pacific Coast have been growing for thousands of years and are still growing today.

Figure 7–6 Growth by accretion (a) and intussusception (b). The copper sulfate crystal grows by adding more $CuSO_4$ on the outside. A chick takes in feed, assimilates it, and uses its components to synthesize new proteins and other structural substances.

REPRODUCTION AND MUTATION

That a dog should give birth to puppies, that a cat should bring forth kittens, and that an oak should give rise to a stand of young oaks seems only natural. And yet the phenomena of reproduction are among the most miraculous and exciting to be found in the universe. The unbelievable complexity of a living organism is reproduced with an exactitude that is far beyond the present (and currently foreseeable) capabilities of human technology.

The term *reproduction* implies the production of something similar to that which already exists. Just how this comes about is the essence of the reproductive process. For thousands of years man believed that life could arise spontaneously from the soil or from inanimate objects. Even Aristotle believed such tales as that mice might arise from decaying wheat, that the mud of a stagnant pond could give rise to frogs, and that maggots and worms might miraculously appear from rotting meat. Of course, today such notions seem silly. But such things *were* observed. Frogs do appear to arise from the mud of ponds, maggots will appear to emerge from a mass of rotting meat. Today we know that frogs hibernate over winter in the mud at the bottom of lakes and ponds, and that flies lay eggs in meat, which subsequently hatch out into wriggling maggots. But the ancients were not experimenters. Appearances seemed to satisfy their curiosity.

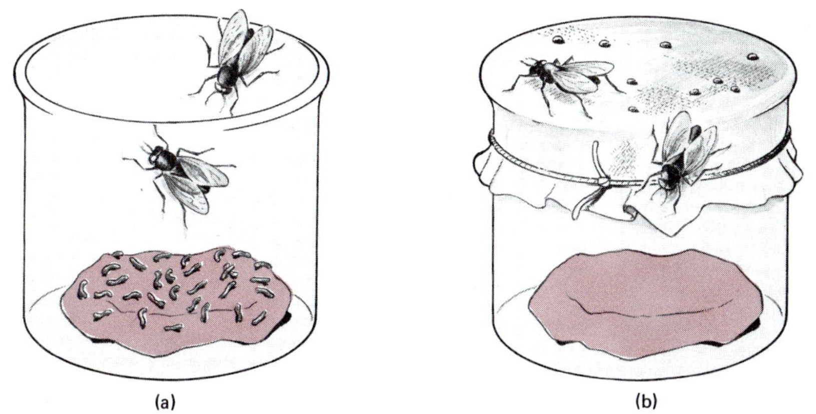

Figure 7–7 The Redi experiment. Maggots appeared on meat in an open jar to which flies had free access (a). Another piece of meat, placed in a covered jar, remained free of maggots (b). Fly eggs could be observed on the gauze cover.

During the Renaissance, however, men began to examine with new eyes what had previously been taken for granted for thousands of years. They began to probe more deeply with the refined tools of the emerging scientific method. Thus, by the seventeenth century the belief in the **spontaneous generation of life** was being actively attacked by a number of scientists. A major breakthrough was achieved in 1668 when the Italian scientist Francesco Redi performed a series of brilliant experiments clearly demonstrating that maggots do not arise spontaneously from meat (Fig. 7–7). Redi prepared several wide-mouthed jars containing quantities of meat. He divided these jars into two groups. The mouths of one group of jars were covered with a fine gauze, which prevented flies from entering. The mouths of the second group were left open and readily accessible to the flies. In time, maggots developed in the second group but none were observed in the first group. Redi observed, however, that tiny eggs were sometimes deposited on top of the gauze covering the wide-mouthed jars. Subsequent researchers confirmed Redi's work, and the idea of spontaneous generation for macroscopic organisms languished and died.

But with the expansion of the use of the microscope, and the discovery of microscopic organisms, the notion of spontaneous generation was reborn and flourished anew. It was argued that these newly discovered creatures could arise out of specks of dust or other nonliving material. After all, they often were found swarming where none were seen to be before.

In 1765 another Italian scientist, Lazzaro Spallanzani, performed a series of experiments which seriously discredited this belief (Fig. 7–8). Spallanzani prepared two sets of vessels containing nutrient broth. One set was permitted to stand with their mouths open to the air, while the contents of the second set were boiled in an attempt to destroy any microbes which might be present within the broth. These vessels were then sealed. Although the exposed set of containers seethed with microbial activity, the sealed vessels were completely devoid of life.

Surely this was sufficient evidence to sound the death knell for the proponents of spontaneous generation. But this was not to be the case. Instead, they insisted that the act of boiling had destroyed some "vital force," perhaps present in the air, which was necessary for spontaneous generation. The final blow came, however, in 1862, nearly one hundred years later, when the French scientist Louis Pasteur performed a simple yet conclusive experiment. Pasteur repeated the experiments of Spallanzani, but he used a specially shaped flask for his experiments. Instead of sealing the flask after boiling, he drew out a long S-shaped tube with its end permitted to remain open to the air (Fig. 7–9). The configuration of the tube was such that air could freely enter and leave the flask,

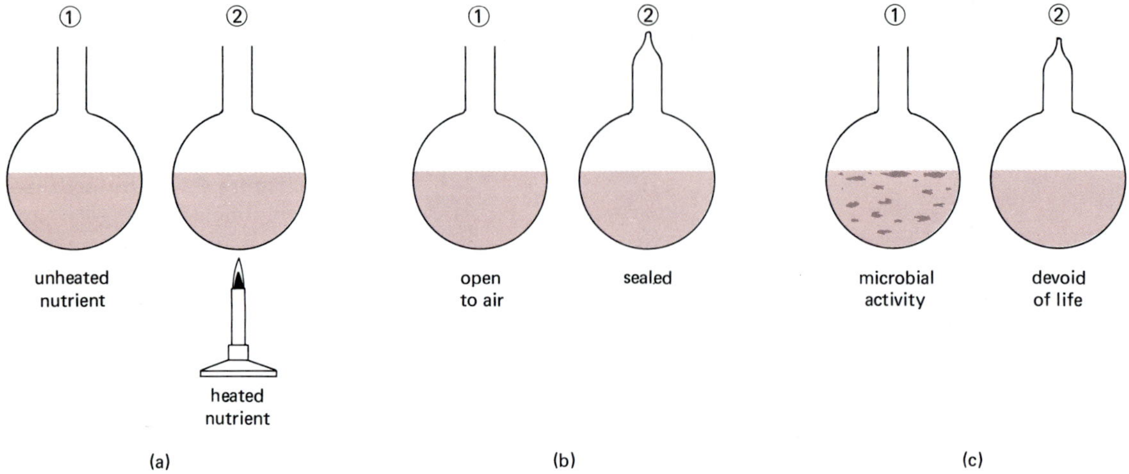

Figure 7–8 The Spallanzani experiment. The broth left exposed to air soon teemed with microorganisms. The broth that was heated and then sealed off remained sterile.

but dust particles and microorganisms were prevented from entering because of the sharp bends, which effectively served as traps. When the flask cooled and was permitted to remain undisturbed, no microorganisms were seen after days, weeks, and even months. However, when the S-shaped tube was broken off, permitting dust to enter, the broth soon teemed with microorganisms. Thus Pasteur effectively discredited once and for all the myth of spontaneous generation.

But is the concept of spontaneous generation really a myth? If the statement propounded by Virchow in 1855, "Omnis cellula e cellula" ("All cells arise from cells"), is to be taken as true and we are to assume that all life has arisen from previous life, then how are we to explain the appearance of the *first* life on earth without invoking either a strict interpretation of the Bible or the theory of spontaneous generation? This question will be discussed in greater detail in Chapter 18; suf-

Figure 7–9 The Pasteur experiment. The S-shaped neck protected the nutrient broth from exposure to dust particles, although air could enter freely. Under such conditions, a sterilized broth remained sterile. When the S-shaped neck was subsequently broken off, the solution quickly became inoculated with microorganisms.

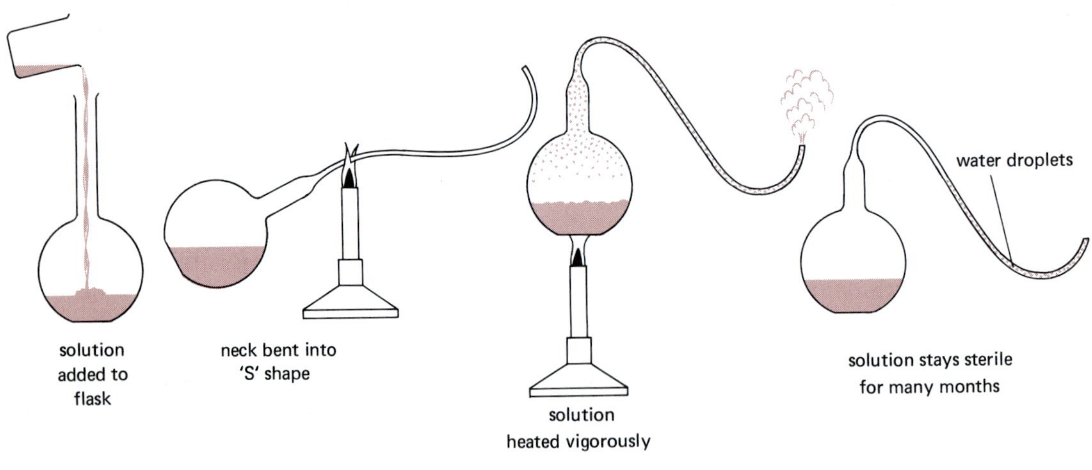

fice it to say here that the vast majority of biologists today believe that life arose on earth some billions of years ago from inanimate chemicals by a process of "chemical evolution."

The processes of reproduction are quite varied among the living species of this world. But two basic types of reproduction may be distinguished: **asexual** and **sexual.** Many types of asexual reproduction are to be found among life forms, ranging from a simple splitting in two **(binary fission)** to more complicated mechanisms in which **spores** first are formed and in turn give rise to new individuals. In sexual reproduction, on the other hand, two individuals are required to consummate the process. Each produces specialized cells, **gametes,** which play an integral role in the reproductive process. Two gametes —a sperm from the male and an egg from the female—combine to form a new cell, a **zygote,** which develops into a new individual.

Thus, in sexual reproduction each parent donates a part of itself to the offspring, and in these parts are carried the blueprints which together determine the ultimate plan of the newborn individual. But occasionally something goes amiss. Some trait or feature appears in the offspring which is nowhere found in the parents. Such an anomaly is called a **mutation,** which may be defined as a sudden change in a trait that in turn is passed on to the succeeding generation of cells. The most important mutations are those occurring in the gametes, sometimes resulting in the inheritance of the change by the offspring. But mutations can also occur in the body or **somatic cells.** Most mutations are either deleterious or have no particular obvious effect. The effects of some deleterious mutations are quite severe: two-headed calves, legless dogs, eyeless children, and countless other tragedies. But mutations have also been a powerful force in the evolutionary development of life on earth, and without their influence, doubtless man himself would never have emerged.

IRRITABILITY AND MOVEMENT

Sometimes a moth will flutter directly into a flame, singeing its wings once or twice until some apparently compelling force draws it to a final smoldering death. Periodically hordes of lemmings smother the Scandinavian countryside and stream toward the sea to a seemingly pointless destruction. A veldt full of wildlife may suddenly be transformed into a riot of kaleidoscopic motion at the crack of a twig or the scent of a man. The family cat arches its back beneath a stroking hand.

Each of these movements is in response to a stimulus. Sometimes such stimuli are clear and understandable; at other times they are obscure and the object of much debate among scientists. But whatever the stimulus, the ability of an organism to respond to it is termed **irritability.** Frequently this response takes the form of gross movement, as indicated above. But on other occasions this response may assume many other forms. The odor or even the mere thought of food may result in salivation and the onset of other digestive processes in both man and many lower animals.

In plants, response to various stimuli may result in gross movement, as when a vine climbs up a trellis, or it may take some other form, such as the whitening of leaves when kept in the dark. Generally, movements in the plant kingdom are quite slow, though some exceptions do exist (for instance, the Venus flytrap, a carnivorous plant that snaps shut on an unwary insect that has lighted upon it).

In the animal kingdom, complex systems have arisen for the perception of, analysis of, and response to stimuli. The light receptors (photoreceptors) in the eyes, the sound receptors (phonoreceptors) in the ears, and the chemical receptors (chemoreceptors) in the nasal passages and on the tongue, as well as the complex and integrated nervous system and the muscular system—indeed, essentially all the systems in the organism—play a role in its response to stimuli.

ADAPTATION

An important characteristic that sets man apart from the other living organisms of the earth is his ability to adapt his immediate environment to suit his own needs in a multitude of ways. If he is cold, he dons clothing; if he is hot, he resorts to air-conditioning;

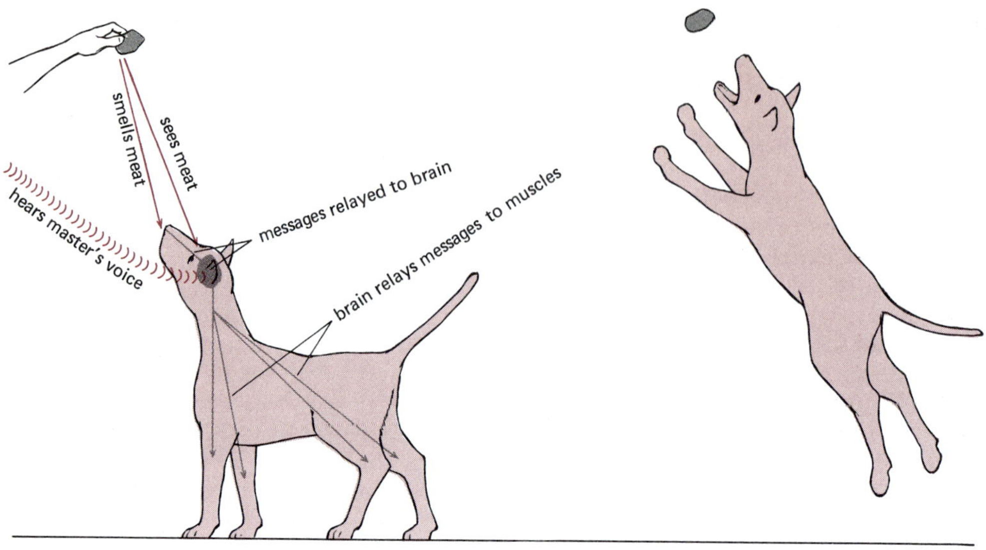

Figure 7–10 Living organisms perceive and respond to stimuli.

with the flick of a switch he can make his surroundings dark or light or even rainbow-hued. He can even change his internal environment: with pep pills, tranquilizers, aspirins, or any number of other drugs, he can lift his spirits, quiet his "nerves," or cast himself adrift into a peaceful sleep.

But the creatures of the world (and man himself, of course) are, to a large extent, at the mercy of the whims of Nature. For billions of years, through the long processes of evolution, countless numbers of creatures have survived and thrived, to be replaced in turn by species which were better able to

Figure 7–11 Life forms have adapted to a variety of environmental conditions.

adapt to the always potentially hostile environment. For example, the mighty dinosaurs, which were the dominant forms of life on earth for more than a hundred million years, gave way to more adaptable small, furry, ratlike creatures, the ancestors of the modern mammals.

A fantastic variety of life forms have adapted to the diverse spectrum of conditions of our world. Fish swim effortlessly under crushing pressures in the depths of the seas—they literally explode if brought up to the surface. The frozen wastes near the poles have their characteristic animals and plants, and so do the steaming jungles near the equator and the arid deserts of both Western and Eastern hemispheres.

Plants and animals have adapted not only to the physical conditions of the earth, but to one another as well. Often intricate interdependent relationships among different species of living organisms have evolved, so that the demise of one species may adversely affect another or even an entire community. The lives of those that eat and those that are eaten, and of those that compete for food or sunlight or shelter, are intricately interwoven in a delicate balance.

Man, in his efforts to reshape his own environment, has sometimes unwittingly—and sometimes purposely—caused dislocations in the delicate balance of nature, the ultimate effects of which we have yet to comprehend. But as new balances are struck, it is those species best able to adapt that survive.

Criteria for the Classification of Living Things

It is the year A.D. 2092. The spaceship *Santa Maria II* has just landed on an alien planet. The ship's biologist descends the gangplank and peers out into a strange, unearthly world. Numerous objects never before seen by man move about before him. Others, just as strange, show no signs of movement. His job is first to determine whether "life" exists on this planet and, if so, to classify this life into some meaningful and useful system.

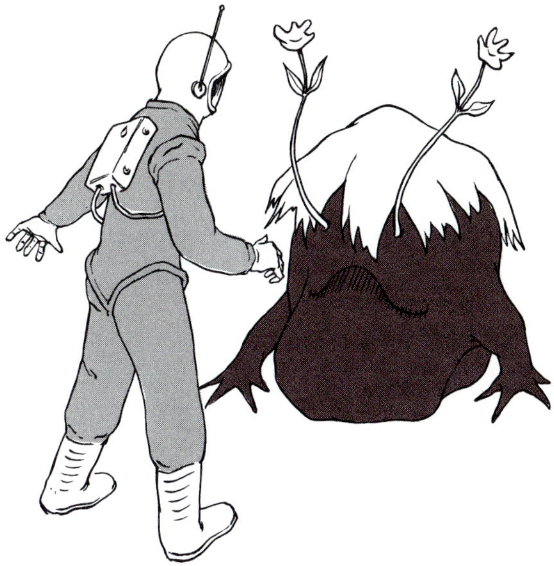

Figure 7–12 Are you a plant or an animal?

Today this scene seems like a flight of science fantasy, but scientists are actually making plans and devising instruments for such an eventuality. We have already discussed the criteria by which we determine whether an object is "alive" or not. Here we shall consider criteria by which we may classify such living organisms.

Assuming we have decided whether an organism is a plant or animal or protist, our next task is to attempt to classify it into a series of progressively smaller groups so that we can then (1) find it and material related to it rapidly and painlessly and (2) appreciate its relationship to the other organisms and, perhaps, (3) infer additional and as yet uninvestigated structural and physiological characteristics from its position in the classification scheme (much in the manner that a chemist can often predict some of the chemical properties of an element or compound before it has been discovered or synthesized).

Such basic characteristics as similarities and differences in structural, physiological, developmental, ecological, and genetic aspects of the organism are among the major factors which are considered in the classification of living things.

CLASSIFICATION OF PROTISTS

Within the highly diverse group of organisms comprising kingdom Protista, the presence of either a single cell, a loose aggregation of cells (colony), or an integrated complex of cells differentiated into tissues and organs is an important criterion of classification. The mode of nutrition, whether **autotrophic** (manufacturing their own food) or **heterotrophic** (feeding upon other living organisms or upon decaying organic matter), is another important criterion.

Other factors are also considered. Thus, the animallike protozoa are commonly distinguished by the presence of such locomotor organelles as pseudopods, cilia, and flagella (or by a sessile or passively floating existence). A main factor in differentiating the different types of true fungi is the mode of reproduction, while an important criterion for distinguishing the different groups of algae is their most conspicuous pigment.

CLASSIFICATION OF PLANTS

A consideration of such structural aspects of higher plants as the reproductive organs and tissues plays a vital role in determining the exact taxonomic position of a particular species. For instance, the presence of spores or seeds, and the types of seeds, flowers, and fruits, are important considerations. Naturally, such vegetative structures as the types of roots, stems, and leaves (or the absence of one or more of these structures), the types and quantity of conducting and supporting tissues (or the absence of one or the other), along with various other structural aspects, aid in the classification of plants. Physiological, genetic, ecological, embryological,* and geographic factors also play a part in determining the various relationships among plants. Recently, biochemical analysis has become an important tool toward this end.

The mosses, liverworts, and hornworts, which have no conducting tissues, are thus distinguished from the remaining members of the plant kingdom, the vascular plants. These in turn are categorized either as ferns (and other types that reproduce by the formation of spores) or as seed plants. The latter are further subdivided according to the type of seed, whether naked or enclosed in an ovary. Within each large group, then, the factors enumerated above are studied to achieve finer and finer distinctions.

*Unlike the photosynthesizing and other plantlike protists, all of the true plants form embryos and are often collectively termed Embryophyta.

CLASSIFICATION OF ANIMALS

The fact that there are three or four times as many animal species as plant species on earth seems to imply that the classification of animals may be a more difficult task. But this is not necessarily true. Taxonomists have evolved a number of criteria on the basis of which decisions are made in the process of classification of animal species.

One of the primary criteria is the number of cells and the level of organization of the animal. The phylum Porifera includes those animals which possess a **cellular level** of complexity—i.e., the individual cells are relatively uncoordinated in their activities. Coelenterates have a primitive nerve network which permits a fairly high level of coordination of these animals (and thus we refer to this group as having reached the **tissue level** of organization or complexity). Flatworms, with brain, eye, and various digestive organs, are the first group at the **organ level** of complexity.

The type of structural symmetry (or presence or lack of it) is one of the most important criteria in the classification of animals. We might say that all objects in the universe fall into one of two categories: symmetrical or asymmetrical. Many types of symmetry are observed in this world. Basically we may define **symmetry** as a similarity or correspondence of shapes on opposite sides of a midline or central point (Fig. 7–13). Thus, a circle shows what is termed **radial symmetry.** The spokes of a wheel, a tire, an asterisk (*), a hydra, and a starfish all exhibit radial symmetry. Man is a good example of **bilateral symmetry:** if an imaginary line is drawn down the midline of his body through his forehead,

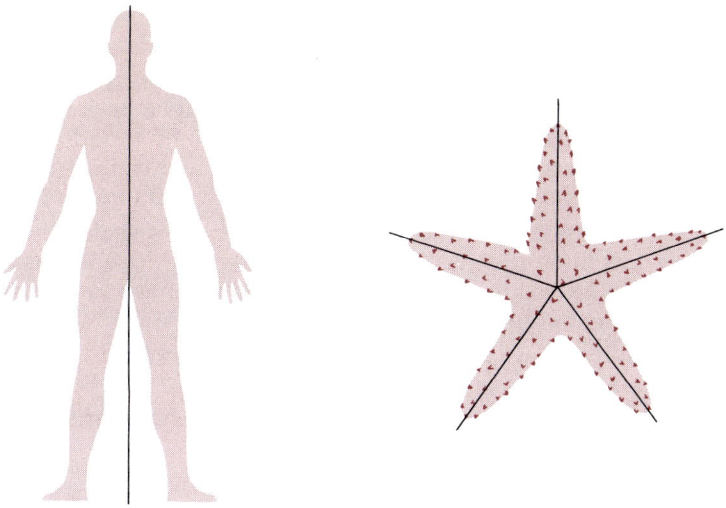

Figure 7–13 Symmetry is an important criterion in animal classification. Most higher animals are characterized by bilateral symmetry. (Curiously, the radially symmetrical starfish has a bilaterally symmetrical immature form.)

nose, mouth, and navel, the two halves would be essentially similar. Most of the higher forms of animals exhibit this type of symmetry.

The presence of **segmentation** of the body is a criterion used in the classification of animals, particularly the various worms. For example, the earthworm is distinguished from many of the lower worms by the fact that its body contains a series of repetitive units or segments. Segmentation is found also in arthropods and chordates.

Many other structural characteristics are also considered by taxonomists. The skeletal, digestive, circulatory, nervous, excretory, and other systems play a role in the classification of animals.

Aside from the structural aspects, which of course play a dominant role in the classification of animals, other factors also are considered. The physiology, ecology, and genetics of organisms are some of these. Biochemistry has shed some additional light upon taxonomic groupings, and while as yet comparatively little work has been done in this area, the potentialities are enormous. As chemical methods of the analysis of macromolecules improve, we can expect to see a significant upsurge in the use of biochemistry as a taxonomic aid.

QUESTIONS FOR REVIEW AND THOUGHT

1. Is a comprehensive, generally accepted system of classification of living organisms really necessary?
2. Outline the essential points of Linnaeus's binomial system of classification.
3. Why are there such discrepancies in reporting the total numbers of animal and plant species?
4. Discuss the criteria for distinguishing separate species in light of the variations among breeds of dogs and successful matings between horses and donkeys.

5. What are the criteria for distinguishing between plants and animals? In what cases are these criteria inadequate?
6. Why is taxonomy not considered to be an "exact" science?
7. You are chief biologist of the spaceship *Santa Maria II.* You have just set down on an unexplored planet, and scouts have brought back samples of velvety purple, pincushion-shaped structures which they found associated with rocks and boulders. You have to determine whether they are alive. What criteria will you use, and what sort of tests will you employ?
8. Your tests of the structures described in Question 7 have revealed that these structures are in fact life forms. What criteria will you now use in classifying them among the kingdoms of life? Could other planets have life forms that would not fit into any of the kingdoms with which we are familiar? Explain.
9. Describe the "landmark" experiments in the controversy on spontaneous generation.
10. Is the ability to adapt to environmental conditions a characteristic only of living organisms? Discuss.

CHAPTER **8**

Panorama of the Protist and Plant Kingdoms

The Borderline of Life

If a tiny blue crystal of copper sulfate is placed in a beaker containing a saturated copper sulfate solution, it will slowly grow in size. With each passing day, its dimensions will increase. It will continue to grow (if adequate solution is available) for months, years, and decades. Is this life?

If a tiny piece of sodium metal is dropped onto the surface of a beaker of water, it will fizzle and will begin to skid crazily over the surface of the water. If a tiny drop of oil is placed on a solution of glycerol and alcohol, it will send out "pseudopods" and slither in a manner resembling an amoeba. Are these examples of life?

If a tiny crystal of tobacco mosaic virus is placed on a tobacco plant, the leaves will shortly become speckled in the mosaic pattern characteristic of the disease. Is this crystal alive?

Most people would agree that in the first three instances, although characteristics of life were superficially manifested, neither the crystal of copper sulfate, nor the piece of sodium metal, nor the drop of oil was really "alive." However, the case of the tobacco mosaic virus is a matter of sharp dispute among scientists. These "organisms" can be crystallized in chemically pure form. In this form they might be mistaken for ordinary table salt. They cannot reproduce unless they are within a living organism. They have neither a nucleus, cytoplasm, mitochondria, a cell membrane, nor most of the other structures associated with a normal cell. These invading viruses are essentially nothing more than DNA or RNA molecules. And yet they do possess many of the characteristics normally associated with life: they respond to stimuli, they mutate, and they grow and reproduce (albeit normally only in a living organism). Because viruses lie at the borderline of life, they usually are not given a place in classification systems.

VIRUSES

Viruses are unbelievably small. If the period at the end of this sentence were considered as a miniature raft, it could hold more viruses than there are people on this earth. One of the

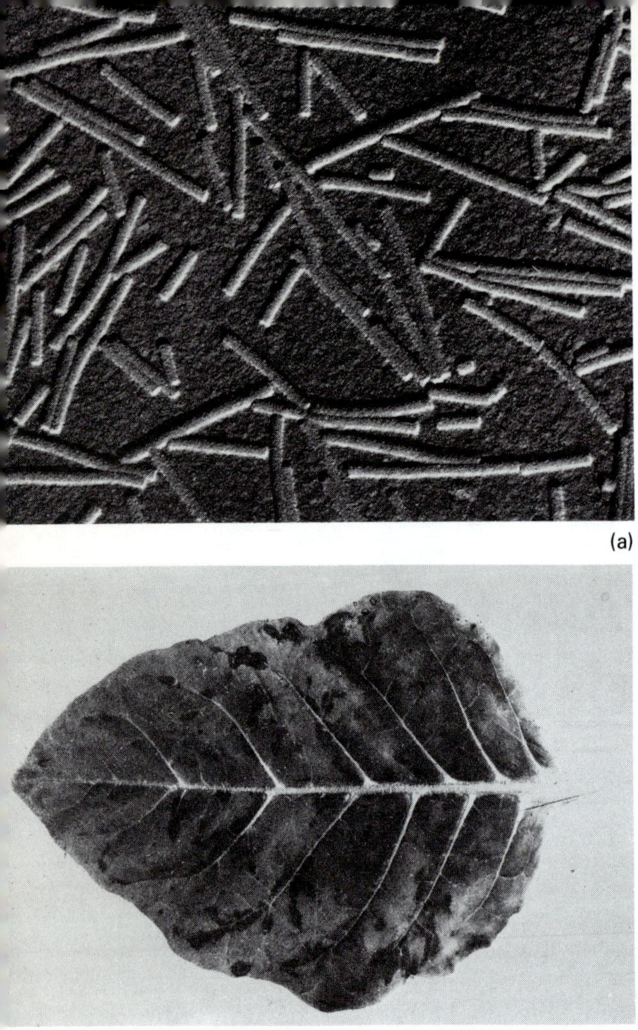

(a)

(b)

Figure 8–1 Tobacco mosaic virus. (a) Tobacco mosaic virus, magnified 94,600×; (b) leaf of a tobacco plant infected with TMV.

smallest viruses is the influenza virus (9–10 mμ). But the largest viruses (e.g., ornithosis virus, ca. 500 mμ) are almost as large as some of the smallest bacteria.

Viruses are comprised essentially of an outer protein coat and an inner nucleic acid core. Viruses that infect plants (e.g., the tobacco mosaic virus, or TMV) generally contain RNA as the nucleic acid. There are also a number of animal-infecting viruses which have RNA as their nucleic acid (e.g., poliomyelitis, influenza, and mumps viruses). The DNA viruses (except for bacteriophages) usually infect animals. Such diseases as smallpox and yellow fever are caused by DNA viruses.

Viruses may assume a great variety of shapes, which may be classified according to their symmetry (Fig. 8–2): **spherical** or **cubic** symmetry (e.g., adenovirus, chicken pox), **helical** or screw symmetry (TMV, influenza), or **complex** symmetry (vaccina, bacteriophages).

Some DNA-containing viruses specifically attack bacteria. These viruses are called **bacteriophages** (or **phages**). They have been of immense importance to geneticists and pathologists in elucidating the mechanisms of life.

Among the bacteriophages most commonly used in genetics research are the "T-even phages." T2 and T4 bacteriophages infect *Escherichia coli*, a bacterium commonly found in the human colon. These bacteriophages resemble a tennis racket, with a hexagonal head and a taillike appendage (Fig. 8–3).

MYCOPLASMAS

Mycoplasmas are the smallest living things capable of independent existence. These organisms are completely invisible under the light microscope and may reach dimensions as small as 0.1 mμ (ca. 1000 Å) in diameter. The surprising fact is that these tiny organisms have a unit membrane approximately 100 Å thick, which is of about the same dimensions as in normal-sized cells. More than thirty different types of mycoplasmas are known, and additional ones are being discovered each year.

Some mycoplasmas have been implicated in respiratory and other diseases afflicting such animals as cattle, sheep, goats, chickens, turkeys, mice, and rats. Recently, evidence implicating them as causative factors in pneumonia has been reported.

In the 1940s, M. D. Eaton isolated a mycoplasma from the sputa and lungs of pneumonia patients. Hence mycoplasmas as a group are sometimes referred to as pleuropneumonialike organisms (PPLO). It was not until recently, however, when human volunteers were administered a tissue-culture strain of the PPLO, that its causative relationship to pneumonia was firmly established.

Mycoplasmas are truly remarkable organisms. They apparently represent the ultimate

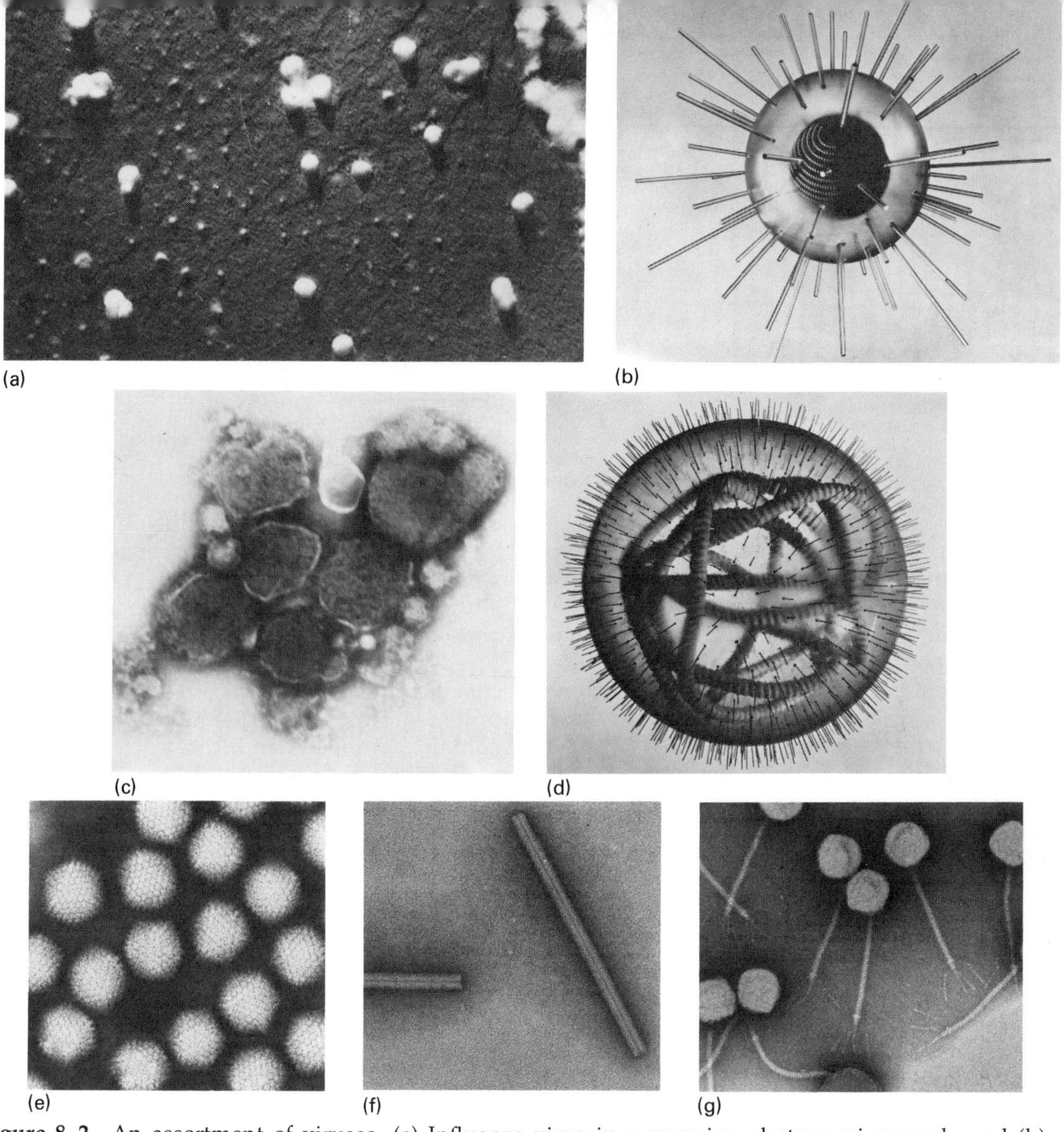

Figure 8–2 An assortment of viruses. (a) Influenza virus in a scanning electron micrograph, and (b) a model designed from electron micrographs; (c) electron micrograph and (d) model of measles virus; electron micrographs of (e) adenovirus, 92,6000×, (f) tobacco mosaic virus, 117,000×, and (g) T5 bacteriophage, 72,000×.

Figure 8–3 (a) T4 bacteriophage; (b) a bacterium, *Escherichia coli*, infected with phage T2. Note the phage protein coat attached to the surface of the bacterium in the upper left.

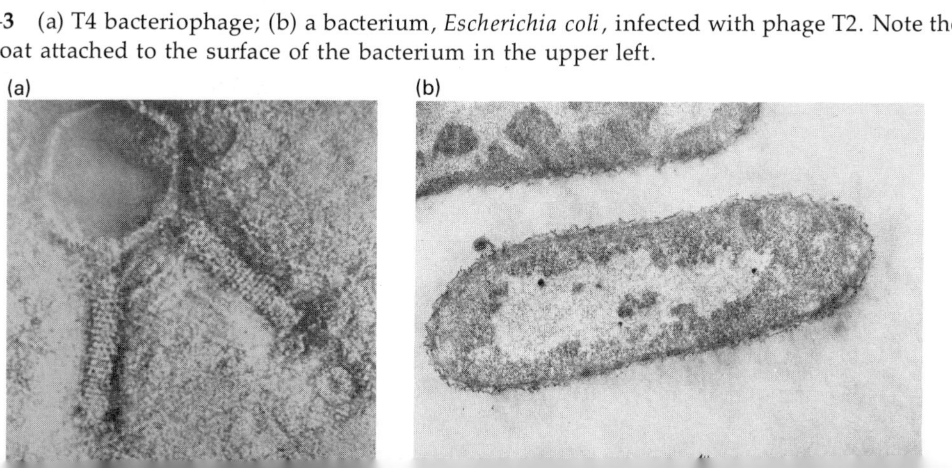

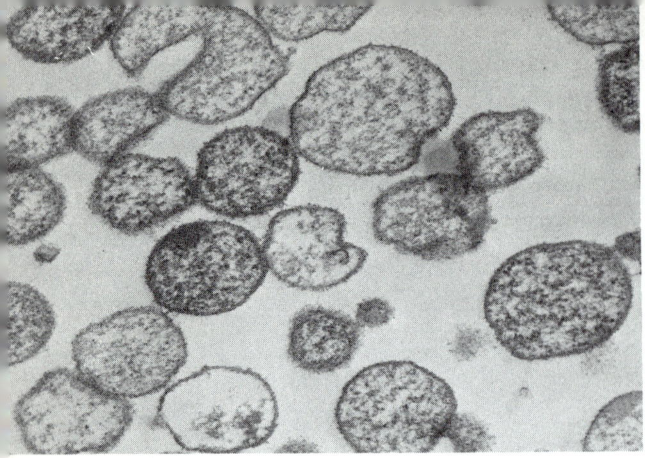

Figure 8–4 Mycoplasmas isolated from a goat (31,740×).

in diminutive size as far as an independent, definitely living entity is concerned. It has been calculated that each of the smaller organisms can contain no more than twelve hundred macromolecules. Considering the number of different activities carried on by these "midgets" (i.e., growth, reproduction, and other essential metabolic activities), this would permit little more than a single molecule to function in each category. This is indeed a rare creature. In living systems, redundancy is the general rule: usually thousands of identical molecules and often several alternate mechanisms for a particular process are available to the cell. But it appears that in the mycoplasmas Nature has designed an "economy model."

RICKETTSIAS

In 1910 Howard T. Ricketts, after whom the **rickettsias** are named, died from a disease caused by these organisms. In 1909 Ricketts described the rickettsial organism responsible for Rocky Mountain spotted fever. In the following year he identified the rickettsia responsible for typhus by examining the blood of Mexicans afflicted with this disease. It was there, in Mexico, that he contracted typhus and died shortly thereafter. In honor of his work with these organisms his name was affixed to the genus which includes them.

Rickettsias are intermediate between viruses and bacteria both in size and constitution. Some rickettsias are rod-shaped, and others are spherical (Fig. 8–5). Like viruses, essentially all of the rickettsias are incapable of metabolic activities outside a living organism. But rickettsias have a cell wall and contain both DNA and RNA.

Rickettsias are responsible for a number of diseases, which include typhus, Rocky Mountain spotted fever, trench fever, and Q fever. These diseases are transmitted to rats, rabbits, dogs, cattle, sheep, goats, and man via such **vectors** as lice, ticks, fleas, and mites.

Rickettsias can grow and reproduce only in living hosts. No sexual reproduction has been observed in these organisms, reproduction occurring by simple binary fission.

Figure 8–5 Rickettsias of (a) epidemic typhus, (b) endemic typhus, (c) Rocky Mountain spotted fever, and (d) American Q fever.

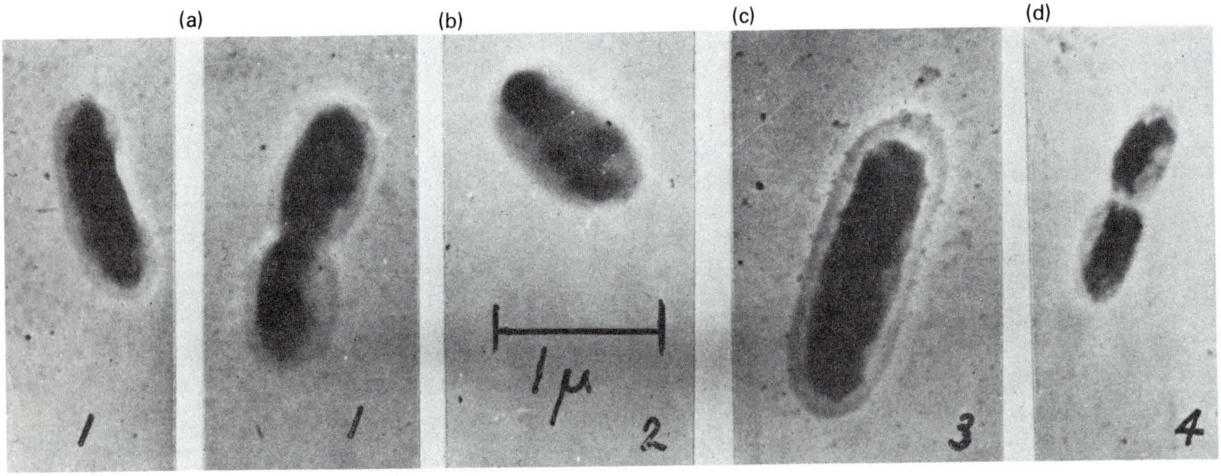

Bacteria: Phylum Schizomycophyta

Our bodies are covered with millions of invisible microorganisms. Most of these "germs" are **bacteria** (phylum **Schizomycophyta**). Even after a thorough washing with soap and water, bacteria have been found clinging to the skin.

Hordes of bacteria can also be found swarming in our foods and drinking water (unless it has been treated with chlorine for the express purpose of killing them), in the soil by the millions per gram, in the air that we breathe, and even within our bodies, warm and protected, living upon the food we ingest and the by-products of our metabolism. Some of these unseen hordes are dangerous pathogens, whose uncontrolled multiplication may bring disease and death. Others are beneficial, or even vital to our lives. Ammonifying and nitrifying bacteria convert the nitrogen of the atmosphere to forms usable by plants, and ultimately by animals and man. The decay bacteria of the soils and waters save the world from inundation by an otherwise ceaseless pileup of dead organisms, and they return the constituents of their bodies to the pool of usable substances. Vitamins are produced for us by the microscopic guests that inhabit our intestines. The bacterial flora of the human alimentary canal are so numerous that it has been estimated that from a fourth to a third of the dry matter of human feces is composed of the bodies of bacteria.

This fact is even more astonishing when we consider the size of an individual bacterium. The period at the end of this sentence is large enough to cover as many as two hundred fifty thousand bacteria spread out in a layer one bacterium thick.

Bacteria are often classified according to their shape into three distinct types (Fig. 8–6): rod-shaped **(bacillus),** spherical **(coccus),** and spiral or corkscrew **(spirillum).** The most common are the bacilli, the least common (only a hundred species or so) are the spirilla. Although bacteria are unicellular organisms, they are often found in groups of two, three,

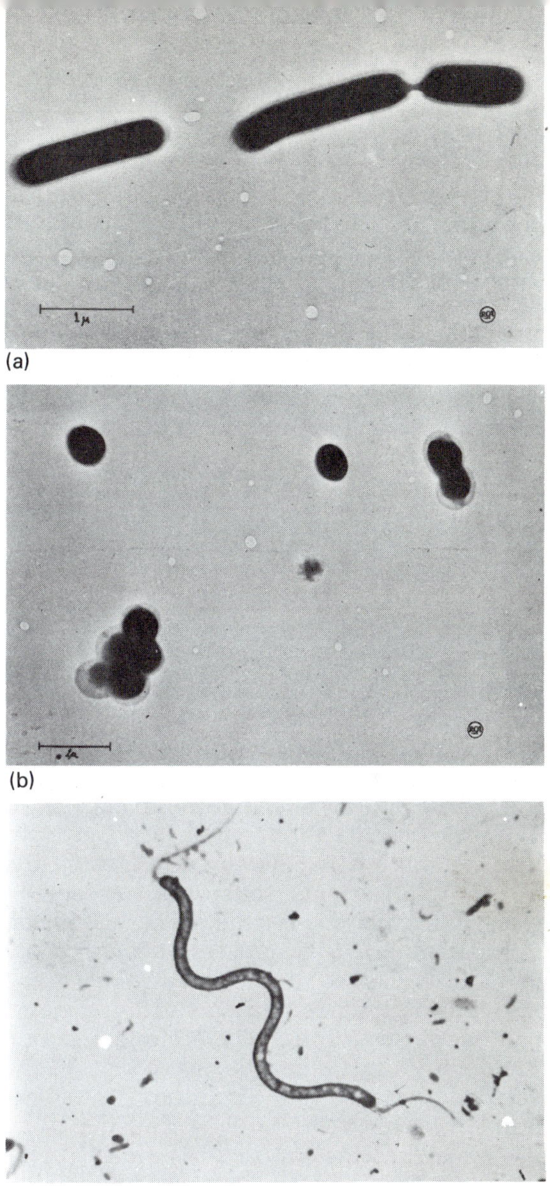

(a) (b) (c)

Carolina Biological Supply Company

Figure 8–6 Typical bacteria: (a) bacilli, (b) cocci, and (c) spirillum.

or more (ranging up to scores) in the form of chains or clumps.

Bacteria lack a recognizable nuclear body, but certain areas respond to nuclear stains showing localized areas containing DNA. Bacteria have been found to be devoid of mitochondria, Golgi bodies, and even an endoplasmic reticulum. But the bacterial cell may contain vacuoles, deposits of reserve foods, and pigments. Many bacteria possess whiplike cellular outgrowths (flagella), which permit them to move about freely in a liquid environment. The bacterial cell wall contains no cellulose, but is a complex structure consisting of proteins, polysaccharides, and sometimes lipids.

Reproduction in bacteria usually occurs by simple fission. The two equal daughter cells may separate immediately, or they may adhere together, forming chains or loose clusters. Some bacterial species can divide every twenty minutes. At such a rate of reproduction, under ideal conditions a single bacterium could give rise in a single day to more than 1,000,000,000,000,000,000,000 descendants weighing more than 2000 *tons*! Fortunately, such "ideal" conditions are not to be found on this earth. Long before any such astronomical figure could be approached, the growth rate would slow down drastically as a result of depletion of food supplies, crowding, and the accumulation of toxic waste products.

Bacteria have also been found to engage in a form of sexual reproduction. In mixed cultures containing two strains with differing growth requirements, offspring are sometimes found that possess the characteristics of both parent strains. A process of conjugation has been observed, in which two bacterial cells lie side by side and part of the genetic material passes through a tube from one cell (the **donor**) to the other (the **acceptor**).

Under unfavorable conditions, in some forms of bacteria (primarily bacilli) the cytoplasm may become enclosed in a small spherical or oval **spore**. Here the organism remains dormant until conditions are again suitable for germination. Bacterial spores are extremely resistant, even to such extreme influences as detergents and high temperatures. However, they can be destroyed by sterilization with steam under pressure.

Bacteria show great variety in the conditions they require for optimum growth. Although most grow best at intermediate temperatures, such as room temperature or human body temperature, some "cold-loving" bacteria can thrive in foods in a refrigerator, close to the freezing point of water, while other, "heat-loving" bacteria can live in the waters of hot springs. Some species and strains of bacteria have highly specific requirements for certain salts, vitamins, or other nutrients. These varied growth requirements of bacteria have been widely utilized in studies by microbiologists and microbial geneticists.

Since most bacteria do not possess chlorophyll, most of them are **parasitic** (infect a living host) or **saprophytic** (live on dead organic matter). Others are independent of organisms, living or dead, and satisfy their energy needs by metabolizing such inorganic sources as nitrogen, sulfur, nitrates, nitrites, and ammonia. Many of the latter organisms play an instrumental role in making the soil habitable for plants. The saprophytes are important in varied ways. They include the decay bacteria and numerous fermentative bacteria. Man has long utilized the activities of the latter bacteria in the manufacture of butter and cheese, wine, vinegar, sauerkraut, and in such processes as the tanning of leather and the retting of flax. In recent years the industrial applications of microbiology have been expanding dramatically. Now, in huge temperature-controlled tanks, specially bred strains of bacteria busily metabolize carefully selected media, producing valuable products for man.

Despite the many useful (or at least harmless) species of bacteria, the most publicized bacteria are those pathogenic forms that produce diseases in man, animals, and plants. These organisms, and methods of controlling them, will be discussed in Chapter 23.

Protozoa: Phylum Protozoa

About three centuries ago an obscure Dutch lensmaker announced the discovery of a new world. Unlike the famed explorers who braved the open seas and the treacherous jungles of distant lands, Anton van Leeuwenhoek sighted his new world in a drop of rainwater. From an open pot which had stood outside for a few days, he removed a small sample of liquid and examined it under one of his newly ground lenses. To his surprise he saw a circus full of life. The drop of water was swarming with tiny creatures swimming and crawling busily about, all too small to see with the naked eye. Slipperlike "animalcules" glided gracefully through the water, propelled by numerous tiny "feet." In this swarming world in a drop, these slipperlike creatures sometimes bumped into snakelike organisms, and the two swam their separate ways. Among the debris in the water drop, amorphous globs slithered slowly about, first sending forth little "horns," then oozing forward. Leeuwenhoek spent many, many hours peering at his newfound world.

For the most part, this world that Leeuwenhoek observed was the world of the protist. And most of the motile organisms he observed were protozoa.

The phylum **Protozoa** is a diverse group of microscopic unicellular organisms. The largest protozoan may reach a millimeter in length. The smallest may be only a few microns long, just barely visible through Leeuwenhoek's lenses. Most of the protozoa are motile, free-living water dwellers, inhabiting fresh or salt waters. They may move about with the aid of numerous rapidly beating cilia or one or more whiplike flagella. Still others may ooze along, sending forth pseudopods, projections of their jellylike bodies. But some protozoa are sessile, spending much or perhaps essentially all of their lives anchored to some object. Other protozoa may live in colonies, loose associations of a number of single-celled organisms. Within these colonies, such as those of *Volvox,* there may even be the beginnings of specialization of function among the individual members.

Some protozoa possess chloroplasts and can synthesize their own food in the presence of sunlight. But most (and many of the photosynthetic varieties, as well) take in food from the medium. This food may be in the form of either organic and inorganic nutrients or other microscopic organisms. Some protozoa live as parasites within the blood or other tissues of multicellular organisms. Such are the dysentery amoeba and the flagellate trypanosome that causes sleeping sickness.

Protozoa do not possess a cell wall, such as that which characterizes the cells of plants and bacteria. The single-celled bodies of some protozoa, such as the amoeba, are surrounded merely by a thin plasma membrane, which encloses but gives no definite shape to the jellylike protoplasmic contents. Others, such as the paramecium, have a definite shape, imparted by a stiff but flexible outer covering called a **pellicle.** Some protozoa, such as radiolarians and foraminiferans, are encased in tiny shells or skeletons constructed from particles of inorganic substances such as silica or calcium carbonate. Those huge deposits of chalk, which make up the famous white cliffs of Dover, in England, were laid down in an eon-long process of gradual sedimentation of incalculable numbers of such tiny skeletons.

The minute size of the protozoa might lead one to think that they are extremely simple organisms. Yet this could hardly be further from the truth. The protozoan may possess only a single cell; yet within this single cell it packs the apparatus to perform nearly all the major functions performed by multicellular organisms. It senses and responds to its environment; it obtains (or manufactures) food, which it digests and metabolizes for energy and building materials; it eliminates its wastes; it grows and reproduces. With but a single cell, the protozoan has no specialized tissues and organs; but it does have specialized structures within the cell, called **organelles.** In addition to a **nucleus** (or sometimes more than one), which directs the activities of the cell and carries the genetic information of the organism, a protozoan often contains various **vacuoles,** globular structures containing food or water. The **contractile vacuole** found in many protozoa serves as a sort of hydrostatic organelle, pumping out excess water and some waste materials. Locomotor organelles found in this diverse group include numerous hairlike **cilia,** the beating of which is synchronized, thus effectively propelling the organism through its watery medium; whiplike **flagella,** which lash back and forth with the same effect; and **pseudopods** ("false feet"), protoplasmic regions which are temporarily extended from the body of the orga-

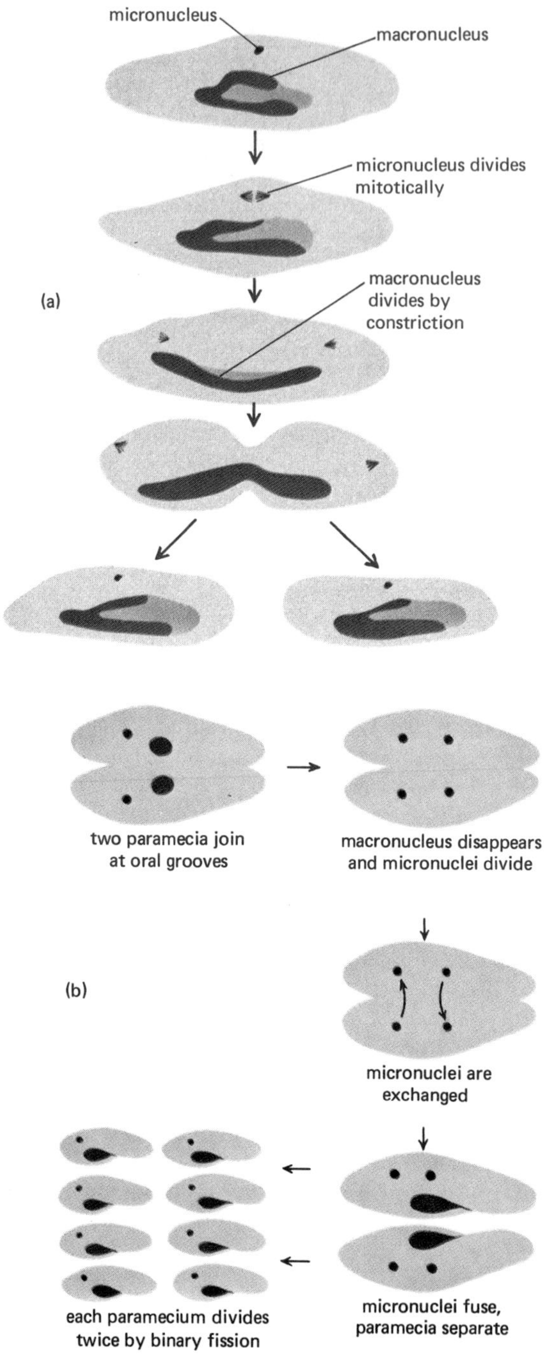

Figure 8–7 *Paramecium* can reproduce either by binary fission (a) or by conjugation (b). In binary fission, one parent yields two daughter cells that are exact copies of the parent. In conjugation, two parents yield eight daughter cells, each of which combines hereditary traits of both original parents.

nism to produce the creeping motion characteristic of the amoeba. Sessile protozoa may secrete a stalk with which they attach themselves to the solid bottom or to some other stationary object, such as a rock or the stalk of a water plant. Some, like *Stentor*, may detach themselves and swim away, to settle down elsewhere, if conditions prove unfavorable.

Protozoa generally reproduce by **binary fission:** when an organism reaches a certain size, it divides in two, reproducing its genetic material and producing two smaller but otherwise identical copies of itself (Fig. 8–7a). In some species there may also be a primitive form of sexual reproduction, called **conjugation** (Fig. 8–7b). For example, two paramecia of different mating types (they are morphologically indistinguishable to the observing scientist, but the paramecia can tell the difference) temporarily join together. During this intimate union, the two partners exchange some of their nuclear material. After they separate, each proceeds to divide twice by binary fission. Each paramecium is thus replaced by four tiny replicas that incorporate traits both of the original parent and of its mating partner.

When environmental conditions become unfavorable, as when a pond or roadside puddle dries up or freezes over, its protozoan population may enter a dormant condition. Each tiny organism forms a protective covering about itself, called a **cyst.** Protozoa can survive thus encysted for months or even years and then, when conditions are again favorable, resume their activities again.

The thousands of species of protozoa are classified in five major groups (Fig. 8–8). Class **Sarcodina** includes the amoebas and other organisms which move about with the aid of pseudopods. An amoeba looks like a shapeless blob of jelly, but it can creep about in self-directed motion, capture and eat other tiny organisms, and reproduce itself (by binary fission). Some organisms of this class are encased in protective shells, with openings through which tentaclelike pseudopods protrude for feeding.

Class **Flagellata** consists of protozoa that move about through the lashing action of whiplike flagella. A common example of this class is *Euglena,* a strange blend of "plantlike" and "animallike" characteristics. It is free-swimming and may take in nutrients from its surrounding medium, but it also normally possesses chloroplasts and manufactures its own food in the presence of sunlight. Other common flagellates are the sleeping sickness trypanosome and the dinoflagellates—free-swimming, photosynthesizing organisms that are among the most numerous life forms of the oceans. (The dinoflagellates are usually classified as algae, though their status is in dispute among taxonomists.)

The third important class of phylum Protozoa, class **Ciliata,** includes the one-celled organisms that move about through the synchronized beating of numerous hairlike cilia. These cilia may be studded over the entire outer surface of the body, as in the common ciliate *Paramecium,* the "slipper animalcule," or they may be arranged primarily in a row around a rim, as in the trumpet-shaped *Stentor* and cup-shaped *Vorticella.* The cilia may be used not only for locomotion, but also to set up feeding currents, which continuously swirl the water of the medium, along with food particles, into the feeding orifice of the organism.

Another class of protozoa, **Suctoria,** includes species that are completely sessile as adults. They spend their adult lives attached to some solid object by stalks and feed with long, protruding, sucking tentacles. The larval stages, formed by a process of internal budding, have cilia, but these disappear when they settle down to a sedentary life.

The members of the last major class of protozoa, class **Sporozoa,** have no special means of locomotion. The members of this class are all parasites and reproduce by the formation of numerous spores. *Plasmodium,* the sporozoan responsible for malaria in man, is an important representative of this class. This parasite spends part of its life cycle within the body of the anopheles mosquito and is transmitted to humans through the bites of this insect pest.

Amoeba, Euglena, and *Paramecium,* representatives of the three most important classes of the phylum Protozoa, are often studied as "type organisms."

Carolina Biological Supply Company

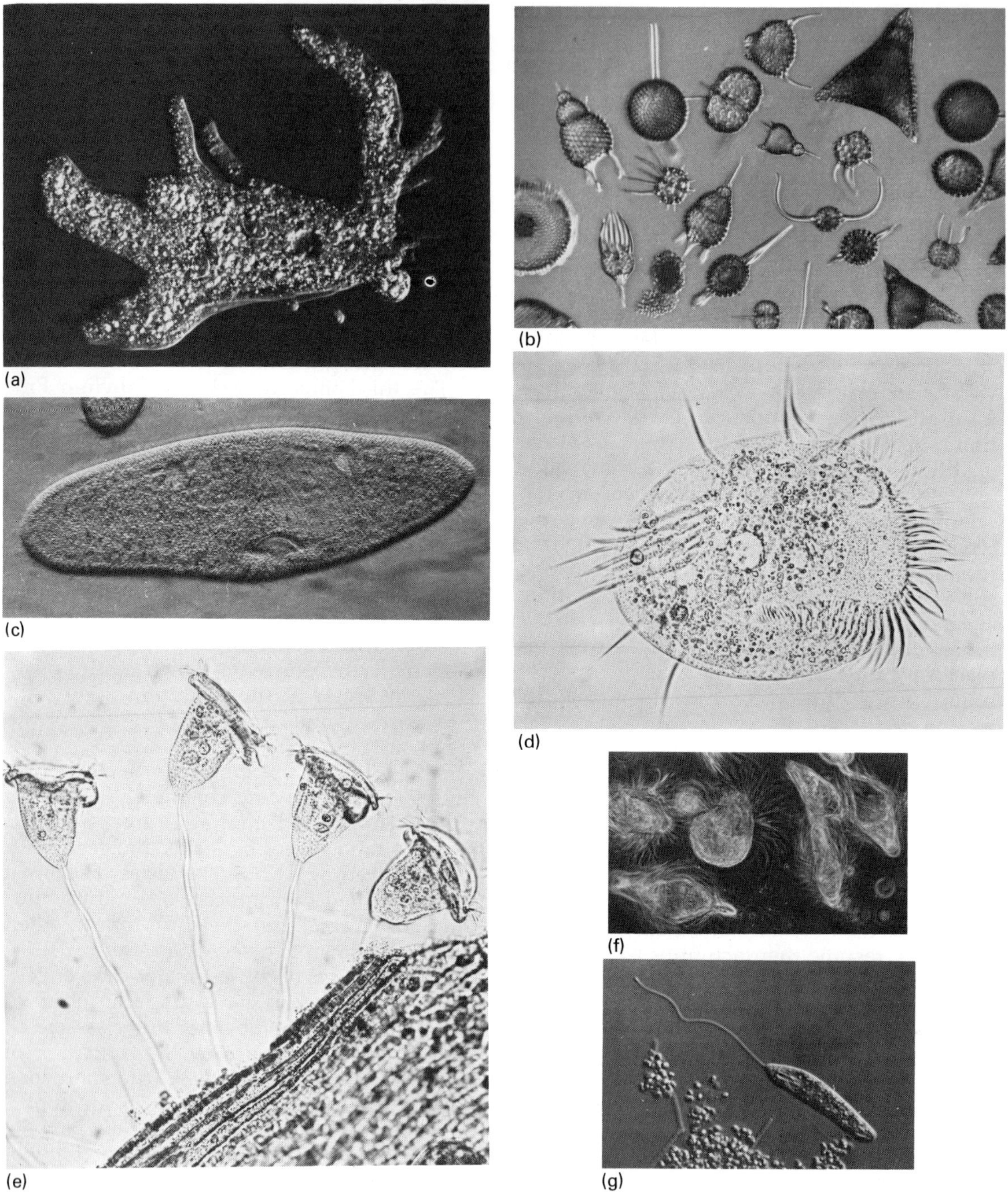

Philip L. Feinberg, Omikron

Philip L. Feinberg, Omikron

Figure 8–8 Some representative protozoans. Sarcodina: *Amoeba proteus* (a) and radiolaria (b); Ciliata: *Paramecium multimicronucleatum* (c), *Euplotes* (d), and *Vorticella* (e); Flagellata: flagellates from the intestine of a wood-feeding roach, *Cryptocercus punctulatus* (f) and *Peranema trichophorum* (g).

Algae: Phyla Cyanophyta, Chlorophyta, Euglenophyta, Chrysophyta, Pyrrophyta, Phaeophyta, and Rhodophyta

The surface waters of seas occasionally turn a deep red over an area that may extend for thousands of square miles, as though stained by the blood of a cataclysmic battle. These outbreaks or "red tides" are usually caused by a sudden proliferation of countless numbers of red-pigmented algae called dinoflagellates.

Algae range in size from microscopic cells to giant kelp, which may extend for more than six hundred feet. There are about eighteen thousand known species of algae distributed throughout essentially all of the watery areas on the surface of the earth: lakes, ponds, rivers, seas, from the cold Arctic to the hot equator. Algae have been found even in hot springs where temperatures exceed 85 °C. Some forms of algae are found in the soil, on trees, on stones, and even on fence posts and growing on the long fur of sloths.

Although almost all algae contain chlorophyll, the range of colors among the various species is quite extensive: red, brown, yellow, blue-green, etc. This is so because many species contain a second pigment present in such excess that it masks the green color of the chlorophyll. For a long time, the algae were classified in a single phylum, divided into four classes (blue-green, green, brown, and red algae) according to the most conspicuous pigments. More modern classification systems recognize a number of different phyla in the group, differentiating them according to various structural and reproductive characteristics and the nature of the stored food, along with a consideration of the prominent pigments (Fig. 8–9). These phyla include the **blue-green algae (Cyanophyta),** which are the most primitive algae; the **green algae (Chlorophyta),** including some marine and many fresh-water forms, such as the common pond scum; the **animal-like algae (Euglenophyta),** flagellated forms such as *Euglena,* usually with no true cell wall;* the **yellow-green algae, golden-brown**

*We have already discussed these organisms under the phylum Protozoa.

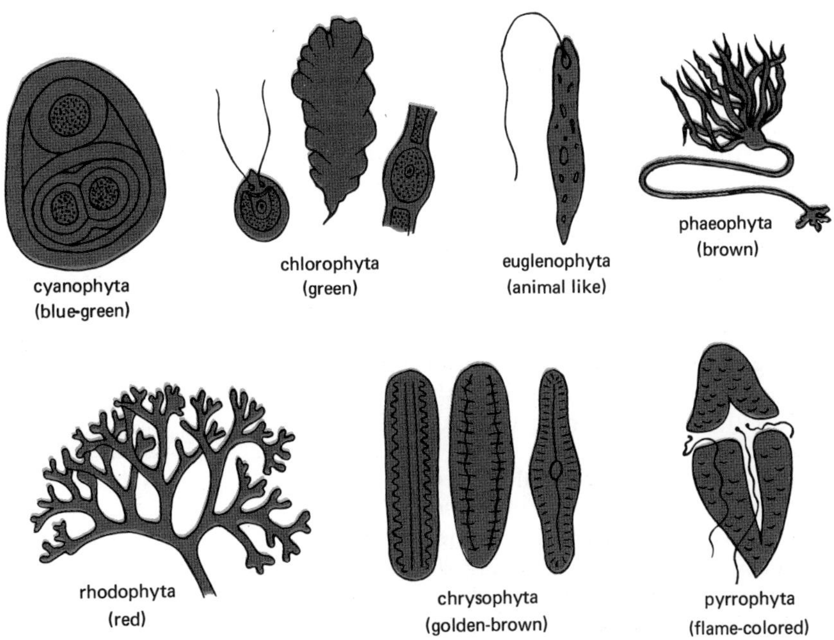

Figure 8–9 Typical representatives of the major groups of algae (not drawn to the same scale).

algae, and **diatoms (Chrysophyta),** a varied group, including both motile and nonmotile, unicellular and colonial forms; the **flame-colored algae (Pyrrophyta),** chiefly the dinoflagellates; the **brown algae (Phaeophyta),** mainly marine organisms, including the rockweed *Fucus,* the seaweed *Sargassum,* and the giant kelps; and the **red algae (Rhodophyta),** also mostly marine organisms, many of which have delicate, feathery branches.

Algal cells may occur singly, or in thread-like strands **(filaments),** or in clusters, depending on the species (Fig. 8–9). Many of the filamentous types have **holdfasts,** which anchor the filament to some immobile object, such as stones or twigs. A number of forms possess flagella. In some of the colonial forms, in which various numbers of cells are loosely aggregated, there may be the beginnings of specialization among the members of the colony, with a differentiation of the vegetative and reproductive functions. Some of the brown algae are quite advanced, possessing structures resembling stems, roots, and leaves.

The filamentous forms of algae may reproduce by fragmentation: the filament breaks up into many smaller sections, each one giving rise to a new filament. The simpler species may reproduce by fission; but in general, asexual reproduction takes place by the formation of spores, which are formed within the cell and liberated into the surrounding medium, thus giving rise to new individuals (Fig. 8–10). Some of these spores are flagellated and capable of independent movement **(zoospores).** Sexual reproduction occurs in many species of algae. This is accomplished by the production of sex cells **(gametes),** with their subsequent union, forming a **zygote.**

Algae are of fundamental importance to the continuation of life on earth in that they provide a vital link in numerous food chains; that

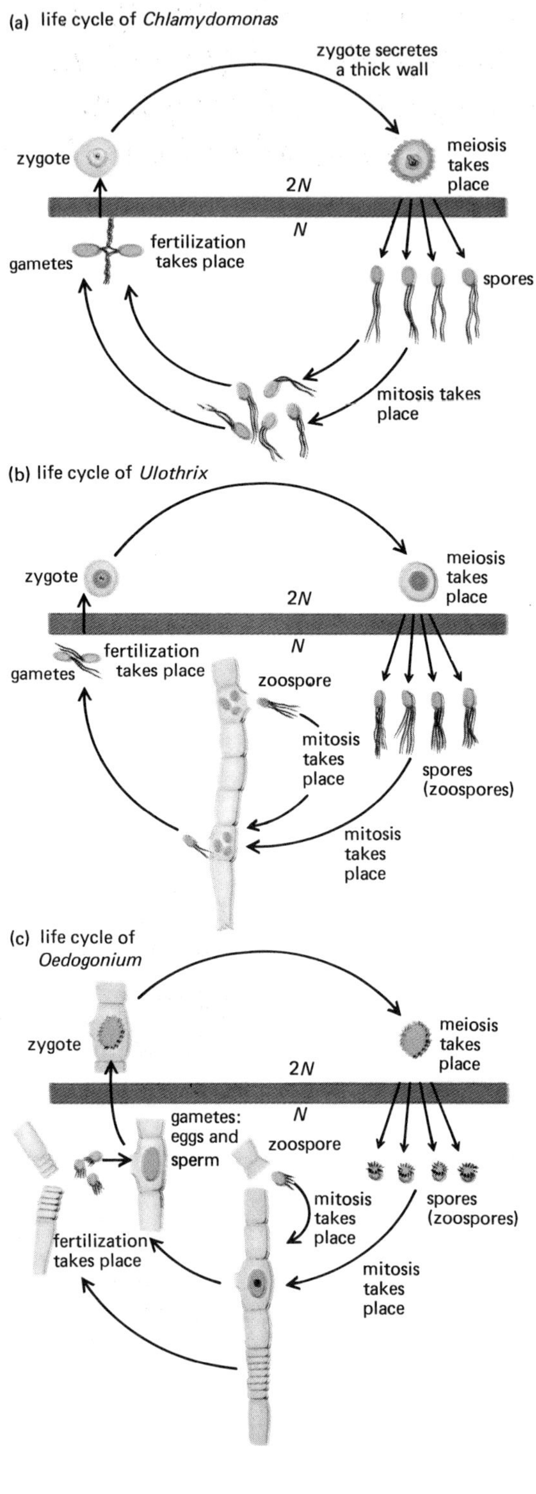

Figure 8–10 Life cycles of three species of green algae: (a) *Chlamydomonas;* (b) *Ulothrix;* (c) *Oedogonium*. Notice that each alga can reproduce both asexually (zoospores are discharged and develop into a new individual) and sexually (fusion of gametes, followed by production of spores).

is, they are the sources of food for many animals throughout the world. The upper few hundred feet of the waters of the seas are a veritable soup of **plankton,** mostly microscopic single-celled algae that drift passively along with the current. A large fraction of these are diatoms, encased in fragile frameworks of silica crystals of varied and delicate beauty. These plankton organisms also include other microscopic forms, such as dinoflagellates, some manufacturing their own food, others feeding upon other plankton. Tiny animals, such as minute crustaceans called **copepods,** feed upon the microscopic forms of life and in turn provide food for other, larger organisms—and so on up the chain of life. Algae also serve as important sources of iodine, and they are used industrially to produce glue, jellies, and culture media (in the form of **agar,** a jelling agent extracted from red algae). A widely used food additive, **carrageenan,** is made from a type of red algae called Irish moss. This ubiquitous substance reacts with proteins, particularly the casein in milk. Carrageenan is used to prevent the cocoa in chocolate milk from settling, to keep puddings from "weeping," and to add bulk and "mouthfeel" to soups, sauces, syrups, and cottage cheese.

True Fungi: Phylum Eumycophyta

Such seemingly divergent forms as bread molds, yeasts, mushrooms, and *Penicillium* belong to the same group, the **true fungi** (Fig. 8–11). Despite their widely differing appearances, they have certain characteristics in

Figure 8–11 Some representative fungi. Phycomycetes: *Rhizopus nigricans*, bread mold (a); Ascomycetes: *Penicillium chrysogenum*, a mutant form of which now produces nearly all of the world's commercial penicillin (b); Basidiomycetes: *Amanita muscaria*, the fly amanita mushroom (note the gills on the underside of the cap) (c) and *Lycoperdon sp.*, puffballs, growing on a rotting stump (d).

common. All true fungi are chlorophyll-free, plantlike protists that lack vascular tissues and true roots, stems, and leaves.

Most of the more than forty thousand species of true fungi known are multicellular saprophytic or parasitic organisms. The vast majority of the members of this group possess long, slender, filamentous, branching, multinucleated structures called **hyphae.** The hyphae may be grouped into masses collectively termed **mycelium.** Yeasts, an exception to the general rule, are single-celled, nonfilamentous organisms.

The true fungi (phylum **Eumycophyta**) are grouped into four classes. Class **Phycomycetes,** the **algal fungi,** is comprised of organisms that show considerable resemblance to some of the green algae. This class includes both aquatic forms (water molds) and such terrestrial forms as the bread mold fungus *(Rhizopus nigricans)* and the late blight of potatoes *(Phytophthora infestans),* which caused the great potato famine in Ireland in the mid-nineteenth century.

The **sac fungi,** class **Ascomycetes,** represent an extremely large group of fungi, including the cup fungi, the powdery mildews, and various other organisms that cause scab, rot, leaf spot, and numerous other common and destructive diseases of plants. The yeasts are other important representatives of the sac fungi.

Class **Basidiomycetes,** the **gill fungi,** includes mainly such saprophytic, fleshy fungi as mushrooms, toadstools, puffballs, and bracket fungi, along with the rusts and smuts that cause plant diseases resulting in substantial economic losses every year.

The last group of true fungi, class **Deuteromycetes** or **Fungi Imperfecti,** is a catchall category into which are grouped the fungi that do not fit into any of the other three classes. In contrast to the first three classes, only asexual forms of reproduction have been observed. Indeed, all fungi are classified mainly on the basis of their reproduction and life cycles. These often intricate cycles usually include both asexual forms of reproduction, such as budding, spore formation, and fragmentation, and various types of sexual reproduction. Whenever a sexual form of reproduction is discovered for a fungus species temporarily classified in the Fungi Imperfecti, the species is then moved into the appropriate one of the other three classes.

In addition to causing numerous diseases among plants and animals, true fungi are responsible for such human diseases as athlete's foot and ringworm. But the picture is not all negative. Some edible fungi are considered great delicacies. Various saprophytic fungi, together with the decay bacteria, are vital links in the chain by which the bodies of dead organisms are reduced to their constituents, which can then be reused by other, living organisms. And important large-scale industrial processes, including the production of life-saving antibiotics, are based on the culturing of specially selected and bred strains of fungus organisms.

Slime Molds: Phylum Myxomycophyta

The growing acceptance of the three-kingdom classification system of protists, plants, and animals has made the minds of many conscientious taxonomists rest much easier with regard to a number of "borderline groups." One such group was the euglenoids, which we have already discussed. The correct placement of another group, the **slime molds,** has traditionally made biologists even more mentally uncomfortable. For these strange creatures, classed by themselves in phylum **Myxomycophyta,** possess a number of characteristics of both plants and animals. They creep slowly across damp surfaces and ingest food in amoeboid fashion. Surely these are animal characteristics. But eventually their movement ceases, and, plantlike, they produce **spores,** which, after further development, combine in pairs giving rise to **zygotes.** These, in turn, develop into multinucleated, slimy, unprotected masses known as **plasmodia** (Fig. 8–12). These plasmodia (i.e., slime molds) are widely distributed on the earth. One can find them in almost any moist place containing rotting

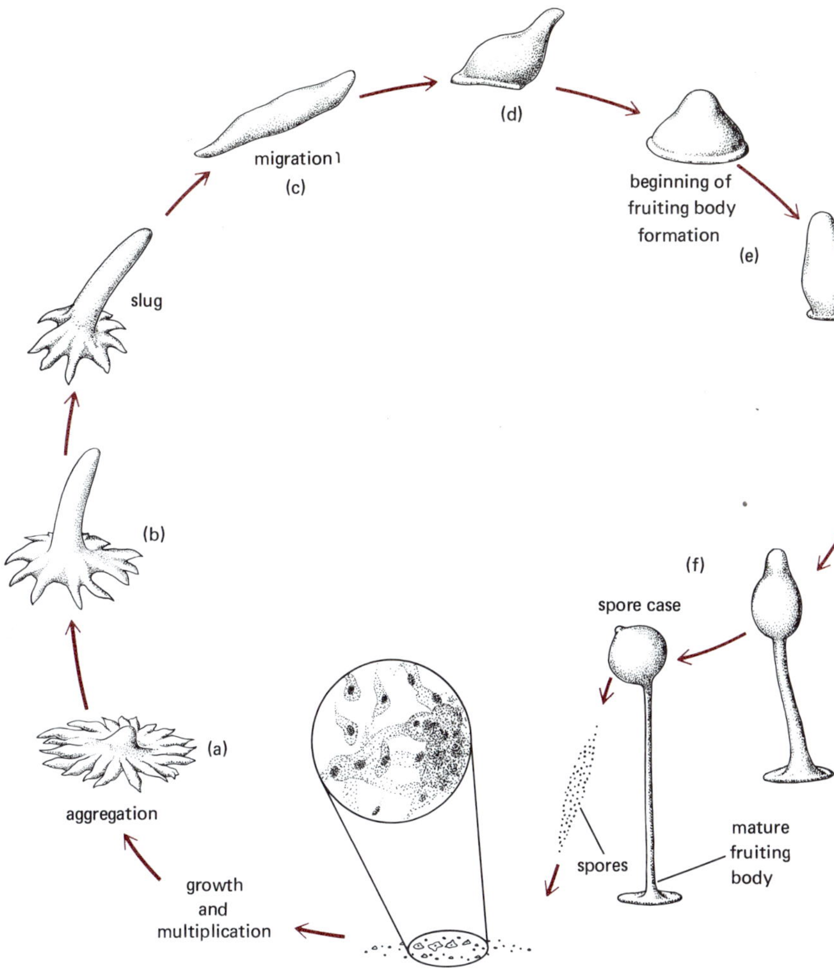

Figure 8–12 Life cycle of a slime mold. Spores germinate and form amoeboid cells, which aggregate to form a star-shaped mass (a). This mass develops into a moving sluglike form (b,c), which is transformed (d,e) into a fruiting body (f) consisting of a spore case, stalk, and base. Spores are discharged from the spore case and can begin the cycle again.

stumps and other decaying organic matter. Some plant diseases are attributed to these organisms.

The Plant Kingdom

Plant life in its varied forms is to be found over virtually all the lands of the earth: from the flowers and crops growing in gardens and fields planted by man, to the stands of mighty sequoia trees of California, to the water hyacinth choking waterways in the southeastern United States, to the grassy pampas of Argentina, to the moss-covered rocks of the Arctic region. Even in the parched and seemingly uninhabitable deserts, strange plant forms such as the cactus and the agave tenaciously cling to life.

Scientists believe that the first plants were relatively small, simple forms, which developed from plantlike protists. Gradually systems of conducting tissues evolved, which permitted a great expansion both of the size of the individual plants and the types of habitats they could utilize.

Mosses and Liverworts: Phylum Bryophyta

A continuous tapestry of green velvet carpets vast stretches of northern forests and tundras. The major components of this covering are **mosses** and **liverworts** (phylum **Bryophyta**). Despite the extensiveness of their growth, the individual plants are rather small. Many of the twenty-three thousand species that make up these two groups are quite tiny, requiring a magnifying lens to discern details. Few mosses and liverworts exceed a foot in size.

These plants are the most primitive land plants, lacking true roots, stems, and leaves and a vascular system (Fig. 8–13). It is the lack of a vascular system, which would provide a means of conducting water and other substances from one part of the plant to another, that limits the size to which mosses and liverworts can grow. (In a land plant, mineral nutrients and metabolic products cannot simply diffuse in and out, in an interchange with the medium, as in the case of water plants. The plant must possess some sort of efficient system of transport if it is to grow to any significant size.) The rootlike structures **(rhizoids)** of mosses and liverworts perform the functions normally attributed to roots, such as anchorage and mineral absorption, but they have a relatively simple design.

Mosses and liverworts are distinguished from algae (which they resemble in many respects) in that their gametes are enclosed in a single layer of cells, while those of algae are merely enclosed by a cell wall. They possess specialized reproductive organs and achieve a form of sexual reproduction. Sperms are produced in an **antheridium,** while egg production takes place in an **archegonium.** Furthermore, fertilization occurs *inside* the archegonium, with the sperm swimming to meet the egg.

All mosses and liverworts exhibit an **alternation of generations** (Fig. 8–14). The conspicuous plant body belongs to the **gametophyte** generation, the form that produces the gametes. The other generation, the **sporophyte,** is generally a tiny structure, at least partly dependent on the gametophyte.

Some liverworts, such as the commonly studied *Marchantia,* have a small, simple green body called a **thallus** (Fig. 8-13). Delicate rhizoids grow down from the underside of the thallus and absorb mineral nutrients from the moist soil. Even more numerous than the thallose liverworts are the leafy liverworts,

Figure 8–13 Typical liverwort (*Marchantia*) and moss (*Polytrichium*).

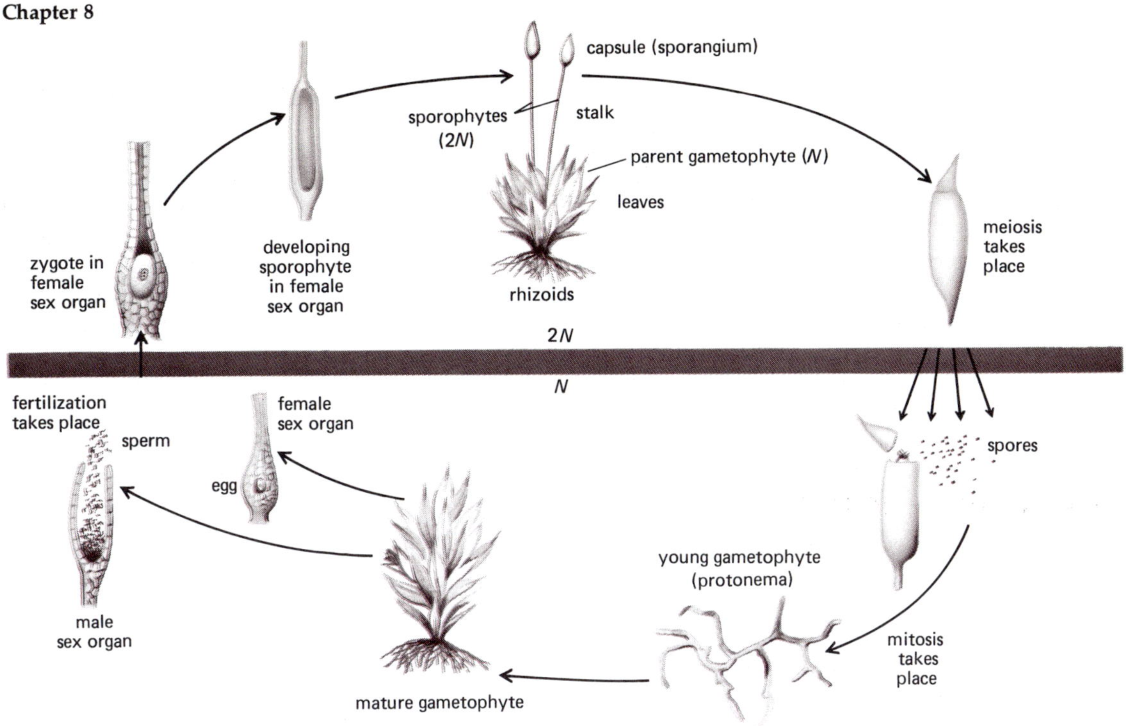

Figure 8–14 Life cycle of a moss. There is a distinct alternation of generations: the sporophyte and gametophyte.

tiny green plants that may easily be mistaken for mosses. Their leaves do not show the internal differentiation of tissues characteristic of vascular plants. The two groups can be distinguished by the fact that the "leaves" of mosses are arranged in radial symmetry, attached all around the stem; in the leafy liverworts, on the contrary, they are arranged in three rows along the plant body.

The group of mosses includes the true mosses (most of the common species), the sphagnum or peat mosses (all in the single genus *Sphagnum*), and the rock mosses which are small dark plants that grow in Arctic and alpine regions. The leaves of most mosses are extremely tiny, and except along the midrib are only one cell thick. Like the liverworts, the mosses have no vascular transport system, and materials pass from one cell to another by diffusion. Mosses and liverworts thus are normally limited to moist and shaded habitats where absorption is facilitated. (Some mosses are able to survive periods of drought in a dormant state.)

Plants like the mosses and liverworts are thought to have been the first invaders of the land. But though they had some success, the bryophytes represent an evolutionary dead end. Further development came in the group of vascular plants, the tracheophytes.

Vascular Plants: Phylum Tracheophyta

When the land areas were invaded by green plants descended from various types of algae, two main groups developed. One line gave rise to the nonvascular bryophytes, the mosses and liverworts. This was to prove a blind alley. For these plants, though able to live on land, were chained by their inherently inefficient transport systems to regions providing ample moisture. Moreover, the same inherent disadvantage imposed severe limitations on their possible size: mosses and liverworts are characteristically tiny plants, frequently only an inch or so in size.

The other line, the vascular **tracheophytes,** showed a much greater potential for versatility. With an efficient transport system, the modest ancestral forms evolved into a variety of shapes and sizes, ranging all the way up to towering trees. Many of the ancient tracheophytes are now either extinct or on their way to extinction. But the ferns, the conifers, and the profusion of flowering plants, from violets to redwood trees, are all tracheophytes, the ruling plant group of our world.

FERNS: CLASS FELICINAE

Ferns once covered the lands of the earth. The lush jungles of vegetation that flourished in the Carboniferous period, some 350 million years ago, were mainly ferns and fernlike plants. Their growth was so extensive that their remains, piling up over the ages, formed the major source of our huge deposits of coal and petroleum.

Today the **ferns** (class **Felicinae**) have to a large extent given way to their successors, the seed plants. But they have not vanished by any means. They are to be found from the tropical rain forest to rocky areas beyond the Arctic Circle. They range from minute water ferns to the great tree ferns of the tropics, which may reach a height of sixty feet.

The body of a fern is divided into roots, stems, and leaves. The stems are often **rhizomes,** underground stems that may creep along through the soil, branching and giving rise to new shoots. The roots normally grow down from the stems. The leaves of a fern, commonly called **fronds,** consist of a slender stalk and the leaf **blade.** The blade may be simple, but more often is compound, consisting of many smaller leaflets. These may in turn be divided, producing the feathery appearance of the common fern fronds, familiar as "greens" in a florist's bouquet.

The internal structure of the fern leaf, characterized by specialized conducting tissues, is quite similar to that of a flowering plant. It is in the leaves that most of the photosynthetic activity of the plant takes place.

The life cycle of the fern, like that of bryophytes, is characterized by a pronounced alternation of generations (Fig. 8–16). But in contrast to the mosses and liverworts, in the case of the fern the conspicuous generation is the sporophyte. The large, leafy sporophyte plants produce asexual spores, each of which develops into a small green nonvascular, heart-shaped **prothallus,** the sexually reproducing gametophyte.

Figure 8–15 Some common ferns: maidenhair fern (a), cinnamon fern (b), and the royal fern, showing a fertile frond with sporangia (c).

(a) (b) *Lynwood M. Chace, Omikron* (c)

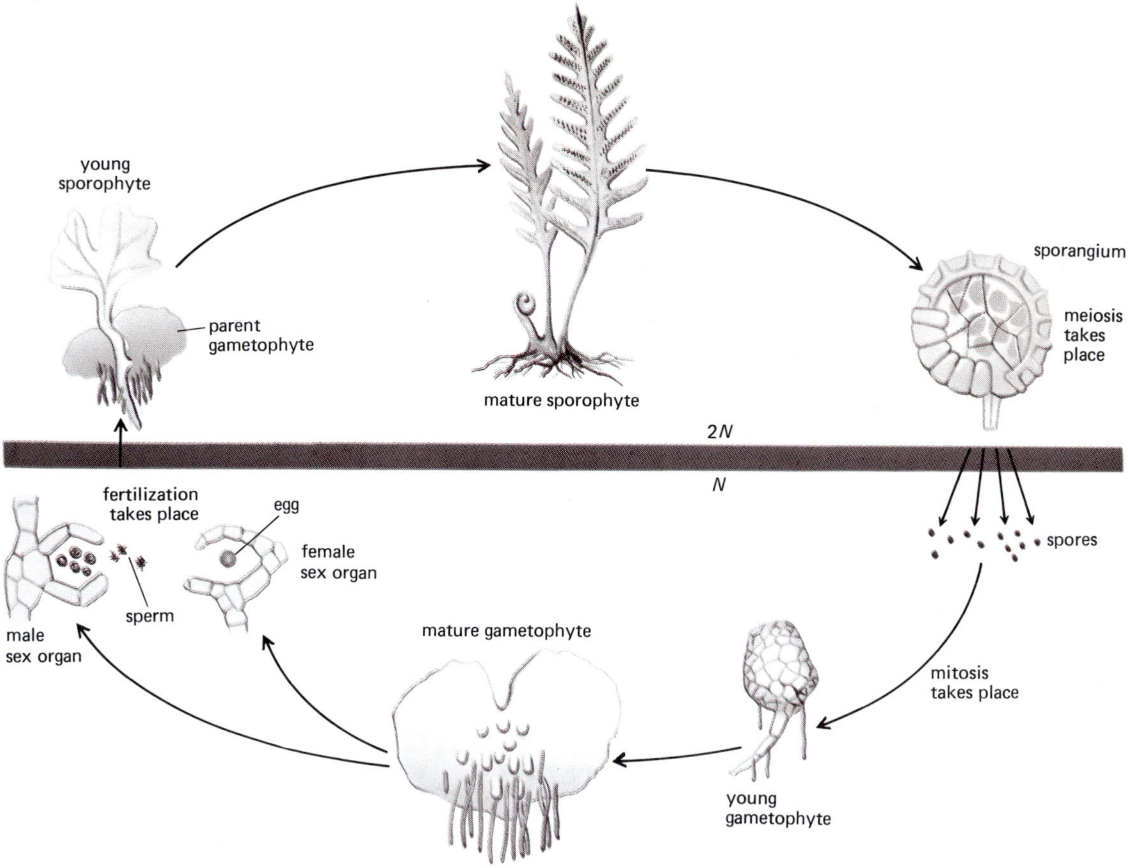

Figure 8–16 Life cycle of a fern. Alternation of generations is observed, but the more conspicuous form is the sporophyte.

SEED PLANTS: CLASSES GYMNOSPERMAE AND ANGIOSPERMAE

Seed plants are the most abundant as well as the most beautifully colored members of the plant kingdom. Their distribution is literally worldwide, ranging from the ice-covered expanses of the polar regions to the humid tropical jungles along the equator. Representatives of this group are among the largest living things on earth. The sequoia trees of California frequently exceed 250 feet in height, with trunk diameters in excess of 20 feet. Other members of the group are less than 1/16 inch across. Still others are creepers, aquatic or desert plants, woody shrubs, herbs, etc.

Seed plants assume an endless variety of shapes and forms (Fig. 8–17). They are usually subdivided into two classes: **gymnosperms,** which include the conifers (pines, firs, cedars, spruces), and **angiosperms,** which comprise more than two-hundred-fifty thousand flowering plants. The angiosperms may be further subdivided into two subclasses, **monocotyledons** (the grasses, including corn, wheat, and rice, as well as some flowering plants, such as the lily) and the **dicotyledons** (most flowering plants—poppy, buttercup, tomato, etc., and trees such as oaks, maples, and willows).

In the seed plants, the trend in the alternation of generations seen in the ferns is taken a

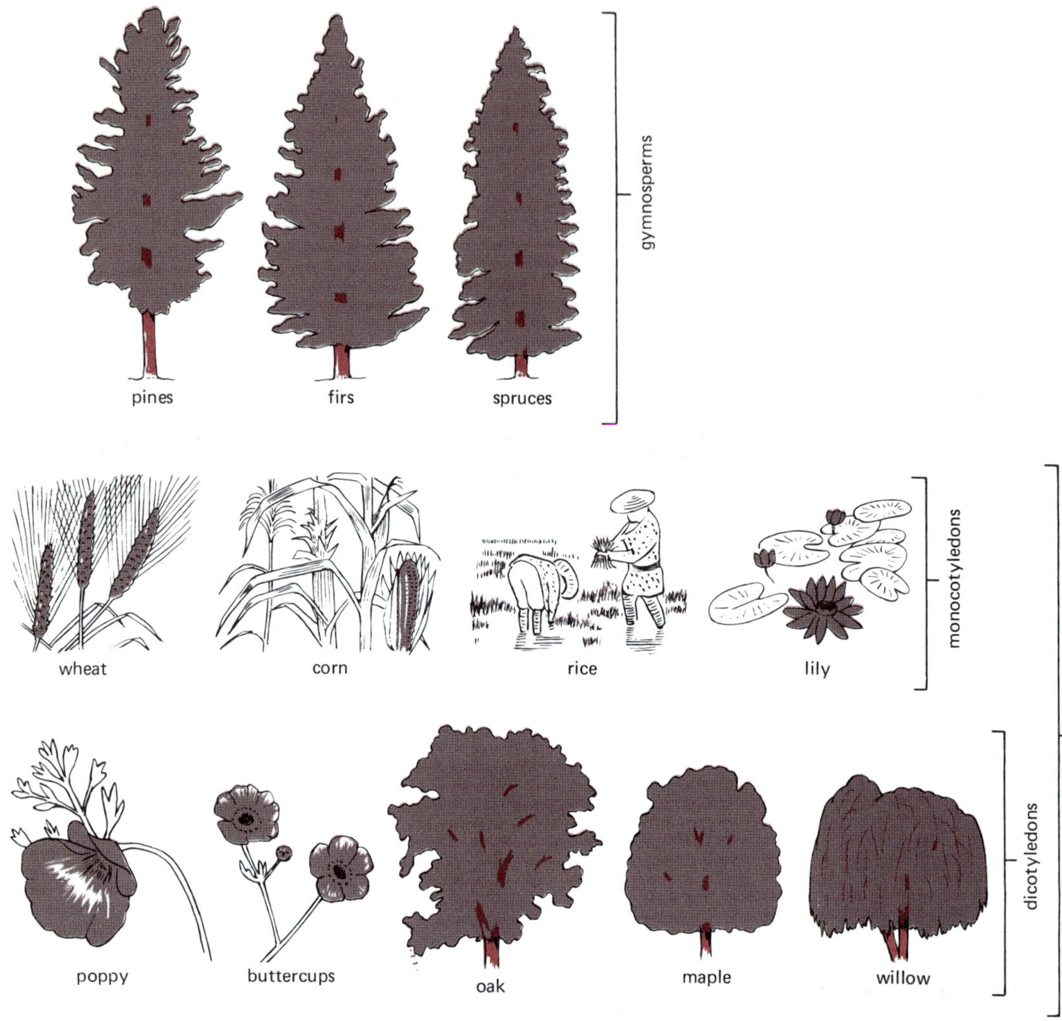

Figure 8–17 Seed plants.

step further. Not only is the sporophyte the most conspicuous form, but the gametophyte is reduced to such an extent that it is no longer a free-living, photosynthesizing form but merely a specialized portion of the sporophyte.

In both gymnosperms and angiosperms, reproduction is sexual. The male germ cells, in the form of **pollen,** are transferred to the female reproductive organs, and within the ovary the sperms fertilize the eggs (Fig. 8–19). Each resultant zygote develops into an **embryo** which, together with stored food material, is enclosed within a protective coat. The embryo, together with its food reserves and outer covering, is called a **seed.** The embryo is protected from drying out by its **seed coat,** and it can remain dormant for long periods if conditions are not favorable for **germination,** the process of explosive growth and development that ultimately gives rise to a new sporophyte.

The term *gymnosperm* means literally "naked seed," and comes from the fact that the seeds of gymnosperms are not enclosed in

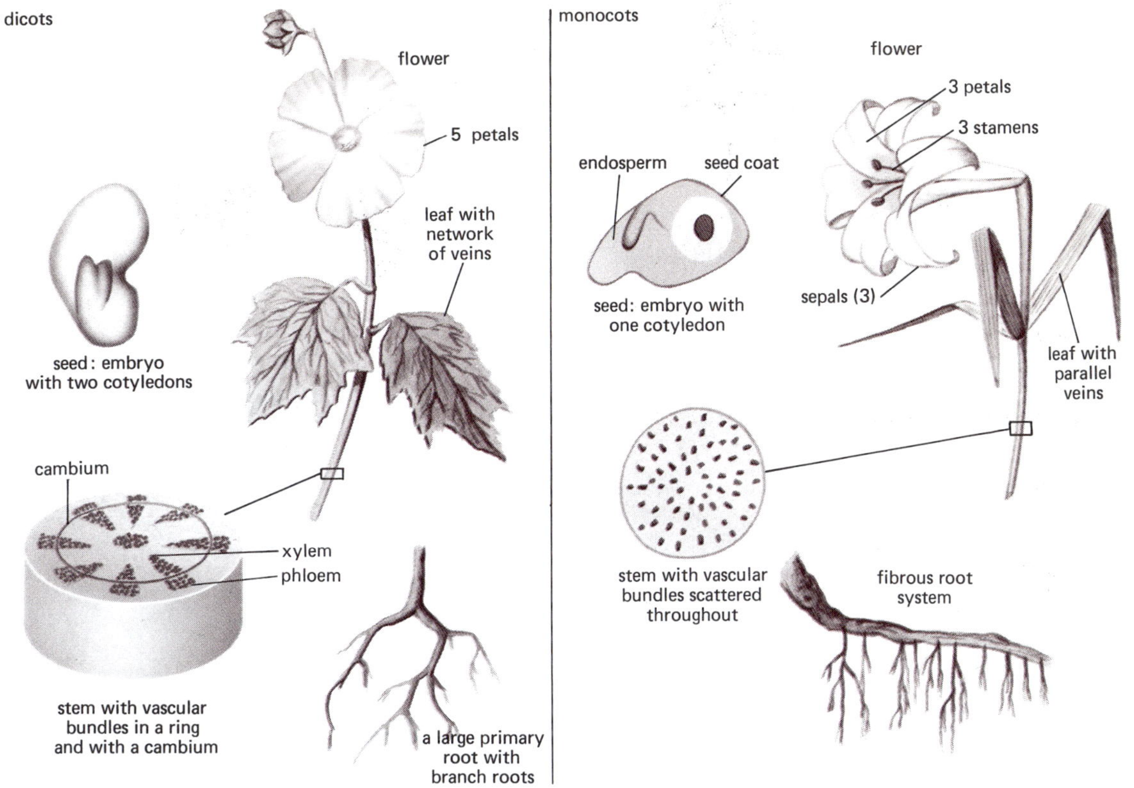

Figure 8–18 Characteristic distinctions between monocots and dicots.

a special covering of tissue. In angiosperms ("covered seeds"), on the other hand, the seed or seeds of a plant are encased in a covering of tissue called a **fruit.** The fruit is often fleshy and may be attractively colored and flavored; it develops from the ovaries of the plant and associated structures.

Gymnosperms also differ from angiosperms in a variety of other ways, the most obvious of which is in the reproductive structures. The reproductive organs of gymnosperms are contained in characteristic **cones** (Fig. 8–20a). The angiosperm reproductive organs, on the other hand, are contained in **flowers;** in a typical flower, the male reproductive organs **(stamens)** and female reproductive organs **(pistils)** are surrounded by a ring of **petals** (which may be colored and/or scented) and an outer ring of **sepals** (Fig. 8–20b).

The vegetating seed plant (the sporophyte) may be divided into three major regions: **roots, stems,** and **leaves.** These major organs of the plant differ both in general structure and in functions.

ROOTS

We usually think of **roots** as underground structures, but in many instances they extend for considerable distances through the air before entering the soil. The banyan trees of tropical Asia, which sometimes reach heights of more than a hundred feet, make numerous contacts with the ground by means of **aerial roots.** The roots of some other plants never even touch the ground. For instance, the aerial roots of the orchid are entirely aboveground and are provided with special cells that absorb moisture from the atmosphere.

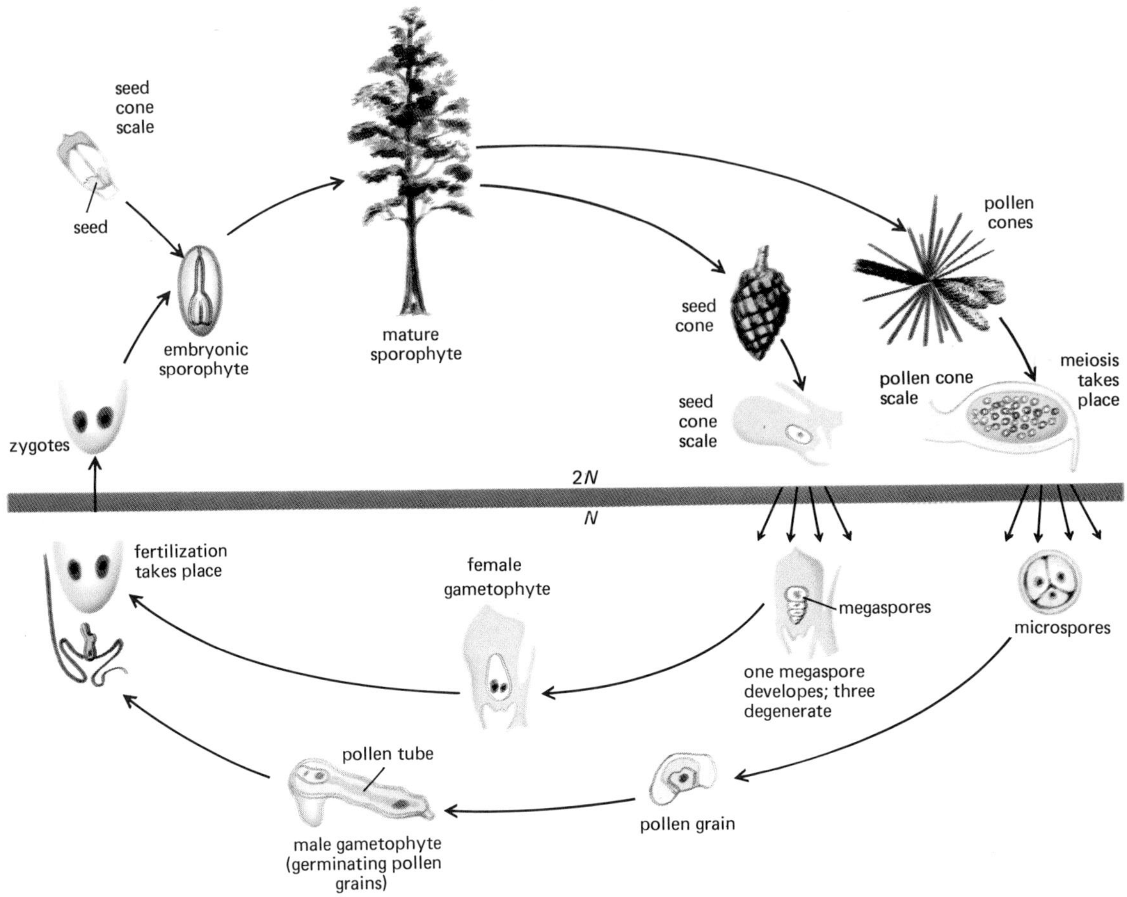

Figure 8–19(a) Life cycle of a gymnosperm (pine).

Notwithstanding these and other exceptions, the vast majority of roots are found underground and perform a number of vital functions. Perhaps the most obvious and certainly one of the most important of these functions is **anchorage.** The buffeting winds and even the mere weight of the superstructure (stem and leaves) might dislodge a plant with an inadequate root system.

A second and equally vital function is the **absorption** of minerals and water from the soil. Not only are these ingredients necessary for photosynthesis, but they are also required for many of the vital metabolic functions carried on by the plant's tissues. The absorption of moisture (and dissolved minerals) by the roots occurs through **root hairs,** long, slender, delicate lateral projections from the root, each of which is derived from a single epidermal cell. The absorption of water and minerals by the root occurs by various mechanisms, including diffusion, capillary action, and "transpiration pull"—the tendency of water to flow upward in the xylem vessels to take the place of water lost by evaporation from the leaves (Fig. 8–21).

The third major function of the root is of less immediate importance, but serves the long-term needs of the plant. This is **food storage.** The photosynthetic process results in the production of excess carbohydrates. These and other unused materials are transported from the leaves, where they are manufactured, to the roots, where they are stored for future use. Man has taken advantage of this function of roots by using carrots, beets, radishes,

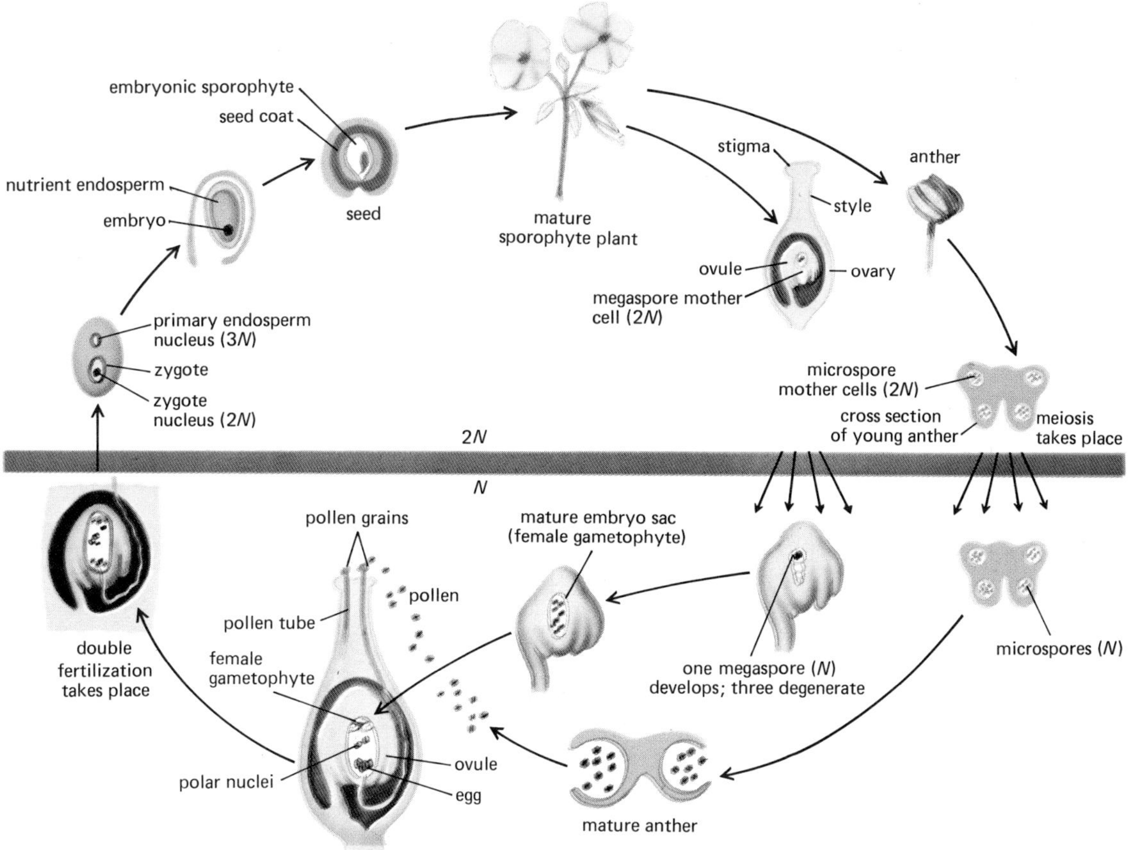

Figure 8–19(b) Life cycle of an angiosperm. In both cases the conspicuous form is the mature sporophyte. The gametophytes are much reduced structures, enclosed within the tissues of the sporophyte.

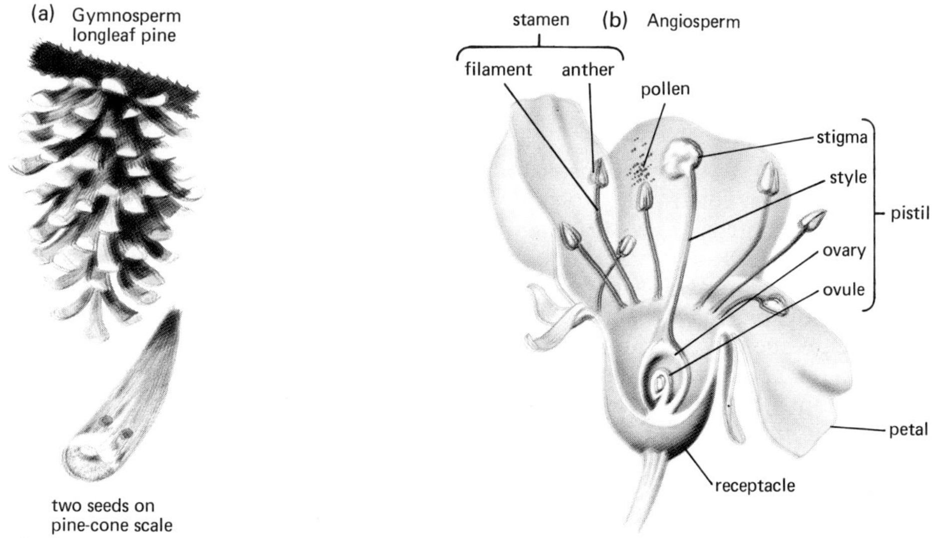

Figure 8–20 Reproductive organs of the gymnosperm (a) and angiosperm (b). The seeds of gymnosperms ("naked seeds") develop on an exposed surface, such as a pine-cone scale. The seeds of angiosperms ("covered seeds") develop inside the ovary of the flower; the ovary and associated structures develop into a fruit.

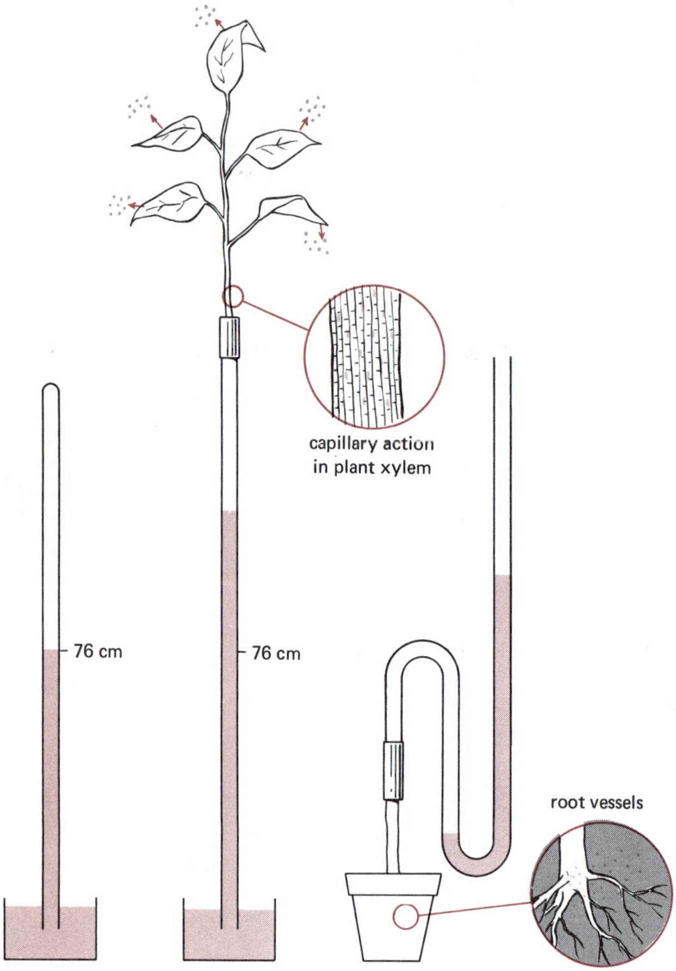

Figure 8–21 Evaporation from the leaf surface, capillary action in the xylem vessels, and diffusion of water into the roots all contribute to the movement of fluids in plants.

turnips, and other root crops as basic foods.

Plants may have various types of roots, depending on their species and stage of development, as well as environmental conditions. The development of a root system can be watched by placing a bean seed in a jar, secured to a backing of moistened blotting paper (Fig. 8–22). Within days, a tiny root emerges from the seed. This first root is known as the **primary root.** After a few days, branches begin to extend horizontally from the primary root: these are the **secondary roots.**

Occasionally roots extend from stems or even from some types of leaves. An example of the latter is the African violet. Roots which arise from structures other than the primary root or one of its branches are known as **adventitious roots** (Fig. 8–23). The prop roots of corn and the aerial roots of poison ivy (which attach the stems of this vine to some object for support) are examples of adventitious roots.

The root system of a plant consists of its entire mass of roots. The two most common types of root system are the **diffuse** or **fibrous root system** and the **taproot system** (Fig. 8–24). The fibrous root system is composed of numerous relatively slender roots of approximately equal size. The various grasses, such as corn, wheat, and barley, exhibit fibrous root systems. In a taproot system, the primary

Figure 8–22 Germination of bean seeds. Root hairs are beginning to appear on each primary root. Note that all the roots grow downward.

Figure 8–23 Adventitious roots of poison ivy anchor the vine to supports such as a tree trunk.

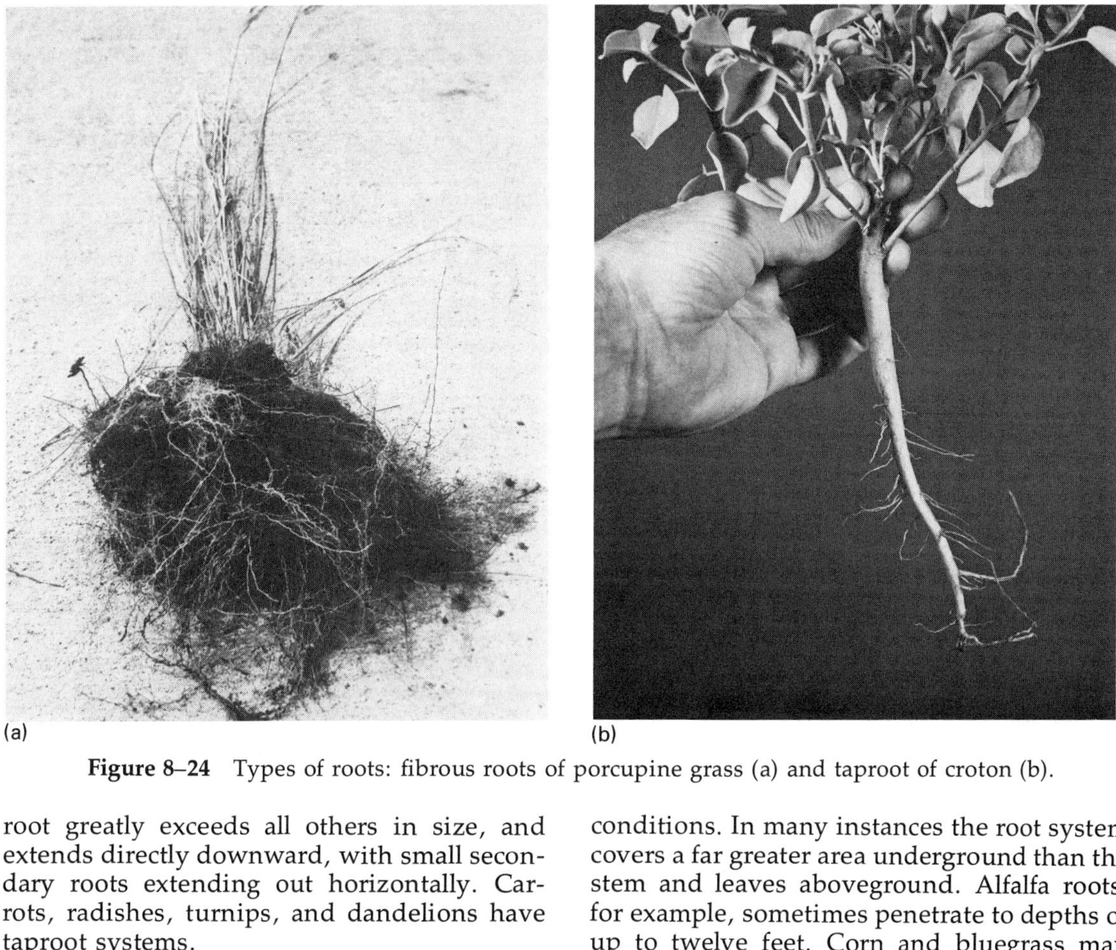

(a)

(b)

Figure 8–24 Types of roots: fibrous roots of porcupine grass (a) and taproot of croton (b).

root greatly exceeds all others in size, and extends directly downward, with small secondary roots extending out horizontally. Carrots, radishes, turnips, and dandelions have taproot systems.

The extent to which root systems penetrate the surrounding soil varies considerably with different species and different environmental conditions. In many instances the root system covers a far greater area underground than the stem and leaves aboveground. Alfalfa roots, for example, sometimes penetrate to depths of up to twelve feet. Corn and bluegrass may have rather shallow root systems, in which most of the roots are confined to the upper ten or twenty inches of soil but extend horizontally for several feet in each direction. This is an adaptive device permitting these plants to grow in regions of scanty rainfall, where the rain rarely penetrates to the deeper layers of the soil.

The extent and complexity of some root systems is staggering. H. J. Dittmer, studying winter rye, counted about 14 *million* roots with a total length of about 380 miles on a single plant! An elm tree may have roots spreading over an area 120 feet in diameter. Why is such an extensive root system necessary? The answer lies in the enormous amounts of water lost by evaporation from the leaves of the plant, in a process called **transpiration.** A single corn plant may lose up to two quarts of water each day in this manner, and a large apple tree may lose almost 20 tons of water during a single growing season. Since all this water must be replaced, it is taken into the plant mainly by absorption through the roots.

STEMS

When we think of a stem we often have in mind a slender, delicate but flexible structure a few inches long. But the mighty three-hundred-foot trunks of the redwoods of the West Coast region, some of which have circumferences in excess of ninety feet, also qualify as stems. A **stem** may be defined as that portion of the plant which is aboveground, with the exception of the leaves and reproductive structures (i.e., flowers and fruits, or cones). But as in the case of roots, we find a number of exceptions. The food-storing **tubers** of the potato, the **bulbs** of the onion and hyacinth, and the extensive **rhizomes** of quack grass all represent special types of underground stems.

The most obvious functions performed by the stem are its provision of mechanical **support** for the leaves and its role as a physical **link** between the roots, where water and minerals are absorbed, and the leaves, where these materials are utilized in photosynthesis in the manufacture of organic materials and thence returned to the roots for storage. One might wonder why such a link is needed—why the leaves of plants cannot grow directly out of the roots. This would be possible, but the presence of a stem provides certain advantages to the plant by raising the leaves up to some height above the ground or permitting the formation of lateral extensions. In the constant competition with other plants for sunlight and growing room, this arrangement makes for greater flexibility and adaptability.

Many stems also serve as important areas of **food storage.** Examples are the stalks of sugar cane, which store large amounts of sugar, and the thickened underground tubers of potatoes, in which starch is stored.

In some plants, such as the cactus, photosynthesis is restricted to the stem, the leaves being reduced to small spines. Although this is an extreme case, it is not unusual to find photosynthesis taking place in the newly formed stems of trees and shrubs.

The stems of such creeping plants as strawberries (known as **runners**) have a **reproductive** function: they grow horizontally along the ground, giving rise to new plants at various points that come in contact with the ground.

Figure 8–25 Stems may be modified for special functions: the sweet potato tuber is an organ of food storage (a); strawberry runners help to propagate the plant (b); and the succulent stems of cactus store water and carry on photosynthesis, while the leaves are reduced to slender spines (c).

(a) Lynwood M. Chace, Omikron (b) (c)

The spiral stems **(twiners)** of the morning glory and the clinging stems **(tendrils)** of Boston ivy, on the other hand, provide support in climbing.

Seed plants may be classified into two main groups, according to the basic type of stem: **herbaceous** and **woody.** Most herbaceous plants have relatively smooth, soft green stems and are **annuals** (i.e., they complete their life cycles in a single growing season). They are often killed with the first frost. Other herbaceous plants (such as carrots and beets) store food in their roots. With the frost the stem and leaves die, but they are replaced the following spring, and go on to produce seeds. Then the plants die, and the seeds start the two-year cycle over again. These plants are known as **biennials.**

In contrast to the herbaceous plants, woody plants have a much thicker, harder, and tougher stem and are covered with a layer of cork, which gives them a roughened exterior. In general, woody plants are **perennials** (i.e., they live for longer periods than the biennials). (Some living sequoia or redwood trees are more than four thousand years old.) A **tree,** which is a woody perennial, has a single main stem, called a **trunk.** This usually gives rise to numerous **branches,** which are generally narrower than the trunk. On the other hand, a **shrub** is a woody perennial that usually gives rise to several stems of approximately the same size. In general, shrubs are smaller than trees.

The gymnosperms are woody-stemmed plants. Among the angiosperms, most of the monocotyledons are herbaceous, while many of the dicotyledons are woody.

One of the outstanding features of the internal structure of stems is the bundles of

Figure 8–26 Cross sections of stems: (a) corn (monocot herbaceous stem); (b) sunflower (dicot herbaceous stem); (c) pine (young woody stem); (d) pine (tree trunk showing annual rings).

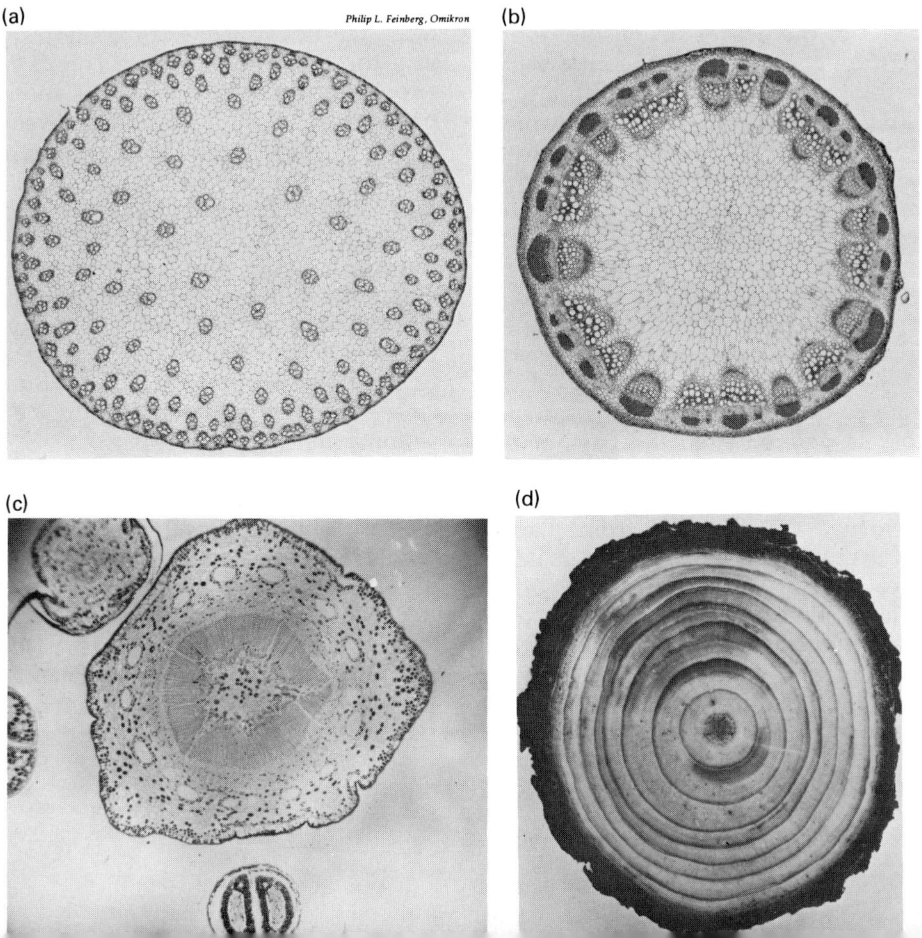

conducting vessels (xylem and phloem), called **vascular bundles.** Through these vital connecting links water is transported from roots to leaves and various materials are transported up and down the plant.

During the first year of growth, the woody stem bears a close resemblance to a typical dicot herbaceous stem (Fig. 8–26). But with each succeeding year the cambium gives rise to an additional layer of xylem (on the inside) and a much thinner layer of phloem (on the outside). When the stem is viewed in cross section, the yearly deposits of xylem are clearly distinguished as **annual rings.** This is due to the fact that the xylem cells formed in the normally wet spring are much larger and have thinner walls than those formed during the dry, hot summer (when growing conditions are less favorable). The inner layers of xylem cease conducting fluid and eventually die, but they continue to add strength to the tree. These inner layers are known as **heartwood.** The outer layers of xylem, known as **sapwood,** are active in carrying water and minerals up to the leaves from the roots. Because of the dependence of the size of the xylem layers on the amount of precipitation, it is possible to determine the approximate seasonal rainfall (in a given region), for hundreds and even thousands of years in the past, by examining the annual rings on the cut surfaces of felled trees and fossils.

Stems and products derived from them are extremely useful to man. Ample evidence of this statement can be found in any home or classroom. The books and newspapers, the floors and the furniture, are only a few of the items produced from the wood of trees. Commercial cork is derived from the bark of the cork oak. Rubber comes from the sap (conducting fluid) in the stems of rubber trees. Numerous drugs are obtained from plants. A number of these chemicals have a striking physiological effect on humans and other animals. Many other useful products, such as maple sugar and turpentine, are derived from the stems of plants. And of course, various stems, including potatoes, onions, and sugar cane, serve as sources of food.

LEAVES

The verdant beauty of a leaf-laden forest contrasts sharply with the bleak drabness of the same forest in midwinter. The delicate traceries of green leaves are not only decorative, however: they also play a vital role in the life of the plant. The most important and universal function of **leaves** is the **manufacture of food** materials. This is accomplished through the process of photosynthesis. With the aid of the light-sensitive pigment chlorophyll, contained in organelles of the leaf cells called chloroplasts, some of the energy of sunlight is captured in chemical bonds, as simple inorganic chemicals—carbon dioxide from the air and water brought up from the soil via the roots and stems—are transformed into organic compounds—first simple sugars, then more complex compounds.

Leaves are extremely well adapted for their photosynthetic function. The broad, thin leaves of most plants afford a maximum surface area to catch the sunlight. (The total surface area of the leaves of some trees exceeds fifty thousand square feet. This is more than enough to cover the surface of an entire football field.) In addition, the leaves of many plants turn to present the maximum possible surface to the rays of the sun in its apparent movement across the sky during the day.

Carbon dioxide for photosynthesis enters the leaf through minute openings called **stomata.** These stomata are usually located on the underside of the leaf, and there may be many thousands of them on a single leaf. The size of their apertures is regulated by a pair of crescent-shaped guard cells (Fig. 8–27). The stomata usually gape open during the day, when active photosynthesis is taking place, and are partly closed at night. Transpiration, the evaporation of water from the plant into the surrounding atmosphere, occurs through the stomata. Thus, their closing at night helps to conserve water for the plant.

The leaves of some plants exhibit special adaptations. For instance, the Venus flytrap is provided with sensitive hairs on the surface of its leaves (Fig. 8–28). If an insect alights on a

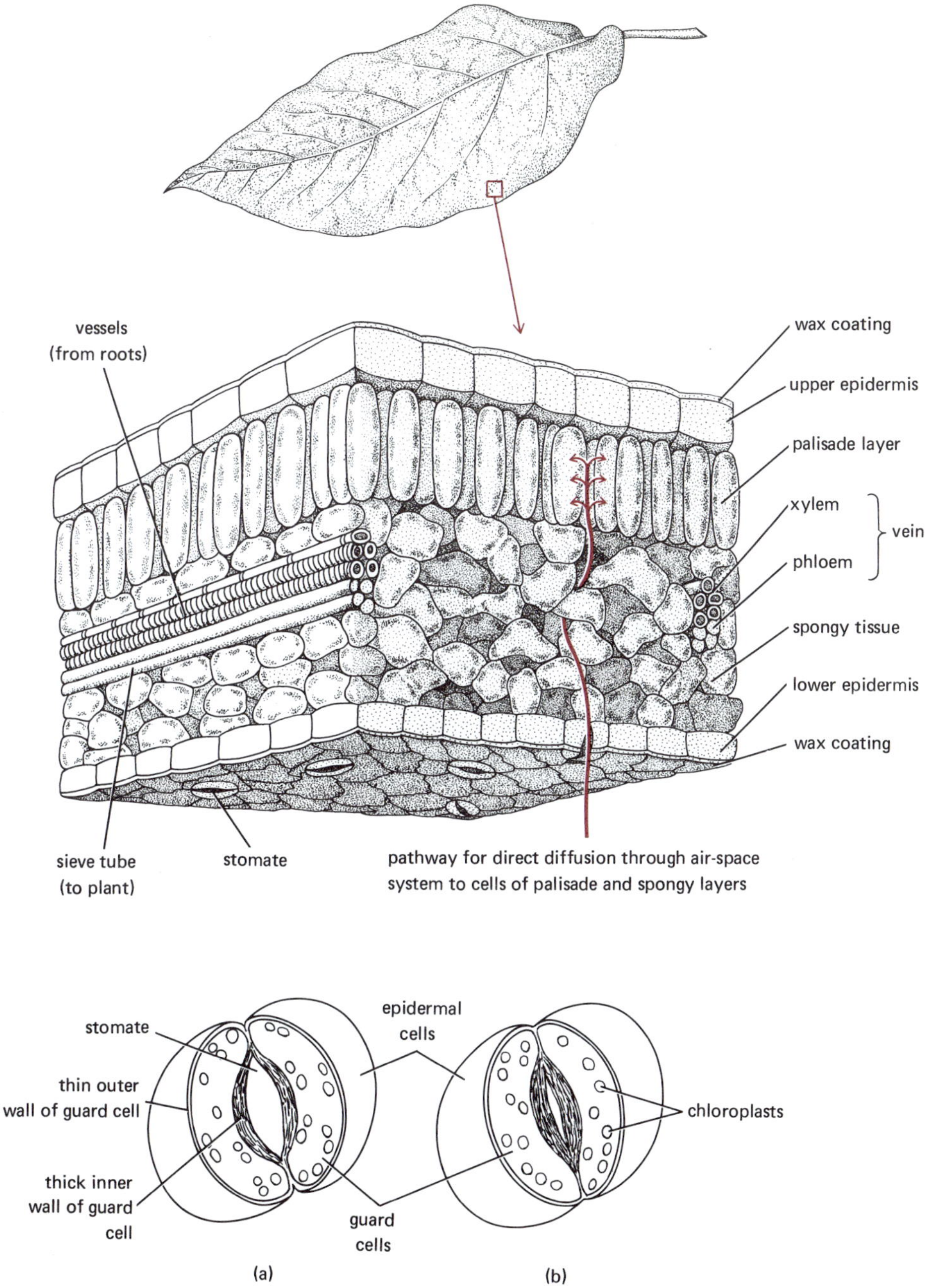

Figure 8–27 Structure of a leaf. The opening and closing of the stomata (a and b) regulate gas exchange with the atmosphere.

Figure 8–28 The leaves of the Venus fly trap can snap shut to trap and digest an unwary insect.

leaf, it will trigger one of these hairs, causing the leaf to fold in. The trapped insect is then killed and slowly digested by a fluid secreted by the plant.

In some desert plants, the leaves serve as a storage area for water. The agaves of the deserts of the U.S. Southwest and Mexico have succulent, fleshy leaves of this type. In other desert plants the leaves have almost disappeared. The spines of the cactus (which serve as a protective device, warding off animals that might try to eat the plant) are the only evidence of leaves in this plant. This adaptation serves as a means of reducing the loss of water by transpiration. The stem of the cactus takes over the function of food production.

The leaves of plants vary enormously, both in size and in shape. The leaves of some small

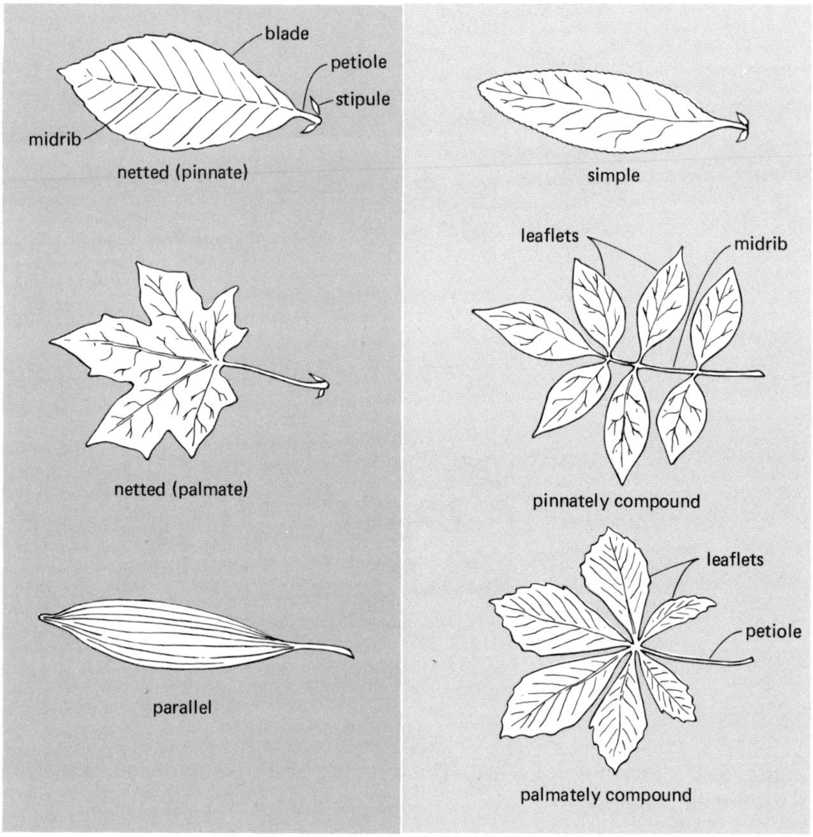

Figure 8–29 Leaves are distinguished according to patterns of venation: parallel or netted (pinnate leaves have veins branching out from a central midrib; palmate leaves have several principal veins diverging from a single point near the base of the blade). The leaf blade may be simple; or it may be compound, consisting of a number of leaflets, which may be arranged either pinnately or palmately.

herbaceous plants measure barely an eighth of an inch across, whereas those of some palms may be fifty to sixty feet long. Some order can be traced in the seemingly infinite variety of leaf shapes by classifying them according to their patterns of **venation,** that is, the arrangement of vascular bundles in the form of **veins.** There are two main types of venation in leaves: **parallel** and **netted** (Fig. 8–29). In the first type, the veins run parallel to one another. This arrangement is typical of monocots, such as corn and lily plants. Netted venation appears as a network of progressively smaller interwoven veins. Dicots such as the oak and maple typify two distinct types of netted venation. In oak leaves, smaller veins branch out from a single large **midrib.** In maple leaves, several relatively large veins form the framework from which smaller veins arise.

Although most leaves occur as single **blades** (known as **simple leaves**), some are divided into various numbers of separate **leaflets** (Fig. 8–29). Leaves of the latter type are referred to as **compound leaves.** Examples include clover and walnut.

In the autumn in temperate climates, the brisk winds scatter tons of dried and shriveled but beautifully colored leaves among the forests and meadows. The conifers retain their needles, but most of the other trees and shrubs lose their leaves. Soon the green of the summer gives way to the bleak, leafless winter. At the base of each leaf stalk, a layer of parenchyma cells grows across, weakening its connection to the stem. This layer is called an **abscission layer;** toward the end of the growing season, the abscission cells begin to break down; a brisk wind may sever the leaf's last tenuous ties to the stem, and it falls. The bright colors of the autumn leaves are due to the presence of yellow, orange, red, and purple pigments in the leaves. When a leaf dies, its chlorophyll quickly breaks down, and now the other pigments can be seen.

QUESTIONS FOR REVIEW AND THOUGHT

1. How do mycoplasmas and rickettsias compare with viruses in size and complexity? Is there any reason to consider them more "alive" than viruses?
2. Would the destruction of *all* bacteria on our planet be helpful or harmful to human life? Explain.
3. Compare and contrast a typical protozoan to: (1) a human cell; (2) a multicellular organism.
4. Give examples of common fungi and their impact on human life.
5. What advances in morphology do mosses and liverworts show over the algae and fungi?
6. Describe the adaptive advantages the tracheophytes have over the bryophytes.
7. Why do you suppose the ferns, which once dominated the lands of the earth, gave way to the gymnosperms and angiosperms?
8. Trace the concept of alternation of generations through the phyla of the plant kingdom. Is there any comparable phenomenon in the animal kingdom?
9. What functions of the stem overlap those of the root? Which functions are unique to each?
10. Are leaves really a necessary part of the plant? Explain.

camel
bat
mountain
sheep
hawk
snake
scorpion
lizard
dragonfly
pelican
deer
rabbit
seagull
seal
bear
butterfly
mouse
copepod
crayfish
rotifer
lobster
fox
ant
beetle
turtle
frog
jellyfish
spider
bass
elephant
grasshopper
whale
sea urchin
barnacle
sponge
starfish
shark
monkey
earthworm
crab
rhinoceros
clam
crocodile
snail
octopus

CHAPTER 9

Panorama of the Animal Kingdom

The Animal Kingdom

It has been estimated that more than $1^1/_2$ million different animal species are to be found on the earth. An incredible multitude of species, from minute insects to ponderous elephants, from timid mice to roaring lions, inhabit the lands. Worms burrow in the ground, and birds fly through the air. The seas, too, teem with animal life. Starfish crawl along the ocean bottom, while jellyfish float in the surface waters. Shrimps and squids, a variety of fishes, and sea mammals such as seals and whales are among the creatures that swim in the seas.

A close examination of the numerous species within various animal groups reveals that they are not discrete entities, totally divorced from the rest. Instead, they show varying degrees of similarity to other groups, even to distant groups. Basic sequences and patterns of development seem to emerge. Within and among the animal phyla there appear to be progressions toward greater and greater differentiation and specialization of cells, tissues, and organs.

Figure 9–1 The animal kingdom.

Evolutionists have found convincing evidence in support of their theories on the origin and development of life on earth in just such comparisons of the different groups. Presumed protozoanlike ancestors, it is believed, gradually were grouped together into associations of cells. In time, groups of cells in such cell "colonies" became specialized for various functions, such as food getting, digestion, and protection from the environment. Such loose colonial associations of specialized cells could well have been the ancestors of the **metazoa**—multicellular organisms that we now classify as animals.

The metazoa, with their complex organizations of highly specialized cells, have many advantages over the single-celled protists. First of all, **differentiation** and **specialization** of cell structure and function generally result in greater efficiency for the organism as a whole. Each specialized cell in a multicellular organism is often modified for optimum functioning in a particular task and therefore can perform it better than a single cell that has to "divide its attentions" among many different functions, each requiring diverse structural adaptations. Multicellular organisms can also grow much larger than single-celled forms, thereby increasing their chances of survival. The metazoan organization of differentiated organs has permitted exploitation of land habitats, which would be impossible for protists, which are dependent upon an exchange of materials through the cell membrane by liquid diffusion.

But with all their advantages, differentiation and specialization of cells also have some inherent disadvantages. Interdependence is a two-edged sword. It provides many mutual benefits for the associated cells, but also means that the death of any significant fraction of one type of cell may spell the end for the entire organism. Each cell, moreover, is surrounded by other cells, all of which have similar needs for food, water, and oxygen and have waste materials that must be disposed of. Networks of distribution, unneeded by protists, are therefore vital for metazoa.

Increased differentiation also seems to be associated with a reduced ability for **regeneration**—replacement of lost parts. In the development of the individual metazoan organism, just as in the evolution of the metazoan phyla, there is a progressive increase in differentiation and a corresponding progressive decrease in potential for regeneration. If the individual cells of a developing embryo at a very early stage are separated, each generally can regenerate all the missing parts and eventually give rise to a whole new organism. This is essentially the basis of identical twins, triplets, and so on. Similarly, if a relatively unorganized animal such as a sponge is cut into parts, each part can regenerate a new sponge. But among the more highly differentiated organisms, the regenerative powers gradually decrease: a flatworm or a starfish can be regenerated from a small piece of the original organism, but a segmented worm usually can regenerate only a certain highly specific portion of its head or tail region. And in such highly organized and differentiated animals as man, only a few types of tissue, such as skin and liver tissue, can be regenerated.

With these general patterns in mind, we shall now discuss in greater detail some of the major animal phyla and the characteristics and habits of some of their most important representatives.

Sponges: Phylum Porifera

With the rise of the modern plastics industry, the natural **sponge,** which was a household fixture literally for millennia, has been largely superceded by brightly colored, fine-textured synthetic substitutes. Nonetheless, natural sponges are still preferred by many, and are

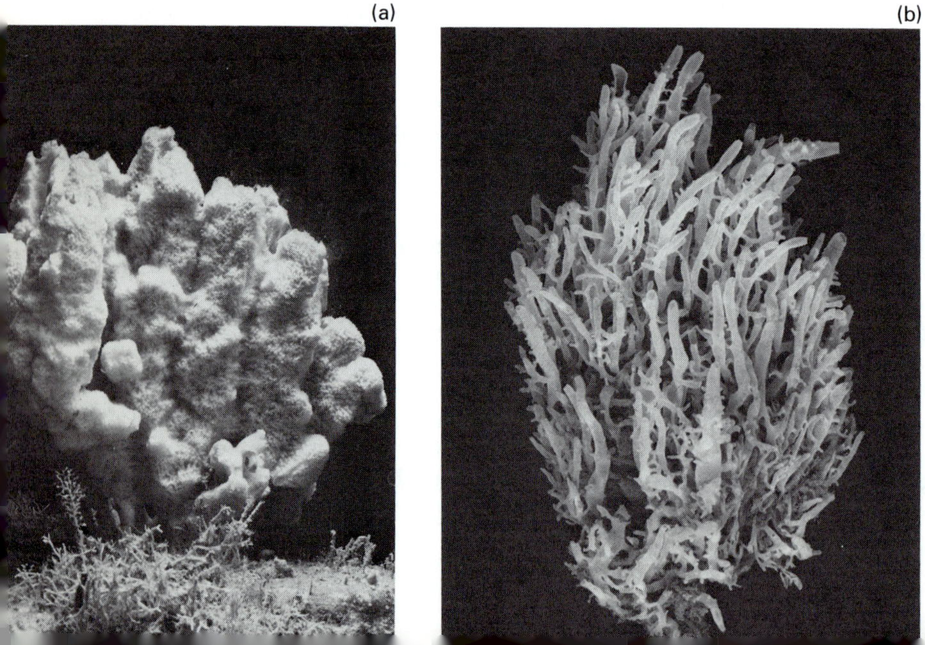

Figure 9–2 The slimy sponge, *Desmacidon fruticosum,* growing on a rock in deep water (a) and a branched calcareous sponge, *Leucosolenia complicata*, (b).

put to uses quite similar to those for which they were employed by the ancient Greeks and Romans. Two thousand years ago or so, just as today, sponges were used as bathing aids and in scrubbing floors and furniture. The ancient Greeks also used them as padding in their shields and armor (a prototype of today's foam rubber padding), and the Romans used them as mops and paintbrushes. In fact sponges were even used as "drinking cups" in ancient times: the drinker dipped his sponge into liquid and then squeezed it into his upturned mouth.

The bath sponge with which we are familiar bears little resemblance to the living creature from which it is made. Actually, the bath sponge is only the dried skeleton of the sponge, made of a horny, fibrous material called **spongin.** Not all sponge species have spongin skeletons. Some have skeletons formed of calcareous or limy little spikes. Other species have silicious skeletons, which, though of no particular commercial use, may be very beautiful when dried, resembling delicate traceries of spun glass.

In a number of respects, sponges are quite different from other animals. In fact, they represent an evolutionary dead end. Their body organization is quite rudimentary: they have no organs, movable parts, or appendages. In fact, the sponge rather resembles a colony of protozoa, and it has been called a "republic of cells." The extreme lack of organization of a sponge is manifested in its amazing abilities for regeneration: not only can a sponge readily regenerate lost parts (or even a whole sponge from a single small cut part), but this organism can even be entirely reconstituted from its separated individual cells. If a sponge is squeezed through a piece of fine silk cloth, through which only individual cells can pass, and then these cells are suspended in water, they reorient themselves, reunite, and grow back together into another whole sponge.

Sponges are quite plantlike in some respects. The adult sponge is sessile, living quietly in the water, attached to a rock or to the sea or stream bottom. (Most of the approximately three thousand sponge species are salt-water dwellers, with a few fresh-water representatives.) Sponges grow much like plants and, unlike all the other metazoa, have no nervous tissue; thus impulses cannot be conducted from one part of the body to another. Only the free-swimming larvae (immature forms) of the sponges possess the means of locomotion.

A living sponge rather resembles a piece of raw liver. The fresh-water forms are usually a dull brown, but the marine forms are often brightly colored. The branched and channeled irregularly shaped bodies of the sponges often serve as refuges for small fish, worms, crustaceans, and other water animals.

The basic body plan of a typical sponge is a vaselike structure enclosing a central cavity (Fig. 9–3) with a single opening at the top. Under a microscope, the body wall of the sponge is found to contain numerous tiny openings or pores. The formal name of the phylum to which the sponges belong, **Porifera,** means "pore-bearing." The inner surface of the body wall is lined with flagellated cells, which set up currents in the water. These currents result in a continuous flow of water, which is filtered through the pores in the walls, bringing oxygen and food (minute animals, plants, and protists) into the inner body cavity. The flagellated cells absorb the filtered food particles, and the current, with accumulated waste products, is then directed out through the "mouth" of the sponge.

Sponges may reproduce asexually by various means. A common form is budding, in which small new individuals are budded off the parent sponge, just as among many plant and protist species. Fragmentation and regeneration of new sponges from the fragments is another mode of asexual reproduction.

Sexual reproduction in sponges involves the production of eggs and sperms, either both in the same organism or each in organisms of either sex. In the latter type, sperms from a male sponge are swept along in the currents and drawn into a female sponge, where they fertilize the eggs. The zygote develops into a free-swimming flagellated larva, which leaves the parent sponge, swims

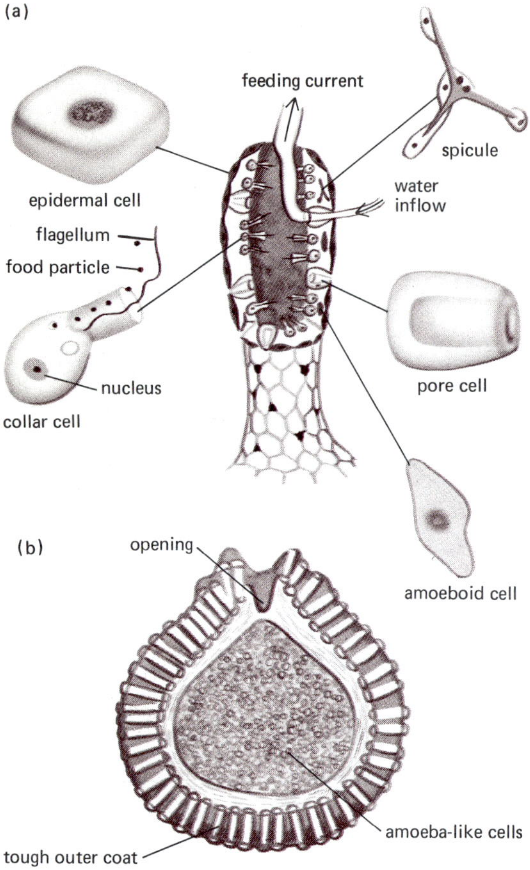

Figure 9–3 (a) A member of a colony of sponges. The top part is cut away to show the internal structure. There is some specialization among this "republic of cells." (b) A gemmule, formed by some freshwater sponges under threat of unfavorable conditions such as freezing temperatures or drought. When conditions again become suitable, the amoebalike cells flow out of the opening and form a new sponge.

about for a few hours, then settles down to the bottom, to take up a sedentary life like that of its parent.

Coelenterates: Phylum Cnidaria

The Portuguese man-of-war is a lovely sight. A brightly colored balloonlike creature, blue, purple, red, or greenish, floats upon the warm ocean waters, sometimes seeming to sink beneath the surface, only to bob up again. But the wise swimmer regards the man-of-war at a distance. For beneath the gaily colored, sometimes foot-long balloon trail long tentacles, equipped with deadly stinging cells. Contact with even a single tentacle will raise an excruciatingly painful red welt on the flesh of an unwary human swimmer. Man-of-war stings may produce temporary paralysis or even death. Such is the result for the fish upon which this well-armed floating dreadnaught feeds. Yet one tiny fish, *Nomeus,* seems to be miraculously immune to the man-of-war's deadly poison. In fact, the little fish makes its home within the tentacles of the Portuguese man-of-war. If, on its ventures out from its strange home, *Nomeus* is menaced by a larger fish, it darts quickly back into the sheltering tentacles. The larger fish, following, is paralyzed and seized by the stinging tentacles,

Figure 9–4 A small fish, *Nomeus*, seeks sanctuary among the deadly tentacles of the Portuguese man-of-war.

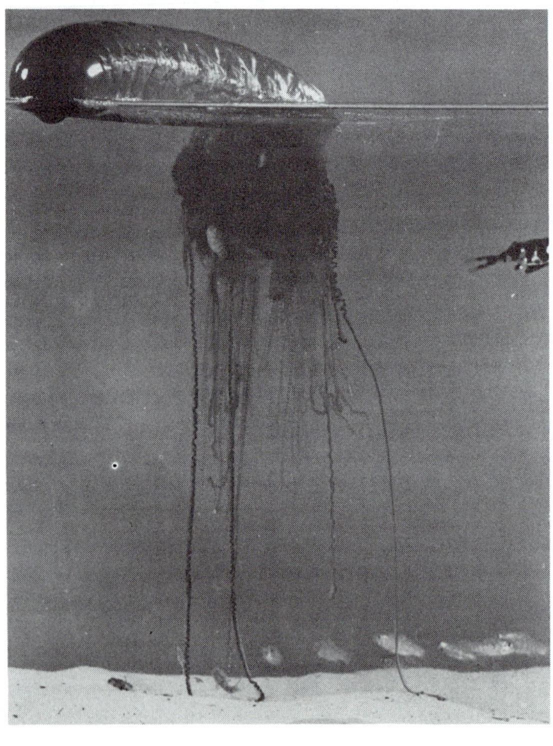

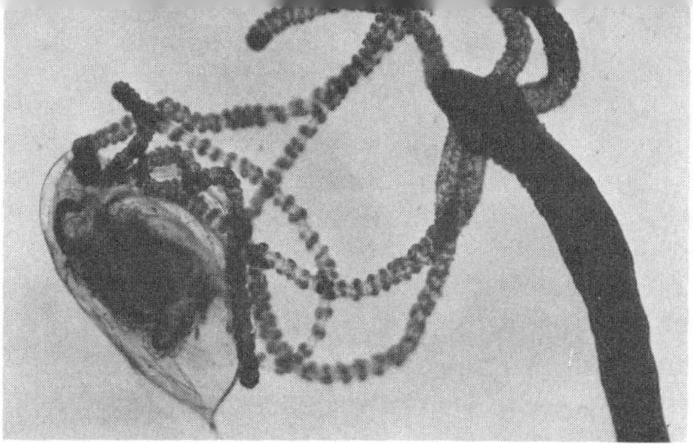

(a)

(c)

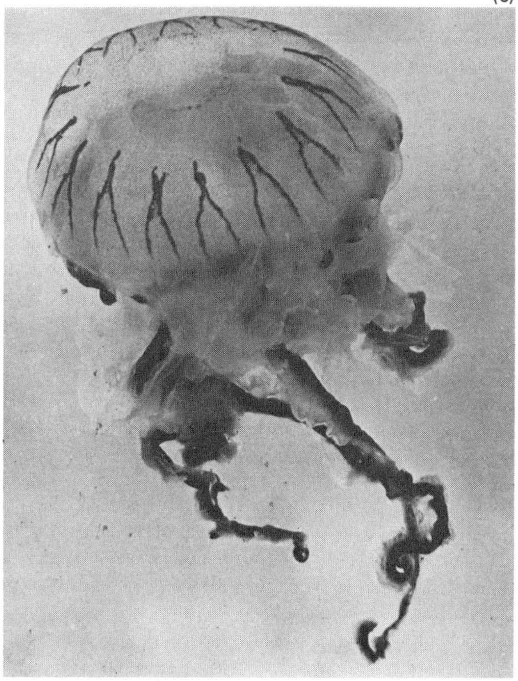

(b)

(d)

William M. Stephens, Omikron

Figure 9–5 Cnidarians include the hydra, shown here catching a waterflea (a), the sea anemone (b), jellyfish (c), and coral (d). The living coral polyps shown here have the typical white limy skeleton; their color comes from algae that live within the tissues of the polyps.

and then becomes a meal for the man-of-war and its little guest.

The Portuguese man-of-war belongs to the group of **coelenterates,** now known as the phylum **Cnidaria** (named from the **cnidoblasts,** or stinging cells possessed by the members of the group). Other members of this water-dwelling phylum are the tiny, stalklike hydra, with its crown of waving tentacles; various types of jellyfishes; the sea anemones, "animal flowers" of the ocean; and the corals, with hard, calcareous skeletons, colonies of which may gradually build up whole atolls, reefs, or even islands of coral (Fig. 9–5). Taxonomists place the more than nine thousand coelenterate species in three classes: **Hydrozoa,** including *Hydra,* the colonial form *Obelia,* the Portuguese man-of-war, and some jellyfish; **Scyphozoa,** exclusively marine forms, including the large oceanic jellyfish; and **Anthozoa,** including sea anemones, corals, sea fans, and sea pens.

The typical coelenterate body consists of a radially symmetrical body wall surrounding a central body cavity (Fig. 9–6). This cavity, within which food is digested, has but a single opening, the mouth. The mouth is surrounded by tentacles, which convey food to it; undigested materials and waste products are ejected through the mouth.

Unlike the sponges, the cells of coelenterates are differentiated into definite types, forming tissues, including digestive, muscular, nervous, and sensory tissues.

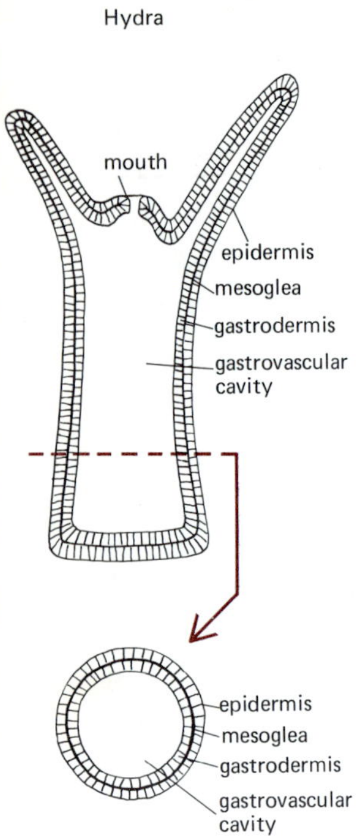

Figure 9–6 The hydra (longitudinal and transverse sections).

There are two basic morphological types found among the coelenterates: the **polyp** and the **medusa.** The polyp is a sessile* form, usually column-shaped. The base of the column is generally attached to a substrate; at the other end of the cylindrical stalk is the oral (mouth) end, surrounded by tentacles. The medusa is a bell- or umbrella-shaped free-swimming organism. The tentacles dangle from the outer margin of the bell. From the center of the underside of the bell projects a tubular structure, at the free end of which is the mouth. (Despite the apparent structural difference, the medusa may be regarded as basically an upside-down polyp, as is evident from an examination of Fig. 9–7.)

**Hydra,* though primarily sessile, can move about in various ways. It can creep along the bottom by amoeboid movements of its basal cells. It can let go and float along in the currents. It may move along the bottom in a looping manner, somewhat like a caterpillar, bending over to cling to the bottom with its tentacles, then advancing its base, clinging with that while it moves its tentacled end forward, and so on. The most amusing type of *Hydra* locomotion to watch is a type of somersaulting motion. The hydra bends over, clings to the bottom with its tentacles, lets go entirely with the base, and flips it over to assume a new position on the other side, thus completing a full cartwheel.

Some coelenterates, such as *Hydra,* sea anemones, and corals, exhibit only the polyp forms; others, such as some jellyfish, are exclusively medusae. Some coelenterates, such as *Obelia,* show an alternation of generations, with the medusae reproducing sexually to give rise to polyps; the latter in turn reproduce asexually, yielding medusae. In species where either the polyp or the medusa form is absent, the same individual may reproduce either sexually or asexually.

Some polyps, such as *Hydra,* are found singly. But a number of polyp species are colonial, with many individuals living attached to one another. There may be some specialization of function in the colony. For example, *Obelia,* a typical colonial coelenterate, has a number of specialized feeding polyps and reproductive polyps, growing in a branchlike arrangement (Fig. 9–8). In the Portuguese man-of-war, a number of polyplike individuals are suspended from a medusalike float, which can be expanded with a self-generated gas or collapsed at will. The polyplike individuals are specialized into defensive, nutritive, and reproductive forms. The stinging polyps dangling from the float may be as much as sixty feet long, and are extremely dangerous even for some time after the death of a man-of-war washed up onto the shore.

The stony corals are close relatives of the flowerlike sea anemones. But these polyps are not soft-bodied: they secrete a hard calcium carbonate skeleton. The coral polyps live in interconnected colonies, mainly in warm, shallow waters, and are especially abundant in tropical areas. The addition of new members of the colony slowly builds up characteristic reefs or atolls.

Within and among the branching growths of corals live thousands of other species of animals, some associated with the corals in cooperative relationships of varying intimacy. These species include certain shrimps and crabs, which actually live within the coral growths. Such cooperative relationships, like that of the little fish *Nomeus* with the Portu-

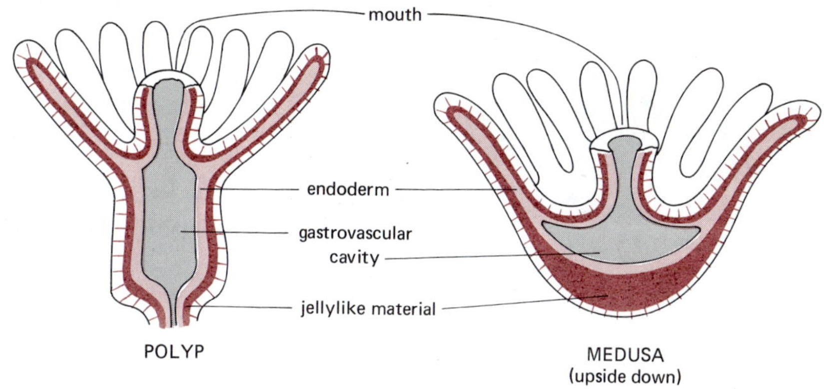

Figure 9–7 Despite their apparent dissimilarity, the polyp and medusa forms of coelenterates have the same basic structure—as becomes evident if the medusa is turned upside down.

guese man-of-war, are common in nature and are referred to by the term **symbiosis** (literally, "living together"). Other examples of symbiotic relationships will be described in Chapter 24.

Figure 9–8 Life cycle of *Obelia*, a typical coelenterate. *Obelia* is a colonial hydrozoan, in which the predominate stage is a polyp that forms a branched colony by budding. There is some differentiation in the colony, into feeding and reproductive polyps. The reproductive polyps bud off free-swimming medusae, which reproduce sexually.

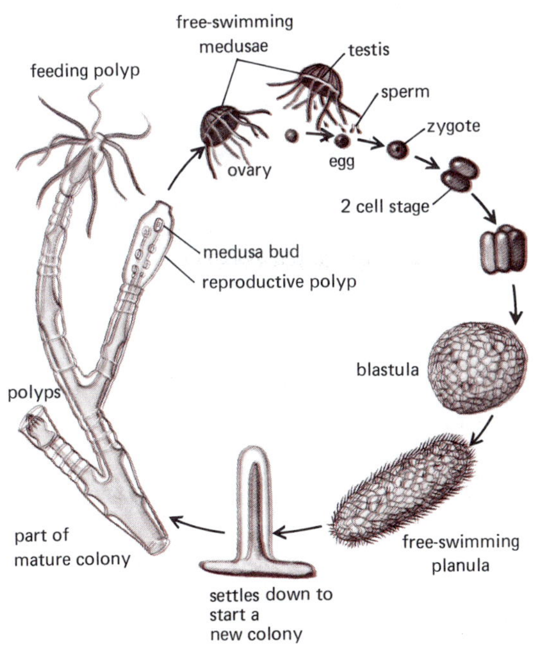

Flatworms: Phylum Platyhelminthes

In recent years, the tiny, ribbonlike flatworms have achieved a quite disproportionate degree of fame in scientific circles. Simple to raise and care for, the tiny planarian (one kind of flatworm) had long been a favorite laboratory specimen in view of its unusual facility in regenerating lost parts. When it was announced that planarians could be conditioned to make simple choices and solve simple mazes, scientists were excited at the interesting prospects of study opened up, and the lowly flatworm even became the exclusive subject of a scholarly journal, *Worm Runners' Digest*.

The **flatworms,** phylum **Platyhelminthes,** are widely distributed throughout the world. The more than six thousand species are placed by taxonomists in three major classes: **Turbellaria, Trematoda,** and **Cestoda.** The turbellarians are free-living, soft-bodied forms that abound in both fresh and salt waters. The best-studied representative is the common freshwater planarian *Dugesia,* found in ponds and streams. The trematodes, or **flukes,** are all parasitic (either internally or externally with regard to the host organism off which they live). They show a number of adaptations to their parasitic life, such as special suckers for attachment to the tissues of their hosts, and some species have complex life cycles, requiring a highly specific series of successive hosts.

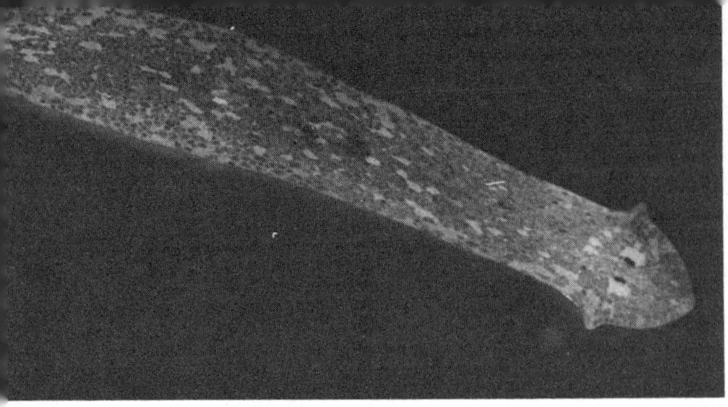

(a)

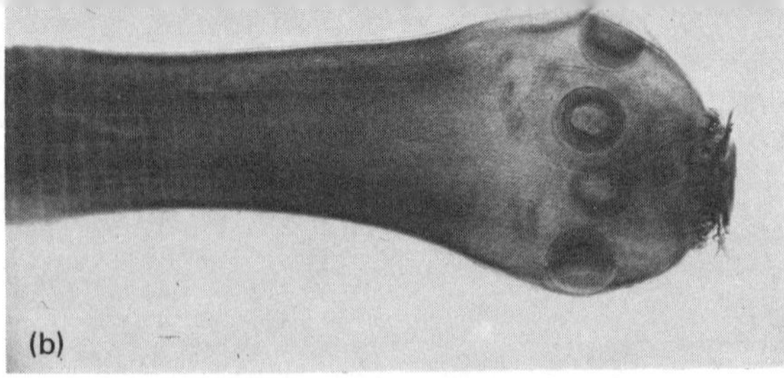

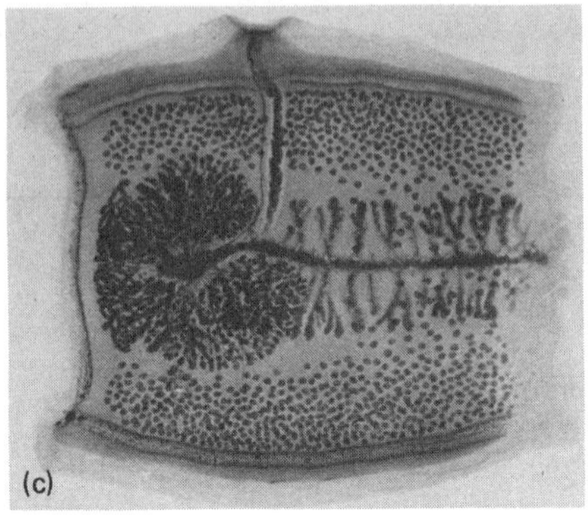

Figure 9–9 Two flatworms: the free-living planaria, *Dugesia tigrina* (a), and the parasitic tapeworm *Taenia pisiformis*, the head and a segment of which are shown in (b, c).

An example is *Fasciola hepatica*, the sheep liver fluke. The cestodes, or **tapeworms,** are another important class of internal parasites, with bodies divided into segments. Common representatives are *Taenia saginata*, the beef tapeworm, and *Taenia solium*, the pork tapeworm. These parasites are acquired by man largely by eating raw or improperly cooked meats.

The flatworms exhibit a number of significant advances in complexity of structure and function in comparison with the sponges and coelenterates. They are bilaterally symmetrical, with a flattened overall body shape, well-distinguished **dorsal** (back) and **ventral** (belly) surfaces, and a distinct head and tail. They have reached the organ-system level of development, with specialized nervous system, digestive system, excretory system, and reproductive system, and with definite copulatory organs, the penis and vagina. The organ systems are often greatly reduced or modified in the parasitic flatworms, which possess characteristic special adaptations to their way of life.

In the planarians, bilateral symmetry permits improved locomotion. They can glide along an underwater surface by first secreting a layer of mucous slime, and then propelling themselves in this layer by the beating of numerous cilia attached to the ventral body surface. They can also crawl along by attaching the **anterior** (front) end to the surface of the substrate with slime, then bringing the **posterior** (back) end forward by muscular contractions (a sort of looping motion). Planarians can also swim freely in the water by undulating movements.

The planarian meets life literally "head on": with its characteristic locomotion and bilateral symmetry, the anterior end of the worm is the part that comes into contact with changes in the environment first. Appropriately, there is a concentration of sensory receptors, which perceive light, touch, and vibration, in the head region (Fig. 9–10a). The light receptors, two dark spots on the dorsal surface of the head, are the first examples of eyes in the animal kingdom.

The planarian's mouth, which is in the middle of the ventral surface of its body, leads into a dead-end digestive system. After digestion, the waste products are sent back out through the mouth.

Like most of the flatworms, planarians are **hermaphroditic**—reproductive organs of both sexes occur in the same animal. Though bisexual, the planarian cannot fertilize its own eggs—it must have a mate. When two planarians mate, the penis of each is inserted into the genital pore ("vagina") of the other, and sperm are deposited in both. (Each flatworm thus acts as both "mother" and "father.") After fertilization of the eggs, capsules are formed around the zygotes and yolk cells. The capsules are later attached to some object, and the eggs eventually hatch into small planarians.

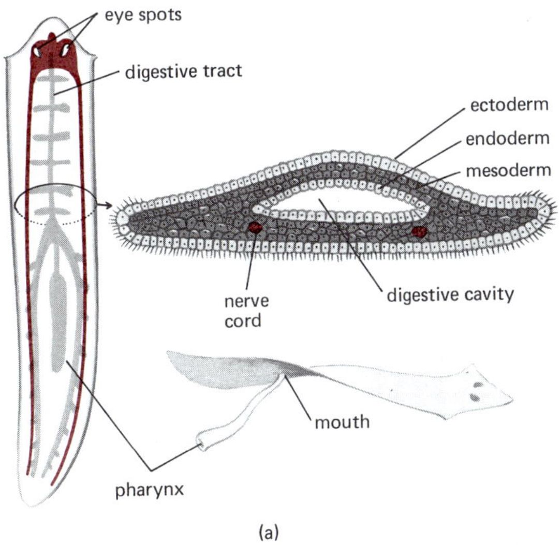

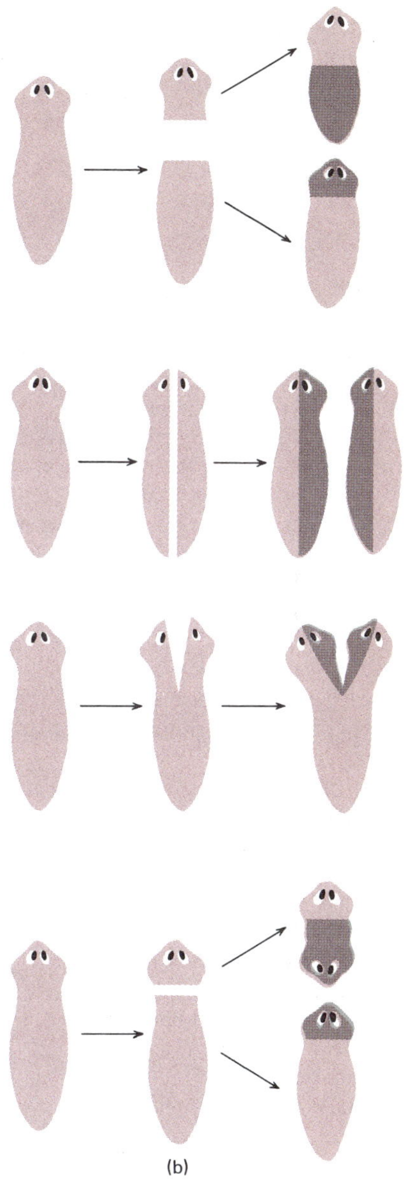

Figure 9–10 (a) Structure of a planarian flatworm. (b) The flatworm can regenerate lost body parts.

Planarians may also reproduce asexually by simple fission, the animal dividing into two parts by a transverse split just behind the pharynx. Each of the two segments then regenerates the missing parts.

The regenerative powers of the planarian are well known and are commonly studied in the laboratory. With proper handling, even a relatively small segment of a planarian can regenerate all the missing parts, yielding a complete new worm. Regeneration will occur whether the worm is cut longitudinally or transversely. Two-headed or two-tailed worms can be produced by splitting a planarian longitudinally only halfway (Fig. 9–10b). And if a flatworm is cut transversely just behind the sensory lobes, the head segment regenerates not a new body and a tail, but a new head, resulting in a strange double-headed monster.

Roundworms: Phylum Nematoda

Roundworms, members of the phylum **Nematoda,** may well be the most ubiquitous of all animals. Nearly every sample of soil or natural water has its nematode inhabitants: nematodes are found living within the roots, stems, and leaves of plants, and as parasites living within members of nearly all animal species. By far the majority of the more than eight thousand nematode species so far catalogued (it is believed that there are many more species not yet identified) are harmless saprophytes, or tiny organisms that prey on soil microflora. But a number of species are dangerous parasites upon man and his domestic animals and plants.

The trichina worm, *Trichinella spiralis,* lives as a parasite on hogs and is most prevalent in areas where hogs are fed on garbage. The immature larvae form capsules or encyst within the muscle tissue of the hog, and as such may be eaten by humans. A single ounce of badly infested meat may contain as many as a hundred thousand encysted trichina larvae. If pork is not cooked properly, the trichina cysts that it may be carrying are not killed. Instead, the digestive juices free the larvae, which complete their development within the intestines of their human host (Fig. 9–11). Here they mate and produce fantastic numbers of eggs, which are carried through the bloodstream to the voluntary muscles. There the embryos develop into larvae and encyst. The larvae die within a few months, but the cysts become calcified and remain. It has been estimated that nearly 10 percent of the United States population has at least a mild case of trichinosis, caused by these parasitic roundworms.

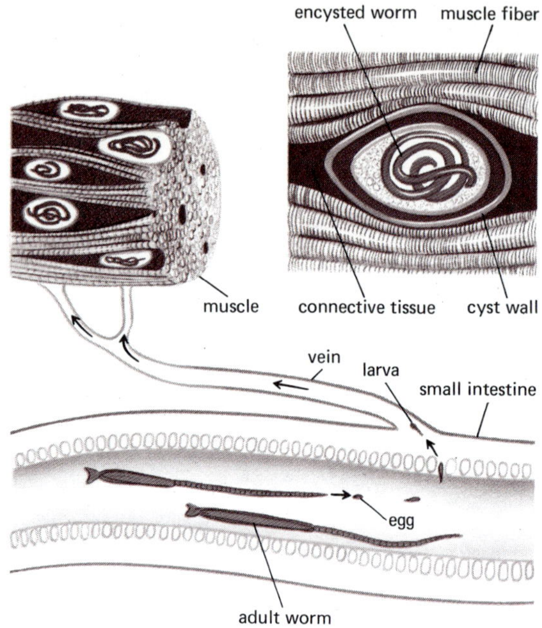

Figure 9–11 Life cycle of the trichina worm, *Trichinella spiralis.* Eating raw or improperly cooked infected muscle tissue results in digestion of the cyst wall, freeing the encysted worm to develop in the intestines. Eggs laid by adult worms develop into larvae, which migrate to muscle tissue and encyst.

Man as a host is rather a dead end for *Trichinella spiralis,* since there is generally no way for the cysts to be transferred to a new host (except perhaps among cannibals). But this offers little solace to the human trichinosis victim. In fact, in the case of extremely heavy infestations, trichinosis can be rapidly and quite painfully fatal.

Roundworms are bilaterally symmetrical, cylindrical worms, tapering at the ends like a spindle. Their resemblance to tiny threads is the rationale for the phylum name, Nematoda, which comes from a word meaning "thread." They have a basic **tube-within-tube body plan,** with the tubular digestive system situated within the tubular body (Fig. 9–12). At the anterior end of the digestive cavity is a mouth opening, and at the posterior end an anus. The roundworms thus have a one-way digestive tract, far more efficient than those

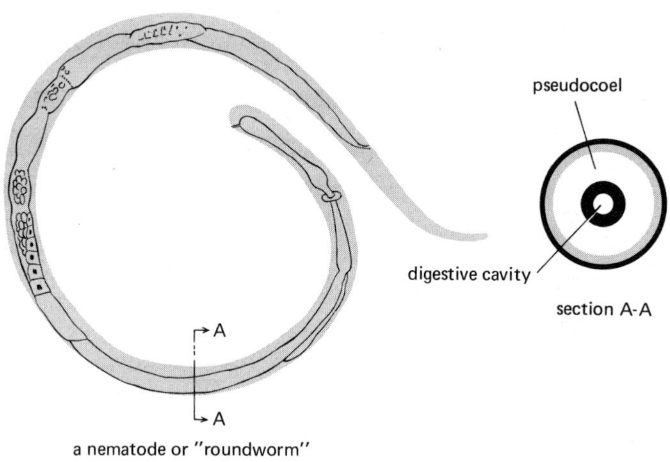

Figure 9–12 A typical roundworm. The cross section emphasizes the tube-within-tube body plan.

of the coelenterates and flatworms, in which the ingestion of food and the egestion (elimination) of undigested materials both occur through a single opening.

The nematodes are covered by a thick and rather rigid **cuticle,** which severely limits their movements. In liquid media they are capable only of whiplike lashings, but they manage to wriggle along fairly effectively over solid surfaces and within the tissues of hosts. Nematodes may also be carried about from place to place by the wind, in dust, and within animals and seeds.

The sexes are usually separate in nematode species, the first phylum in which this condition is the rule rather than the exception. The probabilities of suitable matings are enhanced by the tremendous fecundity of the roundworms, literally thousands of eggs being produced at a time. Thus, the likelihood that both male and female larvae will be taken in by a new host is considerable.

In addition to trichina worms, the roundworms include other important parasites of man and his domestic animals. Infestations by hookworms was at one time one of the most serious health problems in the rural areas of the southeastern United States. The roundworm *Ascaris* causes great damage as it burrows from one site in the body to another. The filarial worms, which are a serious health problem in tropical areas and require an intermediate host to complete their life cycles, are also nematodes. One of these worms is responsible for the condition of elephantiasis (Fig. 9–13). The Guinea worm, which coils beneath the skin, may reach a length of several feet. It is estimated that man is subject to infestation by at least fifty different species of nematodes; the dog can harbor thirty-six known roundworm parasites, the cat thirty-three, cattle more than fifty, and horses nearly seventy.

In addition, there are nematodes that parasitize many cultivated plants, sapping the

Figure 9–13 Parasitic worms blocking the lymphatic vessels cause the grotesque swellings of elephantiasis victims.

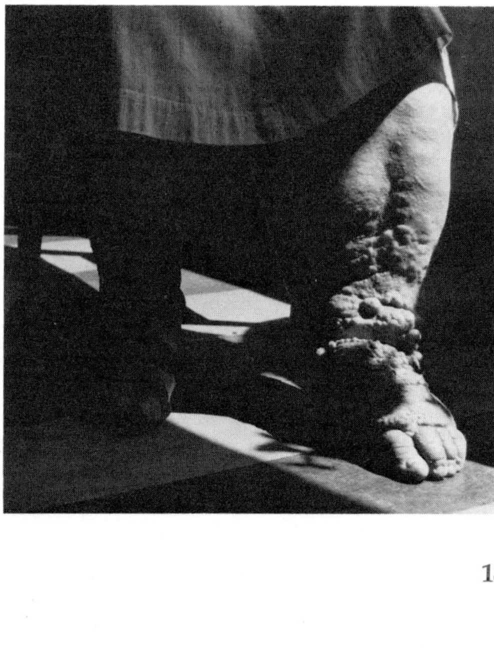

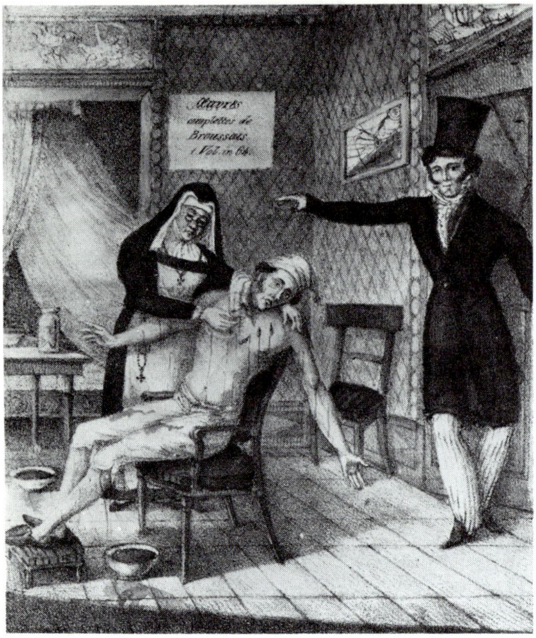

Figure 9–14 Bleeding with leeches was once a standard medical treatment.

vitality of the infested plants and severely reducing crop yields. Some of these are amazingly persistent. The wheat eelworm, *Anguina tritici,* which may produce as many as ninety thousand individual roundworms in a single gall (the characteristic root knot produced at the site of infestation), can remain dormant up to twenty years in a dry wheat kernel.

Segmented Worms: Phylum Annelida

Under the warm sunlight a large flock of sheep browses on the lush green grass of a pasture. Yet *beneath* this quiet scene feeds another "flock," far more numerous than those that can be seen. Literally millions of earthworms churn through the soil. Indeed, it has been estimated that the combined weight of all the earthworms living within the soil of such a pasture may be even greater than the total weight of the sheep that the pasture supports.

The common earthworm, *Lumbricus terrestris,* lives within the soil and has probably done more for human agriculture than all man's inventions and innovations. The earthworm literally eats its way through the soil. Earth is taken in at the anterior mouth. After passage through the worm's digestive tract (in the course of which food materials are absorbed by the worm), the earth, now mixed with fallen leaves, worm body wastes, etc., is ejected from the posterior anus. This "enriched soil" is discharged either underground or at the surface in piles of castings (worm feces). As the earthworm burrows through the soil, it makes the soil more porous and easier to cultivate. Charles Darwin, who made an extensive study of the habits of earthworms, estimated that earthworms, with their castings, can bring as much as eighteen tons of topsoil up to the surface of one acre of ground in a single year. In a period of a hundred years, this could amount to a layer of several inches of topsoil.

Not all **annelids—segmented worms**—live in the soil, by any means. In fact, most of the seven thousand or more species are marine or fresh-water dwelling animals. Such species as the sand worms, bristle worms, tube worms, and clamworms live in burrows or under rocks in the sea. Phylum **Annelida** also includes an important class of parasitic forms, the leeches or bloodsuckers. These possess special sucker modifications: the posterior sucker anchors the leech temporarily to its host, while the anterior sucker is used to suck the host's blood. The most famous of these segmented worms is *Hirudo medicinalis,* the "medicinal leech," which was commonly used by physicians as a cureall for common ills as late as the nineteenth century. The leech secretes a substance, hirudin, which inhibits blood clotting in the host and keeps the blood flowing freely, even continuing to ooze from the wound after the leech has been removed. Until the comparatively recent development of more effective anticoagulants, hirudin itself was used for medicinal purposes.

Like the other worms, the annelids have bilateral symmetry and well-developed organs and organ systems. They show a high degree of **cephalization**—concentration of nerve tissue in the head region. One of the most dis-

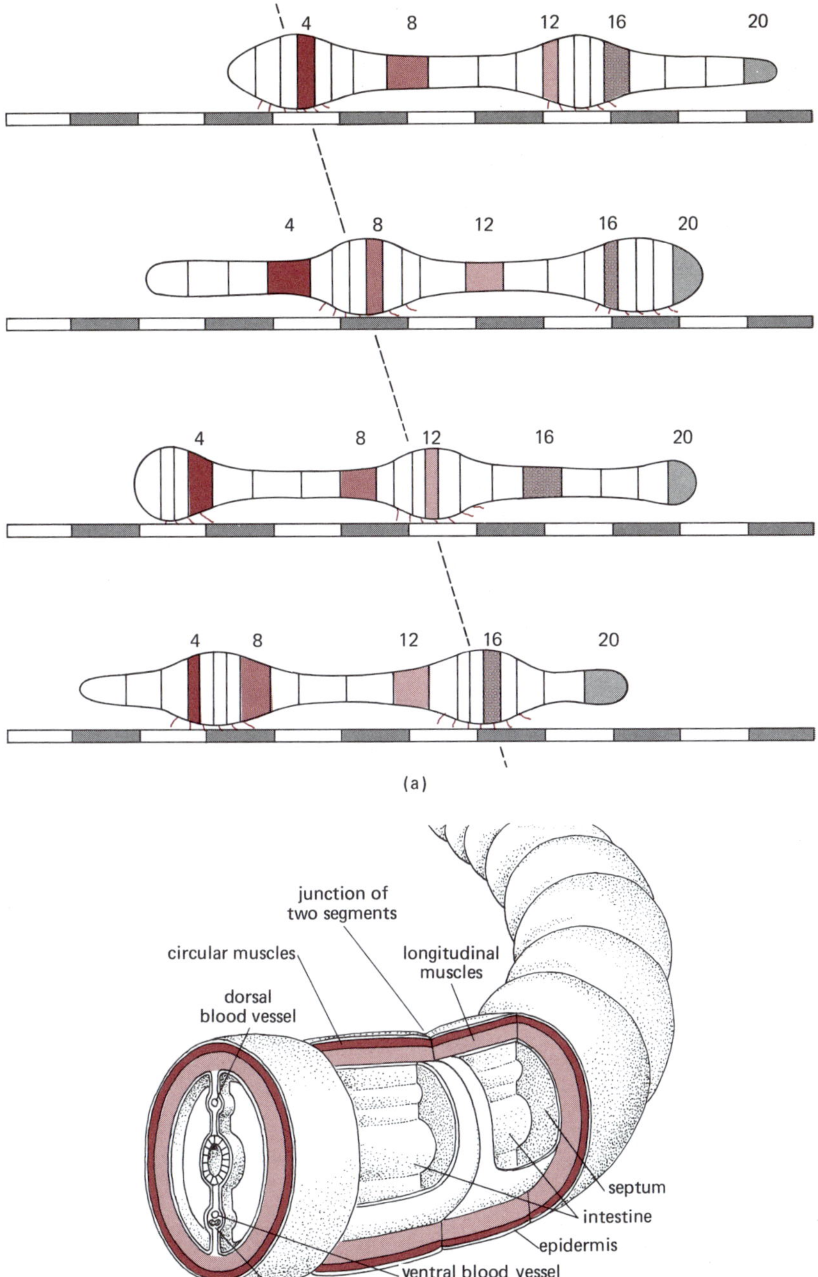

Figure 9–15 (a) Earthworm locomotion. (b) Cutaway view of earthworm body structure. Each segment of the earthworm's body is equipped with its own set of circular and longitudinal muscles and is capable of considerable stretching or contraction. The earthworm moves by successive waves of contractions first of circular then of longitudinal muscles, extending the anterior portion of its body and then effectively consolidating its gain. The bristlelike setae help to wedge the worm against the solid wall of the burrow. The setae protrude when the longitudinal muscles contract and are withdrawn when the circular muscles contract.

tinctive characteristics of the annelids is their **segmented body plan:** the body is partitioned into a series of separate compartments, with serial duplication of many of the organs (Fig. 9–15a). In some species there are paired appendages on nearly every segment; on others, such as the earthworm, there are small projections or bristles called **setae.** These appendages are used in locomotion, permitting the worm to get a better grip on its burrow. The segmentation of the earthworm and its well-developed system of circular muscles permit great flexibility. The earthworm can contract to a fraction of its usual length (with a concomitant thickening of its body, permitting it to wedge itself against the walls of its burrow) or it can stretch out to a much greater length than normal, moving through loose soil like a battering ram (Fig. 9–15b). The giant earthworm of Australia *(Megascolides australis)* is about three or four feet long and about two inches thick when fully contracted, but can stretch out to a maximum length of ten to twelve feet.

In the marine worms, the sexes are usually separate; but earthworms and leeches are generally hermaphroditic. Earthworms usually come out of their burrows to mate; copulation always occurs at night,* and usually during warm, moist weather. During copulation, the two earthworms align themselves with heads pointing in opposite directions and cling tightly together with special setae and sticky secretions, as each worm deposits sperm into the seminal receptacle of the other (Fig. 9–16). Following copulation, the worm forms a cocoon around the segments containing the reproductive organs. As the worm withdraws from the tubular cocoon, it deposits first eggs and then the sperm from its seminal receptacle. Fertilization occurs within the cocoon, and the fertilized eggs develop there directly into tiny earthworms.

Despite the fabled regenerative powers of the earthworm, they are generally inferior to those of the flatworm. Regenerative capacity varies among earthworm species, most being able to regenerate only a few head and/or tail segments.

*The habit of earthworms of coming onto the surface at night to feed on fallen leaves and other debris and to copulate is responsible for their common name of "night-crawlers." It is during these nocturnal sallies out of their burrows that earthworms are generally caught for fishing bait and for preserved specimens to be studied in the biology laboratories of schools and colleges. Combined sales in North America for these two thriving businesses run into the millions annually.

Mollusks: Phylum Mollusca

A garden snail inches along the ground, gliding on a trail of slime and carrying its coiled house upon its back. A soft-bodied, shell-less slug hides under a damp rock in the same garden. A torpedolike squid jets along through the water, occasionally loosing a cloud of inky fluid that masks its activities, permitting it to capture its confused prey and escape from baffled pursuers. A clam lives quietly near the low tide line, opening its double shell to feed when it is covered by the ocean waters and closing tightly or burrowing deeply into the sand when the waters retreat. Oysters are assiduously cultivated by man, to be served "on the half shell" to please his palate or to yield up glowing pearls to delight his eyes. Shipworms bore into the hulls of vessels and the wooden pilings of wharves and piers, causing great annoyance and financial damage to man. And a timid octopus, unmindful of his ferocious reputation, hides in a rock crevice, turning pale at the sight of a skindiver.

Figure 9–16 Earthworms come up to the surface to copulate.

William M. Stephens, Omikron

Figure 9–17 Mollusk shells make a beautiful and varied picture (a). Living gastropods and pelecypods often have a far different appearance: (b) garden snails and (c) the *Lima* clam or file shell. The primitive chiton (d) and the intelligent octopus (e), which has no external shell, are also members of the phylum.

The diverse organisms enumerated above represent a heterogeneous array indeed. Yet they are all members of the same phylum, **Mollusca.** Mollusks are a quite ancient group, and we have an excellent fossil record of them, extending back as far as 550 million years ago. The six classes of the phylum include about a hundred thousand living species and even more fossil forms. In numbers of representatives, therefore, this phylum is second only to the arthropods.

Mollusks are generally soft-bodied, unsegmented animals, characterized by a **mantle**—a glandular mass which hangs down over the body mass somewhat like a cloak (Fig. 9–18). In many of the molluscan species, the mantle secretes a hard, limy shell (single or double). In some species, such as clams and snails, the shell is external; in others such as the squid, a modified shell serves as an internal supporting structure. The cuttlefish, a close relative of the octopus, has an internal

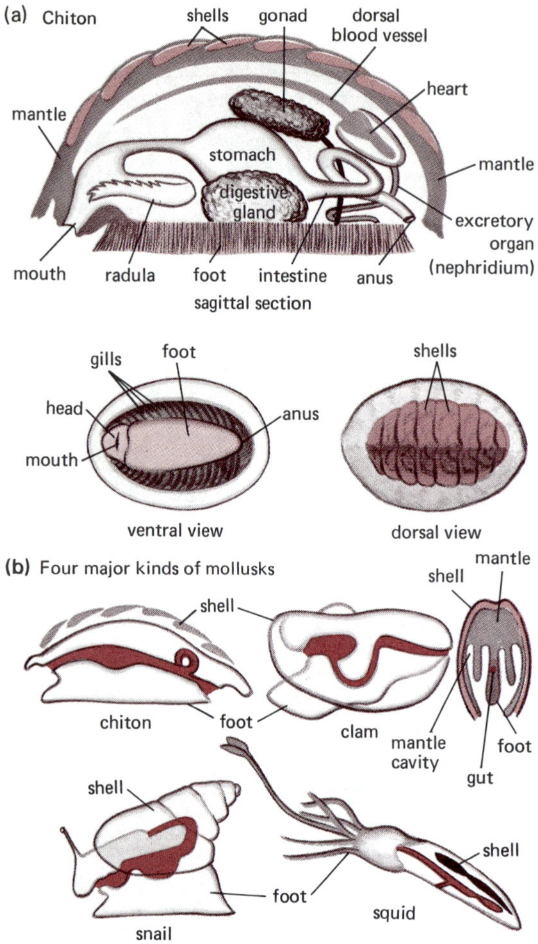

Figure 9–18 Molluscan body plan. (a) The structure of the chiton is repeated, with modifications, throughout the phylum. (b) Representatives of the four major classes of mollusks.

"cuttlebone," which is prized as a canary food supplement in view of its high lime content. Many mollusks also have a characteristic tonguelike rasping organ, the **radula,** which is used to rasp off bits of plant life so that they can be more easily swallowed.

Mollusks are characterized by bilateral symmetry, but in some forms, such as coiled snails, the internal organs are twisted into an asymmetrical arrangement.

The limy shells of the mollusks are quite varied and distinctive, as any shell collector knows. They may be single or double (the two halves of double shells being held tightly together with amazing strength by powerful muscles); mollusk shells may be flat, cuplike, tubular, or intricately coiled. The shelled species usually have a muscular foot that is used for locomotion. (In the octopus and squid, the foot is modified into tentacles.)

The most primitive class of mollusks, **Monoplacophora,** until 1952 was thought to consist only of fossil forms. But in 1952 ten living specimens were dredged up from the Pacific Ocean, near the west coast of Mexico, and revealed some interesting anatomical relationships among phyla. In general appearance, these creatures were clearly mollusks, but the internally segmented construction of their soft body parts showed close resemblances to the annelids.

Class **Amphineura** includes mainly the rather primitive chitons, tiny sand-colored organisms that bear an eight-plated coat of armor. They normally crawl along rocks on the seacoast, feeding upon algae. If pried from their firm hold on the rock, they curl up into tight little armor-encased balls.

Class **Scaphopoda** consists of a small group of mollusks that live encased in long tusk-shaped shells, open at both ends. A common example is the tooth shell, *Dentalium,* which spends its life buried deeply in the sand, with only the tip of its shell protruding into the water. Mollusks of this class are generally commercially unimportant, with one exception: the shells of the genus *Dentalium* were frequently used by West Coast Indians as money.

The large class **Gastropoda** (literally, "stomach foot") includes all the snails and their shell-less relatives, the slugs. They are often called **univalves,** because of their single shell. Except for the soft-bodied slugs (including maritime forms called nudibranchs), the bodies of gastropods are encased in hard, coiled shells. The pattern of coiling is usually quite characteristic of the species. Snails and other gastropods have broad, flat, muscular feet (one to an organism) and well-developed heads with two eyes, often at the ends of raised stalks. Except for a few carnivorous forms, such as the oyster drill, which incurs the ire of man by feeding upon economically valuable oysters, most snails and slugs are herbivorous. Snails are prized as gourmet delicacies by some humans, but this economic value is far overshadowed by their nuisance value in many areas, since they may do extensive damage to vegetable crops and also serve as intermediate hosts for some dangerous parasites upon man and his domestic animals.

The class **Pelecypoda** (literally, "hatchet foot") includes the **bivalves,** so called because the soft body is encased in two shells, each called a valve, held together by a ligament. The bivalves range in size from tiny species a few millimeters in diameter to the giant bear's paw clam, *Tridacna,* of the tropical Pacific waters, which ranges up to six feet long and may weigh up to five hundred pounds.

Bivalves generally live quietly at the bottom of seas or streams, often just at or below the low tide line, subsisting on food particles filtered from the water that they draw in under their folded mantles, between the two halves of the shell. Bivalves have virtually no cephalization and generally have few and rather inefficient sense organs. (An exception is the scallop, which has a row of eyes and tactile organs along the edge of its mantle.) Locomotion is generally slow and laborious among the bivalves, and is accomplished with a thick, muscular foot that is extended between the valves. The oyster does not even have such a foot; although the larval form is free-swimming, the adult oyster is entirely sessile. Again the scallop is an exception in the class: it can scoot along rapidly through the water by clapping its two valves together, resulting in a rather crude form of jet propulsion.

The various bivalves, especially oysters and clams, are highly valued by man, not only as food, but also for the pearls that they sometimes secrete. Few bivalves secrete the perfect spherical pearls that are used in expensive jewelry; even among the best "pearl oysters," found off the coasts of India, Japan, and northern Australia, hundreds of wild oysters must be opened to obtain a single good pearl. Pearls

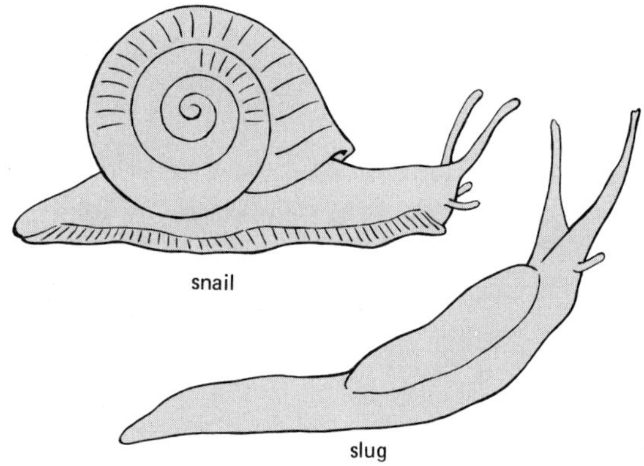

Figure 9–19 Snails and slugs, both gastropods, are extremely similar in structure, except that the slugs lack the coiled shells characteristic of snails.

are formed by the secretion of nacre (mother-of-pearl) around a foreign irritant that has somehow been introduced inside the shell, such as a grain of sand or some boring organism. The Japanese have set up a thriving industry growing "cultured pearls" in captive oysters. An irritant such as a grain of sand or a fragment of mother-of-pearl is introduced into an oyster, which is then lowered in a wire cage back into the oyster bed, there to start upon the five- to seven-year process of producing a good pearl. The mother-of-pearl that lines the inside of the shells of bivalves is also used extensively for making buttons and other ornamental objects.

Another common bivalve, with a negative economic significance, is the shipworm, *Teredo,* which is not really a worm at all. This bivalve uses its hard shell to bore into the hulls of wooden boats and wharves, causing much damage.

The last class of mollusks, **Cephalopoda** (literally, "head foot"), includes the octopus, squid, cuttlefish, and chambered nautilus. Only the latter has an external shell; the other members of the class rely for protection not upon armor, but upon a well-developed nervous and sensory system, with adaptations permitting keen alertness and speedy locomotion. They all have well-developed heads

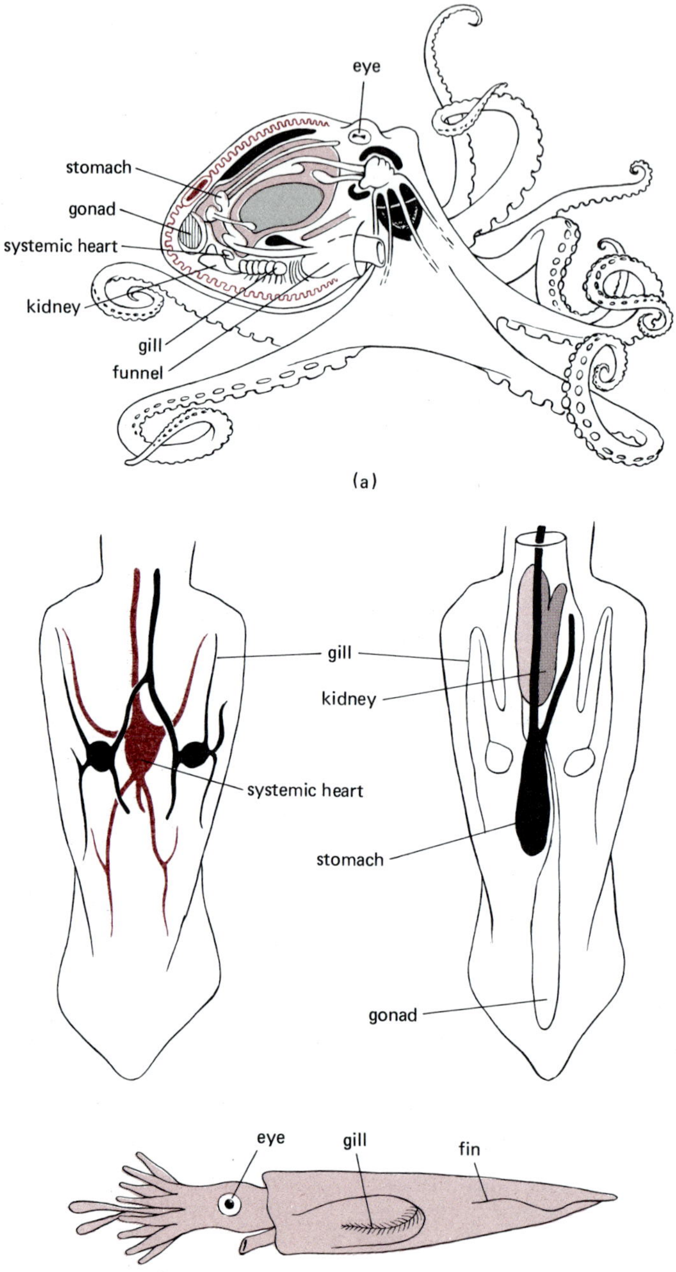

Figure 9–20 Structure of the octopus (a) and squid (b).

with large, efficient eyes. The mouth, equipped with powerful beaks or jaws and a strong radula, is surrounded by a ring of tentacles furnished with suction cups. These tentacles are used to grasp prey and bring it to the mouth. The octopus also uses its tentacles to clamber over the rocks of its home. But the octopus and squid also have a much more rapid mode of locomotion—by forcefully ejecting a stream of water from under the mantle. Squids and octopuses have rather a sophisticated adaptation for offense and defense: they can release at will a charge of inky particles that spread out to form a diffuse "smoke screen," as mentioned earlier.

Another interesting characteristic of cephalopods is their ability to change color according to their "moods." The body surface of the cephalopod contains many pigment-bearing cells, chromatophores. When the animal is alarmed, it changes color rapidly by changing the alignment of the chromatophores.

The ferocious reputation of the octopus is quite exaggerated. There is no evidence to indicate that they ever attack humans without provocation. (Though a human attacking an octopus might have a rough time of it, since the octopus outclasses him eight arms to two, and may in addition deliver a venomous bite.) Although giant octopuses up to twenty-eight feet in diameter have been reported, these are rare; the average octopus is a timid creature only a few inches wide.

Forms of reproduction vary greatly among the mollusks. In most species the sexes are separate, and even in hermaphroditic forms, such as some snails, cross-fertilization is the general rule. Some species eject clouds of sperm and eggs into the water; fertilization takes place externally, and there are no provisions for care of the enormous numbers of young produced, most of which perish. In other species, fertilization is internal.

In many gastropods and most cephalopods, there is some form of courtship preceding the actual mating. Care of the young is provided for by the more highly developed species, such as the mother octopus, who stays with her eggs all during their development, carefully cleaning and aerating them. She takes her responsibility so seriously that she usually does not eat at all during this period, which may be as long as six weeks.

In some species, such as the octopus, the eggs develop directly into small duplicates of the parents. In other species, there are intermediate larval forms. The sessile species especially are generally characterized by free-swimming larvae.

Echinoderms: Phylum Echinodermata

To a bather chancing to encounter a starfish stranded in a tidal pool on the beach, this creature is an object of beauty and fascination. Yet to an oysterman, starfish are bitter enemies. For starfish, like many humans, consider oysters and other bivalves a delicacy and destroy commercially valuable beds of such bivalves. Various methods have been used to control the starfish pests, the simplest being to collect them by trawling and destroy them. Up to a few generations ago, oyster fishermen practiced what they thought was an effective method of destruction: they chopped captured starfish into pieces and threw the parts back into the ocean. But later study of the habits of the starfish revealed that it has marvelous powers of regeneration: even a single arm may be able to regenerate all the missing parts and produce a new organism. The ill-advised oystermen therefore, far from destroying the pests, were actually rather effectively propagating them. Now other means of destruction are used, including chemical treatments of starfish habitats.

The six thousand or so **echinoderms** (their name means literally "spiny skin") are all marine animals. They include the familiar five-pointed starfish (or sea stars, as they are sometimes more aptly called), the brittle stars, sea lilies, sea cucumbers, sea urchins, and sand dollars. In addition to their disastrous effects upon commercial oyster, clam, and other bivalve beds (a scallop will be observed

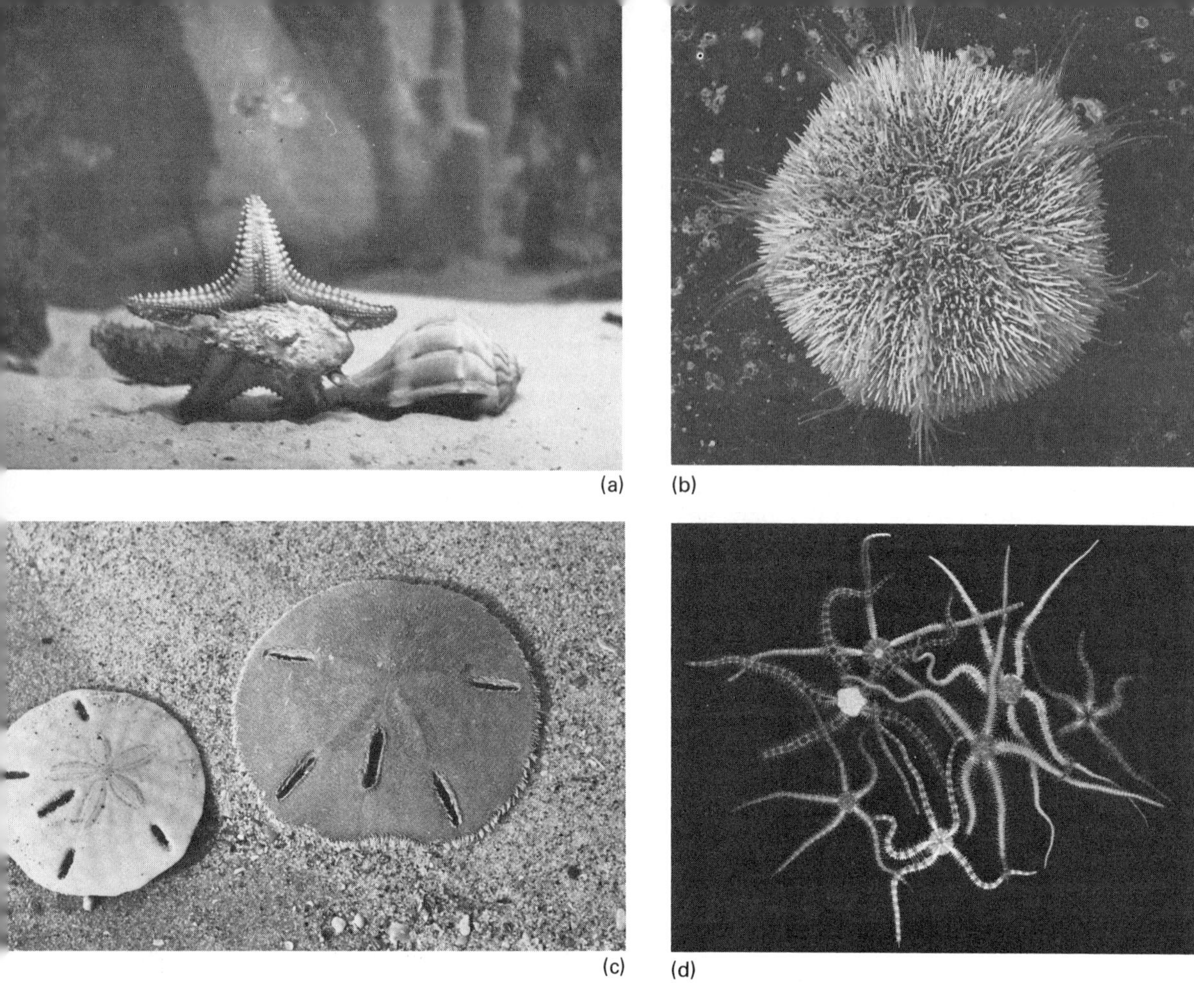

Figure 9–21 Echinoderms: a starfish feeds on a dead spiny boxfish (a); a sea urchin, viewed from above (b); sand dollar skeleton and living sand dollar (c); common brittlestars (d).

to swim rapidly away at the approach of a starfish), starfish serve as useful scavengers along the ocean bottom, and keep the populations of various mollusk forms down. The brittle stars, also abundant, are less well known because of their tendency to hide under rocks and inside sponges, coming out in search of food generally at night. They have long slender arms, and move by thrashing them about. When in danger, the brittle stars can detach their arms—hence their name. Sea urchins and sand dollars are round, without the characteristic raylike arms. But the sea urchin especially lives up to the phylum name of "spiny skin," rather resembling an animated pin cushion. Sea cucumbers, on the other hand, have soft, leathery bodies, rather than the spiny skin typical of their relatives. The sessile, plantlike sea lilies are also members of the phylum.

Echinoderms are generally bottom dwellers. Some are sessile, but most move about slowly with the aid of an unusual **water vascular system,** in which water is forced in and out of tiny **tube feet** by hydraulic pressure (Fig. 9–22).

One might at first be inclined to think that the taxonomists have made a mistake in considering the headless, brainless, radially symmetrical echinoderms as *higher* invertebrates. Yet an examination of the typical echinoderm life cycle reveals the justification for such a classification. The free-swimming larvae of echinoderms are bilaterally symmetrical and in fact so closely resemble the larvae of the acorn worm, the most primitive chordate, that

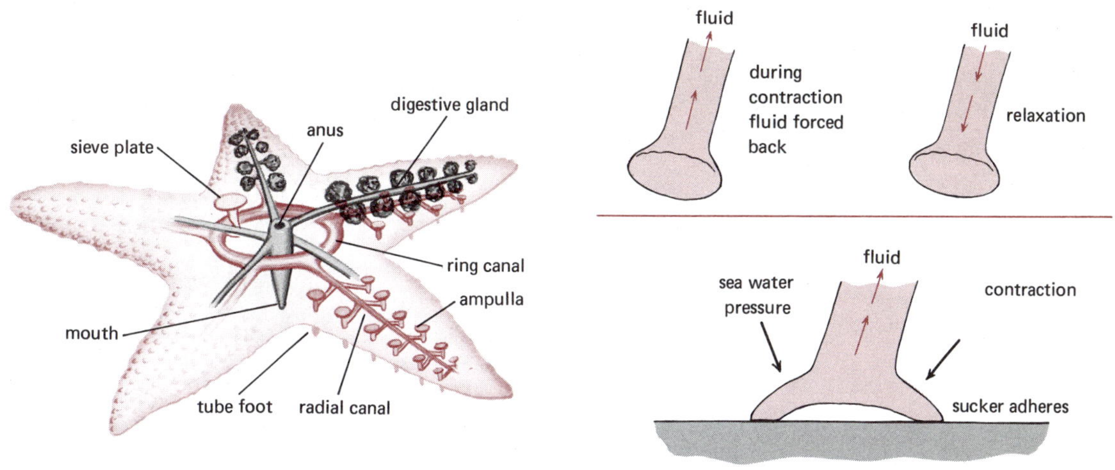

Figure 9–22 The starfish moves about with the aid of a water-vascular system. Sea water enters through the sieve plate and passes down the radial canal of each arm into the hollow tube feet. The ampulla, a muscular sac at the top of each tube foot, contracts to force water into the tube foot. The digestive system of the starfish, shown in black, includes a ventral mouth and a dorsal anus. In feeding, this echinoderm can evert its stomach through its mouth opening to digest clams and other prey in their shells.

it is difficult to distinguish between the two. The prevailing view among evolutionists today is that both the chordates (including man) and the present-day echinoderms are descended from some bilaterally symmetrical free-swimming form something like the echinoderm larvae. The characteristic radial symmetry of the adult echinoderms and the very different characteristics of the chordates probably represent the evolutionary results of adaptations to very different environments.

Most echinoderms have both an **exoskeleton** (external skeleton), formed of calcareous projections, varying from small lumps to long spines, and a calcareous **endoskeleton** (internal skeleton), formed of separate plates.

The tube feet of starfish and other echinoderms are used not only for locomotion, but also for grasping. They can exert incredible force: by attaching its suckerlike feet to the shells of bivalve mollusks and exerting a slow and steady pull, a starfish can pull the two shells apart. It then proceeds literally to turn its stomach inside out, extending it into the shell of the mollusk and digesting its contents.

Sexes are generally separate in the echinoderms. A mature starfish discharges millions of eggs or sperms into the water. Fertilization takes place externally. Because of the large numbers and easy accessibility of echinoderm eggs and sperms, as well as certain interesting characteristics of their early embryonic development, they have been extensively studied and experimented on by embryologists. The bibliography of research conducted on a single sea urchin species fills an entire volume.

Arthropods: Phylum Arthropoda

If earth were visited by an unbiased extraterrestrial observer, this observer might at first have difficulty determining the dominant form on the planet. After rejecting the abundant but passive plants, the alien might search among the animals. Likely his criteria of dominance might include not only sheer numbers of organisms, but also the extent and diversity of their adaptations to the varied environs of this world. Quickly his attention would be riveted upon the ubiquitous **arthropods.** In sheer numbers they overwhelm the earth. (Nearly 80 percent of the known animal species of this world are arthropods.) They

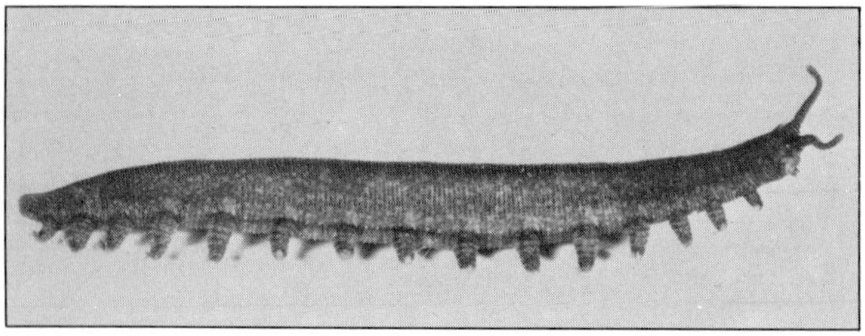

Figure 9–23

(b)

Figure 9–24

(c)

(a)

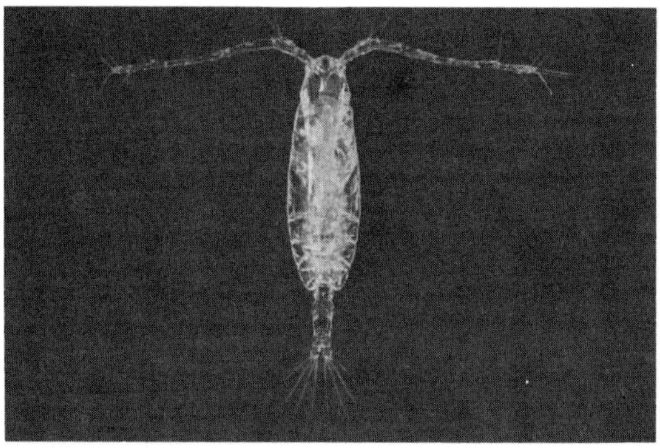

(d)

Figure 9–25

(a)

(b)

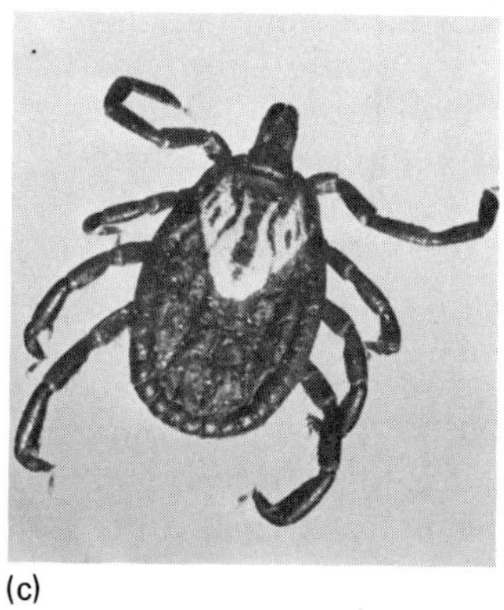

(c)

(d)

(a)

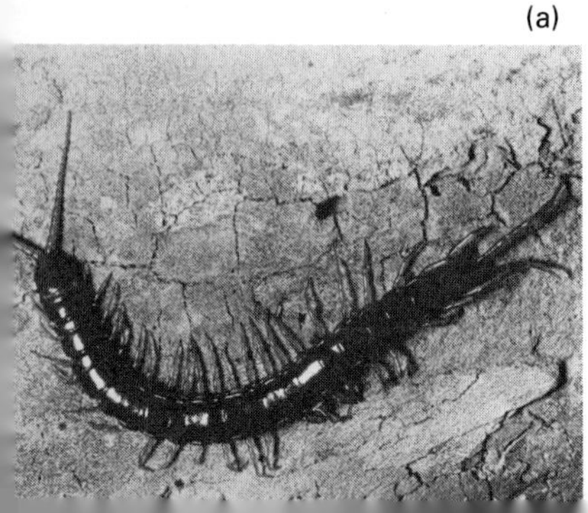

(b)

Figure 9–26

(a) Lynwood M. Chace, Omikron

Figure 9–27

(b)

(c)

Figure 9–28

(a)

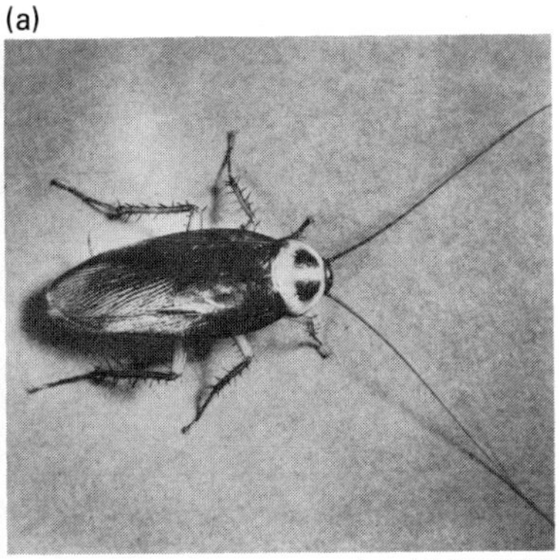

Lynwood M. Chace, Omikron

(b)

(c)

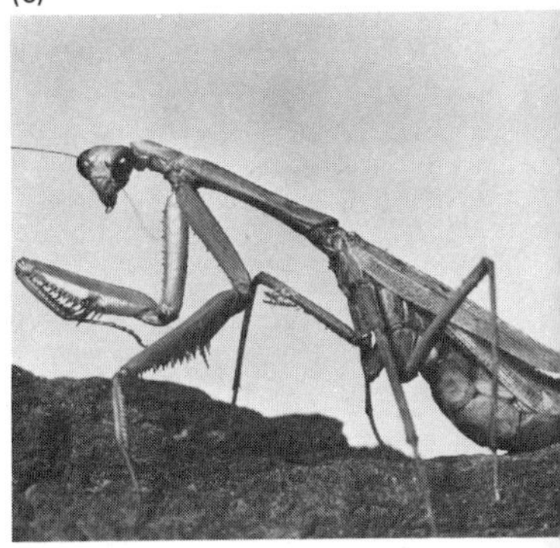

Figure 9–29

Figure 9–30

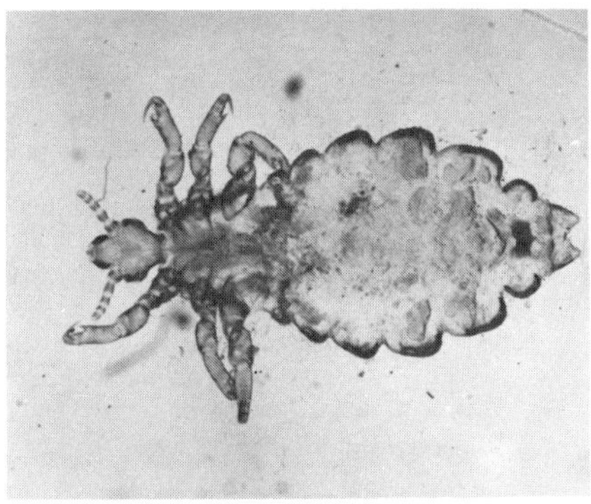

Figure 9–31

Figure 9–32

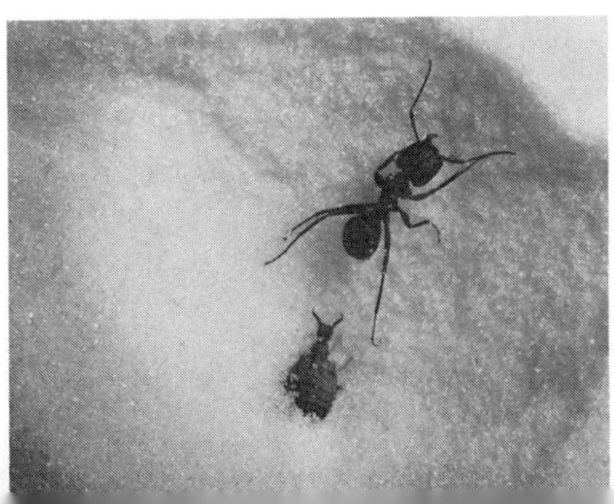

(a) (b) *Lynwood M. Chace, Omikron* (c)

Figure 9–33

(b) (c) *Lynwood M. Chace, Omikron*

Figure 9–34

Lynwood M. Chace, Omikron (a)

Figure 9–35

(a) (b)

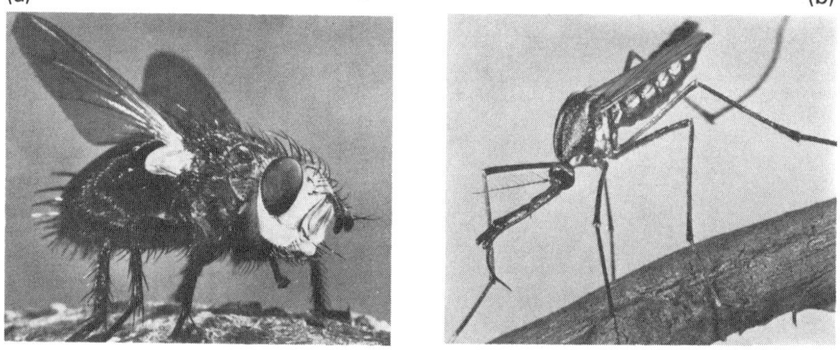

Figure 9–36

(a) (b) (c) *Lynwood M. Chace, Omikron*

Figure 9–23 The "walking worm," *Peripatus*.
Figure 9–24 Crustaceans: a copepod, *Acartia clausi* (a), from the plankton, a key link in the food chains of the sea; buoy-making barnacles, *Lepas fascicularis* (b), settled down to a sedentary life; a ghost crab (c) with eyes on stalks and the right claw larger than the left; a crayfish (d) with a pair of sensitive antennae.
Figure 9–25 Arachnids: the orange garden spider, *Argiope aurantica*, biting a viceroy butterfly that has just been wrapped in silk (a); the deadly scorpion, *Centruroides sculpturatus* (b); the dog tick before feeding (c) and a female tick engorged with blood (d), who has dropped off her host and is about to lay eggs.
Figure 9–26 The centipede (a) has one pair of legs per segment, while the millipede (b) has two.
Figure 9–27 Insects: the primitive silverfish (a); the mayfly (b) will live just one day in its adult form; the dragonfly (c) forms a basket with its legs to catch its prey on the wing.
Figure 9–28 Orthoptera: cockroach (a), grasshopper (b), and praying mantis (c).
Figure 9–29 Termites live in communities, with a division of labor among different structural types. Note the two soldier termites in the picture, with large heads and pincer jaws. The insects' pale bodies will never be exposed to light unless by accident if their nest is opened by a predator or natural disaster.
Figure 9–30 The human body louse.
Figure 9–31 This giant water bug has captured a fish.
Figure 9–32 An ant slides helplessly down the sand as the ant lion that dug the pit waits below.
Figure 9–33 Coleoptera: the stag beetle (a), potato beetle (b), and cotton boll weevil (c).
Figure 9–34 Lepidoptera: (a) at left a silver-spotted butterfly and at right a monarch butterfly; (b) a luna moth; (c) close-up of the head of the tomato sphinx moth, showing the long coiled proboscis used to sip nectar from deep-throated flowers.
Figure 9–35 Diptera pests: the fly (a) and mosquito (b).
Figure 9–36 Hymenoptera: harvester ant (a), bumblebee (b), and mud wasp (c). The mud wasp builds a nest of mud, then captures and paralyzes spiders, which she seals in the nest as food for the young wasps that will hatch from her eggs.

occupy every conceivable niche and pursue incredibly diverse modes of life. Arthropods are to be found in the depths of the sea, high in the air above, and on all the lands of the world. Even their closest rival, man, might be looked upon as their slave. He has built houses to shelter and feed cockroaches, termites, and other insects. He has planted fields of crops for beetles, weevils, locusts, and other arthropods to feed upon; he has even improved these crops to make them more nutritious. Man's attempts to throw off the yoke of slavery have been in vain. He has tried noxious sprays and other means in desperate efforts to break his bondage to his arthropod overlords. But to no avail. The masters of the planet have learned to tolerate man's puny tantrums. Individual arthropods might perish, but Arthropoda live on.

The nine hundred thousand or so known arthropod species fall into many classes. **Onychophora** is a small class consisting mainly of the "walking worm," *Peripatus* (Fig. 9–23). This caterpillarlike creature lives exclusively in the Southern Hemisphere and seems well on the way to extinction. The species is significant because it combines a strange mixture of annelid and arthropod characteristics, and thus is a kind of "missing link" between the two phyla, living evidence of their evolutionary relationship.

A much larger and more important class is **Crustacea,** which includes the lobsters, crabs, crayfish, shrimps, barnacles, and water fleas (Fig. 9–24)—generally aquatic forms (an exception is the pill bug, which lives under moist logs and stones). Many of the large crustaceans are eaten by man; some of the smaller ones are important links in the food chains of salt- and fresh-water fishes and whales.

Class **Arachnida,** which includes spiders, scorpions, ticks, mites, and king crabs, is another arthropod class of great interest to man (Fig. 9–25). Spider silk is, weight for weight, the strongest known material on earth. Spiders and some other arachnids also possess poison glands with which they paralyze their prey. Spiders aid man in killing many harmful insects, but they can also inflict painful stings if he gets in their way. In the case of the black widow spider, the venom is so potent and is injected in such large amounts that death may result. Scorpions can also be dangerous, especially to children. Mites and ticks are frequently annoying parasites, and ticks may also transmit dangerous diseases, such as Rocky Mountain spotted fever.

The stereotype of the many-legged crawler is well fulfilled by **Chilopoda,** the centipedes, and **Diplopoda,** the millipedes (Fig. 9–26). There are several important differences between the two classes. Centipedes ("hundred legs") have a single pair of legs attached to each segment of their long bodies; in the case of millipedes ("thousand legs"), each body segment is actually two fused segments, and has two pairs of legs. Centipedes are generally carnivorous and prey upon various insects; millipedes are herbivorous, and some species do damage to cultivated plants.

The largest arthropod class is **Insecta,** the insects. This class, with well over six hundred thousand living species, is generally divided into about two dozen orders, including: **Thysanura,** such as the silverfish to be found under old logs in the forest or dining luxuriously on the paste holding the wallpaper to your walls; **Ephemeroptera,** the mayflies that sometimes create a nuisance to man with their swarming but are valuable sources of food for fish; **Odonata,** dragonflies and damsel flies, which help to keep the mosquito population down; and **Dermaptera,** earwigs, which—contrary to popular belief—do not crawl into people's ears and bite them with their powerful pinchers.

Another important insect order is **Orthoptera** (Fig. 9–28), a diverse group including such friends and foes as cockroaches (some species of which lead an idyllic existence as nonpaying guests of man); grasshoppers and locusts, which feed on man's crops, as well as mantids, which prey upon them; camouflaged walkingsticks and leaf insects, which look amazingly like parts of the plants upon which they feed; and the musical crickets.

Isoptera, the termites, live in complex societies and feed on cellulose; they are of great interest to man, in a negative sort of way, because of their great predilection for the wood frames holding up his buildings (Fig. 9–29).

Anoplura, the sucking lice which feed on man and his domestic animals, and **Homoptera,** including leafhoppers, cicadas, aphids, and scale insects, many of which feed on man's crops, are equally disliked. **Hemiptera,** including water boatmen, back swimmers, giant waterbugs, bedbugs, chinch bugs, and stink bugs, is another insect order containing a number of "public enemies" (Fig. 9–31).

In somewhat better repute in human circles are the members of **Neuroptera,** including lacewings and ant lions. The larvae of lacewings are called aphis lions because of their practice of voraciously gobbling up all the aphids in their path. Ant lions are also predacious: they dig pits in the loose sand and lurk at the bottom, waiting for an unwary ant to fall into their traps.

Another important order, **Coleoptera,** includes beetles, weevils, and fireflies, among others (Fig. 9–33). This enormous order contains about 40 percent of all the known insect species. Many beetles and their larvae feed voraciously on plant leaves and other plant parts, causing great damage to crops. Other species perform the useful functions of scavengers, feeding on decaying material. Some beetles, such as the ladybug, are predacious, feeding on harmful insects such as aphids and scale insects.

Many humans collect varied and beautiful specimens of the order **Lepidoptera,** the butterflies and moths (Fig. 9–34). A number of these insects are also of considerable economic importance: the larvae of some feed on crop plants, trees, and fabrics; the silkworm, on the other hand, produces a useful product, silk. Many flowering plants are pollinated by butterflies or moths.

Siphonaptera, the fleas, and **Diptera,** including flies and mosquitoes, are two more classes of great concern to man (Fig. 9–35). In addition to their painful and annoying bites, these insects may also be carriers of disease. Bubonic plague, the dread black death that periodically swept through Europe during the Middle Ages, sometimes decimating whole areas, is transmitted by fleas parasitic on rats. Flies may transmit anthrax, tularemia, and typhoid fever and may play host to various protozoan and roundworm parasites. Mosquitoes may pass on malaria, yellow fever, encephalitis, and filariasis.

Another major insect class, **Hymenoptera,** includes ants, bees, and wasps, many of which live in complex social organizations (Fig. 9–36). These insect societies generally consist of a queen, a large fertile female that lays eggs for the colony; males, whose sole function is to mate with the queen; and various castes of specialized workers, sterile females who care for the all-important queen and for the young, gather and process food, and perform all the other necessary tasks for the group. All the ants are social insects; but among the bees and wasps, there are many solitary species—in fact, the social species are actually in the minority. Bees perform an important function in the balance of nature, pollinating many important flowering plants as they travel about gathering nectar and pollen, which they transform into honey and other products.

The arthropods share a number of structural features with many of the other invertebrates. They possess bilateral symmetry, organ systems, a segmented body, and a well-developed head with efficient sense organs.

But arthropods also possess a number of unique features. The most important of these is their **chitinous exoskeleton:** the arthropod body is enclosed in a tough, jointed case made of the organic polymer **chitin.*** The presence of an exoskeleton has a number of important ramifications in the structure and function of arthropods. The presence of a tough outer covering provides a number of advantages. Chitin is a material with great mechanical strength and is impervious to numerous chemical agents. It protects the arthropod from bruises and infections and also helps to restrict evaporation of body fluids, thus permitting successful adaptation to a terrestrial environment. The chitinous exoskeleton provides support for the body organs and permits more efficient action of the muscles, which are attached to the inside of the exoskeleton.

But as the knights of old discovered, a suit of armor possesses a number of disadvantages along with its obvious advantages. First of all, it is heavy. There is a definite physical limit to the weight of armor that a creature of a given size can carry around. Moreover, even with jointed armor, there is a certain loss of maneuverability. (Medieval knights in full armor had to be lifted onto their horses. Once unhorsed in the heat of battle, they could not remount unaided and usually had to continue the fight on foot.) The chitin armor of the arthropods thus places definite limits on the size that they can attain. Aquatic forms, such as lobsters, crayfish, and shrimp, can grow somewhat larger than their terrestrial relatives, because the buoyancy of the water helps to support some of the weight of their exoskeletons. But even these rarely exceed a length of two feet or so. Insects, the major terrestrial forms, are never larger than eleven inches, and most are an inch long or less. Flying insects are usually even smaller than the crawling forms. (So perhaps we need not ever worry about giant wasps and other insect monstrosities dreamed up in science fiction horror movies!)

A tough exoskeleton also has another important disadvantage. It does not allow for growth. Since the size and shape of the exoskeleton remain essentially fixed, if an arthropod is to increase in size, it must **molt**—shed its old exoskeleton and grow a new one. Arthropods generally molt from four to seven times before they reach adult size. During the transition periods an arthropod is soft-bodied and extremely vulnerable to its environment.

Another unique and important feature of the arthropods is the presence of segmented

*The crustacean armor is additionally reinforced by deposits of lime in the chitin.

appendages. (The name of the phylum, Arthropoda, literally means "jointed feet.") In fact, to the casual observer, a major characteristic of the "crawly creatures" seems to be the many mobile legs. Insects have three pairs of jointed legs, spiders four pairs, crustaceans five pairs, and some of the multilegged species such as centipedes and millipedes may have up to 115 pairs. The segmented appendages of the arthropods show a great variety of adaptations. In addition to the legs, there are usually antennae and sometimes wings. The various appendages, including the legs, may be adapted not only for locomotion, but also as sensory organs ("feelers"), offensive and defensive weapons (as in the claws of lobsters and beetles), and tools (such as the bee's specialized legs, used as baskets for pollen gathering).

Another uniquely arthropod characteristic is the **compound eye.** The compound eye is multifaceted, each tiny facet contributing a view from a slightly different angle of vision, overlapping to some degree with the image provided by the facets around it, all together making up a mosaic image. Arthropods generally have a pair of eyes, each of which may consist of thousands of units.

The sexes are generally separate among the arthropods, and bisexual reproduction is common. But **parthenogenesis,** the development of young from unfertilized eggs, is also sometimes observed. In most species, fertilization is internal: the sperm are generally transferred to receptacles within the abdomen of the female and retained there until the eggs are about to be laid. A few species retain the developing eggs within their bodies and give birth to living young. But most lay eggs, usually in enormous numbers, and embryonic development takes place outside the body of the mother. In some species, such as grasshoppers, cockroaches, and aphids, the young that emerge from the eggs are miniature replicas of their parents. But in many species, such as bees, ants, moths, and flies, there are intermediate larval forms, often quite different in appearance from the adults (Fig. 9–37). The radical changes from larva to adult comprise the phenomenon of **metamorphosis.** Frequently most of the active growth occurs in the larval stage.

Chordates: Phylum Chordata

If the hypothetical extraterrestrial visitor who decided that arthropods rule the earth were actually to try to contact these "rulers," he would be in for a surprise. For he undoubtedly would soon be noticed by a member of a quite different phylum and politely (or not so politely) informed that he had made a mistake.

For numbers are not the whole story. Though outnumbered by far, the forty thousand or so species of **chordates** have a number of characteristics to recommend them as the dominant phylum on earth. On the average, the individual members of this phylum are much larger than the individual invertebrates. Even the giant squid, which sometimes reaches a length of 60 feet, is hardly a match for the great blue whale, which is more than 100 feet long and weighs a good 150 tons. And even a mouse is much larger than most insects.

Chordates can also challenge the arthropods in diversity of habitats and niches. Chordates are to be found in the depths of the oceans, in the midst of parched deserts, burrowing in the ground, and soaring through the air. In addition, one chordate, man, not only has shown an ability to adapt to a diversity of environments, but has demonstrated an ability to adapt a diversity of environments to his own needs. With a superior ability to reason and solve problems, man is winning dominance not only over the physical world about him, but also over the other species that share this world with him. His is the power to shape, to change, and to destroy.

Like some of the invertebrates, the chordates are bilaterally symmetrical and possess some degree of segmentation. But a number of unique features set them apart. All chordates at least at some time in their lives possess a flexible, cartilaginous rod running longitudinally down the dorsal portion of the body. This is the **notochord,** from which the name of

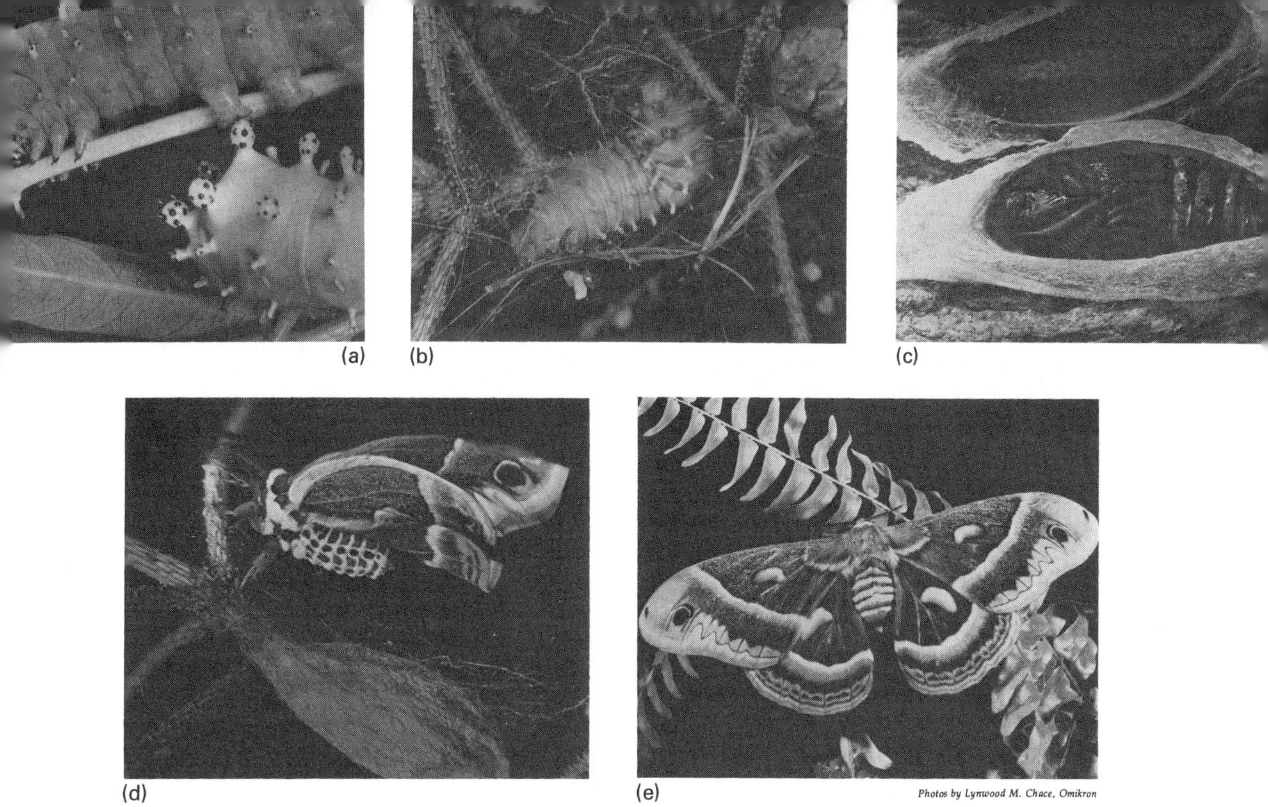

Figure 9–37 Life cycle of a moth: (a) the caterpillar of the Cecropia moth clings to a twig; (b) in the fall the caterpillar spins a silken cocoon about itself; (c) a cocoon cut open to show the pupa: protected from the cold of winter, the caterpillar was metamorphosing into a moth and would have emerged in the spring; (d) when the moth first emerges from the cocoon, its wings are soft and very fragile; it clings to a twig until they dry and harden; (e) an adult Cecropia moth.

the phylum is derived. Some primitive chordates retain the notochord throughout their lives. But in the vertebrates, including fishes, amphibians, reptiles, birds, and mammals, the notochord fuses into the bones of the vertebral column or backbone long before the animal reaches maturity (Fig. 9–38).

The **endoskeleton** made of living tissue, be it the cartilaginous notochord of the primitive chordates or the bones of the vertebrates, is a key characteristic of this phylum and one of the major reasons for its success. A light and flexible yet strong endoskeleton, providing a framework of support for muscles and body structures and permitting growth without shedding, allows for much greater body size than the heavy and restrictive arthropod exoskeleton.

Another typically chordate characteristic is the possession, at some stage of development,

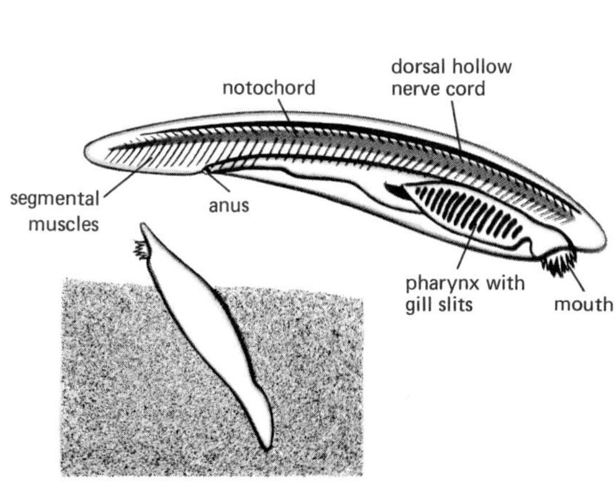

Figure 9–38 Amphioxus, a primitive chordate, leads a sedentary life as a filter feeder, half buried in the bottom sand. It is frequently studied in the laboratory as a key "step" on the evolutionary ladder.

of pairs of **gill pouches,** outpocketings of the lining of the throat passage and matching inpocketings of the body surface. In most aquatic forms, these break through to form gill slits, which provide for an efficient respiratory system: water passes in through the mouth, over the gill filaments, where gas exchange takes place, and out through the gill slits. In terrestrial chordates, the gill pouches are vestigial, appearing as such only during embryonic development. The adult forms breathe with lungs.

The chordates possess a dorsal tubular nervous system, enlarged at the anterior end to form a **brain.** This system has proved to be the most efficient nervous system among all the animal phyla.

Phylum Chordata includes several subphyla in addition to the familiar Vertebrata, which includes man. At the boundary line, and somewhat in dispute among taxonomists as to whether it should really be included among the chordates, is **Hemichordata.** This small group includes the primitive acorn worms, which combine a number of chordate characteristics with other characteristics that seem more appropriate to the echinoderms (Fig. 9–39).

Definitely chordates, but also quite primitive, are the members of the subphyla **Cephalochordata** and **Tunicata.** The former is represented by the tiny fishlike creature amphioxus, which is a perfect "model chordate" and is commonly studied as such in biology laboratory courses.

The tunicates are sessile marine animals, which are named for the celluloselike covering that surrounds their bodies like a tunic. Tunicates are frequently called sea squirts, from their habit of squirting streams of water from external body openings when they are disturbed. The baglike adult tunicates seem to resemble primitive sponges far more than chordates. But their larvae are much more characteristic of the phylum, rather resembling frog tadpoles and possessing a well-developed notochord in the tail (Fig. 9–40).

It is the subphylum **Vertebrata** which contains most of the animals with which we are familiar. Tiny goldfish and fierce sharks, hopping frogs and toads, crawling snakes, turtles and lizards, chickens, ducks, and songbirds, and all the mammals, from pekinese to tiger, from opossum to elephant, from mouse to man—all are vertebrates.

All of the thousands of vertebrate species share unique features setting them apart from all the "invertebrates." All have a **vertebral column,** a segmented backbone that has replaced or surrounded the primitive notochord. This is made of flexible cartilage in the lower vertebrates and of bone in the higher forms. Through the vertebral column runs the dorsal nerve cord, the **spinal cord.** At the anterior end of the spinal cord, the enlarged brain is protected within a bony cranium, or **skull.**

Another typically vertebrate characteristic is an internally jointed endoskeleton of bone or cartilage, which serves as a supporting framework, to the outside of which the muscles are attached. Vertebrates have a well-developed ventral heart, with two, three, or four chambers, which pumps red blood, carrying hemoglobin in red blood cells, through a circulatory system of arteries, veins, and capillaries. Other important characteristics are an **integument,** consisting of dermis and epidermis, along with scales, feathers, or hair; two pairs of appendages; and a tail, extending beyond the trunk. This tail is absent in some species, such as apes, men, and guinea pigs.

Figure 9–39 An anatomical model of the acorn worm, *Dolichoglossus.*

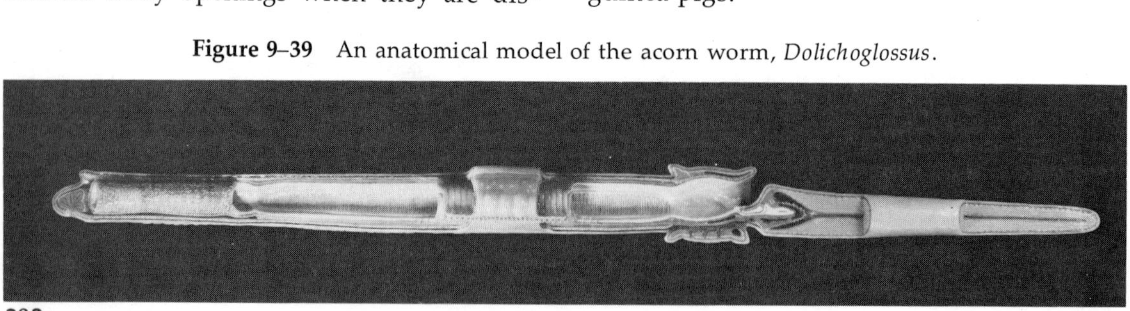

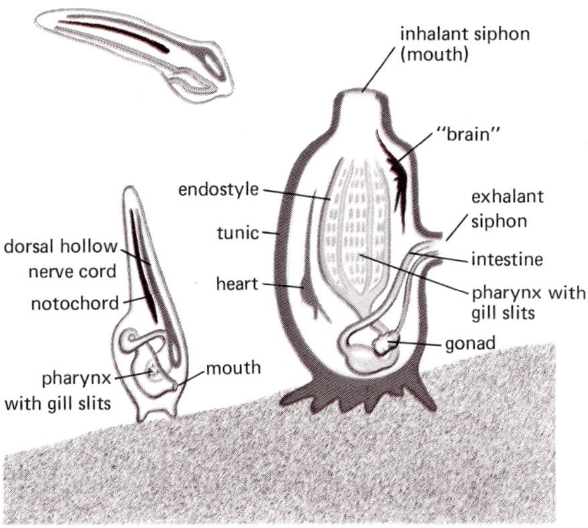

Figure 9–40 The tunicate, *Molgula manhattensis*, pictured in the photograph, seems to have an invertebrate structure. But the baglike adult develops from a tadpolelike larva with a notochord.

FISH: SUPERCLASS PISCES

It has been predicted that the great farms of the future will be in the sea, and among the great "crops" that are raised and harvested will be the fish that live therein. For this and other reasons, scientists and food industrialists are showing increasing interest in the habits and habitats of the fish that live in the oceans and other waters of the world.

Fishes show a number of important advances over the invertebrates. Backbones appear for the first time in fishes. Jaws and paired fins are important portents of things to come in the evolution of the vertebrates. The vertebrate kidney appears in the fish, which also possess a highly developed nervous system and a high degree of cephalization. Even nostrils and air-breathing lungs appear in a few fish species, along with various specialized adaptations such as organs that generate electricity.

The fish are a varied group, comprising several classes. The living (i.e., nonextinct) representatives of class **Agnatha** are the cyclostomes, jawless elongated, eellike fish with neither paired fins nor scales. These include the hagfishes, exclusively marine, which are mainly scavengers but may also attack living fish, and the parasitic lampreys. The latter are jawless creatures with a round sucking mouth equipped with a rasping tongue. They attach themselves to other fish by their suction-cup mouths, rasp a hole in the fish's flesh, and suck its blood. These unpleasant creatures spend most of their adult lives in the sea, but migrate into fresh water to spawn. The eggs hatch into amphioxus-like larvae, which for a few years live quietly in nests in the bottom mud, feeding on organic debris. Then they metamorphose into adult lampreys and make their way back to the sea, wreaking havoc upon the fish population as they go. Although lampreys are eaten as a delicacy in some areas, they destroy far more food sources for man than they provide.

After the construction of the Welland Canal around Niagara Falls in 1829, the Great Lakes were opened to the sea lamprey *Petromyzon marinus*. Gradually the insidious pests began to move in upon this rich new territory. More and more often the lake trout and other commercially important fish were found with lamprey scars. Soon the fish yield began to drop precipitously, and the once mighty lake trout industry was virtually wiped out. Studies of the lamprey's habits have indicated that it is most vulnerable in its upstream migra-

Figure 9–41 Fish: (a) These fish bear the scars of bloodsucking lampreys, primitive jawless fish. The tiger shark (b) and roundel skate (c) are both cartilaginous fishes. The bony fishes show great variety: (d) a school of herring streams past a silver pompano; (e) the gaudy lionfish comes from the South Pacific; (f) the tiny neon goby lives with the huge moray eel and swims fearlessly into the big fish's mouth, scavenging parasites and food particles lodged around the eel's teeth; (g) the sea horse combines an unusual appearance with an off-beat life style: the female deposits her eggs in a pouch in the male's abdomen and he incubates them and cares for the young.

tions to its spawning grounds. Dams and electrified fences have been constructed in an effort to stop the lampreys in their migration, but these have proved expensive and not very effective. Recent discoveries in the area of chemical control offer hope that selective chemical agents can be used to kill off the young lampreys without harming the other, more desirable fish species.

The second major class of fish, **Chondrichthyes,** is an extremely ancient class that includes the sharks, skates, rays, and chimaeras. These fish have no bones: their entire skeletons are made up of cartilage. (This group is therefore frequently called the **cartilaginous fishes.**) They have five or more pairs of flat, platelike gills, and their gill slits are visible from the outside. Their characteristic gills are the source of another alternative name for this group, the **elasmobranchs** (derived from words meaning "plate" and "gill"). Unlike the cyclostomes, they do possess jaws and paired fins, and their bodies are covered with characteristic toothlike scales.

The elasmobranchs are mainly marine animals; the few fresh-water species live in areas that were originally part of the oceans but were cut off by geological upheavals. Elasmobranchs are carnivorous. Sharks especially are renowned for their rapacious habits and are frequently called the "wolves of the sea." They are strongly attracted to blood in the water, and there have been many authenticated reports of fatal attacks on humans. Sharks are eaten by humans in many areas (usually under another name, with more innocuous connotations), and their large livers are valued for their high content of oil and vitamin A.

The flattened skates and rays generally feed on mollusks and crabs. Some have peculiar modifications, such as the stingrays, with a sting at the base of the tail, and electric rays, with specialized electric organs.

Most of the fish with which we are familiar belong to the third, and presently largest, class of fishes, **Osteichthyes.** These fish are frequently referred to as **bony fishes** (the literal translation of their name), in view of the fact that their skeletons are constructed mainly of bone. One feature which readily distinguishes them from the cartilaginous fishes is the presence of the **operculum,** a cheeklike plate that covers the gills. Another distinguishing feature is the presence of overlapping flat scales that cover their bodies. Most of the bony fishes possess a **swim bladder,** which is used to regulate their buoyancy so that they can remain at any particular depth in the water without a constant expenditure of effort.

The bony fishes are a diverse group indeed. They are found both in fresh water and in the seas. They vary in size from tiny fish less than half an inch long to ocean giants weighing more than a ton. Some are torpedo-shaped, others slender and eellike, or flattened like a flounder, or resembling round balls. Some of them possess unique adaptations. Luminous fish are common, with varied colors and arrangements of "living lights." Flying fish can leave the water and glide through the air for short distances. The swordfish carries with him an enormous spearlike battering ram, which can even impale a boat. Electric organs are frequently observed. Some fishes use them to deliver powerful electric shocks that stun their prey; others generate weaker electric currents and appear to utilize them for orientation as a sort of electrical radar system. A curious developmental adaptation is observed in the flounder. This animal starts life as a quite normal-appearing fish, but as it grows, it begins to assume a sidewise orientation in swimming, with one side gradually becoming the dorsal surface, and the other the ventral surface. The eye on what is now becoming the underside of the fish then migrates to the dorsal side, and the adult flounder is able to lie flat on the bottom, with both eyes surveying the water above for prospective meals.

A small but important group of bony fishes developed nostrils and lungs and once were able to spend short periods on land. Modern representatives of this group are the lungfishes and the coelacanths. (The latter were thought to have been extinct for more than 70 million years until a living specimen was caught off the east coast of Africa in 1938. Since then more than a dozen specimens have been caught in the deep waters north of Madagascar.) It was this branch of the bony fishes that is believed to have developed into the first land vertebrates, the amphibians.

AMPHIBIANS: CLASS AMPHIBIA

With the population explosion looming more and more threateningly upon the horizon, increasing attention is being paid to research on ways to better utilize the resources of our planet. More and more interest is being centered upon the seas, which cover a far larger fraction of the globe than the inhabitable land areas. Man has even begun to dream that he himself may some day make his home in the sea, observing its wonders with his own eyes and harvesting the rich yields of vast farms within the ocean depths.

If man and other land animals do some day succeed in inhabiting the sea, they will be reversing a process that is believed to have occurred some 350 million years ago. At that time, a period of intermittent drought, it is believed that some primitive lungfish-type creatures crawled and dragged themselves from pools that were drying out, over the land, to pools that still retained some water,

Figure 9–42 Life cycle of the frog.

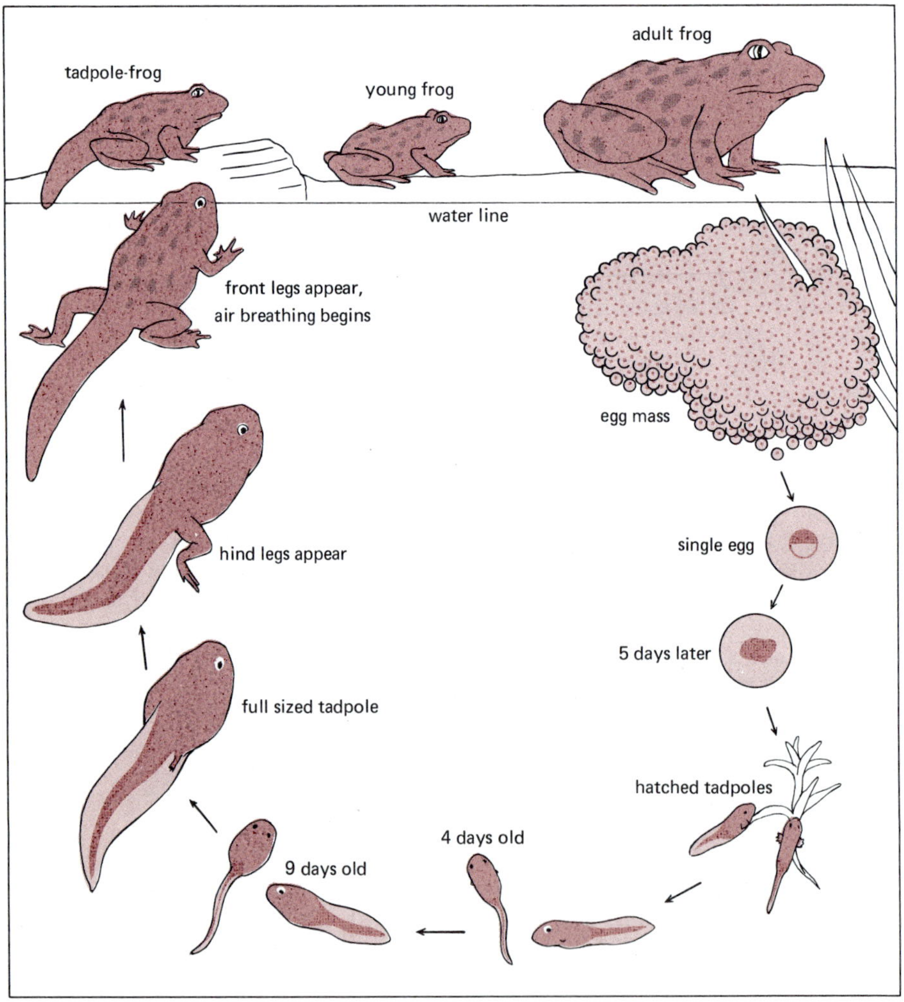

(a) (b) (c)

(d)

Figure 9–43 Amphibians: (a) bullfrog leaping; (b) green tree frog; (c) male Fowler's toad singing at the edge of the breeding pool; (d) spotted salamander.

and thus survived. These then gave rise to the first true **amphibians,** animals with habits partly aquatic and partly adapted to life on land.

The modern amphibians generally begin life as aquatic tadpoles, which breathe with gills, swim with a **caudal** (tail) **fin,** and are herbivores. These fishlike tadpoles eventually change or metamorphose into four-footed carnivores, which breathe with lungs and can live on the land (Fig. 9–42). Yet these adult amphibians are not entirely free of the water, by any means. Their naked skin (with no fish-like scales or other protective covering) is generally soft and moist, and body fluids are readily lost through it. Most amphibians therefore remain close to a water source or in moist locations, where there is no danger of the body drying out. (This need to guard against dehydration also explains the nocturnal habits of many land-based amphibians.)

Amphibians still retain another tie to the water: they must return to a body of fresh water to spawn.* Fertilization is generally external, requiring an aqueous medium.

The approximately three thousand species of presently living amphibians are placed by taxonomists in three orders: **Urodela,** the salamanders and newts; **Anura,** the frogs and toads; and **Apoda,** a small group of caecilians —legless, wormlike forms that burrow in the damp soil of tropical and subtropical forests (Fig. 9–43).

*A few amphibian species lay their eggs in damp places, rather than in actual bodies of water.

The salamanders and newts retain a tail throughout their lives. They have two pairs of tiny, inefficient legs, which cannot support the full weight of the animal. Each leg ends in a clawless foot; the forelegs are four-toed and the hind legs five-toed. There are two main types of salamanders. One type, typified by the mud puppy *(Necturus)* of the eastern United States, is an exception to the usual amphibian pattern and never undergoes the metamorphosis from gills to lungs. These salamanders live an exclusively aquatic life. The other type does undergo metamorphosis, exchanging lungs for the gills of the immature form and spending the adult life on the land. A typical example of the latter type is the tiger salamander, *Ambystoma tigrinum.*

The frogs and toads—tailless amphibians—comprise the other important order of class Amphibia. The tails of the gill-breathing tadpoles are lost in the course of development to the adult forms. The bodies of frogs and toads are much shorter and more squat than those of salamanders. They generally have well-developed hind legs and much smaller front legs. In the water, frogs propel themselves along with the characteristic "frog kick"; on the land they generally move about by hopping. Frogs and toads usually sit on their haunches rather than standing as four-footed animals do, thus taking the weight off their rather inadequate forelegs. Like the salamanders, they have clawless feet, with four toes on each foreleg and five on each hind leg.

Frogs and toads are important in keeping down the number of insect pests, which they capture by lightning thrusts of their adhesive, protrusible tongues.

REPTILES: CLASS REPTILIA

The first invaders of the land, the amphibians, were successful only to a limited degree. Lungs permitted them to take in life-sustaining oxygen from the air, and paired legs permitted them to move about on the solid surface of the land. But the amphibians still retained many ties to the water. A line of their descendants, the **reptiles,** were much more successful in their conquest of the land. In fact, at one time reptiles, especially dinosaurs, were masters of the earth. Giant herbivorous reptiles browsed on the lush vegetation and were preyed upon by equally titanic reptilian carnivores. Reptiles radiated into every available habitat and niche on the land; some even took to the air, and others reinvaded the seas.

A number of structural and physiological developments in the reptiles helped to eliminate some of the drawbacks inherent in the amphibian pattern and effectively emancipated the reptiles from their dependence on the water. Lungs and legs, the important amphibian innovations, were retained (except in some legless forms, such as the snakes). The three-chambered heart of the amphibian was modified into a four-chambered heart, which made for an even more efficient circulatory system. But among the most important reptilian innovations were the ones in the area of water retention. Instead of a soft, moist, water-permeable outer covering, the reptile has a dry, scaly, watertight skin that greatly retards water loss by the body. Another water-saving improvement in the reptilian design is in the physiology of excretion. In fish, nitrogen-containing wastes are excreted in the form of ammonia, a rather poisonous substance, which must be greatly diluted with water to avoid ill effects to the organism. This presents no great problem for the fish, which lives totally immersed in water. In amphibians, nitrogenous wastes are disposed of in the form of urea. This substance is not toxic, but it is highly water-soluble, and a great deal of water is needed to wash out the urea. In most reptiles, however, nitrogenous wastes are excreted as uric acid. This compound is relatively insoluble in water. Thus, water can be resorbed and the wastes neatly deposited as little piles of solid uric acid. (This solution to the water problem is also utilized by the birds, and was independently arrived at by other animals living in dry habitats, such as insects and land snails.)

A third important breakthrough on the water-emancipation front is connected with the mode of reproduction in reptiles. First of all, fertilization takes place internally, so that no external watery medium is needed for the mobile sperm. Furthermore, the eggs are provided with a rich supply of yolk, which serves as an energy source for the developing embryo, and after fertilization the eggs are encased in a leathery, waterproof shell that prevents their drying out. Such eggs, therefore, need not be laid in the water, but can instead be deposited in sand or soil.* The eggs develop into miniatures of the adult forms, without an intermediate gill-breathing larval form.

The sixty-five hundred or so species of living reptiles are a small remnant of the class that once ruled the earth. The living species are classified in three major orders: **Chelonia,** the turtles; **Squamata,** the lizards and snakes; and **Crocodilia,** the crocodiles and alligators (Fig. 9–44).

Turtles and tortoises are generally toothless, and their bodies are enclosed in a protective shell. The ribs and vertebrae of these

*Among some reptiles the eggs are retained in the body of the mother until they hatch.

Figure 9–44 Reptiles: (a) box turtle; (b) scarlet king snake (c) a ctenosaur, or false iguana; (d) an American alligator.

reptiles are greatly enlarged and incorporated into the hard, bony shell. The heavy shell has resulted in great modification of the basic body plan and has meant a definite sacrifice of speed and mobility. But the turtle or tortoise can generally entirely withdraw its head, legs, and tail into the cavity of its rigid shell and thus present an armored fortress to any enemy.

The term *turtle* is usually applied to aquatic forms and *tortoise* reserved for land forms, but there is much confusion in the usage. Marine turtles have fins instead of feet, and some have lost the ability to withdraw their appendages into the shelter of their shells. Both marine and land forms must lay their eggs on the land. The eggs are generally buried in sand or soil and left unattended to hatch.

Various forms of turtles are found in fresh- and salt-water habitats and in moist woods and gardens. The common garden turtle *Terrapene* feeds on insects and vegetation; the snapper *Chelydra* is a rather vicious carnivore. The largest turtles are the sea turtles *Dermochelys,* the leatherbacks, which reach a length of seven feet and may weigh over a thousand pounds. The largest land forms are the giant Galápagos Island tortoises, *Testudo,* which sometimes weigh as much as five hundred pounds and may exceed four feet in length. Turtles are noted for their longevity (and a concomitant slow rate of growth). Ages of over 150 years have been recorded for some turtles.

The lizards and snakes are built for speed rather than armor. Most lizards have well-developed legs; they are superficially similar to the salamanders, but differ externally from them in that their bodies are covered with scales and their feet are clawed. Snakes, with their slender, streamlined bodies, can slide swiftly through grasses and underbrush. Lizards differ from snakes in the possession of movable eyelids and external ear openings. Snakes, on the other hand, possess sensitive tongues that serve as olfactory organs.

At first one might wonder how such different-seeming creatures as lizards and snakes could be classed in the same order. The rationale for this apparent paradox becomes more evident if we consider the species of legless lizards and some primitive forms of snakes with vestigial hind legs.

Snakes and lizards may be found in varied habitats, from dry deserts to moist forests, fresh-water ponds and streams, and oceans. Most of the lizards are beneficial to man, since insects are their principal diet. Only the Gila monster of the southwestern United States and Mexico is dangerous to man. Some lizards have rather easily broken tails, and can readily leave them behind in the jaws of a predator, twitching violently, while the rest of the lizard beats a hasty retreat. The tail is then regenerated.

Despite their sinister reputation, most snakes are harmless. Legless, they move along by undulating movements of their bodies, either over solid ground or in the water. They feed on insects, small rodents, or fish, depending on the species. Snakes can swallow prey much thicker than their own slender bodies by means of an unusual jaw that can be dislocated temporarily, permitting an enormous gape. Some snakes, such as the pit vipers (copperhead, water moccasin, and rattlesnake) and the coral snake can inflict poisonous bites; a dose of the venom may even be fatal.

Figure 9–45 A snake can temporarily dislocate its jaw to swallow prey thicker than its own body. Here a corn snake is swallowing a white-footed mouse.

The largest living reptiles are the crocodiles and alligators. They are primarily aquatic forms, quite awkward on land but agile in the water. The crocodile and alligator are quite similar, but can be distinguished by the shape of the head: the alligator has a broad head with a rounded snout, whereas the crocodile has a triangular head, a more pointed snout, and protruding teeth. In addition, the crocodile tends more to prefer salt or brackish water.

Two representatives of the order Crocodilia are found in the United States. The American alligator, *Alligator mississippiensis,* is found from North Carolina to the Texas coast. The American crocodile, *Crocodilus americanus,* is found in southern Florida and the Florida Keys.

Though most alligators and crocodiles are timid creatures, living quietly on fish and water birds, some species, such as the African crocodile, are said to be dangerous man eaters.

Alligators and crocodiles, like most other reptiles that have returned to an aquatic life, still retain the reptilian adaptations to life on the land. They must surface periodically to breathe, and they return to the land to lay their eggs.

BIRDS: CLASS AVES

Man has always been fascinated by birds. To be able to fly—to soar through the air, free of the bonds of his everyday life—has been a human dream far longer than the span of recorded history. In the Icarus legend, an audacious man fashioned birdlike wings in an attempt to unlock the secrets of flight, but paid dearly for this knowledge by the loss of his beloved son. The Renaissance genius Leonardo da Vinci, who also dreamed that some day man might take to the air like the birds, left us detailed sketches of prototype aircraft.

(a) (b) (c)

(d) (e)

Figure 9–46 Birds: (a) brown pelicans on their nesting island; (b) a mallard duck swims with her brood; (c) screech owl; (d) black-capped chickadee; (e) male ostrich.

In addition to the sheer miracle of flight, other aspects of birds have long fascinated and intrigued man. The regular departure and return of bird species, timed to the seasons; their uncanny accuracy as navigators, traversing sometimes thousands of miles, always along precisely the same route, returning to precisely the same destination; the melodious songs of birds, which may have inspired man's first musical efforts; and their often brilliant plumage, which man appropriates to adorn himself—all these have been objects of wonder.

Birds, class **Aves,** are believed to have evolved from forms of bipedal reptiles that took to the air some 125 million years ago. Like their reptile ancestors, birds reproduce by internal fertilization and lay shell-encased eggs, within which an amnion encloses a watery medium and a rich supply of yolk—the energy source for the developing embryo.

But birds show a great number of far-reaching modifications from the reptilian pattern, modifications basically associated with adaptation for flight. Except on the legs, the reptilian scales have been modified into feathers, light and strong. These provide larger surfaces for support by air currents, yet minimum weight, and in addition serve as effective heat-conserving insulation. The bird skeleton, too, is a masterpiece of design for maximum strength with minimum weight. The hollow, tubular bones are strategically braced, strutted, and honeycombed with bony fibers. Instead of heavy jaws, the bird has a horny bill, which can serve simultaneously as part of the eating apparatus and for fine manipulations. The skeleton of a bird is incredibly light for its size: for example, the skeleton of a frigate bird, which has a wingspread of seven feet, weighs a mere four ounces—even less that the weight of its feathers!

The limbs of a bird are also highly modified. The anterior pair is modified into feather-covered wings, operated by powerful breast muscles. The posterior pair of scale-covered legs, with clawed feet, is variously modified for perching, swimming, or running, depending upon the species.

Even the birds' streamlining of body shape and stripping away of all conceivable excess body weight would be ineffective in flight if it were not for their amazing system of energy conversion—a system of unparalleled efficiency. Birds consume high-calorie foods and can transform an unusually high fraction of the energy taken in into usable forms. They must feed voraciously to keep their metabolic mills churning busily, and yet can lay down stores of high-energy body fat that can last them through long nonstop migratory flights.

Unlike the reptiles, the birds are "warm-blooded," or, **homoiothermic**—they maintain a constant body temperature independent of the environment. The body temperature is maintained at 107° to 113°F, depending upon the species. The mechanisms for heat manufacture and conservation—the insulating feathers and amazingly efficient energy metabolism, together with an extremely rapid heartbeat, exceeding 1500 per minute in some small species—are balanced by the most efficient known natural cooling mechanism. In addition to the usual lungs, birds possess a number of air sacs in various parts of the body, and a large fraction of the inspired air is circulated for cooling purposes. (The air sacs also greatly enhance the efficiency of the respiratory gas exchange, thus further promoting an efficient energy metabolism.)

The internal organs of birds are unusually compactly arranged within the body cavity, again with a pruning of all possible excess weight. There is no bladder for temporary urine storage (nitrogenous wastes are excreted in the form of solid uric acid), and feces are not retained in the lower intestines, as they are in mammals. The four-chambered heart pumps blood through an extremely efficient circulatory system.

The sensory organs of birds are also highly modified for a life of flight and the special problems of spotting potential food sources while soaring through the air. The olfactory apparatus is greatly reduced, but the eyes are enormous. In some birds, the eyes weigh even more than the brain. The head is freely movable; some species can rotate their heads nearly full circle. The eyes of birds can focus efficiently on both near and distant objects, and they have excellent color vision. In addition to two outer movable eyelids, a thin third eyelid, the **nictitating membrane,** can be drawn over the eye, permitting the bird to follow objects across the background of the sun and providing protection for the eyes during dust storms.

Reproductive practices among the birds are far more complex than among the reptiles. Elaborate, instinctive courtship rituals characteristically include a highly vocal song. There is a considerable degree of parental care for the young, from the preparation of some sort of nest in which to lay the eggs to incubation of the hard-shelled eggs and feeding of the helpless young from the time they hatch until they are ready to function on their own.

Though the major structural and physiological adaptations of birds are associated with their ability for flight, bird species have radiated into nearly all conceivable habitats. Some live on the ground and have entirely lost the ability to fly. The puny wings of the ostrich, for example, are quite inadequate to support its massive bulk. Other birds have come to live on or even in the water. The penguin, for example, has wings modified into finlike structures for swimming; but penguins return to land to lay their eggs and raise their young.

Bird species have quite varied dietary habits. Some feed mainly on seeds and green shoots (as any farmer who has ever erected a scarecrow can testify). Many bird species are important agents of biological control, helping to keep down the population of numerous insect pests. Birds of prey, such as eagles, hawks, and owls, feed on various small animals such as rodents and, in some cases, other birds.

Man also finds some birds (and their eggs) good eating. Various species of domesticated fowl such as chickens, turkeys, ducks, and geese are raised as a major source of food for man.

MAMMALS: CLASS MAMMALIA

About 150 million years ago, the earth was dominated by ponderous reptiles. Dinosaurs of all description lived, ate, fought, and laid their eggs in the most varied habitats—in the marshes, plains, and forests of the world. Giant reptilian sea serpents ruled the seas, and other reptiles were soon to take to the air.

Figure 9–47 Mammals: (a) the flying squirrel—really a glider rather than a flier; (b) an African elephant calf nursing; (c) the star-nosed mole, adapted for a life underground; (d) the harbor seal, adapted for a life underwater; (e) an English setter with her pups; (f) a rhesus monkey stores food in two cheek pouches; if danger threatens, it can flee to the trees and finish its meal in safety.

It seemed that the age of the dinosaurs would last forever. But even at the height of their rule, tiny creatures of a new type were beginning to appear. Minute, mouselike creatures scurried amid the grasses, shivered in terror behind the rocks, and sought refuge in the trees as the ponderous dinosaurs rumbled by. Who would have thought that these inconspicuous little creatures, the first mammals, would be the ancestors of a group that would one day supercede the dinosaurs and dominate the earth?

The principal features that distinguish the **mammals** from the other classes of vertebrates are the **mammary glands** (specialized structures that secrete milk used to feed the young) and the presence of hair on the body. Most mammals have a full coat of **body hair.** Even man, who generally considers himself rather hairless, lacks hair only on the palms of his hands and the soles of his feet. Adult whales are essentially hairless, but their embryos have a full coat of hair at certain stages of prenatal development.

All mammals are air breathers (as are all reptiles, birds, and mature amphibians). The operation of the lungs is facilitated by the presence of a **diaphragm,** an internal muscular partition separating the chest cavity from the abdominal cavity. Movement of the diaphragm expands or contracts the chest cavity, in coordination with the expansion and contraction of the lungs.

The blood, containing nonnucleated red blood cells, is pumped through the closed circulatory system by a four-chambered heart.

Thus, as in birds, there are actually two separate circulatory routes: one carries the oxygen-poor blood to the lungs, where gas exchange takes place, and then back to the heart. The other route carries the oxygen-rich blood around to the various structures of the body and returns the oxygen-poor blood to the heart.

Mammals, like birds, are homoiothermic—the only other class with a system of body temperature regulation independent of the environment. Precise temperature control is aided by the insulating action of the fur and in some cases by evaporation of water from the **sweat glands,** located in the skin. The skin also contains other glands, such as the mammary glands already mentioned (usually situated in specific portions of the abdomen) and the **sebaceous glands,** which secrete lubricating oils.

Perhaps the greatest advances of the mammals are in the nervous system. The mammalian nervous system is the most efficient in the animal world, and the highly developed brain permits a degree of adaptability, through learning, unmatched by the members of any other class.

Reproduction and care for the young are highly advanced among mammals. Fertilization is always internal; the male possesses a copulatory organ, the penis, which transfers sperm to the female vagina. The eggs are generally minuscule, with extremely little yolk. The eggs develop within the body of the mother, except in one small subclass (which has only two living species), the monotremes. In all the rest of the mammals, with the exception of another small subclass, the marsupials, the developing embryos are nourished through a modification of the embryonic membranes known as the **placenta,** which is attached to the inner uterine wall. The young are usually relatively helpless at birth and are cared for for long periods by the parents, especially the mother.

The approximately fifteen thousand species of living mammals are a diverse group. They range in size from tiny two-inch shrews to giant blue whales, which frequently exceed a hundred feet in length. Mammals are to be found in plains and forests, in swamps and deserts, swinging from the trees and burrowing in the ground. Some mammals, such as the seals, dolphins, and whales, have invaded the seas (though they must rise to the surface periodically to breathe). Some, such as the bats, have taken to the air and can fly as well as any bird. The eating habits of mammals vary from a strictly vegetarian diet to a diet composed exclusively of insects or of the flesh of other mammals, birds, and fish. Some mammals, such as mice, pigs, and man, are **omnivorous,** subsisting upon a variety of foods, and thus are even more highly adaptable to varied habitats.

The various mammals show numerous structural and physiological adaptations to their way of life. The limbs, especially, are adapted to characteristic modes of locomotion—walking and running (on four feet or two), leaping, swimming, and flying. Moles, which spend most of their time burrowing underground, show a number of special adaptations for more efficient digging. Monkeys, which spend most of their time in the trees, have hands with well-differentiated fingers and opposable thumbs, well adapted to grasping and manipulation. The New World monkeys of South America even have a "fifth hand," a prehensile tail with which they can grasp tree branches as they swing along.

The teeth of mammals are differentiated into specialized types for cutting, tearing, or grinding. These are developed to various degrees in various species, depending upon their way of life. Desert mammals have highly developed powers of water retention. The gerbil, for example, can generally obtain sufficient water from the foods that it eats, with no need for a free source of water. Special metabolic adaptations permit these mammals to gain much of the water needed from fat metabolism.

There are three subclasses of living mammals: **Protheria,** the **monotremes; Metatheria,** the **marsupials;** and **Eutheria,** the **placental mammals.**

Mammals similar to the monotremes are believed to have been the first to develop from their reptilian ancestors. The only two species

(a) (b)

Figure 9–48 Living representatives of the ancient mammals: (a) the duckbilled platypus, an Australian monotreme; (b) the opossum, an American marsupial.

of monotremes now surviving, the duckbilled platypus and the spiny anteater, are both native to Australia. Both have a full coat of mammalian fur, but lay shelled eggs like those of reptiles. The young, after hatching from the eggs, are fed by the mother with milk secreted in a broad milk field on the ventral body surface. There are no nipples; the young lap up the milk that oozes from their mother's abdomen.

The marsupials are believed also to have appeared early in the evolution of mammals and to have been the dominant form of animal life on earth at one time. The living marsupials include the numerous marsupials of Australia, including the kangaroo, and one American species, the Virginia opossum *(Didelphis virginiana).* A major characteristic of the marsupials is a pouch or **marsupium,** situated on the abdomen of the mother. There is no placenta; the young develop only to a limited extent within the uterus of the mother and then are born. The offspring, tiny, hairless, and helpless, are still really at a fetal stage of development in comparison with the young of other mammals. These tiny creatures somehow grope their way up the mother's abdomen and into her pouch. (The mother makes no effort to help them, and will not even pick one up if it should fall.) There in the pouch the tiny marsupial young attach themselves each to a nipple, through which milk is pumped into their mouths, and there they complete their early development.

In the members of the largest mammal subclass, Eutheria, the young are retained within the body of the mother to a fairly advanced degree of development, and are nourished during their intrauterine development through a placenta.

The placental mammals are usually classified in fifteen major orders, which are summarized in Table 9–1.

The various organ systems and physiology of the mammals will be discussed in much greater detail in the following chapters. Special emphasis will be placed on the biology of the mammal that man in his egocentric way generally tends to think of as the most important—man himself.

TABLE 9-1 Orders of living placental mammals

Order	Examples
Artiodactyla	Even-toed ungulates: cattle, deer, camels, pigs, giraffes, hippopotamuses
Carnivora	Dogs, cats, bears, weasels, seals, sea lions
Cetacea	Whales, dolphins, porpoises
Chiroptera	Bats
Edentata	Armadillos, sloths, anteaters
Hyracoidea	Hyraxes
Insectivora	Primitive insect-eaters: moles, shrews
Lagomorpha	Rabbits, hares, pikas
Perissodactyla	Odd-toed ungulates: horses, zebras, rhinoceroses
Pholidota	Pangolins
Primates	Lemurs, tarsiers, monkeys, apes, man
Proboscidea	Elephants
Rodentia	Rats, mice, squirrels, beavers, gophers, porcupines, guinea pigs
Sirenia	Dugongs, manatees
Tubulidentata	Aardvark

QUESTIONS FOR REVIEW AND THOUGHT

1. Are multicellular organisms more efficient than protists? Discuss from the standpoints of individual cells and of the whole organism.
2. On what basis are sponges, sea anemones, oysters, and other sessile animals classified in the animal rather than the plant kingdom?
3. Compare the alternation of generations in coelenterates to the types of alternation of generations found in the plant kingdom.
4. Give examples of radial and bilateral symmetry in the animal kingdom, citing at least two phyla for each.
5. Describe some examples of mollusks, illustrating the diversity of this phylum.
6. A visiting extraterrestrial might believe that the dominant forms on earth belong to the group of arthropods. Discuss.
7. Discuss the advantages and disadvantages of a chitinous exoskeleton.
8. Distinguish between warm-blooded and cold-blooded animals and give examples of each. Are any invertebrates warm-blooded? Is the blood of cold-blooded animals always colder than that of warm-blooded animals?
9. Discuss the advantages of reptiles over amphibians in adaptation to life on land.
10. Discuss the advantages of placental mammals over egg-laying animals in provision and care for the young. Are there any disadvantages?

PART III

BASIC PROCESSES OF LIFE

CHAPTER **10**

Matter and Energy for Life—I: Input and Output

Foods for Life

A living organism exists precariously poised in a state of dynamic equilibrium. There must be a constant flow of materials and energy into the system, to be utilized in the ever-turning metabolic mills of the cells. Some materials are used as building blocks and deftly assembled into the intricate and highly organized constituents of life. Other materials are oxidized and broken down to simpler units, with a release of chemical energy that is used to power the unceasing reactions of metabolism. By-products and waste materials build up and must be disposed of somehow, if the cell and the organism are not to be choked with wastes and poisoned.

Living organisms have developed a variety of methods of obtaining **foods**—the basic organic raw materials for life. Some (the plants and many protists) make their own food, by chemical synthesis from simple inorganic compounds, utilizing energy from the sun's radiations. Other organisms must eat, deriving their nourishment from complex biochemicals synthesized by other living things. But here more problems are encountered. Chunks of leaves or fruit or flesh are far too large to penetrate to their ultimate destination, the individual cells. Even the biochemical constituents of food are in most cases too gross to breach the barrier of the cell membrane. Thus, the intake of food is only the first step in a chain of processes of digestion, in which the food is then successively broken down into increasingly more manageable units, finally reaching the stage of relatively small molecules that can be utilized by the individual cells. Many protists and other small organisms can get along with a minimally organized, loose flow of materials into and out of the cell. But with increasing size, complexity, and specialization of function, increasingly complex and highly organized systems of organs are needed for the digestion of food materials, its assimilation into the body substance, and the elimination of wastes.

(a)

(b)

Figure 10–1 With the energy of sunlight, plants can produce all the food they need.

Modes of Nutrition in Living Organisms

What would happen if you never ate, but merely sat outside taking sunbaths, drank somewhat salty water, and breathed in plenty of fresh air? You would starve to death, of course. Yet plants do not eat (with a few rare exceptions); from such simple ingredients as sunlight, water and salts, and gases from the air their bodies are able to synthesize thousands of intricate substances that provide sources of fuel and building materials for their growth, reproduction, and other essential life processes. They perform this daily miracle through a series of complicated chemical reactions collectively known as **photosynthesis.**

Organisms that can synthesize the organic molecules required for their vital activities from carbon dioxide, minerals, and water by using a source of energy (such as sunlight) are called **autotrophs.** The autotrophs can be classified in two main groups, according to the source of energy that they utilize. **Photosynthetic autotrophs** derive their energy from the sun. The plants and various pigmented protists belong to this group. **Chemosynthetic autotrophs,** on the other hand, utilize the biological oxidation of inorganic substances as their source of energy. This is a much smaller group, consisting exclusively of bacteria. The substrates used by these microorganisms include a wide variety of inorganic substances, such as ammonia, nitrites, nitrates, hydrogen sulfide, and ferrous salts. The individual bacteria are highly specific in their utilization of a particular substrate. For example, nitrite bacteria *(Nitrosomonas)* oxidize ammonia to nitrites. These nitrites may then be oxidized to nitrates by entirely different species: nitrate bacteria *(Nitrobacter)*. Each type would starve on a substrate that the other would find richly bountiful. Together the various types of nitrogen-utilizing bacteria play an important role in the nitrogen cycle of the earth, converting various sources of nitrogen in the soil to forms more usable by plants. The sulfide bacteria oxidize hydrogen sulfide and other sulfides to sulfates, while iron bacteria oxidize ferrous (Fe^{++}) salts to ferric (Fe^{+++}) salts.

Organisms that require a continual intake of organic molecules preformed by other organisms are classified as **heterotrophs.** Members of this group include many protists, such as bacteria, fungi, and protozoa; a few nongreen plants (e.g., some orchids); and all the animals. Heterotrophs are subdivided into several groups, depending on the form in which organic materials are incorporated into the organism.

Most animals ingest organic materials in the form of solid masses, which usually undergo a process of digestion before they are finally absorbed into the tissues of the organism. This mode of nutrition is called **holozoic nutrition.** Some holozoic heterotrophs subsist almost exclusively on grasses and other plant life, a diet high in carbohydrates. These are the **herbivores,** a group that includes horses,

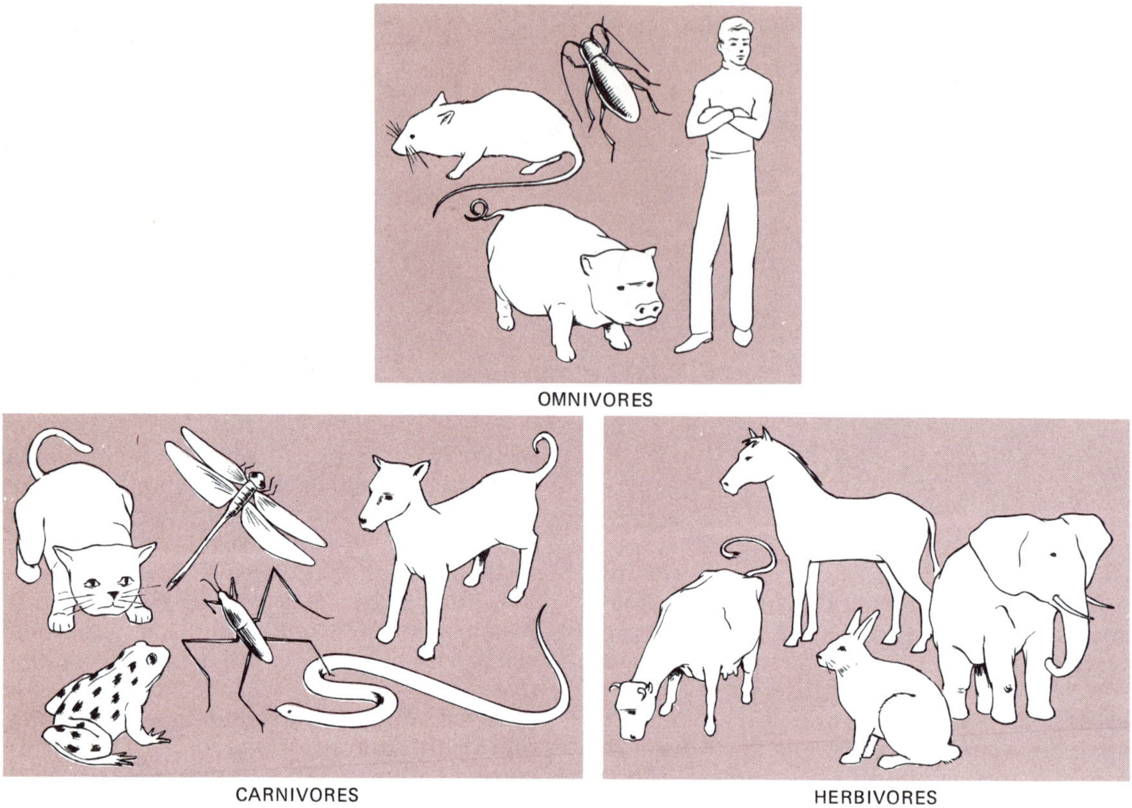

Figure 10–2 Diet requirements vary among holozoic heterotrophs.

cows, elephants, and rabbits. Many insects, from caterpillars to grasshoppers, are also restricted to a herbivorous diet. Other animals live almost solely on animal flesh, a food source high in fats and proteins. These are known as **carnivores** and include members of the cat and dog families. A variety of other animals, from waterbugs and dragonflies to frogs and snakes, are also meat eaters. Man, on the other hand, is one of the relatively few animals that habitually consume a wide variety of food substances: the **omnivores.** Man shares this trait with such species as the mouse, the pig, and the cockroach.

Most fungi and bacteria are unable to utilize food materials in the intact form. Instead, they subsist on the products of living protoplasm or the decomposition products of dead organisms. **Saprophytes,** which feed on the decomposition products of dead organisms, are found in great abundance in the soil and in various other media of the earth. This group includes mainly bacteria and fungi. Some of these organisms can subsist on a limited assortment of organic compounds, together with salts and water. For example, the bacterium *E. coli* will thrive in a solution of glucose, salts, and water. With these meager raw materials, utilizing the sugar as a source of energy, this bacterium can produce a vast array of complex chemicals to maintain its normal metabolic processes. Other saprophytes have quite specialized nutritional requirements, which have been used to great advantage in the investigation of the genetic basis of life (see Chapter 15).

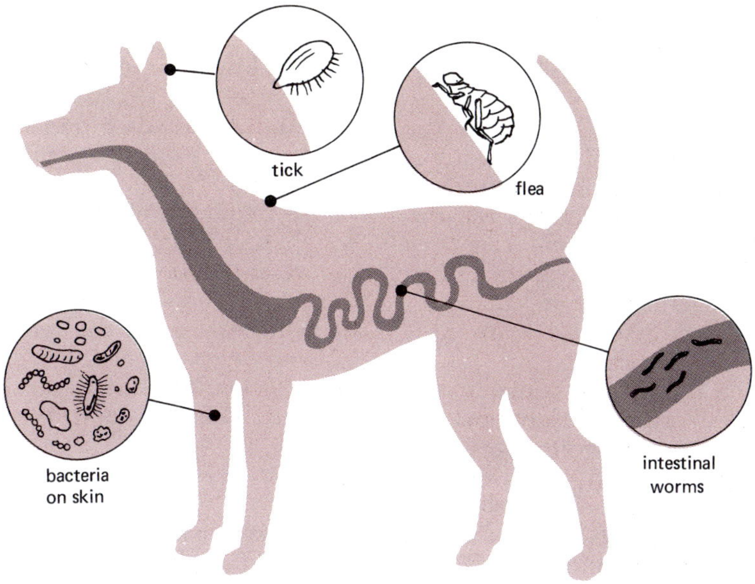

Figure 10–3 Animals often play host to a variety of parasites, adapted to live off their host in various ways.

Many species of protists and a few plants and animals have evolved modes of nutrition in which they have become dependent on a living host for a supply of food materials. These organisms, known as **parasites,** infest virtually all forms of life, from bacteria to man. Many parasites are quite particular in their host preferences and can live on or in only a single species or closely related species. A cotton aphid, for example, would die on a pea plant; pea aphids can also survive on bean plants but would die if placed on a carrot seedling.

Some parasites may live in relative harmony with their hosts, merely withdrawing enough material to sustain their own needs and causing no serious harm. But others may cause varying degrees of damage, including the death of the host. These disease-producing or pathogenic organisms include many viruses, bacteria, fungi, protozoa, and other forms. The common ailments of mankind, such as colds, tuberculosis, measles, and cholera, are caused by bacterial and viral parasites that attack and destroy human tissues. The ceaseless search for treatments and cures for diseases is a dramatic chapter in the history of man.

Photosynthesis

Scientists have long dreamed of being able to make food synthetically in the laboratory. With two-thirds of the world's human population underfed, a simple, inexpensive process for manufacturing food on a large scale would be a boon to mankind. Although a basic energy source, glucose, has already been synthesized in the laboratory, the yield has been disappointingly low and the process far too expensive for practical purposes. Yet this process, and others even more complicated, are performed continually by plants and many protists. In this daily miracle, cheap, simple, and abundant raw materials—the carbon dioxide of the air and the water and minerals of the soil (and of the seas, ponds, and other bodies of water for aquatic forms)—are combined in the presence of solar energy to produce glucose and other organic materials. This is the essence of the process of **photosynthe-**

sis, which results in the conversion of solar energy to chemical energy, stored in the form of the chemical bonds of organic compounds.

EARLY INVESTIGATIONS

The growth of plants has long been an intriguing mystery to man. The growth of animals is clearly associated with the food they eat. But plants do not "eat"—how, then, can they grow? A natural assumption was that they obtain their nourishment from the soil. Although this is partly true, we know now that this is not the whole story.

A dramatic step toward the realization of the true nature of the source of plant nourishment was taken in the seventeenth century by the Flemish physician Jan Baptista van Helmont. He planted a 5-pound willow shoot in a pot containing exactly 200 pounds of soil (Fig. 10–4). After five years, during which van Helmont carefully watered the young tree with only rainwater or distilled water and protected the soil from contamination, he uprooted the tree and carefully weighed both it and the soil. He found that the tree had gained 164 pounds 3 ounces, while the soil had lost only 2 ounces! Obviously the soil was only a minor (albeit indispensable) source of nourishment. Van Helmont theorized that the weight gain of the plant must have come from the water that he added. This was only part of the answer, however. Ironically, van Helmont had the rest of the answer at his fingertips, but he never realized it. For another of his major scientific studies dealt with the gas evolved when wood is burned: carbon dioxide.

The first inkling that gases are involved in the growth of green plants came over a century later, in the experiments of the English scientist Joseph Priestley. In a series of brilliant investigations in the early 1770s, Priestley

Figure 10–4 The van Helmont experiment. After five years, a five-pound willow shoot had gained 164 pounds 3 ounces, while the soil in which it was planted had lost only 2 ounces.

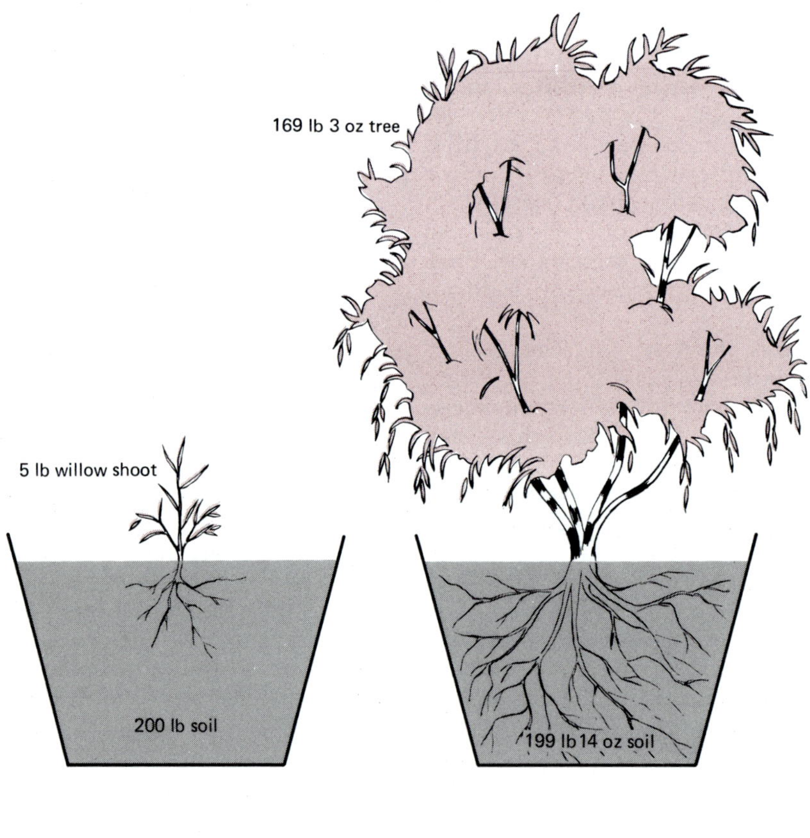

demonstrated that a gas (later identified as oxygen) is given off by growing green plants and can support the life of a burning flame or of a tiny breathing animal, such as a mouse. In one experiment, Priestley placed a mouse under a bell jar and a sprig of mint growing in water under another bell jar. The mouse soon died, but the mint continued to grow. When, however, a mouse and a sprig of mint growing in water were placed under a single bell jar, both remained alive (Fig. 10–5). Priestley further demonstrated that a green plant could "freshen" air in which a candle had burned out or a mouse had suffocated.

Priestley's experiments sparked further work in this area. Gradually it was discovered that light is essential for the production of oxygen by plants and that only the green parts of the plants can produce this oxygen. Carbon dioxide was found to be absorbed by photosynthesizing plants, and the utilization of water was established and measured. The role of the pigment chlorophyll was discovered, and the chemical reactions of photosynthesis were worked out, first overall and then in increasingly fine detail.

PHOTOSYNTHETIC STRUCTURES

Until 1954 it was believed that photosynthesis could take place only in an intact organism—or at least in a substantial portion of one, such as a leaf and stem immersed in water. But then the American physiologist Daniel Arnon and his coworkers perfected a method for extracting tiny subcellular structures called **chloroplasts** from spinach leaves. They discovered that a suspension of these particles in a test tube can carry out the entire photosynthetic process when it is exposed to light. No source of energy other than visible light, and no unusual conditions (such as high temperature) are required; the photosynthetic process takes place in a manner essentially identical with that in the intact cell, with the absorption of carbon dioxide from the air and water from the medium and the production of carbohydrates and evolution of oxygen.

Each cell in the leaf of a green plant usually contains from twenty to a hundred chloroplasts. Within the chloroplasts, the light-sensitive pigment **chlorophyll** is found concentrated within stacks of double-membraned disks **(lamellae).** The stacks of chlorophyll-containing lamellae are known as **grana** and are imbedded within a colorless granulated medium called **stroma** (Fig. 10–6).

The chlorophyll molecules in the lamellae are closely packed in thin layers, alternating with layers of proteins, lipids, and carotenes and other pigments, an arrangement which enhances contact with various enzyme systems and facilitates energy transfer among the pigment molecules.

Figure 10–5 The Priestly experiment. A plant placed under an air-tight bell jar ultimately died. So did a mouse under similar conditions. But if a mouse and a plant were placed together under the same bell jar, both survived.

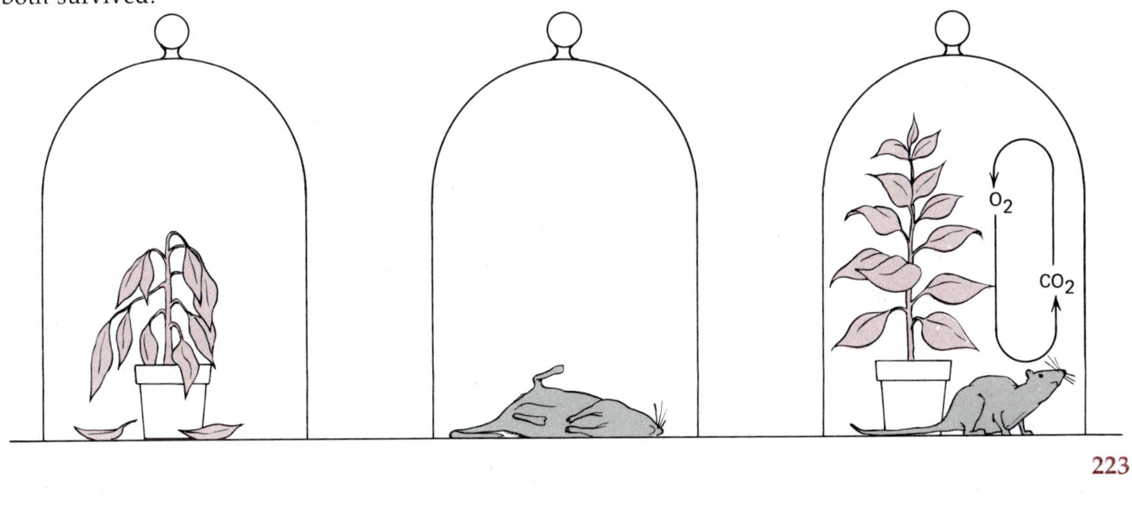

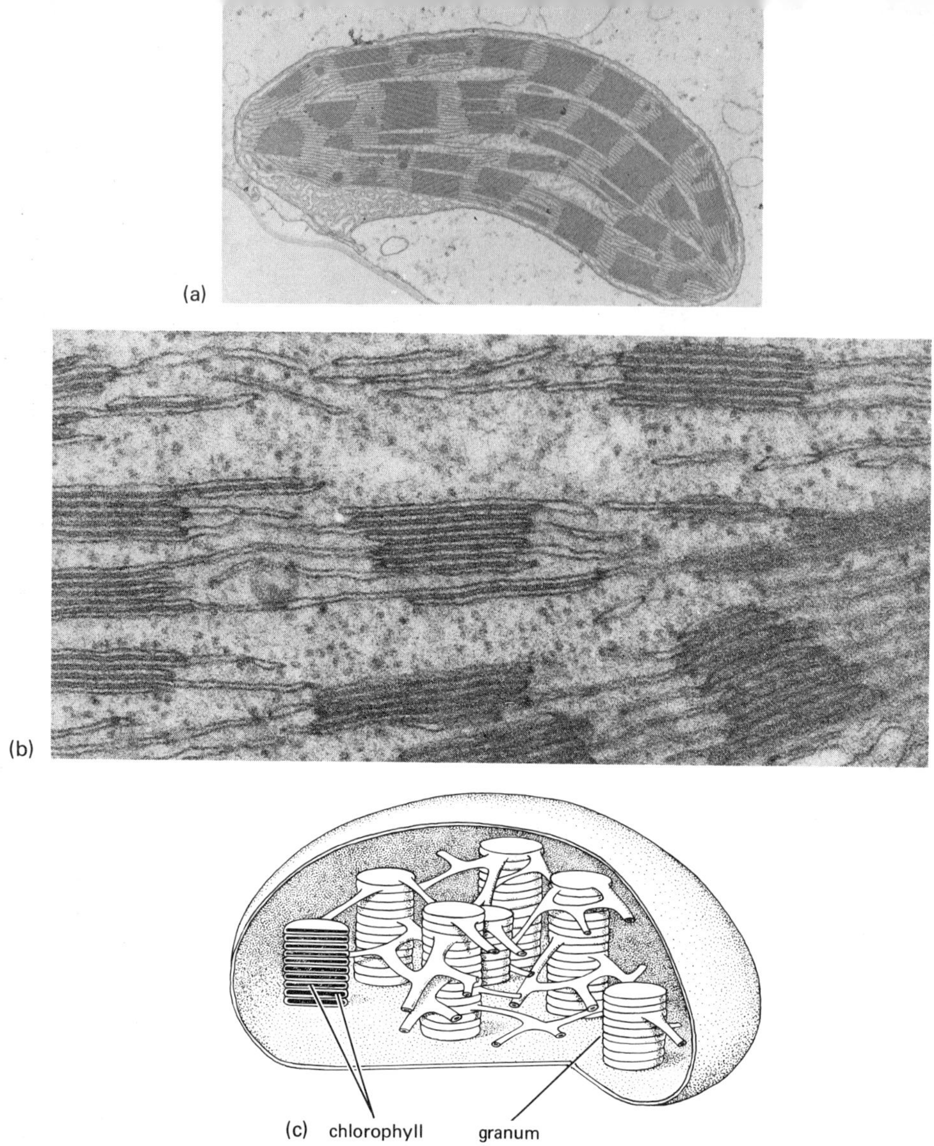

Figure 10–6 (a) Electron micrograph of a chloroplast of a corn leaf cell, (10,120×). (b) An electron micrograph taken at higher magnification (62,400×) shows the details of an individual granum. (c) Schematic diagram of the structure of a chloroplast.

The chloroplasts arise from tiny colorless structures called **proplastids,** which reproduce themselves in the cytoplasm of the cell. When they are exposed to light, the proplastids produce chlorophyll and develop into **chloroplasts,** which are also capable of self-duplication. Sprouts grown in darkness do not have the characteristic green color and are referred to as **etiolated** (Fig. 10–7). Etiolated sprouts soon turn green when they are exposed to light, and chlorophyll is formed in their chloroplasts.

In protists, the raw materials for photosynthesis (water and carbon dioxide) are supplied by simple diffusion through the plasma membrane. In higher plants, on the other hand, specialized structures function to supply the needed raw materials to the photosynthesizing cells. Water and minerals are absorbed from the soil or water medium

Figure 10–7 The bean seedlings in the pot toward the front were grown in the light; those to the rear were grown in darkness.

through tiny root hairs into the root system, from which they are delivered through a xylem network into the leaf and finally to the photosynthesizing cells. Carbon dioxide from the air is admitted to the leaf through tiny openings in the leaf surface called **stomata.**

THE PROCESS OF PHOTOSYNTHESIS

It is estimated that about a trillion tons of organic matter is manufactured each year by the photosynthesizing protists and plants of the earth. This figure is well over one hundred times the weight of *all* the products produced industrially by man within the same period. Yet with this fantastic total, the great preponderance of the light that strikes plants is "wasted": frequently less than 1 percent of this light is actually stored in organic materials.

Before light energy can be stored in chemical compounds, it must be "trapped" in some way. In the plant, the incident light energy is captured by light-sensitive pigments, the best known of which is chlorophyll.

The chlorophyll molecule* bears a striking resemblance to the heme portion of the hemoglobin molecule (Fig. 10–8). Both these compounds are metal porphyrins, with the metal ion (magnesium in the case of chlorophyll,

*Green plants actually contain two slightly different varieties, chlorophylls a and b.

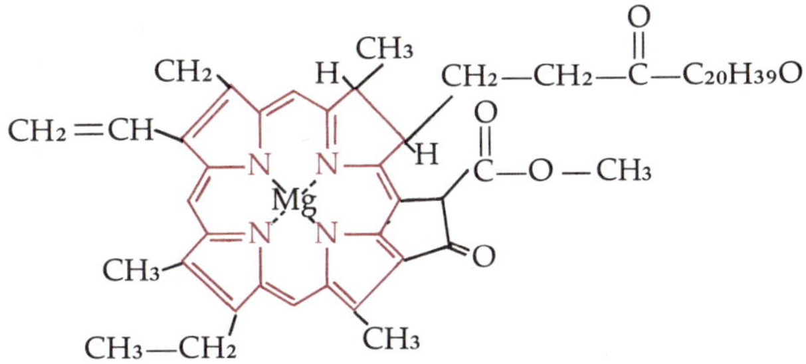

Chlorophyll a

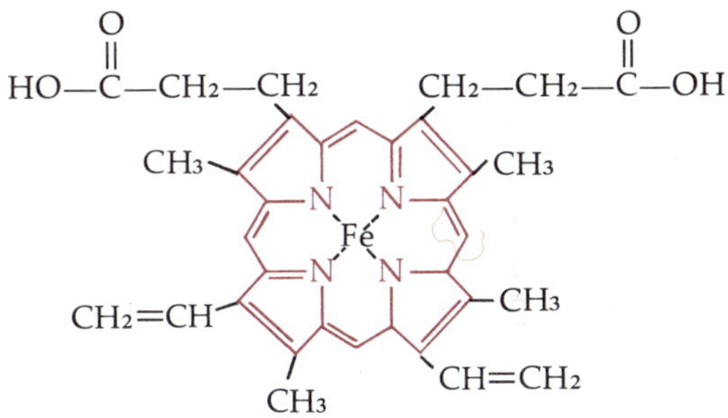

Ferroheme

Figure 10–8 Chemical structure of chlorophyll compared with heme.

iron in the case of heme) in the center of the porphyrin ring system. The latter is a rather intricate and symmetrical nitrogen-containing organic molecule, with many conjugated double bonds that facilitate energy transfer.

The chlorophyll molecule is thus uniquely suited to its role, although it is by no means a perfect "energy trap." Sunlight is represented by a whole spectrum of wavelengths, from the infrared, through the visible spectrum, down to the ultraviolet. But the chlorophyll molecule absorbs most efficiently in two very specific wavelength regions (Fig. 10–9): in the violet-blue region (400–500 mμ) and in in the orange-red region (600–700 mμ). In the green-yellow region (500–600 mμ) the efficiency of light absorption by chlorophyll is quite low. We see evidence of this fact whenever we look at a green plant, for its green color is the result of *reflected* light, i.e., that

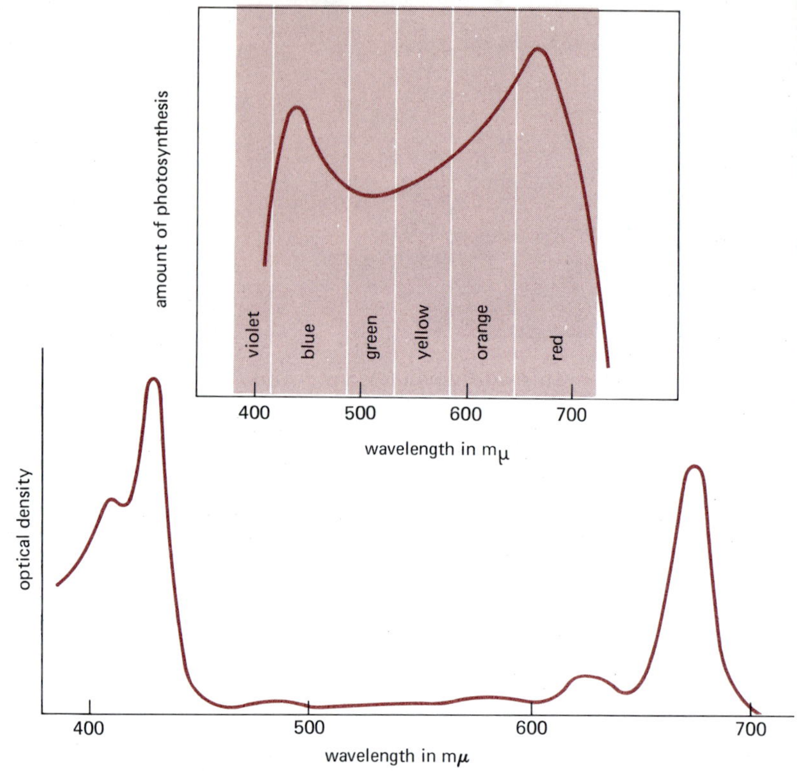

Figure 10–9 (a) The light absorption spectrum of chlorophyll a. The absorption peaks fall in the blue and red regions. (b) The action spectrum of photosynthesis. The peaks approximately correspond to the absorption maxima of chlorophyll a; the photosynthetic activity in the middle region of the visible spectrum reflects the contribution of other light-absorbing pigments.

portion of the visible spectrum which is not absorbed.

The light in the green-yellow region is not entirely wasted by green plants, since the chloroplasts also contain accessory pigments: the yellow and orange carotenes and xanthophylls, which absorb more efficiently in the green and blue-green regions. These accessory pigments aid in absorbing light energy and transfer it to chlorophyll, a process that is facilitated by the layered packing of the chlorophyll and other pigment molecules in the chloroplasts.*

Photosynthesis cannot begin in total darkness. But the striking of a single match may be sufficient to initiate this process in a plant. When the light intensity increases, the rate of photosynthesis increases up to a certain point, known as the **light saturation point,** beyond which a further increase in light intensity will not increase the rate of photosynthesis. The light saturation point varies with different plant species, and it is generally about a tenth to a third of the intensity of bright sunlight for plants that normally live in bright sunlight. Hence, photosynthesis is not usually interrupted by a blockage of sunlight by a passing cloud.

Although light is a necessary factor in the overall process of photosynthesis, it has been found that this process actually consists of two separate but interrelated sequences of reactions, one which occurs only in the presence of light and another sequence that does not require light and may even proceed in total darkness. Radioactive tracer studies, using radioactive isotopes of oxygen and carbon,

*Certain algae contain blue, red, or brown pigments, in addition to chlorophyll a. These accessory pigments absorb energy and transfer it to chlorophyll. They are particularly effective at depths of 50 to 300 feet below the water surface, where only blue, green, and yellow wavelengths of light penetrate.

have revealed many details of the light and dark reactions of photosynthesis. These are schematically summarized in Fig. 10–10.

The overall equation of photosynthesis may be written as:

$$\underset{\text{carbon dioxide}}{6CO_2} + \underset{\text{water}}{6H_2O} \xrightarrow[\text{chlorophyll}]{\text{sunlight and}} \underset{\text{glucose}}{C_6H_{12}O_6} + \underset{\text{oxygen}}{6O_2}$$

The light reactions of photosynthesis result in the formation of a reducing agent, NADPH (reduced nicotinamide adenine dinucleotide), and an energy source, ATP (adenosine triphosphate), which are utilized in the conversion of CO_2 (carbon dioxide) to carbohydrate in the "dark reactions" of photosynthesis.

Paradoxically, the "dark reactions" of photosynthesis usually take place in the light. They are called "dark reactions" because they do not *require* light and may take place, at least for short periods, in total darkness. (Since the dark reactions depend on a continual supply of products of the light reactions of photosynthesis, they soon come to a halt in the dark as these products are used up.)

Much of our present knowledge of the dark reactions of photosynthesis was gathered by Melvin Calvin and his associates at the University of California, in studies that eventually won Calvin a Nobel Prize. In studies of photosynthesis by the green alga *Chlorella* in the presence of $C^{14}O_2$ (carbon dioxide containing a radioactive isotope of carbon), Calvin discovered the various intermediate products in the conversion of CO_2 to glucose and other organic substances and worked out the sequence of reactions. This was accomplished by exposing *Chlorella* preparations to light for brief periods, ranging from seconds to minutes, then stopping the reaction and subjecting extracts of the cell contents to paper chromatography. By applying the technique of radioautography to the chromatograms, Calvin and his associates determined the compounds into which the radioactive carbon was incorporated after various periods. Five-carbon, six-carbon, and three-carbon molecules are involved in the sequence of reactions, which ultimately yields the six-carbon sugar glucose.

The glucose produced in photosynthesis is converted to various other substances, such as other sugars, starch, organic acids, and amino acids, and thence into more complex biopolymers, such as proteins and nucleic acids, as well as special substances such as lignin, gums, resins, and alkaloids. Starch is the major form of stored food in plants. It may be deposited in the roots or stems or in specially modified storage organs, depending on the plant species. Reserves of stored food are also laid down in seeds, for the use of the growing embryo until its own leaf system is well enough established to provide food through photosynthesis. Whenever a plant calls upon its reserves of stored foods, the high-molecular starch must first be digested into smaller glucose units before it can be utilized, just as animals must digest their food.

Nutrition

Today, when two-thirds of the world barely subsists at the edge of starvation, the problems of adequate nutrition continue to play a major role in shaping the policies of nations.

Figure 10–10(a) Electron flow in the light reactions of photosynthesis results in the synthesis of ATP and NADPH.

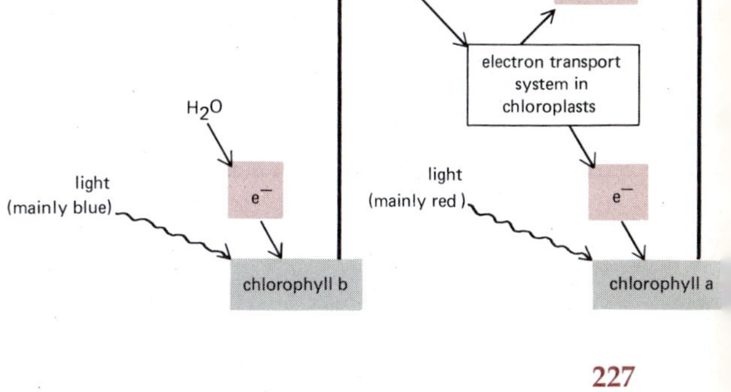

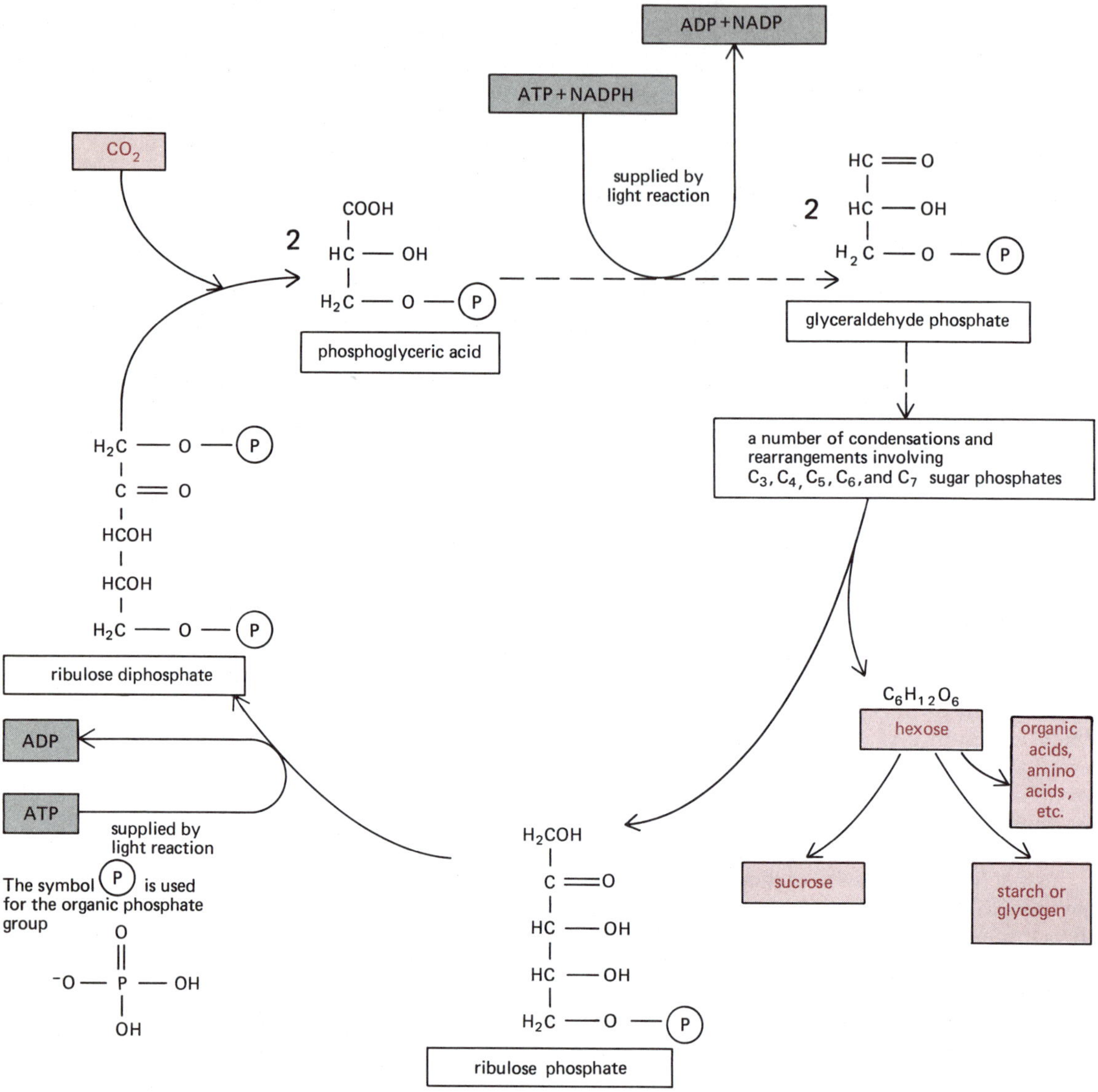

Figure 10–10(b) In the dark reactions of photosynthesis, the carbon in carbon dioxide is used to build glucose and ultimately other organic products. The ATP and NADPH produced by the light reactions of photosynthesis (a) supply the energy for the energy-requiring dark reactions.

In the "well-fed" nations, problems of nutrition are also an important factor. Here, in contrast to the low food consumption of the deprived nations, food intake may be so high as to cause problems of overweight, and even then adequate nutrition often remains an elusive goal. For though the quantity of food may be sufficient, the proper combination of necessary food factors may not be present.

An adequate diet for man should consist of appropriate proportions of each of the following: (1) carbohydrates, (2) fats, (3) proteins, (4) water, (5) minerals, and (6) vitamins. Substances containing some or all of these items in edible form are called **foods.** Foods are used in numerous ways in the body, which may generally be grouped in the following categories: (1) as a source of energy for the various metabolic processes; (2) for the building, maintenance, and repair of the tissues of the body; and (3) as regulators of metabolic activities.

MEASUREMENT OF BODY ENERGY REQUIREMENTS

Today America is a "calorie-conscious" country. Low-calorie foods—even "no-calorie" foods—abound on supermarket shelves. The energy content of foods is commonly measured in terms of a unit more accurately called the "large calorie" or **Calorie.** This is actually a kilocalorie (kcal), the amount of energy required to raise 1000 grams (1 kilogram) of water 1°C (1 degree centigrade).

A certain minimum amount of energy is required to maintain the basic cellular and body functions, such as the heartbeat, breathing, and maintenance of the body temperature. This basic rate of energy expenditure is called the **basal metabolic rate (BMR).** An individual's BMR is measured after a night of restful sleep with a lapse of least twelve hours after the last meal. The person must recline for at least thirty minutes immediately before the test with a minimum of physical or psychological distractions. The basal metabolic rate has been found to be proportional to the surface area of the body, and also depends on the age and sex of the person. For most adult males, the BMR ranges from 1600 to 1800 Calories per day; for adult American women, the normal range is 1300 to 1500 Calories per day.

The simplest activities of everyday life, such as the digestion of food, sitting, standing, walking, and so on, require additional energy expenditures. Additional Calories are also required for the performance of chores or work activities, the amount depending on the type of activity and its duration (Fig. 10–11). A sedentary worker (e.g., a clerk or typist) weighing about 150 pounds would have a total daily energy expenditure of about 2200–2500 Calories. A person engaged in more strenuous activity (e.g., a metalworker or furniture painter) has a daily energy expenditure of between 2800 and 3200 Calories. Lumberjacks working in cold weather may use as many as 8000 Calories in a day.

Another quantity used in measuring energy relationships in the body is the **respiratory quotient (R.Q.):**

$$\text{R.Q.} = \frac{\text{Volume of } CO_2 \text{ eliminated from body}}{\text{Volume of } O_2 \text{ absorbed by body}}$$

The respiratory quotient can be computed theoretically for the basic types of food substances, according to the equations of their oxidation in the body. For carbohydrates, the volume of carbon dioxide formed is equal to the volume of oxygen consumed:

$$C_6H_{12}O_6 + 6O_2 = 6CO_2 + 6H_2O$$

Thus, for carbohydrates the R.Q. would be 1. Because proteins represent a more reduced state than carbohydrates (i.e., they contain proportionately less oxygen per molecule), more oxygen is required to oxidize them to carbon dioxide and water. Hence, the R.Q. for proteins is lower than that for carbohydrates, about 0.8. Fats represent an even more reduced state, and, as expected, their R.Q. is lower still, about 0.7.

The respiratory quotients measured experimentally for various organisms and for various tissues of an organism can be compared with the theoretical values for food substances to yield some interesting information.

Figure 10–11 The amount of food that must be consumed to maintain a healthy body is proportional to the amount of energy expended in daily activities.

For example, brain tissue exhibits an R.Q. of about 1.0. This factor seems to indicate that carbohydrates are used almost exclusively in brain tissue metabolism. For most other tissues, the R.Q. is about 0.85, indicating a more balanced use of carbohydrates, fats, and proteins.

Shortly after a meal high in carbohydrates, the respiratory quotient of the body approaches 1.0. However, about ten hours later, the R.Q. approaches 0.7, the value for pure fat metabolism. The interpretation of this observation is that after the body has consumed its ready source of carbohydrate, it then begins to use its fat reserves for energy. If the fat reserve is exhausted (as during starvation), the R.Q. will then approach 0.8, indicating that the proteins of the body are now being metabolized.

Another good correlation with the theoretical respiratory quotients is observed in the measured R.Q.'s of various types of animals (Fig. 10–12). Typical herbivores, such as horses and cows, have R.Q. values ranging from 0.90 to 0.95, as might be expected from their high-carbohydrate diet. Carnivores, whose diet is high in fat and protein, have an R.Q. of about 0.75. Omnivores such as man show an intermediate value in the neighborhood of 0.85.

FOODS FOR THE BODY

In some respects the human body resembles a machine. To be sure, it is a far more complicated "machine" than an automobile or an airplane or a locomotive. It has the qualities of self-analysis, self-repair, and self-reproduction; but like machines it has a major limitation: to continue functioning, the human body requires an external source of energy. For machines this source of energy may vary widely—electric power, steam power, nuclear energy, or chemical fuels. The human body, on the other hand, is limited to a single source: chemical fuels. And this source is further restricted to organic chemicals.

The organic fuels used by the body fall into three major groups: carbohydrates, fats, and proteins. The energy released in the body by the oxidation of a gram of carbohydrate or protein is about 4 Calories. Fats, on the other hand, are more economical fuels, delivering about 9 Calories of energy for each gram consumed.

Very few foods are *pure* carbohydrate, protein, or fat. Take a roast beef sandwich, for example. The bread is largely carbohydrate, but the grain used to make the flour also contains some proteins and oils (fats);

Figure 10–12 The respiratory quotient is correlated with the type of food being metabolized.

the shortening used in the dough is high in fat, and the eggs used are rich in fats and proteins. The bread also contains moisture and small amounts of various vitamins and minerals. The roast beef in the sandwich is mainly protein, but it is marbled with fat and also contains some amounts of carbohydrates, vitamins, minerals—and the ubiquitous water.

Carbohydrates. For most of the human population of the world, carbohydrates make up the bulk of the diet, in the form of wheat, corn, rice, or potatoes. These are the most abundant and cheapest of foods, and they consist mainly of starch. Plant sugars, such as those found in fruits, are also an important carbohydrate food source.

Carbohydrates, in the form of the simple sugar glucose, are the first line of energy reserves for the body. The blood contains a certain amount of glucose at all times. This amount, called the **blood sugar level,** is usually about 90 mg percent (i.e., 90 mg of glucose per 100 ml of blood), but after a meal it may rise to about 120–140 mg percent. If the blood sugar level falls, we feel hunger; if it drops too low (below 50 mg percent) and remains at that level, the delicate brain cells will be starved of energy for their metabolic activities, and irreversible damage may ensue.

Fortunately the body possesses a number of mechanisms that help to maintain the blood glucose at a normal level, even under the most adverse conditions (Fig. 10–14). The activity of the liver is a key factor. This organ is capable of storing glucose in the form of **glycogen,** or animal starch, a high-molecular polysaccharide. The liver removes excess glucose from the blood and converts it to glycogen; glycogen can later be reconverted to glucose and released into the blood if an energy need is signaled by a drop in the blood sugar level.

This activity of the liver is influenced to some extent by the hormone insulin, a small protein that is secreted by a gland called the pancreas. At high insulin concentrations, the liver tends to store glucose as glycogen. Insulin also stimulates the use of glucose by most of the nonhepatic cells of the body, which effectively reduces the blood glucose level still further. A low rate of insulin secre-

Figure 10–13 The roast beef sandwich pictured contains all of the major types of nutrients: protein, fat, carbohydrate, water, vitamins, and minerals.

Esias Baitel, Omikron

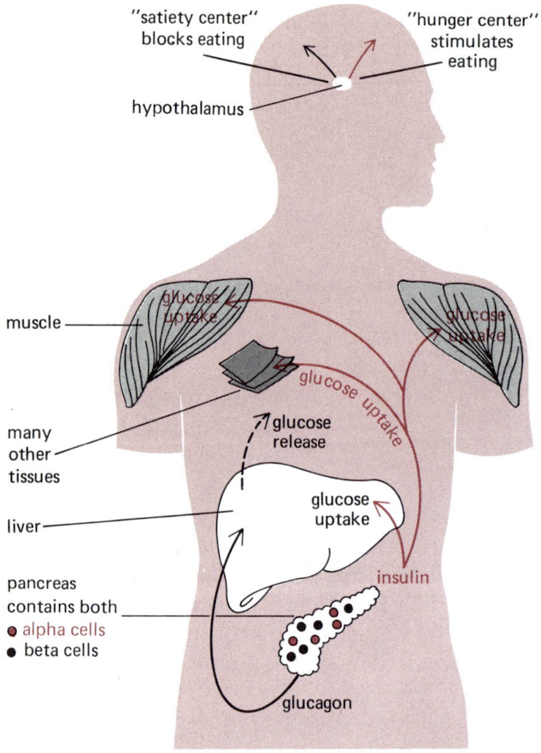

Figure 10–14 The blood sugar level is regulated by a number of interrelated systems. A key organ is the pancreas, which secretes two antagonistic glucose-regulating hormones, insulin and glucagon. The hypothalamus in the brain monitors blood glucose levels and can stimulate or inhibit eating in response to a drop or rise in blood glucose levels.

tion, on the other hand, will allow the blood sugar level to rise as high as 200 to 300 mg percent before the liver will act to convert blood glucose to glycogen. Other hormones also help to regulate the blood glucose level, in a manner that is not yet fully understood.

Fats. In the industrialized countries, particularly the United States, fats have become a major constituent of the diet. It is estimated that in the United States, as much as 40 percent of the daily caloric intake is comprised of fats. Major sources of fats are butter, margarine, the various cooking and salad oils, milk, meats, eggs, and nuts.

Hardly a month goes by without the appearance of some scientific report implicating fats in some form as a causative or contributing factor in heart disease. This continual barrage of incriminating evidence has tended to give the public the impression that fats are "bad." And yet, the intake of fats (at least some forms of fat) is essential to life.

Fats are triglyceryl esters of fatty acids, and when hydrolyzed in the intestines are absorbed into the blood as fatty acids and glycerol. The absorption of the fat-soluble vitamins—A, D, E, and K—is greatly facilitated by the presence of fat. In addition, certain fatty acids, including the unsaturated fatty acids linoleic and linolenic acids, are essential nutrients for the maintenance of good health.

The products of fat digestion are transported to the liver, where cholesterol and other body fats are synthesized. These fats may be used as a source of energy, or they may be stored in adipose tissue distributed through various parts of the body (Fig. 10–15). Such storage depots account for about 15 percent of the total weight of an adult male and

Figure 10–15 Fat is deposited in characteristic depots scattered over various parts of the body.

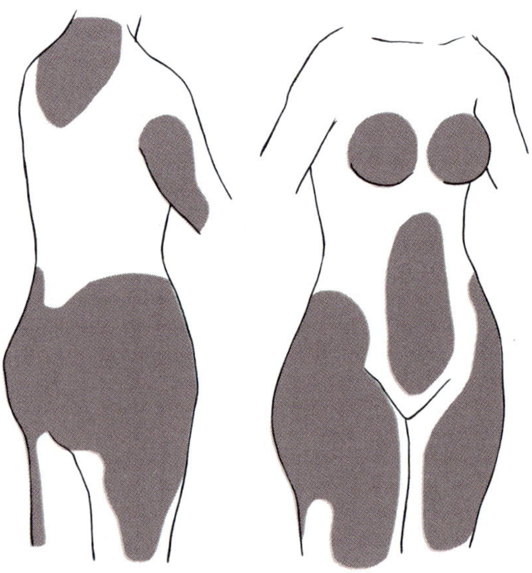

20 percent of the total weight of an adult female. Because of this quantity and because the deposits of adipose tissue are strategically located, fats serve as excellent insulators and shock absorbers. These fat deposits may be augmented by the intake of excess carbohydrates, which are converted to fats and stored. Fats serve as important constituents of the various cellular and tissue membranes of the body and are also included in lipoproteins and other key body chemicals.

Fats may play a negative role in the body, however, through the formation of deposits within the walls of the arteries. This fat build-up, a condition known as **atherosclerosis,** results in a gradual narrowing of the arteries. The heart must work harder and harder to force blood through the narrowed arteries, and a clot may plug them entirely, resulting in a heart attack or a stroke. As a result, physicians are urging Americans to eat less fats and more proteins.

Proteins. Proteins are the major building materials for the body, and an adequate supply is absolutely vital to life. A severe protein deficiency results in a wasting disease known as **kwashiorkor,** an African word that can be translated approximately as "the disease the child gets when the new baby is born." For when the new baby is born and the older child no longer receives his mother's milk, he is plunged from a relatively high-protein diet to one that may supply enough calories but is woefully deficient in proteins. This condition can result in stunting of physical growth and mental development, and ultimately in death (Fig. 10–16).

The primary protein sources in the Western diet are meats, milk, eggs, and fish (animal sources); and grains, peas, beans, and nuts (vegetable sources). The digestion of proteins leads to the formation of amino acids. The proteins of living organisms are composed mainly of twenty different amino acids. The adult human is capable of manufacturing twelve of these amino acids in sufficient quantities to meet virtually all normal needs. The other eight amino acids, called **essential amino acids,** cannot be made by the body

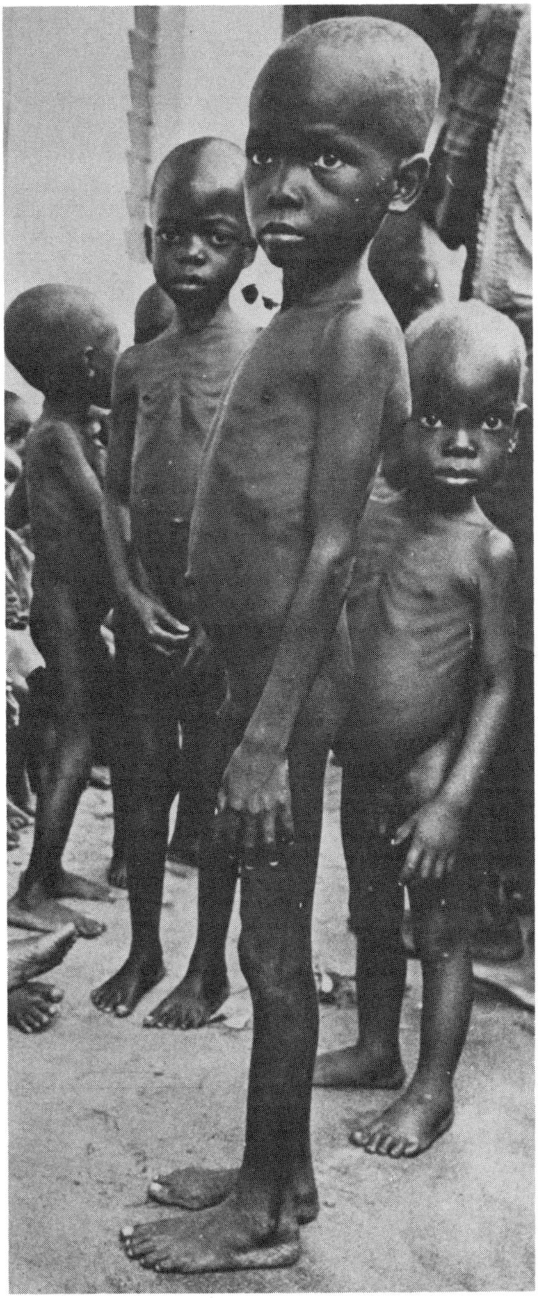

Figure 10–16 The swollen bellies and light-colored hair of these Biafran children are typical symptoms of the protein-deficiency disease, kwashiorkor.

and must be ingested. Thus, to say that a certain quantity of protein should be eaten for good health is somewhat misleading. Even more important, the proteins eaten should be balanced so that sufficient amounts of all of the essential amino acids are obtained. Foods vary widely in their amino acid composition. Corn protein, for example, normally lacks two of the essential amino acids (tryptophan and lysine). By eating a variety of protein sources, one can be more assured of obtaining a proper balance of amino acids.

Amino acids serve many functions in the body. They may be assembled at the ribosomes of the cells to form a great variety of proteins. Excess amino acids may be deaminated (i.e., the amino group is removed) in the liver and used eventually in cellular respiration as a source of energy. Or excess amino acids can be converted into fats and stored as such to serve as an energy reserve.

In the normal adult, about 30 grams of body proteins are degraded each day into amino acids, which are then deaminated and oxidized. This amount must be replaced if serious consequences are to be avoided. To be on the safe side, physicians usually recommend a daily protein consumption of about 70 grams for a 150-pound man.

One of the end products of protein metabolism is ammonia. But this compound is potentially poisonous, even in relatively low concentrations. Fortunately, the liver converts the ammonia to urea, a rather nontoxic substance, which is then carried by the blood to the kidneys, where it is eliminated from the body in the urine.

If laboratory animals are fed a synthetic diet consisting solely of pure carbohydrates, pure fats, and pure proteins (or amino acids), they soon die. If water is added to the diet, a similar group of animals will live for a longer period of time, but soon they, too, will sicken and die. The further addition of either small quantities of various mineral salts or vitamin supplements will not be of much help in sustaining the life of the animals. But a combined diet incorporating all of these substances in appropriate ratios will maintain health and vigor.

WATER AND MINERALS

About two-thirds of the total body weight is composed of water. Its critical importance to life is indicated by the fact that a person may live for as much as a month or two without food, provided water is available; but if water is also withdrawn, death will follow within four to six days.

The vital functions of water in the body include its role as a "universal solvent" for chemical reactions; its participation in hydrolytic reactions; its service as a carrier vehicle for various other substances in the blood, lymph, and intracellular fluid; and its role in the body's heating and cooling systems.

Mineral salts comprise about 5 percent of the total body weight. Some inorganic substances act as important structural constituents, such as the calcium and phosphorus in the bones and teeth. Substantial quantities of chlorine, sodium, potassium, and magnesium are also present in the body, participating in various body functions such as muscle contraction, nerve impulse conduction, and the maintenance of the acid-base balance in the blood. Small quantities of iron, zinc, and fluorine, along with minute amounts of such **trace elements** as iodine, copper, cobalt, manganese, selenium, boron, aluminum, arsenic, silicon, and nickel are also found in the tissues of living organisms. Some of these inorganic substances have key functions in the metabolic reactions of the body—for example, in organometallic complexes and as cofactors for enzymes. For some other inorganic constituents, no useful role has yet been established.

VITAMINS

The sale of vitamin supplements is a multimillion-dollar business in the United States. And yet the question, "Are vitamin supplements really necessary?" is still controversial. Of course, physicians regularly prescribe vitamins for young children, pregnant women, and people with specific vitamin-deficiency diseases. But if you ask your physician whether vitamin supplements are advis-

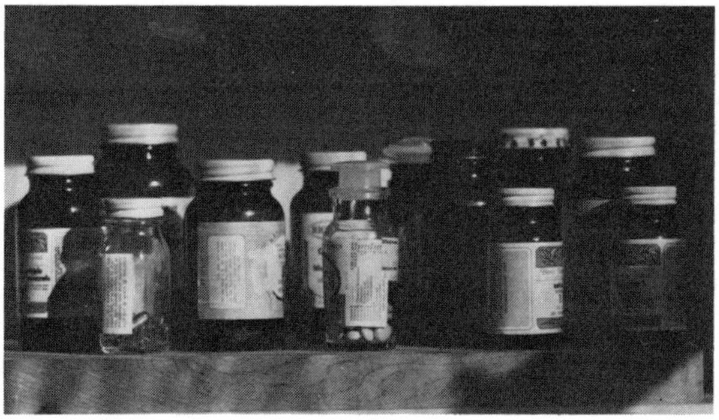

Figure 10–17 Vitamin supplements: are they necessary in the modern diet?

able, he may very well answer: "Vitamin supplements are not necessary if you eat a balanced diet." This is not necessarily a relevant answer. For American nutritional habits include the ever-dieting mother, the salad-scorning, meat-and-potatoes father, the fad-following daughter, and the teenage son who seems to subsist on a steady diet of frankfurters and soft drinks.

Even if the conscientious housewife makes a determined effort to provide the family with a "well-balanced diet," they still may not receive the proper proportions of the essential vitamins. Modern techniques of processing foods often tend to remove or destroy vitamin constituents. For example, the pasteurization of milk destroys its natural vitamin C content, and the manufacture of flour, involving the removal of the hulls from the grains, strips off many of their vital ingredients, especially the B-complex vitamins. Grain-product manufacturers reintroduce perhaps a dozen of the scores of vital nutrients that have been removed and call their product "enriched." Meanwhile, the discarded hulls provide a highly nutritious fodder for livestock.

The efforts of manufacturers to supplement processed foods with vitamins introduce still another factor, which may distort the concept of a balanced diet. Milk, bread, cereals, and numerous other food products are commonly enriched with various vitamins in varying amounts. Whereas this may seem to be a beneficial practice, the cumulative effect of this practice may in fact be harmful. It is known that excessive amounts of vitamins A and D, which are stored for long periods in the body, are toxic to the human system and may cause permanent damage. One of the nationally sold breakfast foods advertises that each serving provides a complete adult minimum daily requirement of the major vitamins. When one considers the cumulative effects of vitamins provided by natural foods, artificially "enriched" foods, and possibly vitamin pills in addition, it is easy to see why dangerous levels of certain vitamins may build up.

Vitamins are organic substances that are needed in very small amounts for the normal metabolic processes and cannot be manufactured by the body itself. Most of the vitamins act as cofactors or prosthetic groups for enzymes in the body.

The existence of vitamins was not realized until the twentieth century, although the **deficiency diseases** resulting from their lack had been observed for thousands of years. Now dozens of vitamins are known, with widely varying structures, including steroids, alcohols, organic acids, amines, and esters. Of this large number, little more than a dozen have been definitely shown to be needed in human nutrition, though numerous others are suspected to be essential to good health. What may be termed an essential vitamin for one species is not necessarily such for another. For example, although vitamin C is required for human nutrition, the rat and mouse can synthesize this vitamin in adequate amounts. (The guinea pig is one of the few animals that shares the human need for vitamin C, and as a result has been a valuable experimental animal for nutrition studies.) Plants, of course, can synthesize all the vitamins considered essential to man, as well as all the other vitamins that have been discovered. Indeed, plant substances provide major sources of vitamins in the human diet.

Vitamins are grouped into two basic categories according to their solubility. Vitamins A, D, E, and K are found to be soluble in fats and oils, and are therefore called the **fat-soluble vitamins.** These vitamins tend to be stored in the body for relatively long periods,

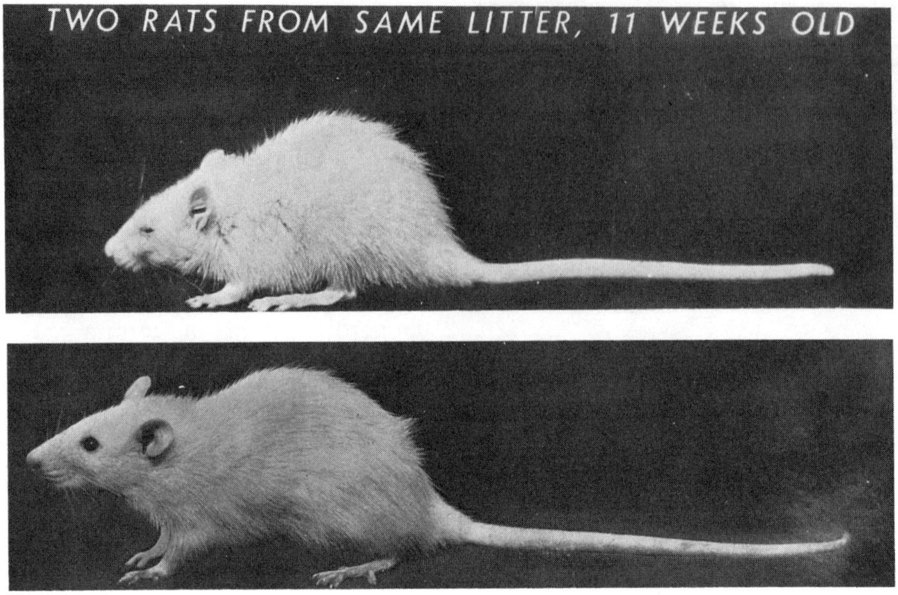

Figure 10–18 Two rats from the same litter show the effects of vitamin A deficiency. (a) A rat raised without vitamin A has an infected eye, rough fur, and a generally sick appearance. It weighs only 56 grams. (b) A littermate raised on an adequate diet has bright eyes, sleek fur, and appears alert and vigorous. It weighs 123 grams.

and thus a steady daily intake is not vital. But accumulation may become a factor in the case of high levels of intake.

The vitamin B complex and vitamin C are soluble in water, and are known as the **water-soluble vitamins.** The B-complex vitamins are, in reality, a large group of vitamins that are frequently found together in various foods. Among this group are thiamine (vitamin B_1), riboflavin (vitamin B_2), niacin, pantothenic acid, pyridoxine (vitamin B_6), biotin, folic acid, cobalamin (vitamin B_{12}), inositol, choline, and para-aminobenzoic acid. The water-soluble vitamins are not stored in the body; excesses are eliminated rapidly. As a result, cumulative overdoses do not present a problem, but a steady daily source of these vitamins must be ensured.

Several problems arise in the determination of whether a particular vitamin is essential for man. One of the standard methods of determining whether a vitamin is essential to a particular species is to prepare a special diet whose individual ingredients are known exactly and to exclude a specific vitamin from the diet fed to a particular group. If the animals of this group fail to grow normally (as compared to a suitable control group fed the same diet to which the test vitamin has been added) or if they show signs of disease, it may be concluded that the vitamin is essential. Reversal of the effects of vitamin deficiency by the timely administration of therapeutic doses of the vitamin provides corroboration of such a conclusion.

Such an experiment could not ethically be pursued using human subjects. Hence, this otherwise fruitful avenue is usually closed to the physiologist attempting to study human nutrition. Instead, he must seek out cases of deficiency already present in the population. In a number of cases this is rather difficult, for some vitamins are required in such small quantities and are so widespread among common foods that cases of deficiency are extremely rare. Vitamin E deficiency is such a case. Matters are further complicated by the fact that some vitamins are actually produced

TABLE 10-1 Some important vitamins

VITAMIN	FOOD SOURCES	CONDITIONS CAUSED BY DEFICIENCY
Fat-soluble		
Vitamin A (retinol)	Butter, eggs, fish liver oils,green and yellow vegetables and fruit	Dry and brittle skin, susceptibility to infection, night blindness, xerophthalmia (an eye condition)
Vitamin D (calciferol)	Butter, egg yolks, irradiated milk, liver oils; also synthesized in skin exposed to sunlight	Rickets, soft bones, retardation of bone and tooth formation
Vitamin E (alpha-tocopherol)	Vegetable and animal oils, meat, eggs, green vegetables	(Demonstrated in animals: sterility and death of embryos, degeneration of muscles and nerves)
Vitamin K	Green vegetables; also synthesized by intestinal bacteria	Slow blood clotting and hemorrhage
Water-soluble		
Vitamin B complex:		
Vitamin B_1 (thiamine)	Whole-grain cereals, nuts, yeast, liver	Beriberi (degeneration of nerves and muscles)
Vitamin B_2 (riboflavin)	Milk, cheese, eggs, liver, meat, yeast, wheat germ, leafy vegetables	Cracks in the corners of the mouth, sores on lips and tongue, stunted growth, eye disorders
Vitamin B_6 (pyridoxine)	Meat, liver, eggs, nuts, whole-grain cereals, fresh vegetables	Stunted growth, anemia, atrophy of lymphoid tissue and susceptibility to infection
Vitamin B_{12} (colbalamin)	Liver, meat; also synthesized by intestinal bacteria	Pernicious anemia
Niacin	Meat, yeast, fresh vegetables, whole wheat, beer	Pellagra (skin inflammation, diarrhea, mental disturbance)
Pantothenic acid	Meat, eggs, peanuts, sweet potatoes, fresh vegetables, whole-grain cereals	Stunted growth, dermatitis, gray hair, numbness and pain in feet
Biotin	Liver, yeast, egg yolk, molasses	Skin and eye inflammations (raw egg contains avidin, which can combine with biotin and prevent its absorption)
Folic acid	Leafy vegetables, liver	Anemia, stunted growth, impaired antibody synthesis
Vitamin C (ascorbic acid)	Citrus fruits, tomatoes, green peppers	Scurvy (bleeding gums, anemia, painful joints, slow healing of wounds); large doses of vitamin C have been used to prevent or alleviate the symptoms of the common cold

in the intestines by bacteria present there and are absorbed into and utilized by the body. This is true of vitamin K in rats and to some extent of vitamin B12 in man. The problem of the physiologist studying humans is further aggravated by the fact that some vitamins, especially the fat-soluble vitamins, are stored in the body. A diet deficient in one of these vitamins, therefore, might not be accompanied by any various symptoms of deficiency for some time.

Despite these limitations, a wealth of information regarding the physiological impact of a number of vitamins has been accumulated over the years. Information on the major vitamins, their food sources, their role in the human body, and the conditions caused by deficiencies of them, is summarized in Table 10–1.

Digestion

All living organisms share a need for organic raw materials to provide sources of matter and energy for their metabolic processes. The means by which different organisms obtain these raw materials vary widely, as we have seen. But once they have been obtained, in one way or another, these raw materials must be reduced in size and complexity before they can be utilized by the cells. This is accomplished through processes of **digestion.**

The basic chemical processes associated with digestion are remarkably similar throughout the kingdoms of life (Fig. 10–19). Proteins must be broken down to their constituent amino acids. Starch and other polysaccharides must be split into their constituent simple sugars. Fats must be hydrolyzed to glycerol

Figure 10–19 In processes of digestion the body breaks down complex organic compounds into simpler molecules, which can then be used to synthesize new complex organic compounds needed by the body.

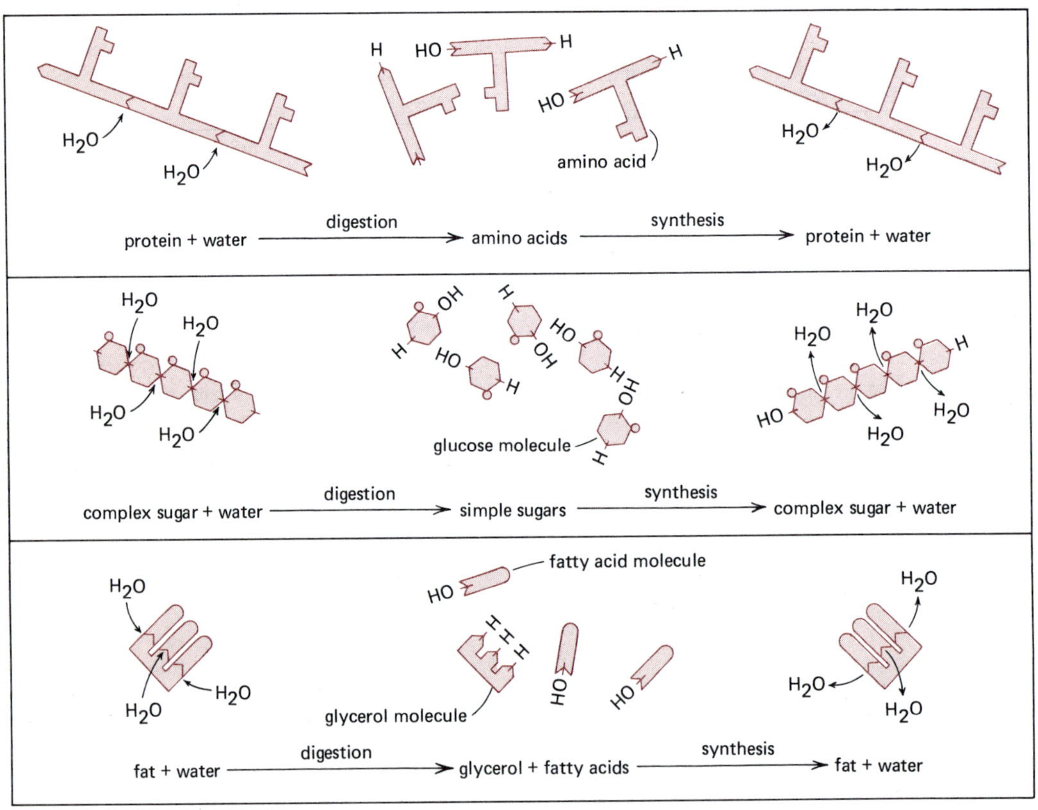

and fatty acids. These chemical reactions of digestion are catalyzed by enzymes, which show striking parallels among the most diverse groups. Proteolytic (protein-cleaving) enzymes extracted from amoebas greatly resemble the human digestive enzymes pepsin and trypsin. The plant enzyme papain is widely used to soften or "predigest" meat for human consumption.

Though the basic chemical processes of digestion are similar throughout the kingdoms of life, there is wide variation in the structures of the digestive apparatuses of the various groups. In plants, in most protozoa, and in sponges, food materials penetrate the cell membrane, and digestion actually takes place within the cells (or cell) of the organism. This mode of digestion is termed **intracellular digestion.** In higher animals, however, although digestion usually takes place within the animal's body, the digestive processes are localized within a specially modified tube (gut), and food materials do not actually penetrate the epithelial membranes of the body until after they have been digested into simpler constituents. This type of digestion is called **extracellular digestion.**

INTRACELLULAR DIGESTION

Natural waters contain about 2 to 20 mg of dissolved organic matter per liter. Aquatic protists can therefore obtain some nourishment by simple diffusion through the cell membrane. But most protozoa have additional means of obtaining food and the rudiments of a "digestive system," even within their single cell. The amoeba is a predator that engulfs its prey (small protozoa and algae) by a process of phagocytosis (Fig. 10–20). After the amoeba's pseudopods have engulfed the potential food source, they merge, effectively enclosing the food within a tiny membrane-surrounded "bubble" or vacuole within the cell. This **food vacuole** moves slowly through the amoeba's cytoplasm, as lysosomes pour digestive enzymes into it. As the food is digested, usable organic materials diffuse out through the vacuole membrane into the cytoplasm, and waste materials remain. The vacuole gradually drifts to the outer surface of the cell, where it opens, discharging its indigestible contents into the surrounding waters.

Ciliates such as the paramecium characteristically have a more organized digestive system (Fig. 10–21). An **oral groove** is surrounded by a border of waving cilia, which set up currents in the surrounding water. Bits of food (bacteria, algae, or bits of detritus), brought in by the feeding currents, pass down the **gullet,** which comes to a dead end. At the end of the gullet, the cell membrane pinches off, surrounding the food particles in a food vacuole. Digestion proceeds in a manner similar to that in the amoeba, except that the undigested waste material is egested (eliminated) through a permanently fixed opening, the **anal pore.**

Figure 10–20 The amoeba engulfs its prey and then digests it inside a vacuole. In this sequence, an amoeba is capturing a paramecium.

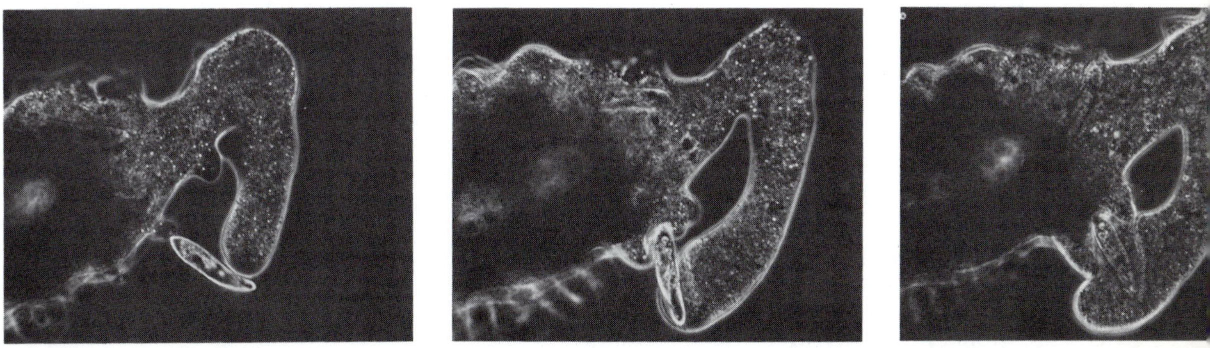

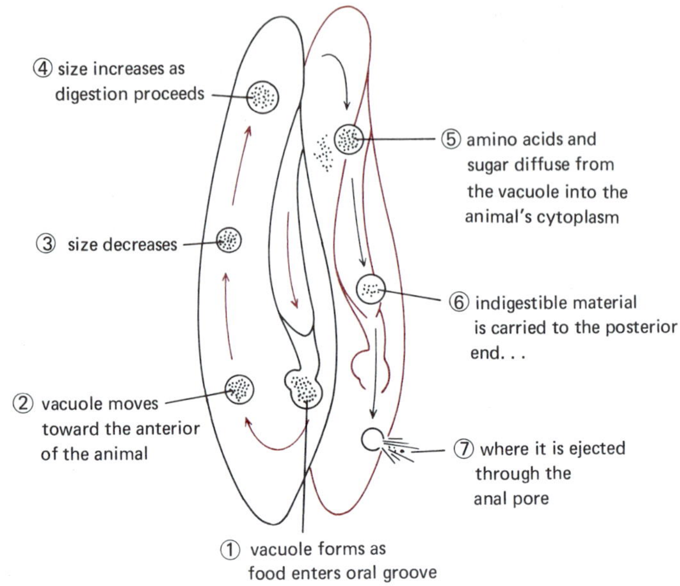

Figure 10–21 Ingestion, digestion, and egestion in the paramecium are basically similar to the corresponding processes in the amoeba but begin and end in fixed structures.

The sessile sponges also eat with the aid of feeding currents, which are set up in the water by flagella lining the inner surfaces of the thousands of narrow, branching channels or pores that extend through the walls of the sponge. The feeding currents direct sea water, laden with microscopic organisms and bits of organic matter, through the channels into the large central cavity of the sponge, the **spongocoel,** and thence out through a large aperture, the **osculum,** at the top of the sponge's body (Fig. 10–22). The spongocoel is lined with flagellated columnar cells known as **collar cells.** The base of each flagellum is surrounded by a delicate transparent upgrowth, the "collar." Movements of the flagella force particles of food against the collar, where they are ingested by phagocytosis and digested. Many sponges also have amoebalike wandering cells, called **amoebocytes,** which digest food materials and carry nutrients to various parts of the animal.

EXTRACELLULAR DIGESTION

In the coelenterates, for the first time in the evolutionary "family tree," extracellular digestion takes place within a digestive

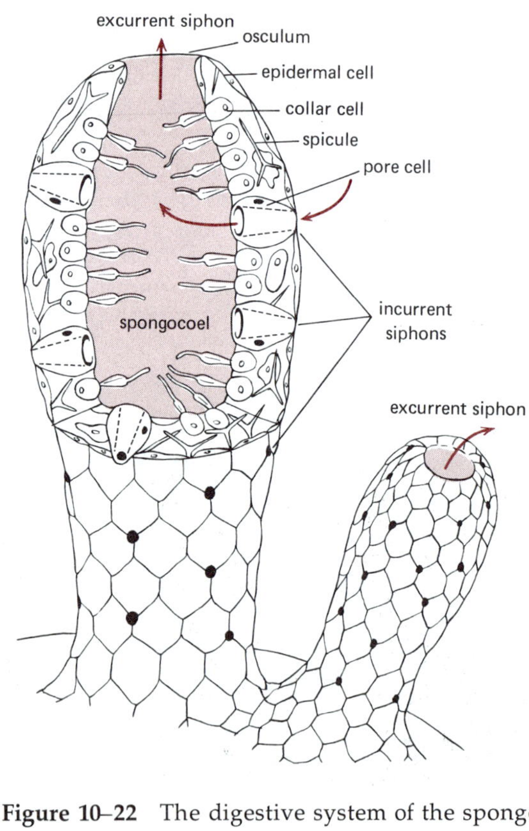

Figure 10–22 The digestive system of the sponge.

cavity or gut. The common pond-dweller *Hydra* is shaped like a slender cylinder, open at one end, with the opening surrounded by a fringe of waving tentacles. If a hapless microorganism touches one of these tentacles, stinging cells called **cnidoblasts** discharge minute "poison darts" that pierce and paralyze the prey (Fig. 10–23). The hydra's tentacles then grasp the prey and stuff it into the body cavity, the **coelenteron.** There digestive juices secreted by cells lining the cavity dismember the prey into smaller pieces. But the hydra is not entirely committed to extracellular digestion. For the half-digested bits of food are then grasped by pseudopods extending from other lining cells, and digestion is completed intracellularly. Waste materials pass out again through the single opening to the coelenteron.

The digestive system of the free-living (as opposed to parasitic) flatworms, typified by the planaria, is more complicated than that of the coelenterates, but it has a basically similar plan. A muscular **pharynx,** located on the underside of the planaria's body, protrudes through the animal's **mouth** and sucks off bits of food—small crustaceans or the remains of dead organisms. The

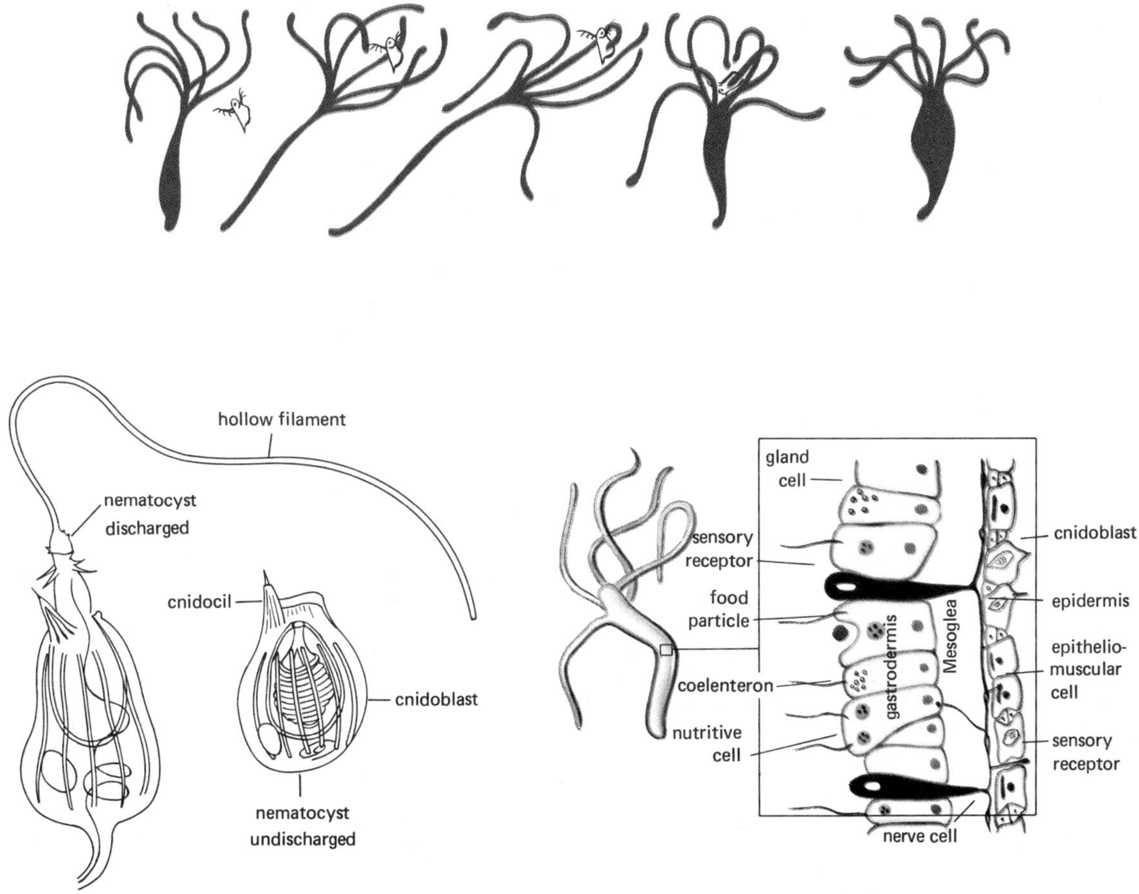

Figure 10–23 The hydra catches its prey with tentacles equipped with stinging cnidoblasts. The cells that line the body cavity digest the prey partly extracellularly (gland cells) and partly intracellularly (nutritive cells).

pharynx leads into the branching **intestine,** the ramifications of which effectively distribute food materials to all parts of the body. Here too, digestion is both intracellular and extracellular: some lining cells secrete digestive enzymes into the digestive cavity, while others send out pseudopods to engulf partly digested fragments and complete the process in food vacuoles. As in the coelenterate, the planaria's digestive system is a two-way system: waste materials leave the body through the same mouth opening through which the food was initially taken in.

ADVANTAGES OF A ONE-WAY DIGESTIVE SYSTEM

In the higher animals, the commitment to extracellular digestion becomes complete. Greater efficiency of the digestive process is promoted by a basic body plan, which is repeated with modifications in all the higher animals, from the earthworm to man. This is the **tube-in-tube body plan.**

A glance at a diagram of an earthworm (Fig. 10–24) will reveal that it looks rather like a piece of thick-walled tubing. The walls of the tubing represent the body of the animal, and the cylindrical cavity within corresponds to the digestive tract. The inner cavity begins at an opening in the front end of the animal (the **mouth**) and ends at another opening in the rear, the **anus.** Among higher animals, the outer contour of the body is often modified from a simple cylinder into complex shapes, with a variety of appendages. The inner cavity also is modified from a simple cylinder to an elaborate series of bulges, pouches, and tubes. The latter may in addition be intricately coiled and packed within the body, so that the "inner tube" is actually far longer than the "outer tube" that encloses it. But the basic design is still there: a mouth at one end of the body opens into a roughly tubular cavity that runs continuously to another, terminal opening, the anus.

The tube-in-tube body plan confers some important advantages in carrying out the function of digestion. Instead of the inefficient "two-way traffic" of the coelenterate and flatworm digestive tract, it permits a continuous one-way flow of materials from the mouth through the digestive tract to the anus. Specialized organs along the way can

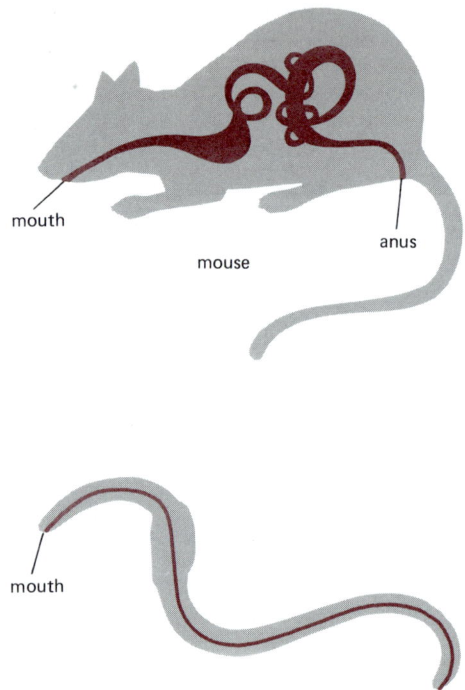

Figure 10–24 Both the mouse and the earthworm have a basic tube-within-tube body plan, though the tube of the mouse's digestive system is elaborately looped and coiled.

work on the food materials in turn, digesting specific components and preparing the mass for the action of the next "stations" along the route. Digested food materials can be removed en route, and waste materials can continue unimpeded on the way out of the body, without any danger of their getting mixed up with new batches of incoming food.

FROM WORM TO MAN

The earthworm literally eats its way through the soil in which it lives. Soil, laden with leaves and decaying organic matter, is sucked in through the earthworm's mouth by the muscular **pharynx.** From there it passes through the **esophagus** into a thin-walled chamber, the **crop.** The walls of the crop are expandable, permitting the soil-food mixture to be stored there temporarily. Next the ingested material passes into a hard, thick, muscular-walled chamber, the **gizzard.** Here the food is ground up into a fine mass by the combined action of the muscular walls and pebbles that have been taken in along with the soil. The material then passes into the long, straight **intestine,** where digestive enzymes break down the food materials, and the products of digestion are absorbed through the lining cells into the body. Undigested material, along with the soil, passes out of the **anus** in the form of small pellets, called **castings,** which are often found in heaps next to earthworm burrow entrances. The action of earthworms helps both loosen the soil, making pockets for air to be entrapped, and to enrich the soil with organic matter. Both these effects are extremely beneficial for the growth of plants.

A curious variation of digestion is found in the starfish. Here digestion is not only extracellular, but it often takes place outside the body! The starfish has a rather small mouth, located on the underside of its body. Small prey animals can be taken in through the mouth, but feeding on large mollusks or fish presents a problem.

The manner in which a starfish copes with a large clam or oyster is a masterpiece of ingenuity. The muscles of the bivalve can hold its two shells so tightly together that human strength could not pry them apart. The starfish grasps both shells with numerous suckers on the bottoms of its tentacles and pulls, slowly and steadily. Eventually the muscles of the bivalve relax sufficiently to permit the shells to gape open a minute fraction of an inch. Immediately the starfish everts its stomach (literally turns it inside out) and inserts it into the gap between the shells. It then pours out digestive juices, digests the soft body of the bivalve right within its own shells, and sucks in the resultant soupy mass.

In many animals, including man, digestion begins with a mechanical process, in which food is ground or chopped into smaller pieces. This is followed by a chemical process, in which the bits of food are further decomposed by enzymes into simpler chemical constituents, which can then be utilized by the body cells for building materials and energy.

In humans the mechanical breakdown is begun by chewing the food with teeth. Many other species of vertebrates are also equipped with teeth, but not all of them are equipped for chewing. Herbivores typically have broad, flat teeth that are excellently suited for grinding plant materials. But carnivores such as lions and tigers, many fish, and seals typically use their teeth only for holding and tearing their prey. They do not chew their food, but swallow large chunks of it whole. The mechanical breakdown is accomplished largely by grinding in the muscular stomach.

Insects are characterized by a variety of types of digestive systems, depending on the type of food eaten by the particular species and the manner in which it is eaten. The mouths of some insects are modified into long, sharp-pointed sucking tubes, like the needle of a hypodermic syringe. These insects, such as mosquitoes and aphids, eat only liquid food—blood of animals or juices of plants. No mechanical breakdown is needed to prepare such food for the chemical processes of digestion. Other insects are

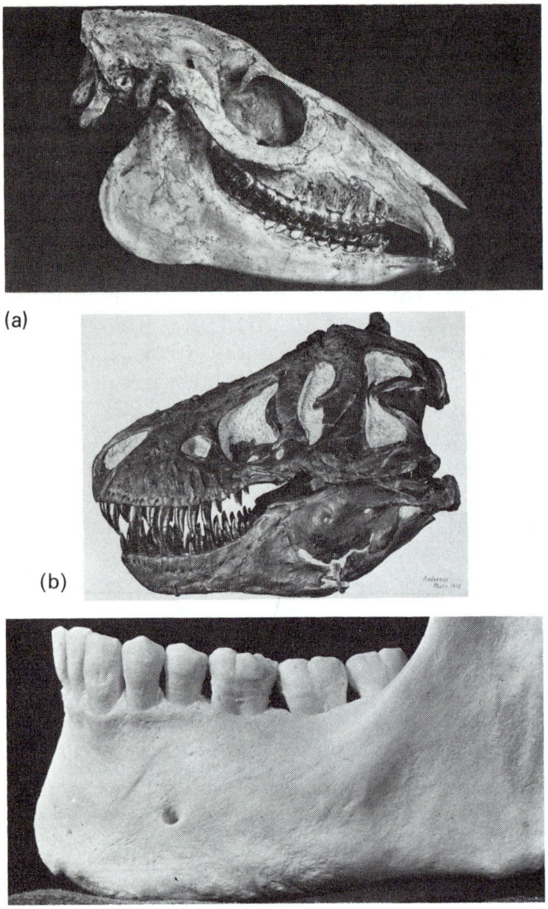

(a) (b) (c)

Figure 10–25 Teeth are adapted to an animal's way of life: (a) skull of *Mesohippus*, an extinct herbivore; (b) skull of *Tyrannosaurus*, an extinct carnivorous dinosaur; (c) lower jaw of man, an omnivore.

equipped with hard mouth parts, which function analogously to teeth in biting and chewing. In addition, some mechanical breakdown may occur by grinding in a muscular gizzard.

The digestive system of the grasshopper (Fig. 10–27) is typical of the latter group. The grasshopper feeds primarily on plant leaves, biting off pieces with its hard mouth parts. As food enters the **mouth** or **oral cavity,** it is mixed with **saliva** secreted by a pair of **salivary glands.** The food then passes through a short **esophagus** and is temporarily stored in a thin-walled **crop.** The food is ground and pulverized in the muscular **gizzard.** This leads into the **stomach,** where most of the digestion and absorption takes place. Further absorption occurs in the **intestine;** waste materials are stored temporarily in the **rectum,** and are finally eliminated through the **anus.**

In some ways, the digestive system of the bird is strikingly similar to that of the grasshopper. The bird also bites off chunks of food with its hard mouth parts (its **bill**) and swallows them whole. The ability to store food temporarily in a distensible **crop** is a useful adaptation, which permits the bird to alight briefly, gobble down large quantities of food, and then take to the safety of the air again, to digest its food at leisure. The chunks of food are broken down in the bird's **stomach,** by a combination of churning and the action of digestive juices. Among seed eaters, whose food requires more strenuous measures for effective decomposition, part of the stomach is modified into a thick-walled, muscular **gizzard** with a hard, horny lining. The grinding action of the gizzard is aided by the curious practice of these birds of swallowing small pebbles and stones. These stones remain in the gizzard of the bird and help to grind its food. (Curiously, seals, which also swallow their food whole, have the same habit of swallowing small stones, which help to grind the food in their stomachs.)

The digestive systems of most mammals are basically similar to that of man, with modifications from the general plan that suit the dietary habits of the particular species. Let us turn now to an examination of the parts of the human digestive system and how they work.

The Human Digestive System

It seems astonishing to consider that within each of our bodies there is a tube some 30 feet long. The actual distance from mouth to anus, measured along the outer surface of the body, may be only a few feet. But the digestive tract that connects these two openings turns and twists and coils. The small intestines, in particular, which make up

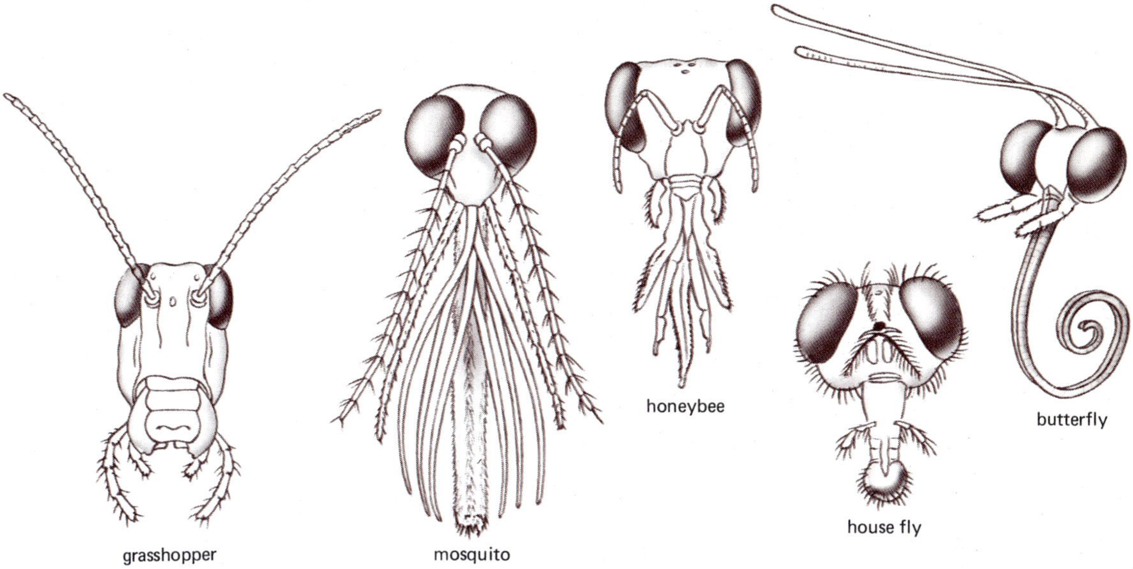

Figure 10–26 An assortment of insect heads. The mouth parts show a great diversity of modifications corresponding to the type of food and the manner in which it is eaten.

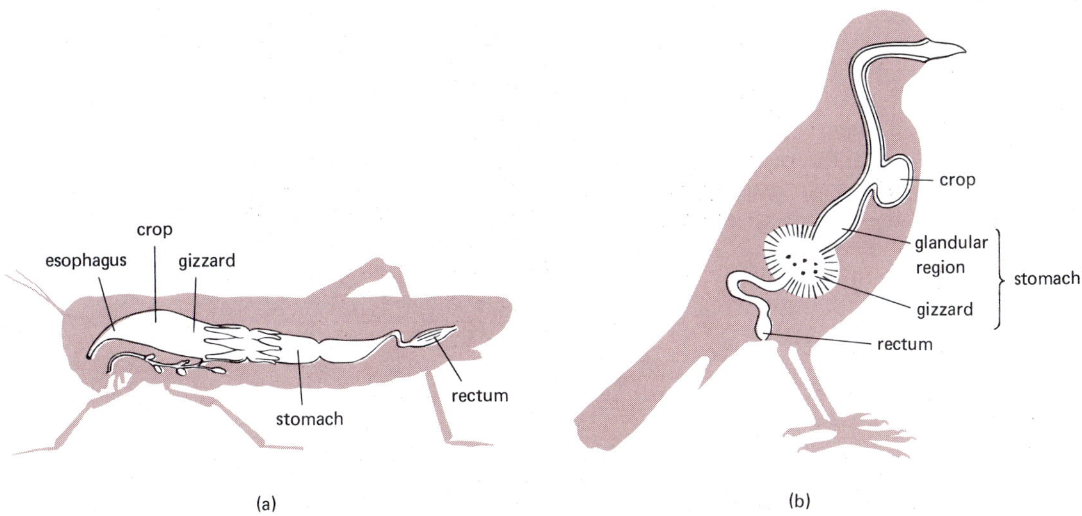

Figure 10–27 Digestive systems of the grasshopper (a) and bird (b). Note that both utilize a crop for temporary storage of food and a muscular gizzard for grinding food. These similarities are independent adaptations to similar modes of feeding.

more than 20 feet of the total length of the digestive tract, are packed like loops of spaghetti into the abdominal cavity.

THE MOUTH

The entrance to our one-way digestive tract is through the mouth. This opening is guarded by movable, fleshy **lips** (a mammalian innovation, possessed by no other group of vertebrates or invertebrates), and within them a border of **teeth.** These are two sets of hard, bonelike structures, embedded individually in the fleshy **gums** that cover the lower and upper **jaws.**

The mouth opening leads into a cavity with a vaulted roof, formed by the **hard** and **soft palates.** Attached to the floor of the **mouth** or **oral cavity** is a movable, fleshy **tongue.** An interlacing of muscles permits the tongue to be manipulated in virtually any direction. Its surface is studded with **taste buds,** cup-shaped collections of specialized sense cells that are sensitive to particular chemical configurations that are perceived as characteristic tastes. The sense cells of the tongue are capable of responding to four basic types of tastes: sweet, sour, salty, and bitter. The taste buds for each particular type of taste are concentrated in specific areas of the tongue surface (Fig. 10–28). Few food substances and flavoring agents stimulate only one particular type of taste receptor. (One of the major disadvantages of certain artificial sweetening agents, for example, is their tendency to leave a bitter aftertaste.) The varied flavors that humans can distinguish, therefore, are, to a substantial degree, the result of combinations of stimulation of the four types of taste cells.

In addition to its functions in the digestive process, the tongue also aids in modulating sounds emerging from the mouth cavity and thus assists in the articulation of speech.

Adult humans possess 32 teeth, distributed in two corresponding sets, half in the lower jaw and half in the upper jaw. The teeth on one side of the mouth are essentially a mirror image of those on the other

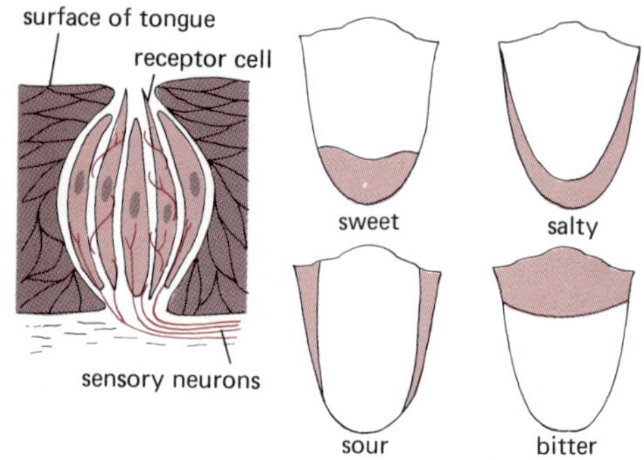

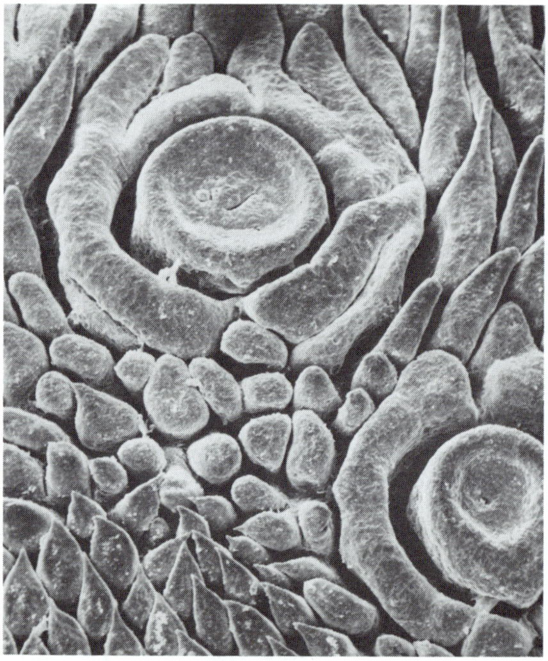

Figure 10–28 The sense receptors of the tongue, grouped into structures called taste buds, are sensitive to four basic types of tastes: sweet, sour, salty, and bitter. The four types of taste buds are distributed nonuniformly over the surface of the tongue, as is shown in the small maps of the tongue surface. The variety of tastes we perceive are combinations of stimulation of the four types of taste buds, together with sense impressions from the olfactory receptors of the nose. The scanning electron micrograph depicts taste receptors in the tongue of a dog.

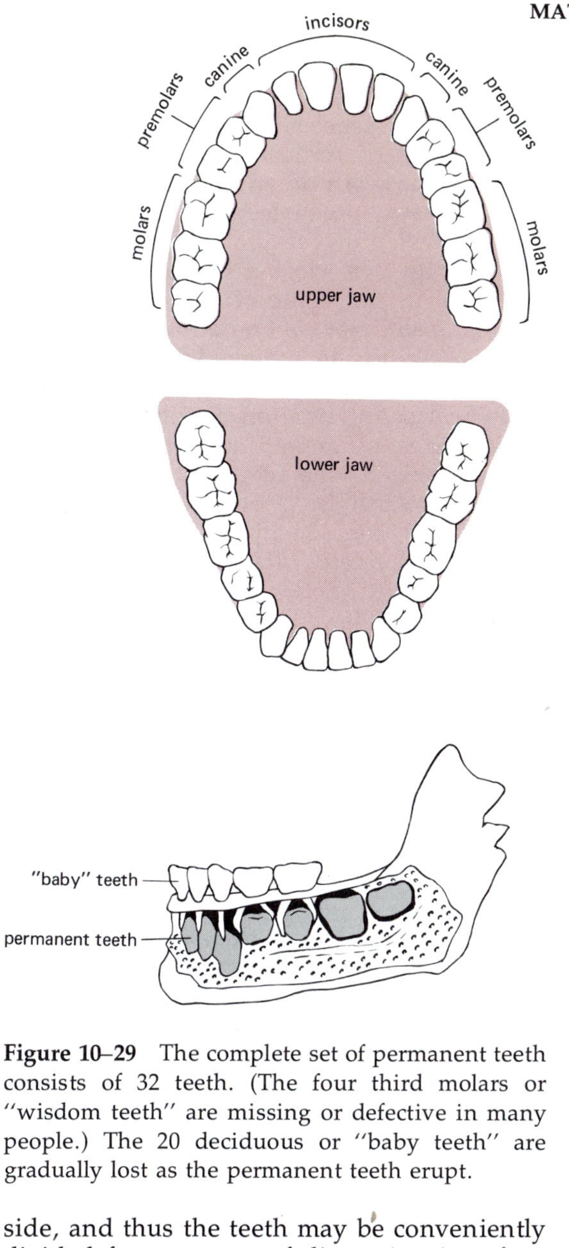

Figure 10–29 The complete set of permanent teeth consists of 32 teeth. (The four third molars or "wisdom teeth" are missing or defective in many people.) The 20 deciduous or "baby teeth" are gradually lost as the permanent teeth erupt.

side, and thus the teeth may be conveniently divided for purposes of discussion into four essentially identical quadrants (Fig. 10–29a). Proceeding from the middle of the jaw back toward the hinge, we find first in each quadrant two chisel-shaped teeth, the **incisors.** These are used for biting and cutting food. Next is a pointed, cone-shaped tooth, which resembles the fangs of a dog; from this resemblance is derived its name, **canine.** The canines are used for tearing food. Posterior to the canine are two broad and flattened teeth, the **premolars** or **bicuspids,** which are adapted for crushing and grinding. The last three teeth in each quadrant, the **molars** or **tricuspids,** are somewhat larger than the bicuspids but are constructed on basically the same plan and are used for the same purposes. The 32 teeth of the adult human thus include a total of 8 incisors, 4 canines, 8 premolars, and 12 molars.

The jaws of a child would not be large enough to accommodate the full set of adult teeth. This problem is solved by the fact that a normal human has two complete sets of teeth during his lifetime (Fig. 10–29b). The first set of 20 **milk teeth** or **deciduous teeth** begins to appear in the first year after birth, being completed generally by the end of the second year. The milk teeth are smaller than the permanent teeth, and they do not include any molars. At about five to seven years of age, the milk teeth begin to fall out, and the permanent teeth begin to erupt. By about twelve years of age a person usually has his first 28 permanent teeth. The last 4 molars (the **wisdom teeth**) usually erupt at about eighteen years, but in many people they are undersized and decay readily or do not appear at all.

The outer covering of the teeth, called **enamel,** is the hardest substance in the body. It contains only about 4 percent organic matter and is largely composed of calcium apatite with small amounts of calcium phosphate. Beneath the layer of enamel is a layer of **dentin,** which resembles bone. Within the dentin is a region of soft tissue interlaced with connective tissue, blood vessels, and nerves; this is the **pulp.** The portion of the tooth protruding above the gum into the oral cavity is designated as the **crown.** The **neck** of the tooth is enclosed by the gum, while the **root** fits snugly into a socket in the jawbone. **Cement** affixes the tooth to the jawbone in the region of the root.

If teeth are not cleaned regularly and thoroughly, they become coated with a deposit called **plaque.** Bacteria thrive in the bits of decomposing food and tissue debris

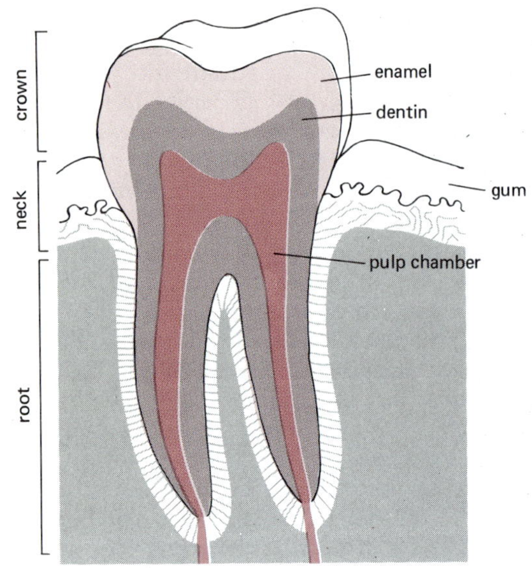

Figure 10–30 Structure of a tooth.

contained in plaque, particularly in the crevices between teeth. They secrete acids that can eat through the hard enamel coating of teeth, causing **dental caries** (tooth decay). A diet high in sugars promotes the growth of decay bacteria, particularly if they are allowed to remain on the teeth after eating.

The teeth of fish are quite different from human teeth, being rootless, undifferentiated pointed cones. In many fish, including sharks, the teeth are continually replaced as they are worn through, a characteristic that many false teeth wearers must wish we shared.

Although the teeth of all the mammals are basically similar, the distribution of the various types shows modifications according to the dietary habits of the species. Among herbivores, such as the horse and cow, the molars are large and flat, well suited for grinding grass. The lower incisors are sharp and well developed, permitting them to crop vegetation by pulling it with the tongue and upper lip across the sharp edges of the lower incisors; the canines are often absent. Flesh-eating animals, on the other hand, usually have large canines, suitable for tearing meat into chunks that can be swallowed. Gnawing animals have especially prominent incisors. The incisors of rodents grow continuously and must be worn down by ceaseless gnawing. (If a rat is prevented from gnawing, his incisors will grow several inches long and prevent him from closing his mouth.) Sometimes, as in the tusks of the elephant and walrus, teeth may be modified for special purposes. With the extreme specialization of mammalian teeth, a zoologist can gain a great deal of information about an animal's habits merely by examining its teeth. Indeed, paleontologists, who reconstruct the history of life on earth by a study of bones and other fossils, can sometimes build up a surprisingly complete picture of an animal's appearance and mode of life from an examination of a few teeth (see Fig. 10–25, p. 244).

DIGESTIVE PROCESSES IN THE MOUTH CAVITY

The first part of the digestive process may begin before you even take a bit of food into your mouth. The sight and smell of food—sometimes even the thought of a favorite dish—is often enough to start a flow of **saliva** flooding into the mouth cavity. This enzyme-rich fluid is secreted by three pairs of **salivary glands** (Fig. 10–31).

There is actually a continuous flow of saliva, occurring throughout the day, even when no digestion is taking place. This secretion of fluid (totaling about 1 to $1^1/_2$ quarts a day!) keeps the mucous membranes of the mouth cavity moist.

When you take a bite of food—let us say, a roast beef sandwich—into your mouth, it immediately is subjected to a variety of physical and chemical actions (Fig. 10–32). The chewing motions of the teeth break down the large chunks of solid material into smaller bits. These are continually moved about by the tongue and become more or less thoroughly mixed with saliva, depending on how long the mouthful is chewed. Gradually the food mass is shaped into a ball, called a **bolus.**

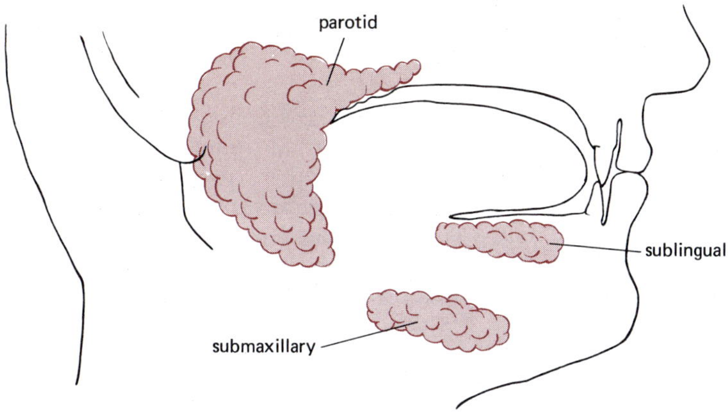

Figure 10–31 The salivary glands.

Figure 10–32 On its way through the digestive system, a roast beef sandwich is gradually broken down to its constituent nutrients and absorbed into the body.

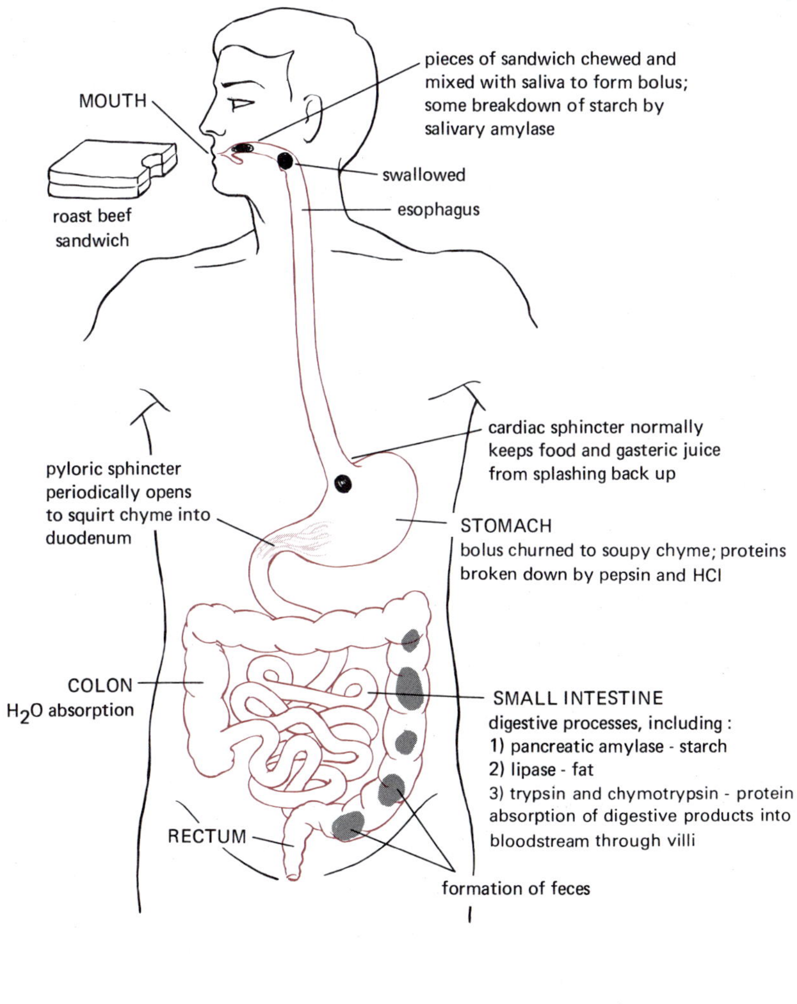

Saliva performs a number of important functions as the food is being chewed. It moistens and softens the food, aiding in the formation of the bolus. Saliva acts as a solvent for food materials, permitting the taste buds on the tongue to be activated (for these sense cells operate only in the presence of moisture). Saliva also contains an enzyme, **salivary amylase,** which begins the chemical processes of digestion. As can be guessed from its name, salivary amylase catalyzes the hydrolysis of amylose, a major constituent of starch, ultimately yielding many units of the disaccharide maltose. Farther along the digestive tract, another enzyme, maltase, works on the maltose, adding water to it and splitting it into two molecules of glucose (Fig. 10–33). Thus, in the mouthful of roast beef sandwich, the meat and fat would be only softened and physically broken down into smaller pieces in the mouth. The starch in the bread, however, would be chemically decomposed as well, the extent of the process depending on the total time of chewing. The pH of saliva (6.3–6.8) is the optimum value for the effective action of the salivary enzyme.

After the food has been shaped into a bolus, it is ready to leave the mouth cavity for the next way station in the digestive process, the stomach. With the aid of the tongue, the bolus is manipulated into the back of the mouth cavity and initiates the **swallowing mechanism.** This is a series of events which, once begun, proceeds without any further voluntary control. The **uvula,** a pendant, fleshy lobe at the back of the mouth cavity, rises, shutting off the air passage up into the nose (the nasopharynx). This act prevents any solids or liquids from entering the nasal cavities. Simultaneously, muscular contractions cause a valvelike fold of tissue, the **epiglottis,** to close the top of the air passage leading to the lungs like a trap door. Thus, food is prevented from passing into the respiratory tract. The position of the tongue prevents the food from returning to the mouth, so that when the muscles in the walls of the passageway leading down from the mouth cavity (the **pharynx**) contract, the bolus is pushed down the only remaining accessible pathway, the **esophagus.**

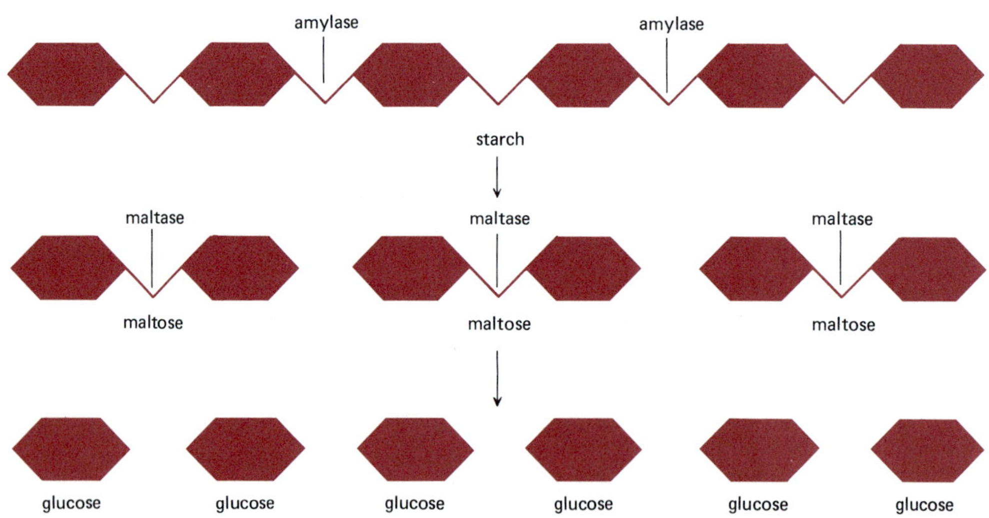

Figure 10–33 Two successive enzymatic reactions digest starch to its glucose constituents.

The swallowing mechanism requires precise coordination and timing. Nearly always it operates perfectly, but occasionally things go a bit awry. Failure of the uvula to completely block off the nasopharynx may result in the swallowing of air while eating. (Babies, who are novices at the job of eating, are particularly prone to swallow air in this way and therefore must be carefully "burped" after meals to dislodge air bubbles from the stomach.) Or occasionally some liquid may be splashed up into the nose. Failure of the epiglottis to close the windpipe results in the introduction of solid or liquid particles into the air passages. This initiates an automatic mechanism of coughing and sneezing, which may continue until the offending particles are expelled.

The **esophagus** is a straight, thick-walled, muscular tube, which extends for about ten inches and connects the pharynx with the stomach. A succession of contractions and relaxations of circular muscles in the walls of the esophagus propels the bolus down to the stomach within about five to six seconds. Before the first astronauts were sent out into space, physicians were worried about whether food could be swallowed properly under conditions of weightlessness, without the action of gravity to help the bolus or liquid down. But no difficulties were encountered. The wavelike contractions of the esophagus and other portions of the digestive tract, referred to as **peristalsis,** carry the food along in a quite normal manner even at zero gravity.

THE STOMACH

The stomach is a thick-walled muscular pouch, situated somewhat to the left side of the body just beneath the diaphragm. At the top of the stomach, demarcating the boundary with the esophagus, is a ring of smooth muscle, the **cardiac sphincter.** A similar valve of circular smooth muscle, the **pyloric sphincter,** marks the end of the stomach, separating it from the small intestine.

The stomach has rather elastic walls, and can expand to hold as much as two to three

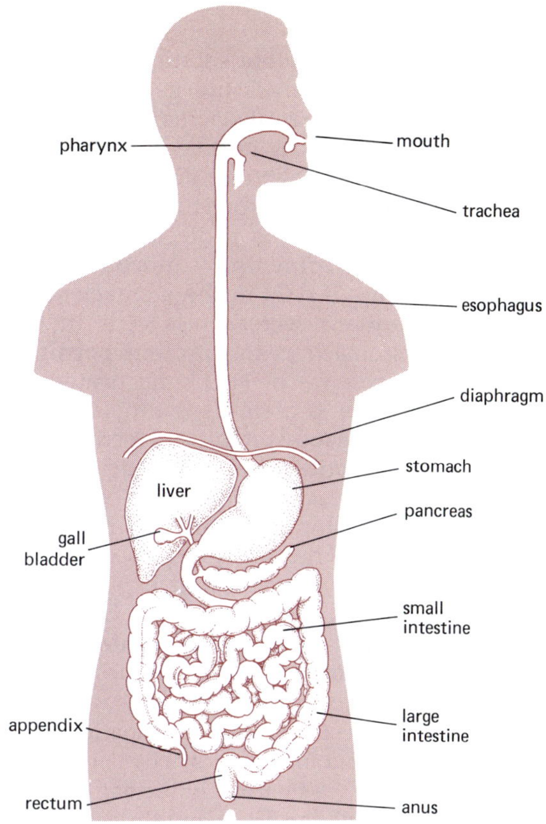

Figure 10–34 Structures of the human digestive system.

quarts after a large meal. When the stomach is empty, its walls contract to the point that their inner surfaces nearly touch one another. This condition initiates contractions of the stomach muscles, which stimulate sensory nerve endings, giving rise to hunger pangs. When food enters the stomach, at first virtually all peristaltic activity ceases. But as additional food is taken in, mild contractions begin and travel in a wavelike motion toward the pyloric region. As time passes, the contractions become more frequent and violent. After an hour or so (depending on the quantity and type of food), the entire stomach is churning and grinding the food, which is transformed into a thick, soupy mixture known as **chyme.**

Several kinds of glands in the stomach wall secrete digestive juices into the cavity.

One type secretes a dilute solution of hydrochloric acid, which gives the gastric juice a pH between 1 and 2. It seems astonishing that the stomach of a normal human contains an acid solution strong enough to burn or decompose any unprotected tissue. The lining of the stomach is protected from the action of its acid contents by mucous secretions of a group of cells in the stomach wall. Another important secretion of cells in the wall of the stomach is the protein **pepsinogen.** In the presence of HCl, this protein is converted to the protein-digesting enzyme **pepsin,** which cleaves protein molecules into smaller units (peptides and amino acids).

Much of the early knowledge of what goes on in the stomach came from a very curious accident, the potentialities of which were immediately realized and exploited by an alert physician. In 1822 a young Canadian trapper, Alexis St. Martin, accidentally shot himself in the stomach. He was successfully treated by an American army surgeon, William Beaumont. But when the wound healed it left a fistula (a permanent opening) leading from the stomach to the outside through the abdominal wall. For the next eight years, Beaumont supported the young trapper, his wife, and their growing family and used St. Martin as a living laboratory for the study of digestive processes. The fistula permitted samples of gastric juices to be drawn off after various types of meals, and samples of food could also be inserted directly into the stomach and then pulled out again at various stages of digestion.

Let us return for a moment to the bolus formed from a mouthful of roast beef sandwich and follow its fate in the stomach. After the bolus is propelled through the esophagus, it passes through the cardiac sphincter, which then closes above it, somewhat as the shutter of a camera closes in. (Occasionally the cardiac sphincter mal-

Figure 10–35 One of the Beaumont experiments. A shooting accident left a young Canadian trapper with a fistula leading into his stomach. His doctor, William Beaumont, conducted a series of experiments on the functioning of the stomach. In the one depicted here, pieces of meat in a mesh bag were inserted into the stomach through the fistula and then withdrawn after various periods of time.

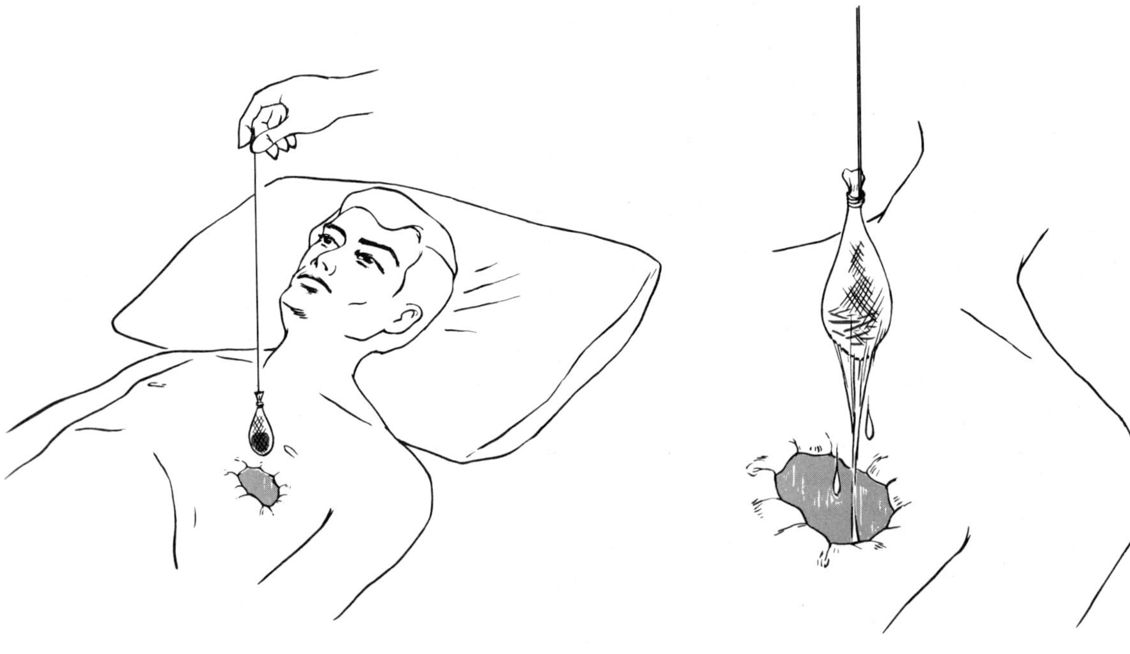

functions and opens accidentally, permitting a bit of the acid gastric juice to splash out, creating the sensation of "heartburn." Accidental openings of the cardiac sphincter are quite frequent among infants, causing them to "spit up" some of their milk or food.)

When the bolus hits the acid gastric juice in the stomach, its pH is abruptly lowered, thus rendering the salivary amylase inactive. The churning action of the stomach stirs the bolus material into soupy chyme, and various components of the gastric juice begin to act upon the food substances. The animal proteins of the meat and the vegetable proteins of the bread are hydrolyzed by pepsin, forming peptides and amino acids. The hydrochloric acid present hydrolyzes disaccharides into monosaccharides, thus continuing the digestion of starches initiated by the salivary enzyme. Hydrochloric acid also acts as a germicide, killing most of the bacteria that have entered the stomach with the food.

If the food had spoiled, or if something happens to upset you while the food is in your stomach, or if you have an attack of motion sickness, the **vomiting mechanism** may be initiated. Impulses are sent out from the vomiting center of the brain, located in the medulla, and deep contractions of the lower stomach wall are induced. Simultaneously the cardiac sphincter relaxes; the esophagus then relaxes, and the diaphragm and abdominal muscles contract strongly. Food is propelled forcefully from the stomach up the esophagus and out through the mouth. The vomiting reflex can serve as a protective mechanism, ridding the body of harmful substances that have been taken into the stomach. The reflex is lacked by some mammals, including the rat. The effectiveness of some rat poisons is due in part to the inability of these animals to vomit out the poison.

If the food is wholesome and digestion proceeds calmly and quietly, after a time the churning of the stomach raises the pressure of its contents higher than the pressure prevailing in the upper portion of the intestines (the **duodenum**). The pyloric sphincter then opens, and a portion of chyme is squirted out of the stomach and into the duodenum. (A few substances, such as glucose and alcohol, are absorbed rapidly and directly through the stomach wall into the bloodstream.)

A normal meal takes about three to four hours to pass through the stomach. Some foods, such as fats, have a tendency to depress the peristaltic movements in the stomach and thus reduce the rate at which food enters the small intestine. A high fat meal may thus remain in the stomach for more than five to six hours. The ingestion of extremely cold liquids or substances such as ice cream also tends to delay the action of the stomach.

THE SMALL INTESTINE

The small intestine is a tube about 2 inches in diameter in the region near the stomach, narrowing to 1 inch as it coils and twists for more than 20 feet before joining the large intestine. The first 10 to 12 inches of the small intestine is called the duodenum. It is held rigidly in place by a transparent membrane, the mesentery, which supports most of the visceral organs. Beyond the duodenum, the small intestine continues for 8 to 10 feet as the **jejunum.** The last 10 to 12 feet constitute the **ileum.**

Leading into an opening in the wall of the duodenum are two small tubular ducts through which secretions flow from two large digestive glands, the **pancreas** and the **liver.**

A variety of substances are secreted by the pancreas into the digestive tract. One, the salt sodium carbonate, neutralizes the stomach acid and rapidly raises the pH of the intestinal contents to a value of about 8. This is the optimum value for the functioning of the various digestive enzymes that operate in the intestines. In addition, several enzymes are secreted by the pancreas, including (1) an **amylase,** which hydrolyzes starch to maltose; (2) a **lipase,** which hydrolyzes lipids (fats) to glycerol and fatty acids;

(3) two proteinases, **trypsin** and **chymotrypsin,** which hydrolyze peptide bonds that were not cleaved by pepsin (these two enzymes are actually secreted in the inactive forms, trypsinogen and chymotrypsinogen, which are converted to the active enzymes in the intestines); (4) a **carboxypeptidase,** which removes, one by one, the amino acids at the acid end of peptide molecules, thus aiding in the task of protein digestion; and (5) **nucleases,** which hydrolyze RNA and DNA to their constituent nucleotides.

The secretion of the liver, **bile,** is not an enzyme but plays an important role in the digestion and absorption of fats. Bile is secreted continuously by the liver; between meals it accumulates in a storage organ, the **gallbladder,** and then is expelled into the duodenum when chyme enters the intestines. Bile is composed mainly of water, inorganic salts, organic bile salts, and bile pigments, all by-products from the breakdown of hemoglobin in the liver. The bile salts act as emulsifying agents, breaking large fat globules down into tiny droplets that are more readily acted upon by lipases. They also form complexes with fatty acids and facilitate their passage through the intestinal walls. Bile stimulates peristaltic movements in the intestines and aids in the absorption of the fat-soluble vitamins.

The cells lining the intestines also secrete an **intestinal juice,** which contains a variety of enzymes. The major digestive enzymes and other secretions, their sources and roles in digestion are summarized in Table 10–2.

From the time food passes through the pyloric sphincter until it leaves the small intestine through another valve, called the **ileocolic sphincter,** the small intestine re-

(a)

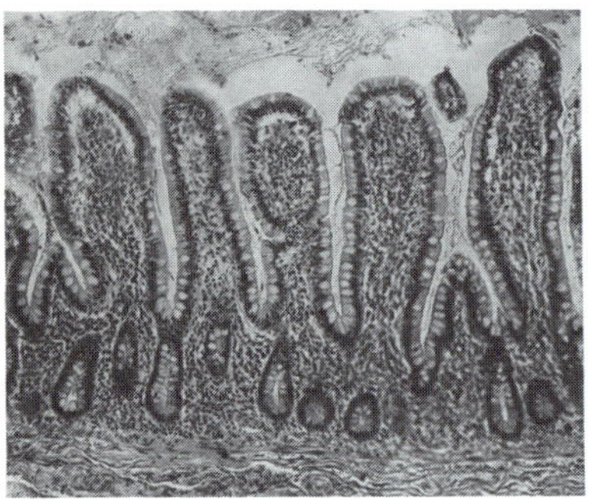

Figure 10–36 (a) Photo of villi of the human intestines, 66×. The villi that cover the intestinal surface provide an enormous increase in surface area. A fold of the human intestine is shown in (b) at the left. The structure of a villus is depicted in the center, while the small blowup at the right shows the fine structure of the epithelial cells of the villus. Including the additional surface provided by the microvilli, it has been computed that one square inch of intestinal wall has 600 square inches of absorptive surface.

(b)

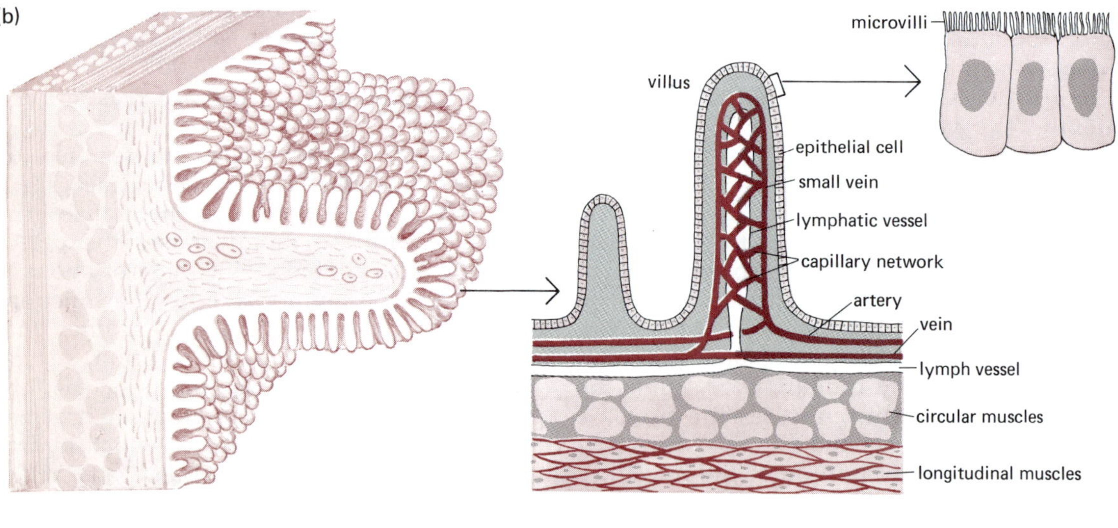

mains in almost constant motion. Combinations of rhythmic waves of contraction, repeated alternations of contraction and relaxation, and to-and-fro motions combine to mix the chyme thoroughly, bring it into intimate contact with the intestinal walls, and move it along toward its ultimate destination, the anus.

ABSORPTION IN THE SMALL INTESTINE

The lining of the small intestine, particularly the ileum, is studded with enormous numbers of tiny fingerlike projections, called **villi,** about 1 mm long. Each villus contains a network of blood capillaries and a tubular projection of the lymphatic system, called a **lacteal** (Fig. 10–36). The electron microscope reveals that the surface of each villus is in turn covered with thousands of smaller projections, the **microvilli.** The net effect is a tremendous increase in the effective surface area of the small intestine. It has been computed that the lining of the small intestine has a total surface area of about 100 square feet, or about five times the total surface area of the skin.

This vast surface area of the small intestine represents a major gateway through which the products of digestion are taken into the body for ultimate use by the body cells. Amino acids, sugars, nucleosides, glycerol, fatty acids, and other small molecules and ions pass through the intestinal membrane by processes of diffusion and active transport.

Amino acids, sugars, and various other water-soluble materials pass into the network of capillaries that surround the lacteal and are carried through the bloodstream to the liver. There they are stored temporarily and then passed on through the bloodstream to the body cells. Fatty substances pass into the lacteals and are carried into the lymphatic system, eventually emptying into the bloodstream.

If any poisonous substances have penetrated into the digestive tract and have reached this point of the digestive process, they are absorbed into the bloodstream and carried to the liver, which is the body's major chemical factory. In addition to its many functions already enumerated in this chapter, and various others, the liver is the major detoxifying organ of the body, capable of converting various poisons to less toxic products. These then pass out in the bile and eventually find their way out of the body through the intestines along with the undigested food materials.

THE LARGE INTESTINE

Just below the sphincter that separates the small and large intestines, the large intestine balloons out into a blind pouch, the **caecum.** At the bottom of the caecum is a long, narrow, fingerlike projection, the **vermiform appendix.** Serving no known function in man, this outpocketing of the caecum may become a focal point for bacterial infections, leading to a condition of inflammation called **appendicitis.** Then the appendix becomes a danger to the rest of the body, for it may balloon out and ultimately burst, spewing bacteria out into the abdominal cavity. The surgical removal of an inflamed appendix to prevent such an outcome is referred to as **appendectomy.**

From the caecum the **large intestine** extends up the right side of the body to the top of the abdomen; this portion is called the **ascending colon.** The large intestine continues horizontally across to the left side, as the **transverse colon,** then turns sharply downward (the **descending colon**) and loops over toward the midline of the body in an S-shaped curve (the **sigmoid colon**). At the midline of the body, the large intestine turns sharply downward again and continues for about six inches as the **rectum,** a straight tube that terminates in an opening known as the **anus.** The anal opening is guarded by two sphincters, an inner ring of smooth muscle and an outer ring of skeletal muscle. The latter valve permits voluntary control over the retention or elimination of the **feces,** the contents of the rectum.

Most of the water in the nearly five quarts of digestive juices and mucous secretions emptied into the digestive tract each day is reabsorbed in the large intestine. The undigested and indigestible remains of the in-

TABLE 10-2 Digestive Secretions

SECRETION	TYPE	SOURCE	EFFECT
Active in mouth			
Salivary amylase	Enzyme	Salivary glands	Starch → maltose
Mucin	Carbohydrate-protein	Salivary glands	Binds food particles together
Active in stomach			
Gastrin	Hormone	Stomach	Secretion of gastric juice (HCl + enzymes)
Hydrochloric acid	Mineral acid	Stomach	Pepsinogen → pepsin; disaccharides → monosaccharides; kills bacteria
Pepsin	Enzyme	Stomach	Proteins → peptides
Rennin	Enzyme	Stomach	Coagulates milk proteins
Active in small intestine			
Sodium bicarbonate	Salt	Pancreas	Neutralizes HCl
Amylase	Enzyme	Pancreas	Starch → maltose
Lipase	Enzyme	Pancreas	Fats → fatty acids + glycerol
Proteinases: trypsin + chymotrypsin	Enzyme	Pancreas	Peptides → smaller peptides
Carboxypeptidase	Enzyme	Pancreas	Peptides → amino acids
Nucleases: RNase +DNase	Enzyme	Pancreas	Nucleic acids → nucleotides
Bile salts	Steroids	Liver	Emulsify fats
Aminopeptidase	Enzyme	Intestinal glands	Peptides → amino acids
Enterokinase	Enzyme	Intestinal glands	Trypsinogen → trypsin
Maltase	Enzyme	Intestinal glands	Maltose → glucose
Sucrase	Enzyme	Intestinal glands	Sucrose → glucose + fructose
Lactase	Enzyme	Intestinal glands	Lactose → glucose + galactose
Active in pancreas			
Secretin	Hormone	Duodenum	Secretion of sodium bicarbonate
Pancreozymin	Hormone	Duodenum	Secretion of digestive enzymes
Active in gallbladder			
Cholecystokinin	Hormone	Duodenum	Release of bile

gested food (largely cellulose), along with nitrogenous wastes, some salts, fats, white blood cells, and cells from the intestinal lining, make up the bulk of the weight of the feces, the contents of the large intestines. Bacteria may constitute 10 percent or more of the weight of the feces. (These are generally not disease bacteria, but normal inhabitants of the digestive tract, which live quietly on bits of undigested organic matter and may incidentally contribute some vitamins to the welfare of their host.) The fecal material becomes increasingly concentrated as it moves through the colon.

The human rectum is normally empty. However, the presence of chyme in the stomach or duodenum may trigger a reflex resulting in strong peristaltic waves coursing through the entire colon (Fig. 10–37). This action forces the contents of the descending and sigmoid colon into the rectum. The presence of material in the rectum stimulates nerve endings that produce a desire to defecate. In infants, the stimulus of a full rectum triggers the **defecation reflex,** which results in involuntary contractions of the muscles in the intestinal walls and a relaxation of the anal sphincters. With increasing maturity, the child learns to ignore the defecation reflex if this seems desirable at the moment, and defecation becomes a voluntary act. The action of the reflex can be augmented voluntarily by taking a deep breath, closing off the glottis, contracting the diaphragm and abdominal muscles (straining), and inhibiting the external anal sphincter.

If the desire to defecate is ignored, the rectum becomes acclimated to its new volume, and a new desire to defecate does not come until there is a new influx of material from the colon. Repeatedly ignoring the desire to defecate may result in the rectum becoming permanently insensitive to distension. A consequence of this condition is **constipation,** difficult and infrequent defecation. Delay in defecation results in excessive drying out of the fecal material, adding further to the difficulties of defecation and setting up a vicious cycle. A diet high in bulk (indigestible cellulose), as provided by fruits and vegetables, and regular defecation habits tend to minimize the incidence of constipation.

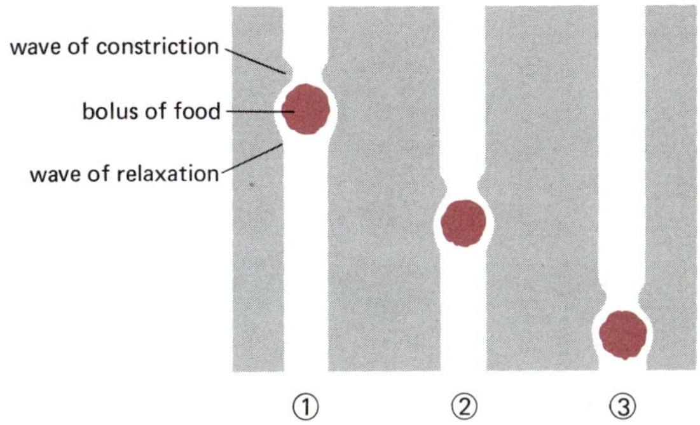

Figure 10–37 Food is moved through the digestive tract by the wavelike contractions of peristalsis.

Another disorder of the intestinal function is **diarrhea,** which is the opposite of constipation. In this case the contents of the colon move along too quickly, preventing sufficient absorption of water and resulting in frequent watery defecations.

Excretion

A growing problem of our modern world is the increasing pollution of air, water, and soil by the waste products of industrial activity. Factories and automobiles pour out varied mixtures of carbon monoxide and dioxide, hydrocarbons, oxides of nitrogen, sulfur, lead, and fluorine compounds, and numerous other substances. More and more it is being realized that we must find effective means of dealing with these waste products, lest they irretrievably poison the biosphere—our living world.

The body of a living organism has its own pollution problems, just as serious though on a far smaller scale. The ceaseless chemical reactions of metabolism result in the production of a variety of unwanted byproducts, which if allowed to pollute the cells and conducting fluids of the body will ultimately accumulate and poison it. Carbon dioxide, a by-product of respiration, is one of the common pollutants of living organisms. (In plants, during photosynthesis the excess oxygen produced becomes a substance that must be disposed of.) One of the most difficult pollutants for the animal organism to cope with is the nitrogenous wastes formed in various aspects of protein metabolism. The deamination of amino acids results in the formation of ammonia, which is a quite toxic substance. This compound must rapidly be converted to less toxic substances, and these must ultimately be eliminated from the organism.

The process of disposal of metabolic wastes is referred to as **excretion.** This term is often confused with several other processes: egestion, defecation, and secretion. **Egestion** is the elimination of undigested matter through the digestive tract. **Defecation** is a special case of egestion, referring to the process whereby feces, consisting of undigested food material, dead bacteria, dead and worn-out cells, are eliminated from a digestive tract. For the most part these materials take no active part in the metabolism of the organism and therefore cannot be considered as metabolic wastes. **Secretion** is the process by which products of metabolism are released, usually from specialized glands, to perform useful functions at a site other than the site of origin. Thus, secretion generally does not involve a function of elimination at all, although occasionally products of secretion may be eliminated (as in the case of the liver's secretion, bile, much of which ultimately passes out through the intestines after performing its functions).

EXCRETION IN PLANTS, PROTISTS, AND LOWER ANIMALS

Plants have no organized excretory system; nor, as a rule, do they have any particularly serious problems of disposal of waste products. Indeed, except for some gaseous effluvia that pass out through the stomata, plants barely excrete at all. There are several reasons for this. First, the rate of catabolism is much lower in plants than it is in animals, so that toxic products accumulate more

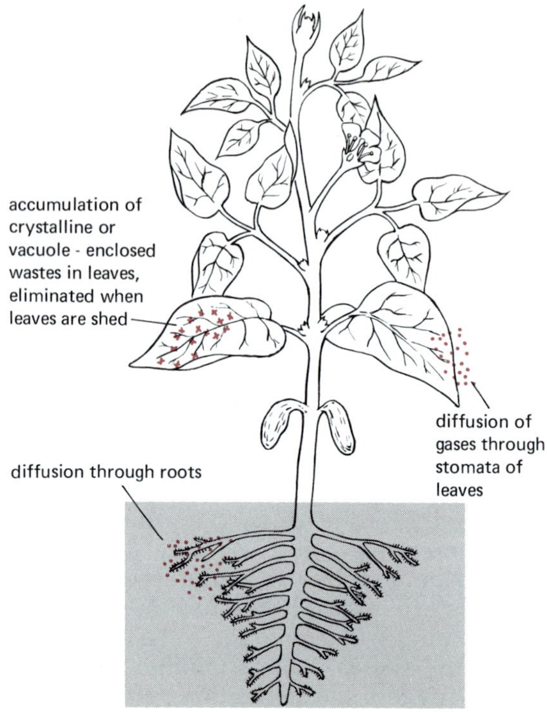

Figure 10–38 The simple excretory mechanisms of plants are sufficient to meet their needs.

slowly. In addition, since a major fraction of plant metabolism is based on carbohydrates, as opposed to the protein emphasis in animals, there is much less production of toxic nitrogenous compounds. Some deamination of amino acids does occur in plants, but before the ammonia produced can reach toxic levels, the plants can reutilize this substance for the synthesis of new proteins. Similarly, the carbon dioxide and water produced in carbohydrate catabolism can be reused as raw materials for photosynthesis.

Plants cope with occasional buildups of toxic substances quite simply. If they cannot be sent out into the atmosphere by transpiration, they may be walled up in vacuoles, where they remain sealed off, unable to harm the living tissues of the plant. Or substances may be accumulated in the form of solid crystals. Substantial quantities of oxalic acid crystals build up in spinach leaves, for example; as the older leaves are shed, the waste product is effectively eliminated.

Marine protozoa, sponges, and coelenterates eliminate their waste products by simple diffusion through the cell membranes. There is no need for elimination of water, because the medium in which the animals live is essentially isotonic to their intracellular fluid. But freshwater organisms live in a hypotonic environment. Water tends to diffuse through the cell membrane; this influx of water, together with the water produced metabolically, results in an excess, which, if unchecked, can cause the cells to swell and explode. Freshwater protozoa and sponges are usually equipped with a **contractile vacuole,** a device that continually pumps water out of the organism.

With increasing structural complexity, the vast majority of the cells of multicellular organisms are no longer in intimate contact with the external environment, and simple diffusion no longer suffices to remove the waste products of metabolism, particularly nitrogenous wastes. Organized excretory systems are needed to cope with the problem. Such relatively unsophisticated animals as flatworms are equipped with specialized **flame cells,** which extract nitrogenous and other wastes from the body fluid and send them out through excretory ducts.

The **nephridia** of annelids resemble the **nephrons** of vertebrates (Fig. 10–39). Both types of structures are in intimate contact with capillary networks and receive waste products from the bloodstream.

THE VERTEBRATE KIDNEY AND EXCRETORY PROBLEMS OF TERRESTRIAL ANIMALS

The common feature in the vertebrate excretory system is the **kidney,** containing tiny coiled, tubelike structures called **nephrons.**

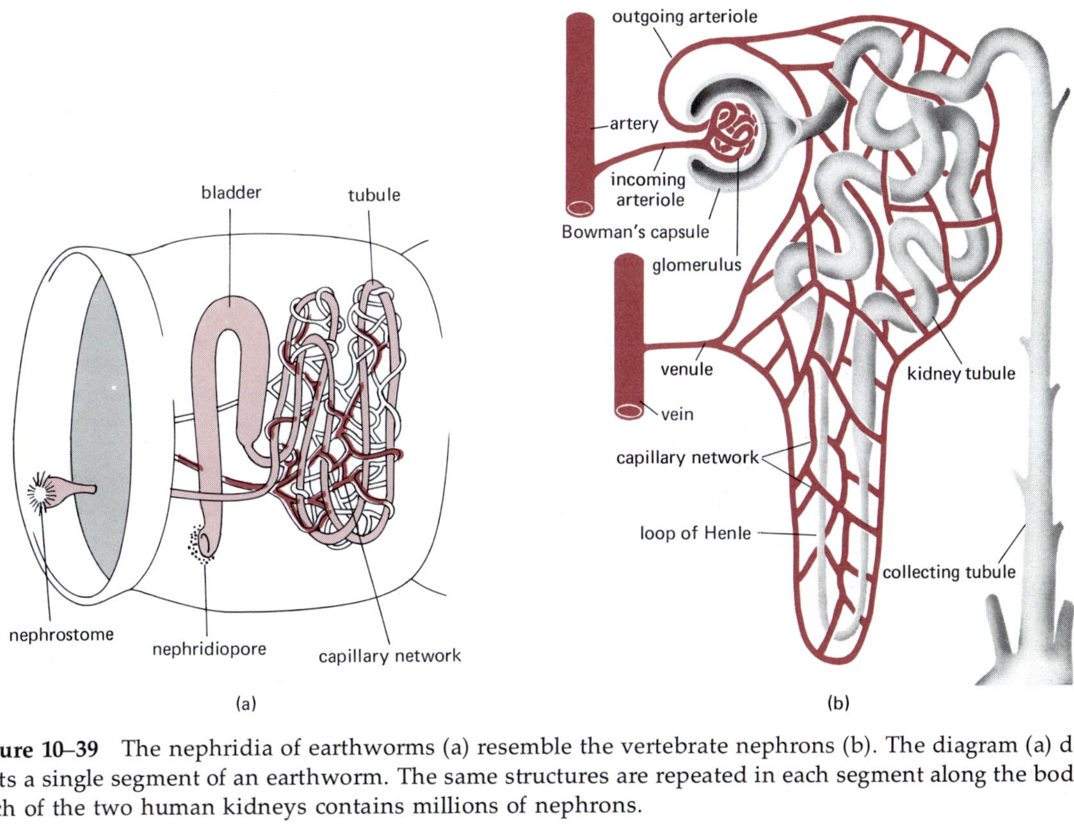

Figure 10–39 The nephridia of earthworms (a) resemble the vertebrate nephrons (b). The diagram (a) depicts a single segment of an earthworm. The same structures are repeated in each segment along the body. Each of the two human kidneys contains millions of nephrons.

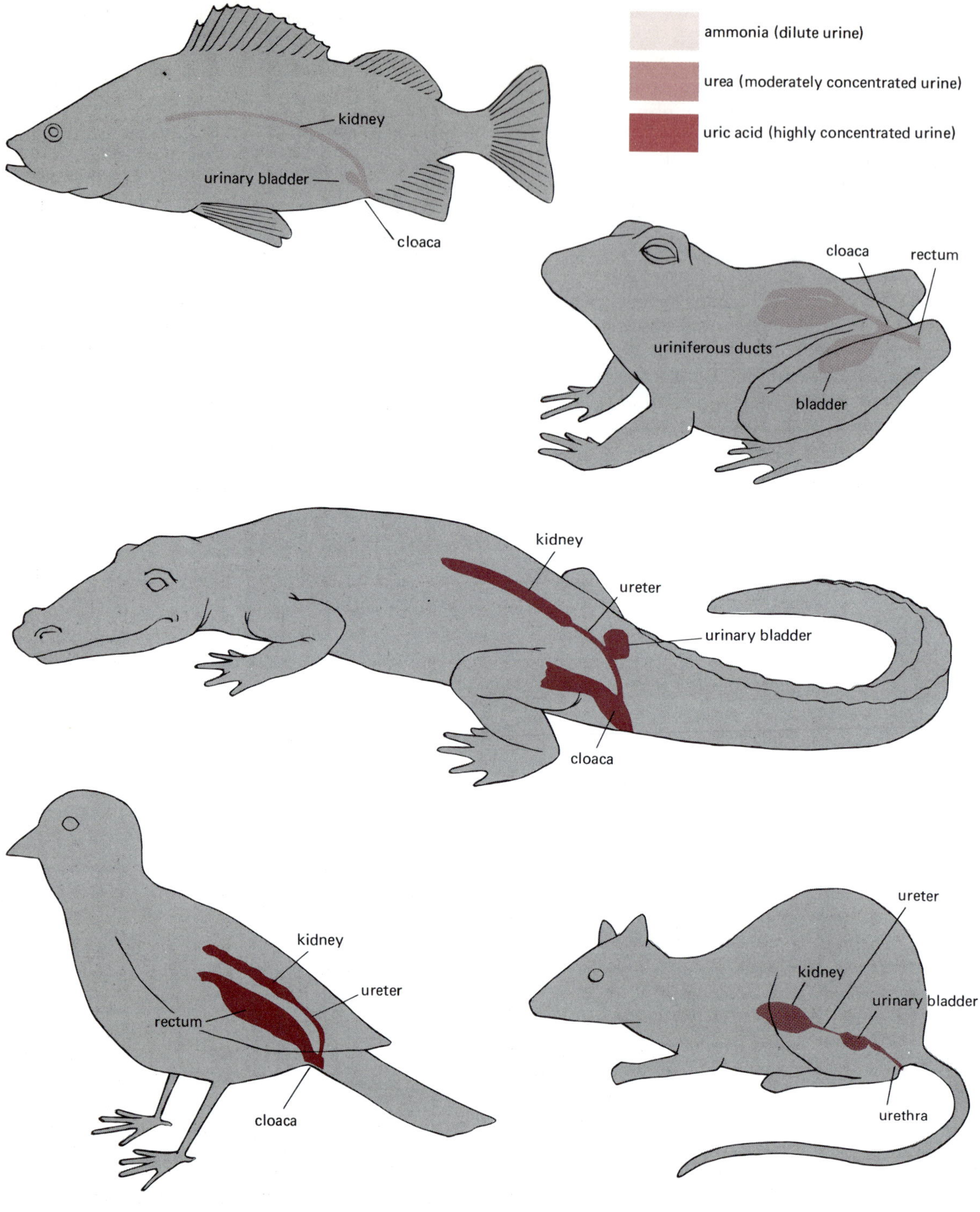

Figure 10–40 The vertebrate excretory system. Freshwater fish and amphibians, with access to a watery medium, excrete a dilute, watery urine, while birds and reptiles excrete their nitrogenous wastes in the form of nearly solid uric acid.

The kidneys of fish, amphibians, reptiles, and birds empty their accumulations of waste products into a common chamber, the **cloaca,** into which flow the products of the digestive, reproductive, and excretory systems (Fig. 10–40). Most mammals do not have a cloaca.

Among fish, where a source of water presents no problem, nitrogenous wastes are eliminated by flushing out ammonia, diluted to a level of low toxicity with copious amounts of water. But land animals do not have unlimited sources of water at their disposal. Their excretory systems are characterized by varying degrees of water conservation. This is accomplished by converting ammonia to other nitrogen compounds that are not as toxic and therefore do not have to be diluted as extensively. The major nitrogenous constituent of mammalian urine is urea (NH_2CONH_2). Reptiles and birds carry the water conservation even farther, excreting a highly concentrated, almost dry urine, containing a high concentration of uric acid:

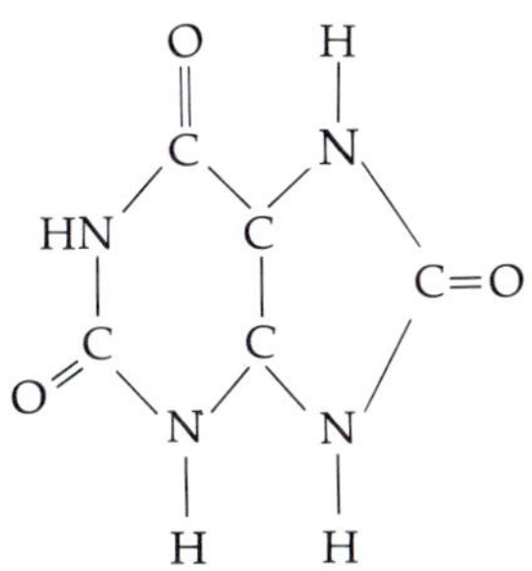

(There are exceptions to this general rule: water birds, such as ducks, excrete rather dilute urine, while some desert mammals, such as kangaroo rats, conserve water by excreting a highly concentrated urine.)

EXCRETION IN MAN

Although the kidneys play the primary role in the human excretory system, a variety of other organs of the body also aid in removing substantial amounts of metabolic wastes (Fig. 10–41).

The **lungs** provide the main avenue for the excretion of carbon dioxide. Each day nearly 300 liters of this gas are eliminated from the human body through this route. In addition, about 350–400 ml of water is excreted through the lungs daily.

The **liver** also functions in an excretory role. Although most of the constituents of bile are reabsorbed by the colon and recycled to the liver, the bile pigments (formed from the de-

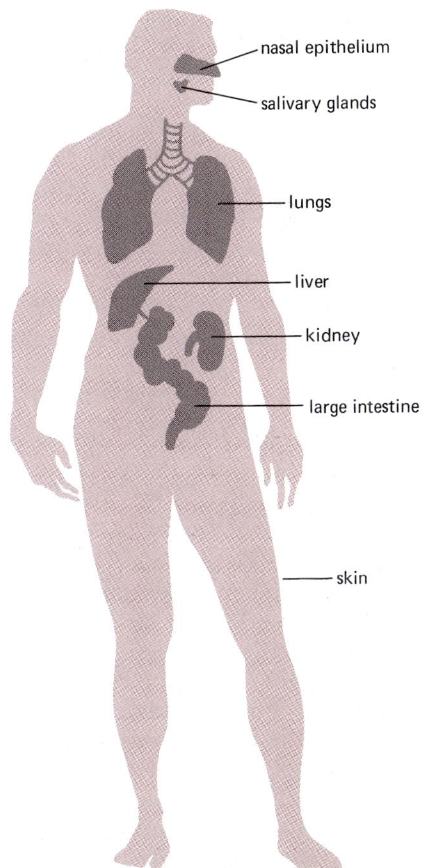

Figure 10-41 Some important organs of excretion in man.

composition of hemoglobin) are eliminated from the body in the feces.

The epithelial cells lining the **intestines** secrete certain salts, including calcium and iron salts, into the lumen (cavity) of the colon; these salts, together with the water that is not reabsorbed, pass out of the body with the feces.

The **skin** is an important organ of excretion. The approximately 2.5 million **sweat glands** embedded in the skin discharge sweat through coiled ducts directly onto the skin surface. Sweat is mostly water (99 percent) and contains about one-eighth as much solid matter as does a corresponding amount of urine. The solid material excreted by the skin consists of essentially the same substances that are found in urine: Na^+, K^+, Ca^{++}, Mg^{++}, $PO_4^{=}$, $SO_4^{=}$, urea, ammonia, uric acid, lactic acid, pyruvic acid, etc. The pH of sweat is usually acid, varying from 5 to 7. The body normally loses 0.5 to 2 liters of perspiration daily. But under strenuous exercise in the sun, the amount lost may rise to as much as 4 liters per hour. (Thus, a baseball pitcher often weighs about 10 pounds less at the end of a game than before it started.) Large as these figures may be, the sweat glands normally excrete only 5 to 10 percent of the nitrogenous wastes of the body.

The **kidneys** are the major excretory organs in man, excreting about 75 percent of the nitrogenous wastes of the body. In addition, they stabilize and regulate the concentrations of many of the substances that are normally found in the blood. They rid the body of many toxic substances that may find their way into the bloodstream, maintain the acid-base balance of the body fluids, and keep the volume of all the body fluids under a well-regulated control. Thus, the kidneys play a key role in the body's continuing struggle to achieve **homeostasis,** a maintenance of constancy of the internal environment.

THE HUMAN KIDNEYS

The human kidneys are a pair of kidney-bean-shaped organs, about four to five inches long, located against the dorsal wall of the abdominal cavity, just below the stomach (Fig. 10–42a). Within the darker outer layer of the kidneys, the **renal cortex,** is an inner, thicker core (the **renal medulla**) and a funnel-shaped chamber (the **renal pelvis**).

The **nephron,** a microscopic, highly coiled structure about half an inch long, is the structural and functional unit of the kidney. Its structure (Fig. 10–42b) provides intimate contact with the bloodstream. Each kidney contains about 1 million nephrons. The open end of each leads into the renal pelvis.

The **ureter,** an extension of the renal pelvis, is a muscular tube about ten inches long. There are two ureters in the body, one leading from each kidney to a storage organ, the **urinary bladder.** This is a hollow, expandable, muscular organ, located in the lower portion of the abdominal cavity. A valve at the point of attachment prevents backflow of urine into the ureters.

Urine, formed in the kidneys and stored temporarily in the urinary bladder, is voided through the **urethra,** a duct that leads from the bladder to the outside of the body by way of the penis in the male and via the **urethral orifice,** located just anterior to the vagina, in the female. A sphincter of smooth muscle at the juncture of the urinary bladder and urethra and another sphincter of skeletal muscle in the first inch or so of the urethra aid in the voiding of urine.

The kidneys perform a prodigious amount of work. About 1.3 quarts of blood is pumped through the kidneys each minute; this represents a quarter of the blood that is pumped through the heart each minute and means that the equivalent of the entire blood supply of the body passes through the kidneys in less than five minutes.

FORMATION OF URINE

The average man voids between 1 and 2 quarts of urine each day. Even if he were wandering on a desert, with no fluids to drink, urine would still be formed (though in reduced quantities and in a more highly concentrated form). How this urine is formed and how the body adjusts urine formation to variations in

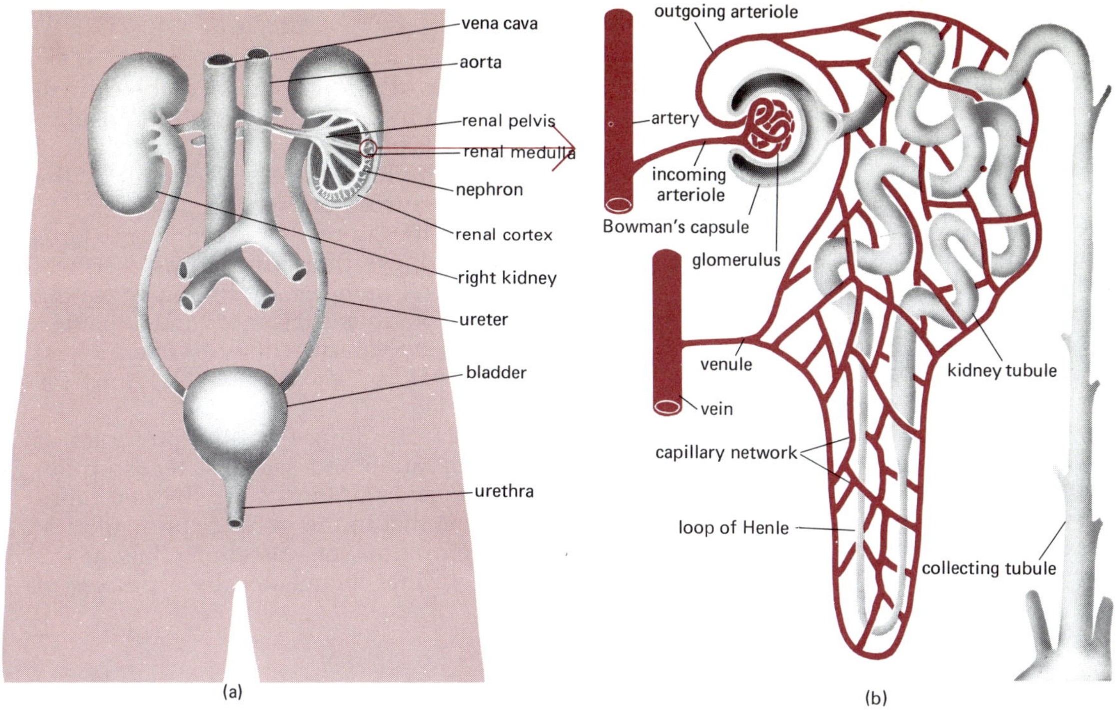

Figure 10–42 The urinary system. A magnified detail of the internal structure of the kidney, depicting its functional unit, the nephron, is shown at the right.

the internal and external environments is an interesting and complex phenomenon.

The formation of urine involves three processes: pressure filtration, selective reabsorption, and tubular secretion or augmentation. The net effect of these processes is to enable the kidney to remove waste products while conserving substances that are of use to the body.

The first stage in the formation of urine is a sort of preliminary screening process called **pressure filtration.** In this process the larger molecules of the blood (primarily proteins) and formed elements (blood cells and platelets) are screened out by filtration under pressure, driven by the contractions of the heart. Water and other small molecules, such as glucose and nitrogenous materials, diffuse out from the **glomerulus,** a cluster of capillaries in the nephron, to the adjacent **Bowman's capsule,** a hollow sac of cells leading directly into the lumen of the nephron. The nephric filtrate at this stage resembles blood plasma, without the blood proteins.

Each day our kidneys filter about 180 quarts of fluid. If all of this material were permitted to pass out of the body, it would mean the loss of about 400 pounds of fluid each day! Fortunately the human kidneys excrete only about 1 to 2 quarts of urine daily. The remainder of the material is reabsorbed by the capillary network surrounding the kidney tubules by a process of **selective reabsorption.** In this way the huge quantities initially lost in the preliminary screening of pressure filtration are reclaimed for the body's use.

The final concentrations of the solutes in the fluid that enters the collecting tubules of the kidneys are vastly different from those found in the initial glomerular filtrate. Large quantities of mineral ions, including Na^+, K^+, Ca^{++}, Cl^-, HCO_3^-, and to a lesser extent $PO_4^{\equiv}$ and $SO_4^{=}$, are reabsorbed by an enzyme- and energy-requiring process of active transport, and are returned to the bloodstream through capillaries that lead out from the nephrons. The selective transfer of solutes to the blood increases the water concentration in the nephric filtrate and results in a passive process of osmosis, transferring water to the blood. By the time osmotic equilibrium is restored, about 98 to 99 percent of the water in the nephric filtrate has been reabsorbed.

The amounts of ions and other substances reabsorbed in the tubules depend on the levels of the substances in the bloodstream. If the content of a particular substance in the blood is unusually low, larger amounts of it than usual will be reabsorbed, until the level in the blood is readjusted to the normal value. If blood levels of a particular substance are unusually high, smaller amounts than normal will be reabsorbed, again tending to readjust the level in the blood to a normal value. In the latter case, unusually high levels of the substance will be excreted in the urine, and analyses of urine can thus be used to check indirectly on the metabolic state of the body. For example, the condition of diabetes mellitus, caused by poor function of the insulin-secreting cells of the pancreas, is characterized both by high blood glucose levels and by the presence of glucose in the urine.

Through the process of selective reabsorption, the kidneys thus function as a sort of quality-control monitor, continuously testing the levels of various substances and adjusting them when necessary. They can even adjust the blood pH, by exchanging Na^+ ions for H^+ ions.

The urine resulting from the processes of pressure filtration and selective reabsorption has a urea concentration more than 60 times greater than that in the initial nephric filtrate. The concentrations of waste materials in the urine are further augmented by a process of

Figure 10–43 Some mechanisms of urine formation in the nephron. Filtration of blood in Bowman's capsule yields a filtrate isotonic to the blood. As the filtrate passes down the proximal convoluted tubule toward the loop of Henle, water diffuses out into the surrounding tissue fluids, and sodium ions diffuse in, resulting in a hypertonic fluid in the tubule. Passing through the ascending loop of the distal convoluted tubule, the filtrate loses Na^+ by diffusion and becomes hypotonic; during the flow down the collecting tubule water diffuses out into the salty surrounding fluids, resulting in the final hypertonic filtrate, urine. Processes of active transport also contribute to urine formation, adding certain substances to the filtrate against the concentration gradient and removing others.

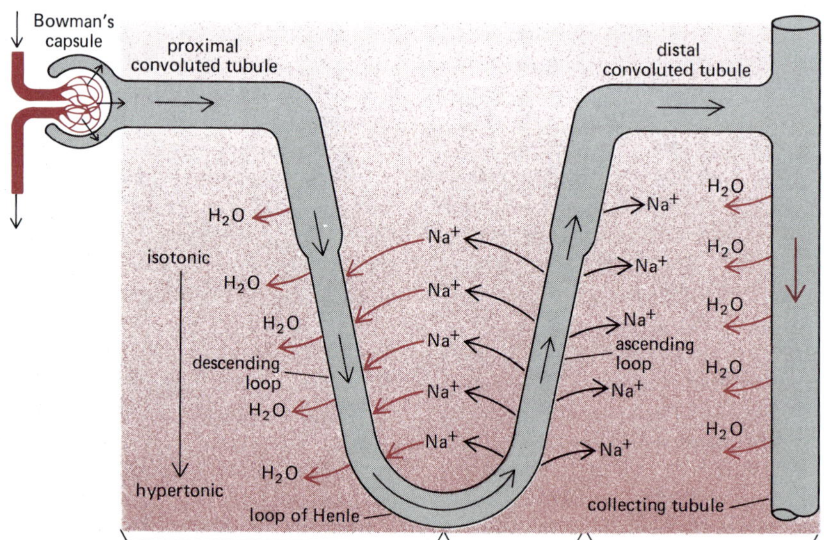

tubular secretion. The cells of the kidney tubules remove substances from the blood in the surrounding capillaries and secrete them directly into the lumen of the tubules. Tubular secretion involves processes of active transport. Although this mechanism is normally considered to play a relatively minor role in the human kidney, it apparently is significant in the removal of certain foreign substances, such as dyes and drugs (including penicillin), from the blood.

The work of the kidneys is monitored indirectly by sense receptors in the brain that are sensitive to changes in the osmotic concentration of the body fluids. If these fluids become too concentrated, signals from the receptors trigger the release of **antidiuretic hormone (ADH)** from the posterior pituitary gland. ADH makes the lining of the kidney tubules more permeable to water; thus more water is returned to the blood and the urine becomes concentrated. Ultimately an optimum concentration in the body fluids is restored, and the flow of ADH ceases. Alcohol interferes with this homeostatic mechanism by inhibiting one of the links in the ADH-secreting system. As a result, less water is removed from the urine, and a person who has had a few drinks notices an increased need to urinate.

URINATION

Urine formation is a continuous process. As this fluid accumulates in the collecting tubules of the nephrons, it trickles through the renal pelvis and then is carried to the urinary bladder via the ureters, with the aid of peristaltic contractions of the ureter walls. The muscular walls of the bladder readily expand to accommodate the accumulating urine. When the volume of urine reaches 250–300 ml, the sensory nerve endings in the walls of the bladder send impulses to the brain, giving rise to a sensation of fullness and a desire to urinate. Simultaneously, impulses are sent to a control center in the spinal cord and relayed to cause relaxation of the sphincters and contraction of the walls of the urinary bladder, resulting in an expulsion of urine through the urethra. In infants the voiding of urine is an involuntary process; gradually voluntary control over the **voiding reflex** is acquired, and the urge to urinate can be inhibited if desired.

QUESTIONS FOR REVIEW AND THOUGHT

1. Discuss the basic modes of nutrition in living organisms and give examples of each.
2. The chlorophyll molecule absorbs light most efficiently in two narrowly specific wavelength regions. Is the rest of the incident light "wasted"?
3. Outline the basic processes of photosynthesis.
4. Discuss the six basic types of constituents of foodstuffs. Should "no-calorie" foods be classified as foods?
5. Are vitamin supplements (*a*) necessary? (*b*) desirable?
6. Distinguish between extracellular digestion and intracellular digestion. Does intracellular digestion ever occur in man?
7. Discuss the advantages of a one-way digestive system.
8. Does the ability to taste foods play any role (*a*) in the digestive processes? (*b*) in the survival of the organism?

9. Trace the progress of a frankfurter in a bun through the digestive tract, from mouth to anus. Include a discussion of the enzymes that act on the food materials, their effects, and the organs that secrete them.
10. Why is a well-developed excretory system essential for the well-being of higher animals, while plants can generally get along without any organized excretory system at all?

CHAPTER **11**

Matter and Energy for Life—II: Circulation and Utilization

Introduction

In some ways the human body can be likened to a large modern nation. Complex factories (the metabolizing cells) turn out a vast array of products, both for internal consumption and for export to other cells of the body. These cells are linked to one another and to major sites of body activity by the intricate network of highways (or rather, waterways) of the circulatory system. Along the channels of the bloodstream float red blood cells, carrying materials to and from the cells like laden barges. Through the waterways of the body move the "law-enforcement agents," the white blood cells, policing the body cells and defending the body from "foreign invaders" —bacteria, viruses, and other microbes. The major ports of entry into the body—the digestive tract and the respiratory tract—are concerned with the passage of materials—food, fuels, and gases from the atmosphere—into and out of the body. The activities of the body are linked by an extensive communications network: the nervous system, with its extensions the sense organs, providing links to the outside world. The body's activities are controlled and coordinated by a variety of "government agencies"—the control centers of the brain and spinal cord and the glands of the endocrine system, which mediate body functions through messages transmitted over the nerves and through chemical messengers that travel through the bloodstream.

Thus, the circulatory system is vital in the functioning of various body systems. It participates in the transport of food and fuel materials from the digestive system to the body cells and in the removal of metabolic wastes from the cells to the waste-disposal or excretory system. It plays a key role in the protection of the body from disease and in the coordination of body activities. The circulatory system is closely interlinked with the functioning of the respiratory system, which functions to bring oxygen into the body and remove the waste carbon dioxide that is produced in the cells' energy metabolism.

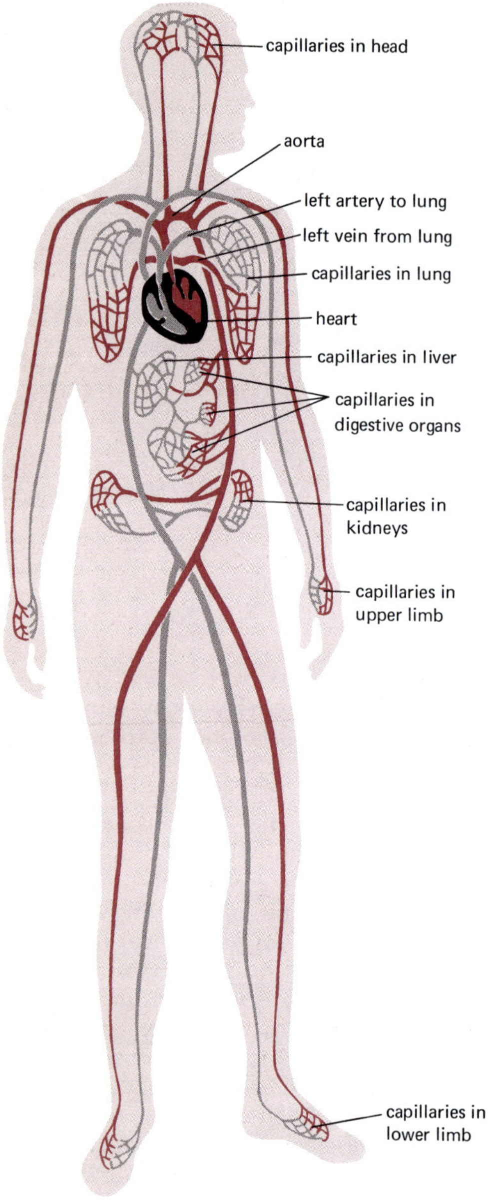

Figure 11–1 The circulatory system provides all the body organs and tissues with nutrients and oxygen and carries away waste products of the cells.

Circulatory Systems

For one-celled organisms there is no real need for an organized circulatory system. Oxygen and nutrients can be taken in and waste products disposed of through simple exchange with the environment. Substances move about through the cell carried by currents in the cytoplasm, in a sort of circular motion referred to as **cyclosis** or **cytoplasmic streaming.** Similar unorganized arrangements also suffice for the nutritional and respiratory needs of many of the lower animals, including sponges, coelenterates, and even flatworms. None of the cells of their bodies is too far from the external and internal media for convenient and simple exchange of materials.

But with increasing structural complexity and specialization, more and more cells become buried within the animal's body, shielded from contact with the environment by layers of other cells. An organized circulatory system is needed to reach these cells, to bring them supplies of oxygen and nutrients and carry away their waste products.

The earthworm has a circulatory system surprisingly similar in basic design to that of the higher animals (Fig. 11–2). Red blood, colored by the presence of hemoglobin, an oxygen-carrying pigment, is pumped by five pairs of "hearts" —actually contractile, thickened blood vessels. The earthworm's blood flows from the "hearts" into a thick blood vessel, which branches and gradually narrows

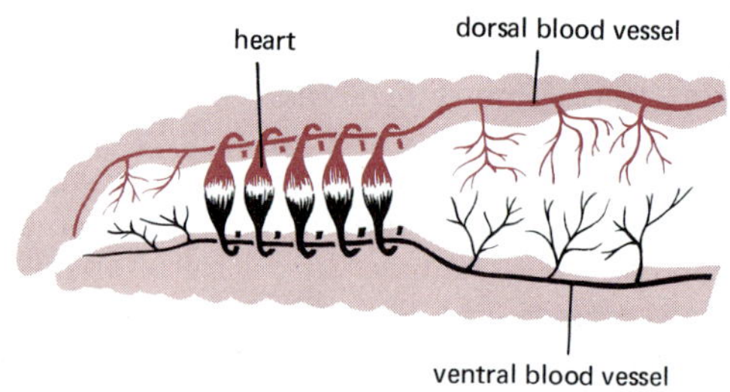

Figure 11–2 Circulatory system of the earthworm: a closed circulatory system.

into fine vessels called capillaries. Through the fine walls of the capillary network, gases and other materials are exchanged with the tissues of the worm's body. The capillaries widen into larger blood vessels, which return the blood to the "hearts." Because the earthworm's blood is confined to a closed system of blood vessels, it is said to possess a **closed circulatory system.**

Many invertebrates, on the other hand, have what is referred to as an **open circulatory system.** One of the simplest examples of this type is the circulatory system of the grasshopper (Fig. 11–3). A series of tubular contractile hearts pump blood into a wide blood vessel called the aorta, thus sending it the length of the insect's body. But then the blood empties into a series of large cavities called sinuses, through which it moves sluggishly back toward the hearts. The grasshopper's blood is thus confined to closed blood vessels only through a brief portion of its trip around the body. The insect's blood is generally colorless; there is no need for an oxygen-carrying pigment because the insect has a separate system of branching air tubes that distribute gases from the atmosphere through the whole body. This separate provision for the respiratory needs of the body permits the grasshopper to get along satisfactorily with its rather inefficient open circulatory system.

Both open and closed circulatory systems are found among various groups of invertebrates. In general, blood flows very sluggishly through an open circulatory system, and this type of circulation is quite inefficient. (Indeed, the clam's heart beats just six times a minute.) In many groups, aside from the insects, the blood aids in respiratory functions, transporting oxygen and carbon dioxide, and it is supplied with a respiratory pigment capable of combining with these gases. In some cases this pigment is the blue hemocyanin, instead of hemoglobin. The lobster, for example, is a true "blue-blood."

Among the vertebrates, the closed circulatory system is the rule. The sequence of groups, from fish to amphibians to reptiles to birds and mammals is marked by a steady increase in sophistication of heart structure and a corresponding increase in circulatory efficiency. The fish, for example, has a very simple heart (Fig. 11–4). A single receiving chamber, the atrium, leads into a single pumping chamber, the ventricle. From the ventricle the fish's blood is sent coursing strongly to a network of fine capillaries in the gills. There, an exchange of gases takes place: oxygen dissolved in the water passing through the gills diffuses through the thin capillary walls into the blood, while carbon dioxide that has been carried in the blood from the body cells diffuses out into the water. But as the fish's blood flows through the intricate capillary bed, it slows down drastically; the oxygen-rich blood that emerges from the gill capillaries flows very sluggishly through the body, eventually back to the heart. Obviously, for a more efficient delivery of oxygen-rich blood to the body, some means would be needed to speed up the blood flow again after it has picked up a new load of oxygen.

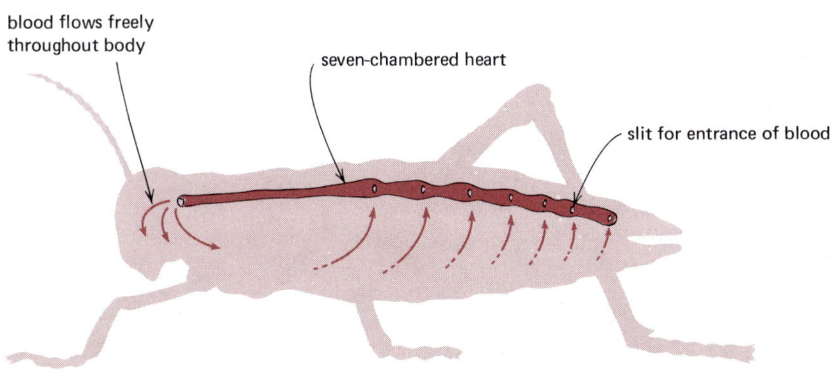

Figure 11–3 Circulatory system of the grasshopper: an open circulatory system.

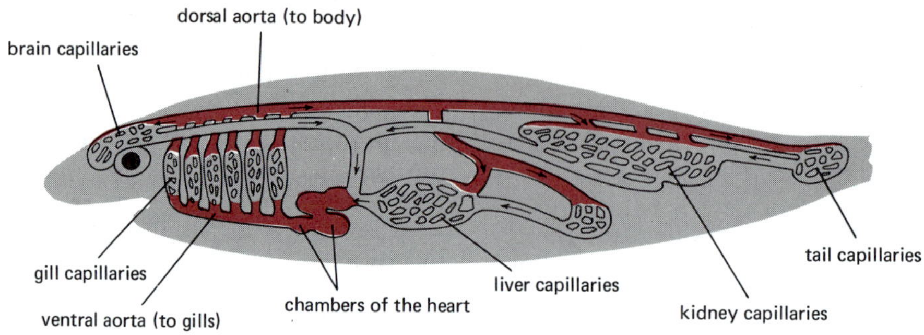

Figure 11–4 Circulatory system of the fish: a closed circulatory system with a two-chambered heart. (Heart and arteries shown in color.)

This is accomplished in an ingenious manner in the amphibian circulatory system, in which the heart is made to do double duty. The blood vessels are arranged in two main loops: one carries oxygen-poor blood from the heart to the capillaries in the lungs, where gases are exchanged, and then the blood flows back to the heart; from there the blood, now oxygen-rich, is sent out again, around a circuit through the body, eventually to return, oxygen-poor, to the heart. Two receiving chambers (atria) are provided (Fig. 11–5), thus effectively separating the incoming oxygen-rich and oxygen-poor blood flows. But this gain in efficiency is somewhat offset by the fact that both atria empty into a single ventricle. Mixing of the oxygenated and deoxygenated blood is minimized by partial partitions within the heart; these act as baffles, directing the blood flow toward the proper blood-vessel openings.

The partitions in the heart are better developed in reptiles, and a septum or partition divides the ventricle into two nearly separated compartments. In birds and mammals the separation is complete: the heart is divided into four separate chambers, two atria and two ventricles, and the blood flows through two complete systems of blood vessels: from the heart to the lungs and back again (the **pulmonary circulatory system**) and from the heart through the body and back again (the **systemic circulatory system**) (Fig. 11–7).

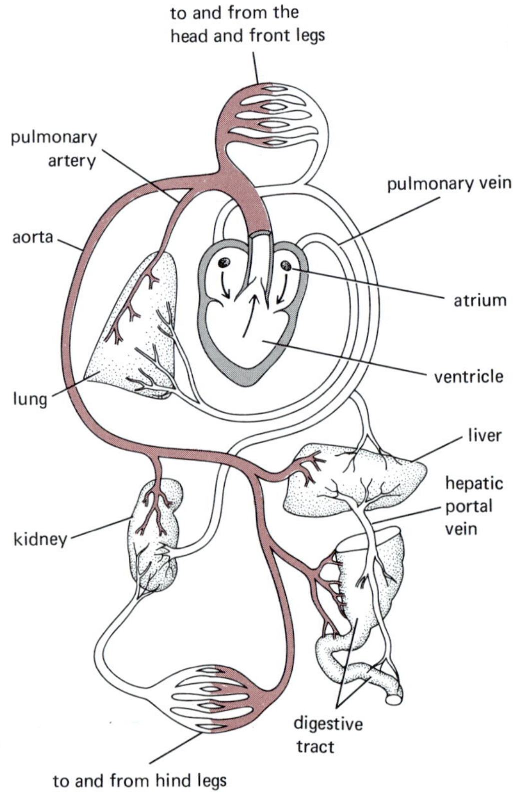

Figure 11–5 Circulatory system of the frog. Arrows indicate the direction of blood flow in the three-chambered heart; valves prevent backflow.

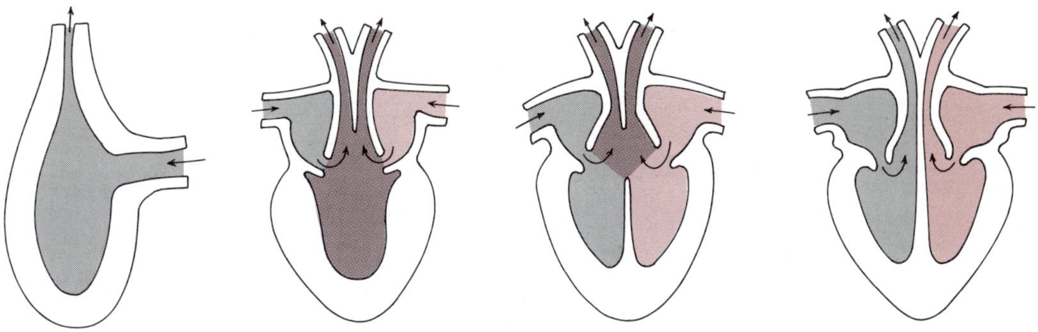

Figure 11–6 Stages in the development of the vertebrate heart. Oxygenated blood is denoted by color. Note the increasing efficiency in preventing mixing of oxygenated and deoxygenated blood, culminating in the completely partitioned mammalian heart.

The Human Circulatory System

THE HEART

The human heart may be visualized as a busy intersection, at the center of the intricate highway network of the bloodstream. Incoming blood from the lungs, rich in oxygen, is routed through the heart and out again into blood vessels carrying it to various parts of the body, while oxygen-poor blood from the body cells is routed through the heart and out through blood vessels that carry it to the lungs. The two blood flows cross but do not mix; a "traffic jam" in the heart is averted much in the same manner as is provided by an overpass or underpass at a highway intersection. The two flows move unimpeded through two completely separate pathways, provided by the four compartments of the heart.

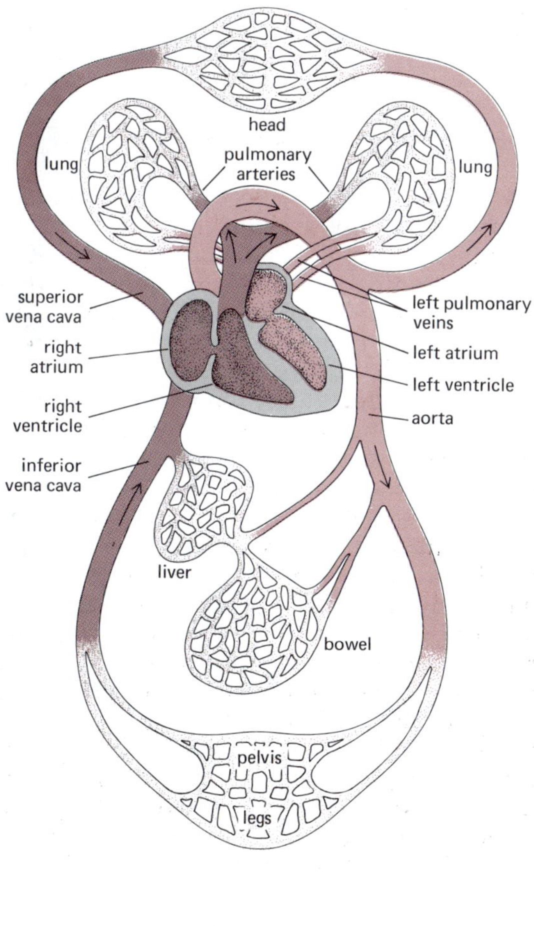

Figure 11–7 The human circulatory system. A four-chambered heart, typical of mammals and birds, provides for an efficient separation of oxygen-rich and oxygen-poor blood. The blood vessels form two basic loops: the pulmonary system, which carries oxygen-poor blood to the lungs for oxygenation and oxygen-rich blood back to the heart; and the systemic system, which carries oxygen-rich blood to various parts of the body and returns oxygen-poor blood to the heart.

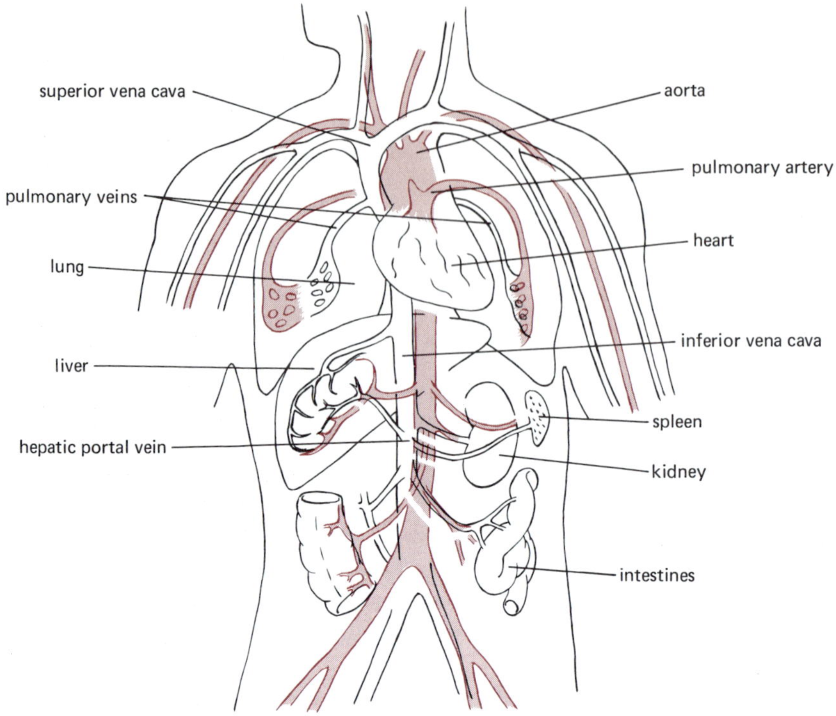

Figure 11–8 Structures of the human circulatory system (arteries in color).

The heart is a somewhat cone-shaped organ, situated roughly in the middle of the chest cavity, tilted slightly to the left (see Fig. 11–8). The rounded portion at the top consists of two chambers, the **atria.** The heavy muscular cone at the bottom is partitioned into two chambers called **ventricles.** The term *atrium* comes from a Latin word meaning "antechamber," and the atria are actually the receiving chambers of the heart. The ventricles are the pumps of the heart and send the blood coursing out through the blood vessels.

Blood from the lungs flows into the left atrium through four blood vessels called **pulmonary veins.** At the bottom of this atrium is an aperture guarded by a double-cusped muscular valve, the **mitral valve.** Blood passes through this valve into the left ventricle and then, by a powerful muscular contraction of the ventricle walls, is expelled through another valve into a very large blood vessel called the **aorta** (Fig. 11–9). The aorta is the largest blood vessel in the body; after arching toward the back and downward, it ramifies into a number of smaller branches, **arteries,** which carry blood throughout the body. The arteries leading from the aorta ramify further into smaller and smaller blood vessels, ultimately terminating in extremely fine-bore (about 8–10 μ in diameter) **capillaries.** These are in intimate contact with the body cells, and an exchange of materials occurs through their thin walls. The capillaries lead into gradually enlarging blood vessels, the **veins,** which carry the blood, now poor in oxygen and laden with carbon dioxide, back toward the heart. The veins from the head region ultimately empty into a large vein called the **superior vena cava,** while the veins from the rest of the body merge to form the **inferior vena cava.** Blood from both these veins flows into the right atrium and thence, through a three-flapped valve called the **tricuspid valve,** into the right ventricle. A strong contraction of this ventricle sends the blood rushing out through the **pulmonary trunk,** a large blood

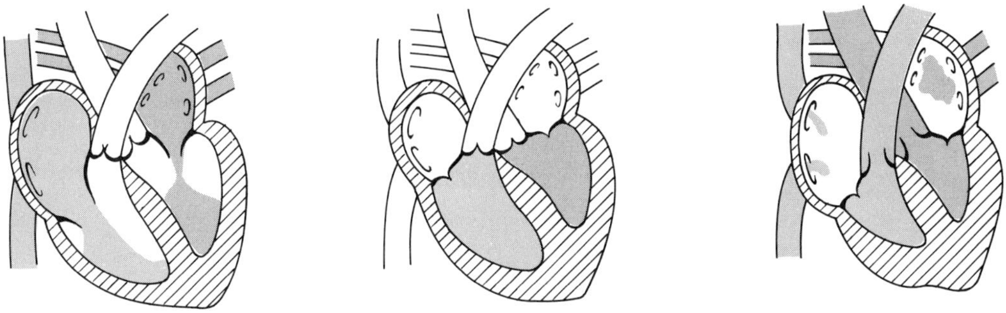

Figure 11–9 Operation of the valves of the heart during the filling and emptying of the ventricles. Note that the two ventricles relax and contract together.

vessel that branches into the right and left **pulmonary arteries,** each going to one of the lungs. In the capillary network in the lungs, there is a gas exchange, with carbon dioxide diffusing out through the thin capillary walls into the tiny air sacs of the lungs (and thence out of the body) and oxygen from the inhaled air passing into the bloodstream. From the lungs the blood returns to the heart through blood vessels culminating in the four pulmonary veins.

The human heart is a remarkable structure. Although it weighs only 9 to 12 ounces, it pumps more than 400,000 tons of blood in the course of a normal life span—enough to completely fill a cylindrical tank 100 feet in diameter and taller than the Empire State Building. The heart beats about 70 times a minute, or a total of about 2.5 billion times in a lifetime. Although the contractions of the left and right ventricles were described separately, they actually occur simultaneously. The heartbeat consists of a series of events, which together make up the **cardiac cycle:** As blood enters the atria from the pulmonary and systemic circulatory systems, the atria are in a relaxed state. Both atria then contract, forcing blood through the open tricuspid and mitral valves into the respective ventricles. Now the ventricular walls contract, producing a rise in the pressure within the ventricles and forcing the tricuspid and mitral valves shut with a characteristic sound. The pressure within the ventricles now exceeds that within the arteries; the **semilunar valves** (half-moon-shaped valves opening out into the pulmonary trunk and aorta) are forced open, and blood is propelled out of the heart. The ventricles then relax; their pressure falls below the arterial pressure, and the semilunar valves close with a sharp thump. The two sounds of the valves closing are the characteristic "lub-dub" that can be heard through a stethoscope. Injury to the heart valves produces characteristic changes in the heart sounds, which the trained physician can detect readily. The phases of relaxation of the compartments of the heart are referred to as **diastole,** the phases of contraction as **systole.** The complete cardiac cycle usually takes about a second in an adult, but an infant's heart may normally beat twice as rapidly.

The rate of the heartbeat is controlled through the nervous system, through a center called the **pacemaker,** located in the wall of the right atrium. Various chemicals, particularly the hormone **adrenaline,** can also affect the rate of the heartbeat. At times of great emotional stress, larger-than-normal amounts of adrenaline are secreted, and the heart beats faster. Injections of adrenaline can also stimulate a heart that has stopped to resume beating. Other substances, such as the poison muscarine found in poisonous mushrooms, tend to slow or even stop the heartbeat. Exercise, temperature variations (e.g., fever), and various other conditions can also affect the heartbeat.

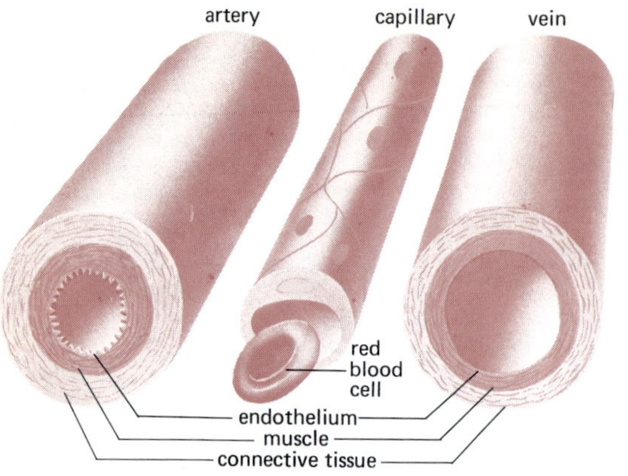

Figure 11–10 Structure of the walls of three types of blood vessels. Both arteries and veins have three-layered, muscular walls (note the greater thickness of the arterial walls). Capillary walls consist of a single layer of epithelium and are permeable to gases and liquids.

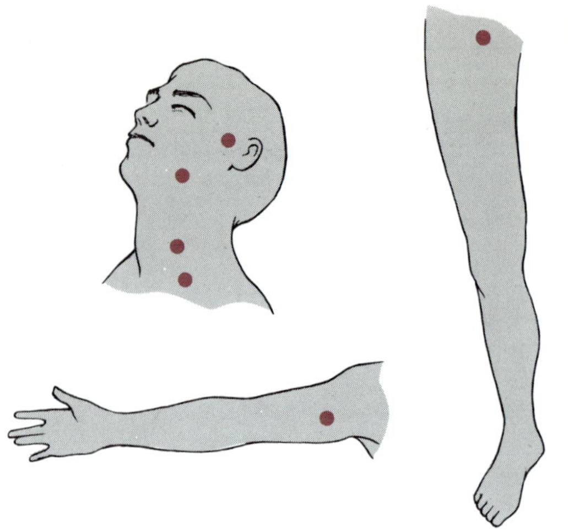

Figure 11–11 The pressure points. In the regions indicated, major arteries lie close to the surface of the body. Pressure on these points can be used as an emergency first aid measure to stop hemorrhaging. (A pressure point situated between the wound and the heart is used.)

BLOOD VESSELS

The blood vessels of the circulatory system are long, tubular structures of three main types (Fig. 11–10). The **arteries** are large, thick-walled vessels that carry blood away from the heart. They ramify into smaller and smaller branches (the smallest arteries are called **arterioles**) and finally become continuous with minute **capillaries,** which are usually less than a millimeter long and are microscopic in cross section. The capillaries are arranged in networks or "beds," arranged so that most of the cells in the body are adjacent to a capillary. Although an individual capillary is minute, there are so many of them that their total combined length in an adult human is estimated at about 60,000 *miles!* The tiny capillaries coalesce into larger blood vessels (**venules,** or literally, "little veins"), which in turn merge to form larger and larger **veins.** These carry blood back to the heart. Veins are thinner-walled and more delicate than the arteries.

In addition to the initial impulse imparted to the blood by the contractions of the heart, contractions of the artery walls also help to keep blood flowing along. These contractions can be felt as the **pulse** in various parts of the body where an artery lies close to the surface. The pulse is conveniently taken on the wrist, but the characteristic beating of the arteries can also be felt in a number of other parts of the body, such as the temples, in the hollow of the collarbone, and the groin (Fig. 11–11). These are the locations of the "pressure points," areas where pressure applied as a first aid measure can check severe bleeding from a torn artery.

The pressure in the arteries is usually measured in the arm, with an instrument called a **sphygmomanometer.** Two different values are usually obtained, for there is a regular variation in the arterial pressure that is coordinated with the cardiac cycle. Immediately following the ventricular systole (contraction), there is an upsurge in the arterial pressure (systolic pressure). The pressure then subsides, reaching a low point during the ventricular diastole (relaxation). The latter value represents the diastolic pressure. Typical normal blood pressure values for an adult are a systolic pressure of about 120 mm Hg, diastolic pressure of about 80 mm Hg.*

*Pressure is usually measured in mm Hg—the number of millimeters a column of mercury will rise in a closed tube connected to the system. Normal atmospheric pressure is about 760 mm Hg.

Blood rushes out of the aorta at a rate of about a foot per second, but it slows down steadily and is traveling at only about a millimeter per second by the time it reaches the capillaries. This slowdown is due to a physical principle: in general, when a fluid flows in a confined channel, its velocity is inversely proportional to the cross-sectional area of the channel. Although the aorta is about an inch in diameter and an individual capillary has a diameter of perhaps 10 μ, there are enormous numbers of capillaries. Their *total* cross-sectional area is hundreds of times greater than the cross section of the aorta.

The nature of the blood flow also changes as it reaches the capillaries. In the arteries the blood flows rapidly in spurts, propelled by the contractions of the heart and the artery walls. By the time it reaches the capillaries it has diminished to a slow and steady flow.

The physical relationships of the blood-vessel diameters provide for some speedup of the blood flow as it passes into the veins, but the efficient return of the blood to the heart presents some problems. This is so particularly for the return of blood from the legs and feet in a human being standing upright. For the thinner, less muscular walls of the veins cannot contract to speed the blood along. The blood is aided in its flow by the normal contractions of the body muscles, which squeeze the blood vessels in a "milking" motion as a person walks and moves about. The leg and arm veins are also equipped with valves (Fig. 11–12), which permit blood flow toward the heart but tend to keep the blood from slipping backward. When a person stands motionless for a long time (as in the case of soldiers standing at attention), the efficiency of the return of blood to the heart is reduced. Blood begins to pool in the legs and feet, thus reducing the amount of circulating blood available to be sent to the other parts of the body, particularly the head region. But the brain requires a large and steady supply, and if this is reduced, damage to the brain tissue may result. An efficient automatic corrective mechanism has been evolved: the person faints. His body immediately assumes a horizontal position, and now that blood does not have to be returned against gravity the circulation is rapidly improved. As normal circulation in the brain is restored, the person revives. Experienced soldiers learn to avoid pooling of blood in the legs by inconspicuously tensing and relaxing their leg muscles. Lecturers often adopt the protective mechanism of pacing or swaying while they speak, thus keeping their leg muscles moving.

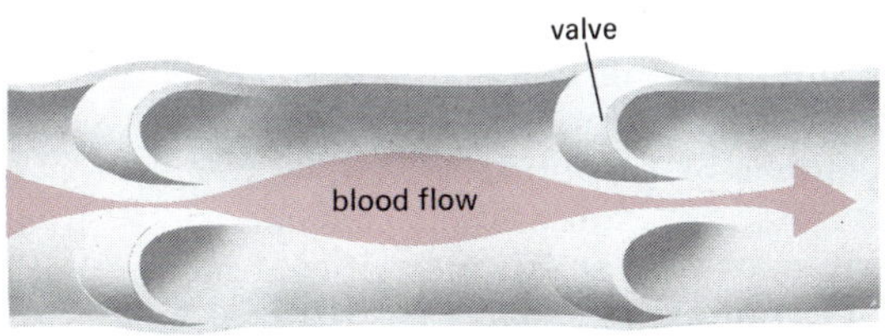

Figure 11–12 Valves in the veins regulate the direction of blood flow, preventing backflow. This action is particularly important in the return of blood from the legs, against the pull of gravity.

The Blood

Human blood is a red, watery liquid. Although it appears to be relatively homogeneous, it is actually a suspension of solid particles in a watery liquid in which various other substances are dissolved.

If a sample of blood is placed in a test tube and spun in a centrifuge, the suspended solids will settle to the bottom of the tube to form a red sediment, leaving a clear, straw-colored liquid at the top (Fig. 11–13). The sediment, which comprises about 45 percent of the volume of the blood, consists of blood cells and other particles that are together termed the **formed elements.** The liquid comprising the other 55 percent of the blood is the **blood plasma.** A healthy adult has about five to six quarts of blood flowing in his circulatory system.

Blood serves a number of important functions in the body. It serves as a medium for gas exchange in the lungs and in the tissues of the body. It transports food, oxygen, wastes, and other materials to the parts of the

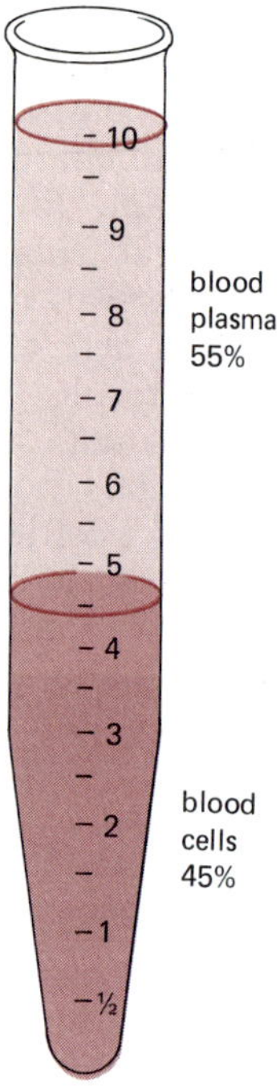

Figure 11–13 Centrifuging a blood sample reveals that blood contains about 55% liquid and 45% solid formed elements.

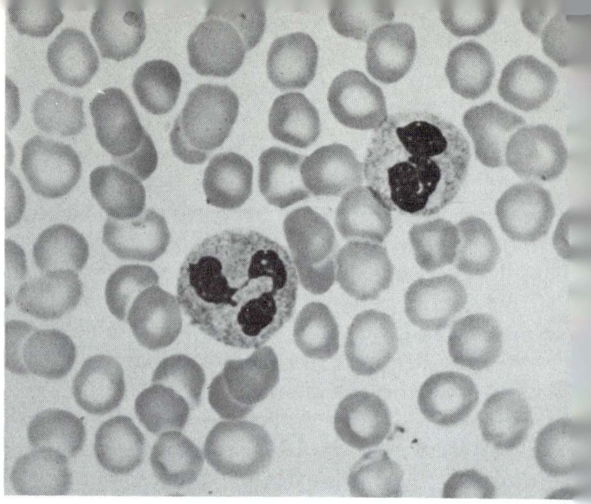

P. Feinberg, Omikron

Figure 11–14 Human blood: red and white blood cells, 1095×.

body where they can be most effectively used or eliminated. It aids in regulating the salt, acid, and base balance in the body and functions in temperature regulation by aiding in the redistribution of heat. Components of the blood are key factors in the body's system of defense against invading microbes; the blood serves as a vehicle for transporting these "germ fighters" and also for transporting chemical substances (hormones) that help to regulate the body's activities.

RED BLOOD CELLS

There are about 5 *million* red blood cells in each cubic millimeter of blood flowing through the blood vessels of an adult human. The total number of red blood cells in the body is thus estimated at about 25–35 trillion.

Red blood cells or **erythrocytes** are reddish yellow flattened discs with thickened rims, which look something like miniature doughnuts whose centers have been plugged (Fig. 11–14). They are about 7 to 8 μ in diameter (i.e., about the same as the diameter of the smallest capillaries) and are encased in a flexible membrane that permits them to bend and squeeze their way through the capillary network. Mature erythrocytes have no nucleus (the nucleus present in the immature forms degenerates); they live in the mature form for about 120 days and they are destroyed without ever dividing.

A major constituent of red blood cells (about a third of their weight) is the red pigment **hemoglobin.** This is an iron-containing protein that can combine with either oxygen or carbon dioxide, thus greatly enhancing the red cells' ability to transport these gases. Each red cell contains about 300 million hemoglobin molecules, each of which can carry four molecules of oxygen.

A deficiency of iron in the diet or various diseases may result in red blood cells being lost or destroyed more rapidly than new ones are formed. As a result, the concentration of red cells in the blood decreases, and the person suffers from a condition known as **anemia.** Chronic fatigue is a typical symptom of anemia (the body and brain cells are not receiving an optimum supply of oxygen and removal of carbon dioxide). Many forms of anemia are successfully treated with liver extract or iron salts, often in conjunction with B vitamins.

Some individuals—especially those from African and Mediterranean countries—are susceptible to having red blood cells with a sickle or half-moon shape (Fig. 11–15). Sickle cells have a tendency to clump together and

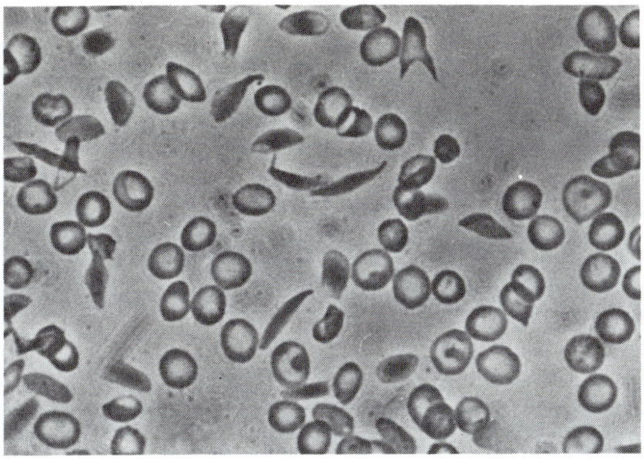

Figure 11–15 Blood of a sickle cell anemia patient, 462×.

may block small blood vessels, causing severe pains and tissue damage. These cells are also more fragile than normal red blood cells. They rupture easily, spilling out their hemoglobin, and can produce an anemia so severe as to lead to an early death. It has been found that **sickle-cell anemia** is a hereditary disease, caused by a difference of a single amino acid in each half of the hemoglobin molecule. (The protein portion of hemoglobin consists of two identical halves, each containing about three hundred amino acids.)

WHITE BLOOD CELLS

The red blood cells for the most part are carried passively along in the blood flow. But blood also contains another type of cells, which can move actively, even swimming against the current. These are the white blood cells or **leukocytes.** They are generally larger than the red cells but far less numerous: each cubic millimeter of blood contains about five to ten thousand white blood cells—i.e., there is about one white cell for every thousand red cells. Unlike the erythrocytes, white cells have a prominent nucleus and their shape is rather amorphous. Indeed, both in appearance and in behavior the white blood cells somewhat resemble amoebas.

There are a number of different types of leukocytes (Fig. 11–16), classified according to their size and their reaction to various dyes used in cytological studies.

The white cells not only can move about through the bloodstream, but they can also squeeze readily through pores in the capillary walls in a process called **diapedesis** (Fig. 11–17). In this manner they slip in and out of the bloodstream and meander through the body tissues on a sort of "police patrol."

The white blood cells make up the major defense force of the body. If a white cell encounters an invading bacterium, it sends out pseudopods and literally engulfs and then digests the invader, much as an amoeba captures and "eats" its microscopic prey. This is the process of **phagocytosis** (Fig. 11–18). White cells migrate in droves to sites of infection of inflammation in the body, apparently alerted by chemical "messages" sent out by damaged body cells. A white cell cannot go on ingesting bacteria indefinitely: each dying bacterium releases toxins, which gradually accumulate in the white cell, poisoning it. After a white cell has taken in a certain number of bacteria, perhaps a dozen or two, it too succumbs to the cumulative effect of the bacterial toxins. When a white cell dies, its place on the fighting line is taken by others. In raging infections, many bodies of dead white cells accumulate, forming the semisolid yellowish white substance known as pus.

White blood cells in the lining of the liver and spleen inspect the red blood cells that pass through these organs and engulf and decompose those that are old or damaged. Much of the hemoglobin from these red cells is reused as such, but some is broken down to pigmented components, which are secreted in the bile and ultimately give the feces their characteristic color. The white blood cells themselves are much shorter-lived than the red blood cells, living for less than two weeks on the average.

Some white cells are also active in the production of antibodies, substances which help further in the body's defense, as will be described later in the chapter.

Certain diseases and conditions, such as allergies, worm infections, and acute viral and bacterial infections, result in a sharp increase

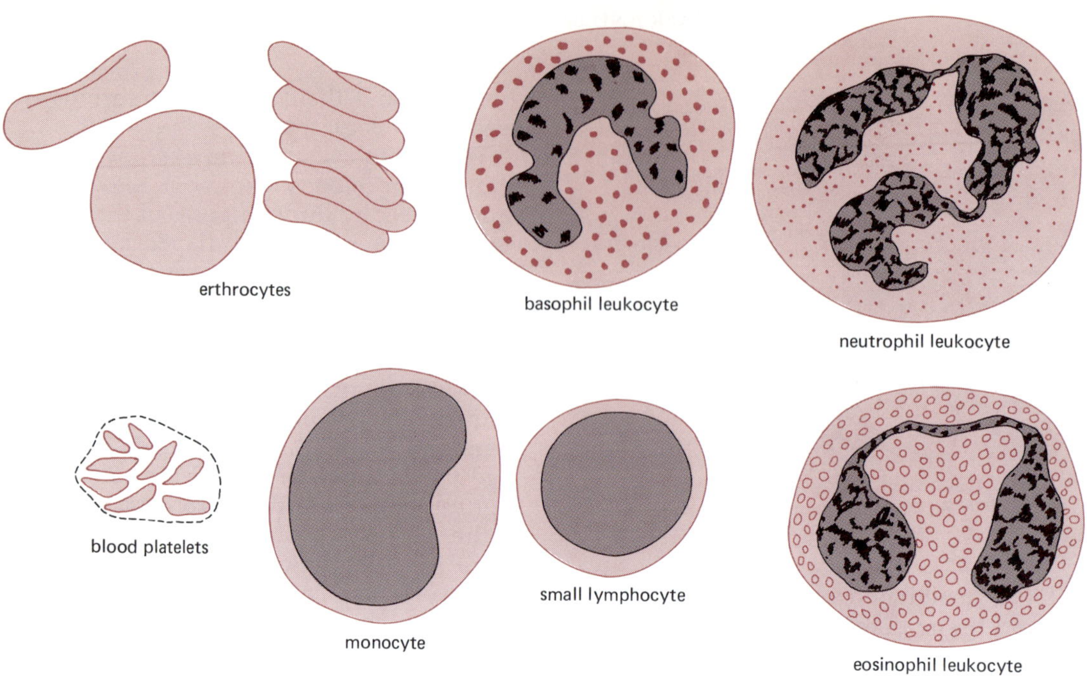

Figure 11–16 Types of blood cells. Both red cells (erythrocytes) and white cells (leukocytes) are derived from undifferentiated **stem cells**, produced in the bone marrow.

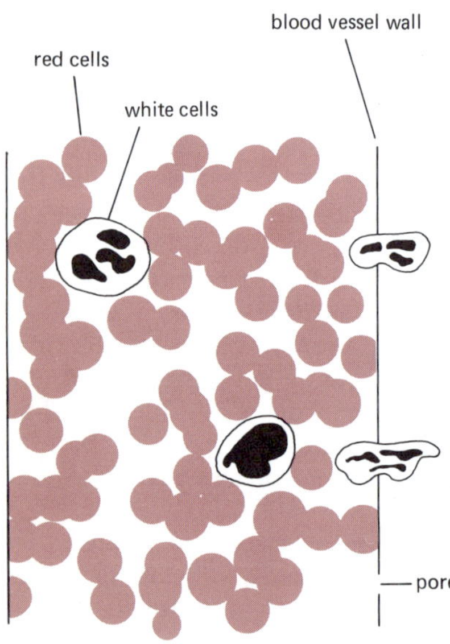

Figure 11–17 Patrolling white cells squeeze in and out of pores in the blood vessel walls in a process known as diapedesis.

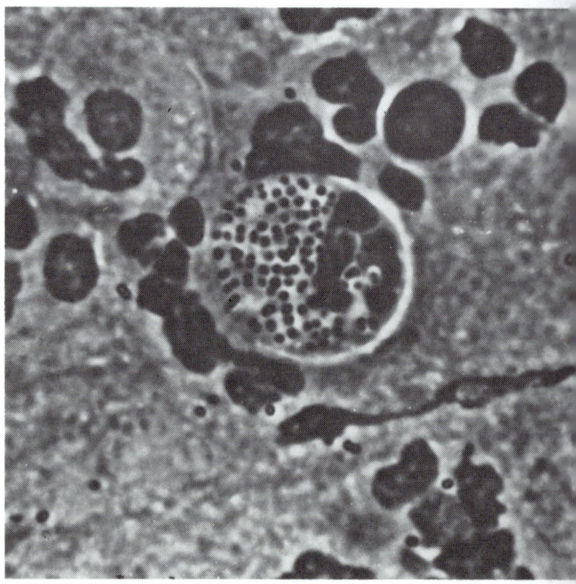

Figure 11–18 The large leukocyte in the middle has ingested numerous bacteria (**Neisseria gonorrhoeae**), 1000×.

in the population of one particular type of leukocytes, the type affected depending on the nature of the infection. A **differential white cell count** is thus a valuable diagnostic aid.

BLOOD PLATELETS

The third component of the formed elements actually represents only fragments of cells. These are the **platelets,** colorless, anucleated, disklike structures with a diameter of only about 2 to 4 μ. They are formed in the bone marrow by the fragmentation of large white blood cells. Each cubic millimeter of blood normally contains between two hundred and four hundred thousand platelets. They are delicate structures that rupture easily and generally live only about a week. The blood platelets play a key role in the mechanism of blood clotting, as we shall see shortly.

BLOOD PLASMA

Blood plasma is an extremely complex solution containing hundreds of different substances. These include water (90–92 percent), proteins (7–8 percent), carbohydrates, lipids, hormones, enzymes, vitamins, antibodies, amino acids, fatty acids, glycerol, dissolved gases, salts, and nitrogenous wastes. Some of these substances play important roles in the specific functions of the blood; others are being transported in the blood flow from the site of their manufacture to the site of their intended use in the body (or to the excretory organs for ultimate removal from the body).

Most of the proteins in the blood may be classified as **serum albumins, serum globulins** (alpha, beta, and gamma globulins), and **fibrinogen.** The gamma globulin fraction contains numerous antibodies, and this fraction is frequently extracted from blood and administered as a protective measure to persons exposed to serious diseases in order to provide a temporary immunity. Fibrinogen plays an important role in blood clotting.

The blood proteins are **amphoteric** (i.e., have both acid and basic functional groups) and as such can act as buffering agents, helping to maintain the blood pH within a narrow range. They also aid in the transport of various other substances in the blood, such as lipids (which would normally not be water-soluble) and carbohydrates. The blood proteins also include minute concentrations of a variety of enzymes and some hormones. Unlike the other components of the plasma, blood proteins do not diffuse through the capillary walls.

The salt content of the blood plasma is maintained at a relatively stable value (0.9 percent). The most abundant ions found in the plasma are Na^+ and Cl^-. Potassium, calcium, bicarbonate (HCO_3^-), and phosphate ($PO_4^{\equiv}$) ions are also major constituents of the plasma salts. Bicarbonate ions are one form in which carbon dioxide is transported in the blood, and they also act as buffering agents. Largely due to their action, the blood pH is kept at a relatively stable level of 7.3–7.4.

BLOOD CLOTTING

The human body possesses a remarkable and complicated means of preventing excessive loss of blood due to minor injuries. Within the confines of the blood vessels, blood remains in a fluid state. But if a sample of blood is withdrawn with a syringe and placed in a test tube, it will be observed to turn into a gelatinous mass within five to eight minutes (the normal clotting time). On closer examination (Fig. 11–19a), the mass or **clot** is found to consist of a network of microscopic, needlelike threads in which red and white cells are enmeshed. The same thing happens whenever a blood vessel is broken, as by a cut or rupture: within minutes, a gooey clot forms and plugs up the hole in the vessel wall, stopping the leakage of blood.

The reactions of blood clotting involve the blood platelets and several of the blood proteins (Fig. 11–19b). When a blood vessel is damaged, the delicate blood platelets in the blood flowing by are caught on the rough edges of the lacerated vessel wall. They at first adhere and then disintegrate, setting off an intricate chain reaction by spilling large quantities of the enzyme **thromboplastinogenase** out into the blood. This enzyme works

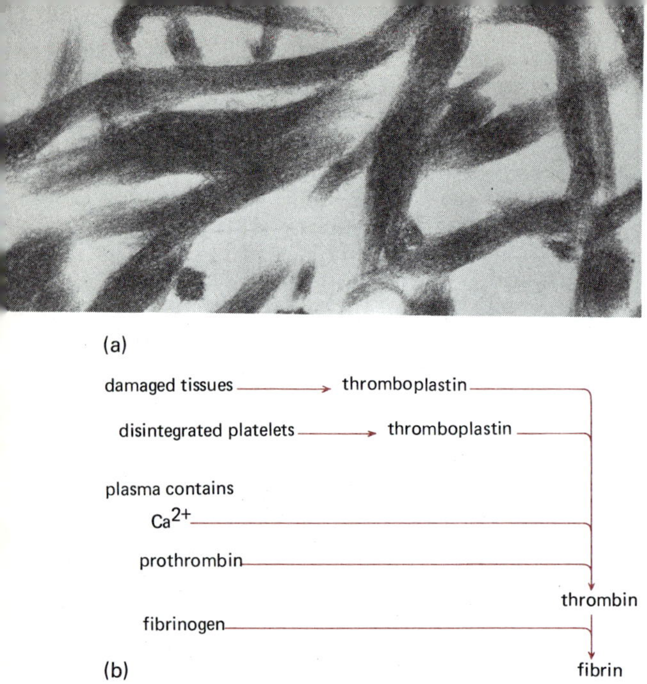

Figure 11–19 (a) Electron micrograph of a clot of fibrinogen that formed when thrombin was added (59,500 X). (b) The sequence of chemical reactions leading to blood clotting.

on one of the plasma proteins, **thromboplastinogen,** converting it to another protein, **thromboplastin.** In the presence of calcium ions, thromboplastin reacts with still another plasma protein, **prothrombin,** converting it to **thrombin.** Thrombin in turn acts on the plasma **fibrinogen,** forming **fibrin.** The latter polymerizes into a threadlike network of fibers, which form the framework of the clot. Passing white and red blood cells are caught in the sticky fibers and held there, blocking up the gaps in the network and eventually forming a solid mass.

The buildup of fat deposits within the artery walls roughens their inner surfaces and increases the incidence of platelet rupture. As a result, a clot may also form within a blood vessel. This clot may become dislodged and carried to other parts of the body. It may eventually lodge in a smaller blood vessel and cut off the flow of blood through the vessel. If this occurs in one of the main vessels supplying the heart or the brain, a heart attack or stroke may result.

Under normal body conditions, there is a continual breakdown of blood platelets, resulting in a continual outpouring of thromboplastinogenase into the blood. However, unless the level of the enzyme exceeds a certain level, the formation of clots is prevented by the presence of **heparin,** a mucopolysaccharide that blocks the conversion of prothrombin to thrombin. Physicians frequently administer heparin or other **anticoagulants** (anticlotting agents) to patients with damaged hearts or blood vessels to reduce the danger of clot formation within the vessels. The opposite situation occurs in the disease **hemophilia,** in which there is a deficiency of thromboplastin formation. Persons suffering from this condition bleed profusely and uncontrollably from the slightest cuts, because their blood does not clot properly.

Accessory Organs of the Circulatory System

One normally tends to think of the circulatory system as consisting of the heart and blood vessels. But actually a number of other organs play important roles in the circulatory system.

The **liver** and **spleen** have large spaces called sinuses, which can serve as reservoirs for blood. In times of physical or emotional stress, or in case of hemorrhage, the muscles in the walls of the liver and spleen contract, forcing more blood out into the general circulation. The blood of the spleen is especially rich in red blood cells. (In some animals as much as 25 percent of the red blood cells are normally stored in the spleen.) Large phagocytic cells in the liver and spleen, called **macrophages,** filter out the damaged and worn-out cells as the blood passes through these organs. In addition, the spleen has been found to be an active center of antibody production. But it is not necessary for the survival of an adult human; if it must be removed, its functions can generally be taken over by other organs of the body.

The **bone marrow** is another key accessory organ of the circulatory system. This is a spongy tissue found within the cavities of the bones of the body, particularly the short flat bones, such as the ribs, sternum, clavicle, vertebrae, skull, and pelvic bones, and the ends of the long bones, such as the humerus and femur. In the adult, all the red cell production and hemoglobin production take place in the bone marrow—specifically, in regions whose richness in red cells gives them the name of **red bone marrow.** Some of the white blood cells and platelets are also manufactured here. A second type of marrow, the **yellow bone marrow,** is found in the shafts of the long bones and abounds in fat cells. Under emergency conditions, this bone marrow can also manufacture blood cells.

If cancer (a disease characterized by uncontrolled growth of cells) develops in the bone marrow, abnormally large numbers of white blood cells may be produced. This disease, **leukemia,** strikes victims of all ages and may be accompanied by severe anemia. Although neither the cause nor a cure have yet been discovered, doctors are now having some success through the use of complex drugs, transfusions, and other forms of therapy. In many cases an apparent normalization or remission can be achieved.

The Lymphatic System

The thin walls of the capillaries are quite permeable to small molecules, such as water molecules. As a result, some of the fluid of the blood continually leaks out into the tissues. This colorless fluid, called **lymph,** is simply blood plasma without the proteins (which for the most part cannot diffuse through the capillary walls). The lymph bathes the tissues, providing additional opportunities for an exchange of materials with the body cells. Most of the fluid lost from the capillaries eventually diffuses back into other blood capillaries, but a substantial portion does not. This lost lymph eventually passes into a separate circulatory system, the lymphatic system (Fig. 11–20). Diffusing first into the blind-ended **lymphatic capillaries,** the lymph is drained into successively larger and larger **lymphatic veins,** the largest of which are equipped with valves to prevent backflow of liquid. Ultimately two large lymph vessels return the lost fluid to the veins of the blood circulatory system.

Scattered through the lymphatic system, particularly in the region of the groin, abdomen, neck, and in the armpits, are oval-shaped masses of lymphoid tissue called **lymph nodes.** These nodes contain sinuses, or cavities, into which the lymph flows. Abundant phagocytic cells in the lining of the walls of the sinuses filter out bacteria and other contaminants that have penetrated into the lymph. During serious infections, the lymph nodes become especially active and may become enlarged, producing painful "swollen glands."

Defenses against Disease

It has often been remarked facetiously that doctors and drugs do not really cure patients. To some extent this is true, for in most cases the aim of the physician is to do what he can to improve the odds in the body's ceaseless fight against disease. Although drugs may help, the final cure is effected by the body itself.

The body has a number of lines of defense against the constant bombardment of potentially inimical microbes. The skin (as long as it remains unbroken) provides an effective barrier to the entry of microscopic invaders. The acid in the stomach kills many of the bacteria that are taken in with food. The mucous membranes lining the air passages and lungs are equipped with cilia, which help to expel some invading microbes and foreign particles breathed in with the air. But hordes of bacteria, viruses, and fungi manage to breach these preliminary defenses and are met by the body's second line of defenses, concentrated in the circulatory system and associated organs: the liver, spleen, thymus, bone marrow, and lymph nodes. These organs together make up a disease-fighting complex

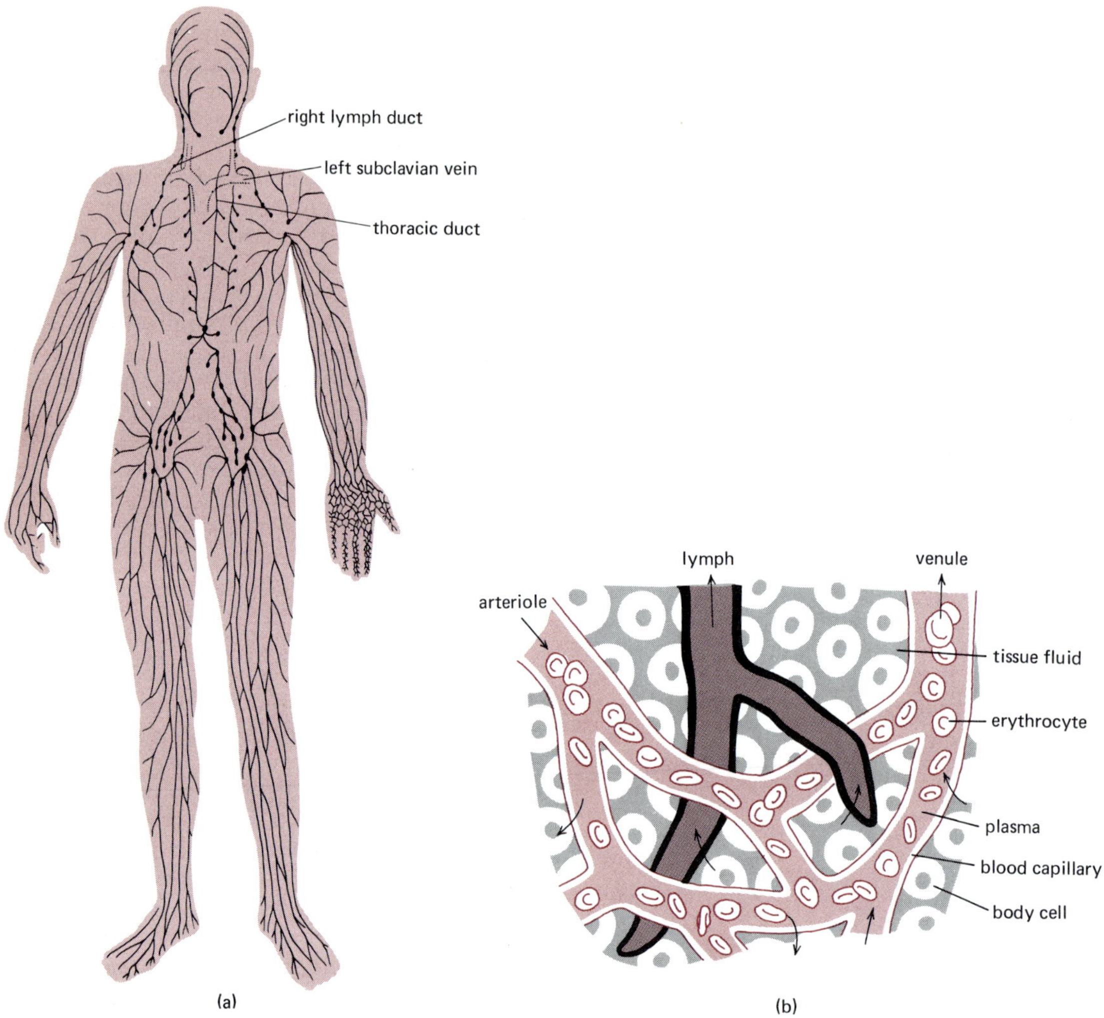

Figure 11–20 (a) The human lymphatic system. (b) Fluid diffuses out of blood capillaries into the surrounding tissues. Some of this fluid diffuses into blind-ended lymphatic capillaries and is ultimately returned by the lymphatic system to the blood circulatory system.

collectively referred to as the **reticuloendothelial system (RES).**

We have already seen how some of these defenses operate. In the case of local infections, the damaged cells secrete a substance (the exact nature of which is not yet known) that attracts phagocytes. These white blood cells engulf and kill the invaders. Phagocytic cells in the linings of the lymph nodes and macrophages in the sinuses of the liver and spleen also help to screen out potentially dangerous microbes. The phagocytes are effective against most bacteria but generally ineffective against viruses. The latter may be controlled by a substance called interferon, which is produced and released by virus-invaded cells. (The mechanism of interferon's production and action will be discussed in Chapter 23.) But some microorganisms multiply too rapidly to be kept in check by these measures, and some types of bacteria are resistant to destruction by the phagocytes.

The deciding factor that often finally brings success in the body's fight against an invading microbe is the production of **antibodies.** Each type of antibody is a protein with a structure that is believed to correspond, physically and chemically, to the structure of a particular **antigen**—a foreign substance that has penetrated into the body. The normal human body has an amazing ability to distinguish between "self"—the hundreds of thousands of chemicals that make up its own tissues—and "nonself"—specific chemicals from other organisms of the same species or different species.

Antibody production has been the object of intensive study in recent years, but researchers are not yet entirely certain exactly where and how it occurs. The process is believed to begin when an invading microbe or other antigen is ingested by a phagocyte, perhaps in a lymph node. The phagocyte apparently uses the antigen as a pattern and begins to turn out complementary antibodies (Fig. 11–21). The information is passed on to other white cells and to intensively protein-synthesizing **plasma cells** in the tissue spaces

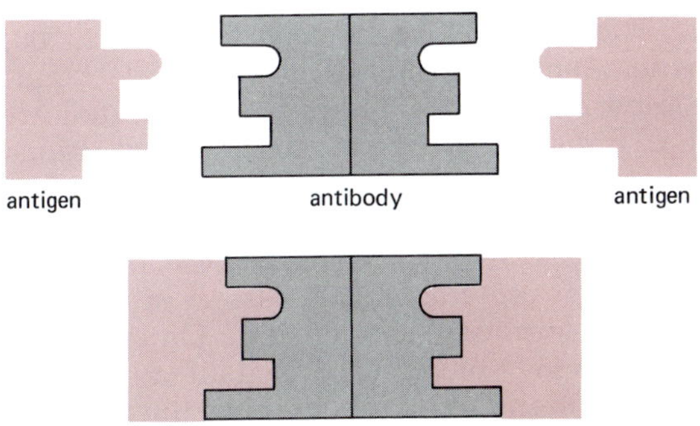

Figure 11–21 Antigen-antibody reaction. There is evidence indicating that antibody molecules are proteins consisting of two identical halves. Each contains an active site, exactly complementary to a specific antigen. When antibody and antigen combine, the antigen is inactivated.

and lymph nodes, and these cells also begin to produce the new antibody.

The antibodies react with particles of the antigen and inactivate them. The manner in which this occurs varies according to the particular types of antibody and antigen involved. **Antitoxins** are antibodies that render bacterial toxins harmless. **Opsonins,** another type of antibody, alter the surface of bacteria in such a way as to make them more easily engulfed by phagocytes. Often the production of large numbers of antibodies marks the turning point in a serious illness; gradually the body's defense forces gain control, and the invading microbes are destroyed.

After all traces of the antigen have disappeared from the bloodstream, the massive production of its specific antibody slows down. But a small pool of the antibodies still remains—the pattern is kept permanently on file. If the body is subsequently invaded by a similar antigen, it will be prepared to respond with a virtually immediate explosion of antibody production that will snuff out the incipient infection before it has a chance to grow to dangerous proportions. The person is then said to have acquired an **immunity** to the disease associated with that antigen.

The immunity mechanism has been successfully utilized in the treatment of diseases by the development of **vaccines,** artificially introduced antigens that stimulate the formation of antibodies, which in turn will confer immunity against dangerous disease-causing microbes. The first experimental vaccination was performed in 1796, long before the immunity mechanism or even the causes of disease were understood. The English physician Edward Jenner noticed that milkmaids who had had cowpox, a mild disease, acquired a protection from smallpox, a major scourge of the time. He injected material from the sores of a cowpox patient into a human subject and then demonstrated that smallpox immunity had been acquired. We know now that the technique works because the cowpox virus is similar enough to the smallpox virus that the antibodies whose production it stimulates are also effective against the more dangerous pathogen. Vaccines are now available against a number of bacterial and viral diseases, including polio, measles, influenza, typhoid fever, yellow fever, and bubonic plague. In the case of diphtheria and tetanus, the vaccines stimulate the production of antibodies against the bacterial toxins, rather than the bacteria themselves.

Invading microbes usually have an outer coating of protein or polysaccharide, and bacterial toxins are generally proteins. As might be expected, the body's immunologic mechanism characteristically responds not only to invading microbes, but also to any sort of foreign proteins and polysaccharides that happen to penetrate into the bloodstream. This property has created great problems for the rapidly developing medical technique of organ transplantation, as will be discussed in Chapter 23. The body recognizes the transplanted organ as "foreign" and ultimately rejects it unless special precautions are taken.

Sometimes the immunologic mechanisms may go awry, or may function to the ultimate harm of the body instead of its aid. Allergies are cases in which the body "overreacts," responding to the invasion of a minor irritant, such as dust or ragweed pollen, with an enormous mobilization of defensive forces more suitable to a major threat to life. Some diseases, such as rheumatoid arthritis, are believed to be autoimmune diseases, in which the body has mistakenly "recognized" some of its *own* chemicals as foreign bodies and has begun to manufacture antibodies against them. Some modern theories hold that cancer is due to

Figure 11–22 Gerald Edelman with a model of the antibody whose structure he delineated.

an opposite kind of defect of the immunity system: the body has failed to recognize the cancer cells as foreign and therefore has failed to manufacture antibodies against them. The key to an understanding and possibly a cure for all these conditions may lie in a better understanding of the immunologic mechanism. Encouraging progress is being made. One of the most exciting recent achievements in the field has been the delineation of the complete amino acid sequence of an antibody, as well as the general contours of its structure (Fig. 11–22) and some insights into the manner in which its active sites operate.

Blood Types and Blood Transfusions

Blood transfusions to replace the blood lost in severe hemorrhages were first tried back in the seventeenth century, using animal blood for the transfusions. The patients all died. Later, better results were obtained by using human blood for transfusions, but still a large proportion of the patients died. No one could understand why, or predict which patients would live and which would die.

The mystery was finally solved in 1900, when an Austrian biologist, Karl Landsteiner, discovered that cases of death following transfusions were caused by reactions among the blood proteins. Landsteiner discovered two characteristic proteins in red blood cells, which seemed to be paired with two other proteins in the plasma, in a sort of antigen-antibody relationship. Of course, a person's own blood never contained both of a particular antigen-antibody pair. But if it was mixed with blood from another subject that did contain the appropriate protein, an immediate reaction occurred, and the blood "clumped" or coagulated. If this occurred within the body of a patient who had just received a transfusion, the result would be massive blood clots, which could lead to blockage of the blood flow and ultimate death.

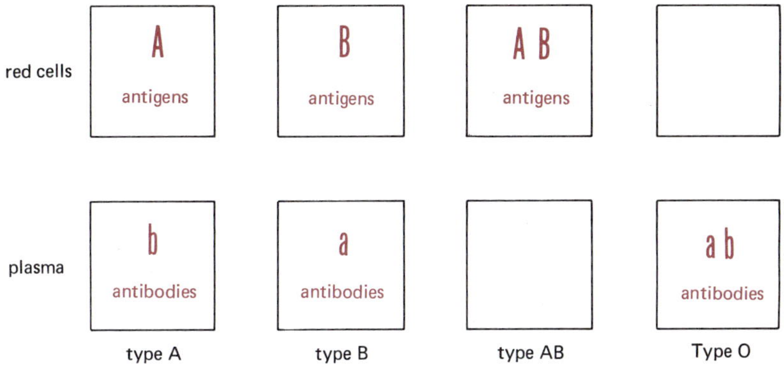

Figure 11–23 Red blood cells and blood plasma contain proteins that are linked together in antigen-antibody relationships. (No blood type contains *both* of a corresponding antigen-antibody pair.)

Landsteiner's red cell proteins were labeled **A** and **B,** and the plasma protein factors were labeled **a** and **b.** Four different combinations were theoretically possible: (1) a person could have A proteins in his red cells and b proteins in his plasma; or (2) B proteins in his red cells and a proteins in his plasma; or (3) both A and B proteins in his red cells and no corresponding proteins in his plasma; or (4) both a and b proteins in his plasma and no corresponding proteins in his red cells (Fig. 11–23). Investigations of blood samples taken from numerous subjects revealed that all four cases occur. Case 4 or type O is the most numerous in the United States population (about 45 percent), with case 1 or type A a close second (about 42 percent). Smaller groups of the population are found to have blood corresponding to cases 2 and 3 (types B and AB; 10 and 3 percent, respectively). The blood factors have been found to be inherited, by a mechanism that will be described in Chapter 14.

Clumping reactions in blood transfusions can be avoided by giving the patient blood from a donor whose blood has matching antigen and antibody proteins. Blood samples can be "typed" (i.e., their blood group determined) by taking a few drops and adding them to samples of known type A and type B blood, as diagramed in Fig. 11–24. In emergency cases, when an exact match cannot be obtained, blood of a different type may be

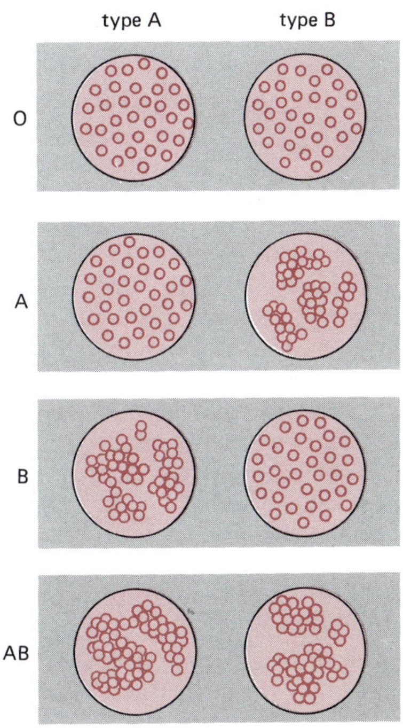

Figure 11–24 Blood typing. Blood samples are added to each of two standard specimens, type A and type B. Then the presence or absence of "clumping" is observed.

used in certain specific cases. For example, a person with type A blood (with A antigens in his red cells and b antibodies in his plasma) can donate blood to persons of types A and AB, but can receive blood only from persons with types A and O. Type O (no antigens in the red cells, both a and b antibodies in the plasma) is called the "universal donor" because such blood can be donated to any other group, but persons with type O blood can *receive* blood only from other type O's. In each case the general rule is that the red cells of the *donor* must contain no antigens corresponding to the antibodies in the plasma of the *recipient*. The presence of antibodies in the introduced blood against the recipient's own red blood cells is not too serious because the plasma introduced is quickly diluted by the patient's own plasma.

Since Landsteiner's work, various other blood protein factors have been discovered, resulting in a rather complicated "blood type." One of the most important of these is the **Rh factor,** named for the rhesus monkey, in which it was first observed. This is an antigen (or rather a number of closely related substances) found in the blood of most Americans. About 15 percent of Americans do not have the Rh factor in their blood; they are referred to as Rh-negative. Persons who do possess the factor are Rh-positive. Normally, blood does not contain any antibodies corresponding to the Rh factor: Rh-positive persons would not have any antibodies for their own proteins, and Rh-negative persons would never have been sensitized to the antigen. But if an Rh-negative person receives a transfusion of Rh-positive blood, he will develop the corresponding antibodies. The volume of red cells introduced is usually so minor that there will be no serious consequences for the person who received the transfusion. But if the person happens to be a woman married to an Rh-positive man, there may be serious consequences to her future children.

The Rh factor is inherited as a "dominant" factor; therefore, there is a great likelihood that any children will be Rh-positive if either of the parents is Rh-positive. When an Rh-negative mother whose blood has been sensitized to the Rh factor (i.e., has been stimulated to produce antibodies to it) is carrying an Rh-positive child, there may be some leakage of blood across the barrier of the placenta (Fig. 11–25). The mother's antibodies may find their way into the child's blood, resulting in clumping of his red cells. Depending on the degree of mixing, the baby may die before birth or may be born with a condition of anemia called erythroblastosis fetalis.

In addition to mismatched transfusions (which occurred quite often before the Rh factor had been discovered), an Rh-negative mother's blood can also be sensitized in another way. If she is carrying an Rh-positive child, some of its blood may leak into her bloodstream, either late in the pregnancy or, even more likely, during the actual birth. The child born in such a pregnancy will generally be unharmed, because he will be safely born before the antibodies in his mother's blood have built up to a dangerous level. But any subsequent Rh-positive child the woman conceives may be in jeopardy.

In recent years, erythroblastosis fetalis has generally been treated by giving the baby a complete exchange transfusion either at birth or even while still in the womb, replacing his damaged blood with new whole blood. In 1968, however, a new vaccine became available, the widespread use of which may ultimately bring the disease under complete control. The treatment is given to an Rh-negative mother whose blood has not yet been sensitized, within a day or two after she has given birth to an Rh-positive baby. Paradoxically, the woman is injected with Rh antibodies. These react with any antigen that has leaked into her blood during the birth, thus removing the stimulus for her body to make antibodies of its own. After a short time, both the antigen and the injected antibodies are eliminated from the mother's blood, and she is still antibody-free, capable of having another pregnancy in safety. Unfortunately, the treatment does not help women whose blood has already been sensitized.

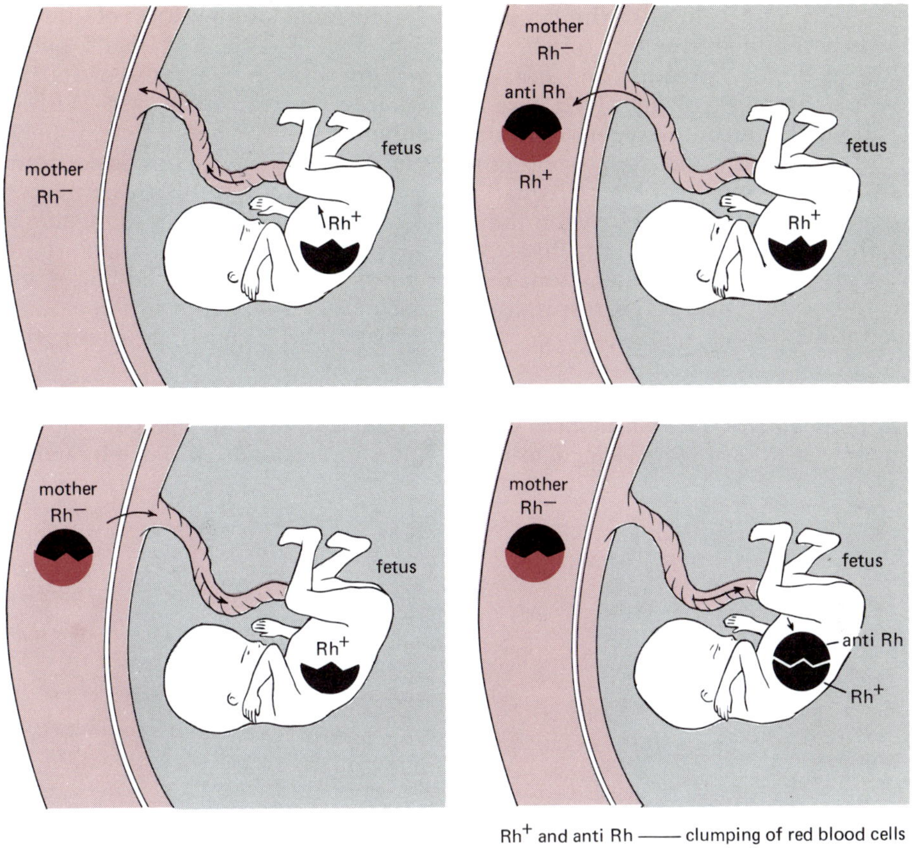

Figure 11–25 Rh conflict. When a mother with Rh negative blood is carrying a fetus with Rh positive blood, there may be some leakage of fetal blood through the placental barrier. The mother's blood, thus sensitized, produces antibodies against the Rh^+ antigen. Transfer of the antibodies to the fetus produces clumping of its red cells, resulting in the serious disease, erythroblastosis fetalis. If sensitization occurs at all, it usually does so at the time of delivery. Thus the first Rh^+ child is usually unaffected, but a hazard may be created for future Rh^+ offspring.

Respiration

One of the most important functions of the circulatory system is to carry oxygen to the cells, where it is used in the oxidation of food materials to provide energy for their activities, and to carry away the carbon dioxide that is produced as a waste product of this oxidation. The organism obtains its oxygen from the environment. Single-celled organisms and animals with a very simple body structure in which essentially all the body cells are in contact with the environment (e.g., sponges and flatworms) do not need any special structures or organ systems to bring oxygen into the body. They obtain their supplies of this vital gas by simple diffusion. But higher organisms require more intricate means of getting oxygen to all their body cells, and most animals have an organized respiratory system, which works in conjunction with the circulatory system to accomplish this function.

The water of ponds, streams, and oceans contains various amounts of dissolved gases, including about 1 percent oxygen derived both from the atmosphere and from the photosynthetic activity of aquatic plants. Aquatic animals are able to obtain all the oxygen they need from the water in which they live. A variety of forms derive this oxygen from the water with the aid of specialized respiratory organs called **gills.**

The most familiar form of gills is that of the fish (Fig. 11–26). The fish's gills are found within slits on the sides of its body, somewhat to the rear of its eyes. Within these **gill slits** are many rows of intricately branched plates, which have a pink color from their rich supply of blood capillaries.

A fish "breathes" in the water by opening and closing its mouth regularly. The water that is drawn in is forced backward when the fish closes its mouth. At the same time, its throat muscles constrict, leaving the water no-

Figure 11–26 Respiratory system of the fish. Cutaway sections in (a) show the gills and heart. Gas exchange with the water flowing over the gill filaments occurs in the profuse bed of gill capillaries.

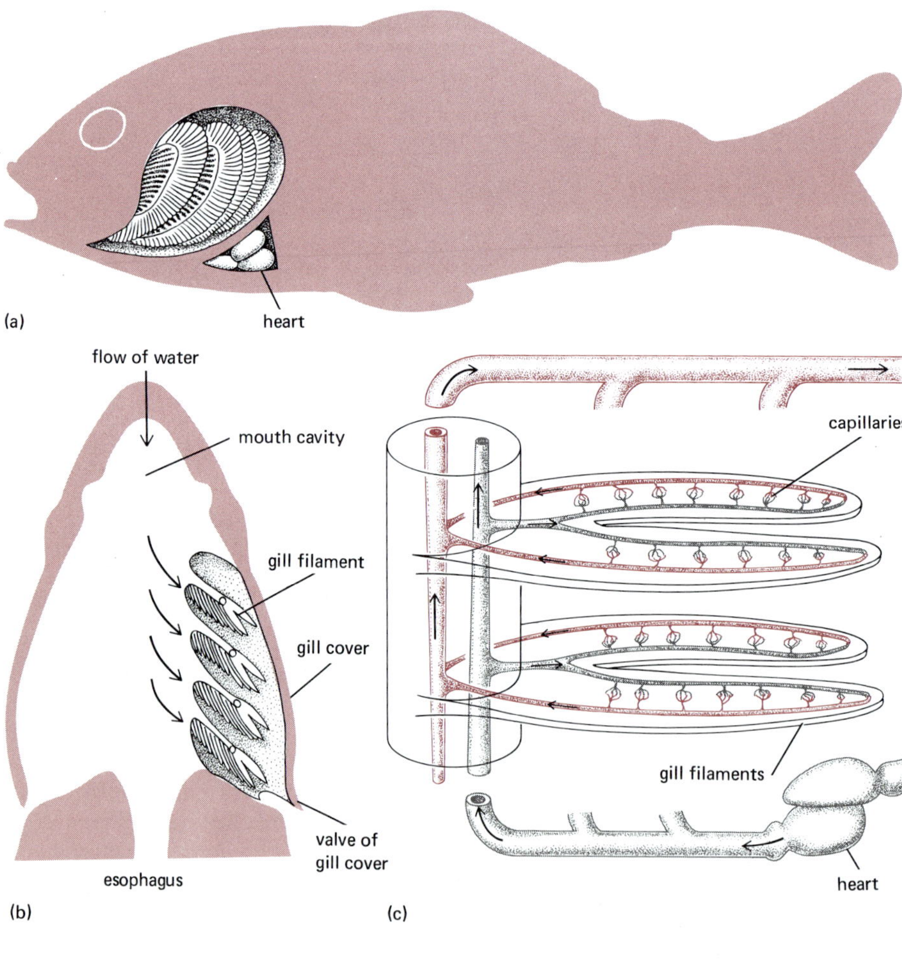

where to go except back through channels leading over the gill plates. As the water bathes the gill plates, oxygen diffuses through their thin membranes into the capillary bed. At the same time, carbon dioxide diffuses out from the capillaries into the water and is carried away when the water is swept out through the gill slits.

A number of other water animals, such as clams, squids, and even some aquatic insects, also breathe through gills. Researchers have recently been trying to devise plastic-membrane artificial gills that could be used by humans. The prototypes produced thus far are rather unwieldy, but if compact models can be developed, man may some day be able to swim in the water without having to carry an oxygen supply with him.

Some terrestrial animals that live under moist conditions are able to breathe through modified gills or even by diffusion through their moist skin. The earthworm, which lives in damp soil, gets its oxygen and eliminates its carbon dioxide in the latter manner. The respiration of frogs and other amphibians is augmented by a gas exchange through their moist skin.

But animals firmly committed to dry land are faced with a respiratory paradox: they do not have access to sufficient water to keep the outside of their bodies moist, but the diffusion of gases, to proceed effectively, must occur through at least a thin film of moisture. Terrestrial animals generally solve the problem through a respiratory system consisting of passages leading into the body. There, shielded from the drying effects of the atmosphere, the vital gas exchange can take place.

The respiratory system of most terrestrial insects consists of an intricately branched system of air ducts, called **tracheal tubes,** which lead into the body from openings called **spiracles** (Fig. 11–27). Bristles at the spiracles help to filter out dust particles; muscular valves permit the insect to regulate the opening and closing of the spiracles. The spiracles have been observed to open and close much more frequently when the insect is moving about actively than when it is at rest. In many insects, contractions of the body muscles help

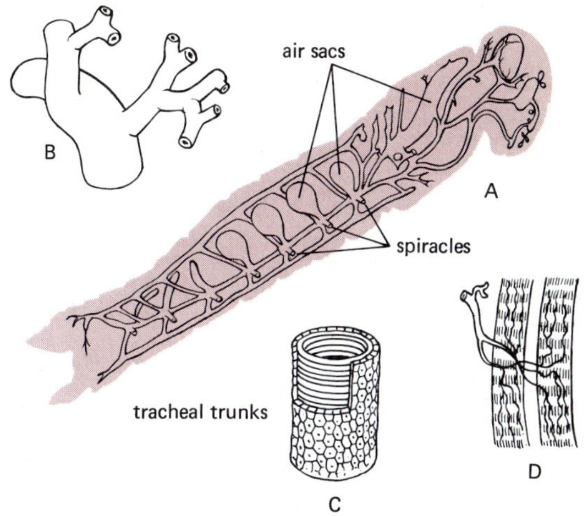

Figure 11–27 Respiratory system of the grasshopper, a typical insect.

to pump air in and out of the air passages.

The numerous branches of the tracheal tubes reach virtually all the cells of the insect's body, and an exchange of gases occurs directly. Thus oxygen and carbon dioxide do not have to be carried actively by the insect's circulatory system. Indeed, as has already been mentioned, the blood of insects generally lacks any special gas-carrying pigment.

Spiders, like insects, generally have a system of tracheal tubes. But they usually possess in addition special respiratory structures called **book lungs.** This curious name comes from their structure: they are constructed very much like gills, and the membranes that comprise them are folded like the pages of a book.

The terrestrial vertebrates—amphibians, reptiles, birds, and mammals—have respiratory systems consisting of air passages terminating in two thin-walled air sacs called **lungs.** These are generally subdivided into many tiny sacs, thus greatly increasing the effective surface area over which diffusion can take place.

The flight of birds requires enormous expenditures of energy, and birds have correspondingly large oxygen requirements. These are satisfied in part by supplementary air sacs protruding from the lungs (Fig. 11–28), which

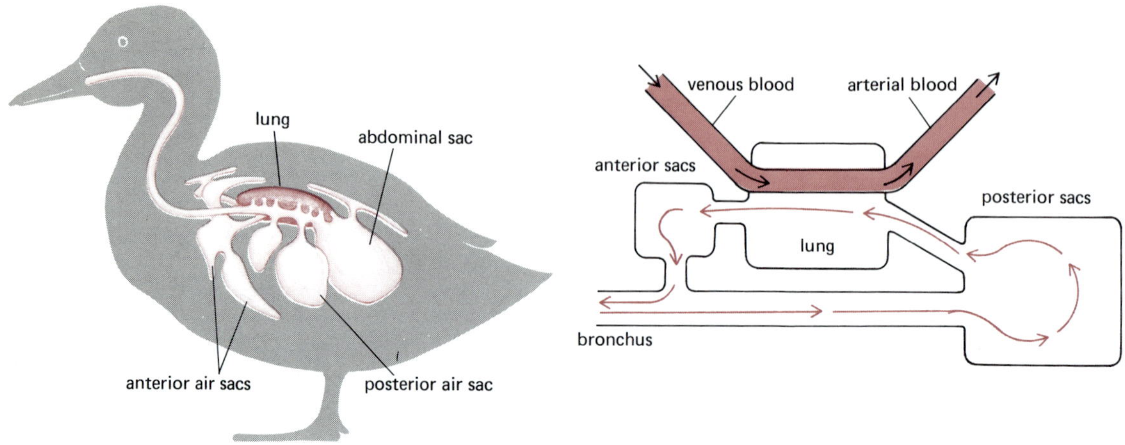

Figure 11–28 Respiratory system of the bird. The bronchi divide to form many small tubes, called parabronchi, which function like the alveoli of mammals. The air sacs, which are connected to air spaces in the bones, help to increase respiratory efficiency and add buoyancy to the bird's body structure. The flow of air in the lungs is opposite to the direction of blood flow, providing for a highly efficient countercurrent exchange that permits birds to fly at altitudes of 20,000 feet or more.

help to increase the efficiency of respiration. The air sacs serve the additional function of reducing the bird's overall body weight, since air weighs much less than the corresponding amount of body tissue.

The Human Respiratory System

Breathing is as vital and as automatic a body function as the beating of the heart and the digestion of food. You do not have to continuously think about breathing and can do it quite efficiently whether you are asleep or awake. Yet, unlike the beating of the heart and the digestion of food, if you do make a conscious effort, you can vary the rate and volume of your breathing. Indeed, you can even stop breathing entirely—for about a minute. Then, try as you might to continue holding your breath, you will suddenly gasp and breathe in again.

In normal human respiration, air is inhaled through the nose and then passes back and down, through a chamber called the **pharynx,** through the **larynx** or voice box, down a passage called the **trachea,** which branches into two **bronchi,** and then through a succession of narrower and narrower tubes (the **bronchioles**), which ultimately terminate in minute air sacs, the **alveoli** (Fig. 11–29). There an exchange of gases occurs through diffusion into and out of the surrounding capillaries, and then the air is exhaled again, retracing the same route in the opposite sequence.

The nostrils, the openings to the nasal cavities, are guarded by a fringe of bristly hairs that screen out any large particles from the entering air. The nasal passages themselves, and the respiratory passages into which they lead, are lined with a moist mucous membrane, equipped with ciliated cells. The net effect of these structures is to warm and moisten the incoming air and further screen out foreign particles. When the nose is blocked, during a cold or allergy, a person can breathe through his mouth instead. But this is not as efficient. Air passing through the mouth is not warmed as effectively as in the nasal passages, and the lining of the mouth and throat is rapidly dried out by the continual flow of air.

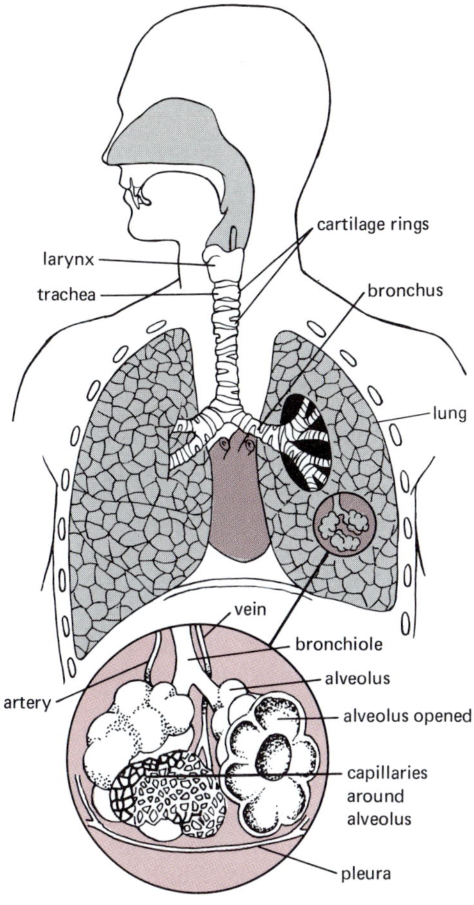

Figure 11–29 The human respiratory system.

The pharynx leads into two separate passageways, the **trachea** (windpipe) and esophagus (food pipe). The manner in which the trapdoorlike epiglottis guards the entrance to the trachea during swallowing was described in Chapter 10. If, however, the epiglottis malfunctions, or if solid or liquid particles are breathed in through the nose, an automatic cough and/or sneeze mechanism is set off and continues until the offending particles are forcefully expelled from the air passages.

The walls of the trachea are reinforced with firm, elastic rings of cartilage. Two folds of tissue, the **vocal cords,** stretch across the **larynx,** the enlarged chamber at the top of the trachea. When air passes through the larynx (usually during exhalation), the vocal cords may be set vibrating. A corresponding vibration is induced in the column of air in the pharynx and mouth cavity, and a sound is produced. The tension of the vocal cords can be varied with the aid of special muscles, and the pitch of the resultant sound can thus be changed.

Particles that have not been filtered out in the nose are trapped in the mucus-covered lining of the trachea, bronchi, or bronchioles. The moving cilia on the lining cells sweep the mucus upward, and it finally reaches the mouth as sputum, which can be spit out or swallowed.

The alveoli of the lungs look rather like a bunch of grapes. The surface area provided by these tiny clustered sacs is enormous: the two lungs of a normal adult human contain about 300 million alveoli, with a total surface area of between 400 and 800 square feet. (By contrast, an average adult male has a total skin surface area of only 20 square feet.) The alveoli are surrounded by a network of capillaries, and the outer membranes both of the alveoli and of the capillaries are extremely thin. Gases diffuse readily through the membranes in both directions.

THE GAS EXCHANGE

The inhaled air drawn into the lungs is relatively rich in oxygen (about 21 percent) and poor in carbon dioxide (0.04 percent). The blood delivered by the capillaries, on the other hand, is rich in carbon dioxide and poor in oxygen. Processes of diffusion in the direction of the concentration gradients therefore result in a net flow of oxygen from the alveoli into the capillaries and a net flow of carbon dioxide from the capillaries into the alveoli.

The lungs are not entirely emptied during exhalation; a reserve of about 2.5 liters of "used air" remains in the alveoli. Considering that an average man at rest breathes in and out only about 500 ml of air with each breath, the amount of new air brought into the lungs is a relatively small fraction of the air already there. (The amount of air breathed in and out can be increased substantially—to several liters—by making a special effort, and may rise to about 5 liters during strenuous exercise.) As a result of this relatively inefficient turnover of air in the lungs, the air within the alveoli usually contains more carbon dioxide and less oxygen than atmospheric air (about 13 percent oxygen and 5 percent carbon dioxide).

Diseases that reduce the surface area of the alveoli lower the efficiency of the gas exchange with the bloodstream. For example, in pneumonia many of the alveoli fill with fluid and cannot function in the gas exchange. In a condition known as emphysema, the partitions that divide the clusters of alveoli are broken down (Fig. 11–30), thus reducing the area in which gas exchange can take place. The heart must work harder to pump more blood to the lungs, so that the body will have a sufficient supply of oxygen; heart failure may ultimately result. Emphysema is a growing problem in our modern society, particularly among older people. Studies have revealed considerable evidence linking the incidence of emphysema both with smoking and with air pollution.

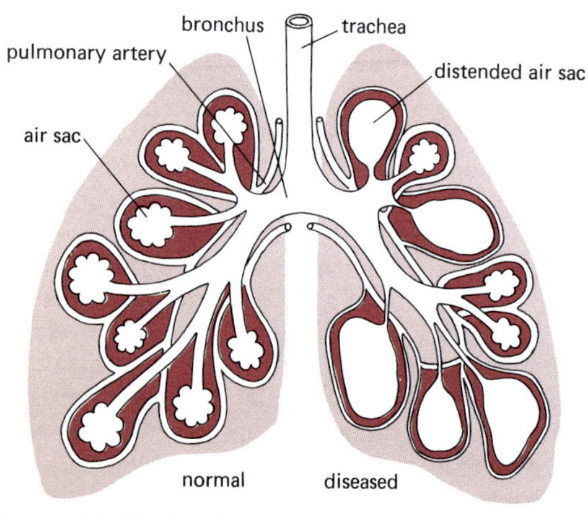

Figure 11–30 Emphysema causes a breakdown of the alveolar membranes, converting clusters of tiny alveoli to large distended sacs. The surface area for gas exchange is thus reduced, decreasing the efficiency of respiration. Emphysema also causes a reduction of the elasticity of the alveolar walls and a constriction of the bronchi, limiting air movement.

Indeed, smoking has a variety of damaging effects on the respiratory and circulatory systems. Comparisons of lung tissue of smokers and nonsmokers show a striking difference: a smoker's lungs are filled with tiny particles of deposited soot (Fig. 11-31). Air pollution causes a similar "lung pollution," but even an urban nonsmoker has far less soot in his lungs than a smoker. Inhaled smoke paralyzes the ciliated cells in the lining of the respiratory passages that would normally sweep the soot particles out and thus thwarts the normal clearing mechanism of the lungs and bronchi. Components of cigarette smoke cause changes in cells that can lead to cancer. (Pipe and cigar smoking has not been linked with lung cancer simply because their smoke normally is not inhaled; they have been implicated in cancers of the lips and mouth.) Smoking also increases the risk of heart and circulatory disease in several ways. Nicotine in cigarette smoke speeds up the heart rate and increases the pressure in the arteries, increasing the needs of the heart muscle for oxygen. Yet another major constituent of the smoke, carbon monoxide, can compete with oxygen in combining with the hemoglobin of the red blood cells, thus reducing the amount of oxygen they can carry to the heart and brain cells. Cigarette smoke has also been found to increase the tendency of blood to clot inside the blood vessels. All these factors can contribute to heart attacks and strokes.

The oxygen that passes into the bloodstream from the alveoli combines with the hemoglobin in the red blood cells in a reversible reaction, forming the compound **oxyhemoglobin.** This compound is a bright scarlet red and imparts a characteristic bright red color to oxygenated blood. (Hemoglobin itself is a dark, purplish red.) The oxygenated blood is carried through the heart and thence to the body cells. There the oxyhemoglobin gives up its bound oxygen to the cells. Carbon dioxide from the body cells diffuses into the blood. Some of it dissolves in the plasma and is carried mainly in the form of bicarbonate ions; some of the carbon dioxide is loosely bound to hemoglobin in the complex **carbaminohemoglobin.**

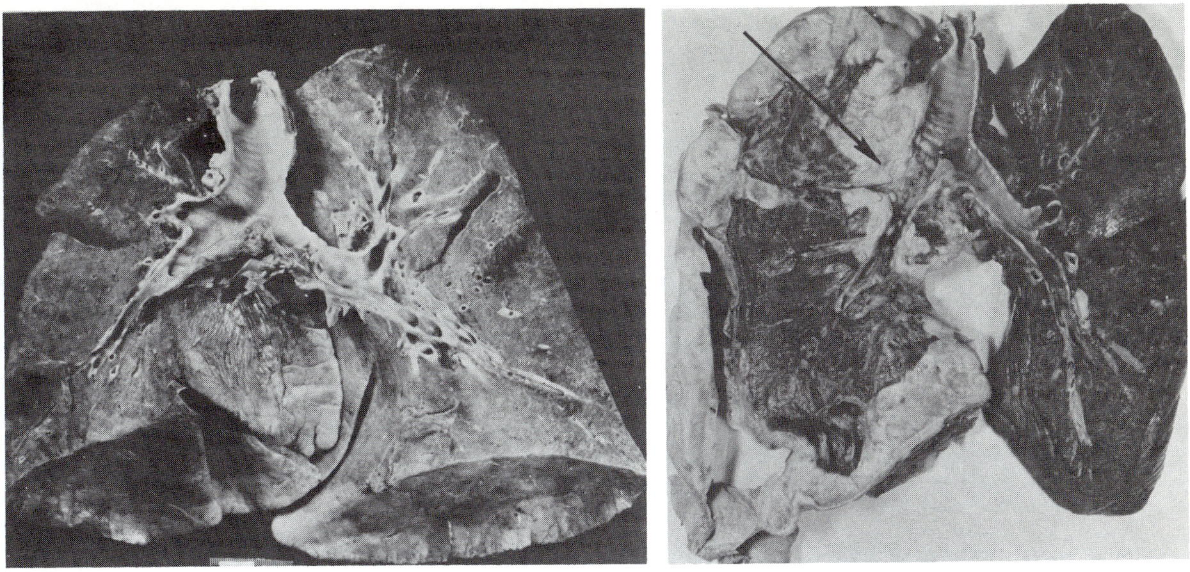

(a) (b)

Figure 11–31 Smoking produces tar deposits in the lungs, causes an enlargement of the alveoli (with a consequent reduction of the effectiveness of gas exchange), and may stimulate the development of cancer. (a) Lungs of a nonsmoker; (b) lungs of a heavy smoker with lung cancer (cancerous area indicated by the arrow).

MECHANISMS OF BREATHING

The lungs are surrounded by a thin membrane, the **pleura,** and the thoracic or chest cavity is lined with a similar membrane. The two pleural membranes adhere tightly together, "cemented" together by a film of moisture. Because of the tight seal between the two membranes, any action that produces an expansion of the chest cavity results in a corresponding expansion of the lungs. In man the chest cavity is expanded by moving the enclosing rib cage up and outward, while at the same time the muscular diaphragm, which forms the floor of the cavity, is contracted (Fig. 11–32). When the volume of the lungs is increased by such expansion, the air pressure in them drops considerably below atmospheric pressure. Since the lungs communicate with the external atmosphere through the air passages of the bronchi, trachea, and nose, air from the atmosphere rushes in to equalize the pressure. Inward movement of the ribs and relaxation of the diaphragm produce a contraction of the lung volume. The pressure within the lungs is correspondingly increased, and air rushes out through the air passages, again equalizing pressures with the atmosphere.

An adult human normally breathes in and out regularly every four or five seconds. The rate of breathing and the intricate muscular coordination involved are automatically controlled by a special respiratory center in the brain, in the region of the medulla oblongata. This region of the brain has been found to be extremely sensitive to the carbon dioxide content in the blood. When the carbon dioxide level in the blood reaching the medulla rises above a certain value, the respiratory center automatically increases the rate of respiration until the excess carbon dioxide is eliminated and its level in the blood has returned to normal. It might seem strange that the rate of respiration is controlled by the carbon dioxide level, rather than the oxygen level, when oxygen is the vital constituent in the inhaled air. The apparent paradox is resolved when we consider that these two gases are intimately associated in the processes of cellular respiration.

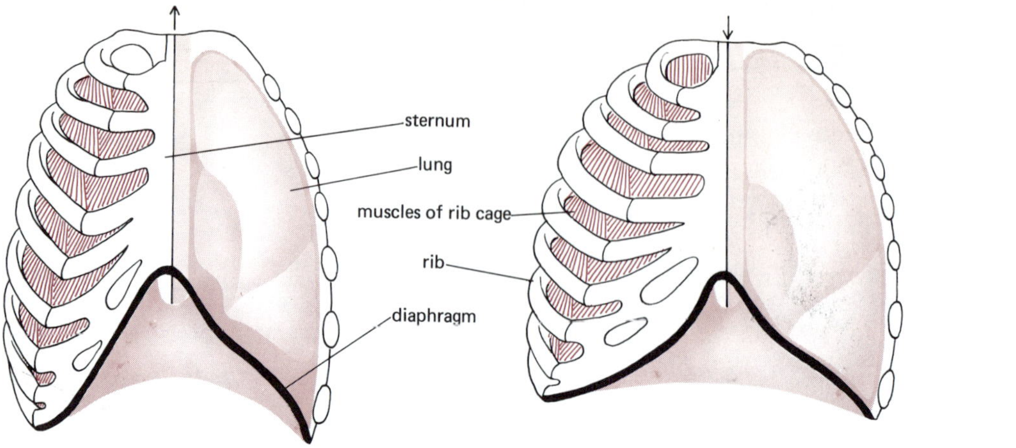

Figure 11–32 The mechanism of breathing. On the left: exhalation. The diaphragm is relaxed, volume of the chest cavity reduced, air is forced out of the lungs. On the right: inhalation. The diaphragm is contracted, volume of the chest cavity expands, and air is sucked into the lungs.

Cellular Respiration: The Energy Cycles

The sequence of chemical reactions that release energy for the body's activities has much in common with the burning of a bonfire. Chemical "fuel" is oxidized by oxygen in both cases, releasing the energy locked inside the chemical bonds and yielding simpler compounds as oxidation products. But there are also a number of important differences. A fire burns rapidly, and the energy released is given off in the form of heat and light. The oxidation reactions of the body's energy metabolism, on the other hand, proceed in a series of small, controlled steps; much of the energy released in each step is recaptured and stored in the high-energy phosphate bonds of ATP. Only a little of the energy released in metabolic reactions is given off as heat, and occasionally light (in bioluminescent organisms).

Throughout the kingdoms of life, there is a striking similarity of the basic processes by which energy is released from organic fuels and converted to a readily available form. Generally these processes begin with **glycolysis**, an energy-releasing process that can take place in the absence of oxygen and therefore is referred to as the **anaerobic** ("airless") **pathway**. In glycolysis, the six-carbon sugar glucose is subjected to a sequence of reactions ultimately yielding the three-carbon compound pyruvic acid. Two hydrogen atoms are eliminated in the process; they are taken up by an intermediate hydrogen acceptor, and the energy released is stored in the form of ATP.

At this point the respiration processes show some divergence, depending on the species involved. For example, in yeasts and some bacteria (and in higher plants under anaerobic conditions), the pyruvic acid is decarboxylated (i.e., loses carbon dioxide) to yield acetaldehyde; then two hydrogens are enzymatically added to form ethyl alcohol (Fig. 11–33). This process is known as **alcoholic fermentation.**

In some bacteria (and in animal cells in the absence of oxygen), the pyruvic acid may be enzymatically reduced by the addition of two hydrogens, forming lactic acid. This process, known as **lactic acid fermentation,** and the alcoholic fermentation mentioned above do not yield any further useful energy. Moreover, the end products of fermentation are toxic to the cell, and if allowed to accumulate may kill the organism that produced them. Fermentation processes in higher plants and animals therefore are generally emergency measures, which

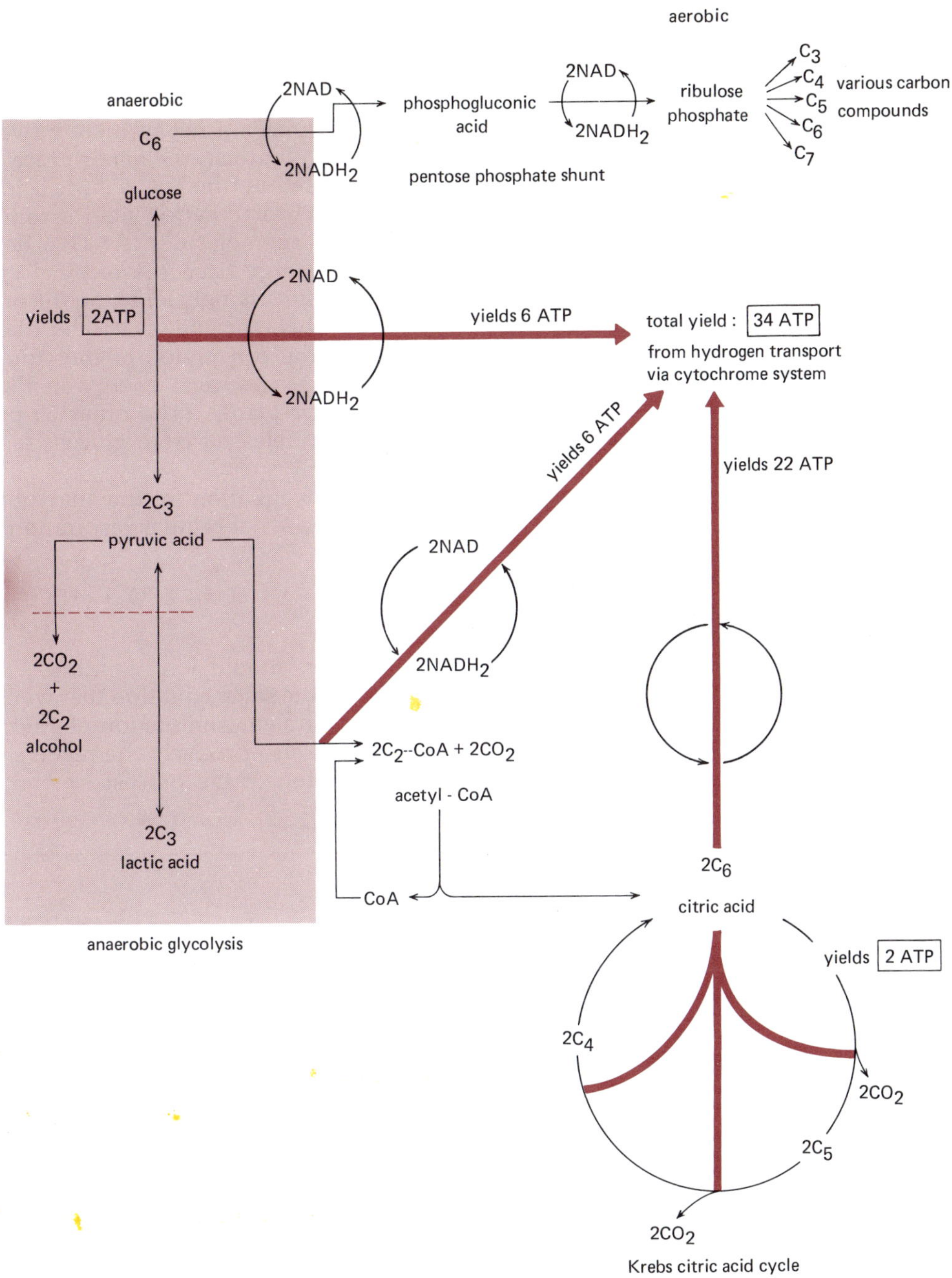

Figure 11–33 The reactions of cell respiration. The energy stored in sugars is converted to the more readily available form of ATP, and the sugars are ultimately degraded to carbon dioxide and water. The oxygen-consuming reactions of aerobic respiration yield a combined total of 36 molecules of ATP, as contrasted with a yield of only 2 molecules of ATP from anaerobic respiration.

permit the glycolytic processes to go on, with their release of energy, when the cell has no access to atmospheric oxygen.

If oxygen is present, most organisms, including higher plants and animals, can further metabolize the pyruvic acid* and hydrogens yielded by glycolysis in a series of aerobic (oxygen-requiring) steps comprising several closely interrelated cycles, ultimately resulting in the production of carbon dioxide, water, and a number of energy-rich ATP molecules. These reactions, comprising the **Krebs citric acid cycle** and the **cytochrome cycle** are diagramed in Fig. 11–33. The **pentose phosphate shunt** provides an alternate pathway of energy metabolism, as depicted. All these reaction sequences are collectively termed **aerobic respiration.** Although only the last reaction in this integrated complex actually requires the direct participation of oxygen, a lack of this vital substance would soon bring the entire aerobic phase of respiration to a halt. This is due to the accumulation of intermediate products, which act as feedback inhibitors, effectively halting the earlier reactions in the series as well.

*Pyruvic acid is a pivotal compound in cell metabolism. In addition to the respiratory pathways, the pyruvic acid yielded by glycolysis can also be converted to amino acids or fats, in processes of anabolism.

The interlocking cycles of aerobic respiration are far more efficient in providing usable energy in the form of ATP molecules than the anaerobic glycolytic cycle. The entire process of glycolysis yields a net of merely two ATP molecules. By contrast, the major processes of aerobic respiration (the Krebs citric acid cycle and the cytochrome cycle) yield a combined total of thirty-six molecules of ATP. The overall yield of energy from the original glucose molecules is surprisingly large, on the order of 40 percent. This efficiency is greater than that achieved in the conversion of the stored energy of coal to electrical energy in the best modern power plants. (The other 60 percent of the energy released from glucose is dissipated as heat.)

The overall equation for the anaerobic and aerobic processes of cellular respiration is:

$$C_6H_{12}O_6 + 6O_2 \rightarrow 6CO_2 + 6H_2O$$

This equation should look familiar, for two reasons. It is the same equation that would describe the complete combustion of glucose by burning. It is also precisely the reverse of the overall equation of the process of photosynthesis.

QUESTIONS FOR REVIEW AND THOUGHT

1. Distinguish between open and closed circulatory systems. Why is the open system called "open" when it is actually enclosed in the animal's body?
2. What are the main functions of a circulatory system?
3. Trace the path of a red blood cell through the heart, through the pulmonary circulatory system, through the heart, and through the systemic circulatory system back to the heart.
4. How do white blood cells fight disease germs?
5. What are the functions of the lymphatic system?
6. Describe the mechanism of Rh conflict in a pregnant woman and the manner in which the new Rh vaccine can help protect future children.
7. Compare respiration in the gills of a fish with respiration in human lungs.
8. Breathing polluted air over long periods leads to deposits of soot in the alveoli of the lungs. How can this impair the respiratory function?
9. Describe the gas exchange that occurs in the lungs and that which occurs between the body cells and the circulating blood cells.

CHAPTER **12**

Interaction with the Environment

Introduction

Man has already walked on the moon. Some day he will set foot on more distant worlds. There he will search for other forms of life. This task may not be as simple as it seems, for life as we know it, and as we cannot yet imagine it to be, may take many forms. But no matter what form they may take, living things must possess a certain capacity. They must have some awareness of their environment and respond to changes in that environment.

A number of separate yet interrelated systems are involved in the interaction of living organisms with their environment. First of all, characteristics of the environment and changes in them must be perceived. This entails not only the receiving of stimuli, but also their meaningful interpretation. Light may impinge on a man's eyes, but if the connections of eyes to brain have been severed, he will not "see" anything.

Furthermore, living things, unlike nonliving matter, have an ability to respond to their environment, to adjust conditions so that they are more favorable to survival. This response in many cases is expressed in some form of motion, which may be toward or away from the stimulus that elicited the response, depending on its nature and its relationship to the responding organism. A plant will turn its flowers and leaves so that the greatest possible surface is exposed to the rays of the sun, while a cockroach will flee from light.

Response to stimuli also requires a certain degree of coordination of the activities of the organism. The more sophisticated and highly differentiated the response, the greater the degree of coordination required. The single-celled *Paramecium*, for example, after bumping into an obstacle, will respond by backing off, turning at an angle, and then proceeding ahead in a new direction (Fig. 12–1). This is surprisingly sophisticated behavior for a single cell, but closer observation reveals the limited nature of the response. The *Paramecium* always turns at precisely the same angle after meeting an obstacle, no matter what the external conditions. Its new advance may bring it smack up against the same obstacle again; it responds to this contretemps by repeating exactly the same strategy. Eventually

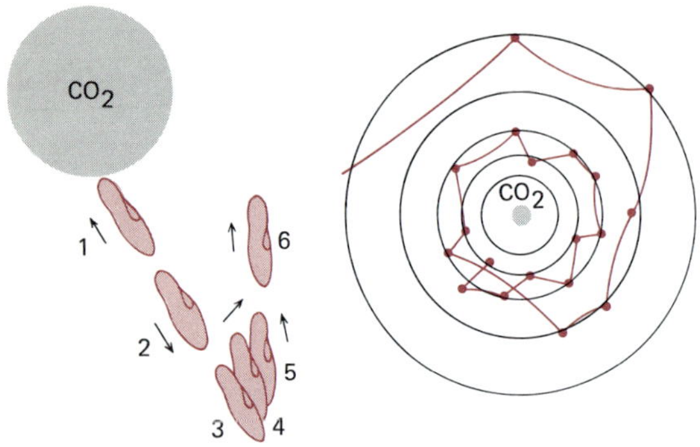

Figure 12–1 The paramecium reacts to an obstacle with a highly stereotyped set of responses which are not modified by experience.

(if no other pond creature has come along to eat it in the meantime) the thwarted *Paramecium* will blunder its way through, though it may have missed many "obvious" escape routes in the process. By contrast, the sophisticated coordination provided by the human brain and nervous system permits a man not only to analyze many aspects of a situation, but also to think ahead and project probable future consequences of alternative responses, and act in accordance with a specific choice of behavior.

In multicellular organisms, the activities of the various cells, tissues, and organs and of the organism as a whole may be coordinated both chemically and electrically. Plants seem to depend most upon chemical control, through compounds called hormones, although some signs of bioelectric activity have been detected in plants. The activities of animals are controlled and coordinated not only chemically, through hormones, but also electrically, through a network of nerve cells and associated structures that make up the nervous system. Electrical messages can flash along the nerves much more rapidly than chemicals can be transported from one site to another. As a result, animals as a group are able to respond much more rapidly to changes in the environment than are plants. These differing capacities are reflected in the typically sedentary life of most plants and the typically mobile existence of most animals.

The responses of living organisms to their external environment are directed mainly toward maintaining constant and optimum conditions in their internal environment. Thus, the mechanisms for perceiving and responding to stimuli are an important part of the system of homeostasis.

Sense Organs

Some fascinating experiments have recently been conducted to determine the psychological and physiological effects of prolonged isolation. In some of these experiments the subject has been suspended in a body-temperature fluid, in total darkness, in a sound-proof environment, constructed so that even the echoes of the subject's own voice and breathing are muffled. Imagine for a moment what it

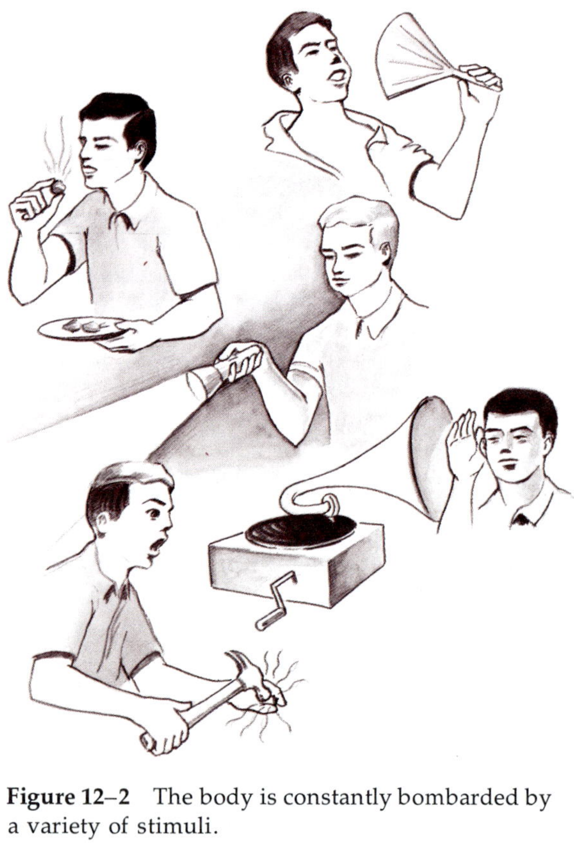

Figure 12–2 The body is constantly bombarded by a variety of stimuli.

would be like to be in an environment almost totally free of external stimuli. No light and color, no ticking clocks, or rustle of leaves, or traffic noises, no sensation of heat or cold. It is small wonder that the prolonged absence of noise and other stimuli has been found to be shattering to man's psychological equilibrium.

In normal life, living organisms are constantly bombarded by a variety of stimuli. A spectrum of electromagnetic radiations streams down upon us. Physical vibrations are transmitted through the air and through the earth. An incredible diversity of chemical substances float through the medium, whether the air in which we live or the water that is the environment of aquatic organisms. Objects are encountered with varying degrees of force, have diverse surface configurations or textures, and offer varying degrees of resistance to pressure.

Living organisms on earth are equipped to receive and evaluate a wide range of stimuli: **chemical receptors** perceive chemical substances, and they are interpreted in the form of taste and smell; **photoreceptors** receive electro-magnetic radiations within the limited region of "visible light" and function in the sense of sight; **phonoreceptors** are sensitive to the vibrations of sound waves and function in the sense of hearing; **touch receptors** respond to contact and changes in pressure; and **thermoreceptors** are sensitive to changes in temperature.

All these receptors are tuned to stimuli from the external environment. These receptors may thus be called **exteroceptors.** But a living organism also has an internal environment, and a continuing flow of information on the condition of the various body systems is of key importance for an organism's survival. **Interoceptors** provide information on the relative positions and states of various parts of the body and produce such sensations as hunger, thirst, and fatigue. Some interoceptors relay their information directly to the appropriate effectors, without giving us any conscious awareness of their activities. Thus, the body temperature is kept regulated within a narrow range by the temperature-sensitive receptors found within the hypothalamus. Receptors continuously monitor the state of the circulatory system and automatically adjust the blood-vessel bores to maintain optimum functioning. A variety of receptors found in the inner ear, the soles of the feet, and distributed through various parts of the body provide information on the relationship of the body to the earth's gravity and thus aid in maintaining equilibrium.

The particular sense organs present and the degree of dependence of the organism upon specific ones vary with the organism and its way of life. We humans are highly sight-oriented. We have ears that bring in sound data, organs of taste and smell in our mouth and nose, and a variety of touch and temperature receptors in our skin. But we tend to rely mainly on the evidence of our eyes. ("Seeing is believing.") The areas of the brain involved in coordinating and interpreting optical data are correspondingly highly developed in humans—far more so than in most other animals, except for our monkey relatives and birds, which also rely heavily on sight. Many animals, including rodents and dogs, depend more heavily on their senses of smell and hearing; correspondingly, the olfactory (smell) and auditory (hearing) areas of their brains are relatively highly developed. This is particularly true of animals that live in darkness or are active mainly at night. An extreme case is the bat, a large segment of whose brain is devoted to the analysis and coordination of auditory data. In a water environment, the chemical senses (mainly taste) are often the most important.

CUTANEOUS RECEPTORS

The human skin contains a variety of sense receptors, responding to a number of different types of stimuli. These can be mapped by marking off a checkerboard in ink on the skin, for example, of the palm of the hand, and then testing each small square with a sharp pointed rod, with a thin bristle, and with heated and chilled glass rods (Fig. 12–3). Such experiments reveal that different types of receptors, each with its own characteristic pattern of dis-

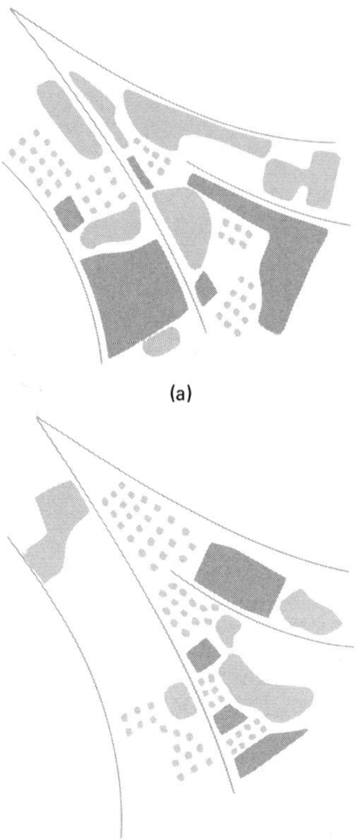

(a)

(b)

Figure 12–3 A map of cold spots (a) and hot spots (b) on the palm of the hand. The sensations were most intense in the colored areas, less intense in the black areas, and least intense in the dotted areas. There was no particular sensation in the blank areas. A comparison of the two maps reveals the nonuniform and dissimilar distributions of the hot and cold receptors in the skin of the palm.

tribution, are responsible for the sensations of pain, touch, heat, and cold.

Touch receptors are of varied types, including free nerve endings and specialized encapsulated structures. The body hairs also are extremely sensitive to touch stimuli. Touch receptors are especially abundant in the fingertips. (It is said that some safecrackers make their touch even more sensitive by sandpapering off some of the dead outer skin on their fingertips.) The lips also contain an unusually large number of touch receptors. This undoubtedly has some relationship to the popularity of kissing. Some areas of the body, however, are quite sparsely supplied with touch receptors. Although there may be as many as a hundred touch receptors per square centimeter in the fingertips, in some regions of the back the individual touch receptors are as much as two inches apart.

Pain receptors, which end as bare fibers beneath the skin, represent the most numerous of the cutaneous receptors and are in fact distributed throughout the body in essentially all tissues (except the brain). Pain receptors are sensitive to a wide variety of stimuli, including chemical, thermal, mechanical, and electrical.

Although pain is often looked upon as a scourge of mankind, it actually serves as an important protective mechanism: pain arises in many situations in which tissues of the body are being damaged and may lead to an appropriate response to remove the pain-causing stimulus. Often the response is automatic, performed without conscious thought. Even a small child who has never encountered a hot stove before will immediately draw his hand away if he touches one. The "uncomfortable feeling" one gets after sitting in the same position for a long time results from a lack of blood flow to the compressed areas and is automatically relieved by an often unconscious shift in position.

Sometimes the brain may be fooled as to the exact origin of pain. For instance, a pain in the heart region may be felt as a pain in the left shoulder, or a pain in the stomach may be felt between the shoulder blades. This "referred pain" is probably due to the complexities of the neuronal circuitry of the body (i.e., there may sometimes be a "crossing of the wires").

The "feeling" of pain apparently is comprised of two major components, the physiological and the psychological. The sight of a hypodermic needle in the doctor's office, for example, may be far more anguish-producing for a child than the actual injection. There also seems to be a distinction between the perception of pain and the response to it, the latter component being centered in the brain. Certain anesthetics and operations severing the appropriate neural connections to the brain can eliminate the response to pain without affecting the awareness of it. The patient is still conscious of the pain sensation, but it no longer causes him any anguish. The surgical technique is sometimes used to provide relief in terminal cancer cases.

Pain receptors are apparently associated with the sensations of tickling and itching. A mild, barely threshold stimulation of the pain receptors is believed to be the cause of tickling, while slightly stronger stimuli presumably cause itching. There is a great deal of individual variation of the threshold or minimum level of strength at which a stimulus is actually felt as pain.

There are three basic types of thermal receptors: cold, warm, and pain receptors. The pain receptors respond to extremes of heat and cold (Fig. 12–4). (That is the reason why intense heat and intense cold give essentially the same sensation, and a piece of ice may feel "burning" to the touch.) The thermal receptors do not give an absolute picture of the thermal environment, but rather furnish sensations of relative heat and cold. The thermal receptors can also respond to changes in temperature. When the temperature of the skin is falling, a person feels colder than if the temperature had remained steady at the lower level. Eventually the thermal receptors adapt to the new temperature environment. That is why a hot bath feels burning hot when one first steps in, and a cold pool initially feels freezing cold, but after a while both begin to feel comfortable.

CHEMORECEPTORS

On a platter before you is a thick, juicy, charcoal-broiled steak, smothered in onions and mushrooms. With a sigh of anticipation, you raise the first forkful to your mouth, and chewing with obvious satisfaction, you exclaim: "Mmm, that tastes good!" This statement exemplifies a popular misconception: much of the "taste" of the steak is in reality due to stimulation of smell receptors located in the nasal cavities. One can readily appreciate this fact during a cold, when mucus accumulations block the nasal cavities leading to the pharynx and most foods seem "tasteless."

The sense receptors for taste and smell are classified as **chemoreceptors** because each is activated by direct chemical stimulation. But these chemicals must be in solution to be effective as stimulators. Thus, a dry tongue (which houses the taste receptors) will not respond to powdered sugar. But normally the tongue is moistened somewhat, and the sensation of sweetness will be readily appreciated as the sugar quickly dissolves. The nasal membranes also contain a moist covering, which

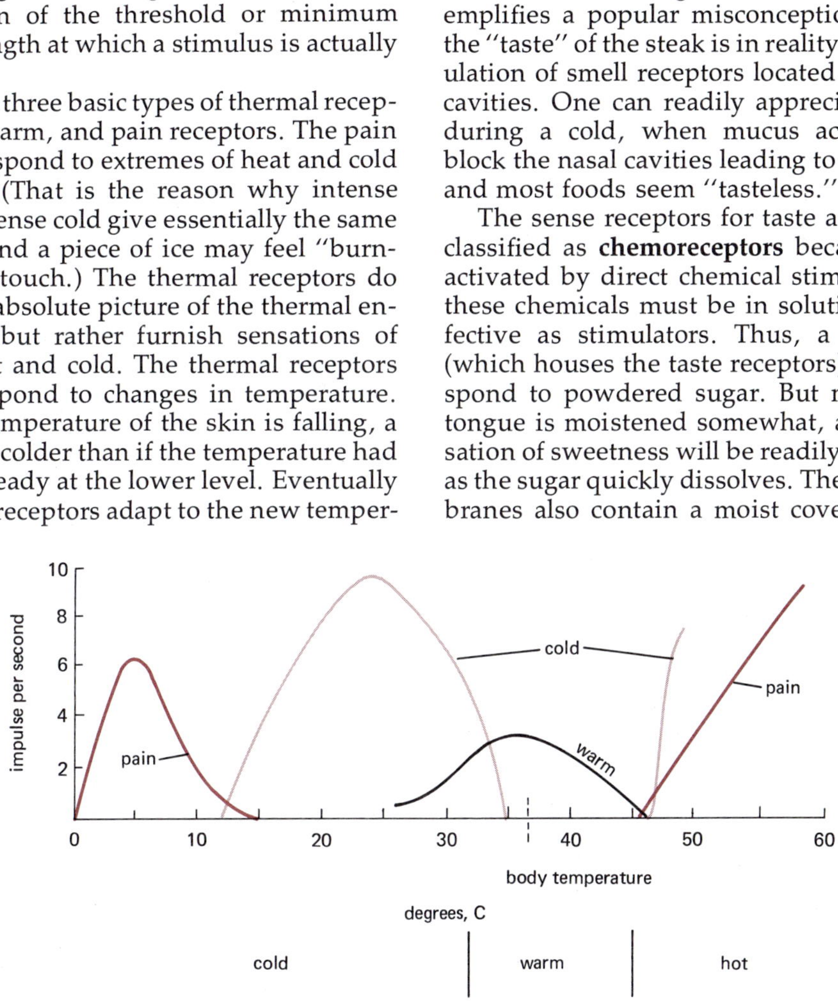

Figure 12–4 Response of pain, cold, and warm receptors, plotted against the temperature. Note the conflicting sensations aroused by exposure to extreme heat or cold.

not only prevents desiccation of the delicate underlying tissue, but also facilitates the dissolution of potential chemical stimulators of the smell or **olfactory receptors.**

The receptors for taste and smell are extremely sensitive and can detect stimulants in proportions of just a few parts per million (in the case of taste) and even of a few parts per billion (for the sense of smell). The characteristic pungent odor of escaping oven gas, for example, is due to a foul-smelling chemical, methyl mercaptan. This chemical is added to the odorless natural gas in such minute amounts that a single cubic yard of methyl mercaptan would suffice to treat enough gas to provide for the cooking needs of all the families in New York City for an entire year.

The olfactory receptors, which occupy a total area about the size of a postage stamp in the epithelial tissue of the nasal cavity (Fig. 12–5), are quite sensitive, but they also are subject to rapid fatigue. An odor perceived quite sharply at first soon ceases to impinge upon a person's awareness. A famous chemist relates that he and a colleague were once thrown off a bus for this very reason. They had been working in the laboratory, studying some particularly pungent and volatile compounds. Their olfactory senses had adapted to the odor; they were totally unaware that their

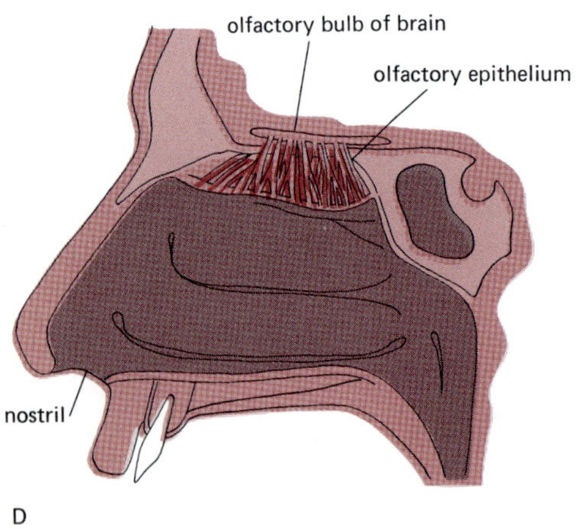

Figure 12–5 Chemicals in the inhaled air excite receptors in the olfactory epithelium of the nasal cavity; messages are transmitted from the receptors to the olfactory bulb of the brain.

clothing was permeated with the compounds, and they literally reeked.

To be perceived by the olfactory cells lining the nasal cavity, chemical substances must possess several properties: they must be volatile and they must be soluble both in the watery mucous material coating the nasal membranes and in the lipid-containing membranes of the olfactory cells. There appears to be some correlation between the three-dimensional configuration of a molecule and its capacity to invoke the sensation of a particular odor. Two quite different chemicals may smell alike if they have the same overall spatial configuration.

Man's ability to smell is quite rudimentary in comparison with that of many other members of the animal kingdom. Deer and other animals detect the presence of predators by their odor carried by the wind, while predators may detect the presence of potential prey in the same way. Some male insects can detect the characteristic odor of females of the same species at a distance of several miles. Plants produce odoriferous attractants which entice insects and ensure pollination, while affording the insects a source of food.

In Chapter 10 we saw that the tongue contains four different kinds of taste receptors, concentrated in specific areas and sensitive to the four basic types of taste: sweet, sour, bitter, and salty. The tastes of most foods are due to varying combinations of these basic tastes. The concentration of a substance may also greatly affect the emotional response to it. For instance, very low concentrations of sweet-tasting substances may be unpleasant. (It may be for this reason that highly diluted fruit juices sold as "fruit drinks" in the summertime are sometimes quite unpalatable.) Highly concentrated sweets usually evoke very pleasant sensations. Sour, salty, and especially bitter tastes, on the other hand, evoke the opposite response: they may be pleasant at low concentrations, but unpleasant at high concentrations. These principles are important guidelines in the culinary arts.

It is a well-known fact of common experience that certain foods (for example, liver, spinach, caviar, and olives) are considered highly tasty by some people, while others find them unappetizing. While it is true that a person can often be "educated" to develop a liking for foods that he formerly disliked, individual variations in food likes and dislikes may not be entirely capricious. There is some evidence to indicate that substances do not taste the same to all people. For example, a chemical known as PTC (phenylthiocarbamide) is readily detected as a bitter-tasting substance by the majority of people, but is quite tasteless to between 15 and 30 percent of the population. The tendency to be a PTC "taster" or "nontaster" is hereditary.

Taste or **gustatory receptors** are found in many animals, particularly those that live in the water. Although the taste receptors of humans and most other higher vertebrates are confined to the mouth and pharynx, some organisms have similar receptors in odd locations. Chemoreceptors are distributed over the bodies of many fish, and butterflies have taste receptors in structures that would correspond to the feet of higher animals.

PHOTORECEPTORS

The vast majority of organisms on earth are receptive to light in the "visible" region of the spectrum. (Visible light, i.e., that which is perceived by the human eye, is the narrow band of electromagnetic radiation with wavelengths from about 400 mμ—violet light—to about 750 mμ—red light.) In plants, energy in the visible region of the spectrum is essential to the process of photosynthesis. Radiations of shorter wavelengths (ultraviolet) and longer wavelengths (infrared) play a key role in such fundamental processes of plant physiology as germination and flowering.

Although plants seem to have a generalized sensitivity to light, many animals and even many protists have specialized structures for light reception. In the simplest cases, such as the red-pigmented eyespot of the single-celled *Euglena* and the cup-shaped ocelli of the flatworm, the specialized structures are sensitive merely to the presence or absence of light or to its relative intensity. *Euglena* swims toward light of a certain, optimum intensity and away from light that is too bright; the flatworm tends to avoid light, but it can be trained to override this natural preference for the sake of a food reward. Most higher animals are equipped with true eyes, which are capable of perceiving patterns of reflected light rays, and suitably developed regions of the brain that integrate these patterns into meaningful images.

Figure 12–6 The arthropod compound eye is made up of hundreds of thousands of individual ommatidia or simple eyes. (a) The head of a hornet; (b) close-up of a fruit fly's eye; (c) an image of Anton van Leeuwenhoek, photographed through the eye of a crab, 144×.

(a) (b) (c)

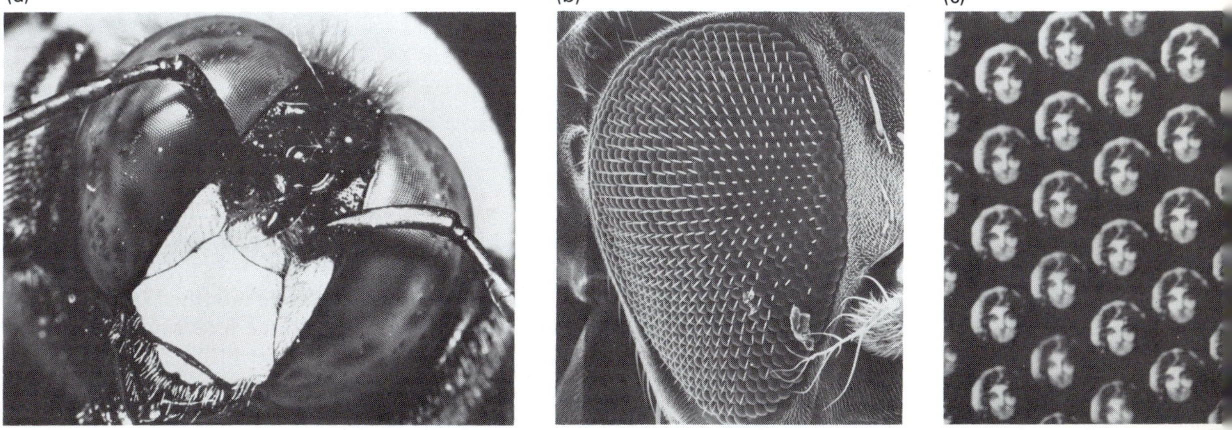

A number of insects have compound eyes, consisting of hundreds or even thousands of tiny identical units, called **ommatidia** (Fig. 12–6). Each ommatidium registers the relative light intensity reflected from a particular segment of the visual field. The result is a mosaic picture, which is integrated into a meaningful whole by the insect's brain. The process is analogous to the printing of halftone illustrations. If you examine a halftone illustration in a newspaper with a hand lens, you will observe varying patterns of black dots against a white background and white dots against a black background. The relative sizes and concentrations of these dots form mosaics which result in the total picture. The resolution of the image depends on the number of dots in a given area. Similarly, the number of ommatidia in a compound eye determines the sharpness of the image perceived by the insect. The eye of a bee, containing a relatively large number of ommatidia, can resolve objects to a much finer degree than the eye of a grasshopper, with relatively few ommatidia. But the vertebrate eye, constructed on a different principle, has far greater resolving power than even the eye of a bee.

The major parts of the human eye are shown in Fig. 12–7, together with a diagram of the manner in which light rays are focused in

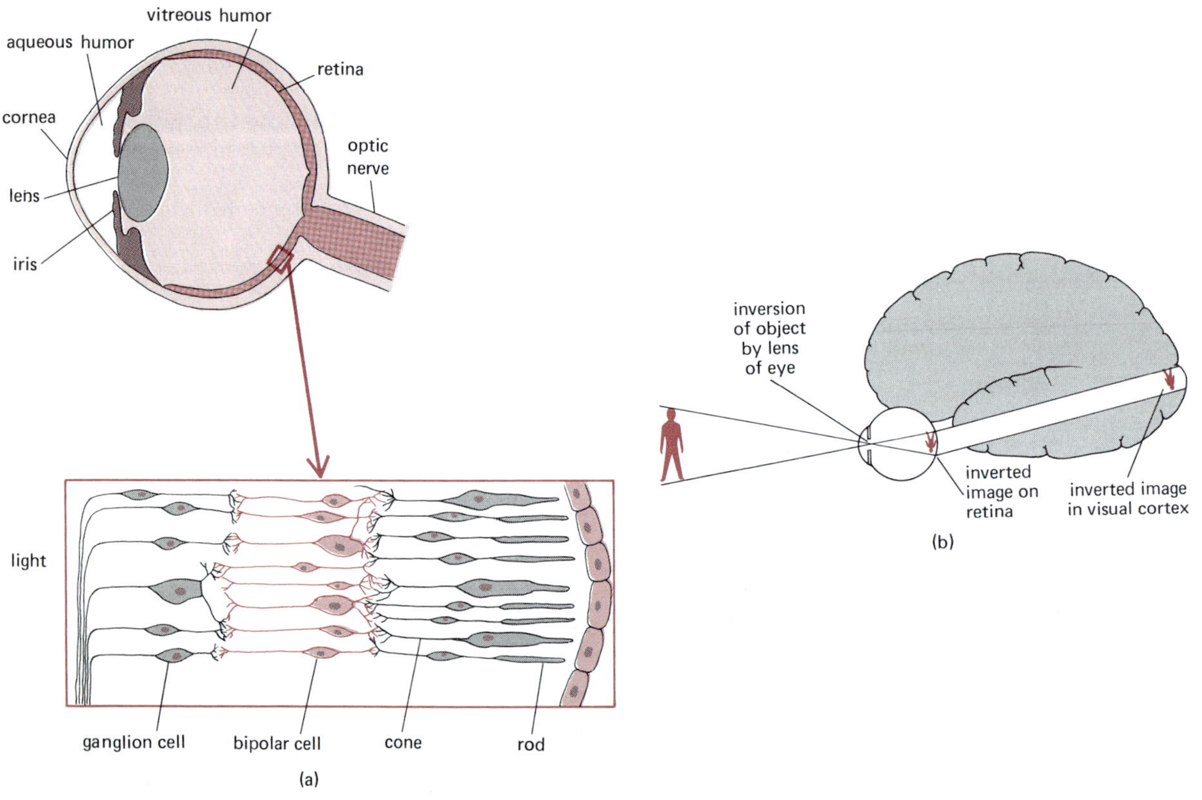

Figure 12–7 (a) The human eye. (b) The lens of the eye inverts the image of an object. The inverted image projected onto the sensitive cells of the retina is transmitted via the optic nerve to the visual center of the brain. The brain uses various cues of orientation such as touch and gravity to interpret the image as rightside up. The brain's interpretation of what the eyes see is learned and can be modified in the light of experience. In experiments in which subjects continuously wear inverting spectacles, after several days they suddenly see things "rightside up" again. If the spectacles are removed, the world looks upside down until the brain again readjusts.

it. The human eye works very much like a camera. The tough outer coat of the eyeball, the **sclera,** has a transparent anterior portion, the **cornea,** through which light is admitted. The **choroid,** a darkly pigmented layer just inside the sclera, acts like the black inner surface of a camera to prevent internally reflected light from blurring the image. The **iris diaphragm** controls the amount of light admitted through an aperture, the **pupil,** to a **lens.** Muscles attached to the lens can change its shape, so that objects at various distances can be brought into focus. The light rays from an object are focused on the **retina** at the back of the eyeball. (Because of the way the light rays are focused by the lens of the eye, the image produced on the retina is inverted, as can be seen in Fig. 12–7.) The light-sensitive retina, on which the light rays are focused, is a complex structure consisting of several layers including an outer pigmented layer and inner layers of specialized photoreceptors and other nerve cells. These nerve cells send data about the image on the retina to the brain via the **optic nerve;** the brain interprets these data, and we see.

Two basic types of photoreceptors are found in the retina: the **rods** and **cones.** The rods (about 125 million per eye) are distributed over nearly the entire retina, while the cones (about 7 million per eye) are concentrated primarily in one small region in a direct line with the center of the cornea. The cones are sensitive to color and have a rather high threshold: they will not respond when the light intensity drops below a certain level. The rods have a considerably lower threshold. Thus, in a dimly lit room you would not be able to see colors (since the cones would not be stimulated), but you would be able to make out the general form and other characteristics of objects in the room (due to stimulation of the rods).

The rods contain a light-sensitive pigment known as **rhodopsin,** or **visual purple.** This substance is synthesized in the retina from a protein and vitamin A. (Adequate supplies of this vitamin are therefore essential for good night vision.) Stimulation of the rods (which is ultimately passed on through nerves to the brain) involves chemical conversions of rhodopsin, initiated by light.

The cones occur in three different varieties. Each type contains its own specific pigment, sensitive to light in a particular wavelength region: red, blue, or green. Though the cones are stimulated independently, their collective action results in the perception of a kaleidoscopic variety of colors. In all, the human eye is capable of distinguishing more than seventeen thousand different hues.

In some people, one or more of the three types of cones is absent. About 8 percent of the male population and less than 0.5 percent of the female population suffer from colorblindness. They are unable to distinguish between certain types of colors—for example, red and green. This can create difficulties for colorblind drivers when traffic lights deviate from the common pattern of red on top, green below. In rare cases, no cones are present, and the person sees everything in patterns of blacks, whites, and grays. This is the case for most of the mammals, with the exception of man and the other primates. Most birds have excellent color vision, however, and many reptiles, fishes, and insects can distinguish colors.

The size and shape of the eyeball is maintained by a filling of fluid, which also aids in refraction of the light rays. Normally the eyeball is just the right size and shape that the focus of the image coincides with the retina. But some people are born with eyeballs that are too long or too short. If the eyeball is too long, the image is formed before the retina is reached, no matter how hard the lens system tries to accommodate. As a result, only very near objects can be seen clearly. This condition is called **myopia** or nearsightedness. It can be corrected by placing a diverging lens in front of the eye, which increases the focal length of the system as a whole (Fig. 12–8a).

If the eyeball is too short, on the other hand, the distance to the retina is not long enough for an image to be formed, except for very distant objects. This condition is called **hypermetropia,** or farsightedness. As a person ages, his lens system loses its ability to accommodate the lens, and as a result he may be-

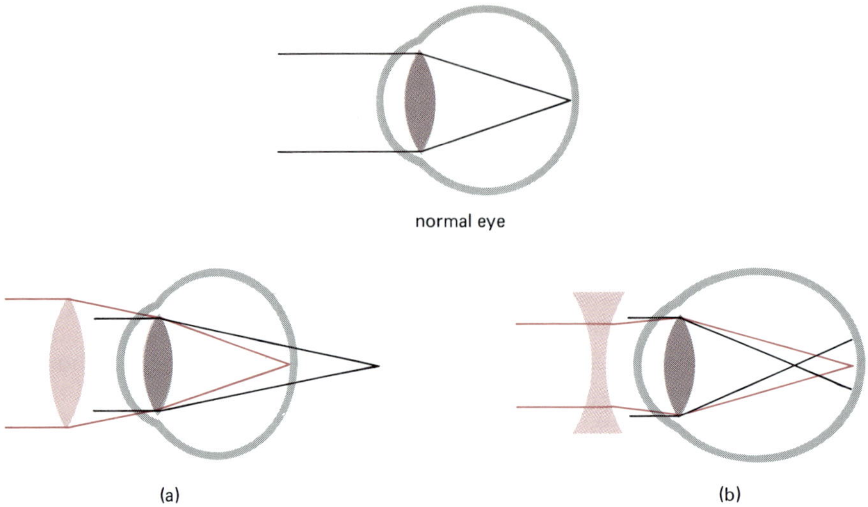

Figure 12–8 Correction for farsightedness with a converging lens (a) and for nearsightedness with a diverging lens (b). In each case, the lens helps the eye to focus light rays onto the retina, producing a clear image.

come farsighted even with a normal-sized eyeball. Hypermetropia is corrected by the use of converging lenses, which shorten the overall focal length of the system (Fig. 12–8b).

Another common eye defect results from aberrations in the front surface or cornea of the eye. If this is not perfectly spherical, rays coming through certain parts of the cornea will be focused differently from rays coming through other parts, and the person will be unable to get a distinct image of the whole. The constant strain of attempting to accommodate can lead to headache and eye fatigue. This condition is called **astigmatism.** It is corrected by the use of cylindrical lenses oriented to adjust for the aberrations of the cornea. Actually, corrective glasses often are made to adjust for several eye defects simultaneously, and as a result they are complex combinations of cylindrical and spherical surfaces.

Figure 12–9 A test for astigmatism. Do all the lines in the pattern appear equally sharp to you?

PHONORECEPTORS

The human ear is an amazingly sensitive organ. The faintest sounds that it can detect are perhaps one-trillionth as loud as the loudest audible sounds. Little would be gained by a significant increase in the efficiency of the human ear; in fact, life might be rather uncomfortable, for our ears would pick up a perpetual buzzing sound due to the constant collisions of the molecules of air in the ear passages.

The human ear consists of several different parts, each with its own functions (Fig. 12–10). The first part of the ear that a sound vibration reaches is the **outer ear.** This acts as a tube that channels and focuses the sound, concentrating it in one direction by reflection. Actually the human outer ear is greatly reduced in comparison with the ears of some other animals. Dogs and horses, for example, not only possess more distinctively tube-shaped outer ears, but they can also use them as "directional antennas," turning them to receive sounds more effectively.

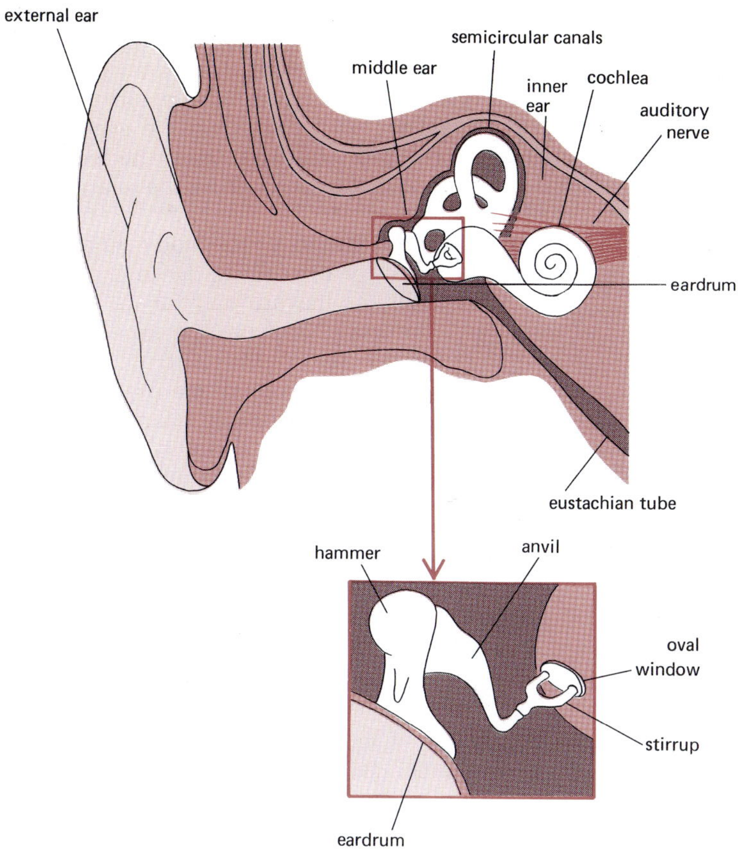

Figure 12–10 Structures of the human ear.

At the end of the channeling tube of the outer ear is an elastic membrane, the **eardrum,** which is set to vibrating like a miniature drumhead when sound vibrations reach it. Following is a system of three small bones, the **malleus, incus,** and **stapes** (their names, derived from their shapes, mean "hammer," "anvil," and "stirrup"). Sound vibrations are transmitted by these bones to another membrane, the **oval window,** which marks the end of the middle chamber (the **middle ear**).

The **inner ear,** containing a structure coiled like a snail shell (the **cochlea**), is filled with liquid and contains the auditory nerve endings. Another elastic membrane, the **round window,** moves in and out as vibrations are transmitted through the fluid of the inner ear, thus permitting expansions and contractions of the otherwise rigid cavity.

The vibrations in the canal of the inner ear are translated into impulses, which are ultimately transmitted along the auditory nerve to a specialized region of the brain. There they are interpreted as meaningful sounds.

In addition to its auditory functions, the human ear is the major seat of another important function—the maintenance of balance. Within the inner ear, in addition to the cochlea, are three **semicircular canals** and two small sacs, the **utricle** and the **saccule.** Each of these sacs contains a sense organ consisting of hair cells covered with a gelatinous layer, in which tiny bony masses called **otoliths** are embedded. When the head is tilted, the position of the otoliths in the gelatinous mass is changed (Fig. 12–11a). The shift in weight is detected by the hair cells, and signals are transmitted to the brain.

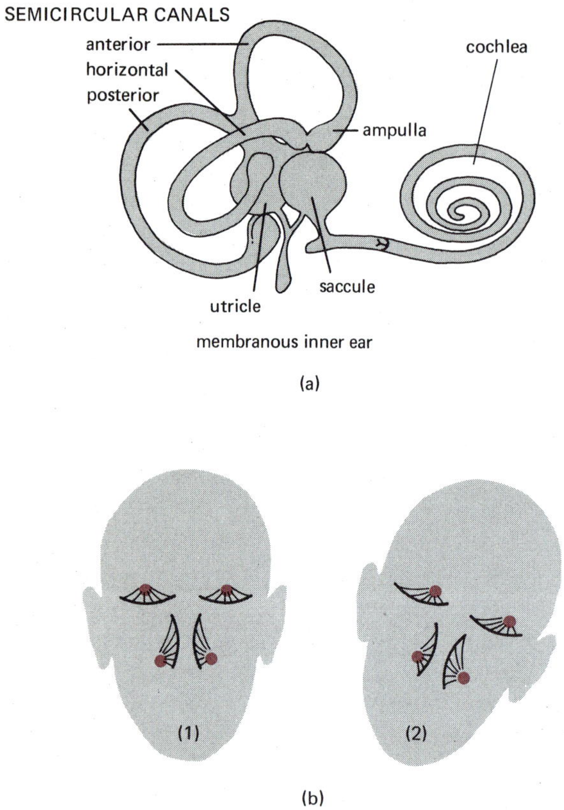

Figure 12–11 (a) Organs of balance in the inner ear. (b) Inside the saccules and utricles, crystals of calcium carbonate (otoliths) rest on sensitive hair cells. When the position of the head is changed, the crystals are shifted and exert a different pattern of pressures on the hair cells. Signals transmitted to the brain inform it of the change, which is correlated with other positional cues.

The semicircular canals also provide information on the position of the head. They lie approximately at right angles to one another, two vertical and the other horizontal. One end of each fluid-filled semicircular canal expands into an enlarged chamber containing a group of sensitive hair cells (Fig.12–11b). When the head is bent or moved, the fluid in at least one of the semicircular canals will be set in motion and stimulate the hair cells, stimulating impulses that are relayed to the brain. The brain, having received and coordinated all the messages relating to changes in the position of the head, together with data from the eyes, from pressure receptors in the soles of the feet, and from **proprioceptors** indicating the relative positions of other parts of the body, directs appropriate adjustments of the body muscles to keep the person from falling down.

Sometimes, as in various forms of motion sickness, the impulses from the inner ear become random and intermittent, due to frequent changes in the position of the head. Dizziness (after being rapidly spun in a circle) or nausea (in seasickness) are typical symptoms of this condition. Inflammation of the semicircular canals and the utricle may result in vertigo, a severe loss of equilibrium: the floor seems to heave as one walks, the walls may appear to tilt, and judgment of spatial relationships may be impaired.

The Nervous System

The sense organs of animals bring in a constant flow of information about the environment. Some of these data deal with sources of potential danger to the organism—the sight or sound or smell of a predator, increasing cold or heat, the presence of harmful chemicals in the air or water. Some data deal with potentially promising situations, which can aid in the organism's survival or well-being if it can take advantage of them—the presence of food or shelter or a possible mate. Often rapid and decisive action is required if a situation is to be turned to advantage or saved from disaster. A mouse that lingers to examine a bobcat at leisure will soon be a mouthful; a dawdling bobcat will find that his mouse has disappeared before he can pounce. A female firefly that fails to blink the proper recognition signals will not be joined by any male.

A high degree of control and integration of body activities and a great diversity of response to external stimuli is provided by a

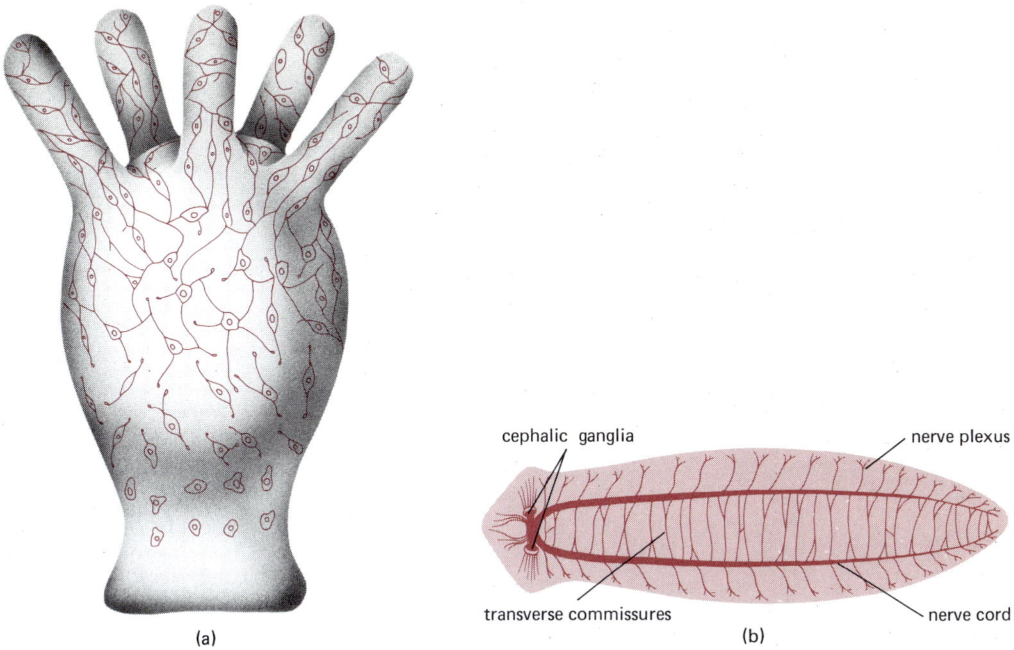

Figure 12–12 (a) The nerve net of a hydra. Stimulation of one part is transmitted uniformly over the whole body. (b) Ladderlike nervous system of a flatworm: the beginnings of a central nervous system.

nervous system, involving both chemical and electrical mechanisms of message transmission.

An early stage in the evolution of the nervous system is the primitive **nerve net,** found in coelenterates (Fig. 12–12a). Stimulation of one portion of the body of a hydra or jellyfish results in a generalized reaction that spreads throughout the entire organism.

The beginnings of a central nervous system appear for the first time in the flatworms. Conglomerations of nerve cells in a large **ganglion** take on the characteristics of a primitive brain; a pair of nerve cords extends backward in a ladderlike fashion, giving rise to branching peripheral nerves (Fig. 12–12b).

Further refinement of the brain occurs in the higher invertebrates: the octopus has been found to exhibit a sophisticated degree of learning capacity, and many insects have rather well-developed brains, with a large area devoted to the integration of visual data. Insects, however, are generally characterized by a high degree of stereotyped behavior: a specific stimulus almost invariably elicits a specific response, with little modification in the light of experience.

In the vertebrates, the brain increasingly assumes a dominant role, and a single, hollow, dorsal nerve cord, the **spinal cord,** acts as a secondary control center. The brain, a straight-line arrangement in the lower vertebrates, becomes increasingly more complex and folded upon itself (Fig. 12–13). Learning —modification of responses in the light of experiences—becomes more important, and the forebrain, which houses the centers of "intelligence," becomes enormously enlarged until (in the primates) its volume exceeds that of all the other sections combined. It is the brain of man that sets him apart from the other creatures of the earth. It is this structure which has given him the audacity (and the ability) to attempt to change his world.

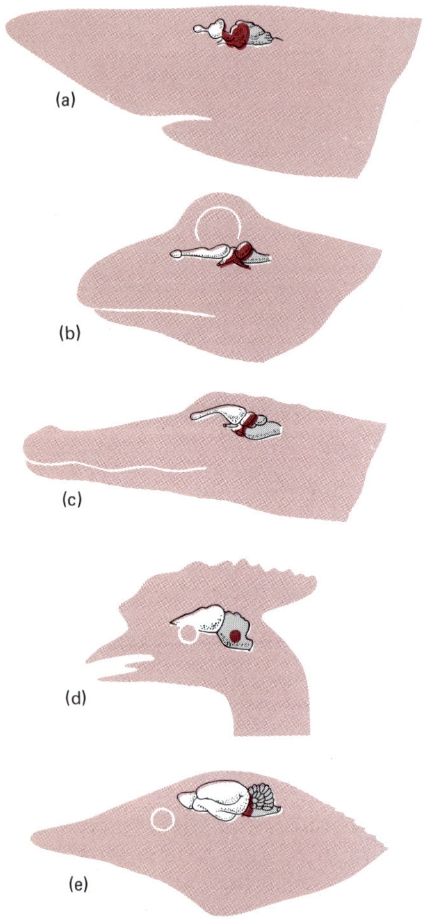

Figure 12–13 Evolution of the brain: (a) fish, (b) frog, (c) reptile, (d) bird, (e) shrew (a primitive mammal), (f) man. Note the progressive changes in the relative size of the midbrain (colored region) and the forebrain (white region).

THE NEURON

The nerve cell or **neuron** is the fundamental structural and functional unit of the nervous system. It consists of a nucleated **cell body,** from which one or more fibrous extensions branch out (Fig. 12–14). The **dendrites** are the extensions or **processes** carrying impulses *to* the cell body; they are frequently extensively branched. The **axon,** which carries impulses *from* the cell, may be extremely long—up to a yard or more in the legs of a man and more than eight feet in the neck of a giraffe. It seems strange to think of a microscopic cell being many feet long, but the nerve cells are extremely slender. The axons of nerve cells frequently branch out into numerous **terminal branches.**

The dendrites and axon of many nerve cells are enveloped in one or two protective sheaths, which may be considered analogous to the insulating sheathing on electric cables.

Individual neurons communicate with other neurons through a nonprotoplasmic gap known as a **synapse.** Large numbers of nerve fibers (axons) are often found together in bundles and are called nerves. Neurons that carry messages to the brain or spinal cord are called **sensory** or **affector neurons.** (Many of them bring data from the various sense organs.) Neurons that carry impulses from the brain or spinal cord to muscles or glands are called **motor** or **effector neurons.** (Many of them carry messages that stimulate muscle contractions resulting in motion.) **Connector** or **association neurons,** which are found only in the brain and spinal cord, permit communication between sensory and motor neurons.

THE NERVE IMPULSE

Impulses are transmitted along a neuron according to an "all-or-none-law." If the stimulus is too weak, the neuron will not "fire" (i.e., carry an impulse). But if the stimulus is above some threshold value, an impulse will be initiated and then will travel along the neuron at a speed that is entirely independent of the intensity of the original stimulus. Moreover, although many different kinds of stimuli can in-

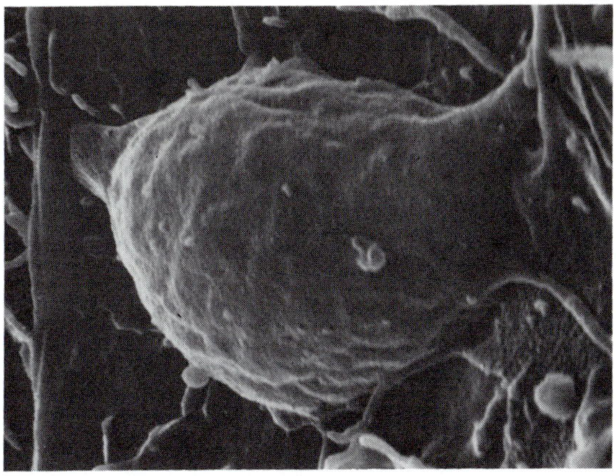

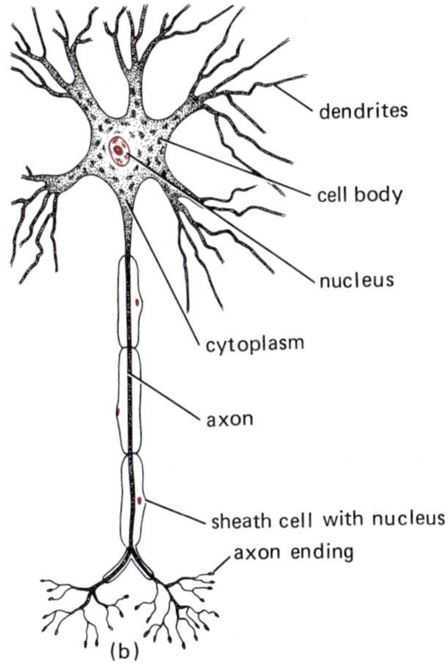

Figure 12–14 The neuron. The photograph is a scanning electron micrograph of a neuron from the rat cerebellum, maintained in culture (7,150 X). (b) Diagramatic view of a typical motor neuron.

itiate the impulse—heat, electricity, a sharp blow—once the impulse begins, it is transmitted at a uniform rate throughout the neuron, and this rate is independent of the nature of the original stimulus. (Strong stimuli do, however, tend to give rise to a larger number of impulses per unit time than weak ones.)

A nerve impulse travels along a nerve fiber as a result of a series of electrochemical reactions (Fig. 12–15). In the resting neuron, there is a continual, energy-consuming process of active transport of Na^+ (sodium) ions from within the neuron out into the extracellular fluid and of K^+ (potassium) ions in the opposite direction. As a result, substantial concentration gradients are built up across the nerve cell membrane. The concentration of sodium ions outside the nerve cell is ten times as large as that within the cytoplasm, while the potassium-ion concentration within the neuron is thirty times that in the extracellular fluid. In each case we would expect a tendency for diffusion to equalize the concentrations on both sides of the membrane. Some K^+ ions do diffuse out, but the membrane of the resting neuron is practically impermeable to the passage of Na^+ ions. As a result of the unbalanced flow of ions out of the cell, the membrane acquires a slight positive charge on the outside and a corresponding negative charge on the inside. An equilibrium is established between the tendency of K^+ ions to diffuse out and their attraction for the negatively charged interior of the neuron.

Certain stimuli, however, including electricity, heat, mechanical deformation, and certain chemicals, can selectively alter the permeability of the nerve cell membrane to sodium ions. When this occurs, Na^+ ions move into the neuron. If the alteration of permeability is great enough (i.e., if the stimulus exceeds the threshold), sodium ions rush in and eliminate the potential difference entirely or even momentarily reverse the polarization of the membrane. This process results in an increase in the permeability of adjacent portions of the membrane, and the depolarization is successively repeated in a chain reaction all along the length of the neuron.

After depolarization, processes of diffusion and active transport reestablish the normal resting potential difference, and the neuron is ready to "fire" again.

The mechanism outlined accounts for the transmission of an electrical impulse along the length of the neuron to the terminal branches of the axon. But how does the impulse get across the gap of the synapse to the dendrites of the next neuron?

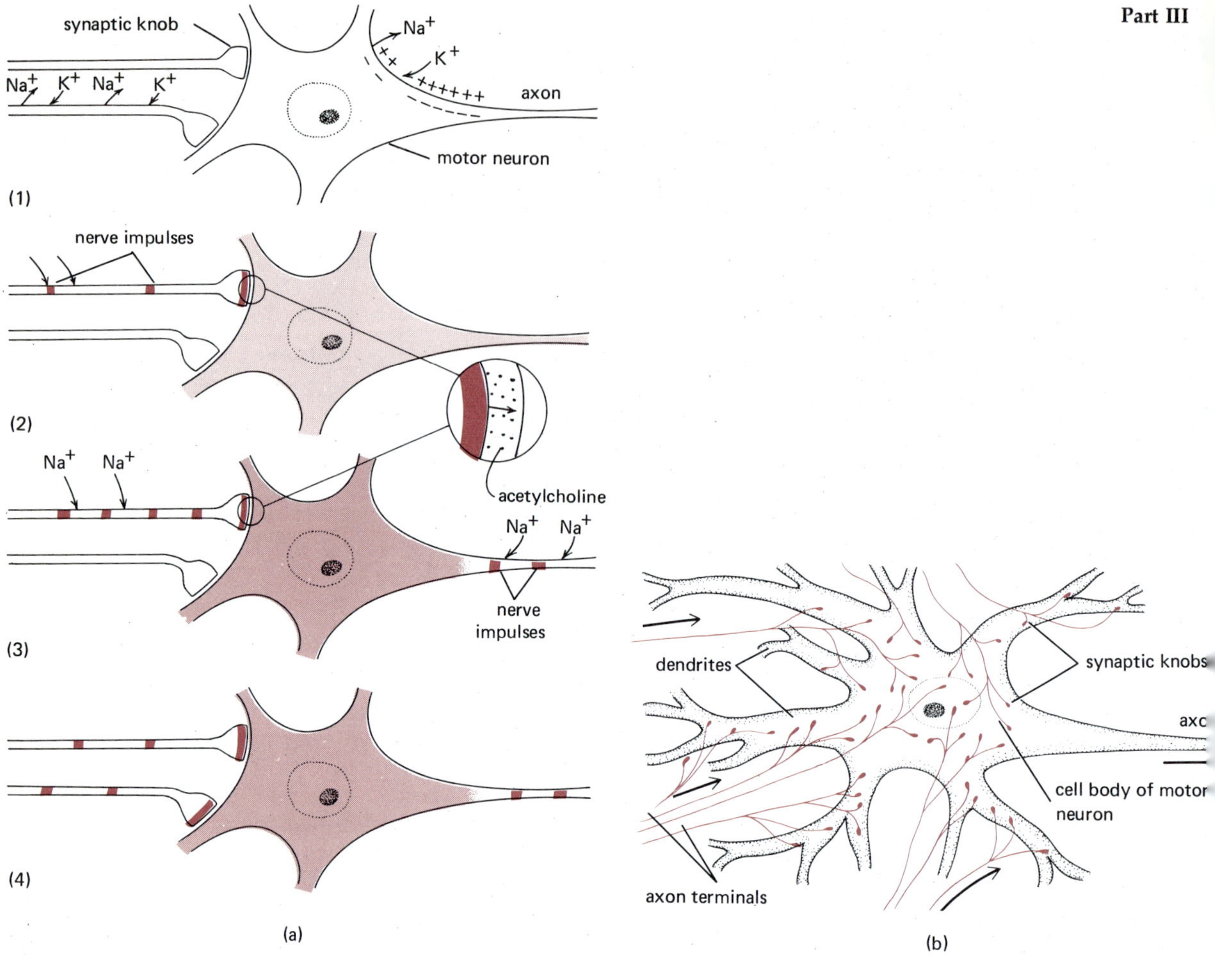

Figure 12–15 (a) Schematic depiction of the excitation of a neuron. In the resting neuron (1) there is an active transport of Na^+ out of the neuron and K^+ into it, resulting in a polarization of the cell membrane. Transmission of an impulse involves successive depolarizations along the neuron and the secretion of neurohumors into the synapse. If the stimulation does not exceed the threshold of excitation (2), the next neuron in the sequence will not "fire." Greater stimulation at one synapse (3) or excitation at more than one synapse (4) can spark depolarization of the next neuron and continue the transmission of nerve impulses. In (a) the terminal branches and synaptic knobs are disproportionately magnified and simplified for clarity; the actual neuron with its synapses looks more like (b).

The situation is analogous to a line of railroad tracks running down to the edge of a river and resuming at the other side, with no apparent means of connection between the two ends. A train approaches the river and stops at the bank. It is met there by a ferryboat, which conducts it slowly across to the tracks on the opposite bank.

In a similar manner, messages are conducted along the neuron sequences in the body. When an impulse reaches the terminal branches of a neuron, it liberates one or more chemicals (**neurohumors**), which then diffuse across the synapse to the dendrites of the following neuron in the sequence. The neurohumors (including the substance **acetylcholine**) act as a stimulus initiating an impulse in the second neuron. After the acetylcholine has done its job, it is removed from the scene by the enzyme **acetylcholinesterase,** which hydrolyzes the molecule into inactive fragments. If this did not occur, the acetylcholine would continue to stimulate the second neuron, causing it to fire over and over again. Some effective insecticides are acetylcholinesterase inhibitors; they are extremely dangerous to use for they can interfere with the nervous activity of man and other mammals as well as insects. Some chemical warfare agents called nerve gases are also acetylcholinesterase inhibitors.

Figure 12–16 The human nervous system.

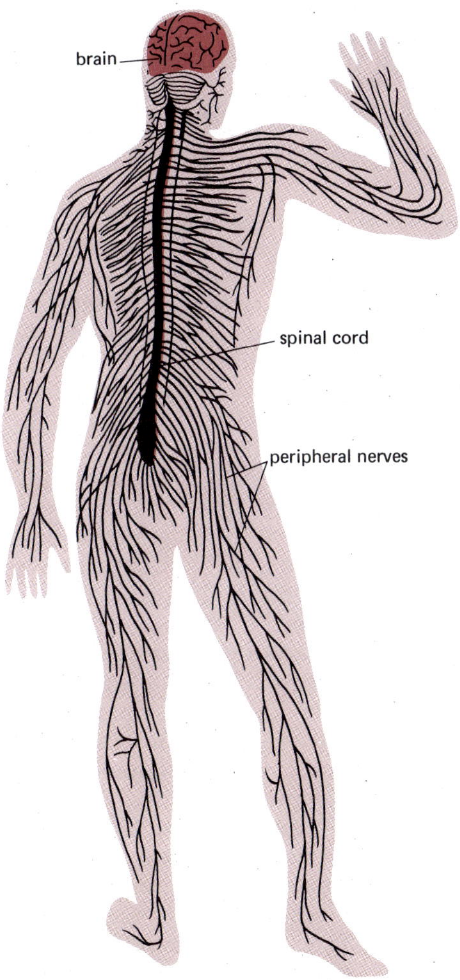

THE CENTRAL NERVOUS SYSTEM

The human nervous system is comprised of more than 10 billion neurons. Through this intricate network run impulses which are the foundations of hopes, dreams, desires, creative thoughts, and innumerable bits of data that continually flood the body's sensory organs. These data pass from the sensory organs along the **peripheral nervous system** (Fig. 12–16) to the brain and spinal cord, which comprise the **central nervous system.** It is the central nervous system that screens the incoming impulses, shunting and directing them to the appropriate regions of the body via connector neurons and efferent spinal and cranial neurons (the latter being parts of the peripheral nervous system).

The delicate central nervous system is well protected from the constant jarring that is an inevitable result of normal body movements. The brain is enclosed in a hard, bony **skull (cranial bones),** while the spinal cord is enclosed within the **vertebral column.** In addition, three separate fibrous membranes, the **meninges,** provide added protection for the brain (Fig. 12–17). Lymphlike cerebrospinal fluid, bathing the meninges and filling the chambers of the brain and channel within the spinal cord, provides a cushioning effect.

THE SPINAL CORD

The spinal cord, comprising a mere 2 percent of the weight of the central nervous system performs two vital functions. It serves as a pathway for the two-way conduction of nerve impulses between the brain and the peripheral nervous system. In addition, it serves as an independent control center for the automatic or reflex actions of the body.

Many actions protecting the organism from possible harm must be performed too rapidly for conscious thought if they are to be effective. The automatic withdrawal of the hand from a hot stove has already been mentioned. The blink of your eye as a mosquito zooms in toward it is another automatic protective mechanism. Such actions are called **reflex actions.** They are mediated directly through the

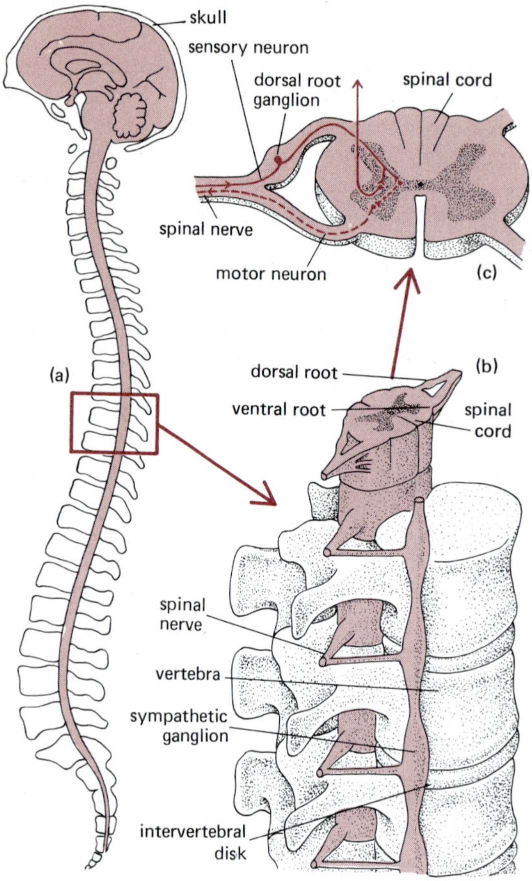

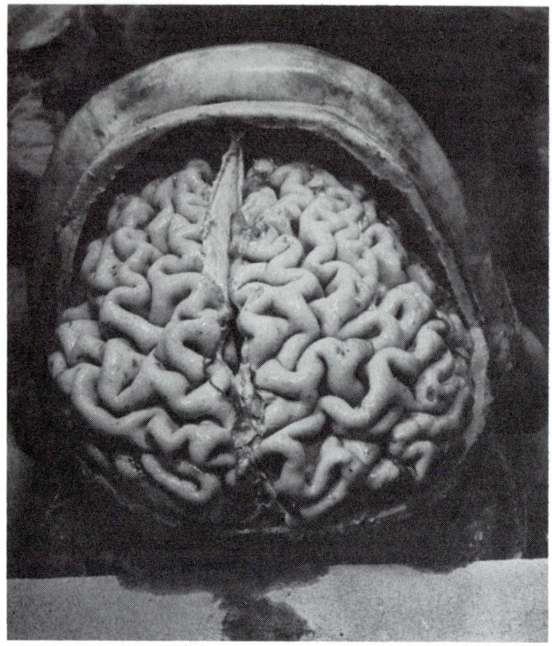

Figure 12–17 The organs of the central nervous system are well protected: the spinal cord is encased in a flexible sheath of bony vertebrae, while the brain, wrapped in several layers of protective membranes, is housed in the bony skull. The photo shows an open skull, viewed from above, with the meninges removed to show the surface of the brain.

spinal cord or through reflex centers in the lower portion of the brain, without first relaying messages to the higher brain. (The brain is often informed of the event afterward.)

REFLEX ACTIONS

A simple **reflex arc** generally includes a receptor, which is usually a specialized peripheral ending of a sensory neuron (which may be associated with a special sense organ); a sensory or afferent neuron, which conducts impulses to the spinal cord; a connector or association neuron in the spinal cord, which synapses with the incoming sensory neuron and the outgoing motor neuron; a motor or efferent neuron, terminating in an effector; and an **effector,** consisting of a muscle or gland which performs the actual response (Fig. 12–18).

Simple reflex actions include the rapid removal of a burnt finger from a hot stove, the blinking of the eyes upon the unexpected approach of an object, the constriction of the pupil in response to a bright light, and the knee jerk when the tendon just below the knee cap is tapped. All these actions occur before the brain is informed of the initial stimulus.

Once the brain is brought into the picture, the reflex action may be complicated by further refinements. Thus, removing your burnt finger from the hot stove is mediated directly through the spinal cord. But when you exclaim "Ouch!" and shake the finger or put it in your mouth, this is the work of the brain.

Conditioned reflexes represent a further increase in the complexity of reflex action. This group of reflexes represents learned responses to specific stimuli. The great Russian phys-

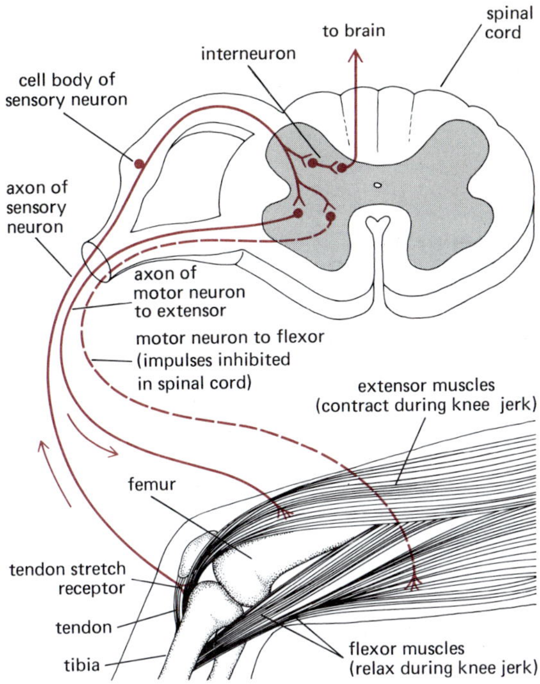

Figure 12–18 Nerve pathway of the knee-jerk reflex.

iologist Ivan Pavlov was a pioneer in the field of conditioned reflexes. In one series of experiments he utilized the reflex action of salivation in response to the presentation of food. (You may have noticed that your mouth starts to water when you sit down to an appetizing meal, even before you have taken a bite.) Pavlov rang a bell each time he presented meat to dogs. After a certain number of such associations, he omitted the meat and merely rang the bell. The dogs salivated on cue.

After the discovery of the conditioned reflex, there was an initial wave of attempts to reduce all human learning and experience to terms of conditioned reflexes. Although it soon was realized that this is a gross oversimplification, conditioned reflexes nevertheless play an important role in our daily lives. The sudden jamming down of the brake pedal when the car in front stops unexpectedly, the response elicited upon hearing one's name, and the mad dash for the door by students when the class bell rings are common examples of conditioned reflexes.

In addition to the reflexes mediated through the spinal cord, there are reflex centers in the brain (particularly in the lower region adjacent to the spinal cord) that control particular reflex actions. These include such automatic mechanisms as breathing, the heartbeat, the adjustment of the blood vessel bore, swallowing, sneezing, vomiting, and coughing, as well as temperature control, sleep and wakefulness, and various emotional centers. Conditioned reflexes are to a large extent controlled by a higher division of the brain, the cerebral cortex.

THE BRAIN

The brain is the major clearing center that integrates, regulates, and controls many of the activities of the body. It is the seat of consciousness and sensation. We are aware of external stimuli—lights, sounds, tastes, pain, and temperature variations—because the brain receives and interprets sensory impulses arriving from the external and internal environment. The brain is the source of the esthetic and creative and intellectual activities that make man unique among the living creatures of the earth. Violent emotions, as well as tender feelings, drives, and urges of various kinds, are also derived from the activity of the brain. This versatile organ is an intricate, unbelievably complex, compact marvel weighing a mere three pounds.

The brain may be considered to consist of three major parts: the cerebrum, the cerebellum, and the brain stem. The brain stem may be further subdivided into the medulla oblongata, pons, midbrain, thalamus, and hypothalamus (Fig. 12–19).

The regions of the **brain stem,** as has already been mentioned, house the control centers for a number of vital reflexes as well as for the interpretation of various sensory data and the control of certain emotional states.

The **medulla oblongata,** which might be regarded as an enlarged extension of the spinal cord, houses the reflex centers for breathing, the rate of the heartbeat, and the dilation and constriction of blood vessels, as well as for sneezing, coughing, vomiting, and swallow-

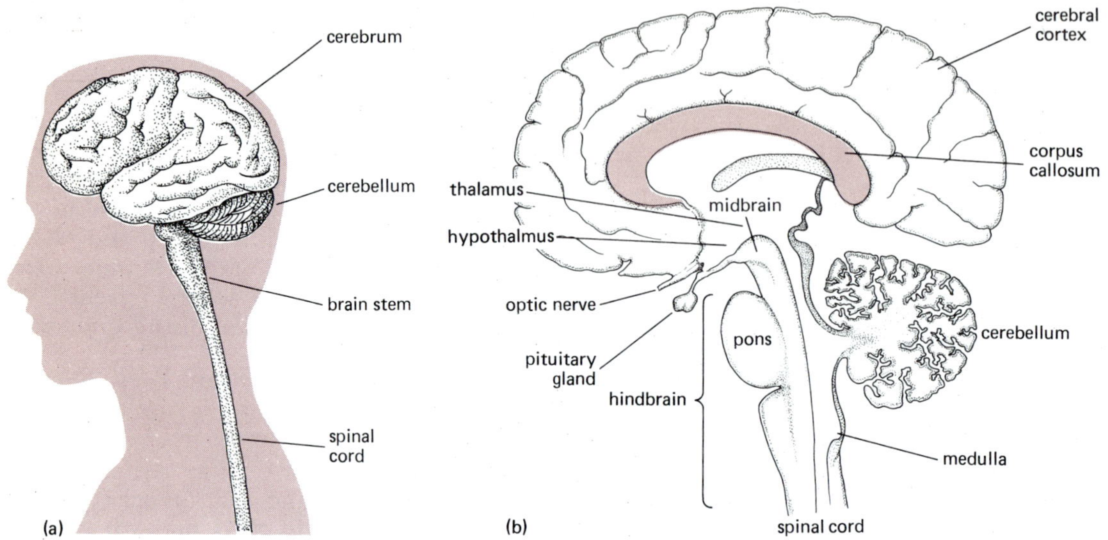

Figure 12–19 The human brain. (a) External view. (b) Longitudinal section through the midline of the brain, showing the inside of the right hemisphere.

ing. A skilled practitioner of karate utilizes a knowledge of this vital area in a paralyzing blow to the back of the neck. Such a blow, properly aimed, can even cause death.

The **pons,** the "bridge" of the brain, houses a respiratory center, while the **midbrain** contains auditory and visual centers and also plays a role in the control of equilibrium, muscle tone, and postural reflexes. (The last are the continual small adjustments of the body muscles that help to keep a person standing upright.)

Running through the midbrain, from the medulla and pons up through parts of the hypothalamus, is a cone-shaped network of neurons called the **reticular formation.** This nerve network houses the **reticular activating system** or **RAS.** The RAS acts as a sort of central clearing house for the flood of sense information that comes into the brain. It filters out stimuli that are weak or that form familiar patterns, letting only the particularly strong or novel signals pass up to the higher brain for conscious perception. It is this action of the RAS that permits you to concentrate on a particular thought or activity, disregarding background noises and other potential distractions.

In addition, there is a continual flow of signals down from the cerebrum to the reticular formation. Through this continual dialogue, the RAS works to keep the higher brain awake and alert. At night, when you compose yourself for sleep and the influx of information into the RAS is at a minimum, this activating system shuts down the conscious mind and permits you to sleep. But an annoying noise or a distracting worry may keep the exchange of signals flowing between the midbrain and the cerebrum and keep you tossing and turning far into the night.

The messages carried by such sensory neurons as those concerned with pain, temperature extremes, and rough body and skin contact are initially appreciated in the **thalamus.** Such sensory impulses are eventually relayed to the cerebral cortex, where their more delicate aspects are perceived.

The **hypothalamus** houses a number of important centers (Fig. 12–20), including those for the regulation of body temperature, appetite, water balance, the heartbeat, blood pressure, smooth-muscle activity, sleep, carbohy-

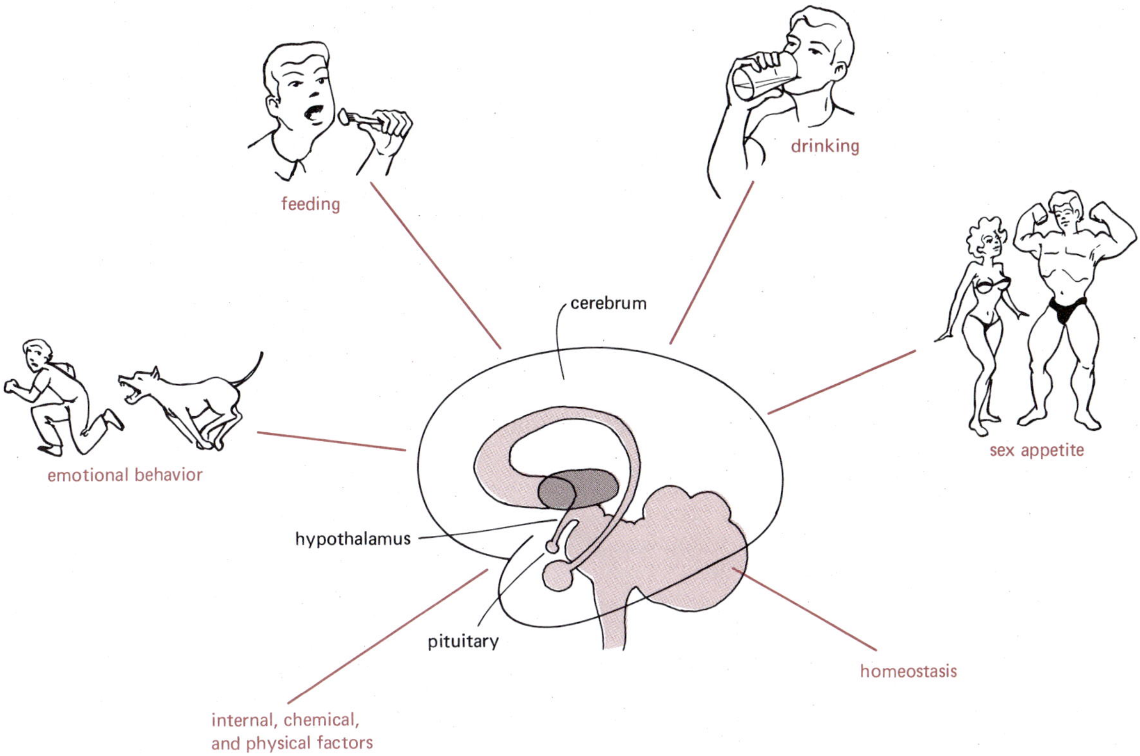

Figure 12–20 The hypothalamus contains control centers that regulate a variety of emotional states and crucial body functions.

drate and fat metabolism, and such emotions as fear and rage. (An impulse through an electrode implanted in the appropriate region of a cat's hypothalamus can make it quiver in fear at the sight of a mouse.) Centers controlling the sexual drive, hunger, and thirst are also located in this region. The hypothalamus is also linked with the hormonal and reproductive mechanisms of the body.

Scientists have speculated about the awesome possibilities that are unfolding as knowledge of the hypothalamus is gathered. Some have envisioned a world populated by robotized humans with electrodes implanted in their pleasure and pain centers, enslaved by unscrupulous dictators who send out radio signals to "reward" or "punish" them and thus control their actions.

Such fantastic prospects are quite within the realm of scientific possibility. A "pleasure center" has been mapped in the hypothalamus. Rats that have been taught to press a lever that transmits a stimulus to this pleasure center have been observed to press the lever again and again until they collapse from exhaustion. In their singleminded pursuit of self-stimulation, these rats ignored food, water, and members of the opposite sex. An apparently similar area has been mapped in the human brain. One human subject with an ex-

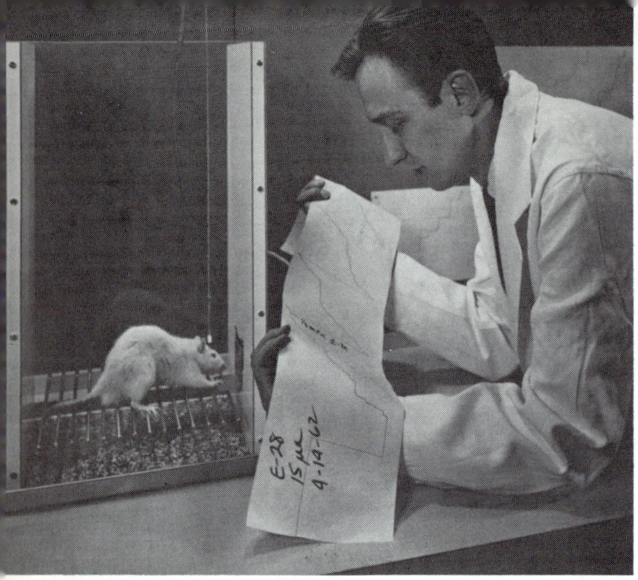

Figure 12–21 This rat has an electrode permanently implanted in a reward center of its brain. Trained to press a lever to gain an electrical reward, it is being used in drug evaluation studies.

perimental electrode implanted in his pleasure center pressed a button to stimulate this center three thousand times in ninety minutes before the circuit was disconnected by observing physicians.

Although **electrostimulation of the brain (ESB)** through implanted electrodes holds alarming potentials for misuse, this technique is already showing promise for positive uses. In a series of experiments with monkeys, for example, researchers mapped about two hundred different points in the brain stem that control movements of various parts of the arms and legs. A computer that was programmed with these brain maps was then connected to electrodes implanted in the brain stem of a paralyzed monkey. Computer-controlled patterns of stimulation enabled the monkey to feed itself, to climb, and to scratch its back. The monkey could even control its own actions, through the computer, by means of a set of switches. Techniques such as this could be used to help victims of strokes and head injuries to regain the use of their paralyzed bodies. Other ESB experiments are aimed at the use of implanted electrodes to feed sense impressions from a television camera directly into the brains of blind people, to permit them to "see" again.

The thalamus, the hypothalamus, and structures of the midbrain are linked together with parts of the cerebrum to form a looping network called the **limbic system** (Fig. 12–22). It is the limbic system of the brain that colors thoughts with emotions. It functions to recognize disturbances of equilibrium—due to hunger or a sex drive, a danger or threat, or a less tangible worry or disappointment. The emotional tag that the limbic system places on such upsets helps the higher brain to recognize the problems and take steps to restore equilibrium.

The **cerebellum** and **cerebrum** are the "higher" parts of the brain. In man they are quite large and well developed. The cerebellum acts in concert with the cerebrum to regulate and control skeletal-muscle activity. As a result of this control, the body is able to execute smooth, steady, efficient, and well-coordinated movements. In addition, the cerebellum maintains muscle tone. Through its control of muscular contraction, the cerebellum helps to maintain equilibrium and posture.

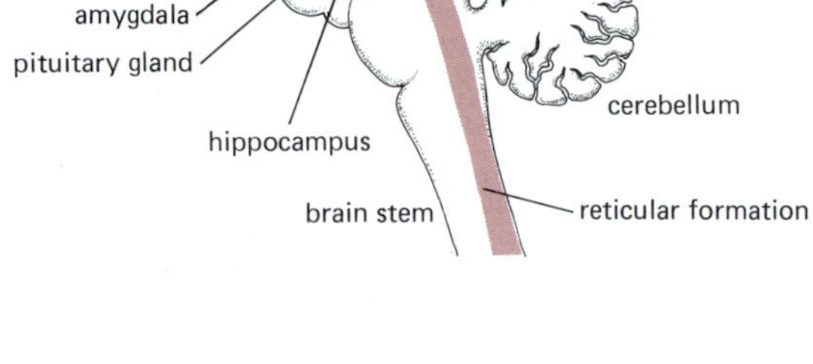

Figure 12–22 The limbic system is an integrated network that includes the hippocampus, the septal area, the amygdala, the hypothalamus, the thalamus, parts of the reticular formation, and the limbic region of the cerebral cortex.

The centers of memory, learning, reasoning, and many other faculties, such as sight, hearing, and touch are found in the cerebrum. This structure is the largest part of the human nervous system, comprising more than 5 billion of the 10 billion neurons that constitute the nervous system. A longitudinal fissure separates the cerebrum into two halves or hemispheres, which communicate through a broad band of white fibers. The outer layer of the cerebrum, the **cerebral cortex,** has a wrinkled appearance due to the presence of deep grooves, shallow depressions, and extensive folds or convolutions.

The cerebral cortex has been extensively mapped by a variety of techniques, including: direct stimulation by electrical impulses and observation of subsequent responses; deliberate destruction of definite areas in the brain (in humans, in the treatment of epilepsy) and observations of victims of brain damage; study of infants born with little or no cerebral cortex;

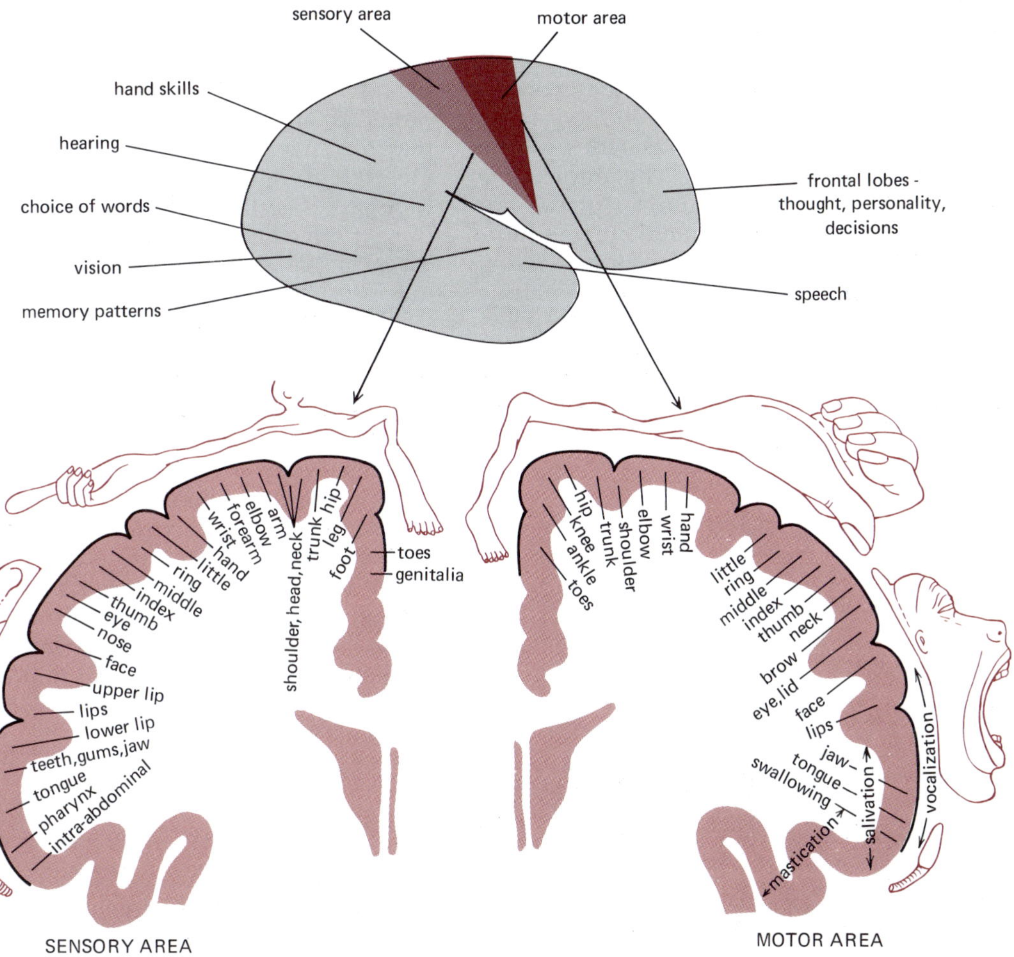

Figure 12–23 Maps of the brain. Various functions have been traced to specific regions of the cerebral cortex. The sensory and motor strips (relating to the reception of messages from sense receptors and the stimulation of motor activity in the body, respectively) have been mapped in particular detail. Note that the representation of various parts of the body in the sensory and motor areas is proportional not to their size but to their function.

and study of the electrical activity of the brain. Using such techniques, three major areas of the cerebrum have been identified: the motor, sensory, and association areas (Fig. 12–23).

The **motor area** of the cerebrum is associated with controlling the contractions of the skeletal muscles. In the **sensory area,** the brain interprets such sensations as heat, cold, touch, and pressure. Interestingly, the nerve fibers connecting with the sensory and motor areas of the brain are twisted in such a way that each hemisphere of the brain is associated with the opposite side of the body. Usually one hemisphere is dominant, resulting in the phenomenon of "handedness"—greater dexterity of one side of the body. The majority of people are right-handed (i.e., the left hemisphere is dominant). The unfortunate minority of left-handers must contend not only with desks, scissors, and other equipment designed for the convenience of right-handed people, but also with numerous unpleasant connotations of the word *left,* as it is expressed in various languages, such as *sinister* (Latin) and *gauche* (French). (The positive term *dexterity,* of course, comes from a Latin word meaning "right.")

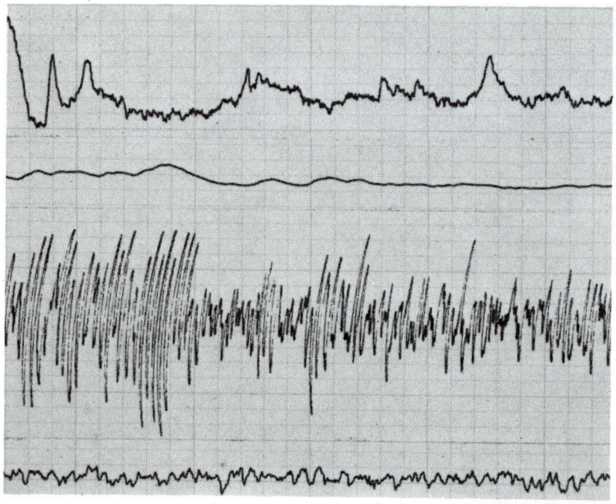

Figure 12–24 A typical EEG. Electrodes pasted to various regions of the face and scalp pick up minute electrical currents generated in various parts of the brain, which are amplified and recorded by moving pens on long sheets of graph paper.

The largest portion of the cerebral cortex is occupied by the **association areas.** The integrative functions carried on in these areas culminate in the unique characteristics of man: imagination, insight, and other attributes of creativity. In addition, this highly developed area gives man an unrivaled potential in such faculties as memory, reasoning, and learning capacity. Visual and auditory centers are also located in this region. Electrical stimulation of parts of the visual centers may not only cause the sensation of light, but may trigger sensations of past visual experiences. Damage to the association centers of the brain may result in aphasia: an inability to recognize certain kinds of symbols. For example, the person may hear people speaking but be unable to remember or speak the names of objects, although he recalls and understands their functions perfectly well. Damage to the frontal lobes of the cerebrum (the anterior portion) may result in pronounced personality changes.

The cerebral cortex has been observed to generate minute pulses of electricity. The electrical activity of the brain, referred to as **brain waves,** has been studied by placing electrodes on the subject's scalp and amplifying the resultant electrical activity. Several types of brain waves have been recorded (Fig. 12–24). Their nature varies according to whether the subject is at rest with his eyes closed or looking at something. Characteristic waves are observed during sleep. The nature of the brain waves is influenced by excitement and is markedly changed by certain brain disorders, such as epilepsy.

The Autonomic Nervous System

Each moment the heart beats, the stomach and intestines rhythmically contract, numerous glands secrete varied chemical substances into the cavities and blood vessels of the body. All this occurs, and yet we are not aware of it.

The portion of the peripheral nervous system which innervates cardiac muscle, smooth muscle, and glands is known as the **autonomic nervous system** (Fig. 12–25). The body nor-

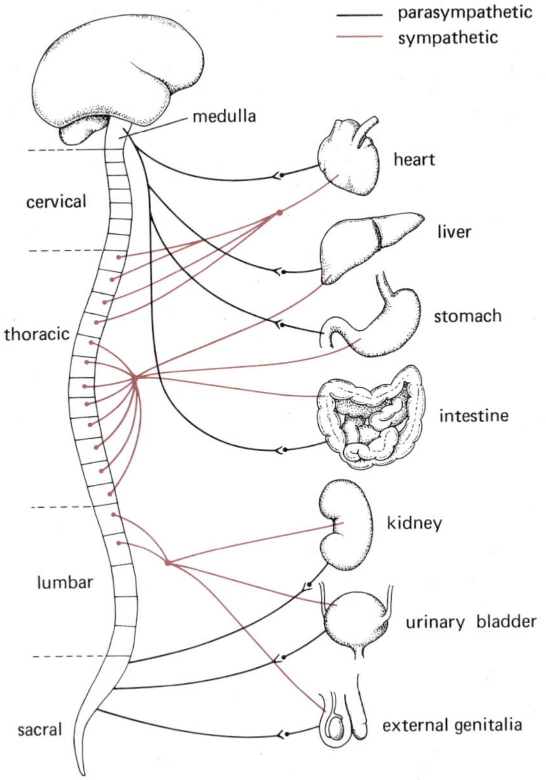

Figure 12–25 The autonomic nervous system and some of the organs it innervates. Most organs are innervated by both parasympathetic and sympathetic fibers, which work antagonistically.

mally exercises no voluntary control over the organs innervated by the autonomic nervous system, including the heart, blood vessels, thoracic and abdominal organs, iris, ciliary muscles, hair muscles, and all the glands of the body.

Most of the body structures are supplied with *two* pairs of autonomic nerve fibers, resulting in antagonistic action. The **sympathetic neurons** carry impulses which result in an activity opposite to that of the **parasympathetic fibers.** If one initiates or speeds up a particular activity, the other inhibits it or slows it down. Both types of fibers, however, are only of the motor variety.

In general, the parasympathetic nervous system exerts its influence during periods of rest and helps to maintain a constant internal environment. Its regulation of the visceral effectors results in such actions as a slowing of the heart, stimulation of peristalsis and secretion of digestive juices, and a constriction of the bronchioles.

The sympathetic nervous system prepares the organism for emergency conditions where rapid mobilization of the body resources and a maximum of energy production are often required. The regulation of the visceral effectors by the sympathetic nervous system leads to such results as an elevated heart rate accompanied by an increase in blood pressure, an increase in blood sugar (due to the breakdown of glycogen in the liver), dilation of the bronchi (facilitating breathing), and an increase in the rate of sweating. The "fight-or-flight" response, stimulated by the hormone adrenaline, is mediated through the sympathetic nervous system.

The functions of the autonomic nervous system were previously thought to be completely beyond conscious control. Yet tales of yogis and religious mystics who survived burial alive, walked over beds of sharp nails, and performed other extraordinary feats—if they could be believed—seemed to indicate that some degree of conscious control of such "automatic" functions as breathing, heart rate, and blood clotting might indeed be possible. Finally, in the 1950s, controlled studies of yogis revealed that these men can indeed exercise an unusual degree of control over their breathing, heart rate, and other autonomic functions. Meanwhile, in laboratory experiments, rats were trained to control their heart rate, blood pressure, the contractions of their intestines, and the rate of formation of urine. A rat could even be taught to dilate the blood vessels in one ear and not the other, producing a one-sided "blush."

Using similar techniques of **biofeedback conditioning,** providing the subject with constant monitoring and information on any changes in the target function, physicians are now training heart patients to correct irregularities of their heart function and lower their blood pressure. It is hoped that such studies may also provide clues to the puzzling field of **psychosomatic illness** (diseases caused by ef-

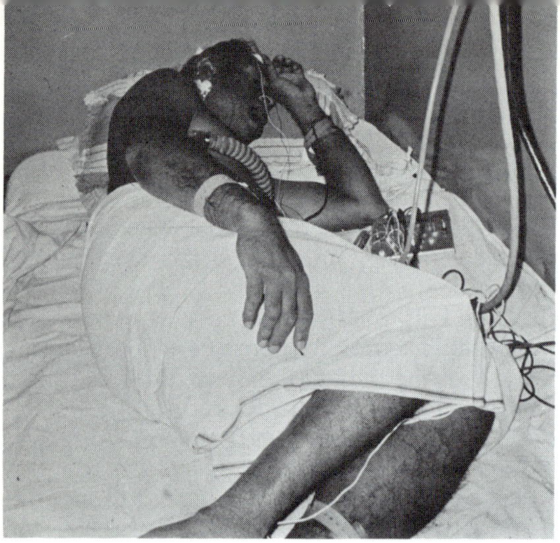

Figure 12–26 In a controlled experiment in New Delhi in 1970, Ramanand Yogi entered a trance in an airtight box.

fects of the brain on the body, rather than by a specific pathogen), and that man may ultimately be able to harness the powers of the brain to cure the body's ills.

The Endocrine System

The nervous system provides for the control and coordination of an enormous variety of responses to changes in the internal and external environment. The time of responses mediated by this system is frequently measured in seconds or even fractions of a second. But the body's activities are also controlled and coordinated by another system, the effects of which are more subtle though equally pervasive. This is the **endocrine system.** Its control over the activities of the cells and organs is exerted through the manufacture of chemical substances called **hormones;** these are released or secreted by specialized cells that may be grouped into structures called **glands** and are carried through the bloodstream to the site of their action. The direct and indirect effects of the endocrine system on the organism are frequently profound, but they become manifested much more slowly than those of the nervous system. An electrical impulse can travel along a nerve much more rapidly than a chemical hormone can be transported in the blood. The effects of hormone secretion are usually detected within minutes rather than seconds, and sometimes days, months, and even years may pass before the full effects of a particular hormonal activity become evident.

In some cases the activities of the nervous and endocrine systems are coordinated. For example the adrenal medulla will secrete its hormones only in response to the appropriate neural stimulation.

Some hormones, such as growth hormone and thyroxin, may affect all or nearly all of the cells in the body. Others affect only a specific organ. ACTH, for example, is secreted by the pituitary gland and is carried through the bloodstream to the adrenal gland, where it stimulates the adrenal cortex to produce its own secretions. The specific tissue affected—in this case, the adrenal cortex—is called the **target tissue.**

Chemically, hormones do not belong to any one class of substances: they may be proteins (e.g., insulin), polypeptides (secretin), or steroids (progesterone).

The endocrine glands are sometimes referred to as the "ductless glands," since their secretions pass directly into the body fluids. This is in sharp contrast to the **exocrine** or **duct glands,** which discharge their secretions through tubelike ducts into specific regions of the body, where they exert their action. The salivary glands, for example, are exocrine glands that empty their secretions into the mouth cavity through salivary ducts. The hormone gastrin is secreted by endocrine cells in the stomach wall into the blood, in which they are carried to exocrine glands in the stomach. The latter are stimulated by the endocrine hormone to secrete hydrochloric acid through ducts into the stomach.

In man the major endocrine glands are the **thyroid,** the **parathyroids,** the **pancreas,** the **adrenals,** the **gonads (testes** and **ovaries),** and the **pituitary.** The arrangement of these endocrine glands in the body is shown in Fig. 12–27. The hormones secreted by each of the major glands and their functions are summarized in Table 12–1.

The basic life processes, including growth, development, maturation, and reproductive functions, are controlled and regulated by hor-

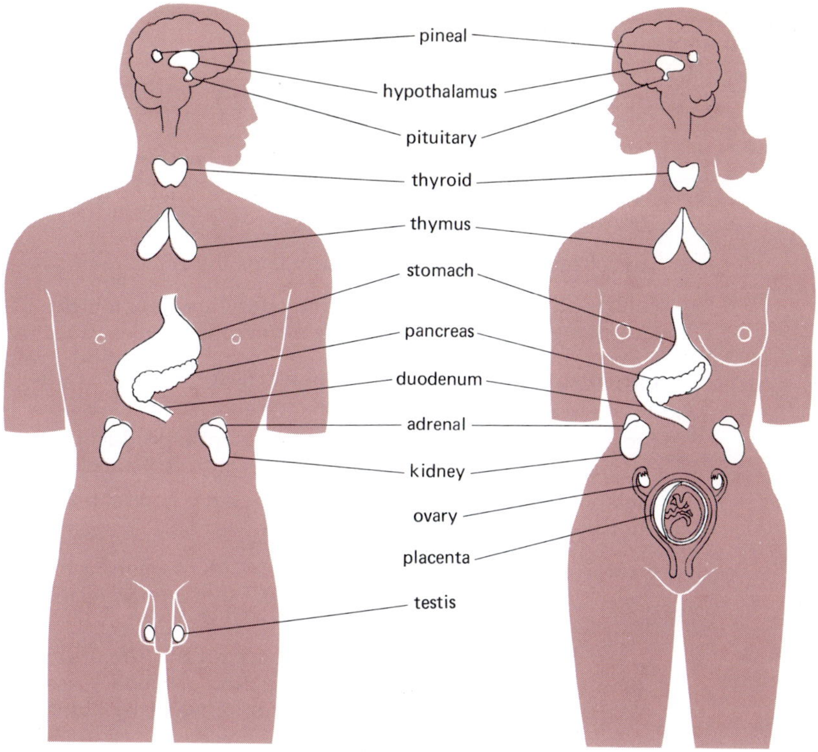

Figure 12–27 The major glands of the human endocrine system.

mones. Frequently the quantities of hormone involved are quite small. An injection of as little as 0.001 mg (about 0.00000003 ounce) of adrenaline will speed up the heart rate of a dog.

Exactly how hormones exert their action in the body has been a subject of lively speculation in recent years. There is evidence to indicate that hormone action is mediated through the sequence of specific DNA activation (or inhibition of protein suppressors of specific genes), RNA synthesis, and ultimate enzyme production. But increasingly, attention has been focused on events that take place at the cell membrane. Studies along this line, involving a compound called **cyclic AMP,** won a Nobel Prize for E. W. Sutherland in 1971.

The name *cyclic AMP* is an abbreviation for cyclic adenosine monophosphate, and this compound is closely related to ATP (adenosine triphosphate), one of the key links in the "energy currency" of living cells. Inside the cell, ATP can be converted to cyclic AMP through the action of an enzyme, **adenyl cyclase,** which seems to be fixed in the cell membrane. According to the latest hypotheses, a hormone* does not actually enter the cell on which it acts. Instead, it reacts at the surface of the cell membrane with a specific receptor site. This reaction may stimulate the enzyme adenyl cyclase on the inner surface of the membrane, thus sparking an increase in cyclic AMP synthesis, which in turn provides the starting signal for the specific effect on the cell that the hormone produces. Or the reaction with the receptor site may inhibit another

*The cyclic AMP model of hormone action seems to apply only to protein and proteinlike hormones. Apparently the steroid hormones, such as the sex hormones, have a different mode of action and may actually penetrate into the target cell, perhaps in combination with a special receptor chemical.

TABLE 12-1 Important hormones and hormonelike substances

GLAND	HORMONE	MAIN EFFECTS
Anterior pituitary	Growth hormone	Stimulates growth
	Adrenocorticotropic hormone (ACTH)	Stimulates adrenal cortex
	Thyrotropic hormone	Stimulates thyroid
	Follicle-stimulating hormone (FSH)	Stimulates growth of follicles in ovary and seminiferous tubules in testis
	Luteinizing hormone (LH)	Stimulates secretion of sex hormones by ovaries and testis
	Prolactin (LTH)	Stimulates milk production by mammary glands; controls "maternal instinct"
	Melanocyte-stimulating hormone	Controls skin pigmentation in lower vertebrates
Hypothalamus (stored in posterior pituitary)	Oxytocin	Stimulates uterine contractions and release of milk by mammary glands
	Vasopressin (antidiuretic hormone, ADH)	Stimulates contraction of smooth muscles, constriction of blood vessels, and reabsorption of water by kidneys
Pineal	Melatonin	Inhibits ovaries; may help to regulate pituitary
Thyroid	Thyroxin, triiodothyronine	Stimulate oxidative metabolism (increase BMR)
	Calcitonin	Regulates calcium metabolism (antagonist of parathormone)
Parathyroids	Parathormone	Regulates calcium and phosphorus metabolism
Thymus	Thymosin	Stimulates maturation of immunologically competent lymphocytes
Stomach mucosa	Gastrin	Stimulates secretion of gastric juice
Duodenal mucosa	Secretin	Stimulates secretion of pancreatic juice
	Cholecystokinin	Stimulates release of bile by gallbladder
Pancreas (alpha cells)	Glucagon	Stimulates conversion of liver glycogen to blood glucose; raises blood sugar level

enzyme, **phosphodiesterase,** which normally breaks down cyclic AMP. This effect also results in a net increase in the amount of cyclic AMP in the cell. Still another type of specific receptor site may be involved in a reaction with a hormone at the cell membrane. This type of reaction may inhibit adenyl cyclase, and reduce the amount of cyclic AMP in the cell.

In each case, the "first messenger" (the hormone) must have its message translated by a "second messenger" (cyclic AMP), which works inside the cell to produce the ultimate effect (Fig. 12–28). The variety of specific re-

TABLE 12-1 (Cont.)

GLAND	HORMONE	MAIN EFFECTS
Pancreas (beta cells)	Insulin	Stimulates glycogen storage and glucose utilization; lowers blood sugar level
Kidney (plus blood)	Hypertensin	Stimulates vasoconstriction, causing blood pressure rise
Adrenal medulla	Adrenaline	Stimulates "fight-or-flight" reactions, reinforcing action of sympathetic nerves; stimulates glycogen breakdown
	Noradrenaline	Similar to adrenaline; constricts blood vessels
Adrenal cortex	Glucocorticoids (cortisol, cortisone, hydrocortisone, etc.)	Inhibit incorporation of amino acids into proteins; stimulate conversion of proteins to carbohydrates; help maintain normal blood sugar level
	Mineralocorticoids (aldosterone, deoxycorticosterone, etc.)	Regulate sodium-potassium metabolism
	Cortical sex hormones (adrenosterone, dehydroepiandrosterone, etc.)	Stimulate development of secondary sex characteristics, especially male
Ovaries	Estrogen	Stimulates development and maintenance of female sex characteristics and behavior
	Progesterone	Stimulates female sex characteristics; maintains pregnancy
Placenta	Chorionic gonadotropin	Acts with other hormones to maintain pregnancy (its detection is basis of pregnancy tests)
Testes	Testosterone	Stimulates development and maintenance of male sex characteristics and behavior
Seminal vesicles (and others)	Prostaglandins	Variety of effects including stimulation of uterine contractions, vasoconstriction, raising and lowering of blood pressure
Damaged tissues	Histamine	Increases capillary permeability
Various body cells	Chalones	Work with adrenaline to inhibit cell division.
Various specialized glands	Pheromones	Produce effects on other members of the same species (sex attractants, resorption of embryos, territorial markers, food trails, etc.)

ceptor sites and the chemical makeup of the cell itself help to ensure that each hormone will act precisely on the appropriate cells and produce its own characteristic effect.

Hormones frequently influence one another's activity. Some hormones stimulate the secretion of a second hormone or reinforce its action. Others inhibit the synthesis, secretion, or activity of a second hormone. Thus, both positive and negative feedback mechanisms are important in hormone action (Fig. 12–29). The pituitary gland, in particular, secretes a wide variety of hormones, some of which affect virtually all the body's cells, while others

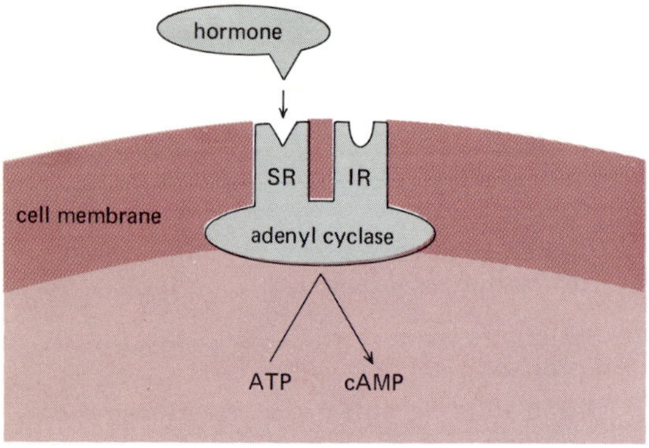

Figure 12–28 Two-messenger mechanism of hormone action. The hormone (first messenger) reacts at the surface of the cell membrane with a specific stimulatory (SR) or inhibitory (IR) receptor site. In the case of stimulation, the enzyme adenyl cyclase then catalyzes the synthesis of cAMP within the cell. cAMP, the second messenger, then stimulates the cell to synthesize its characteristic products. Reaction at an inhibitory site inhibits the action of adenyl cyclase and causes a drop in cAMP levels within the cell.

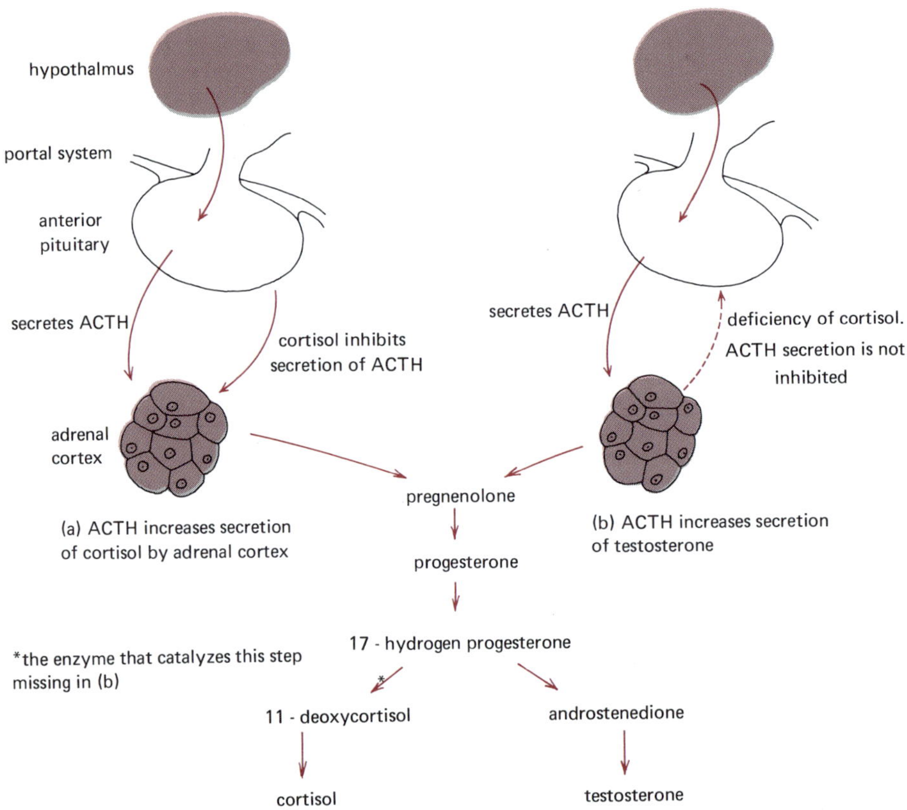

Figure 12–29 Feedback mechanisms regulate secretions of the endocrine glands. In a normal human (a), a releasing factor from the hypothalamus stimulates the pituitary to secrete its hormone ACTH. ACTH acts on the cells of the adrenal cortex, stimulating them to secrete the hormone cortisol. Cortisol in turn inhibits the secretion of ACTH and thus indirectly helps to regulate its own secretion. In a certain type of hereditary adrenal cortical virilism (b), a key enzyme in the synthesis of cortisol is lacking, and the intermediates are instead converted to the male sex hormone testosterone. ACTH secretion is not inhibited, and the sequence of reactions leading to testosterone production continues unabated, resulting in masculinization.

stimulate or control the functioning of other glands. For this reason the pituitary is often called the "master gland."

Although similar hormones show some chemical variation from species to species, they exhibit a surprising degree of interchangeability in their functioning. Beef, sheep, and hog insulin can be used successfully to treat diabetes (caused by an insulin deficiency) in humans. In frogs, the hormone of the thyroid gland (thyroxin) has been shown to stimulate the changes that result in metamorphosis from tadpole to adult frog. In mammals the functions of thyroxin are quite different: it regulates the rate of cellular oxidation. Yet the addition of beef thyroid extract to water in which tadpoles are swimming will cause them to metamorphose immediately into adult frogs. Some plants produce substances that are surprisingly similar to insect hormones. These substances disrupt the endocrine metabolism of insects feeding on the plants and thus help to protect the plants from excessive insect depredations.

The functions of the endocrine glands are so vital to the normal growth and health of the organism that deficiencies of specific hormones can lead to a variety of serious conditions or even death. Diabetes, resulting from a deficiency of insulin, has already been mentioned. This hormone, secreted by the pancreas, controls the rate of utilization of glucose from the blood. Deficiencies of other endocrine hormones may result in dwarfism or giantism, grotesque obesity, and mental retardation.

The interactions of the sex hormones, both with one another and with hormones of the pituitary and adrenal glands, are extremely complex. The sum total of these interactions determines both the formation of the sexual organs and their fertility and the formation and maintenance of a whole constellation of secondary sex characteristics, including the growth of hair under the armpits and in the pubic region and (in men) on the chest and face, the breasts and rounded contours of women, and the differences in the structure of the larynx that result in men's having lower-pitched voices than women.

It is curious that both men and women normally have supplies of both male and female sex hormones. It is the ratio between the concentrations of the two groups of hormones that determines whether a person will seem masculine or feminine. At one time, female sex hormone therapy was tried in an attempt to minimize the possibility of heart attacks in middle-aged men. The therapy produced disturbing side effects—many of the men began to develop enlarged breasts and other feminine characteristics. Similarly, in women with adrenal tumors or other conditions that result in a lowered production of fe-

Figure 12–30 Endocrine gland disorders can give rise to various disturbances of growth and metabolism: (a) acromegaly (hypersecretion of the pituitary gland); (b) cretinism (hypothyroidism from birth, resulting in dwarfism and mental retardation if untreated); (c) goiter (hypothyroidism often due to lack of iodine.

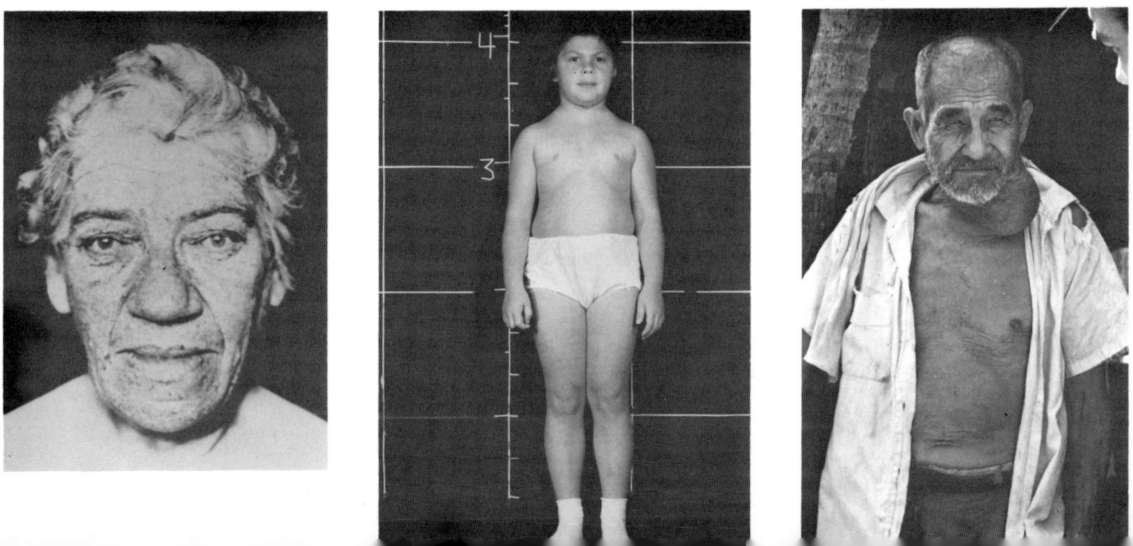

Figure 12–31 A bearded lady, suffering from a hormone imbalance.

male sex hormones or increased production of male sex hormones, a masculinization may result, including growth of a beard and deepening of the voice.

The administration of sex hormones can even produce a complete sex reversal. An egg-laying hen can be transformed into a virile rooster by removing her ovaries and treating her with the male sex hormone, testosterone. Rudimentary testes within her abdomen develop, and "she" not only grows a comb and wattles, but also crows and can father chicks. An analogous shift of the hormone balance can convert a rooster into a hen. (One wonders if that is what happened accidentally in an Italian town back in the sixteenth century: court records of the town relate that a rooster was condemned to die at the stake because he allegedly laid an egg.)

A variety of female sex hormones govern the menstrual cycle, in which a mature ovum is released each month and makes its way through the female reproductive organs while the lining of the uterus is prepared for the possible implantation of a fertilized egg. (If this does not occur, the lining breaks down and is expelled from the body in the menstrual flow; then the cycle begins all over again.) Intensive research in recent years has been centered on gaining insight into the hormonal interactions during the menstrual cycle and synthesizing artificial sex hormones, so as to control the process and either prevent or enhance the possibility of pregnancy. Numerous "birth-control pills" have now been devised. These are typically taken, one a day, either during the entire month or for a certain period of days in a stretch. They act to regulate the hormonal level so as to inhibit the release of a fertile ovum. Similar research, which has not yet come to fruition, is aimed at altering the male hormone levels to inhibit the production of fertile sperm.

In addition to the major endocrine organs enumerated, hormones are also produced by several other structures. During pregnancy, the **placenta** (the structure within the uterus that nourishes the unborn child) is a powerful regulating organ for the body of the mother, producing sex hormones that augment those produced by the more permanent organs. Curiously, many animals (even those normally vegetarians) eat the placenta immediately after it has been expelled following the birth of the young. This provides the exhausted mother with nourishment, and the hormones of the placenta may also aid in bringing her uterus back to its normal condition rapidly.

Hormones that stimulate various digestive glands and organs are secreted by endocrine cells in the stomach and duodenum.

Still somewhat of a puzzle is the **pineal body,** a small pea-sized structure situated between the two halves of the cerebral cortex, just above the cerebellum. Although no endocrine function has yet been discovered for this structure in man, a hormone called melatonin has been extracted from the pineal bodies of cattle and other mammals. This hormone has been found to affect the pigment granules of the skin of tadpoles, causing their skin color to lighten. (It does not have a similar effect on mammals.)

In various animals, including frogs, rats, and birds, the pineal gland has been shown to be sensitive to light, impulses from which reach it through the retina of the eye. In the

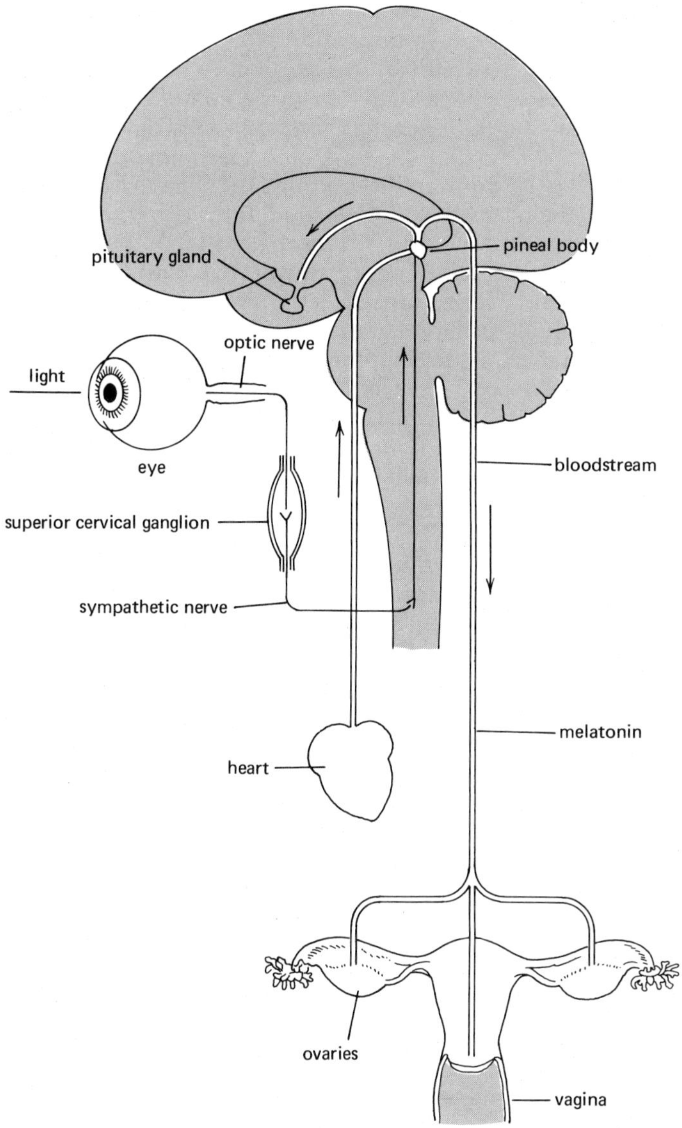

Figure 12–32. Changing day length can affect the functioning of the female reproductive organs through hormone production by the pineal gland.

rat, continuous exposure to light has been found to lead to a decrease in the size of the pineal gland, with a decrease in the production of melatonin and a concurrent increase in the size of the ovaries and acceleration of the estrous cycle (the cyclic alternation of fertile and infertile periods). Injections of melatonin had the opposite effect: a decrease in the size of the ovaries and a slowing of the estrous cycle. The pineal gland is believed to control the circadian rhythms of the body, twenty-four-hour cycles in the body temperature and various other body functions. It is believed to be linked with the "biological clocks" that regulate the annual breeding and migration patterns of birds and other animals.

Movement and Locomotion

The word *life* calls many pictures to mind. In nearly all of these there is one common element—**movement.** Thinking further, one might be inclined to object: "But only animals and protists can move, and plants certainly are alive as well." Yet plants also exhibit movement. Some types of plant movement can be observed only over a lapse of time: plants grow, turn to follow the path of the sun, and send out roots in search of moisture. Another type of plant movement can be observed only under the microscope: there one can see that the cytoplasm within the cells themselves is undergoing a constant streaming motion. This type of motion is also observed among animals and many protists: the cell itself (or the organism itself) is capable of self-initiated, self-propelled motion from one place to another, or **locomotion.**

The major modes of locomotion utilize the property of **contractility.** This property is shared, to some degree, by all animal cells. In some protists and in animals composed of a relatively small number of cells, the whole organism may undergo a paroxysm of motion as a result of the contractile nature of the protoplasm itself. In more complex, highly multicellular organisms, the cells have become specialized; some have nearly lost the property of contractility, while others have developed it to such a degree that they constitute the basic sources of motion for the body (Fig. 12–33). These highly developed contractile cells are called **muscle cells.** They are grouped together in larger units, known as **muscles.**

In Chapter 6 the three basic types of muscles—skeletal, smooth, and cardiac—were described. The muscles concerned with movement of the body parts and with locomotion are skeletal muscles, whose contraction is under voluntary control.

In locomotion, the muscular system works in conjunction with the skeletal system—the supporting framework of the body. In vertebrates, the skeletal system is comprised of an arrangement of bones, jointed or articulated at

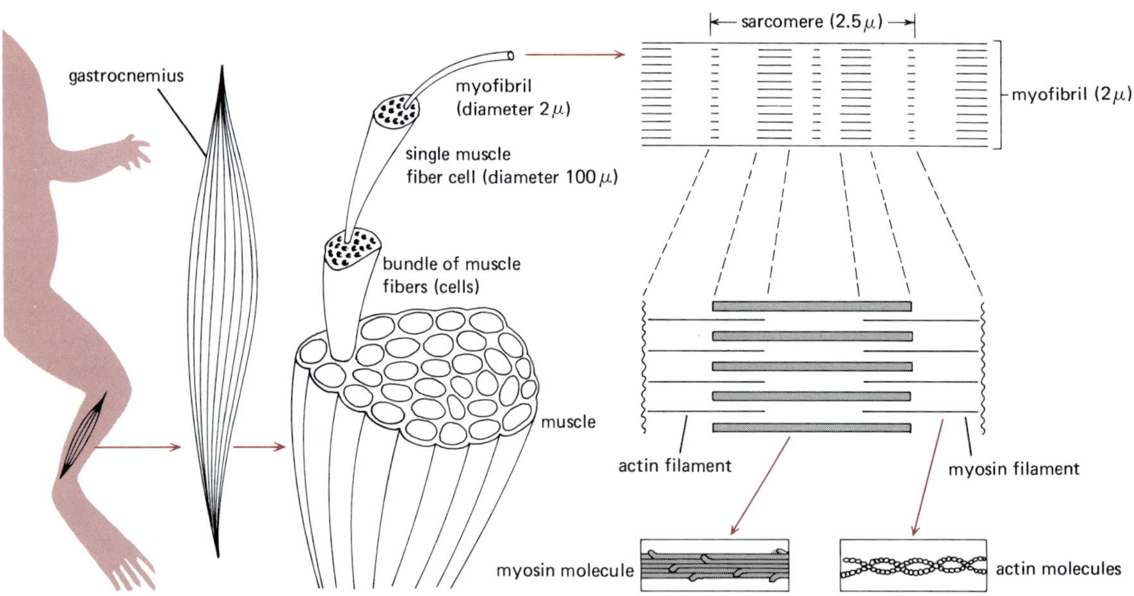

Figure 12–33 A skeletal muscle is made up of bundles of muscle fibers, surrounded by a sheath of connective tissue. Each single fiber (i.e., muscle cell) consists of a number of tiny myofibrils, each made up of two types of protein filaments, actin and myosin. During contraction, opposite actin filaments slide toward one another along the thicker myosin filaments in a manner that is not yet completely understood.

intervals in such a way that legs, necks, and other body parts can be bent or rotated while still being supported. The bones are embedded within the soft tissues of the body, and thus the vertebrate skeleton may be referred to as an **endoskeleton.**

Many soft-bodied invertebrates do not have a supporting skeleton. Some, such as clams and snails, are encased in hard, limy shells. This "armor" provides good protection, but it does not afford any flexibility for locomotion. The arthropods, as a group, are provided with flexible, jointed skeletons that provide for great versatility of locomotion, but their skeletons are constructed on a quite different plan from the vertebrate skeleton (Fig. 12–34). Insects and other arthropods are characterized by an **exoskeleton,** a supporting superstructure that covers the animal's entire body like a coat of armor. This is commonly made of chitin. The muscles of the arthropod are attached to the inner surface of its exoskeleton, much as muscles of vertebrates are attached (either directly or through tendons)

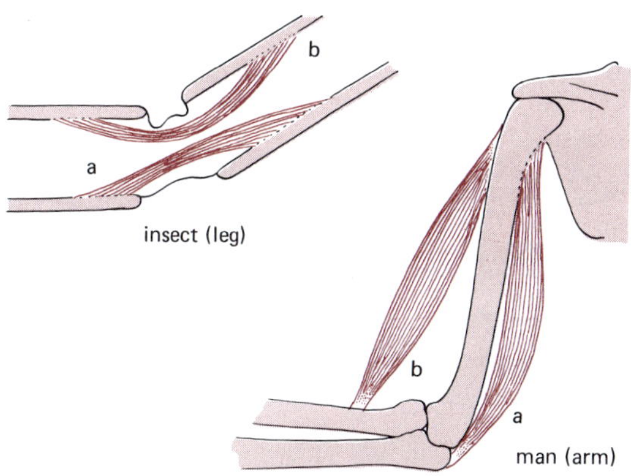

Figure 12–34 Muscles and skeleton work together for support and movement. In the insect, the muscles are attached to the inside of the exoskeleton; in man the muscles are attached to the outside of the endoskeleton. In each case movement is produced by pairs of antagonistic muscles. In both joints pictured, muscle *a* is an extensor muscle (when it contracts, the leg or arm extends) and muscle *b* is a flexor muscle (when it contracts, the leg or arm is flexed or bent).

to the outer surfaces of bones. Although the joints of the arthropod's exoskeleton provide flexibility, the sheer weight of the exoskeleton puts an effective upper limit on the potential size to which the animal can grow; if it is too large, the exoskeleton will be too heavy to lug around. Terrestrial insects rarely grow more than an inch or two long. Aquatic arthropods, such as lobsters and crabs, can grow considerably larger because the buoyancy of the water helps to support the weight of the exoskeleton.

Modes of locomotion vary widely among the forms of life. The mode of locomotion employed by a particular species is determined generally both by its basic structure and by the medium through which it must move. Water-dwellers solve the problems of moving about in quite different ways from those species that live on the land. Still different mechanisms have been evolved by the birds and insects that soar through the air. Some animals can employ several modes of locomotion, depending on the circumstances. A human, for example, first learns to creep on his belly, then to crawl on hands and knees, and then to walk and run upright. If he wishes, he may hop or leap. (The first astronauts to "walk" on the moon discovered that a two-legged hop, covering about 6 to 8 feet in a hop, was the most efficient means of locomotion to take advantage of the moon's lower gravity.) When man takes to the water, he swims, using a variety of techniques, and a sort of swimming motion is also an appropriate form of locomotion in the cabin of a spaceship during the weightless portion of a space flight.

All the types of locomotion employed by living organisms are governed by the laws of physics dealing with motion through media of various densities, with velocity, resistance, and with the intricate force interrelationships between the medium and the "machinery" of the body—the articulated system of levers that comprises the skeletal system, driven by the motive force supplied by the muscles.

LOCOMOTION IN THE WATER

Protists, as a group, are water-dwellers, and so are a large proportion of the animal kingdom. Every one of the dozen or so animal phyla has some representatives that inhabit the waters of the world; only four phyla have members that move about on the land areas, and only two phyla have representatives that fly through the air. This observation seems to reinforce the theory that life arose in a watery medium and thence spread to the land areas.

Organisms use a variety of methods to move freely through the water, or **swim.** Many protists, typified by *Paramecium* and *Euglena* move through the water propelled by the motions of short, threadlike cilia or long flagella. In *Paramecium* the beating of the cilia is rhythmic and is coordinated by interconnecting fibers that join in a sort of "control center" in the region of the gullet. Flagella lash back and forth in a whiplike manner.

The mechanisms of the action of cilia and flagella are not well understood. Actinlike and myosinlike proteins, similar to those that operate in muscle contraction in vertebrates, have been extracted from these organisms and may play some role. The presence of ATP as an energy source has been found to be required, just as it is for muscle contraction.

It is interesting to note that *Paramecium*, *Euglena*, and many other protists that swim with the aid of cilia or flagella have a streamlined, ovoid body shape, basically similar to the general outlines of a fish (Fig. 12–35a). This shape is excellently suited to moving through the water, a medium characterized by both resistance and buoyancy. The smooth, rounded contours minimize turbulence: the water parts and passes smoothly along the swimming organism's sides.

The fish moves through the water by sweeping its tail sideways against the resistance of the water; the forward component of

Figure 12–35 Most swimming animals have streamlined body outlines, regardless of their means of propulsion. The paramecium (a) swims by rhythmic beating of the short cilia distributed over its body surface (b). Another microscopic pond dweller, the euglena (e), is propelled through the water by lashing of a single long flagellum. The whale (c) and fish (d) (not drawn to scale) get their forward thrust from sweeps of a leverlike tail. The whale's horizontal tail flukes move up and down, while the fish's vertical tailfin moves from side to side.

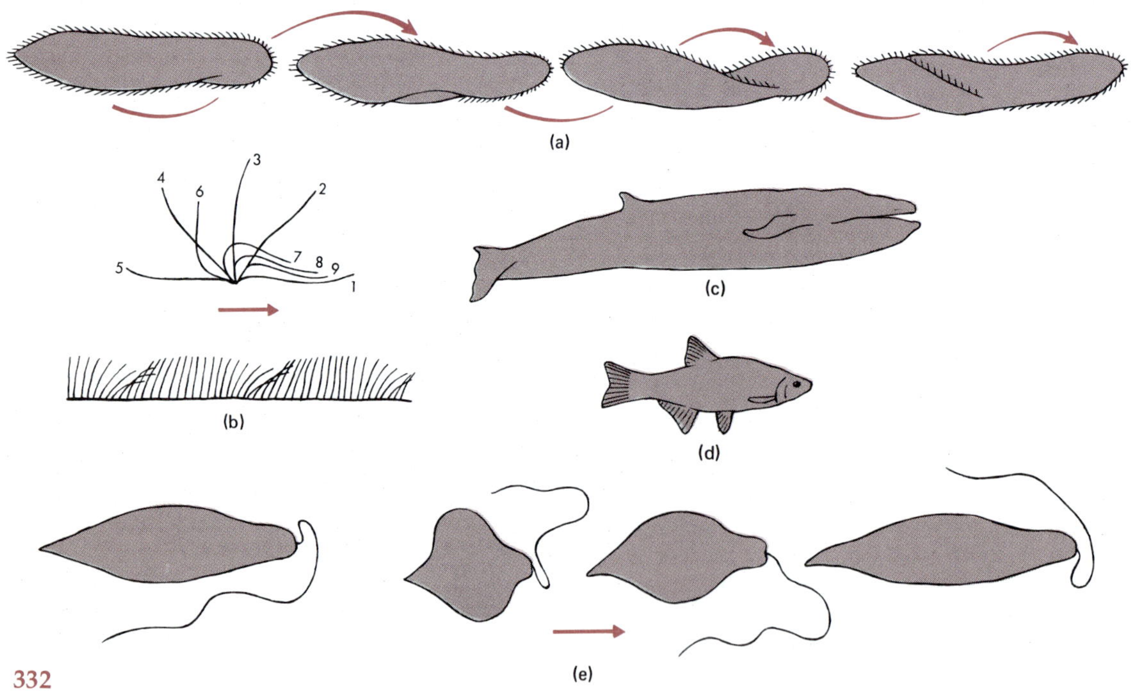

the force is transmitted along the hinged vertebrae of the backbone and propels the fish forward. The forward thrust imparted to the fish body by the leverlike tail is comparable to that transmitted to a ship by its propeller (Fig. 12–35d). The fins aid in balancing, automatically correcting the orientation of the fish if it should accidentally roll, yaw, or pitch.

The fishlike shape, with propulsive muscles and tail, is found not only among the fish, but also among the water-dwelling mammals—the whales, porpoises, and dolphins. In the mammalian groups, the streamlined fish shape was evolved independently from a more typical mammalian body plan, and the system of propulsion is slightly different: the vertebral column is flexible up and down, rather than from side to side, and the tail flukes are horizontal, rather than vertical as in the fish (Fig. 12–35c).

In both fishes and the fishlike mammals, speed generally increases with body size, up to some limit, at which point water resistance increases to such an extent that all the increased muscle capacity is required just to overcome it, rather than to increase the propulsive efficiency. Thus, the maximum cruising speeds of the porpoise and the giant whale are about the same, approximately 20–25 miles an hour.

Squids and a few other marine species swim by a variation of **jet propulsion** (Fig. 12–36). By squirting a jet of water out in front of its body, the squid can propel itself backward. This is a clear illustration of Newton's third law of motion, which states that to every action there is an equal and opposite reaction. This principle is also utilized by the scallop, which claps its two shells together forcefully to produce a jet.

Turtles and some other species get about in the water by **paddling.** The four limbs of the turtle are flattened, very much like the paddles of a canoe, and they are used in much the same manner.

The Portuguese man-of-war and related coelenterate species skim about on the surface of the water very much like a sailboat. An inflated, balloonlike crest, which projects above the surface of the sea, catches the air currents like a sail. This gas-filled "sail" can be inflated or deflated, thus varying its wind resistance.

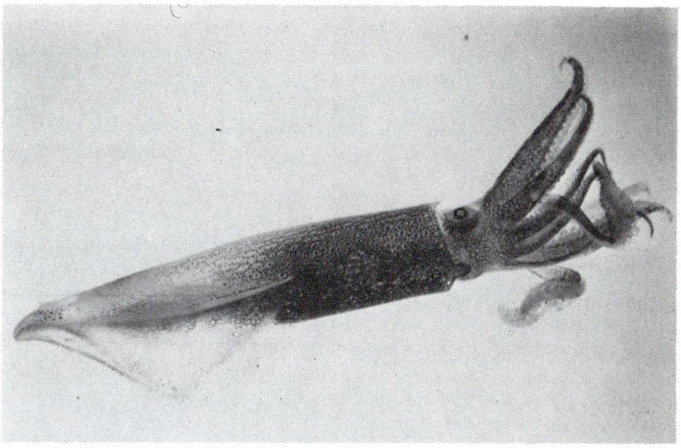

Figure 12–36 The squid moves through the water by a form of jet propulsion.

Some water-dwellers do not move freely through the water, but instead creep along the bottom. The pond-dwelling *Amoeba* oozes along in typically **amoeboid movement.** A bulge protrudes from the flexible cell, and the rest of the cell seems to flow into it. Amoeboid motion appears to be associated with reversible sol-gel transformations within the cytoplasm. Some contractile proteins similar to those found in higher organisms have also been isolated from the *Amoeba,* but their role in its locomotion is not yet understood.

A number of higher invertebrates creep along the ocean bottom by modifications of **hydraulic systems.** The water vascular system of the starfish permits it to move by forcing seawater into and out of tiny tube feet. These tube feet grasp the rocky bottom like suction cups and pull the starfish along. The clam also uses a hydraulic modification of locomotion (Fig. 12–37). It thrusts its hatchet-shaped, muscular "foot" forward; blood then flows into the foot, which swells and serves as an anchor. By contraction of the retractor muscles, the clam inches forward. The blood is then forced out of the foot, which shrinks back from its attachment to the sand, only to thrust forward again and repeat the process.

LOCOMOTION ON LAND

Some land-dwellers also crawl along or burrow within the solid surface. The earthworm

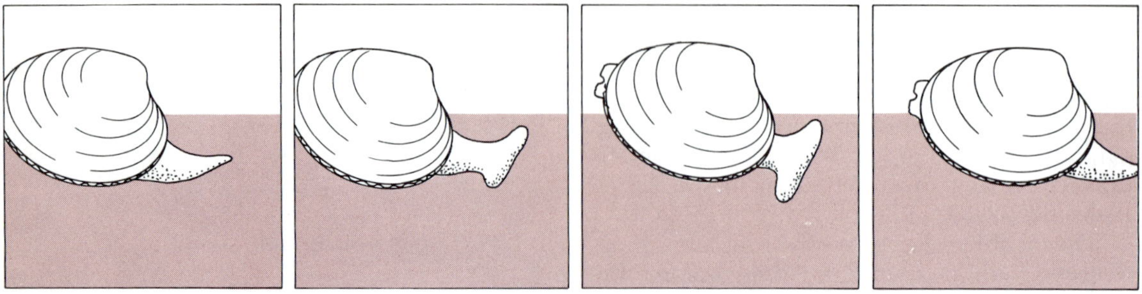

Figure 12–37 The clam's muscular foot swells and contracts by means of an internal hydraulic system to move the clam slowly along.

moves through its cylindrical burrow with the aid of two sets of muscles: circular and longitudinal. When the outer, circular layer of muscle fibers is contracted and the inner, longitudinal layer is relaxed, the earthworm's body becomes long and thin. When the outer layer is relaxed and the inner layer contracted, the body becomes short and thick. Waves of alternate and opposed contractions and relaxations of the two muscle layers swell or stretch one portion after another of the earthworm's body, permitting it to stretch forward then wedge itself against the sides of the burrow, thus consolidating its advance. Paired bristles along the worm's sides help it to brace itself against the walls of the burrow.

Most of the animal species that live on the land practice some form of **walking** and/or **running** (Fig. 12–38). In this type of locomotion, the body must be supported clear of the ground for any degree of speed and efficiency. (Reptiles as a group are rather inefficient in this respect, tending to drag their bellies along on the ground.)

Support and balance are important factors in walking and running. In the insect, the support is provided by the chitinous exoskeleton. At rest, the insect is well balanced on its six legs, with the head and abdomen serving as counterweights. In walking, the insect supports itself on three legs at a time: the front and back leg of one side and the middle leg of the other, which function as a tripod. The other three legs are carried forward and set down in turn, moving the insect forward. This tripod mechanism permits insects to walk and run quite rapidly—relative to their body size. Many insects have alternative modes of locomotion, such as flying or leaping, which they can use in emergencies when especially rapid covering of greater distances is required for escape from potential predators.

In the four-legged vertebrates, the same factors of support and balance are primary. Support is provided by the articulated endoskeleton. Balance is provided by the interplay of the muscular system and the skeletal system, coordinated by the nervous system.

In walking, a four-legged animal lifts one leg at a time in a regular pattern (leaving a tripod on the ground at each step). In running the same pattern is followed, but the second leg is lifted before the first one is back on the ground. This permits great speed, but requires a finely developed balance and highly efficient skeletal system. The horse can run for sustained periods at 20 miles an hour, and can sprint briefly at speeds up to 45 mph. The cheetah, the fastest animal, has been clocked at speeds up to 70–75 mph, but only for extremely short periods (about 15 seconds). In running we again observe a general increase in speed with increasing body weight, but here, too, a limit is reached as the overwhelming bulk of the animal becomes a factor. The elephant is incapable of true running. Its body is so massive that the task of supporting its weight on only two legs during rapid motion would place an insupportable strain on its skeletal system.

In four-legged species, the animal's weight is distributed in such a way that its center of gravity falls somewhere between the forelegs and hind legs. In many animals, such as the

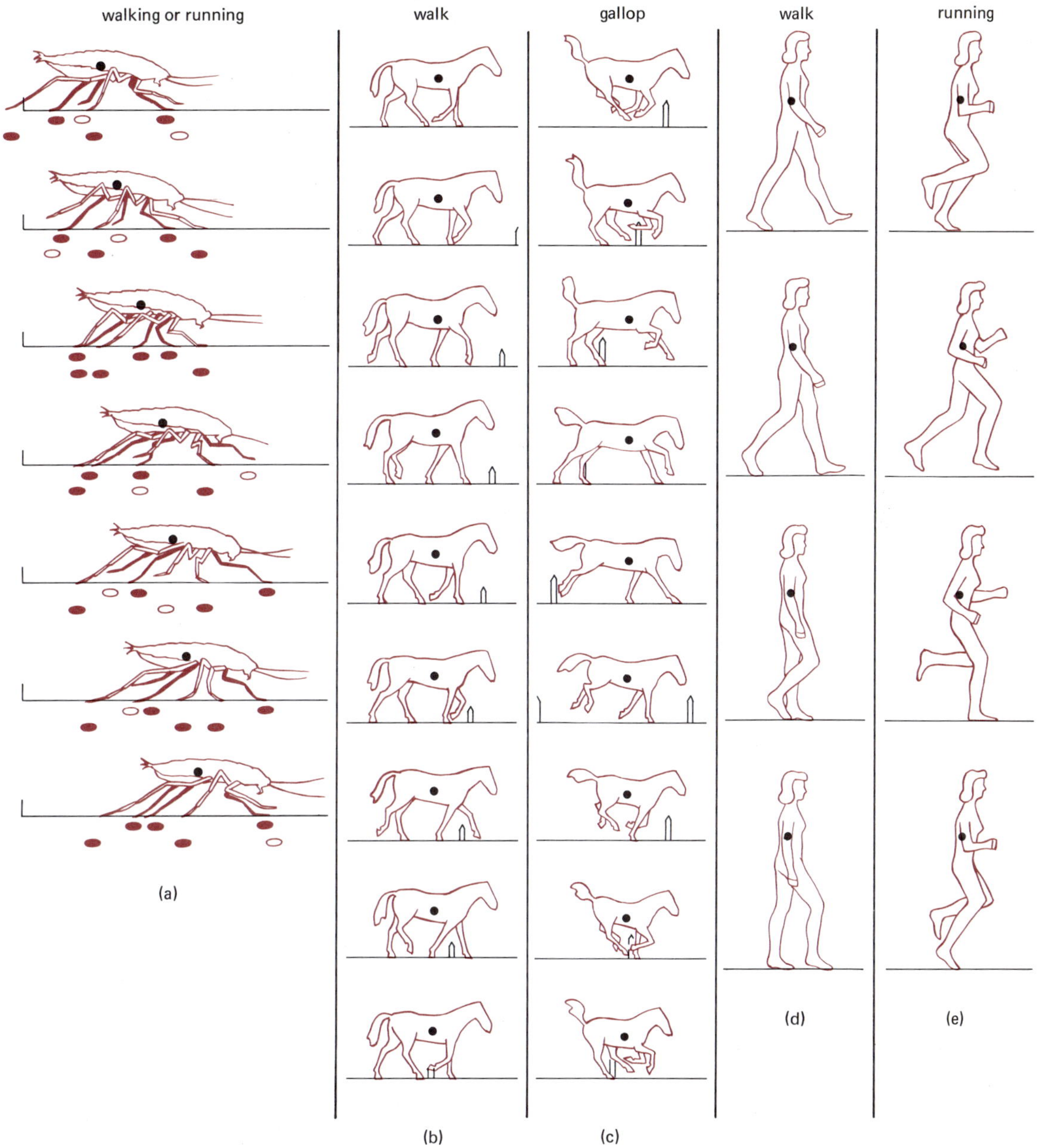

Figure 12–38 Hexapodal (a), quadrupedal (b, c), and bipedal (d, e) locomotion. With fewer legs there is decreasing inherent stability, and balance and coordination become increasingly important. The colored dots indicate the approximate location of the center of gravity.

bear, the center of gravity is shifted so far back toward the hind limbs that the animal can readily lift its forelimbs and walk on its hind limbs unsupported. This tendency is even more pronounced among the primates, particularly man, where **bipedal locomotion** is the dominant form. But in two-legged locomotion, an efficient system for maintaining balance is particularly important. In walking, one leg is lifted and the body remains supported on only the single remaining leg; in running, both legs may be momentarily off the ground at the same time. The development of bipedal locomotion in man has resulted in considerable structural modification of the skeleton and muscular system, in comparison with his

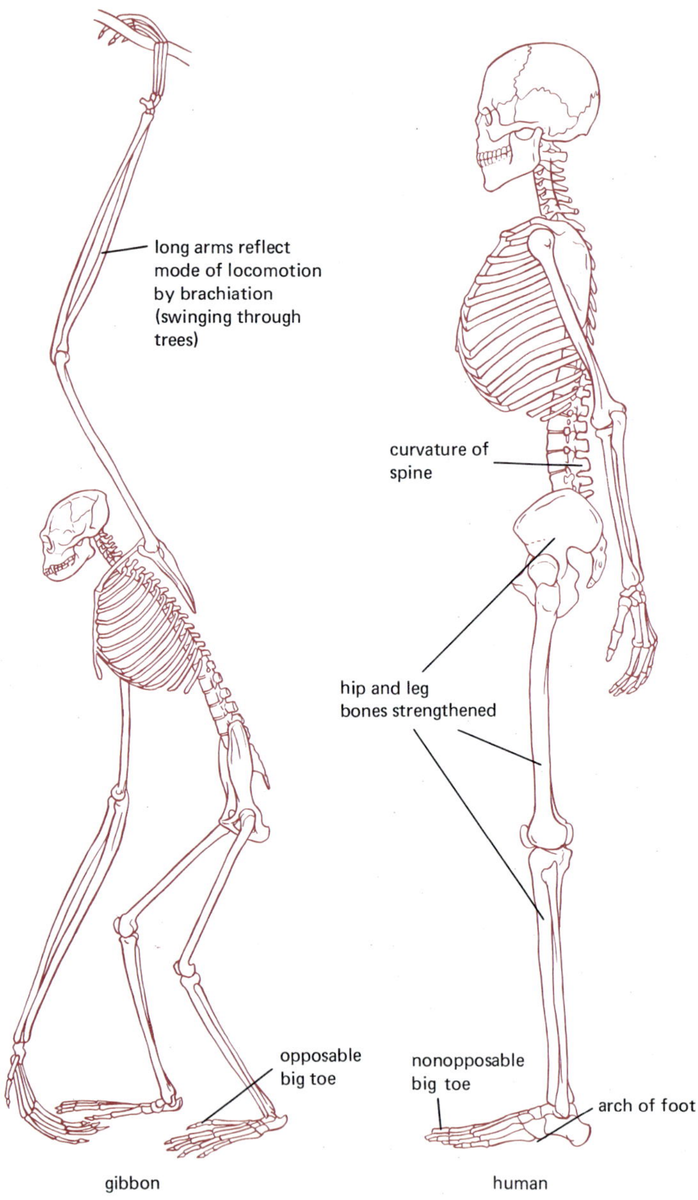

Figure 12–39 The development of an upright posture and bipedal locomotion resulted in considerable modification of the primate skeleton.

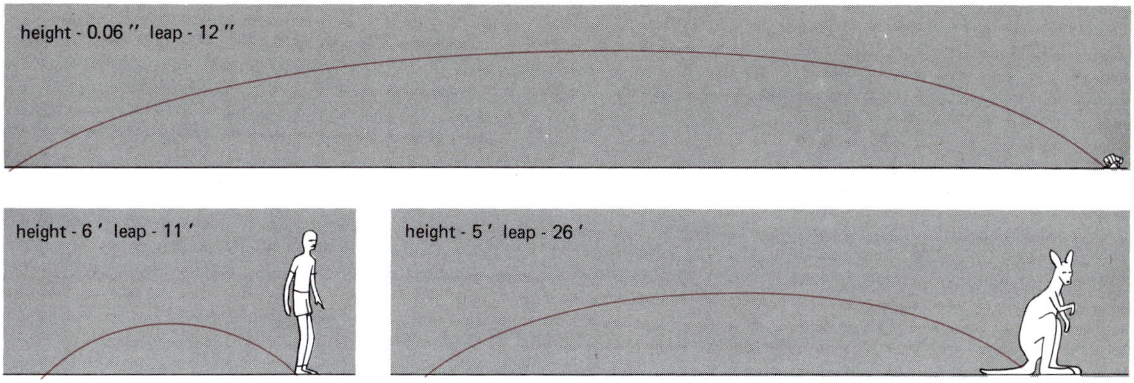

Figure 12–40 Champion leaper: the flea.

primate relatives (Fig. 12–39). The frequent occurrence of such skeletal disorders as fallen arches and "backaches" is an indication that evolution has probably not yet reached an optimum in this respect. The problems of maintaining balance during bipedal locomotion are aggravated by the wearing of high-heeled shoes and during pregnancy. Usually the curvature of the spine must be modified to compensate for the shift in the body's center of gravity in such cases.

Many animals, such as the kangaroo, frog, grasshopper, and flea, are especially adept at an alternative method of locomotion: **leaping.** Such species generally have one pair of legs that are much larger and stronger than the others and are capable of propelling the animal powerfully through the air for huge distances. To appreciate the capabilities of such animals, let us compare the relative jumping abilities of man, the kangaroo, and the flea. From a stationary position, a champion broad-jumper can leap about 11 feet (approximately twice his height). A kangaroo can cover a distance of 26 feet, or about 5 times its height. Although a flea can leap only about a foot, this represents a distance 200 times its body length!

LOCOMOTION IN THE AIR

For countless centuries man has marveled at the apparently effortless ability of birds to fly and has yearned to duplicate this miracle of nature. We find evidence of such yearnings in the Icarus myth of the ancient Greeks and in the prophetic drawings of aircraft sketched by Leonardo da Vinci about four hundred years before the Wright brothers flew the first successful airplane at Kitty Hawk.

Watching a bird effortlessly flapping its wings in flight, one might think that the mechanism of bird flight is rather simple. And yet nothing could be farther from the truth. Despite years of intensive investigation, the various mechanisms involved in the flight of a bird have not been entirely unraveled. A great deal of light has been shed on these mechanisms by extensive use of high-speed photography.

Figure 12–41 High-speed photography has revealed some of the secrets of bird flight. Shown here: a flock of white ibis.

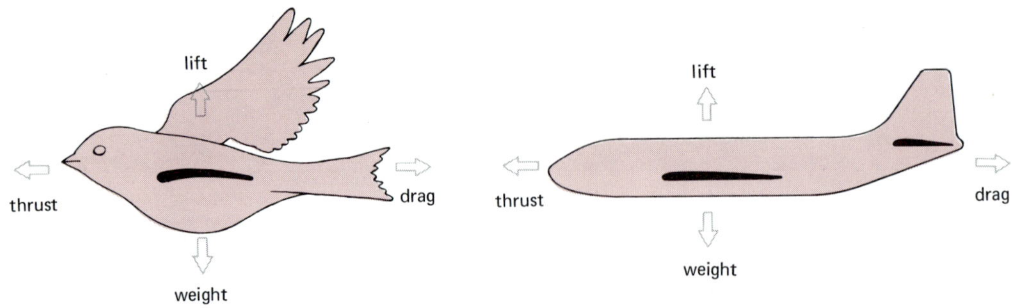

Figure 12–42 The flight of a bird and that of an airplane show many basic similarities.

There are many similarities between the flight of a bird and that of an airplane, but there are important differences, too (Fig. 12–42). In both, the wings provide lift, using the buoying effect of the air flowing past their surfaces. But unlike the airplane, in which thrust is provided by a propeller or a jet, the wings of a bird must also provide the propelling force. The basic elements of the flapping of a bird's wings are a downward stroke, which provides lift, and an upward stroke, providing both lift and forward thrust. The exceptional lightness of birds, due to their hollow-boned skeletal structure, their numerous inflatable air sacs, and their coating of feathers, further aids in their overall flight performance.

Aside from birds, only one other vertebrate, the bat, is capable of flight. The mechanisms involved are similar to those of bird flight. In addition, numerous insects, including butterflies, bees, and houseflies, are capable of flight. Usually, the smaller the insect, the more rapid is the beating of its wings. The swallow-tailed butterfly beats its wings about five times a second, the horsefly two hundred times, and the tiny midge more than a thousand times a second. This extreme muscular activity requires an enormous amount of energy, and the insect must feed frequently to sustain such flight.

In addition to active flight, characterized by the flapping of wings, many aerial organisms are capable of sustained passive flight, or **gliding.** The most proficient gliders are birds, some of whom may soar through the air for hours at a time. In such flight the wings are held in an extended rigid position, and the bird may make extensive use of rising air currents to achieve the longest possible glide. The best gliders travel long distances horizontally with the smallest possible loss of altitude, but the force of gravity ultimately produces a gradual but inevitable descent.

Some terrestrial and even aquatic animals also utilize the principle of gliding, although they are not nearly as proficient as birds. Flying fish, for example, can glide for 10 to 50 yards or more, remaining in the air for periods of several seconds. Among mammals, in addition to the bats, which can glide for long distances, a number of species can glide from tree to tree. Well-known examples are the flying squirrels, which have membranes running between their wrists, bodies, and hind legs, giving them the appearance of animated kites.

Responsiveness in Plants

One would not expect to see a plant walking down the street and pausing at a busy intersection, carefully looking both ways before crossing. We smile at such a flight of fancy because we recognize that plants not only lack means of locomotion comparable to those of animals, but furthermore do not possess a nervous system to coordinate such a sophisticated series of activities. Nor do they have specialized sense organs.

Yet plants do respond to external and internal stimuli, and they do manifest some degree of coordinated activity. A marked difference between this activity and that exhibited by animals is in the rate of response. An animal, with an intricate network of nerve fibers connecting highly sensitive sense receptors with efficient effectors, through integrating control centers, can react almost instantaneously to stimuli. Plants, lacking such a nervous system, are coordinated by a system of chemicals resembling the hormonal systems of animals. This necessarily results in a much slower response, which may become manifest in a matter of minutes, hours, or even days.

Plants respond to stimuli largely by differential growth. For example, if growth occurs to a greater degree on one side of the stem tip, the stem will gradually bend in a direction away from the side of greater growth (Fig. 12–43). Such a growth response is referred to as a **tropism.**

Many types of tropisms are exhibited by plants; they are classified according to the stimulus that elicits the response. For example, geotropism is a response to the earth's gravity. Growing seeds strongly exhibit this tropism; no matter how they are oriented when planted, their roots grow downward and their shoots grow upward. The roots are said to exhibit **positive geotropism,** and the shoots exhibit **negative geotropism** (i.e., with and against the direction of the gravitational force, respectively). Such a tropism has obvious survival value; if each seed that falls on the ground had to be oriented in a very specific manner in order to germinate and grow normally, a large percentage would certainly fail to do so.

Figure 12–43 Phototropism: the corn seedlings on the left are bent toward a side light source.

Other tropisms include **phototropism** (response to light), **hydrotropism** (response to water), **thigmotropism** (response to touch or contact), and **chemotropism** (response to some chemical). For example, a morning glory vine climbs up (geotropism) and entwines about (thigmotropism) a trellis as its leaves and flowers turn toward the sun (phototropism). Meanwhile, its roots burrow downward in the soil (geotropism), inching toward a source of subterranean moisture (hydrotropism).

Without specialized sense receptors, how do plants "perceive" external stimuli? And without a coordinating nervous system, how do they respond to stimuli through the tropisms enumerated above?

Beginning about 1880, the famous Charles Darwin and his son Francis performed a series of experiments that paved the way to an understanding of the mechanisms underlying plant tropisms. Noticing that the shoots of grass seedlings bend toward a source of light, the Darwins covered the tips of some seedlings with black paper caps, covered all but the tips of other seedlings with black paper cylinders, and left a third group completely uncovered. When the bed of seedlings was illuminated from the side (Fig. 12–44), the completely uncovered shoots bent toward the light. So did those with only the tips uncovered. But the shoots with their tips covered by black caps continued to grow straight up. The Darwins concluded that some "influence" somehow controlled the plant's activities. But for more than a generation, the nature of this "influence" remained a mystery.

Another landmark in the study of plant tropisms came in the years 1910–1913, with the experiments of Boysen-Jensen. This Danish biologist worked with oat coleoptiles. (The coleoptile is a hollow, cylindrical sheath that protects the primary leaf of a germinating grass seed as the young shoot pushes its way up through the soil.) Boysen-Jensen found that if an oat coleoptile is decapitated, and then a layer of gelatin is applied to the stump and the cut-off tip is replaced on top of the

Figure 12–44 The Darwin experiment. An uncovered seedling bends toward the light in a differential growth reaction. So does a seedling with all but its tip covered. But light has no effect on a seedling with its tip covered with a black paper cap. Conclusion: Some "influence" in the tip of the seedling controls the plant's activities.

gelatin (Fig. 12–45), the coleoptile will show the same phototropism as an intact coleoptile. If a layer of gelatin is applied to the stump but the tip is not replaced, no bending toward a light source is observed. Similarly, no phototropism is noted if the stump and its cut-off tip are separated by a sheet of impermeable mica instead of the gelatin layer.

Taking his investigations a step farther, Boysen-Jensen discovered that if the tip is cut only halfway across, and a sheet of mica inserted, phototropism will be observed normally if the cut is made on the side on which light falls, but not if the mica is inserted on the side away from the light source. Evidently the Darwins' "influence" is transmitted from the growing tip to the rest of the plant, and it is transmitted along the dark side of the shoot.

An important step was taken in the late 1920s by F. W. Went, who coined the term *"growth substance"* for the controlling influence. He found that if oat coleoptiles are decapitated and the cut tips are placed on blocks of agar for a time, then the agar blocks alone, without the tips, can stimulate growth of the decapitated coleoptiles (Fig. 12–46a). In addition, if the agar block is placed eccentrically on the decapitated coleoptile, differential

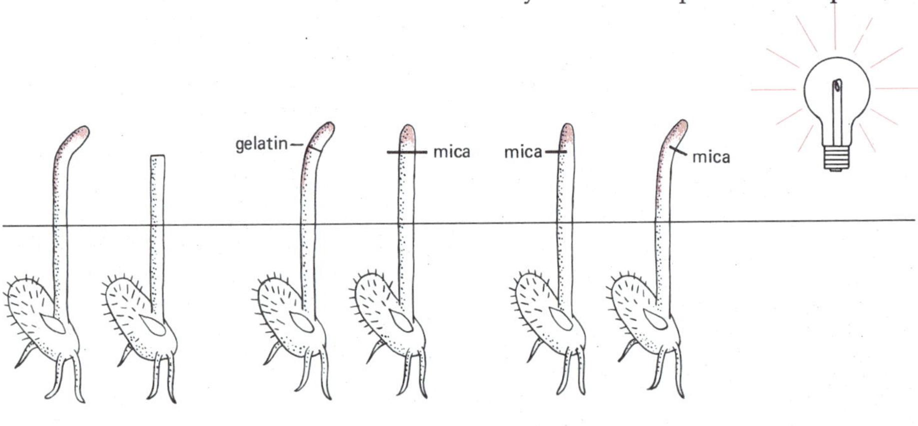

Figure 12–45 The Boysen-Jensen experiments. Light has no effect on a decapitated oat coleoptile. But if a layer of gelatin is applied to the stump and the cut-off tip is replaced, the coleoptile bends toward the light. If a sheet of impermeable mica is placed between the cut tip and the stump, there is no reaction to light. The reaction to cutting the coleoptile halfway across and inserting a sheet of mica in the cut depends on the side on which the cut was made: no reaction to light with mica inserted into the dark side of the shoot, normal reaction with a mica sheet in the lighted side. Conclusion: The growth hormone is transmitted from the growing tip down the dark side of the shoot.

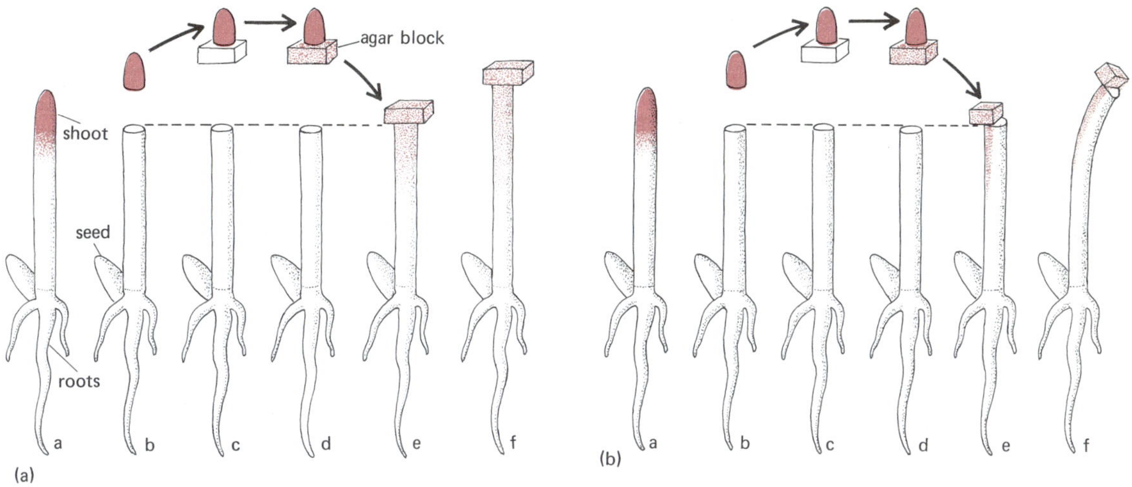

Figure 12–46 The Went experiments (conducted in darkness). (a) If the tip of an oat coleoptile is removed, the stump stops growing. Went placed the cut-off tip on an agar block, then placed the treated block on the stump, which resumed growth. Conclusion: A growth hormone diffused from the tip into the agar block, and then into the stump. (b) When the experiment was repeated, placing the agar block off-center on the cut surface of the stump, the resultant growth was uneven: the stump bent away from the side on which the agar block was placed. Conclusion: The hormone is transported vertically down the shoot but not laterally across it. The cells directly below the block received the hormone and grew faster than the cells on the other side.

growth will result, with the coleoptile bending in the direction away from the agar block (Fig. 12–46b). In fact, the amount of curvature is so precisely related to the amount of the "growth substance" that had diffused into the agar that this reaction is used as the basis of an extremely sensitive biological test for the presence of plant growth hormones. This bioassay is far more sensitive than any chemical test yet devised: a single milligram of a plant growth hormone, an amount approximately comparable with a single grain of sugar, would produce a 10-degree curvature in about 50 *million* coleoptiles.

The growth substance (or substances) found in the coleoptile tips was named **auxin,** and it was soon found in many other sources, including corn germ oil, malt, and—surprisingly—human urine! It was from the urine of humans fed a diet rich in wheat germ that enough of the auxin was finally isolated to determine its nature (in 1931): it was found to be indole-3-acetic acid, structurally related to the amino acid tryptophan.

Since then, other chemically related substances have been synthesized, and many have been found to regulate plant growth. One synthetic auxin, β-naphthaleneacetic acid, is widely used to prevent premature dropping of apples. Another synthetic auxin, 2,4-dichlorophenoxyacetic acid, better known as "2,4-D," is such a potent stimulator of plant growth that it is paradoxically used as a weed killer. Dicots, including many of the most common weeds, are especially susceptible to its action, and 2,4-D stimulates them to metabolize so wildly that they literally consume themselves and die.

Auxins may either stimulate or inhibit growth, depending on their concentration. Low concentrations of the natural auxin, indole-3-acetic acid (IAA), generally stimulate plant cells, whereas high concentrations inhibit growth. The exact levels that stimulate or inhibit depend on the portion of the plant involved. Quite low IAA concentrations, of the order of one part in 10 billion parts of water, maximally stimulate growth of the roots,

Figure 12–47 The rows on the right were treated with 2, 4-D six weeks before the photo was taken. Pre-emergence spraying with this herbicide controls weeds such as crabgrass and foxtail.

but do not affect the stems and buds. At somewhat higher concentrations—one part in 50 million or so—root growth is inhibited, but the buds are stimulated. At still "higher" concentrations, perhaps one part per hundred thousand, both the roots and buds are inhibited, but stem elongation is stimulated. Growth of the stems is not inhibited until the IAA concentrations reach approximately one part in five hundred.

Auxin is produced in the growing tips of the plant: in the tips of the stems and roots and in the terminal buds. It diffuses from these sites of embryonic tissue to the older, differentiated portions of the plant, sometimes working against gravity, as in the case of the roots.

The directed flow of auxins, in conjunction with their differential effect on different parts of the plant, explains plant tropisms. For example, if a plant is placed on its side, the normal downward flow of auxin from the tip of the stem will be uneven (Fig. 12–48), resulting in a pooling of auxin on the underside of the stem. This auxin will stimulate greater growth of the lower side of the stem, outstripping that of the upper side and causing the stem gradually to turn upward. In the roots, on the other hand, the gradual accumulation of auxin on the lower side brings its concentration up to inhibitory levels, and the upper side experiences differentially enhanced growth, resulting in a net turn downward. When a vertical position of the tips is reestablished, the flow of auxin again becomes uniform, and all sides of the stem and root grow at an equal rate, maintaining a vertical orientation.

The differential sensitivity of various parts of the plant to auxin also explains the puzzling phenomenon of apical dominance. As long as the terminal bud of a plant remains intact, the lateral buds do not develop. But if the terminal bud is removed, the plant begins to send out lateral shoots from below and becomes bushier. In this case, the inhibiting influence of auxin diffusing down from the terminal bud is removed, permitting the lateral buds to develop.

Auxins work in conjunction with other plant growth regulators. One class of these is the **gibberellins,** which were first discovered in Japan, in long, spindly rice plants infected with a fungus disease called "foolish seedling disease." The gibberellins are a chemical family including gibberellic acid and its derivatives. In contrast to the auxins, which stimulate a local increase in growth, gibberellins exert a more general growth-stimulating effect, characterized mainly by pronounced elongation of the stems (Fig. 12–49).

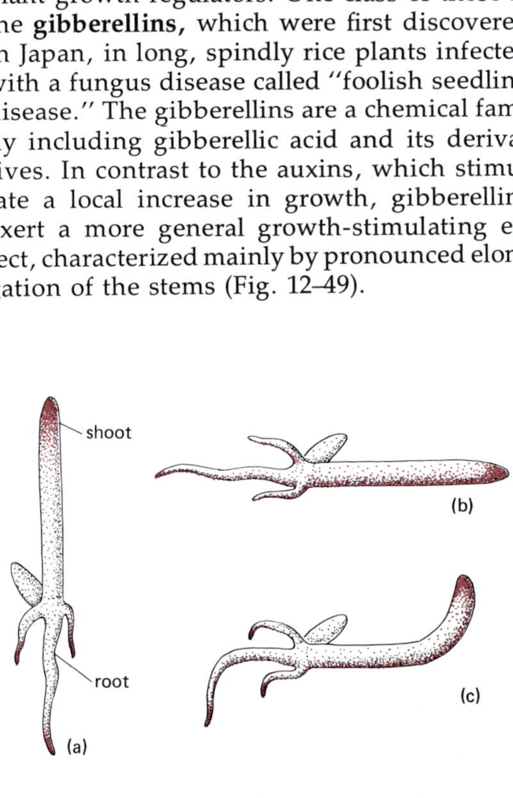

Figure 12–48 Auxin effects account for geotropism. (a) In a normal seedling, shoots grow upward and roots grow downward. Turning a seedling on its side (b) results in a pooling of auxin in the lower side. The increased auxin concentrations stimulate growth in the shoot and inhibit it in the root. As a result, the tip of the shoot turns upward and the tips of the roots turn downward (c).

Figure 12–49 The plant on the right was treated with gibberellin; the plant on the left was untreated.

A third class of plant growth regulators is the **kinins.** The most important of these is kinetin (6-furfurylaminopurine). In contrast to auxins and gibberellins, which promote cell enlargement, the kinins stimulate cell division. Rapidly dividing cells in plant cancers have been found to produce abnormally large amounts of kinins.

The kinins appear not to be transported through the plant, but instead are produced and utilized in localized regions. They are used commercially to retard the yellowing of cut leaves and flowers.

Another plant hormone, the existence of which has been postulated but which has not yet been isolated in pure form, has been named **florigen.** This substance is believed to be involved in the stimulation of flowering (Fig. 12–50).

A curious pattern has been observed in the flowering of plants. Some species begin to bloom as soon as they have reached a particular size and stage of development. But other species bloom only at particular times of the year, no matter how large and well developed they may be at the time (Fig. 12–51). It has been found that for the latter species, flowering is triggered by seasonal fluctuations in the length of the day (or night—evidence gathered thus far seems to indicate that the actual determining factor for blooming is the length of exposure to uninterrupted darkness). Such plants may be either "short-day" or "long-day" plants, depending on the species. Plants whose blooming is indifferent to the length of the day are termed "day-neutral."

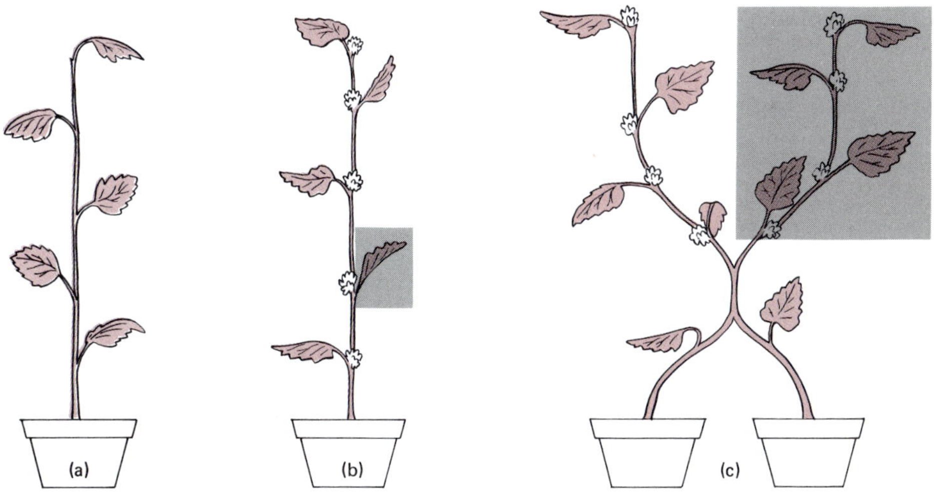

Figure 12–50 An experiment demonstrating that the stimulus for flowering is transmitted by the leaf to the flower buds. (a) The nights were too short to initiate flowering. (b) Only part of a single leaf (shaded area) was exposed to a suitable photoperiod for flowering, while the rest of the plant was on a short-night schedule. The entire plant bloomed. (c) Plant (a) was on a short-night schedule; the shaded portion of plant (b) was exposed to a suitable photoperiod for flowering. Both plants bloomed.

(a) (b) (c)

Figure 12–51 The flowering of some plants depends on the length of the days and nights. (a) Radish, a long-day plant; (b) poinsettia, a short-day plant; (c) tomato, a day-neutral plant.

This phenomenon, known as **photoperiodism,** explains why certain plants are restricted to rather narrow bands of latitude, even though growing conditions outside the band would otherwise seem to be perfectly satisfactory for them. For example, the short-day plant ragweed will not grow in the northernmost parts of the United States (hay fever sufferers take note!) because by the time the days are short enough (or the nights long enough) to initiate flowering, the autumn frosts have killed off all the plants. The long-day plant spinach, on the other hand, cannot be grown in tropical regions, because the days never become long enough (or the nights short enough) to initiate its flowering. Since individual plants, even among the same species, may differ greatly in photoperiodism, it has often been possible to breed strains that can be grown outside their normal latitude distribution.

The flowering of plants has also been found to be associated with a light-sensitive pigment, **phytochrome,** which exists in two forms, absorbing in the red and far red regions of the spectrum, respectively.

QUESTIONS FOR REVIEW AND THOUGHT

1. How do the sense organs bridge the gap between the organism and the environment?
2. Describe the various types of sensory receptors in the skin. Are they all uniformly distributed?
3. What do we mean by "visible light"?
4. The human ear contains two different types of sense organs. Explain.
5. Would you expect the afferent portion of the nervous system of an animal to function if the animal were placed in an isolation chamber completely devoid of external stimuli? Explain.
6. What are the basic components of the nervous system of higher-animal organisms? What advantages does this type of organization have over the coelenterate nerve net?

7. What are the advantages, if any, in having two different systems for control and coordination of body activities, the nervous and endocrine systems?
8. Messages are transmitted through the nervous system electrically and through the endocrine system by chemical means. Is this statement true or false? Explain.
9. Discuss swimming, walking, and flying from the standpoints of speed, energy requirements, and efficiency of adaptation to the requirements of the medium in which the animal lives.
10. How can plants respond to changes in their environment without possessing any specialized sense organs with which to perceive these changes?

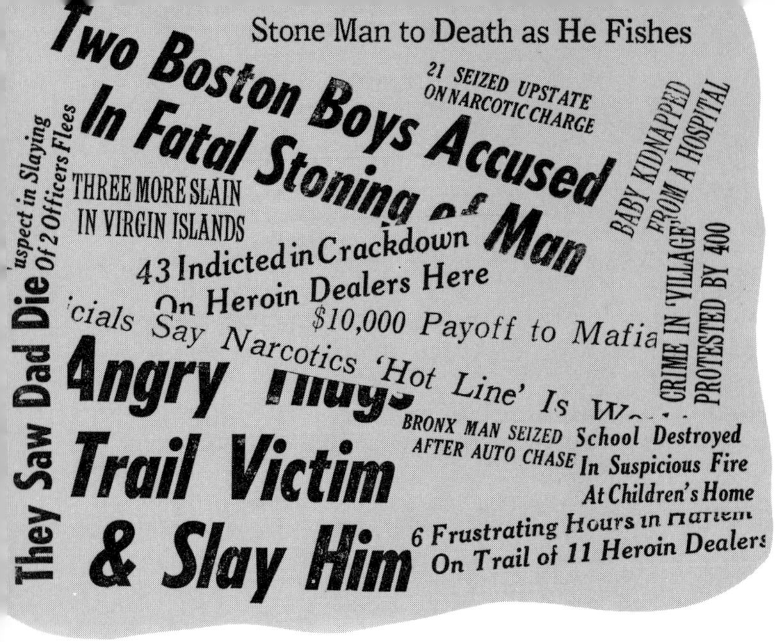
Stone Man to Death as He Fishes
Two Boston Boys Accused In Fatal Stoning of Man
21 SEIZED UPSTATE ON NARCOTIC CHARGE
BABY KIDNAPPED FROM A HOSPITAL
Of 2 Officers Flees
THREE MORE SLAIN IN VIRGIN ISLANDS
43 Indicted in Crackdown On Heroin Dealers Here
$10,000 Payoff to Mafia
CRIME IN 'VILLAGE' PROTESTED BY 400
Narcotics 'Hot Line' Is
They Saw Dad Die
Angry
Trail Victim & Slay Him
BRONX MAN SEIZED AFTER AUTO CHASE
School Destroyed In Suspicious Fire At Children's Home
6 Frustrating Hours in Harlem On Trail of 11 Heroin Dealers

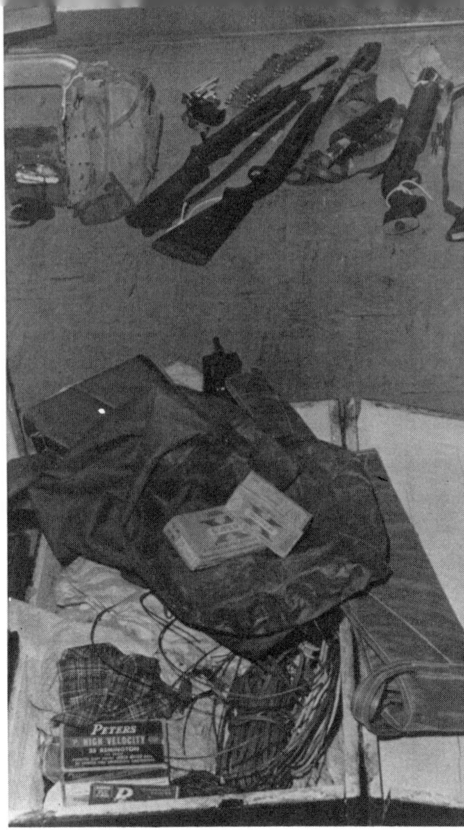
PETERS
HIGH VELOCITY

CHAPTER **13**

Behavior

Introduction

In American cities today, more and more, people are barricading themselves in their homes behind double-locked doors and grilled windows. Only the brave walk the streets at night. Newspapers are splashed with stories of assassinations, murders, plane hijackings. Are we truly becoming a "sick society"? Or has man's behavior been consistent since his appearance on earth? What makes people act the way they do?

The study of behavior has exploded into the public view in recent years. A number of popular books have attempted to examine the nature of man and his society in the light of his animal origins and to explain the patterns of human behavior in terms of prototypes in the behavior of animals. The concepts of territory, courtship rituals and family life, dominance, and aggressiveness have taken on new dimensions.

Observations of animal behavior per se are nothing new—indeed, they undoubtedly date back far beyond the beginnings of recorded history. Naturalists have traditionally interspersed their descriptions of animal species with anecdotes of their observations of animal behavior in the wilds or in captivity. Frequently human motivations were read into these incidents and human attributes ascribed to ants, lions, and monkeys alike. However, **ethology,** the modern science of animal behavior, may be considered to date back to the work of the pioneering ethologists, Konrad Lorenz and Niko Tinbergen, beginning about the 1930s. The importance of their work was recognized by the 1973 Nobel Prize in Physiology and Medicine, which they shared with Karl von Frisch, who studied communication among honey bees.

The **behavior** of an animal may be defined as its outwardly directed activities in response to a stimulus (the origin of which may be either external or internal). Behavior is only part of the overall activity of an animal and is distinguished from the multitude of activities which occur within the organism. The digestive processes, the con-

Figure 13–1 Today's troubled world.

tinual adjustment of the blood vessel bores, and the secretion of hormones by the various endocrine glands are activities vital to the organism, but they cannot normally be detected by an observer without the use of special instruments. The activities classed as behavior, on the other hand, are frequently quite visible even to a casual observer. The sniffing of a dog at an elusive scent on the ground, the immediate rush of a troop of howler monkeys to rescue an infant dropped from a tree, and the sucking response of a newborn human infant as it nestles close to its mother's breast are all examples of behavior.

Behavior may be elicited by a variety of stimuli, both external and internal. Hunger pangs may elicit a generalized restlessness, and then an apparently aimless roaming about, before purposeful activities directed toward securing food are actually begun. A change in the hormonal balance might cause a bullfrog to suddenly seek a mate after he had been ignoring the female frogs around him all spring. A cat may seek a shady spot under a bush and curl up to sleep out the noonday heat. The tiny protist *Euglena* will swim toward a light source of moderate intensity, but it will turn around and swim away if the light is too bright (Fig. 13–2). A trained watchdog barks and lunges menacingly at the approach of a stranger.

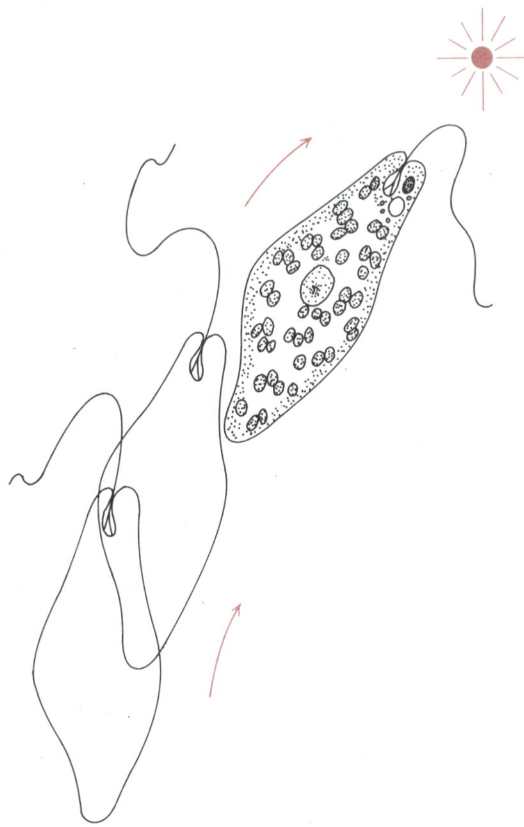

Figure 13–2 *Euglena* will swim toward a light source of moderate intensity.

From the examples cited, it is clear that behavior frequently results in an adaptation to changed environmental conditions. The responses of *Euglena* to light, for example, show definite adaptive advantages and help to promote its survival. Moderate illumination provides energy for photosynthesis in *Euglena's* chloroplasts, but light of too great intensity may cause irreparable damage to the organism.

Early observations and studies of animal behavior had a number of shortcomings. First of all, many of them were conducted on animals in captivity. Recent studies have shown, however, that an animal in captivity does not behave normally. A pampered pet, particularly one that has been raised by humans from infancy, may come to identify with human beings more than with its own kind, and as a result may not interact normally with others of its own species. In zoos and other collections, the frequent crowding and the unnatural surroundings, devoid of the normal sights and sounds of the natural habitat and lacking opportunities for interaction with other species, may give rise to abnormal behavior patterns. Deprived of the daily stimuli of seeking food and avoiding predators, an animal in captivity may become bored and restless. He may spend excessive amounts of time engaging in sexual activities; if he is also deprived of the companionship of others of his species, he may become abnormally preoccupied with his own body, sometimes even habitually engaging in self-destructive acts out of sheer frustration at the lack of normal outlets for his energies. Some interesting parallels have recently been drawn between the effects of crowding and confinement on zoo animals and the frustrations of modern man living in a crowded urban environment.

Studies of animal behavior in a natural environment also frequently suffered from serious shortcomings. First of all, in many situations it is extremely difficult for a naturalist to get close enough to an animal community to make meaningful observations without significantly changing the environment through his very presence. But even when such practical difficulties are overcome, the observations of the naturalist are still filtered through his own perceptions and background of preconceived opinions. In a complex natural environment, an animal is ceaselessly bombarded by a variety of stimuli, both internal and external. The observing naturalist may be aware of only a small fraction of these stimuli. His selection of the relevant data and his interpretation of them become matters of subjective choice, colored by his own personality and training.

Interpretations of the early studies were often beclouded by anthropomorphism, a tendency to interpret the behavior of other creatures, especially animals, from the standpoint of human feelings and attributes. It was thought that a bird sits upon her eggs because she wants to be a mother, and then feeds her voracious babies because she "loves" them. Similarly, it was suggested that insects fly up to a light because they are curious.

In addition, early descriptions of behavior in the living world were frequently anecdotal, consisting mainly of descriptions of single instances of unique and striking behavior, often related by untrained observers. These behavioral anecdotes might or might not be representative of the behavior of the species as a whole. Combined with anthropomorphism, such anecdotes resulted in labeling of species with manlike characteristics—the "sly" weasel, the "crafty" fox, the "king of the beasts," the "greedy" pig.

Toward the end of the nineteenth century, the pendulum swung to the opposite extreme. The pioneers in the emerging science of comparative psychology held that examples of behavior should not be explained by complicated, anthropomorphic mechanisms, imputing emotions and purposiveness to the organisms involved, if the same behavior can be explained by a simpler series of responses. The application of the principle of parsimony to animal behavior was phrased by C. L. Morgan in 1894, and is now known as **Morgan's canon:** "In no case may we interpret an action as the outcome of the exercise of a higher psychical faculty, if it can be interpreted as the outcome of the exercise of one which stands lower in the psychological scale." Like Occam's razor, Morgan's canon can be used to cut away much of the beclouding anthropomorphism in the study of behavior. The sudden closing of the Venus flytrap when a fly alights upon it, but not when its leaves are stirred by a passing breeze, may be explained on the basis of a simple reflex response to the brushing of specialized sensitive filaments by a foreign object. There is no need to ascribe to it any feelings of hunger, or recognition of the fly as potential food. Similarly, the bird may sit upon her eggs because certain sensitive areas on her breast are satisfied by the sensation. It has

Figure 13–3 This herring gull is trying to brood a wooden egg that is twenty times as large as the normal eggs for her species.

been found that brooding birds are just as willing to sit upon pebbles of the appropriate size and shape, even though these can never hatch.

But Morgan's canon, applied indiscriminately, can lead to an equally absurd extreme. As we approach man on the evolutionary scale, we find animals with sense organs and nervous systems more and more approaching those of man himself. With this increasingly manlike apparatus, it is reasonable to assume that such organisms might also exhibit increasingly manlike processes of reasoning and even emotion. Is the dog or cat that trots out to meet its returning master exhibiting merely a simple reflex? Can the behavior of a chimpanzee who carefully fits a short pointed stick into the hollow on the end of a longer stick and then uses the resultant "tool" to reach a piece of fruit outside its cage, beyond reach of its arms, be satisfactorily explained without bringing in some "higher psychical faculties"?

Since the early twentieth century, the emphasis in the study of behavior has been increasingly upon objectivity of observation and interpretation. Attempts have been made to eliminate variables, to conduct experiments under controlled conditions. To some extent controlled laboratory situations are unreal situations, but they can be of great aid in determining the specific aspects of the natural environment to which the observed behavior is a response. An effort is made toward balance of interpretation: an observed activity is not explained in anthropomorphic terms, but neither is it necessarily reduced solely to the basis of a series of simple reflexes.

Many techniques are used to study behavior in lower organisms. Surgical techniques are often used to set up certain internal conditions within an animal in order to study the role of the nervous system, sense organs, and endocrine system in behavioral response. Experimental psychologists compare the learning capacities and modes of response of different species through the use of mazes, puzzle situations of various types, and devices presenting the subject with choices. The naturalist, too, still has his place in the study of behavior. Here, too, improved techniques of sampling and observation have been introduced.

Unlearned Behavior; Instincts

All living organisms exhibit behavior in one form or another. A light flashes on and a cockroach dashes for the nearest dark hole. But a pill bug in the same situation would crawl toward the light.

We have already seen that plants characteristically exhibit similar orientation responses to various stimuli. They grow with their shoots pointing "upward" and their roots pointing "downward"; with their shoots bending toward the light and the roots away from it; and so forth. As we saw in Chapter 12, these highly specific responses of plants, elicited by highly specific stimuli, are called **tropisms.** The orientation is accomplished by a slow process of differential growth, rather than by locomotion or the relatively rapid movements possible in animals.

Animals also may exhibit "tropisms." (The term *tropism* is generally reserved for plant behavior, and the term **taxis** is used for the orientation responses of animals.) Light is a stimulus eliciting fixed responses—taxes—among many species. The organism may characteristically move toward the stimulus, away from it, or may orient itself at some fixed angle to the stimulus. For example, the grayling butterfly *(Eumenis semele),* when menaced by a predator, flies straight up toward the sun. The cockroach, on the other hand, characteristically avoids light. Foraging honeybees and ants find their way back to the colony with the aid of a "light compass": they orient themselves at a fixed angle to the sun.

Taxes, like tropisms, are named according to the stimuli that elicit them: phototaxis, geotaxis, chemotaxis, and so forth. The upright orientation of the fish, belly downward, is a product both of phototaxis (effect

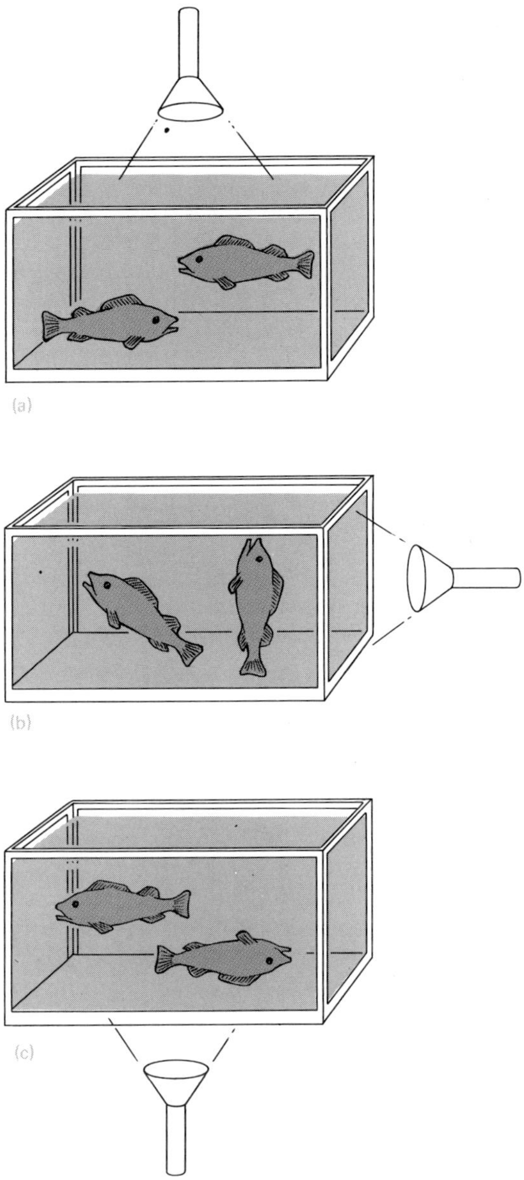

Figure 13–4 The normal fish at the left in the tank maintains a normal upright orientation regardless of the position of the light source. The fish at the right has had the labyrinths of its inner ears destroyed. Unable to perceive gravitational forces, it assumes an orientation appropriate to the position of the light source.

of light) and of geotaxis (effect of gravity). If the labyrinth of the inner ear of a fish is destroyed, it can no longer perceive gravitational forces. Then the fish will orient itself solely on the basis of the direction from which the light source is shining (Fig. 13–4). If the light shines in from the side of the tank, the fish takes a position with its back to the light, even if this necessitates assuming a vertical position. If the light shines in from the bottom, the fish again places its back to the light, this time swimming upside down.

Another type of relatively simple unlearned behavior is **kinesis,** a general increase in activity in response to a stimulus. Lampreys are characteristically more active in the presence of high light intensity, while cockroaches show the greatest activity at low light intensity. Hens exhibit greater egg-laying activity in the light, a kinesis that poultrymen use to advantage: increasing the day length by the use of artificial lights results in an increase in the egg yield.

Another type of unlearned, or **stereotyped behavior** is the **reflex.** Like taxes, reflexes are relatively fixed responses elicited by specific stimuli, which are innate in that they are based upon inherited characteristics of the nervous system. Unlike taxes, reflexes generally involve only a part of the body, rather than the whole body, and they do not necessarily involve an orientation to the stimulus. (In fact, a taxis may involve a number of specific reflexes.) The knee-jerk reflex, elicited by a sharp rap just below the kneecap, is a well-known example of a reflex response. But, especially among higher organisms, even simple reflexes are subject to some modification. For example, it is possible, through anticipation of the rap on the knee and conscious concentration upon the muscles involved, to inhibit the knee-jerk reflex.

Many of the behavior patterns of lower organisms are explainable largely on the basis of combinations and chains of reflexes and other simple stereotyped responses. Even among organisms whose behavior is based essentially upon stereotyped responses,

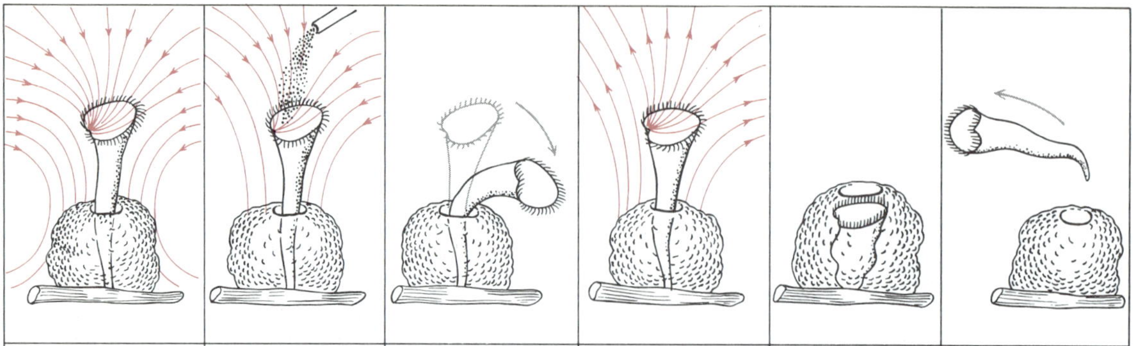

Figure 13–5 The protozoan *Stentor* has a repertoire of several alternative responses to an irritating stimulus.

there may be several alternative responses to a particular stimulus. A ciliated protist, *Stentor,* is a trumpet-shaped single-celled creature that secretes a rigid tube around the lower portion of its stalk (Fig. 13–5). It normally lives anchored within its tube, quietly feeding by drawing in a continuous stream of water by the coordinated movements of its cilia. If a stream of carmine particles intrudes upon its peaceful existence, *Stentor* exhibits a varied repertoire of responses. First it bends over from one side to another. If the irritating particles are still impinging, it reverses the movements of its cilia, thereby reversing the feeding current and directing water *out* of its cone rather than inward. If the stimulus continues, *Stentor* withdraws into its tube for half a minute or so. If the stream of carmine particles continues, we find that even at the protist level on the evolutionary scale, there can be some modification of behavior in the light of experience: if *Stentor* tentatively ventures forth from its tube and finds the stimulus still there, it does not try bending or reversing its feeding current; it immediately withdraws into its sheltering tube again. If the stimulus still persists after a number of trial ventures, the *Stentor* detaches itself, abandons its tube, and swims away, ultimately settling down and building a new tube in a new location free of unfavorable stimuli.

INSTINCTS

The most complicated types of unlearned behavior are the **instincts.** The term *instinct* has been badly misused. The layman frequently speaks of "instincts" as connoting some mysterious "wisdom of nature," or to refer to automatic responses, whether learned or unlearned ("I did it by instinct"). Even biologists have often misused the term to refer to any pattern of behavior the underlying mechanism of which is unknown. For example, before anything was learned about the natural "sonar" system of the bat, it was frequently stated that bats fly with such uncanny accuracy "by instinct." As a result of this misuse and overuse, the term *instinct* is still in rather questionable repute among many comparative psychologists. But biologists frequently find the term quite useful and meaningful, provided its definition and limitations are clearly understood. The term *instinct* is now used to refer to unlearned behavior (either performed without previous experience or unmodified by previous experience) that is characteristic of the species (rather than the individual) and is

Figure 13–6 Male mosquitoes cluster around a tuning fork emitting the frequency of the female mosquito's characteristic whine.

adaptive (but is not consciously directed by the individual exhibiting the behavior).

Instincts include many highly complicated behavior patterns. The spider spins a web exactly characteristic of its species, even though it may never have encountered any others of its own kind. A male mosquito, with no sexual experience, immediately recognizes the characteristic whine of the female of his species, and will even fly directly to a tuning fork that is emitting the appropriate frequency. The caterpillar spins a perfect cocoon in the first and only attempt of its life. A young bird, which has never taken part in a migration, navigates perfectly to the new nesting grounds of its flock, perhaps half a continent away, even if it is held back from migrating with the experienced birds.

It is frequently extremely difficult to determine whether a particular behavior pattern is truly innate and unlearned, or whether it is learned or modified by previous experience. For example, it was long thought that cats chase and kill mice instinctively. But more intensive studies revealed that cats generally do not kill mice unless they have learned to do so as kittens by watching and imitating adult cats. Away from such influences, kittens have been raised to live quite peaceably with mice, birds, and other species of their "natural prey."

Generally, as in most questions in biology, the "instinct vs. learning" controversy is not really a true either-or question, but rather a mixture of some of each—the innate and the acquired—in varying proportions, depending upon the situation. For example, birds instinctively sing, with a song characteristic of their species. A cuckoo or cowbird, reared by foster parents of a different species, does not copy their melodious song, but instead develops the harsh and limited repertoire of its true parents, which it has never seen or heard. A chaffinch, if raised in isolation, will sing a recognizable chaffinch song, although a much simpler one than that sung by a chaffinch raised with adults of its own kind.

The question of the relative roles of innate behavior and learning is further complicated by the fact that many types of stereotyped behavior require a definite level of maturation of the individual before they can be manifested. Before the appropriate neural circuits are connected, the stimulus will evoke no response; after they have been established, the response will be exhibited, regardless of the organism's experiences in the meantime. Even such a relatively simple reflex response as the blinking of the eye at the approach of a foreign body is not exhibited by the newborn human baby. At a later stage of development, the blinking reflex is observed. One might reason that this is an acquired response, learned through unpleasant experiences with foreign bodies. But blinking has been found to be a true reflex, which is established when the appropriate neural circuitry is ready. Various modes of locomotion are also examples of innate behavior patterns (modifiable to some degree) that are not established until a definite stage of maturation has been reached. Baby rats, for example, are unable to walk until some days after birth. Before they have reached the walking stage, they cannot be taught to walk. Baby guinea pigs, on the other hand, which have spent a much longer gestation period within the uterus of the mother and emerge much more fully developed in many respects (with open eyes, a full coat of fur, etc.), are up and about within minutes after birth. Similarly, the swimming behavior of young salamanders has been found to be innate, exhibited after a definite stage of maturation has been reached, rather than a learned activity.

Studies of instinctive behavior have revealed that definite symbols appear to act as **"releasers"** of appropriate behavior patterns. These releasers are frequently not the entire stimulus, but only a specific part of it, a symbol or **"sign stimulus,"** often in the form of a specific color, shape, sound, taste, or odor. Among organisms with good color vision, the color red in an appropriate situation frequently serves as a releaser of instinctive behavior. For example, the male stickleback, studied by Niko Tinbergen, a fish with an intricate and interesting pattern of courtship and mating behavior, vigilantly defends his territory against other males of his species. The male stickleback has a red abdomen, and the defending male viciously attacks all intruding males. The pattern of territorial defense seems to be released by the red belly: a male stickleback will just as vigorously defend his territory against a model of a fish of a different species with a red-painted abdomen, or even a stylized oval with a red underside, but may ignore a perfect model of a stickleback without a red spot (Fig. 13–7). Similary, a male English robin will defend his territory not only against other red-breasted males of his species, but also against an intruding bunch of reddish feathers. Yet he will rarely attack a complete stuffed male robin whose breast has been painted brown or gray.

The color red is also an important sign stimulus for baby seagulls. The adult gull has a red spot on its bill, and the gull chicks peck at this spot when "begging" for food. The adult then opens its bill and offers food to the chicks. But if the red spot is painted out or painted on a different part of the adult bird's head, the hungry chicks do not peck at the bill, but merely look puzzled. Incubator-reared baby seagulls that have never seen an adult will peck at a red disc for a day or so after hatching. But if the disc never responds by providing them with food, the behavior pattern gradually dies out.

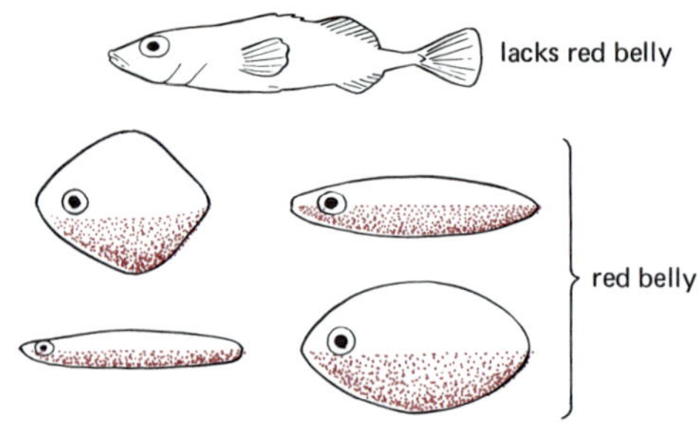

Figure 13–7 The male stickleback will vigorously defend his territory against crude models with a red-painted underside but will ignore a perfect copy of a male stickleback lacking the red belly.

The effect of characteristic shapes as releasers of instinctive behavior is interestingly demonstrated by an experiment in which a wooden model with the shape depicted in Fig. 13–8 is drawn along above the heads of young ducks, geese, or game fowl. If the model travels in the direction corresponding to the left in the figure, the young fowl pay little attention. If the same model is drawn in the opposite direction, they scurry away in fright. Evidently the model traveling with the long "neck" in front resembles a goose or waterfowl in flight; in the opposite orientation it looks like the silhouette of a short-necked, long-tailed bird of prey, such as a hawk. Further studies have suggested that it is the relative unfamiliarity of the latter silhouette, rather than any specific association with a predator, that triggers the fear response.

Odors and tastes also play important roles as releasers of innate behavior patterns in the animal world. The male dog or cat becomes sexually excited when it scents odors (imperceptible to man) given off by a female of the species in the estrous state.

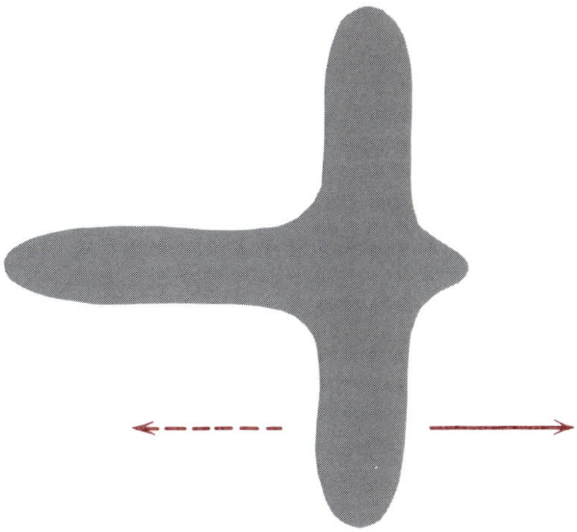

Figure 13–8 The characteristic shape of a bird flying overhead elicits appropriate responses from young ducks and geese, as demonstrated with the wooden model pictured. When the model was moved to the left, simulating the flight of a waterfowl, there was no response; but movement toward the right, simulating a long-tailed bird of prey, produced fear and escape reactions among the young fowl.

Such secretions, called **pheromones,** play an important role in the mating activities of such varied animals as lions and mice, cockroaches and gypsy moths. (A number of insect sex attractants have been isolated and used as lures in pest control devices.) Pheromones also serve other important functions in the animal world. The free-swimming larvae of barnacles are believed to be attracted to settle down by chemicals released into the water by sessile adult barnacles. Rabbits mark off their territories by rubbing secretions of special chin glands on objects along the boundary. Bees, ants, and other social insects communicate partly through a complex of biochemical secretions that are scented and tasted.

Instincts are commonly referred to as inherited behavior patterns. But it should be clearly understood that behavior as such cannot be inherited. Instead, one inherits certain potentialities, which may or may not be expressed, depending upon the interaction of both genetic and environmental factors. Behavior patterns are the end result of a long developmental process determining the structure and physiology of the organism. Absence or deviation of a key gene in the intricate complex of genes involved in this developmental process might block the ultimate expression of a behavior pattern: a caterpillar whose genes controlling the development of silk-secreting glands were impaired could not spin a cocoon. The environment, too, must provide opportunities for full development of the potential if a behavior pattern is to be expressed. For example, a female rat in the last stages of pregnancy typically builds a crude nest out of rags, paper, wood shavings, or similar materials. This behavior is not learned: a young female rat that has never seen any other rats building nests will construct a characteristic nest for her first litter. But if the rat has been raised in an empty cage and has never been given any opportunities for practice in carrying things, she will not build a nest prior to the birth of her litter.

Learning

If we were to rate the members of the animal kingdom on their ability to learn from experience, the dunce cap would surely go to the protists and sponges. For despite repeated attempts, no one has yet successfully demonstrated an instance of "learning" in these organisms. However, among the metazoans, a spectrum of gradations of learning capacity has repeatedly been demonstrated in the laboratory.

Figure 13–9 This Norway rat made a nest of sawdust for her litter in the corner of a storeroom.

Learning may be defined as an enduring (possibly even permanent) change in behavior in the light of past experience. The process involves a change in the functioning (and in some instances in the structure) of the nervous system. For example, if a bird's wing becomes crippled and it can no longer fly, this is not an example of learning, even though there has been a permanent change in behavior resulting from experience. Transitory changes in behavior are also ruled out: a hungry animal behaves quite differently from a satiated one, but as soon as its hunger is appeased it goes back to the typical behavior pattern of the satiated animal.

Biologists have some difficulty in agreeing on the definition of learning, especially regarding the application of this definition to concrete cases. Is the behavior of *Stentor* described earlier in the chapter a case of learning? Can a planarian learn? The latter question is of more than academic interest. Some of the landmark studies on the mechanism of learning have been conducted on flatworms.

The available evidence indicates that the ability to learn increases with increasing complexity of structure and functioning of the organism, particularly its nervous system. Flatworms and annelid worms (such as the earthworm) can be taught to solve problems involving simple alternative choices, such as avoiding a rough spot in the culture dish or solving a Y- or T-shaped maze (Fig. 13–10). These organisms have rudimentary nervous systems, but little central control. In fact, if an earthworm is taught to solve a T-maze, and then its head (containing its brain) is cut off, after a temporary disturbance the headless worm shows little loss of its ability to solve the maze.

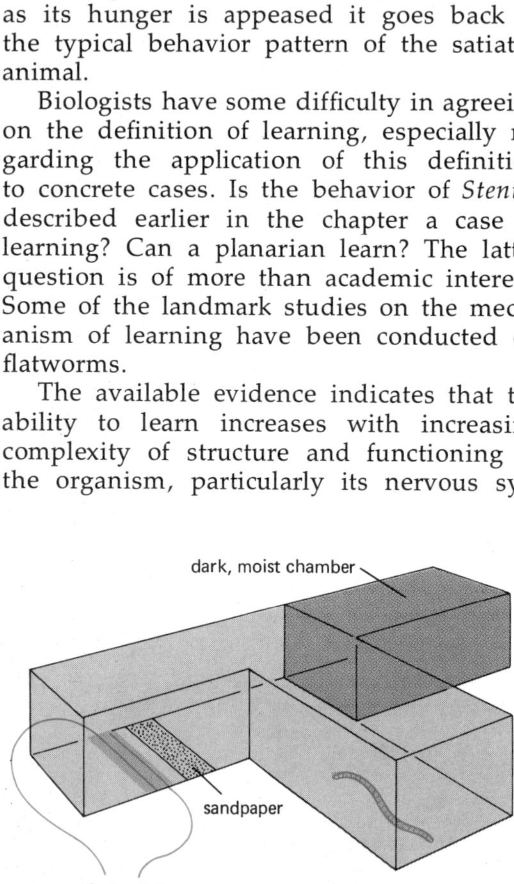

Figure 13–10 In the simple T-maze pictured, the dark, moist chamber represents a reward for the earthworm, while the sandpaper represents punishment. Eventually the worm will learn to choose the right turn at the fork nearly all the time.

In higher animals, centralized control becomes more and more important, with the brain playing an increasing role in learning capacity, as well as in the control of behavior in general. Among the mollusks, snails have shown some ability to learn T-mazes; but not all snails can be successfully trained, and even those that do learn the maze are not always consistent in their performance.

Octopuses, on the other hand, have excellent vision and a rather well-developed brain. An interesting series of experiments has established that the octopus can readily learn to make distinctions between stimuli and to modify its behavior on the basis of past experience.

An octopus was placed in an aquarium containing a small pile of bricks at one end. The octopus promptly retired behind the bricks. When a crab was dangled in the aquarium at the end of a string, the octopus came out, captured the crab, and went back to its bricks. Then a crab was presented attached to an electrified white plate. When the octopus tried to seize the crab it received an electric shock. Turning pale (contraction of its pigment cells is under nervous control) and blowing on its shocked tentacles, the octopus retired to its lair. After about twelve trials, in half of which the crab was presented alone and in half with the electrified plate, the octopus began to show reluctance to approach a crab accompanied by the plate. By twenty-four trials it consistently made the correct choice. Later it was found that octopuses could be

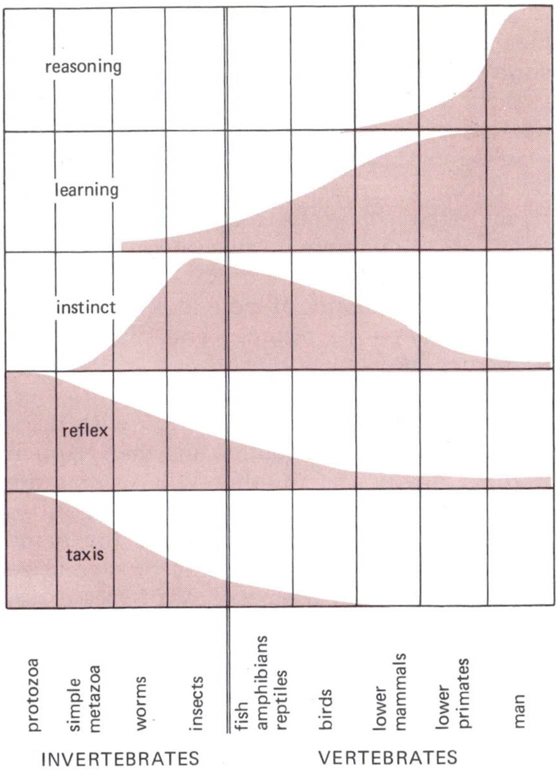

Figure 13–11 The components of behavior vary in proportions among the different groups of animals. Instinct reaches a peak among the most advanced invertebrates (insects), while learning and reasoning play an increasing role among the vertebrates.

trained to even finer distinctions: to come out to feed when the crab was attached to a small plate and to remain in the nest when it was attached to a larger (electrified) plate; to discriminate between different-shaped plates; and so forth.

Learning has also been demonstrated among the insects. Bees can be trained first to settle upon a square of blue paper upon which a dish of sugar water is placed, and then to settle upon the blue paper alone. They can discriminate between the blue paper and other squares of different colors, even a gray of equal intensity, and no matter what the relative positions of the squares. Ants and cockroaches can be taught to solve simple mazes. But by and large, behavior in the invertebrates is based chiefly upon stereotyped responses, and learning plays a relatively minor role.

In the vertebrates, on the other hand, particularly among birds and mammals, we find evidence of highly developed learning capacity. Studies of the learning capacities of fish, reptiles, and amphibians indicate that they can master mazes and detour problems generally better than invertebrates.

Birds learn even more rapidly, and can master problems of surpising complexity. The canary, for example, can be trained to grasp an abstract concept: to choose consistently the odd member of a group of objects (Fig. 13–12). For example, a canary may be trained to look under the one screw in a row of aspirin tablets, no matter what the position of the screw with respect to the row. Then the canary can transfer the concept of oddness to new situations. For instance, it will successfully pick up its food pellet from under one aspirin tablet placed in a series of half a dozen screws, or under a single chess man in a series of screws.

Figure 13–12 A canary can be trained to pick out the odd item in an assortment.

Among the mammals, considerable learning capacity is exhibited by rats and mice, even more by horses, dogs, and cats. Studies of dolphins are revealing a surprisingly high degree of "intelligence." But by far the greatest and most varied potential for learning is shown by the primates: monkeys, apes, and particularly man. The degree of learning ability is closely correlated with the neural and motor capacities of the organism. A dog could not even attempt to master a problem situation calling for fitting a short pointed stick into the hollow end of a longer one to reach a piece of food outside its cage; it does not have the requisite motor equipment. Its paws are not prehensile, and its jaws are adapted for holding and carrying objects, but not for any delicate manipulations. A chimpanzee, on the other hand, solves such problems with ease. The ability to learn and to solve problems of increasing complexity seems also to be correlated with brain development, particularly the cerebral cortex.

THE MOLECULAR BASIS OF MEMORY AND LEARNING

The ability to learn—to modify actions in the light of past experiences—rests upon another important ability: the ability to remember these past experiences and the lessons learned from them.

The concept of memory is an intriguing one for scientist and layman alike. Basic to the learning process, it is usually correlated to some extent with "intelligence" and reasoning ability. The capacity of the human memory is fantastic: a person can train himself to recall flawlessly hundreds of stanzas of poetry or have at his fingertips thousands of chemical formulas and the meaningful relationships among them. The ability to acquire new memories seems to deteriorate somewhat with age, although it is a frequent observation that a very old person, on the verge of senility, may have very vivid and accurate recollections of his early childhood, even though he is unable to recall what he ate for breakfast that morning.

There seem to be two basic types of memory: a temporary storage of data, or **short-term memory,** and a relatively permanent recording of information that may persist for many years—**long-term memory.** One's own name, address, and important experiences are indelibly fixed in the mind. Even when conscious recollections of them fade from the immediate awareness, memories may be triggered by a suitable association. On the other hand, if you look up an unfamiliar telephone number and place a call, it is very likely that by the end of the conversation you will have forgotten the number. If you have no further association with it for some time, perhaps months, and then happen upon it again, the number will seem quite unfamiliar. Similarly, if one is introduced to a stranger, one frequently forgets the name unless a special effort is made to memorize it or unless some dramatic event creates an association that indelibly fixes it in mind. "Cramming" for an exam similarly leaves transitory patterns of thought that soon fade away with barely a trace (short-term memory). But a regular program of study pursued over a long period of time builds up long-term memories of the subject that can be drawn upon for a lifetime.

The functioning of memory, in humans, seems to be associated with the cerebral cortex. Some interesting and revealing results have been obtained in operations on epileptics, when various portions of the cerebral cortex have been stimulated electrically. Patients have reported vivid recollections of past experiences: a voice of a friend not seen for many years, the young face of a mother long since dead. Patients seem to relive parts of their past while still retaining an awareness of the present.

Numerous theories have been offered to explain the nature of memory. The recent emphasis has been upon attempts to explain memory on a molecular basis, and such studies are yielding some promising leads.

The present ferment of research into the molecular basis of memory and learning was initially sparked by experiments with the lowly flatworm. Planarians were trained to react to an external stimulus. They were then cut in two. Since flatworms are capable of complete regeneration, the tail section grew a new head and the head grew a new tail. Despite the fact that the "brain" of the planarian is in the head region, both newly formed animals retained much of their previous learning experience. However, if both severed sections were permitted to regenerate in water containing RNase (an enzyme that deactivates RNA), only the portion that contained the original head retained the "learning" experiences and showed behavior similar to that of the original uncut planarian.

More startling still, when trained flatworms were ground up and fed to untrained ones, some transfer of "learning" seemed to have taken place. The "cannibals" that had eaten trained planarians were able to be conditioned in a much shorter time than planarians that had been fed untrained animals. Thus, the experiments seemed to indicate a transfer of some sort of chemical memory factors, and the brain RNA seemed to be implicated.

Initially researchers had difficulty in repeating these landmark experiments, and some challenged the validity of conducting learning experiments with planarians at all. But then experiments with rats and mice showed similar results: there seemed to be some sort of transfer of learning experiences when RNA extracts from the brains of trained rats were injected into untrained animals. These experiments sparked a furor of controversy. It seemed illogical that specific "memory molecules" would be able to pass through the various barriers of the body and exert an influence on the brain of the recipient. Moreover, there was again an initial difficulty in repeating the experiments. Eventually, however, a number of other researchers reported corroborating findings. Indeed, cross-species memory transfers were accomplished: RNA extracts from the brains of hamsters trained to respond to a light flash by going to their feeding box conferred a similar response on rat recipients (Fig. 13–13).

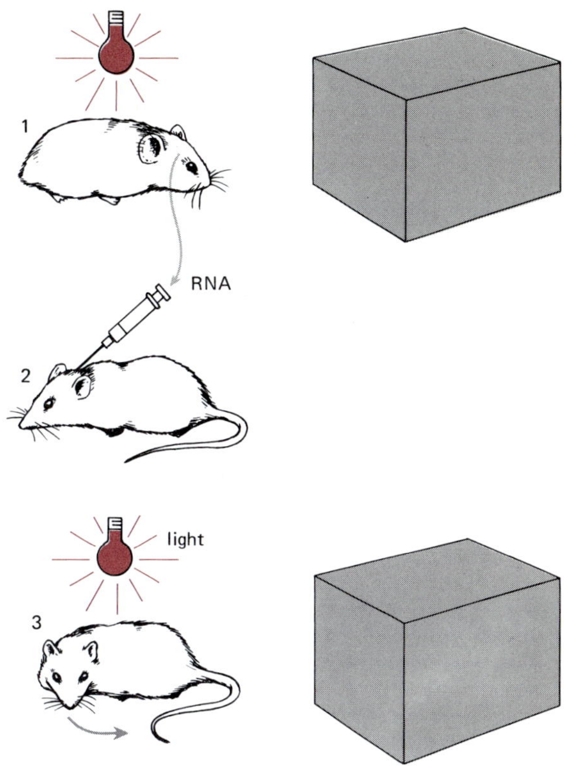

Figure 13–13 Memory transfer by injection. The hamster (1) was trained to enter its feeding box when a light flashed. Then an RNA extract from its brain was injected into an untrained rat (2). The rat then responded to a flashing light by turning toward its feeding box (3).

The question of whether "memory molecules" could penetrate the various barriers of the body has been resolved by radioactive tracer studies revealing that chemicals from portions of organs removed from animals and injected into recipients do penetrate rather rapidly to precisely the same organs of the recipients. And in other studies, injections of *liver* extracts from trained rats were found to produce some apparent transfer of learning experience! It is conceivable that specific memory molecules, while concentrated in particular parts of the brain, might also find their way to other parts of the body.

With the flood of experiments indicating the possibility of chemical memory transfer, research into the nature of chemical memory storage took on a new pitch of intensity. It was found that the neurons in the brains of trained rats contain as much as 35 percent more RNA than those of untrained rats. The protein content of the neurons of rats' brains also rises after intensive mental activity. Injections of yeast RNA into aged humans suffering from memory impairment produced a marked improvement in their ability to recall recent events. (Side effects preclude the routine administration of RNA as a treatment for senility, but intriguing and promising results have been obtained in tests of drugs, such as Cylert, that stimulate the body's own RNA production.)

RNA's key role in memory seems to lie in its role in protein synthesis. In one series of experiments, goldfish taught to jump over a hurdle in the middle of their tank to avoid an electric shock were unable to retain the memory of this training if they were given an injection of a drug that inhibits protein synthesis. Recently, researchers have succeeded in isolating and synthesizing an actual "memory chemical" that keys a specific memory: fear of the dark. This chemical, called "scotophobin," proved to be a polypeptide.

The exact mechanisms of memory storage in the brain are still controversial. But putting together the various pieces of evidence, modern researchers have worked out a reasonably coherent theory:

A new experience, after analysis and correlation in various parts of the brain, produces a complex pattern of pathways among millions of brain neurons, linked together by repetition or association with related memories or ideas, the neurons along the same sequences are stimulated to fire again and again, and a series of changes is set off. RNA is produced, and it in turn forms specific proteins. The latter may work to strengthen the connections of certain synapses along the pathway, or they may even stimulate the growth of new branches of the nerve cells, forming new synapses. Gradually a particular pattern of firing neurons is fixed; this pattern is called a **memory trace** or **engram** (Fig. 13–14). Each of the neurons in a particular engram may be interconnected with numerous other memory traces. And from these interconnections the brain builds up its intricate "filing system," full of "cross-references" that help in retrieving particular memories.

The growing understanding of the molecular basis of memory and learning has encouraged brain researchers in their quest for ways of correcting abnormalities of mental development and even of improving the mental capacity of normal people. The search for chemical "mental stimulators" is being avidly pursued. The potential impact of such studies on our society is enormous. Most of the major discoveries throughout history have been made by a surprisingly small segment of the population. If this base of "genius" could be significantly broadened, the fruits of the resultant quantum jump in human activity would be beyond the power of our imagination.

HABITUATION

One of the simplest types of learning observed in living organisms is **habituation.** When an organism is repeatedly exposed to a stimulus without any significant results associated with it, either positive or negative, its natural response tends to die out. A cow brought out to pasture in a field near a busy highway will at first turn her head to look at each car whizzing by. But as cars, trucks, and buses continue to pass the field, the cow gradually becomes accustomed to the periodic noises and ceases to turn her head. With each passing day of exposure to the stimulus, the cow pays less and less attention to the stimulus, and is less likely to show any response. Similarly, a human infant gradually becomes habituated to the usual noise and bustle of the household, and may sleep peacefully even if the radio is playing, his parents are talking, or his older brothers or sisters are squabbling.

If, however, a continually repeated stimulus is associated with some consequence of

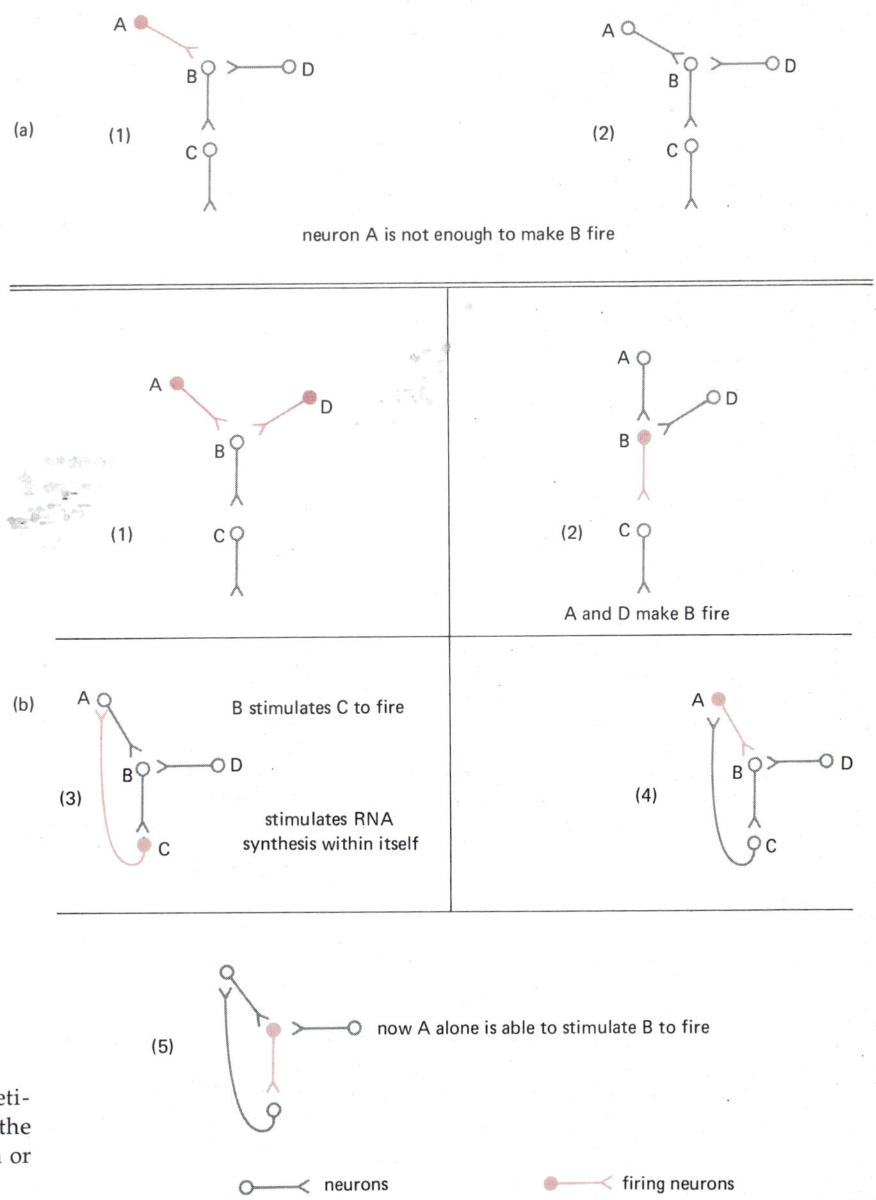

Figure 13–14 A theoretical mechanism for the formation of an engram or memory trace.

significance for the organism—either positive, like the provision of food, or negative, like an electric shock—then the organism shows an appropriate response. With each repetition of the stimulus and of the response, resulting either in the avoiding of an unpleasant situation or the promoting of a positive effect, the association between stimulus and response becomes stronger. It seems that the more often a neural circuit, stimulus → response, is used, the less resistance this circuit offers to subsequent impulses. This phenomenon is known as **facilitation,** and plays an important role in the learning process. Conversely, failure to use a particular neural pathway for long periods of time ultimately results in extinction of the response, and plays a key role in the processes of habituation and forgetting.

CONDITIONING

These are the principles underlying the classic experiments of the Russian psychologist Ivan P. Pavlov, discoverer of the **conditioned reflex.** Pavlov's landmark studies were performed on dogs. A dog begins to salivate at the sight of meat. Pavlov discovered that if another stimulus, such as the sound of a buzzer, was repeatedly offered at the same time as the meat, and then the second stimulus was presented alone, the dog would salivate at the sound of the buzzer just as though it had been offered a piece of meat. The primary stimulus (in this case, the piece of meat) is called the **unconditioned stimulus,** and the secondary stimulus, to which the subject learns to respond, is termed the **conditioned stimulus.** If the conditioned stimulus is repeated many times without simultaneous reinforcement with the unconditioned stimulus, the salivation response gradually fades out, a process termed **extinction.** After a long rest away from the test situation, however, the conditioned response often reappears spontaneously. It seems that at least some aspects of the learning are permanent. In fact, pigeons trained in a simple response and then removed from the training situation before extinction has an opportunity to occur have been found to exhibit the conditioned response as long as ten years after the original training.

Pavlov's experiments revealed a number of other aspects of conditioning. For example, the dog at first tended to generalize in its responses to conditioned stimuli. If the original conditioned stimulus was a bell, then the dog would salivate in response to bells of various tones. But by reinforcing only a particular tone and allowing all other tones to become extinguished, the experimenter can train the dog to **discriminate** among the different tones, responding only to the one selected. (This technique is valuable in investigations of animal behavior not only in studies of the learning process, but also in determining innate capacities of animals to distinguish among different sounds, colors, etc.)

Pavlov's experiments dealt with reflex behavior. It was therefore important to determine whether the general principles that he discovered are also applicable to voluntary behavior. This was accomplished by the American scientist B. F. Skinner, who devised the "Skinner box" for experiments with rats, pigeons, and other laboratory animals. In the special cage devised by Skinner, a lever is provided; if the animal presses the lever, a small food pellet is released into an accessible trough (Fig. 13–15). A hungry rat placed in such a cage soon, by random activity, presses the lever.

Figure 13–15 A trained albino rat presses a lever to obtain food.

When the rat discovers that a bit of food has been provided, it quickly learns to press the lever for more. If pressing the lever results in a reward of food only when the cage is lighted, the rat soon stops pressing the lever when the light is off. It thus learns to discriminate between the conditions of light and darkness.

HABITS

When a stimulus is repeated over and over and the same response is given, gradually a **habit** is formed. Habits are learned responses, which, through repetition, we come to make consistently and automatically to a given situation (stimulus). A key factor in the formation of a habit is the repetition of the association of the specific stimulus and the response. This process is akin to the establishment of a conditioned reflex; indeed, habits are often conditioned reflexes, or series of conditioned reflexes, in which the response to one stimulus in turn acts as the stimulus for the following conditioned responses in the sequence.

The normal activities of our day are studded with a surprising number and variety of habits—actions that we perform literally without a thought. The alarm clock rings, and we respond with a characteristic routine. The sequence of washing, toothbrushing, and dressing is performed without any conscious attention. In fact these actions may be performed to the accompaniment of the news on the radio or deep thoughts about the coming events of the day. Through breakfast, on the way to work or school, and so on through the day a person performs many more "routine" actions, which we refer to as habits.

Habits are not innate behavior. They are persistent, but not permanent, and can be modified or superseded by other habits. If habits are not repeated relatively often, they tend to die out, just as knowledge tends to be forgotten if it is not used or referred to for long periods of time. The **forgetting process** is promoted by the acquisition of new or conflicting knowledge or habits.

Esaias Baitel, Omikron

Figure 13–16 A habit with some element of reward is difficult to break.

Habits may be highly desirable. They permit the performance of a variety of routine tasks with a minimum expenditure of time and effort. When one tries to teach a young child how to tie a shoelace or to hold a pencil in the appropriate position for writing, one realizes with surprise how automatic these procedures have become after long repeated experience, and how cumbersome each operation is if conscious attention must be given to each phase of it.

People also generally accumulate a collection of undesirable habits. These frequently involve some element of reward, and therefore tend to be self-reinforcing and rather difficult to break. Often the most effective way of breaking a habit is to substitute another habit for it—this is the technique used by the person who substitutes gum chewing for cigarette smoking.

PROBLEM SOLVING

The experimenter carefully places a guinea pig in a maze and sits back, notebook in hand, ready to record the number of correct and incorrect responses. The guinea pig retreats into the nearest corner and cowers there, but nothing further happens. The experimenter has overlooked an important factor. A strong stimulus or drive is needed in this situation to overcome the guinea pig's natural timidity. One such drive is hunger. If the guinea pig were starved before the experiment, it would be more inclined to explore the maze in search of

Figure 13–17 A monkey will work a puzzle for its own sake, without the motivation of any additional reward.

food. Indeed, hunger is a common source of **motivation** utilized by the experimental psychologist. Other sources of motivation are thirst, sex drive, fear, and curiosity.

Many drives appear to originate in the hypothalamus. They need not be consciously felt. They arise from a disturbance of homeostasis and result in patterns of behavior that continue until the internal equilibrium is restored. A drive or need may be satisfied by instinctive behavior; or the behavior patterns may be partly or wholly learned.

There is frequently a direct relationship between the rapidity of learning and the strength of motivation. For instance, the hungrier a rat is, (up to a point), the more rapidly it can learn to run a maze. Of course, no matter how strong the motivation, an organism cannot be impelled to learn an activity that is beyond its innate capacities. A rat, for instance, no matter how hungry, cannot be taught to speak: it does not have the appropriate vocal and neural apparatus.

Skinner discovered another interesting facet of motivation: stronger motivation appears to be developed in a problem situation when rewards are irregular than when a given response is always rewarded. If he adjusted his box so that a food pellet was delivered every time the lever was pressed, the rats would stop working as soon as Skinner discontinued the food. But if the food pellets were delivered only part of the time, at irregular intervals, the rats went on working much longer when the supply was cut off. Apparently they had greater difficulty in distinguishing whether the cutoff was permanent or not. Skinner suggests that a similar mechanism is responsible for the popularity of slot machines and other forms of gambling among humans.

Among some species, particularly cats, monkeys and apes, and man, curiosity seems to be a strong source of motivation: actions are frequently performed not from any desire for reward or fear of punishment, but apparently for their own sake. A chimpanzee provided with a hook and eye or other simple lock mechanism will work the puzzle of its own volition. In a Butler box (Fig. 13–18), a cage with a small window that can be opened by the animal inside but closes automatically in a few seconds, a monkey will spend hours opening the window and

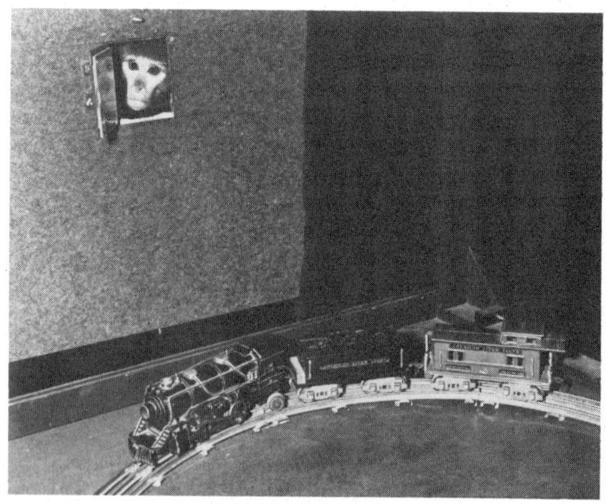

Figure 13–18 A monkey in a Butler box looks out at an electric train.

peering out. It is found that monkeys will open the window more frequently the more "interesting" the sights are outside: a toy electric train chugging around its track will elicit many more window openings than a pile of blocks. Apparently the most interesting and motivating sight for a monkey, judging by the number and frequency of window openings in a Butler box, is—another monkey.

Various techniques are used by experimental psychologists to study the problem-solving behavior of animals. We have already described the Skinner "serve-yourself" box. Mazes are another common type of test, utilizing the normally greater activity of a hungry animal to determine its ability to memorize a series of appropriate responses. (In studies of ant behavior, motivation is frequently achieved by placing the maze between the ant and its anthill.) Depending upon the innate capabilities of the animal being tested, the maze may range from a simple Y- or T-shape to the intricacies of the Thorndike puzzle box depicted in Fig. 13–19.

Another standard type of test situation is the detour problem, in which the animal is tied in such a way that it must first detour back away from its desired food reward in order to reach it. In yet another type of test situation, the animal is required to make a choice between alternatives: a lighted door and a dark door; two different colors; two different shapes; left and right. These test situations can be made more complex: for example, the animal can be taught that a triangle is the correct answer when the background is solid white but that a circle is correct when the background is striped. Or there may be a definite and repeated sequence of correct answers—for example, RRLLRRLL, and so forth. A cat can master a double alternation series (after lengthy training) up to the point RRLL, but evidently cannot go any farther. Raccoons can master the sequence RRLLRR. Monkeys can extend the series to RRLLRRLL, but only man seems able to extend the series indefinitely.

"TRIAL-AND-ERROR" LEARNING

How does an animal learn to solve problems? A classic experiment was performed in 1898 by the American psychologist E. L. Thorndike, who was strongly influenced by the implications of Morgan's canon. Thorndike defined a problem situation as one in which the goal sought cannot be directly attained by the performance of a simple act included in the animals' repertoire; either a novel action or a new integration of the available actions is needed. Thorndike placed a cat in a cage, the door of which could be opened by pulling on a

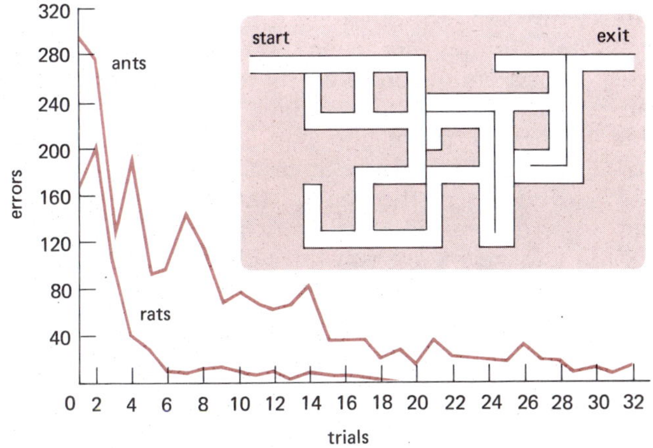

Figure 13–19 The maze shown was used to test the learning abilities of rats and ants. (A food reward was used to motivate the rats, while for ants the maze was placed in the route to their nest.) As is shown by the graph, rats learn faster than ants, but both eventually master the intricacies of the maze.

latchstring hanging down inside the cage. The cat responded at first by running through its standard repertoire. It meowed piteously. It tried to squeeze between the bars, and clawed and bit at them. Finally, one of its random motions resulted in a pull on a loop in the latchstring, and the door opened, permitting it to leap out. The next time the cat was placed in the box, it again went through its standard repertoire. But this time it reached the latchstring a little sooner. In successive trials, the number of incorrect responses gradually decreased, until the cat eventually was able to go to the latchstring almost directly. Thorndike theorized that through successive sequences of "trial and error," and eventual success, the response that was rewarded by escape from the cage was gradually reinforced, while the responses that were not rewarded (meowing, squeezing, biting at the bars) were gradually extinguished. There was no need to assume any understanding of the situation on the part of the cat, he argued; the solution to the problem was learned by a process of conditioning.

INSIGHT

The learning theories of Thorndike, Pavlov, and others were unchallenged for some years. But then some psychologists began to assert that although trial and error and conditioning were sufficient to explain some types of learning, they did not give the whole story. Another classic series of experiments was carried out by the psychologist W. Kohler while he was isolated on Tenerife in the Canary Islands during World War I. Kohler worked with chimpanzees. In one experiment, for example, he placed a chimpanzee in a cage with a banana hanging outside the cage, just beyond arm's reach. The chimpanzee tried futilely to reach the banana by thrusting his arm through the bars of his cage. Then he noticed a stick lying on the floor of the cage. In one swift motion, he picked up the stick, poked it out through the bars, and knocked in the banana. In other experiments, chimpanzees fitted two sticks together, or piled up boxes and climbed up on them to reach high-dangling food. In all cases, there was no prolonged period of trial and error. After an abortive attempt to solve the problem through an obvious simple response (and sometimes not even that), the chimpanzee apparently perceived the difficulty and what was required to overcome it.

Figure 13–20 In Kohler's experiments, chimpanzees solved problems of reaching food rewards using tools at hand without any preliminary trial-and-error attempts.

Such problem-solving behavior is called **insight learning.** Some consider it a subdivision of the general category of trial-and-error learning, in which the initial trials and errors are run through mentally and the successful response is the one that is overtly manifested. In any case, insight involves the integration of past experiences into a new combination, yielding a response appropriate for the new situation. Insight is based upon the development of concepts, or a certain degree of abstract reasoning. For example, two basic thought processes were involved in the chimpanzee's solving of the banana-out-of-reach problem: *inductive reasoning,* involving the learning of a general principle (for example, that a stick or similar device can be used to extend one's reach to cover a greater distance) from past experiences with specific situations, and *deductive reasoning,* applying the general principle learned to a new specific situation.

Many lower animals cannot learn to abstract at all: a rat, for example, cannot master the concept of the oddity principle. If it learns to receive food by picking a square from a series of two circles and a square, it cannot then transfer this learning to a new series of two triangles and a circle, but must learn the new combination all over again by a process of trial and error. Monkeys and apes, and especially man, on the contrary, show a high degree of ability for such abstract reasoning.

Biological Clocks and Rhythms

Today, with giant jets whizzing through the sky, carrying people across continents and oceans, we are faced with problems undreamed of by our forebears. Some of these problems are sociological, but others have a physiological basis. One of these, known as jet exhaustion, is becoming more and more common among businessmen, diplomats, and others who travel frequently around the cities of the world. The well-known radio and television commentator and writer Lowell Thomas described a frightening series of experiences that dramatically brought home to him the physiological dangers inherent in fast-paced modern travel through the airways. In a period of about a year, he had flown on a number of trips to various parts of the world, crossing all the twenty-four time zones at least twice, and resting very little between trips. Then he began to notice disturbing signs. Several times he experienced blackouts, and finally, one day, after jetting halfway around the world again, he had the feeling that a sheet of plate glass separated him from the people around him. Movement and speech were difficult, and his hands were shaking so that he could hardly hold a pen. Alarmed, he called a doctor, and was informed that he had been abusing his biological clocks too long.

Within us, numerous biological clocks "tick away." Our heart rate, body temperature, metabolism—each has its own rhythm. And the various rhythms of the body are precisely adjusted not only to the time of the world outside, but also to each other. Before the age of the airplane, and especially the jet, travel from one time zone to another was a relatively slow process. Thus, the body had time to adjust all its rhythms gradually to the new time schedules. But now, when one can eat supper, then hop aboard a jet, and embark from the flight several hours "later" in the early *afternoon,* the rhythms of the body are no longer in tune with their surroundings and may fall seriously out of phase with one another as well.

A recent Federal Aviation Agency study, attempting to determine the effects of rapid changes of time zones upon jet pilots, yielded some curious results. One test group was flown through ten time zones from Oklahoma City to Manila, another group was flown east from the same starting point to Rome, and a third group was flown south to Chile (Fig. 13–21). Immediately upon landing, the members of all the groups were run through a test of reaction times, involving the punching of a telegraph key in response to a light flash. Both groups flown across time zones showed a significant lengthening of reaction times in comparison with similar tests before the flight; but the group flown a similar distance south, crossing no time zones en route, exhibited no significant change in reaction time.

Endogenous rhythms are by no means confined to man: they are to be found virtually universally throughout the kingdoms of life. Plants show daily rhythms in their

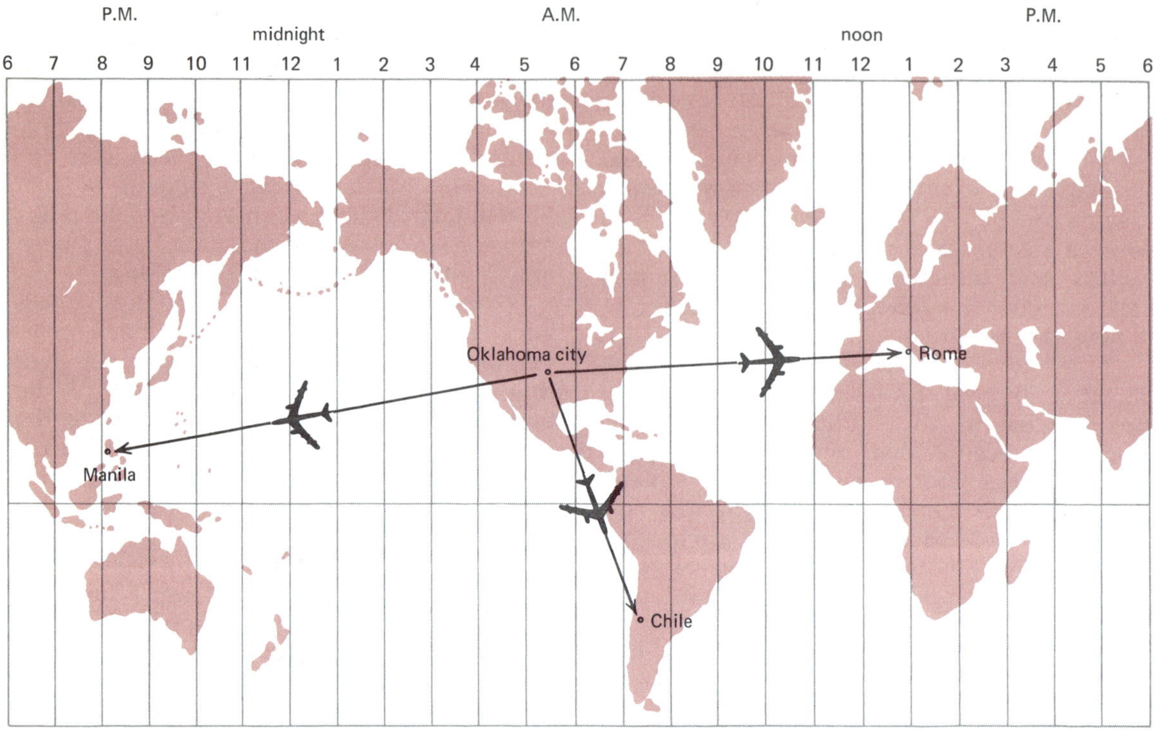

Figure 13–21 In an experiment on jet pilots, flying through time zones was found to lengthen reaction times, while a southward flight over the same distance did not produce any significant effect on reaction time.

movements and physiological processes (Fig. 13–22), and many species show characteristic seasonal patterns as well. Circadian rhythms—or rhythms timed to a period of approximately twenty-four hours—are also common among animals. Like man, many animals show spontaneous periods of activity and repose that are correlated with such body cycles as body temperature, hormone levels in the blood, excretion of salts in the urine, and susceptibility to drugs and toxic agents. Some animals are characteristically nocturnal, exhibiting their greatest activity during the night; others (like man) are diurnal, most active during the daytime; other organisms are crepuscular, most active at dawn or dusk.

Some marine animals, especially those that live along the shore, exhibit pronounced tidal rhythms. Some are characteristically active at high tide (oysters and clams open their shells when the tide is in), others (such as fiddler crabs) at low tide. Even when such organisms are transferred to the unchanging conditions of a laboratory tank, they continue to exhibit the same rhythms, in tune with the tides that no longer wash over them.

Some rhythms observed in plants and animals are of longer duration, apparent "biological calendars." Rhythms may be timed to the lunar month, an average of 29.5 days. Marine animals and plants, subjected to the moon-synchronized patterns of the tides, frequently show monthly rhythms in their reproductive cycles; curiously the human menstrual cycle is also of about the same duration. Although this cycle is not timed to any particular phase of the moon, it is thought that there may once have been

Figure 13–22 Plant sleep movements: the wood sorrel, *Oxalis grandis,* in the daytime (a) and at night (b).

such a relationship. Annual rhythms are also observed, such as the yearly cycles of reproduction in some animals and the rhythms of germination, growth, flowering, and fruiting in plants.

Some organisms, such as bees and migratory birds, have an extraordinary innate time sense, a constantly consulted "inner clock" that permits them to navigate accurately by sun or stars, automatically correcting for changes in their positions as the day (or night) progresses.

Research into the fascinating subject of biological clocks and calendars has been intensified in recent years, as the growing problem of "jet lag" has focused more and more attention upon the phenomenon. It has been found that some biological clocks can be reset. For example, for a long time little was known about the reproductive activities of the guinea pig, a common and much-studied laboratory animal with normally nocturnal habits, because the estrous cycle of the female usually has its peak in the wee hours of the morning. Finally one ingenious researcher put a group of guinea pigs on an artificial day, turning the lights on at night and off during the day. After a few weeks of this artificial schedule, the internal rhythms of the guinea pigs adjusted, and their mating behavior could be studied at a more convenient time for the researcher.

A number of studies have been conducted to determine how fixed the twenty-four-hour cycle is in man. In one study, conducted in Spitzbergen, Norway, during the summer season of continuous daylight, groups of student volunteers were given "cheating" wristwatches. These watches appeared to be normal, but actually compressed the "day" to twenty-one hours or lengthened it to twenty-eight hours. Many of the students could not adapt, but some began to show twenty-one- or twenty-eight-hour body temperature cycles. Studies were also made of their body chemistry, and analyses of urine samples revealed an interesting phenomenon. Among the students who had adapted best to the new schedule, and were urinating at the appropriate times of "day" and retaining urine during the "night," the cycles of excretion of sodium and calcium were also adapted to the new schedule. But the potassium excretion was still on a twenty-four-hour cycle—out of phase with the others! The implications of this finding are disturbing: what other cycles of the body might also be out of phase, and what might be the long-term consequences of shifts of the circadian rhythm?

Despite intensive research, scientists are still not certain of exactly what factors are responsible for the characteristic rhythms observed among living organisms. Some have suggested that the organism, quite independently of its environment, possesses natural biochemical cycles. Through a long process of natural selection, these cycles have come to match closely the natural geophysical frequencies and now are passed on genetically. But this hypothesis fails to explain some puzzling characteristics of the biological clocks and calendars. These rhythms are strikingly independent of factors that normally modify the metabolic rate, such as temperature and various drugs. The fiddler crab, for example, exhibits characteristic color changes timed to the solar and lunar cycles (Fig. 13–23). When the crab is transferred to constant laboratory conditions, these regular color changes continue. Moreover, the cycle remains constant over a 20°C temperature range, independent of the changes in metabolic rate produced by such a temperature variation. Many plant seeds similarly show an annual rhythm of germinating capacity when stored at constant conditions, no matter whether the temperature of storage is −22°C or +45°C.

These peculiarities of the biological rhythms seem to be explained by another hypothesis, which holds that the endogenous rhythms are actually exogenous, depending upon a continuing response of the organism to pervasive geophysical rhythms. Until recent years, this hypothesis could not have been tested. For no matter how "constant" the laboratory conditions—light, temperature, motion of "tides"—a test organism was always subjected to the rhythms of the earth upon which it exists. But now, as man is venturing out into space, away from the influences of geophysical rhythms, more may be found out about the basic causes underlying the rhythms of life.

SLEEP

The average person spends more than twenty years of his life asleep. Yet until the past few decades, this most familiar and yet mysterious aspect of the human circadian rhythms had received little serious scientific study. Now the space program has galvanized research into the mechanisms of sleep, and with the aid of refined techniques of measuring the state of body functions, particularly electroencephalography (the study of the minute electric pulses from the cerebral cortex), much valuable information has been gained.

Although many species do not "sleep" in the sense that humans do, regular periods of a temporary suspension of many normal activities have been found to be characteristic of such diverse animals as ants, fish, and mammals.

Among humans, patterns of sleep vary considerably both among individuals and among cultures. Modern man tends to take his sleep in one continuous stretch of about eight hours, once a day, generally during the hours of darkness. It is thought that primitive man may have taken his sleep in a series of "catnaps," as circumstances warranted, and some people claim to do well on such a sleep schedule in our modern world.

Much of the modern research into the nature of sleep has been pioneered by

Figure 13–23 The fiddler crab periodically changes color from a light to a dark phase.

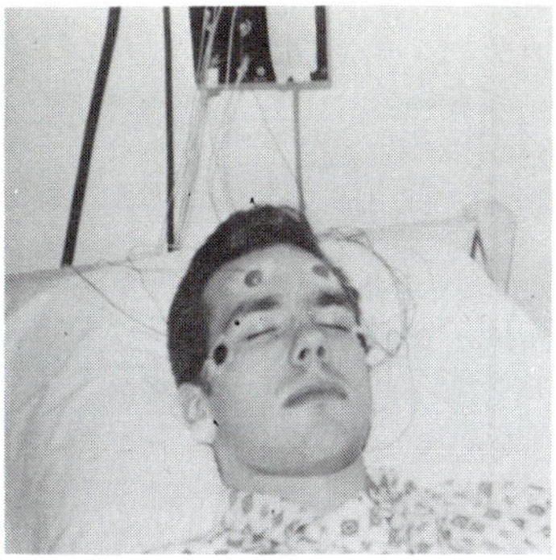

(a)

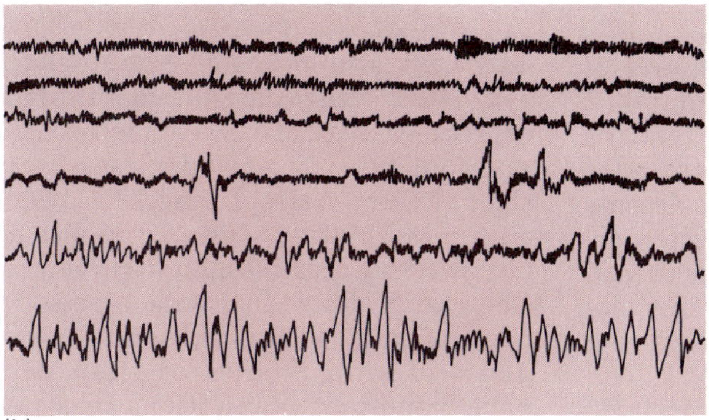

(b)

Figure 13–24 (a) A volunteer sleeps in a sleep laboratory, with electrodes pasted to his head. (b) Typical EEG tracings during the waking state and various stages of sleep.

Nathaniel Kleitman at the University of Chicago. The subject of a modern sleep experiment lies down to sleep in a bed in a darkened laboratory room, festooned with a surrealistic array of taped-on electrodes, thermistors to measure his body temperature, and various other sensors. The sleep researcher in an adjoining room fights his own fatigue as he monitors the electronic equipment that ultimately produces thousands of feet of EEG tracings in a single night's observations. Such experiments have yielded a general picture of normal sleep, as well as various significant patterns of abnormality that are already proving valuable in medical diagnosis.

During sleep there is reduced activity in a number of areas of the body. The muscles are relaxed and the breathing is slower and more regular. The heart rate and blood pressure are significantly lower, although during periods of dreaming both may increase dramatically. The body temperature during sleep is usually somewhat lower than normal (by about 1°F); the eyes are usually rolled up or unfocused, and pupils are constricted.

Although there is considerable variation among individuals, a general pattern of sleep is usually observed when the subject is allowed to sleep for a normal span of time (Fig. 13–25). After a gradual process of relaxation, an extremely deep sleep is usually attained between the first and second hour. This deep sleep seems to be essential to well-being; if a person is deprived of it for several days in a row (by being disturbed slightly—not enough to wake him entirely—each time he passes into the deep-sleep state) and then allowed to sleep undisturbed on the following night, he will spend more time in deep sleep than usual, as though making up for lost time.

Gradually, toward the middle of the sleep period, the depth of sleep decreases, until the person may be on the verge of waking. The second half of the sleep period is generally characterized by episodes in which the person's eyes move rapidly back and

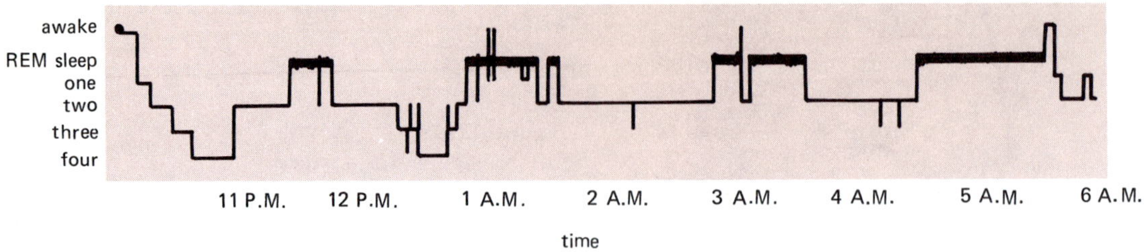

Figure 13–25 Graph of a night's sleep. The sleeper shuttles back and forth from one stage to another. As the night progresses, he spends increasing time in REM (dreaming) sleep and decreasing time in stage 4 (deep) sleep.

forth, as though he were watching a movie. By waking people during periods of rapid eye movements (REM), experimenters have found that these are correlated with episodes of dreaming. It has been found that dreams may last from several minutes up to an hour. Contrary to popular belief, they occur in real time; that is, they unfold at the same rate at which events would normally occur in the real world.

REM sleep—i.e., dreaming—has been found to be essential to mental health. If people are prevented from dreaming by awakening them at the first sign of rapid eye movements, they become fretful, irritable, and prone to hallucinations. This occurs even if the person receives a full eight hours of sleep each night. Sleep studies show that all people dream every night, even those who claim that they do not dream at all. Apparently people differ widely in their ability to recall their dreams. The vividness of recall also depends on the phase of sleep in which the person awakens and the suddenness with which he awakens. (Dreams are recalled most vividly when a person is awakened very rapidly right in the middle of an REM episode.)

Sleep deprivation can have serious consequences. Volunteers who have spent prolonged periods ranging up to a week or more without sleep have shown a gradual and then accelerating deterioration of awareness and alertness, losses of memory, progressive inability to perform various mental and physical tasks, instability, impairment of judgment, and ultimately vivid hallucinations and even schizophrenic manifestations. In most cases these alarming manifestations clear up spontaneously after a long session of restful sleep. Enforced sleeplessness has been used as a form of "brainwashing" in police and military interrogations; when deprived of normal sleep for prolonged periods, a person tends to become highly suggestible and shows a breakdown of will. Studies of sleep deprivation have important implications in a number of areas of modern life in which persons are placed under stress for prolonged periods. A sleep-deprived soldier may disastrously misinterpret an order; a groggy intern on night duty may make an error of judgment that costs the life of a patient. Indeed, some studies indicate that mortality among hospital patients after heart attacks and operations may be correlated to a large extent with inability to sleep for long periods due to the unfamiliar surroundings, anxiety, and rigid hospital routine, which may be poorly timed to the patient's normal body rhythms.

A byway of sleep research that has been pursued particularly in the USSR is the electrical induction of sleep with the aid of a device that delivers extremely mild, periodic shocks to the head region. Such electrical stimulation induces a deep sleep that may be sustained for hours and even days. Soviet physicians using this technique have claimed cures for a number of neuroses, as well as for such conditions as severe and recurrent headaches and cases of hypertension that do not have an organic basis.

Territorial Behavior: Migration and Homing

The phrase, "A man's home is his castle," is not at all restricted to humankind. Instead, this is an axiom to which many species subscribe. Among fish, for instance, as was mentioned earlier, the male stickleback stakes out a territory and ferociously guards it against all intruders (except for a gravid female). As soon as he sights the red belly of another male approaching his boundary line, he dashes off to the challenge. Literally bristling with indignation, he chases the other stickleback back into his own territory. But as they near the nest of the former intruder, the roles are reversed; the angry defender is now the aggressor, and is promptly chased back into *his* own territory. Finally the two fish glare at each other across their invisible boundary line; then both turn away, still glowing their intense emotion, and nibble at the sand at the bottom.

TERRITORY

A **territory** may be defined as the part of an animal's home range that it will defend against others of its species. The territory may include only the breeding or nesting area; or it may include part or all of the feeding territory, or even the entire range of the animal. The size of the territory varies with the species and, within species, with the age and sex of the organism (females and younger individuals often establish smaller territories), with the time of year, breeding season, etc.

Bird species are noted for their territorial behavior. The song of a male bird frequently serves notice to other males of the species that he has established a territory and is prepared to defend it. (In addition, the song may announce his readiness for courtship. The choice of a mate is characteristically made by the female bird, and she does not choose a male who does not have a territory staked out.)

Figure 13–26 A cardinal displays on his territory.

A bird's territory is usually respected by other birds; often the sudden attack by a bird defending his territory is enough to frighten the intruder into retreating without bloodshed. So strong is the territorial instinct that even a caged male robin, placed in an area that he had established as his territory before his capture, will flutter about his cage, break into song, and put on a typical aggressive display if another male robin encroaches upon his boundaries. Yet the same caged robin, placed in the territory of another bird, is silent and timid, frequently huddling in a corner of his cage.

Territories are also defended by various mammals, such as squirrels and chipmunks, which aggressively defend their homesites and sometimes feeding areas as well. Deer frequently establish a large mobile territory, the boundaries of which are established not in fixed land areas, but rather with respect to the location of the females of the harem. Male fur seals stake out territories on the beach during the breeding season and only then establish their harems (of a size proportional to their territory).

Invertebrates, too, may establish and defend territories. The fiddler crab vigorously defends an area of a yard or two around the entrance to its burrow. Ants and crickets also exhibit territorial behavior.

The establishment and defense of a particular territory has various adaptive advantages for an organism. The longer it remains in a particular territory, the more familiar it becomes with various natural refuges and avenues of escape. Outside its familiar territory, it is somewhat at a loss and much more vulnerable to predators. The establishment and mutual respecting of territories keeps the population density down to a set level, preventing overcrowding and helping to ensure sufficient food and living space for all. In addition, excessive fighting is minimized, and there is less aggressiveness in the immediate vicinity of the breeding site, thus minimizing chances of trampling or injuring eggs or young. The general link of territorial defense with mating practices helps to ensure that the strongest members of the species will leave the most progeny.

The existence of home ranges and territories implies some ability of the organism to recognize landmarks and find its way back to its specific nest or burrow, or a specific source of food or water. Many mammals with a highly developed olfactory sense leave scent markers in their home territories. Wild boars and bison, for example, rub against trees; bears roll in their urine and rub against "frontier trees" of their home range. Some species, such as deer and lions, have special scent glands, which release secretions (pheromones) upon trees, foliage, or other parts of their environment. These scents serve not only to aid in orienting an animal in its own home range, but also to inform others of its species of its presence and territorial claims. Bears, for example, generally will not venture into an area in which they scent another bear's urine. Male lions are attracted by the scent sprayed upon tree trunks by special glands of a female lion in the estrous state.

HOMING AND MIGRATION

In addition to scents, various visual landmarks aid in the sometimes marvelous exploits of homing behavior observed in animals. Many of the tales of cats, dogs, and other family pets finding their way back to their owners or former owners can be attributed to visual and other clues picked up by the animal while it was being taken away from its home range, or to chance or some search pattern bringing it back into an area in which it is familiar with enough landmarks to orient itself. Other cases may have been cases of mistaken identity.

But some species do have some very special ability to orient themselves and find their way to a specific area. Homing pigeons are noted for their ability to return to their home roosts. Migratory species often return year after year to exactly the same nesting and feeding grounds. The annual flights of the whooping crane have been carefully plotted and are breathlessly watched twice a year by nature lovers and conservationists all along their route. Each spring, the northern fur seals leave their feeding grounds in Pacific Ocean waters at about the latitude of California, to converge upon the bleak Pribilof Islands in the North Pacific. Their feats of navigation are even more amazing when we consider that the older females, pregnant with the young conceived in the previous breeding season, usually give birth within a day or so after their arrival.

Man has long been fascinated by the amazing navigational feats of migratory birds. Their semiannual flights may be up to thousands of miles in length, and sometimes include long stretches of nonstop flight over open sea, out of sight of any landmarks. The latest evidence seems to indicate that birds can navigate by taking sightings from the sun and stars, perhaps with corrections for the earth's magnetic field. Navigation during migration is unlearned: young birds, which have never made the flight before, frequently migrate before or after the parent flock, yet find their way to the habitual feeding grounds. Caged birds flutter against the bars of their cages during the migratory season, orienting themselves in the proper direction for the flight (if they have a view of the sky—Fig. 13–27). If they are then released after the flock has gone, they too can find their way alone to

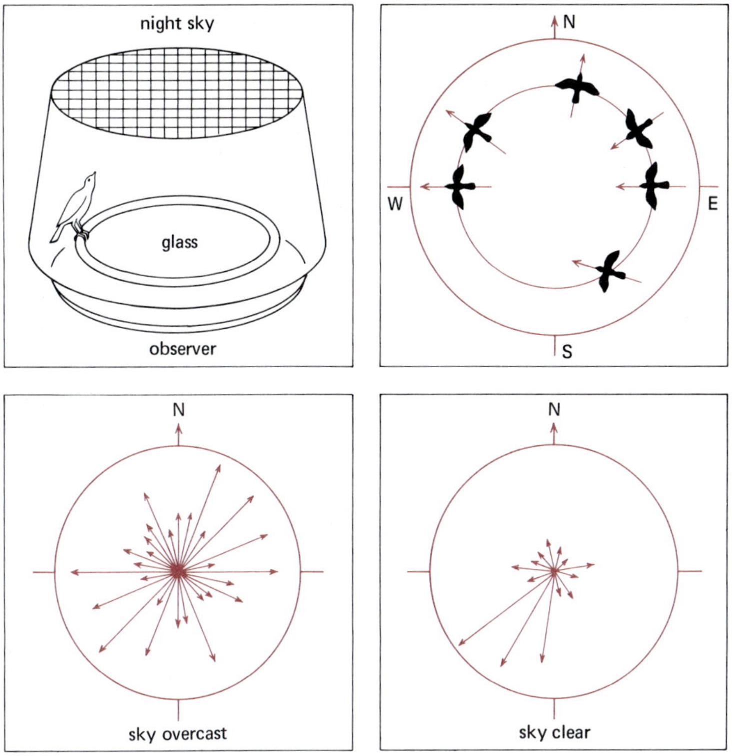

Figure 13–27 During the migratory season, an observer beneath the glass-bottomed cage (a) records the bird's flutterings (b). If the sky is overcast (c), the bird just flutters randomly. When the sky is clear, the bird tends to orient itself on an appropriate migratory route.

rejoin the flock. Experiments with warblers in Germany seem to add further support to the theory that birds use the positions of the stars to obtain navigational "fixes." Caged warblers placed in a planetarium oriented themselves according to the apparent night sky. When the star positions were changed to those visible above various parts of Germany through which the migratory warblers pass in their flights, the caged birds assumed the appropriate direction, exactly following the migratory route. Migratory birds seem to have a constantly consulted inner clock, by which they can make adjustments for the changes in star positions according to the time of day or night.

The "why" of migration is perhaps even more intriguing than the "how." Even less is known about this facet of migration. Research is gradually yielding bits of the puzzle; although the evolutionary origin of migration is still cloaked in mystery, more and more is being learned about the physiological bases of migration. It has been found that subcutaneous fat deposits are greatly increased in birds before long migratory flights. The birds burn this subcutaneous fat as fuel on their journey. Some studies seem to indicate that the duration of the periods of daylight may have an influence as a triggering factor in migration. Increasing the hours of light by artificial illumination, no matter what the temperature, has been found to result in an increase in the size of the gonads and in a tendency to fly north. It is hypothesized that the light may act through the pineal gland, which in turn stimulates the gonads and other glands.

Whatever the mechanism of bird migrations, they have definite adaptive advantages for the species. The winter nesting grounds, in the tropics or subtropics, are generally crowded, and there is great competition for the available food and living space. Migration takes the birds to their summer nesting areas, farther north, where it is less crowded and food and space in which to breed and live are more abundant. But the approaching cold of winter kills off much of the available vegetation and other sources of food, and the birds go winging off to their southern refuge.

Other species may migrate specifically for reproductive purposes. A fascinating case is the life story of the fresh-water eels of the genus *Anguilla.* When certain tiny fish were first discovered out in the Atlantic Ocean southeast of Bermuda, it was thought that these were a separate species of fish, which was given the name *Leptocephalus brevirostris.* Careful studies by the Danish scientist J. Schmidt revealed that these tiny salt-water fish are really larvae of the fresh-water eels, now referred to as leptocephalus larvae. Both the American and the European eels share the same spawning grounds (Fig. 13–28). After hatching, the tiny ocean-dwelling larvae begin their journey to the continental mainland, gradually developing into elvers, eellike larval forms capable of living in fresh waters, and finally into the adult eels. The American eels take about a year to travel from the spawning grounds to the mainland, and complete their transfor-

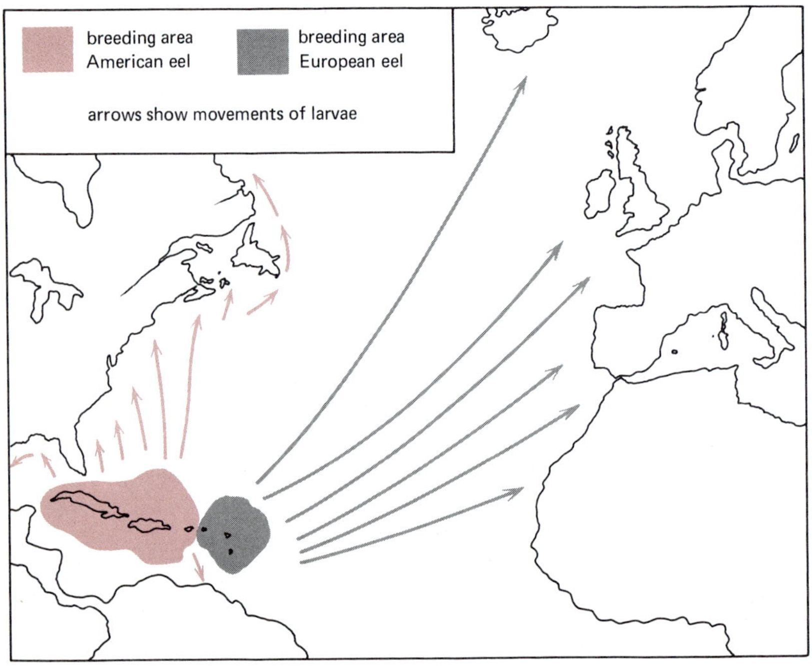

Figure 13–28 Migrations of the eel. Both American and European eels spawn in the Sargasso Sea region, then travel across the ocean in opposite directions and ultimately take up a fresh-water existence. Inset photos show the leptocephalus larva (a), elver (b), and adult eel (c).

(a) (b) (c)

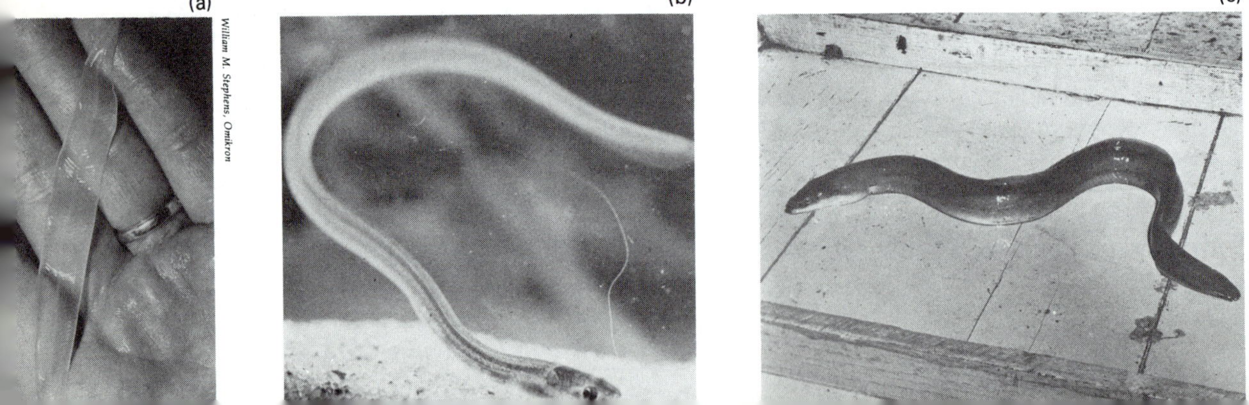

William M. Stephens, Omikron

mation into elvers around the time they arrive. The European eel larvae, on the other hand, have a three-year period of development into elvers, which matches their longer journey to the European mainland. The adult eels spend their lives in fresh-water streams until they head out into the ocean, and by as yet undetermined means, find their way to the spawning grounds of their birth. There they spawn and die.

Imprinting and Affection

War brings many tragedies. Among the most pitiful of these are the thousands of orphans, left without homes, without loved ones to care for them. Psychologists are just beginning to realize the full impact of an early life without love. Children raised from infancy in impersonal institutions where the employees are too busy to give them much individual attention and "mothering," show a lower survival rate than children raised in a normal family atmosphere. And those who do survive are frequently emotionally warped. They have difficulty in establishing normal relationships with friends and later with marital partners and their own children.

Some light on early emotional development has been cast by studies of rhesus monkeys, conducted for some years by H. F. Harlow at the University of Wisconsin. These studies grew out of an accidental discovery. In attempting to produce and maintain a healthy, disease-free colony of young rhesus monkeys for study, Harlow separated infant monkeys from their mothers a few hours after birth and placed them under controlled conditions of nurture and physical care. A much higher survival rate was achieved, and testing procedures were much more convenient. But as the monkeys grew, it was discovered that though they were physically healthy, they were mentally disturbed. They would sit in their cages and stare into space, or hug themselves and rock back and forth for hours at a time. Some developed compulsive habits, and some even exhibited fits of self-destructive behavior. When placed together with other monkeys, either those raised similarly or normal monkeys, they did not socialize, and were unable to bring any sexual behavior to the point of copulation.

Disturbed and fascinated by these unexpected results, Harlow was soon launched upon a series of studies of the emotional development of monkeys. In one series of experiments, infant monkeys were raised with two types of "mother surrogate": one a wire frame with an attached feeding bottle, the nipple protruding from the "chest" of the frame; the other type a terrycloth-covered model roughly resembling a mother monkey, either with or without feeding bottle (Fig. 13–29). All the experimental monkeys, even those fed from the wire frame, soon showed a distinct preference for the terrycloth "mother surrogate," spending much time clinging to it, evidently pleased by the sensation of bodily contact. If the monkeys were frightened, as by a mechanical teddy bear beating a drum, they ran to the artificial "mother" for comfort.

Figure 13–29 A young monkey clings to his mother surrogate. Note the wire frame with feeding bottle in the background.

When placed in strange surroundings, they at first huddled close to the terrycloth "mother," then boldly explored, returning from time to time for reassurance. Monkeys without such a "mother surrogate," on the other hand, cowered in a corner when placed in strange surroundings, and would not explore.

The early development of monkeys raised with terrycloth "mother surrogates" appeared normal. But later emotional disturbances emerged. They had difficulty relating to other monkeys. If, however, infant monkeys were raised with cloth surrogates, and, in addition, permitted to play in a playpen with other baby monkeys for a set period each day (twenty minutes or more), they made relatively normal social adjustments. This finding was contrasted to the emotional maladjustment observed in monkeys raised with their own mothers but kept isolated from any social contact with other baby monkeys.

Harlow concludes that both mothering (especially including extensive physical contact—holding, grooming, etc.) and a certain amount of socialization with members of the peer group are vital in the process of normal emotional development, at least among rhesus monkeys.

Figure 13–30 Konrad Lorenz surrounded by a quartet of imprinted goslings.

There have been many indications, in the work of Harlow with monkeys and in the work of others with many other species, that there are key periods in the formation of normal affectional relationships For example, soon after hatching a gosling begins to follow its mother about. This response was at first thought to be instinctive. But careful experimentation by Konrad Lorenz and others revealed that this early identification of young animals with their parents is actually a *learned* response, which has been given the name of **imprinting.** In one of the early experiments, Lorenz divided a clutch of eggs laid by a grayleg goose into two lots. The first was brooded over by the goose; the second lot was hatched in an incubator. The naturally hatched group of goslings began to follow their mother about as soon as they were able to walk; the incubator goslings, however, were exposed soon after hatching not to the mother goose, but to the investigator Lorenz. They took to following Lorenz about just as though he were their mother.

Imprinting has been observed among many other species. Ducks, for example, are imprinted by anything nearby that emits a "quack" or otherwise attracts their attention when they are in a certain, highly specific impressionable stage soon after hatching. Under normal circumstances, this is the mother duck. But in laboratory experiments with incubator-hatched ducklings, they have been imprinted to people imitating duck calls, male duck decoys (ducklings so imprinted later tend to follow male ducks instead of females), and even moving inanimate objects, such as mechanical floor polishers. The imprinted object is considered as the young animal's mother, and then, when it grows to adulthood, it may even make the imprinted object the object of its mating behavior.

Biological researchers studying animals and people raising young animals as pets often inadvertently imprint them, sometimes with comic results. Stories of dogs and cats who "think they are people" and identify with humans instead of their own kind are common. Lorenz became the object of the affections of a jackdaw that had become imprinted to him soon after hatching. The jackdaw persisted in attempting to feed Lorenz caterpillars, forcibly if necessary.

It has been suggested that failure to imprint appropriately may lie at the basis of some emotional and social maladjustments among humans, especially those among orphans, indicated above. Thus, normal individual development includes a whole complex of factors, both external and internal, both innate and learned. And in these processes of individual development may lie the bases of social organization. Despite the difficulties inherent in the measurement and interpretation of emotions and their interaction with behavior, strides are being made in research, and part of the story is already beginning to unfold.

Figure 13–31 The herons' stretch display—prelude to mating.

Courtship and Family Life

The green herons have arrived at their northern breeding grounds. Gone is their drab winter plumage; the males especially are striking, with their yellow-orange legs, their blue-black patch of skin just anterior to their eyes. Their personalities have undergone an equally striking change, too. Each male strives to stake out a large territory, aggressively defending it against all comers. Round and round he flies, patrolling his boundaries, alighting now and then upon trees along the frontier to emit loud, harsh *scaws*. Any intruder of his species, male or female, is greeted by a menacing ruffling of his back feathers, then an opening of the bill and rasping warning, and finally—if necessary—a furious onslaught. The female herons at first remain quietly at feeding grounds along the edge of the nearby river, stream, or lake. But after a number of days or weeks, they are attracted by the males' harsh cries. As soon as a female enters a male's territory, she is furiously chased off, just like any other trespasser. But the attraction remains: undaunted, she keeps returning. After a few days, the male may allow her to stand upon his boundary line. Now he begins to build a nest, or to repair a nest used the previous year. Now the female is allowed to enter the territory—but not too near the nest! The male flies round about his territory; the area that he defends against the ardent female becomes smaller and smaller. Soon they begin to fly about together, and at last the female is permitted to alight upon the nest tree.

Now the ritual of courtship begins in earnest. The male begins a stretch display, in which, with his bill stretched upward, he ruffles his head and neck feathers, and his entire body sways from side to side. The female joins him in the nest, and for hours they stand quietly side by side. Then comes a period of mutual caressing of heads and bills and nibbling of feathers. Finally, the female takes an active role, with a stretch display of her own. Copulation follows, and now the herons are a pair. No longer is it "his" nest, but now "their" nest, vigorously defended by both.

Among birds, fish, insects, and many other species, courtship is a strictly prescribed, highly species-specific ritual. This ritual consists of a series of stereotyped acts, each serving as a sign stimulus, releasing the appropriate stereotyped response by the other partner. Generally it is the female that chooses her mate, and the males during the breeding season vie to present a pleasing display for her. Birdsongs are not only announcements of territorial claims, but also often announcements that a male of a certain species is an eligible bachelor. The melodious song of the male nightingale is probably just as pleasing to the female nightingale as it is to human ears (though the female heron of course much prefers the harsh *scaw* of her chosen male). The gorgeous tail feathers of the male peacock are displayed in a courtship ritual for the benefit of the female. The drab little bower bird is without any bright display of plumage. Instead, he builds an elaborate nuptial bower of plant fibers, grasses, and twigs, which he decorates with flowers, shells, colorful pebbles, and so on. Some bower birds even paint their bowers with the juices of berries. The bower, sometimes requiring months of meticulous labor to prepare, is used only for lovemaking. After mating, the female then builds her own nest high up in a tree and there lays her eggs and raises her family alone. The male meanwhile maintains his beautiful bower, and sometimes, when the baby birds are grown, the female brings them back to their father's bower and the members of the family perform a sort of ritual display for one another.

Among the mammals, display is usually a less prominent part of the courtship ritual, but is often equally important. Especially among nocturnal species, the tactile and chemical senses frequently play a more important role than the visual senses, and sniffing and licking play an important role in the courtship patterns.

The stereotyped, instinctive nature of courtship and mating rituals in many cases serves to prevent interbreeding. If at any stage of the ritual the appropriate sign stimulus is not given or fails to elicit the appropriate response, the ritual is broken off, and copulation does not follow. The patterns are inherited. In one interesting experiment, in which two species of ducks were with great difficulty bred together, the offspring were characterized by courtship activities that differed from those of both parents. In some cases the behavior was a combination of the traits exhibited by the two parent species; in other cases, the courtship rituals resembled those of neither parent, but did resemble those of other species of ducks.

Patterns of family life are equally varied and characteristic of the species. Some species, such as the foxes and many birds, are monogamous, a pair remaining together during the entire season or even for life. Other species, like the fur seal, establish and maintain large harems. In some species, the act of mating is performed in a single encounter, as in the nuptial flights of winged ants and the mating of the spider (which often has disastrous results for the male; after his function has been performed his mate may kill him). Reproductive behavior among the howler monkeys is quite a casual affair; a female in the estrous state approaches the nearest male in the vicinity, with no apparent selection by either sex.

There is also great variation in patterns of care for the young. In some cases there is no care whatsoever: the female merely deposits a clutch of eggs in some unprotected location and leaves. In some cases there is a rudimentary amount of care, as seen among the reptile species that bury their eggs in the warm sand to incubate and hatch. In other species, the parent cares for the young all through the prenatal period (either by brooding upon or otherwise attending the eggs, or by actually retaining the eggs within the body until a fairly advanced stage of development has been reached), and even afterward. The latter pattern is quite common among mammals, where, with increasing emphasis upon learned (as opposed to stereotyped) behavior, a lengthy protected period is required for the young to acquire the knowledge and skills necessary for survival.

The parent that cares for the young also varies with the species. In some species it is the male. Among seahorses, for example, the mother's role is completed when she has deposited the fertilized eggs in the male's

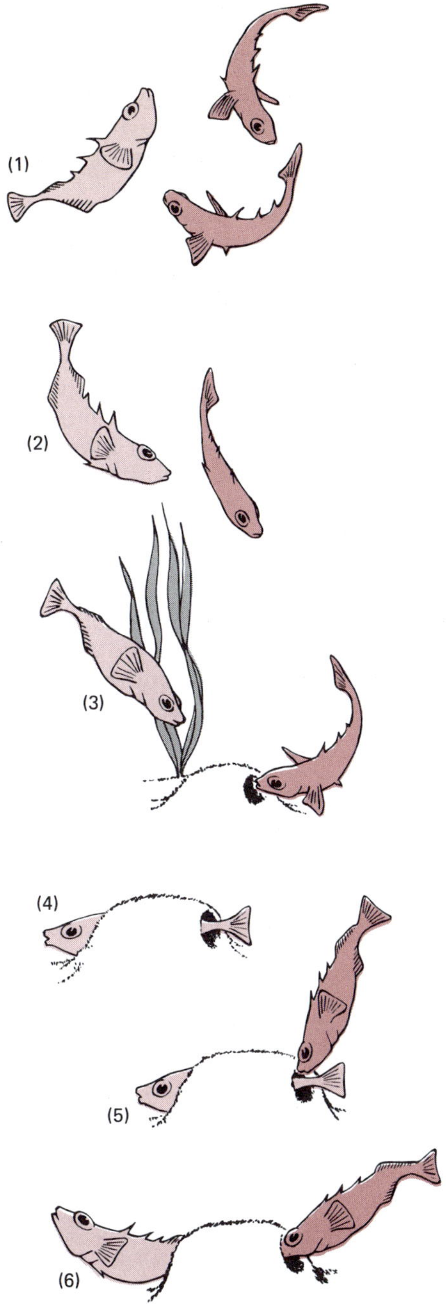

Figure 13–32 The male stickleback welcomes a gravid female into his territory (1) and escorts her to the nest he has prepared (2, 3). After she has entered the nest (4), he prods her at the base of the tail (5), stimulating her to deposit her eggs. Then the male drives the female from the nest (6), enters it himself, and deposits his sperm.

pouch; the father then broods the eggs until they hatch. Among sticklebacks, the male builds a nest into which he lures the female (Fig. 13–32). After she has laid her eggs and the male has deposited his sperm, the female leaves, and the male cares for the eggs. The stickleback is a very conscientious father. He remains by the nest and fans with his fins, sending currents of water through the nest and keeping the eggs well aerated. If the oxygen content in the water drops, the male stickleback fans faster.

In other species, such as many birds, the parents take turns caring for the young. The male penguin broods over the eggs, living off his stored body fat, while the mother is off at sea, fattening up on fish. Then she returns to take over the child care duties, and the male goes off to fatten up. Pigeons also take turns caring for their young. The baby pigeons are fed with a sort of milk, secreted by special milk glands possessed by both parents.

In some species, particularly (but not exclusively) in polygamous species and species where mating relationships are transitory or promiscuous, the female has the sole care of the young.

The amount of care given to the young is to some extent determined by the physical and physiological mechanisms of the reproductive process. But among the various species of wasps, for example, although the basic reproductive process is similar, the provision for the young varies. The female may simply deposit the eggs or build a more or less protective structure to house them. In some cases a fat caterpillar or other juicy morsel is paralyzed and neatly laid away for the larvae to eat when they hatch. In some wasp species, eggs are laid by a queen, and the young are cared for by the group, with an elaborate social structure; castes of workers perform the necessary tasks.

The interactions of male, female, and young form one type of social behavior. Indeed, in some species, the family unit is the basic social unit. But in some species, such as the wasps just mentioned, there are more elaborate forms of social organization.

Social Behavior

Man is often called a "social animal," since he lives in groups or communities whose members interact in numerous and complex ways. Other species, too, exhibit varying degrees of social behavior. Some fish swim together in enormous schools; some birds migrate together in huge flocks; and some insects swarm together in incredible hordes. In contrast to man's social structures, these societies are not highly organized. Fish, birds, or swarming insects tend to remain together, often at exactly prescribed and species-specific distances from their nearest neighbors, but—except among social insects such as bees and ants—there are no hierarchies, no orders of dominance, no division of labor.

Many animal societies are more complexly organized. Many mammals live together in social groups. Among carnivores such as wolves, small bands or packs, often consisting of an old male, one or several mates, their young, and a few young adults, are common. Herbivores such as sheep, deer, and elephants often band together in large herds. Such associations provide greater protection from predators and involve a certain amount of cooperation. While grazing, the herd is often arranged with the younger, stronger adult animals around the outside of the group, with the immature and weaker individuals in the protected center of the herd.

LEADERSHIP AND DOMINANCE

Frequently there is a leader of the group whom all the members of the group tend to follow when moving on to new territories or in times of danger. In some species, such as goats and howling monkeys, the leader may vary from one instance to another. For example, among the howlers, whichever male finds a good route through the trees makes a clucking noise, and the entire group veers over to follow him. In other species, as in sheep, there is a well-established leader of the flock or herd, usually an old female. This leadership arises from the tendency of the young lambs to follow their mothers. The tendency persists to some extent even after the lambs have grown to adulthood and have lambs of their own following them.

In species that commonly fight among themselves, orders of dominance-subordinance are frequently established. Dominance may be, but frequently is not, synonymous with leadership in the group. Among deer, for instance, the stags round up harems of females during the breeding season, fighting among themselves for females and butting the females of their little flocks if they try to get away. But the true leaders of the herd, who determine the direction of the group, are generally old females, who exert their sway without the use of force.

One of the best-known examples of a social hierarchy organized on the basis of dominance-subordinance relationships is the "pecking order," first observed among barnyard hens by the Norwegian scientist Schjelderup-Ebbe (Fig. 13–33).

A casual observer might readily notice that there is normally much squabbling in a group of hens, with one pecking at another a fre-

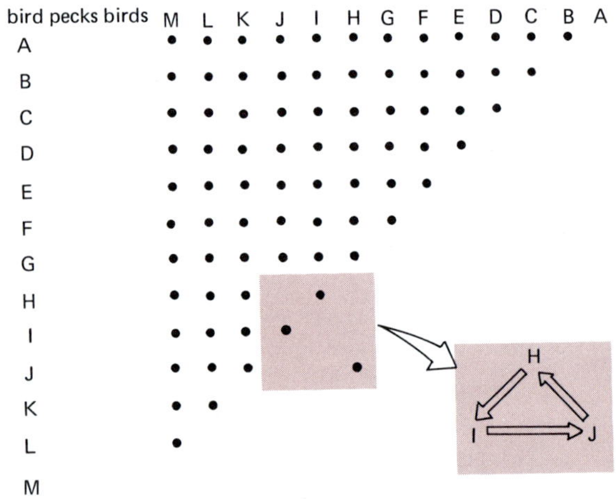

Figure 13–33 A pecking order in a flock of brown leghorn hens. The otherwise regular pattern is interrupted by a triangular relationship among birds H, I, and J. Note that bird A pecks all other birds in the flock and is pecked by no one; bird M at the bottom of the pecking order is pecked by all and may peck no one.

(a) Omikron (b)

Figure 13–34 (a) A section of honeycomb with worker bees busy at their various tasks. At left nurse bees are feeding young larvae in the open cells. In center a bee is entering a queen cell to feed a queen larva a special food called royal jelly. At right are capped cells in which worker bee pupae are metamorphosing into adults. (b) Older worker bees forage for the hive, gathering nectar and pollen.

quent occurrence. Schjelderup-Ebbe observed the situation more closely, identifying the individual hens, and discovered a pattern in the pecking. One hen had a well-established position of first at the feed, and could peck at any of the other hens without their retaliating. The others would only run away, dodge, or crouch beneath her onslaughts. In fact, a mere threat of a peck was frequently enough to result in a hasty retreat by one of the subordinate hens. Within the rest of the group there were similar relationships: hen number 2, for example, felt free to peck any other hen *except* hen number 1, without fear of retaliation. Similarly, hen number 3 could peck any other hens except hens numbers 1 and 2, and so on, down to the hen at the bottom of the pecking order, whose lot it was to be pecked by all the other hens without being able to peck any of them.

Such a pecking order is established gradually, as a form of learned behavior. When two strange hens are brought together, they generally fight, vigorously flapping and pecking. At last one hen gives up and flees. At their next meeting, the fight is repeated, but generally the same hen gives up, usually much sooner. Gradually the winning hen acquires the "habit" of success in conflicts with the other hen, while the losing hen acquires the habit of defeat, and assumes a subordinate role.

An established pecking order has definite adaptive advantages for the group. When each hen "knows" its roles with respect to each other hen of the group, destructive fighting within the group is minimized. It has been observed that a flock of hens organized into a hierarchy of dominance eats more and lays more eggs than a group of strange hens recently brought together.

Pecking orders have been observed among a number of other species, including mice and cows. The strength of the hierarchy is often dependent upon the relative abundance and distribution of the food supply. A herd of goats grazing in a field where the food sources are scattered, for example, do not fight over the available food supply. But in a group of goats kept in a farmyard, where position at the feeding trough is important, a dominance order is soon set up, with the strongest animals at the top, enjoying ready access to the food supply.

INSECT SOCIETIES

Paradoxically, aside from man, the most complex and stratified societies are found among insects. In fact, a number of insect species are commonly known as "social insects." These are mainly species of bees, wasps, ants, and termites. All the species of ants and termites now known are social, whereas among the bees and wasps, younger groups from the evolutionary standpoint, there are also many solitary species.

Among the social insects, roles in the society are strictly and biologically determined. There are workers of various types, and sometimes soldier castes as well (these are all females); queens (usually a single mature queen to an entire group), who are assiduously tended by the workers, and whose sole function is to lay eggs for the next generation of the colony; and males whose sole function generally is to fertilize the queen. The caste of a particular individual is determined even before it hatches. The queen bee, for example, stores the sperms deposited by a male in their nuptial flight and has the ability either to release portions of the sperm-containing fluid to fertilize the developing eggs or to withhold the sperms, resulting in the depositing of unfertilized eggs. Such unfertilized eggs develop into haploid individuals, which are males; the fertilized eggs develop into females. But even then the story of caste determination is not complete. For, depending upon the nutrients fed to the developing larvae by the workers, they may grow into sterile workers or soldiers. Or, if the larvae are fed a special hormone-rich "royal jelly," they will become mature queens. The latter are able to be fertilized by a male. The new queens then leave the group to found a new colony, as is common among ants, or continue the old colony, if the old queen has left with a swarm of followers to establish a new home, as is the general practice among bees.

The duties of a particular worker may change during its lifetime. For example, it has been found that during the first twenty-one days of its life, a worker bee passes successively through the stages of "janitor" bee, cleaning the cells of the hive; then a nurse bee, charged with feeding and caring for the developing larvae; then a building bee, working on the erection and repair of the wax combs of the hive; then as a honey maker, evaporating nectar by fanning with its wings and performing the necessary chemical conversions, and a pollen stamper, storing pollen away in the form of bee bread; then as a guard bee, standing guard at the entrance to the hive; and finally, from the twenty-first day of life onward, a forager, roving the countryside in search of nectar and pollen.

The responses of a social insect are basically stereotyped, determined by its genetic constitution, modified by hormonal influences, and released by specific sign stimuli. Even the different roles played by worker bees are generally correlated with the stage of development of specific wax glands, nursing glands, etc. The dependence of each individual upon the group as a whole and the degree of adaptation to a specific role within the group are so extreme that the beehive, the anthill, and the termite colony each might also be thought of as a "superorganism," the constituents (i.e., members) of which are incapable of independent existence.

Insect societies are frequently compared to human societies (sometimes to the detriment of the latter: it may be said that insect societies have no crime, unemployment, or neglect of children). Yet in many respects the resemblance is only superficial. As we have seen, insect societies are rigidly stratified by inherent biological factors. Among humans, on the other hand, as witnessed by the great diversity of social structures that have prevailed in different parts of the world and at different times during man's history on earth, it is evident that there are no such inherent biological limitations. Although hierarchies of status exist (and have existed) in virtually all human societies, they are by nature relatively ephemeral, as, in fact, are the particular forms of the societies themselves.

Communication and Language

Man is searching the heavens for alien species which he hopes may one day communicate with him. For on this planet he is alone. With no other terrestrial species can he communicate with any degree of meaningfulness. Yet in the sea there is a species which may, to some degree, end man's isolation. For years now, experiments have been conducted with the bottle-nosed dolphin *(Tursiops truncatus)*.

It was long known that the friendly dolphin is quite intelligent and easily trained, and it has long been a favorite at aquariums and water shows (Fig. 13–35). But when it was discovered that dolphins apparently communicate with one another in a manner rather similar to some forms of human speech, an intensive research effort focused upon this species was begun. Scientists working with bottle-nosed dolphins since the 1950s report that the animals can solve visual problems, show a high degree of social cooperation, and can emit meaningful sounds. Communications between dolphins have been recorded and analyzed, and they show great similarities to human conversations. A dolphin characteristically transmits, in the form mainly of whistles and clicks, and then waits for a reply from the other dolphin before continuing. In addition to the typical whistles and clicks, dolphins also emit various high-pitched sounds, described as quacks, wails, bleats, barks, buzzing, etc., some quite resembling sounds produced by the human voice. These sounds are frequently elicited upon hearing human speech, and dolphins have been taught to speak and (so the scientists working with them maintain) to understand some human words. It seems that this may turn out to be man's most successful attempt thus far to communicate with another species.

Previous attempts to communicate with other animal species have generally ended in failure, usually because of innate inability of the animal species to imitate human sounds or inadequacies of brain development. A parrot, for example, can repeat words but apparently has no understanding of what it is saying. In one classic attempt, a couple, Cathy and Keith Hayes, actually raised a chimpanzee, Viki, as their own child. With great difficulty, Viki was taught to pronounce a recognizable "mama," "papa," and "cup," and to use these words in a meaningful way. But though her vocabulary was laboriously increased to seven words by

Figure 13–35 Trained dolphins go through their routine.

the time she was six years old, her foster parents report that she was apparently reaching her limit at that point. Chimpanzees and other mammals seem to have very little vocabular control over their larynx and other motor apparatus. Sounds are usually associated with emotion, producing such patterns of communication as warning and distress cries. In subsequent studies, however, chimpanzees have been successfully taught to communicate more extensively in a more congenial medium: sign language. Chimpanzees trained to use their hands to form standard sign language words have been observed to invent new symbols for themselves to use in conveying messages to humans and have even spontaneously begun to use sign language to communicate with other chimpanzees.

COMMUNICATION AMONG ANIMALS

In addition to his (so far rather unsuccessful) attempts to teach animals to speak his language, man has also been fascinated by the study of how animals communicate with one another.

Birdcalls have received much study. It is usually found that any particular species has at least six or seven distinguishable calls. These can generally be interpreted as indications of danger, distress, hunger, presence of food, announcement of territorial claims, etc. Modern techniques of recording and oscillographic analysis of birdcalls are yielding new information on nuances of difference imperceptible to the human ear. These techniques have also had some practical applications. A Pennsylvania State University professor rid a small Pennsylvania town of its troublesome population of starlings by driving past their roosts with a sound truck, blaring out a much-amplified recording of the starling distress cry. The starlings all left posthaste.

In addition to the characteristic and rather stereotyped calls, communication among birds also includes typical postures and movements. These "cues" are even more important among mammals, which generally are more silent than birds (with the exception of a few species, such as the howling monkeys).

Modern brain research has spun off an intriguing new avenue for the investigation of communication and other aspects of behavior. Communication among groups of monkeys is now being studied with the aid of devices similar to those used to monitor the vital functions of astronauts during space flights. For example, radio signals to an electrode implanted in a specific area of a monkey's brain stimulate him to emit a certain sound. Observers note the reaction of the rest of the group to that particular sound and deduce its significance. Other aspects of behavior can similarly be provoked and studied. Indeed, when a set of switches is made accessible to the monkeys, they can even learn to use electrical brain stimulation to calm down a disturbed member of their group.

Much work has been done in observing communication among honeybees. For years the common observation that a rich source of nectar (or sugar water) discovered by a lone scout bee will soon be swarming with foragers, even before the returned scout has left the hive again, seemed inexplicable. Painstaking research by the German zoologist Karl von Frisch revealed some of the answers. He marked individual bees so that they could be readily distinguished and observed behavior in the beehives through red glass windows (since bees are blind to red light, they are unaware that they are being watched under such conditions, though human observers can see quite well in the red light).

Von Frisch discovered that a foraging bee that has discovered a good food source will, upon returning to the hive and discharging most of her honey, perform a kind of dance. The bees in the hive gather round and, in the darkness of the hive, "observe" the dance, apparently by a combination of tactile, taste, and smell perceptions. If the food source is close (the exact limit of "close" varies with the species), the scout bee performs what is known as the **round dance** (Fig. 13–36a). She crawls about on the vertical surface of the honeycomb in a tight little circle, dancing alternately clockwise, then counterclockwise. A feeling of excitement and the scent of food is communicated to the bees massed around her, which begin to follow her movements and then fly en masse out of the hive, searching randomly about the nearby area for the reported food source.

If the source of food is farther away, the scout bee performs a more complicated dance, which von Frisch named the **waggle dance.** In this dance, the returned scout bee performs a figure eight motion on the vertical surface of the honeycomb, wagging her abdomen vigorously during the straight run along the center of the figure (Fig. 13–36b). It is amazing how much information can be packed into the waggle dance. The frequency of the runs conveys to the massed workers in the hive the distance of the food source from the hive. Among von Frisch's bees, for example, a frequency of eight turns in fifteen seconds corresponded to a distance of 200 meters, six in fifteen seconds to 500 meters, between four and five to 1000 meters, two to 5000 meters, and only a single run in fifteen seconds to denote a distance of 10,000 meters. These relationships were found to be amazingly constant on different days, and even with different colonies of bees (but they do vary among different bee species). The waggle dance also gives informa-

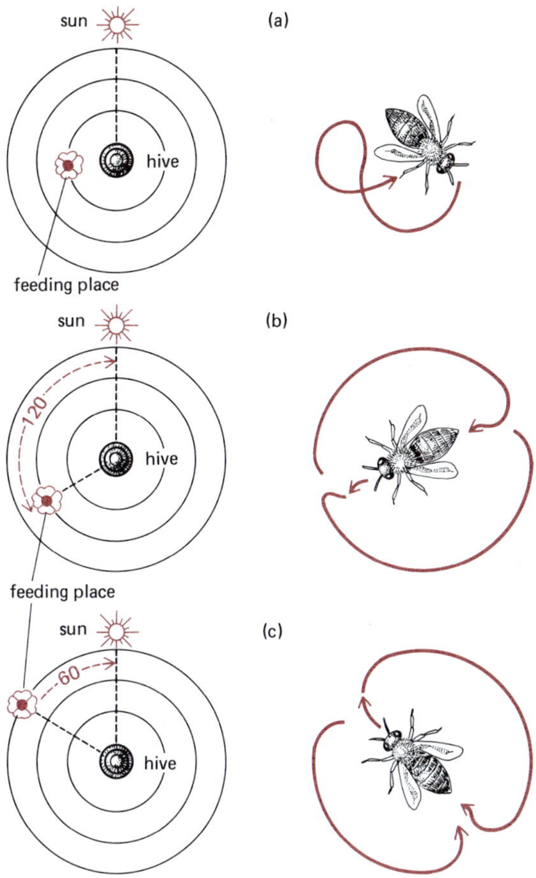

Figure 13–36 Honeybees communicate by variations of two basic dances, performed in the hive: the round dance (a) and the waggle dance (b, c). The round dance signals the presence of a food source nearby. The waggle dance contains navigational directions to a more distant food source. The run down the middle of the figure eight gives the direction: the angle from the vertical is the angle from the position of the sun. Distance is indicated by the number of turns in a unit time. The bee scout dances faster for a closer food source (b).

tion on direction of the food source from the hive. Von Frisch discovered that a waggle dance oriented so that the straight run is directed straight up along the vertical surface of the hive corresponds to a location in the direction away from the sun. Various angles with respect to the sun (along the horizontal) are translated into the appropriate angles from the vertical. The built-in clocks of the bees also operate in their interpretation of the information reported by the scout bees. Not only can the group of foragers fly directly to the location indicated by the scout bee, but, if held in the hive, they still fly to the correct location hours later, having corrected for the change in the sun's position.

The waggle dance also provides some information on the type of food source: a richer food source elicits faster waggling by the scout as she performs her dance. In addition, it has been found that the waggle dance is accompanied by a whirring noise produced by slight movements of the scout bee's wings. This whirring is followed by a short "beep," which appears to come from the surrounding workers and seems to indicate that they have heard and understood the message. Intensive analysis of recordings indicates that the sounds, too, carry information on the distance and type of nectar supply. How the bees can hear the message is still a puzzle.

Using a knowledge gained by careful observation and experimentation, man has been able to "talk to" bees. Artificial models of bees, attached to oscillators and impregnated with an appropriate sweet scent, have been introduced into beehives, and, through an appropriate bee dance, large numbers of foraging bees have been correctly directed to food sources at various distances and directions from the hive.

But it should be emphasized that such conversations, whether between bee and bees or between man and bees, are highly concrete. Neither bees nor any other terrestrial species (insofar as we know), with the exception of man, has developed a system of communication including the use of versatile symbols and permitting the communication of abstract ideas. Thus, in his search for another species with which to hold a meaningful dialogue, man is still alone.

QUESTIONS FOR REVIEW AND THOUGHT

1. Devise an experiment which might indicate whether a mother hen sits on her eggs as a result of instinct or as a form of learned behavior.
2. Do you believe that there is a qualitative difference between learning in bees, for example, and in man?
3. Discuss the implications to our society of a drug that could substantially raise the intelligence of every human on earth.
4. Discuss the probable physiological effects of the common practice of rotating work shifts in certain professions (e.g., policemen, hospital employees) from the standpoint of biological clocks.
5. Does man exhibit territorial behavior?
6. Discuss the advantages of migration and the special problems of navigation that a migratory animal faces.
7. Occasionally one hears a story of a human child raised by a family of wolves. Discuss the plausibility of such stories in the light of the phenomenon of imprinting.
8. From the evolutionary standpoint, what selective advantages are there in some of the elaborate patterns of courtship that are frequently observed among organisms?
9. Compare the beehive to a human community, citing the important similarities and differences.
10. If we received "intelligent" signals from outer space, how could we go about communicating with the senders of these signals?

PART IV

FROM EGG TO OLD AGE

CHAPTER 14

Heredity

Introduction

Any child knows that a cat will give birth only to kittens, not to puppies, or rabbits, or chipmunks. At first glance a litter of kittens seems to be a set of tiny copies of the mother, with the same number of ears, eyes, legs, and a tail. The shape of the face is essentially the same, the body is covered with the same type of fur. And yet closer examination reveals numerous differences among the kittens, and between individual kittens and their mother, which may become more and more accentuated as they mature. Often the color of the fur is quite different. More subtle differences may be detected in the shape and color of the eyes, specific features of the face, body size, and even personality.

The same basic pattern is repeated from species to species: although there is a great overall similarity between parents and their offspring (and to a somewhat lesser degree among all members of the same species), there are numerous minor variations. It is a common observation that children generally have some traits just like those of one parent, other traits exactly like those of the other, some traits intermediate between the two parents, and some traits shown by neither parent.

The branch of science concerned with the study of the inheritance of traits is known as **genetics.** In the past few decades, the field of genetics has been one of the most intensively researched areas of scientific endeavor. A significant proportion of the recent Nobel Prizes in chemistry and in physiology and medicine have been awarded to scientists whose work

Figure 14–1 Tiny copies of their mother.

has been directly or indirectly related to genetics. This is a clear recognition of the fact that genetics is not only of academic interest, but is also of extreme practical importance to man. Nearly all aspects of an organism are to some degree genetically determined—from the gross anatomical features to the microscopic subcellular structures, to the very chemicals of the body.

It seems almost incredible that the amazing diversity and complexity of a higher organism can be transmitted by way of two virtually invisible gametes, the egg and the sperm. An understanding of just how all this information could be transmitted via such tiny vehicles has become possible only through massive research efforts essentially begun at the beginning of this century and culminating in the post–World War II period in intensive studies of the molecular basis of heredity (a field now referred to as **molecular genetics**).

The era of modern genetics is frequently dated from the year 1900. But it actually began a generation before, with the experiments of Gregor Mendel.

Gregor Mendel, the Father of Modern Genetics

In nineteenth-century Austria, in the town of Brünn (now Brno, Czechoslovakia), an obscure monk by the name of Gregor Mendel quietly puttered about in his monastery's garden and performed a series of experiments that now stand among the greatest scientific achievements of man. By simple yet cunningly devised crosses of the garden pea, combined with mathematical methods of analysis, Mendel established the basic laws of heredity. Yet at the time he was completely ignored. An eminent biologist of the time, to whom Mendel diffidently sent a summary of his results, dismissed him with an icily polite acknowledgment; the only recognition that Mendel was ever to receive for his biological work in his lifetime was the publication of some papers in the transactions of the local natural history society in 1866 and 1869. For a

Figure 14–2 Gregor Mendel, the "father of modern genetics."

generation these papers languished, unread.

In 1900, sixteen years after Mendel's death, one of the most remarkable coincidences in the annals of scientific investigation occurred. Three scientists* in three different countries simultaneously discovered, in the course of their own researches, the basic principles of heredity. In searching the literature to see whether anything had been published on this promising new line, *each* happened upon the obscure journal in which Mendel published and realized with a shock that his "new" findings had been anticipated more than a generation before. In the best traditions of scientific integrity, *each* scientist now published the important new discovery, giving full credit to Gregor Mendel and reporting his own independent findings only as corroboration. Now the climate was right for acceptance of Mendel's theories; they sparked a wave of active research on the principles of heredity, and the era of modern genetics had dawned.

Why is Gregor Mendel considered the "father of modern genetics"? Man has known

*Hugo de Vries in the Netherlands, Karl Correns in Germany, and Erich von Tschermak in Austria.

for centuries and likely for thousands of years that traits are passed on from parent to offspring. The expression, "He looks just like his father," has probably been a commonplace since the days of the cavedwellers. What new insight did Mendel contribute?

Mendel quantized genetics. Before Mendel, biologists had attempted to fathom the secrets of heredity by considering the organism as a whole. Their efforts to consider simultaneously the inheritance of numerous traits frequently led to utter confusion. A tendency to consider small numbers of offspring also frequently obscured the basic principles.

Mendel was able to achieve a significant advance because in any particular experiment he traced only a single or at most a few distinct traits. Furthermore, his selection of garden peas and his choice of traits to follow were fortunate for several reasons. First, the pea plant gives rise to a large number of offspring, far more than can be achieved from the mating of two humans or even two mice. Second, since the pea plant is capable of either self-fertilization or cross-fertilization, Mendel was provided with a great flexibility in his experimentation. Third, the pea plant has a number of quite distinct and easily observed traits, all of which may be inherited independently of one another. Still another factor in Mendel's success was his painstaking tabulation of the results of his crosses. With the masses of data that he compiled, he was able to subject his findings to a meaningful statistical analysis and to discern underlying patterns.

While puttering in his garden, Mendel had noticed that some plants always "bred true" for a specific trait—whenever they were self-pollinated, their offspring always exhibited exactly the same characteristics as the parent—while others when self-pollinated gave rise to a variety of offspring, some differing quite radically from the parent. In searching for the underlying principles responsible for these differences, Mendel concentrated on seven distinct traits, among them stem length, color of seeds, shape of seeds, and color of flowers. He bred pure strains for each trait—for example, a strain of peas that always bred true for yellow pods, another that always produced green pods—and then crossed them with one another.

In one typical experiment, Mendel took plants from a true-breeding strain of tall peas and carefully cross-pollinated them with plants from a true-breeding strain of dwarf peas. He gathered all the seeds from the cross-bred plants and planted them in turn the next season. All the offspring of the first generation were tall; there was not a dwarf among them!

Following up this lot of peas, Mendel carefully self-pollinated all the first-generation offspring, gathered the seeds that they produced, and planted those the following season. Among the second-generation offspring, although the majority of the plants were tall, there was a substantial number of dwarf plants, greatly resembling their "grandparents." Careful tabulation of the numbers of plants of each type revealed that there were approximately three tall plants for every one dwarf (Fig. 14–3).

For all seven pairs of traits that Mendel studied, this interesting pattern was repeated: if a plant from one strain was crossed with a plant from the complementary strain, the offspring of the first generation were all alike—they exactly resembled *one* of the parents. No offspring resembling the other parent were obtained at all. Yet if these first-generation offspring were crossed with one another or if they were self-pollinated, the second generation always contained individuals resembling each of the original parent strains. Moreover, the two types were always related in a definite and constant whole-number ratio: there were three of the type that appeared in the first generation offspring to every one of the other type.

From this finding Mendel deduced that traits are inherited in distinct units or factors, and that each adult organism has two such factors for each trait, one derived from each of its parents. During the formation of the gametes, each pair of factors separates; any single gamete receives a half-sized set of factors including only one from each pair. When two gametes unite to produce a new organism, the paired relationship is restored. This concept of separation or segregation of the pairs of he-

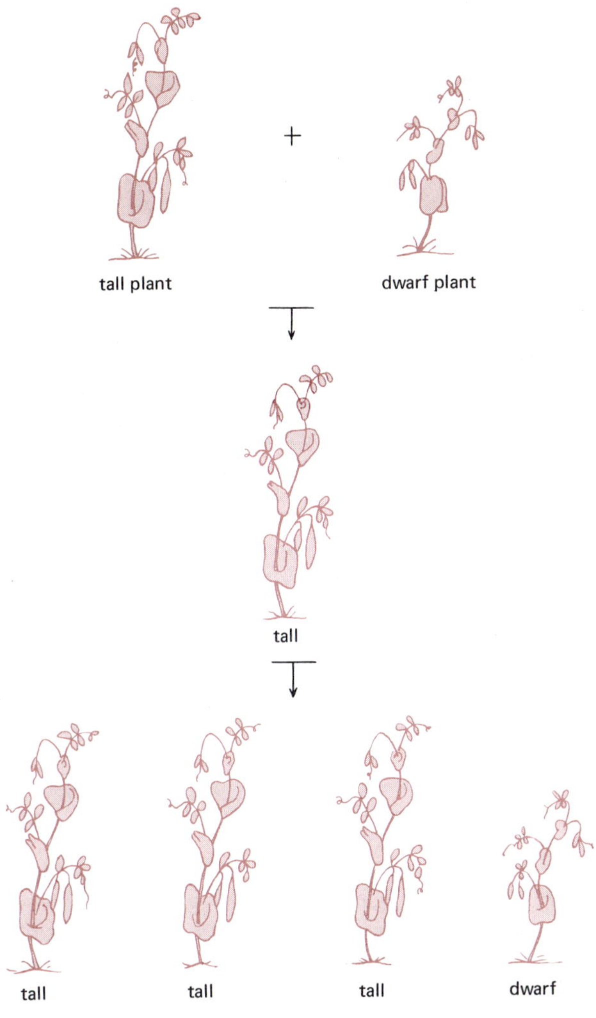

Figure 14–3 When Mendel crossed tall pea plants with a dwarf strain, all the progeny were tall. But when these tall progeny were self-pollinated, their offspring included both tall and dwarf plants, in about a 3 : 1 ratio.

redity factors is commonly referred to as Mendel's first law, or the **law of segregation.**

In any particular individual, the two heredity factors of a pair may be the same, as in the strains that always breed true for a given trait; or they may be one of each, as in the first-generation offspring resulting from a cross between two different pure strains. In such **hybrid** or cross-bred progeny, one factor seems to possess the ability to "dominate" the other (e.g., tallness in Mendel's pea plants), and only its effects are apparent in the observed trait. Yet the fact that the "dominated" factor is still present becomes apparent in the next generation, when the hidden trait reappears in some of the offspring.

In reporting his work, Mendel instituted a system of notations and terminology that is still used today. The parent strain is denoted as **P.** The first generation of offspring obtained from the parents is denoted as $\mathbf{F_1}$ (standing for first filial generation), the second generation as $\mathbf{F_2}$, etc. A trait that is always expressed in the first generation of offspring if at least one of the parents possessed it is referred to as **dominant.** A trait that does not appear in the first generation if crossed with a dominant, but then may crop up among the members of

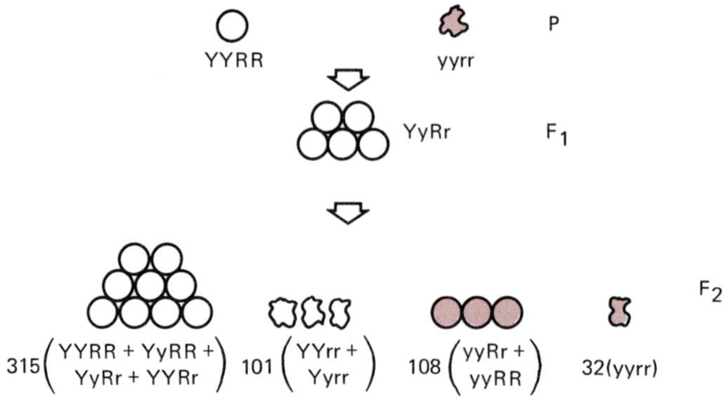

Figure 14–4 A dihybrid cross. The first-generation progeny of a cross between peas with round yellow seeds and peas with wrinkled green seeds all had round yellow seeds. In the second-generation progeny a ratio of approximately 9 round yellow : 3 wrinkled yellow : 3 round green : 1 wrinkled green was observed, indicating that the traits of round vs. wrinkled seeds and yellow vs. green seeds are inherited independently.

the second generation offspring, is called **recessive** (e.g., dwarfism in Mendel's peas). In abbreviated notation, the dominant trait is symbolized by a capital letter, the recessive trait by a small letter. Each offspring is denoted by a pair of letters, indicating the two heredity factors that it possesses with respect to that particular trait. Thus a tall plant that always breeds true when self-fertilized is denoted as TT; a true-breeding dwarf plant would be tt; a tall plant that will give a mixture of tall and dwarf offspring is denoted as Tt.

In addition to single crosses (i.e., the crossing of two variants of the same trait, Mendel also studied the results of crossing variants of two distinct and different traits at the same time. For instance, he crossed true-breeding pea plants with long stems and round seeds with plants that bred true for short stems and wrinkled seeds; pea plants from a pure strain producing round yellow seeds with a pure strain with wrinkled green seeds; and various other combinations. Figure 14–4 summarizes the results of one of Mendel's experiments on such **dihybrid crosses,** as they are called. In this case, just as in the other cases that he studied, the first generation of offspring showed only the dominant forms of both traits, while the second generation had exactly the ratio of types that would be obtained if the two traits were inherited independently of one another. This formed the basis for Mendel's second law, the **law of independent assortment:** any two or more pairs of factors of heredity may assort at random during the formation of the gametes, and each pair is distributed independently of all the other pairs.* As a result, all possible combinations of traits are ultimately obtained.

From today's vantage point, it seems almost incredible that Mendel's contemporaries virtually ignored his findings. But we might to some degree excuse them for their shortsightedness if we consider that at the time of his work there was not an inkling of the chromosomal and molecular basis of genetics. This lack of basic background only makes Mendel's insight all the more remarkable.

With the burst of activity following the rediscovery of Mendel's work in 1900, geneticists began to delve more deeply into the phenomena and mechanisms of heredity. From statistical ratios on the macro-scale they

*A similar independent assortment is observed in the dealing of a (well-shuffled) deck of cards. It is unlikely that any particular hand will be all clubs, or all spades or hearts. Unlike a hand of cards, however, which might contain, say, two or even more sixes or Queens—each gamete will receive only one of each pair of factors of heredity—such as height or seed color.

progressed to the physical basis of heredity within the cell itself, and gradually the role of the chromosomes, the genes, and ultimately the molecular basis of heredity began to crystallize.

Chromosomes and Genes

In the early twentieth century, the role of the chromosomes in heredity was finally recognized. Mendel's basic hereditary unit was equated with units on the chromosome strand, which were visible in micrographs as periodic beadlike structures. The chromosome was thus visualized essentially as a string of beads. In this analogy, each bead could be thought of as a different color, symbolizing the different traits carried by the chromosome. But since, as predicted by Mendel's theory, chromosomes occur in **homologous** (corresponding) pairs, there must also be a possibility of additional differentiation between corresponding beads on the two chromosomes of the pair, symbolizing the variants of a particular trait. For example, one might visualize two strings of beads with the same sequence of colors, in which the crimson bead on one is made of plastic, while the crimson bead on the other string is cut glass; or one of a pair of turquoise beads is smooth, while the other is rough.

When the existence of actual beadlike structures on the chromosomes (called chromomeres) was first detected under the microscope, biologists assumed that they were equivalent to the functional units of heredity, which the pioneering American geneticist T. H. Morgan named **genes.** With added research and greater insight into the mechanisms of heredity, it was realized that this picture is actually an oversimplification. But although there is not an exact one-to-one correspondence between the chromomere and the gene—indeed, the exact physical relationship of the gene to the chromosome is not yet clear—a great deal has been worked out about the way the genes function in heredity.

In fact, modern research has carried studies beyond the level of the gene, to the molecular basis of heredity. The chromosomes have been found to consist of huge coiled molecules of DNA in association with molecules of protein. The nature of these nucleoproteins and their function in carrying and transmitting the information of heredity has unfolded largely in the years since the early 1950s and will be discussed in the next chapter.

Chromosomes are found within the cell nuclei of virtually all organisms. Within a given species, the number of chromosomes per cell is constant and is characteristic of the species. (See Table 5–1 in Chapter 5.) Despite such extremes as the roundworm *Ascaris,* with only a single pair of chromosomes, and a fernlike plant species of the genus *Ophioglossum,* with a total of 768 chromosomes, the vast majority of the species have chromosome numbers between 10 and 80. Man has a chromosome number of 46 (23 pairs). His closest relatives, the gorilla, the orangutan, and the chimpanzee, all have 48 chromosomes. The dog has 78, the cat 38, and the horse 66 chromosomes. The garden pea, which Mendel studied, has a complement of 14 chromosomes. (It is an interesting coincidence—or more likely a result of shrewd observation and experimental design—that the seven traits that Mendel selected for study were later found to be situated one on each of the seven separate pairs of chromosomes of the pea.)

The fruit fly *Drosophila melanogaster* has become one of the most valuable subjects for genetic studies largely because of its unusually small chromosome complement (four pairs, or a total of 8 chromosomes; see Fig. 14–5). An additional advantage is provided by the occurrence of giant chromosomes, up to a hundred times as long as normal, in the salivary glands of the larvae of this organism. By delicate techniques of micromanipulation, these giant chromosomes can be stretched out and their finer details discerned. Detailed chromosome maps have already been drawn up for the four pairs of chromosomes of *Drosophila,* relating influences on specific traits to specific sites on the chromosomes.

The characteristic chromosome number is maintained, throughout the successive mitotic divisions of the cell, through a process of re-

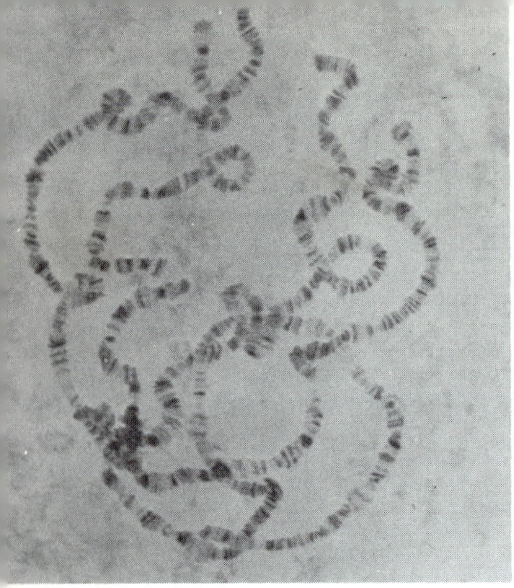

(a)

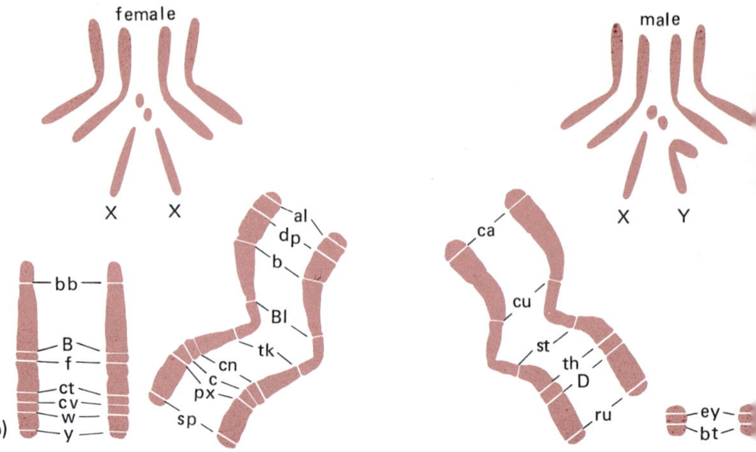

(b)

Figure 14–5 (a) When salivary gland preparations from *Drosophila melanogaster* are squashed, micrographs show the structure of the chromosomes in clear detail. (b) Schematic depictions of the *Drosophila* chromosome set, sorted into pairs. The positions of some of the mutations that have been mapped on these chromosomes are indicated by letters.

duplication and subsequent separation of the original chromosomes. The processes of reproduction and transmission of genetic information to new generations also involve a different kind of cell division, in which the characteristic chromosome number is not maintained. This process is called **meiosis.**

Meiosis

A seventeenth-century Irish theologian, Archbishop Ussher, estimated that man was created in 4004 B.C. If this assumption is correct, then there would have been approximately two hundred generations of human beings in the six thousand years since man first appeared on earth. Now, we know that a new human being is produced by the union of a pair of sex cells or gametes, one from each parent. If this whole process occurred solely by mitosis, each gamete of Adam and Eve would have contained 46 chromosomes. Such gametes, combining, would form cells with 92 chromosomes. The children of Adam and Eve would therefore have had 92 chromosomes in each of their cells, including their gametes. Their children, in turn, would have had 184-chromosome cells, the next generation 368, the next generation 736, and so on, until we today would have cells with 46×2^{200} chromosomes each. Yet a single set of such a number of chromosomes would take up more space than the entire universe. Since the latest estimates of the age of man indicate that Archbishop Ussher's figure was far too low—the first appearance of *Homo sapiens* probably dates back well beyond 100,000 years ago—the situation becomes even more absurd. Obviously something must happen in the process to change the number of chromosomes in the gametes. Such a "happening" is the special kind of cell division called meiosis.

Meiosis is a series of two consecutive cell divisions in the course of which the chromosomes are duplicated only once. The process starts with a normal **diploid** cell, possessing the full chromosome complement for the species (an assortment of homologous pairs), and ends with four cells, each of which contains half the number of chromosomes of the original cell. This halved chromosome number is called the **haploid** number, and the haploid complement consists of one of each pair of the diploid set. Because of the reduction of the number of chromosomes that occurs in meio-

Figure 14–6 (*Opposite*) Comparison of mitosis and meiosis. For clarity, only one pair of chromosomes is pictured (one member of the pair shown in color) and nuclear membranes and nucleoli are omitted. Note that in mitosis the diploid state is maintained in the two daughter cells. Meiosis results in the production of four haploid daughter cells. Fertilization restores the diploid state.

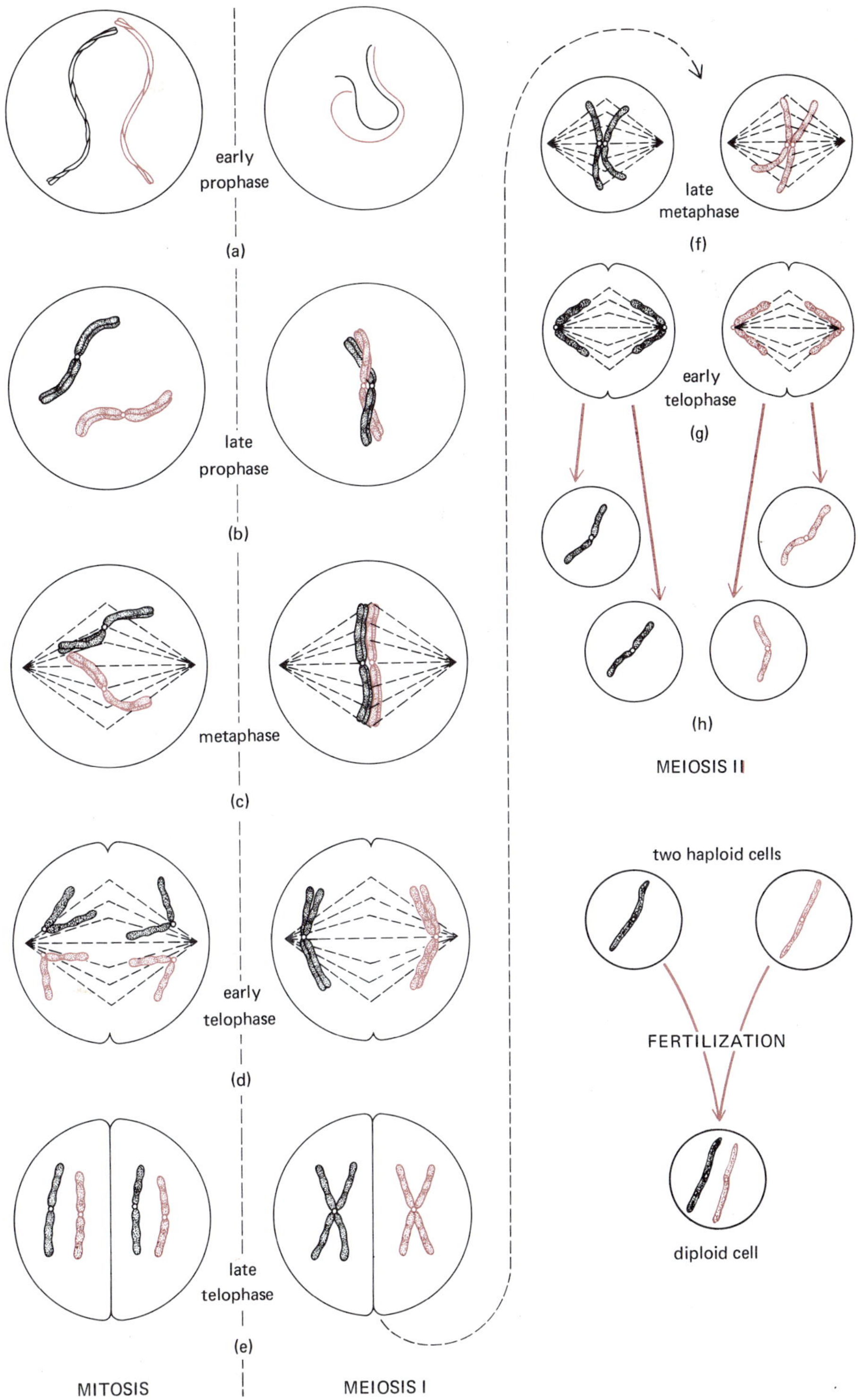
early
prophase
(a)
late
prophase
(b)
metaphase
(c)
early
telophase
(d)
late
telophase
(e)
MITOSIS
MEIOSIS I
late
metaphase
(f)
early
telophase
(g)
(h)
MEIOSIS II
two haploid cells
FERTILIZATION
diploid cell

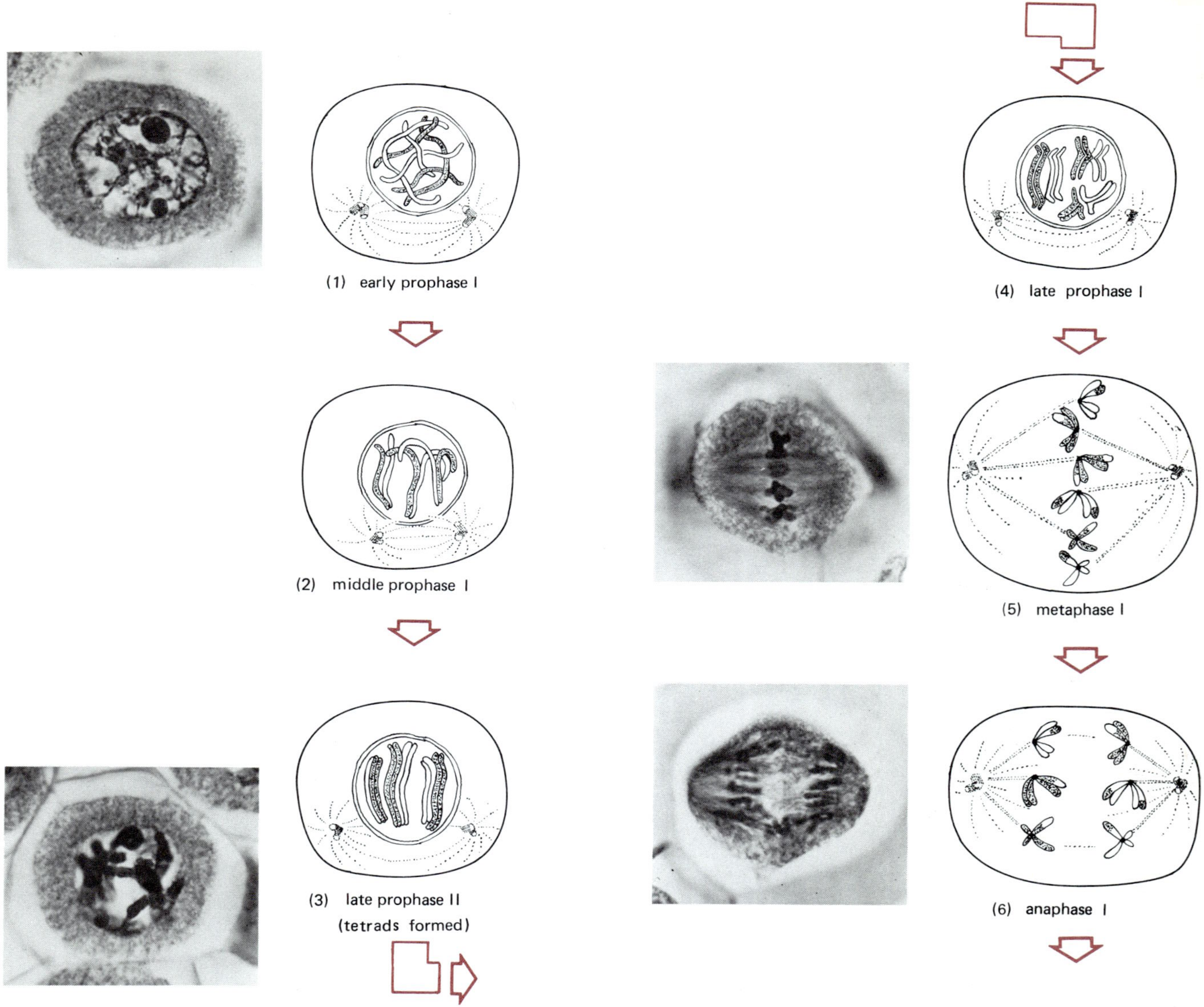
(1) early prophase I
(2) middle prophase I
(3) late prophase II
(tetrads formed)
(4) late prophase I
(5) metaphase I
(6) anaphase I

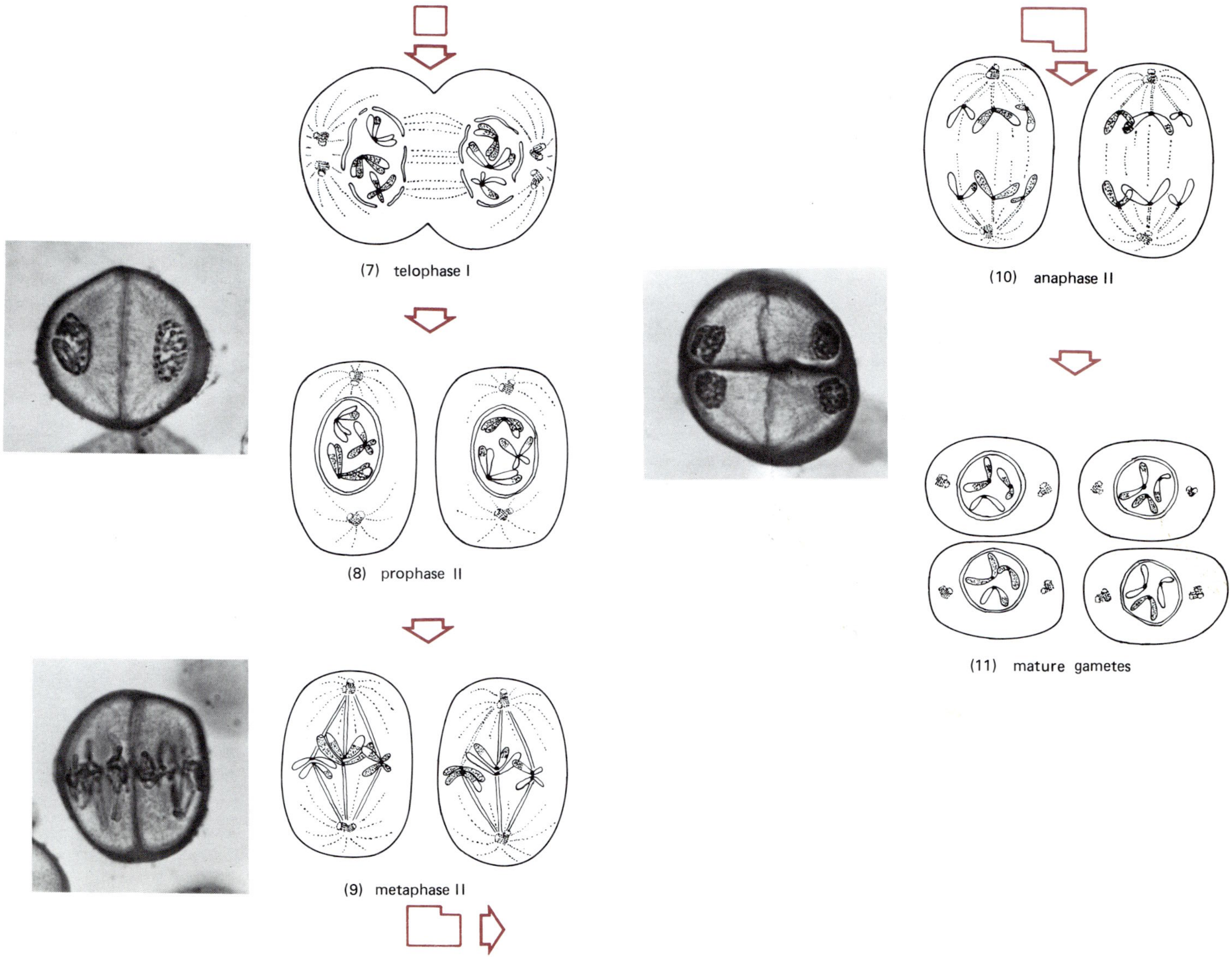

Figure 14–7 Meiosis: the diagrams outline the process in simplified form for an animal with a chromosome number of six; the photos illustrate stages of meiosis in the lily (chromosome number 24).

sis, the process is often called **reduction division.**

Meiosis might be thought of as a process opposite to fertilization: meiosis results in halving of the chromosome number, while fertilization results in doubling. The two processes are key links in the cycle of sexual reproduction: without meiosis, successive fertilizations would result in successive doublings of the chromosome number, ad absurdum. Without fertilization, successive meioses would be unfeasible.

In many organisms, the two divisions of meiosis follow quickly one after the other, with no intervening pause. In others, the process halts temporarily at one phase or another. In vertebrates, for example, the egg normally proceeds through the first meiotic division, then stops at the metaphase of the second division. Only if fertilization occurs (or if the egg is artificially activated by pricking or by some other physical or chemical means) does it continue on to complete the second division. (The sperm, itself already haploid, meanwhile waits quiescently within the egg until the egg has finished its process of meiotic division.) The timing of meiosis and fertilization in the life cycle also varies greatly among species. Among sexually reproducing protists, such as various algae, meiosis frequently occurs soon after the zygote is formed (or after an intervening period of dormancy), and the bulk of the life cycle of the organism is spent in the haploid state. In higher animals, the situation is the reverse. Fertilization usually occurs soon after meiosis; only the gametes are haploid, and during the major portion of the life cycle the cells are in the diploid condition. In various other species, the two key phases of sexual reproduction may be spaced at different points of the cycle, with either haploid or diploid forms being dominant. In any case, in any organism that reproduces sexually, meiosis and fertilization must alternate, with a regular halving and doubling of the chromosome number, thus maintaining the same basic complement from generation to generation.

In the interphase preceding the first division of meiosis, the chromosomes are reduplicated, just as in the interphase preceding a mitotic division. As meiosis proceeds, just as in mitosis, a prophase, a metaphase, an anaphase, and a telophase can be distinguished. But distinct differences begin to become apparent and are intensified as, without an intervening interphase, the chromosomes of the daughter cells proceed into the prophase of *another* division. Since there has been no second reduplication of the chromosomes, the separation that occurs in the second division of meiosis results in a halving of the normal diploid set, and the daughter cells formed from this second division receive only one of each pair of homologous chromosomes.

The successive stages of the two divisions of meiosis are diagramed in Fig. 14–7. An important point to note, which holds the key to an understanding of the transmission of inherited characteristics, is the fact that the homologous chromosome pairs are assorted randomly in meiotic division. Thus, if it were possible to mark the chromosomes in a particular cell to show which one of each homologous pair originally came from the father and which came from the mother,* it would be found that the haploid cells ultimately obtained from meiosis each contain a complete set of a random assortment of paternally and maternally derived chromosomes. This principle provides an insight into the patterns of heredity that we see about us, and helps to explain why a person may have some characteristics resembling those of his maternal grandmother, others of his maternal grandfather, and still other traits inherited from his paternal grandmother and grandfather.

Since the human cell contains 23 pairs of chromosomes (Fig. 14–9a), and each pair separates randomly and independently during the formation of gametes, a fantastic number of different possible assortments of chromosomes can be found in the gametes formed by a single person. Thus, it becomes highly unlikely that any two gametes would be exactly alike. Since, further, the combination of ga-

*This has effectively been done in certain cases, by studying the transmission of chromosomes with an unusual shape that can readily be spotted on micrographs of dividing cells (Fig. 14–8).

(a) (b)

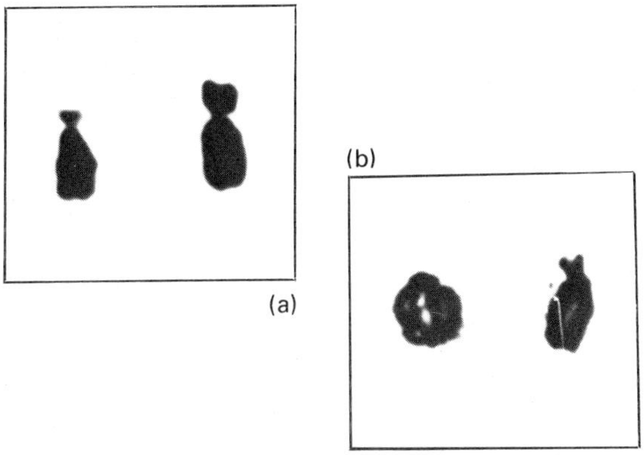

Figure 14–8 Some mutations produce visible changes in the chromosomes that can be followed readily through successive generations. (a) Short arm in a human chromosome 5, shown with its normal counterpart; (b) ring chromosome 15 in a human male, shown with a normal chromosome 15 from the same cell.

metes in fertilization is also a random process, the probability that two children of the same parents would be exactly the same is vanishingly small. (An exception, of course, is the case of identical twins, in which two duplicate cells, formed by mitosis, somehow become separated and develop into two independent organisms.)

Occasionally in meiosis the paired chromatids fail to separate properly, and in the subsequent division one gamete receives both members of a particular pair while the other does not receive any. This meiotic error is referred to as **nondisjunction.** When such a gamete combines with a normal gamete to begin the life of a new organism, the cells of that organism will thus wind up either with only one of the chromosome pair in question or with three instead of the normal two. These conditions can give rise to various anomalies. Down's Syndrome or mongolism—characterized by severe mental retardation and a typical complex of physical traits, including short stature, short stubby fingers and toes, unusual finger and palm prints, and especially a round face with folds of the eyelids very similar to those characteristic of Asian races (Mongoloid)—has been traced to nondisjunction: the

Figure 14–9 (a) A normal human female chromosome set, with 46 chromosomes, including 2 X chromosomes. (b) The chromosome set of a girl with Down's Syndrome shows an extra 21st chromosome as a result of non-disjunction. (c) A mentally retarded boy with Down's Syndrome with the typical physical characteristics of the condition.

(a) (b) Omikron (c)

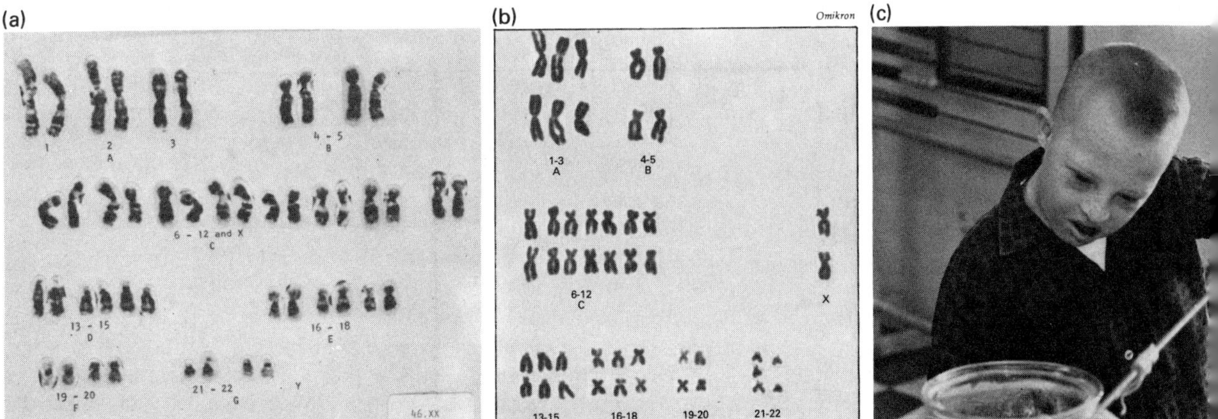

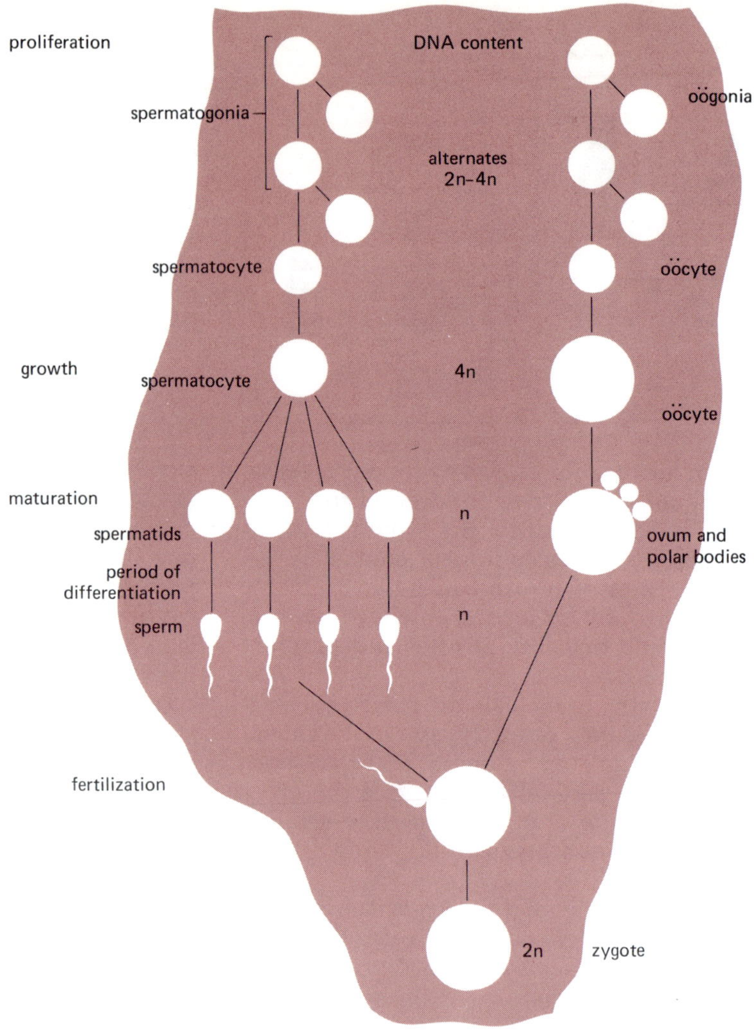

Figure 14–10 Sequences of steps in the formation of eggs and sperm. (n is the haploid chromosome number.)

affected individual has 47 chromosomes instead of the usual 46, with a triplet of chromosomes corresponding to the normal 21st pair (Fig. 14–9b).

Meiotic division is the basis of the formation of gametes, a process that is specifically referred to as **gametogenesis.** In humans and a number of other species, this process occurs somewhat differently depending upon the sex of the organism in which the gametes are formed.

In the formation of the male gametes—the process of **spermatogenesis**—meiosis results in the formation of four haploid sperm (Fig. 14–10a). In **oögenesis**—the formation of female gametes—however, only one of the four meiotic daughter cells is ultimately functional as a reproductive cell; the other three disintegrate (Fig. 14–10b). At first glance, this seems to be a wasteful process, but it has important functional significance in the process of reproduction.

The amount of stored food in the egg cell is of critical importance in the survival of the potential early embryo. If oögenesis consisted of a meiotic cycle directly analogous to spermatogenesis, then four equal-sized eggs would be produced, and each of these would contain only one-fourth the amount of stored food possessed by the original cell. Instead, the meiotic divisions of oögenesis are unequal; in each case one daughter cell receives half the chromosomes but *nearly all* the cytoplasm, while the other receives half the chromosomes but practically no cytoplasm. The meiotic divisions thus result in the production of one haploid **egg cell,** which has retained nearly all the cytoplasm of the original parent cell, and three haploid **polar bodies,** which have been so sorely shortchanged of cytoplasm that they are practically bare nuclei (Fig. 14–11). These polar bodies disintegrate, while the single egg cell remaining is ensured of an optimum food supply for the potential embryo.

In other species, notably higher plants, an unequal meiotic division, resulting in the formation of egg cells and polar bodies, is utilized in somewhat different ways for the same ultimate end: the provision of the maximum amount of stored food for the developing embryo.

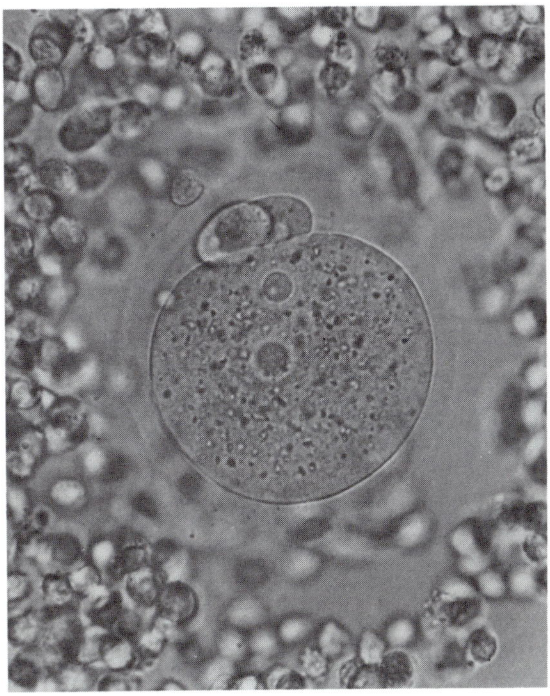

Figure 14–11 A living fertilized mouse ovum, taken from the oviduct 20 hours after copulation. Two polar bodies can be seen at the top of the ovum.

Some Basic Concepts of Genetics

After the dramatic triple rediscovery of Mendel's work in 1900, the biological world was in a ferment. Throughout the world, biologists plunged into this new area of research. An American researcher working at Columbia University, T. H. Morgan, was especially fortunate in hitting upon an almost ideal experimental animal. The tiny fruit fly, *Drosophila melanogaster,* has an unusually convenient combination of qualities. It is large enough to examine and handle easily, yet small enough to raise by the hundreds in a space easily accommodated in a standard laboratory. It requires no exotic or expensive food—a little mashed bananas with yeast will

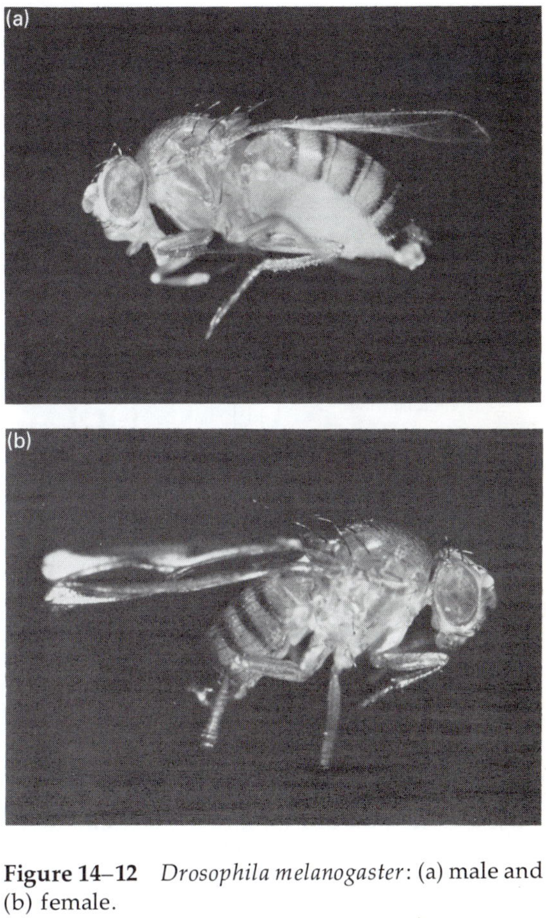

Figure 14–12 *Drosophila melanogaster*: (a) male and (b) female.

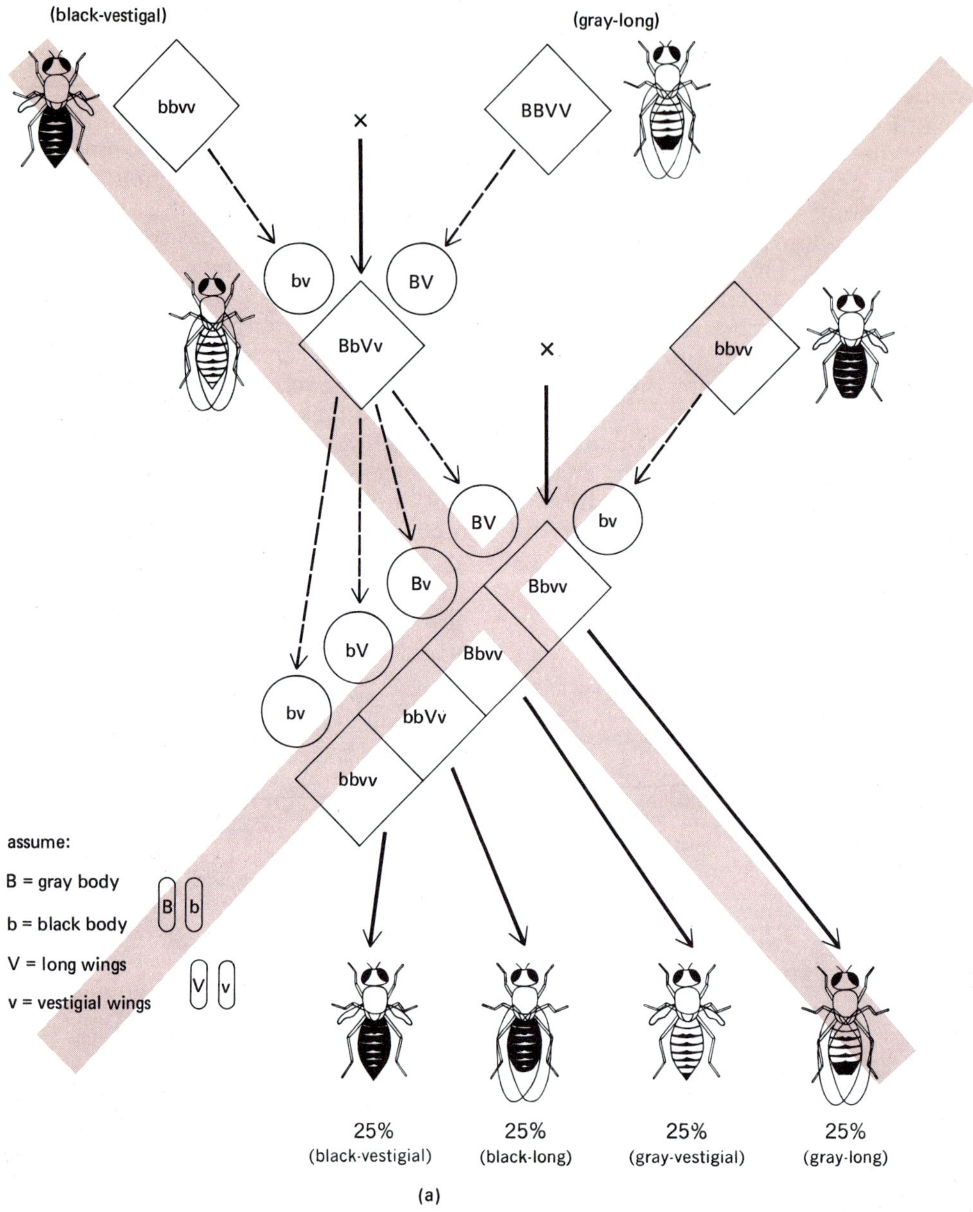

Figure 14–13 One of Morgan's experiments. Black-bodied vestigial-winged fruit flies were mated with wild-type flies (gray-bodied normal-winged). As expected, all the first-generation progeny had the external appearance of the wild type. Their mixed heredity was revealed in a test cross: a mating with pure-bred black-bodied vestigial-winged fruit flies. But the second generation progeny did not show the distribution expected from an independent assortment of the body color and wing size traits (a). Instead (b, opposite), among most of the F_2 fruit flies the black body–vestigial wing and gray body–normal wing combinations were inherited together—a linkage group. Yet the presence of a consistent and significant proportion of flies in which the traits *were* assorted randomly indicated that crossing-over is also involved.

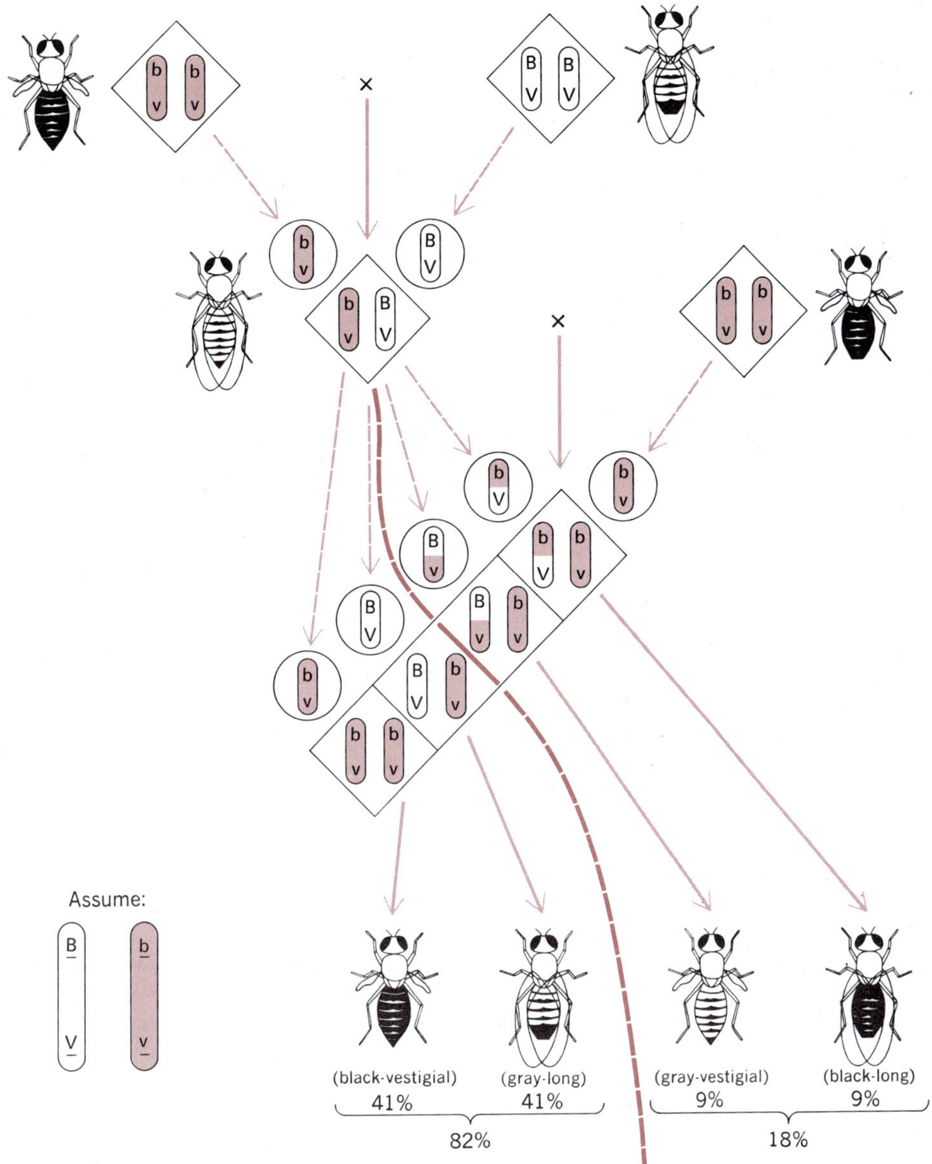

(b)

do—and can be raised easily in small milk bottles at room temperature. Fruit flies are easily handled: a whiff of ether will immobilize them for examination, and they soon recover unharmed. They show extremely characteristic sex differences; an experienced observer can rapidly separate the flies according to the contour of the body and distinctive markings, and thus ensure that the sexes will be kept separate until precise matings are arranged. *Drosophila* is not at all shy about mating in captivity; a cross of a pair of adult flies promptly results in a brood of a couple of hundred eggs. In about ten days these offspring grow into adults and in turn are capable of producing progeny of their own. The fruit fly has only four pairs of chromosomes, and the larval form has giant chromosomes, so large that the details can easily be discerned under a microscope. With all these advantages, it is small wonder that progress in genetics research was especially rapid once intensive use of the fruit fly was begun.

Morgan was first introduced to *Drosophila* in 1906, and there began a fortunate partnership, which was ultimately to bring to Morgan a Nobel Prize and to the fruit fly a degree of renown accorded to few species on earth. Working with this tiny fly, Morgan embarked upon a series of studies that opened up new avenues of approach to genetics, the consequences of which have not yet been fully realized. In the early 1900s, W. S. Sutton had proposed a chromosome theory that related Mendel's hereditary "factors" to the chromosomes. In this theory, Morgan's insight revealed an explanation for some puzzling results of his fruit fly studies. Some of the traits that he studied followed Mendel's law of independent assortment quite well. But other pairs of traits did not seem to assort independently at all; instead, they seemed regularly to be inherited together. For instance, if a pure line of fruit flies carrying the recessive traits of black body and vestigial wings (Fig. 14–13) is crossed with a pure line of the normal or **wild type,** possessing the dominant gray body and normal-sized wings, the 9:3:3:1 ratio expected from the law of independent assortment is not obtained in the F_2 generation. Instead, approximately three gray-bodied, normal-winged fruit flies are observed for every one black-bodied, vestigial-winged fruit fly, with practically no black-bodied, normal-winged or gray-bodied, vestigial-winged individuals in the F_2 progeny. The characteristics of body color and wing size seem to be passed on as though they are linked together. Many other pairs of traits show similar behavior. In fact, Morgan discovered four groups of "linked" characteristics that generally are inherited together—a perfect correspondence to the four pairs of chromosomes observed under the microscope.

But Morgan's further studies of the newly discovered *Drosophila* **linkage groups** revealed another perplexing fact: the linked traits did not *always* stay together. Although the bulk of the F_2 generation from a cross of a pair of linked traits fell into a simple 3:1 ratio, there were still some individuals that showed the independent assortment of the traits. Somehow, in a small but consistent percentage of the offspring the chromosome sets were being shuffled.

There was by now too much evidence in support of the chromosome theory to discard it entirely because one set of data stubbornly refused to fit. Yet scientific scruples could not permit the troublesome data to be shoved into a dark corner and conveniently forgotten. In 1916, in a brilliant flash of insight, Morgan slashed through the apparent paradox. He proposed that portions of the chromosomes containing part of the set of genetic factors might be able to be exchanged between homologous chromosomes during meiosis (Fig. 14–14). Although definitive proof was not to come for many years, this indeed has proved to be the case, and the phenomenon has been descriptively named **crossing-over.**

Now Morgan had the basis for a comprehensive theory of the units of the chromosome —the genes—which would explain both Mendel's theories of the inheritance of characteristics and his own new findings on inheritance in the fruit fly.

Morgan visualized the chromosome as consisting of a series of discrete units of heredity, the **genes.** These are arranged in a specific linear sequence, each gene occupying a definite position or **locus** on the chromosome strand. Two homologous chromosomes each contain corresponding genes at each particular locus. A pair of corresponding genes on homologous chromosomes both govern the inheritance of the characteristic, but variations upon this characteristic might arise through a slight change or **mutation** in a gene. For example, the two genes governing body color of the fruit fly are situated on corresponding loci of homologous chromosomes of the fly. But one gene might be a gene for the normal gray body color, while the other gene might be one for the unusual black body color. Two contrasting variations of a trait are called **allelomorphs,** more commonly shortened to **alleles.** The terms *gene* and *allele* are often

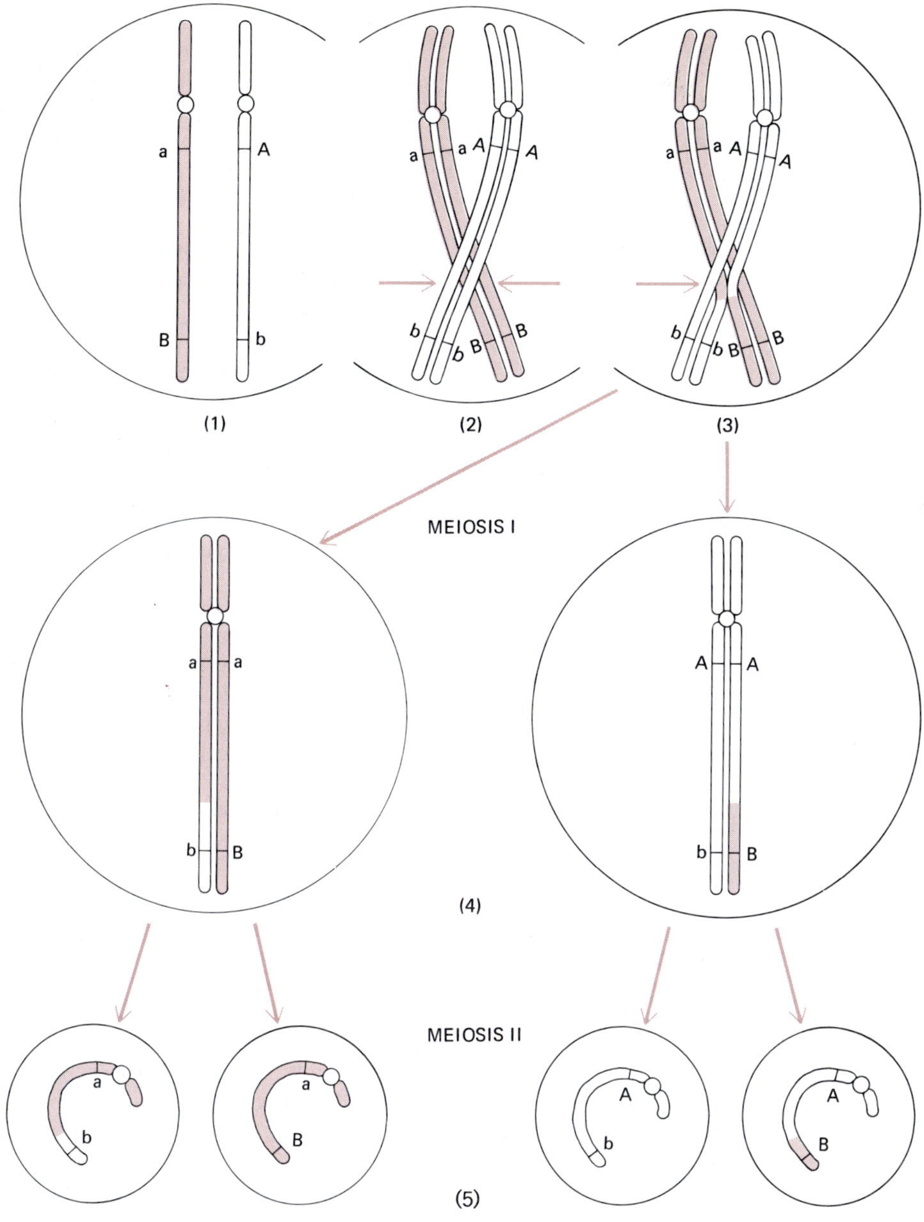

Figure 14–14 In crossing-over, portions of homologous chromosomes are exchanged during meiosis, resulting in a redistribution of linked genes.

used interchangeably, but the term *allele* emphasizes the fact that there are two (or more) alternative forms for a gene controlling a given trait. Thus, it is said that the gene for black body in *Drosophila* is an allele of the gene for gray body.

An individual fruit fly possessing both alleles, one for black and one for gray body, has a gray body, since the gene for gray body is dominant over its mutant allele. But if this fruit fly is mated to a black-bodied individual, half of the progeny will be black-bodied, half

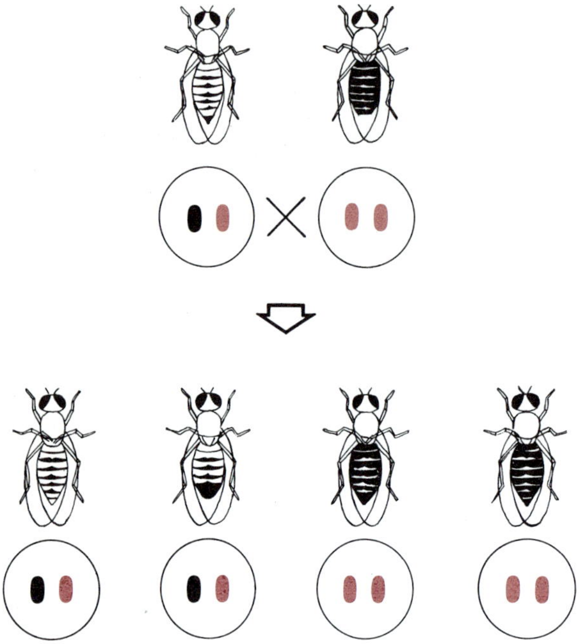

Figure 14–15 A cross of a heterozygous gray-bodied fruit fly with a homozygous black-bodied fruit fly reveals that the gray-bodied fly actually carried both body-color alleles. (Otherwise no black-bodied progeny could have been obtained.)

gray-bodied (Fig. 14–15). This mating clearly indicates that the gray fruit fly possessed both alleles, even though only one was overtly expressed.

The overt expression of a particular trait, such as blue eyes or brown hair, is referred to as the **phenotype,** while the assortment of genetic factors that the individual possesses with respect to the given trait is called the **genotype.** The phenotype is a description of appearances; the genotype can be determined only by inference, by a comparison of the characteristics of the individual with those of his parents and his offspring, in the light of knowledge of the principles of inheritance of the trait in question.*

*With the current prevalence of hair dyes, wigs, colored contact lenses, false eyelashes, etc., the human phenotype is fast becoming almost as difficult to discern as the genotype.

Since the chromosomes occur in pairs, a normal individual will have two genes controlling each trait. These may be the same or different, depending on the combined set he received from his parents. When both genes at a particular locus are the same, the condition is referred to as **homozygous,** and the individual carrying the identical genes is called a **homozygote** with reference to that trait. If the two genes at a particular locus on homologous chromosomes represent different alleles, the condition is known as **heterozygous,** and the individual is a **heterozygote** with reference to that trait.

Because one allele is generally dominant to the other alternatives for the trait,* it is not usually possible to determine whether an individual is homozygous or heterozygous by merely looking at him. Individuals with apparently identical phenotypes can have quite different genotypes. A mouse with a glossy dark-brown coat, when mated with an albino mouse, might have all brown offspring. Another brown mouse, indistinguishable from the first by the most careful examination, might in a similar mating produce a litter with some white offspring, some dark brown, and some a light amber brown resembling the fur color of neither parent. The genotype can thus be determined in this case only by observing the results of suitable matings.

On the other side of the coin, identical genotypes might not produce the same phenotype. For the genes determine only the direction that development will take, the ranges within which a given trait may be expressed. The end result of development is determined

*There are some exceptions to the dominant-recessive pattern. In cases of incomplete dominance, a heterozygote will have a phenotype in which *both* alleles are expressed. A common example is the inheritance of flower color in certain plants (fortunately for Mendel, not in the garden pea). In such cases, for example, crossing of a red-flowered variety with a white-flowered variety would yield F_1 offspring with pink flowers, instead of the red flowers that would be expected if simple dominance were the rule (Fig. 14–16). Another common example is the inheritance of blood type factors in humans, such as M and N, which are equally expressed, and A and B, which are also equally expressed and in addition are both dominant to the alternative recessive allele O.

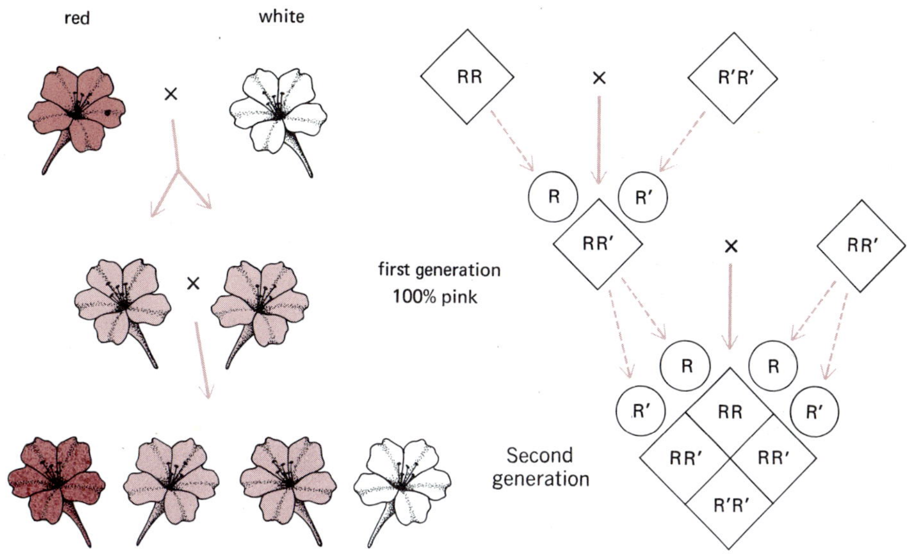

Figure 14–16 Incomplete dominance in the four-o'clock. Neither the allele for red flowers R nor that for white flowers R′ is dominant. First generation hybrids show the effect of both genes (pink flowers). In the second generation progeny, the parent types reappear, showing the expected distribution of the alleles.

not only by the basic genetic information, but also by the interaction of this information with the other genetic factors and with the complex of environmental factors at each stage of development. At each stage, key reactions take place that might be blocked or turned in another direction counter to the effect normally produced by a particular gene. A child with the genetic constitution of a creative genius might succumb to a case of phenylketonuria, which, unless promptly diagnosed and treated, can lead to mental retardation. Or an accident involving a severe blow to the head might damage his brain and similarly thwart the full phenotypic expression of his genotype.

Some Basic Techniques of Genetics on the Macro-Scale

From Mendel onward, many of the basic principles of genetics were worked out from a study of the results of selected matings or **crosses.** One of Mendel's most fruitful techniques was a cross between pure (homozygous) lines carrying different alleles, followed by crosses of the heterozygous or hybrid first-generation offspring with one another (or with themselves by self-pollination).

For example, Mendel crossed peas with yellow seeds and peas with green seeds (Fig. 14–17a). In the first-generation progeny, all the seeds were yellow. With what we now know about the chromosomes and genes and their behavior during gamete formation and fertilization, we can deduce that all the first-generation offspring were heterozygous and that the yellow allele is dominant to green. Thus the genotype of the F_1 peas can be represented as Yy.

Now what should happen if two Yy pea plants (with a yellow seed phenotype) are crossed? Each plant will form gametes, half of which carry the Y allele and half of which carry the y allele. Since the combination of the gametes is random, we should expect all possible combinations in the genotypes of the offspring: YY, Yy, and yy. Moreover, there should be twice as many Yy combinations as either YY or yy—i.e., a YY:Yy:yy ratio of 1:2:1. Thus, since the yellow allele is dominant, we can expect three-quarters of the F_2

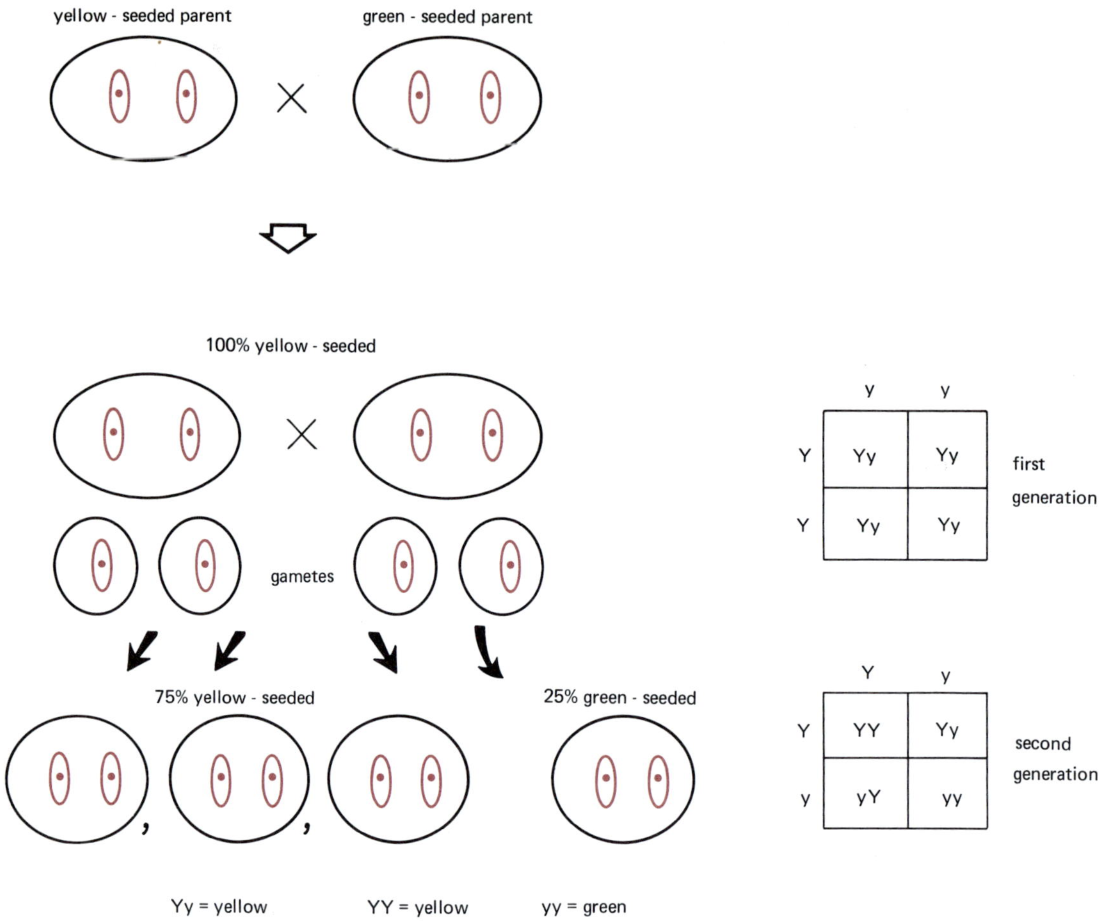

Figure 14–17 The Punnett square provides a quick and simple way to determine the predicted results of a cross.

progeny to have yellow seeds (1 YY + 2 Yy) and the other fourth to have green seeds (1 yy). This is almost exactly what Mendel obtained: the F_2 offspring consisted of 6022 yellow and 2001 green, for a ratio of 3.01 to 1.

The line of reasoning by which the expected ratio of a cross is determined can be succinctly summarized in a diagram, called the **Punnett square,** in honor of the English geneticist R. C. Punnett who first proposed it. All the possible gametes from each parent are arranged along two sides of a square, as depicted in Fig. 14–17b. At the projected intersections of each gene in each subdivision of the Punnett square, the appropriate combination of letters is filled in, to show how the gametes would combine in fertilization. The genotype ratio can then easily be determined by adding up the genotypes denoted in the subdivisions of the square. For a simple cross involving only one trait, the Punnett square has only four subdivisions. But expanded squares can be used similarly to work out the expected ratios in crosses where two or more independently assorting characteristics are considered—for example, the mating of a black guinea pig with a rough coat with a white, smooth-furred guinea pig (Fig. 14–18).

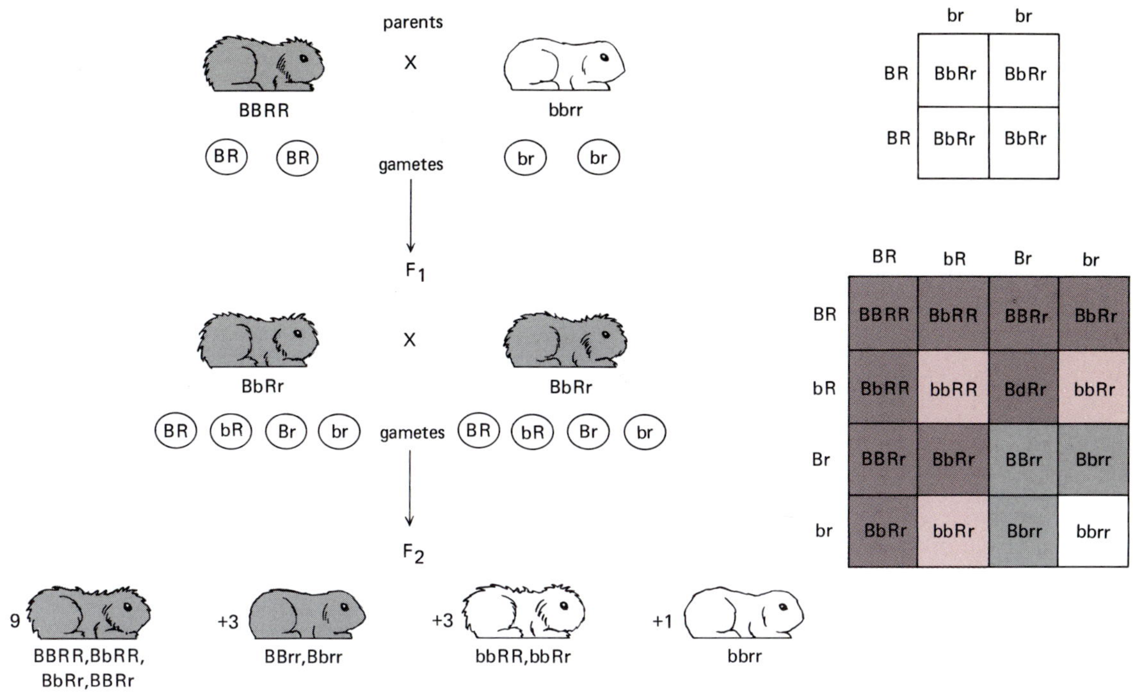

Figure 14–18 A dihybrid cross. Among guinea pigs, black fur is dominant to white fur, and a rough coat ("Abyssinian" variety) is dominant to a short, smooth coat.

Once the dominant-recessive relationships are known for a particular trait, the genotype of a particular individual can be worked out by a suitable **test cross,** also called a **backcross.** For example, suppose you purchased an unusual curly-haired mouse from a pet shop and wanted to know whether it was homozygous or heterozygous. (The curly-haired gene in mice, called *Rex,* is dominant over straight hair, just as it is in humans.) You could mate your curly-haired mouse with a straight-haired mouse (since straight hair is recessive, you would know that the latter mouse was homozygous). If the mouse in question were pure *Rex* (RR), then all the offspring of the cross RR × rr would have the genotype Rr, and would therefore all have the curly-haired phenotype. If, on the other hand, the curly-haired mouse were heterozygous, with only one *Rex* gene (Rr), then the test cross would yield a litter with (ideally) half curly-haired and half straight-haired mice (Fig. 14–19). By comparing the actual results of the mating with the expected ratios, worked out with Punnett squares, you could readily determine the unknown genotype.

The alert reader, particularly if he has ever raised animals or has tried to trace the inheritance of characteristics in his family, may have picked up an apparent flaw in the material presented thus far: in the world about us, characteristics are very seldom inherited in simple Mendelian ratios. In humans, brown eye color is dominant to blue eye color, and brown hair is dominant to blond. But a heterozygous brown-haired, brown-eyed man married to a homozygous blond-haired, blue-eyed woman may have *all* blond-haired, blue-eyed children, or all brown-haired, brown-eyed children, or any combination in between, including offspring with brown hair and blue eyes or blond hair and brown eyes. Two different litters from the same pair of mice similarly might show quite different

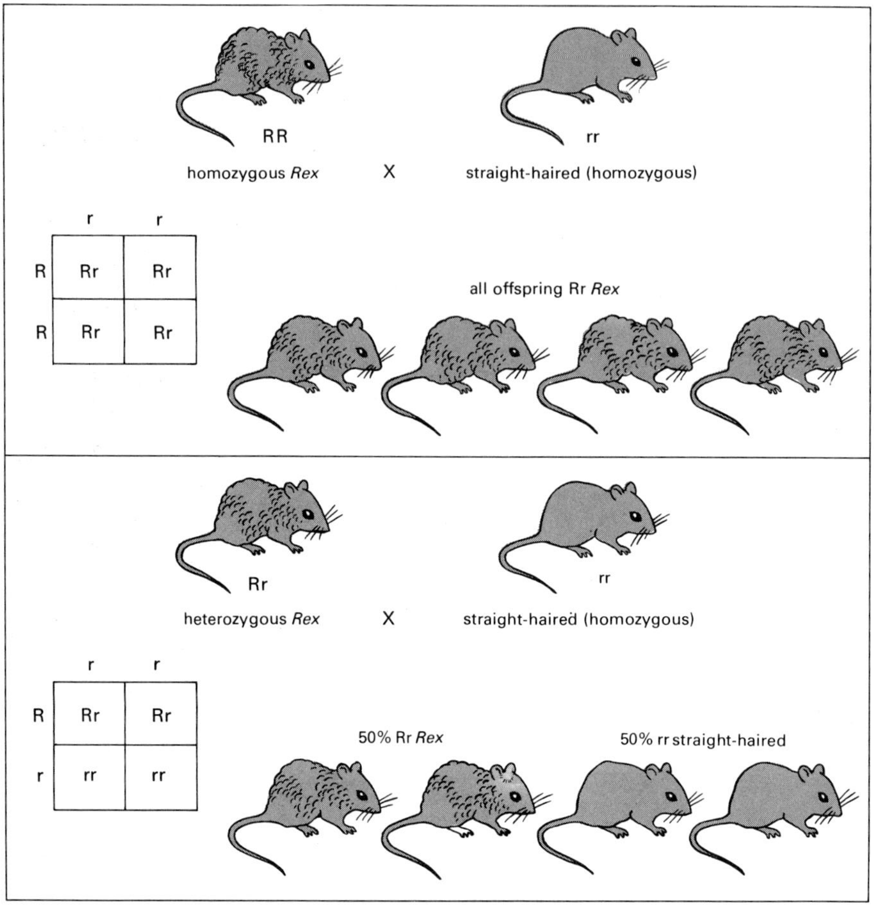

Figure 14–19 A test cross of a mouse with the *Rex* phenotype reveals whether it is homozygous or heterozygous.

ratios of fur colors, and neither of them corresponding to the expected ratio worked out on a Punnett square. Indeed, even the ratio of pea progeny in the Mendel experiment cited at the beginning of the section was not *exactly* 3:1. Are Mendel's laws invalid after all?

The answer to this disconcerting paradox lies in the fact that the laws of genetics are basically laws of *probability*: they describe the statistical probability that a given mating will produce a given result. And it is well known from considerations of statistics that in general the larger the size of the sample, the better are the laws of probability fulfilled. For example, if you toss a coin once, there is a 50 percent chance that it will come up heads and a 50 percent chance that it will come up tails. Let us say that you toss a coin and it comes up heads. Now, if you toss it a second time, you *again* have the same 50 percent chance of obtaining heads and 50 percent chance of obtaining tails. Anyone who has ever tossed a coin enough times knows from experience that in the first few tosses (unless the coin is loaded) almost any imaginable result might be obtained—half of each, all heads, all tails, or anything in between.* But

*The author just took a moment out to try it and obtained the following sequence in the first ten tosses: H T H H T H H H H T, or 70 percent heads and only 30 percent tails.

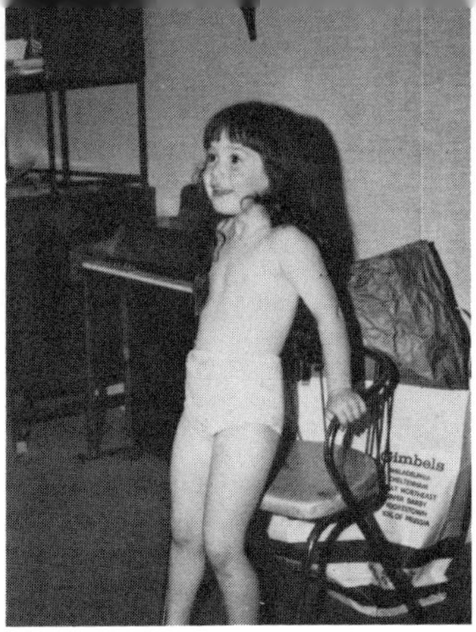

Figure 14–20 Multiple genes determine the inheritance of human skin color. The resultant differences in the amount and distribution of the dark pigment melanin produce a continuous gradation of skin tones.

with enough patience, as the series of tosses is extended, the ratio obtained more and more approaches the expected 1:1.

The same situation is observed in the case of genetic crosses. Although a single litter of mice may have a phenotype ratio far different from that expected according to the laws of heredity, a dozen litters obtained from similar crosses will show better correspondence, a hundred litters will show still better agreement, a thousand such litters even better, and so on.

Sometimes a small but consistent discrepancy from the expected ratios is observed and fails to disappear as the sample size is increased. Such cases have pointed up areas in which Mendel's laws do not apply entirely or are in need of modification, and have led to important breakthroughs in unlocking the secrets of the genetic mechanism. We have already encountered one such example, in Morgan's discovery of the principle of crossing-over.

In cases where a gross deviation from the expected Mendelian ratio is obtained, more than one set of alleles may be involved in the expression of the trait. For example, a cross of black guinea pigs with albino guinea pigs might give a ratio of 9 black to 3 brown to 4 albino among the F_2 progeny. Two different sets of alleles are involved here. One determines whether pigment will be produced or not (the recessive homozygote is an albino). The other allele determines the fur color—*if* pigment is produced at all. (Try working out the results of the cross, assuming the original black parent had the genotype CCBB and the albino parent ccbb, where C and c are the alleles for pigment production and albinism, and B and b are the alleles for black and brown fur color.)

The effects of the basic pair of alleles that produce blue or brown eyes in humans are modified by a variety of other genes, producing such variations as gray, green, and hazel eyes. In the case of skin color and height, it is believed that an undetermined number of multiple pairs of alleles all contribute to the expression of the trait, thus producing a continuous gradation.

The field of Mendelian genetics and the refinements introduced by Morgan and later researchers have provided a wealth of insight and detailed information about the nature of heredity in man and the other organisms of our planet. Space would not permit a detailed exposition of this material here, but before passing on, in the next chapter, to the most actively researched area of genetics today—the molecular level—let us briefly consider two more aspects of "macrogenetics" of great interest and practical importance to man: sex determination and sex linkage.

Sex Determination and Sex Linkage

In view of the overwhelming preponderance of similarities between the human male and the human female, in contrast to the few (important) differences, it is not surprising that when we examine the chromosomes we find that 22 of the 23 pairs of chromosomes are held in common by both male and female. These 22 pairs, the same in both sexes, are called the **autosomes.** The 23rd pair is referred to as the **sex chromosomes,** for here the only obvious difference lies. The human female has two homologous sex chromosomes, called the **X chromosomes.** The human male has one X chromosome, quite indistinguishable from the X chromosomes of a typical female. But the other member of the 23rd pair in the male does not "match." It is a different chromosome that has no exact counterpart either among the autosomes or among any of the chromosome complement of the female. The odd chromosome in the human male is called the **Y chromosome,** and it is somewhat shorter than the X chromosome (Fig. 14–21).

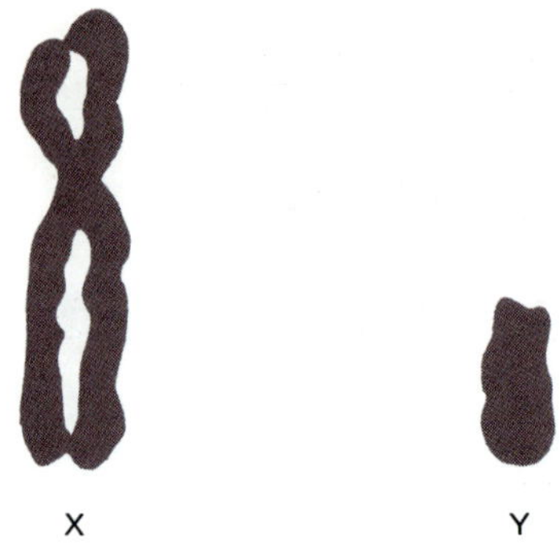

Figure 14–21 The human sex chromosomes.

During meiosis, the chromatids are segregated in such a manner that each gamete receives a set of 22 autosomes (one from each homologous pair) and *one* sex chromosome. Each gamete (and polar body) produced by meiosis in the female receives an X chromosome, since there was a homologous pair to start with in the diploid cell. But in gametogenesis in the male, the four haploid sperms produced do not have equivalent sets: each of the four receives a set of 22 autosomes, but two of them each receives a copy of the single X chromosome, while each of the other two receives a copy of the single Y chromosome. When a male gamete (sperm) fertilizes a female gamete (egg), the resulting zygote (fertilized egg cell) may have either an XX sex-chromosome pair or an XY chromosome pair, depending on which sperm united with the egg (Fig. 14–22).

In any particular fertilization, the probability of an XX genotype for the resulting offspring is 50 percent, and the probability of an XY genotype is also 50 percent. Each fertilization is entirely uninfluenced by the last—if one has had three girls in a row, the probability of having a boy the next time (based on purely genetic considerations) is still 50 percent. And, as in the case of the simple crosses discussed earlier, although the expected 1:1 ratio holds quite well when large samples are considered, great discrepancies may be observed for the extremely small sample that is the average human family. Nearly everyone is acquainted with a family that has longed for a boy (or girl) but has had two, three, four, five, or even more girls (or boys) in a row.*

*The problem of the ratio of the sexes is not quite so simple as a consideration of the XX and XY genetic combinations would indicate. Other factors apparently influence which sperm will fertilize an egg. For example, the Y-carrying sperm can swim slightly faster than the X-carrying sperm and thus reach the egg sooner more often than its X counterparts. Whatever the cause of the nonrandomness of the combinations, it has been estimated that the sex ratio at fertilization is over 130 males to each 100 females. The death rate of males before birth is proportionately higher, for reasons that will be discussed later in the section, and so by the time of birth the ratio has dropped to 107:100. The disproportionately high death rate of the males continues after birth, and by fourteen–eighteen years, the beginning of the reproductive age, the ratio has evened out to 100:100 (Fig. 14–23).

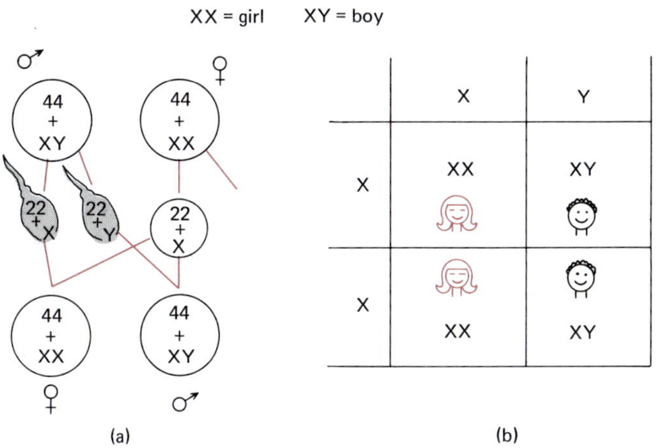

Figure 14–22 Genetic determination of sex in humans.

What is there about an XX combination of sex chromosomes that results in a female and an XY sex-chromosome combination that results in a male? An obvious conclusion would be that the Y chromosome must carry some sort of dominant gene for "maleness." Although there was for some time considerable uncertainty and controversy on this seemingly simple point, it is now generally agreed that, at least in man and the other mammals, this is indeed the case.

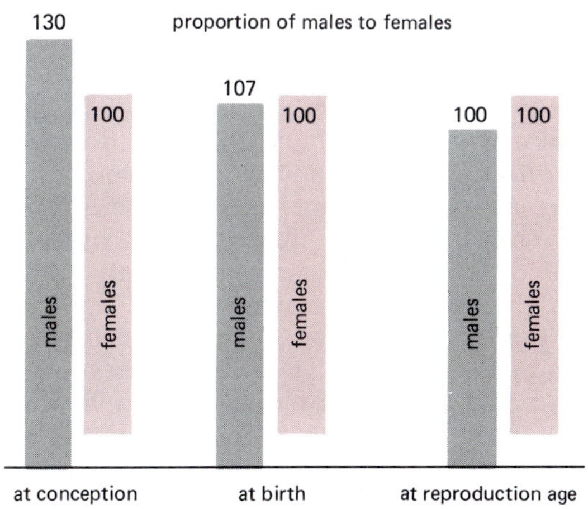

Figure 14–23 Human sex ratios in various age groups.

Interestingly, as so often occurs in the study of life, light was cast on the normal process of sex determination by a study of certain abnormalities. Sometimes, by an error of meiosis (nondisjunction), a gamete fails to receive a sex chromosome. If such a gamete combines with a gamete carrying an X chromosome, the resultant zygote has the genotype X0. Such a zygote is generally viable and grows into an adult that is morphologically female, although her sex organs do not mature fully. By a similar error of meiosis, a gamete sometimes receives an extra sex chromosome. Combination with a normal gamete will produce the genotype XXX, XXY, or XYY. The first will be a female, whereas the other two will have a recognizably male phenotype. An XXY male has male sex organs, but his body contours, particularly the breasts, are somewhat feminized. The XYY genotype has been in the news, on the basis of studies showing a significantly greater incidence of the genotype among criminals and unusually aggressive men than in the general population. (However, the fact that the genotype also occurs in some individuals who are apparently quite normally adjusted citizens emphasizes that the extra dose of "maleness" is not a cause of criminality, although it may be a predisposing factor.)

The confusion on the mechanism of sex determination in man was due to the rather conflicting results obtained on other species. The geneticist's favorite, *Drosophila,* also has a set of autosomes (three pairs) and a pair of sex chromosomes, XX for the female and XY for the male. In the case of the fruit fly, the Y chromosome is not shorter than the X, but it is readily distinguishable under the microscope by the characteristic hook on its end. If you were to jump to the "obvious" conclusion that the Y chromosome is responsible for maleness in *Drosophila,* you would be quite wrong. For an X0 fruit fly produced by nondisjunction is a male, even though it has no Y chromosome at all. On the contrary, a fruit fly with the genotype XXY—or even XXYY—is a female. Obviously, in this case the Y chromosome is not the sex-determining factor at all, although it is necessary for fertility (the X0 male is sterile).

Further studies of chromosome abnormalities in *Drosophila* have revealed that in this species sex is determined by the ratio of the X chromosomes to the autosomes. The X chromosomes apparently determine the traits of "femaleness," while the autosomes apparently carry the traits for "maleness."

The mode of sex determination varies considerably among species. In birds, moths, butterflies, and some fish and amphibia, the *male* has a pair of homologous sex chromosomes XX (sometimes denoted as ZZ to avoid confusion), and the female has a nonmatching pair XY (or WZ). Other species have an XX, X0 pattern of sex determination, in which there is no "odd" sex chromosome, and sex determination works according to the *Drosophila* pattern.

The existence of a pair of chromosomes specifically associated with sex raises some interesting questions. Do these chromosomes also carry some non–sex-related genes? And since in the male (in humans and many other species) the two sex chromosomes do not match, what happens to the expression of traits determined by the genes with no corresponding allele?

It is found that the sex chromosomes do indeed carry a great amount of nonsexual information, and in the male a unique situation is created: a recessive gene carried on the X or Y chromosome may be expressed even though it is present only in a "single dose" because there is no homologous allele to dominate over it. Such was the case for the first instance of sex linkage ever observed: the white-eye gene of *Drosophila* discovered by Morgan. The mutation appeared in a male who, as was later deduced, was carrying it on his single X chromosome. All the first-generation offspring of this white-eyed male, as expected, were red-eyed. When these heterozygous flies were mated with one another, a 3:1 ratio of red- to white-eyed second-generation offspring was obtained, but closer examination revealed that *all* the female granddaughters were red-eyed, while half the males were red-eyed and half white-eyed. White-eyed females were ultimately obtained by breeding white-eyed males with heterozygous females. The inheritance of the sex-linked white eye character in *Drosophila* is diagramed in Fig. 14–24.

The concept of sex linkage has explained many puzzling observations. We have already noted that human males show higher death rate than females at all ages and even before conception. This situation could be explained by the occurrence of various deleterious recessive genes on the X chromosome. A female would have to inherit a double dose of such a gene to be affected. In the male, however, with no allele to balance such a gene, its deleterious effects would be expressed in full strength, resulting in lower vitality and a higher death rate at various stages of development.

The concept of sex linkage also provides an explanation for certain diseases and disabilities, such as hemophilia and colorblindness, that are much rarer among females than among males. Characteristically such conditions are inherited in crisscross fashion: an affected male apparently transmits the condition to none of his children, but then the condition reappears among his grand*sons*—specifically his *daughter's* sons. Assuming that the condition is transmitted by a gene situated on the X chromosome, the puzzling pattern of inheritance is easily explained: none of the sons receives the defective gene from the father, since they inherit only a Y chromosome from him; all the daughters, outwardly normal, are carriers of

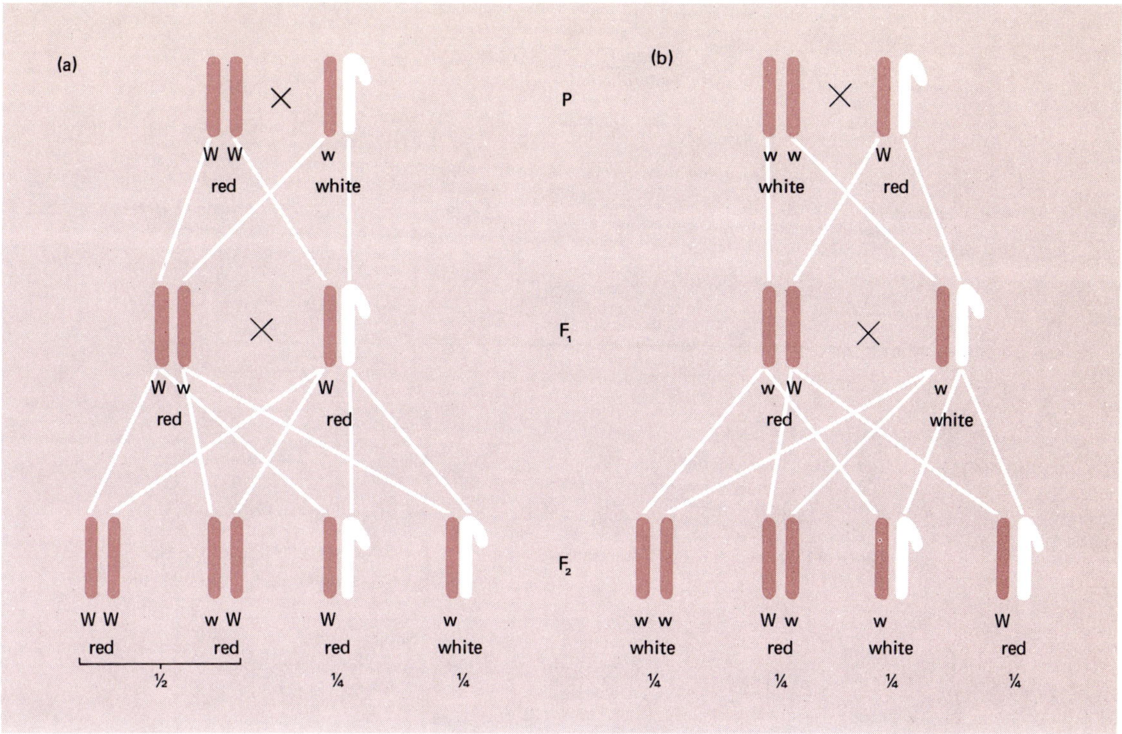

Figure 14–24 Inheritance of the white-eye gene in the fruit fly. The distribution observed in the F_2 progeny of the original male mutant (a) can be explained by assuming that the eye color gene in question occurs on the X chromosome, with no corresponding allele on the Y chromosome. White-eyed females were ultimately bred from white-eyed males (w) and heterozygous females (Ww). The results of their matings (b) confirmed the hypothesis of sex linkage of the white-eye gene.

the defective gene on the X chromosome that they received from their father. Half of their sons will inherit that defective X chromosome.

Over a hundred sex-linked genes have been identified in *Drosophila*. In humans, the study of pedigrees has revealed the existence of a number of sex-linked conditions, carried on both the X and the Y chromosomes. The most famous of these is the "royal hemophilia pedigree," which started with Queen Victoria and is diagramed in Fig. 14–25. Queen Victoria had eight children, four boys and four girls. One of her sons suffered from hemophilia, a defect of the blood-clotting system. The other children *seemed* unaffected. But as the further pedigree reveals, two of her daughters had also inherited an X chromosome with a gene for hemophilia and were carriers of the disease. (The other five children, including the ancestor of the present queen of England, were normal.) Hemophilia cropped up again and again among Victoria's male descendants, and indeed this "royal" gene may have changed the course of world history. For one of these descendants was the Tsarevitch Alexis, the only son of Tsar Nicholas II of Russia, and the sickliness of this much-beloved only son is thought to have been one of the indirect causes of the Russian Revolution.

In addition to the typical sex-linked characteristics inherited on the X or Y chromosome, there are some other characteristics that definitely seem to be associated with sex but do not show the same pattern of inheritance. These are the **sex-influenced** or **sex-limited** traits. Such traits are inherited on the autosomes, but their expression is altered by the hormone balance of the individual. For example, in mammals both males

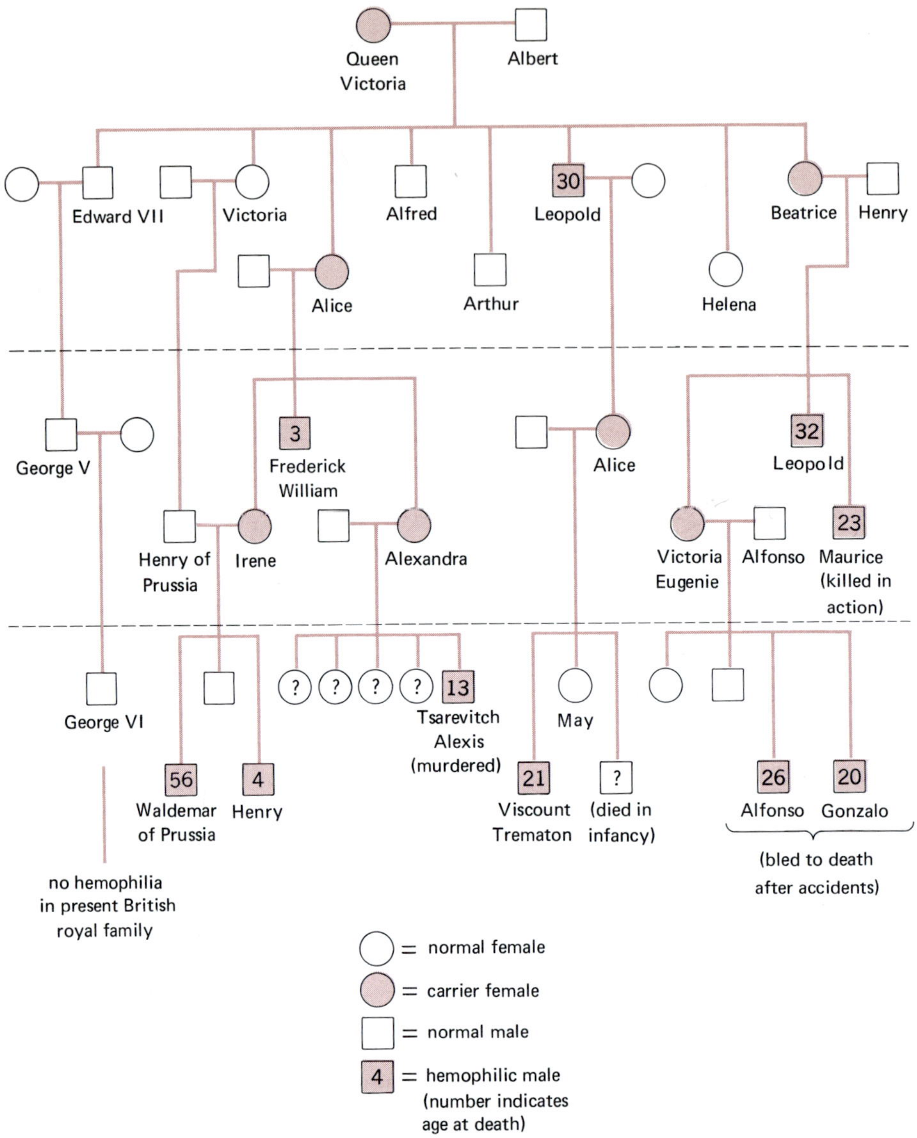

Figure 14–25 The hemophilia pedigree among the descendants of Queen Victoria. (Many of the grandchildren and great-grandchildren who did not inherit the recessive allele have been omitted.)

Figure 14–26 Like father, like son—pattern baldness is inherited by a sex-limited mechanism.

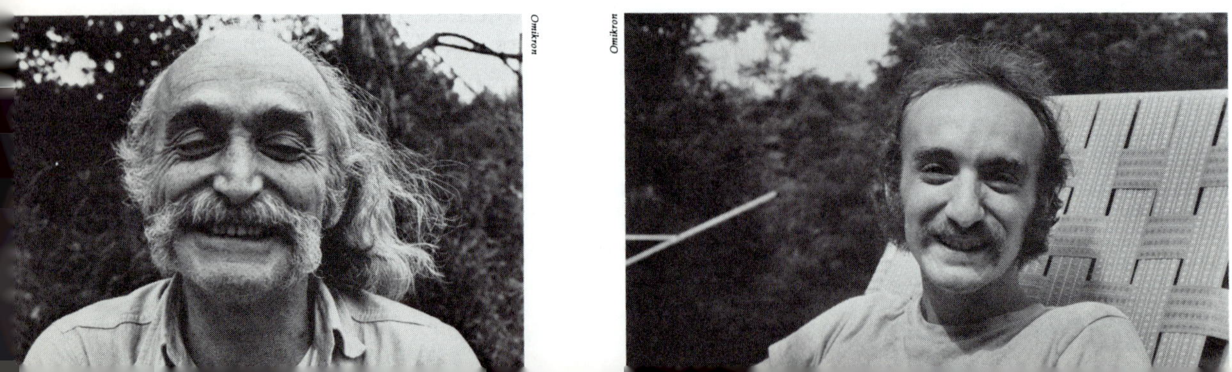

and females inherit genes for milk production, but only females have the appropriate hormones to permit expression of the genes. The fact that males, too, inherit the genes, is utilized in animal husbandry in the practice of progeny selection: champion bulls are selected for siring milk cows on the basis of the milk yields of their daughters, after preliminary selection from among the sons of champion milk cows. Thus the milk-yield genes are transmitted from champion milk cow to son (who gives no milk at all) to milk cow granddaughter.

A common sex-limited characteristic in humans is pattern baldness (Fig. 14–26). Bald men are quite common, whereas bald women are rare. A study of the pedigrees of pattern baldness indicates that the baldness gene seems to act as a dominant in males and a recessive in females. The key role of hormones in determining the expression of this gene is indicated by the lack of baldness among castrated males and its appearance in women with an excess of male hormone.

QUESTIONS FOR REVIEW AND THOUGHT

1. In the light of the explosion of genetics research at the gene level, is Mendel's contribution to genetics still valid and significant?
2. The number of chromosomes in a typical diploid set is directly related to the degree of development of a species. Is this statement true or false? Discuss.
3. Compare and contrast mitosis and meiosis.
4. Discuss the advantages of employing the fruit fly in genetics research.
5. Discuss the relationship between the chromosome set of an organism and phenomena of linkage.
6. Distinguish between gene and allele, genotype and phenotype, homozygous and heterozygous, mutant and hybrid.
7. Two blond parents can have only blond children. Two brown-haired parents may have brown-haired or blond children. Pure red hair often tends to "skip a generation," but many descendants of red-haired people have reddish blond or reddish brown hair. Characterize the brown, red, and blond hair alleles with respect to dominance or recessiveness (complete or incomplete).
8. A brown female mouse, mated with a brown male, gives birth to a litter consisting of six brown mice, two amber-colored mice, and two albino mice. Devise an experiment to determine the fur-color alleles that each parent is carrying.
9. A couple has had three daughters in a row. What is the probability that their next child will be a son? Explain.
10. Why do sex-linked disorders most frequently affect males? Can there be any sex-linked characteristics that affect females only?

CHAPTER 15

The Molecular Basis of Genetics

Redefining the Gene

For thousands of years man has observed the effects of gene action. But only in the past generation has he studied the gene itself. In the early decades of modern genetics research, the gene was looked upon as a discrete unit, the "atom" of heredity. Further studies revealed inconsistencies in this concept even as they cast more light on the ultimate nature of the gene and its action.

Meanwhile, suspicions were growing that a specific biochemical substance was the determiner of heredity. And strong presumptive evidence was accumulating to indicate that this substance, the stuff of which the genes are made, is deoxyribonucleic acid, DNA. Cytochemical studies with Feulgen's reagent, a dye that specifically stains DNA and only DNA (Fig. 15–1), revealed that this substance occurs in the chromosomes and, with few exceptions, nowhere else in the cell. Moreover, body cells were found to have generally twice as much DNA as the gametes—an excellent correlation with the known diploid and haploid number of chromosomes in somatic cells and gametes, respectively. These and other findings were intriguing evidence—but not proof—that DNA, intimately associated with the chromosomes, is the "stuff" of heredity.

More conclusive evidence was obtained in a historic series of experiments on the

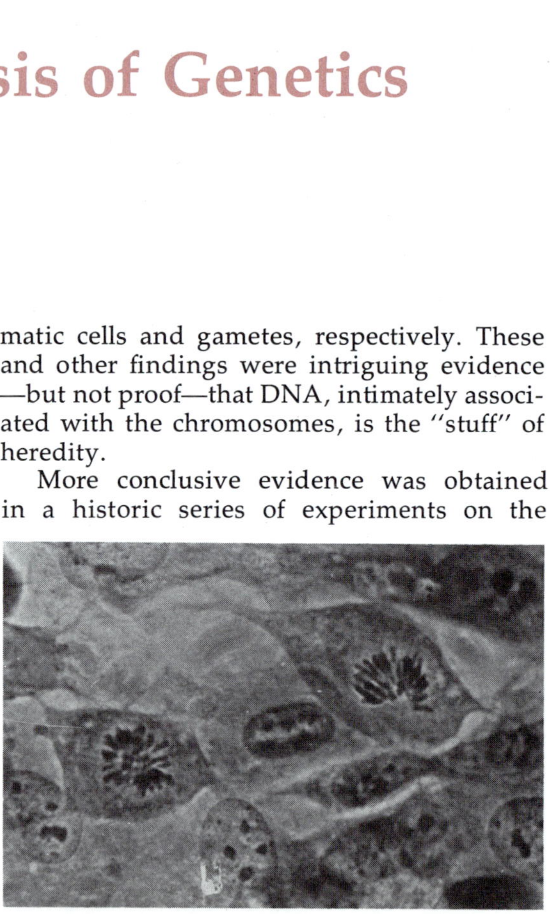

Figure 15–1 These fibroblast cells, cultured in vitro, have been stained with Feulgen's reagent, which makes the DNA in the chromosomes stand out sharply, dark against a paler background of unstained cytoplasm.

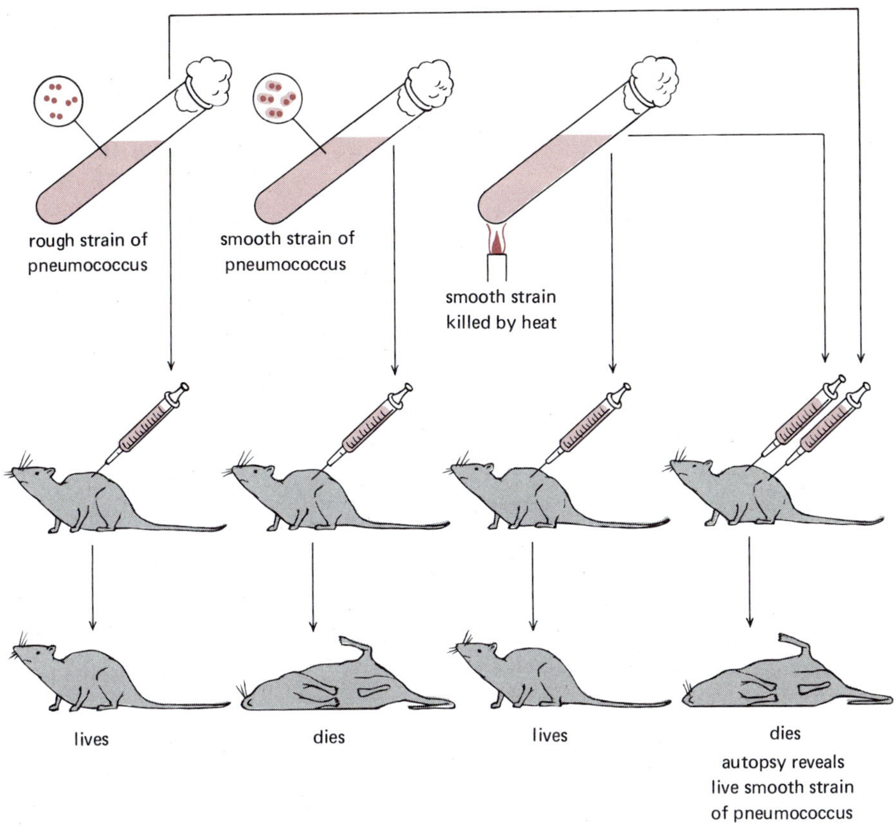

Figure 15–2 The Griffith experiment. The smooth strain of *Pneumococcus* was found to be virulent, while the rough strain was nonvirulent. The virulent properties of the smooth strain were lost when the bacteria were killed by heat. But if *both* the killed virulent bacteria and the nonvirulent rough strain were injected, the rats died. Somehow the nonvirulent strain was "transformed."

genetic material of the pneumonia bacterium. In 1928 the English researcher F. Griffith discovered that the virulent properties of a killed pneumococcus strain seemed to be transmitted to live pneumococci of a nonvirulent strain if the live and killed bacteria were injected simultaneously into mice (Fig. 15–2). Somehow the hereditary information from the killed virulent bacteria had been transferred to the living nonvirulent strain, and the latter had been transformed into a virulent variety.

The search for the "transforming principle" was on. In 1944 the research team of O. T. Avery, C. M. McLeod, and M. McCarty finally found the answer. Reproducing Griffith's experiments in a test tube and separating and testing various chemical constituents of killed bacteria, they discovered that only the DNA (deoxyribonucleic acid) could modify the heredity of living bacteria of a different strain.

The announcement by Avery and his associates that the "transforming principle" was actually DNA met with some surprise among geneticists. It had been thought that only proteins were complex enough to be determiners of heredity. Could the DNA molecule, too, provide sufficient variety and complexity? It consists, after all, of building blocks (nucleotides) each containing a mole-

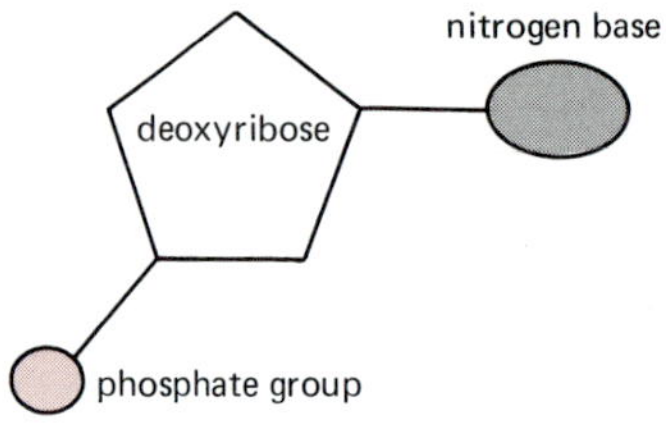

Figure 15–3 The structure of a DNA nucleotide. The 5-carbon sugar deoxyribose is joined to a phosphate group and a nitrogen base (purine or pyrimidine).

cule of the five-carbon sugar deoxyribose, a phosphoric acid residue, and an organic base, either a purine or a pyrimidine (Fig. 15–3). Only four different organic bases, the purines adenine and guanine and the pyrimidines cytosine and thymine, are generally found in DNA molecules. Can an "alphabet" of only four "letters" provide enough scope of variation for the thousands of different genes controlling the development of the varied traits that make up a complex living organism? Many eminent scientists seriously doubted that the nucleic acid molecule could provide sufficient flexibility.

These doubts were dramatically washed away with the announcement by James D. Watson and Francis H. C. Crick in 1953 of their model of the DNA molecule and the subsequent avalanche of research supporting their findings. This epic announcement was an outgrowth of the new influx of physicists and physical methods of analysis into biochemistry and the growing interdisciplinary field, **molecular biology.** On the basis of the known properties of DNA, both as a chemical entity and as a unit of heredity, and of detailed X-ray diffraction studies of the DNA molecule conducted by the physicist Maurice H. F. Wilkins, Watson and Crick suggested that DNA exists in the form of a double helix (Fig. 15–4), linked together by a series of paired nitrogen bases. The adenine of one spirally coiled chain is linked to a thymine in the corresponding position of the other, the guanine in one chain to a corresponding cytosine of the other, a thymine to an adenine, a cytosine to a corresponding guanine, and so on down the line.

A glance at the structural formulas in Fig. 15-4 reveals that the Watson-Crick model is perfectly in accord with the general structure of the nitrogen bases. They are of different sizes and possess different reactive groups, and only with the pairing of the bases indicated could a double helix of uniform width be obtained. Furthermore, an obvious consequence of such a scheme of base pairing is that the number of adenine units must be equal to the number of thymine units, and the number of guanines must be equal to the number of cytosines—exactly what is actually observed.

Another important consequence of the Watson-Crick model is that each strand of the DNA molecule is complementary to the other. For example, if the base sequence along a portion of one DNA chain is A T G C C T T T A A G G C, then the nucleotides on the corresponding portion of the other chain will exhibit the base sequence T A C G G A A A T T C C G.

The complementary nature of the two DNA chains suggests a mechanism for the

(a)

Figure 15–4 (a) James Watson and Francis Crick with a model of the double helix; (b, opposite) a schematic depiction of the DNA structure they proposed.

important self-replicating activity of DNA. If the two linked helical strands of the DNA molecule were somehow "unzipped" into two separate strands, and each of these strands served as a pattern or template upon which free nucleotides could be assembled to form a new polymer molecule, then the nucleotides on strand I would pair with the complementary free nucleotides, and the new polynucleotide assembled would be exactly complementary to it—in other words, *an exact replica of strand II.*

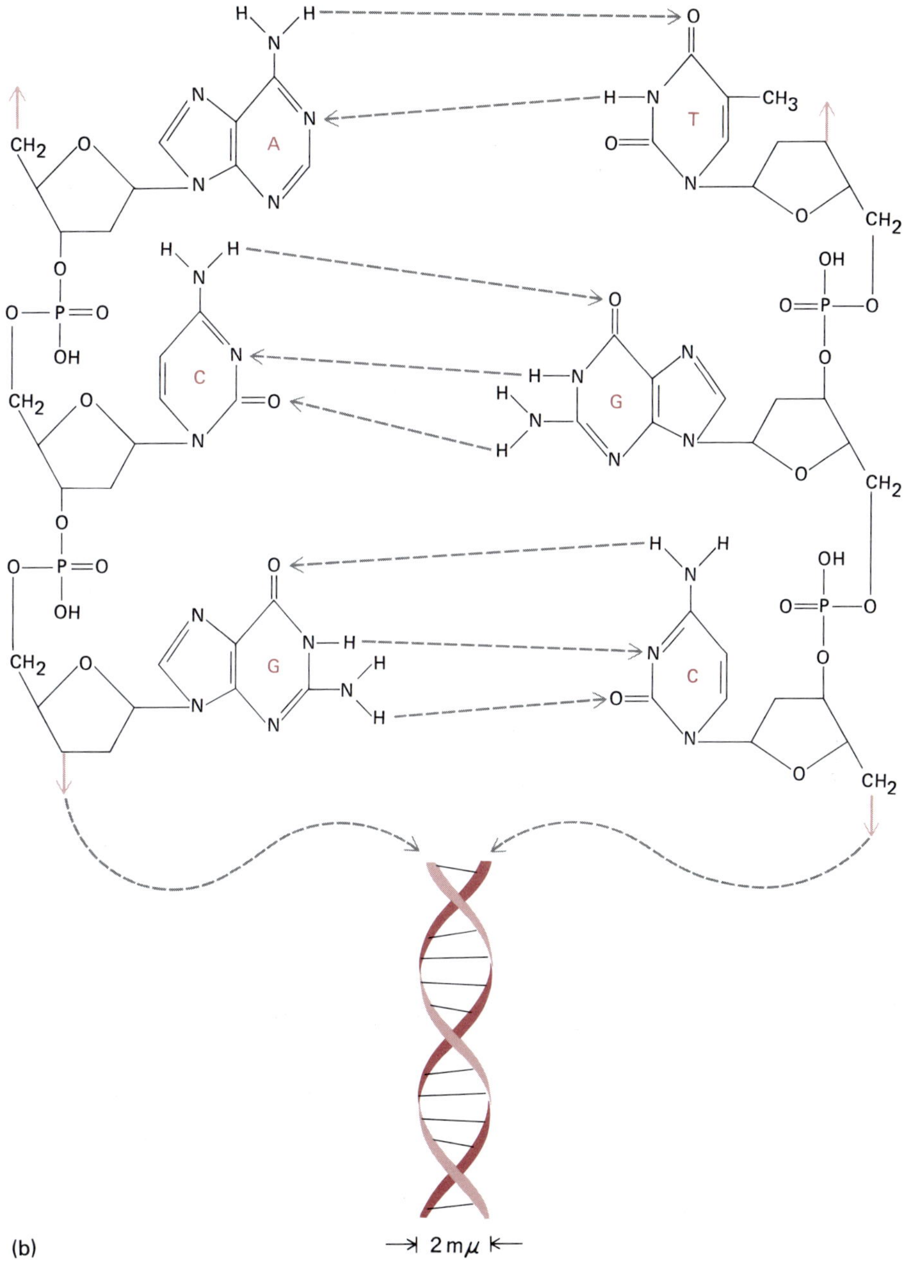

(b)

The DNA molecule as depicted by Watson and Crick thus satisfies an important criterion for genetic material: the ability for self-replication. Can such a molecule also serve as a unit of function, directly or indirectly controlling the production of all the diverse molecules of the cell and their finely coordinated activities? At first it might seem that a simple assortment of four different nitrogen bases would not possibly offer sufficient flexibility.

If we attempt to arrange the four letters A C G T in all possible linear sequences, we find a total of factorial 4 (4!) or $4 \times 3 \times 2 \times 1 = 24$ different combinations. But if we also consider all possibilities in which one or more of the letters in the four-letter "word" is duplicated—e.g., A A A C, A A C A, A T T G, G G G G, etc.—we wind up with a total of 4^4 or 256 permutations. Similar considerations for a twenty-letter message, spelled out in a four-letter alphabet, yield a total of $4^{20} = 1{,}099{,}511{,}627{,}776$ possible permutations. But the DNA molecule is not a mere four or even twenty nucleotides long—it usually consists of millions of nucleotides. For such a number of units, the possibilities for permutations become virtually infinite. Thus, the four-letter alphabet provided by the four DNA bases actually is quite ample to spell out any conceivable message or series of messages required for the activities of the cell.*

Watson and Crick's ingenious proposal,† along with the furious research into the genetic material that it sparked, has resulted in a new look at the gene (Fig. 15–5). The classical view of the gene as a single unit of function (determining a particular phenotype), of transmission (passing on the genetic information to the next generation), and of mutations (with inherent potential for giving rise to change in a trait) has gradually been superseded. S. Benzer, a leading proponent of the new view, has proposed that the unit of function be termed the **cistron.** Each cistron would presumably carry the blueprints for the synthesis of a single protein or polypeptide (the "one gene—one protein" hypothesis). The cistron may contain many sites of possible mutation; each such site is termed a **muton,** the smallest subunit of a cistron that can be modified to give rise to a mutant form. The third former meaning of the gene, the unit of transmission, is now to be called the **recon,** the smallest subunit that is interchangeable through genetic recombination, or a measure of the distance between adjacent mutons. The latest evidence seems to indicate that the muton consists of a single nucleotide pair. We shall now take a closer look at this muton and at the mechanisms of mutation at the molecular level.

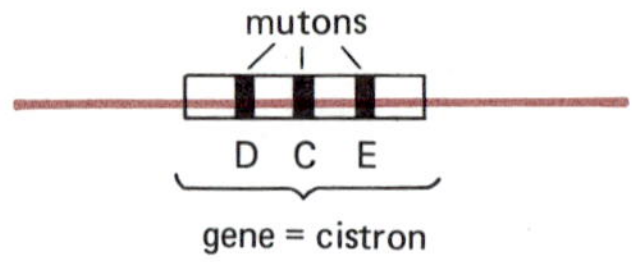

Figure 15–5 The cistron, the functioning gene, may contain numerous sites of possible mutation.

*Indeed, a surprising degree of diversity is provided by an "alphabet" consisting even of a mere two letters. The Morse code, made up of various combinations of "dots" and "dashes," can be used to transmit any message originally spelled out in the twenty-six-letter English alphabet. The language of the electronic computer likewise is based upon only two symbols, + and −. Yet computers are among the most versatile creations of man, capable not only of solving mathematical problems involving numbers of any size desired, but also of other tasks of various degrees of sophistication, including the translation of words frcm one language to another. Compared to these cases, the four-base alphabet of DNA seems positively profligate.

†For which they shared the Nobel Prize in physiology and medicine with Wilkins in 1962.

Mutations

The DNA molecule as envisioned by Watson and Crick has an incredible capacity for storing information. It has been estimated that the genetic information contained in the 46 chromosomes of a single human zygote, if transcribed in print, would fill a thousand books of 500 pages each. Let us assume that a team of expert typists recopied these 500,000 pages about 2^{45} times (the number of divisions the zygote and its daughter cells undergo in forming a human being), producing a total of more than 17 quintillion pages of genetic information. How many typographical errors might we expect the typists to make in recopying the material? One a page? One in 10 pages? If they were really good, one in 20, 30, even 50 pages of typing? The copying of genetic information in the body is fantastically accurate. "Typographical errors"—mutations—are made in no more than the equivalent of once in every *million* pages!

Most of these "errors" are minor. For example, a typist might leave out the "i" in "heredity" or perhaps misspell it with an "e." The resultant "heredty" or "heredety" would look rather strange, but these forms are not entirely nonsense. An intelligent reader would quickly realize what was meant. Similarly, a minor change in a gene might result in the development of a trait that was slightly changed but still functional in the larger overall plan of the organism as a whole. For example, a slight change in the development of pigment in the eyes of a fruit fly would produce a quite distinct change in its phenotype, but would not be intrinsically detrimental to the survival of the fruit fly as an organism.

But the "typographical error" might be a major one. Suppose, for example, that a typist copying the word "heredity" accidentally shifted her hands over on the keyboard. The word would then come out "gwewsurt." The word would be meaningless, and the puzzled reader would be forced to skip over it. Such major errors also occasionally occur in the copying of genetic information, and may result in a failure to synthesize a key enzyme or in the malformation of an important structure. Generally such major mutations are deleterious; in some cases they might have little effect for the organism as a whole, or in a very few cases might even be advantageous.

Mutations—sudden, inheritable changes in the structure of the genetic material—had long been observed and often capitalized upon by farmers and selective breeders. For example, the ancon breed of sheep (Fig. 15–6), characterized by unusually short legs, arose from a mutation that occurred in a New England flock back in the eighteenth century. The shrewd farmer who owned the flock noticed the new trait in one of his lambs in 1791 and realized its potential advantages: sheep with short legs cannot jump over pasture fences. Using the mutant animal as the breeding ram for his flock, he eventually derived a pure strain. Navel oranges, seedless grapes, and various other abnormal forms that man has found useful also arose from mutations that were utilized long before geneticists had any workable idea of what caused them.

Figure 15–6 Shrewd breeders have taken advantage of occasional useful mutations, such as short-legged sheep that cannot jump over fences.

Around the turn of the century, scientists finally began to catch up with practical breeders in this important area of genetics. The Dutch botanist Hugo de Vries, whose name we have already encountered in the amazing story of the triple rediscovery of Mendel's theories, was the first geneticist to study the phenomenon of sudden inheritable changes in characteristics. In the 1880s he became intrigued by striking variations in the evening primrose, a plant that had been introduced from America and was growing wild in European meadows. He collected a number of variant specimens and cultivated them. Some of the changes were stably passed on to subsequent generations. De Vries published a collection of his accumulated data and generalizations based on them in 1901 in a book called *The Mutation Theory*.

The study of mutations took a quantum jump when T. H. Morgan announced in 1910 that he had observed a new type of fruit fly with white eyes instead of the normal red type. An intensive search was launched for more mutations of *Drosophila*. New forms did not appear very often, but such enormous numbers of fruit flies were grown and examined that soon there were lines of fruit flies with eyes of varied hues and all shapes and sizes (and even no eyes at all); flies with bent wings, curved wings, dumpy wings, miniature wings, vestigial wings, and no wings; flies with black bodies, yellow bodies, ebony bodies; hundreds of new strains were bred from mutations that appeared in single individuals (Fig. 15–7).

Although in general the genetic information is duplicated and passed on with fantastic faithfulness, the processes of mitosis and meiosis do provide numerous opportunities for errors to creep in.

For example, the reductive division of meiosis may fail to occur, leaving the gamete with a doubled set of chromosomes. Or a mistake in pairing or a failure of the paired chromosomes to separate during the formation of gametes may result in a zygote possessing only one chromosome of a particular "pair," or perhaps even three. These changes in the number of chromosomes can result in an altered phenotype and may be passed on by heredity to subsequent generations.

Changes may also occur within the chromosomes themselves, through accidents at various stages of chromosome reduplication (Fig. 15–8). A chromosome may break, in one or more places, and the broken ends may heal together—but not necessarily in exactly the same manner in which they were originally attached. A portion of the broken chromosome might be omitted, or one segment might be turned around before rejoining, resulting in a chromosome with the same contents as the original one but a different base sequence. The segments might be rejoined in the wrong order, or a fragment from one chromosome might become attached to the end of another. Such chromosome mutations can often be observed in electron micrographs, and many such changes have been useful in marking chromosomes for the correlation of gene maps.

All these gross chromosome changes can cause modifications, sometimes profound, in the phenotype of an organism. But phenotypic changes can also be produced by finer alterations within the chromosome, changes in the chemical constitution of the genes themselves—often a single change in one of the DNA bases.

Gene mutations can be induced by a variety of physical and chemical influences. The first of these to be studied was the mutagenic effect of X rays, first discovered by the geneticist H. J. Muller working with *Drosophila*. In some of Muller's experiments, in which fruit flies were exposed to high doses of X rays, mutations appeared at a rate a hundred times greater than the spontaneous ("natural") mutation rate. The types of mutations that appeared were the same as those that

(a), (b), (c), Carolina Biological Supply Co.

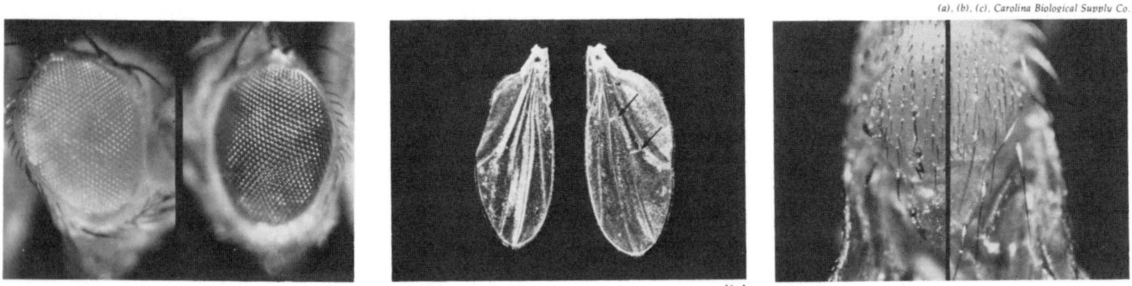

(a) (b) (c)

Figure 15–7 Some *Drosophila* mutations: (a) white eye and wild type, (b) crossveinless wings and wild type, (c) forked bristles and wild type.

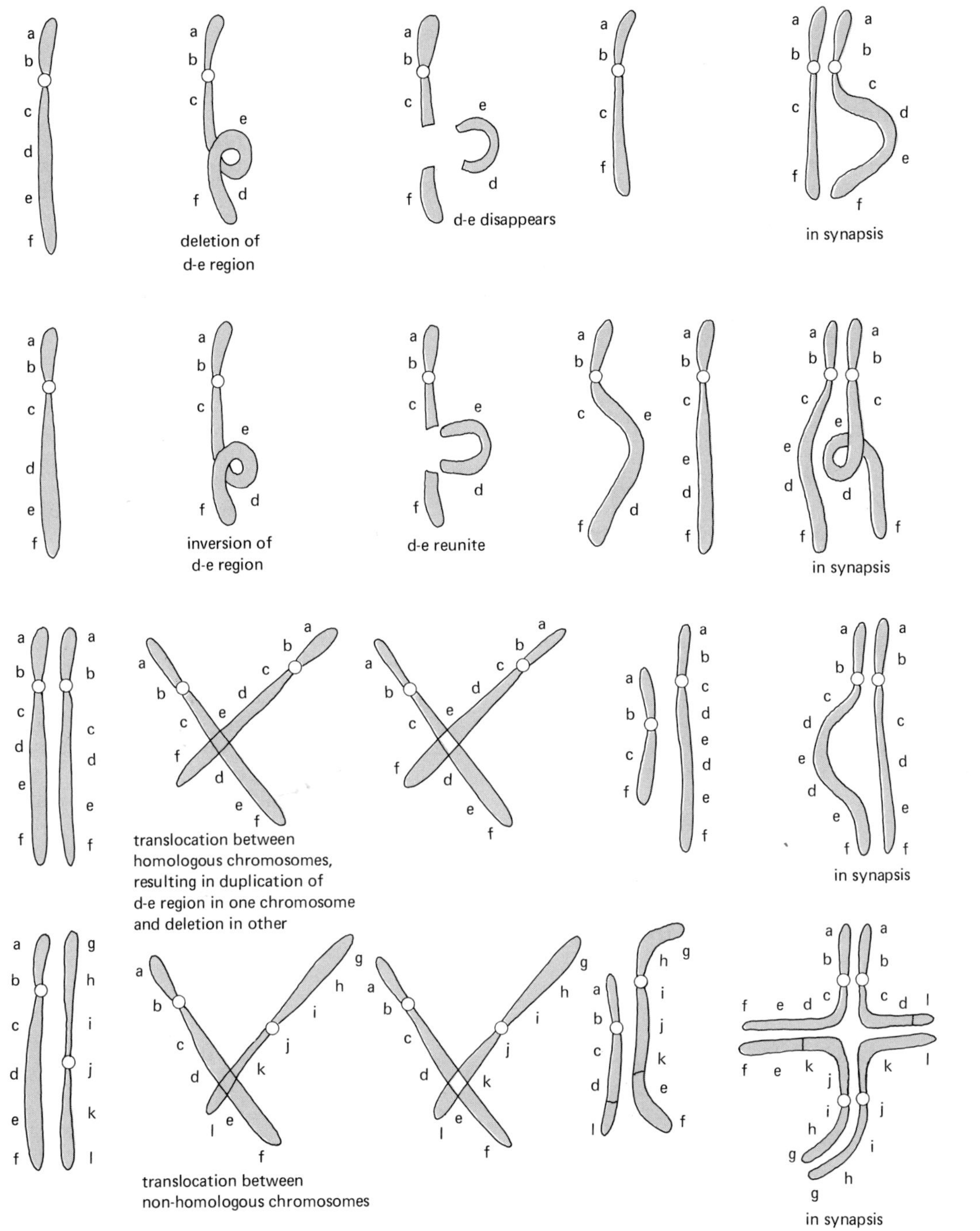

Figure 15–8 Various types of chromosome mutations. In synapsis (the pairing of homologous chromosomes), each type presents a characteristic picture in electron micrographs.

appeared spontaneously, only there were many more of them in the irradiated populations.

Subsequent studies established that X rays induce mutations in a variety of species of the animal, plant, and protist kingdoms. Soon it was learned that other types of radiations, including gamma rays, ultraviolet radiation, and neutrons, can also induce mutations in living organisms. The effects of these radiations seemed to be nonspecific, increasing the numbers of both chromosomal and gene mutations. It was assumed that the radiations cause some disturbance in the DNA molecule, resulting either in the destruction of some genetic unit, or in some chemical rearrangement culminating in a change in the genetic information.

In addition to radiations, many other factors have also been found to induce mutations. One of these is high temperatures. For example, fruit flies raised at 27°C show a mutation rate 2.5 times as great as that in flies raised at 17°C. Chemical substances, such as mustard gas and its nitrogen analogs, ethylurethane, phenol, nitrous acid, various analogs of the nitrogen bases of the nucleic acids (e.g., 5-bromouracil), and even such a widely consumed chemical as caffeine, have been found to significantly increase the mutation rate in various organisms.

Chemical mechanisms have been worked out for the action of some of these chemical mutagens. For example, the base analog 5-bromouracil is structurally similar to thymine (Fig. 15–9). If 5-bromouracil (BU) is incorporated in place of a thymine molecule in one of the DNA chains, then the base pair A–T is changed to A–BU. But 5-bromouracil is not *exactly* like thymine, and in the next DNA replication on the template containing the BU molecule, a copy error may be made: instead of the appropriate adenine molecule, the new DNA chain may contain a molecule of guanine. The base pair is now G–BU. In the next replication, using the new guanine-containing chain as a template, a molecule of cytosine will be lined up as the appropriate mate for guanine in the site where the original substitution of 5-bromouracil for thymine occurred. We now have a complete change in a letter of DNA "code" on both chains: in successive replications we have gone from the original A–T to A–BU, then G–BU, and finally G–C.

A mechanism along somewhat similar lines has been worked out for the mutagen nitrous acid. But there is an important difference here: nitrous acid is not an analog of one of the purine or pyrimidine bases. Instead it seems to act upon the bases within the DNA molecule, changing their pairing properties and resulting in subsequent copy errors. Mutations have been induced in such diverse specimens as viruses, bacteriophages, bacteria of various kinds, molds, and salmon sperm by treatment with nitrous acid. Yet this chemical (in the form of sodium nitrite) is used regularly as a preservative for meats and other foodstuffs and is found commonly in such preparations as frankfurters, bologna, and salami. Though no evidence has as yet

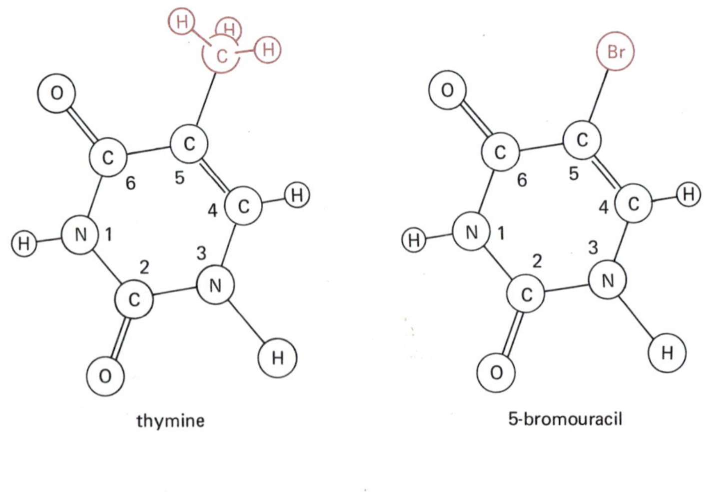

Figure 15–9 The base analog 5-bromouracil is extremely similar in structure to the normal base thymine. Its insertion into a DNA chain can result in a copy error.

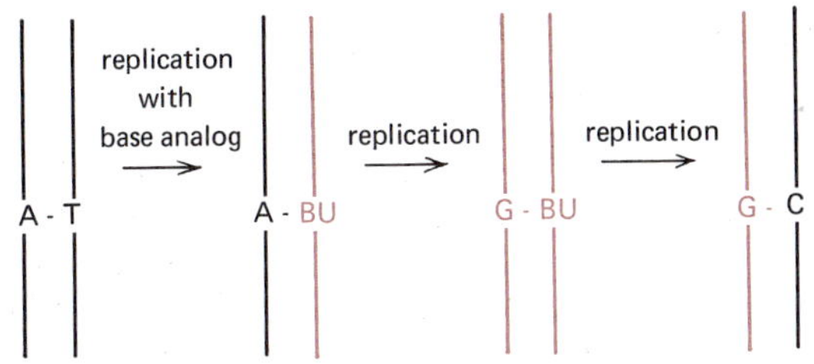

Figure 15–10 The mechanism of 5-bromouracil mutagenesis.

been obtained to indicate that nitrous acid would cause any mutagenic effects in humans, there are some indications that nitrite food additives may hold further dangers. In the body, nitrites can be converted to nitrosamines, chemicals that are known carcinogens (cancer causers). It might be wondered whether the full story of nitrites and their effects on man is yet to unfold.

Figure 15–11 Focuses of genetics research: (a) *Neurospora crassa,* (b) *Escherichia coli.*

(a)

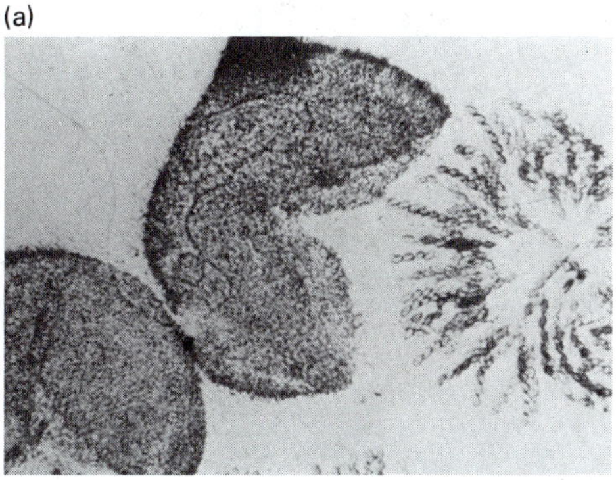

(b)

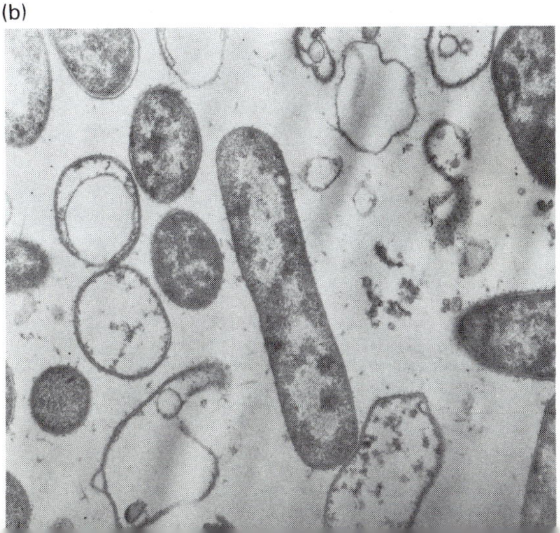

Experiments in Microbial Genetics

The discovery of *Drosophila* as a species for genetics experiments early in this century played an important role in the explosive growth of this field of biological science. With the aid of the fruit fly the gene theory was worked out, and such basic principles as sex linkage and the mutagenic effects of radiations were discovered. The second explosion of research in the field, in the second half of the century, has been based in large part on the growing use of another type of organism, in some respects even more suitable for genetic studies. Microorganisms—including the bread mold *Neurospora crassa,* the intestine-dwelling bacterium *Escherichia coli,* and the bacteriophages that live on this bacterium—are now supplanting the fruit fly in geneticists' favor.

Microorganisms offer enormous advantages in the laboratory. The fruit fly, which produces two hundred or so offspring in an interval of about ten days and can be stored away by the hundreds in little bottles on the laboratory shelf, was a quantum jump in convenience over the mouse, which produces a litter of perhaps ten at intervals of a couple of months and requires racks of space-consuming cages. In the same way, the bacterium, which can double its population in twenty or thirty minutes and will grow nicely in the millions in compact culture dishes and test tubes, is infinitely more convenient to work with in comparison with the fruit fly.

But amazing rapidity of multiplication, enormous population size, and convenience of laboratory cultivation are not the only features to recommend microorganisms as subjects of genetic studies. In multicellular organisms, a change in genetic information and its ultimate phenotypic expression are separated by a long chain of developmental interactions, which can be modified in various directions by a great diversity of environmental and other extraneous factors. In unicellular organisms, much of the complexity of interaction and interdependence is short-circuited, and the expression of the genes can in many cases be studied much more directly.

Most important, the available evidence indicates that with the exception of minor differences, the basic principles of genetics apply equally to *all* organisms, whether protist, plant, or animal. Thus the findings obtained on microorganisms can be utilized to gain a valid insight into the genetic processes of higher organisms.

Because of the extreme smallness of the individual organisms and the tediousness of trying to discern traits on such tiny and numerous individuals, the examination of bacterial phenotypes is usually performed on colonies. These are macroscopic growths, each containing many individual organisms, which form on a culture dish with suitable solid medium after a portion of the bacterial culture has been inoculated onto it (Fig. 15–

Figure 15–12 These three Petri dishes were inoculated with three different strains of *Leuconostoc mesenteroides,* isolated from sugar cane juice. Note the difference in colony size. The colonies in dish 1 were so transparent that print could be read through them, while those in dish 3 were opaque.

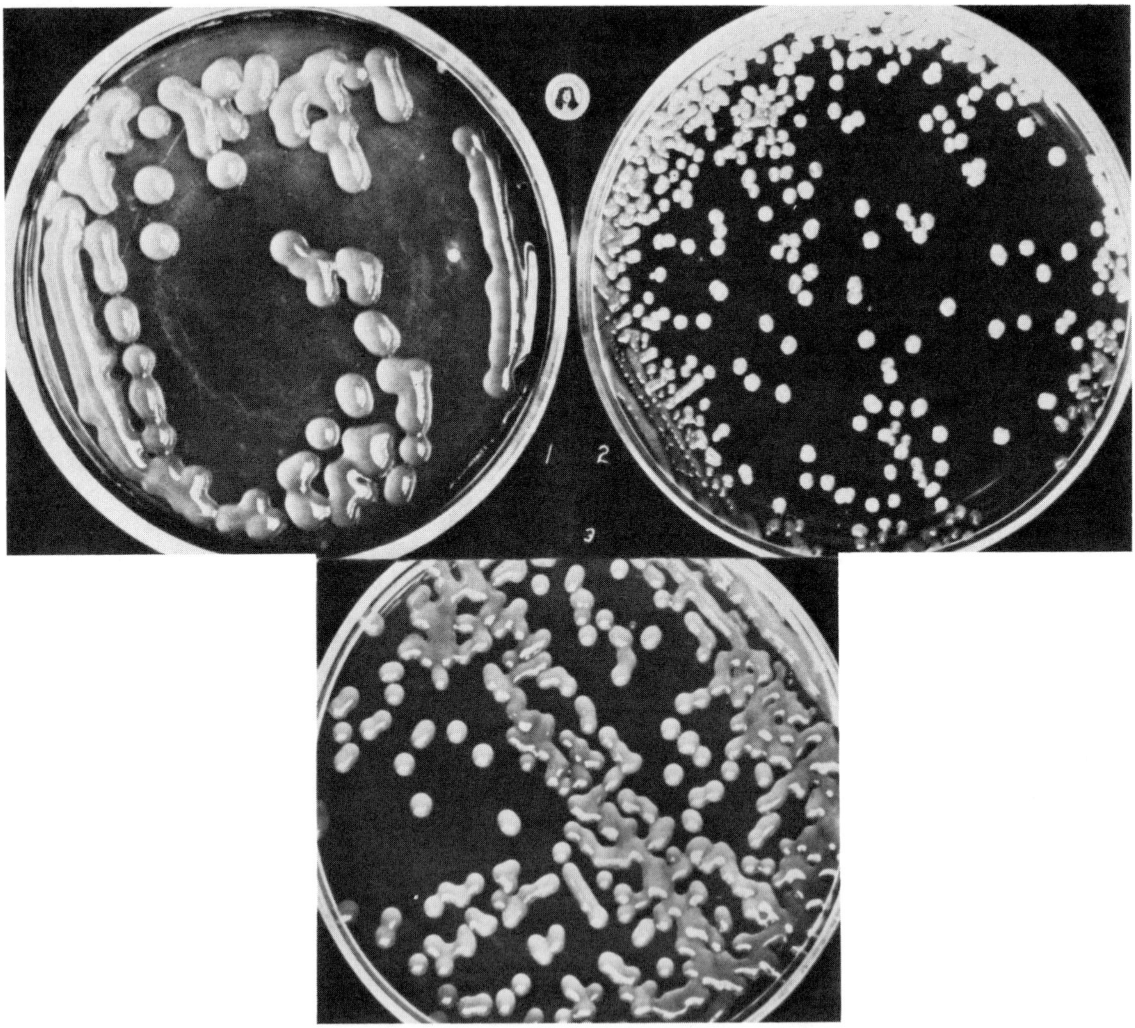

12). The procedure is genetically valid, since each colony is a **clone,** a genetically uniform population derived by mitotic division from a single cell. Various hereditary characteristics of the colony may be studied, including size and morphology, resistance to antibiotics, and dependence on specific nutrients.

Although bacteria normally reproduce by simple binary fission, they can also engage in a form of sexual reproduction called conjugation, thus providing opportunities for an exchange of genetic material. Gene recombination can also occur through the mediation of an infecting bacteriophage, in a somewhat similar manner to the transformation of pneumococci that led to the discovery of DNA as the material of heredity.

Much valuable genetic research has also been conducted with bacteriophages. These tiny viruses consist basically of DNA surrounded by a protein coat, with a sheathed tail that serves to attach the phage to a bacterial cell (Fig. 15–13). The DNA of the phage then penetrates into the bacterium (the protein coat remains outside) and literally takes over the genetic mechanisms of the bacterium, directing it to produce more and

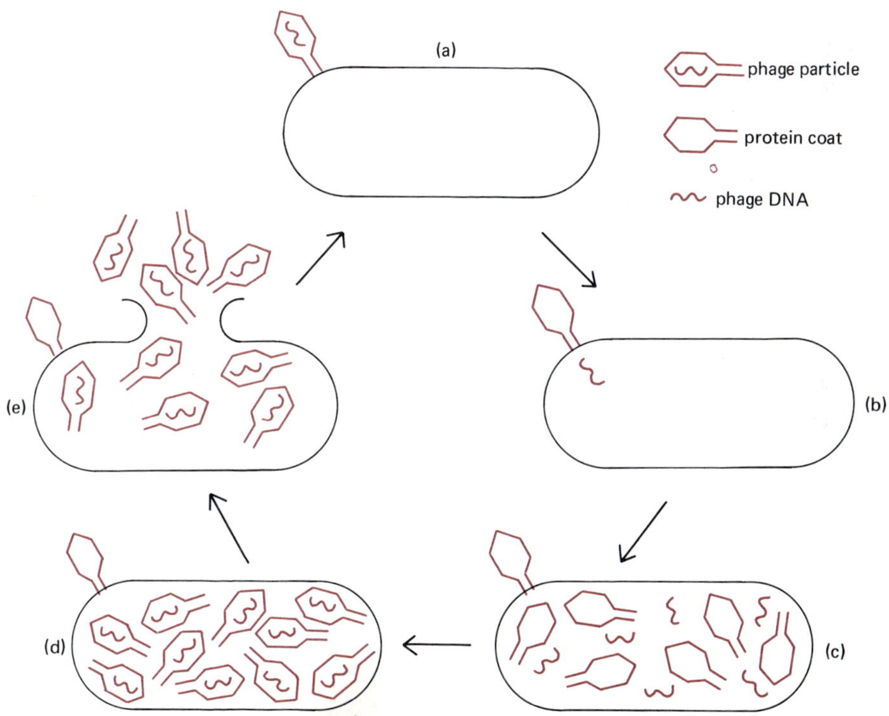

Figure 15–13 The life cycle of a bacteriophage. The phage attaches itself to a bacterial cell (a) and injects its DNA (b), leaving its protein coat outside. The phage DNA then directs the synthesis of new phage components inside the bacterial cell (c), using the resources of the bacterium. The new phage particles are assembled (d), and then the bacterium bursts (e), spewing out new phage particles, each of which can begin the cycle again on a new host cell (a).

more new phage particles. Eventually the bacterial cell is lysed and bursts, spewing forth a crop of new phages.

In a way, bacteriophages can thus be regarded almost as "naked genes," and as such they offer a number of unique advantages for genetic studies. Simple as they are, phages can mutate and even exhibit a form of genetic recombination like that of bacteria. Since phages are so minute (visible only under electron microscope magnification) and morphologically simple, their genetics is studied at the level of the colony rather than the individual. Major characteristics usually considered are the size and morphology of **plaques.** These are clear patches formed on a solid layer of bacterial growth as a result of lysis of the host bacteria (Fig. 15–14). Host range is another inherited trait that can be followed: most bacteriophages can infect only certain specific bacterial hosts.

Viruses that affect plants and animals are also being profitably studied by geneticists. Most popular has been the **tobacco mosaic virus** (TMV), a rod-shaped form consisting of a central rodlike core of RNA (which acts like the DNA of bacteriophages), surrounded by a thick cylindrical protein covering. The tobacco mosaic virus has been crystallized; the structure of its protein has been delineated, and the nucleotide sequence of its RNA has been partially determined. The protein covering and the RNA core have been taken apart and put back together again. In fact, researchers have even been able to "reconstitute" a viable virus from the protein covering of one strain and the RNA core of another. (In such cases, the resultant virus always exhibited the characteristics of the strain from which the RNA was taken). Various mutations have been induced in TMV with such mutagens as nitrous acid, and a start has been made toward discerning the relationships between specific changes in the RNA nucleotides and corresponding changes in the TMV protein.

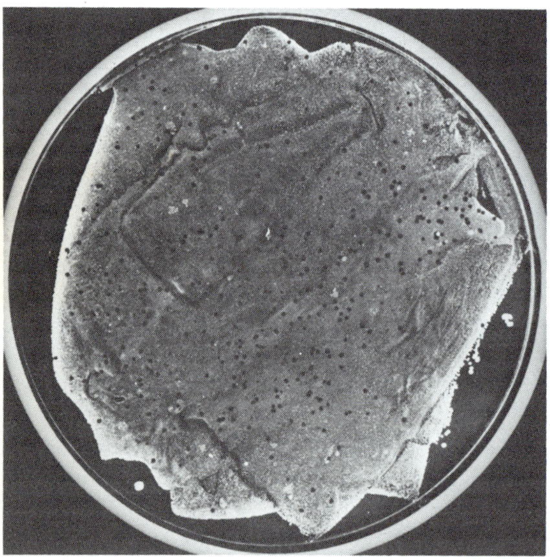

Figure 15–14 A mixture of a young culture of *Staphylococcus aureus* and staphylococcal bacteriophage was spread over the agar on this dish. A total of 381 phage plaques can be seen scattered over the film of bacterial growth.

DNA–The Master Chemical

DNA is sometimes referred to as "the code of life." It contains the blueprints for the fantastic diversity of structures and biochemical entities that make up a living organism. Just what is this code? How is it "read"? How is this information actually translated into the "hardware" of the cell? The mechanisms by which DNA controls the activities of the cell are not clearly understood. But it is known that DNA produces messenger RNA, which in turn directs protein synthesis at the ribosomes. DNA also produces two other types of RNA. Transfer RNAs collect the activated amino acids, the building blocks of the proteins, and bring them to the site of protein synthesis. Ribosomal RNA constitutes the bulk of the cellular RNA and "reads off" the message from the messenger RNA (Fig. 15–15).

The DNA of each cell in our bodies contains the information required for the building of all the structures of the body and the functioning in all its constituent ele-

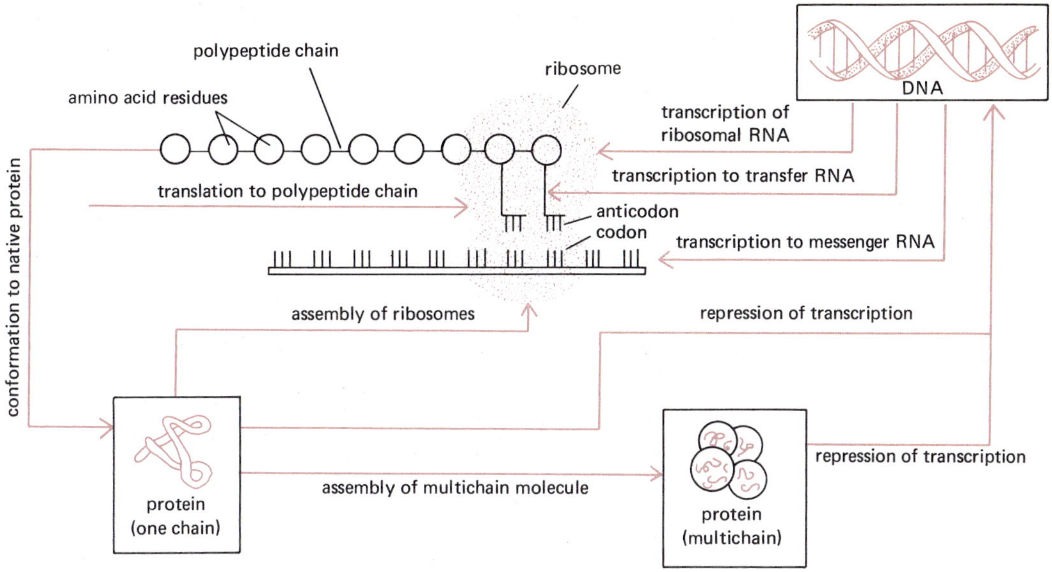

Figure 15–15 Flow of information in the cell. Parts of the DNA template are transcribed as three types of RNA: ribosomal, transfer, and messenger RNA. The transfer RNA's collect amino acids and deliver them to the ribosomes, where they are assembled into peptide chains according to the instructions inscribed in messenger RNA.

ments. Yet most of our body cells are highly specialized, performing only their own characteristic functions. It is clear that most of the genes in any particular cell are "turned off" most of the time, although they can be activated under special circumstances. Just how a particular gene on a DNA molecule is "turned on" and begins to manufacture RNA is still a matter of lively interest and debate in scientific circles.

Studies of bacteria by the French researchers François Jacob and Jacques Monod have cast new light on the problem of what genes are and how they work. Jacob and Monod have discovered **regulator genes** that regulate the activity of the structural genes that carry the actual blueprints for protein synthesis. A regulator gene may be situated at some distance along the chromosome away from the gene or gene complex that it regulates. Next to the structural gene, or genes is another type of controlling gene, the **operator gene.** It acts like an on–off switch, controlling the protein-synthesizing activities of the adjacent genes. The structural genes controlled by a single operator gene are usually associated with the synthesis of enzymes or other proteins with closely related functions. Together with their operator gene, they form a complex called an **operon** (Fig. 15–16). For example, the *lac* operon studied by Jacob and Monod consists of an operator gene and three structural genes, each corresponding to a different enzyme involved in the bacterium's lactose metabolism.* Each gene (and the regulator gene) can mutate independently, and a mutation in each would have a different effect on the chain of biochemical reactions of lactose metabolism. In the view of Jacob and Monod, who won a Nobel Prize for their work, the regulator and operator genes interact with substances in the cytoplasm that can induce or repress their activity.

The theories of Jacob and Monod effectively describe the operation of genes in **procaryotes** such as bacteria—organisms with essentially naked DNA as the material of

*The *lac* operon has been isolated from *E. coli* phages by a Harvard Medical School research team headed by Jonathan Beckwith.

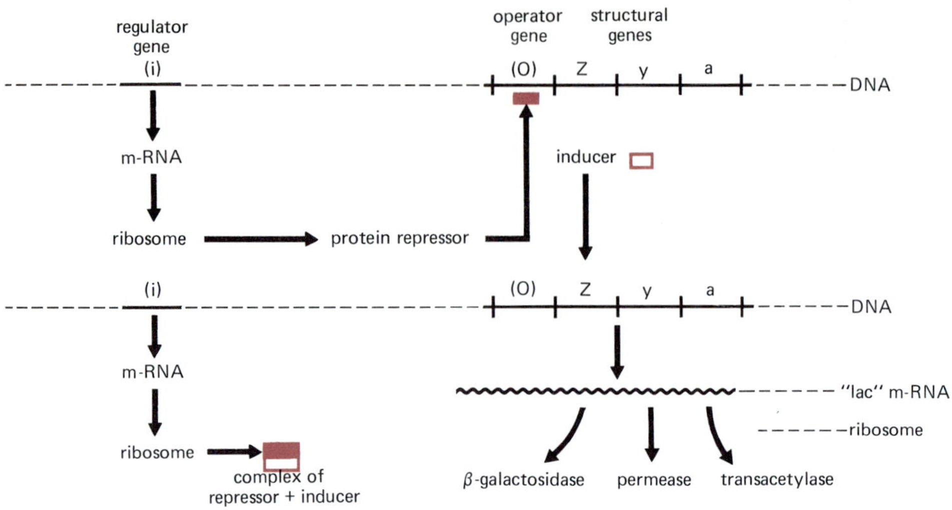

Figure 15–16 The lac operon. Each of the three structural genes controls the synthesis of a specific enzyme involved in bacterial lactose metabolism. All three genes are turned on and off by the adjacent operator gene, which in turn is regulated by a more distant regulator gene.

heredity. But in higher organisms, **eucaryotes,** the DNA of the chromosomes is associated with proteins, and the regulation of gene activity seems to work somewhat differently. According to one hypothesis, the **histones,** basic proteins associated with DNA in the nucleoprotein, form a protective sheath that keeps most of the DNA molecule inactive. Hormones in some cases have been found to be involved in the activation of DNA. Much experimental evidence seems to corroborate the presence of specific repressors, which keep the genes inactive unless they are removed. Other equally convincing evidence points to the existence of specific activators as key factors in gene action. New insights may ultimately result in a complete synthesis of the two viewpoints.

The explosion of genetics research in the past generation has resulted in an elucidation of many aspects of the mode of action of DNA—how it reduplicates itself, how it produces RNA, and how RNA functions in the synthesis of proteins.

DNA REPLICATION

The initial stages of DNA replication have sometimes been likened to the opening of a zipper. The two strands break apart at one end, and, one after another, the hydrogen bonds holding the base pairs together give way. Before the unzipping process has traveled the full length of the coil, an alignment of free nucleotides from the medium has begun at the free, single-stranded ends (Fig. 15–17). One strand—let us say, with the base sequence A C C G T T G A—attracts to itself free nucleotides containing the complementary bases—in this case, T G G C A A C T. These are lined up in the appropriate places, the sugar phosphate portions of the nucleotides join together into a continuous chain, and, one nucleotide after another, a complementary strand is assembled on the exposed template. Since this strand is complementary to the DNA strand on which it is being formed, it is exactly identical with the other DNA strand, which has just been unzipped from the template.

Meanwhile, the same process is occurring on the other freed DNA strand, resulting in the assembly of a new strand of DNA complementary to it and identical with the first original DNA strand. As the process of unzipping and assembly of new complementary strands on the exposed DNA templates

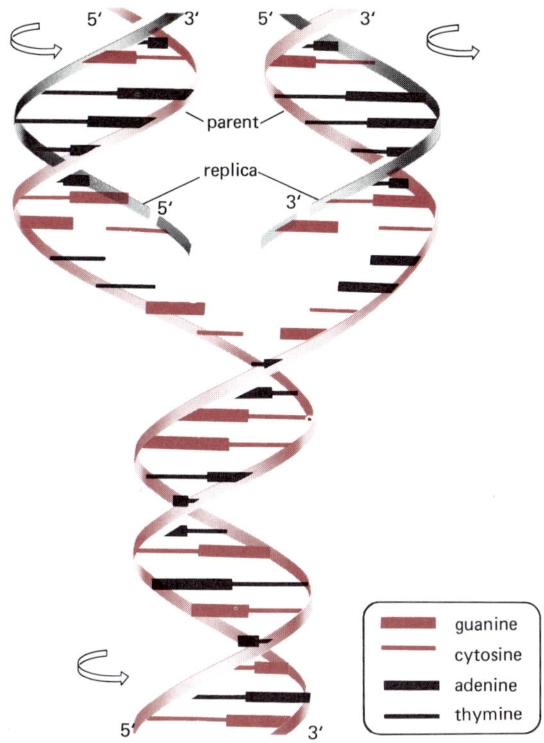

Figure 15–17 DNA replication according to the Watson-Crick model. Assembly of daughter strands proceeds on the exposed templates of the unwinding DNA double helix. When the process reaches the end of the double helix, there will be two complete, separate DNA molecules. Barring mutations, the two daughter helices will have an identical base sequence; each will contain one strand of "old" DNA and one newly synthesized strand.

reaches the end of the long helix, replication is complete. As can be seen from Fig. 15–17, we now have two daughter helices, each consisting of one strand of the parent DNA molecule and one strand of newly assembled DNA.

This is the scheme of DNA replication that follows logically from the Watson-Crick model of DNA. The reaction in scientific circles after its announcement was: It is elegant and intellectually satisfying, but does it work? Is this really what happens in the cell? Numerous subsequent experiments indicated that this is precisely the case. One corroboratory experiment involved the use of DNA containing the heavy nitrogen isotope N^{15}. In other experiments, conducted by Arthur Kornberg and associates, DNA synthesis was conducted in a test tube and proceeded precisely as predicted by the model.

DNA AND PROTEIN SYNTHESIS—THE GENETIC CODE

The cracking of the code of life, the transcription of genetic information in the DNA molecule, stands out in history with other great code-breaking achievements, such as the discovery of the Rosetta stone, which led to the decipherment of the hieroglyphic inscriptions of ancient Egypt. Before describing the exciting discoveries that provided insight into how the four-letter alphabet of the DNA molecule is translated by the cell into the twenty-letter amino acid alphabet of the proteins, let us briefly discuss the present knowledge of the process of protein synthesis itself and the role of DNA in directing this process.

It is believed that the process of protein synthesis is initiated when a portion of the DNA molecule is activated. In the presence of a specific enzyme—**RNA polymerase**—nucleotides are now assembled on one strand of the exposed DNA, forming a complementary polymer chain. This process is similar to DNA replication, but this time the nucleotides assembled possess the five-carbon sugar ribose instead of deoxyribose, and the nucleotide thymidylic acid (containing the purine thymine) is replaced by another nucleotide with similar pairing properties, uridylic acid (containing the purine uracil). As a result, the polymer synthesized is a chain of RNA, rather than the complementary DNA that is produced in DNA replication. For example, a portion of the DNA strand with the base sequence G T T C C A A C G would synthesize the complementary RNA strand: C A A G G U U G C.

This polymer RNA molecule, known as **messenger RNA** (mRNA), travels to the ribosomes, small subcellular particles (Fig. 15–18).

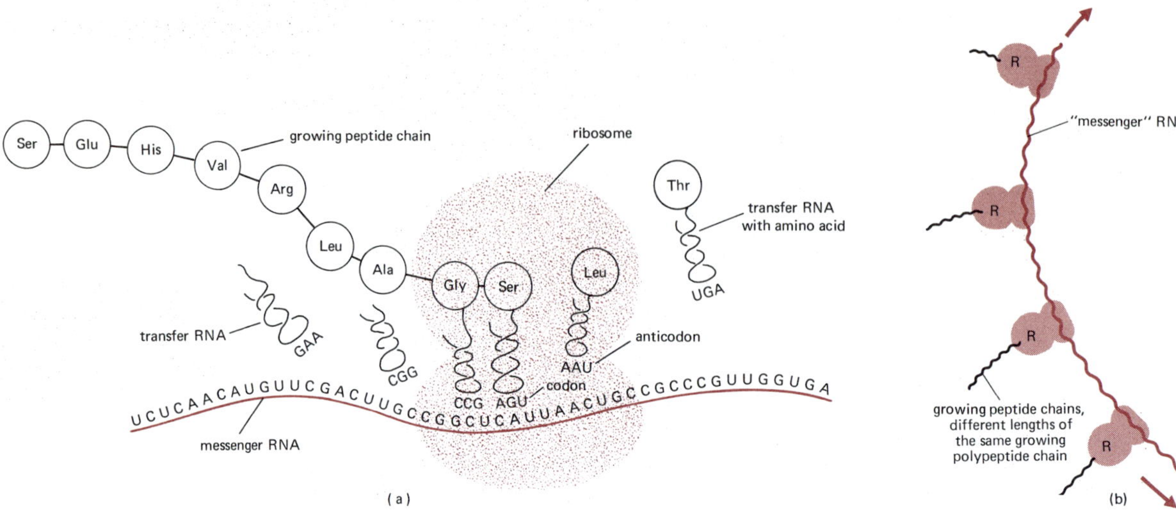

Figure 15–18 Protein synthesis at the ribosomes. (a) Amino acids are delivered by transfer RNA molecules and assembled into peptides according to the "code" inscribed in messenger RNA. (b) A messenger RNA molecule often threads itself through a series of several ribosomes, at each of which a peptide chain is synthesized. The resultant cluster is called a polysome.

Meanwhile, elsewhere in the cell, other RNA molecules of a different type, known as **transfer RNA** (tRNA), have been gathering up activated amino acid molecules. Each amino acid molecule has been activated by reaction with ATP and a specific activating enzyme, and it is picked up by a small and highly specific transfer RNA molecule. Each amino acid–tRNA combination is also carried to the ribosomes, which contain their own type of RNA, **ribosomal RNA.**

The molecule of messenger RNA, which provides a template complementary to the portion of the DNA template on which it was synthesized, forms a transitory complex with a ribosome and some tRNA molecules. The ribosome now moves down the mRNA chain, lining up appropriate tRNA molecules with their activated amino acid residues on the mRNA template. As the ribosome moves along, it gradually accumulates a dangling "tail"—a growing polypeptide molecule—as more and more activated amino acid residues are added according to the blueprint provided by the mRNA. Meanwhile, the messenger RNA may thread its way through several ribosomes, linking them together in a cluster called a **polysome.** (Such clusters were observed on the early electron micrographs of cells, and at that time were quite a mystery.) When the ribosome reaches the end of the mRNA molecule, it drops off, and the completed polypeptide or protein molecule is released.

Although the transfer RNA molecule contains several score nucleotides, three of these are of particular significance. For it is these three specific nucleotides which are complementary to a triplet of nucleotides on the mRNA template. This mRNA triplet is known as a **codon,** and its complementary triplet on the tRNA molecule is known as the **anticodon.**

It is the codons that form the essence of the code of life. For they provide the "dictionary" that permits the cell to translate from the four-letter alphabet of DNA (through the complementary four-letter alphabet of mRNA) to the twenty-letter amino acid alphabet of the cell proteins.

The initial breakthrough in cracking the code of life came in 1961, when Marshall W. Nirenberg discovered that a synthetic ribonucleic acid, polyuridylic acid (. . . U U U U U . . .), could serve as a template for protein synthesis in vitro, without the action of living cells. But the long polypeptide formed contained *only* phenylalanine, no matter what other amino acids were present in the medium! In the light of prior and subsequent evidence indicating that the mRNA codons consisted of triplets of nucleotides, it seemed evident that the codon for phenylalanine is UUU.

Almost immediately an explosive surge of research began to yield the codons for the remaining amino acids. At first only the nucleotide content of the codons was determined. But soon new techniques made it possible to ascertain the exact sequence of the bases in each codon.

The "dictionary" of the genetic code is presented in Table 15–1. From an examination of this table it is evident that most of the amino acids can be coded by at least two different codons. A particularly interesting case is that of the codons UAA and UAG, sometimes referred to as "nonsense" codons because they have not been found to correspond to any amino acid. There is evidence to indicate that these codons serve as signals to terminate the polypeptide chain.

The fact that the genetic code is read in triplets of nucleotides provides a basis for a molecular interpretation of some types of mutations. For example, experiments of Crick and associates on phage T4 indicate that certain types of induced and spontaneous mutations are caused by the deletion or addition of single base pairs. Yet in some instances, if there were *three* additions or deletions in the same chromosome, or if a single chromosome contained both a deletion and an addition, the wild-type phenotype was obtained again; that is, no mutation was observed. A consideration of the triplet code gives us an idea of how this finding might be explained.

Suppose the DNA nucleotide sequence corresponding to the wild-type phenotype is: C A T C A T C A T C A T C A T C A T, which is correctly "read" in triplet codons: C A T / C A T / C A T / C A T / C A T / C A T. If one of the base pairs were deleted from the DNA molecule, then since there are no intrinsic "punctuation marks" in the code (i.e., no "commas" to separate one meaningful triplet from another), the resultant shift of the bases would produce a new "message," spelled out in a different series of triplets: C A T /$_{\ominus}$A T C / A T C / A T C / A T C / A T. The effect of addition of an extraneous base pair would be simi-

TABLE 15-1 The genetic code

Codon	Amino acid	Codon	Amino acid	Codon	Amino acid	Codon	Amino acid
AAU	Asparagine	GAU	Aspartic acid	CAU	Histidine	UAU	Tyrosine
AAC		GAC		CAC		UAC	
AAA	Lysine	GAA	Glutamic acid	CAA	Glutamine	UAA	Stop signals
AAG		GAG		CAG		UAG	
AGU	Serine	GGU	Glycine	CGU	Arginine	UGU	Cysteine
AGC		GGC		CGC		UGC	
AGA	Arginine	GGA		CGA		UGA	Stop signal
AGG		GGG		CGG		UGG	Tryptophan
AUU	Isoleucine	GUU	Valine	CUU	Leucine	UUU	Phenylalanine
AUC		GUC		CUC		UUC	
AUA		GUA		CUA		UUA	Leucine
AUG	Methionine	GUG		CUG		UUG	
ACU	Threonine	GCU	Alanine	CCU	Proline	UCU	Serine
ACC		GCC		CCC		UCC	
ACA		GCA		CCA		UCA	
ACG		GCG		CCG		UCG	

lar: C A T /$_{\oplus}$G C A / T C A / T C A / T C A / T C A / T. But if both an addition and a deletion happened to occur, and they were close enough on the chromosome so that the incorrect portion of the "message" was small enough to be insignificant, the original reading might be virtually restored: C A T /$_{\ominus}$A T C/ $_{\oplus}$G A T/C A T/C A T/C A T.

Although much of the research on cracking the genetic code was done on microorganisms, there is considerable evidence that this code is universal throughout the kingdoms of life. It provides powerful insights that are already aiding in the understanding of hereditary diseases. For example, sickle cell anemia has been traced to an alteration of a single amino acid in each chain of the hemoglobin molecule: a glutamic acid is replaced by valine. A glance at the "dictionary" of the genetic code reveals that glutamic acid is coded by the nucleotide combinations GAA and GAG, while the codons GUU, GUC, GUA, and GUG cause the incorporation of valine into a growing peptide chain. A change in a single nucleotide A $\rightarrow$ U converts the glutamic acid codon to that of valine, and thus gives rise to pervasive changes in the functioning of the blood system.

With a basic understanding of the workings of the genetic code, scientists are now seeking ways to use this knowledge to modify and perhaps to improve the heredity of living organisms. A research team led by H. G. Khorana has already synthesized genes in a test tube without the intermediate agency of living organisms or substances derived from them. The growing field of genetic engineering is striving to learn how to selectively alter the genetic mechanisms of the cell, ultimately to correct or prevent such hereditary defects as phenylketonuria and sickle cell anemia. This exciting new field will be discussed in Chapter 22.

QUESTIONS FOR REVIEW AND THOUGHT

1. Has it been conclusively proven that DNA is the transmitter of heredity? Explain.
2. Describe the Watson-Crick model of nucleic acid structure.
3. How can the DNA molecule contain all the information necessary for the cell's activities in chemical sequences consisting of variations of just four nitrogen bases?
4. Distinguish among: gene, cistron, muton, and recon.
5. Discuss mutations from the molecular standpoint.
6. List and discuss factors that have been found to induce mutations. Do they all induce the same type(s) of mutations?
7. Progress in genetics in the past century has been marked by three major stages, centered upon studies of three principal organisms: the garden pea, the fruit fly, and *Escherichia coli*. Discuss.
8. Discuss the similarities and differences of DNA replication and RNA synthesis.
9. Discuss the working of the genetic code in protein synthesis.
10. Discuss mutations in the light of the genetic code.

CHAPTER 16

Reproduction and Embryology

Introduction

One of the major inventions of the future, as envisioned by some imaginative science fiction writers, will be the "matter duplicator." This device would be capable of analyzing, atom by atom, the structure of any material object—a stone, a television set, or even a man—and producing a virtually identical copy. The existence of such a device would have a profound impact upon our civilization, changing all our basic concepts of economy, government, and even personal relationships.

For various reasons, the matter duplicator as envisioned may never become a practical reality. Yet to some extent, our world already contains uncountable multitudes of functioning matter duplicators. Each living organism holds the potential for reproducing more of its kind. Through the process of reproduction the living species of the earth are perpetuated. For though each individual organism is mortal, life itself is provided a form of immortality through the succession of generations.

Reproduction, as its name implies, is a process in which one or more cells or organisms give rise to one or more new individuals of essentially the same type. Unlike the "matter duplicator" envisioned by science fiction writers, the reproduction of fine details may not be entirely faithful: although the basic patterns are preserved, the permutations of heredity may result in considerable individual variation among the offspring.

Two basic categories of reproduction are generally recognized. In **asexual reproduction** a single organism gives rise to one or more offspring. **Sexual reproduction,** on the other hand, generally involves the participation of two individuals. Each of these parents contributes part of its genetic material in the form of a sex or germ cell known as a **gamete.** The two gametes are of two different types (sexes); the male gamete is generally referred to as a **sperm,** while the female gamete is called an **egg** or **ovum.** Each of the gametes is generally haploid, containing just half the normal complement of chromosomes for the species. They

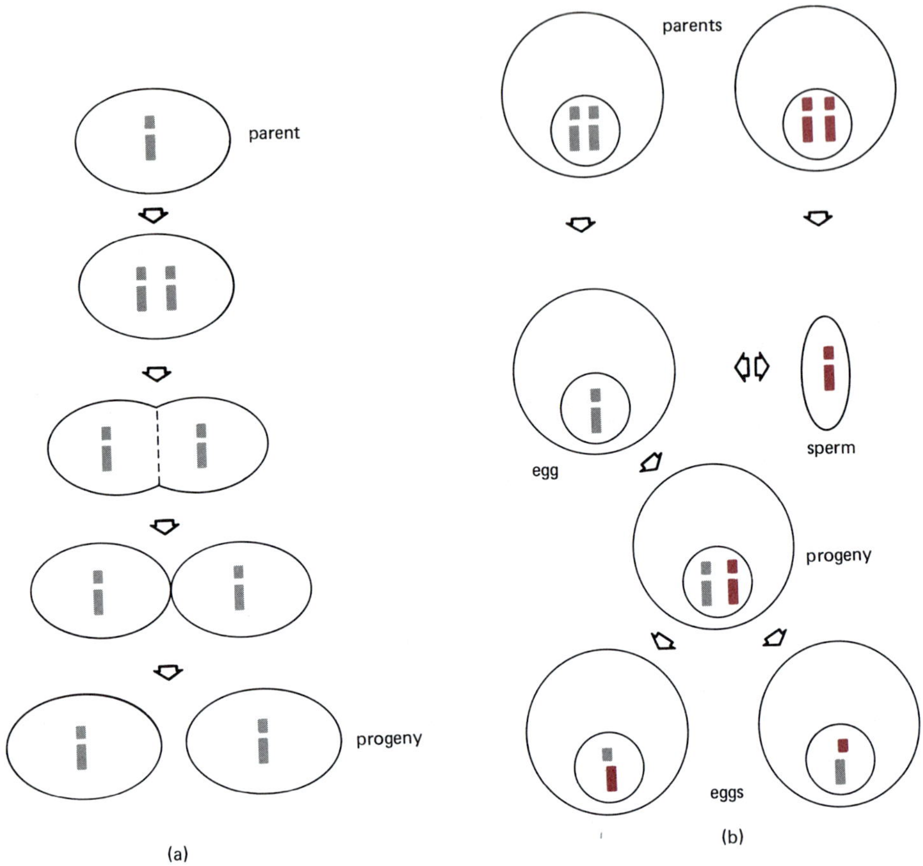

Figure 16–1 Asexual reproduction (a) results in progeny with the same genetic heritage as the original parent. Sexual reproduction (b) provides an opportunity for mingling of different hereditary lines and the assortment of chromosomes and genes into new combinations.

fuse to form a fertilized egg or **zygote.** The zygote resulting from their union has the normal diploid set of chromosomes and represents the beginning of the life of a new organism. Embarking upon an explosive process of cell division and differentiation, the zygote is ultimately transformed into a functioning organism, possessing the normal characteristics of its species.

Sexual reproduction has some advantages over asexual reproduction. The offspring of asexual reproduction carries a set of chromosomes identical with that of its parent (except for any random errors of duplication). An asexually reproducing species would therefore tend to be a rather static line, showing extremely little variation of characteristics from one generation to another. In sexual reproduction, on the other hand, there is scope for considerable individual variation as two sets of chromosomes, one each from two different parents, are combined. Sexual reproduction inherently provides greater flexibility for adapting to changing conditions of the environment and thus furnishes the raw materials for the evolution of life.

Asexual Reproduction

In **binary fission,** the "mother cell" literally gives her life for her children: two identical (half-size) daughter cells are formed by an equal division of the very protoplasm of the original cell. After the division is complete, the original cell no longer exists as an entity, although its substance is preserved in the form of the two daughter cells.

Binary fission is the most common form of reproduction among the protists. It is essentially a process of mitosis, much like the manner in which the cells of higher organisms are multiplied for growth and repair of the body. The full chromosome set is duplicated, and each of the daughter cells receives one set, together with approximately half the cytoplasm of the original cell.

Budding is a form of asexual reproduction which, like binary fission, produces two cells (or organisms) in place of one; but in this case one of the new individuals is far smaller than the other. Thus, the process results in recognizable "parent" and "offspring," the latter being a miniature copy of the parent.

Budding is observed both in single-celled organisms and in multicellular organisms (Fig. 16–2). In the single-celled yeast, budding involves an unequal kind of cell division, in which each of the resultant individuals receives a full set of chromosomes, but one of them (the "bud") receives considerably less than half of the available cytoplasm. In the multicellular hydra, a bud appears as a knob of undifferentiated cells on the side of the parent's body. Cell division and differentiation in the bud eventually result in the formation of a miniature copy of the parent, which may detach and go on to live independently, grow to full size, and reproduce itself in turn. Many plants can similarly reproduce asexually through the formation of buds or other conglomerations of undifferentiated tissue which can give rise to new individual plants. But both for the hydra and for the plants, budding is generally not the sole means of reproduction: there is usually an alternative, sexual form of reproduction as well.

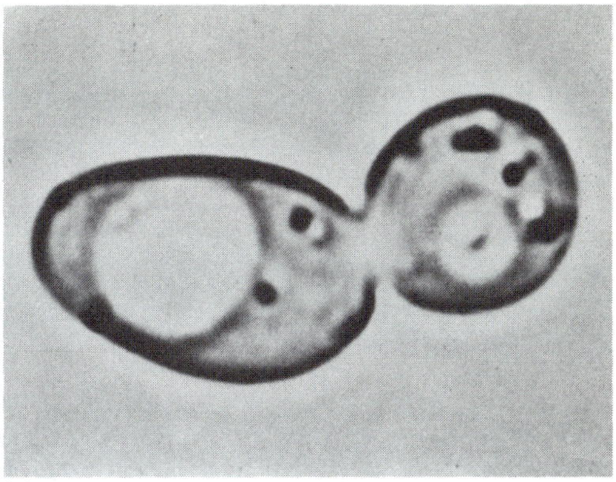

(a)

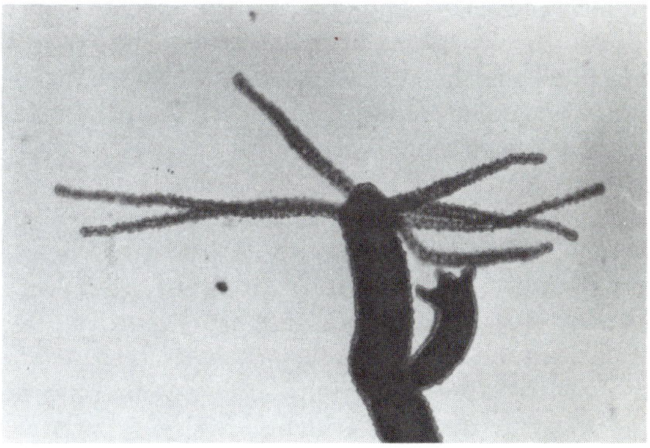

(b)

Figure 16–2 Budding: (a) single-celled yeast and (b) multicelled hydra. In each a single parent produces a single offspring genetically identical to itself.

Mitotic processes can result in reproduction in several other ways as well. Algae, for example, frequently grow in long filaments—actually colonies of loosely associated single cells. The action of turbulent waters often tears and shreds the algal filaments into fragments. Each of these grows, by a process of mitosis, into a new individual filament. Thus, **fragmentation** can serve as a mode of reproduction among some species.

The **regeneration** of lost parts observed among some of the lower animals is in some respects very similar to fragmentation, and this process, too, can serve a reproductive function. Oystermen, incensed at the depredations of oyster-eating starfish, used to chop up the pests into pieces whenever they caught

Figure 16–3 The long-arm starfish, *Luidia ciliaris*, readily breaks off its arms. All the arms on this specimen have been broken at least once; the regenerated portions are lighter in color than the remnants of the original body.

one. But so great are the regenerative powers of the starfish that the oystermen were unwittingly helping to propagate their foe: a starfish can regenerate new arms readily (Fig. 16–3), and even a single arm can regenerate the rest of the organism as long as part of the central disk is attached. The flatworm, also well known for its regenerative ability, is frequently used in the laboratory in studies of the phenomenon. Not only can cut pieces of flatworms generally regenerate a whole worm, but this organism has also been observed spontaneously reproducing in this manner. The head and tail of the worm suddenly begin to pull in opposite directions, in a tug-of-war that ultimately results in the worm's splitting apart in the middle. The head then grows a new tail, while the tail grows a new head, and there are two flatworms from the original "parent."

Many plants and protists regularly reproduce asexually through the production of **spores.** The latter are usually tiny, well-protected cells, frequently capable of surviving harsh conditions of drought and temperature extremes in a dormant state. When environmental conditions are again congenial, the spores germinate, ultimately developing into normal adult individuals. Spore formation (called **sporulation**) is quite common among fungi, mosses, and ferns. Members of these groups produce clouds of invisible spores, which are released from the parent plant and carried for hundreds and even thousands of miles by air currents. The air we breathe is seldom free of spores.

The forms of asexual reproduction discussed thus far have mainly been characteristic only of the protists and "lower" animals and plants. The more highly developed members of the animal and plant kingdoms have generally evolved sexual modes of reproduction, either in addition to asexual forms, or exclusively. However, even among the higher organisms there are some instances in which reproduction actually occurs through an anomalous asexual variation on a basically sexual pattern.

In some species, for example, an egg cell may spontaneously develop into a new individual without being fertilized. This process, known as **parthenogenesis,** is quite common among pond-dwelling rotifers and in certain orders of insects. In aphids there is typically an alternation of normal sexual reproduction and parthenogenesis: during the spring and summer months, parthenogenic reproduction prevails, and only females are produced (Fig.

Figure 16–4 Each female aphid can produce a series of parthenogenic young, without the appearance of males.

Omikron

16–4). In the fall, as the weather is growing cool, males appear, sexual matings occur, and the fertilized eggs are tough, resistant forms that overwinter and hatch the following spring. Under laboratory conditions, aphids have been raised for more than fifty consecutive generations without the appearance of males, reproducing solely by parthenogenic means. Eggs of a variety of species, including frogs, turkeys, and rabbits, have been stimulated artificially to develop parthenogenically, yielding relatively normal-appearing adults. (The parthenogenic offspring, however, have usually been smaller and less vigorous than their conventionally reproduced counterparts.)

Another asexual variation on the theme of sexual reproduction is **polyembryony:** the development of two or more identical offspring from a single fertilized egg. Identical twins are formed when the zygote divides and then instead of remaining together, the two resultant cells separate and continue developing as though each had been a zygote (Fig. 16–5). If the split occurs at a later stage, a larger number of identical offspring will result. Since the zygote has a normal diploid chromosome complement and its divisions occur by mitosis, each of the cells that it yields has the same set of chromosomes as all the others. Among humans, multiple births are a relative rarity, and not all of them are cases of polyembryony. (Nonidentical or fraternal twins, triplets, etc., result from the simultaneous release and fertilization of more than one egg cell.) But in some species, multiple identical births are the rule, rather than the exceptions. In the armadillo, for example, four identical offspring are invariably produced at each birth. Among some parasitic insects, a single egg will split up during the course of development, resulting in the formation of several hundred identical offspring.

Sexual Reproduction

Throughout most of the animal kingdom (and the higher plants as well), the concept of sexual reproduction is virtually synonymous with the existence of *two* sexes, and their interplay has given rise to an enormous variety of structural and behavioral modifications. But in the protist paramecium, sex life becomes complicated to a staggering degree: paramecia occur in as many as eight different mating types, each of which can function as a unique "opposite sex." This organism, like some other types of protists, can participate in a special kind of sexual reproduction called **conjugation** (Fig. 16–6). During this process, two paramecia, each representing a different mating type, temporarily join. Nuclear divisions take place within each of the mating paramecia, and there is an exchange of some of the nuclear material, leaving each with a complete, but somewhat modified set of chromosomes. The two organisms then separate and embark upon mitotic divisions that yield offspring with added vigor and vitality.

Apart from the exotic mating behavior of the multisexed paramecia, organisms that practice sexual reproduction generally exhibit two distinct types of reproductive organs: male and female. The male reproductive organs produce **sperm,** which are usually tiny, motile sex cells. The motile sperm characteristically seek out the much larger female sex cells **(eggs** or **ova),** which are produced in female reproductive organs. The reproductive organs in general are referred to as **gonads;** specifically, the male gonads are termed **testes,** while the female gonads are called **ovaries.**

Figure 16–5 Identical twins are formed from a single zygote.

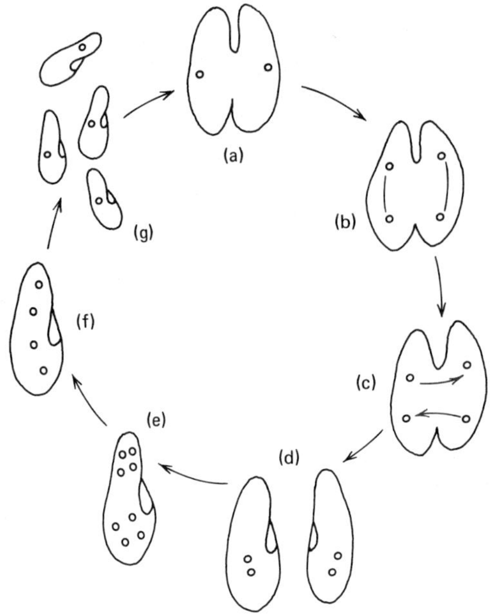

Figure 16–6 Conjugation in *Paramecium*. Two individuals of appropriate mating types join (a). The macronucleus of each disappears, and the diploid micronucleus passes through two meiotic divisions, producing four haploid micronuclei in each organism. All but one of the micronuclei in each *Paramecium* disintegrate, and the remaining micronucleus divides mitotically, yielding two haploid micronuclei (b). Then there is a transfer of genetic information: one of each pair of micronuclei migrates into the other paramecium (c). When the conjugating *Paramecia* separate, the two haploid micronuclei of each fuse to form a diploid zygote micronucleus. Each mating partner then undergoes a series of mitotic divisions (e,f), during which the macronuclei reappear, and each ultimately yields four daughter cells (g). The diploid micronucleus of each of the eight daughter cells of the conjugating pair contains a combination of genetic information from both original parents.

A single organism may have sex organs of both types; or the sexes may be distinct, with a single organism possessing either testes or ovaries (or in the case of plants, either male or female flowers or cones) but not both. Each pattern is found in both the plant and animal kingdoms. Animals possessing both male and female organs in a single individual are referred to as **hermaphrodites;** they are found among various invertebrate species, including the earthworm, parasitic tapeworms, oysters, snails, and slugs. The majority of flowering plants are hermaphroditic, but in the animal kingdom a separation of sexes is the predominant pattern. Even among the hermaphroditic animal species, self-fertilization is rare. It is prevented by a variety of mechanisms. For example, in oysters the testes develop before the ovaries. In no case are both gonads functional in the same oyster at the same time. In the earthworm, both sex organs are functional simultaneously, but mating occurs in pairs, with sperm from each worm fertilizing the eggs of the other.

The substantial size difference between eggs and sperm is correlated with a difference in their basic functions. In general, the functions of the sperm are twofold: to carry a (haploid) set of chromosomes and to meet and penetrate the egg. A set of chromosomes does not take up very much space, and the sperm's additional apparatus of a motile tail-like flagellum and various packaging materials, energy supplies, and chemicals to react with the protective coating of the egg do not add very much to its total bulk. As a result, a sperm characteristically resembles a microscopic tadpole (Fig. 16–7).

The egg, in turn, also has two basic functions: to carry a (haploid) set of chromosomes and to provide nourishment for the early development of the new organism that will result from its fertilization by a sperm. It is the latter function that accounts for the far greater bulk of the egg, in the form of stored food material called **yolk.** The human egg is approximately forty thousand times as large as the human sperm; the differential represents stored food. Yet, compared with other species, the human egg contains relatively little yolk. The largest amounts are contained in the eggs of reptiles and birds, since here the developing offspring remains within the eggshell for an extended period, during which the yolk provides its only source of nourishment. The human ovum is about 0.14 mm in diameter, minuscule in com-

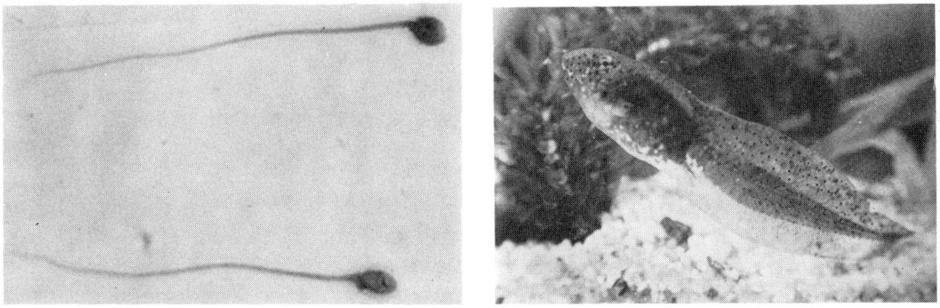

Carolina Biological Supply Co.

Lynwood M. Chace, Omikron

(a) (b)

Figure 16–7 A human sperm (a), like a tadpole (b), has a distinct head and a long, motile tail.

parison with a hen's egg, and especially in comparison with a 200-mm-long ostrich egg. The main difference is yolk.

Fertilization, the meeting and fusion of sperm and egg, is a critical stage in the overall process of sexual reproduction. The motile sperm is adapted for movement in a watery medium. Among aquatic organisms, the sex cells are usually released into the water by both parents, and fertilization takes place externally (i.e., outside the body of the female). The inherent inefficiency of external fertilization is compensated for in various ways. Organisms that employ this mode of reproduction generally release enormous numbers of gametes. A single American oyster, for example, can liberate millions of eggs at one time. At the same time, other oysters liberate motile sperm in such profusion that the waters turn milky with them. Chemical signals released into the water frequently alert an organism of one sex to the presence of members of the opposite sex (and same species) in the vicinity, and the release of sperm into the water may trigger the release of eggs, or vice versa.

In some species, external fertilization results from an actual act of **mating,** in the usual sense of pairing of two individuals of opposite sex. When frogs are ready to mate, the male climbs upon the back of the female and tightly clasps her about the abdomen with his forelegs. They may remain in this close embrace for as long as several days; then pressure exerted by the male on the female's swollen abdomen helps to force eggs out of her body into the surrounding water. At the same time the male is stimulated to release his sperm, which flows out over the eggs.

A considerable gain in efficiency was achieved through the development of types of internal fertilization, in which the male deposits his sperm directly into a specially modified cavity or receptacle in the female's body and fertilization occurs in that protected environment. In this case the motile sperm move through a medium of fluid secretions within the female's body. Frequently the male is equipped with a specialized copulatory organ, called a **penis,** with which he can effectively insert his sperm some distance into the appropriate cavity of the female. Thus, a loss of sperm is minimized. Male copulatory organs are found among many representatives of the animal kingdom from flatworms up to man. They are not universal even among

Figure 16–8 The female wood frog is in the act of laying eggs. The male, clasping her from above, prepares to release his sperm, which will fertilize the eggs externally.

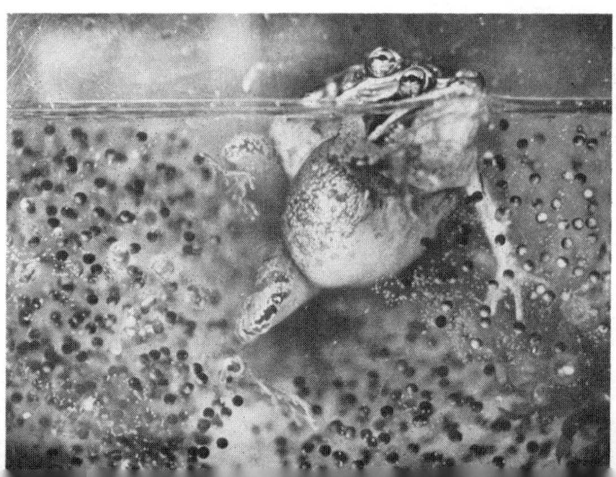

Lynwood M. Chace, Omikron

higher animals, however; birds, for example, do not have penises even though they practice internal fertilization. (Sperm are transferred to the female bird by a form of suction.)

Another important phase of the sexual reproductive process is the early development of the new organism, after the egg has been fertilized. Among species that practice external fertilization, the early development of the offspring—from zygote to embryo, and then perhaps through one or more immature forms—occurs in the water where fertilization took place. Characteristically, the developing young of these species are ignored by their parents and are unprotected against the depredations of predators and destructive forces of the environment. (The solicitous stickleback father is a notable exception). Generally, the main factor promoting survival against these unfavorable odds is the enormous numbers of offspring produced by such species. When a single spawning yields millions of eggs and sperm, the chances are enhanced that at least a few individuals will survive.

The species that practice internal fertilization are characterized by a wide range of degrees of care for their young. Some mothers merely deposit their eggs and leave; usually, however, at least some rudimentary protection is provided, such as a scooped-out hollow in the sand or a protective sac attached to a twig. Among many species, even when the mother "abandons" her young, some provision is made not only for their protection but also for their food supply after they hatch from the eggs. Many insects, for example, deposit their eggs on or in the leaves, stems, or fruits of plants or in the bodies of animals that will provide suitable nourishment for the larvae that emerge from the eggs. Among the higher vertebrates, the practice of remaining with or near the developing eggs becomes increasingly frequent. Some reptiles remain near the nest and may even guard the young for some time after hatching. Birds are noted for building elaborate nests (Fig. 16–9), incubating their eggs by sitting on them, and feeding, protecting, and instructing their fledgling young. The ultimate in postfertilization care is found among the placental mammals, in which the developing young are retained within the body to a fairly sophisticated stage of development and are nourished with food materials derived from the mother's tissues. (There are some scattered instances of internal development among other groups as well, notably among some insects and fish. But although the young of such species are "born alive," they generally are not directly nourished by the tissues of the mother while they are within her body; instead, they subsist on a reserve of yolk, just as in related species that lay eggs.) Among the mammals, parental care usually extends for some period after birth, as the offspring complete their growth and development and are instructed in necessary skills for survival.

Figure 16–9 Birds care for their young: (a) a mother swan broods her eggs on a nest in a mound of sticks and grass built on the ground; (b) eggs in a red-eyed vireo nest fastened to a fork of tree branches; (c) a wood thrush brings an earthworm to the hungry nestlings.

Lynwood M. Chace, Omikron (a) (b) (c)

The trend to internal fertilization and to increasing parental care for the developing young is paralleled by a progressive decrease in the number of offspring produced in a single mating. As dangers from the environment are minimized, enormous numbers of offspring are no longer needed to ensure survival, and quality of care is enhanced by having fewer individuals among which attention must be divided. Moreover, the size of the mother's body sets an upper limit on the amount of living matter that can be accommodated: either number of offspring or degree of development must ultimately be sacrificed, and the higher vertebrates have taken the path of increasing degree of prenatal development. Thus few mammals have more than a dozen offspring in a litter, and many species have only one or two at a time.

The Human Reproductive System

In more reticent times, when it was felt that a young man or woman was reaching an age when he or she would need some practical knowledge about sexual reproduction, he or she was given a lecture on a subject euphemistically entitled "the birds and the bees." Although this approach did correctly stress the basic similarities in patterns of sexual reproduction through the animal kingdom, the choice of birds and bees as illustrative examples was singularly inapt. As has already been noted, birds lack a specialized male copulatory organ, although they do engage in internal fertilization. And the sexual practices of bees are quite inhumane, from the human standpoint: among the social bees the only males are drones, haploid individuals whose sole useful function is to mate with the queen bee. Although the majority of workers in the hive are females, they do not engage in any sexual behavior at all. This privilege is reserved for a special caste, the queen bees (usually only one to a hive). For the queen bee, mating is strictly a once-in-a-lifetime affair; after her nuptial flight, she stores the supply of sperm that she has received from her mate and settles down to the serious business of laying eggs. (The drone who won her favors meanwhile has usually fallen dead from the nuptial flight.)

Obviously there are great disparities between reproduction in birds and bees and that in humans. However, human reproductive structures and functions do have a great deal in common with those of the more closely related groups of placental mammals, and much of our knowledge of human reproduction has come from insights gained in suitable animal studies. Like the other placental mammals, humans come in two varieties: male and female. Although the male and female sex organs of humans are derived from similar embryonic structures, by about the seventh week of prenatal development they have already begun to differentiate into recognizable forms, and the sex of a baby is thus established long before birth. For convenience, we shall discuss the male and female human reproductive organs separately.

THE MALE REPRODUCTIVE SYSTEM

The main functions of the human male reproductive system are twofold: to produce male germ cells (sperm) and to deliver them to the female in a manner that will enhance the chances of fertilization.

The male gametes are produced in the **testes** (Fig. 16–10), a pair of organs about the size and shape of pigeon eggs. The testes are suspended in the **scrotal sac,** a loose pouch of skin which forms a continuation of the abdominal cavity. In the developing embryo, the testes are formed within the abdominal cavity proper; at about the time of birth they descend into their final position in the scrotal sac. Occasionally one or both of the testes fails to descend, a condition called cryptoorchidism. Unless this condition is corrected with drugs or through surgery, sterility will result, for no viable sperm will be produced.

The position of the testes in the loosely hanging scrotal sac seems rather vulnerable, but this inherent disadvantage is counterbalanced by a definite survival advantage.

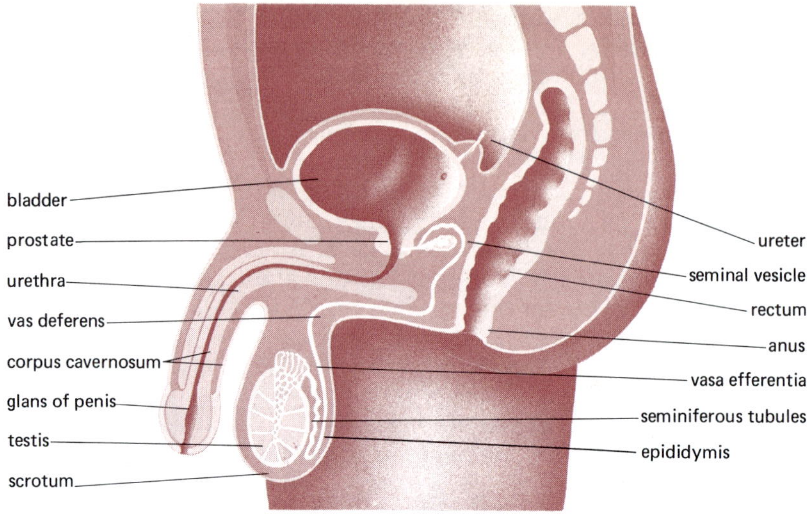

Figure 16–10 The human male reproductive system.

Heat retention in the scrotal sac is not as efficient as in the rest of the body, and as a result the temperature in the testes averages about 3–5°F below normal body temperature. This lower temperature is apparently a necessary prerequisite to the normal production of sperm. Since temperature is also a factor in the induction of mutations, the position of the testes probably also contributes to genetic stability. (It has been computed that the temperature rise due to the wearing of trousers has produced a definite increase in the human mutation rate.)

Sperm are not produced until **puberty.** In early adolescence an increased flow of hormones from the pituitary stimulates development of the testes. These organs in turn produce not only sperm, but also a hormone of their own: the male sex hormone **testosterone.** This hormone produces varied effects throughout the body, promoting a deepening of the voice, the growth of hair on the face and body, and development of the muscles.

The sperm are actually produced in numerous tiny, coiled tubes found within each testis. These tiny tubes, called **seminiferous tubules,** ultimately empty into a large convoluted duct known as the **epididymis.** Here the sperm may be stored for a time. The epididymis leads into the **vas deferens,** a duct that swings upward in a loop into the abdominal cavity, then around and out into the penis, ultimately merging into the **urethra,** a duct that also carries urine out of the body. Glands called the **seminal vesicles** empty secretions into the vas deferens during its loop through the abdominal cavity. Two more sets of glands, the **prostate** and the **Cowper's glands,** also contribute their own secretions (Fig. 16–11). These secretions join with the sperm to produce a white fluid called **semen.**

As in many other animal species, the human male copulatory organ is a structure called the **penis.** It is composed of erectile tissue, and the skin covering the tip is particularly sensitive.

THE FEMALE REPRODUCTIVE SYSTEM

In the human male, both the gamete-producing organs and the copulatory organ are situated in appendages that extend beyond the limits of the actual body cavity. In the human female, on the other hand, the reproductive organs are all tucked away inside the abdominal cavity (Fig. 16–12). The main functions of the female reproductive system are also twofold: to produce functional female gametes and to provide a site for the fertilization and subsequent development of the embryo. The female reproductive organs

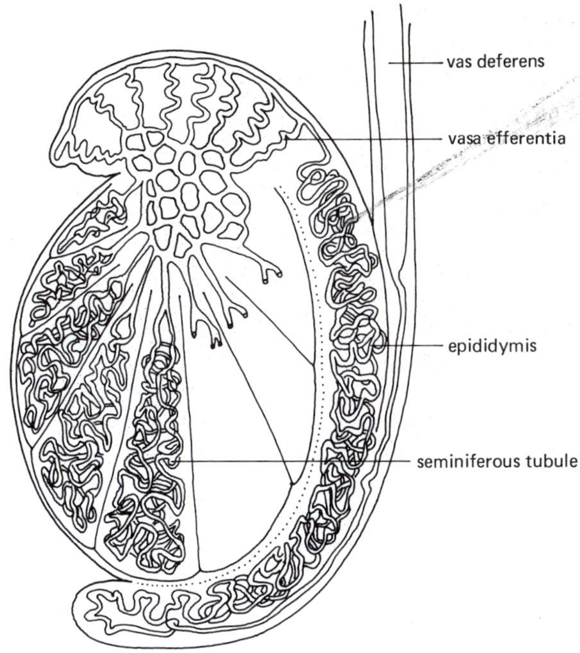

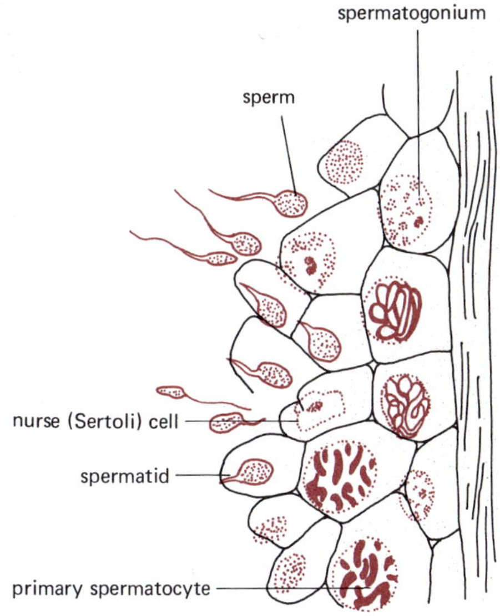

Figure 16–11 Sperm are produced in the seminiferous tubules, which are lined with spermatogonia (derived from primordial sex cells) and nurse cells, which nourish the sperm as they develop through a series of stages (spermatogonium, spermatocyte, spermatid) into mature sperm.

also act as endocrine organs, secreting hormones that facilitate their major functions. (The male reproductive organs perform a similar endocrine function.)

Eggs or ova are produced in the **ovaries,** a pair of oval-shaped structures about the size of olives. At birth, the ovaries of a human female contain nearly half a million immature egg cells. These remain dormant until some time between ten and fifteen years of age. At this period of puberty, a complex cycle of changes in the endocrine system triggers

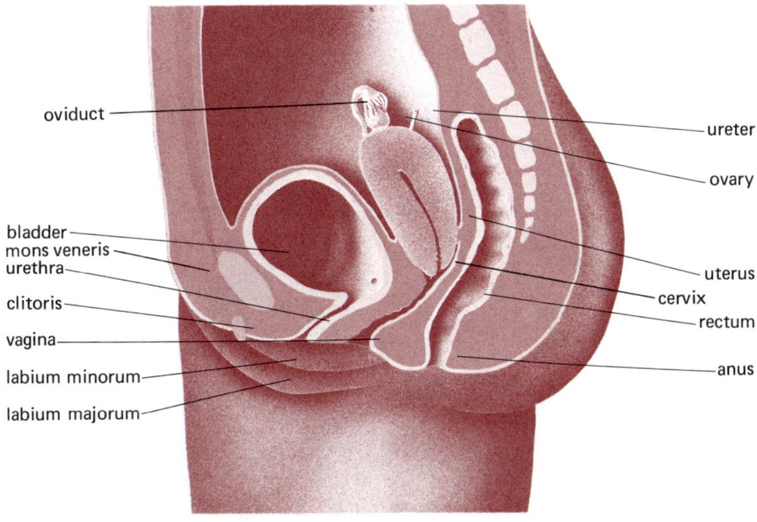

Figure 16–12 The human female reproductive system.

the development of various secondary sex characteristics, such as growth of the breasts and the appearance of characteristic patches of body hair. Changes are also produced in the sexual organs themselves, the ovaries, which become important endocrine glands secreting sex hormones. Now the ova begin to mature, usually at a rate of one a month. (The ovaries alternate, each one thus producing one mature ovum every other month.)

The growth of the maturing ovum and the completion of its first meiotic division occur within a fluid-filled sac called a **follicle.** The follicle secretes sex hormones called **estrogens,** but when the ovum is mature, the follicle ruptures, releasing the ovum into the body cavity. The mature ovum quickly passes into the open end of the nearby **oviduct** or **fallopian tube.** It is slowly conveyed, by a combination of ciliary movements and peristaltic contractions, down the tube into a pear-shaped structure called the **uterus.**

Meanwhile, the old ruptured follicle has degenerated into a yellow body called the **corpus luteum.** This structure begins to secrete a new sex hormone, **progesterone,** which produces several effects, the most important of which are a stimulation of a buildup of the lining of the uterus and an inhibition of the formation of a new follicle. If the ovum has been fertilized by a sperm by the time it reaches the uterus, it will soon become implanted in the uterine wall and will embark upon the development leading to the formation of an embryo and ultimately a child. If the egg has not been fertilized, however, it degenerates and passes out of the uterus and out of the body through a channel called the **vagina.** (Unlike the male, the human female does not have a single passageway serving to carry both urine and germ cells; the vagina is entirely separate from the urethra, and the opening of the latter is just anterior to the opening of the vagina.)

After about ten days, if fertilization has not occurred, the secretion of progesterone by the corpus luteum ceases. The corpus luteum shrinks, and the lining of the uterus begins to break down, ultimately being discharged from the body through the vagina in a process called **menstruation.** A new follicle begins to develop in the other ovary, and the cycle begins over again. Usually there is a cycle of twenty-eight to thirty days between the onset of successive menstrual flows; **ovulation** (the release of a mature ovum from a follicle) generally occurs on about the fourteenth or fifteenth day after the beginning of the preceding menstruation.

The external area into which the vagina and urethra open is enclosed in two sets of fleshy folds called **labia.** Near the anterior junction of the labia is the **clitoris,** a pea-

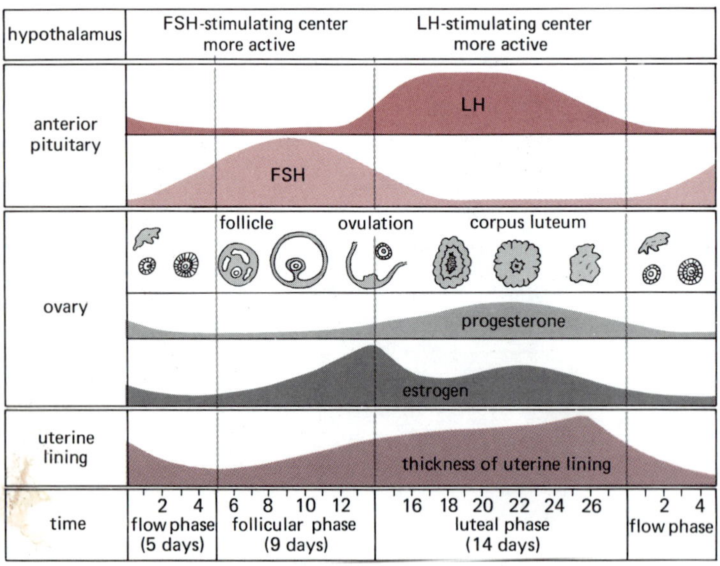

Figure 16–13 The human menstrual cycle.

sized, highly sensitive structure, which is homologous to the male penis (i.e., has a similar structural derivation). Both the clitoris and the hair-covered mass of fatty tissue anterior to it (the **mons veneris** or "mound of Venus") may become extremely sensitive during sexual excitement and serve as organs of sexual stimulation.

Copulation and Fertilization

A human egg cell is fertilized by only a single sperm. Yet in a single act of sexual intercourse, about 300 to 500 *million* sperm are transferred from the male to the female. Nature's apparent profligacy is not as illogical as it might seem, however, as will become clear as we learn more about the actual process of fertilization.

One of the most critical aspects of the reproductive process is to get the sperm and egg together so that fertilization may take place. This occurs during the act of sexual intercourse or copulation. During sexual excitement, the penis, which normally hangs down limply, becomes erect and hardened as the erectile tissue becomes engorged with blood. Insertion of the penis into the vagina of the female, followed by rhythmic movements, eventually brings on heightened feelings of sexual excitement known as an **orgasm.** At the same time, there is an **ejaculation,** in which 3–4 ml of semen is forcibly expelled from the penis into the vagina.

During copulation, the action of the penis against the clitoris, the vaginal walls, and the cervix (the entrance to the uterus) greatly enhances the sexual excitement of the female and may lead to an orgasm comparable to that experienced by the male. During female orgasm, the fallopian tubes and the vaginal and uterine walls undergo rhythmic contractions. Although the female orgasm is not a necessary prerequisite for fertilization, it does aid in the movement of the sperm in the direction of the uterus and fallopian tubes.

Fertilization, when it occurs, takes place in the upper third of the fallopian tube. It rarely takes place in the uterus, for by the time an unfertilized egg reaches the uterus—about four days after ovulation—it has already degenerated. In fact, the human egg is normally fertilizable for only about a twenty-four-hour period of time after ovulation. It then begins to degenerate. (Human sperm are equally delicate, degenerating within a day or two after copulation. One almost wonders how so many babies manage to be conceived under such inauspicious conditions.)

If fertilization does occur, the fertilized egg or zygote continues its trip along the fallopian tube in the direction of the uterus, dividing and redividing as it proceeds. Eight to ten days after ovulation, the developing embryo becomes implanted in the lining of the uterine wall. At this stage it already consists of several hundred cells.

After this brief description of copulation and fertilization, it may still seem unclear why it is advantageous for so many sperm—300 to 500 million—to be provided by each ejaculation, since there is usually only one egg available for fertilization at any particular time. The apparent wastefulness actually provides a number of adaptive advantages. First of all, many of the sperm go astray and do not enter the fallopian tubes at all, and approximately half of those that do find their way into the tubes enter the tube that does not contain an ovum. Even when a num-

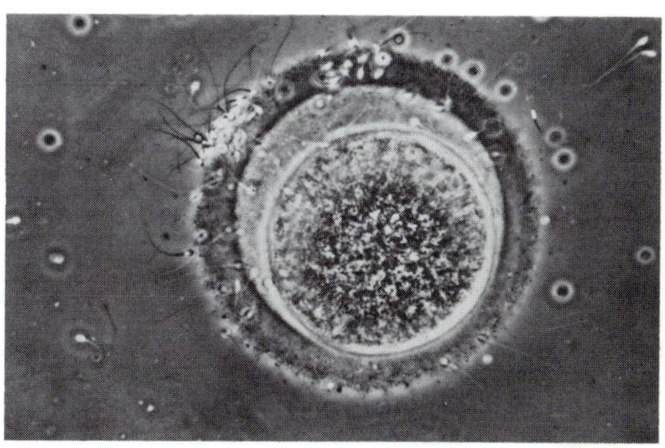

Figure 16–14 Sperm bombard the outer envelope of a human egg.

ber of sperm actually do encounter a mature ovum, it is not necessarily the winner of the race that wins the prize—i.e., fertilizes the egg. The egg itself is enclosed in a jagged covering of cells, called the **corona radiata** as a result of its resemblance to the corona of the sun (Fig. 16–14). The corona radiata effectively prevents the union of any sperm with the egg and must be dissolved away before fertilization is possible. The head of each sperm is equipped with a small structure that literally explodes like a miniature bomb when it hits the egg, releasing a tiny charge of the enzyme **hyaluronidase.** This enzyme hydrolyzes hyaluronic acid polymers, which are a major component of the cementing material that binds the cells of the corona radiata together. Each sperm carries such a small amount of hyaluronidase that it is estimated that fully 35 million sperm are needed to impact with the egg before a large enough hole is breached in the corona radiata. When this is accomplished at last, the next sperm to reach the egg fuses with it. Immediately the outer membrane of the egg thickens and effectively prevents the entrance of additional sperm. Now the ovum completes the second division of meiosis, resulting in a single haploid egg cell and three polar bodies. The nuclei of egg and sperm fuse, and after a time the fertilized egg begins to divide, doubling and redoubling its number of cells by a process of mitosis. A new individual life has been conceived, and the female is now **pregnant.**

As might be guessed from the description of the intricate sequence of processes and interactions that lead to fertilization, far from all copulations actually result in the conception of life. In this situation, humans are somewhat unusual. In most species, reproduction is seasonal, and a male and female do not engage in copulation unless conditions are optimum for the promotion of conception. (Indeed, in some species such as rabbits, the act of copulation actually triggers ovulation in the female.) Among a number of mammal species in which reproductive cycles are not strictly seasonal, it is often the case that although the male can be sexually stimulated at virtually any time, the female is sexually receptive only for fairly brief periods, which occur in regular cycles determined by the ebb and flow of hormone secretions. These cycles of receptiveness are referred to as **estrous cycles.** They characterize the sexual behavior of cats, dogs, rodents, and various other mammals. If a female is in a period of receptiveness or "heat," she will excrete characteristic hormones (pheromones) that serve to attract males; any resultant sexual activity will very likely result in pregnancy. If the female is in a different portion of her estrous cycle, she will not produce the characteristic sex attractants, and will resist any sexual advances of males.

The human sexual cycles have been the subject of intensive research efforts aimed along two quite contradictory lines. Studies are being made toward developing more effective means of contraception—the prevention of pregnancy following sexual activity. Equally vigorous efforts are being directed toward finding means of enhancing

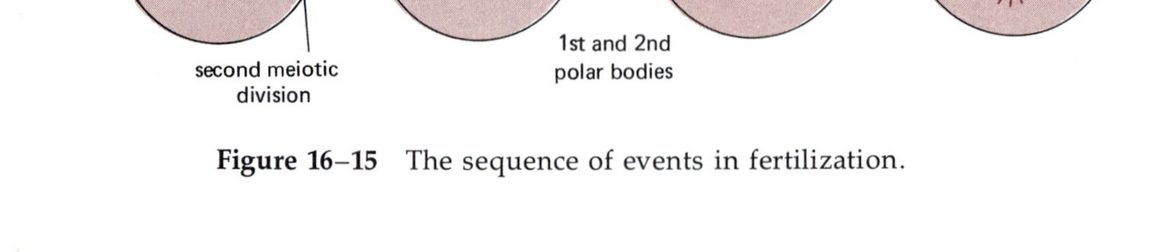

Figure 16–15 The sequence of events in fertilization.

fertility, to help couples who want children but have been unable to have them. Considerable progress has been made on both fronts. For example, a number of drugs have been developed to treat infertility due to failure of the female to ovulate. One of these has rather spectacular side effects: the chemical tends to stimulate multiple ovulation and commonly results in the conception of twins, quadruplets, and in at least one case already recorded, septuplets. (Unfortunately, none of the latter survived.)

Efforts at contraception take several basic pathways. Mechanical barriers to the passage of sperm, such as condoms, have been used for a long time. Spermicidal liquids, foams, and jellies have also been employed. The "rhythm method" attempts to utilize calculation of the natural female menstrual cycle to minimize chances of conception. Since ova and sperm each have an effective lifetime of only a day or two, it would be expected that only copulations occurring at or near the time of ovulation will result in conception. Allowing a few days on either side of the probable ovulation date to take various irregularities into account, this still leaves a substantial portion of the month comprising a theoretical "safe period," in which sexual intercourse should not result in pregnancy. Unfortunately, some women tend to be so irregular in their monthly cycles (some may even ovulate twice or more in a month) that the rhythm method is not very effective. Efforts to enhance its accuracy by charting the monthly temperature rise that occurs on the day of ovulation are not as helpful as they might be, since the temperature rise is only a fraction of a degree and might be masked by slight fever or a temperature abnormality due to a cold or other body condition.

One of the major contraceptive efforts has centered upon attempts to artificially regulate the monthly cycle by administering hormone preparations in a sequence that artificially maintains the body in a pregnancylike state for most of the month, suppressing ovulation, and then stimulates a menstrual flow (Fig. 16–16). This is the basis of "the pill," probably the most reliable contraceptive device yet devised. The question of long-term side effects, however, has not been definitely resolved. Some statistical evidence, for example, indicates that birth control pills may slightly increase the risk of thrombophlebitis, an inflammation of the veins combined with blood clotting within the blood vessels, and pulmonary embolism, the formation of blood clots within the lungs. Diabetes and other hormonal imbalances may also be triggered by "the pill," and thus women whose family history indicates a higher-than-average susceptibility to such conditions are usually advised to avoid this method of contraception. (Of course, pregnancy and childbirth can also result in these and other equally dangerous side effects, so it is often a question of weighing the relative potential hazards.)

The other major contraceptive measure now in practical use is the insertion of an intrauterine device, a bit of stainless steel or plastic made in various shapes and designs (Fig. 16–17). For some reason that is not yet entirely understood, the presence of such a foreign body in the uterus prevents implantation of an embryo in the uterine wall, and thus pregnancy cannot result even if an egg is fertilized.

Vasectomy, the surgical ligation of the vas deferens in the male, and a similar surgical sterilization of the female by ligation of the fallopian tubes have been gaining in popularity in recent years. However, at present these techniques are usually irreversible steps and thus can be used as contraceptive measures only by those who are sure they will not want to have additional children in the future.

Meanwhile research on contraception continues, and the development of safe and reliable male contraceptives, "morning-after" pills, and other variations may be expected in the future.

Early Embryonic Development

The early development of the embryo is surprisingly similar among a variety of animal species—so similar, in fact, that valuable insights into human development have

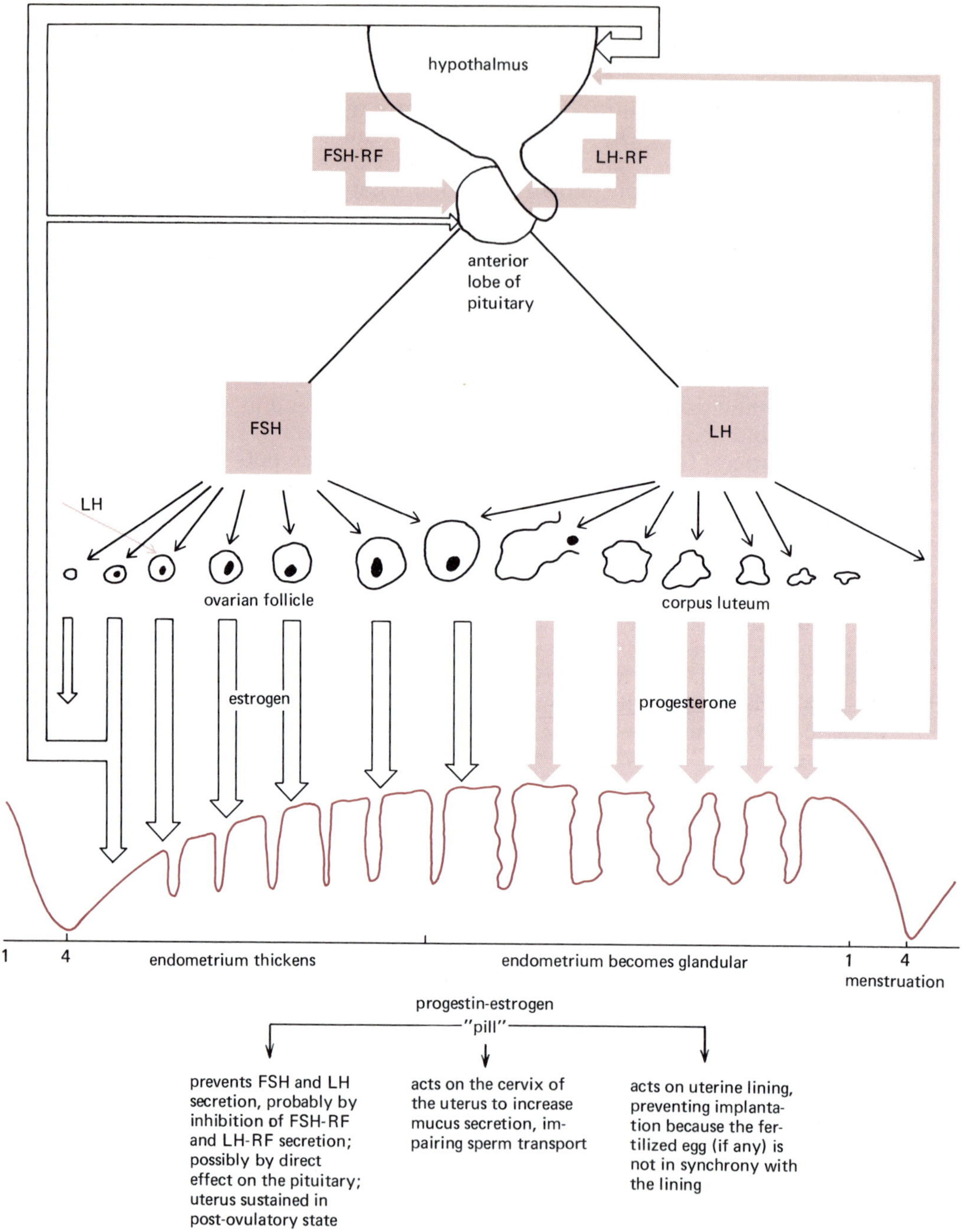

Figure 16–16 The female reproductive cycle is normally controlled in a complex feedback interrelationship with the hypothalamus and the endocrine glands. The "pill" artificially regulates hormone levels, resulting in a feedback inhibition of ovulation.

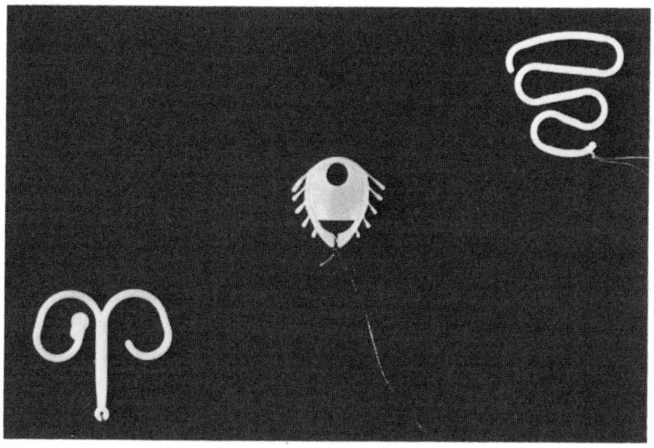

Figure 16–17 An assortment of IUD's (from left to right): the Safe-T-Coil, the Dalkon Shield, and the Lippes Loop.

been gained through studies of fertilized eggs of such diverse species as frogs, chickens, and even sea urchins. Such eggs are much easier to obtain in large numbers and much more convenient to observe: frog and sea urchin eggs, for example, develop in water, where they can be continuously observed; a "window" carefully cut into the shell of a fertilized chicken egg provides an effective means of observing the early development of that species.

Human fertilized eggs, on the other hand, develop within the body of the mother, and our knowledge of the stages of early human embryonic development comes largely from observations of embryos taken from pregnant uteruses removed for medical reasons in the operation of hysterectomy, and from aborted embryos and fetuses. Efforts have been made to maintain human embryos under artificial conditions for varying periods of time (usually a few weeks at most), but the "test tube babies" envisioned in Huxley's *Brave New World* are still an achievement for the distant future. Indeed, it was not until 1968 that researchers actually brought about a fertilization of a human ovum in vitro. By the time you read these words, a child may have been born as a result of fertilization in a "test tube." But it will have completed its embryonic development in the womb of a real mother.

The egg and sperm that join to form a zygote are each haploid, carrying half the characteristic number of chromosomes for the species. The zygote is thus diploid, with a full set of chromosomes—a pair of each type. After the nuclei of sperm and egg fuse, for a time there is no apparent activity. But the zygote is doubtless a center of considerable behind-the-scenes biochemical activity, for then a furrow appears, and the single cell divides into two smaller duplicates of itself. This is the first mitotic division in a series of divisions comprising the process of **cleavage.** This process continues, the new cells remaining clustered together, until gradually a hollow ball of cells is formed. This ball is called the **blastocyst.**

At this stage of development, the ultimate plan of the organism has not yet revealed itself. A biologist looking at such a ball of cells could not be sure what species it was going to develop into. Even at this stage, however, he could eliminate some possible groups from observations of the general pattern of cleavage (Fig. 16–18). In eggs with a great deal of yolk, such as those of birds and reptiles, the yolk tends to be concentrated at one edge or "pole" of the sphere, and the mitotic activity is concentrated chiefly in the region of the opposite pole. In eggs with relatively little yolk, on the other hand, early cleavage occurs relatively uniformly through the sphere, giving rise to similar-sized daughter cells. Both sea urchin and human eggs are of the little-yolk type. The eggs of insects and some other invertebrates follow a rather different plan of early development: the nucleus and cytoplasm of the unfertilized egg are in the center, surrounded by yolk. As cleavage proceeds, the daughter nuclei with adhering cytoplasm migrate to the surface and form a continuous layer of protoplasm about the yolk.

The cells of the tiny blastocyst continue to divide and redivide. Soon an indentation appears in the ball of cells and deepens, until the tiny organism-to-be, now at the **gastrula** stage of development, resembles a double-walled cup, or an old tennis ball into which someone has pressed a thumb.

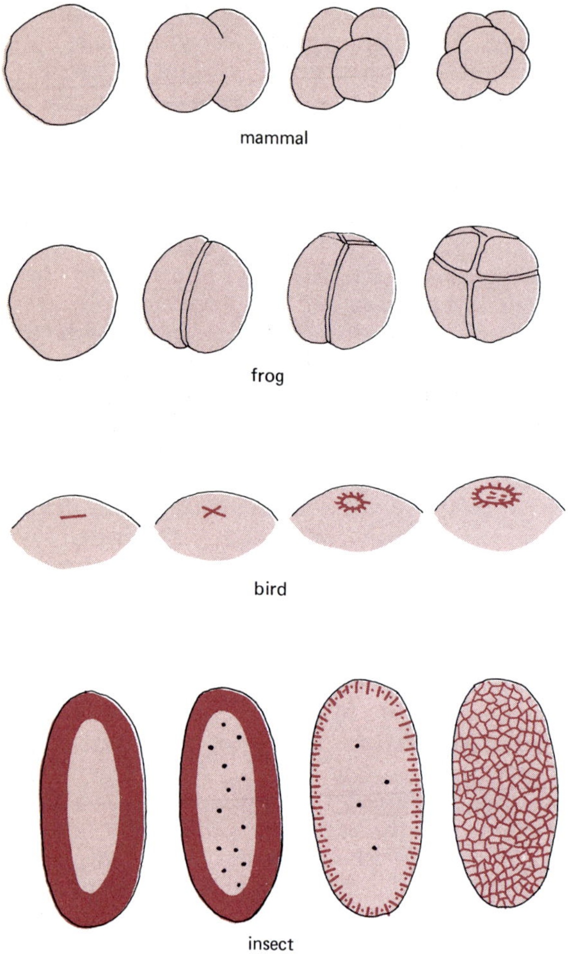

Figure 16–18 Typical cleavage patterns illustrating variations in the relative proportions of yolk.

With all this furious development, little if any actual growth has taken place up to this point in the developing organism or embryo. Thus far, at least overtly, development has taken the form chiefly of an increase in the number of cells by mitotic division.

But soon the embryo does begin to grow, to increase in size and total mass. And soon another major aspect of development is manifested: **differentiation.** Cells begin to become specialized. An outer layer, called the **ectoderm,** becomes distinguished from the inner layer, the **endoderm.** In time a third basic or germ layer, the **mesoderm,** is formed between them. Gradually these layers differentiate further and give rise to the various tissues, such as nerve tissue, muscle tissue, and connective tissue (Fig. 16–19). Rudimentary organs are formed and actually begin to function before their development is complete. Some organs are formed but do not begin to function until after birth or even beyond.

Even at this late stage, the similarities among embryos of different vertebrate species are striking. But the small differences gradually become accentuated until the embryo becomes recognizably a fish or frog, a reptile, a bird, a mouse, or a man.

Pregnancy and the Placenta

When a human baby is born, it is already about nine months old. It has spent these nine months of life since conception living as a parasite within the body of its mother, drawing nourishment from her tissues and dependent for its well-being upon her physical state. The normal period of pregnancy or **gestation** period (i.e., the time from the fertilization of the egg to the birth of the child) is 266 days for man. In comparison with other mammals, this is a medium-length gestation period, as befits man's body weight and the degree of development undergone by the young before birth. Rodents, for example, are generally small animals and are characterized by relatively short periods of intrauterine development. The young of house rats and mice are born in an extremely immature state after a gestation period of only three weeks; those of guinea pigs, far more highly developed at birth, are born after a gestation period of more than two months. The elephant, by contrast, has a pregnancy lasting nearly two full years.

During its nine months of life before birth, the human embryo passes through a series of startling transformations. From a featureless single cell, it passes first through the blastocyst and gastrula stages; the three germ layers form, and soon recognizable structures begin to take shape.

About eight to ten days after fertilization, the developing embryo becomes established in the uterus. It secretes proteolytic enzymes, which destroy some of the cells in the uterine lining. The embryo then sinks into the uterine wall at the site of the damage, and a rapid growth of tissue soon covers it.

As the implanted embryo continues its growth, it sends out fingerlike projections into the surrounding uterine lining, within which an extensive system of capillaries is formed, connected to the embryo's own developing vascular system. At the same time, spaces in the lining of the uterus form and eventually fill with maternal blood. Thus, although there is no direct contact between the blood of the **fetus*** and that of the mother, there is a diffusion of gases, food, and waste materials across the membranes separating the maternal and fetal systems. The complex of fetal and maternal tissues and blood vessels in which this exchange occurs is called the **placenta.** The embryo's body is attached to the placenta by the tubular **umbilical cord,** which contains the two umbilical arteries and one umbilical vein. The time of implantation of the embryo in the uterine wall and formation of the placenta may be a period of hormonal upset in the mother's system, and it is during this period that she may experience the nausea and other symptoms colloquially known as "morning sickness."

Toward the end of the third week, when the mother has just about realized that her menstrual period is overdue and may soon suffer the first symptoms of "morning sickness," the heart of the embryo is already beginning to beat. At this time, the embryo itself is little longer than 1.5 mm, or the height of the letter *e* in this sentence. A little over a week later, at about the end of a month, the embryo has spurted to about 5 mm in length. The heart is large in proportion to the size of the body and is now divided into four chambers. Around this time a pair of little buds appear on the sides of the body, followed in a day or so by another pair farther back from the already well-distinguished head region. In still another day or two, eyes can be discerned. At about this time the embryo also exhibits a prominent tail and gill pouches. By about the sixth week, the embryo has grown to about 13 mm (about half an inch). The arms and legs have lengthened considerably with the hands and feet distinguished, and the features of the face are beginning to emerge.

At the end of the second month, when the embryo is about an inch long, its face is definitely humanoid, with distinct eyes, ears, and nose. The embryo is now able to respond to stimuli, such as touch, by the contraction of voluntary muscles, and it is beginning to exhibit spontaneous movements. The tail now begins to disappear, as it is covered by the growing buttocks. At this time the general plan of organs and organ systems has been essentially completed. Virtually all that remains is an increase in size and a "finishing off" of the details of development.

Toward the end of the third month, when the embryo is about three inches long, its sex can be distinguished, and now that it has all the basic human structures it is designated a fetus. Considering all the major developments that are crowded into the first few months of intrauterine development, it is not surprising that this period is particularly sensitive to all sorts of adverse influences upon the mother—effects of drugs (such as the now-banned thalidomide), poor nutrition, and diseases such as German measles. Such influences may result in gross abnormalities or even death of the embryo.

By the end of the second third of its prenatal development, the fetus, now about 13 inches long, has a typical human appearance. Its fingernails and toenails are fully formed, as is a full head of hair, and

*Although there is some difference of opinion as to the exact point at which the human embryo should be called a fetus, this point is often taken at about ten weeks, when the embryo has begun to look recognizably human.

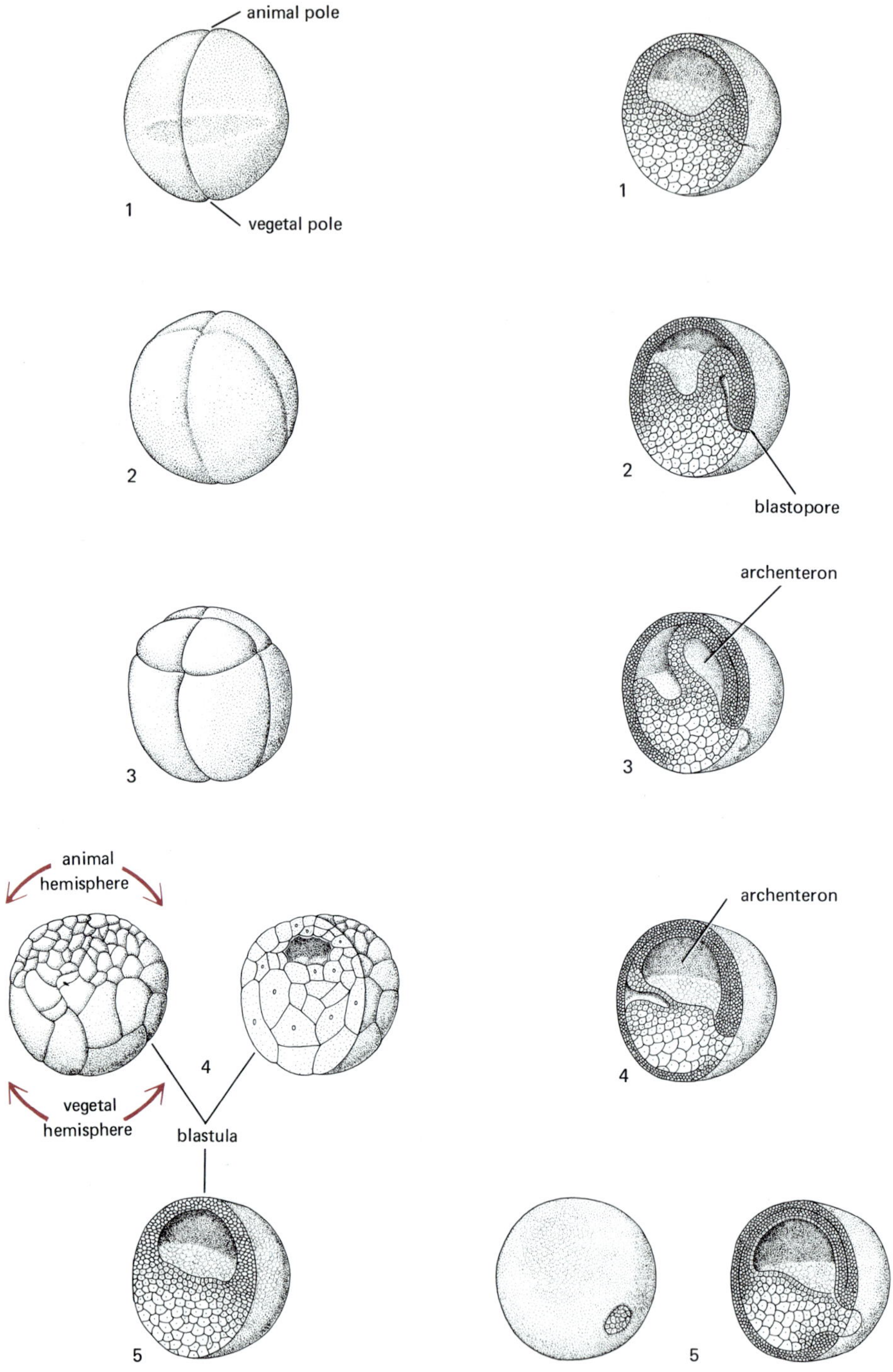

Figure 16–19 Early embryonic development of frog tadpole: from zygote to tissue differentiation.

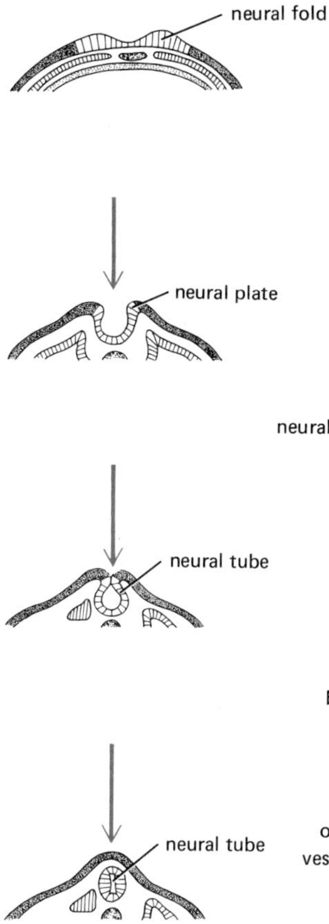

(above) 4 successive cross-sectional views of the development of the neural tube between stages (c) 1-3.

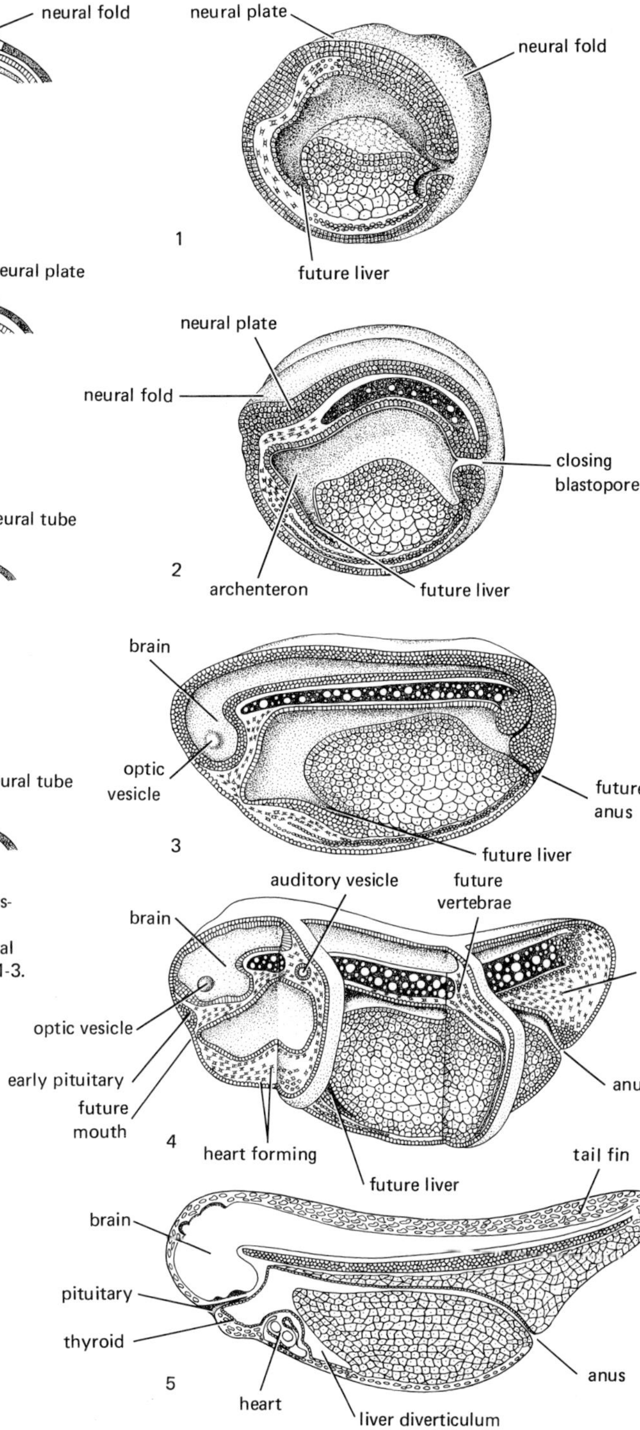

(c)

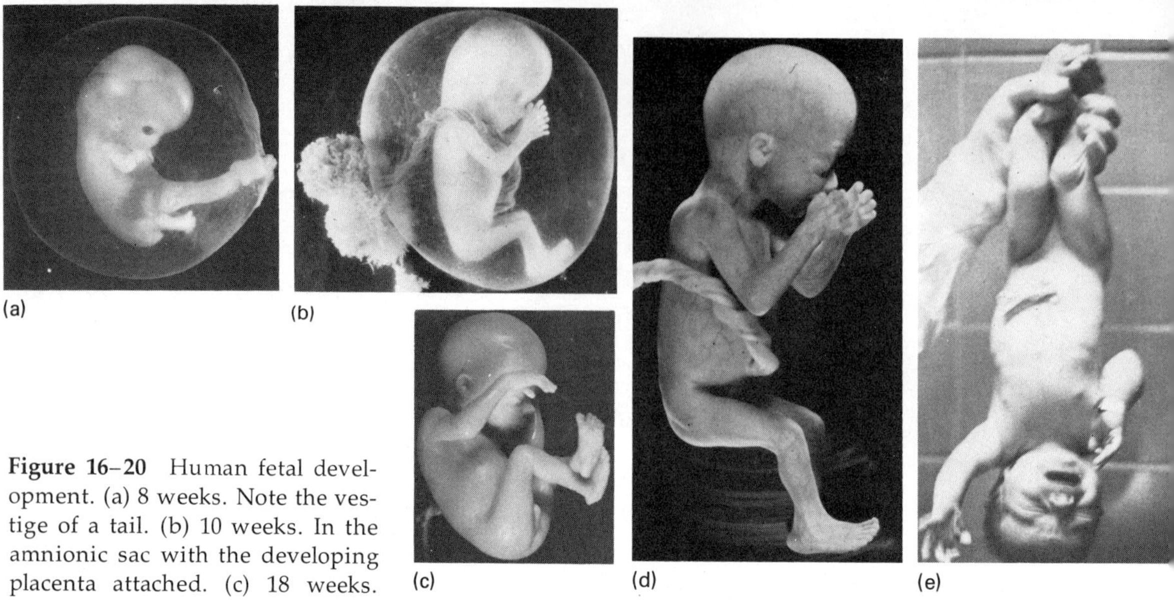

Figure 16–20 Human fetal development. (a) 8 weeks. Note the vestige of a tail. (b) 10 weeks. In the amnionic sac with the developing placenta attached. (c) 18 weeks. Unmistakably human and beginning to make its movements felt. (d) 28 weeks. Old enough for a chance at survival if it were born prematurely. (e) Birth at term. The end and the beginning.

it has just acquired eyelashes and eyebrows. The fetus is now reaching a critical stage. If it were born prematurely at the end of the sixth month, even with modern techniques of incubator care it could not survive for more than a few hours. But by the twenty-eighth week, the surface-active agent in the liquid film lining the alveoli of the lungs is formed, and this seems to be the turning point: a fetus born at this time, usually weighing about three pounds or so, may survive with proper care and develop into a normal adult.

During the final months of prenatal development, the wrinkled skin of the fetus is firmed out as fat deposits are laid down under the skin, and the fetus becomes more and more active. It sleeps and wakes up, kicks, startles at loud noises, may even suck its thumb, and may turn many times in the womb, until, just before birth, it reassumes the head-downward position, ready to be propelled through the birth canal.

Within the uterus, the embryo is enclosed within a pair of membranous sacs (Fig. 16–21). The inner membrane is the **amnion,** and the outer is called the **chorion.** Within the amnion, the embryo floats in an isotonic fluid that serves as a protective shock absorber, cushioning the embryo against jolts and jars.

A third extraembryonic membrane, the **allantois,** is derived from the digestive tract of the embryo. In egg-laying animals, the allantois takes the form of a rather large sac that serves as a receptacle for waste materials and may also function as a breathing organ, exchanging oxygen and carbon dioxide through the egg shell. In man the allantois becomes incorporated into the umbilical cord, and its blood vessels become the two umbilical arteries and the umbilical vein.

Still a fourth extraembryonic membrane, the **yolk sac,** is found in the human embryo, where it is small and nonfunctional. In reptiles and birds, on the other hand, the yolk sac serves as the major source of food for the developing embryo.

During its nine months of prenatal development, the human embryo undergoes an incredible increase in size—from a microscopic single cell to a six- to nine-pound mass of protoplasm comprised of nearly 10 trillion cells.* The uterus undergoes a correspondingly startling increase in size. The virgin uterus

*This astonishing rate of growth is actually rather slow in comparison with those of many other vertebrates. At the other extreme, a blue whale embryo grows to a 20-foot-long, 2-ton infant within a gestation period of little more than nine months!

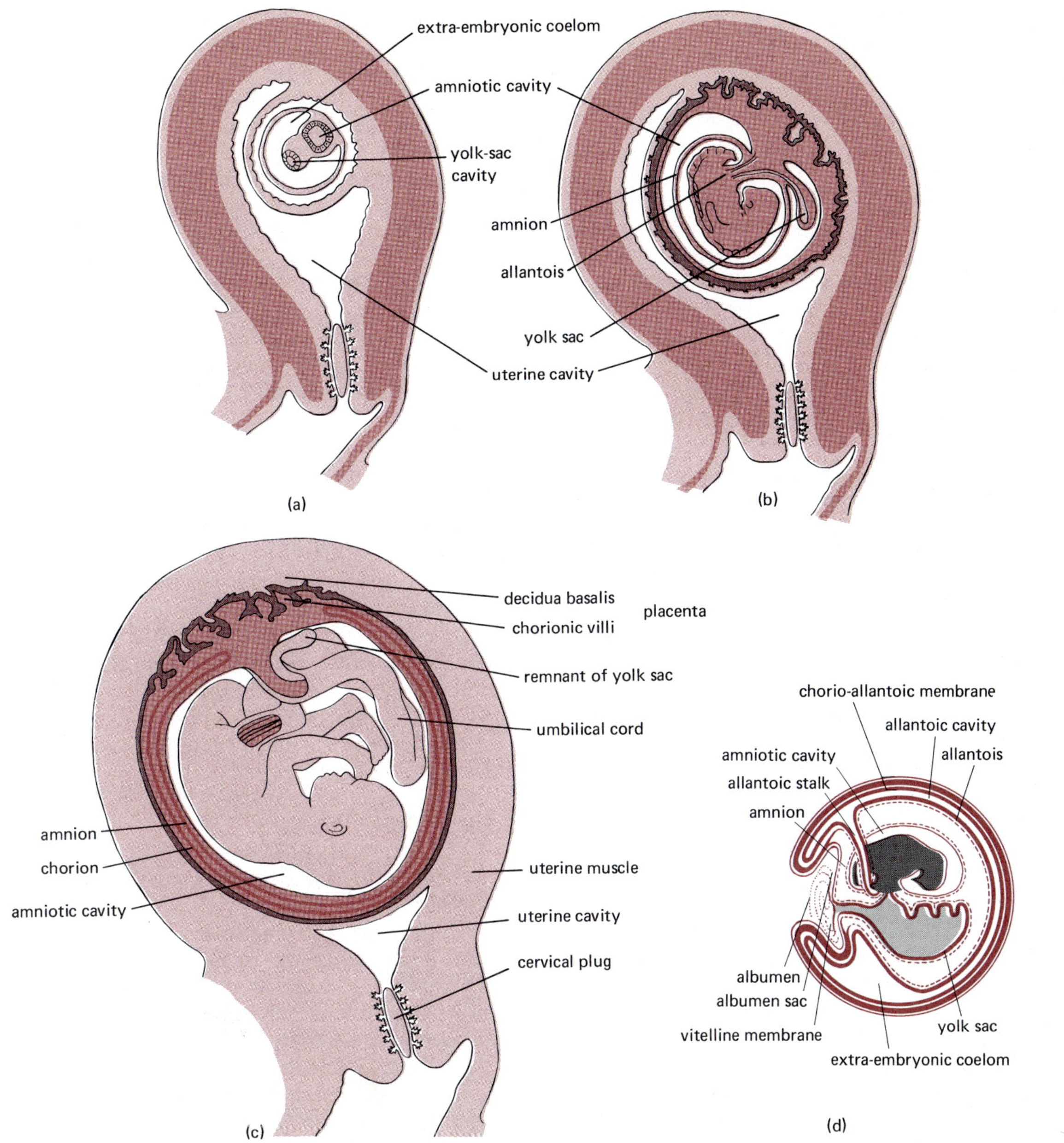

Figure 16–21 Development of the human embryo and its extraembryonic membranes (a, b, c); note the difference in relative size of the yolk sac and allantois of a developing chick (d). The chick utilizes the yolk as its source of nutrients during development, and the yolk sac is richly supplied with blood vessels. The allantois is a receptacle for the chick's urinary wastes. The fused chorioallantoic membrane is a site of gas exchange. In the developing human, the functions of food supply, gas exchange, and waste removal are performed by the placenta. The yolk sac is insignificant, and the remnants of the allantois are incorporated into the umbilical cord.

has a capacity of about 2–3 ml. This capacity increases to about 5–7 liters at the time of birth.

Birth and Lactation

For nearly nine months the unborn child develops in a completely sheltered universe. He rests in gently supporting fluids. No sharp chills or sudden flushes disturb his peaceful equanimity. Darkness reigns. Sounds are muted, and shocks and pressures are dissipated by surrounding fluids. Slowly he grows and takes on a human shape which becomes more and more apparent with the passage of time. Soon he begins to make exploratory movements with his developing body, stretching an arm, kicking a leg, blinking eyes which do not yet see, and sucking a thumb that chances to find his mouth. Though conditions become increasingly cramped as he nearly fills the confining uterine cavity, life is serene in the sheltering watery world of the unborn.

But this serenity comes to an abrupt end for the fetus when the very walls of his world begin to heave and contract. At first the contractions are mild and infrequent. Then they become progressively more frequent and severe, culminating at last in a series of violent upheavals which forcibly propel the infant out into a world of light, noise, and endless problems. The warm, shielding, and sustaining aquatic environment is suddenly replaced by a strange, thin medium: air. At the same time the oxygen supply from the mother is cut off. With a convulsive gasp the infant draws in this strange new mixture of gases and fills his lungs for the first time. Intricate and vital changes now occur within the circulatory system of the newborn, as the old pathways of circulation through the umbilical cord are shut off, the arc of pulmonary circulation is established as a major link, and the heart must adjust to the new blood flows. With the temperature shield removed, the infant's body must learn to operate its system of thermostatic controls. At first this is rather erratic, but it improves with time.

After nine months of comparative calm, what precipitates this sudden upheaval? The contractions that propel the infant out into the world are apparently initiated by a shift in the mother's hormone balance. At the end of pregnancy, the progesterone level of the blood drops. Its inhibition of uterine contractions is thus lifted, and the uterine muscles, sensitized by mechanical pressures and strains, begin to contract. Another hormone, oxytocin, may contribute to the growing wave of uterine contractions. Thus begins the process of childbirth, also known as **parturition** or **labor.**

During labor, the fetus is gradually moved down toward the **cervix** (the opening of the uterus), usually head first, and the cervix dilates, opening to permit its passage. Mechanical stimulation and hormonal secretions interact in a heightening interplay, as the articulation of the pelvic bones loosens, permitting the fetus to be propelled out through the cervix and down the vagina (Fig. 16–22). The entire process, from the first "labor pain" to the emergence of the baby from the greatly dilated opening of the vagina, usually lasts from eight to twenty-four hours in the first pregnancy, but may take only minutes after many pregnancies.

During the third and final stage of labor, which follows a brief lull after the birth of the baby, the uterus resumes its contractions, resulting in the detachment and expulsion of the placenta (now called the "afterbirth"). This process usually takes ten to forty-five minutes and is accelerated if the baby is allowed to suck at the mother's breast soon after birth. The act of sucking precipitates a reflex mechanism which stimulates secretion of the hormone oxytocin. This hormone promotes uterine contractions, leading not only to rapid expulsion of the placenta but to a minimization of bleeding in the uterus as well.

Breast-feeding of the baby is actually an integral part of the birth process and of great physiological benefit to the mother as well as to the baby. In reality, the breasts

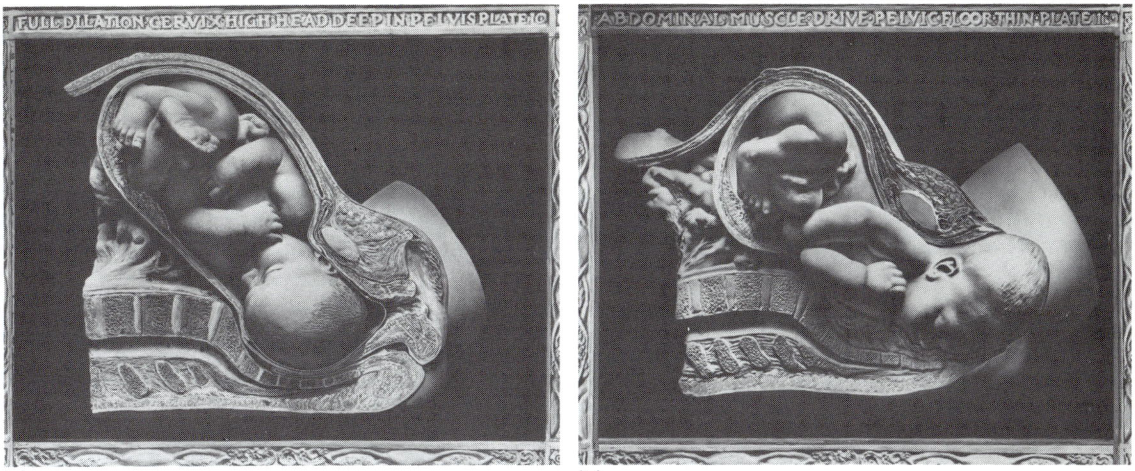

(a) (b)

Figure 16–22 Models of birth: (a) the cervix is dilated, and the infant's head is deep in the pelvis; (b) contractions of the abdominal muscles propel the infant out through the vagina. Note the elongation of the head permitted by the relatively soft cranium. The skull will return to normal shape within a few days.

have been preparing for many months before the birth of the baby, promoted by hormonal secretions. Strangely, true milk is not produced until two or three days after birth. Instead, a fluid called **colostrum,** which contains basically the same amounts of protein and lactose as milk but almost no fat, is secreted at first. The infant receives only small amounts of this fluid, but it is rich in antibodies to diseases to which the mother is immune.

The actual suckling of the baby involves an interplay of hormones and reflexes, resulting in release of the milk that has been continuously secreted and stored in tiny cavities within the breasts. The hormone oxytocin, which triggers the milk release, also stimulates uterine contractions which return the uterus to its normal state more rapidly than if the baby is not breast-fed.

Milk is sometimes called "nature's most perfect food." But since the composition of milk varies from species to species, what might be optimum for one mammal may not necessarily be suitable for another. For instance, cows' milk contains about twice as much protein as human milk and much less sugar. Hence, in preparing formula for human infants from cows' milk, the milk must be diluted with water to lower the protein content, and sugar is frequently added. Cows' milk is deficient in certain components. For instance, it is low in iron and vitamins C and D. Therefore, modern bottle-fed babies are started early on vitamin supplements and are shortly thereafter introduced to solid foods.

Reproduction in Higher Plants

In both the animal and plant kingdoms, the evolution of life to larger and more complex forms has been characterized by an increasing emphasis upon sexual modes of reproduction. But although two sexes are involved in each case, the forms of reproduction have developed along quite different lines in the two kingdoms. The differences have been largely determined by the major differences between plants and animals—the relative immobility of plants contrasted with the mobility, reactivity, and heightened perceptive abilities that generally characterize animals. Whenever two sexes are involved in reproduction, a key factor is that of getting the gametes together for fertilization. This could be accomplished readily by having both produced by the same organism. But survival advantages are usually enhanced by **outbreeding,** a mingling of genetic material from different sources. In most animal species, a variety of sense receptors and locomotor organs permit potential mates to find each other and come together over relatively great distances. But the specializations of plants have not provided such sophisticated systems for location and locomotion. The plants have evolved their own characteristic means of bringing gametes together.

The principal groups of higher plants are the seed plants: the gymnosperms and angiosperms. Gymnosperms, particularly pine, spruce, and other conifers, form vast forests in the northern temperate zone. They provide about three-fourths of our lumber supply. The angiosperms, or flowering plants, provide virtually all our cultivated food crops, as well as sources of oils, drugs, and textiles.

Both gymnosperms and angiosperms evolved from groups of plants characterized by an alternation of generations: a diploid generation of plants (the **sporophyte**) reproduced asexually to form a haploid **gametophyte** generation; the latter, in turn, reproduced sexually to form sporophytes; and so on in an endless circle. In the ferns, the sporophyte is the dominant generation, with the gametophyte an insignificant little plant. Among the seed plants, the gametophyte is reduced still further to a small structure within the sporophyte, totally dependent upon it for nourishment.

In the pine, a typical gymnosperm, the reproductive process begins with the formation of **cones,** usually in the early spring. Two types of cones, male and female, are borne on each tree (Fig. 16–23). The tiny male cones contain cells that undergo meiosis; the resultant haploid spores develop into pollen grains. The large, woody, mature female cones are what one normally thinks of when "pine cones" are mentioned. The female cones are also small at first. They contain ovules within which are cells that will ultimately develop into haploid egg cells.

The pollen cone is relatively short-lived; within a space of a few weeks it grows, produces spores that grow into few-celled pollen grains, releases them, and dies. The pollen grains are carried by the wind, some to reach a female cone, where the next phase of reproduction, **pollination,** occurs (Fig. 16–24).

Some of the pollen grains that have drifted through the air are caught in a sticky secretion in the female cones and are drawn into the ovule. Now the pollen grain absorbs nourishment from the sporophyte and grows a pollen tube, which burrows into the tissue of the ovule and ultimately delivers sperm

Figure 16–23 (a) Pistillate (female) cones of the Austrian pine; (b) staminate (male) cones of the slash pine.

(a) (b)

Rocke, Omikron

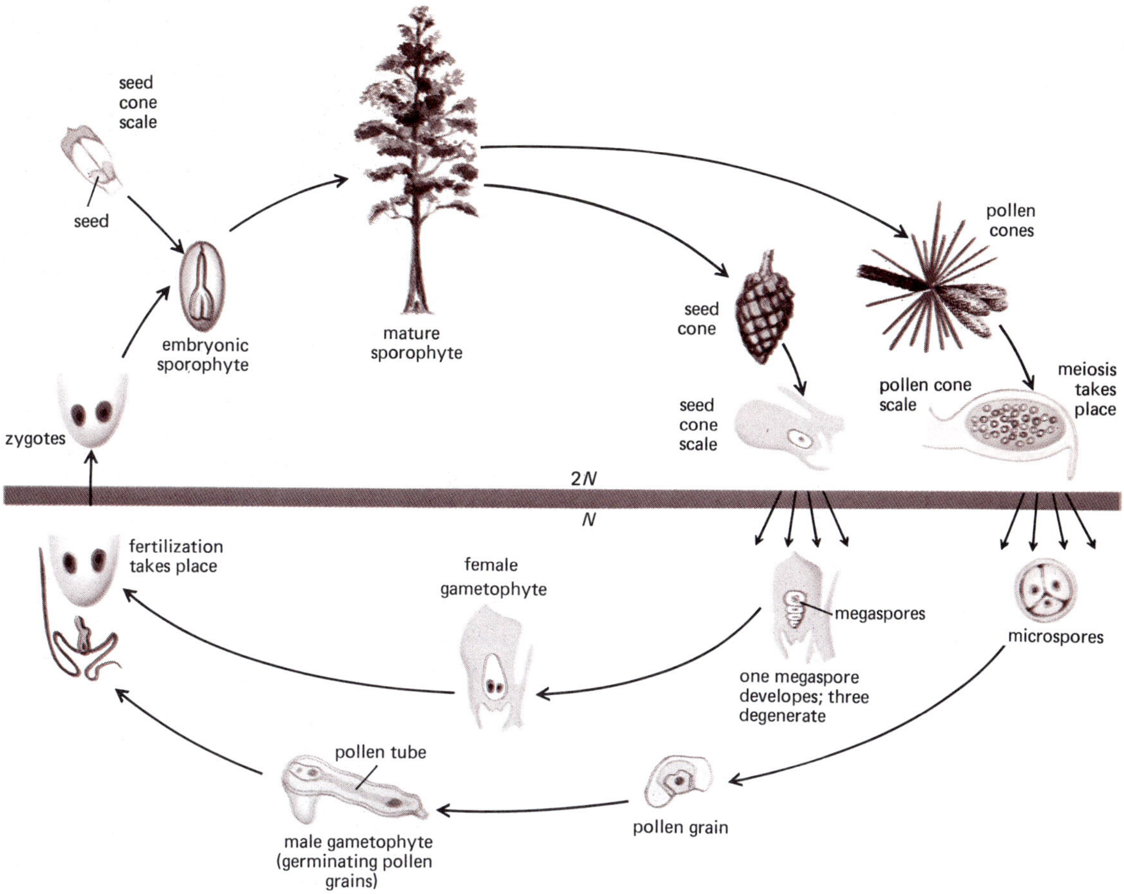

Figure 16–24 The life cycle of a pine tree. The tree is the mature sporophyte. The gametophytes, much reduced, are enclosed within the tissue of the sporophyte. After fertilization the female gametophyte, with its enclosed zygote and a tough seed coat produced by the parent sporophyte, becomes the seed that forms on the seed cone scale.

to eggs. The inexorable advance of the pollen tube over the fraction of an inch to its goal takes nearly a year in the pine, although in some conifers this process is completed in a few weeks. Meanwhile, within the ovule, cell divisions produce mature eggs and a food supply for the later development of the embryo. At last the tip of the pollen tube penetrates to the egg and bursts open; sperm fuses with egg, and a zygote is formed. The diploid zygote divides and differentiates, and its surroundings (the ovule) enlarge, ultimately producing a seed. The pine seed thus consists of food storage tissue, the seed coat, and the embryo. At the end of the second year, the seeds are shed from the cones, eventually fall upon the soil, and after a period of dormancy, germinate in the soil the following year.

Reproduction in the angiosperms has reached an advanced degree of specialization, permitting these species to adapt to a wide range of conditions of life on land, as well as, in some instances, in the water. A major innovation among the angiosperms is the additional protection furnished to the ovule and later to the embryo by its enclosure within the ovary, a portion of the flower. The name *angiosperm* literally means "covered seed," and is derived from this protected state of the ovule. In contrast, in the *gymnosperm* (which literally means "naked seed"), the seeds are borne exposed on the surface of the cone scale. The flower is thus the main distinguishing feature of the angiosperm. Flowers vary enormously in size, shape, color, and odor, but they are present in some form in all members of this class.

The reproductive organ of the angiosperm, the flower, develops from a bud and is essentially a specialized branch, bearing reproductive leaves. Some species produce flowers as soon as they have reached a certain stage of growth; in others the onset of flowering is apparently triggered by the length of the day (or night), as was discussed in Chapter 12. Among biennials (plants with a two-year life cycle), such as the carrot, only vegetative structures are produced the first year; flowering occurs in the second year of growth.

In most angiosperm species, the flowers contain both male and female organs (**stamens** and **pistils,** respectively). Such flowers are termed perfect flowers, in contrast to imperfect flowers, which contain only the organ of one sex (Fig. 16–25). In some species characterized by imperfect flowers, such as corn (*Zea mays*), the same plant bears both male (staminate) and female (pistillate) flowers, located in different regions of the plant. In others, such as the willow (*Salix*), the sexes are separated, and a plant may bear either male or female reproductive organs, but not both.

The overall process of reproduction in the angiosperms is, in many respects, similar to that of the gymnosperms (Fig. 16–26). The male reproductive organs produce haploid spores, which develop into pollen grains; the latter are carried to the female reproductive organs by wind, water, or animals. There a pollen tube grows into the ovule, where meiosis has occurred. A sperm from the pollen tube fuses with an egg within the ovule, fertilizing it. But a second sperm fuses with two polar nuclei that were formed during meiosis, producing a triploid cell that develops into the **endosperm.** This double

(a) (b)

Figure 16–25 (a) A perfect flower: the haw. Each flower contains both male and female organs. (b) Imperfect flowers: corn. The male tassels and female cobs develop on the same plant. Each single strand of "silk" is a combined style and stigma, attached to an ovary on the cob. Pollen grains pollinate each silk individually, and the fertilized ova develop into kernels on the cob.

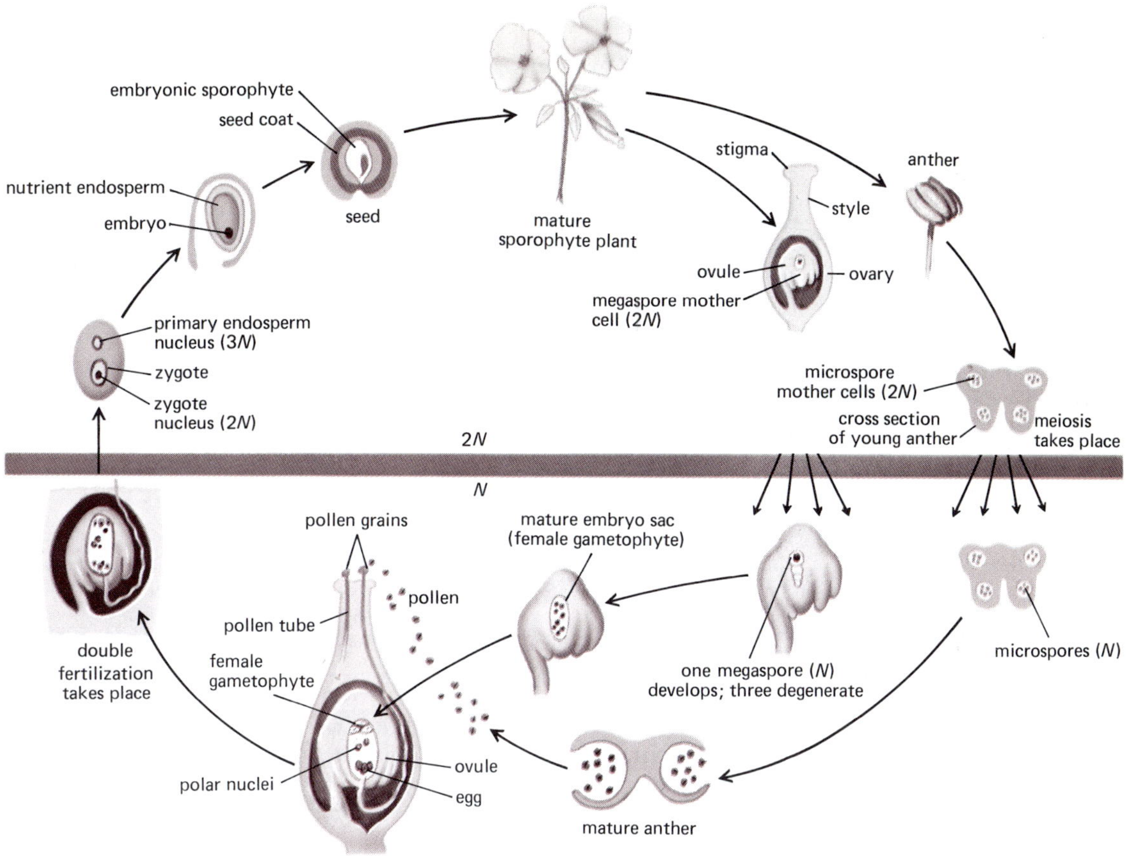

Figure 16–26 Life cycle of an angiosperm. As in the gymnosperm, the prominent generation is the sporophyte, and the gametophytes are reduced structures, enclosed within the tissues of the sporophyte.

fertilization thus produces a zygote and material that will nourish the developing embryo.

As the diploid zygote develops into the embryo of the seed and the triploid endosperm nucleus divides and redivides to produce the food storage tissue of the seed, the ovary increases greatly in size; when it is mature, it is known as the **fruit.** The mature seeds are dispersed in various ways, with or without the enclosing fruit. When they encounter suitable conditions (sometimes after a period of prolonged dormancy, which can be broken only after exposure to extreme conditions, such as the cold of the winter), they may germinate, giving rise to a new sporophyte.

THE FLOWER

Although the gymnosperms depend upon the wind to carry the pollen grains to the female organs of other plants, many of the angiosperm species have evolved more efficient means of cross-pollination. Insects and other animals, attracted to the flowers, pick up a load of nectar and in the process acquire a dusting of pollen. In subsequent trips to other flowers of the same species, the pollen acquired may be brushed off, ultimately fertilizing the eggs of the second flower. Various species of bees, moths, beetles, and other insects, as well as bats and birds, have evolved such mutually beneficial associations with various

Lynwood M. Chace, Omikron.

(a)

(b)

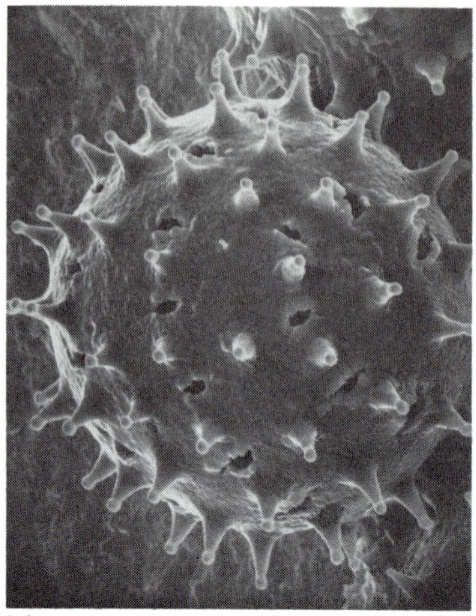
(c)

Figure 16–27 Insects visit flowers to feed on their nectar and pollen but inadvertently transfer pollen grains from one flower to another. (a) a monarch butterfly; (b) a bumblebee. (c) Hibiscus pollen grains, magnified 700 ×. It is easy to see how such grains could cling to the hairs or legs of an insect.

species of flowers. Specialized leaves of the flowers have developed into brightly colored petals, with colors and odors well suited to attract particular animal visitors. The shape of flowers may also be adapted to the characteristics of their pollinators, providing landing platforms for bees, long tubular vases for insects with a long proboscis or birds with a long bill, and even intricate spring traps that dump a load of pollen on an insect's back. Even flower habits are often adapted to the habits of pollinators: flowers pollinated by twilight-flying species often open late in the day, while those pollinated by species active in daylight frequently close up for the night.

With all the great variety of flower forms, the flowers of all the angiosperm species are constructed according to the same basic plan, which is diagramed in Fig. 16–28.

FRUITS AND SEEDS

The mature seed contains both the embryo and its supply of stored food, the endosperm. Why then, one might wonder, does the apple go to all the trouble of laying down a succulent layer of food *around* the seeds, the fruit? The flesh of the apple fruit plays no role in the nourishment of the germinating embryo; does it have any adaptive function?

The apple and other fleshy fruits aid in the dispersal of the seeds. An animal is attracted to the fruit by its color and tempting odor and eats it. It may then spit out the seeds onto suitable ground for germination. Or, protected by their tough seed coats, the seeds may be swallowed whole, pass un-

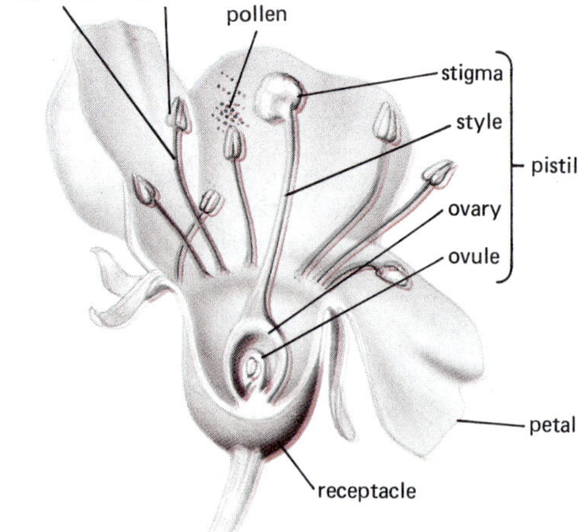

Figure 16–28 The structure of a typical flower.

harmed through the animal's intestines, and ultimately may be deposited at random, sometimes at great distances from the original habitat.

The fleshy apple, of course, is not the only kind of fruit produced by plants. From the viewpoint of the botanist, a fruit is defined as a structure consisting of one or more pistils, with or without adjacent parts of the flower. This definition has no relation to the commonly used distinction between "fruits" and "vegetables." By the botanical definition, many of what we normally consider as vegetables, such as the tomato, cucumber, and green pepper, are actually fruits, whereas many other edible vegetables are actually modified leaves, stems, or roots.

Regardless of the type of modification that eventually produces the fruit (Fig. 16–29), the wall of the ripened ovary is referred to as the **pericarp.** The pericarp may be dry and hard, as in nuts and the grains of the cereal grasses. It may split at maturity, releasing the seeds, as in the milkweed and the legumes, which include peas and beans. Or the pericarp may become fleshy and succulent at maturity, as in the berries (including the grape and the tomato). It may be partly fleshy, with a hard or leathery outer rind (as in watermelons and citrus fruits, respectively), or it may have a fleshy outer part enclosing a hard inner part—the "stone"—which in turn encloses the seed (such as the peach and the cherry). All the fruits just described are simple fruits, formed from a single pistil. Fruits may also be formed from a single pistil with adjacent parts (apple and pear) or from a number of separate ovaries derived from a single flower or several flowers (e.g., raspberry and pineapple, respectively).

Fleshy fruits and nuts are commonly dispersed to various distances from the parent plant by animals. Cockleburs and other fruits with spines or barbs are also carried away (unintentionally) by animals. The winged maple seed and the delicate parachute of the dandelion are well adapted to wind dispersal. Some fruits, such as witch hazel pods, literally explode, shooting their seeds for short distances out of the seed pod. Fruits like coconuts, with large air spaces or corky floats, may be spread from place to place by rivers, streams, and oceans.

The seed, no matter what its overall shape, form, or size, consists of three basic structures: the seed coat, which protects the seed from excessive evaporation, mechanical injury, and damage by parasites; the embryo; and the endosperm, which provides nourishment for the embryo (Fig. 16–30).

The embryo is essentially the new sporophyte in miniature. The cotyledons, or seed leaves, digest and absorb food from the endosperm, and they store food for the growing embryo. Monocot seeds have one cotyledon; dicot seeds have two. The **epicotyl,** the portion of the embryo above the point of attachment of the cotyledons, contains cells that will ultimately grow into the shoot of the new plant. The **hypocotyl,** the part of the embryo below the point of attachment of the cotyledons, will ultimately develop into the primary root.

When conditions are favorable (sufficient moisture; sufficient oxygen; temperature in a suitable range, especially 70–85°F; light of the proper quality, intensity, and duration; suitable soil acidity, carbon dioxide content, etc.) and dormancy, if necessary, has ended, the seed may **germinate** (Fig. 16–31). It absorbs water and swells. As the water content of the protoplasm is restored, the enzymes of the embryo begin to function, and metabolic activities are resumed in a gathering crescendo. Stored food is digested, and there is an enormous increase in respiration. Cell elongation begins, and the development of the embryo takes up where it was suspended with the ripening of the seed. Soon the seed coat bursts, and the hypocotyl emerges. The young root develops rapidly, anchoring the new plant in the soil, and immediately begins to absorb water and minerals. Shortly thereafter the epicotyl emerges from the seed coat.

The pattern of emergence and development of the young seedling varies considerably among different species. In some, such as the bean, the cotyledons are carried above the

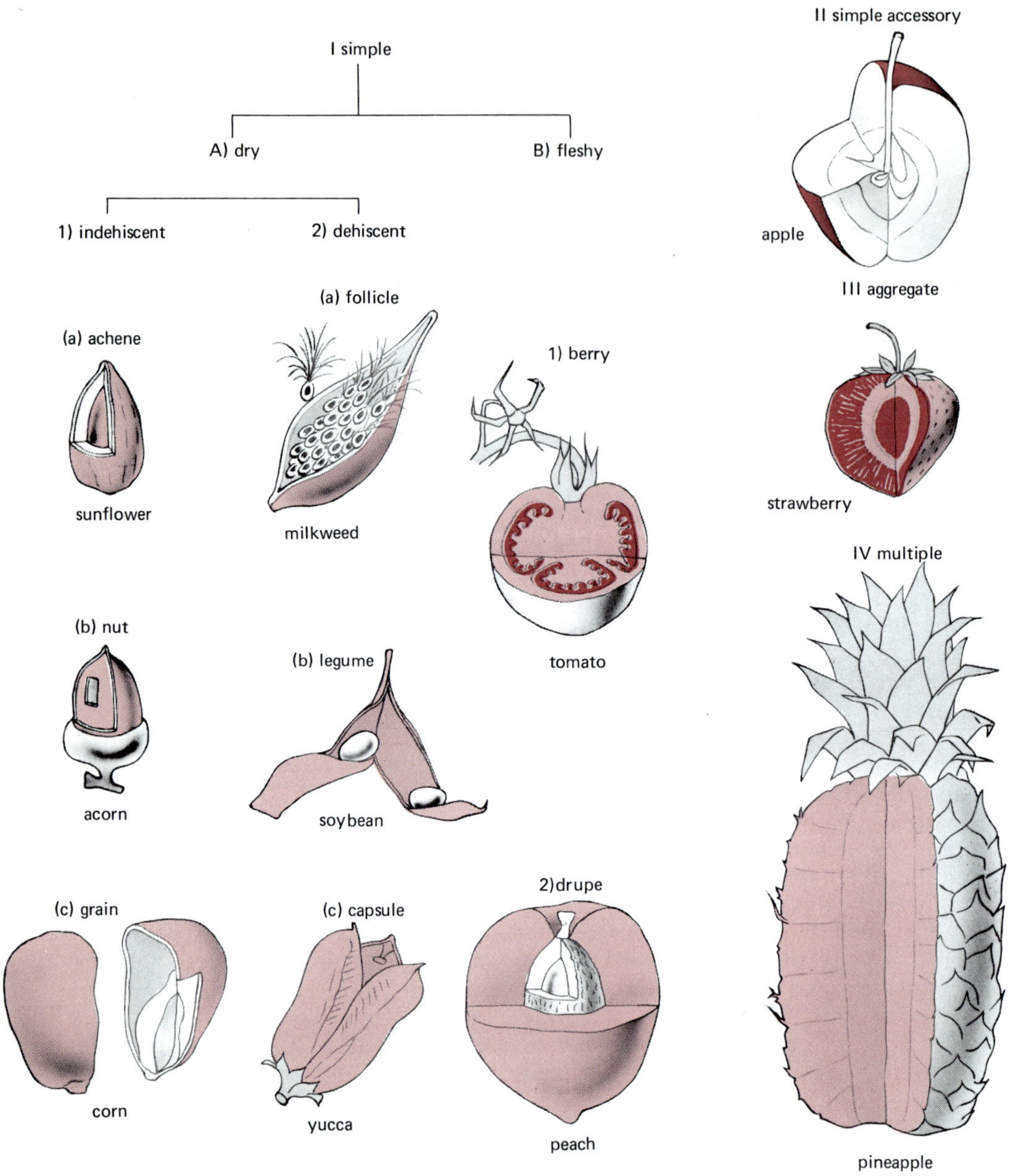

Figure 16–29 Various types of fruits.

ground with the epicotyl, and slowly wither as their cargo of food is consumed. In other species, as in corn and other grains, the cotyledon remains in the soil. At any rate, as soon as the shoot is thrust up into the light, it turns green and begins to manufacture its own food. The young sporophyte is now an independent plant.

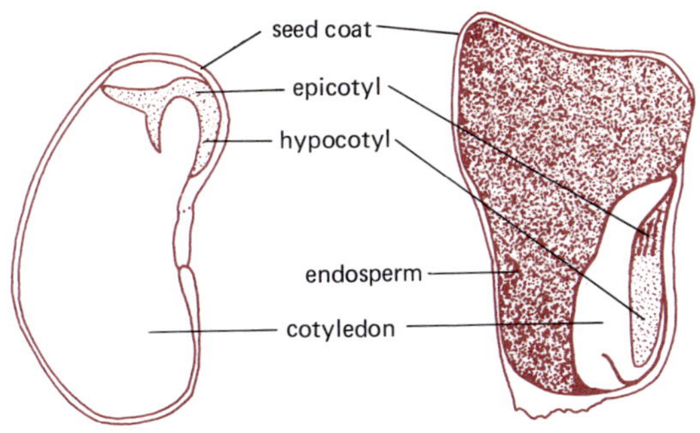

Figure 16–30 Structure of typical dicot (left) and monocot (right) seeds.

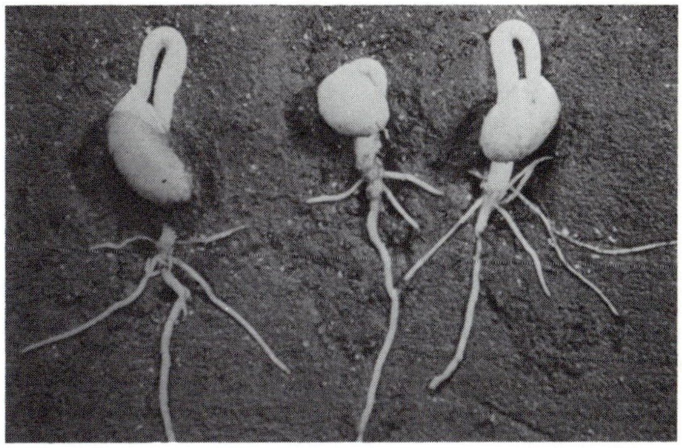

Lynwood M. Chace, Omikron.

Figure 16–31 Soil has been cut away to expose these germinating bean seeds. As the developing roots thrust down into the soil, the hypocotyl elongates. After it has burst through the soil surface, the young seedling will straighten up and pull the cotyledons and plumule up out of the already split seed coat. The seedling, exposed to sunlight, will quickly turn green and begin photosynthesis.

QUESTIONS FOR REVIEW AND THOUGHT

1. Discuss the advantages and disadvantages of asexual and sexual reproduction.
2. What is the significance of the gross size difference between sperm and egg?
3. Discuss meiosis and fertilization from the standpoint of chromosome number.
4. Discuss the significance of hormones in the female reproductive cycle.
5. If the human female had an estrous cycle, how would the evolution of our society probably have been changed?
6. How will our society be changed if fertilization in a test tube and mass duplication of embryos are perfected?
7. At what stage does a human embryo become recognizably human?
8. Breast-feeding is outmoded and unnecessary. Bottle-fed babies grow just as well as breast-fed babies. Are these statements true or false? Discuss.
9. Discuss adaptations in plants that enhance the probability of pollination.
10. What are the advantages, if any, of a period of dormancy before a seed will germinate?

CHAPTER **17**

Growth and Senescence

Introduction

Man loves to compartmentalize. He separates the animals from the plants, male from female, human from nonhuman, and so on. Similarly, when he considers the life cycle of the organism, he cannot resist the temptation to compartmentalize this too. He speaks of embryonic development, postembryonic growth and differentiation, and senescence (aging), as though these were separate and distinct phases. Yet in reality, the life of an organism is a continuous sequence of interrelated events, which begins with the zygote and ends with the death of the organism.

We have already discussed the first part of this continuous process: the changes that occur between conception and birth. In this chapter we shall deal with the remaining aspects of development: the growth of the organism after birth, processes of postembryonic differentiation, regeneration and repair of tissues, senescence, and finally the death of the organism.

Growth

Normally we tend to think of growth, at least in living things, as a finite process. An amoeba grows, reaches a certain size, and then splits into two smaller copies of itself, each of which starts the growth process over again. A tiny human zygote grows and differentiates to form an embryo, which grows and differentiates further to produce a child. The child continues to grow, until his late teens or early twenties, and then growth essentially stops. Now there is a fine counterbalance of catabolic and anabolic processes: there is a turnover of cells and tissues, and those that are broken down are replaced by new cells and tissues, but there is no longer a net increase. (In later years, the balance may even be shifted further: breakdown may exceed replacement, and there may be a net shrinkage of the total cell mass of the body.)

Yet in the living world growth is not always finite. Some plants have the capacity to go on growing, seemingly indefinitely. The giant se-

(a) (b) Lynwood M. Chace, Omikron

Figure 17–1 The sequoias of California (a) can grow continuously for thousands of years; the Galapagos tortoise (b), noted for its 150 year lifespan, also continues to grow throughout its lifetime.

quoias of California are the products of more than four thousand years of continuous growth. Some animals, too, seem to grow extremely slowly, but for inordinately long spans of time. Turtles appear to go on growing throughout their entire lifetimes, and life spans of more than one hundred fifty years have been reported for some individuals of this group. Generally for organisms that go on growing indefinitely, death results from accident or disease, rather than from any intrinsic process of senescence.

Indeed, there are indications that even cells and tissues of organisms that normally show a finite span of growth may have within them the potential to go on growing indefinitely, given optimum conditions. Cells and small groups of cells taken from various plants and animals, including man, have been perpetuated in tissue culture for periods of years, continuing to grow without any signs of cessation. The most famous instance of this experimental "tissue immortality" was that of a small piece of embryo chicken heart, which was removed from an incubating egg in 1912 by the French-American physician and Nobel Prize winner Alexis Carrel and placed in a warm nutrient medium. The tiny heart fragment continued to beat for a while; then it began to slow. Reasoning that toxic products of contraction or other metabolic activity might be accumulating and poisoning the cells, Carrel and his assistant carefully washed the heart fragment with warm isotonic salt solution and placed it in fresh nutrient medium. Beating was resumed almost immediately. For a time everything went smoothly. Bathed in a warm, periodically replaced nutrient medium, the bit of tissue continued to grow actively, nearly doubling in mass each day. But soon the mass of the tissue had increased so much that the bathing nutrient medium could not diffuse effectively to the central portions. Carrel and his assistant eventually worked out a regular routine of changing the nutrient fluid and carefully trimming away part of the new tissue mass every forty-eight hours. World War I came and went; the bit of tissue continued to grow actively. After ten years time, the longest any chicken of that breed might ever be expected to live, the routine still went on: growth, trimming, and change of nutrient fluid. In the early 1930s, Carrel collaborated with Charles

Lindbergh to perfect a perfusion pump, which made the process more effective. In 1939, Carrel returned to France, and a year later became involved with the Vichy government, an association that was to cost him his position after the liberation in 1944. Meanwhile, back in America, the bit of chick heart grew steadily on. Carrel died later in 1944, and still the tissue continued to grow, faithfully tended by his assistant. Finally, thirty-four years after the original bit of chick embryo tissue was first placed in the nutrient solution, after more than three life spans of a normal chicken, the experiment was deliberately stopped.

Recent investigations of cell cultures, using refined techniques, have cast some doubts on the validity of Carrel's findings. The modern researchers suspect that the nutrient solution used to bathe the bit of chick embryo tissue was contaminated with cells from the nutrient medium, and that it was these new cells that provided the growth far beyond the normal life span of the chicken. However, where care has been taken not to introduce new cells, it has been shown that certain types of mouse cells and cancer cells can be grown in cultures indefinitely.

The process of growth in living organisms involves a taking in of relatively nonspecific "raw materials" from the external environment and their assembly into the specific substances of the cell. Plants and photosynthesizing protists manufacture all their cellular constituents from very simple raw materials—carbon dioxide, water, and some mineral salts—binding the radiant energy of the sun in the form of chemical energy. Animals must take in more complex building blocks by eating other living things or their products. But these ready-made components are not incorporated as such: proteins, starches, fats are first broken down (digested) into simpler building blocks (amino acids, simple sugars, fatty acids and glycerol), which are then assembled into new proteins, polysaccharides, and fats highly specific for the organism manufacturing them. Some of the raw materials, in both plants and animals, are "wasted"—lost to the environment in the form of various metabolic wastes. But growth implies a net gain: the total nutrients taken in from the environment exceed the materials returned in the form of metabolic wastes.

Growth may occur by an increase in the size of the individual cells, as it does in protists, or by an increase in the total number of cells (essentially the mode of the explosive growth that transforms a single-celled zygote to a human baby, with a total mass billions of times as great). Frequently both types of increase—in cell number and size of the individual cells—are involved in growth.

In multicellular organisms, growth of the organism as a whole (as opposed to growth of its individual cells) is characterized by a **differential growth** of its individual parts. In the early human embryo, for example, the skin is at first represented by a single surface layer of cells. As the embryo grows in size, the skin layer must grow rapidly if it is to continue to cover the surface of the embryo. This growth occurs at first by division of the cells, each daughter cell then growing to the size of the parent cell and dividing in turn. During this period, the number of cells approximately doubles in each cycle of the cell division, and the skin layer grows rapidly in surface area but not in thickness. Then comes a phase of growth in which the surface layer increases in thickness as well as in surface area. It is found that the skin layer has now become several strata. But only the cells in the original layer are continuing to divide. Some of the cells from the original layer migrate upward to form the upper strata. They cease to grow actively as individual cells (though they do enter upon a distinct process of differentiation or specialization), but growth of the skin as a whole continues. The overall rate of growth, however, is lower, since the cells of one portion of the skin are continuing to double, while the rest are no longer dividing at all.

Differential growth can be observed in many instances on the macro-scale. It has been found that the tails of mice increase in length more than 40 percent more rapidly than does the body as a whole. As a result, though a newborn mouse seems relatively short-tailed, it has a proportionately quite longer tail as an adult.

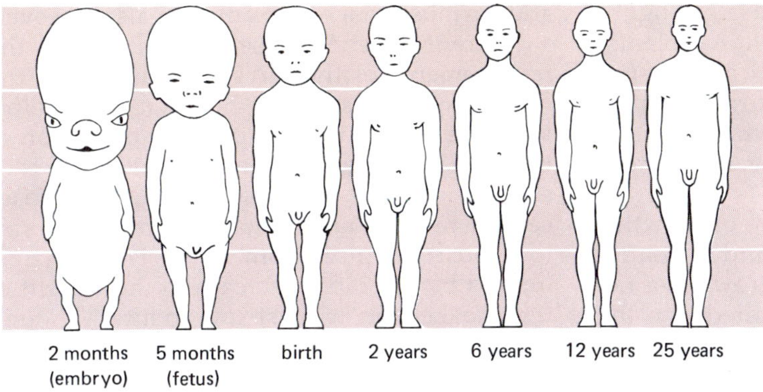

Figure 17–2 Growth is a differential process, resulting in a gradual change in the body proportions.

The opposite relationship is observed for the size of the head. The relative proportions of the human body at various stages of development are depicted in Fig. 17–2. Although the head corresponds to nearly a quarter of the length of the newborn human infant, the head grows less rapidly than the rest of the body, and by adulthood its fraction of the total length is closer to an eighth. Each organ and body part has its own specific pattern of growth. The external reproductive organs seem disproportionately large in a newborn infant, but during the preadolescent period they grow more slowly than some of the other parts of the body and become proportionately quite small. Then, at puberty, there is another period of explosive growth, during which these organs reach approximately their adult size.

In the growth of a plant, several distinct phases can be detected. At first there is a preliminary period during which little or no growth is observed. This period is called the **"lag" phase.** Then growth picks up, and the organism increases in size at a faster and faster rate. Growth at this time is exponential, with a doubling at equal intervals of time. This period is called the **"log" phase,** because of the logarithmic form of the growth curve (Fig. 17–3). But then a point is reached at which growth levels off, until eventually no growth is observed. This period is known as the **stationary phase.**

Curiously enough, it has been found that a

Figure 17–3 The S-shaped growth curves of individuals and populations show striking similarity.

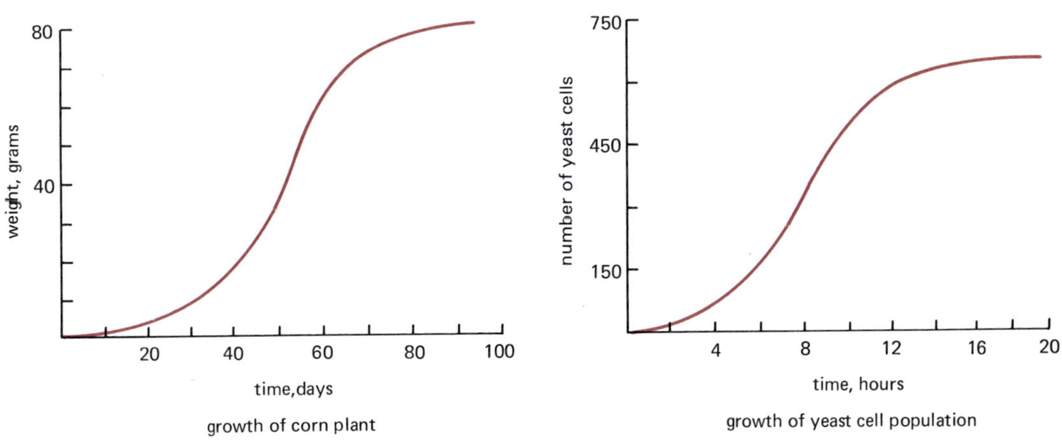

growth curve of the type depicted in the figure is characteristic not only of individual plants, but also of animals, populations of bacteria, and indeed of populations of organisms in general. For all organisms in which growth has been measured, the same overall pattern has been observed. What varies is the length of the growth period and the rate of growth.

Various factors have been found to control growth. First of all, of course, growth—a net increase in protoplasm—is limited by the supply of raw materials. In a famous "landmark" experiment, McCay and associates found that by raising young rats on a diet sufficient in such constituents as vitamins and minerals, but deficient in calories, the growth rate of the rats could be substantially lowered and their life spans correspondingly increased. (When the rats, after periods of retardation of growth up to a thousand days, were placed on a normal diet, they rapidly grew to the adult size, and lived out a relatively normal adult life, even though littermates kept on a normal diet from weaning were long since dead.)

Hormones also play a major role in stimulating and regulating growth processes both in plants and in animals, as we have already seen (Chapter 12).

Growth among the vertebrates is essentially a continuous process. Among the insects and various other groups of animals, however, a different pattern is observed, dictated by the requirements of the physical structure of the organism. An insect's body is encased not in a soft skin but in a rigid coat or **exoskeleton** of **chitin.** This tough substance provides a considerable degree of protection, but it imposes certain restrictions upon its wearer. The size to which an insect can grow is ultimately limited by the rapid increase in the weight of the exoskeleton it must lug about. Although the chitin coat is jointed like a medieval suit of armor, it does impose some limitations on body mobility. And unlike the cellular skin of vertebrates, a chitin exoskeleton cannot expand as its wearer grows. Instead, an insect must shed its "skin" or **molt** in order to grow (Fig. 17–4).

As an insect's exoskeleton begins to become too tight, a soft new coat of chitin is formed beneath it. Then the old exoskeleton splits, and the insect wriggles out, leaving behind a dry, empty shell of its former self. Its new exoskeleton is still soft and pliable. Quickly it breathes in as much air as it can hold, inflating its body to a considerably larger volume. Gradually the new chitin hardens, and its inflated form will provide the insect with additional room for growth. Thus growth in insects occurs by a series of discon-

Figure 17–4 Pale and moist, an adult cicada (a) crawls out of the split outer covering of the immature nymph form; gradually its new exoskeleton hardens and darkens (b) and its wings unfold. The snake also sheds its skin—in this case a true skin or epidermis. This shed snakeskin (c) forms a complete outline of its former inhabitant.

(a) (b) (c)

Lynwood M. Chace, Omikron

tinuous stages, with an abrupt increase in body size after each successive molt.

Just what causes the cessation of growth that sets in at adulthood in most multicellular species (with the notable possible exceptions of certain trees, fishes, and reptiles) is still unclear. Scientists would very much like to know whether the cessation of growth is a necessary concomitant or even a result of the processes of increasing differentiation, or whether they are two different processes of the metabolic activity that normally occur together by mere coincidence.

Scientific attention has also been concentrated upon another puzzling aspect of growth—the occasional instances of growth run wild, cancer. In cancerous growth, the normal controls seem to be suspended. There is neither differentiation of the cells nor any check upon their ceaseless self-reduplication. In one attack on the problems of cancerous growth, the eminent biologist Albert Szent-Gyorgyi has discovered two hormone-like substances, promine and retine, which appear to work antagonistically: promine promotes growth and retine retards it. These substances have been found in a variety of organisms, from clams to mushrooms to humans. (Indeed, an important source of the substances is human urine.) Research into the action and interaction of the two substances is not only a promising avenue of cancer study, but is also bringing new insights into the basic growth processes.

Metamorphosis

The study of nature is full of surprises. An amateur naturalist exploring the coastal waters of the Atlantic might come upon a flock of small tadpolelike creatures swimming about. Wondering what tadpoles could be doing in salt water, he might catch a few in a jar and take them home to see how they would develop. Much to his astonishment, within a few hours the "tadpoles" would cease swimming about and attach themselves to the bottom of the jar with special adhesive glands. The tails would disappear, resorbed into the body, but no legs would develop. No lungs would form to replace the gills; the eyes would disappear. The typically chordate "tadpoles" would turn before his startled eyes not into air-breathing frogs, but into the sedentary, water-dwelling, rubbery, baglike creatures called tunicates.

Of course, the dramatic change of the free-swimming, bilaterally symmetrical tadpole-like tunicate larval form to the rubbery, baglike adult is intrinsically no more astonishing than the transformation of the frog tadpole

Figure 17–5 Metamorphosis—two "tadpole" larvae turn into far different adults: (a) bullfrog tadpole; (b) tunicate larva; (c) baglike adult tunicates; (d) a young wood frog absorbing the remains of its tail.

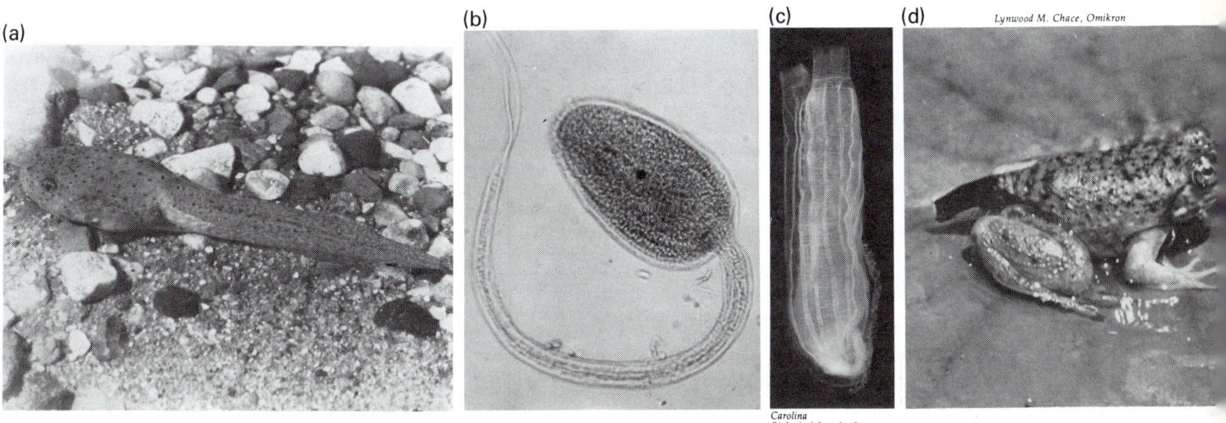

Lynwood M. Chace, Omikron

Carolina Biological Supply Co.

itself (Fig. 17–5). A totally aquatic, gill-breathing, legless animal, which swims about propelled by its powerful finlike tail and eats only plant life and detritus, which it grinds with horny teeth, is transformed into an amphibious creature, at home both on land and in the water, which breathes with lungs, leaps and swims by powerful kicks of its well-developed hind legs (it has no tail), and eats exclusively living animal forms, which it catches in motion by lightning flashes of a "flypaper" tongue. The scope and magnitude of the changes involved may seem no less dramatic than if, for example, an embryo were to develop into a rabbit that hopped about the meadow for a time, contentedly munching the grass, and then suddenly turned into an eagle and flew away.

Such dramatic changes are common in many groups of the animal kingdom. In many instances, the larval form that initially develops from the embryo is so unlike the adult into which it eventually changes that biologists first classed them in totally different species or even different phyla. Such diverse sessile organisms as sponges, echinoderms, mollusks such as oysters and clams, and the tunicates described above all have free-swimming larvae that bear little resemblance to the adults. Among the amphibians also, dramatic changes are observed in the course of postembryonic development. And even more dramatic changes are observed in some insect species. The crawling caterpillar, which feeds voraciously and at length walls itself up in a cocoon, is a far cry from the delicate butterfly that emerges from the cocoon, lives for a brief while, mates, and dies.

The term **metamorphosis** is used to connote the phenomenon of dramatic changes in morphology, physiology, and usually mode of life occurring within a relatively brief period of postembryonic development, without any increase in size (or even with a decrease in size).

Research into the causes and mechanisms of metamorphosis has disclosed a hormonal control of the process. In amphibians, the thyroid gland has been definitely implicated. A normal frog tadpole, for example, increases greatly in size after hatching from the egg, as

Figure 17–6 Axolotl, the salamander that never grows up. Note the feathery external gills.

it feeds upon the abundant microscopic green plant life of the pond or stream. But for a long period (which varies from a few months to more than two years, depending upon the species) the general appearance and proportions of the animal remain basically the same. Then the buds of the hind legs begin to grow and differentiate, and forelegs break through the skin in the breast region. Within the body even more profound changes are occurring, and soon the animal ceases to feed and crawls out of the water. The changes proceed even more quickly now: in a matter of days, the tail has been resorbed, and the tadpole, smaller now, has been transformed into a frog.

This normal course of events can be experimentally disturbed in a variety of ways. If, in a relatively early tadpole stage, the larvae are fed or injected with thyroid hormone or large doses of elemental iodine, they will undergo a precocious metamorphosis, turning prematurely into miniature frogs. Yet if the thyroid glands of tadpoles are removed, they will continue to grow, but will remain tadpoles all their lives, never turning into frogs at all. Curiously enough, their reproductive organs may develop fully, eventually producing ripe sperm or large eggs. There is a natural parallel of this experiment. The Mexican newt axolotl (Fig. 17–6) retains its typical aquatic tadpole form, with the feathery external gills

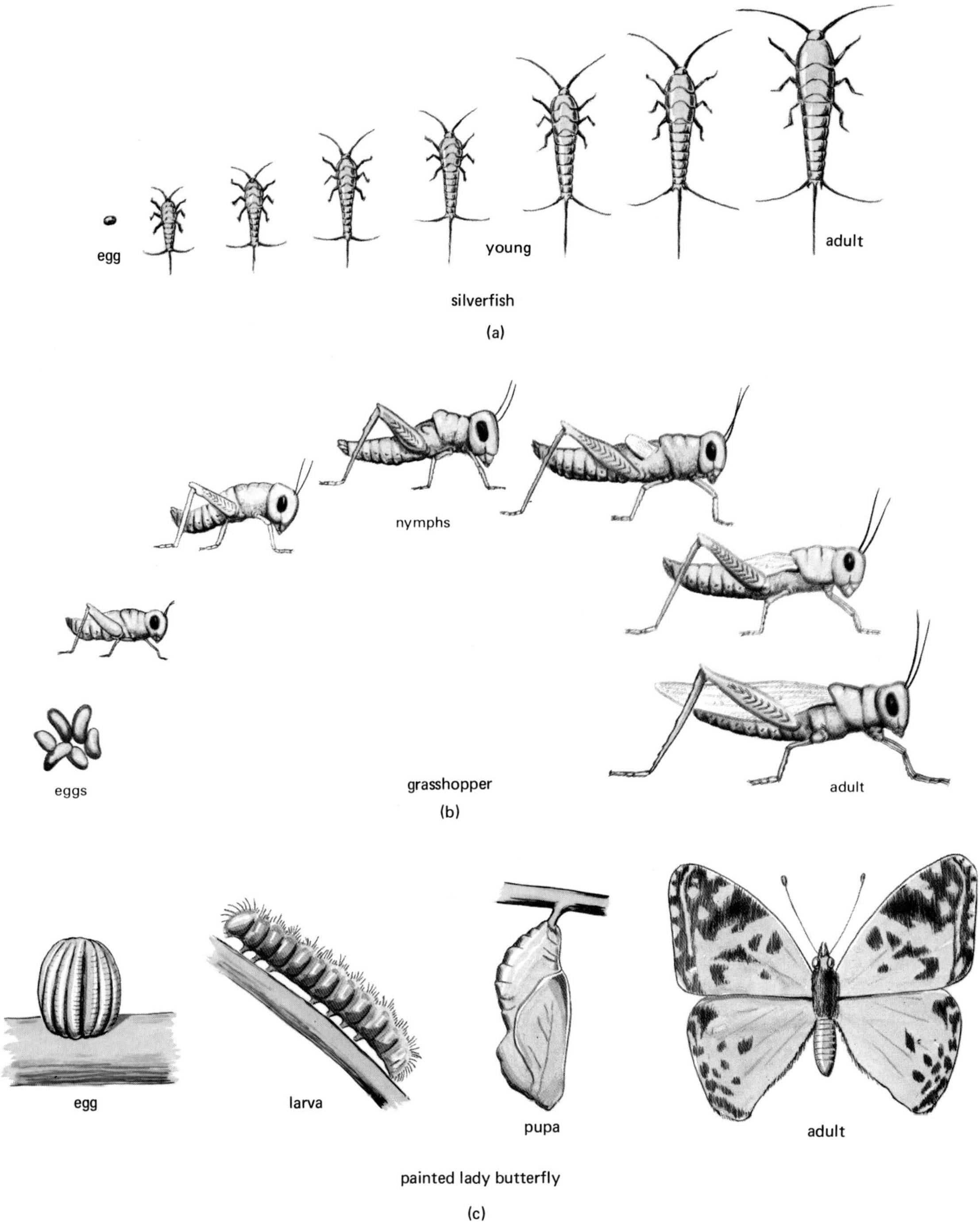

Figure 17–7 Examples of lack of metamorphosis (a), incomplete metamorphosis (b), and complete metamorphosis (c) are found among the insects. A pupal stage occurs only in (c).

characteristic of newt tadpoles, throughout its lifetime, reaching sexual maturity without ever undergoing metamorphosis. Noticing that axolotl normally is found in waters with an unusually low iodine content, experimenters tried feeding thyroid hormone to axolotls. The fat tadpolelike creatures promptly turned into small, black, gill-less newts.

Several different types of metamorphosis are observed in insects (Fig. 17–7). In some species, such as the silverfish, there is no metamorphosis at all: the egg hatches into a tiny replica of the adult form. This insect grows in size with each successive molt, eventually reaching the size of the adult.

In another variety of postembryonic insect development, called **incomplete metamorphosis,** the larval form shows some resemblance to the adult and is transformed into the adult type by a series of gradual, successive changes, resembling the adult more and more with each molt. This type of larval form is usually called a **nymph;** incomplete metamorphosis is characteristic of such species as roaches and grasshoppers. The morphological changes involved in the incomplete metamorphosis of insect species that have aquatic larvae (called **naiads**), such as the dragonfly, are more complex than in cases when both nymph and adult insect are air breathers.

In many insect species, **complete metamorphosis** is observed: the egg develops into a **larval stage,** usually a crawling, wormlike form, which spends most of its time feeding voraciously; the larva then passes into the quiescent **pupal stage,** and the pupa finally develops into the adult **(imago)** stage. In the pupal stage, the insect is usually enclosed in a protective case, and though no activity is visible to the observer, profound morphological and physiological changes are taking place. The **imaginal discs,** present in the body of the larva, develop into the organs of the adult, and most of the specialized cells of the larva die. In the adult stage, the principal activities are concerned with propagation of the species, although the adults of most insect species also feed (usually to a lesser degree than the larvae).

Hormones have also been found to play a key role in insect metamorphosis. A fine balance of several different hormones controls growth, molting, and metamorphosis. A hormone produced in the brain stimulates the secretion of a hormone known as **ecdysone** (the molting hormone) by the prothoracic glands (Fig. 17–8). Ecdysone controls growth and differentiation in interaction with a third hormone, **juvenile hormone,** which stimulates larval development but inhibits metamorphosis. This hormone is secreted by the **corpora allata,** two tiny endocrine glands posterior to the brain. It is the relative proportions of ecdysone and juvenile hormone that determine the course of development of the insect. In the presence of large quantities of juvenile hormone, ecdysone stimulates larval growth through a series of molts. If the amount of juvenile hormone present is lowered ecdysone stimulates formation of the pupa, and when there is no juvenile hormone at all the final stage of metamorphosis results in the adult. (At the end of the larval stage, juvenile hormone ceases to be secreted.)

Experiments have been conducted with juvenile hormone similar to those on the influence of the thyroid gland upon amphibian metamorphosis. It has been found that if the corpora allata are removed from a young larva, it enters into the pupal stage at the next molt, ultimately producing a miniature adult form. If juvenile hormone is administered to a mature larva, ready to undergo pupation, this stage does not ensue; instead, the next molt simply yields a much larger larva.

In recent years, juvenile hormone has been a subject of lively interest among researchers, some of whom have hoped that this substance may hold some of the keys to the aging process. Juvenile hormone has been isolated from noninsect sources as well, including mammals. (What role it plays in these organisms is not yet known.) The structure of this substance has been delineated, and studies of its role are continuing actively.

Ecdysone has also been a center of scientific interest in recent years. This hormone has been linked with the formation of chromo-

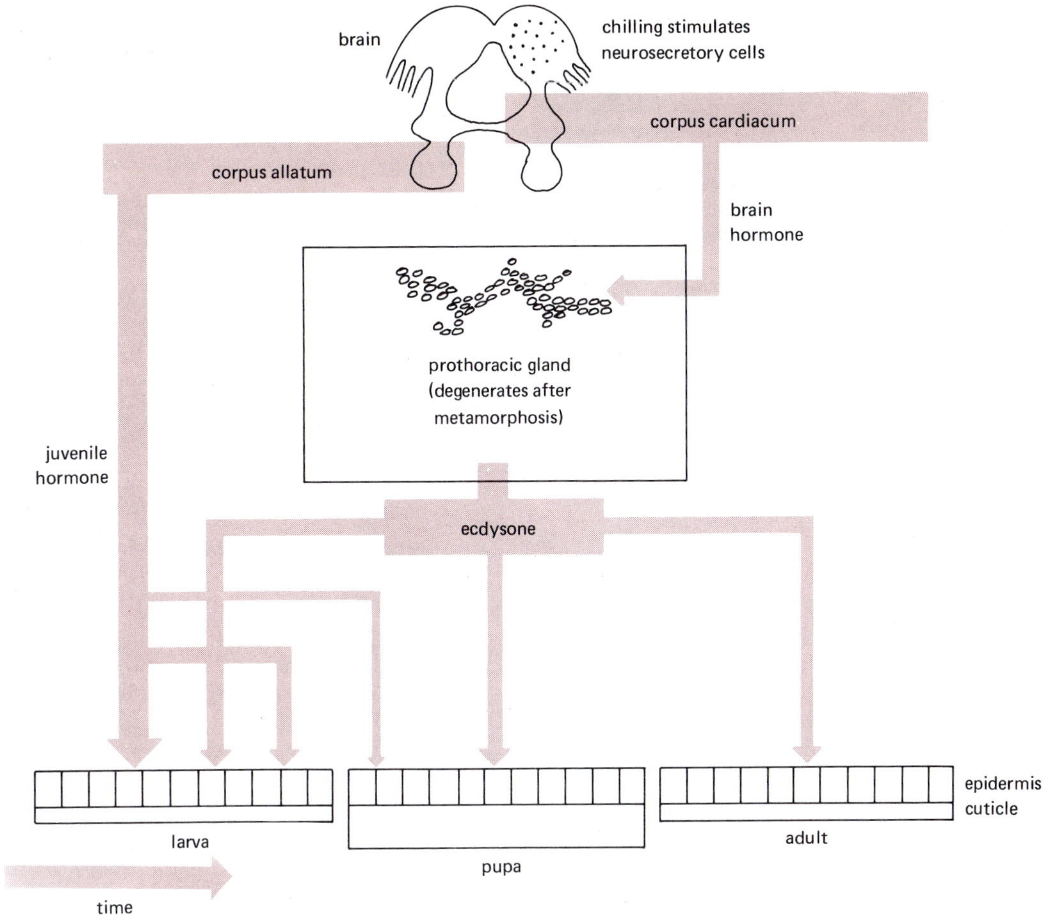

Figure 17–8 Hormonal control of metamorphosis in the silkmoth.

some puffs observed in the giant salivary gland chromosomes of certain insect larvae and thought to be associated with processes of cell differentiation. The puff formation resulting from administration of ecdysone has been taken as evidence that hormones may play a key role in activating certain genes.

Repair and Regeneration

A circling hawk swoops down suddenly upon a small lizard crawling about among the leaves. With a pounce it is clutching the tail of the unlucky reptile. The lizard struggles; there is a snap, and while the attention of the hawk is fixed upon the detached but still wriggling tail, the rest of the lizard scurries inconspicuously away to safety. During the days that follow, an amazing series of events unfolds. A bud appears at the site of the break, resembling somewhat the limb buds that appear on a developing embryo. As the lizard continues about its daily affairs, the bud grows and grows, and gradually begins to take on the appearance of a recognizable tail. Soon it is a full-grown tail (Fig. 17–9). It is not a perfect tail, to be sure (it is reinforced by a rod of cartilage instead of a bony skeleton), but it is a functional tail, and if it should hap-

Figure 17–9 This lizard lost its tail, perhaps in a dispute with a predator. The bud of a replacement can be seen at the end of the stump; ultimately the full tail will be regenerated.

pen to be clutched by another hungry hawk, it will break off just like the old tail, and another new tail will grow back.

During the lifetime of an organism, many accidents may occur. Through disease or mishap, tissues, organs, or whole body parts may be damaged or lost. In a fall, skin may be scraped off or cut open; bones may be broken. To some extent, the body possesses the ability to repair damage to its parts. Wounds may heal; broken bones may knit together. This ability for self-repair varies greatly among individuals. Within a species, younger individuals seem to possess a much greater capacity for healing, which is gradually diminished as they age. The various species also differ greatly in their ability for repair and regeneration. Most of the vertebrates, like man, can replace only certain tissues, such as the skin and subcutaneous tissues that grow in to cover a wound. A lizard can replace a lost tail; if a salamander's leg is cut off, a new one will grow in its place. A crab can grow a new claw. An earthworm can replace a lost tail or even a head. A lowly flatworm can even regenerate all the missing parts for a whole new worm, starting from just a small segment of the body. A starfish can grow a new arm—and an arm can grow a new starfish! And a whole new carrot plant can be grown from a few callus cells. In general, the ability for **regeneration,** the replacement of structures or parts that have been lost by accident or mutilation, is greatest in the relatively nondifferentiated "lower" organisms, and is gradually lost as differentiation increases—both with age and with increasing phylogenetic complexity.

Scientists have long marveled at the regenerative capacities of such organisms as flatworms and salamanders, and wondered: Why is it that a salamander can regenerate a lost limb and a man cannot? Is regeneration a special adaptive process found only in certain species, or a general property of living matter that species such as man have somehow lost? Are there specific substances that prevent large-scale regeneration in higher organisms? Might we some day discover techniques that would permit the human body to regain its lost powers of regeneration and literally heal itself?

It is not surprising that intense interest has been centered upon research into the phenomenon of regeneration. Many fascinating details have been discovered, and some general principles have been derived from them. But much mystery still remains.

An interesting and quite definite series of events occurs during the healing of a wound (Fig. 17–10). First it is covered by a protective scab, formed by a blood clot and the dried residue of serum exuded from the wound. Beneath the scab, which provides mechanical protection for the delicate exposed tissues and prevents them from drying out, undifferentiated cells begin to penetrate into the wound area from the surrounding region. Scientists are not yet in agreement as to the origin of these undifferentiated cells. They may result from dedifferentiation of specialized cells near the wound, or may come from a pool of unspecialized cells that migrate over to the wound area. Whatever their origin, the undifferentiated cells multiply rapidly, invading the wound area, and bridge it with a fibrous growth of connective tissue. As this thickens, it forms "scar tissue," and the protective scab falls off. Gradually adjacent cells begin to multiply and invade the scar tissue, often eventually replacing it entirely with the normal tissue for the site.

In the case of regeneration of a whole body part, such as that of a lost limb by a salamander, the process begins in much the same manner. But after the wound has healed, a

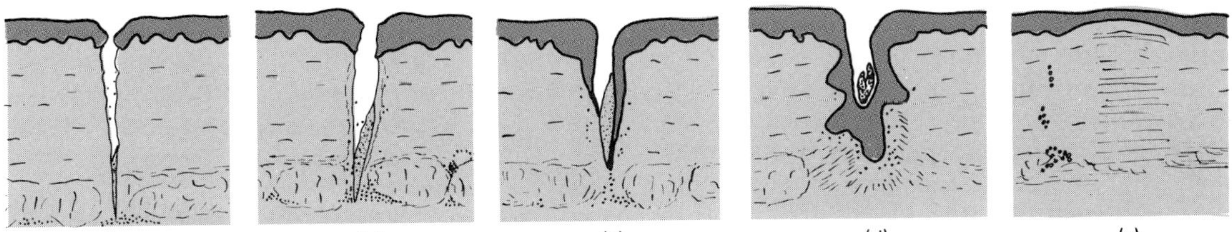

Figure 17–10 Successive stages in the healing of a wound. (a) An early clot has formed, but there is still some bleeding in the subcutaneous tissue; (b) the clot lines the walls of the wound, the epidermis thickens, and the subcutaneous tissues show an inflammatory reaction; (c) epidermis migrates down the walls of the wound and phagocytes remove debris; (d) epidermis has covered the defect and the scab is ready to be sloughed off; (e) a raised, thin epidermis covers a dense maturing scar.

bud of undifferentiated cells appears, rather resembling the limb bud of a developing embryo. Gradually, as the bud grows, its cells become differentiated into bone, muscle, connective tissue, integument, and the various other tissues that comprise the functional leg.

The "why" of regeneration has given rise to lively speculation and has prompted many ingenious experiments. According to one major theory, the organism possesses a system of growth inhibitors that becomes more and more effective as the organism grows to adulthood. If tissues or body parts are removed, the customary inhibitor is no longer supplied from the site of the wound, and there is a proliferation of new tissues moving into the site and restoring the missing part. When restoration is complete, further growth is again inhibited.

In some instances, complete removal of the part is not even necessary: regeneration can be stimulated by an injury to the part, which apparently interrupts the supply of the inhibitor from it. For example, regeneration in a planarian can be induced not only by cutting off a portion of the body, but even by making an incision in the body. Many curious planarian monsters have been experimentally created in this way. If the elbow region of a salamander limb is injured by exposure to ultraviolet radiation, even though the limb is not removed, an accessory, perfectly formed forearm may grow out at the site of the irradiation.

But growth inhibitors do not appear to be the whole story. Regeneration also seems to be correlated with metabolic activity. In a planarian, for example, there is a definite gradient of metabolic activity, with physiological processes most active in the head region and increasingly less active down toward the tail. The rate of regeneration similarly decreases from the head to the tail region. The metabolic-gradient concept provides a possible explanation for the fact that in regeneration, polarity is normally conserved: a segment cut from the middle portion of a planarian will regenerate a new head at the cut surface corresponding to the anterior end of the original worm and a new tail at the cut surface corresponding to the posterior end. Similar observations of conservation of polarity are common among various eggs, embryos, and lower organisms.

In addition to the growth inhibitors, certain growth stimulators also seem to be involved in the processes of regeneration. Hormonal influences have been demonstrated in certain cases. And the nervous system seems to play a key role, at least in regeneration among vertebrates. For example, if the nerves leading to a regenerating salamander limb bud are cut, growth of the limb bud will cease. Eventually, if the nerves are continually cut as they begin to regenerate in the direction of the bud, the limb bud itself will be resorbed, and no new limb will grow. Similarly, if the limb of a tadpole is removed, a new one will grow back easily. Yet removal of a limb from an adult frog will not be followed by regeneration, *unless* extra nerves are moved surgically to the stump of the missing leg. It has been postulated that some vital growth-stimulating substance is released by the nerves and promotes regeneration. Recent experiments in which regeneration of a lost leg of a frog was stimulated by irritating the stump seem to offer some hope that techniques may some day be devised to greatly increase man's inherent potential for healing and regeneration.

Senescence

Must we grow old? Man has asked and re-asked this question through the ages. After a life rich in experience, when a man feels that he is just beginning to gain a wealth of knowledge and an appreciation of life, he helplessly finds that his body and even his mind are deteriorating progressively, bit by bit. When life should be most rewarding, the imminence of death casts a darkening shadow.

Over the centuries man has chased will-o'-the-wisps—anything that might promise a hope of eternal youth. From Ponce de Leon searching for his Fountain of Youth to modern men journeying to a clinic in Bucharest in the hope that procaine treatment will rejuvenate them, man has clung to the idea that aging is not inevitable—that some day, in some way, the inexorable deterioration can be halted and perhaps even reversed.

Aging or **senescence** may be defined as a biological process characterized by a gradual deterioration of the structures and functions of the organism, resulting in increased susceptibility to disease. Organisms generally do not die of "old age" but rather from disease or accidents, which may be a consequence of the deterioration that accompanies aging.

If we look about us in the world of life, we may discover some interesting facets of the processes of aging. Living forms tend to have a characteristic life span. The rotifer *Philodina citrina* (Fig. 17–11), a tiny multicellular animal found in ponds, commonly lives for about three weeks and toward the end of its life shows degenerative changes reminiscent of aging in higher organisms. The rat is old at three years, the dog at fifteen, man at seventy. Each species appears to have a definite span of time during which the normal processes of growth, maturation, and aging take place. Apparently there is a genetic factor in aging. And yet there seem to be exceptions to the pattern. Many of the giant sequoia trees of the West Coast are still alive and growing after thousands of years. It is said that many fish and reptiles similarly continue to grow throughout their lifetime and do not age. In these organisms death usually is the result of accident, disease, or predation. In addition man has been able to experimentally upset the natural pattern of senescence in some animal species.

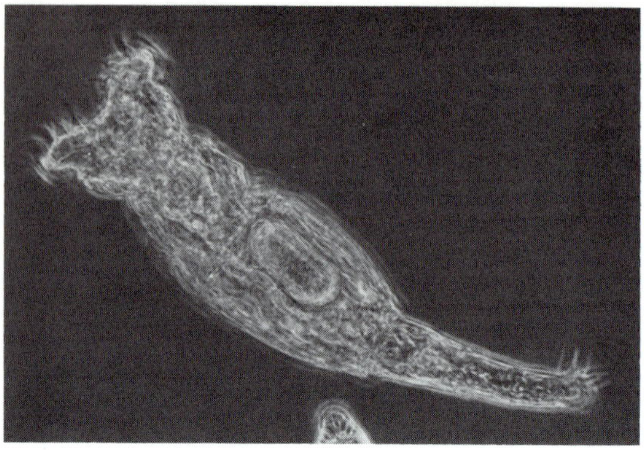

Figure 17–11 The rotifer, a common pond organism, is an intriguing model for aging research. Some species, such as *Philodina citrina*, are born, reproduce, age, and die in just three weeks.

In the 1940s A. I. Lansing performed a series of fascinating experiments with the rotifer. These parthenogenic organisms lay eggs over a period of days. For instance, *Philodina citrina* lays eggs beginning on the fifth day after hatching and every day thereafter for a period of two weeks. Lansing isolated these "first-day eggs" and then, in turn, isolated the "first-day eggs" of these second-generation organisms. The process was repeated for fifty-four generations. Whereas the original parents lived an average lifetime (i.e., 24 days), there was a gradual increase in the life span of successive generations of offspring, until by the fifty-fourth generation the life span was 104 days! A similar series of experiments was conducted with eggs laid by older rotifer mothers. The eggs laid on the seventeenth day gave rise to offspring that lived an average of only 18 days. The third generation of this line lived only 14 days; the eggs of the fourth generation never even hatched. As a result of this and other experiments, Lansing concluded that offspring from younger mothers tend to live longer than offspring from older mothers and that the effects are cumulative. Surprisingly, there seems to be some correlation between Lansing's findings on rotifers and actuarial tables on human life expectancies. According to the records of the Metropolitan Life Insurance Company, on a statistical basis children from younger mothers tend to live longer than children from older mothers.

Earlier in the chapter we described another series of experiments in which lifetimes were artificially lengthened. In the experiments performed by Clive McCay, the life spans of rats were considerably lengthened by underfeeding these rats from the time of weaning. Another series of dietary experiments, which might be considered an outgrowth of the classic McCay experiments, was performed in the laboratories of Monsanto Chemical Company in the early 1960s. Young chicks and weanling mice were fed a diet deficient in the essential amino acid tryptophan. Normal growth and development apparently ceased as long as this amino acid was withheld. At the end of a year, the experimental animals were placed on a normal diet. Just as in McCay's experiments, the normal growth pattern was resumed. The year of "suspended growth and development" was reflected in a year added to the lifetimes of these animals.

We have said that aging is accompanied by a progressive deterioration in the organism. For instance, if we compare a seventy-five-year-old man with a thirty-year-old man, we find that there is an appreciable decrease in brain tissue (a drop of 44 percent), number of taste buds (64 percent loss), and kidney glomeruli (44 percent decrease). In addition, the physiological functions have also deteriorated: the heart output of the older man is only 70 percent that of the younger man, and the strength of his grip is scarcely more than half as great.

Why do we age? Why does this progressive degeneration take place? Is aging inevitable? Many theories have been offered to explain the ubiquitous process of aging. We shall discuss here several of the most important of these theories. The theories discussed below may not necessarily be mutually exclusive. Indeed, various facets of each may play a role in the aging process.

Many scientists feel that aging is a necessary concomitant of life. These scientists state that aging is literally "programmed" into the genetic material of the organism—that in the course of living certain gene-determined reactions occur, independent of the environment, which ultimately lead to a gradual degeneration of the organism. These biologists point to the fact that each species has a characteristic and predetermined life span, and even within a species there are long-lived strains and short-lived strains. For instance, in man it is known that children of long-lived parents tend to live longer than the average for the population.

A variation of this theory of gene-determined life span proposes that each organism inherits a characteristic energy reserve upon which it draws to meet its life-long needs. According to this theory, propounded by Hans Selye and others, repeated stresses, which draw more heavily upon the body's reserves, would tend to shorten the life span of the organism. This theory seems logical, but has some rather obvious weaknesses. For instance, one would expect an athlete to have a foreshortened life span; yet this is not at all the case.

An even more intriguing theory holds that in the life of the organism certain waste products of metabolism accumulate within it. This gradual buildup of poisons interferes with the normal functioning of the organism. There is much experimental evidence to support this view. Since the turn of the century, experimental evidence has been accumulating revealing that pigments accumulate with age in the tissues of virtually all higher organisms. Progressively increasing amounts of yellowish pigment have been detected in nerve, muscle, liver, and other cells of man and other species. Bernard Strehler has observed a linear increase in the amount of the yellow "senility pigment" lipofuscin in human heart tissue, at a rate such that this pigment comprises 6 to 7 percent of the total volume of the heart muscle cells by the age of ninety. It is still a matter of controversy whether the senility pigments are contributing causes or merely concomitants of the aging process.

Other experiments appear to support the thesis of a gradual buildup of metabolic poisons. We have already described Carrel's famous "immortal" embryonic chicken heart tissue, which was seemingly kept alive for decades by washing away the waste products and paring the tissue to keep it small enough so that all parts of the tissue could be reached by the bathing nutrient solution.

In another series of experiments, in the laboratory of Hans Selye, a thin glass tube was inserted beneath the skin of a rat (Fig. 17–12). In time, membranous tissue covered the ends of the tube, and fluid filled the tube. Eventually, a tiny cord of cells grew from each of the two membranes into the tube and finally met and joined at the center. It was found that within a matter of weeks, these cells showed signs of deterioration and died. However, if the fluid in which they were bathed was changed periodically, they remained alive and vigorous for considerably longer periods of time.

Another theory of aging, which has been gaining more and more adherents in scientific circles, is the somatic mutation theory. The proponents of this theory claim that under the influence of various environmental factors, such as sources of ionizing radiation, and such internal mutagens as hydrogen peroxide, mutations tend to accumulate among the somatic cells of the body. Since the vast majority of somatic cells rarely divide during the course of the lifetime of the individual, there is little opportunity for the elimination of faulty (mutation-bearing) cells through the selective processes that operate during cell division. For example, H. J. Curtis has found a progressive increase in the number of aberrations in the liver cells of mice, reaching a minimum of 22 percent of all the

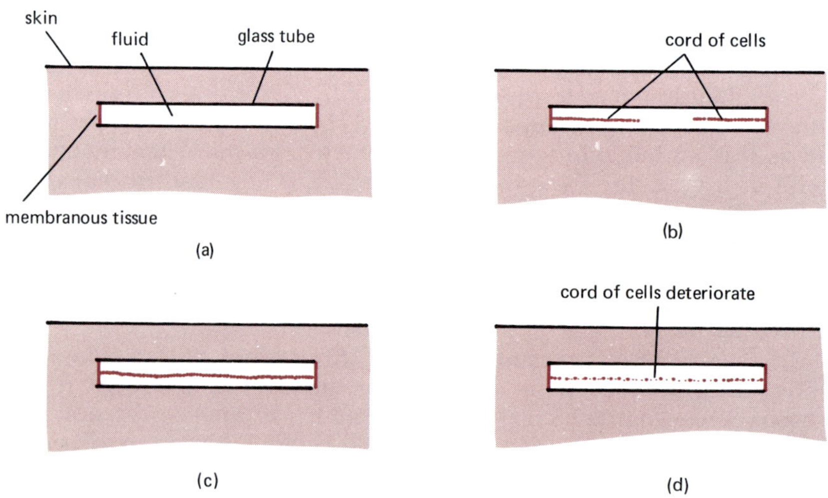

Figure 17–12 The Selye experiment. A glass tube was inserted beneath the skin of a rat. Gradually membranous tissue covered the ends of the tube and fluid filled it. A cord of cells grew into the tube from each end and joined. Then, unless the fluid in the tube was changed periodically, the cells deteriorated and died.

liver cells at the age of twelve months. The proponents of the somatic mutation theory contend that with an ever-increasing percentage of abnormal cells accumulating in a particular organ, that organ becomes less and less efficient and hence may place undue stress upon other organs, hastening their deterioration as well.

Death

Man has been able to adjust to every fact of life except its end—death. He has devised numerous schemes and beliefs to circumvent this seemingly inevitable event. Beliefs in reincarnation and in an eternal life in the hereafter are attempts to negate the concept that the termination of life is an irreversible trip into oblivion. But death occurs all about us and within us. Our white blood cells are continually destroying invading microbes. Every moment, millions of cells within our bodies perish. Throughout the world, storms, fires, floods, earthquakes, and other natural phenomena destroy many billions of organisms each year. In natural communities, in the course of normal cycles of life, countless trillions of organisms are being devoured and destroyed by others. But were an organism to survive all of these dangers, it too would succumb to a "natural" death. Thus death appears to be an inevitable consequence of life.

If the death of organisms in a community is studied on a statistical basis, some very interesting patterns emerge. For instance, if the number of deaths of organisms is plotted against the age of death, we arrive at a curve which varies depending upon the particular species observed and the conditions under which they are living. But nonetheless, certain definite general patterns are observed. For instance, Fig. 17–13 shows the pattern of death in various animal species. In a population of oysters, where death occurs most frequently at young ages as a result of influences imposed by a hostile environment, only an extremely small proportion of the population reaches the potential life span. In the case of a population of fruit flies raised in the laboratory under optimum conditions and protected from all the normal environmental hazards, nearly all the individuals of the population survive to a period somewhat approximating the potential life span; then all die off within a very short period. The human survival curve falls somewhere between these two extremes. Modern medical advances have brought the curve closer and closer to the pattern typified by the fruit fly curve. But an important fact should be noted: in each case there is a definite "potential life span." Despite the actuarial tables, which over the past decades have shown a definite upward trend in human "life expectancy," the advances of modern medicine have *not* lengthened man's potential life span, which remains approximately a hundred years. They have merely increased the proportions of the population that live to greater ages within this life span. They have reduced the relative numbers of individuals

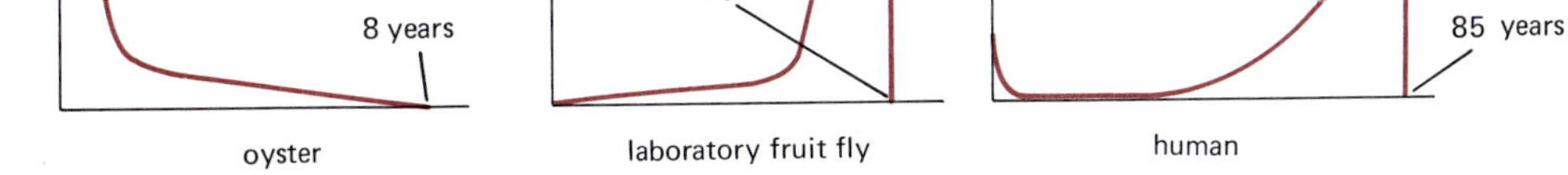

Figure 17–13 Plots of mortality versus age at death yield widely differing curves depending on the species and the conditions of its environment.

that succumb at early ages to accident and disease and increased the relative numbers of those that go on to die a "natural death."

What is the phenomenon of death? It may be defined as the cessation of all metabolic processes. Death of individual cells is common, and appears to be under some sort of genetic control. In some tissues, there is a rapid turnover of cells. It is estimated, for example, that there is a complete turnover of human red blood cells approximately every 120 days: within this time, virtually every red blood cell in the bloodstream dies and is replaced by a new one. Yet some cells, such as certain nerve cells, last throughout the life of the organism. In some cases, death of specific cells is a definite part of the "blueprint of development" of an organism. The cells in the tail of a tadpole die off and are resorbed as it changes into a frog. In the insect, the very process of metamorphosis depends upon death of part of the cells of the organism: as the specialized organs of the larva die, they serve as an energy source for the rapidly growing and differentiating imaginal disks, which develop into the organs of the adult.

But what of the death of the organism as a whole? Is this, too, written into the blueprint of the genes? In multicellular organisms, death usually ensues as a result of the failure of some vital organ to continue carrying on its vital functions. This failure may be the result of damage caused by disease or accident, or by the deteriorations that accompany the process of senescence. When one vital organ fails, the effects of this failure spread more and more widely among the interrelated systems of the body. In man, in many instances the heart or some other part of the circulatory system is the vital organ first affected. But no matter which vital organ is the first to fail, eventually the effects of this failure reach the circulatory system, and the heart stops. Until recently, this was the generally accepted signal of death. For deprived of the food and oxygen delivered by the circulatory system, and deprived of its vital role of removal of potentially toxic wastes, the cells of the body begin to die off one by one. Some types of cells die more rapidly than others. Nerve cells are the most vulnerable: it has been found that irreparable brain damage sets in about four or five minutes after the heart of a man has stopped. Other types of cells and tissues die more slowly; but eventually all succumb. Death of the cells is accompanied by rapid and irreversible changes in the structure of the cell, and ultimately a breakdown and dissolution of the cell contents. The organizing influence is gone.

Must "clinical death"—stoppage of the heart—mean the death of the organism? In recent years, great strides have been made in techniques of resuscitation, the saving first of experimental animals and then of humans that have clinically "died." Large hospitals now have emergency resuscitation teams that often race to the bedside of a critically ill patient whose heart has just stopped. In the crucial few minutes before brain damage might set in, they literally bring him back to life, buying time for surgical techniques and drugs to work and cure the malfunction of some vital organ that caused the "fatal" breakdown. One of the most famous cases of successful resuscitation was that of the eminent Russian physicist Lev Landau, who sustained "fatal" injuries in an automobile crash. His grief-stricken students, colleagues, and friends, including some of the most prominent men in Soviet science, literally refused to let him die. Specialists and equipment were rushed in from all over the Soviet Union, and Landau's friends set up a round-the-clock vigil, bringing him back from clinical death time and time again, as surgeons worked feverishly and at last successfully to repair the damage.

Achievements such as this have become increasingly commonplace. Today, with the use of improved heart-lung machines, it is possible to keep a person's heart beating for hours after all signs of brain activity have ceased. This technique has recently been used extensively to keep organs for organ transplants in good condition until the recipient has been readied for the operation. However, the flurry of activity in the field of heart transplants sparked a controversy over the need for a valid and binding definition of death. If the old definition of clinical death as stoppage of the heartbeat is accepted, then a physician who removes the heart of a person kept "alive" by a heart-lung machine must be guilty of murder. It has been suggested that a cessation of brain wave activity be taken as the definition of clinical death instead—for example, the observation of a flat alpha wave for twenty-four hours. Yet such a definition might render heart and other organ transplants virtually impossible.

QUESTIONS FOR REVIEW AND THOUGHT

1. What factors affect the rate of growth of an organism?
2. Discuss the growth curve and its universal application.
3. Describe some examples of complete metamorphosis.
4. How can one explain the fact that salamanders and some other vertebrates are capable of regenerating limbs, and man is not?
5. Describe the processes involved in the healing of a wound.
6. Is senescence a necessary concomitant of life? Explain.
7. Briefly discuss some of the current theories on aging.
8. Compare the human survival curve with that of the laboratory fruit fly.
9. What changes in our social structure might result if human death were virtually eliminated?
10. Is clinical death reversible? Explain your answer.

PART V

THE UNIVERSE

CHAPTER 18

The Birth of the Universe and the Origin of Life

The Origin of the Universe

When was the beginning of time? *Was* there a *beginning?* And if so, what came *before* the beginning? Is it possible that there *never* was a beginning? The apparent paradox posed by these questions appears to place an answer that we could comprehend and appreciate beyond our reach. And yet today, astronomers are fiercely struggling with answers to these questions.

One of the most provocative (and one of the most difficult to visualize) is a theory posed by the brilliant British mathematician and astronomer Fred Hoyle. His theory, known as the **steady-state theory,** holds that the universe never had a beginning. It has always existed, and in pretty much the same state as it is now: as an ever-expanding universe, with a relatively uniform density throughout. As the galaxies constantly move farther and farther apart, the uniform density is maintained by the creation de novo of new matter, an atom at a time. Ultimately these newly created atoms condense to form huge clouds of gases, then stars and galaxies, which in their turn move outward (Fig. 18–2a).

Figure 18–1 The Crab nebula, seen on a starry night.

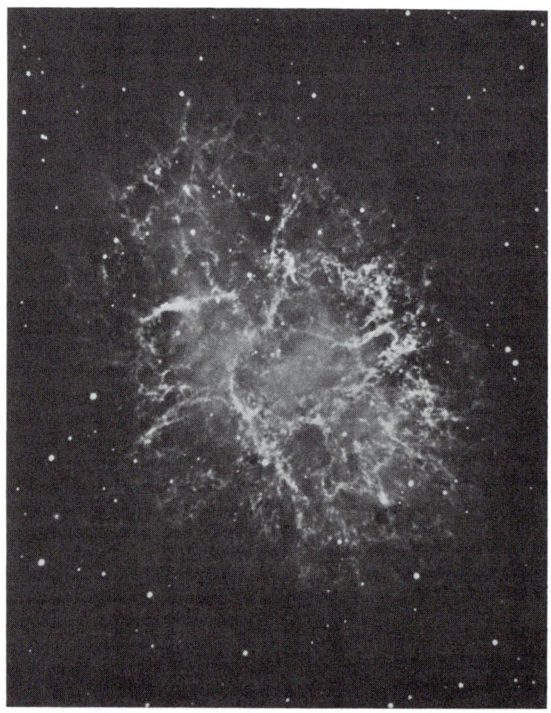

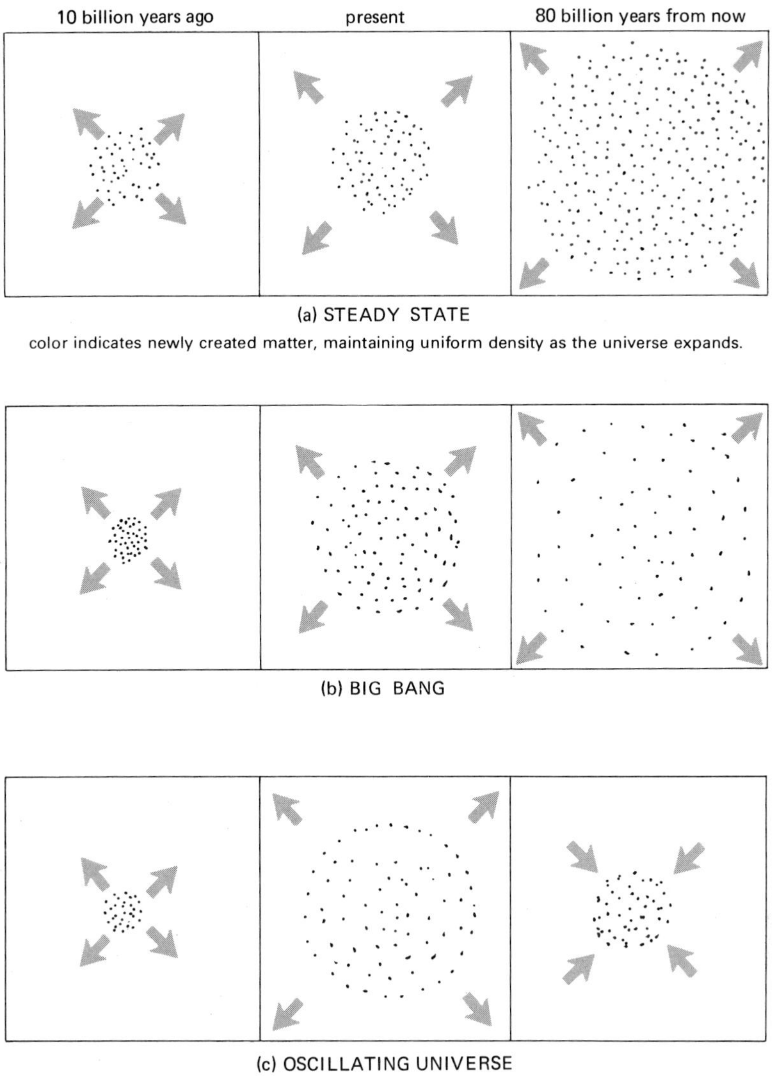

Figure 18–2 Three modern theories on the origin of the universe: the steady-state theory (a), the big bang theory (b), and the oscillating universe theory (c).

A second, more widely accepted theory, and perhaps one which lends itself more readily to visualization, is the **big bang theory.** The proponents of this theory contend that about 10 billion years ago all the matter of the universe was incredibly condensed into a relatively small space. Suddenly, a colossal cataclysmic explosion rent this highly condensed matter, spewing it forth in all directions, with a fantastic burst of energy. This was the "big bang." Since then, the diffuse clouds of matter generated by the explosion have condensed into stars and galaxies. The galaxies of the universe, propelled by the force of this momentous explosion, have been rushing apart, and will continue to rush apart indefinitely (Fig. 18–2b).

An alternate variation of the big bang theory that seems to be gaining an increasing number of adherents is the **oscillating universe theory.** In this conception, the big bang that occurred some 10 billion years ago was not the first; nor will it be the last. Instead, it was but one of an endless chain of oscillations, in which the universe alternately contracts and expands. According to this theory, since the last explosion, about 10 billion years ago, the galaxies of the universe have been expanding and moving apart, and will continue to do so for some time to come. But in time their mutual gravitational attraction will cause them to slow down and ultimately plunge back to another cataclysmic collision, perhaps some 70 billion years from now. The incredible pressures and physical forces generated by such a collision will cause still another colossal explosion. Then another 80-billion-year cycle of expansion and contraction will begin again, with the birth of new stars and galaxies, and perhaps the formation of new worlds, upon which forms of life may evolve again, only to die ultimately in the inevitable conflagration (Fig. 18–2c).

In the past few years some exciting discoveries have begun to cast new light on the relative validity of the theories just outlined. Strange new objects called **quasars** have recently been discovered through their powerful radioemissions. These incredibly bright objects are apparently rushing away from us at fantastic speeds, and represent the fringes of the visible universe. Some of the quasars have been estimated to be billions of light-years away. Indeed, one is estimated to be as far away as 8 billion light-years. This would mean that light left this object barely 2 billion years after the (presumed) big bang. Presently evidence seems to indicate that the quasars are more populous at greater distances, indicating that they were more numerous in the distant past than they are now. These preliminary findings tend to corroborate the big bang and oscillatory universe theories.

Additional confirmation now is coming from observations of another class of new astronomical bodies, quasi-stellar blue galaxies, which resemble the quasars in every way except that they are radio-silent and are about a hundred times as numerous as the quasars.

Some astronomers have recently reported what may be further evidence in support of the big bang theory. Perplexing radio signals have been received that have been interpreted as the lingering effects of the light flash emanating from the cataclysmic explosion 10 billion years ago.

The Origin of the Solar System

Some 10 billion years ago, it is thought, the universe contained many drifting, slowly rotating clouds of dust and gas. Many of them gradually condensed under their own gravitational forces, giving rise to stars. One of these was our own sun. As the cloud of matter condensed, most of it formed a dense core, but about 10 percent remained outside. This material was gradually flattened into a disk as the infant sun and its accompanying matter continued to rotate. As the various parts of the outer material moved past one another at different velocities in their orbital paths, turbulent eddies were created, and concretions began to grow. The concretions, "protoplanets," rotating and revolving about their central star (which at that time had not yet become hot enough to shine as a star), gathered in various stray bits of solid dust particles, such as iron oxides, silicates, and water crystals, that passed within reach of their gravitational forces.

The heavy materials of each protoplanet in time formed a heavy nucleus, which was surrounded by lighter gases. These gases in turn were compressed into a disk by the rotation of the protoplanet, and in the disk satellites formed.

The sun condensed further and was heated by its contraction until it became a seething cauldron of thermonuclear reactions. Its radiations and streams of ejected particles swept away the loosely held light gases and fine dust from around the less massive inner planets. The more massive planets, Jupiter and Saturn, suffered less loss of material,

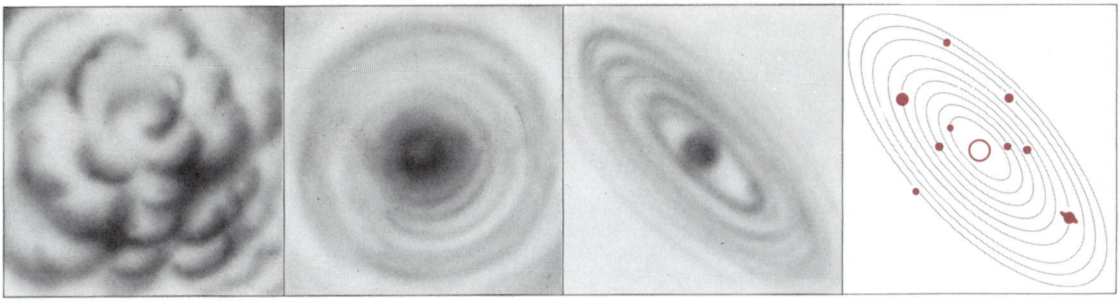

Figure 18–3 The nebular hypothesis on the origin of stars and planetary systems by condensation of dust clouds.

and as a result their atmospheres have a much higher proportion of the light elements, such as hydrogen and helium. (Their specific gravities—ratios of their densities to the density of water—approximately 1, are much lower than the earth value of 5.5.) The loss of mass meant that the planets, especially the inner planets, were less able to hold their satellites. The orbits of the satellites increased in size, and some of them escaped. (A few of these were probably later recaptured, but not necessarily by the planet that had originally lost them.)

The sequence of events just described summarizes the hypothesis currently most widely accepted on the formation of stars and planetary systems. This **dust-cloud hypothesis** or **nebular hypothesis** explains both the general principles of our solar system and many anomalies, such as the unusual rotations of some of the satellites. It is supported by the observation of many similar **nebulae,** clouds of dust and gas, in interstellar space.

This hypothesis has an important implication for the possibility that forms of life may exist on other worlds: according to this hypothesis, the formation of planetary systems is a quite normal event in the evolution of a star, not the unique and improbable accident that some earlier hypotheses pictured it to be.* Considerable indirect evidence is accumulating to indicate that planetary systems may indeed be common in the universe. For instance, the presence of a planetary companion of Barnard's Star has been deduced on the basis of perturbations of its motions. Although the proposed planet has not yet been photographed, its orbit has been inferred, in much the same manner that the planet Neptune of our own solar system was discovered as a result of deductions based on the perturbations of the orbit of Uranus.

*One of the more exotic of these, which was in vogue in scientific circles for some time, was that the planets were formed from huge gouts of matter, torn out from the sun by the enormous tidal forces generated when another star passed close by.

The Origin of Life on Earth

Several billion years ago, our earth, a giant orb, circled the sun. The diameter of the earth was many times its present diameter. With the passage of eons, the earth slowly condensed, and much of its envelope of lighter gases was lost. The very act of condensation created enormous pressures and intense heat. Under the influence of this great heat, which was augmented by radioactive decay, the substance of the earth became fluid enough to permit profound rearrangements of its constituents. The heavy metals, for the most part, migrated to the center, forming a metallic core, and the lighter elements were concentrated in the earth's mantle and crust.

With the passage of time, the earth cooled. But this was a very slow process. Slowly the heat emanating from the interior was radiated out into space. A continual rain fell toward the surface, but for many many millions

Figure 18–4 The primeval earth: raw materials for the origin of life.

of years was vaporized by the extreme heat before it could reach the surface. In time, the earth had cooled sufficiently for waters to remain upon the surface for longer and longer periods of time. At last the oceans formed. Cataclysmic volcanic activity threw up great mountain chains, which in time were worn away by the constant rains and other agents of erosion. Lightning flashed continually through the atmosphere. But this atmosphere did not at all resemble the atmosphere of our earth today. No free oxygen existed in those primeval days. Instead, the atmosphere is thought to have consisted predominantly of water vapor, methane, ammonia, carbon dioxide, and hydrogen cyanide. The lightning flashes scorching through the atmosphere generated numerous chemical reactions among the components of the gaseous atmosphere, the products of which drifted down to the seas below and dissolved therein. Other reactions were generated in the churning seas among the components of the atmosphere dissolved in them, catalyzed by the ultraviolet and other high-energy radiations of the sun that penetrated the cloud cover. Through the frequent volcanic activity of the unquiet earth and long eons of erosion, the oceans of the world gradually became saltier, with a great diversity of minerals. Thus, the warm, soupy seas possessed the ingredients requisite for life. Countless numbers of chemical reactions, simple and complex, were constantly taking place in these seas. Fatty acids and glycerine and then fats were produced. Amino acids were formed, and then linked up into chains, which grew in complexity. Nitrogen bases and sugars were formed, and combined with the phosphates in the sea to form nucleotides. These nucleotides formed chains. Now the requisites for life were here.

Gradually, as the various types of molecules drifted through the seas and collided with one another at random, some tended to stick together, forming transitory complexes. Although these complexes were not yet "living," they began to carry on some activities remarkably resembling those of the living cell. From the warm broth of the seas, they took in simpler molecules of various sizes and compositions, and incorporated them into the complex. Some of the molecular complexes were more successful than others, and these tended to grow and survive: any new change in the molecular aggregate that increased its efficiency in competing for raw materials and enhanced its adaptation to the conditions of the environment tended to persist. Probably some of the complexes developed a sort of enclosing membrane that separated them from the surrounding medium and increased the selectivity of their incorporation of certain molecules and their discharge of others. Some of the portions of the increasingly complex aggregates of molecules had a geometry and chemical properties that proved unusually suited to combine with various types of molecules from the surrounding seas and to change their properties slightly so that they combined more readily with one another. With the aid of these catalysts, further complexity could be achieved. Other portions of the complexes, somewhere along the line, acquired the ability to produce more molecules like themselves, and to interact with and control the activities of the other portions of the aggregation. Gradually, imperceptibly —at exactly what point is but a matter of definition—some of these increasingly complex aggregates of molecules were "alive."

The story just related of course is not a matter of proven fact. For there were no human witnesses to the birth of life on earth. But it does represent a sequence of events that scientists believe may well have led to the origin of life. The foremost proponent of this theory is the prominent Russian biochemist A. I. Oparin. He proposed back in 1924 that the first living things must have arisen from nonliving chemicals present on the primeval earth, and since then he has been at the forefront of speculation and experimentation on the origin of life.

Since the 1950s there has been a blossoming of research on **abiogenesis,** or the origin of life from nonliving matter. Much experimental evidence has been obtained to substantiate various aspects of the theories on the origin of life on earth. The landmark experiment was conducted in 1953. Harold Urey and Stanley Miller, working at the University of Chicago, passed an electric discharge through an "atmosphere" consisting of hydrogen, methane, ammonia, and water vapor (Fig. 18–5). In a week's time, they found in the resultant solution many simple organic compounds—even some amino acids. Other researchers, also attempting to simulate some of the conditions of primeval earth, reported similar findings. Melvin Calvin subjected solutions of carbon dioxide and water to irradiation in a cyclotron and obtained organic acids, with carbon chains of up to four carbon atoms. Other experiments have yielded still more basic organic building blocks—polypeptides, purines, pyrimidines, sugars, and the major energy store of living systems, ATP. Porphyrins—which, like the chemically related chlorophyll, can capture and store light energy—have also been produced.

In further experiments, phosphates like those which must have been present in the rocks of the primeval earth were added, and the amino acids produced in the artificial

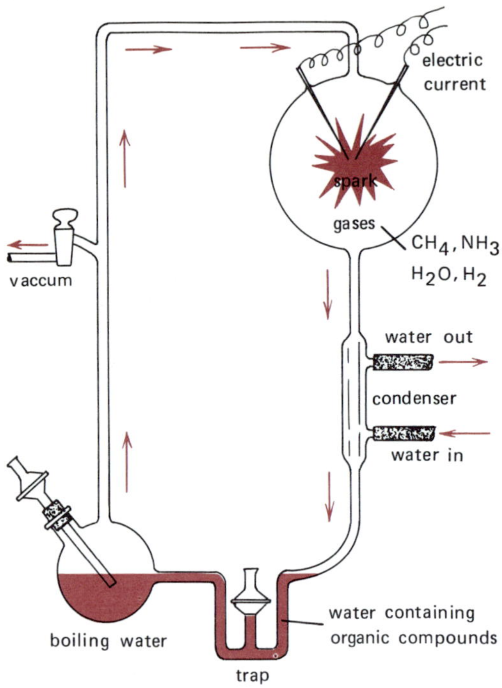

Figure 18–5 The Miller-Urey experiment. A spark in a mixture of gases resembling the primeval earth atmosphere resulted in the production of amino acids and other organic compounds.

atmosphere joined together to form larger, proteinlike molecules. The purines, pyrimidines, sugars, and phosphates have also been linked together to form nucleotides.

Other experiments have been aimed toward reproducing the formation of macromolecular complexes. Oparin mixed two natural substances, gelatin and gum arabic, and obtained a precipitate of tiny coacervate droplets. These droplets showed many properties similar to those that the first simple "protolife" forms must have possessed. When enzymes were added to a suspension of Oparin's droplets, they became bound to the droplets and then catalyzed their characteristic reactions much more rapidly than they did when free of the droplets.

An even more intriguing result was recently obtained by Sidney Fox at the University of Miami. By heating mixtures of dry amino acids to temperatures above the boiling point of water (about 170°C) under conditions such as those existing naturally in volcanic regions, he obtained a variety of proteinlike molecules. These "proteinoids" had many of the properties of proteins produced by living organisms, and they were even found to possess weak enzyme activity.

When the proteinoids were treated with water (as might have happened on the primeval earth when rain fell upon the dry rocks or tides lapped at a volcanic shore), an exciting result was observed. The proteinoids formed numerous tiny spheres. These proteinoid microspheres had many characteristics reminiscent of those of living cells: they were generally about the same size and shape; they tended to shrink or swell when transferred to concentrated or dilute salt solutions, respectively; they showed enzyme activity, which was inactivated by heating, and also some hormone activity; and they were surrounded by a selectively permeable double layer quite analogous to the cell membrane (Fig. 18–6). In general size and appearance, the proteinoid microspheres were similar to certain types of bacteria. And the microspheres were even observed to "reproduce"! When allowed to remain for a week or more in the liquid in which they had been produced, they formed small buds, similar to the buds produced by yeast cells and some bacteria. These buds could be separated from the "parent" microspheres by mechanical, thermal, or electric shock. They then grew to the size of the parents and stopped growing. If they were left undisturbed for another week or so, they, too, produced buds—a new "generation" of proteinoid microspheres.

All this experimental evidence was obtained in small-laboratory apparatuses, under conditions simulating those that are believed to have prevailed during the early life of the earth. Given oceans as reaction media and a billion years or two for the reactions to go on, it seems reasonable to believe that the formation of similar organic substances was a frequent occurrence on primeval earth. Since at that time there were no decay

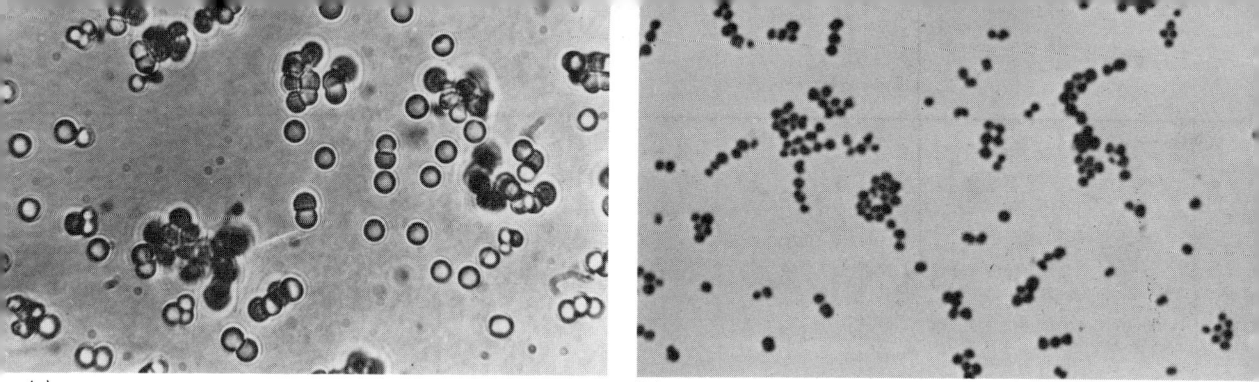

(a) (b)

Figure 18–6 Proteinoid microspheres produced by Sidney Fox (a); *Staphylococcus aureus,* a common bacterium (b)

organisms to degrade them, and (at first) no larger entities to eat them, these organic compounds must have accumulated gradually in the warm seas until they comprised as much as 10 percent of this primeval "broth." Under these conditions, molecular complexes of the type described could well have arisen, grown in complexity, and eventually crossed the invisible boundary line to what we call life. And it is not inconceivable that, given similar conditions, similar chains of events may have occurred many times on many planets circling around numerous stars of the universe—may have occurred, may now be occurring, and may again occur many times in the future.

But for the moment, our attentions are concentrated here on earth. From the first self-replicating chemical complexes to the intricate thinking creatures known as men, a long road still had to be traveled. In the following chapter we shall explore some of the present views of that long and tortuous route, the evolution of life.

QUESTIONS FOR REVIEW AND THOUGHT

1. According to the steady-state theory of the universe, how is a uniform density being maintained?
2. Does the big bang theory explain the origin of the universe? Explain your answer.
3. Is the oscillating universe theory incompatible with the big bang theory and/or the steady-state theory? Explain.
4. Which of the major theories of the origin of the universe seems to be best supported by recent astronomical discoveries? Discuss.
5. According to the modern view of the evolution of the solar system, should the formation of planetary systems be a common or an uncommon event in the universe?
6. Do we have any concrete evidence of the existence of planets revolving about other stars? Discuss.
7. In what ways is the atmosphere of the earth believed to have changed since its formation?
8. What evidence is there that life could have arisen from inorganic chemicals?
9. Could life arise spontaneously under the conditions that now prevail on earth? Explain your answer.
10. What is the most likely constitution of the first "protolife" form?

CHAPTER **19**

The Evolution of Life on Earth

Introduction

Change pervades the universe. Viewed on the cosmic scale, matter is in a constant state of flux. Stars are born and die. Planetary systems appear, are warmed for a time by their parent star, and then either disintegrate in a burst of flame or slowly subside into an eternal chill bleakness.

Earth is presently enjoying the warmth of its parent sun. We have already discussed its early history, its birth, and other events preceding and including the origin of life. Now we shall examine the evidence indicating that life itself has undergone a long process of evolution.

As with many of the major concepts of science, the theory of evolution did not spring full-blown in a sudden flash of inspiration. Through long centuries when the general public and most of the scientific community in the Western world were firmly convinced that the last words on the early history of the earth had been set down immutably in the pages of Genesis, some minds doubted and thought for themselves; some men looked about them with clearer eyes and tentatively voiced their thoughts. Ideas bubbled along like subterranean currents, waiting for a chance to rise to the surface.

This opportunity came at last in 1859, with the publication of Charles Darwin's epic work, and the concept of evolution burst upon the startled world like a bombshell. Darwin, of course, was not the first to propose that present-day life may not have been created in its present forms but instead has developed from earlier forms of life. Greek philosophers, such as Anaximander and Aristotle, speculated on the possibility of an evolution of life. Shortly before Charles Darwin's time, there had been some upwellings of the subterranean ferment in the speculations of G. L. Buffon, Erasmus Darwin (Charles Darwin's grandfather), and Jean Baptiste de Lamarck. But it was Charles Darwin who presented the first comprehensive treatment of evolution and the mechanisms by which it occurs, documented with more than two decades of painstaking observation and correlation.

Victorian society was shocked and alarmed by the new concept, but men of intellect and insight arose to defend and expand Darwin's

Figure 19–1 Charles Darwin.

thesis. New evidence was amassed, some eventually in fields that Darwin had never even heard of, such as genetics and biochemistry, and all fit beautifully into the framework of the theory, rounding it out to even greater strength and validity, until it can now be ranked as one of the major concepts of scientific thought.

Darwin: The Theory of Natural Selection

A voyage whose ultimate impact was to change the patterns of man's thinking about himself and his universe began from Davenport, England, in December 1831 and lasted for five years. During this time *H. M. S. Beagle,* a ten-gun brig, circumnavigated the world on a surveying expedition for the British Royal Navy (Fig. 19–2). A young biologist who had just graduated from college, Charles Darwin, served as its naturalist. On this voyage Darwin observed the similarities and differences among geographically separated animals and plants living on islands and on the nearby mainland. He also noticed similarities and differences among living organisms and fossils of recently extinct forms. These and other observations transformed Darwin from a firm believer in special creation and the immutability of species at the beginning of the voyage to an evolutionist by the time it had ended.

In the years following his great voyage on the *Beagle,* Darwin published extensively on his observations. In the early 1840s he began to formulate his views on evolution, but he confided these views to only a few trusted friends.

One of these men, the British geologist Charles Lyell, had a profound influence on Darwin's thinking and the formulation of his theory of evolution. One of Lyell's major contributions was the division of the various rock strata of the earth's crust into definite periods following a chronological order. The presence of fossils (preserved remains of organisms) associated with many of these strata indicated a similar chronological sequence of appearance of various life forms on earth.

The works of Thomas Malthus also made a deep impression on Darwin. Back in 1798, Malthus had written his *Essay on the Principle of Population,* in which he stated that the human population would eventually outrun the available food supply, and that famine, disease, and war would keep man's numbers in check. The inevitability of overpopulation, in Malthus's view, was due to the fact that the population tended to increase geometrically, while the food supply increased only arithmetically. Darwin happened upon a copy of Malthus's book in 1838 and began to speculate on the applicability of Malthus's reasoning to other species as well. Here was the seed from which a strategic part of the theory of natural selection eventually grew in Darwin's fertile mind.

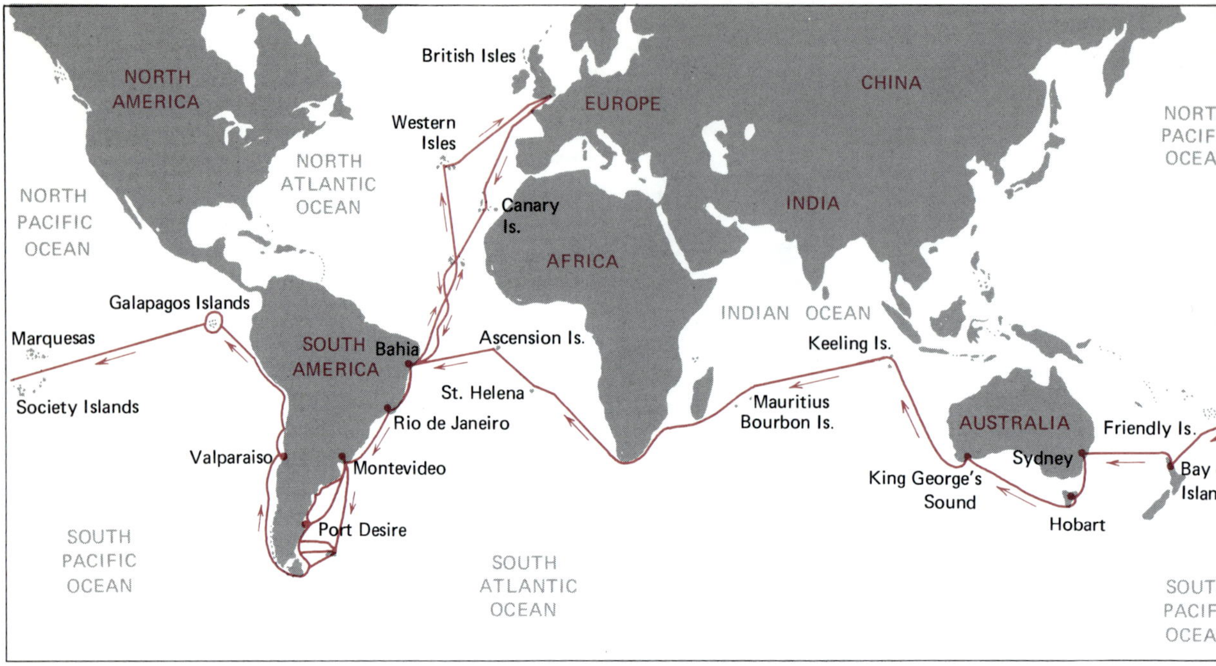

Figure 19–2 The voyage of the *Beagle*.

For more than twenty years following his voyage on the *Beagle*, Darwin continued to accumulate evidence and to expand upon his theory of evolution. He began work on a comprehensive treatise in 1856. But about two years later, when he had completed about half of his book, Darwin received a manuscript from Alfred Russel Wallace, a naturalist then working in the Dutch East Indies (now Indonesia). Darwin was astonished to discover that the manuscript was, in essence, a summary of Darwin's own theory of natural selection. Rather than rush to publication to receive sole credit as the originator of this theory, Darwin showed Wallace's manuscript to friends. One of these was Lyell, who suggested that Darwin and Wallace issue a joint publication. With Wallace's enthusiastic approval, the work of the two men appeared later that year in the journal of the Linnean Society. Meanwhile Darwin redoubled his efforts on his own manuscript, which was published the following year. *On the Origin of Species by Means of Natural Selection, or the Preservation of Favored Races in the Struggle for Life* appeared on November 24, 1859. It was an instant success. The entire first edition was sold out on the very first day.

Darwin's theory of natural selection included the following basic principles:

1. Variation. Variation is a concomitant of life. Within any species, there are wide variations in size, vigor, and a host of other factors. The Great Dane, the fox terrier, and the chihuahua represent huge variations within a single species, the dog. Within a subspecies, there are also enormous variations. Any dog lover will attest that different members of the same breed may show great differences in stature, color, personality, and other traits. Even offspring of the same parents can show wide variation. Since evolution is, in essence, a process involving successive change, the natural variability of organisms provides a pool of such changes,

Figure 19–3 Different breeds, far different appearance and habits, but all *Canis familiaris*: (a) cocker spaniels; (b) Weimaraners; (c) setter.

which play a role in the interaction of organisms with their environment. Darwin, however, did not recognize the true origin of natural variability, i.e., chromosome interactions and mutations.

2. Excessive reproduction. Virtually all the species that exist have an enormous reproductive potential: given sufficient time and ideal conditions, any single species could literally cover the entire earth. The cumulative offspring of a single pair of common fruit flies, if bred under theoretically *ideal* conditions for slightly over a year, would have a total mass greater than that of the entire universe! Even for a species such as man, which usually produces only one offspring at a time and has a lengthy maturation period between generations, the reproductive potential is enormous. If man continues at his present rate of reproduction, it is estimated that in some eight hundred years the world will be so densely populated that there will be one human for every square foot of land surface. And this rate does not even come near to tapping the full potential of human reproduction.

Obviously such unlimited reproduction by any species cannot go on indefinitely. The resources of food and living space on earth are limited. The full reproductive potential of organisms cannot be reached, and many of the offspring that are produced do not survive to produce offspring of their own.

The mechanisms limiting the populations of various species in spite of overreproduction comprise another element of Darwin's theory.

3. Struggle for survival. Frequently, as a result of overreproduction, the number of organisms in a given locality exceeds the capacity of that region to support them. Members of the same species, as well as members of competing species, vie with one another for territory, food, and other available resources. This struggle for survival need not take the form of active battles between individual organisms. (It was Darwin's interpreters and defenders, rather than Darwin himself, who popularized the picture of "Nature red in tooth and claw.") The struggle may actually be quite passive, involving animals living side by side in apparent peace, or even plants quietly competing for the available space, sunlight, water, and other resources.

An illustration may clarify this point. Rabbits are not native to Australia. Some time ago, rabbits were brought over from Europe and adapted successfully to the land. With no natural predators to prey upon them, they adapted so successfully that they literally overran the land, competing with cattle and sheep for the available fodder. In some areas the rabbit population skyrocketed to such an extent that they stripped the plant cover bare

Figure 19–4 Rabbits gathered around an Australian water hole.

and reduced good grazing land to semi-desert. Thousands of valuable sheep and cattle died of starvation, and bounties were placed on the rabbit pests. Finally an insect-borne virus disease, myxomatosis, which is fatal only to rabbits, was specially introduced into Australia in an effort to control the pest.

4. Survival of the fittest. With the wide variations that exist within each species, there are bound to be some characteristics possessed by some individuals that give them an advantage in adjusting to the environment. For instance, among a population of deer, the individuals who can run faster than the rest will have a better chance of outrunning potential predators, such as wolves. In the plant world, a tree that is capable of more rapid growth to greater heights would appear to have an advantage over others in a crowded forest in the constant struggle for a place in the sun. It is these better-adapted individuals—the faster-running deer, the taller-growing tree, the organisms with greater vigor, resourcefulness, and ability to meet the demands of the environment—that have a better chance of surviving to an age when they can reproduce themselves. A particular premium upon even the slightest advantage is imposed by marginal conditions for survival, and especially by changing environmental conditions: only those individuals that can successfully adapt to the new demands of existence—the fittest in terms of those demands—will survive. The constant struggle for survival and the consequent survival of the fittest result in a sort of natural pruning process, in which the weaker lines are weeded out and only the most promising lines (promising from the standpoint of adaptation to the environment) remain and propagate themselves. This process, in conjunction with the next major principle to be discussed, the transmission of adaptive advantages to subsequent generations, makes up the essence of **natural selection.**

5. Inheritance of variations. Those individuals who successfully survive the struggle for existence give rise to offspring that inherit the various traits of their parents. Darwin believed that any new variation which was a factor in the survival of an organism would be passed on to its offspring.* Thus, a deer which happened to develop a musculature that permitted it to better outrun its potential predators would pass this trait on to its offspring. Over many generations of comparable improvements the ultimate descendants might prove to be far swifter than their ancestors. (In the light of modern genetic theory, only a *capacity* for developing highly efficient musculature may be inherited. The actual muscular development of a particular animal would depend on the environmental influences operating within its own lifetime, and not on the muscular development of its parents.)

Another example of the inheritance of variations might be cited from the plant kingdom. Some plants are specifically adapted to pollination by a certain species of insects. Their color, odor, petal shape, and other characteristics are often precisely suited to attract and satisfy the needs of this insect species. A variation in one of these plants that would either lead to greater attraction for potential insect pollinators or that would ensure a greater degree of pollination might result in a significant increase in the number of off-

*This aspect of Darwin's theory on natural selection was one of his weaker points. He and his contemporaries were unaware of Mendel's research on heredity, which was in fact being conducted at the time of the publication of Darwin's theory. The concepts of Mendelian genetics and the role of mutations in introducing new variations into the gene pool did not become a force in biological thought until the twentieth century. Thus Darwin was forced to invent a mechanism of inheritance of variations that was actually a version of the idea of inheritance of acquired characteristics—a concept that has since then been discarded.

Figure 19–5 Darwin's finches. The species 1–6 are ground dwellers, like the ancestor presumed to have migrated from the South American mainland. As the finches became established in the islands, tree-dwelling species (7–13) are believed to have developed.

spring, some of which might inherit the new variation. On the other hand, any plant showing a variation that makes pollination less likely or effective would have fewer offspring, and thus fewer chances to pass on its new trait.

6. Formation of new species. The cumulative effects of many variations, according to Darwin, lead to the formation of new species. On his travels, Darwin saw numerous examples of closely related species that were differentiated by a relatively small number of variations. One of the most dramatic examples was the great number of species of finches on the Galápagos Islands (Fig. 19–5), which differed slightly from island to island and to a somewhat greater extent from the finches found on the adjacent mainland of South America. Darwin recognized that the successive accumulation of variations ultimately leads to the formation of new species, but he overlooked the critical role played by isolation. Without the opportunity to interact with members of the original group, a particular isolated group of individuals of the same species could, over many generations, accumulate a considerable number of unique variations. Eventually they would become so different from the original group that intermating between groups would not lead to viable and fertile offspring—the ultimate test of a new species.

Darwin's publication of the *Origin of Species* set off an explosion of controversy that was to last for generations. Indeed, even today in some isolated parts of the country the topic of evolution still evokes heated controversy and an occasional court battle concerning whether it should be taught in the local public schools.

Due to poor health and a reluctance to enter into heated controversy, Darwin did not actively participate in the debates that followed the publication of his work. But there was no dearth of support, as numerous scientists flocked to defense of Darwin's theory. In particular, Thomas Huxley in England, Asa Gray in America, and Ernst Haeckel in Germany were vocal and effective defenders of his theory.

The Evidence for Evolution

Hardly a day goes by without the uncovering of additional evidence in support of the theory of evolution. Each year thousands and thousands of fossils are uncovered in the course of paleontological excavations, unexpectedly in blasting in mining for mineral ores, by scuba divers, and even by backyard well diggers. Literally thousands of bits of evidence—morphological, physiological, and biochemical—flow from the laboratories each year.

It is hardly conceivable that any intelligent person acquainted with the huge mass of supporting evidence could doubt that the theory of evolution does indeed offer a reasonable explanation for the present state of extant species and those that have become extinct. But despite the overwhelming weight of the evidence, in the true tradition of scientific thinking, one cannot dogmatically say that evolution *did* take place. All one can say with assurance is that the weight of evidence indicates a high probability that evolution did, does, and will continue to take place. This is precisely why we call this concept the *theory* of evolution. It is the best theory yet devised that appears to fit the facts. But Darwin's original concept has already been modified and will doubtless undergo further modification as more evidence comes to light.

THE FOSSIL RECORD

For millions and billions of years, creatures have been living and dying on the surface and under the surface of the earth and in the waters that bathe it. The vast majority of organisms are either eaten by other organisms or die a natural death and are subsequently devoured by scavengers or decomposed by decay microorganisms. A very few are preserved in one form or another as **fossils** by quirks of nature. Even fewer of these are actually found by man. Thus **paleontology,** the study of fossils, is something like working an intricate jigsaw puzzle, most of the pieces of which are still missing and may never be found.

Despite these inherent difficulties, an enormous assortment of fossils have been found and studied, and a surprising amount of information has been pieced together from inspection and analysis of them.

Fossils are found in a variety of forms (Fig. 19–6). Among the earliest fossils discovered were footprints of prehistoric animals. Of course, most of the footprints of creatures that have walked the earth have been rapidly washed away by rains or more slowly obliterated by processes of erosion. But some, left in protected caves or buried by loose soil, have been preserved.

Remains of bones and shells are another common form of fossils. Special combinations of circumstances were needed to preserve

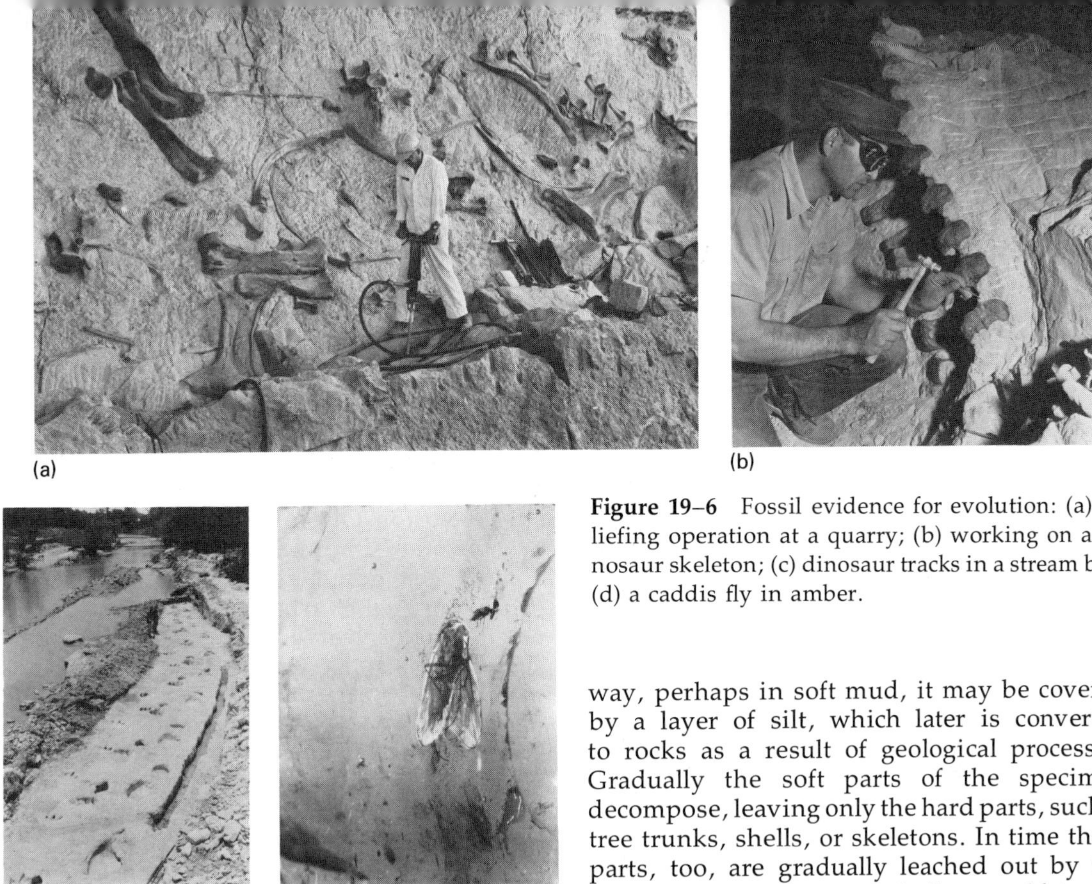

(a) (b) (c) (d)

Figure 19–6 Fossil evidence for evolution: (a) relieing operation at a quarry; (b) working on a dinosaur skeleton; (c) dinosaur tracks in a stream bed; (d) a caddis fly in amber.

whole skeletons or even substantial bone fragments from the general forces of erosion and decay. Rapid burial of an organism in sticky pools of tar, under a searing flow of lava, in quicksand and soft, oozing river bottoms, and beneath thundering avalanches has at times preserved the skeletons of animals from decomposition. Sometimes whole specimens have been effectively preserved. Fossils of insects and spiders embedded in **amber** have been discovered in perfect condition. These small animals were entrapped in the sticky resins secreted by conifers, as long as millions of years ago, and the resinous fluid ultimately penetrated their bodies and hardened. Huge wooly mammoths, buried alive in severe snowstorms thousands of years ago, have been discovered. So well were they preserved in this natural "deep freeze" that their flesh was fed to dogs with no ill effects.

Such a fine degree of preservation is extremely rare. More often, when a plant or animal has been rapidly buried in some way, perhaps in soft mud, it may be covered by a layer of silt, which later is converted to rocks as a result of geological processes. Gradually the soft parts of the specimen decompose, leaving only the hard parts, such as tree trunks, shells, or skeletons. In time these parts, too, are gradually leached out by the ground waters, leaving a hollow mold in the rock that faithfully reproduces the outlines of the long-dead organism. As the ground waters continue to seep through the rock, mineral salts precipitate, forming a **cast** of the mold that is nearly a perfect copy of the original skeleton or shell (Fig. 19–7). The replacement of the original hard parts by mineral salts may be a gradual process, with the ground waters selectively dissolving first the most soluble portions, followed by mineralization, then dissolving of the less soluble portions, and so on. When such replacement, called **petrifaction,** is extremely slow and gradual, minute details of the internal anatomy of the fossilized organism may be preserved, such as the grain of wood, or the pattern of veins on leaves, or the markings of sites of muscle connections on bones. One of the most striking examples of petrifaction is the petrified forest near Adamana, Arizona, extending over an area of more than 100 square miles.

Fossil-bearing layers of rock become covered by layers of new earth deposits, where the cycle of life, death, and occasional preservation is

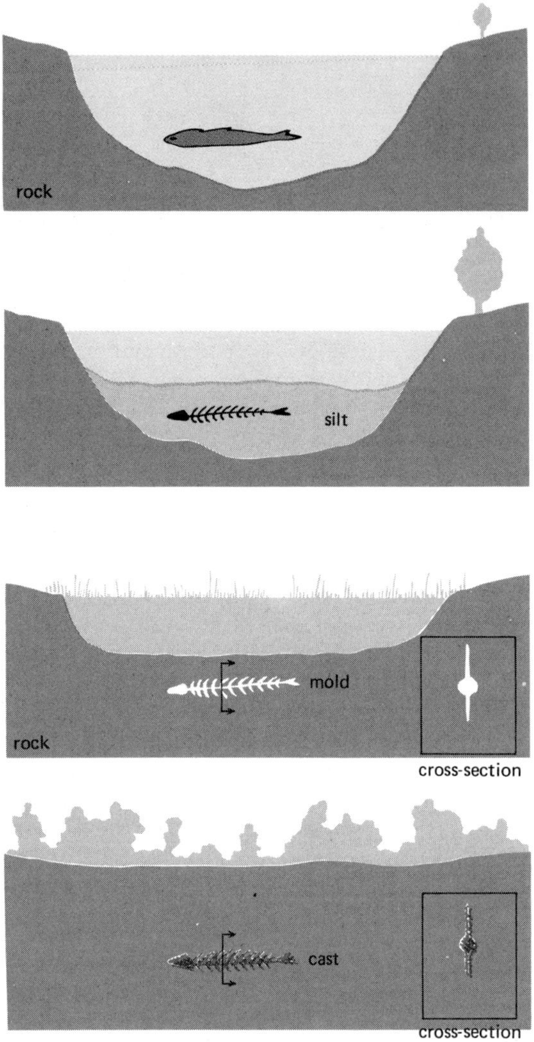

Figure 19–7 Formation of a cast. The soft parts of the fish decompose, and its skeleton gradually becomes enclosed in rock. Groundwaters dissolve the skeleton, forming a mold, and the mold is ultimately filled with mineral salts, producing a cast that is a faithful copy of the original skeleton.

repeated, over and over. The result is a series of rock strata, one above another, each laid down at a different period and each containing a characteristic assortment of fossil remains. In general, the deeper a stratum, the older the rock, and the fewer and less complex the fossils that it contains. But this general pattern is frequently broken by geological upheavals, which may thrust segments of older strata far above much "younger" rocks, and whole strata may be effaced by erosion before new layers are deposited upon the mass.

The fossil record itself, comprising only a tiny fraction of the organisms that have lived and died on the earth, still remains largely unknown. Buried under tons of soil and rock, fossil-bearing strata become uncovered only under rare and unusual circumstances, such as huge geological upheavals, earthquakes, and stream erosion. One of the most vivid examples in the world is the Grand Canyon in Arizona, a mile-deep gorge extending for 280 miles carved out by the Colorado River over a period of millions of years. The walls of the Grand Canyon display a colorful pattern of striations marking the rock strata laid down over the ages (Fig. 19–8). The rich fossil beds contained in these strata give a vivid picture of evolution from primitive algae in the low strata to trees, dinosaurs, camels, and elephants in the upper strata.

Figure 19–8 The record in the rocks in the Grand Canyon.

TAXONOMY

It is one of the strange ironies in the history of science that Carl Linnaeus, a man who was definitely an antievolutionist, devised a system that was later to provide one of the most powerful sources of evidence in support of evolution. Taxonomy, the science of classifying living organisms, is based on the relationships among organisms, including similarities and differences of structure, embryology, physiology, and biochemistry (see Chapter 7). The very fact that the various types of organisms can be arranged in a hierarchy of groups showing varying degrees of similarity and difference seems to imply the existence of a "family" of life forms, with varying degrees of kinship and a common descent. The theory of evolution fits perfectly into the framework of the presently accepted systems of classification and provides a logical explanation for both the diversity of life and the amazing similarities and intergrading characteristics of various species.

COMPARATIVE ANATOMY

It seems surprising that the flipper of a dolphin, the wing of a bat, the foreleg of a horse, and the arm of a man are all constructed in basically the same pattern. Each has approximately the same number of bones, a similar blood and nerve supply, and a similar number and arrangement of muscles. Each has a similar embryological orgin and a similar early pattern of development.

Organs with similar embryological backgrounds and similar anatomical relationships are called **homologous structures** (Fig. 19–9). Such structures may serve similar functions, such as the leg of the cheetah and that of the horse. Or they may take part in such diverse activities as swimming (dolphin's flipper), flying (bat's wing), running (horse's foreleg), or manipulation (man's hand and arm). Their basically similar structure seems to imply a common origin, followed by an evolutionary adaptation to varying demands of the environment.

Homologous structures may be contrasted to **analogous structures** (Fig. 19–10). The latter are structures which perform similar functions but have dissimilar embryological origins and are basically anatomically different. The wings of a bird and the wings of an insect perform the same functions but are not at all related embryologically and are anatomically quite dissimilar. The same is true of the legs, the eyes, and numerous other structures in these two groups.

An advocate of special creation might attempt to argue away the significance of homologous structures as evidence for evolution by

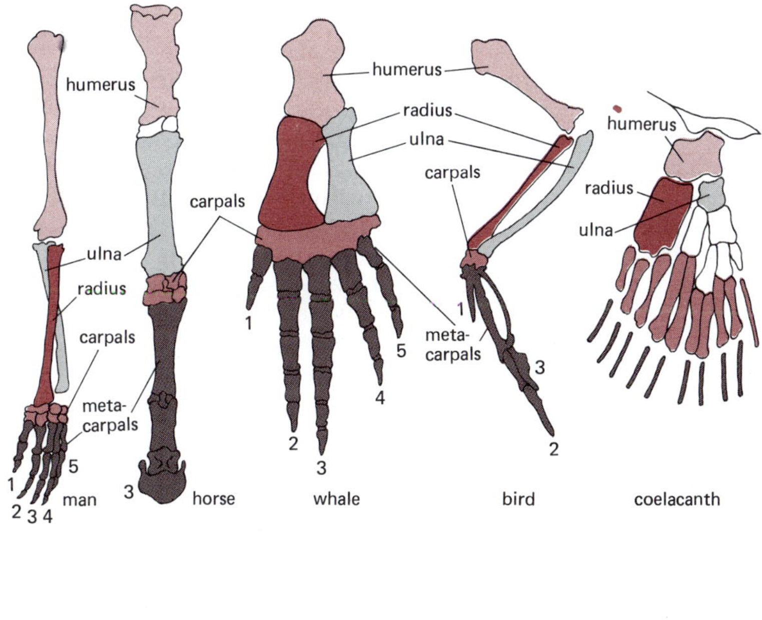

Figure 19–9 Homologous structures. Each forelimb has the same basic internal structure, adapted for a specific set of functions.

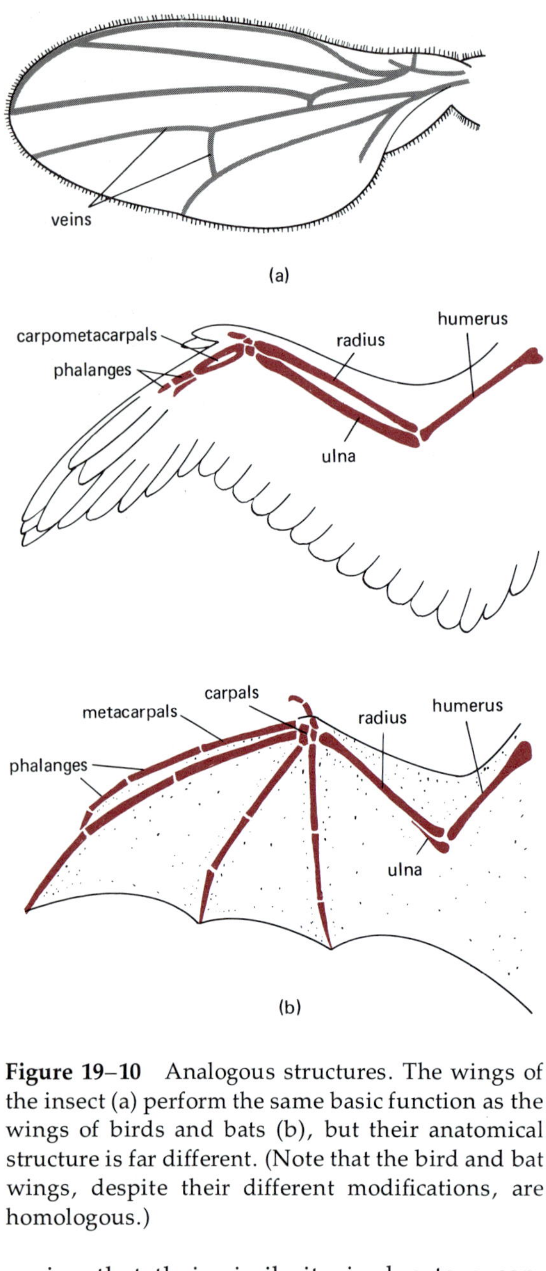

Figure 19–10 Analogous structures. The wings of the insect (a) perform the same basic function as the wings of birds and bats (b), but their anatomical structure is far different. (Note that the bird and bat wings, despite their different modifications, are homologous.)

saying that their similarity is due to a construction according to the best possible plan, rather than to any evolutionary relationship. Yet how could a single structural plan be simultaneously the best for such diverse functions as swimming, walking, galloping, flying, grasping, and manipulating? In addition, the existence of analogous organs indicates that there may be many alternate patterns for the effective performance of similar functions. For example, the eye of the squid and the eye of the fish are derived from quite different embryological origins and are constructed differently, yet both provide for vision under water (Fig. 19–11 a and b). Even more diverse in structure are the eyes of insects and the eyes of birds (Fig. 19–11 c and d).

VESTIGIAL STRUCTURES

Occasionally a human baby is born with a distinct tail. Usually it is discreetly removed by the doctor. This structure has no function in man, and yet virtually all of us bear the *vestiges* of a tail in the form of the four fused bones of the coccyx. Such useless or **vestigial structures** are common throughout the animal and plant kingdoms. In man, for instance, there are more than a hundred such structures or organs. Though they have no known function in modern man, they are nonetheless present and functional in other animals. Among these vestigial structures in man are the vermiform appendix, the third eyelid, muscles of the outer ear or pinna, body hair, and wisdom teeth.

The vermiform appendix in man is comparable to the much larger caecum of herbivores, which in the latter plays a role in digestion. The human appendix has no digestive role and has generally been considered to serve no useful function; recent preliminary studies indicate, however, that it may play a role in the immunity mechanism and the body's defenses against cancer. The nonfunctional vestige of the third eyelid is comparable to the functional nictitating membrane of frogs and snakes, with which they can shield their eyes without entirely cutting off vision. We have all seen horses and dogs pricking their ears and turning them in the direction of a sound. Most humans have lost control over such movements, although they still retain vestigial muscles for the function.

The presence of vestigial structures seems incomprehensible without the assumption of evolutionary relationships, yet within the framework of evolution a mechanism of their

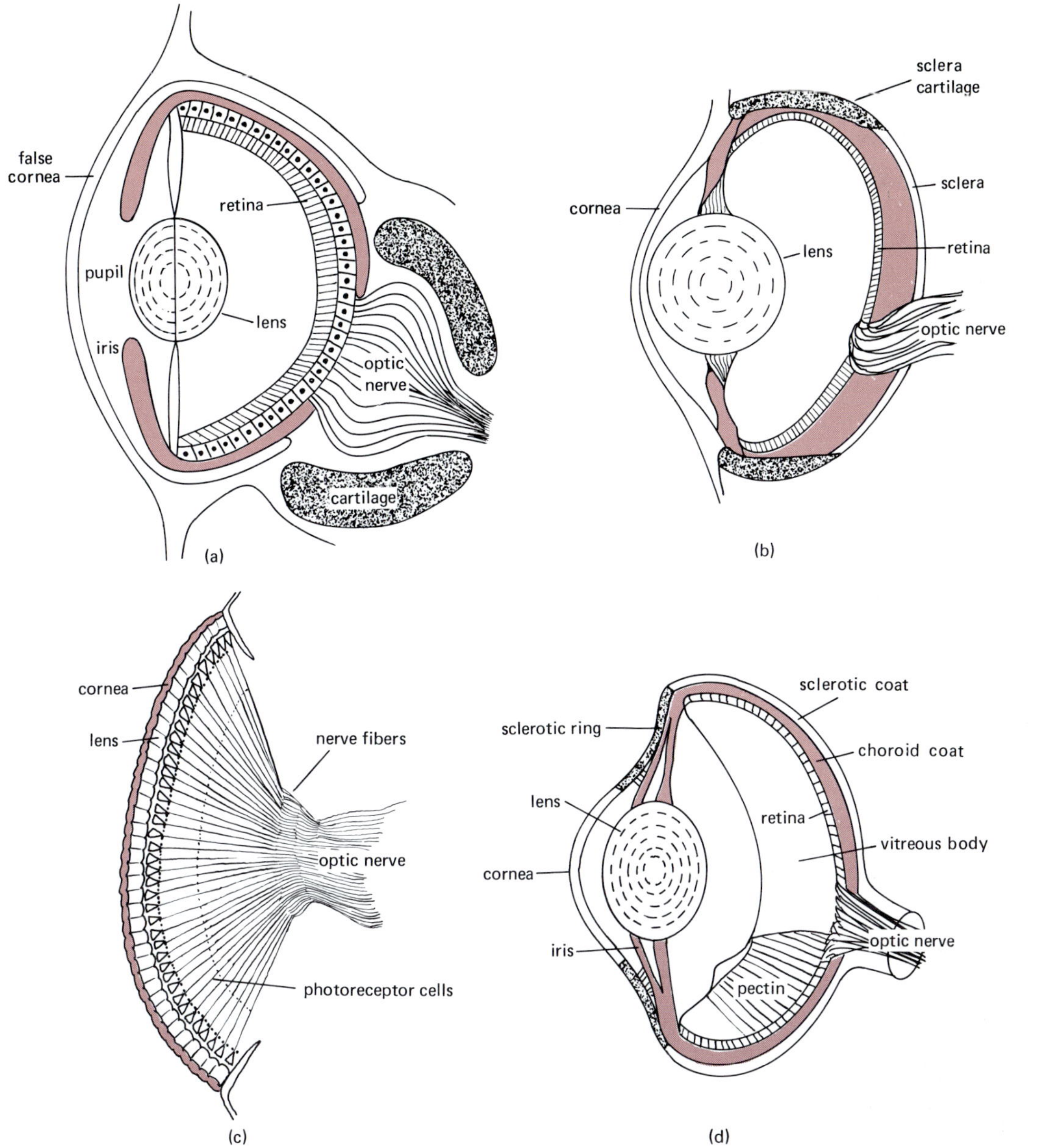

Figure 19–11 Evolutionary solutions to the problem of vision. The cephalopod (a) and the fish (b) take different developmental pathways to reach a cameralike eye structure that provides for underwater vision. The insect (c) and bird (d) use even more divergent structures for seeing.

formation and ultimate elimination can readily be imagined. For example, certain cave-dwelling animals have eyes that are incapable of functioning in vision, while in other cave species the eyes are greatly reduced or virtually eliminated (Fig. 19–13). It may be assumed that for an animal living in the eternal darkness of a cave, mutations that result in impaired vision would not be detrimental to survival of the species and hence

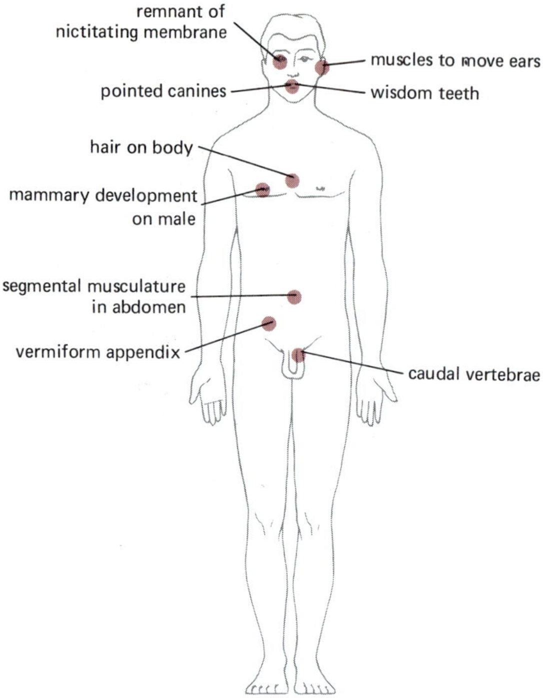

Figure 19–12 A variety of vestigial structures link man to his animal ancestors.

Figure 19–13 Cave dwellers such as this fish have lost their eyes completely in the course of evolution.

might quite likely be passed on. Over many generations, the accumulation of mutations may gradually lead to a reduction and eventually even a disappearance of the eye structures. Vestigial structures may thus be looked upon as way stations in a gradual and continuing process of elimination of structures that have lost their function as a result of changed demands of the environment. In this light, it is interesting to note that many people develop only two wisdom teeth or none at all.

COMPARATIVE EMBRYOLOGY

One of the most dramatic forms of evidence for evolution is seen in the embryological development of species. If we compare the embryos of, say, the fish, salamander, tortoise, chick, rabbit, and man at comparable stages of early development (Fig. 19–14), the similarities are startling. If the figure captions were covered and only the top row of the figure were left exposed, it would be virtually impossible to tell which species is which.

In the animal kingdom, with the exception of the sponges the early embryological development involves practically identical stages, from the zygote to the gastrula. The higher an organism is in the hierarchy of the animal kingdom, the more extensive is its further embryological development. Amazingly, the "higher" organisms pass through many of the same basic stages as the "lower" organisms do, and in basically the same order in which they are believed to have evolved. The human embryo, for example, successively exhibits fishlike, amphibianlike, and reptilelike characteristics, with nonfunctional gill pouches, a tail, and a heart that passes successively through two-, three, and four-chambered stages, ultimately to be modified to the uniquely human pattern. The evolutionary similarities extend even to a succession of basic patterns of biochemical metabolism.

This amazing "evolution" of the embryos of higher organisms was first observed early in the nineteenth century. After Darwin's publication of his *Origin of Species*, the German anatomist Ernst Haeckel proposed the now famous hypothesis, "Ontogeny recapitulates

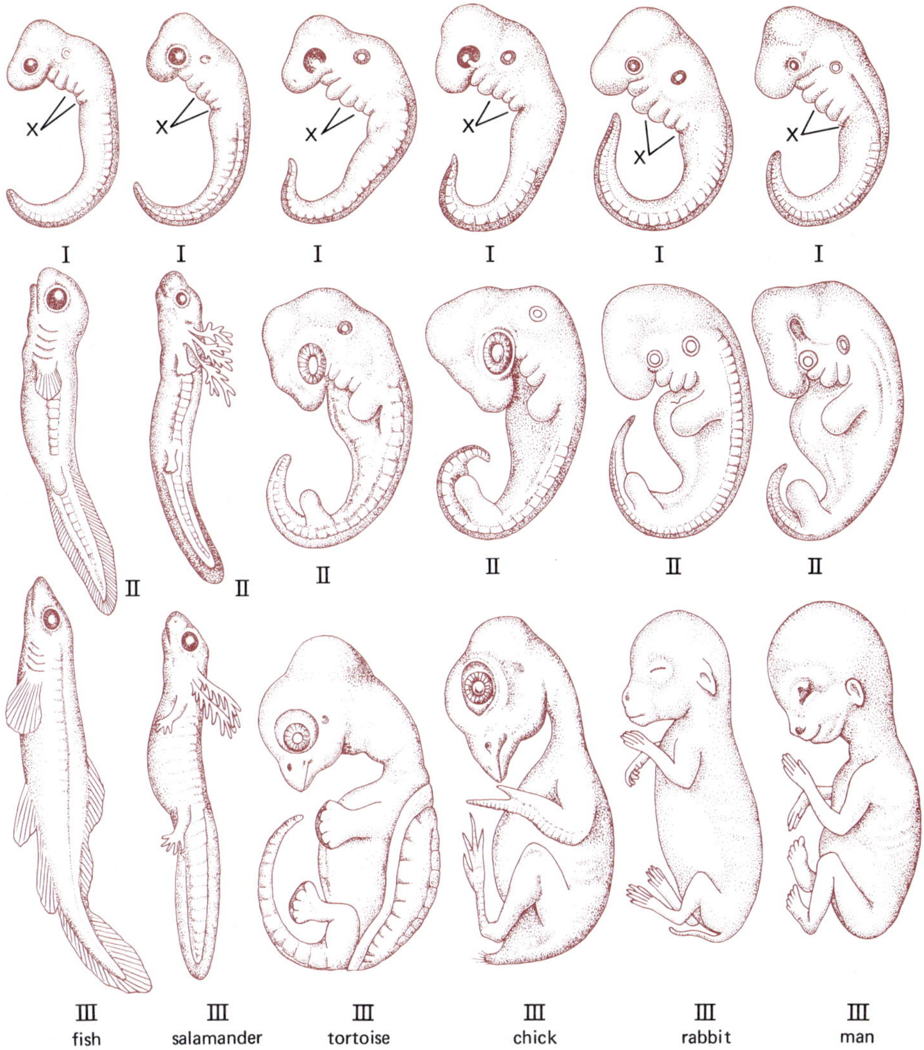

Figure 19–14 Embryos of fish, amphibians, reptiles, birds, lower mammals, and man at corresponding stages of development. Note the similarity of the early stages. (Gill pouches are indicated by x.)

phylogeny." This means that the embryonic development of an individual (ontogeny) repeats or mirrors the evolutionary development of its ancestors (phylogeny); or, as it has been pungently phrased, "During its development an animal climbs its own family tree."

Haeckel's hypothesis was subsequently shown to be correct only in broad outline, since the recapitulation is far from precise and complete. The modern version of the hypothesis, known as the **biogenetic law,** states that *ontogeny is a shortened and modified recapitulation of phylogeny*—that is, the embryological development of higher organisms includes various aspects of the embryological development of their ancestors.

The biogenetic law can be readily understood in the light of modern evolutionary and genetic theory if we assume that the development of species occurred by gradual modification of a basic set of genes. While key mutations have brought new characteristics, the bulk of the gene set was retained, including most of the genes controlling early development. As might be expected, the more closely related two species are, the larger the number of genes in common and the longer the stages of development parallel one another. For instance, even into the seventh month of human development, the human fetus might be mistaken for that of an anthropoid ape.

COMPARATIVE PHYSIOLOGY AND BIOCHEMISTRY

The development of fine techniques of biochemical testing has brought a revolution in the fields of taxonomy and evolutionary theory. New means of establishing kinship among diverse species and, moreover, precise degrees of relationship are now available.

The basic biochemical similarities among the most diverse organisms are astonishing. The DNA-based genetic code is universal, from microorganism up through the plant and animal kingdoms. Virtually all living cells are comprised of proteins containing the same basic twenty amino acids. Cells contain the same basic organic compounds and engage in similar cycles of metabolism with only minor differences among widely divergent species. Certain enzymes found in the digestive tract of man can be found throughout the animal kingdom and even in protozoans. Auxins (growth hormones) from flowering plants will stimulate the growth of pines, ferns, and mosses as well. Sheep, cattle, and hog insulin are successfully used to treat diabetes in humans.

Just as the theory of evolution predicts, biochemical similarities are found to be greater the more closely related the species are. Although virtually any tissue is open to biochemical analysis, studies of the blood are particularly convenient. Using such **serological** techniques, biochemists have been able to pinpoint the positions of organisms in the hierarchy of life.

For example, the long-standing question of the relationship of the rabbit to the group of rodents has been resolved at last through serological studies. In a typical experiment (Fig. 19–15), blood serum from a rabbit is repeatedly injected into a suitable animal, such as a guinea pig. In time the foreign proteins of the rabbit serum (i.e., antigens) stimulate the production of antibodies by the guinea pig's immunity mechanism. These antibodies are specific to the rabbit serum proteins, and if brought into contact with them will cause a rapid precipitation. Now specimens of the sensitized guinea pig's blood are placed in three series of test tubes in standard dilutions. To one series (A) are added samples of rabbit blood, to the second series (B) samples of pig blood, and to the third (C) samples of rat blood. In series A, precipitation is rapid and heavy, even at great dilutions. In series B, the precipitate is lighter, is not observed at the greatest dilutions, and takes somewhat longer to form. In series C, precipitation is even more delayed and occurs only in the more concentrated samples. Thus, it is evident that the rabbit is rather closely related to the pig (an ungulate) and not as closely related to the rat (a rodent). On the basis of tests such as these, along with various morphological studies such as analyses of tooth patterns, taxonomists decided finally that rabbits should not be classified as rodents, as had traditionally been done, and placed them in a separate order, Lagomorpha.

It is interesting that at least some precipitation occurred with all the samples tested. This indicates that there is a direct kinship (of one degree or another) among all the groups involved. Indeed, if this analysis is extended to all the mammals, and even to virtually all the vertebrates, some degree of precipitation will be observed in each case, indicating the underlying relationship of all these species.

GENETICS AND SELECTIVE BREEDING

With modern techniques of microbiology, some of the aspects of evolution can be studied in a petri dish. If the bacterium *Escherichia coli* is cultured under normal con-

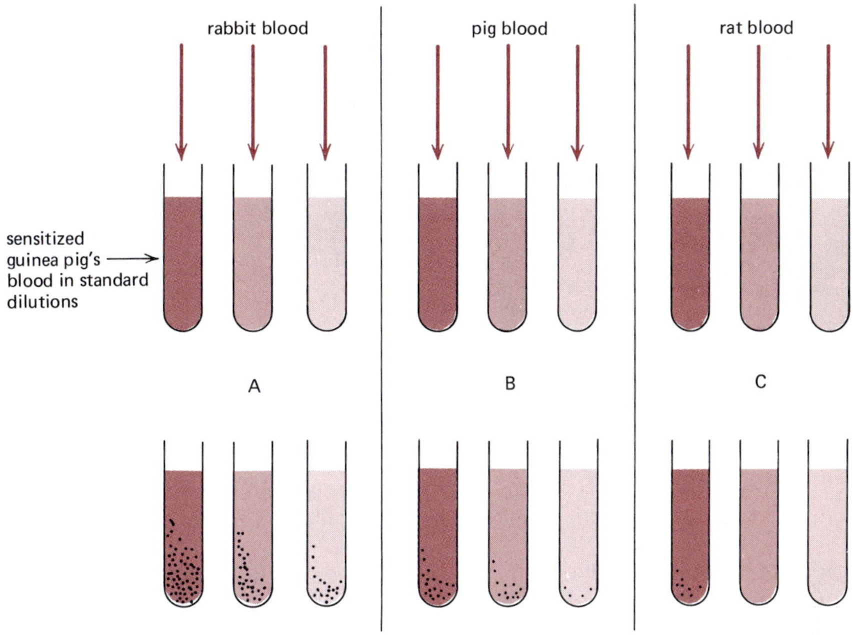

Figure 19–15 Serological tests showed rabbits to be more closely related to ungulates than to rodents. The guinea pig was first sensitized to rabbit antigens by injection of rabbit blood, which stimulated the production of rabbit-specific antibodies.

ditions, variant colonies will occasionally be observed, and one can study the relative survival characteristics of these colonies in comparison with the normal or wild-type colonies. With the use of mutagenic agents, such as X rays or chemical mutagens, the appearance of variant or mutant forms becomes much more frequent. When the conditions are varied to such a degree that the normal type can no longer survive, it is found that one of the sparsely represented forms will now become dominant. It is "fittest" to survive under the changed conditions of the environment. Using this technique, for example, a bacterial variety that can withstand high temperatures may be isolated. By repeating the selection process with the latter variety, strains with even higher heat tolerance may be isolated.

Bacteria are an especially attractive organism to work with since they are extremely prolific, giving rise to a new generation in as little as half an hour, and ultimately to billions of offspring within a day or so. Using the "shortcut" provided by mutagenic agents, scientists have been able to learn a great deal about genetics, selective breeding, and to some degree, the evolutionary processes, working with microorganisms.

But man has been conducting breeding experiments for thousands of years using cattle, dogs, sheep, goats, horses, chickens, and a host of other animals, as well as agricultural plants such as corn, wheat, and rice. As a result he has learned the art of selective breeding and has been able to "breed in" or "breed out" various characteristics, even without a knowledge of the genetic basis behind these attempts. In this way man himself has become a shaping force of evolution and has produced many new forms of plants and animals that differ considerably from their original wild ancestors.

GEOGRAPHIC DISTRIBUTION

Charles Darwin wrote in his autobiography, "The voyage of the *Beagle* has been by far

(a) (b) (c)

Figure 19–16 Man has bred a variety of chickens for eggs, meat, and show: (a) a single-comb white Leghorn male; (b) a white-crested black Polish female; (c) a buff Orpington capon (castrated male).

the most important event in my life." On this famous trip around the world, Darwin took particular note of the patterns of distribution of the fauna and flora throughout the geographic regions that he visited. The explorations of the *Beagle* took him first to the east and west coasts of South Africa, thence to the Galápagos Islands. There he was struck by the gradations of life forms on the various islands and their relationships to those forms on the adjacent South American mainland. It was the relative distribution of closely related species, such as the various types of finches, that provided Darwin with important clues to the nature of the evolutionary process.

A study of the living species found in various parts of the world, as well as an analysis of the fossil remains of past life, reveals patterns of development that dovetail neatly with what geologists have been able to piece together about the history of the earth's crust and its major land masses. Areas that are joined together show a general similarity of life forms in similar types of habitats, while land masses that have been separated for long periods often show widely divergent forms existing under basically similar conditions. One of the most striking examples is the continent of Australia, which is believed to have split off from the supercontinent of Africa–South America–Antarctica some hundred million years ago. The native mammalian life of Australia is unique in the world, consisting entirely of marsupials (pouched mammals such as the wallaby and the kangaroo) and two monotremes (egg-laying mammals: the spiny anteater and the duck-billed platypus) (Fig. 19–17). The only placental mammals found in Australia by settlers are some bat species, which are believed to have island-hopped over, and the doglike dingo and some rodents, which are believed to have been brought over by Malayan aborigines about six thousand years ago. Fossil deposits in Africa and South America indicate that marsupials and monotremes were the dominant mammal forms in these areas, too, precisely at the time when the continental split is believed to have occurred. Later, placental mammals appeared and ramified on the main continent, but apparently the key mutations never arose on the drifting island of Australia, and there the earlier mammal forms remained dominant.

Many puzzling patterns of geographical distribution of life forms are made clear within the framework of the evolutionary concept. For example, various types of camels are found only in Africa and Asia, while the closely related llamas are found only in South America. Yet there are no camellike organisms anywhere in between. An examination of fos-

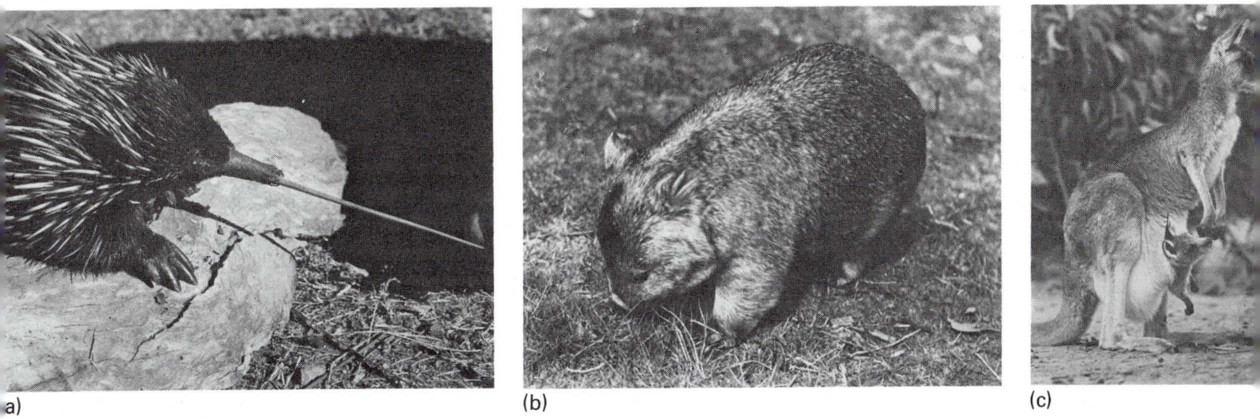

Figure 19–17 The unique mammals of Australia: (a) echidna or spiny anteater (a monotreme), (b) wombat, and (c) kangaroo with young "joey" almost old enough to leave its mother's pouch (marsupials).

sil remains reveals that members of the camel family were once distributed through North America and Europe, as well as in Asia, North Africa, and South America (Fig. 19–18). Indeed, it seems that the camel actually originated in North America and thence spread throughout the world over land bridges existing at the time. Thus, we see that what is now an apparently discontinuous distribution of the camel family at one time represented a continuous spectrum, spread through nearly all the populated land regions of the world.

Thus, the theory of evolution provides highly plausible explanations for the patterns of geographical distribution of living organisms, both present and past. Studies of such

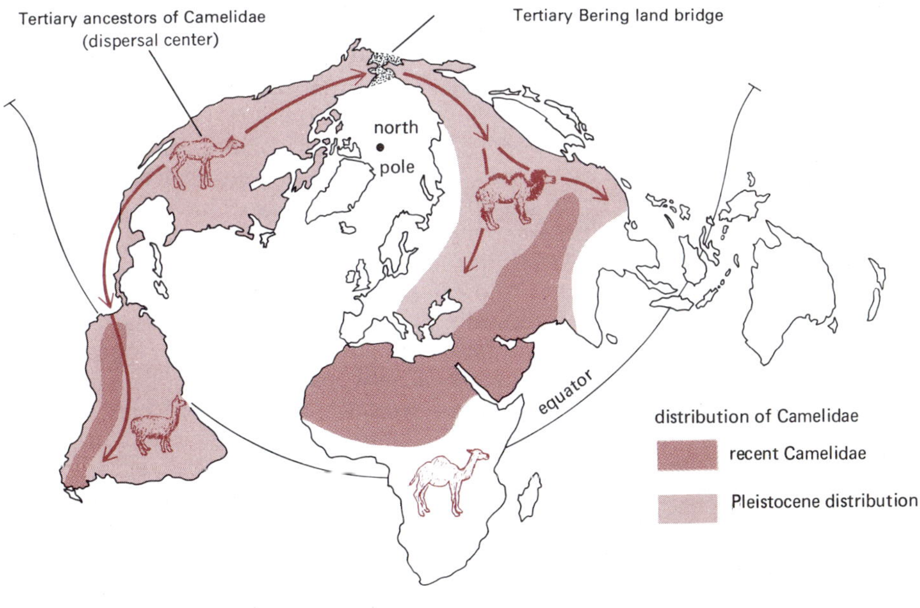

Figure 19–18 The spread of the camel. The distribution of modern members of the camel family is only a remnant of a once continuous distribution.

distributions have furnished scientists with important clues to the workings of the evolutionary process, the mechanisms of which will now be discussed.

Modern Concepts of Evolution

If Darwin were alive today he would doubtless be pleased that the basic aspects of his theory of evolution have survived the test of time and that some of the weaker points of his theory have been strengthened, clarified, and amplified. One of these weak points involved providing a plausible explanation of the source of variations. (As one modern writer has put it, "The only thing Darwin's *Origin of Species* failed to explain was the origin of species.") As indicated earlier, the concept of chromosomal mutations had not yet been introduced to the scientific community during Darwin's lifetime. And because Darwin did not have an accurate understanding of the mechanisms of heredity, he was unable to appreciate fully the vital role of isolation in the process of evolution. Modern studies of population genetics have cast additional light on both factors and have permitted a better overall appreciation of the evolutionary process.

ISOLATION

With the rapid advances in space research and exploration, it seems very likely that some day man will colonize some of the planets and satellites of the solar system. It is even conceivable that he may venture farther to the neighboring stars in search of other habitable planets. By some distant time he may populate the whole of this sector of the galaxy. But the distances between stars are so huge that future colonies of humans might become isolated from the others for thousands and even millions of years. In the intervening time unique mutations would tend to accumulate among the isolated population. In time, if contact with earth were reestablished, the isolated colonists might find that they could no longer have fertile marriages with members of the original human stock. In other words, these isolated humans would have evolved into a new species. A similar process would probably have occurred with any pets and livestock they might have brought along.

Although man's habitat is still his mother earth, nonetheless groups of men have already experienced periods of isolation (for relatively short times, on the evolutionary scale). As a result, various chance mutations have accumulated, giving rise to the races of man (Fig. 19–19). That all human races are readily capable of interbreeding is an indication of the fact that these mutations have not progressed to the point of the creation of new species. With man's ingenious inventions of locomotion permitting greater and greater mobility, beginning perhaps with his primitive dugout canoes and extending to today's modern ocean liners and jets, the geographical isolation of human groups has been effectively ended for virtually all human groups on earth. If free interbreeding were permitted, the varying racial traits would eventually become uniformly distributed throughout the population, and racial differences would literally melt into a common pool.

Throughout the history of the development of life on the earth, many instances of isolation have occurred, extending, in numerous cases, for millions of years. One of the prime examples of this **geographic isolation** (i.e., separation due to geographical barriers) has been the isolation of the Australian continent after its separation from the larger land mass that included Africa, South America, and Antarctica. The barriers that can produce geographic isolation range from vast oceans and impassable mountain ranges down to rivers, ravines, and swamps, each of which may prove to be sufficient barrier to prevent interbreeding between two groups of the same species. Indeed, the barrier may not necessarily be physically impassable; behavior patterns may prevent crossing over barriers that could easily be traversed if an attempt were made. Thus, two populations of meadow butterflies might be effectively separated by a narrow strip of woods.

Once geographic isolation has been prolonged enough to permit the accumulation of a significant number of divergent mutations in two separated populations of the same

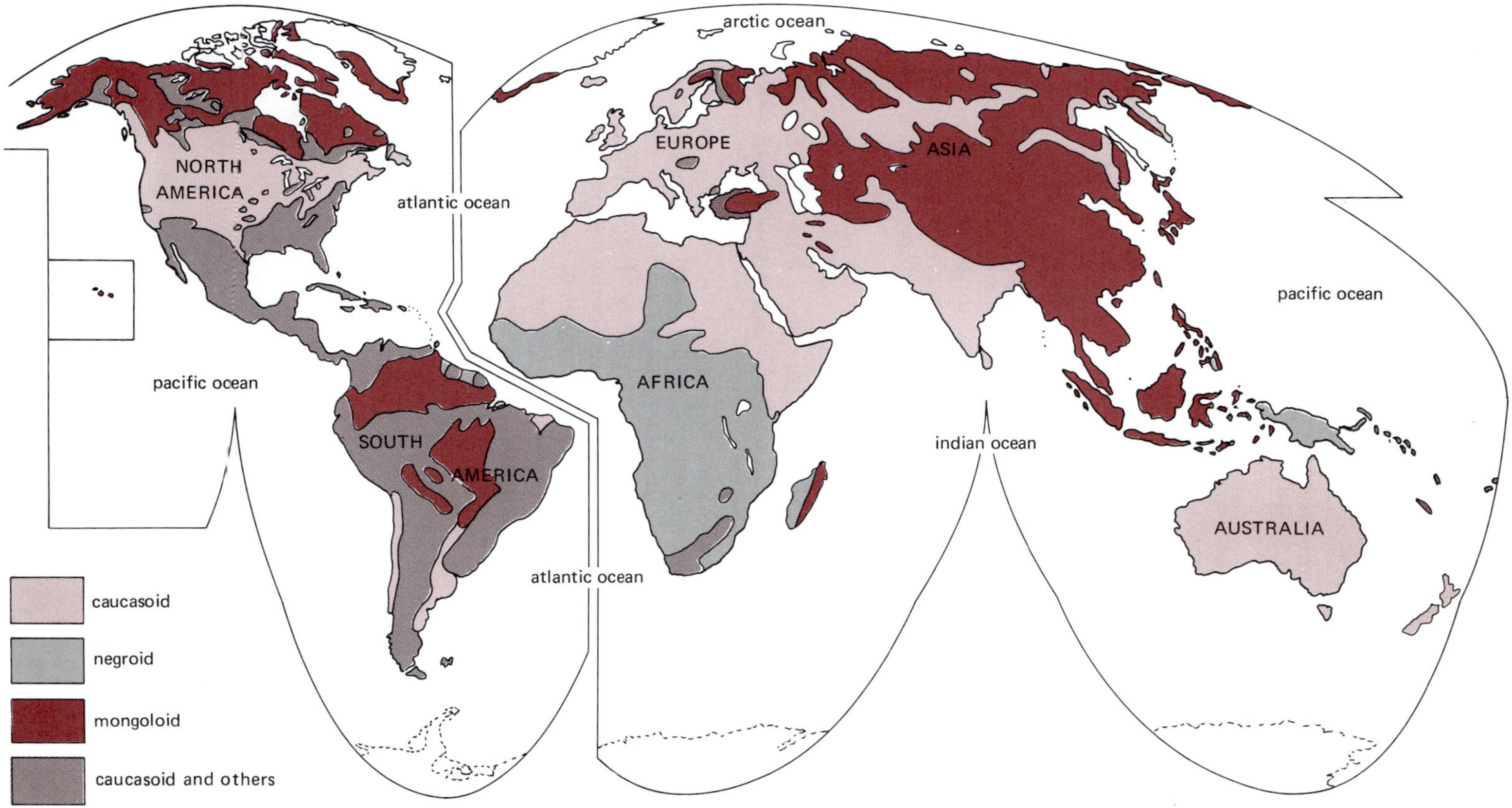

Figure 19–19 The present distribution of the races of man is the result of numerous migrations, some over land (including routes over land bridges that no longer exist) and some over water.

species, a different type of isolation may be superimposed: **genetic isolation.** If the chromosome makeups of the two populations have become so different that no fertile offspring will result from matings between members of the two groups, then they will remain effectively isolated even if the geographical barriers are removed. The members of the two populations will have evolved into two distinct species. Such genetic isolation might be observed in the case of interstellar colonization described at the beginning of the section, and also if the narrow strip of woods separating the two butterfly populations were cut down permitting the butterflies to mingle again.

Other types of reproductive isolation can also have an effect in the evolutionary process. One of these is **ecological isolation,** a condition due to two populations' occupying different habitats in the same area or having patterns of life that prevent the interbreeding even though they live together and might still be interfertile. For example, in the northeastern United States three closely related species of frogs, *Rana clamitans, Rana pipiens,* and *Rana sylvatica* (Fig. 19–20), frequently occupy the same ponds, but they breed at three different, nonoverlapping times of the year. In the Swiss Alps, two distinct races of the moth *Nemophila plantaginis* are found. One occurs exclusively below 5500 feet and the other lives exclusively at altitudes above 8800 feet. There is never any interbreeding at either of those altitudes, but at a height of about 7000 feet a hybrid of the two is found.

MUTATIONS

The increased levels of radiation due to explosions of nuclear and thermonuclear devices are expected to produce an appreciable increase in the rate of occurrence of **mutations,** changes in the genetic material of living cells. Some of these mutations will occur in somatic cells, cells in various parts of the body, and will affect only the cells in which they occur. But others will occur in the reproductive or germ cells, and these changes may be transmitted to future generations.

Although the rate of mutation has probably increased in recent years, the phenomenon of mutation has been a concomitant of life since it first appeared on earth. Our world has always been subjected to a certain level of radiation, from solar and cosmic rays and the decay of radioactive isotopes in the earth's crust, and mutations can also be produced by various other influences, such as certain chemical agents and abnormally high temperatures.

It is understood now, as it was not in Darwin's day, that mutations provide the source of the natural variations upon which the processes of natural selection work. Although mutations have been occurring at random, those changes which gave an organism some definite adaptive advantage in the struggle for

Figure 19–20 Sharing a habitat but ecologically isolated: (a) *Rana clamitans,* the green frog; (b) *Rana pipiens,* the leopard frog; and (c) *Rana sylvatica,* the wood frog.

(a) (b) (c)

existence were likely to spread through the population and eventually become standard characteristics of the species. On the other hand, those mutations which proved deleterious to the organism and put it at a definite disadvantage in the struggle for existence would frequently lead to its early death, often before it had an opportunity to reproduce. Such mutations, therefore, would tend not to be passed along to future generations. Many minor mutations, which possessed neither positive nor negative attributes, would probably be retained within the population gene pool. However, if some change in the environment subsequently made the previously neutral characteristic disadvantageous to survival, then natural selection would probably eventually lead to its elimination from the gene pool of the population.

The mutation in humans that led to sickle-cell anemia has been subjected to two opposing forces of selection. In the homozygous state, the sickle-cell allele results in a crippling and ultimately fatal disease. Heterozygotes (persons carrying one sickle-cell allele and one normal allele) may have only minor symptoms or none at all. About 45 percent of their hemoglobin is of the abnormal variety, and their red blood cells do not usually sickle.

It would seem that negative selective pressures would promote the elimination of the sickle-cell alleles from the population. Yet it has been estimated that the gene is carried by about 20 percent of the U.S. Negro population, and it is also fairly common among people from the Mediterranean region. Further studies have revealed that heterozygotes with one sickle-cell gene have a much-higher-than-normal resistance to malaria. Thus, this gene confers a definite selective advantage in regions of Africa and the Mediterranean where malaria is prevalent. It will be interesting to see whether the frequency of the gene gradually decreases among Americans of African and Mediterranean descent. Modern techniques of genetics counseling may promote its decline. A simple test for carriers of the sickle-cell trait is available. (If a blood sample is deoxygenated, the red cells of a carrier will sickle.) Carriers given this information can intelligently assess the risk of their bearing children with the disease.

By far the majority of mutations are either neutral or harmful. But with fantastic numbers of organisms to work with and the enormous spans of geological time in which to operate the gradual reinforcement and buildup of the small percentage of advantageous mutations through processes of natural selection could result—indeed, has resulted—in an amazing diversity of viable life forms.

In general, except for severely damaging or lethal mutations, when key genes are short-circuited, mutations result in relatively minor changes. Evolutionary development has been largely a process of slow accretion of a long series of small modifications. But one type of mutation, **polyploidy** or a change in the number of chromosomes, provides a mechanism by which, under certain circumstances, the origin of species can occur in a cataclysmic fashion. In polyploidy, an organism may

Figure 19–21 Two Easter lilies: the large one is polyploid, the small one is diploid.

double, triple, or even more increase its chromosome number in a single generation. In most cases such an increase leads to no new species, but rather to an accentuation of the characteristics already present. But occasionally two species will form a hybrid which, after a doubling of its chromosome number, becomes a fertile polyploid.* Such an occurrence was observed in the production of a species of cord grass, *Spartina townsendii* (containing 63 chromosome pairs), which was probably formed from the hybridization of *S. alterniflora* (35 pairs) and *S. stricta* (28 pairs). An indication of the importance of polyploidy to evolution is the fact that, according to some estimates, one-third or more of the species of angiosperms are polyploids.

The concept of mutation provides the key to the evolutionary process. With this concept, it is no longer necessary to assume that acquired characteristics can be passed on to future generations—that, for example, the giraffe evolved a long neck because some short-necked ancestors in the distant past stretched up to reach the succulent leaves hanging down from the tree branches overhead. One need assume only that a mutation promoting longer neck growth occurred at some point in a "normal" short-necked population. The animals bearing the mutation were able to reach high-growing foliage and thus were assured of a plentiful food supply when the lower-growing leaves had been stripped bare. With this adaptive advantage, they prospered and multiplied, passing on the new trait to increasing proportions of the new generations, until at last the long-necked form had become the norm for the species.

Mutations do not occur *because* they are needed. They simply occur, achieve some distribution in a population, and then—if they offer some adaptive advantage—may ultimately become a fixed characteristic of the species. Thus, the new trait usually appears in the population *before* it is really needed. This concept, which is referred to as **preadaptation,** can explain a variety of evolutionary events, from the origin of giraffes to the development of the first air-breathing amphibians from fish stock and the development of penicillin resistance by certain strains of disease bacteria. Mutation provides the raw materials of variability upon which environmental processes operate.

*Interspecies hybrids are usually sterile, since they lack homologous chromosomes. If hybridization is followed by a multiplication of the chromosome set, that problem is solved.

RANDOM GENETIC DRIFT

Sometimes the processes of evolution seem to operate in a manner that has no apparent relation to natural selection or even runs counter to the principle of the survival of the fittest. Characteristics may become fixed in a population that have no particular survival value or even ultimately result in the extinction of the group. This may occur by a mechanism known as **random genetic drift,** which is particularly characteristic of small, isolated populations.

The workings of the mechanism of random genetic drift can be illustrated by a simple example (Fig. 19–22). Suppose that on a particular island there are 1 million members of a certain rodent species. Also assume that a particular trait—albinism, for instance—is carried as a recessive gene by one in a thousand among the population. Under such conditions we would expect to find about a thousand of the rodents carrying the gene for albinism. If some misfortune were to overcome some of these albino-carrying rodents, it is unlikely that such an event would seriously affect the overall frequency of the albino gene in the population. But now assume that on a smaller adjacent island there is a population of only a thousand such rodents, and that here, too, the recessive gene for albinism is carried by about one in a thousand. In this population, only a single animal might be carrying this gene. If this particular animal were to die before reproducing itself, albinism as an allele would disappear from the population of rodents on the smaller island. On the other hand, in the smaller population there would be more intensive inbreeding. If the albino-carrying rat had a large number of progeny and they inter-

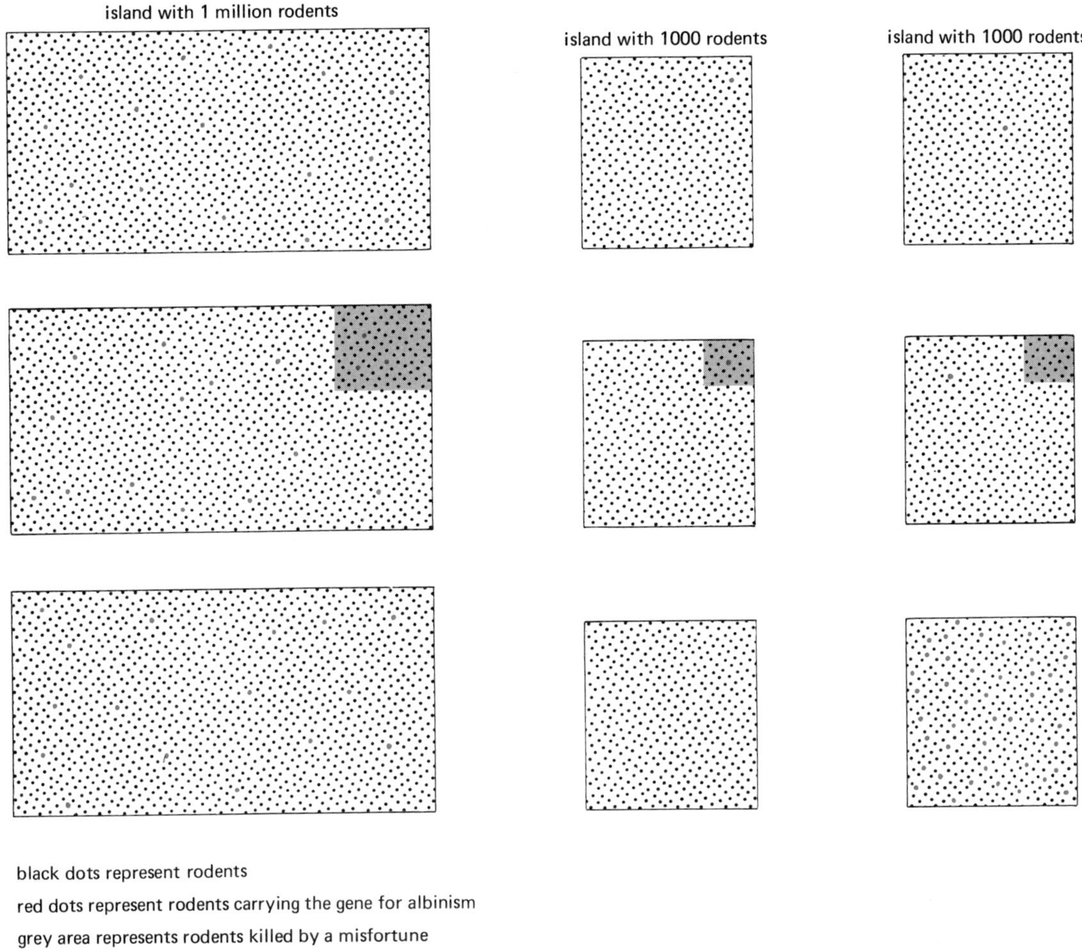

Figure 19–22 Random genetic drift. On the island with a larger population, the loss of a few individuals is unlikely to have a significant effect on the overall frequency of a particular gene. In the case of a small population, an allele may disappear or may undergo a dramatic increase in frequency.

bred to a high degree, the frequency of the albino gene in the population would increase disproportionately, and ultimately the albino gene might spread throughout the local population.

Aside from geographically isolated areas, small populations may also be found under other circumstances. A normal population may be drastically reduced in size by various catastrophic events such as sudden changes in climatic conditions, forest firest, geological upheavals, and epidemics of disease. While some of these events may be selective over a narrow range of traits (such as tolerance to a particular climate, or resistance to a particular disease), they are generally indifferent with respect to the majority of the traits of the species. Hence, most of the gene pool of the surviving population is determined quite by chance, and it may be an atypical set of genes in comparison with the original population. And these may not necessarily represent the best possible gene combinations.

GENE MIGRATION

Through such mechanisms as inbreeding and genetic drift, isolation can have some strange and far-reaching effects on the gene pool of the population. But total isolation in our world is generally a rare and temporary phenomenon. Even Australia, the classic example of isolation, has been invaded by diverse mammalian forms, such as bats, rodents, and primitive man.

If a particular area has been isolated long enough for a unique collection of inherited traits to become characteristic of its animal and plant populations, and then the barriers are removed, individual organisms may begin to move out over the previous boundaries. If the isolation has not been too prolonged and differences within species have not become too pronounced, interbreeding may occur. Such interbreeding can have dramatic effects, both on the local population and on the world at large. An infusion of new alleles from the outside world will bring new elements into the local gene pool; new traits may appear and the proportions of old ones may change. Unique features developed by inhabitants of the previously isolated area can similarly migrate out into the larger population outside the old boundaries.

Figure 19–23 An interspecies mating: the mother, a three-year-old donkey named Jinny, and the father, a five-year-old zebra called Whiskey, shared a paddock at Colchester Zoo in England. Their offspring, who resembles both parents, is classified as a "zedonk."

Another phenomenon that may be observed when barriers of isolation are suddenly removed is an interspecies interbreeding, or **hybridization.** Hybridization can occur only between two closely related species. Although this process is relatively rare in nature, it has been extensively used by man in developing superior strains of agricultural crops. The hybrid form that results from the interbreeding of two species may possess characteristics of either parent, characteristics intermediate between them, or even entirely new traits. In nature, hybrids may prove to have traits that show high adaptation to the environmental conditions. However, most interspecies hybrids are inherently sterile, since they usually lack homologous chromosomes and hence cannot undergo meiotic division in the production of gametes. Occasionally polyploidy may yield fertile forms with double the usual number of chromosomes.

In nature, hybridization may occur between two closely related species growing side by side. The hybrid forms may be sterile in mating with one another but they may show some degree of fertility in matings with the parent species. In such matings, new genes are introduced into the parent population, thus increasing its variability. Such interspecies hybridization is most likely to occur in areas disturbed by man's activities. There environmental conditions have been modified and intermediate habitats are formed that promote the survival of hybrid forms showing intermediate characteristics. A number of examples have been observed, particularly in such intermediate, "hybrid habitats." For example, the sunflower *Helianthus annuus*, an eastern form that has been introduced into the West Coast of the United States by man, grows in a slightly different habitat from the native West Coast form, *H. bolanderi*. In intermediate habitats on the West Coast, hybrid forms have been observed that have backcrossed to varying degrees with both parent species. Now both parental species are exhibiting increased variation, each in the direction of the other.

DIFFERENTIAL REPRODUCTION

Modern genetics has yielded new insights into the workings of Darwin's principle of the survival of the fittest and its role in the process of natural selection. Carried in the chromosomes are numerous inherited characteristics that enhance or detract from an organism's ability to survive—genes for color and size and shape, genes that control the development of muscles and nervous coordination, genes for fertility, and even lethal genes that kill the bearer at an early stage of development. Each organism carries its own particular combination of genes; its premature death, due either to an inherent, genetically determined weakness or to extraneous environmental factors, will remove its genes from potential contribution to the gene pool of the population, while a vigorous life and prolific breeding can help to spread the inherited traits of an organism throughout the population.

It seems obvious that genetic factors conferring greater adaptability, vigor, and fertility will produce more successful individuals in general, which in turn will produce larger numbers of progeny and thus have a greater opportunity to pass on their own characteristics. Conversely, the bearers of less favorable genes will be less likely to survive and will have fewer offspring, and thus their traits may ultimately disappear from the population. But through the generations, the principles of genetics also operate in more subtle ways to result in **differential reproduction** of certain favored groups of characteristics.

For example, not all the potentially fertile members of a population actually mate. Individuals of a wide-ranging species may not be able to find a mate during the breeding season. Among species where there is competition for mates, large fractions of the population may remain unmated. For example, among sea lions, which gather on islands during the breeding season, the actual choosing of mates is prefaced by a fierce competition for territory, in which the males vie with one another for pieces of land on the shore. When the females arrive, each male acquires a harem of a size proportional to the amount of land that he holds. The males who have not succeeded in winning any territory remain apart in a large segregated herd. Only the dominant, "landholding" males are permitted to mate, and thus only their genes are passed on.

Even when mating does occur, the laws of chance and the process of meiosis combine to result in a differential transmission of the genetic material to the offspring. Each zygote contains only half the original set of chromosomes of each of the parents, and certain genes may not be received by any offspring at all. Thus, both random and selective factors contribute to the differential transmission of inherited traits to future generations.

ADAPTIVE RADIATION

Millions of years ago, it is believed, in a pond teeming with life, a fish was born with a pair of seemingly useless characteristics: its fins were stronger and longer than those of its neighbors, and it could breathe air through a crude "lung." For a few minutes each day it

Figure 19–24 Primitive lungfishes gave rise to the first amphibians.

flopped along the muddy shoreline, before slipping back into the warm pool waters. Weeks went by and its uniquely inherited fin structure was strengthened. The days were hot. There was no rain. Steadily the pond waters receded, crowding still more the already dense pond life. Soon little remained of the pond, and life was dying everywhere. One day the pond was no more and all were dead, except for the strong-finned air-breathing fish, who flipped and flopped its way for some distance until it happened to slip into a nearby deeper pond.

That, in a rather simplified form, is the story of evolution of amphibians from fishlike ancestors, which evolutionists have pieced together from the fossil evidence and such "living fossils" as the lungfish and coelacanths. This development occurred some 400 million years ago, during the Devonian period, an age marked by seasonal droughts punctuated by violent rainstorms. Once vertebrate animal forms capable of living for short periods on dry land appeared, further mutations and the processes of natural selection led to a proliferation of new forms better adapted to a life on land. Eventually amphibians, reptiles, birds, and mammals filled innumerable habitats, and their bodies were modified in numerous ways to suit the needs of their lives (Fig. 19–25).

For example, the primitive finlike appendages of the ancestral amphibians were gradually modified into an amazing diversity of forms, ranging from the short, broad, shovel-like limb of the mole (well suited for digging in his subterranean habitat), to the elongated but delicately thin limb of the bat (which supports the wing membrane), to the elongated but heavy limbs of the horse with but a single digit (well suited for running), to the arms of man with his delicately controllable hands (the products of which can be found in the museums, art galleries, and marketplaces of the world).

Thus, a rather primitive structure, the crude finlike appendage, has been modified into numerous specialized structures in the course of evolution, each satisfying the needs of a particular set of environmental conditions for a particular species. This is an example of **adaptive radiation,** sometimes referred to as **divergent evolution.** In this process, related species develop a great external diversity of form, although basic similarities in the fundamental structural plan can be traced.

CONVERGENT EVOLUTION

Another important trend in the evolution of life on earth has been essentially the opposite of adaptive radiation. In **convergent evolution,** two or more relatively unrelated species come to resemble one another, at least superficially.

To the casual observer, the whale, the dolphin, the shark, and the perch may appear to belong to the same family (Fig. 19–26). Each lives in the water, has the same general shape, and swims with the aid of a propulsive tail and a dorsal fin that aids in maneuverability. But a closer examination reveals that these animals do not all belong to the same group after all. The whale and dolphin are warm-blooded, lung-equipped, air-breathing mammals, while the shark and perch are cold-blooded gill-breathers and differ from the whale and dolphin internally in numerous ways. The fossil record reveals that the whale and dolphin are actually descended from four-footed land-dwellers that reinvaded the sea in some long-distant age. In time the descendants of these animals more and more lost the characteristics of their land forebears and more and more assumed the external configuration of the fish.

Why did this occur? It seems that the general shape of the fish is a most efficient configuration for maneuverability and speed for large animals in the water. Hence, every mutation which appeared in the ancestral aquatic mammals, bringing a closer approximation to the general fish shape, provided them with an adaptive advantage and thus had a good chance of being perpetuated in the species.

Implicit in the process of convergent evolution is the premise that there are only a limited number of "best ways" in which some func-

Figure 19–25 Adaptive radiation. A ratlike common ancestor gave rise to a diversity of mammalian forms, each adapted to a particular habitat.

tion may be performed. In the process of natural selection, a characteristic that brings an organism closer to one of these "best ways" has a selective advantage in the struggle for existence. As a result of a cumulative series of such changes, several unrelated species may evolve structurally similar organs, body shapes, or modes of behavior.

THE STORY OF LIFE ON EARTH

Armed with the Darwinian theories of the evolutionary process, supplemented by the concepts of modern genetics and other refinements, paleontologists have studied the fossil record and the evidence of modern species and have pieced together a picture of how life must have developed on our planet. Of course there are many gaps in the picture—considering the vagaries of the processes that contribute to the preservation and discovery of fossils, it is amazing that we have been able to learn so much. But on the whole the inferences and conclusions of the paleontologists have been built up into a rather coherent picture, which accords well both with the available evidence and with our present knowledge of the processes of life.

Table 19–1 summarizes the eras of the evolution of life on earth and their basic characteristics as revealed by the fossil record.

The Precambrian Eras (Ca. 4.5 Billion to 600 Million Years Ago)

It is difficult for us to imagine the earth as it must have existed 4 billion or more years ago.

Figure 19–26 Convergent evolution. Several different lines of animals have adapted in similar ways to life in an ocean environment.

The newly formed crust radiated such intense heat that heavy torrents of rain were quickly vaporized as they neared the hot surface. As time passed the surface cooled slowly, then the quenching rains remained liquid for longer and longer periods of time. Soon the surface was covered with a continuous blanket of water. But huge geological upheavals raised towering mountain chains and formed deep, cavernous, gouged-out basins, which immediately filled with the ubiquitous water. In these early times there could be no life, for the seas were not much below the boiling point.

This turbulent age was part of the **Azoic era,** the time before the appearance of life on earth. How long it lasted is still a mystery, but it probably extended to about 4 to 5 billion years ago, as the crust of the earth was beginning to form. At some point around this time, probably somewhere in the warm soupy seas, a complex organic molecule was fortuitously formed that was capable of self-duplication. Just when this occurred, perhaps sometime between 3 and 4 billion years ago, is impossible to pinpoint precisely, but the results were profound. Within a short span of time (geologically speaking), the seas must have been swarming with these molecules and their mutated or variant descendants. In time, forms which we might call "life" evolved. Organisms appeared that had the ability to bind some of the energy of sunlight into the chemical energy of organic compounds, and other opportunistic forms began to feed upon these primitive "plants," using their ready-made organic substances for energy and building materials.

As the ancestral "plants" and "animals" proliferated in the seas, they changed gradually into new forms. But even as they did so, the planet was changed, too. Oxygen released in the photosynthetic activities of early autotrophs dissolved in the waters of the seas and then, when the saturation point was reached, began to accumulate in the atmosphere. The reducing atmosphere of the early earth was gradually transformed into an oxidizing mixture approximating its present composition, and this oxidizing mixture worked upon the rocks of the earth's crust, modifying their minerals into silicates, carbonates, and other oxidized forms. Some of the oxygen was utilized in the respiration of the proliferating life forms, and carbon dioxide was released in the process. Gradually equilibria were established among the various atmospheric constituents.

The appearance of life on earth marked the onset of the **Archeozoic era,** which lasted until about 1.6 billion years ago. Although a fossil of an algalike form dating back to about 2.6 billion years has been found in a graphitic limestone deposit, the general lack of fossil evidence has tended to restrict our knowledge of the types of life forms that existed during this era. However, we do have some strong indirect evidence that life did indeed exist at this time. Huge deposits of graphite have been observed in the strata of the Archeozoic era, and they are believed to represent the remains of delicate, soft-bodied living forms that have been subjected to the intense pressures and high temperatures of this deep stratum. Extensive calcium carbonate deposits may be the result of the activity of living organisms. If these deposits are indeed the remains of life, then the Archeozoic era must have really abounded in life forms. But their existence must have been constantly jeopardized by the catastrophic geologic upheavals that continued to characterize this era, as in the previous era.

The Archeozoic era has sometimes been called the "era of primitive life." It is separated from the **Proterozoic era,** or the "era of early life," by a colossal series of upheavals, which doubtless caused the extinction of many forms of life then in existence. Few fossils from the Proterozoic era have been found, doubtless because of the difficulties of preservation of soft-bodied life forms. Traces have been found of some protist species, such as primitive algal forms and some protozoa (e.g., Radiolaria) and jellyfish; impressions that resemble worm burrows have also been discovered in rock strata corresponding to the Proterozoic era.

The Paleozoic Era (Ca. 600 to 230 Million Years Ago)

It is one of the great mysteries of geological history that though we have practically no fossil records dating beyond 600 million years ago, the rock strata lying directly above (and hence following in time) are replete with a wide variety of fossil remains. Various theories have been put forward to explain this discontinuity, including the premise that most of the early organisms were soft-bodied and readily decomposed by bacteria, and that the organic matter that escaped such decomposition was transformed by the high pressures and intense heat of the lower regions into graphite. But this would not explain the sharp discontinuity between the barren strata of the late Proterozoic and the fossil-rich early Paleozoic strata. The most plausible explanation for this seemingly sudden burgeoning of life forms is that there was really nothing sudden about it. The colossal geological upheavals separating these two eras were evidently so intense and the unremitting forces of erosion so extensive that they destroyed virtually all fossil remains. And indeed, radioactive dating methods reveal that entire rock strata corresponding to this period of upheaval are actually missing.

Whatever the reasons for the dearth of fossils in the Precambrian eras, as we reach the Paleozoic era we find a virtual explosion of the fossil record. Even at the beginning of the era, every major invertebrate phylum was represented, and before the era came to a close every animal phylum and class, with the exception of birds and mammals, had made its appearance.

The wide variety of life forms that inhabited the seas of the early Paleozoic included numerous algae, protozoa, corals, sponges, brachiopods, primitive echinoderms, gastropods and cephalopods (mollusks), various types of worms, and a group of crustaceans (arthropods) known as **trilobites** (Fig. 19–27a).

TABLE 19-1 The geologic eras

	ERA	PERIOD	TIME FROM START OF PERIOD TO PRESENT (MILLIONS OF YEARS)	PLANT LIFE	ANIMAL LIFE
Precambrian eras	Azoic		~4500	No life on earth; formation of organic molecules from inorganic constituents	
Precambrian eras	Archeozoic		3600	No fossil record; indirect evidence of life from carbon deposits in rocks	
	First great revolution (great erosion and loss of fossils)				
Precambrian eras	Proterozoic		1600	Primitive water plants (algae, fungi)	Marine protozoa; evolution of invertebrates (worms, mollusks)
	Second great revolution (great erosion and loss of fossils)				
	Paleozoic	Cambrian	600	Marine algae (blue-green and probably green algae)	Trilobites, brachiopods dominant; representatives of most modern phyla
		Ordovician	500	Marine algae abundant; probably first land plants	Corals and trilobites abundant; mollusks diversified; first fishes
		Silurian	425	Algae dominant in seas; invasion of land by tracheophytes	Arachnids dominant in seas; rise of fishes; first insects
		Devonian	405	Expansion of land plants; first forests; first gymnosperms	"Age of fishes"; first amphibians; development of insects
		Carboniferous (Mississippian and Pennsylvanian in North America)	345	Great coal forests; lycopods and horsetails dominant at first; then spread of ferns, seed ferns, and gymnosperms	"Age of amphibians"; first reptiles; diversification and spread of insects
		Permian	280	Great forests; decline of lycopods and horsetails	Extinction of many ancient animals; decline of amphibians; expansion of reptiles; mammallike reptiles; modern insects

TABLE 19-1

ERA	PERIOD	EPOCH	TIME FROM START OF PERIOD TO PRESENT (MILLIONS OF YEARS)	PLANT LIFE	ANIMAL LIFE
Appalachian revolution					
Mesozoic	Triassic		230	Dominance of gymnosperms; extinction of seed ferns	First dinosaurs, pterosaurs, and egg-laying mammals
	Jurassic		181	Cycads and conifers common; appearance of angiosperms (dicotyledons)	Dinosaurs abundant; first toothed birds; marsupials
	Cretaceous		135	Gymnosperms decline, angiosperms expand; first monocotyledons; oak and maple forests	Dinosaurs reached peak and became extinct; mammals common; first modern birds; spread of insects
Laramide Rocky Mountain revolution					
Cenozoic	Tertiary	Paleocene Eocene Oligocene Miocene Pliocene	63 58 36 25 12	Rise of monocotyledons, flowering plants; spread and then decline of forests; spread of grasses	Diversification and spread of placental mammals; dominance of land by mammals, birds, and insects; development of manlike apes and first men
	Quaternary	Pleistocene	2	Extinction of many species	Extinction of great mammals
		Recent	0.01	Rise of herbaceous plants	Development of human civilizations

(a) (b)

Figure 19–27 Inhabitants of the early Paleozoic, preserved in stone: trilobites (a) and brachiopods (b).

The trilobites spread rapidly during the early Paleozoic and soon comprised more than half of all the known animal life. The next most abundant group were primitive mollusks with shells, known as **brachiopods** (Fig. 19–27b). Later, reef-building corals appeared in various regions of the world, including areas as far north as Greenland (indicating that these northern regions were warm enough at the time to support the tropical coral). In time the earliest vertebrates appeared: small armored, jawless fishes without fins, called **ostracoderms** (Fig. 19–28a).

The middle Paleozoic is sometimes referred to as the "age of fishes." For as the trilobites began their decline, there was a corresponding explosion in the numbers of varieties of fish. The original ostracoderms gave rise to **placoderms,** fish with jaws and paired fins (Fig. 19–28b). Later, lungfish, lobe-finned fish, the early bony fishes, and sharks evolved (Fig. 19–28c–f). These fishes soon dominated the seas.

During the middle Paleozoic, an important event occurred. Up to this time the lands had been barren, even though the seas were already

Figure 19–28 Fish of the Paleozoic, shown in models: ostracoderm (a), placoderm (b), lungfish (c), lobe-fin fish (d), shark (e), ray-finned fish (an early bony fish) (f).

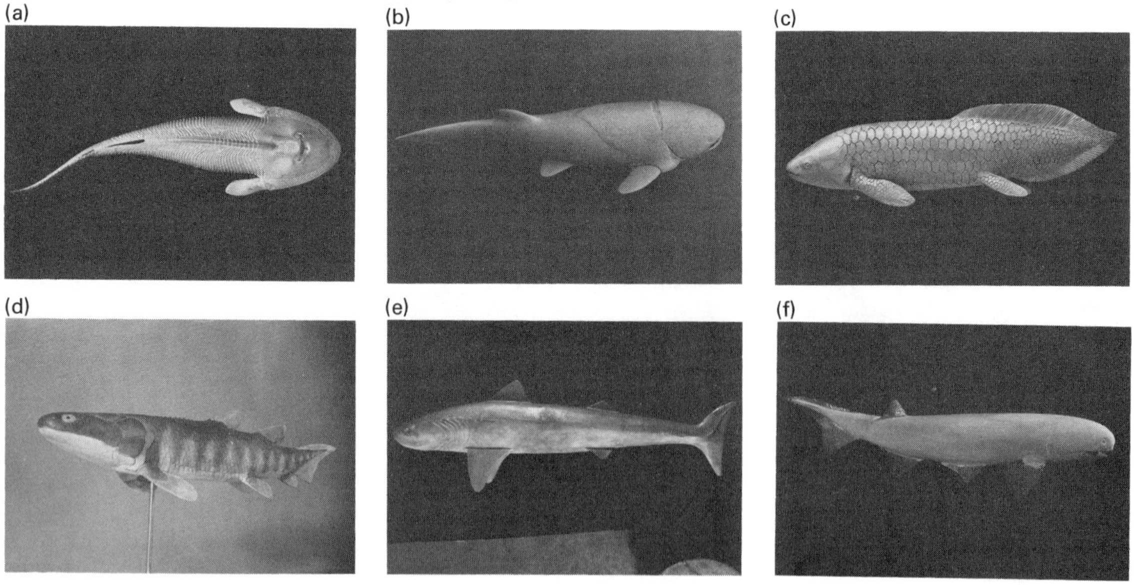

Figure 19–29 A stegocephalian in a typical middle Paleozoic environment.

swarming with life. Now the land areas of the world were rising, and plants began to establish themselves on the land. The first plant species to invade the continents were tracheophytes of the subphylum **Psylopsida,** primitive tracheophytes that are now mostly extinct. These were followed by other vascular plants, including club mosses, horsetails, and ferns. Another historic event was the appearance of the first land vertebrates, primitive amphibians known as **stegocephalians** (Fig. 19–29). These were believed to have evolved from the lobe-finned fish. Air-breathing animals such as spiders, centipedes, and primitive wingless insects may have preceded the amphibians ashore, however.

The late Paleozoic began with the gradual submergence of the land areas and the formation of numerous swampy forests, which were populated by mosses, horsetails, ferns, and giant seed ferns. During this time the dead vegetation gradually gave rise to peat bogs, which eventually became submerged to form the great coal beds characteristic of the strata of the Carboniferous period. Many types of animals roamed the swampy forests of the late Paleozoic. Amphibians dominated for a time (the "age of amphibians"), but then the first reptiles appeared, and before the end of the Paleozoic era they were making serious inroads into the amphibian populations because of definite competitive advantages. One line of reptiles, the therapsids (Fig. 19–30), possessed definite mammalian characteristics; this line ultimately evolved into primitive mammals. Other animal forms inhabiting the forests of the late Paleozoic included winged insects, such as dragonflies. One dragonfly fossil has a wingspan of thirty inches and a body length of fifteen inches. Cockroaches were quite abundant during the late Paleozoic, some giant species reaching three to four inches in length.

During the end of the Paleozoic era the land areas rose considerably, and extensive glaciation led to the demise of many land species. The reduced temperatures probably caused the death of numerous marine forms as well. Many of the inland seas disappeared, further reducing the number of surviving species. Additional loss of life was caused by the huge geologic upheavals associated with the Appalachian revolution, in which enormous areas extending along the eastern coast of the United States and Canada (where the Appalachian Mountains were formed) and other areas of the world, particularly in Europe, were plunged into turmoil as extensive mountain ranges were thrust up. These geological disturbances ushered in the Mesozoic era.

Figure 19–30 Reconstruction of a therapsid, a reptilian ancestor of the mammals.

The Mesozoic Era (Ca. 230 to 63 Million Years Ago)

When we speak of prehistoric life, one of the first images that comes to mind is that of huge dinosaurs waging titanic battles in the swampy forests. But this picture is appropriate to only a small span of the evolutionary past, encompassing about 100 million years out of the more than 2 billion years in which life has existed on earth. This age of the dinosaur transpired during the Mesozoic era. During this era not only did the reptiles achieve dominance on the land, but some reptilian species returned to the sea and others took to the air. The Mesozoic era is therefore sometimes referred to as the "age of reptiles."

Reptiles had first appeared in the late Paleozoic. By the early Mesozoic, they had radiated widely, occupying numerous habitats. The first dinosaurs appeared on the land, but they were rather small. Some of the early dinosaurs and their later descendants were capable of running on their two hind legs, balanced by a long tail, and used their forelimbs to hold food. Others ran on all fours. Many of the

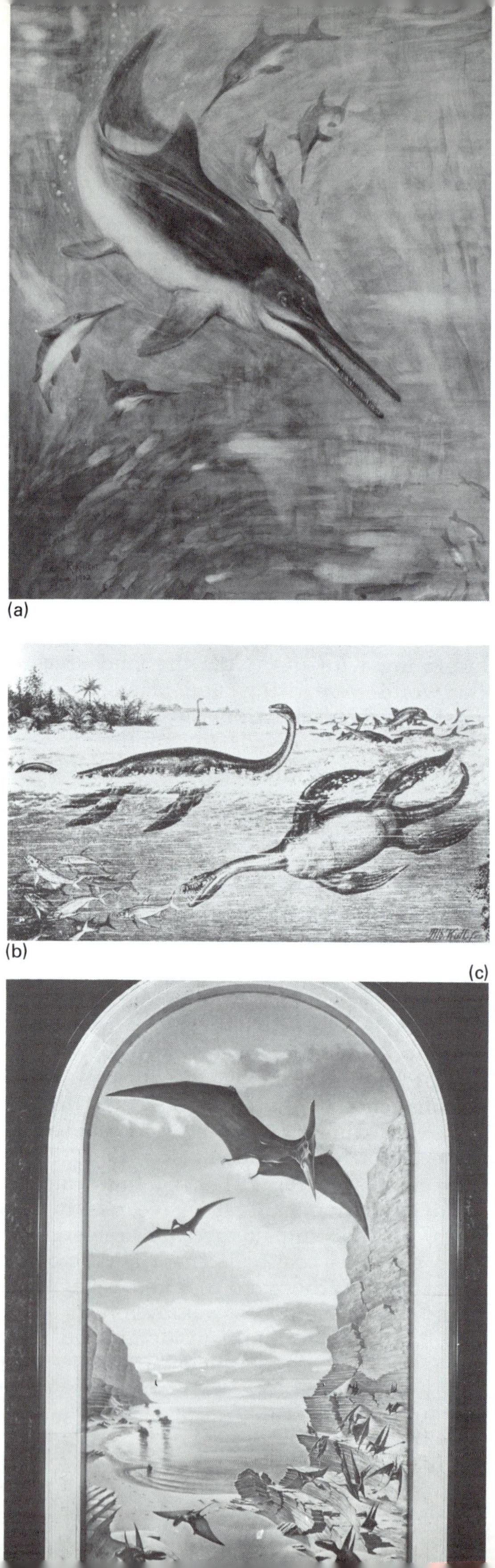

Figure 19–31 (*Right*) Reptiles of the early Mesozoic: ichthyosaur (a), plesiosaur (b), pterosaur (c).

early dinosaurs were carnivorous. Some forms invaded the water and evolved into the fish-like **ichthyosaurs** and whalelike **plesiosaurs** (Fig. 19–31). Others, such as the birdlike pterosaurs, some with wingspans up to twenty feet, took to the air. These flying reptiles had "wings" that consisted of a leathery membrane extending from the elongated little finger of the forelimb to the side of the body and hind legs.

The amphibians declined still further, and many forms became extinct, being unable to compete with the reptilian forms of the time.

One of the most significant events of the early Mesozoic was the appearance of the first mammals. These were rat-sized, egg-laying forms resembling the monotremes. But it was not until the beginning of the Cenozoic era that the mammals were to play an important role.

In the seas, mollusks known as **ammonites** (Fig. 19–32), with intricately coiled shells, became abundant. These were to persist throughout the Mesozoic, and ramified into diverse forms, some as much as ten feet in length.

The climate of the early Mesozoic was generally drier and warmer than at the close of the Paleozoic era. The continents were raised still further, and there were widespread desert

Figure 19–32 Fossil ammonites.

(a)

(b)

Figure 19–33 Survivers from the Mesozoic: cycads (a) and gingkos (b).

areas. The seed ferns had become extinct, and great gymnosperm forests, containing cycads and gingkos, were the dominant plant forms.

In the middle of the Mesozoic era, the climate grew still warmer and much moister. This period saw the appearance and gradual ascendancy of huge dinosaurs (Fig. 19–34a). Enormous amphibious herbivores—such as **Brontosaurus,** which exceeded eighty feet in length and weighed up to thirty-five tons, and **Diplodocus,** which ranged up to ninety feet in length—munched the soft vegetation in the swamps. They in turn served as food for large carnivores, such as **Allosaurus,** which reached a length of about thirty-four feet and was powerful enough to destroy even these lumbering giants.

Figure 19–34 The middle Mesozoic was the Age of Dinosaurs: (a) *Brontosaurus,* (b) *Diplodocus,* (c) *Allosaurus.* This age also produced the first bird: *Archaeopteryx* (d).

Smaller reptiles also abounded: true lizards and crocodiles appeared for the first time in the middle of the Mesozoic and have survived essentially unchanged up to the present day.

Among the amphibians, frogs resembling the present-day forms made their appearance.

An important fossil find from this period is **Archeopteryx,** the first bird, which still retained many reptilian characteristics (Fig. 19–34d). For although it had wings and authentic feathers instead of the reptilian scales, it possessed a full set of teeth in its upper and lower jaws and clawed digits at the ends of its wings. Moreover, its tailfeathers were attached in pairs along a reptilian tail. Unfortunately, we have no way of determining whether this creature was warm-blooded or not.

During the middle of the Mesozoic era, primitive marsupial forms appeared among the mammals. These primitive marsupials were insectivorous and were probably the ancestors of the present-day marsupials.

Most of the modern orders of insects were represented by the late Mesozoic, including termites, grasshoppers, cockroaches, flies, beetles, bees, wasps, and ants. It was at this time that the first angiosperms appeared. These were dicotyledons only, and therefore the great grasslands typical of the Cenozoic era had not yet appeared.

The warm climate of the late Mesozoic encouraged the further development of the giant reptiles. This period marked the appearance of the terrifying flesh eater, **Tyrannosaurus rex** (Fig. 19–35), which measured more than fifty feet in length and had a huge four-foot skull equipped with teeth that were sometimes half a foot long. The primitive marsupials evolved further, beginning in North America; their descendants spread throughout the world. At about this time Australia

Figure 19–35 *Tyranosaurus rex,* king of the carnivorous dinosaurs of the late Mesozoic.

separated from the main continent, carrying with it the early marsupial forms, which eventually came to dominate the Australian continent. Unlike the monotremes which preceded them and which were egg-laying mammals, the marsupials gave birth to live young. However, these young were quite immature and continued their development in an abdominal pouch (a marsupium), where they were suckled until they reached a greater degree of maturity.

On the isolated island-continent of Australia, the marsupials began an extensive adaptive radiation into a variety of habitats, ultimately to produce a diversity of forms ranging from tiny pouched "mice" to huge leaping kangaroos. Meanwhile, on the main continent, a new mammalian form appeared: the placental mammals, which retained their developing young within the body for longer periods, supplying food and oxygen and removing wastes through an organ called the placenta. For the most part the first placental mammals were rather small and subsisted upon insects and worms. But they eventually gave rise to the primitive insectivores, from which, it is presumed, the more advanced forms of placental mammals developed.

Angiosperms prospered and became the dominant plant forms during the late Mesozoic. Among them were vast hardwood forests of maples, oaks, and elms. Monocots, including palms and grasses, made their first appearance in this period. The end of the era was marked by the severe climatic changes brought about by the colossal geological upheavals of the Laramide revolution, which gave rise to the Rockies, Himalayas, Andes, and Alps. The ensuing cold weather, accompanied by the disappearance of the ubiquitous swamplands, led to the demise of many forms of life. The reptiles suffered particularly during this time. Being cold-blooded, the giant dinosaurs were unable to remain sufficiently active to feed themselves. In addition, the marauding habits of the ever-growing mammalian populations doubtless contributed to the demise of the reptiles. Being warm-blooded, the mammals were not as severely affected by the encroaching cold weather. Not only did the relatively small mammals compete actively with the reptiles for the available food, but they frequently feasted upon the reptilian eggs, which were characteristically left unattended. So complete was the demise of the once-mighty dinosaurs that few fossil remains of these creatures have been uncovered in the Cenozoic strata. These climatic changes also led to the extinction of a number of other groups, including numerous species of amphibians and lizards, as well as many plant species.

The Cenozoic Era (Ca. 63 Million Years Ago to the Present)

We are now living in the Cenozoic era. And yet if we could have been present at the beginning of this era, we would hardly have recognized the world. It is true that the huge dinosaurs of the Mesozoic were gone and that mammals roamed nearly everywhere. But the mammals present at the dawn of the Cenozoic era did not closely resemble those of modern times. Instead, the archaic placental mammals had small skulls housing tiny brains, undifferentiated, conical, reptilelike teeth, and five digits in both fore and hind limbs. There were no great grasslands, as exist in the world today, even though the grasses had first appeared in the late Mesozoic. There were, however, small patches of such grass, which augured an explosive expansion that was soon to come. We would have recognized some of the life forms that have remained virtually unchanged since the beginning of the Cenozoic. These forms

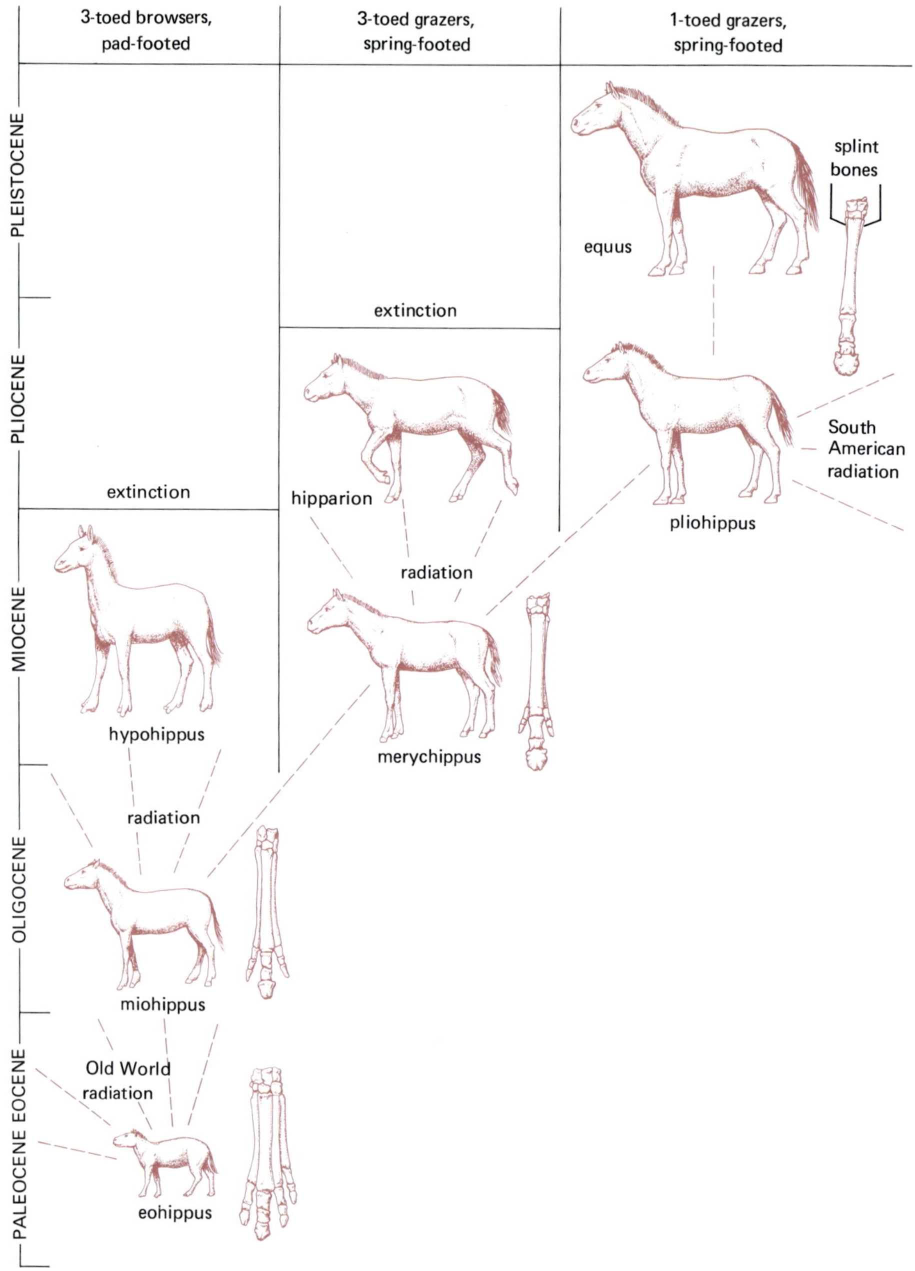

Figure 19–36 The evolution of the horse. The splint bones of the modern horse are vestiges of the additional ancestral toes. According to the fossil evidence, the horse family evolved in North America, then spread across land bridges into the Old World and South America. Horses ultimately became extinct in North America and were reintroduced by the conquistadors (these imported horses gave rise to the mustangs) and by settlers of the continent.

include turtles, crocodiles, frogs, some lizards, cockroaches, ants, and such plant forms as conifers, maples, walnuts, elms, and beeches.

The early Cenozoic saw an explosive growth of mammals. At first the dominant forms were of the archaic type, but these rapidly became extinct. The archaic mammals proved no match for the more intelligent, more adaptive modern mammals that soon began to dominate the land areas of the earth (with the exception of the Australian continent, which was the preserve of the marsupials). Because of this dominance, the Cenozoic is sometimes called the "age of mammals."

But the present (Cenozoic) era is also known as the "age of angiosperms," for beginning in the late Mesozoic and then explosively in the early Cenozoic, the angiosperms literally spread throughout the world. In addition to the large and extensive forests of maple and oak and other dicots, the grasses (monocots) spread through the various land regions of the world. This spread of grasses provided a source of food for the modern herbivores, which were then evolving, and doubtless played a strategic role in their development and spread. In turn, this extensive radiation of herbivorous species encouraged the diversification and development of various modern carnivorous forms. Thus, by the middle of the Cenozoic, most of the modern orders of mammals were not only already in existence, but were rather numerous. These included horses, camels, elephants, pigs, and cattle (ungulates); squirrels, mice, and rats (rodents); tigers, lions, cats, and dogs (carnivores); and monkeys, apes, and lemurs (primates). Indeed, the peak of development of the mammals was reached in the middle of the Cenozoic era, in the period known as the Miocene epoch. From then on to the present there has been some contraction in the number of mammal species. The Cenozoic also saw the spread of the modern birds, which had made their first appearance late in the Mesozoic era.

By the end of the Tertiary period, about two million years ago, the mammals were nearly identical with modern forms. The genus *Homo* (man) had evolved from manlike apes, which in turn had developed from earlier primate ancestors.

The fossil remains of the Cenozoic era are particularly abundant, furnishing excellent records of the evolutionary development of numerous species. This is especially true of the ungulates, which roamed the earth in enormous herds. Fossil remains indicating the development of the horse, camel, rhinoceros, and elephant are particularly numerous. Figure 19–36 summarizes the evolution of the modern horse from a cat-sized dweller of the forests some 60 million years ago.

The Quaternary period of the Cenozoic era began about two million years ago. In the span of these past two million years, four separate glaciations occurred. These glaciations resulted in the extinction of a number of plant and animal species and the redistribution of many more. Such mammals as the giant ground sloths, mammoths, and the saber-toothed tiger died out during this period. *Homo sapiens*, modern man, appeared for the first time. Gradually he established a communal mode of life, which rapidly increased in complexity. The roving bands of hunters domesticated a number of animals, developed techniques of agriculture, and took up a more settled way of life. From that point it was only a short hop, geologically speaking, to the development of civilization, written records, and the series of technological revolutions that ultimately led to our modern way of life.

The Evolution of Man

The Victorian world was shocked to its very foundations following the publication of Darwin's *Origin of Species*. Not that the concept of evolution itself was so appalling, but what of its implications for man! "Do you mean to say that man is descended from the apes?" was the cry. What a blow to man's ego—from the center of the universe to the grandson of a monkey!

Of course this was a misconception. It is now believed that none of the currently living species of monkeys or apes was a direct ancestor of man. Instead, we share a common ancestor. Exactly what this ancestor looked like is not known at this time. There are great gaps in our fossil records of the evolution of man. In fact, the record of man's development is poorer than those of many other species. For example, whereas literally hundreds of separate fossils representing a single species of horse or camel or elephant at a particular stage of development are available, much of our knowledge of some forms of primitive man is based upon perhaps a single jawbone or skull, sometimes augmented by finds of tools and weapons that the primitive man may have fashioned and used, and by the bones of animals he may have killed. The paucity of the human fossil record is probably due to a number of factors, including his rapid and rather recent development (probably within the last couple of million years), the fact that primitive man was never very numerous and probably lived in small groups rather than huge herds, and also man's greater intelligence, which tended to keep him out of such "fossil traps" as tar pits, oozing river bottoms, and glacier freezes.

It is now generally accepted that man shares a common heritage with other members of the primates, a group whose modern representatives include lemurs and tarsiers, the Old World and New World monkeys, and the anthropoid apes: the gibbons, orangutans, chimpanzees, and gorillas (Fig. 19–37). The immediate ancestor of the primate line was apparently a type of insectivore that took to the trees (a habitat that offered greater safety from predators and new sources of food) and by about 60 million years ago gave rise to the first primates of which we have fossil records, the tree shrews. The demands of life in the

Figure 19–37 Primates, our closest relatives in the animal kingdom: (a) squirrel monkey, (b) young velvet monkey, (c) white-handed gibbon, (d) chimpanzee.

(a) (b) (c) (d)

trees were met with some typically primate adaptations, found among all the modern forms, including primitive tree shrews that still survive in limited areas of the world such as Malaya and the Philippines. These include flat-nailed, five-fingered hands, with opposable thumbs (good for moving about in the treetops by grasping branches); eyes both directed forward and focused so that the fields of vision overlap, providing for stereoscopic, three-dimensional vision (good vision, and particularly depth perception, are excellent advantages when an animal depends for survival on its ability to judge its leaps from branch to branch); a proportionately greater brain development in comparison with other mammals (partly associated with visual coordination, but providing obvious additional advantages); and a single pair of mammary glands and a general tendency to give birth to only one or two offspring at a time (when mother is swinging through a tree, too many offspring are a severe encumbrance). The primates are also characterized by relatively long periods of dependence of the offspring, and by a high degree of reliance upon patterns of learned behavior, which must be transmitted to the young, rather than instinctual behavior, which is automatically known at the appropriate stage of development.

The early tree shrews developed into prosimian forms, resembling the present-day tarsiers and lemurs, which spread through most of the world. Geological changes resulted in a submerging of the old land bridges, and the continent of South America became separated from the rest of the world. New World monkeys with prehensile (grasping) tails evolved in South America, while in Asia and Africa part of the prosimian stock developed into Old World monkeys, which are either tailless or have nonprehensile tails. From the ancestral tree shrews through the prosimians to the monkeys there was a progressive shortening of the snout, a significant development of the brain (especially the cortex), an improvement of the focusing of the eyes, and a reduction and modification of the teeth. Most of the monkeys retained the arboreal (tree-dwelling) habits of their ancestors, but the baboons adapted to a life on the ground and developed a form of cooperative communal life that has been extensively studied.

In the Old World a second line of higher primates evolved, the anthropoid apes. The modern chimpanzee and gorilla, in particular, show a marked increase in relative skull size and brain development over the Old World and New World monkeys. They bear striking structural resemblances to man, and serological studies of biochemical similarities indicate that they are man's closest living relatives.

But man apparently is not descended from the apes, at least not from any forms resembling present-day apes. The line that was to lead to man is believed to have branched off from the common anthropoid stock at an earlier, less specialized stage.

In recent years anthropology, the study of man's origins and development, has been in a ferment. Sensational finds have been made in Africa indicating that the crucial steps from man-ape to ape-man may have been taken on that continent. Moreover, studies of animal behavior have yielded increasing evidence that not only the development of man's physical makeup, but also his behavior and perhaps even his patterns of socialization have their roots in an evolution from animal beginnings.

Numerous fossil remnants of an extremely humanlike type of anthropoid ape dating back to some 500,000 to 2 million years ago or even earlier, have now been discovered in various parts of Africa. These "man-apes" have been classed in the genus **Australopithecus;** the fossils found seem to represent two main species, *Australopithecus africanus* and *A. robustus* (Fig. 19–38). The latter, with a heavily developed chewing apparatus adapted for a vegetarian life, was apparently an evolutionary sideline, now extinct. But *A. africanus* is believed to be in the line that led eventually to modern man. *A. africanus* was somewhat smaller than modern man, and his brain was considerably smaller—about 500 cc (the brain capacity of modern man averages about 1350 cc). But his teeth were quite characteristically human, and he walked upright on

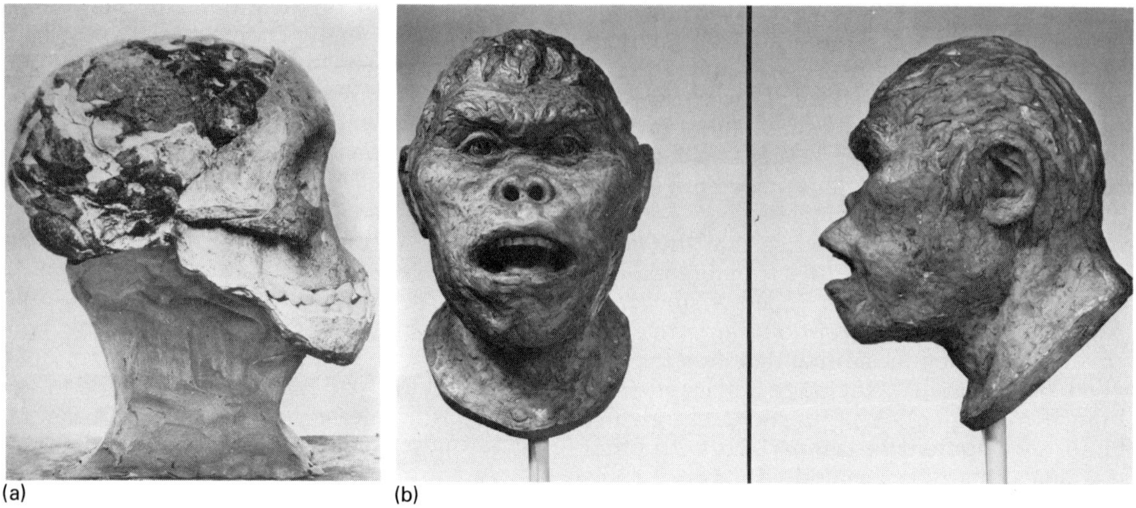

Figure 19–38 From fossil skulls, (a), anthropologists have reconstructed the appearance of Australopithecus (b).

two legs. The first discovery of an australopithecine skull by Raymond Dart in 1925 was greeted by disbelief and derision. At the time it was generally believed that the earliest man-apes or ape-men must have had large brains. Most authorities believed that "Piltdown man" represented a typical early forerunner of the human line. "Piltdown man" was later ignominiously exposed as a fraud, consisting of a modern human skull combined with a suitably modified ape jaw, but the image of a Piltdown-like ape-man lingered in the minds of anthropologists for decades in the face of steadily mounting piles of evidence. Now at last these preconceived ideas have been discarded, and the australopithecine line has been recognized as ancestral to man. Indeed, despite their small brains, the australopithecines apparently achieved quite a degree of sophistication. Remains of animal bones and other artifacts found with these fossils indicate that they lived as hunters, probably in cooperative groups, used carefully selected animal bones as tools and weapons, and may even have fashioned crude stone tools.

The discovery of an increasing number of fossils showing a continuous gradation of characteristics has made it difficult to say precisely at what point *Australopithecus africanus* evolved into the genus *Homo*. Remains of early men with sharply increased brain capacity (about 900 cc) have been found in Java and China and dated at some 730,000 years ago. Similar fossils have since been found in northern and eastern Africa and in Europe. These fossils have been classed as the species *Homo erectus*. This early man still had a smaller brain than modern man, but there is evidence that he used fire and fashioned a varied assortment of stone tools and weapons. In the early 1970s, fossils with a distinctly human appearance were discovered in Kenya, beneath a rock stratum dated about 2.8 million years ago.

The popular caricature of **Neanderthal man** (Fig. 19–39), the best known early representative of *Homo sapiens*, is grossly exaggerated. He did have heavy brow ridges and a rather protruding jaw, but he was not the shambling savage popularly supposed. He walked fully upright, and his brain was as large as or even larger than that of modern man. He lived in caves in small groups, fashioned a variety of tools, and buried his dead with ritual offerings of food and ornaments. It seems likely that the European variants of Neanderthal man, with the characteristic low forehead, large teeth, and heaviest brow ridges, repre-

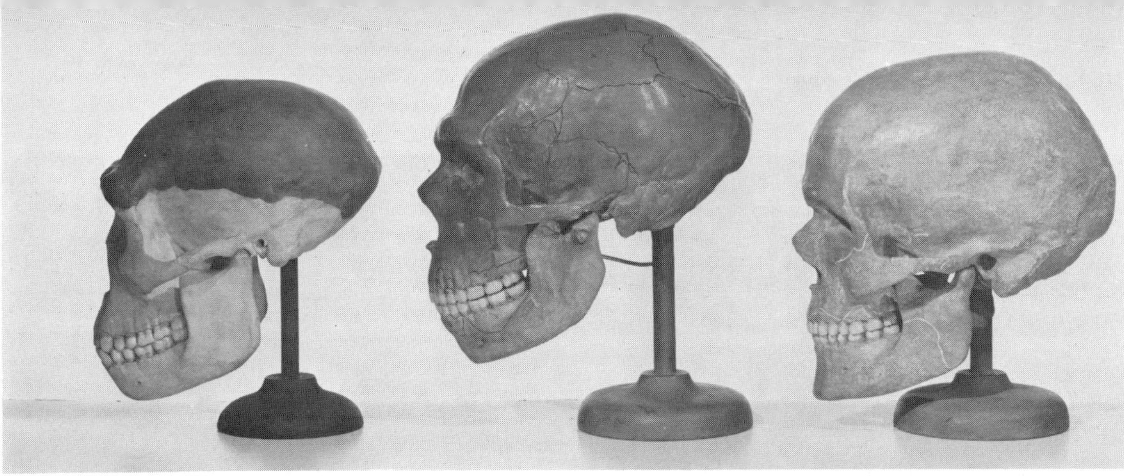

(a)

(b)

Figure 19–39 (a) Skulls of prehistoric men (left to right: Pithecanthropus, Neanderthal, and Cro-Magnon); (b) restorations of these three prehistoric men.

sented an evolutionary sideline and were not direct ancestors of present-day man. Modern *Homo sapiens* is believed to have evolved in other parts of the world and then migrated into Europe, where he supplanted the more specialized Neanderthal race.

The human stock has existed in basically its present form for at least 50,000 years. Human fossils dating that far back have been found in a variety of locations. The early man that they represent has been named **Cro-Magnon man** (Fig. 19–39), after the French cave in which such fossils were first found, but he is classified simply as *Homo sapiens sapiens,* just as is modern man. (Neanderthal man is usually given a subspecies suffix: *Homo sapiens neanderthalensis,* to differentiate him from the modern form.) Cro-Magnon man stood about six feet tall, had a high forehead without eyebrow ridges, a definite chin, and possessed a brain capacity even larger than that of modern man (about 1700 cc). He left many tools, weapons, and utensils, including bone needles, indicating that he probably fashioned clothing for himself. Cro-Magnon man is noted for his cave drawings—vivid

Figure 19–40 Cro-Magnon artists in the cave of Font de Gaume.

murals of mammoths, bison, and other animals of his time (Fig. 19–40). These murals show a precision in delineating anatomical detail that was unequaled until the time of the Renaissance and a high degree of skill in depiction.

Cro-Magnon man and Neanderthal man were contemporaries for many thousands of years. Some anthropologists believe that Cro-Magnon man exterminated his less intelligent rival; others feel that interbreeding between the two groups took place, producing various hybrid forms.

At any rate, the physical form of man had become established, and his further evolution has been a cultural evolution, which has proceeded at greatly different rates in different parts of the world. With the development of agriculture, the domestication of animals such as the dog, the pig, sheep, goat, and cow, and the adoption of a more settled way of life, man had the leisure to fashion new tools, weapons, and implements, develop pottery and weaving, and invent the dugout canoe and the wheel. Further expansion and concentration of population led to the construction of various types of dwellings of increasing strength and permanence; the discovery of the technology of metalworking led to technological revolutions that have helped to shape our modern world.

It is now generally agreed that the varied races of modern man all belong to a single species, *Homo sapiens.* Although the range of variation of numerous characteristics in man is great, the ultimate test of a species is met in virtually all cases: all the different types of present-day man are interfertile. In fact, it has been stated that probably the only "intrahuman cross" that has not yet taken place is that between the Eskimo and the African Bushman—and this has doubtless been only through lack of opportunity.

In view of the basic similarities among men, the definition of "races" of humans is largely a matter of semantics. A **race** may be defined as a population the members of which differ significantly from other populations of the same species in the frequency of one or more characteristics or traits. Generally, several major groups are recognized (Fig. 19–41). The **Caucasoids** are frequently referred to as the white race, but their skin actually ranges from pinkish to dark brown. This race shows the widest diversity and has been classified into a number of subgroups, such as Slavic,

Figure 19–41 The races of man: (a) Caucasoid; (b) Negroid; (c) Mongoloid; (d) Australoid.

Mediterranean, etc. The second major race, the **Negroid** race, has a typically darkened skin, which is believed by some scientists to have been the result of a mutation from an original white-skinned stock some time in the long-distant past.* The **Mongoloid** race represents the third major race. The group includes not only the Oriental stocks but also the North and South American Indians, who are believed to have migrated over to the North American continent via the Alaskan land bridge some time in the late Pleistocene epoch. Finally, some anthropologists view the Australian aborigines as a separate racial stock, the **Australoids.** They show many primitive features, including prominent brow ridges and a small brain (averaging about 1200 cc)* and are thought to have descended with little change from the original *Homo sapiens* stock.

*Another interpretation holds that the light-skinned races arose from dark-skinned ancestors, through a mutation resulting in reduced melanin production. The melanin deposits in the skin of dark-skinned races helps to screen out the strong ultraviolet radiation characteristic of Equatorial regions. The lack of melanin in the skin of fair-skinned inhabitants of far northern regions permits their bodies to make maximum use of the weak sunlight of those regions for the synthesis of vitamin D.

*However, there is no evidence that the Australoids are any less intelligent than other races. Indeed, there is no scientific evidence to indicate any inherent intellectual superiority of any of the races of man in comparison with any others.

The definitions of human races are, to a large extent, arbitrary. It is generally recognized among anthropologists that there are no "pure" races. To a large extent the gene pools of the various groups of men have been intermingled for thousands of years. In view of this, it is ironic and unfortunate that racial tensions are among the most persistent and potentially dangerous problems that plague man today.

QUESTION FOR REVIEW AND THOUGHT

1. If Charles Darwin did not originate the theory of evolution, what was the significance of his contribution?
2. Might you expect two paleontologists working independently on a strange life-containing planet to come up with basically the same interpretation of the fossil record? Explain.
3. Compare man with a mosquito and a rat with respect to analogous and homologous structures found in these three organisms.
4. Considering civilized man's present mode of life and the current trends in this mode of life, what presently functional structures might be expected to evolve into vestigial structures?
5. Explain the statement: "Ontogeny recapitulates phylogeny." Is this statement literally accurate?
6. Discuss and assess the importance of the biochemical evidence for evolution.
7. What is meant by random genetic drift? Is this a positive or negative factor in progressive evolution? Explain.
8. Are there still opportunities for evolution through adaptive radiation in the modern world? Explain.
9. Trace the evolution of the major groups of animals and plants through the various geological eras.
10. Is natural selection operating among humans today? Explain.

PART VI

OUR MODERN AGE

CHAPTER 20

Atomic Energy and Radioactivity

Introduction

For perhaps the first time in history, man has the potential of annihilating both himself and all life on his planet. A miscalculation, a political impasse, or some other unforeseen event might precipitate a nuclear war—a war to end all wars. Yet, paradoxically, the same discoveries that have given man this awesome power have also placed him at the threshold of a new era of ease and plenty, with the potentiality of tapping virtually limitless resources of energy and raw materials.

These vast new energy resources are stored within the nuclei of chemical atoms—the basic units of matter that cannot be broken up by conventional means. It was barely a generation ago that man first managed to split atoms and ushered in a new age, the full implications of which may not be grasped for many years to come.

The atomic age was born in a ball of fire and a swelling, mushroom-shaped cloud. The product of a "crash program" of wartime research, the first applications of nuclear science were destruction of lives and property. But after World War II ended, although research toward the development of new and better nuclear weapons continued, there was a gradual shift toward seeking ways of harnessing the energy of the atoms for peaceful purposes.

Figure 20–1 The atomic bomb.

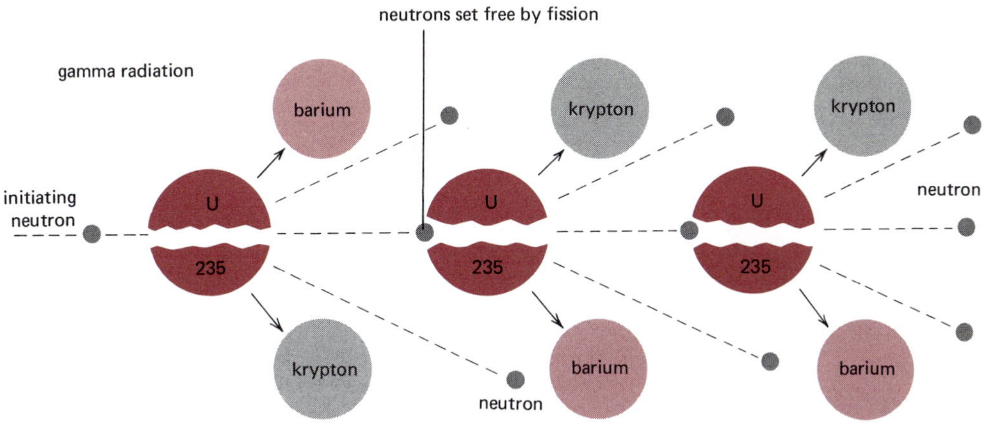

Figure 20–2 A nuclear chain reaction. Each fission of a uranium-235 nucleus liberates neutrons that can spark the fission of others.

Nuclear Fission

Nuclear fission, or the splitting of an atom's nucleus into two more or less equal fragments, may occur spontaneously in the case of the heaviest elements or may be induced by the capture of bombarding particles, such as slow-moving or **thermal neutrons.** Large amounts of energy are released in the process, and the new nuclei formed are referred to as **fission products.**

Not all elements can undergo nuclear fission. Uranium-235 is the only naturally occurring **fissionable material** (material capable of capturing neutrons and then splitting into two particles with great kinetic energy). Another naturally occurring isotope of uranium, U^{238}, is classified as a **fertile material.** Although it does not itself undergo fission, when it captures a thermal neutron it is converted to a radioactive isotope which rapidly decays to a fissionable isotope of another element, plutonium-239.

Nuclear fission can yield a number of different isotopic products, depending on how the nucleus is split. For example, the nuclei of uranium and plutonium can split in forty or more different ways during fission, and as a result the detonation of a nuclear fission bomb results in the production of eighty or more different fission products.

In the fission of uranium or plutonium, between two and three neutrons are released. Each of these can induce a further fission, and thus a **chain reaction** can be started. In such a chain reaction, fission would be induced by neutrons liberated in the previous step of fission, and the nucleus in splitting would yield more neutrons, which would in turn induce further steps of fission (Fig. 20–2). Since each individual fission reaction yields two or more neutrons, and thus can theoretically result in the initiation of two or more fission reactions, it is easy to see that such a fission chain reaction would readily turn into a runaway process, resulting in an explosion. But not all captures of neutrons result in fission. Some merely reduce the kinetic energy of the neutron. Moreover, if the mass of fissionable material is too small, many neutrons will escape from its surface before they have a chance to be captured. Physicists have calculated the **critical mass*** required to keep a nuclear chain

*In a **fission bomb,** the mass of fissionable material is separated into two parts, each less than the critical mass individually, but together exceeding the critical value. They are brought together at the proper moment by an explosive charge of TNT. Only then can the energy-yielding chain reaction start.

Figure 20–3 The nuclear reactor at Oak Ridge, Tennessee.

reaction running smoothly, and through suitable reactor design such self-sustaining fission reactions can be controlled and kept going at a steady rate.

In the **nuclear reactor** designed in the massive World War II research effort (Fig. 20–3), the fissionable material is in the form of cylindrical rods embedded in a large mass of graphite. Graphite (a modification of pure carbon) is a light element that has a negligible tendency to capture neutrons, but collisions with its nuclei slow them down. Graphite thus tends to moderate the energy of the neutrons, and as a result is called a **moderator.** In addition, movable rods of a different substance are also embedded in the graphite block. These are the **control rods,** and they are made from a substance such as boron or cadmium that has a high **neutron-capture cross section**—that is, a high tendency to capture neutrons. These control rods can be inserted to various depths in order to regulate the rate of the nuclear chain reaction by absorbing the excess neutrons and thus determining the number that will be available for captures resulting in fission.

Nuclear reactors are used for a variety of purposes. The energy that is released in the fission reaction can be converted to electrical energy and other useful forms of power. Reactors are also used for the production of a variety of isotopes, many of which do not occur in nature and can be obtained only in fission reactions. These isotopes are then used for research, in the medical therapy of cancer and other diseases, and for numerous other purposes. Moreover, reactors can be constructed with special channels or "holes" in which biological specimens and various other materials can be placed to study the effects of radiations on them.

Nuclear Fusion

A bucketful of water contains enough energy to propel a fleet of ships around the world. In laboratories throughout the world, man is learning to tap this reserve of energy for useful purposes. This energy can be released through reactions of nuclear fusion.

In nuclear *fission* reactions, the nucleus of a heavy atom is split to yield fragments, nuclei of atoms of lower atomic weight. Nuclear reactions essentially opposite to fission are also possible, in which nuclei of light atoms *fuse* to form new nuclei of heavier atoms. For example, two hydrogen nuclei (protons, $_1H^1$*) can fuse to form a deuterium nucleus (deuteron, $_1H^2$). Deuterium nuclei in turn can fuse, liberating still larger nuclei and energy. Other important simple fusion reactions involve lithium nuclei and yield an impressive release of energy.

The energy released in nuclear fusion reactions comes from a conversion of mass as predicted by Einstein in his famous formula, $E = mc^2$ (where E is the energy released, m is the amount of mass that has disappeared, and c, a "constant," is the speed of light —186,000 miles/sec). In each of the reactions depicted, the mass of the product nucleus (or nuclei) is slightly smaller than the sum of the

*In chemical notation, the mass of an isotope is indicated by a superscript and its atomic number by a subscript. Thus, $_1H^1$ has one proton and no neutrons in its nucleus, whereas $_1H^2$ has one proton and one neutron (for a total atomic mass of two). Both are varieties of hydrogen, as evinced by the atomic number one.

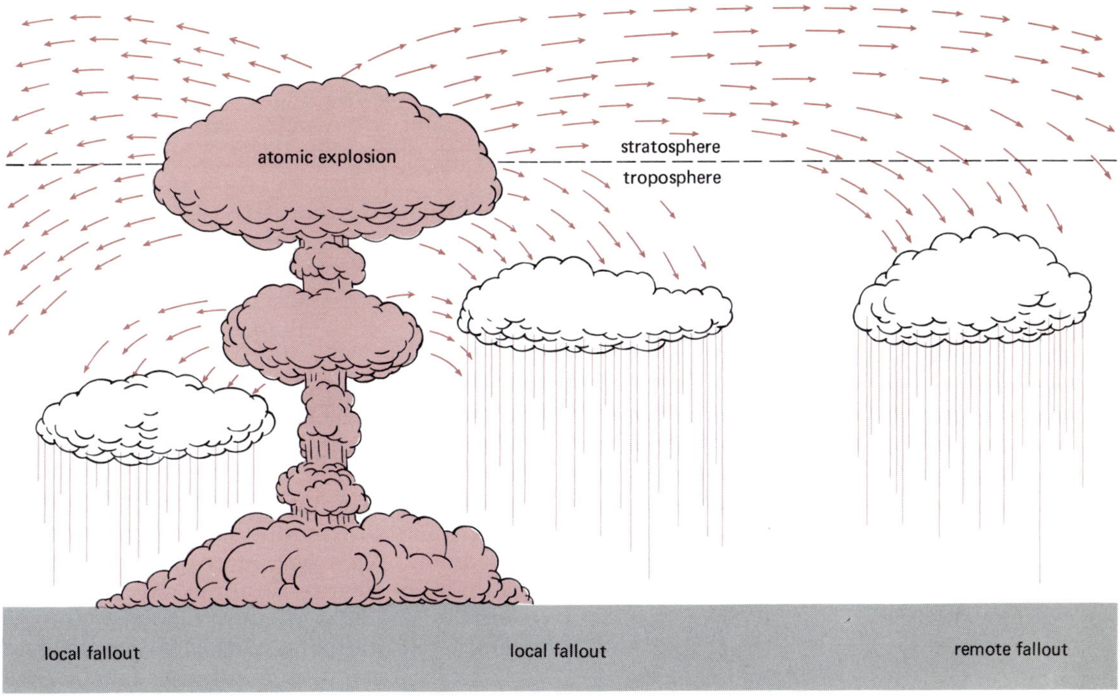

Figure 20–4 An atomic explosion can produce immediate contamination of local land areas (local fallout) and delayed contamination of both local and distant areas resulting from contamination of the troposphere and stratosphere.

masses of the nuclei that fused to form it. The mass difference is minute, but is converted to a substantial yield of energy. Pound for pound, fusion reactions involving deuterium isotopes release three times as much energy as the fission of uranium or plutonium, and more than a million times the energy released in chemical reactions by typical chemical fuels such as gasoline.

It would seem that nuclear fusion reactions hold a great potential as sources of usable energy. But there is a "catch": such reactions require the input of enormous energies, such as are supplied at the superhot temperatures that prevail in the sun and other stars. In fact, it is believed that nuclear fusion reactions supply the tremendous amounts of energy that stars radiate out into space. Because of the high temperatures required—temperatures at which matter is ionized to the special state known as plasma—nuclear fusion reactions are frequently referred to as **thermonuclear reactions.** Until the development of nuclear fission devices, such temperatures could not be produced on earth at all. But the temperatures generated in the fission of uranium are of the order of millions of degrees. Thus, it is not surprising that the first hydrogen fusion reactions were carried out in devices sparked by the detonation of a fission bomb.

After fusion explosions had been achieved, research efforts were concentrated on achieving controlled nuclear fusion that could be harnessed as an energy source for peaceful uses. The problems involved were threefold: (1) to produce the temperature necessary to ignite the fusion chain reaction (a few hundred million degrees), (2) to maintain a suitable plasma density (perhaps 1000 to 10,000 times less than that of atmospheric air), and (3) to contain the reaction system in some way for at least a few tenths of a second (ordinary container materials could not withstand the enormous energies involved). At first scientists and engineers were highly optimistic about the prospects for harnessing fusion power; then as the magnitude of the problems involved were realized more clearly, many experts grew doubtful that this goal could be achieved at all. More recently encouraging progress has been made, and a cautious optimism is growing again.

Radioactive Fallout

The blazing fireball and blasting shock wave of an atomic explosion are followed by a more insidious but equally deadly effect. The dust particles that drift down to the ground from the hovering mushroom cloud have been infected with radioactivity by the fission products released in the explosion. These radioactive particles, together with the fission fragments, are now classified as **local fallout.** Invisibly the radiations from the fallout particles pervade the area. Unlike the other immediate effects of the explosion, their presence can be detected only with radiation detectors such as Geiger counters—and by their subsequent effects on living organisms exposed to them.

The immediate fallout effects, resulting in a rain of radioactive debris down upon the area of the explosion, is not the whole problem, however. Only part of the radioactive particles are rapidly deposited after the explosion. Another portion remains in the troposphere for periods of hours, days, or even months. Other **radionuclides** (radioactive isotopes whose formation is induced by bombardment by neutrons released in the nuclear explosion) are carried up into the stratosphere and may be retained there for as long as five years. Both the tropospheric and the stratospheric radionuclides serve as reservoirs for future fallout. They may be carried over great distances by winds, and then they may gradually settle out or may be collected in cloud formations and then washed down by rain or snow, resulting in heavy episodes of fallout far from the site of the original explosion.

The nature of the fallout radionuclides and the amount of radiation that they emit change with time. From the instant immediately following the detonation to a period twenty-four hours later, the radioactivity drops by a factor of more than 6000. This drop is due largely to the disintegration of fast-decaying isotopes, and the rate of decrease subsequently slows down considerably as they are exhausted. The effects of the longest-lived radionuclides, such as strontium-90, cerium-144, cesium-137, and ruthenium-106, continue to be significant for up to seventy years.

Explosions of nuclear weapons, both during the Second World War and more substantially in subsequent extensive testing of various nuclear devices, have significantly raised the level of background radiation all over the world. Increases in this level have been recorded after every test; after periods of moratorium on nuclear testing, the radiation levels gradually settle down, particularly a few years after the cessation of testing. At various times, in various locations, radioactive fallout from nuclear explosions has resulted in radiation levels high enough to threaten the health and lives of the population. But even where the levels are not great enough to offer a direct hazard to life, the increased radioactive background is still a cause for concern. There is no dose threshold for the effects of radiation on living organisms—that is, a single exposure, no matter how small, may be enough to produce some change in living tissue. No one really knows what the ultimate effects of even the present radiation levels may be. That is why groups of concerned scientists in many countries have banded together in efforts to halt uncontrolled testing of nuclear weapons.

In 1962, the United States, the USSR, and Great Britain agreed to a limited nuclear test ban, barring all but underground tests of nuclear weapons. But the newer members of the "nuclear club," France and China, have stubbornly refused to enter into the treaty agreement.

The radionuclides deposited by fallout may be inhaled by humans from the atmosphere or may be picked up through contact with contaminated objects. But this is not the only or

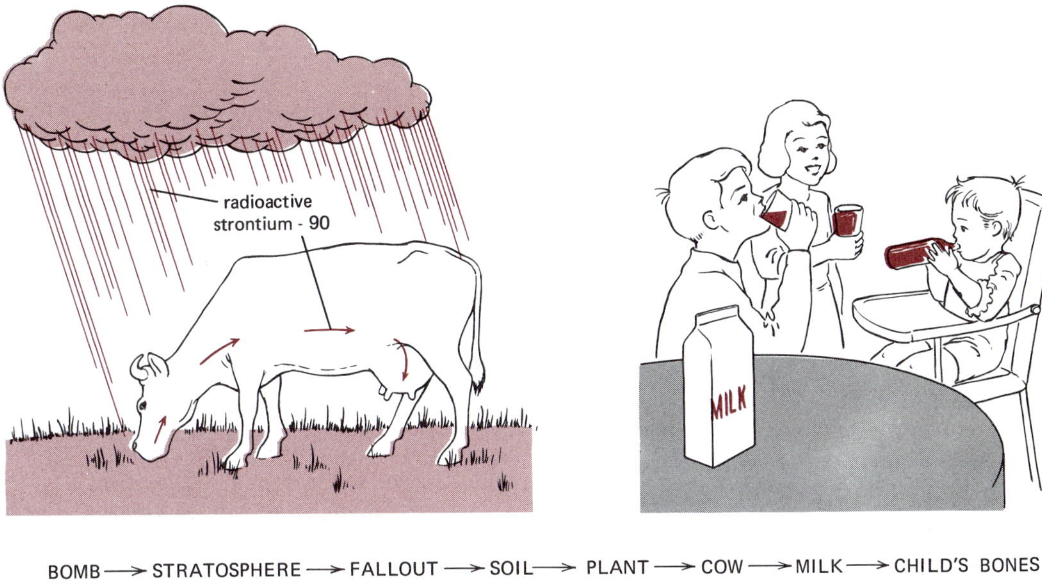

Figure 20–5 Radioisotopes such as strontium-90 can be transmitted along food chains, ultimately to man.

even the major source of hazard for man. Fallout products are distributed through the biosphere along a number of intricate pathways referred to as food chains (Fig. 20–5). Radioactive particles fall upon the soil or the ocean and then are taken in by plants. Animals eat the plants, and then man may eat meat from the animals or consume animal products such as milk. The isotopes transmitted most effectively along the food chains are iodine-131, barium-140, strontium-89, strontium-90, and cesium-137. The iodine-131 hazard, transmitted through food contamination, is most significant during the early period after the explosion; with a half-life of only 8 days, its proportion among the radionuclides rapidly dwindles to insignificance. A serious hazard is presented for far longer periods by strontium-90, both because of its long half-life (25 years) and because of its chemical similarity to calcium. Like calcium, strontium tends to be transmitted in milk in large amounts. Even more important, like calcium, strontium is deposited in bone, a tissue characterized by a relatively low metabolic turnover. There it remains for years, subjecting both the bone and the other tissues of the body to irradiation.

Effects of Radiations on Living Organisms

On March 1, 1954, a thermonuclear device was detonated by the United States at the Eniwetok Proving Ground in the Pacific Ocean. This routine test had an unfortunate sequel that immediately focused world concern on the problem of the advisability of nuclear weapons testing in general. Although warnings had been set up, a Japanese fishing trawler, ironically named *Lucky Dragon,* blundered directly into the path of a "hotspot" of tropospheric fallout. The trawler's catch of fish was seriously contaminated with radioactive particles and had to be destroyed. The twenty-three members of the crew suffered radiation burns and showed various other signs of radiation sickness, such as lowered blood counts. Meanwhile a shift of the wind carried the radioactive cloud over the Marshall Islands, where there was another episode of fallout. A number of inhabitants of the islands received small to moderate doses of radiation and had to be evacuated to a neighboring island until contamination had been eliminated and their homes had been rebuilt. The islanders suffered various symptoms ranging from nausea and vomiting to burns and loss of hair. Follow-up studies by visiting medical teams revealed few lasting effects of the radiation exposure, but these two incidents dramatically focused world attention on the hazards of radiation.

THE NATURAL RADIATION BACKGROUND

With the present concern about the consequences of the increased radiation levels due to the testing of nuclear weapons and possible accidents at nuclear power installations, it is easy to lose sight of the fact that actually a certain amount of radiation is and has always been a quite normal part of the environment of our planet. Radiation is an influence to which life has been exposed ever since it appeared on earth; indeed, it is believed that the origin of life and its subsequent development have been greatly influenced by the effects of radiation.

The radiation naturally present (i.e., that which was characteristic of our planet even before the first nuclear device was detonated) is referred to as the **natural radiation background.** It arises from several sources: from outer space (in the form of cosmic rays) and from the disintegration of naturally occurring radioisotopes found in the rocks of the earth's crust, in building materials, in soils, and in the air of our atmosphere. Common naturally occurring radioisotopes include potassium-40, carbon-14, radium-226 and its decay products (particularly radon, a radioactive gas), and thorium-232 and its decay products (including thoron). The naturally occurring uranium isotopes, uranium-235 and uranium-238, were used as materials in the production of the first atomic bombs. The natural radiation remains relatively constant in time, but it varies from place to place, according to the local conditions of soil, rock, etc.

FACTORS DETERMINING RADIATION EFFECTS

The effects of radiations—whether the radiations of the natural background or those released in nuclear explosions or by other artificial means—upon living organisms depend on a number of factors. First of all, the amount and kind of radiation are important determining factors. Although all types of ionizing radiations, such as X and gamma rays, alpha and beta particles, protons and neutrons, are capable of producing similar effects on biological systems, the particular dose that will produce a particular effect varies greatly from one type to another. For each type of radiation, a **relative biological effectiveness (RBE)** is determined, comparing its effectiveness with that of X rays. Similarly, not only the amount of radiation (the **dose**) received by the organism, but also the length of time of exposure, the **dose rate** (the amount of radiation delivered per unit time), and whether the dose was delivered all at one time or was divided into fractions—all influence the ultimate effects of radiations of the organism. In the latter case, it has been found that the same dose delivered in one session produces a substantially greater effect than the same total dose fractionated into two or more sessions with a lapse of time between them. The difference is greater the longer the intervening time between irradiations.* It has been found that during this time between successive exposures to radiation, various repair processes occur in the irradiated organism, permitting it to recover to some extent from the effects of the first irradiation. But it has also been observed that generally, no matter how long the interval between irradiations is made, there is some residual effect from the first exposure, a certain component of irreversible damage to the organism. This is an important factor in cases

*__Radiation__ is the term used for the actual ionizing rays; **irradiation** refers to the exposure of a biological system or other object to radiations. One speaks, for example, of the "radiation dose" and the "time of irradiation."

of chronic exposure to radiation, when after a long series of exposures effects may be manifested even though the dose received in a single exposure is not great enough to produce any observable effects.

Another important factor in determining the effects of radiations on living organisms is whether the whole organism is subjected to irradiation **(whole-body irradiation)** or whether only part of the body is exposed to radiations—and, moreover, which part (Fig. 20–6). For different organs and tissues differ in their **radiosensitivity** (their susceptibility to the effects of radiations) and in their importance to the life of the organism. For example, a dose sufficient to produce acute radiation sickness if delivered to the whole body may have very little effect if concentrated only on the hand. In mammals, the germ cells and rapidly proliferating cells such as the blood-forming components, the lining of the digestive tract, and the skin are most radiosensitive, while mature bone and muscle cells and the components of the central nervous system are more radioresistant. There are also wide differences in radiosensitivity among different species and among individuals of the same species. For example, warm-blooded animals are generally more sensitive to radiations than cold-blooded organisms, and plants are generally more radioresistant than animals. Extremely young animals are more radiosensitive than mature adults, but after maturation there is usually an increase in radiosensitivity with advancing age. Seeds and plants in the dormant phase are quite resistant to radiations, while germinating seeds and plants in phases of active growth are considerably more radiosensitive. Thus, the plants of an area might survive massive irradiation during the winter whereas damage would be severe if a similar dose were delivered in the spring.

Environmental factors are also important in determining the degree of radiosensitivity. It has been found that the presence of water enhances the effects of radiation (the extreme dehydration of the dormant seed is one of the major factors in its high radioresistance), and these effects are also increased by the presence of oxygen. Radiosensitivity also

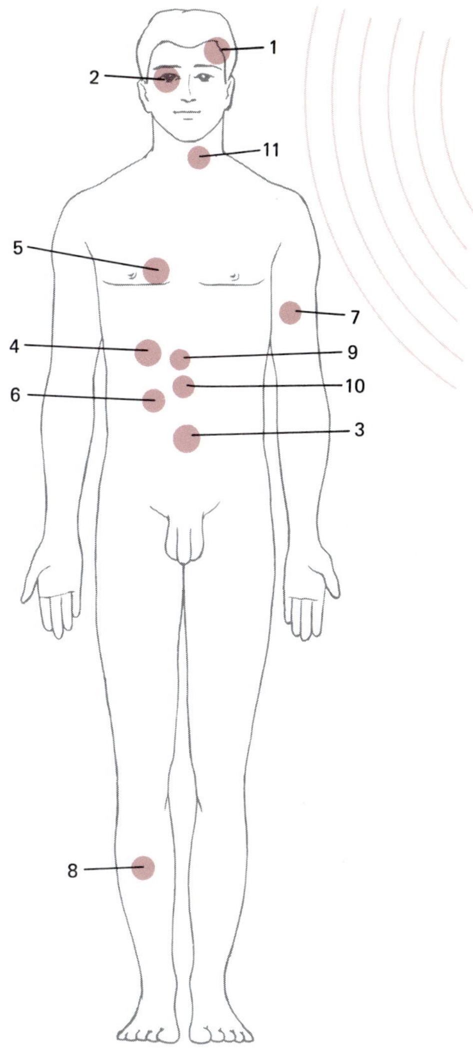

Figure 20–6 If only part of the body is subjected to irradiation, the consequences will depend to a great degree on the individual radiosensitivities of the body structures exposed to the rays. Some key organs and systems: (1) skin and hair; (2) eyes; (3) alimentary canal; (4) liver and gall bladder; (5) lungs; (6) urinary tract; (7) bones; (8) blood and bone marrow; (9) lymphatics and spleen; (10) adrenal glands; (11) thyroid gland.

Figure 20–7 "Hiroshima Maidens," disfigured by the atomic blast, at their arrival at New York where they received free plastic surgery.

increases with increasing temperature, but this seems to be linked with the increased metabolic rate rather than being an intrinsic effect of temperature itself.

BIOLOGICAL EFFECTS

The biological effects of radiation may be divided into two broad categories: somatic and genetic. The **somatic effects** are those manifested in the affected organism itself and range from acute radiation sickness to such chronic effects as premature aging, a nonspecific shortening of the total life span, and the development of cancers of various types. The **genetic effects** of radiation consist of mutations induced in the germ cells and are manifested in future generations.

When humans are subjected to whole-body irradiation at a dose level higher than 100 rads,* they may develop a complex of symptoms known as the **acute radiation syndrome.** The time of onset of the symptoms and their degree of severity are generally determined by the magnitude of the dose. The symptoms of the acute radiation syndrome may include vomiting, reddening of the skin, later a loss of hair, a drop in the white blood cell count, and the appearance of small or massive hemorrhages (at higher doses). The patient generally experiences a great feeling of fatigue, and temporary or permanent sterility may result. At higher doses, the number of deaths mounts sharply; at doses of about 500 rads, 50 percent of untreated patients die.

In contrast to the acute radiation syndrome, the first symptoms of which may appear within hours after exposure, the **chronic effects** of repeated exposure to lower doses may take months or even years to be manifested. They include a general shortening of the life span, lowered fertility, an impaired growth rate, increased incidence of tumors and leukemia, and the development of cataracts and circulatory diseases. In the early years after the discovery of the phenomenon of radioactivity the dangers of radiation were not realized, and little effort was made to minimize the doses delivered, for example, in fluoroscopic

*The **rad** is defined as 100 ergs of absorbed energy per gram of absorbing material. The name of the unit is actually an acronym for "*r*adiation *a*bsorbed *d*ose."

diagnostic procedures, or to protect the operators of such devices from scattered radiations. X rays were even used for such trivial purposes as fitting shoes by fluoroscoping the feet of the prospective purchaser. As a result, a high incidence of leukemia and other chronic effects of radiation exposure was eventually discovered among X-ray technicians. Indeed, Marie Curie, the discoverer of radium, died of leukemia, probably as a result of her researches with radioactive substances. When the dangers inherent in radioactivity were realized, measures were taken to minimize the doses absorbed by the patient and the number of exposures, and focusing cones and appropriate shielding were introduced to protect the X-ray personnel. The fluoroscope machine disappeared from the shoe store. New hazards continue to pop up, however, as in the recent wave of concern provoked by the discovery that certain types of color television sets were emitting potentially dangerous levels of X rays.

The genetic effects of radiation are due to mutations induced in the chromosomes of the germ cells by exposure to ionizing rays. Although the great majority of mutations, radiation-induced or otherwise, are harmful to the organism, some do result in improvements in characteristics of the species. Radiation has therefore been used fruitfully for the selective breeding of improved varieties of agricultural plants.

It should be emphasized that radiation-induced mutations are not a new phenomenon in our world. The natural radiation of our planet, described at the beginning of the section, has been inducing mutations in living organisms ever since life appeared on earth. Indeed, it is believed that such mutations have been a prime factor in the evolution of the varied life forms that now inhabit our planet. However, man's activities in the past century have been adding both to the general radiation background and to the radiation dose received by the human population. Each episode of radioactive fallout, each diagnostic X ray contributes its own bit to a person's lifetime dose. Since a certain portion of the effect of radiation is irreversible and cumulative, and since there is believed to be no threshold for the genetic effects of radiation, this is a factor to be reckoned with. It is estimated that a doubling of the mutation rate in the world's human population would probably occur if the dose were increased by some 10 to 80 roentgens* per generation.

MODE OF ACTION OF RADIATIONS ON BIOLOGICAL SYSTEMS

Ionizing radiations can exert their influence on living organisms in two ways. The organism may be exposed to **external radiation,** as in X-ray diagnosis or therapy, exposure to a nuclear explosion, and irradiation in the biological channels of a research reactor or by an irradiation unit used for cancer therapy. In all these cases there is a single exposure to radiation, and its direct effects last only as long as the direct exposure is continued.

However, in certain cases radioisotopes may be taken into the body. This may occur through the respiratory tract, for example, by the inhalation of air contaminated with radioactive fallout products, or through the alimentary tract, as by eating contaminated foods. Radioactive contaminants may be absorbed through the skin or mucous membranes of the body, and radioisotopes may be injected as a form of therapy for various medical conditions. In all these cases, when radioisotopes are incorporated into the body, these isotopes subject the organism to continuous irradiation until the last traces of radioactivity either are eliminated from the body by normal processes of excretion or decay to nonradioactive products.

The effects of **internal radiation** are greatest in the areas surrounding the site of concentration of the radioisotope (Fig. 20–8). The different radioisotopes tend to be deposited primarily in specific types of organs or tissues, which depend on the metabolic peculiarities of the particular element involved. For example, strontium-90, which has a chemical re-

*A **roentgen** (R) is defined as the quantity of X or gamma radiation producing ions carrying one electrostatic unit of electricity in 0.001293 gram of air.

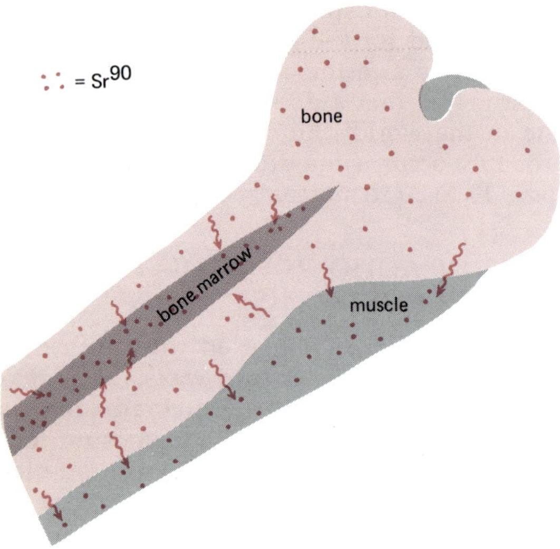

Figure 20–8 Strontium-90, deposited in bone, produces gamma irradiation of body tissues throughout a person's whole lifetime. Effects are strongest in the bone marrow, causing damage to developing blood cells and platelets, resulting in anemia and reduction of oxygen transport in the blood, lowering of defenses against disease, and hemorrhaging due to lack of platelets; and in the bone, causing malignant bone tumors and bone necrosis (or decay). Gamma rays emitted by deposited Sr^{90} may also penetrate into surrounding soft tissues, affecting dividing cells and causing mutations and tumors.

semblance to calcium, is deposited mainly in the bone, while cesium-137, which resembles potassium, tends to be concentrated in muscle tissue. The rate at which a radioactive substance is eliminated from the body depends not only on such physical factors as the size of the particles and their solubility and on the means by which they were taken into the body, but also on the metabolic rate of the tissues in which they were deposited and on the balance of other chemical substances in the diet and in the organism. For example, the rate of elimination of radioactive strontium from the body is influenced by the amount of calcium present. The complex of these factors determines the **biological half-life** of each radioisotope (the time required for half the dose of substance to be eliminated from the body or from the tissue or organ being considered). The biological half-life thus is quite different from the radioactive half-life of an isotope (the time required for half of it to disintegrate by radioactive decay), and an isotope with a long radioactive half-life may have a short effective half-life in the body if it is rapidly eliminated.

The effects of radiations on living systems may be exerted in a number of ways. Ionizing radiations, as their name implies, interact with matter to produce various ionized and excited particles. This action may break down critical molecules and structures in the cell. In addition, the ionized and excited particles may themselves interact with the matter of the cell to produce further changes. Such indirect effects are produced particularly through the effects of radiations on water, which is normally a major constituent of living cells. The free radicals, peroxides, and other highly reactive particles resulting expand the initial effect of the incident radiation. The effects of radiations on living tissues may also result in the production of toxic substances. These **radiotoxins** continue to exert their deleterious effects on the organism for some time. A high concentration of oxygen dissolved in the tissues of the body also enhances the effects of radiation, apparently by contributing to various oxidation reactions, yielding reactive peroxides and other substances.

PROTECTION FROM THE EFFECTS OF RADIATION

As scientists have learned more about the mode of action of radiations on living organisms, they have been able to devise means of providing protection from these effects. One of the most obvious means of radiation protection is the use of **shielding,** which prevents the radiations from reaching the organism or part of it. The type of radiation shield needed depends on the type of radiation involved, particularly its penetrating power. Alpha particles produce very high concentrations of ion pairs in their path, but they dissipate their energy very rapidly. As a result, only those alpha particles with unusually high energy can penetrate even through human skin; thus, external alpha radiation presents a negligible hazard, and shielding is not needed. Beta particles can be stopped by a quarter-inch of plastic or wood. Depending on their energy, X rays may be stopped by lead-coated gloves or apron, or may require a heavy concrete shield. Thick shields of concrete, iron, or lead are also required for gamma rays and neutrons. Even if the whole body cannot be shielded, the protection of certain critical organs can be extremely helpful. For example, recovery from the bone marrow destruction produced by heavy irradiation can be enhanced by shielding the legs. Cells from the protected bone marrow later migrate to other parts of the body and replace the destroyed blood cells. Shielding plays an important protective role not only in the operation of nuclear reactors and X-ray machines, but also in radiation therapy of cancer and other diseases in which the irradiation is strictly limited to the area requiring treatment (Fig. 20–9).

Research has also revealed the efficacy of **anoxia** (a reduction of the oxygen supply) in minimizing radiation effects. One of the most promising areas of study is the use of **chemical protectors.** Certain chemicals, particularly those containing free sulfhydryl groups (—SH) and those with both a free sulfhydryl group and an amine group (—NH_2 or —NR_2), have been found to provide some protection from the effects of radiations, reducing the mortality rate and the severity of symptoms. Most of these chemical radioprotectors are effective only when administered before or during irradiation. But a few seem to offer limited protection when administered shortly after irradiation as well. Other chemicals are used to prevent the absorption of incorporated radioisotopes and enhance their elimination from the organism. An example is the use of aluminum hydroxide, which binds radiophosphorus in an insoluble compound that is not absorbed in the gastrointestinal tract.

Figure 20–9 A betatron in action at the Boston University Medical Center. The instrument can concentrate radiation at the site of a cancerous tumor.

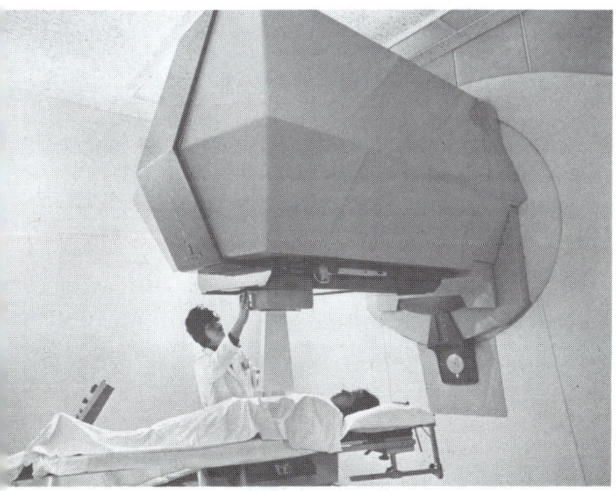

Peaceful Uses of Atomic Energy

The preceding sections dealt mainly with harmful effects of radiations on living systems. However, atomic energy and radioactivity have already been used to provide great benefits, and it is hoped that in future years the peaceful uses of atomic energy will expand far more.

The use of radioisotopes in tracer studies was discussed in Chapter 4. By "tagging" specific portions of molecules with labeled atoms, researchers have been able to study specific aspects of metabolism both in plants and in animals. For example, calcium-45 and phosphorus-32 have been used to study the processes of bone accretion and resorption, the formation of teeth, milk and egg formation and transfer through the placenta, and the function of the kidneys and excretory mechanisms. Such isotopes as tritium, carbon-14, sulfur-35, iodine-131, iron-59, copper-64, chlorine-36, potassium-40, cobalt-57, -58, and -60, sodium-22 and -24, zinc-65, strontium-85 and -89, and so forth have been similarly used, both for metabolic studies, and in some cases for medical diagnosis and radiotherapy. Iodine-131, for example, is used routinely in assaying the function of the thyroid gland, in scanning for thyroid tumors, and in the treatment of certain types of thyroid cancer.

Figure 20–10 Inside the sealed chamber, tobacco plants are being grown in a controlled environment containing an isotope of carbon. The "tagged" plants will be studied to trace certain components of tobacco smoke.

Carbon-14 is particularly useful in a variety of metabolic studies, both on plants and on animals.

In a typical plant cell study, growing plants are placed in a closed growth chamber (an "isotope farm," shown in Fig. 20–10) with an atmosphere containing "labeled" carbon dioxide ($C^{14}O_2$). In the process of photosynthesis, the plants take in the labeled carbon dioxide, and then, as it undergoes various conversions in the plant, the label is incorporated into a variety of chemical compounds. (Indeed, it was the use of C^{14} tracer studies that revealed the basic sequence of biochemical steps in the process of photosynthesis.) As the plant continues to grow in the chamber, the radioisotope is incorporated into such substances as alkaloids (basic nitrogen-containing substances such as caffeine, morphine, nicotine, quinine, and reserpine), some of which are used as drugs. The labeled alkaloids can then be extracted from the plants and administered to animals, and then their effects and metabolic conversions in the body can be followed.

The induction of mutations by ionizing radiations has been successfully utilized in the selective breeding of agricultural plants and of microorganisms used in various industrial processes. Although most mutations are harmful or neutral, occasionally a valuable change is induced. This change can then be enhanced by suitable breeding. Thus, such plants as barley, oats, wheat, rice, beans, peas, peanuts, clover, tobacco, flax, and others have been treated with radiations, sometimes in conjunction with chemical mutagens. Mutants with such desirable characteristics as resistance to disease and insect pests, a size more convenient for harvesting, earlier or later ripening, improved quality and/or quantity of yield, and increased hardiness have been isolated (Fig. 20–11). Some of the new variants obtained have been introduced into agricul-

Figure 20–11 Radiations: tools of the plant breeder. (a) The shade house at the UT-AEC Agricultural Research Laboratory in Oak Ridge, Tennessee, where irradiated plants are examined for useful mutations; (b) research biologists at the Brookhaven National Laboratory examine a new hybrid tobacco plant produced by a technique of cell fusion.

(a)

(b)

tural practice and are rapidly supplanting the traditional varieties.

The use of radioisotopes and radiations in agriculture has not been limited to tracer studies of metabolism and selective breeding of plants and animals. Radioisotopes have also been helpful in the study of soils and the processes of decomposition and exchange that occur in them, in the study and measurement of water sources, in veterinary medicine, and in the study and control of insect pests. In the last area, radiation sterilization of insects has proved to be a promising method of pest eradication without affecting any other organisms in the biological community or leaving any harmful pesticide residues. The procedure was shown to be feasible in a landmark project on the screwworm fly, *Callitroga hominivorax* (Fig. 20–12). Females of this fly lay eggs in cuts or wounds of warm-blooded animals, such as sheep and cattle. The larvae hatch out and burrow into the flesh, feeding on it and killing the animal. Losses due to the depredations of the screwworm fly used to amount to more than $20 million annually in the southeastern United States alone. After field studies and laboratory experiments, a pilot program was tried on the island of Curaçao. Male flies were exposed to a cobalt-60 source at a dose that sterilized them but still left them able to mate. Then these male flies were released on the island over a period of several months.

Figure 20–12 The screwworm fly.

The sterilized male flies competed with the native normal flies and mated with the females. The screwworm fly population rapidly declined, as successively fewer matings resulted in viable offspring. Within five months, screwworm flies had been entirely eradicated from the island. The procedure has subsequently been used in other areas, and studies are extending the technique to other species of pests.

Even nuclear explosions are gradually being harnessed to peaceful uses. The Plowshare Program* of the U.S. Atomic Energy Commission, begun in 1957, has sponsored research on possible engineering and scientific uses of nuclear explosives. Experiments conducted under the auspices of the program have shown that nuclear explosions can be successfully used to excavate craters and ditches. It has been proposed that, using nuclear explosions, a sea-level canal a thousand feet wide could be dug across the Isthmus of Panama at a cost far below that using conventional earthmoving methods.

One of the most important of the expanding uses of atomic energy is the harnessing of nuclear reactions for the generation of energy that can be converted to such useful forms as electrical and mechanical energy and used to provide power for the growing needs of modern industrial life.

Nuclear Power: Energy for the Future

Our world is currently faced with an energy crisis. Electrical "brownouts" and sporadic shortages of fuel oil and gasoline have become commonplace, as homes, vehicles, and industries compete for the limited resources of our planet.

Thus far we have depended mainly on fuels derived from once-living organisms. With the rise of industry, the wide use of such readily gathered fuels as wood and peat gave way to

*Named from the biblical quotation, "And they shall beat their swords into plowshares. . . ."

a utilization of **fossil fuels.** These substances were produced over long geological ages of time from the remains of animals and plants that lived millions of years ago. As sand and soil covered these remains and were gradually converted to rock, the temperatures and pressures generated in the buried deposits converted them to coal and oil. New sources of fossil fuel deposits are continually being discovered and the technology of extracting them improved. But it is becoming evident that our current rate of energy consumption will ultimately exhaust *all* of our planet's fossil fuel resources. Thus, alternative energy sources must be developed if we are to continue the industrialized way of life to which we have become accustomed.

Such alternative energy sources are being actively explored. Hydroelectric power, the generation of electricity by the force of rapidly moving water, has long been profitably used. However, this energy source is inherently limited to areas such as waterfalls and other bodies of moving water—which often are not convenient to the regions where the power is needed. (The construction of dams for hydroelectric installations has far-reaching and often undesirable effects on the animal and plant—and human—communities of the area.)

Figure 20–13 The twin cooling towers at the Rancho Seco nuclear station, near Sacramento, can recycle cooling water at a rate of half a million gallons per minute.

The possibilities of gathering solar energy and tapping the geothermal energy of the depths of the earth are being explored. Probably the most promising energy source for near-term exploitation is atomic energy.

After World War II ended, there were glowing predictions that the control of atomic energy would bring cheap, limitless sources of power for mankind. But it was soon apparent that a great deal of technological development would be necessary before energy from atomic fission would be economically feasible. Indeed, it was more than two decades before nuclear power became generally competitive with conventional power sources. The disposal of radioactive wastes generated by nuclear power plants without contaminating the land and waters of our planet still presents some problems. Dumping radioactive wastes into the ocean may have deleterious effects on the intricate food chains of the marine life forms. Various forms of underground burial constitute the most effective means of disposal of radioactive wastes yet devised, but here stringent precautions must be taken so that there will be no radioactive contamination of

Figure 20–14 Fusion—power for the future? This installation is part of an experiment aimed at building up dense plasma (ionized gas), a step toward controlled thermonuclear fusion.

populated areas, nor any seepage of fission products into underground waters.

Waste disposal problems would be minimal with a further advance in nuclear power generation: the achievement of controlled nuclear fusion. Thermonuclear reactions do not produce significant amounts of radioactive by-products, and they can draw upon the virtually limitless supplies of hydrogen isotopes in the oceans for raw materials. Though formidable problems still remain, it seems likely that by the 1980s nuclear physicists will be able to achieve controlled nuclear fusion in the laboratory. Within a decade or two thereafter commercial nuclear fusion plants should become feasible, and at last the original dream of the 1940s—of a source of limitless, extremely cheap energy for the needs of the world—will become a reality.

What will be the effect on the world if limitless supplies of energy become available? There will be an enormous expansion of industry. Industrial processes that are marginal or unprofitable at present costs of electricity will become highly advantageous. The electrolysis industry, in particular, will receive a boost. Electrolytic methods of producing metals will expand. Cheap electrolysis of water will yield not only high-purity oxygen, but also hydrogen, which can be used in such varied areas as steel production and the production of ammonia, which can subsequently be turned into fertilizers, which will in turn foster an expansion of high-yield agriculture. Cheap sources of electric power will have an even greater effect on agriculture in another way. With such power, large-scale desalting processes will become economically feasible, and water from the seas can then be used to irrigate arid lands, rendering them fertile. It has been estimated that seven times as much land as is presently being used for the cultivation of crops could be used for this purpose if only there were sufficient water. Vast coastal agroindustrial complexes are envisioned, in which nuclear power plants, supplying electrical energy for a cluster of factories of various kinds, would be combined with desalting installations, providing pure water to irrigate the surrounding land.

QUESTIONS FOR REVIEW AND THOUGHT

1. Distinguish between nuclear fission and fusion.
2. What is a chain reaction in general and in the specific case of nuclear fission? Cite an example of a chain reaction in another area of science.
3. What is a moderator and why is it a necessary part of a nuclear reactor?
4. Where does the energy released in nuclear fission and fusion reactions come from? Which type has a higher energy yield?
5. Why is it much easier to produce a nuclear fusion explosion than to achieve controlled nuclear fusion?
6. Discuss the patterns of distribution and peculiar hazards of radioactive fallout.
7. What are the sources of the natural radiation background? Has this been a positive or a negative factor in the history of life on earth? How does its magnitude compare with artificially produced radiation now present in the environment?
8. Describe the biological effects of radiations. Are there any means of protecting against them?
9. Discuss the peaceful uses of atomic energy.
10. What are the particular problems associated with the use of nuclear reactors to generate electric power? Should more nuclear reactors be built?

(a)

(b)

(c)

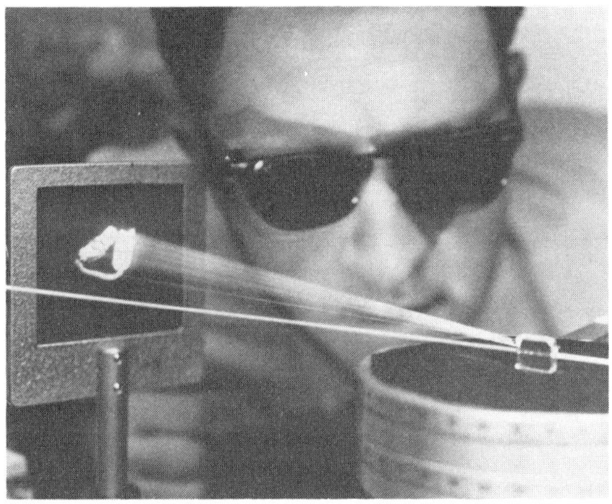

(d)

(e)

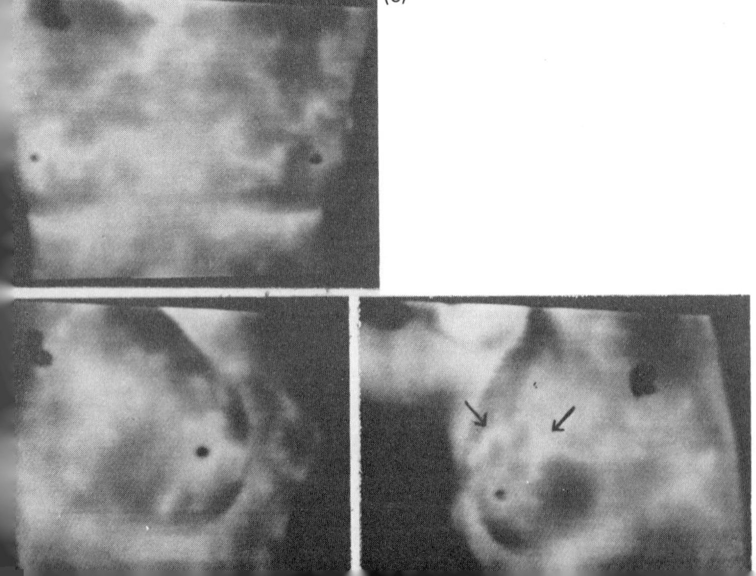

(f)

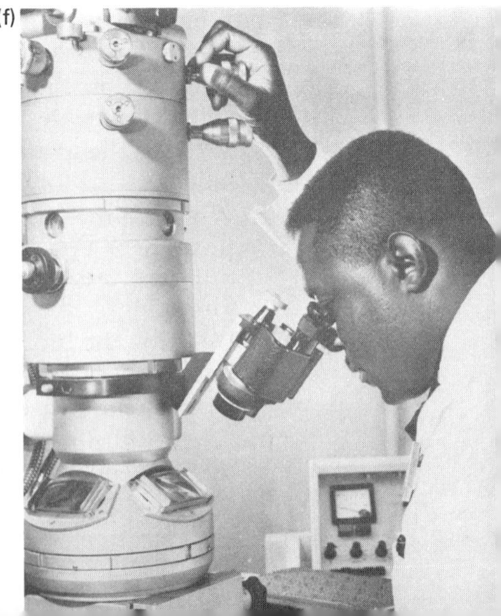

CHAPTER **21**

The Interdisciplinary Revolution

Introduction

We are in the midst of a revolution—a biological revolution. Fundamental discoveries are being made today that will reshape the world of tomorrow. For the first time man is gaining the potential for altering his own genetic constitution—correcting physical and mental defects and perhaps molding his own evolution. Man is even leaving his own planet and searching actively for signs of life on other worlds—the moon, Mars, Venus, and elsewhere.

The seething ferment of activity in the biological sciences is being fed by a growing influx of scientists whose basic backgrounds are in nonbiological fields such as physics, chemistry, and engineering. The fruits of their efforts have been new insights and an interdisciplinary approach that has brought a freshness and excitement to biology, inspiring still more "nonbiologists" to enter the various fields of biological research.

Among the thousands of physicists, chemists, and engineers who have flocked into various fields of biological endeavor during the past few decades are a number of men who have won Nobel Prizes for distinguished work in nonbiological fields. Harold Urey, who won the Nobel Prize in chemistry in 1934 for the discovery of heavy hydrogen, later turned to research in exobiology. Donald Glaser, winner of the Nobel Prize in physics in 1960 for the invention of the "bubble chamber" for the study of subatomic particles, has subsequently performed research on nucleic acids and other areas of molecular biology. Linus Pauling, who spent most of his early years studying the nature of atoms and molecules, with emphasis on inorganic chemistry, won a

Figure 21–1 Scenes of the interdisciplinary revolution: (a) Harold C. Urey; (b) composing computer music; (c) Linus Pauling; (d) a laser beam reconstructs a holographic image stored in a lithium niobate crystal (a thousand different holograms can be stored in the same crystal); (e) thermograms of normal and cancerous breasts (the lightest shades reflect the warmest areas; the arrows point out cancerous "hot spots"); (f) an electron microscope being used in a study of mouse leukemia virus.

Nobel Prize in chemistry for this work in 1954 and has subsequently concentrated his research efforts in molecular biology.

With the infusion of new viewpoints into the biological sciences, the traditional boundary sciences of biochemistry and biophysics are now among the most vital of the scientific fields. Scientists are attempting to learn the chemical nature of biological molecules and the chemical and physical factors that determine their interactions. The molecules of life are being synthesized in the laboratory, and model studies using simpler prototypes are casting new light upon their functions. For example, the intricate reactions of complex metal-containing biochemicals, such as the metallo-porphyrins active in photosynthesis and nitrogen fixation, are being studied indirectly through investigations of complexes of metals with simple compounds such as carbon monoxide and acetylene.

Physical scientists and engineers have brought with them powerful new tools for biology, from the electron microscope to the laser, from radioisotopes to ultrasonics, from cryogenics to thermography. The influence of the computer is felt everywhere, from the coordination of analyses to the simulation of intelligence. In the new hybrid field of bionics, biological systems are studied avidly for the keys they may provide for the improvement of electronic devices. Biologists search for life on other worlds using the hardware of aerospace technology and seek to create life on earth in a test tube.

The beginning of this century was an era of excitement and explosive growth in the physical sciences. In the achievements of that era lay the foundations of our modern technological society. Now it is the turn of biology, and new worlds are on the verge of discovery.

Bionics

The field of **bionics** is quite new. Even the name of the field did not exist before 1960, when it was coined by J. Steele to cover the interdisciplinary science that applies biological principles and techniques to the design of electronic systems. In that same year, the first Bionics Symposium was held in Dayton, Ohio. Bionics has now achieved the status of a major field of scientific endeavor and is continuing to grow at a phenomenal rate.

Modern electronic devices and systems employ a variety of sensors and pickups, generators and modifiers of forms of energy—light, sound, electricity, and so forth. To some extent these devices parallel systems already in existence in the world of living nature, and for the most part manmade devices have a long way to go before they will begin to rival their natural counterparts in compactness and efficiency. This is not too surprising, since man has been developing his electronic devices for only a century or so, while the ultrasophisticated pickups and transmitters and analyzers that animals come equipped with have been through many millions of years of streamlining and refinement in the workshop of natural selection. Thus, it is quite sensible for electronics specialists to turn increasingly to a study of living systems in search of hints for improving their own inventions.

BIOSONAR

One of the first of the electroniclike animal systems to receive the serious attention of physical scientists was the use of sonar systems for the location of external objects.

Bats are generally poor-sighted animals that fly about at night and catch their food on the wing. Their dexterity in flight is amazing: they can flit about with ease through crowded caves without bumping into one another and can locate and catch prey as small as a mosquito at a rate of one every ten seconds. In experiments in a darkened room, with dangling wires a fiftieth of an inch wide suspended randomly at intervals only a wingspan apart, bats were able to negotiate the maze with ease (Fig. 21–2). A bat not only can catch mealworms thrown up into the air in complete darkness, but it can distinguish between mealworms and inert objects of the same size.

Blindfolding a bat has no particular effect on its ability to find its way about in flight. But if its ears are plugged, it blunders about

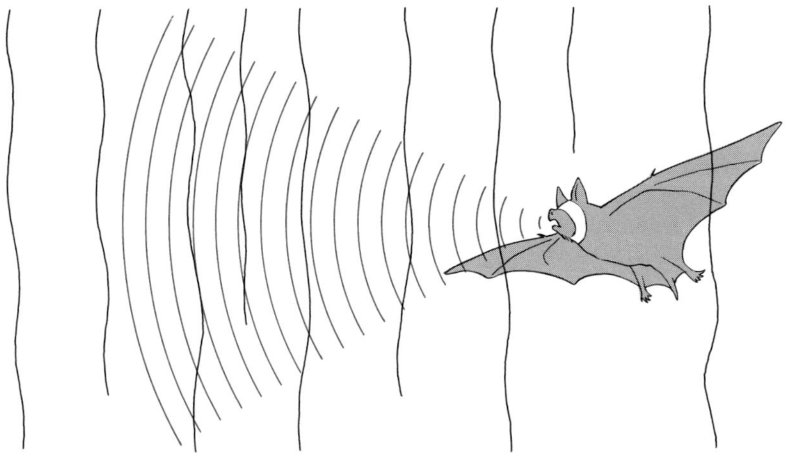

Figure 21–2 A bat's sonar permits it to fly easily through a maze of dangling wires blindfolded. If its ears were plugged, it would hit the wires even with its eyes uncovered.

confusedly, bumping into obstacles in its path. Studies with microphones and sound recording as well as high-speed motion-picture photography have revealed that bats emit (through the nose or mouth, according to the species) a series of high-frequency sound pulses well up into the ultrasonic range. (Bats have been shown to hear sounds as high as 200 kilocycles per second, in contrast to the 20-kilocycle ceiling of adult human hearing.) The large ears of the bats can be turned to act as directional antennas, funneling in the returning echoes and interference patterns produced by the bouncing of the emitted sound waves off obstacles. The auditory portions of the bat's brain are extremely well developed, and the bat is able not only to locate minute objects by an analysis of sound echoes, but also to use the information to plot trajectories of moving objects and to plot an interception course (Fig. 21–3). A flying bat can use its sonar system efficiently even in an artificially produced sound field of high-intensity noise, as long as both of its ears are functioning. Thus, its brain is able to screen out unwanted noise and pick out only the meaningful signals. This ability is analogous to the ability of the human brain to screen out distracting traffic noises in order to concentrate on an interesting conversation.

It has been proposed that the manner in which bats interpret their ultrasonic echolocation signals is similar to the principle of holography, in which a three-dimensional image is reconstructed from the interference patterns produced by reflected signals and a coherent reference signal. In the case of the bat, the reference signal could be supplied from the

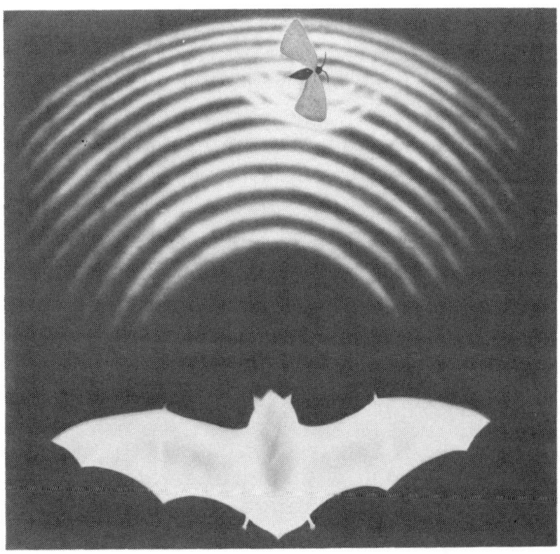

Figure 21–3 How the bat locates its prey: high-pitched sound pulses emitted by the bat bounce off objects in its path; the echoes are interpreted in the bat's brain to form a picture of the environment.

portion of the brain dictating the emitting of a sound of a particular frequency directly to the processing area of the brain. The receiving and processing area of the bat's brain has been found to have a structure greatly resembling an assembly of minute dipoles, which could function in binary digital data processing, much as in a modern electronic computer. A model of the bat's brain has been constructed on this principle. The resulting ultrasonic holograms have been recorded on sound-sensitive photographic plates, and three-dimensional images of objects have been reconstructed.

Sonar systems are used extensively in the water by dolphins, whales, and sea lions. These sea mammals emit patterns of sound and ultrasonic pulses and use them to locate obstacles in much the same manner as the bat does in air. Even man has been shown to have some ability for echolocation. Normally this ability is not very well developed, as humans tend to rely mainly on the sense of sight. But studies with blind people have shown that using a variety of sounds such as hisses and tongue clicks, humans can distinguish differences in size and shape of objects. The subjects studied were able to detect targets as small as 1¾ inches at a distance of 24 inches and a 10-inch target at a distance of 9 feet.

Biosonar systems are fantastically compact and efficient in comparison with man's best electronic efforts to date. Whereas modern radar and sonar sets occupy cubic meters of space, the bat's sonar equipment is neatly tucked away within cubic centimeters. And moths, which are preyed on by bats, have their own sound-receiving equipment packed away in fractions of a cubic millimeter, in "ears" located on the sides of the moth's thorax. Moths can receive and interpret bats' ultrasonic signals at distances greater than the bats' own hearing range. Some species of moth not only are capable of receiving hunting bats' ultrasonic signals and taking appropriate evasive action, but can also send out high-frequency signals of their own that seem to act as warning signals and cause the bats to swerve aside.

"Living lights," such as the winking lights produced by fireflies, have fascinated man for uncounted ages. Written records of glowing creatures date back to beyond 1000 B.C. in Chinese records. This phenomenon, called **bioluminescence,** was well known to Aristotle, who wrote about his observations of the glow of decaying fish and was one of the first to propound the concept of "cold light." A poetic description of the phosphorescent surface waters of the sea was provided in Coleridge's *Rime of the Ancient Mariner:*

> About, about, in reel and rout
> The death fires danced at night;
> The water, like a witch's oils,
> Burnt green, and blue, and white.

This spectacular display of bioluminescence was doubtless caused by large masses of light-producing dinoflagellates, microscopic algae, floating in the surface waters.

The phenomenon of bioluminescence is a subject of lively interest to modern scientists, not only as an important key to an understanding of basic metabolic processes, but also as an extremely efficient conversion of chemical energy directly to radiant (light) energy with virtually no concomitant production of heat.

Bioluminescence has been discovered among diverse groups of living things—among bacteria, fungi, algae (but no green plants), and among varied animal forms in about thirty of the approximately eighty different classes, including sponges, coelenterates, various worms, crustaceans, mollusks, insects, and fishes.

Bioluminescent organisms show a great variety in the type of light emitted. Some species emit light continuously, others only in flashes. The colors of the lights emitted also vary greatly—red, yellow, green, or blue. A fascinating and unusual form is the "railroad worm" of South America. First described in 1647, it was eventually named for the unique arrangement of its colored lights. This two-inch worm (which is actually not a worm, but the larva of a beetle) has a pair of red lights on

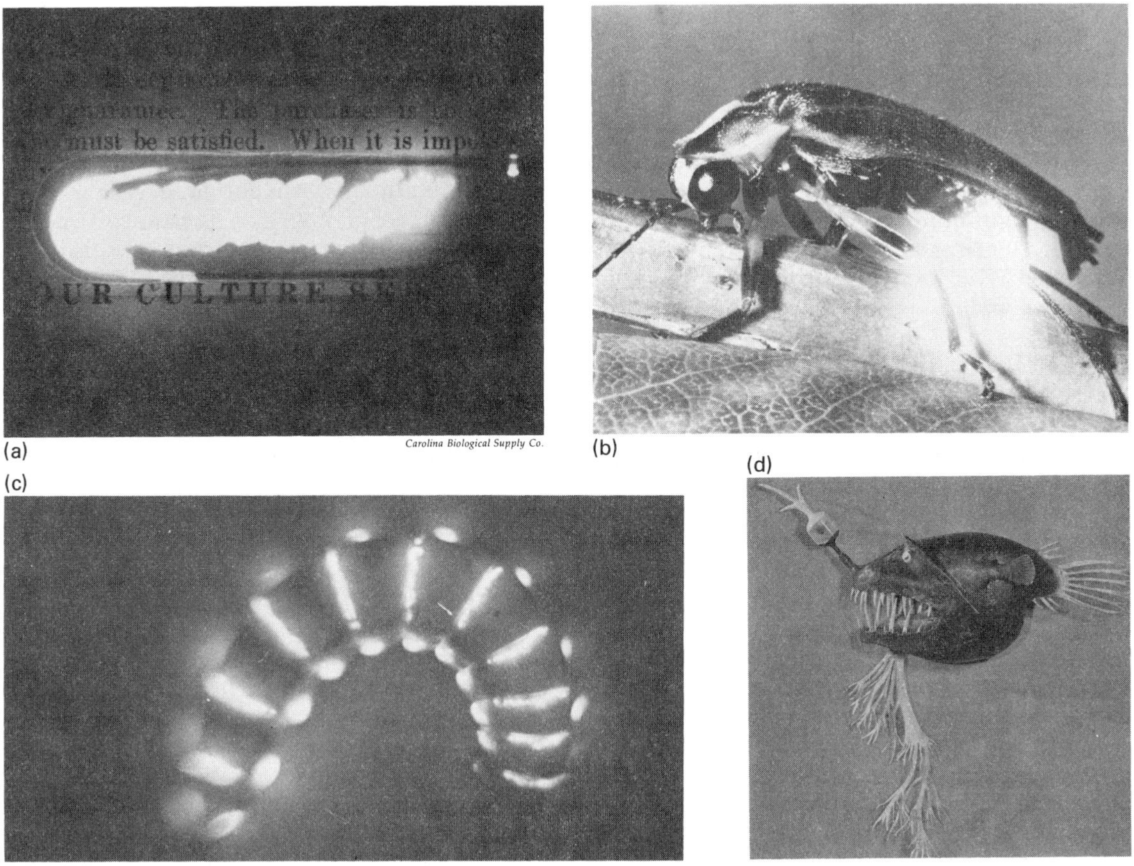

(a) Carolina Biological Supply Co. (b) (c) (d)

Figure 21–4 Bioluminescence: (a) luminous bacteria, photographed by their own light; (b) a firefly with lighted abdomen; (c) a luminous beetle larva, photographed by its own light; (d) the angler fish, whose lighted lantern attracts its prey.

its head and a row of green lights along each side. Studies of this unusual creature have revealed that its red "headlights" can be stimulated to glow by blowing air upon it, while its green side spots can be made to light up by shaking the worm.

The biochemical mechanisms of bioluminescence have been under intensive study, and great strides have been made in recent years, especially in the work of W. C. McElroy and colleagues on the firefly. Bioluminescence has been found to be basically an oxidative reaction, in which a substance called **luciferin** and an enzyme **luciferase,** each nonluminous when separate, produce light when they are mixed under suitable conditions. The reaction requires ATP*; this serves to activate luciferin, which is then oxidized by oxygen, and the energy emitted is utilized for light production. It is believed that the oxidized luciferin, bound to the enzyme, is the light-emitting molecule. The energy conversion is amazingly

*The sensitivity of the reaction to the supply of ATP has been well established. In fact, the bioluminescent reaction has been used as the basis of an analytical reaction for the determination of ATP concentration. As a result, the collection of fireflies has become a profitable sideline for children in the southern United States, who sell fireflies to an ATP manufacturer as raw materials for the quality-control test.

efficient; very little energy is lost as heat, and hence the designation of "cold light."

A comparison of the biochemical mechanisms of bioluminescence in different species shows unusual diversity. Although the same terms—*luciferin* and *luciferase*—are used in each case and the reactions are oxidative, the specific compounds involved are often quite different. For instance, luciferin from the firefly mixed with luciferase from the crustacean *Cypridina* will not yield a bioluminescent reaction.

The color of the light emitted is probably correlated with the type of luciferin molecule. (If several types of luciferin are present, the organism may emit light of more than one color.)

One modern theory on the origin of bioluminescence holds that the mechanism originated early in the history of life on earth, when oxygen was first appearing in the atmosphere. Oxygen was probably toxic to the organisms that lived at the time, and bioluminescence provided an efficient and relatively harmless way of getting rid of it. When organisms appeared that were able to utilize oxygen, bioluminescent mechanisms may have persisted in some cases, either because they did no harm, or because they served some side function of adaptive value.

Bioluminescence does indeed serve definite functions in many organisms today. The flashing lights of the firefly, for instance, are complicated recognition patterns, specific for each species, which help it to find and recognize a potential mate. Many deep sea creatures have lights that help to illuminate their dark watery world. Some species of deep sea fish have lights suspended on stalks, which serve as lures for potential prey or distract potential predators. At least one species of squid, when startled, emits luciferin and luciferase into the water instead of the inky substance used by shallow-water forms; a cloud of luminous water is produced, dazzling the squid's pursuer and permitting it to escape. But for some organisms, the role of bioluminescence is not as clear. For example, a large marine annelid, *Chaetopterus,* spends its entire life buried in the sand at the bottom of the sea in a U-shaped tube, and the light that it emits is unseen by friend or foe. And in many bacterial species, bioluminescence seems to be a mere by-product of metabolism.

BIOELECTRICITY

An unwary swimmer in tropical South American waters might be killed by a surging current of electricity emanating from an electric eel lurking as far away as twenty feet. An electric eel is a nine-foot-long living storage battery capable of discharging more than 550 volts of electricity in a single shock. (Compare this with normal house current, which is usually 120 volts.) As many as four-fifths of the cells in the eel's body are specialized for generating and storing electricity. The electric organs of such an eel are actually modified muscle tissue, each consisting of several thousand

Figure 21–5 The electric eel.

cells arranged in series, in parallel columns within the eel's snakelike body. The series arrangement of the electric cells makes their outputs additive and serves to build up the voltage; parallel arrangement of the long series serves to build up the amperage. As might be expected, the size of the shock that an electric eel can deliver is roughly proportional to its length.

A number of other marine organisms also have powerful electric organs, and some have long been known to man. Electric catfish, which inhabit the Nile, are pictured on ancient Egyptian tombs, and electric rays were used by ancient Romans as treatments for various ills, including some forms of mental illness (and thus may be considered a crude early form of electroshock therapy). Like the electric eel, the electric catfish and electric rays use their electricity to stun their prey, and they exercise remarkable control over the size of the shock that they deliver. (The electric discharge is triggered by a control center in the brain.)

In addition to the strongly electric fishes, there are other species equipped with relatively weak electric organs. The latter group of fish appear to use electricity as a type of location system. The African species *Gymnarchus,* which has been intensively studied by H. W. Lissmann, has a tiny electric organ (at first thought to be "pseudoelectric") in its tail. This electric organ sends out regular pulses of electricity that form a pattern in the water. Any deviations from this pattern produced by various obstacles are perceived by special sense organs in the fish's head (Fig. 21–6). So effective is *Gymnarchus*'s electrical location system that it can capture smaller fishes and avoid obstacles with an effortless precision that would be impossible solely with the use of its small and rather degenerate eyes. Indeed, it can swim backward just as accurately and easily as it swims forward.

This discussion of unusual electric organs should not mislead us. For *all* organisms exhibit some degree of bioelectricity. Virtually every living cell generates at least a minute pulse of electricity, and many organs have characteristic bioelectric activity. The human heart, for instance, generates a current that can be picked up and measured at the surface of the body (0.001 to 0.002 volt). The technique of **electrocardiography** is based on this heart electricity (Fig. 21–7). The brain, too, generates electric currents. Although the potential difference at the surface of the skull, due to the brain currents, is only 0.00002 to 0.0001 volt, this can be picked up by sensitive electrodes affixed to the scalp and recorded by an electroencephalograph. Much has already been learned about the normal brain waves and their deviations during various illnesses. Even plant cells have been found to generate electric currents in the course of their normal activities.

OTHER BIOSYSTEMS OF INTEREST TO ELECTRONICS

The living world is replete with compact, efficient sensor and effector systems that outperform any systems that man has yet devised.

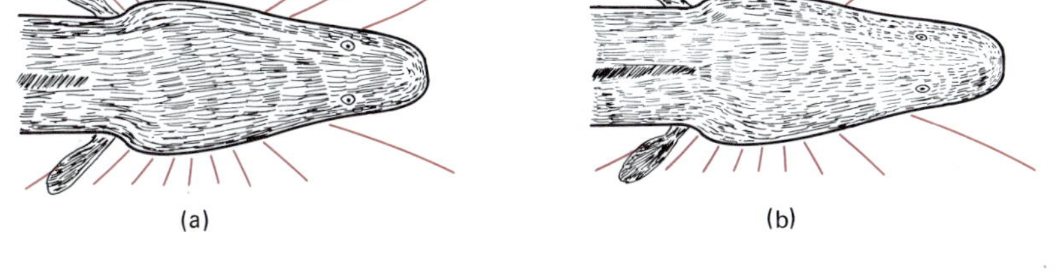

Figure 21–6 *Gymnarchus* finds its way by analyzing distortions of the weak electric field it generates in the surrounding water. Lines of current flow diverge from a poor conductor (a) and converge toward a good conductor (b). Sense organs in the head region detect the effect.

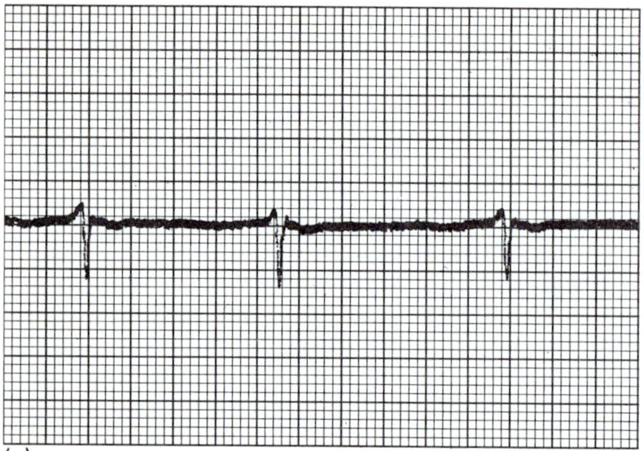

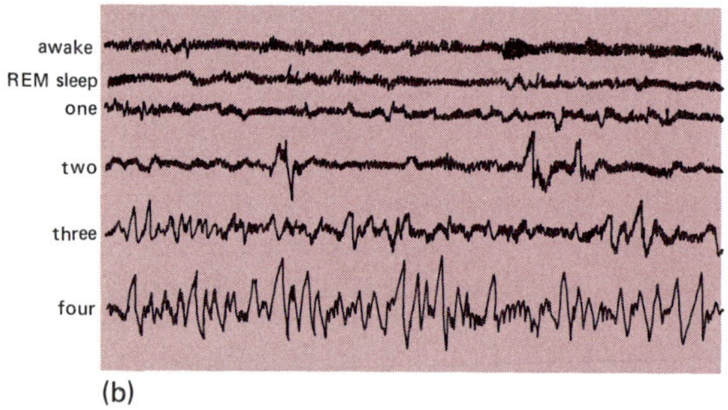

Figure 21–7 Tracings of the body's electrical activity: an electrocardiogram or EKG (a) and an electroencephalogram or EEG (b).

The male gypsy moth, for example, can pick up the scent from a single female of the species located more than two miles away. The best of man-made sensing devices for volatile chemicals (e.g., gas chromatographs) are thousands of times less sensitive than the sensors in these tiny moths. In fact, although the human sense of smell is notoriously poorly developed, a large cosmetics manufacturer found that better sensitivity could be achieved by equipping a gas chromatograph with a human sensor (an experienced perfumer) instead of the usual thermal conductivity detector.

Some animals have extremely sensitive temperature receptors. Pit vipers, for example, are snakes that hunt their prey in the darkness of underground burrows. They locate small warm-blooded animals with the aid of temperature receptors in a pit on each side of the face (Fig. 21–8). By offering such snakes balls at different temperatures (the snake always strikes at the warmer object), it has been found that a pit viper can respond to a temperature difference as small as 0.001°C!

A praying mantis on the ground or a dragonfly in flight can receive a pattern of visual impressions concerning the motion of a small flying insect, analyze and integrate them, strike, and capture its prey, all within 0.05 second. The compound eye of the insect, made up of thousands of individual units called ommatidia, is particularly well adapted for detecting motion. As an object moves across the visual field, ommatidia are progressively stimulated and then turned off, and these signals

Figure 21–8 Pit vipers have sense receptors (pits) sensitive to infrared (heat) radiation.

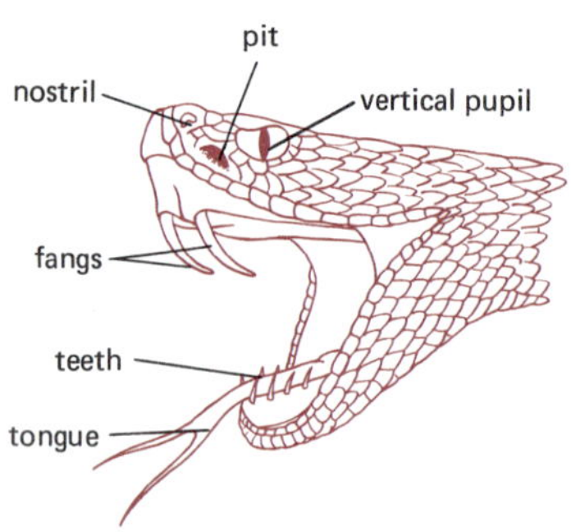

Lynwood M. Chace, Omikron

are transmitted to the insect's brain, where they are interpreted. The frog, which also preys on small moving creatures, is another animal that has an eye-brain system well adapted to ignoring background shapes and responding to small movements. There is an obvious parallel in the development of radar systems for air traffic control, and animal scanning systems are being studied in an effort to improve computerized electronic scanning systems.

Attempts are being made to mimic various biological systems, even those that might be regarded as intrinsically human. For example, a major area of interest is the development of automatic speech duplication. Significant progress has already been made in this field. Perfection of the technique will ultimately lead to considerable improvements in automatic reading machines, remote speech transmission, and remote voice read-out from information storage systems.

Far greater emphasis is being placed on mimicking the neuronal circuitry and information storage and retrieval systems of higher organisms. For example, neuronal circuits made from metals and other inorganic substances have been demonstrated in the laboratory. Some of these systems show surprisingly "lifelike" reactions, even a certain capacity for memory and learning. The use of electronic computers in model studies of learning patterns is a particularly interesting aspect of this field. These experiments, like other areas of bionic study, are of twofold value, since they help to illuminate the processes of life while aiding in the more efficient electronic duplication of its systems.

New Tools for Biology

The cross-fertilization brought by the interdisciplinary revolution has enriched the life sciences with a powerful arsenal of new tools that offer the promise of delving deeper into the inner mechanisms of living organisms than was ever before possible. Technological advances, particularly in various areas of electronics, are providing new avenues both for studying and for manipulating living systems.

LASERS

When the first working laser was demonstrated in 1960, newspapers and magazines abounded with futuristic speculations on deadly ray guns which could wipe out whole populations. Practical ray guns have not yet been developed, but applications of the laser have blossomed in a variety of fields from space technology to tunnel engineering to medicine.

The term **laser** is an acronym, constructed from the initial letters of the phrase, "*l*ight *a*mplification by *s*timulated *e*mission of *r*adiation." The laser itself is a crystal (such as a ruby), a liquid, or a gas in which a large fraction of the atoms or molecules are in an unusually excited high-energy state. If a pulse of energy of just the right wavelength is sent into such a substance, its excited atoms will suddenly lose their excess energy, sending out a pulse of intense radiation. Laser light is coherent light—an extremely narrow beam of light of a single wavelength. (The light from the sun and that of an ordinary lightbulb are random jumbles of many wavelengths and energies, emitted by various atoms at different times.) A laser beam is so narrow and intense that a ruby laser can be used to flash signals from the earth to the moon and back.

Lasers have been described as "a solution looking for problems." They have already been used in varied areas of scientific research. They are convenient tools for studying various physical phenomena; they can produce and probe plasmas and can stimulate photochemical reactions. In experiments where a beam of coherent light is needed, lasers provide an ideal source.

Possible uses of laser beams as carrier waves for communications systems are being explored. One area where lasers would be particularly effective is in communicating with space capsules. During reentry into the earth's atmosphere, the shell of the capsule is heated by friction and becomes surrounded by an ionized gas sheath, which cuts off radio contact between the capsule's crew and the ground. A laser beam, however, can penetrate the sheath.

A laser beam can produce some spectacular effects. Powers of tens of thousands of megawatts can be generated in pulses of the order of a millionth of a microsecond. Such a concentrated light flash tends to heat whatever it strikes with incredible rapidity. Temperature rises of some 20,000,000°F in a few trillionths of a second have been observed. The beam can be finely focused with lenses. Thus, laser beams can be used for precision cutting of sheets of metal (the metal at the site where the beam strikes is virtually instantaneously vaporized) and for such delicate operations as spot-welding a detached retina back into place, drilling decayed teeth, and selectively destroying cancer cells.

When aimed through a microscope, a laser beam can punch a 0.0001-inch hole neatly in the center of a red blood cell and can be used to selectively destroy organelles within cells. Laser beams have been used to stimulate nerve cells without damaging them. It has been suggested that the base sequence in DNA might be selectively altered with the use of laser beams.

HOLOGRAPHY

In 1947 a British electron physicist, Dennis Gabor, suggested a brilliant method for the development of a high-resolution electron microscope. One of the main difficulties in the development of such an instrument lies in the inevitable aberrations caused by the magnetic lenses. Gabor proposed that the lenses be eliminated entirely by exposing photographic film directly to the electron waves scattered from the sample. An unfocused, jumbled picture would result, but this could be reconstructed with visible light, and the image would be greatly magnified because the wavelengths of visible light are much longer than those of electrons. Gabor coined the term **hologram** (literally, "whole message") to refer to the picture that would be obtained, but practical development was not feasible at the time.

With the development of the laser, a source of coherent light, holography became feasible at last. The first holograms were produced by Emmett Leith and Juris Upatnieks of the University of Michigan in 1960. The technique is now used to produce startlingly three-dimensional images of objects ranging from the macroscopic down to the microscopic.

When a beam of light strikes a three-dimensional object—for example, a chess piece (Fig. 21–9a)—light is reflected from each point on the surface of the object. The resultant wave fronts interfere with one another and produce a rather complicated overall shape, reflecting the shape of the object. The amplitude over the surface of each wave front varies, depending on the amount of light reflected by the different parts of the object. A record of the resultant wave fronts would contain enough information for a reconstruction of the image of the object.

In preparing a hologram, a beam of coherent light is sent toward the object to be recorded (Fig. 21–9a). A plane mirror reflects part of the beam directly onto a photographic plate. The mirror is placed between the light source and the object, so that the rays of this part of the beam never strike the object at all. The reflected light is referred to as the **reference beam;** it represents the original undistorted beam of coherent light. Meanwhile the other portion of the original light beam strikes the object and is reflected back toward the same photographic plate. This is the **signal beam,** and in it the wave fronts reflected from each point on the object interfere with one another, producing corresponding changes in resultant wave front. The combination of the reference beam and signal beam impinging on the photographic plate produce an interference pattern, which is recorded as the hologram.

The hologram viewed under ordinary light does not look anything like an image of the object, but it can be reconstructed very simply by shining a laser beam through it from the same direction as the original reference beam (Fig. 21–9b).

Holograms are realistically three-dimensional images. When objects at different

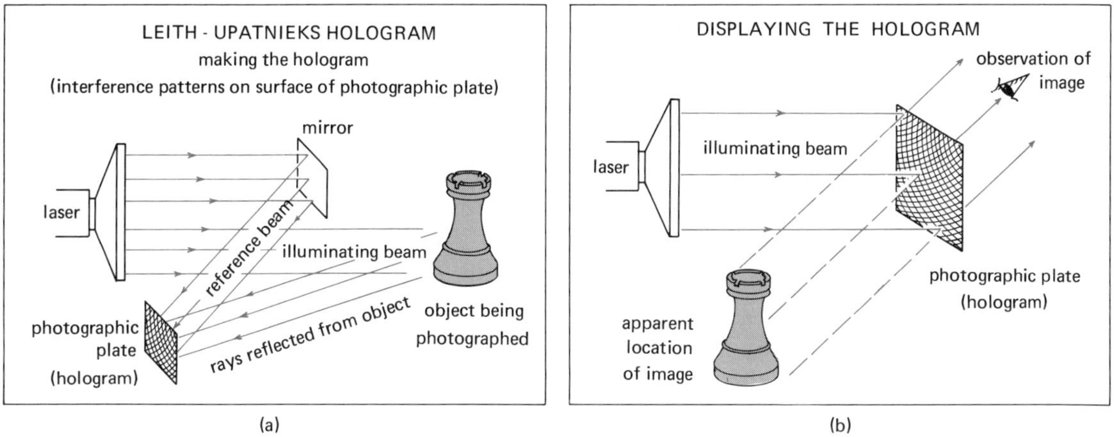

Figure 21–9 (a) Recording a hologram; (b) reconstructing a hologram.

depths are viewed from different angles, they show an apparent relative movement (parallax), just as real three-dimensional objects do. A curious property of holograms is the fact that several different scenes can be recorded on a single photographic plate simply by illuminating the plate at different angles for each recording. This property would be quite useful in the use of holograms to store information. It has already been used to prepare motion pictures of a slowly moving object by changing the angle between the reference beam and the plate after each exposure. But considerable technical difficulties will have to be overcome before large-scale holograms suitable for display in motion picture theaters can be produced. Holography may also have ultimate applications in television.

The implications of holography for biology and medicine are profound. The activities of microscopic life forms and the microscopic structure of specimens can be studied in a manner never before possible. Specimens can be "preserved" in the form of holograms just as they were at a particular instant of time, and then examined at leisure with a microscope. These holographic images can be examined not only with a normal optical microscope, but also by dark-field, phase-contrast, or interference microscopy. Because special techniques for preparing thin sections are not needed, distortions of the specimens are minimized. In addition, Dennis Gabor's original dream of a high-resolution holographic electron microscope is on its way to realization. L. Bartel and C. Ritz of the University of Michigan have recently succeeded in visualizing the clouds of electrons within atoms (resolution about 0.1 Å), using a holographic variation of electron microscopy. The electrons reflected from the nucleus of the atom served as the reference beam.

Holography is useful in the macroscopic world as well. The technique has already been put to work on the industrial assembly line: a hologram of a perfect component is used as a reference standard, to which each new component produced, oriented in the same manner, is compared to reveal flaws. Scientists envision various other uses for holographic techniques, such as in studies of population statistics of deep-water organisms.

The growing practical use of holography is being facilitated by developmental research aimed at making it possible to reconstruct holograms without the use of lasers. Several techniques have already been developed for viewing laser-recorded holograms using white light from the sun or even conventional light sources. A camera that takes films in ordinary light, which can later be processed with a laser to yield holograms, has also been invented.

One of the most exciting new areas of application of holography to biology is the development of holographic theories of memory storage in the brain. If a hologram is broken and an appropriate laser beam is sent through a small fragment of it, the *whole* picture is reconstructed, rather than just a small fragment of it. (Some detail is lost; the amount of detail lost is proportional to the amount of the original hologram that was destroyed.) Similarly, current studies of memory at the cellular and molecular levels indicate that a particular memory is stored not in a single site in the brain, but rather in a large number of sites scattered through the brain. Researchers are coming to believe more and more that the reconstruction of a memory in the brain is analogous to the reconstruction of a hologram.

CRYOGENICS

The world of the ultracold is a fascinating world indeed. Many substances behave strangely, quite unlike their behavior at normal temperatures. Certain metals and alloys known as superconductors are able to conduct electric current almost indefinitely without any loss in strength. At temperatures a degree or two above absolute zero,* liquid helium seems to defy the laws of nature and flows upward against gravity. Exposure to ultracold has been used to open up whole new areas in biology and medicine.

The study of the phenomena that occur at extremely low temperatures and the development of low-temperature techniques is known as **cryogenics.** The temperature range studied is chiefly that below about −200°F. Substances can be cooled to cryogenic temperatures by immersing them in a bath of an extremely-low-boiling liquid. The most frequently used cryogenic liquids include helium, which boils at −452°F (−269°C or 4°K), hydrogen (−423°F or 20°K), nitrogen (−320°F or 77°K), and oxygen (−300°F or 90°K).

The cryogenic liquid may be kept in a container known as a **Dewar flask.** This is a double-walled flask made of silvered glass or highly polished metal, with an evacuated space between the walls. In construction it is therefore very similar to a common Thermos bottle. Liquid air, oxygen, nitrogen, and hydrogen can be kept in a Dewar flask for several hours. Liquid helium, however, will boil off rapidly unless the flask is placed inside a larger Dewar flask containing liquid nitrogen or air.

Refrigeration at cryogenic temperatures is finding increasing use in the food industry. Liquid nitrogen, for example, is used extensively to protect produce from spoilage during shipment by truck over long distances. Cryogenic techniques are also used for preserving biological materials. Sperm of champion bulls, to be used for breeding by artificial insemination, are stored for years at the temperature of liquid nitrogen. Human sperm has been preserved in a similar manner. Cryogenic techniques are also finding increasing use in the preservation of human blood in blood banks. Under normal conditions, stored blood has a shelf life of twenty-one days, but with modern techniques blood can be kept at temperatures below −80°C (−112°F or 193°K) for up to four years with no signs of deterioration. It has been suggested that healthy people might donate blood that could be banked and saved to be retransfused into them in case of future emergency.

It has been suggested that cryogenic techniques could be used for storing organs to be used for transplants. If effective freezing techniques can be developed, there will not be

Figure 21–10 Pouring liquid nitrogen into a cryostatic apparatus.

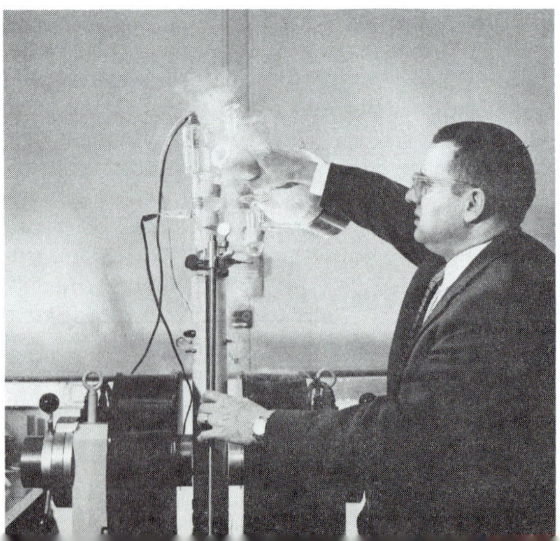

*The theoretical limit at which all molecular motion would cease: 0°K = − 273°C = −459°F.

such a frantic race to correlate a suitable donor and a suitable recipient, and donated organs can be utilized more effectively. But serious problems must be overcome along the way. With the techniques developed thus far, there is a great hazard of damage to the cells during freezing. This can occur either mechanically, through damage by ice crystals formed within the cell, or due to excessive exposure to high salt concentrations, which result when water passes out of the cells to form extracellular ice crystals. Extremely rapid rates of cooling, of the order of several thousand degrees Centigrade per minute, can minimize the effects of high salt concentrations, but beyond a certain value other problems set in. The addition of protective compounds, such as glycerol and dimethyl sulfoxide (DMSO), is of some help. These protectors are generally low-molecular-weight organic compounds that can pass through the cell membrane into the cytoplasm. Their use has already made possible the successful storage of various types of mammalian cells, such as red and white blood cells, bone marrow cells, and spermatozoa, but a small percentage of the cells die. This cell death is not too important in cell suspensions, but can spell failure for an organized tissue or organ.

Cryogenic techniques are producing a quiet revolution in many areas of surgery. Carefully controlled ultracold temperatures are used to destroy cells, the technique being particularly effective in cancer treatment. Lip and skin cancers are treated by freezing the cancerous area; the frozen tissue sloughs off, and normal tissue replaces it, frequently without even leaving a scar. Tumors of the long bones are treated by freezing the bone; the bone and tumor are both killed, but then the dead bone serves as a trellis for the growth of healthy new bone tissue. Brain tumors can be frozen and then lifted out as a solid mass, with minimal danger of hemorrhage. Indeed, cryosurgery in general is so painless and free of hemorrhage risk that it is frequently referred to as "bloodless surgery." It is gaining increasing use in various types of eye surgery—in the removal of cataracts, the fusing of tears in the retina, and in reshaping the cornea.

In 1963 Robert Ettinger published a book, *The Prospect of Immortality,* in which he expressed a growing belief among scientists that some day man will conquer death; if only we can keep ourselves intact until that day comes, we can all be immortal. Ettinger called for the institution of a widespread practice of immediately freezing people who have died of accident or disease, in the hopes that means will eventually be discovered both for curing what killed them and for repairing the damage to cells and tissues caused by the freezing process itself. As a result, a number of societies have sprung up in various cities throughout the world, and more than a dozen people are now in "cryonic" storage under liquid nitrogen.

Cryogenic techniques also hold promise for improving the resolution and efficiency of electron microscopes. At the extremely low temperatures of a cryogenic environment, the amount of volatilization and consequent contamination of the specimen—a very critical factor in electron microscopy—is substantially reduced.

ULTRASONICS

The ultrasonic range of electromagnetic radiation, extending from 20,000 cycles per second up to a billion cps, is a rather peculiar one. The longer waves, just at the upper limit of human hearing, behave like acoustic waves, tending to bend around obstacles. The very short, high-frequency ultrasonic waves are rather similar to light, and tend to rebound from obstacles. Ultrasonic waves, like sonic waves, can be propagated in space only by a carrying medium, solid, liquid, or gas.

Ultrasonic waves can be generated by **piezoelectric crystals.** These are crystals that change their shape when they are subjected to an electric field and generate an electric field when their shape is changed. The crystal can be cut in such a way as to resonate at the desired frequency. Or a large quartz crystal with flat, parallel ends can be subjected on one end to a high-frequency alternating electric field; the resultant alternating mechanical strain is transmitted through the crystal as an ultrasonic wave.

Ultrasonic echo sounders are in growing use in various sonar applications. By sending out high-frequency sound waves and analyzing the patterns of impulses bounced back from various objects, such instruments are being used to explore the ocean floor and to detect schools of fish and other marine life. Sonar systems also have obvious military applications.

Ultrasound has already found a variety of applications in biology and medicine. Focused beams of high-intensity ultrasound can produce a wide range of selective temporary or permanent changes in cells and tissues. Nerve cells are particularly sensitive to the effects of ultrasound. For example, since ultrasound can be focused, changes can be induced in any specific site or depth of the brain. Three-dimensional functional maps can then be obtained by focusing an ultrasonic beam on various tissue structures in animals' brains and observing the correlated changes in responses to stimuli.

The ultrasonic beam can be so finely focused that even such minute structures as the nucleolus found within the nucleus can be selectively destroyed. Ultrasound has also been used to break up microorganisms such as bacteria and viruses into smaller components, which may then be studied. Interestingly, though high-intensity ultrasound destroys microorganisms, low-intensity ultrasound may actually stimulate their growth.

Ultrasonic heat, generated by the absorption of energy and by friction resulting from the effects of the rapid vibrations of the ultrasound on body cells and tissues, has been successfully used in the treatment of arthritis, bursitis, and lumbago.

Ultrasonics has provided a valuable tool for research in diagnosis. The characteristic patterns obtained from ultrasonic bombardment of various types of tissues are under intensive study and are yielding techniques for the diagnosis of malignancy. Such techniques can also be used to diagnose foreign objects within the body, including kidney stones and gall stones, and to diagnose pregnancy and follow the development of the fetus. (Fig. 21–11).

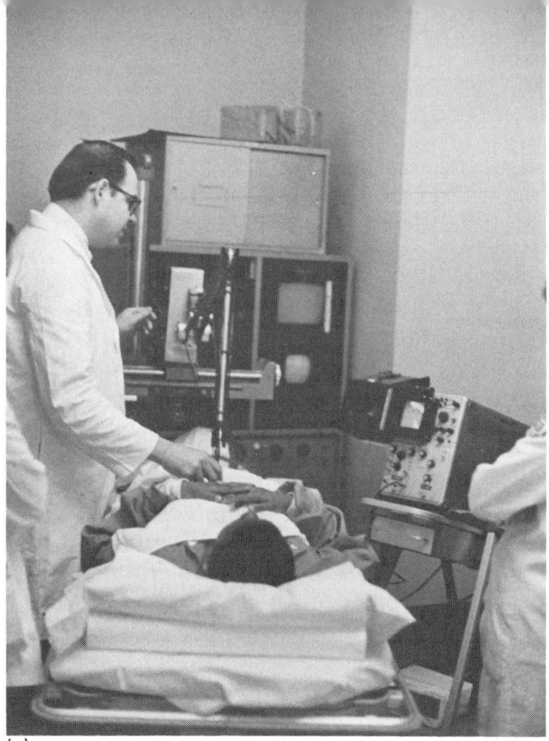

(a)

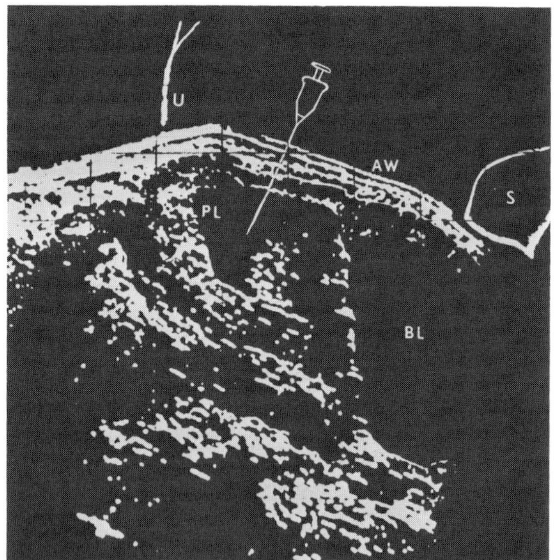

(b)

Figure 21–11 (a) An ultrasonic monitoring unit in action. (b) A sonogram taken by an ultrasonic monitoring machine. U = umbilicus; PL = placenta; AW = anterior wall (of mother); S = symphisis (pubic bone of mother); BL = bladder (of mother). The baby's head is on the left, feet on the right. The hypodermic needle is removing amniotic fluid for examination.

The Computer Revolution

The computing machine has been called the greatest creation of man, and it may well deserve that title. It is hard to conceive of any major segment of our society which is not now dependent in some important way upon computers. Train, plane, and boat schedules are regularly controlled by computers. Bills and checks are processed by computers. Student records in colleges and patient records in hospitals are stored and correlated by computers. Factories are run, sometimes completely, by computers. Planes and rockets are designed and flown through simulated flights by computers. Virtually wherever we turn, we find the computer.

The growth of modern computers from their origin a few scant decades ago has been explosive. The first modern computer, built at Harvard University in 1944, was an electromechanical device, and it was relatively slow in operation. From this original "Mark I" the computer evolved rapidly, aided by the explosive developments in electronics. Vacuum tubes replaced electromechanical mechanisms, to be replaced in turn by transistors, and more recently the ultraminiaturized, solid-state integrated circuits.

From the development of computers in the early 1940s to the present-day "third-generation" computers there has been a phenomenal increase in the speed of their operation. The earliest electronic computers performed perhaps ten or twenty calculations per second. By the late 1950s, the second-generation computers were performing several hundred thousand computations per second. Now Control Data's Star computer can perform more than 100 million operations per second! (A few machines of this type can outperform *all* of the more than fifty thousand computing machines in operation throughout the world by the end of the 1960s.)

What does all this mean in practical terms? Let us assume that the solution of a particular problem in genetics requires 12 million separate calculations. A skilled mathematician, using traditional methods, would require about two thousand *years* to solve the problem. A first-generation computer (say, of 1950 vintage) could have solved the same problem in about a hundred fifty hours of steady work (about six days). Thus, we have already passed from the realm of the impossible to the realm of the feasible. A second-generation computer could solve the problem in about two minutes. The Control Data Star computer could solve half a dozen such problems in less than a single second.

TYPES OF COMPUTERS AND HOW THEY WORK

There are two basic types of computers now in use: digital computers and analog computers.

In a **digital computer,** numerical data are fed into the computer and numerical data are received as the output. These conditions are not as limiting as they might seem at first glance. Although numbers are used as symbols in the digital computer, various codes can be drawn up in which specific combinations of number symbols represent letters of the alphabet or particular words or concepts. Even a picture can be rendered in terms of numbers by arbitrarily dividing the area of the picture into a finely ruled grid and then assigning to each point on the grid specific numbers (coordinates) to signify its location and others to denote the degree of blackening at the point. (This system is used quite successfully in the transmission of pictures by telegraph and television.) Thus, the digital computer has a wide potential applicability, and it is actually the type of computer that is used most widely in business and scientific areas.

The **analog computer,** on the other hand, operates in terms not of numbers but of measurable quantities, such as the distance between two points on a stick, electrical current in a conductor, or the acceleration of a rotating shaft. The input is a series of measurements of an actual system, and the output is in the form of a graph or other pictorial representation. This representation is frequently delivered in the form of a display on an oscillograph screen (Fig. 21–12). An analog computer can yield the results of theoretical variations of conditions, or it can be set up to control a process directly. Its results can be channeled to a

Figure 21–12 Drawing with a light pen on a computer's cathode ray tube display unit.

Figure 21–13 Part of a computer's memory. Wires interconnect the tiny ferrite cores and are used to store and retrieve information.

digital computer that converts them to numerical values. The results obtained by a digital computer can also be channeled to an analog computer, which displays them in a more meaningful form.

How effective a computer will be is determined by the information stored in it and the "program" that tells it what operations to perform. Information may be fed into a computer on punched cards or tapes, magnetic tapes, or even ordinary typed material. Or visual displays may be drawn with an electronic "light pen" on a screen that looks much like a television screen.

The information received by the computer's **input system** is stored in its **memory unit.** This unit may consist of magnetic tapes or discs. A common form of storage of computer memories is on magnetic ferrite cores (Fig. 21–13). These are small, doughnut-shaped rings less than an eighth of an inch in diameter that can be magnetized in either a clockwise or a counterclockwise direction. (The direction can easily and rapidly be switched to the opposite by applying an electric current to a wire running through the hole in the center of the ring.) Each core stores a single **bit** of information, and collections of them can be grouped to store larger amounts of information. Indeed, some of the larger computers today are capable of storing in excess of a million words.

A key part of a computer's functioning is the **program,** which tells it what to do with the information it has received at each stage of the calculation. The computer program must spell out every step in minute detail, for a computer does precisely what it is told to do—no more, no less.

A computer's **arithmetic unit** performs the arithmetical operations of addition, subtraction, multiplication, and division. It operates in a number system that is quite different from the one used in everyday life. The computer uses a **binary code,** in which every number is represented in terms of only two digits, 1 and 0 (Table 21–1). The programmer (a person) feeds information and instructions into the computer in a special language that is automatically translated into the binary-code "machine language" that the computer is equipped to handle. The activities of the computer need not be confined solely to mathematical manipulations. The binary system of symbolism also permits various logical operations, such as a sorting and tabulation of data according to whether they belong to one classification or another, or to both, or to neither. For example, a computer can be used to tabulate and evaluate the results of mass blood tests, automatically selecting which results fall within normal limits and which may indicate a possible disease state.

TABLE 21–1 Binary numbers

Decimal	1	2	3	4	5	6	7	8	9	10	11	12
Binary	1	10	11	100	101	110	111	1000	1001	1010	1011	1100

APPLICATIONS OF COMPUTERS IN THE LIFE SCIENCES

More than any other tool produced by the current cross-fertilization of the sciences and technologies, the use of the computer has already transformed the life sciences and promises to bring about even more profound changes in the future.

A random scanning of the "Applications" section of the magazine *Computers and Automation* yields items like the following:

- Computer-controlled air-monitoring systems are aiding in the enforcement of air pollution measures; in other studies, computers are helping to determine what levels of carbon monoxide can be dangerous to health.
- Computers are being used to study the migratory habits of birds and to map the ecological interactions in natural communities.
- A computer simulation of the action of the inner ear is giving one group of scientists new insights into how the human ear works and how certain types of hearing loss might be remedied. Another group is using computer simulation to study the interaction of hormonal secretion and female reproductive cycles.
- In hospitals, computers are being used in tests for early detection of lung disease, heart disease, and cancer; in statistical studies to determine predisposing factors among stroke victims; and to aid the survival of premature babies.
- Computers can be used to predict the properties of chemical compounds that have not yet been synthesized, and may thus aid in the discovery of important new drugs and other biologically active substances. Computers can scan microscope slides automatically for features of interest far more rapidly and efficiently than a human could. Computers now direct and control the tedious but crucial operations necessary for taking the thousands of readings in an X-ray diffraction study, and thus speed up the determination of three-dimensional structures of complex molecules.
- In another key area, computers are helping to catalog and make available the enormous masses of published results of scientific investigations. These paper mountains have grown to such an extent that it is nearly hopeless for a scientist to be able to keep up to date on all the important developments in his major field of interest, much less related fields and other branches of science. Computer analysis can place the data he needs at his fingertips.

And so it goes—from studies of individual cells to countywide mosquito control programs, computers have become an irreplaceable part of work and research in the life sciences.

MACHINES LIKE MEN?

One of the most exciting areas of computer research is centered upon attempts to duplicate processes of learning and reasoning as they occur in man.

A simple learning machine called *Learm*, demonstrated in 1954, was shaped like a worm and received information through three simple switch receptors in its head, one in the center and one on each side. *Learm* was trained in a T-maze, much like the flatworms used in animal learning studies (Fig. 21–14). At the junction of the T, its central head receptor was activated, causing it to turn either to the right or the left. At first it made either choice with equal probability. But then an object (a "reward") was placed in one arm of the T. When the object was hit by chance, the receptor on that side of *Learm*'s head was activated, putting a charge on its memory capacitor on that side. After receiving a reward on

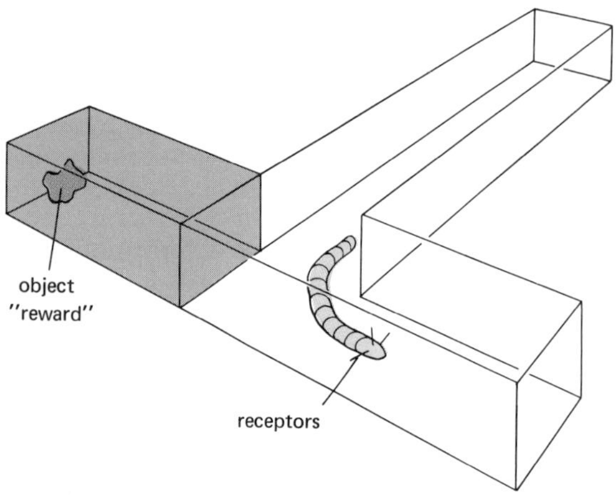

Figure 21–14 Learm, a simple learning machine, could be trained in a T-maze by "rewarding" it through simple switch receptors on its "head."

the right side of the T, *Learm* would tend to turn right with increasing probability. If turns to the left were rewarded instead, the probability of turning would shift to the left. Several successive rewards on one side resulted in consistent turning to that side, with no turns to the other. But if then the rewards were ceased, *Learm* would forget over a period of hours, and gradually the probabilities would even out again.

More sophisticated learning machines have been devised, simulating other aspects of learning in living organisms. Some have been programmed to recognize various patterns and to learn through experience to distinguish and classify new patterns. The Whirling Dervish, for instance, can learn to recognize general characteristics of, let us say, two basic classes of patterns. If a new pattern, never seen before by the machine, is presented, the machine will analyze the characteristics of the new pattern in the light of the remembered classes of patterns and judge which class the new pattern most resembles. The judgment is couched in terms of the relative probability of belonging to a particular class.

The UCLM II, at the University College, London, has been taught to recognize the faces of ten well-known personalities from photographs supplied by the BBC. The learning was a slow process—so slow, in fact, that it proved to be too tedious for human operators to teach the machine to recognize the faces. The job of teaching the "learning machine" was finally turned over to a specially programmed "teaching machine," which patiently displayed the photographs and corrected UCLM II over and over again until it had mastered them. After UCLM II had learned to recognize the set of ten pictures, it was able to identify the personalities when parts of the photographs were obscured, even when only the hairdo or hair and eyes were showing. UCLM II has also been trained to read a set of ten handwritten words that can be arranged into a variety of meaningful sentences and to recognize a set of ten digits, even when they are presented on cards spattered with obscuring inkblots.

In a program called "Student," devised at MIT, a computer was programmed to read statements of algebra problems in ordinary English, derive appropriate equations from the English statements, and solve the problems. When Student runs into difficulty in understanding or solving a problem, it may consult its dictionary files or may request additional information from the computer operator.

In another program developed at MIT, a computer was taught to analyze three-dimensional objects shown in a photograph, resolve them into their simple geometrical component forms, and then reconstruct the objects as they would appear from any requested angle, drawn in the proper perspective.

Relatively simple computers, possessing far less complexity than the human brain, have shown a surprising ability to master and utilize logical patterns presumably similar to those used by man. For example, the JOHNNIAC, an early computer of the mid-1950s, was programmed to prove theorems in symbolic logic. It was then presented with fifty-two elementary theorems from Russell and Whitehead's *Principia Mathematica*. The JOHNNIAC failed to prove fourteen of the theorems, but in one of the thirty-eight in which it was successful it found a shorter and more elegant proof than that originally used by the authors. Later-model machines, with similar programming and larger memories, could successfully prove all three hundred fifty theorems of the book in a matter of seconds.

In efforts to duplicate man's thinking processes, computers have been taught to play games, from tic-tac-toe to checkers and chess (Fig. 21–15). An automatic program suffices for a simple game like tic-tac-toe, but in a sophisticated game like chess the successful player must be able to plan ahead. Computers have mastered this skill. One computer programmed for checkers at IBM beat a former Connecticut State Champion in 1963. In the more complex game of chess, several programmed computers are now playing at the American Chess Federation's Class C level.

Most of the game-playing computers are restricted to a specific game or games and a specific set of programmed instructions. But Newman and Uhr are working on a model that is learning to play a wide variety of games by deriving its own set of independent parameters through its experiences in playing each particular game. This computer has been relatively successful in learning simple games such as tic-tac-toe after brief training periods, and it attempts to generalize what it has learned from one game to another.

Figure 21–15 Playing tic-tac-toe with a computer. This machine can be tied, but it cannot be beaten.

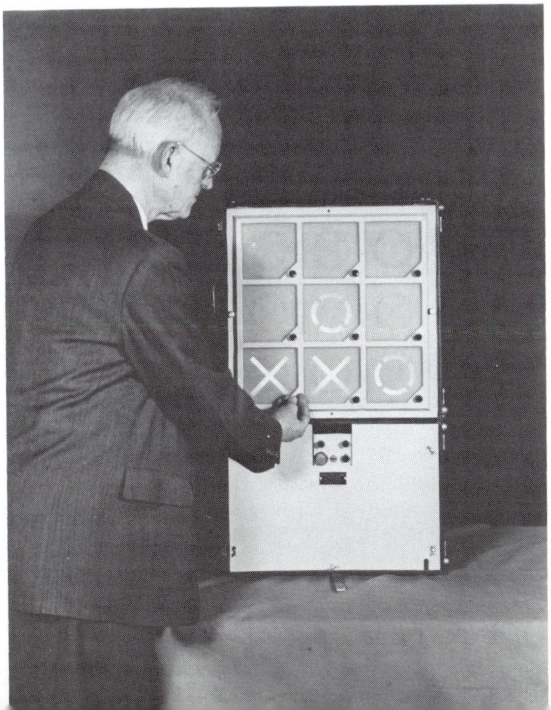

In an interesting offshoot of computer programming, several psychologists and psychoanalysts have been working with computers to design models of personality. A prototype of a neurotic personality has been programmed and studied in interactions with other personality programs. Another model, "Aldous," has been programmed to react to objects in its environment—forming positive and negative attitudes toward them and responding by approach or avoidance or conflict—and to provide simple "subjective" analyses of its responses. The latter are couched in such terms as, "I was slightly attracted, slightly afraid, and not angry. I was not tense."

In other studies, computers have been programmed to write poetry. Sometimes their compositions show an apparent native wit and depth that surprises their human creators. For example, a computer has come up with the following somewhat obscure couplet:

> To belch yet not to boast, that is the hug,
> The high lullaby's bay discreetly crushes
> the bug.

Another computer wrote:

> O panic not to this docile juice
> Finally few of my jackets did distrust the
> goose.

Computers seem to have some way to go before they rank with the best of the human poets. But studies such as these are not only aiding in the improvement of computer technology, but are also giving us a deeper insight into the workings of the human mind.

THE IMPACT OF THE COMPUTER ON HUMAN SOCIETY

The computer has already profoundly affected our world, and its far-reaching influences bring more changes with each passing day. Computers are bringing new benefits never before possible, but they are also bringing new dangers to the fabric of our lives.

In all areas of the business world, computers are taking over various routine tasks involved in the processing and filing of data. Accounts are kept up to date and bills are sent out under the direction of these machines. In fact, it is proposed that the use of computers will soon open the way to a "moneyless society," in which a person's earnings are automatically credited to his account and purchases are automatically debited, all with the aid of a handy, computer-coded credit card. In addition to routine tasks, computers are already being increasingly enlisted to aid in "middle-management-level" tasks—for example, keeping a running count of inventory and automatically ordering new purchases of items that are running low. They can also be used to prepare more effective overall pictures of company operation for the information of top management personnel.

Computers are gaining a growing role in education. Computerized "teaching machines" can provide personalized attention for each student, permitting him to concentrate on the areas of greatest interest and usefulness for him and to progress at his own speed (Fig. 21–16). The computer has infinite patience; it does not grow bored with a slow learner or mind repeating the same point over and over again until the student has mastered it. Small children have shown surprising ability to master the use of a typewriter input console and enjoy playing "learning games" with a programmed computer. Such programs have even shown some success in working with autistic children, those who are so withdrawn that they cannot communicate normally with people.

Figure 21–16 Using a teaching machine.

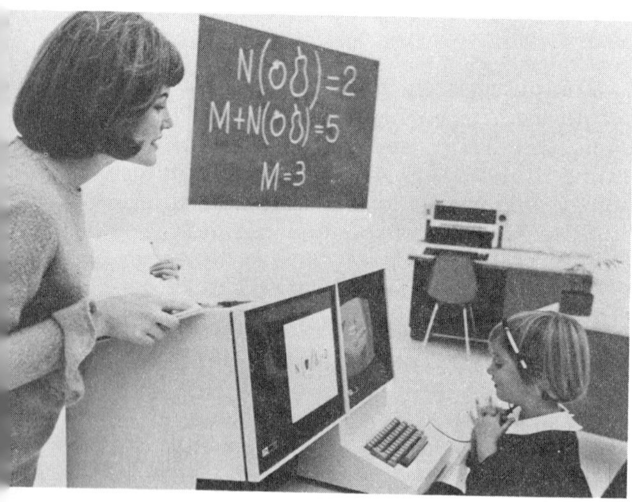

Skillful programming of teaching machines is a crucial factor in their success, and of course computers used in education have their limits. They cannot guide a spirited social studies discussion, for example, nor foster or evaluate manifestations of creativity.

In industry, computers are being used increasingly for the automatic control of machines of various kinds. The growing automation of manufacturing machinery relieves humans of many extremely tedious and repetitive jobs. Particularly in the petroleum, chemical, and metals industries, where production consists mainly of a continuous flow of liquid, dry, or molten substances that can easily be handled automatically, computers are being used to monitor and control production processes from raw materials to finished products. Automation is also increasing in the automobile industry, in mining operations, loading and unloading ships, and in various other areas.

There has been great resistance to automation among many industrial workers who fear that increased automation will result in the elimination of their jobs. It is true that automation does displace human workers, but the computer industry is also creating thousands of new jobs. The main problem is one of dislocation: the jobs eliminated are usually unskilled operations, requiring little in the way of intelligence or training; the jobs created, on the other hand, frequently require a high degree of skill or training. The impetus for headlong expansion of automation, bringing great benefits in increased productivity and efficiency, must constantly be balanced against humanitarian needs for a gradual phasing out of obsolete occupations and intensified job training for displaced workers.

Advances in computer technology over the past decades have resulted in a great reduction in the size of computers with a concomitant increase in their speed of operation. These developments have been accompanied by a steady drop in the cost per unit calculation. The overall effect has been an explosive increase in the usefulness of computers.

The memory unit of an average-sized computer of the early 1950s occupied a volume of 153 cubic feet, or the size of a walk-in closet 5 × 5 × 6 feet. In today's computers (Fig. 21–17), the same amount of information can be stored in a mere 3.3 cubic feet—small enough for a convenient desk-top model and readily fitting into an airplane or spacecraft cabin. Within a decade or two, it is expected that computers will be able to store an equivalent amount of information in far less than a single cubic foot.

Such design improvements have made moderate and large computer facilities accessible to many of the larger companies, but the cost of these computers is quite high. Although small models may sell for less than ten thousand dollars, the moderate- to large-capacity computers are usually leased for rental fees ranging from a thousand to several hundred thousand dollars a month. Sophisticated computer equipment has been placed within the grasp of even small users through a recent innovation called **time sharing.** A single computer is connected to a number of terminals. A fraction of each second of computer processing time is allotted to each of the terminals. The computer works so rapidly, performing so many computations per second, that each user's problem is solved in a reasonable time, while the computer is actually "dividing its attentions" among a number of different users. Time sharing can be effectively used in large organizations, such as research institutions and large manufacturing companies. It can also make computer services available to small consumers. Computations and other computer operations are performed in a large central computer, the terminals of which are rented to users who transmit input data to the computer over direct lines or through telephone services, receive output data in the same manner, and are billed only for the computer time actually used for their business. Extension of time-sharing facilities and expansion of data banks may ultimately make computer facilities available to individual families—one need only lift the telephone, for example, and have the resources of a superlibrary at one's disposal.

Figure 21–17 A modern computer installation.

The expanding use of computers has evoked concern on a number of fronts. The computer is highly efficient but totally impersonal. Anyone who has dealt with a computer-processed book club or department store charge account has probably acquired some insight into the problems of dealing with an impersonal system. Moreover, as the use of computers spreads into more areas of our lives, the potential effects of errors in programming mushroom. It would be a pleasant surprise to get a check for a few hundred thousand dollars accidentally credited to your account by a computer programming error; but what if someone else's purchases were deducted from your bank account, thus damaging your credit standing?

The expansion of data banks containing various items of financial and personal information about citizens is also evoking concern. Such data banks are quite useful to the business community, particularly in the area of credit financing. They could also be useful aids in job placement and in establishing effective systems of medical service. (If you were involved in an automobile accident in a strange town, the local hospital could dial the central data bank, punch in your identifying number, and automatically receive pertinent data on your blood type, drug allergies, and other items that would help in speedy and efficient treatment.) Yet it is feared that such data banks might be a serious invasion of privacy. Too much information in unscrupulous hands could be dangerous. (How much more efficient Hitler's genocide would have been if he had had the facilities of a modern computer-controlled data bank at his command.) With such possibilities in mind, the National Academy of Sciences established a Computer Science and Engineering Board in 1968 with an assignment "to assess the implications of the enormous and somewhat heterogeneous growth of information processing technology as it affects the public and private sectors of our nation."

A less imminent but perhaps ultimately more cogent problem lies in the expanding capabilities of computers. The computers that have been designed and built thus far have been far below the human potential in mentality, experience, and judgment. They can perform computations far more rapidly than a man, of course, but they are still merely tools, extending man's capabilities, rather than self-directed entities. But will there come a time when computers are complex and sophisticated enough to think, plan, and decide for themselves? Can manmade "machines" ultimately evolve into sentient beings? The present feeling among computer scientists is that this is definitely within the realm of possibility. What, then, will be the relationship between man and machine? Will the "robot" retain the role of man's creation or become his successor?

The answers to these questions lie in the future. But they are questions that man will have to face. His survival may depend upon it!

QUESTIONS FOR REVIEW AND THOUGHT

1. Explain why ever-increasing numbers of physicists, chemists, geologists, and engineers are entering biological research and discuss the unique roles they are playing.
2. How would you explain the late start of bionics as an independent science?
3. In addition to the obvious military applications of biosonar studies, what are some of the nonmilitary applications?
4. Describe the basic physiological and biochemical mechanisms of bioluminescence in the firefly.
5. Describe the functioning of *Gymnarchus*'s electrical location system and compare it to the dolphin's biosonar.
6. Discuss applications of holography.
7. What are the present obstacles to practical freezing of spare organs and even entire organisms?

8. Discuss some modern applications of ultrasonics. What animals are capable of perceiving some ultrasonic waves?
9. Computers can already perform calculations millions of times faster than human beings can. Why, then, are they so "unintelligent" in comparison with man? Can computer intelligence ever be raised to a level rivaling that of the human brain?
10. Discuss the impact of computers on modern life.

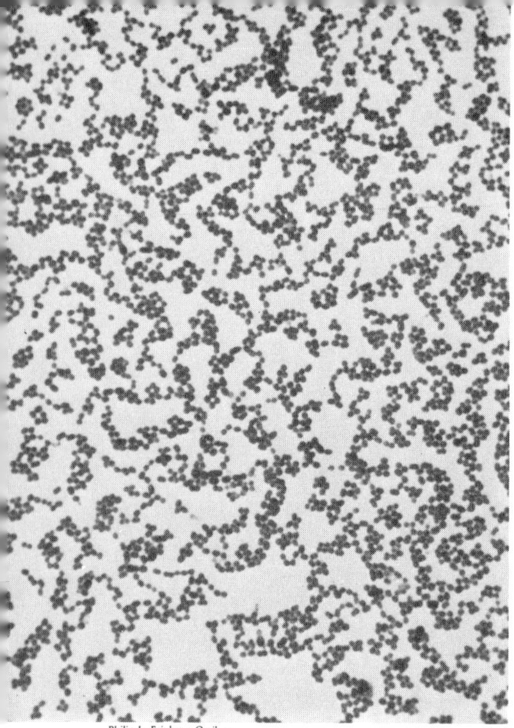

Philip L. Feinberg, Omikron

(a)

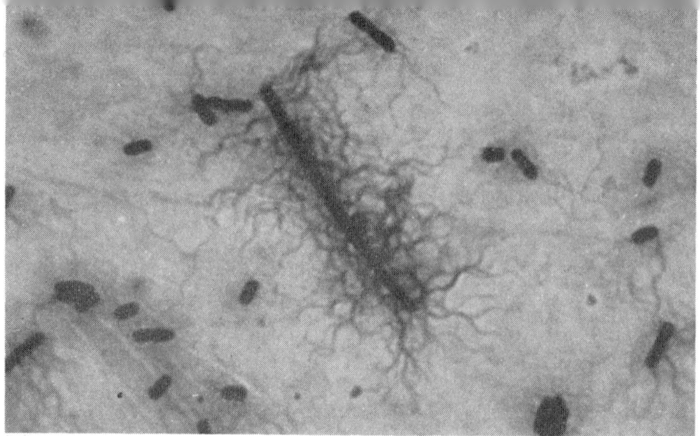

(b)

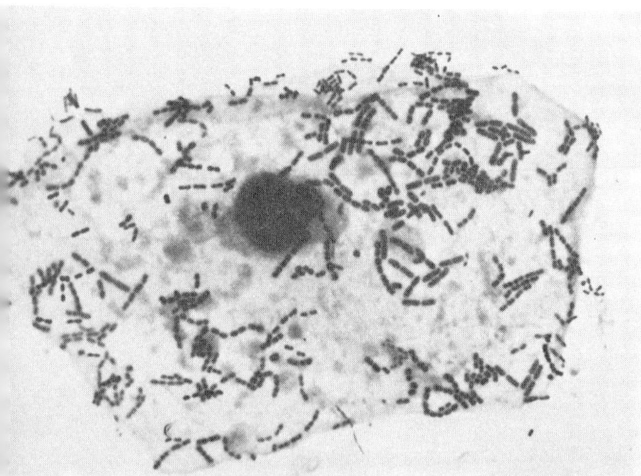

(c)

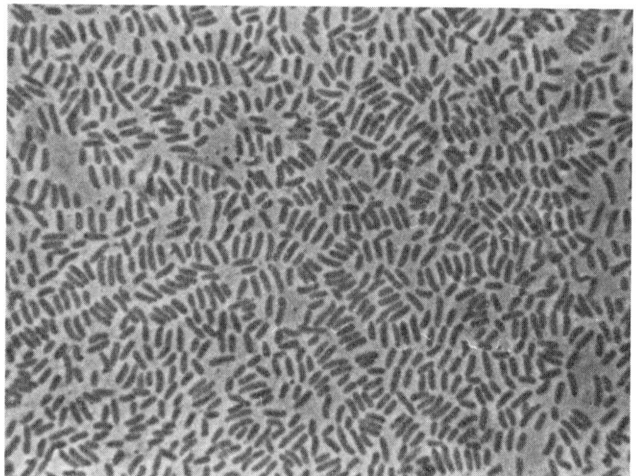

(d)

Philip L. Feinberg, Omikron

(e)

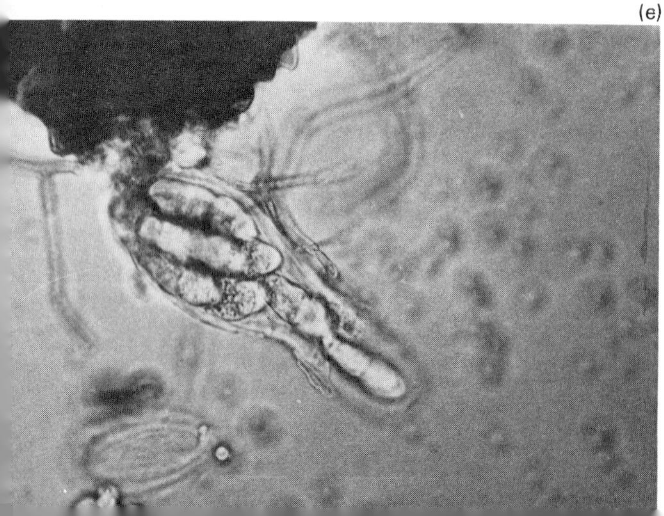

(f)

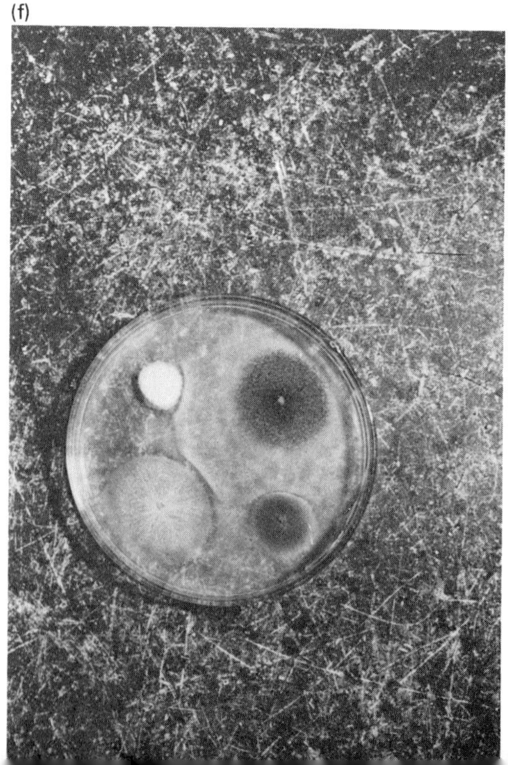

CHAPTER **22**

Frontiers of Biology

Introduction

In past centuries, the growth of the life sciences was closely linked with exciting progress in the exploration of our planet. As new frontiers were opened, and explorers brought back examples of the new life forms they found, theories on the interrelationships of organisms and the history of their development on earth gradually took form. Now the major land masses of our world are largely known, but there are still frontiers to be explored. Unexpected riches are being discovered as man penetrates into the ocean depths, and the first tentative forays out into space have sparked a ferment of speculation on what more exotic life forms may await us beyond the limits of our planet. Just as great a ferment has been stirred by the intensive exploration of man's inner frontiers—penetrating deeper and deeper into the cellular and molecular basis of life. Gradually man is acquiring unprecedented powers to know and even to change himself.

A Germfree World

We live in a world filled with microorganisms. Bacteria swarm on the surface of our skin, no matter how carefully we bathe. The soil we walk on and the water we drink contain swarms of microorganisms. Spores of bacteria and other microscopic creatures drift through the air we breathe, and an assortment of bacteria and viruses is sprayed out in tiny droplets of moisture whenever a person sneezes or coughs. Our bodies, even when we are healthy, are hosts to colonies of microbes, and when we are ill, disease-causing bacteria, viruses, or fungi multiply and spread through our blood or body cells. With its first breath, a newborn baby receives its introduction to our germ-filled world. Though we try to limit the infant's exposure to germs by sterilizing its bottles and disinfecting its diapers, its assortment of **microflora,** microorganisms living in its intestinal tract and other parts of its body, grows steadily.

Figure 22–1 Our germ-filled world: (a) *Diplococcus pneumoniae*; (b) *Proteus vulgaris*; (c) bacteria on an epithelial cell from the mouth; (d) hay bacilli; (e) spores of the potato blight mold, germinating; (f) mold colonies spreading over the surface of nutrient agar in a petri dish.

Just what is the nature of man's relationship to the ubiquitous microorganisms that share his world? Some are disease germs, which destroy body tissues and bring sickness or even death. But other microorganisms can live within the body doing no apparent harm at all. Decay microorganisms cause food to spoil and cloth and wood to rot. But such microorganisms play a vital role in the cycles of nature, breaking down the bodies of dead organisms to their chemical constituents and returning those to the soil, where they can be reused by living plants, which in turn serve as food for animals. Related microbes are responsible for the chemical processes necessary for the production of cheese, sour cream, and alcoholic beverages. Scientists have speculated that man is so dependent on the activities of microorganisms that he could not survive in a germfree world.

The first proposal that such a hypothesis could actually be tested was made by Louis Pasteur in 1885. He suggested that since the interior of a chicken egg is sterile (free of microorganisms), if means were provided to protect the hatching chick from contact with germs, it could be fed sterilized food and raised in a germfree environment. Pasteur believed that an animal so raised could not survive, but he did not have the time or energy to conduct such experiments himself. The decades that followed were marked by sporadic attempts to raise various types of animals under germfree conditions, with spotty success. Isolators were designed, providing means of working with the experimental animals through gloves sealed into the walls of the chamber (Fig. 22–2), but numerous difficulties, such as the problems of preparing adequate sterilized diets,* prevented researchers from maintaining animals in a germfree state for extended periods.

Germfree research finally became practical with the work of J. A. Reyniers and associates at Notre Dame University in the 1930s. Large, workable stainless steel isolators were built. (One was so large that a researcher, wearing a special diving suit, could enter through a tank of germicide and move about freely inside, attending to the experimental animals.) Problems of diet and rearing were laboriously solved, and a variety of animals were eventually born, reared, and kept through many successive generations without any contact with microorganisms. It was definitely established at last that life in a germfree world is possible for virtually any higher animal, as long as sufficient nutrients are provided. (For it was also definitively established that the normal microflora contribute various B and K vitamins to the nutrition of their hosts.)

Figure 22–2 A steel germfree isolator unit of the type Reyniers designed.

Despite the successes of the Notre Dame group, germfree research at first remained a scientific curiosity. For the most part this was due to the great expense of building isolators and the cumbersome procedures associated with maintaining and monitoring the germfree condition. But with the gradual improvement of techniques and materials and the steady acquisition of knowledge from these researches, scientists more and more began to realize the broad potentials that such studies could afford. As a result, germfree laboratories began to spring up throughout the world. Today there are several hundred such laboratories. In no small part, this explosion of research on germfree life has been due to two major factors: the develop-

*In those days it was not known that heat treatment destroys certain vitamins.

ment of low-cost, highly efficient plastic film isolators (Fig. 22–3) and the commercial availability of strains of germfree organisms that can be shipped to virtually any part of the world.

A wide variety of organisms have been raised and studied under germfree conditions, including algae, fungi, protozoa, numerous plants, planarians, insects, chickens, quail, turkeys, mice, rats, rabbits, guinea pigs, dogs, cats, sheep, goats, pigs, cattle, and monkeys. Some of these have been maintained under germfree conditions through dozens of successive generations. Human babies have even been delivered under germfree conditions and maintained in a germfree state. In one case, it had been feared that the infant suffered from a hereditary disorder of the immunological system, and physicians planned to give her massive transplants of immunogenic tissue. But tests revealed that the baby did not suffer from the condition after all; she was removed from her germfree plastic isolator, adjusted satisfactorily to the germ-filled world, and is now a normal, healthy child. A second child has been maintained under germfree conditions for well over a year. He was born without a thymus gland and hence lacked immunological defenses against disease.

Raising germfree organisms requires a meticulous attention to detail. Constant tests must be conducted to be sure that the "biological barrier" has not been breached and that the chamber has not become contaminated with microorganisms. Supplies and food are sterilized carefully, and then passed into the isolator through a sterile lock, which works on a design similar to the "airlock" used in submarines and spacecraft (Fig. 22–4).

Protists and other tiny organisms can be brought into **axenic** (germfree) conditions by isolation with a micropipette, followed by repeated washing with a sterile medium and reisolation. Selective treatment with combinations of antibiotics may be used before and during the process. In the case of higher vertebrates, it has been found that conditions within the egg (for birds and reptiles) or within the womb (in placental mammals) are generally sterile. Therefore, in the case of egg-laying animals, the external surface of the egg is thoroughly decontaminated with germicides, and then the egg is introduced into the germfree environment and hatched there. In the case of placental mammals, the young are delivered by caesarian section and transferred directly to the germfree environment. When these animals have grown up in the germfree environment, they then can mate, reproduce, and rear their young in the usual manner.

Numerous fascinating studies have been conducted on organisms raised in a germfree environment, and many important facts of their existence have been revealed. Organisms

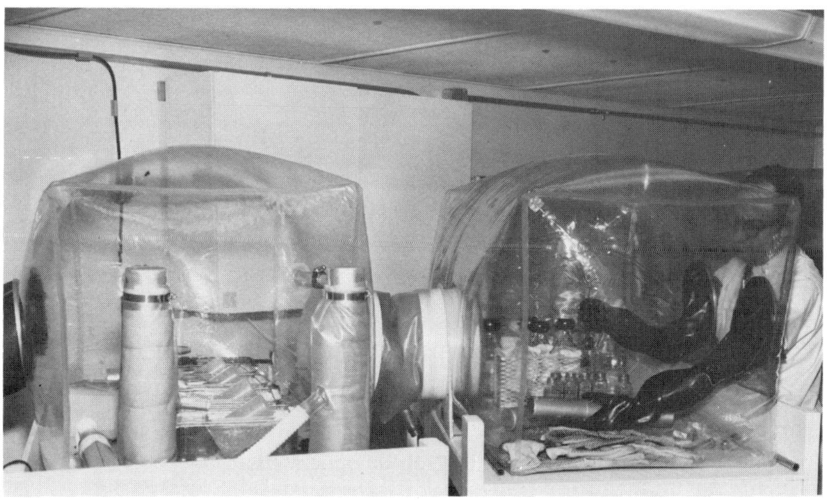

Figure 22–3 Plastic film isolators revolutionized the field of germfree research by making germfree setups accessible to all researchers.

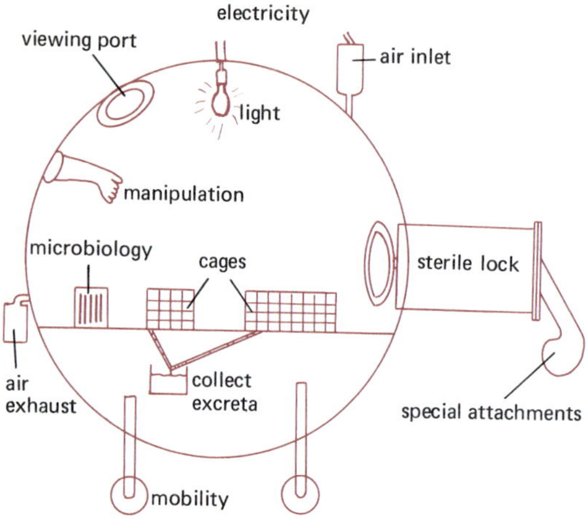

Figure 22–4 Basic parts of a germfree chamber. Numerous variations on the design all serve the essential function of maintaining a barrier against the entry of bacteria.

raised under germfree conditions appear relatively normal. They are just as vigorous as their contaminated counterparts, and show no special nutritional requirements other than for vitamins normally supplied by the intestinal microflora. They mature at the same rate as conventional animals and produce the same numbers of offspring. However, there are certain morphological differences: the weight of the reticuloendothelial system, lymphatic tissue, and the connective tissue in organs normally in contact with microflora is somewhat reduced. The feces of germfree animals are practically odorless, and when such animals die there is no decay.

The field of **gnotobiology** includes not only studies of the germfree state, but also investigations in which germfree animals have been deliberately exposed to one or more strains of microorganisms, the identity of which is strictly known. Such studies have already yielded a number of interesting and useful results. For example, rats raised under germfree conditions *never* suffer from tooth decay, even if they are fed a diet high in sugars. Rats contaminated with various strains of bacteria, on the other hand, do suffer from dental caries under similar conditions, and the nature of the tooth decay depends on the strain of bacteria used. A link has also been established between certain protozoan-caused diseases and accompanying bacteria. When large doses of the amoeba responsible for amoebic dysentery are administered to germfree animals, there are no signs of illness, even though the amoebic parasites multiply vigorously within their bodies. But if bacteria are injected simultaneously, all the characteristic symptoms of the disease appear.

Germfree animals are also being used for studies of the immunological apparatus, the toxicity and fate of new drugs, the origin and growth of tumors, the effects of irradiation, and nutritional studies, including the role of the normal intestinal microflora. They provide unique advantages as "controls," since studies can be conducted with them without distortion of the results produced by the effects of extraneous microorganisms. They also can be used in the bioassay of microorganisms. For example, samples of moon rock brought back by the *Apollo* astronauts were tested for the presence of microorganisms not only by inoculating them onto standard culture media, but also by exposing germfree mice and other animals to them (Fig. 22–5). (All of the tests were negative.)

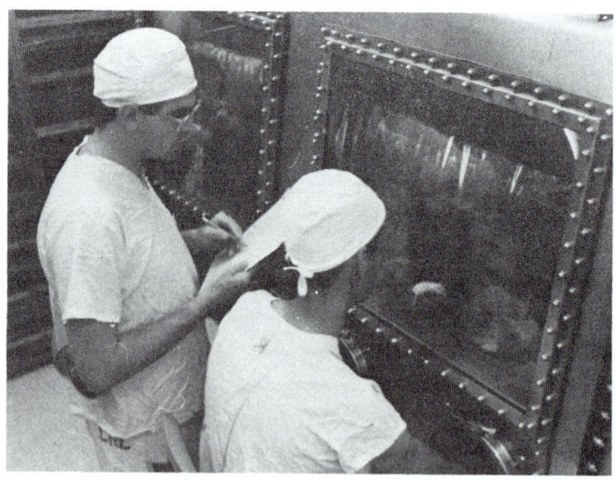

Figure 22–5 NASA technicians injected samples of moon rock material into germfree mice as a test of sterility of the specimens.

Today germfree life is a commonplace phenomenon in the laboratory. But what of tomorrow? Will there be dome-enclosed cities in which the inhabitants breathe sterilized air, eat sterilized food, sleep on sterilized beds, and drive in sterilized vehicles? At first thought, this would appear to be a logical and desirable aim. Communicable diseases could thus be effectively eliminated. But studies with animals have already given reason to doubt that this is indeed a desirable goal. For example, the immunological defenses of germfree animals are considerably lowered. (They still retain an immunological potential, but it must be developed by a gradual exposure to microorganisms. If a germfree animal is suddenly exposed to germ-filled conventional animals, its weak immunological defenses may be overwhelmed, and it may die.) Thus, if man did come to live in axenic cities, he could not travel out into the world beyond the dome without taking stringent precautions. For in so doing, he might seriously jeopardize his life by exposing himself to pathogenic organisms. And once contaminated he would have to be carefully decontaminated before he could return to the domed city. Thus, with a tenuous gain in security, man would lose a considerable degree of a most precious possession—his freedom of mobility.

Genetic Engineering and the Creation of Life

In September 1965, Charles C. Price, incoming president of the American Chemical Society, urged that the scientific community be mobilized to undertake a second *Apollo*-type program. This time the national goal would be the synthesis of life. This exciting proposal was met with almost universal apathy. Apparently neither the scientific community nor the general public believed that such a goal could be achieved in the twentieth century.

Since then, however, several dramatic events have occurred that have brought us much closer to the creation of life and in addition promise to give scientists the ability to manipulate the genetic apparatus of life. For the first time man is gaining the ability to remake himself. This exciting new branch of science, known as **genetic engineering,** is fast becoming one of the most actively researched areas of biology.

At a meeting of the American Association for the Advancement of Science at the end of 1968, noted geneticist Robert Sinsheimer spoke about the prospects for genetic engineering: "All that seems needed is the technology to transfer what we already know to be feasible in bacteria, carrot cells, or frogs to man. I feel strangely akin to the physicists who pointed out in the 1930s that the principles required for the release of the energy locked in the atomic nucleus were understood. Here, too, the principles seem in hand. All that seems really needed is optimism, sustained effort, and support commensurate with the importance of the problem."

Medical science has already tried a variety of approaches to cope with diseases and impairments caused by genetic defects. Eugenic measures such as genetics counseling are aimed at discouraging persons known to be carrying the genes of hereditary diseases from marrying and thus transmitting such conditions to their children. For persons afflicted with conditions due to hereditary defects, remedial measures may be effective. Diabetics, for example, are enabled to function relatively normally by regular doses of insulin, replacing a normal body product that their bodies are unable to manufacture as a result of a genetic defect. In the case of phenylketonuria, in which the genes manufacturing the normal enzymes for phenylalanine metabolism are defective, the use of a suitable diet can avert mental retardation and other consequences if the condition is recognized early enough in infancy. Hereditary defects of the immunological system have been successfully treated by transplants of bone marrow tissue and even by the transplant of a thymus gland. But scientists have long dreamed of repairing the genes themselves or controlling their action.

For example, the condition of diabetes results from the failure of the beta cells of the pancreas to produce enough insulin for the body's needs. This may be the result of degeneration of the beta cells, or the appropriate genes for insulin synthesis may be repressed. Theoretically, such a condition could be treated by adding an agent that removes the repression of the genes in question, either in the beta cells or in other cells of the body. (For the DNA of each body cell contains the complete blueprints for *all* of the syntheses carried out in the body. Each cell "reads out" only a small portion of its genes—i.e., those appropriate to its own particular functions. The rest of the genetic instructions remain repressed.) Alternatively, new genes for insulin synthesis that are not subject to repression might be introduced. Not too long ago, such genetic manipulation would have seemed to most scientists like a far-out, visionary approach. But recent discoveries have pointed the way not only to means of "making genes to order" but also to several avenues of introducing them into cells.

The first major event in this area was the in vitro synthesis of viral DNA by Arthur Kornberg, announced late in 1967. The DNA of the small virus ϕX174 consists of a single strand about six thousand bases long joined in a circle. Kornberg used the viral DNA as a template for the synthesis of a complementary strand that contained the heavier base analogue bromouracil in place of the normal thymine. The rings were broken apart and separated by centrifuging in a density gradient. (The heavier fraction, containing bromouracil, was the "artificial" DNA.) The complementary DNA, synthesized on the natural template, was found to be biologically active. But there was still a chance that some of the natural DNA might have slipped through the separation procedure. Therefore, the artificial DNA was used as a template for the synthesis of a "second generation" of artificial DNA. Since this new DNA was complementary to the form first synthesized, it therefore was an exact duplicate of the original, natural DNA (Fig. 22–6). The new artificial DNA, synthesized entirely in vitro, also proved to be biologically active.

Kornberg's achievement was hailed by President Johnson as "an awesome accomplishment," and was described by some enthu-

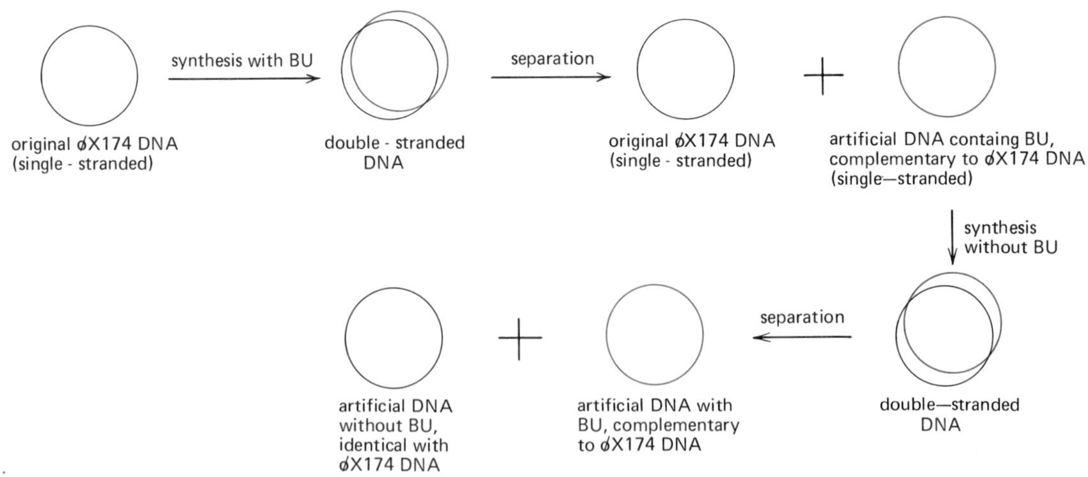

Figure 22–6 The Kornberg synthesis. Arthur Kornberg synthesized an artificial single-stranded viral DNA in vitro, using a natural viral DNA as a template. The use of the base analog BU made the first synthesized form heavier than the original DNA, so that the two could be readily separated by density gradient centrifugation.

siastic scientists as "essentially the synthesis of life." An awesome achievement it was, indeed, but Dr. Kornberg did not actually synthesize life. Rather, he provided appropriate conditions for a normal biosynthetic process to operate in a test tube. But a giant step forward occurred in 1970, when Nobel Prize laureate Har Gobind Khorana synthesized a gene from its component nucleotides without the intervention of components of living systems.

Khorana's achievement was made possible by his development of a new technique of nucleotide synthesis. Conventional methods of chemical synthesis have yielded only short polynucleotides, up to about twenty units long. Yet the synthesis of a gene would require the precise linkage of hundreds of nucleotide units. Khorana's technique, which he calls the "sticky end" technique, has made this possible. The method utilizes an enzyme called DNA-joining enzyme. This enzyme repairs broken strands of DNA in the living cell: it can join the broken ends of two pieces of a DNA strand together, provided that the complementary strand (or an appropriate segment of it) is present (Fig. 22–7). The complementary strand acts as a sort of specially shaped "anvil," onto which the two broken ends are fitted together and joined. Khorana uses the DNA-joining enzyme to join two polynucleotides that together are longer than the "anvil" chain. Dangling at the end of the assembled double strand, therefore, are a few units of single-stranded DNA for which there is no complement. This is the "sticky end." Another nucleotide sequence, with an end complementary to the "sticky end," can be joined to the shorter strand with the aid of more DNA-joining enzyme. This leaves a new "sticky end," on which the process can be repeated, and gradually a whole gene can be built up.

The gene that Khorana synthesized carries the code for alanine transfer RNA in yeast cells. This gene is extremely small, only

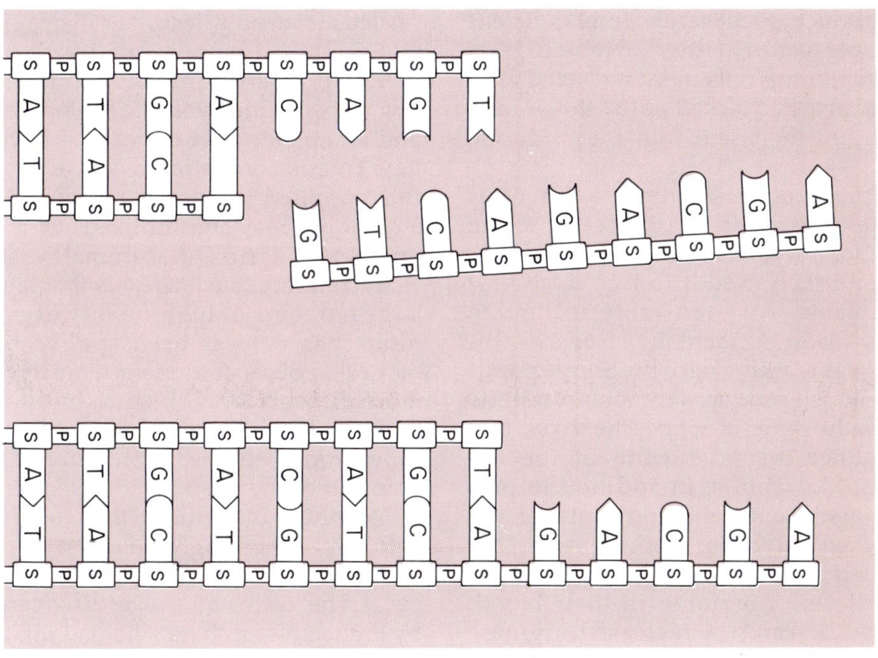

Figure 22–7 Khorana's "sticky-end" technique was used to synthesize a gene. In each step along the way, the complementary strand added was longer than the space being filled, leaving a dangling "sticky end" free for further reaction.

seventy-seven nucleotides long. But the techniques used in Khorana's landmark achievement can also be used to make large nucleotide chains—including the genes of human cells.

If we are rapidly gaining the ability to make genes to order, what about means for introducing them into cells? One method that has been proposed is the use of viruses. When virus nucleic acid enters a cell, it inserts itself into the cell's system of information flow. The viral nucleic acid may direct a replication of itself and synthesis of viral protein from raw materials supplied by the cell. Or it may attach itself to one of the cell's own genes and remain there dormant and "hidden." In experiments with bacteria, it has been found that an invading virus can transfer parts of the genetic material from one bacterial cell to another, adding the genes from its former host to the gene complement of its new host. Thus, viruses specially bred for desirable characteristics, such as the ability to induce the synthesis of a key enzyme, could be used to introduce such genes into deficient cells. Desirable genes, either obtained from normally functioning cells or synthesized artificially, could also be "tacked onto" the normal virus genes and then sent into the body via vaccination.

Exciting progress has been made along this line. Stanfield Rogers, at the Oak Ridge National Laboratory, has modified viral DNA chemically and then used it to introduce new genes into plants. An even more intriguing finding was made accidentally. Rogers's research team was working with the Shope papilloma virus, which produces tumors in rabbits but is not pathogenic in man. The virus was found to induce the production of the enzyme arginase in rabbits, in addition to producing tumors. Routine blood tests of researchers who had been working with the virus revealed that there was a significant drop in the level of arginine in their blood, apparently as a result of increased arginase production. Thus, the typical question that arises in the study of any new biological phenomenon, "Is it applicable to man?" seems to have been answered almost before it was raised!

Various techniques of cell culture seem to hold promise for the repair of genetic errors and other disorders. If an organ is damaged, for example, normal cells could be extracted, cultured, and then reimplanted to take the place of diseased or defective portions. During the culturing process, the genetic makeup of the cells could be altered, through the use of viruses or chemical or physical mutagens, and then the cells with desirable genetic properties could be selectively multiplied. An incredible product of new cell-culture techniques has been the growth of complete organisms from somatic cells (in contrast to the normal growth of the embryo from fertilized germ cells). Using this technique, called **cloning,** cytologists have been able to grow whole carrot plants from bits of undifferentiated callus tissue, and thriving trees have been raised from a few somatic cells. Even normal, living frogs have been grown according to the blueprint provided by frog somatic cells! In such experiments, the nucleus is removed from a frog egg cell, and then the nucleus from an intestinal cell or skin cell of an adult frog, stripped of its cytoplasm, is delicately inserted into the denucleated egg cell (Fig. 22–8). Embryonic development begins, and an apparently normal frog develops. This new frog is a genetic "twin" of the adult frog that supplied the donor nucleus. The cloning technique may be utilized to breed large numbers of agricultural animals with desirable characteristics, such as hens that lay unusually large numbers of high-quality eggs. The technique has not yet been applied to mammalian cells, because these are much smaller and more difficult to work with, but it is expected that it will ultimately be feasible even to grow human "twins" with genetic constitution determined by the nuclei of somatic cells.

Another intriguing line of research in cytology that might have applications in genetic engineering is the production of hybrid cells. The technique was discovered in 1960 by French researchers who had mixed together cultures of two different mouse cancer cells. After a while, cells of a new type were observed; they had resulted from the fusion of pairs of cells of the two types. Isolation of these new cells yielded pure cultures of

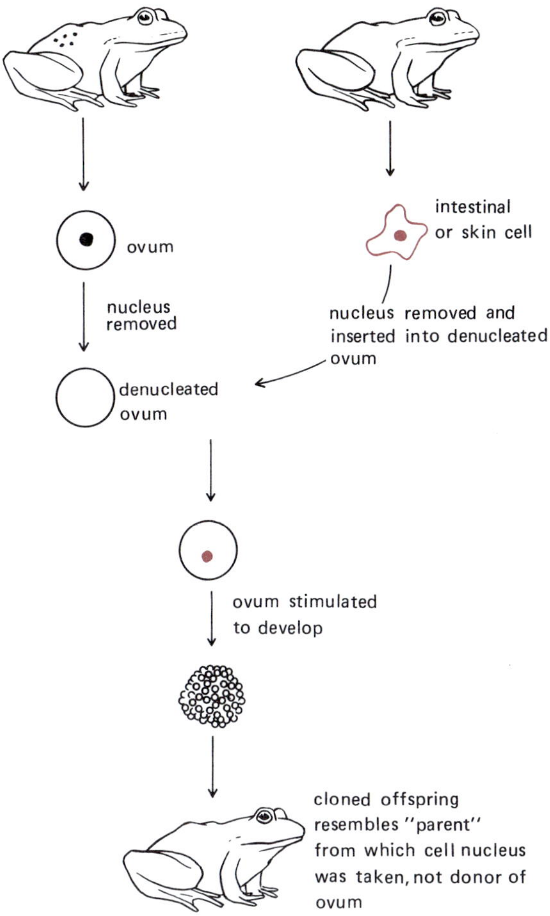

Figure 22–8 The technique of cloning can yield a virtually unlimited number of progeny genetically identical with the original "parent."

hybrid cells. Since then, a variety of hybrids of somatic cells has been produced, including such interspecies hybrids as mouse-rat, mouse-hamster, and mouse-human combinations. In the last case, the hybrid cells have been found to retain most of the mouse chromosomes, but they lose most of the human chromosomes. Studies of such cells are proving extremely valuable for determining the location of particular genes on the human chromosomes remaining. Cell fusion might also provide a means of introducing desired genes into the human body. In addition to hybrids of somatic cells, hybrid embryos, incorporating the characteristics of both sets of parents, have also been produced.

The exciting advances in the infant field of genetic engineering are viewed by some scientists and laymen with alarm. It has even been suggested that research in the field should be deliberately slowed down until man acquires the wisdom to use his knowledge wisely. But geneticists such as Sinsheimer, Kornberg, and Joshua Lederberg contend that the idea of applying new genetic knowledge to modify human characteristics is neither so new nor so horrifying as its opponents imply. "I would suggest," says Sinsheimer, "that those who feel this way are not among the losers in that chromosomal lottery that so firmly channels our human destinies. This response does not come from the 250,000 children born each year in this country with structural or functional defects. . . . It does not include, for example, the 50 million 'normal' Americans with an I.Q. of less than 90."

Oceanography

About 71 percent of the surface of our planet is covered by ocean. We normally speak of individual oceans—the Atlantic, the Pacific, the Arctic, the Indian. But if you look at a globe, you will see that these bodies of water are all interconnected into a single great ocean. We have made great strides in exploring the lands of our earth, and now we are reaching out into space to explore the nearest heavenly bodies. Yet we remain shockingly ignorant of more than two-thirds of our own world that is covered by water. We know more about the surface of the moon than we do about the bottom of the sea. We have achieved great expertise in sending spacecraft through precise trajectories at thousands of miles an hour, while oceangoing merchant ships still plow along at a scant 25 miles an hour. We are beginning an active search for life in outer space, while unknown life forms still await discovery in the depths of the oceans of our own world.

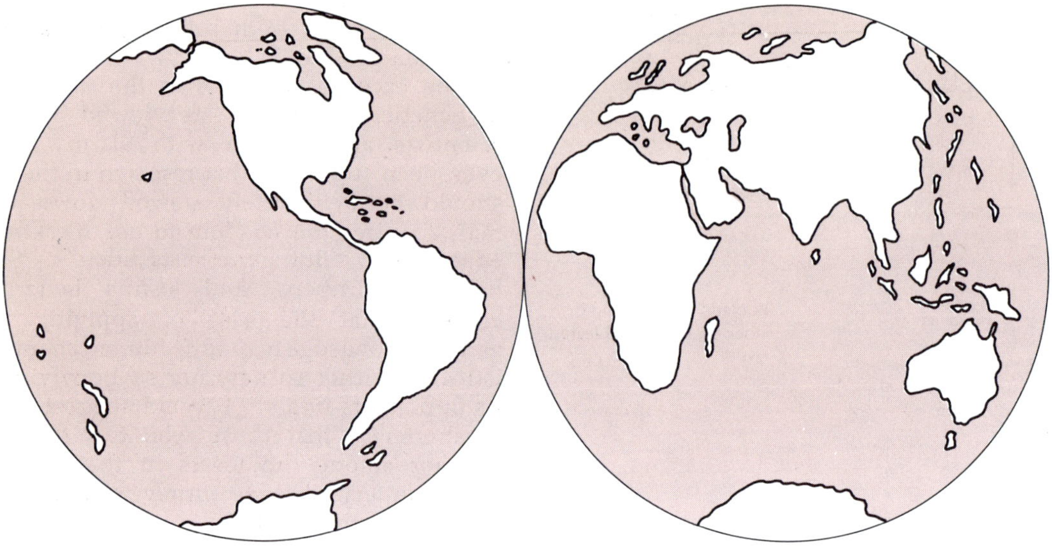

Figure 22–9 About 71 percent of the surface of planet Earth is covered by ocean.

Keen interest in the exploration of the world ocean is belatedly stirring. For the ocean holds a potential of great wealth. About 80 percent of all the animal life in our world lives in the sea. It is believed that the world ocean could yield 200 million tons of fish a year—high-protein food to feed the undernourished of the world, at a cost of perhaps a few dollars per person per year. Mineral riches, too, abound in the world ocean. The ocean waters contain some $30 quadrillion worth of gold—enough to make every man, woman, and child in the world a multimillionaire. On the shallow continental shelves underlying the ocean at the edges of the continents lie rich deposits of tin, diamonds, thorium, and magnetic iron. And beneath the shelves lie half the world's remaining resources of petroleum, along with huge quantities of natural gas, sulfur, and salt. Immense areas of the ocean floor are thickly strewn with large nodules of manganese, which also contain substantial quantities of other valuable metals, including nickel, cobalt, copper, lead, zinc, zirconium, molybdenum, and several rare earths. Underwater phosphorite deposits could prove a valuable source of phosphorus for fertilizer. Nuclear desalination of sea water could serve the multiple purpose of providing water for the irrigation of arid lands and concentrating valuable minerals, as well as furnishing cheap power to the surrounding area.

In addition to their food potential, the animals and plants of the ocean may some day be profitably harvested for a variety of other useful products. Some ocean creatures concentrate minerals from sea water to an amazing extent. A gram of iodine can be obtained from only 200 grams of certain types of dried seaweed, while the iodine concentration in the ocean water is only one gram in about 80 *tons* of water. Lobsters can similarly concentrate cobalt, oysters accumulate copper, and sea slugs concentrate vanadium. Sea creatures also produce a variety of physiologically active substances. Antibiotics have been isolated from such ocean denizens as marine algae, sponges, clams, conchs, squids, oysters, abalones, and sea snails. The sea cucumber, a curious creature that tends to spew out its viscera when it is disturbed, has yielded an extremely toxic substance called holothurin, which may have promise in the treatment of

(a)

(b)

(c)

(d)

Figure 22–10 Treasures of the deep. Some ocean dwellers selectively concentrate chemicals: (a) seaweed bladderwrack; (b) Sargassum sea slug. Others produce antibiotics and powerful toxins: (c) queen conch; (d) puffer fish.

certain heart conditions, as an anticancer agent, and to ease the pain of amputees suffering from destroyed nerve endings. Other potent poisons from octopuses and various types of fish are under intensive study for a variety of effects, such as slowing of the heartbeat, lowering of blood pressure, inhibition of blood clotting, and blocking of nerves. One type of cone shell (a gastropod that stings and eats fish) produces a powerful muscle relaxant, while another type produces a toxin that stimulates muscle contraction. The effects of both these cone shell toxins are reversible. The electric eel has yielded an effective antidote for certain types of pesticide poisoning, while a powerful hormone from the proboscis of a marine worm may be a cancer inhibitor.

A better knowledge of the ocean itself can bring great benefits to man. The ocean currents and the Arctic ice packs are powerful factors in the regulation of weather and climate on our planet. Studies of the ocean floor can yield new information on the nature of the geological forces that are still working to change the face of our planet.

With all the potential benefits to be gained from the exploration and exploitation of the world ocean, why has it taken so long for man to become seriously interested in such studies? The ocean, for man, is basically an inhospitable environment. Even aside from the hazards posed by sharks and other hungry inhabitants of the deeps, there is a basic insecurity about being in a medium to which man must bring his own supply of air to breathe. Gradually, however, the problems of living and working underwater are being overcome.

Various types of diving machines, such as submarines and bathyspheres, have been built. (It is said that Alexander the Great was lowered to the sea bottom in a diving bell to observe the underwater life.) These vehicles provide considerable protection against the deep-sea environment, but there is a concomitant loss of mobility. Mobility can be achieved by sending down an individual diver, but there are complications in this case too.

As the depth below the surface increases, so does the pressure of the water environment, rising to many times the value of atmospheric pressure at the surface. The body withstands the increased pressure well, but problems arise

Figure 22–11 The two-man deep submersible Alvin. The mechanical arm on the outside of the bathyscaph is used to gather specimens from the ocean floor.

when the diver comes up again. At increased pressures the solubility of most gases in the body fluids increases. As the diver comes up, and the pressure on his body is thus lowered, the nitrogen and other gases dissolved in his blood bubble out of solution. If the diver rises too rapidly, this release of gases can produce an excruciatingly painful and potentially fatal condition called **the bends.** Thus, the returning diver must take a long and cautious trip up to the surface, stopping at various stages to allow his body to adjust gradually to the reduced pressure. In the case of work at great depths, far more time may be spent in this process of **decompression** than in actual work at the bottom.

The increased pressures deep in the ocean bring other problems too. Too high an oxygen content in air inhaled at increased pressure may produce oxygen intoxication, which can lead to convulsions, brain damage, and death. Breathing large amounts of nitrogen at depths below about a hundred feet may lead to a form of nitrogen intoxication called "rapture of the deep." The victim has a false sense of well-being, a loss of judgment, and difficulty in performing even simple tasks. Generally mixtures of helium with oxygen (from about 2 to 10 percent oxygen) are delivered to divers' breathing apparatus in place of normal air. But even these measures are not ideal. Breathing a helium mixture tends to distort the speaking voice of the diver, so that he sounds like a "Donald Duck" recording, making communication difficult. Helium also seems to upset the body's thermostat mechanism, making divers feel the chill of the deep waters even more acutely than normally. Helium mixtures also become toxic below about fifteen hundred feet. Heated diving suits are now being tested to solve the problem of cold, and experiments to devise better breathing mixtures, such as oxygen-neon or even oxygen-hydrogen, are under way.

The problem of decompression is minimized to some extent by new techniques of saturation diving. It has been found that after a certain point, the time spent at high pressure does not increase the decompression time necessary. Now a diver can board a pressure chamber aboard the mother ship and be raised to the pressure of the depth at which he will be working. Then he rides a diving bell down to the work site. If he needs to rest, he is brought up, still under pressure, and remains in the pressure chamber on the ship until he is ready to return to the depths again. When his work is complete, he is gradually brought back to atmospheric pressure within the pressure chamber.

An even more revolutionary solution is the installation of "bottom habitats," underwater structures that can be used by divers as a home base without need to come up to the surface (Fig. 22–12). The underwater chamber contains air under pressure. The diver enters and leaves by a floor hatch; water will not enter because its pressure is counterbalanced by the pressure of the air. Decompression is unnecessary upon entering and leaving, because there is no change in pres-

Figure 22–12 Aquanauts leave their underwater home fifty feet below the surface of the Caribbean to begin a scientific experiment. They are participating in a Tektite II program to study the behavior of small groups working and living in an isolated stressful environment analogous to that of astronauts in space flight.

sure. Diving teams have already lived and worked at various depths underwater for periods of several weeks or more in several projects, including the French Conshelf, headed by oceanographer Jacques Cousteau, and the American Sealab projects.

During the Sealab projects, another promising avenue of undersea research has been explored: the potential of cooperation with some of the inhabitants of the sea. Dolphins have been experimentally trained to home in on a distress signal transmitted by a lost diver and to carry a rescue line. Work has also been done with sea lions and seals. It is hoped that such animals can be trained to carry messages and aid in rescue work undersea.

The Exploration of Space

Today we stand on the threshold of an adventure far more exciting than that of Christopher Columbus many centuries ago. For he opened up for his world merely a new continent or two. Our astronauts are now opening our vistas to new *worlds.*

Our space programs have progressed from the first suborbital rocket flights to the launching of artificial satellites and then to unmanned and manned landings on the moon and unmanned missions to our neighbor planets. Future plans for the coming decades include manned missions to Mars and perhaps beyond. Much experience has already been gained, but many problems still remain to be resolved. The development of space exploration involves some intrinsically biological problems vital to the success of the program.

The fields of space biology and space medicine deal with two major aspects of man's new adventure: the effects of conditions of the space environment on living organisms and the devising of workable life-support systems to maintain the health of astronauts during prolonged space flights.

A variety of experiments have been conducted to determine the effects of space conditions, both in actual flights and under simulated space conditions in laboratories on earth. In one ingenious series of experiments, rats were raised under an artificial gravity 4.7 times earth gravity, created by a centrifuge. The reactions of the rats to a return to normal gravity were tested as a key to potential reactions of astronauts to the reduced gravity of the moon and to the weightless conditions of space flight. A curious by-product of the research was the finding that the rats kept continuously at 4.7 G ("gees") lived longer than their normal counterparts, possibly as a result of the depletion of their body fat at the higher gravity.

Other laboratory simulations have tested the effects of isolation and confinement on human volunteers. But the most significant results on the effects of space conditions obtained thus far have naturally been from actual space flights. Orbiting laboratories containing exotic menageries ranging from wheat seedlings and pepper plants to vinegar gnats and frog eggs have been sent up and retrieved from orbit. Mammals including dogs and monkeys have been sent on space flights of varying durations. A variety of telemetry instruments, consisting of sensors and transmitting equipment, have made it possible to continuously monitor the heartbeat, respiration rate, sleep patterns, and other physiological parameters of animals and human astronauts during space flights.

A major factor of the space flight environment is the condition of weightlessness. The effects of weightlessness on living organisms were found to be largely what might be expected. Plants, whose direction of growth is normally determined by geotropism, seemed disoriented. The leaves of pepper plants turned downward from their normal horizontal position, and the roots of wheat seedlings, although apparently normal in structure, twisted about instead of growing downward. Animals showed the effects that might be expected from a sharp reduction on the body's systems of support and circulation: signs of decalcification of the bones, a reduction of muscle tissue, and deconditioning of the heart (a reduction of its ability to function effectively under stress). A disturbing finding was the discovery that bacteria multiplied more rapidly under weightless conditions than under normal earth conditions. If human cells also show an increased turnover rate, deleterious effects could result from prolonged space flights unless special precautions are taken.

Figure 22–13 Astronaut Ronald E. Evans takes a space walk during the Apollo 17 flight.

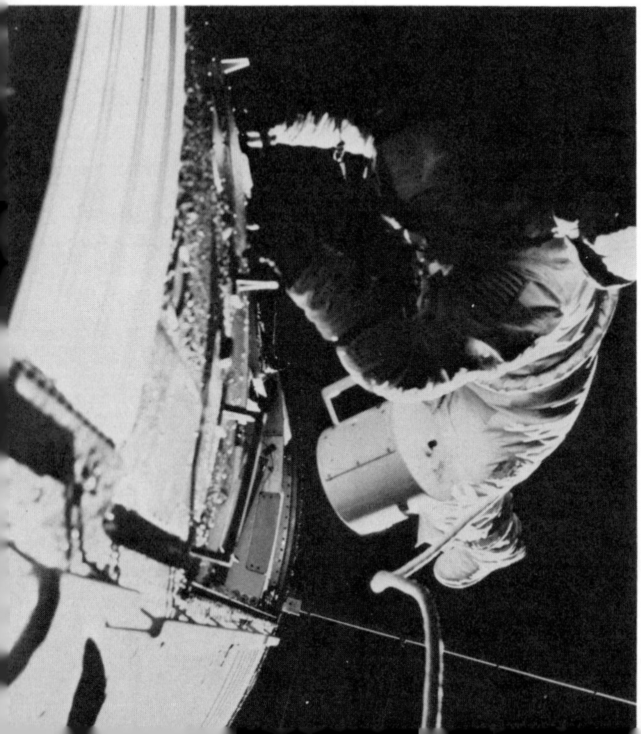

Before the first astronauts were launched into space, it was wondered whether man would be able to adjust to the weightless state and perform normal body functions such as swallowing and urination. As it turned out, these functions produced no difficulties, and the astronauts generally adjusted readily to a weightless condition and seemed to enjoy the sensation, although some instances of motion sickness were experienced. Cardiovascular deconditioning did require a more difficult adjustment to earth gravity after the flight was over, however.

One of the real hazards of space flight is the increased exposure to radiations, particularly during periods of increased sunspot activity. Biological specimens sent up in satellites have shown an increased mutation rate as a result of exposure to radiations. A number of dogs and also human astronauts have had normal offspring after flights in space, but the numbers have been too few and the time too short to draw any definite conclusions. We are still far from a sure understanding of the consequences of long-term exposure to the space environment.

Life-support systems for space flight must provide for the maintenance of a breathable atmosphere, the disposal or reutilization of wastes, and the provision of adequate food supplies and medical treatment. An atmosphere consisting of 100 percent oxygen at a pressure about a third of atmospheric was used in many of the early space flights. This has been recognized as unsatisfactory, both in view of the fire hazard (the *Apollo* tragedy that claimed the lives of three astronauts actually occurred in an atmosphere of pure oxygen at much higher pressure, however) and in the light of an observed loss of red cell mass in such an atmosphere. Apparently the nitrogen in our atmosphere is not as physiologically inert as had previously been thought, for during the first two *Apollo* flights, when a nitrogen-oxygen mixture was used during the beginning of the flight and the oxygen content never subsequently exceeded 95 percent, the loss was insignificant. (In the later flights in which the spacecraft cabin was decompressed during extravehicular activity and then refilled with pure oxygen, significant red cell mass losses were again observed.) Filters remove carbon dioxide from the exhaled air, permitting it to be reused to some extent.

On the *Apollo* flights, body wastes were jettisoned in space, but systems are being devised in which wastes can be recycled to yield pure drinking water and various other useful by-products.

Astronauts thus far have eaten foods packed back on earth and carried along with them. These have been mainly in the form of cubes of dried food that can be eaten in bite-size pieces, pastes in tubes, and freeze-dried foods in plastic moist-packs, which are rehydrated and kneaded before eating. Researchers are experimenting with a variety of foods that can be raised on board spaceships during prolonged flights. The possibility of using algae such as *Chlorella*, both as a food source and as a source of oxygen and a carbon dioxide scavenger, is being explored. Techniques of cell culture might also provide easily harvested sources of high-protein food. Spaceship "farms" of the future might also include such crops as lettuce, tomatoes, and sweet potatoes.

Once the problems of space travel are solved, man will probably attempt to colonize the moon and other bodies of the solar system. A whole host of new problems will arise: What will be the sources of vital oxygen, water, and food for the colonists? What will be the long-term effects of the reduced gravity on the moon, Mars, or the satellites of Jupiter? If moon colonists raised children there, would there be any genetic changes due to the decreased gravity? Would a child born and raised on the moon be able to adapt to the earth environment if he returned later in life?

How will these problems be resolved? Only time can tell.

Exobiology: The Search for Life on Other Worlds

Are we alone? Or may there be life out there—in outer space, on the planets visible to us or on some too distant to be seen? And if so, have these alien creatures been visiting us these past few thousand years? Are flying saucers real, after all?

If scientists had been asked a generation ago whether life exists on other worlds, most would have answered: "Probably not!" But today most scientists believe that there very well may be life on other worlds.

Why this about-face in the attitude of the scientific community? No single answer can be given. But a series of important new discoveries and theories have come to light in the past generation. Scientists have recreated in their laboratories conditions which probably existed here on earth billions of years ago. Under these primitive conditions they have succeeded in forming many of the basic building blocks of life: a staggering array of organic chemicals.

In addition, there has been a ferment of new ideas and theories on the origin of the universe and the birth of stars and planets. Now most scientists agree that there are probably more than 10^{20} (that is, 100,000,000,000,000,000,000) planets in the universe. Each of these worlds revolves around a star, much as our earth revolves about the sun. Scientists believe that many of these planets have passed through (or are passing through) processes similar to those that earth has gone through and at some stage of development have had an atmosphere similar to that of the ancient earth. Given these similar conditions, life may have appeared and evolved on many of these planets, as it did on earth.

The possibly inhabited planets of our universe doubtless have an incredible range and variety of forms of life. Probably on some, life has just begun to emerge from soupy seas of organic chemicals. Others may already have a great variety of life forms. Some of these forms might resemble some of our plants and animals, while others may be akin to nothing we can imagine.

Most scientists would agree that once life has been established upon a world, then, given time, it will evolve into more adaptable and ultimately more intelligent forms. A world-renowned astronomer, Harlow Shapley, has estimated that there may be as many as 100 trillion planetary systems which harbor life at "our level of sentiency."

But is "our level of sentiency" necessarily the limit in the evolution of intelligence? Or does man merely represent an intermediate step along the way?

Many of the inhabited worlds of the universe are doubtless millions (and perhaps billions) of years older than ours. In other words, their inhabitants have had *millions* (or even *billions*) of years to advance to levels far beyond ours. We cannot begin to comprehend just what direction this further evolution might take. There might be beings in the universe whose technology has advanced to a degree of sophistication that would make our greatest achievements seem like children's toys in comparison. There may be beings in the universe who have abandoned technology and developed the powers of mind to an extent that we could not imagine.

But do we have any *evidence* that life forms exist elsewhere in the universe? And if life is so ubiquitous throughout the universe, why have we not been visited by some of these life forms already?

Perhaps we have. Perhaps the flying saucers *are* real. During the past two decades, literally tens of thousands of UFOs (unidentified flying objects) have been reported. Many of these reports have come from responsible and competent observers, such as airline pilots, radar operators, and law-enforcement officers. Most of the sightings can readily be explained by natural phenomena, such as high-flying weather balloons and mirages due to temperature inversions. But hundreds of well-documented sightings remain unexplained.

However, a sobering fact is that no one person has yet offered concrete and incontestable evidence of having met an extraterrestrial, despite several sensational reports of meeting with Venusians, Martians and other aliens.

There are reasons for believing that a visit from aliens, if they do exist, would not be a commonplace event on our world. The distances between stars are so incredibly great that it would take more than twenty-five thousand years to travel to the *nearest* star, even traveling at the great speed of 100,000 miles an hour. In addition, our solar system is tucked away in a near-isolated edge of our galaxy (Fig. 22–14). And there are very few stars even within the incredible distance of 100 trillion miles. Hence, the chances of our being visited by any sentient beings would be quite small.

As to concrete evidence for the existence of *intelligent* beings "out there," at present we have none. However, some recent findings seem to indicate that at least primitive forms of life may be found elsewhere than on earth. A number of scientists have reported finding

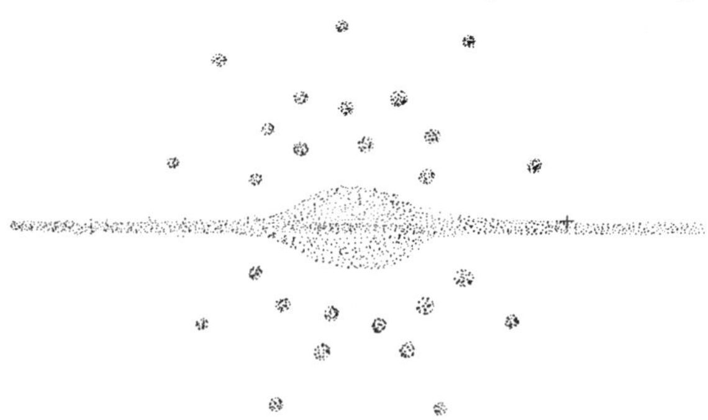

Figure 22–14 An edgewise view of the Milky Way Galaxy. The + marks the position of Earth.

complex organic materials in meteorites which have fallen to the earth. Others have even reported what appear to be fossils of primitive life forms. These findings at present are highly controversial. Some critics have asserted that these "traces of life" are merely contaminants from the earth itself. Others have stated that they may have arisen from inorganic or nonliving sources.

As man moves out into space, to explore the moon and the nearby planets, he will seek life wherever he goes. An important part of the *Apollo* missions was the collection of samples to be returned to earth for a meticulous analysis. Samples of moon rock were inoculated onto culture media and used to "challenge" germfree animals, in order to detect any possible forms of microscopic life. Rigorous quarantine precautions were imposed to minimize the danger of the returning astronauts' infecting the rest of the earth's population with possible exotic "moon bugs," to which humans might not have any natural immunity. For our lifeless moon, these precautions turned out to be unnecessary. But the experience gained in this "dress rehearsal" may ultimately be of life-saving value to earth if our space explorers do some day encounter a world of alien life.

The first explorations of Mars and other planets will probably be by unmanned missions. Ingenious devices have been developed for collecting samples and testing them for the presence of simple forms of life (Fig. 22–15). The Wolf trap, named after its inventor, Wolf Vishniac, consists of an evacuated hollow tube terminating in a fragile probe. When the tip of the probe is shattered by striking the surface, dust will be sucked up into the tube, which contains exposed culture media in which bacteria might grow. The Gulliver life-detection device contains three sticky strings with bullets attached to the ends, which are shot out of the device by a propulsion mechanism and then reeled back into a chamber containing a uni-

(a)

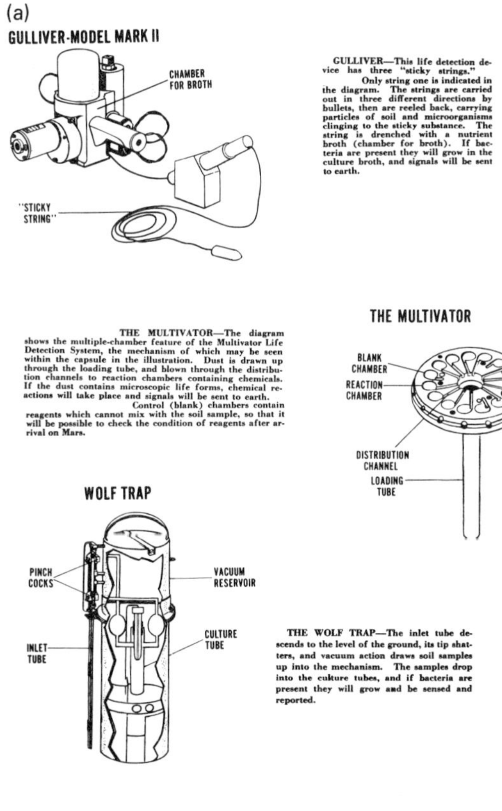

Figure 22–15 The search for life on other worlds: (a) three types of life-detection devices to be carried on unmanned space probes: the Gulliver, the multivator, and the Wolf trap; (b) an artist's conception of the Gulliver in action.

(b)

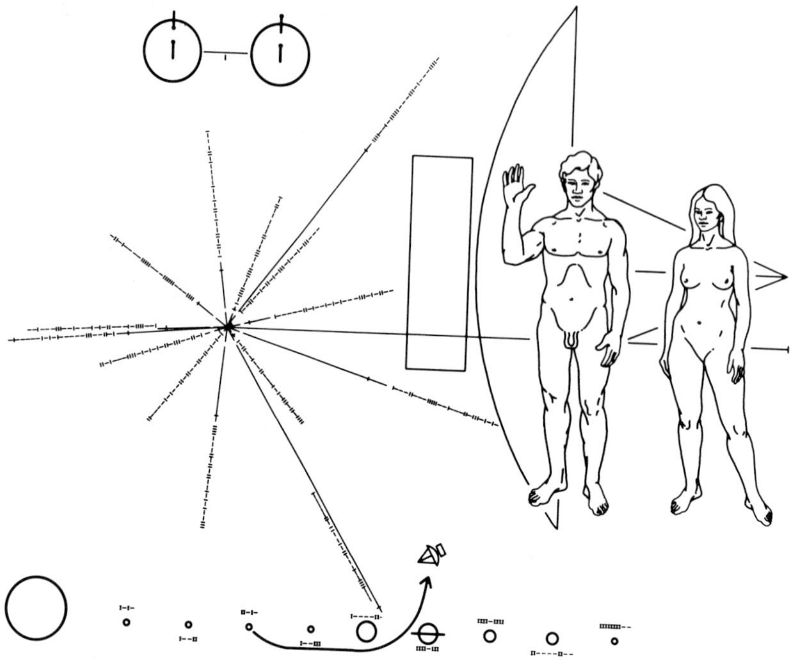

Figure 22–16 An attempt at interstellar communication: this plaque was affixed to the Pioneer 10 space probe.

versal culture broth. In each case, if there is any growth in the culture media, the increase in turbidity will be detected by sensors and signals will be sent back to earth.

If there is life "out there"—on other planets, in our solar system or on planets of other stars—what would it look like? It is generally believed among scientists that the conditions of evolution on the various planets would be specific enough that any highly developed extraterrestrial life forms probably would not look very much like earth organisms in fine details, although there might be some basic similarities in general form. (There are, after all, only a limited number of workable possibilities for locomotion, light reception, and so forth.) Individual differences would be dictated to some extent by special conditions of the planets on which life evolved. The inhabitants of a planet revolving about a red star, for example, might well have vision extending well into the infrared region. The existence of sensors of a type that we do not possess is entirely conceivable. (Even on our own planet there are creatures that can perceive forms of radiation to which we are "blind." For example, bees can see ultraviolet light, and certain species of fish have specialized receptors of electricity.) Although science fiction tales of extraterrestrial humanoids so like humans that they can even mate with our stock are pure fantasy, it is believed that protoplasmic forms of life elsewhere most likely are also based on the intricacies of carbon chemistry. (More exotic speculations have pictured stonelike silicon-based creatures and ethereal "beings" consisting wholly of organized energy patterns.) Concrete plans for the search for life outside our planet have thus far been predicated mainly on the idea that only very simple organisms, such as bacteria and molds, will be encountered. There have been a few attempts to devise a "universal language" to be used in attempts to communicate with intelligent beings (though we have not yet gotten very far in learning to communicate with the "intelligent aliens" on our own planet—the dolphins). Such attempts, however, do not seem particularly serious or urgent . . . yet.

QUESTIONS FOR REVIEW AND THOUGHT

1. Why is the raising of germfree organisms such a difficult task?
2. Summarize man's progress thus far toward the synthesis of a gene. When genes are readily manufacturable, what will be the impact upon our society?
3. Man may someday have the power to shape his own evolution. What organization would you suggest to ensure that man makes rational use of this power? How can it be determined which trends of artificial evolution are desirable?
4. Do you think that Price's proposal of an all-out effort to create life is realistic and desirable? Explain your answer.
5. Why is the ocean considered to be an important frontier of the future?
6. How can sea organisms be used in the future to help mine the mineral resources of the ocean?
7. What are the unique problems involved in exploring the ocean bottom, and what solutions have thus far been worked out?
8. Some scientists look upon the moon as an ideal laboratory. Explain.
9. What physiological effects do you think might result from a visit of a fourth- or fifth-generation lunar colonist to earth?
10. If in truth intelligent life exists on many other worlds, why have we not already been visited by alien delegations?

CHAPTER 23

Frontiers of Medicine

Microbes and Disease

Through the ages man has known and feared the merciless scourges of disease. Yet until the past century or so he has had little inkling of the causes of disease. This lack of knowledge added to his fear. How can one hope to gain safety from a foe whose very identity is unknown? Man searched desperately for causes, as well as cures. Evil spirits were often blamed for the effects of disease, and folk remedies were to be found everywhere (indeed, many still persist today). Some of these, discovered empirically or accidentally, were quite effective. Scientists are more and more turning to an investigation of ages-old herbal remedies, and some of their active components, such as digitalis and reserpine, are used as potent drugs today. But many of the folk remedies, as well as those used by physicians, were of little avail, and frequently killed more patients than they cured. Since the causes of disease were unknown, there was no scientific basis for searching for means of prevention and cure.

But through pioneering efforts of such scientists as Pasteur and Lister, the attention of the scientific world was riveted upon the microbial world. Here were the villains. Contagious diseases were caused by germs, microscopic parasites that were transmitted from one infected organism to another.

With the passage of time, disease after disease was associated with a particular microbe or group of microbes. With the advent of sulfa drugs, and then penicillin and the other antibiotics, marked advances were achieved in the control of various bacterial diseases. Improved techniques of vaccination led to equally dramatic advances in the prevention of viral diseases.

Other important advances have been made in identifying and to some degree controlling numerous parasitic diseases that afflict man in various regions of the world. Some progress in this field has been achieved through the development of improved methods of treating instances of the diseases. But even more important has been the progress in tracing the life cycles of the organisms responsible and discovering key vulnerable stages in which the chain of infection can be broken.

One of the great dreaded diseases of our time is cancer. Its cause in man is unknown.

Granger Collection

M. PASTEUR DANS SON LABORATOIRE

LES EXPÉRIENCES DE M. PASTEUR SUR LA RAGE

Dessins d'après nature de M. Renouard.

Figure 23–1 Louis Pasteur in his laboratory, working on a treatment for rabies. (Engraving from a French newspaper of 1884.)

Indeed, it is widely believed that there is not a single cause, but many causes. For it is believed that cancer is not a single disease, but instead a multitude of diseases with one common element: uncontrolled growth of cells.

Bacterial Diseases

The smell of death was in the air. And more terrifying still, its next victims could be anyone. The townspeople furtively eyed each other, watching for the first telltale signs. Even a suspicion of a symptom could lead to immediate quarantine, isolation from the group, even abandonment. Children lay dying—alone—in boarded-up houses. The stricken huddled in pitiful groups beyond the walls of the town. And through it all the acrid odor of burning flesh pervaded the air, as the bodies of once-loved ones were quickly and unceremoniously thrown upon the flames. Often there were none to mourn them, as whole families were swept away in the tide of the Black Death.

This scene was repeated again and again, for more than a generation, throughout the towns of Europe. For this was the great plague that swept through Europe in the fourteenth century. In many areas this terrible pestilence killed more than three-quarters of the population in the first outbreak alone. Throughout the centuries, outbreaks of the plague have brought death and terror to large segments of the world's population.

Until the end of the nineteenth century and the early years of the twentieth, the origin and transmission of the bubonic plague remained a mystery. Then, in 1894, this disease was traced to a bacterium, *Pasteurella pestis* (also called *Bacillus pestis*), during the virulent Hong Kong epidemic. Intensified research on the transmission of the disease to man implicated the rat as part of the cycle, and in 1898 it was suggested that the disease was spread among rats and humans by rat fleas. Definitive experimental proof of this mode of transmission of the disease was reported in 1908.

With the identification of the phases of the life cycle of the plague organism, sanitation and other prophylactic measures have brought this dread disease under control. In addition, the derivation of nonvirulent strains of the plague bacillus in the laboratory has led to the production of successful vaccines against bubonic plague.

We have briefly traced the history of but a single bacterial disease that afflicts man. But there are literally hundreds of different bacteria which can cause illness and even death of humans. For instance, in addition to *Pasteurella*

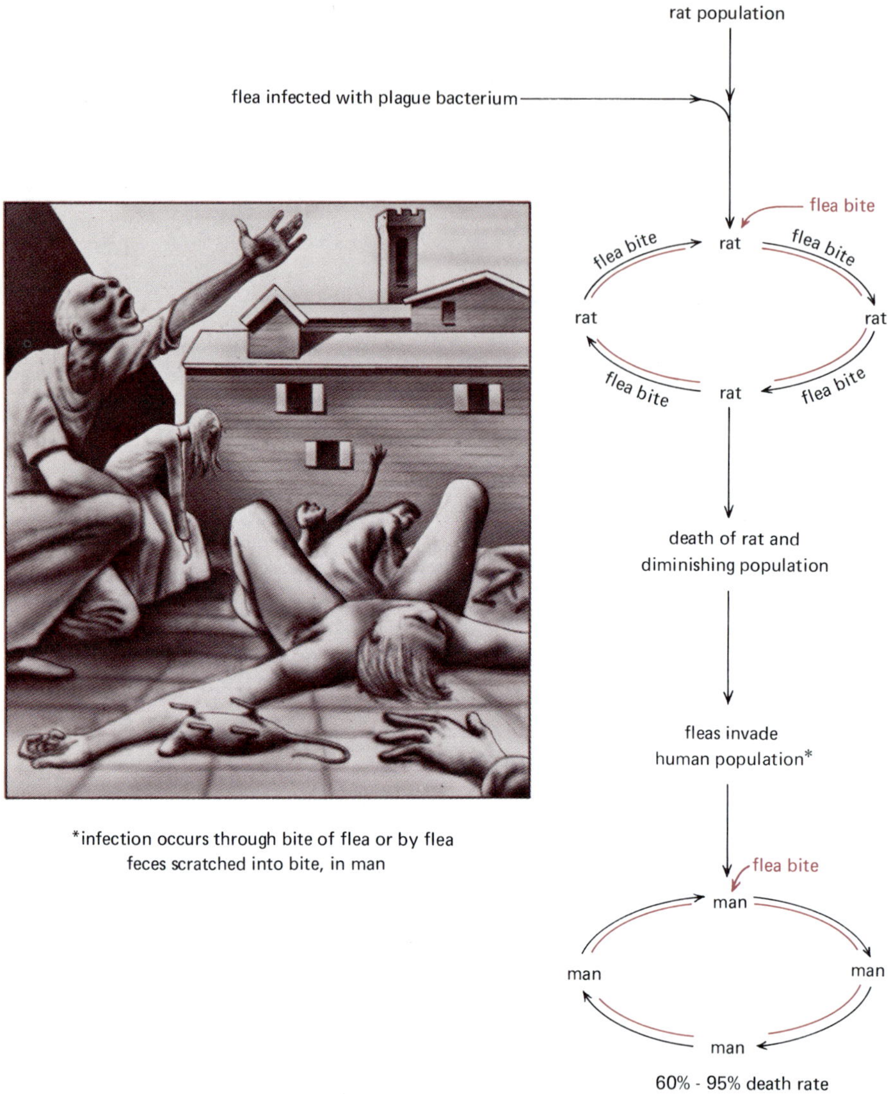

Figure 23–2 The Great Plague of the middle ages was a bacterial disease, transmitted to rats and humans by a flea intermediate host.

pestis, which causes the dread bubonic plague, bacteria cause such varied and serious diseases as typhoid fever (*Salmonella typhosa*), tuberculosis (*Mycobacterium tuberculosis*), cholera (*Vibrio comma*), syphilis (*Treponema pallidum*), pneumonia (*Diplococcus pneumoniae*), scarlet fever and rheumatic fever (*Streptococcus pyogenes*), and food poisoning (many species of the genus *Salmonella*).

The causes and means of transmission of these and other diseases that had plagued man through the ages remained obscure until the nineteeth century. The epic work of Pasteur in 1865 in the isolation of the tiny parasite causing a disease that was killing silkworms and threatening the existence of the French silk industry led to the **germ theory of disease.** According to this theory, com-

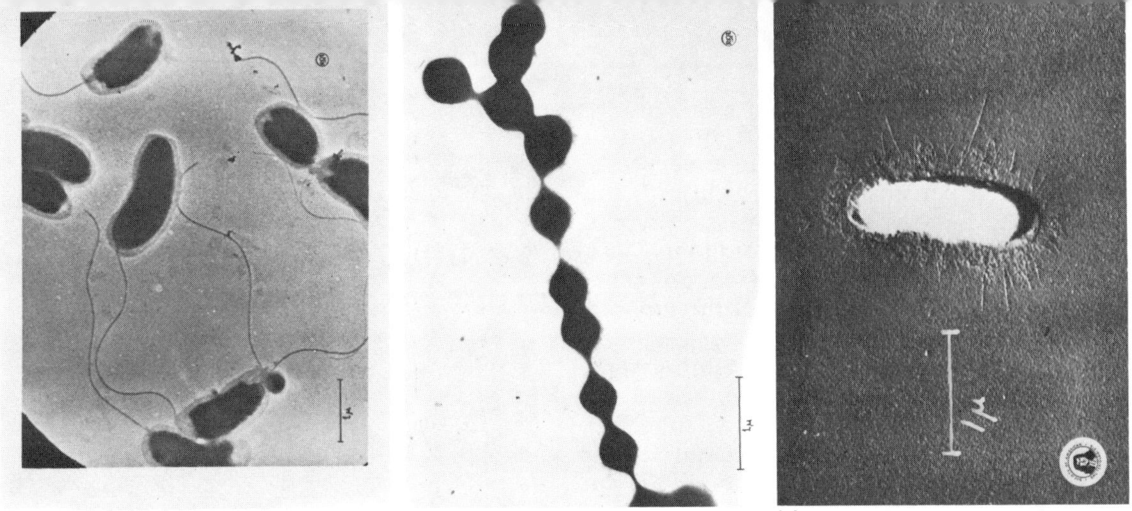

(a) (b) (c)

Figure 23–3 Disease bacteria: (a) *Vibrio comma*, the cholera pathogen; (b) *Streptococcus pyogenes*, pus-forming bacteria; (c) *Salmonella typhimurium*, a food poisoning bacterium.

municable diseases are caused by microscopic parasites, that are transmitted (directly or indirectly) from one infected organism to another.

That this theory is also pertinent to diseases in man was definitively established through the experiments of Robert Koch, a German physician who first isolated the bacterium responsible for the cattle disease anthrax and then focused his attention on the dread human disease tuberculosis. In 1882 he discovered the causative factor in tuberculosis, the tubercle bacillus, *Mycobacterium tuberculosis*, and formulated rules for identifying the causative agent of a disease that are still used today. According to this set of rules, now called **Koch's postulates,** (1) the organism must be found to be *always* present in the diseased organism; (2) the suspected organism must be isolated from the diseased plant or animal and grown in pure culture;* (3) the isolated organisms, after growth in pure culture, must produce the symptoms characteristic of the disease when reinoculated into healthy animals or plants; (4) the newly diseased animals or plants must yield the same microorganisms as those found originally.

With the firm establishment of the germ theory of disease, scientists had many new avenues for the study, prevention, and cure of disease. Intensive investigations of more and more diseases revealed the microscopic villains at work. And with the causative agents identified, scientists could examine more intelligently the various means by which they could be transmitted.

In general, infectious diseases can be transmitted in various ways. The disease-causing pathogens may be transferred by direct contact, as may occur while cleaning or feeding a sick person or—in the case of the bacteria of syphilis and gonorrhea—during sexual intercourse. Transmission may also occur by less direct routes. Bacteria may be breathed into the respiratory tract, either carried on minute dust particles or borne on the droplets expelled by a sneeze. They may be taken into the alimentary tract, sometimes in milk from cows whose udders are infected with pathogens (such as the tubercle bacillus), but more often by a chain of contamination involving various numbers of steps. Disease-producing bacteria are often to be found in the feces of patients and carriers* of the disease. If appropriate sanitary measures are not taken, bacteria may be transferred from hands or objects that have come in contact with these feces and other excreta either directly to the mouth, or to objects or food that is eventually

*In the course of his work, Koch developed techniques of culturing microorganisms on solid media that have formed the basis of modern microbiology.

*Persons harboring active colonies of disease-producing bacteria without themselves showing signs of the disease are called **carriers** of the disease.

taken into the mouth. There is a particular danger when persons who are caring for the sick or who are carriers of a disease prepare food for others. The term "typhoid Mary" as a synonym for *carrier* is derived from a famous case in which a known carrier of the typhoid fever bacterium, *Salmonella typhosa,* worked as a cook for many New York families and institutions for a number of years before she was discovered and forced to change her occupation. Flies and other insects that alight on excrement and then may alight on food are another important source of contamination. Infections may also occur by way of wounds or skin abrasions. And they may be borne by insect, either by direct injection during the insect's bite or by deposition on the skin while the insect is alighted there. (In the latter case, the infectious agent may be rubbed in as a result of scratching of the site of the insect bite.)

After the infectious agent has gained access into the body by one means or another, an **incubation period** characteristically follows. During this time there are no outward signs of infection. But within the body, the invading pathogens are multiplying at the expense of their host. Eventually, after a period that varies from a few hours to a matter of weeks or months, depending upon the disease, the numbers of the pathogens are so great that they cause a disturbance of the normal life processes of the host, and the outward signs of disease become manifest. Then follows a series of pitched battles within the body, with the body defenses pitted against the invaders.

The harm to the host organism caused by pathogenic bacteria is usually produced by **toxins,** poisons manufactured by the invading bacteria. These are of two types, exotoxins and endotoxins.

Exotoxins diffuse out from the living bacterial cell into the surrounding medium—either culture medium (when the bacteria are grown artificially) or tissues of the host organism. A filtrate from a medium in which exotoxin-producing bacteria have been growing will cause the same symptoms as a culture of the bacteria themselves. In fact, the rapid lethal effects of eating foods contaminated with

Figure 23–4 Bulging cans may mean danger. The growth of *Clostridium botulinum* in canned foods creates a gas pressure that usually causes the can to swell.

Clostridium botulinum are due not to multiplication of the bacteria within the host, but rather to the action of exotoxins accumulated during anaerobic multiplication of the bacteria within the stored food before it is eaten. Exotoxins are proteins and include some of the most powerful poisons known. The exotoxins of tetanus and botulism are far more powerful than any snake venom. It has been estimated that a single ounce of the exotoxin produced by *Clostridium botulinum* would be sufficient to kill 400 million humans—nearly the entire population of the Western Hemisphere!

Endotoxins, on the other hand, are produced within the bacterial cell and are released only upon the death and disintegration of this cell. They are less toxic than the exotoxins, and are lipoprotein-polysaccharide complexes. The bacteria that cause typhoid fever and Asiatic cholera are among those that produce endotoxins. (Generally bacteria produce either exotoxins or endotoxins, but not both.)

Virus Diseases

One of the most spectacular military victories ever won by United States armies was our victory over Spain in the Spanish-American War at the close of the last century. Less than 400 Americans were killed on the battlefield. Yet what the Spanish bullets could not accomplish, a far more insidious foe did. By the battalion, United States soldiers fell victim to a number of tropical diseases. In all, the greater part of the more than two thousand noncombat deaths were due to disease. A far larger number of soldiers were put out of action by the debilitating effects of disease.

It was in an attempt to reduce this appalling toll that a commission, headed by the prominent military surgeon Dr. Walter Reed, was sent to the battle area. There the commission did much valuable work on the cause and spread of typhoid fever. After the war was over, in 1899, Reed headed another commission sent to Cuba, a newly designated United States protectorate, where the dread yellow fever was raging.

Yellow fever presented something of a puzzle at the time. It was generally believed to be caused by a bacillus and transmitted through contaminated bedding and clothing. But control measures taken on this basis were of no avail in checking the epidemic. Reed traced the causative agent to a **filterable virus,** the same type of microorganism discovered (in a plant disease) and named by the Dutch botanist Martinus Willem Beijerinck just a few years before. Furthermore, Reed contended that the disease is transmitted by the aedes mosquito.

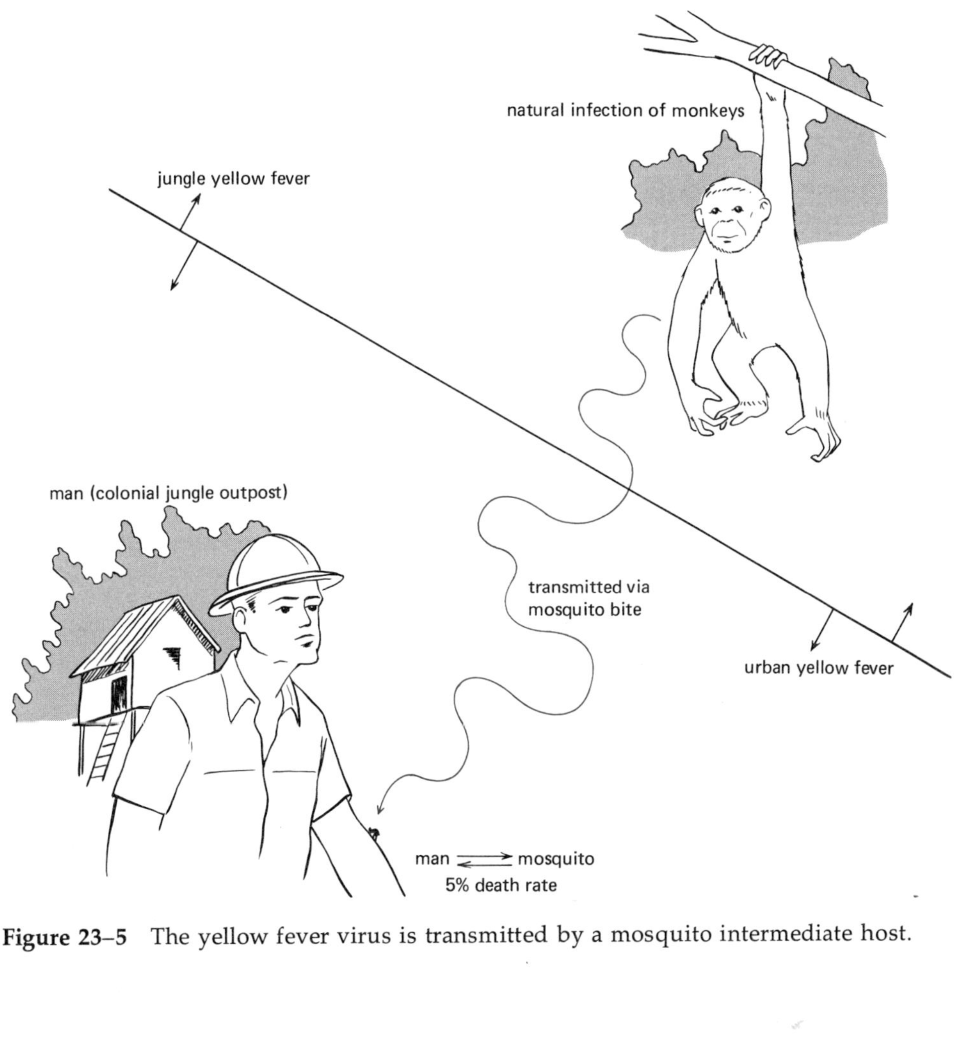

Figure 23–5 The yellow fever virus is transmitted by a mosquito intermediate host.

The authorities were skeptical: what proof did Reed have for such a contention? The disease could not be transmitted to animals, and there was no alternative but to experiment on humans. The physicians of the commission valiantly risked their lives by allowing themselves to be bitten by aedes mosquitoes. Many of them developed the disease, and one died. At great cost, Reed had proven his point. Though he did not live to see all the fruits of his labors, eventually yellow fever was virtually eliminated in Cuba and along the Atlantic seaboard of the United States. Using the principles that Reed had established, the digging of the Panama Canal was made possible.

Yellow fever was one of the first of the human diseases to be attributed to a filterable virus, a microorganism so tiny that it can not be seen by the most powerful optical microscope and passes through a filter so fine as to trap the smallest bacteria.

Paradoxically, virus research was conducted many years before scientists realized that such organisms existed. At the end of the eighteenth century, the English physician Edward Jenner performed his epoch-making experiments establishing that inoculation with material from the blisters of a person suffering from the mild disease cowpox (vaccinia) conferred immunity to the dreaded scourge smallpox. Nearly a century later, Pasteur used similar techniques of inoculation to provide immunity to rabies, dramatically saving the life of a boy bitten by a mad dog in 1885.* In this case, since there was no naturally occurring mild form of the disease to use in making up a vaccine, Pasteur made his own by passing a rabies infection through a series of different animals until its virulence was attenuated. Pasteur was unable to isolate the germ responsible, but the technique worked.

As the years went by, more and more diseases, including poliomyelitis, influenza, measles, and the common cold were traced to viruses. Progress in the study of virus diseases was slow until techniques were developed for growing the viruses outside the living organism, first in chick embryos, and then in mashes of tissues, bathed in blood (with the addition of penicillin to stop the growth of bacteria in the cultures). The latter innovation, for which the American microbiologist J. F. Enders and his colleagues T. H. Weller and F. C. Robbins shared a Nobel Prize, made the polio virus readily accessible to scientific study for the first time. (Prior to Enders's discovery, the polio virus could be grown only in living human or monkey nerve tissue.) This discovery paved the way for the development of successful antipolio vaccines by Salk and Sabin.

The demonstrable mutability of viruses has been an important factor in the efforts of scientists to derive vaccines against virus-caused diseases. On the positive side of the ledger, a virus strain can be readily adapted to a new host, such as a tissue culture or chick embryo; in the process it often loses most of its virulence for its original host,

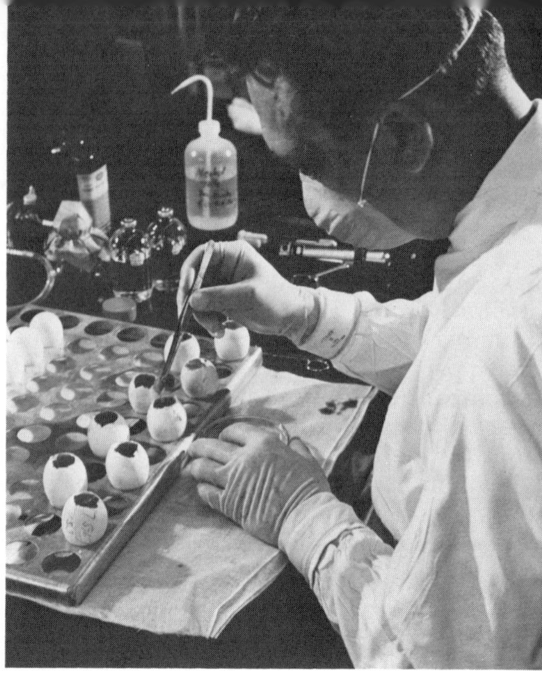

Figure 23–6 Viruses are often cultured in chick embryos for the production of vaccines. The researcher here is wearing a mask and gloves to avoid contaminating the specimens.

*Rabies has an incubation period that can range up to seven months. In the vaccine treatment, the patient's immunity to the virus is gradually built up through a series of injections administered *before* the life-threatening explosive multiplication of the disease microbe can take place.

but may still retain enough similarity to the original form to produce immunity to the original disease. Such an "attenuated" virus can then be used as a vaccine. On the negative side, the mutability of viruses sometimes results in the appearance of a new, more virulent strain, to which a person vaccinated against the original form has no immunity. During the Asian flu epidemic of 1957, and again in 1968, the United States had splendid demonstrations of the speed with which concerted efforts of medical research and drug companies produced and distributed a new vaccine to counter a new strain of the influenza virus, which was reaching epidemic proportions.

In addition to the numerous disease-causing viruses that afflict man, there are also viruses that cause diseases in other vertebrates, in insects, in plants, and even in bacteria. The latter viruses, called bacteriophages, have become extremely valuable subjects of study for researchers in genetics.

Fungus Diseases

Air, soil, and water teem with a host of potential enemies to man and his domesticated animals and cultivated plants. In addition to the hordes of invisible microbes—bacteria and viruses—that we may take into our bodies literally with every breath, there are countless tiny spores of fantastically prolific fungi. Like many of the microbes of our world, many of these fungi are normally harmless to us. Some are saprophytes, which may cause us little concern (such as the gaudy collection of molds that grow on bits of discarded garbage) or considerable annoyance (such as various mildews which may damage damp clothing, or dry rot, which may undermine the wood frames of houses). But some fungi are dangerous parasites that may attack plants, animals, and man himself.

Fungus diseases of man range from common skin infections such as ringworm and athlete's foot to severe and sometimes fatal systemic **mycoses,** caused by less "well-adapted" parasites which are not normally found in living organisms. These microbes normally live a quiet saprophytic life in the soil and may wreak havoc upon the human system if they are accidentally introduced by inhalation or through a wound.

Far more widespread than the fungus diseases that affect man and animals are those that attack various species of plants (Fig. 23–7). It is estimated that each year more

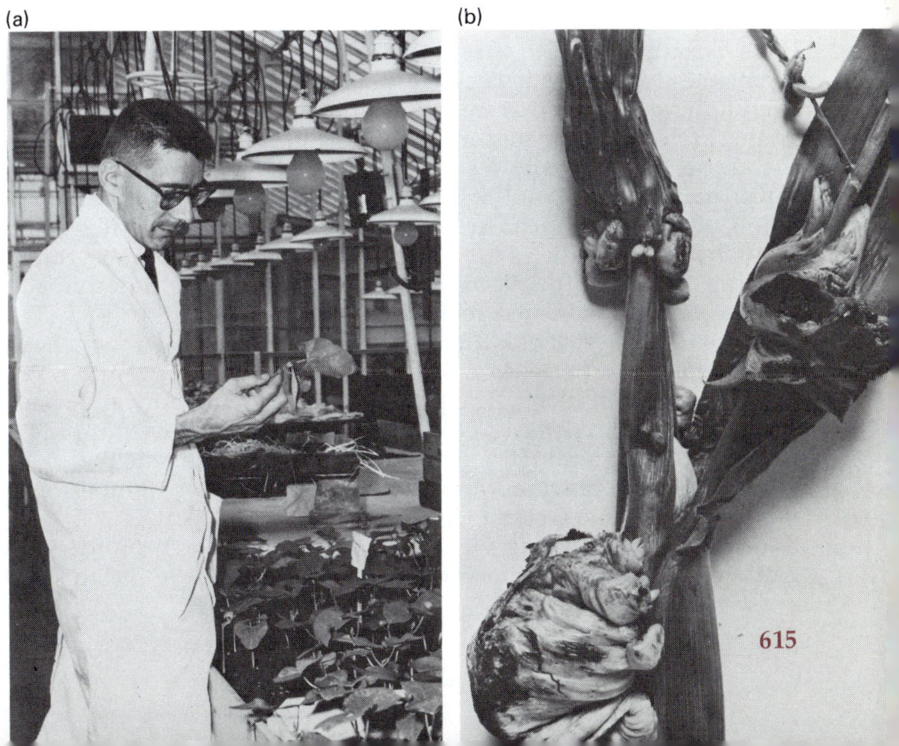

Figure 23–7 Plants are affected by a multitude of fungus diseases. (a) A plant pathologist examines a bean plant infected with fusarium root rot caused by a soil-borne fungus. (The experiment pictured revealed that adding coffee grounds to the soil makes it more fungistatic than normal.) (b) The blister-like growths on this corn plant contain masses of spores of a smut fungus.

than \$3 *billion* worth of crops and forest trees are lost in the United States alone as a result of plant diseases, most of which are caused by fungi. Fungus spores may be borne on plant seeds or transmitted through the soil. Fungus spores may be carried from plant to plant through the air, coming to rest upon a leaf, and in the presence of moisture they may germinate, sending tubes down into the tissues of the plant, finally forming hyphae between or within the plant cells.

The presence of a fungus invader may have dire effects upon the normal development and activities of the infected plant. Photosynthesis may be hampered, either through destruction of chlorophyll or through destruction of active leaf tissue. Cells may be stimulated to cancerous growth, or cell walls may be dissolved away by enzymes. Translocation of water and nutrients may be hampered by the spreading hyphae, and they may secrete toxins that poison the plant. Symptoms of fungus diseases of plants are common and readily recognized: wilt, rot, mildew, spots, scab, rust, blight, smut.

Other Parasitic Diseases

Vast areas of Africa, among the most fertile of the world, are barred to man. In this region lurks a deadly foe: the tsetse fly. Ironically, it is not the fly itself that is so deadly, but an even tinier organism that lives within it. This is the protozoan trypanosome which causes the dread sleeping sickness. Trypanosomes infest much of the local wild animal population of tropical Africa. In these hosts they play the role of the perfect, well-adjusted parasite, causing little harm. Their life cycle, too, is perfectly adjusted to tropical African conditions: tsetse flies, biting an infected animal, take in a hefty dose of trypanosomes with their meal of blood. The protozoan parasites may be transferred to another host immediately, by a mingling with the infected blood from the previous bite. Or they may remain in the body of the fly for twelve to fifteen days, multiplying vigorously and migrating from the alimentary canal of the fly to its salivary glands. Once there, they are ready to be injected into a new host, as the fly bites and injects a dose of anticoagulent-containing saliva.

In the wilds of Africa, this is a smooth, well-running cycle. But when man and his livestock move in, trouble ensues. The tsetse flies take heartily to their new victims, and so do the trypanosomes that they inject with their bites. But without the long ages of mutual evolution, the fine mutual adjustment is missing. In a man's body tyrpanosomes run wild. Multiplying rapidly in the bloodstream, they destroy blood cells and produce anemia; then they move on to the nervous system, destroying the tissues of the brain and spinal cord, producing a profound lethargy (the "sleeping sickness"), and often death. In cattle and horses, the fatal wasting sickness nagana results from trypanosome infection. Drastic measures have been taken; whole villages have been given regular prophylactic doses of drugs or even moved out of areas where sleeping sickness is endemic. Since the discovery of the vector of this disease, the tsetse fly, massive efforts have been concentrated upon the extermination of the pest. Some success has been achieved, but in large areas of tropical Africa the trypanosome still reigns undisputed king.

Another important insect-borne parasitic disease caused by protozoa is malaria. This disease, also common in the tropics, is caused by plasmodia, which are transmitted to man and other animals (principally apes, monkeys, birds, and reptiles) by the bite of an infested mosquito. Malaria has been a scourge of mankind at least since the dawn of recorded history. The symptoms of the disease were precisely described by Hippocrates in the fifth century B.C. (The causes of the disease were not known, however, until the beginning of the twentieth century.) Great progress in the treatment of malaria and in the extermination of its mosquito vector was made during the Second World War, but malaria still remains the number-one disease in total world morbidity.

A protozoan parasite with a simpler life cycle is the amoeba that causes amoebic dysentery. This parasite lives in the alimentary canal of its host, where it damages the intestinal wall and ingests red blood corpuscles. The parasitic amoebas encyst and pass out in the feces. When sanitation facilities and practices are poor, reinfection may occur by eating contaminated food and drinking contaminated water.

In addition to the protozoan parasites, many diseases that plague man and other animals are caused by various roundworms and flatworms, some with complicated life cycles involving a series of successive hosts.

The second most prevalent disease in the world, infecting about 200 million people, mainly in tropical regions of Asia, Africa, and South America, is schistosomiasis. This parasitic disease is caused by several species of blood flukes (schistosomes), flatworms that live in the blood vessels of their hosts. Schistosomes have a complex life cycle (Fig. 23–8). The male and female worms mate in the blood vessels of their human hosts, usually the veins leading from the urinary bladder or intestines. Their eggs pass out in the urine and feces (as many as a million a day from a single human host) and thence find their way into the rivers and streams. The eggs develop into ciliated swimming forms, which then infest an intermediate host, aquatic snails. Within the snail's body, the parasitic larvae develop into forked-tailed swimming larval forms called cercariae, which pass out into the water and then can penetrate the skin of any human who comes in contact with the infested water. In the body of the human host they migrate through the bloodstream to their final home, where they may remain for more than twenty years.

In addition to the debilitating effects on the host due to the strain of supporting the parasites, the spiny eggs of the schistosomes cause severe damage to the walls of the bladder or intestines as they migrate from the blood vessels, and the disease can be fatal. Recently advances against schistosomiasis have been made along two fronts: the development of drugs that kill the parasites in a single dose and the development of chemical molluscicides to eradicate the intermediate host snails from streams and canals.

Many parasites, particularly the various worms, generally do not kill their victims. But in living off their tissues and body

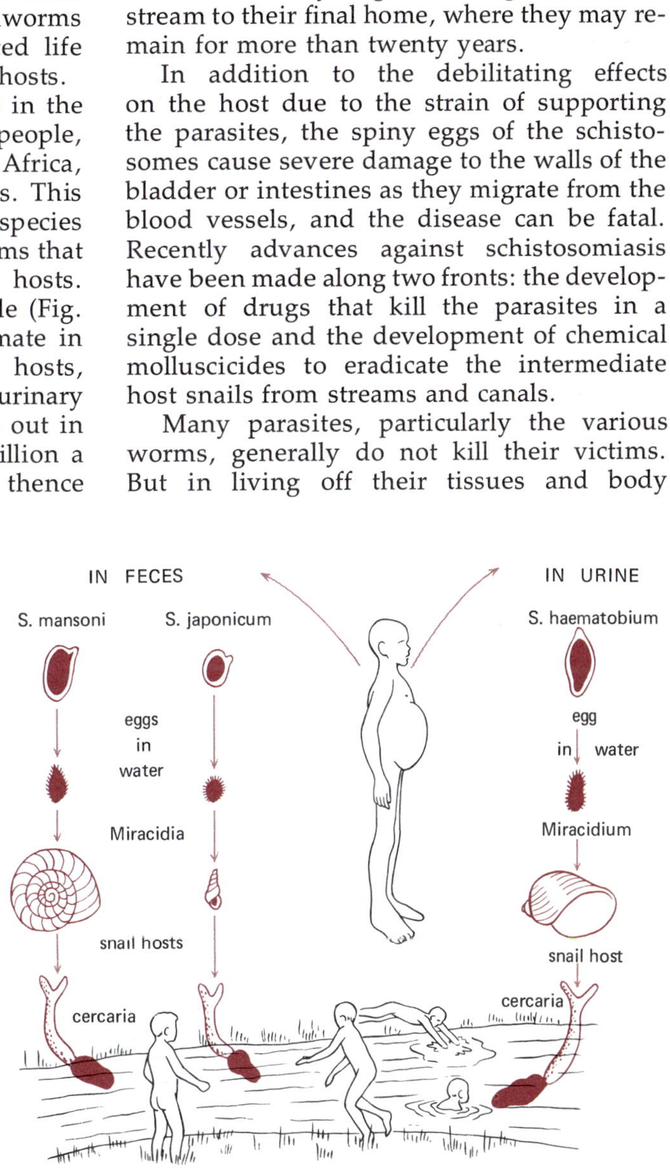

Figure 23–8 Life cycles of several varieties of schistosomes. All are carried by intermediate snail hosts.

products, the parasites frequently sap the energies of their victims, leaving them in a debilitated state. Then the host can fall prey much more easily to bacterial and viral infections that may in turn prove fatal.

Treatment of Infectious Diseases

Any internally widespread bacterial infection spelled almost certain death in the early 1930s. But with the discovery and development of the sulfa drugs and the subsequent development of the **antibiotics,** beginning with penicillin, in the early 1940s, bacterial infections came more or less under man's control. Viral infections for the most part have proved far more stubborn, however; our best weapon against these is still prevention. **Vaccines** have been prepared against numerous viral diseases, including smallpox, rabies, influenza, polio, measles, mumps, rubella (German measles), and even some varieties of the common cold.

The mechanisms by which the body defends itself against invading bacteria and viruses have already been discussed in detail in Chapter 11. In addition to the **active immunity** produced by the reticuloendothelial system, modern medical practice now utilizes the principle of **passive immunity:** the injection of specific antibodies produced in another organism (Fig. 23–9). For instance, in the case of tetanus, a recipient animal such as a horse is injected with a sublethal dose of tetanus toxin. In time the animal responds by the formation of specific antibodies capable of neutralizing the effect of the tetanus toxin. When the antibody titer has reached an optimum level, quantities of blood are withdrawn and the antitoxin-containing serum is extracted. This **antiserum** can then be administered as an emergency treatment for humans who have been exposed to the tetanus microbes in a puncture wound. The immunity provided by treatment with antitoxin is quite transitory. Moreover, there may be some element of danger in administering a mixture containing foreign proteins, to which an

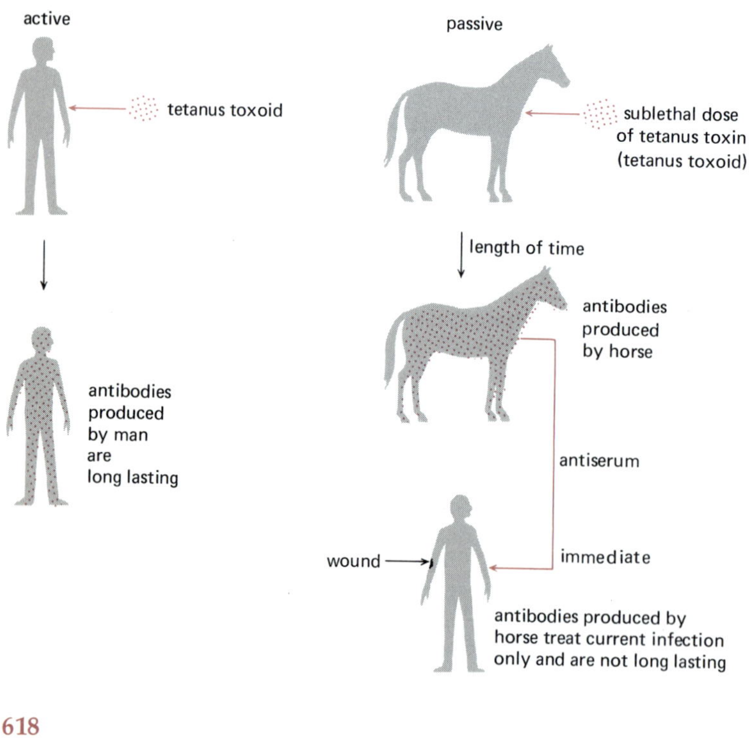

Figure 23–9 Protection from tetanus can be provided by the production of either active or passive immunity. An inoculation of antiserum provides only an immediate, emergency protection; long-lasting protection is provided by an inoculation of tetanus toxoid, producing active immunity.

allergic reaction might be developed. A far more desirable approach is a preventative approach. Administration of tetanus toxoid before there has been any actual infection mobilizes the body's own immunity mechanism and creates a far longer-lasting active immunity.

Not everyone is so farsighted as to keep up a thorough immunization against all the diseases against which vaccines are available. It is therefore fortunate that antisera can be used, at least as emergency measures. Indeed, for some diseases effective antisera were available long before vaccines had been developed. For example, the gamma globulin fraction of the blood of adult humans generally contains a rich assortment of antibodies against many common diseases. Gamma globulin injections are routinely used to provide temporary immunity for unprotected children exposed to polio or measles and for pregnant women exposed to German measles.

Great strides have been made in recent years in the field of chemical attacks upon disease. Chemical approaches generally take two main avenues: killing the invading organisms (the **bactericidal** approach) or stopping their growth until the body's own defenses can be mobilized sufficiently to cope with the problem (the **bacteriostatic** approach). Often the distinction is only a matter of degree, depending upon the dose and other conditions. Penicillin, for example, may exhibit either bactericidal or bacteriostatic effects (Fig. 23–10).

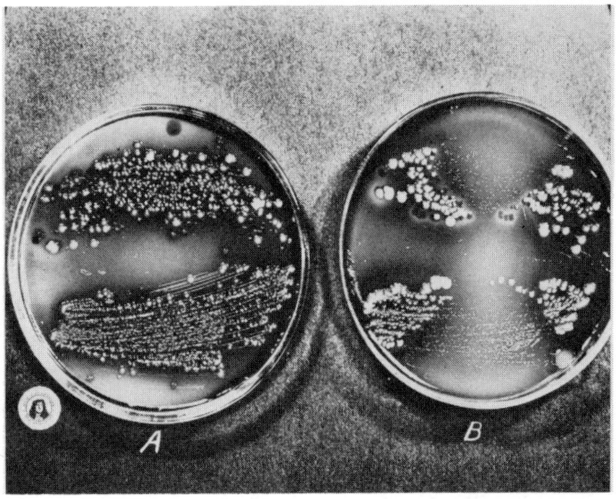

Figure 23–10 Both plates were streaked with a nasopharyngeal swab from a whooping cough patient. A drop of penicillin solution was added to plate B, while the control plate A was similarly treated with a drop of sodium chloride solution.

The field of **chemotherapy** was born in the researches of the German bacteriologist Paul Ehrlich. Ehrlich had done some valuable work on the development of stains for the microscopy of bacteria and white blood cells, and then moved on into the field of immunity and serum therapy. Ehrlich theorized that since stains for bacteria colored them specifically, they must react with some substance in the bacterium and therefore would probably kill it. His goal was to find a "magic bullet" that would selectively kill a pathogen without killing the organism invaded by it. He chose trypanosomes, for this microorganism is not only large and easy to observe, but also destroys one of its victims, the mouse, in 100 percent of the cases. So if he could save just a single mouse, he would know that he was on the right track.

For eight long years Ehrlich struggled, testing hundreds and hundreds of compounds. He had some success with the 606th chemical, dihydroxydiamino-arseno-benzene hydrochloride. But he continued the search, trying to find a better compound. In 1909, when he and his associates were already in the 900s, they happened to try compound 606 again, this time on the syphilis spirochete, for a report had recently appeared describing the spirochete and commenting on its "relationship" to trypanosomes. (This resemblance has since proved quite superficial—but it is fortunate for the world that that bit of misinformation reached Ehrlich at that particular time.) In preliminary tests on animals, the results were dramatic, and in trials on human syphilis sufferers in 1910, compound 606 proved equally effective. Thus, salvarsan, the first of the "miracle drugs," was born.

The next important landmark in the history of chemotherapy was the discovery of the antibacterial properties of the orange-red dye Prontosil by Gerhard Domagk in 1932. Domagk

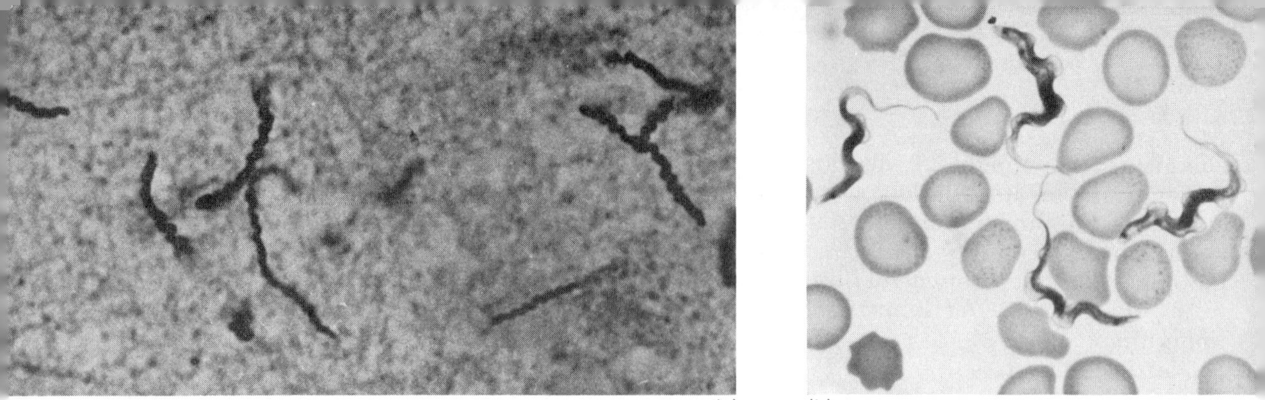

(a) (b)

Figure 23–11 *Treponema pallidum,* the syphilis spirochete, 2010× (a) and *Trypanosoma gambiense,* the parasite of African sleeping sickness, 1200× (b). A fortuitous resemblance led to the discovery of salvarsan, the first of the "miracle drugs."

was testing dyes produced by the German dye company I. G. Farbenindustrie for possible antibacterial activity. He had just obtained promising results with Prontosil on mice when his young daughter fell ill with a serious streptococcal infection. She did not respond to any conventional treatments; in desperation, Domagk tried his experimental compound—and his child recovered dramatically. Further tests of Prontosil revealed that a small portion of the molecule is just as effective as the whole compound. This active portion was sulfanilamide, the first of the sulfa drugs. Sulfanilamide bears a striking chemical resemblance to para-aminobenzoic acid (PABA), a compound needed by bacteria for their metabolic activities (Fig. 23–12). Bacterial enzymes that normally combine with PABA may join with sulfanilamide "by mistake" and be rendered unable to complete their normal reactions. Sulfanilamide thus is also a selective agent—a "magic bullet" that strikes at the bacterium through its special metabolic requirements while leaving the host organism virtually unharmed.

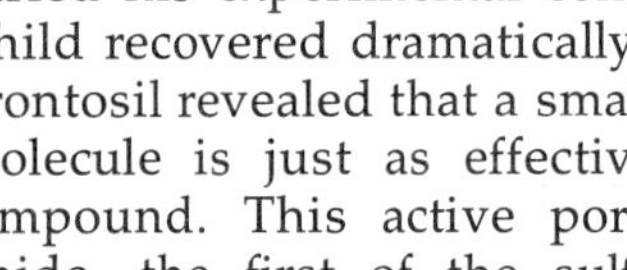

Figure 23–12 There is such a strong chemical resemblance between sulfanilamide (left) and PABA (right) that bacteria mistake the drug for their normal metabolite.

Another important step in the development of chemotherapeutic remedies was the discovery (or rather, rediscovery) of penicillin, a secretion of a mold of the genus *Penicillium.* Other antibiotics have since been discovered among the metabolic by-products of various bacteria and fungi. Their production, both by chemical synthesis and by large-scale culturing of carefully selected strains of microorganisms, is now an important aspect of the drug industry.

But the antibiotics, important—indeed lifesaving—though they may be, have an important drawback. They generally are effective only against bacteria and are powerless to stop the onslaught of viral infections. Indeed, once a viral infection has become established, there is usually little that one can do, except to make the patient comfortable and hope that the body's own defenses will do the rest.

Recently some significant strides have been taken toward the ultimate chemotherapeutic control of virus diseases. For instance, in 1962 H. E. Kaufman discovered that 5-iodo-2-deoxyuridine is effective against herpes simplex, a viral disease of the eye. This drug offers little hope for general use, however, since it is too toxic to be taken internally. Later, researchers at Du Pont announced the discovery of amantadine hydrochloride, a drug effective in preventing the development of a strain of influenza (Asian A) in persons already exposed to the disease. Methisazone (N-methylisatin-β-thiosemicarbazone) has been used successfully for the prevention of smallpox and is being tested for other viral diseases as well. Other potential antivirals, some showing promise of activity against a variety of different viruses, are now being developed.

One of the most interesting stories in the rapidly expanding new field of antiviral research concerns the development of rifampicin. This drug is an antibiotic initially developed as an antibacterial agent and is now extensively used in the treatment of tuberculosis. Studies of the drug's action revealed that it kills bacteria by selectively inhibiting their RNA polymerase, an essential enzyme for RNA synthesis, while leaving mammalian RNA polymerase relatively untouched. Meanwhile, it was discovered that vaccinia virus, a large, DNA-containing virus related to the smallpox virus, contains its own RNA polymerase in addition to its DNA and protein coat. Scientists in several laboratories put the two pieces of information together and reasoned that rifampicin should be active against viruses that have their own RNA polymerase. Such proved to be the case. Rifampicin has been found to be effective against such viruses as vaccinia, smallpox, and an adenovirus.

Meanwhile, other promising studies have proceeded along the line of stimulating the body's own defenses against virus invaders. It had been known for several decades that infection with one virus may produce a temporary, generalized immunity against infection by other, unrelated viruses. This effect is called the **interference phenomenon.** In 1957 it was discovered that the effect is mediated by a small protein called **interferon,** which is produced by virus-infected cells and protects other cells of the body from viral infection. The resistance induced by interferon is general, rather than specific for a particular virus, and therefore it is hoped that the interference mechanism can be utilized against diseases,

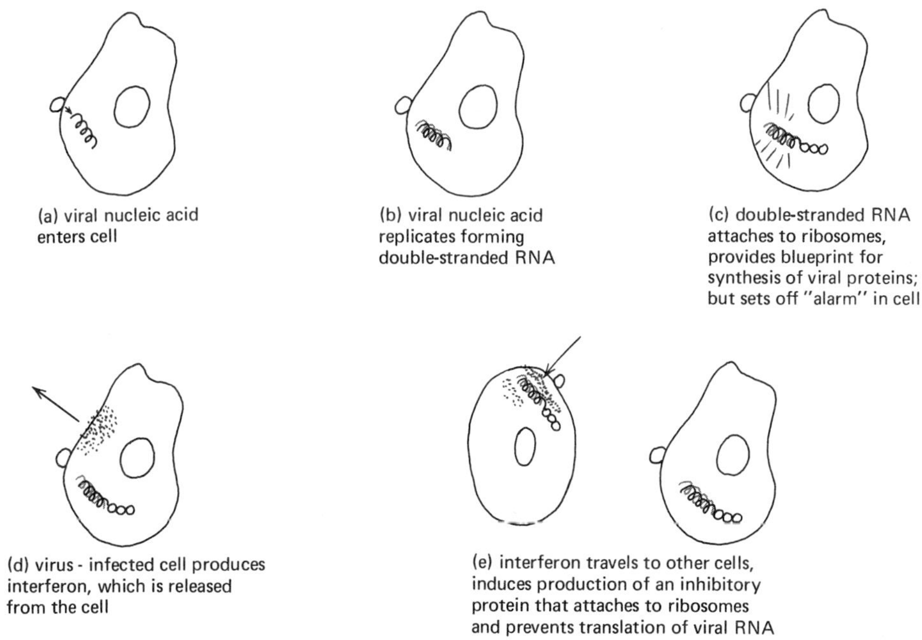

Figure 23–13 Induction of interferon by an invading virus. Researchers are trying to stimulate this mechanism by artificial means to mobilize the body's defenses against virus diseases more rapidly and effectively.

such as the common cold, that are caused by such a variety of different virus strains that conventional vaccines would be ineffective.

Two avenues of approach are being explored in the possible utilization of interferon: harvesting interferon produced in a living organism, then giving it to a person under viral attack; and chemically inducing the body to produce its own interferon. Unfortunately, interferon is highly species-specific, and thus any interferon to be used to treat humans would have to be produced in human cells. Various bacteria, viruses, and chemical substances have been tried as interferon inducers. Such studies received a boost when it was discovered that double-stranded RNA is a highly active interferon inducer, and a tentative mechanism of interferon production was worked out. When the viral nucleic acid penetrates into a cell, it replicates, producing a double-stranded RNA (or in the case of a DNA virus, a double-stranded DNA-RNA hybrid). The new RNA acts as a messenger RNA, attaching to the host cell's ribosomes and inducing the synthesis of viral protein instead of the host cell's own proteins. But it is believed that the double-stranded RNA acts as an "alarm" stimulus for the cell, causing it to produce interferon. The interferon is released and travels to other cells of the body, where it induces the formation of a special translation inhibitory protein. The latter attaches to the ribosomes and prevents translation of the viral RNA message while permitting the synthesis of the host cell's proteins to proceed normally. Thus, the viral infection is effectively contained.

A number of forms of double-stranded RNA have been tested as interferon inducers, including a synthetic cytidylic acid complex, called "poly I:C." Work on these substances is proceeding cautiously because most of them have toxic side effects.

Another promising new line of antiviral research is the synthesis of compounds that work according to the same mechanism as interferon. The industrial dye aurintricarboxylate (ATA) has been shown in the test tube to block the attachment of viral messenger RNA to the ribosomes. ATA itself cannot be used as an antiviral drug because it does not penetrate the cell membrane in living organisms. But it is hoped that suitable chemical modification of the compound, combining the active portions of the molecule with another molecule that will carry it into the cell, will yield a broad-spectrum antiviral agent.

Cancer

The awarding of the 1966 Nobel Prize in medicine and physiology to two cancer researchers (American pathologist Peyton Rous and surgeon Charles B. Huggins) refocused the world's attention upon this dread disease. Each year in the United States alone more than three hundred thousand Americans die of cancer. This disease strikes the young as well as the old. But it is not at all restricted to man; members of virtually every animal phylum exhibit instances of cancerous growth. Indeed, cancer is found among numerous members of the plant kingdom as well. Yet despite the universal incidence of this disease, its basic nature and causes remain unknown. Wildly conflicting theories have been offered to explain the origin and characteristics of cancerous tissue. We remain basically ignorant of the fundamental nature of cancer, despite the fact that the United States has already spent more than $2 billion on studies of this disease.

It has been suggested that cancer is not a single disease, but rather a hundred or more different diseases. Cancer may strike in virtually any tissue of the body: lung, brain, breast, skin, bone, blood, reproductive, intestinal, liver, etc. But in essentially all cases there is at least a single basic common characteristic: the cells exhibit a wild, uncontrolled growth. Indeed, it has been recently demonstrated that whereas normal cells "recognize" their neighbors and cease growing upon contact with them, cancer cells show no such recognition and contact inhibition. Instead, cancer cells continue wild growth, piling over one another in a disorderly array. They literally choke out their normal neighbors.

Recent studies of cultures of cancer cells have shown these cells to be connected by thin bridges of cytoplasm called **cytopons** (Fig. 23–14). When a cell of the cancerous mass happens to divide spontaneously, it is observed that adjacent cells connected to it by short cytopons undergo mitosis almost immediately thereafter. Cells connected by longer cytopons also divide, but after a long time, and division of cells connected to these dividing cells follows. Thus, it seems that some sort of mitotic stimulus is transmitted along the cytopons, triggering a chain reaction of successive cell division among the cancerous mass.

It is not yet known whether cytopons are a general feature of human tumors. If they are, their unique nature suggests several avenues of possible approach to treatment. It has been found that cytopon-connected cancer cells sometimes spontaneously coalesce into giant cells. These giant cells are unstable, usually disintegrating rapidly. If the mechanism of giant-cell formation could be discovered, agents might be devised to stimulate this process, and thus the cancer could be induced to destroy itself. Chemicals that inhibit cytopon formation might also be effective inhibitors of tumor formation. Such approaches would be particularly valuable because they would be specific for cancer cells, leaving normal cells (which do not form cytopons) unaffected.

Figure 23–14 This scanning electron micrograph of a malignant cancer cell shows its spreading branches.

Another common characteristic of cancerous tissue, and a critical factor responsible for the deadliness of this disease, is the phenomenon of **metastasis.** Cancer cells are capable of dislodging from their original site of growth, entering the bloodstream, and being dispersed to other regions of the body. At these new sites, these metastasizing cancer cells continue the growth patterns characteristic of the original cancerous tissue.

Recent Harvard University studies have revealed that most solid tumors produce a protein called **tumor angiogenesis factor** (TAF). TAF stimulates nearby capillaries to grow branches to the tumor, thus providing it with a nourishing blood supply. Without TAF and the effects it produces, the tumor does not grow beyond a tiny size and remains in a dormant, harmless state. Active searches are now underway for means of blocking TAF and thus controlling the growth and spread of cancerous tumors.

The experimental evidence seems to indicate that in some manner DNA plays a fundamental role in cancer. The nature of this role is obscure. Some have theorized that a mutation in the genetic code causes a cell to continue multiplying without regard either to normal replacement requirements or to the normal mechanism of contact inhibition. Others have speculated (without necessarily contradicting the previous theory) that a virus plays a strategic role in the onset of cancer. Indeed, it has been definitively established that more than thirty different viruses cause cancer in animals. The first of these cancer-causing viruses was discovered by Nobel Prize winner Peyton Rous in 1909. This virus causes a type of cancer in hens, which was named Rous sarcoma. Although there is a mounting crescendo of evidence implicating viruses as a causative agent in human cancer, as yet there has been no irrefutable proof of this.

One of the difficulties in viral research on cancer is that even in some animal cancers that have been demonstrated to be virus-caused, whole virus particles generally are not found in cancerous cells after the onset of infection. It has been theorized that the viral nucleic acid becomes incorporated into the cellular DNA. Moreover, the finding of virus particles in cancer cells does not constitute proof of their causative role—they might be fortuitous companions rather than causative agents.

Compounding the difficulties of establishing a viral cause of cancer is the possibility that other agents can serve as contributing factors to a supposedly virus-induced cancer. In such a hypothetical case, cancer would not result from invasion by a virus unless the other factor (such as hormones) or other carcinogenic substances (such as coal tars) are also present.

The exact role of known **carcinogens** (cancer-causing chemicals) such as aromatic hydrocarbons, aromatic amines, azo dyes, nitrosamines, and mycotoxins (produced by fungi, such as the molds that grow on peanuts) is unclear. But C. B. Huggins, who shared the 1966 Nobel Prize with Rous, has stated that he believes these substances may act directly upon the cellular DNA, inducing changes in the genetic material.

For years the standard treatments for cancer have been surgery, radiation therapy, and chemotherapy. If the cancer is caught early, before metastasis, frequently a cure can be effected* by cutting out the cancerous tissues and any surrounding tissues suspected of harboring cancer cells. Many cancers, particularly many deep-seated cancers which are not readily accessible to surgery, may be destroyed by exposure to radiation. Finely focused beams of X rays, rods of radioactive cobalt, needles or beads of radioactive gold, and other radiation sources are used. For instance, in the case of cancer cells that originated in the thyroid gland and have metastasized, a solution of radioactive iodine can seek out and destroy the wide-spread foci of cancerous growth.

Recently, exciting new avenues to the surgical approach have been innovated. Laser beams, which can be focused far more sharply than a beam of X rays, have been used to destroy pigmented cancer cells known as **melanoma** (Fig. 23–15). It is found that these pigmented cells selectively absorb laser beams. Scientists are now attempting to introduce into nonpigmented cancer cells dyes that would permit selective destruction by laser beams.

Another avenue in quite the opposite direction is the use of ultracold temperatures (e.g., the temperature of liquid nitrogen) in "cold surgery." This type of surgery, known as cryosurgery (discussed in Chapter 21), permits a bloodless excision, thus minimizing the chance that stray cancer cells will be carried off by the bloodstream into other areas.

Of the three major approaches to the treatment of cancer, chemotherapy is the area in which the least progress has been made. Paradoxically, this is precisely the area which may hold the greatest promise for ultimate success in the fight against cancer. By far the largest single amount of money spent by the United States government on medical research on cancer has gone toward the testing of enormous numbers of chemical compounds for possible effectiveness against cancer. Only a very small number of these have

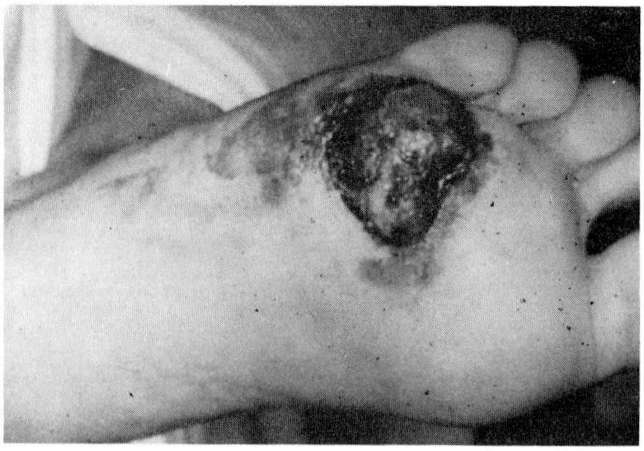

Figure 23–15 Melanoma, a pigmented cancer.

*Cancer specialists are usually guarded in their use of the word *cure*. When it is used, it is generally meant that the patient has been free of signs of cancer for a period of five years or more since treatment.

thus far shown any real promise in limiting the ravages of the disease; up to now, none have shown any promise of a true cure of any of the main types of cancer.

One of the promising approaches to the cancer problem is based upon the thesis that the cancer cell is originally "different" from the normal cell. It is theorized that the cancer cell is so different that the normal body builds up antibodies that effectively lead to the ultimate demise of the cancer cells. But it is thought that in the cancer victim the immunity mechanism has somehow failed to recognize the cancer cells as "foreign." It is felt that if the body of the cancer victim can be made to "recognize" the cancer cells as foreign, it might be made to build up antibodies to these cancer cells and so destroy them. A number of promising experiments are being conducted along these lines.

Support for the possibility of a link between the immunity mechanism and cancer has been obtained in some recent studies. It has been found that recipients of organ transplants who have been treated with heavy doses of immunosuppressive drugs show a significantly higher probability of developing malignant tumors. Comparisons of the immunological responses of cancer patients and normal controls have also shown a correlation between a failure of the immune mechanisms and susceptibility to metastasizing cancer.

Some researchers concentrating on the immunity approach to cancer therapy are attempting to combine cancer cells from human patients with foreign proteins, such as rabbit blood serum gamma globulin, to produce a vaccine that will stimulate the patient's immunity system to reject his own tumor cells. Another approach is the injection of factors from the serum or white blood cells of healthy people, who apparently have built up an immunity to various forms of cancer. The blood of cancer patients who have experienced lasting **spontaneous remissions**—sudden complete recovery from cancer for no apparent reason—is being intensively studied in the hope of preparing transfer serums that will confer passive immunity to cancer.

Another active front in the fight against cancer is the search for differences between cancer cells and normal cells, with the hope of utilizing these differences to selectively fight the cancer cells. Some cancer cells have been found to require certain metabolites, such as particular amino acids, that most normal cells can synthesize for themselves. One of the most promising of the drugs developed on this principle was L-asparaginase. This enzyme breaks down the amino acid asparagine, required by some leukemia cells and some other cancer cells in order to survive. L-asparaginase in high doses has cured mice of leukemia, and after much arduous research and development, methods were devised for producing the drug in large quantities at reasonable cost. Clinical trials proved disappointing, however; the drug showed higher toxicity than expected, for it was discovered that some normal tissues, such as bone marrow cells, also require supplies of asparaginase.

Another difference between cancer cells and normal cells is in their energy metabolism. In normal cells, respiration begins with the anaerobic pathways of glycolysis, then proceeds to the far more efficient aerobic pathways of the Krebs citric acid cycle and the cytochrome cycle. Cancer cells generally do not take that second step. Their major source of energy comes from the inefficient anaerobic breakdown of glucose to lactic acid, which yields only two molecules of ATP per glucose molecule. The lactic acid produced by the cancer cells travels to the liver and kidneys, where it is synthesized to glucose, at a cost to the body of at least six more ATP molecules. In addition, tissue proteins are broken down and converted to glucose to feed the voracious cancer cells. The heavy energy drain put upon the body by the growing cancer ultimately leads to weight loss and debilitation. A new approach to the fight against cancer seeks to inhibit **gluconeogenesis,** the synthesis of glucose in the liver and kidneys. (Blocking glycolysis in the tumor cells directly would block this key pathway of energy metabolism in normal cells as well.) It has been found that various compounds block a key enzyme in gluconeogenesis. One of these, hydrazine sulfate, has shown promising preliminary results in trials on human terminal cancer patients.

Despite discouraging setbacks, there is definite progress on the cancer front. Large-scale testing programs are disclosing a steadily growing arsenal of drugs that are effective against specific forms of cancer. Tests are being devised for earlier detection of malignancy, and earlier detection in turn results in a higher percentage of successful treatment by conventional methods of surgery and radiotherapy. Meanwhile, each year brings more basic knowledge of the nature of normal and cancer cells, new insight into the cause or causes of cancer, and new avenues of fighting it.

Heart Disease

If you asked a person what disease he dreads most, his answer would probably be, "Cancer." Yet the chances are three times as great that he will ultimately die not of cancer, but of heart or circulatory diseases. In the United States today, heart and circulatory diseases are the number-one killer, causing more than a million deaths each year. More Americans die from cardiovascular diseases in any one year than the total American fatalities in all the wars the United States has fought since its founding.

The most common diseases of the heart and blood vessels are generally classified as **degenerative diseases.** They seem to be caused not by the attack of some specific germ pathogen, but rather by a progressive deterioration of key organs of the body, a gradual reduction of their effectiveness.*

As a person grows older, fat deposits gradually build up in the walls of his arteries. The **lumen** or space through which the blood flows is thus gradually narrowed. The heart must work harder to pump the same volume of blood through the narrowed arteries, which increases the wear and tear on this vital organ. Moreover, if a small clot should form, it could lodge in some particular narrowed spot, effectively blocking all further flow beyond that point. Organs "downstream" of the blockage are thus deprived of their supply of essential oxygen and nutrients; if the flow is not restored, cells and tissues die.

Brain cells are particularly vulnerable to a cutoff of oxygen supply. If a clot blocks one of the arteries in the brain, it can result in a stroke, producing paralysis, a loss of the ability to speak, an impairment of other faculties, or death. Even if the outcome is not so dramatic, **atherosclerosis** (the buildup of fat deposits in the artery walls) in the vessels of the brain can lead to a progressive loss of memory, peculiar behavior, and other concomitants of senility.

The heart is another heavy oxygen user in the body. This tireless muscle is supplied with oxygenated blood by two coronary arteries, which branch off from the aorta (Fig. 23–16). Blockage of one of these key arteries can result in a heart attack, causing irreparable damage to part of the heart muscle. The extent of the damage determines whether a heart attack will be fatal or not. If the patient recovers, ultimately other, smaller blood vessels are enlarged, producing an effective **collateral blood supply** to the heart, taking the place of the blocked artery.

What causes this gradual deterioration of the circulatory system? Is it a natural concomitant of the aging process? Some studies seem to indicate that it may not be. Occasionally an extremely old person is found to have "the arteries of a young man"; autopsies of G.I.'s killed in Korea in their late teens and early twenties, on the other hand, revealed that many of them already had significant buildups of fat deposits in their artery walls.

Despite extensive research, a clear picture of the cause of heart and circulatory disease has yet to emerge. But many suggestive studies indicate that a complex of factors is implicated, many of them linked with our modern way of life. Increasing affluence has characteristically brought with it an enormous increase in heart disease mortality—far greater than could be accounted for by the increase in the proportion of older people in the

*Some infectious diseases, such as rheumatic fever, can also produce damage to the heart.

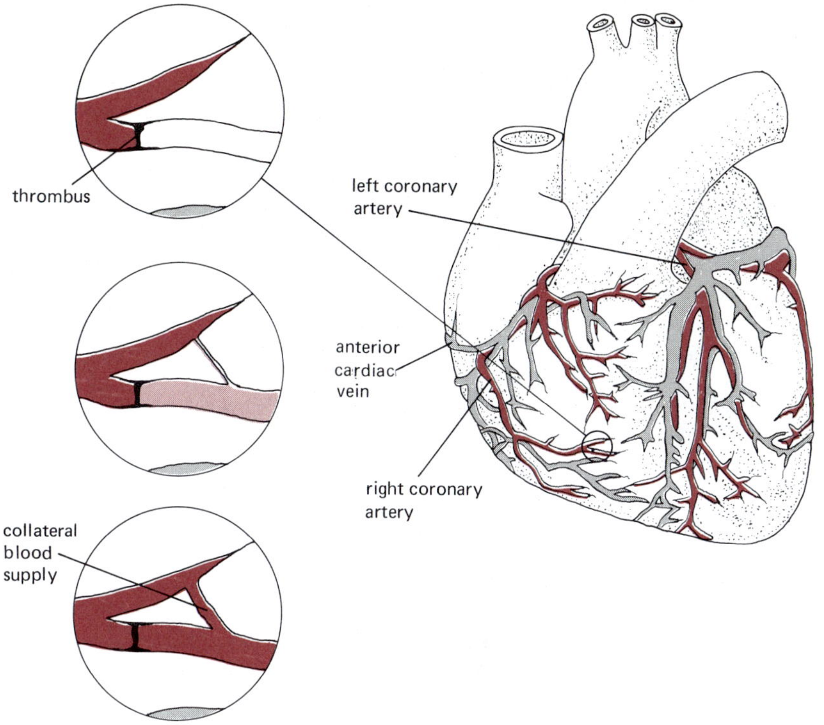

Figure 23–16 Coronary circulation (arteries in color). A thrombus (blood clot) can cut off part of the blood supply to the heart muscle. Ultimately collateral blood vessels may form or enlarge, bypassing the blockage and reestablishing the blood supply.

population. And the age at which heart attacks strike is progressively decreasing among the affluent.

One factor that seems to be definitely linked with a high incidence of heart disease is the "typical American diet," high in carbohydrates, saturated fats, and cholesterol. Another contributing factor is the modern sedentary life. (A regular program of vigorous exercise not only helps to control the buildup of fat deposits in the arteries, but also promotes the establishment of an extensive network of collateral circulation, which can maintain the vital supply of blood to the heart in case of a heart attack.) Many researchers believe that tension is another key factor in promoting heart attacks. The fact that heart disease is far rarer in women before menopause than in men of comparable ages, but that after menopause the rates tend to even out, indicates that female hormones may play a key protective role.

Modern treatments of heart disease are aimed along several major lines. One of the most important is prevention: the stressing of a prudent diet, avoiding excesses of carbohydrates, saturated fats, and cholesterol; the establishment of a regular program of exercise, carefully geared to the physical condition of the individual; the search for a "life style" that cuts down on unnecessary tension and stress. Patients who have been found to be suffering from atherosclerosis are often prescribed carefully regulated doses of an anticoagulant, so as to minimize the chances of a clot forming inside a blood vessel. The development of heart-lung machines (Fig. 23–17) that can temporarily take over the task of pumping and oxygenating the blood has made possible dramatic advances in surgery on the heart itself. Heart defects can be repaired, and collateral blood vessels can be grafted into place to bypass a dangerously blocked coronary artery. Heart surgeons have taken

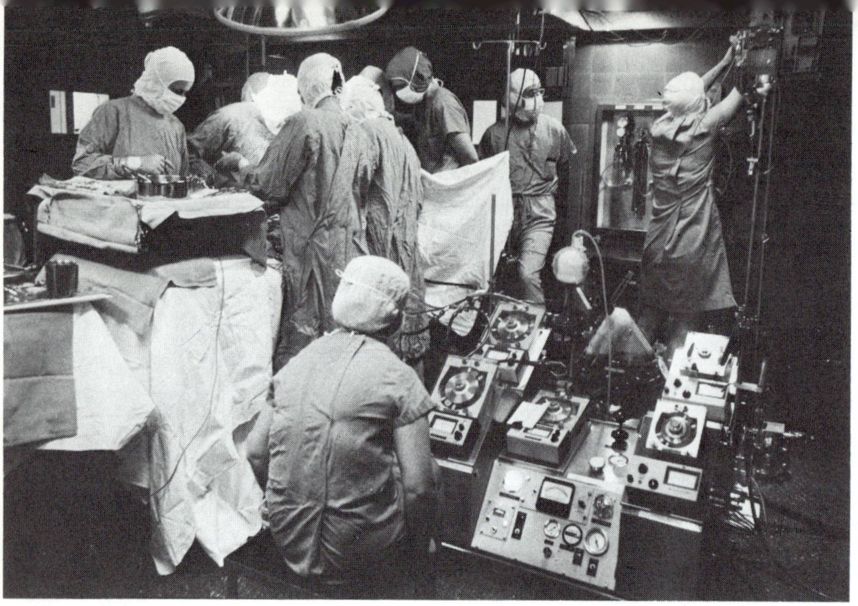

Figure 23–17 A heart-lung machine in action.

the first steps toward actually replacing hopelessly damaged hearts.

Organ Transplants–A Promising Dead End?

The first successful heart transplant by South African surgeon Christiaan Barnard in 1967 focused the attention of the world on a sensational new medical technique. The irreparably damaged heart of a man was removed from his body, and the healthy heart of a young man who had died of a sudden stroke was substituted for it. The first successful heart transplant was followed by a rash of other heart and multiple-organ transplants, and although the rate of failure was at first discouragingly high, more refined procedures are constantly increasing chances for survival.

The transplantation of organs poses a number of unique problems, both medical and ethical. The first transplants to be tried on a large scale were kidney transplants. A normal human is born with two healthy kidneys; if one is removed, he can still function quite normally as long as the other remains healthy. Therefore, living donors can be used to provide a new kidney for a patient whose own kidneys have been damaged beyond repair. Organs for transplants can also be obtained from cadavers. But they must be removed almost immediately after death, before deterioration has set in. And of course, the donor must have died from some unrelated cause—one that did not damage the organs to be used.

Organ transplantation is not merely a matter of removing a diseased organ and carefully connecting up a healthy replacement in its place. The body's immunity mechanism soon comes into play. For the human body normally has the ability to recognize each of its own cells and their characteristic chemical products, and to implacably reject any "outsiders." The first successful kidney transplants were performed between identical twins, whose cells are virtually identical in chemical makeup. But the vast majority of the hundred thousand or so Americans who die each year of kidney ailments do not have convenient identical twins to give them a spare kidney. If organ transplantation was to become anything more than a medical curiosity, means had to be found to temporarily suppress the body's immunity mechanism until the foreign tissue had been present long enough to produce its own "recognition patterns" and thus escape rejection. Successful transplantation would also be aided by careful cross-matching of donor and recipient, so that they would start out with as many tissue factors in common as possible.

In addition to the normal blood types, it was found that there are a number of other tissue compatibility factors, and continuing studies are making possible finer and finer degrees of matching between prospective donors and recipients. Exposures to radiations or the use of immunosuppressive drugs are used to temporarily knock out the recipient's immunity system just before and for some time after the transplant. Special "anti-antibody" sera have also been developed to specifically inactivate the factors responsible for the rejection of foreign tissues.

The use of immunosuppressives carries with it a new set of problems. If too low a dose is used, the transplanted organ will be rejected. If too high a dose is used, the patient will be helpless before the attack of invading microbes; indeed, many of the transplant patients who do not survive die of postoperative pneumonia.

Recently, still another discouraging side effect has emerged: transplant patients have been found to have a higher than average risk of developing malignant tumors. It has been theorized that minute tumors are continually formed in the normal body and are quickly destroyed as their changed properties are discovered by the lymphocytes—the same immunity mechanism responsible for rejection of transplanted organs. Immunosuppressive treatment inhibits the body's defenses against cancer.

Despite the problems and risks, organ transplants are saving the lives of growing numbers of patients. Thus far, kidney transplants have been the most successful. Already hundreds of patients are in their second decade of life after a successful kidney transplant. Several thousand such operations are now performed each year. If a transplanted kidney is rejected, it can often be successfully replaced with another.

After the wave of disillusion that followed the initial surge of heart transplants, the operations continued at a greatly reduced rate. Now that some heart transplant patients have survived for more than five years, the technique is again growing in acceptance.

The mechanical problems of organ transplantation are gradually being solved. Not only kidneys and hearts, but lungs, spleens, livers, and various other organs have been transplanted, with varying degrees of success. But legal and ethical problems still persist.

By what criteria is the physician to decide that a patient's condition is hopeless and that his organs can therefore be used to aid others? If he turns off the heart-lung machine and other elaborate life-support systems that are effectively prolonging the patient's existence (even though his brain may long have ceased functioning effectively), is the physician committing murder? If a distraught relative refuses permission for "mutilation" of the body of the deceased, and thus denies access to an organ that might have saved the life of another dying patient, is he or she committing murder? If a surgeon removes the heart of a patient and transplants a new heart into his body, and then the patient dies, has the surgeon *caused* his patient's death? The legal and moral ramifications of organ transplantation have not yet been resolved.

Meanwhile, the medical profession realizes that organ transplantation, while rapidly becoming an effective treatment for a number of serious conditions, is still a stopgap measure at best. In most cases, for a patient to live, another must die, and the supply of suitable organs for transplantation is thus far from unlimited. Indeed, the available supply cannot be utilized at all efficiently. A successful transplant requires that a suitable donor and a suitable recipient, carefully cross-matched, both be in the same hospital at the same time. The recipient must still be in sufficiently good condition to have a hope of recovering from the operation (yet in bad enough condition to warrant taking the extreme measure of organ transplant in the first place), and the dying donor must be otherwise healthy and must die at a convenient time. This favorable conjunction of circumstances obviously does not happen very often. There will be considerable improvement when technical expertise in cryogenics has progressed to the point where organs can be successfully frozen and stored for long periods without deterioration, but still there is a necessary limitation of supply.

Some hope is provided by developing techniques of tissue culture. It is hoped that some day it will be possible to remove some of a person's own normal cells and from them grow a new organ to be transplanted into his body. Or perhaps studies of regeneration may reveal ways to induce a damaged organ to repair itself. But these are remedies for the distant future. More immediate hope is offered by the use of artificial organs and body parts—prosthetic devices.

When Does a Man Cease To Be a Man?

The patient is finally discharged from the hospital. As he leaves the building, the rays of the sun glint from his stainless steel skull, and as he squints, cutting down the light passing through his transparent plastic corneas, he notices his bus pulling up to the corner ahead. He strides swiftly toward the bus with his Vitallium-supported legs, which are attached to Cerosium hip joints. He can feel the air rushing through his silicone-framed nose and down his silicone trachea into his silicone lungs. With the unaccustomed effort, his brain sends impulses along his tantalum cervical nerve to his electronic pacemaker, which then speeds up the pumping of his artificial silicone heart. The blood, gushing out of his Teflon aorta, courses through smaller Dacron arteries. He shouts to the bus driver, using his new Cerosium jawbone. At last the bus driver hears him and opens the bus door to greet him, as the man (?) steps up into the bus.

Of course, an incident such as this has never really happened—yet. But each of the artificial structures mentioned either has already been used in man or is presently under experimental development in the laboratory.

It is not at all inconceivable that the man pictured above may be a "living" member of our society some day. But will such a being still be a *human* being? At what point does a man cease to be a man?

One might argue: As long as the brain is made of flesh, then we still have a man. But what if in some distant time, when technology is far advanced beyond what it is today, we are capable of replacing the brain, too? Do we still have a man?

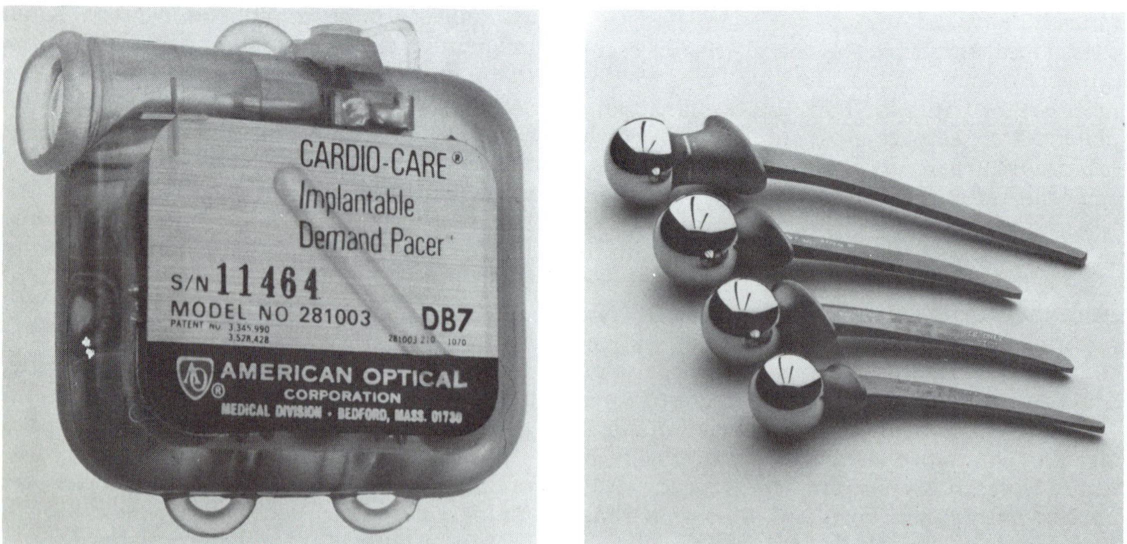

(a) (b)

Figure 23–18 Spare-part surgery: (a) an implantable cardiac pacemaker and (b) an assortment of hip-joint prostheses.

The prospect of replacing human organs with **prosthetic devices** raises moral issues which are not merely fanciful dilemmas to be faced in some far distant future time. Thousands of people are able to live and work today because of electronic pacemakers that control the rhythm of their heartbeats. Active research is underway in the development of artificial hearts, lungs, kidneys, livers, skin, and virtually all other important organs of the body. Artificial limbs are being designed to operate powered by the body's own electricity.

We can expect breakthroughs on many of these fronts in the near future. In August 1966, a team of surgeons led by Michael De Bakey successfully used a heart-assist pump known as a left ventricular bypass, which assumed as much as 60 percent of the pumping action of the patient's heart during a ten-day period. At the end of this time, the device was removed, and the patient, whose heart had repaired itself in the interim, convalesced normally. Dr. De Bakey has predicted that a complete artificial heart replacement for humans may be available before 1980.

In 1969 a complete artificial heart was inserted for the first time into a human patient whose own heart could no longer sustain life while surgeons waited desperately for a suitable donor. After the artificial heart had functioned for sixty-five hours, a donor was found, and a heart transplant was made. But the patient died. A storm of controversy swirled about the attending surgeon, who was criticized for using a device that had not been sufficiently perfected and tested. But it seems only a matter of time before improved models of artificial hearts will be used for longer periods and ultimately permanently implanted.

Will laws be passed to prevent an artificial heart from being inserted? This seems unlikely. Then what about an artificial lung, when one is ultimately perfected? A liver? A kidney? If each is acceptable individually, then what about combinations of them? And ultimately, what about the wholly artificial man? Does this "artificial man" receive the same rights under the law that the "natural man" holds?

As artificial organs are perfected and "artificial men" become a reality, some rather bizarre incidents are bound to arise. Picture a distraught family seeking an annulment of their daughter's marriage on the grounds that "He isn't human!" What effect would there be on the food industry, as more and more people have less of themselves to "feed"? Whole new sets of calorie charts would have to be drawn up for diet-conscious weight watchers.

In the beginning, having an artificial organ may even become a status symbol. (Of course, after a peak is reached, *not* having any artificial organs might then become a status symbol.)

Up to this point we have been discussing the "artificial man" as one arising from the gradual replacement of the organs and other body parts in a living human being. But if we gain the technological know-how to produce adequate replacements for the human heart, lungs, kidneys, liver, and ultimately perhaps even the human brain, then we will have the capacity to build a truly artificial man. Such a "man" (or shall we say robot?) would be able to think and move and perhaps to experience emotions such as love and hate, much as we do. But would such a creature be *alive?* Would such beings be treated as equals under the law? Could they be citizens? Could they vote? Could they own property? Would they be permitted to marry "real" humans?

Some might argue that since these creatures were manufactured, they are merely machines and would have no such rights.

As man's technology improved, he would tend to make his robots more and more intelligent. How long would these "artificial men" submit to a life of slavery? One day these robots might rise up in rebellion, and men might find themselves the slaves of their own creations.

Medical Electronics

A revolution in medical treatment and care has been brought about by the use of computers and a variety of other electronic devices.

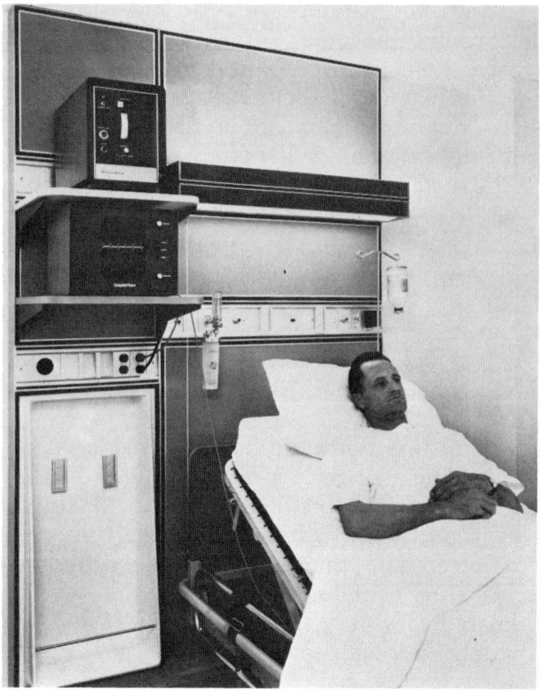

Figure 23–19 The computer-controlled bedside monitor keeps watch over the patient's condition.

The field of medical electronics is a new and fast-growing interdisciplinary area. Each year hundreds of millions of dollars worth of medical electronic instruments are sold by over four hundred different concerns. The field of medical electronics covers literally thousands of items, including mainly electronic and electromechanical instruments and devices which are used on a patient by a physician or nurse, or on a human volunteer or an experimental animal by a biomedical researcher. Such devices may be used for diagnosis or treatment, or to continuously monitor the patient's condition.

The need for instrumentation and automatic analysis is growing by leaps and bounds. For instance, each year nearly a billion tests are performed in the clinical laboratories of the nation. Already the United States faces a shortage of a hundred thousand medical laboratory technicians, and each year the shortage grows worse. To partially fill this gap, laboratories are installing devices such as the Autoanalyzer, produced by Technicon Instruments. Already more than ten thousand of these devices have been sold since they were first introduced in 1957. One line of these devices can simultaneously perform twelve different analyses on a single sample. For instance, a single sample of blood can simultaneously be analyzed for its content of sodium, potassium, chloride, carbon dioxide, calcium, alkaline phosphatase, glutamic oxaloacetic transaminase, bilirubin, total protein, albumin, urea, and glucose. In all, a single machine operated by a single technician can perform more than twenty-five hundred tests in a single day!

Progress in microminiaturization has played an important role in the development of medical electronics and biomedical engineering. Compact monitoring devices have been designed that are so small that they can easily be worn on the body during normal activities. (Some can even be implanted within the body.) Many of these devices are designed to report their readings to radio receivers, which may be some distance away. Thus, heart rates, nerve impulses, and various secretions of experimental animals and human volunteers have been monitored during the performance of a variety of activities, from sleep to strenuous athletics. Such microminiaturized instruments can even transmit their reports from out in space to receiving stations back on earth, permitting continuous monitoring of many facets of the condition of orbiting astronauts.

In the past, there was very little communication between physicians and engineers. However, this situation has been changing rapidly. An encouraging sign of the new attitudes is the fact that since 1960, scores of universities have set up graduate programs in bioengineering.

The growing dependence of modern medicine on the use of computers is dramatic. Many hospitals routinely store medical records, including X rays, other diagnostic data, and family medical histories, in computer systems, where they are available for instant retrieval. Computer-controlled monitoring devices are used in intensive-care units of a number of hospitals to monitor the heart beat, blood pressure, respiration rate, and other indicators, which are constantly compared with data on normal recovery cycles stored in the computer memory bank. Any deviation beyond preset limits results in an alarm or warning flash that may be observed by a nurse who remains at a central monitoring station. Thus, a single device can monitor the progress of a number of patients.

Other computers have been utilized in routine medical examinations. For instance, one computer is regularly used to process automated tests on as many as four thousand patients a month. The computer method is as much as two to three times faster than the standard procedure, and four times as many tests can be performed on the patient for the same cost.

A promising new computer diagnostic program has recently been instituted. Electrocardiogram recordings can be sent over a special telephone line to a central computer in Washington, D.C. Within fifteen seconds, the signal is analyzed and its analysis is sent back to the hospital from which it was originally sent. Similar methods could readily be adapted to other types of medical diagnoses.

Indeed, an important step in that direction has already been taken. A group of doctors working in conjunction with IBM specialists have already developed a Clinical Decision Support System. This system is designed as a comprehensive compendium of medical knowledge that may be useful in the diagnosis and treatment of disease, in a readily retrievable form. The computer acts as a sort of "superconsultant."

Reality is fast approaching some of the most imaginative inventions of the science fiction writers!

QUESTIONS FOR REVIEW AND THOUGHT

1. Devise an experiment which would indicate whether a particular disease affecting the lungs of rats is of microbial origin.
2. Are all diseases caused by microbes? Explain.
3. Why are viral diseases more difficult to control (once they have been contracted by the victim) than bacterial diseases?
4. It has been said that perhaps half of the deaths in the world are directly or indirectly caused by insects. Can this be true? Explain.
5. Why is it said that cancer may in reality be more than one hundred *different* diseases?
6. Briefly describe the standard methods of treating cancer and mention some of the newer approaches which have met with some success.
7. What are some of the drawbacks of the use of antibiotics on man and animals?
8. Why are organ transplants considered by some scientists to be a dead end?
9. At what point in the replacement of body parts by prosthetic substitutes does a man cease to be a man and to enjoy full rights under the law?
10. One important problem besetting those who are involved in research in medical electronics is that there are few men in medicine who know a great deal about electronics and even fewer men in electronics who know a great deal about medicine. What suggestions can you make to alleviate this problem?

PART VII

MAN IN HIS WORLD

Lynwood M. Chace, Omikron

CHAPTER 24

Ecology: The Interrelatedness of Life on Earth

Introduction

The earth teems with trillions of creatures. Each is an entity unto itself, but is capable of affecting and being affected by its neighbors. Life in its multitudinous forms is found throughout the vast expanses of the earth: from encysted microorganisms in the upper reaches of the atmosphere to the mysterious creatures inhabiting the depths of the oceans.

Turn over a large stone, and scores of creatures will scatter before your eyes. Take a sample of air, and hundreds of microbes will wiggle beneath the lens of your microscope. Scoop out a jar of pond water, and more creatures will be seen. Even the desert sands will be found to house numerous life forms. The sea, the soil, and the air provide homes for earth's creatures.

The lives of widely divergent and geographically distant organisms are inextricably intertwined. They may be related by direct physical contact, by competition for the earth's limited resources, or by a complex series of intricate physical, chemical, and biological interactions. The study of the interrelationships between these organisms and their environment is called **ecology.**

Figure 24–1 The diversity of earth's habitats.

The Physical Environment

The lives of earth's creatures are shaped by a ceaseless interaction of numerous factors. One of the most important of these is the nature of the medium in which they live and the physical conditions that prevail in it.

WATER

As we have seen, nearly three-quarters of the surface of our planet is blanketed with water. It is believed that there, in this vast expanse of water, life first appeared and flourished.

In many respects, water is a hospitable and sheltering medium for life. Temperature conditions are relatively stable: whereas changes of 30–40°F within a span of hours are not uncommon on land, such drastic and rapid temperature changes very rarely occur in any particular region of the hydrosphere (the waters of the earth). The water that continuously bathes the bodies of aquatic organisms facilitates an exchange of materials. In the early days of life's evolution on earth, the water of the seas was probably roughly isotonic to the protoplasm of the simple organisms that inhabited them. Now, when the seas have grown more salty and life has also migrated to the low-salt fresh waters of rivers, lakes, and ponds, specialized systems for pumping out water or eliminating salt have been evolved and aid today's aquatic organisms in coping with the problems of maintaining homeostasis.

The currents in the water aid in bringing food to the many sedentary water-dwellers; they also aid in the dispersal of the spores and seeds of plants and the germ cells and immature forms of animal life (Fig. 24–2). Ocean currents have been known to carry such material hundreds and even thousands of miles from their point of origin. The Kuroshio or Japan current, for example, flows eastward nearly five thousand miles across the Pacific Ocean and then swings southward, passing only a few hundred miles from the American mainland.

The buoyant effect of water aids in locomotion and support. Forms of plant and animal life have evolved to great size—far larger than would be possible under the full force of gravity on land—with a minimum of internal structural bracing. Giant squids, some reaching a length of more than fifty feet, glide through the oceans without any supporting skeleton at all. Whales, the largest present-day animals known, manage satisfactorily in the oceans with a skeletal system that would be wholly inadequate on land: when a large whale is occasionally accidentally stranded on a beach, it is totally incapacitated and eventually dies, crushed under the huge body weight that its skeleton cannot support without the aid of water's buoyancy.

Figure 24–2 These cup corals, like other sessile ocean bottom dwellers, rely on ocean currents to bring food to them and disperse their progeny.

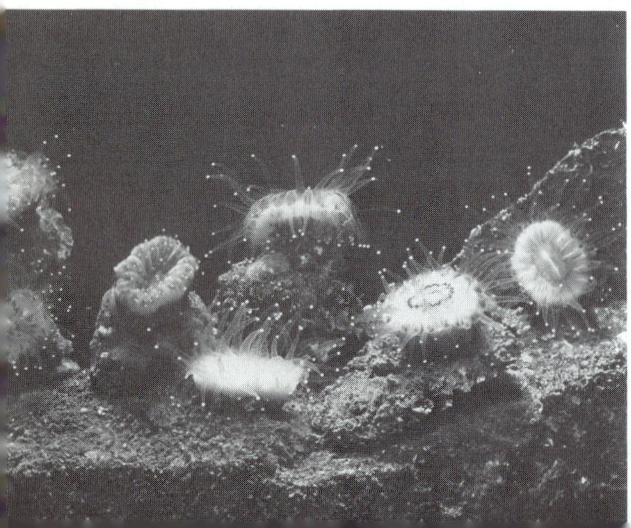

Figure 24–3 The carcass of a beached whale sags lifelessly without the buoyant effect of the ocean water.

AIR

In comparison with the sheltering waters, the lands of the earth present a relatively inhospitable environment for life, although they do offer some compensatory advantages. Both seasonal and daily temperature variations are often quite drastic, and without the buoyant effect of water an organism must possess an effective means of support if it is to grow to any significant size. The vagaries of weather—including winds and storms—bring problems and challenges to land life, but they are also used to advantage by many modern organisms. The wind carries spores, seeds, and insects to far-off places and helps to propagate many species. In areas where rainfall is infrequent, a storm may be followed by a bursting of life: dormant seeds soak up the life-giving moisture and grow rapidly into an oasis of

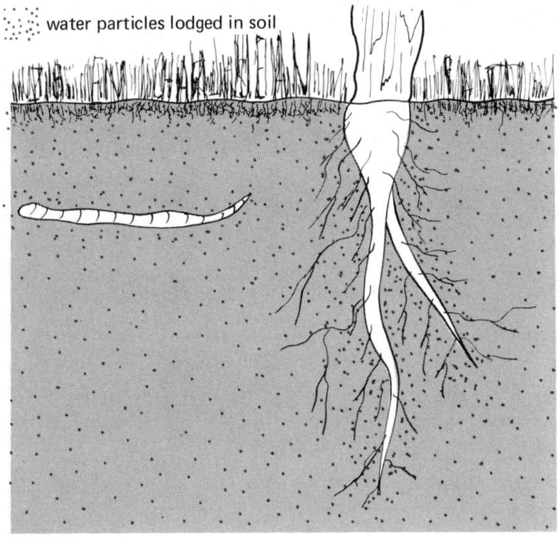

Figure 24–4 Soil-dwelling animals and the roots of plants are actually surrounded by a moist environment.

lush vegetation that soon dies away again as the moisture ebbs away, leaving new seeds or other dormant forms to last until the next rainfall.

Life in an atmosphere of air offers some other advantages over life in water as well. Since air is more transparent to sunlight than water, photosynthesis can take place more efficiently. The higher oxygen concentration in the air (21 percent vs. about 0.4 percent in sea water) boosts the efficiency of respiration among terrestrial animals (especially in birds and mammals) far above that in aquatic forms. The lower resistance that air offers to motion provides a potential for more rapid locomotion.

We should not be misled by the concept of "dry land" as the environment for terrestrial life. Actually the microorganisms, worms, and other small animals that inhabit the soil* are effectively living in water: in the moisture trapped within and among the soil particles (Fig. 24–4). Similarly, the roots of land plants are enclosed in a film of water, though their stems and leaves may sway in a cool dry breeze.

The migration of life from sea to land has necessitated numerous adaptations, many of which have been centered around means of conserving water. For the economy of protoplasm is essentially a water-based economy.

Among plants, the spread to increasingly arid regions has resulted in a succession of water-retaining and water-obtaining adaptations. These include mainly a progressive expansion of the root system and a concomitant reduction of the total surface of the exposed areas aboveground. The gradual adaptation of plant life to a variety of terrestrial environments has sometimes led to extremes: the water lily, a plant that grows in ponds and swamps, has only a minimal root system, whereas the cactus, a typical desert plant, has an extensive root system but leaves reduced to mere spines and a plump, water-storing photosynthesizing stem covered with a thickened cuticle to minimize water loss (Fig. 24–5).

Among terrestrial animals, one of the most important of the problems of survival is similarly the retention of a sufficient amount of water to permit metabolic processes to continue as usual. Amphibians and such soil-dwellers as earthworms are relatively inefficient in their adaptation to life on land. Their soft, thin skins lose water extremely rapidly, and if they are away from a source of water for too long, death from desiccation may result. Such animals are thus limited to land areas readily accessible to sources of water. Reptiles, birds, and mammals (along with most insects) have effectively extended their range of potential habitation by developing an integument that is relatively impermeable to water. Secondary mechanisms for saving water have also evolved in many of these forms. For example, solid feces containing uric acid are excreted by reptiles, birds, and insects. In contrast, amphibians, fish, and mammals excrete their nitrogenous wastes in the form of a dilute urine (mostly an ammonia solution in fish and a urea solution in amphibians and mammals).

In addition to the physical medium of the environment, the energy factors, particularly

*A single acre of soil may house thousands of earthworms, millions of insects, and as many as 20 *quadrillion* (20,000,000,000,000,000) microorganisms.

(a)

(b)

Figure 24–5 Plants adapted to widely differing environments: the water lily (a) and saguaro cactus (b).

light and heat, also have a vital influence on the life of our planet.

LIGHT

Picture the world without light. With the sudden darkness comes an ever-deepening chill. Within a short time frost covers the entire surface of the earth. Temperatures continue to plummet as the earth radiates its precious remaining heat out into the cold stillness of space. All higher forms of life on earth perish. Only dormant seeds and the spores of microorganisms persist. A cold, bleak planet hurtles unseen through space.

A substantial fraction of the sunlight that strikes the surface of our planet is transformed into heat. And since our atmosphere is relatively transparent to light in the visible range and relatively opaque to heat energy, much of the heat is retained. But in addition to its interrelationship with heat, light has a number of unique functions of vital importance to the continued existence of life on this planet.

The use of light energy by plants for photosynthesis is an important factor in determining their mode of existence. Tall trees and climbing vines stretch upward in silent competition for maximum exposure to the sunlight. Some ground-dwelling plants, whose evolution has precluded their participation in this struggle for light exposure, have survived through adaptations permitting them to photosynthesize efficiently at moderate or even low light intensity. Thus, ferns, mosses, violets, and numerous other wild plants develop best in a moderate degree of light, where sunlight filters down to them through the overhanging trees (Fig. 24–6). Indeed, prolonged exposure to direct sunlight can prove fatal for many of these plants.

Many aspects of plant and animal life are influenced by seasonal variations in light exposure. This influence of periodic fluctuations in the duration of daylight upon the lives of animals and plants is known as **photoperiodism.** The migration of birds provides an excellent example of the influence of the length of daylight on the life cycles of animals. At the end of the breeding season, with autumn approaching, the days become progressively shorter. This causes a decrease in the size of the gonads, and at the appropriate time the changed daylight period somehow activates a "trigger" mechanism resulting

Figure 24–6 Trailing arbutus, a shade-tolerant plant that thrives on the forest floor.

in the migration of the birds southward. In time, as the days lengthen, the gonads are stimulated by the increased light, which once more triggers the migratory mechanism for the return flight northward.

Photoperiodism also plays an important role in the life cycles of many plants, as has already been discussed (Chapter 12). Although some species flower as soon as they reach a certain stage of development regardless of the season, other plants will flower only when the periods of daylight (or of uninterrupted darkness) have reached a certain length.

Most animals possess some type of light receptors. In higher animals this takes the form of eyes and is of critical importance in the constant struggle for survival. With the ability to see, color and distinctive patterns begin to play a key role in evolutionary processes through their use in camouflage and recognition signals. Some organisms also produce their own light through processes of bioluminescence, and in a number of cases these "living lights" are of adaptive value.

HEAT

The heat-distributing effects of the atmosphere shield our planet from the extremes of temperature that are observed out in space and on airless worlds such as the moon. The temperature range most compatible with growth and normal development in the majority of earth organisms is between 0°C (32°F) and 50°C (122°F). But this range does not exhaust the survival potential of life. Numerous microorganisms, spores, and seeds have been kept for long periods at temperatures approaching absolute zero (−273°C or −460°F). Although no signs of life are visible during this "deep freeze," normal development often follows after the organisms are thawed. At the other extreme, some algae have been found to be active in the hot springs of Yellowstone National Park in Wyoming, at a temperature of 85.2°C (about 185°F). Active bacteria have been found elsewhere in hot springs with a temperature of 88°C (about 190°F).

Under certain conditions, organisms can adapt to an environment that is normally in-

Figure 24–7 Hot springs: the Devil's Doorways at Yellowstone. Even here some life forms can be found.

tolerable. If the transition to the new environment is gradual enough, the chances of survival are greatly enhanced. For example, many bacteria are capable of forming spores that can withstand a wide range of temperatures, even up to 205°C (about 400°F). This resistant form is achieved through a gradual, partial desiccation of the organism. The common practice of hardening young plants in the early spring by exposing them to progressively decreasing temperatures is another example of adaptation of an organism to a potentially hostile environment. Placing these young, delicate plants directly out in the cold spring ground would, in many instances, prove fatal. In certain regions, the practices of "hardening" are also extended to humans. Repeated exposure to cold or to extremes of heat and cold (such as taking a steam bath and then rolling in snow) presumably makes the body healthier and more resistant to colds and other infections.

The extensive seasonal variations in temperature, particularly in the earth's temperate zone, have given rise to various adaptations among forms of life. These include the thickening of the fur and of fat deposits beneath the skin of animals and the production of thick-walled, cold-resistant seeds, spores, eggs, or other dormant forms by numerous plants, animals, and protists. One of the most outstandingly successful adaptations to seasonal temperature fluctuations on our planet has been the development of mechanisms for maintaining the internal body temperature.* Among warm-blooded or **homoiothermic** animals (mammals and birds), the constant and relatively high body temperatures ensure optimum functioning of metabolic systems in all seasons. This is in sharp contrast to cold-blooded or **poikilothermic** organisms (e.g., fish, amphibians, reptiles, and all the invertebrates), in which the internal body temperature corresponds roughly to the temperature of the environment. When the weather turns cold, chemical reactions in the bodies of these animals slow down, and they grow increasingly sluggish.

Temperatures approaching and dropping below the freezing point of water (0°C or 32°F) often prove lethal to organisms. This result is due both to an upset in the temperature-sensitive enzyme-mediated reactions in the body and to the physical effects of freezing (since protoplasm is a water-based suspension). If the temperature is lowered slowly, large ice crystals may form, often causing extensive cell damage. Rapid freezing yields finer crystals, and damage is minimized.

Higher organisms have evolved various mechanisms permitting them to withstand or avoid freezing temperatures. Evergreens and other trees growing in the cold regions of the earth survive seasons of extreme cold by converting free water into a bound form. This reduces crystal formation, in addition to lowering the freezing point of the protoplasm. Another mechanism for adapting to a cold environment involves the suspension or marked reduction of normal physiological functions. This state, known as **hibernation,** enables a wide variety of animals, including insects, amphibians, reptiles, and mammals, to endure long periods of almost complete inactivity (Fig. 24–9). During hibernation,

*The temperature-control systems, coordinated through the nervous system, include heat production by muscle action (shivering), mechanisms for dilating and constricting the blood vessels close to the surface of the skin so as to enhance or minimize radiation of body heat, and sweating: water volatilized from sweat pores in the skin carries off excess body heat. (Animals without effective sweating mechanisms may accomplish the same end by panting on hot days.)

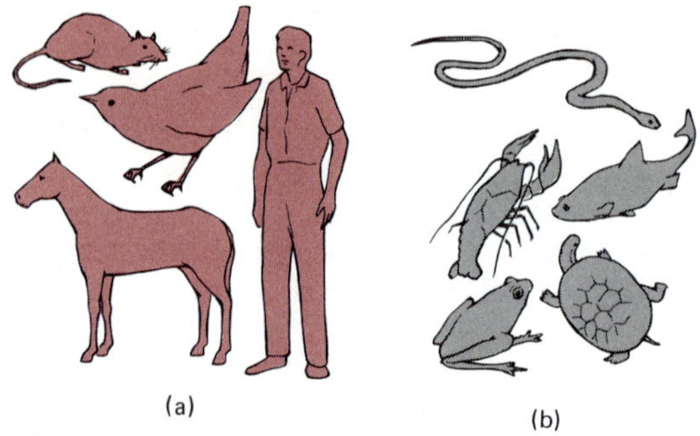

Figure 24–8 Some typical homoiothermic (a) and poikilothermic (b) animals.

metabolism is reduced twenty- to a hundredfold. The body temperature drops to essentially that of the surrounding air (even in hibernating mammals such as ground squirrels, woodchucks, bats, and hamsters), and the slowdown in heartbeat is dramatic (to only a few beats per minute). Breathing is slow and irregular. Some hibernating animals wake periodically and may eat and drink from stored supplies before returning to their hibernating state. The long winter sleep of the bear is not truly hibernation, for its body temperature remains rather high (31–34°C or about 88–93°F).

The mass migrations of birds and of various mammals, such as seals, are also adaptations to temperature fluctuations brought by the changing seasons. By spending the winter in latitudes closer to the equator and summering in regions closer to the poles, an animal not only avoids temperature extremes but also takes advantage of the brief periods of abundant food supply in the less crowded higher-latitude regions.

A number of animals escape the hot dry summers by entering a period of dormancy called **estivation.** (The terms *hibernation* and *estivation* come from Latin roots meaning "winter" and "summer," respectively.) Estivation is practiced by some species among the insects, fish, snakes, and mammals. The ground squirrel, for example, may remain dormant in its burrow and thus escape from periods of drought and food shortage, in addition to avoiding the hot summer weather. Some fish species get through the dry, hot

Figure 24–9 Curled up in its underground nest, a chipmunk hibernates through the cold winter months.

summers by estivating in the caked mud that remains in the bottom of dried-up pools.

The effects of temperature on living organisms have been put to practical uses. Temperature treatment of dormant seeds and buds often leads to a termination of the dormancy period. For example, apples, peaches, and pears require a period of low temperature before blooming. On the other hand, rice and cotton require temperatures exceeding 27°C (i.e., above 80°F) before flowering will occur.

The exposure of germinating seeds or young plants to low temperatures may similarly hasten the onset of flowering. Such treatment, called **vernalization,** is used to practical advantage in the USSR. In many regions of that country, the growing season is so short that winter wheat will be killed by the autumn frosts before it has yielded a harvest if it is planted in the normal manner. Vernalization shortens the flowering time and thus makes the raising of this crop practical.

Biomes

To an orbiting astronaut, the earth appears as a kaleidoscopic panorama: rain forests, grasslands, deserts, and rolling ocean waters are spread out beneath him like patches in a patchwork quilt. Each region is given its distinctive appearance by its characteristic growths of vegetation (or lack of vegetation), and each is characterized by a particular assortment of animal life as well. These regions, each with its own distinctive plant and animal communities, are known as **biomes.** From space it is evident that no sharp line of demarcation separates adjacent biomes. Instead, a gradual blending of adjacent regions is evident, with a transitional area containing vegetation and wildlife common to both.

Several main factors influence the distribution of plant and animal species: temperature (both the range and the rate of change of the temperature), the duration and intensity of sunlight, and the patterns of annual rainfall. Thus, not only the distance from the equator, but also the altitude, as well as particular weather conditions influenced by the major ocean currents, are important in delimiting the biomes of the earth. It is not surprising that the slopes of tall mountains exhibit a succession of biomes, with the higher-altitude regions characterized by patterns of life more typical of regions nearer the poles (Fig. 24–10).

The major terrestrial biomes are the tropical rain forests, grasslands, temperate deciduous

Figure 24–10 An astronaut's view of Earth, taken during the flight of Apollo 17.

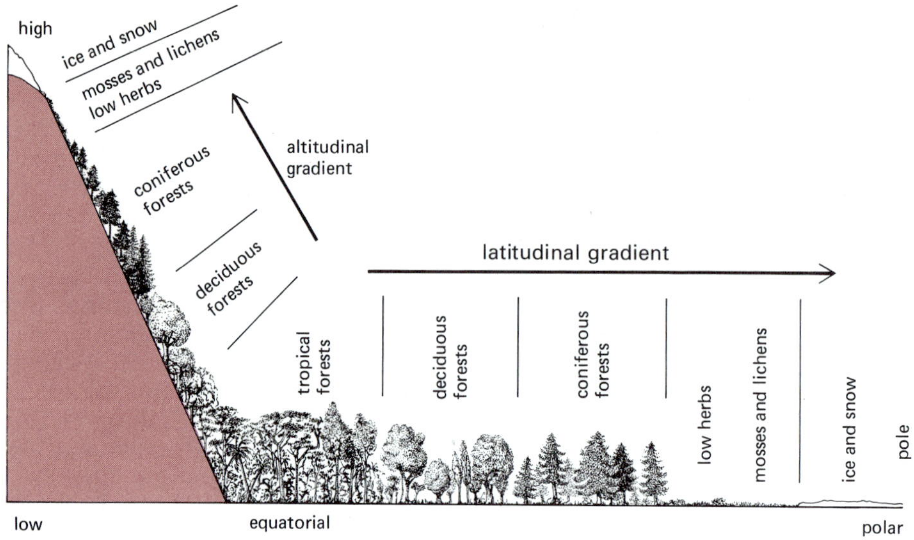

Figure 24–11 Earth's biomes show a typical distribution both according to latitude and according to altitude.

forests, northern coniferous forests, tundras, and deserts. The waters of our planet, too, have their characteristic patterns of life, including the various fresh-water biomes and marine biomes.

The terrestrial biomes became the object of intensive study by scientists of fifty-seven nations when a five-year International Biological Program was launched in 1967. In this cooperative effort, studies were made of the interactions of the animal and plant species of typical environments. The results are being incoporated into complex mathematical equations and fed into a computer for use in predicting the ultimate effects of proposed changes such as the elimination of a particular pest species.

TROPICAL RAIN FOREST

Straddling the equator are tropical rain forests, vast regions of dense jungle containing a bewildering variety of vegetation and animals (Fig. 24–12). The high temperatures prevailing throughout the year and the abundant rainfall (often exceeding 90 inches annually) support the growth of such a dense canopy of tall trees and tangled vines that sunlight rarely reaches the forest floor. Within this canopy, shielded from the violent force of the torrential winter rains and gusty winds, live a varied animal population, including monkeys, sloths, lizards, snakes, and brilliantly colored birds. Tropical rain forests are found in Central America, the Amazon basin of South America, Central Africa, and Malaya and New Guinea in the Pacific area.

GRASSLANDS

In many areas of the globe the annual rainfall is between 10 and 30 inches. This is insufficient to support extensive forests but

Figure 24–12 A typical rain forest community.

Figure 24–13 Grasslands: the Kansas prairies.

does maintain the growth of lush grasses of a number of different types. That grasslands are widely distributed throughout the earth can be deduced from the numerous local terms referring to this type of environment. These include the *plains* and *prairies* of the United States, the *steppes* of Siberia, the *savannas* and *veldts* of Africa, and the *pampas* of Argentina. The grasslands (Fig. 24–13) support an abundance of animal species. The dominant vertebrates are the swift-moving antelopes, zebras, and bisons, which graze on grasses, and numerous burrowing animals, such as gophers, rabbits, and ubiquitous mice. Upon these animals prey lions, vultures, and hawks. The grasslands form the bulk of man's "breadbasket," yielding both cultivated grains (including wheat, corn, and rye) and meat from domesticated sheep and cattle.

TEMPERATE DECIDUOUS FOREST

The land areas in the temperate zones that receive a relatively abundant (30–40 inches), evenly distributed rainfall, with warm summers and cold winters, are often covered with forests of deciduous trees, which shed their leaves in the autumn and remain dormant over the winter. Among the more common varieties are beech, elm, walnut, maple, hickory, chestnut, sycamore, oak, and ash. In comparison with the tropical rain forests, the number of plant species is more restricted and the individual trees are spaced more thinly. This arrangement leaves ample space for the growth of such intervening vegetation as mushrooms, slime molds, and a dazzling array of colorful flowering plants, which are scattered over the leaf-covered forest floor. The variety of animal inhabitants of the deciduous forests include deer, squirrels, mice, raccoons, foxes, opossums, timber wolves, salamanders, and (in the spring and summer) numerous migratory birds (Fig. 24–14).

Figure 24–14 Temperate deciduous forest.

NORTHERN CONIFEROUS FOREST

Stretching across North America, northern Europe, and Siberia, just north of the northern deciduous forests and grasslands, is an area known as the northern coniferous forest or **taiga** (Fig. 24–15). The winters in this region are often extremely cold and severe, with snow covering the ground for months. But the summers are warm, permitting the growth and development of large forests of such conifers as spruce, hemlock, fir, and pine trees. The forests are inhabited by such animals as grizzly bears, moose, wolves, squirrels, and many birds. A number of birds migrate to more southerly regions before the onset of the cold

Figure 24–15 Northern coniferous forest.

Figure 24–16 Snow-covered mountains rise above the tundra in Mt. McKinley National Park, Alaska.

weather, while many of the mammals hibernate through the severe winters.

TUNDRA

The northernmost terrestrial biome, the **tundra,** encompasses the northern regions of North America, Europe, and Siberia. Most of the tundra lies within the Arctic Circle. As one would expect, the winters are long and severe, and the summers are exceedingly short, with temperatures rarely going above 75°F. Just beneath the surface the ground remains in a perpetually frozen condition **(permafrost),** which in some areas extends downward for several hundred feet. Trees do not grow in this forbidding climate; lichens, mosses, and some grasses predominate, with some occasional low-growing shrubs (Fig. 24–16). Surprisingly, many animals inhabit this region during the summer months, as the warmth and sunlight bring a brief spurt of plant growth. Large herds of caribou and reindeer leave the protective forests of the taiga and graze in the tundra. Snowy owls, arctic hares, arctic wolves, arctic foxes, lemmings, and ptarmigans are also found in the tundra, along with swarms of mosquitoes and flies and flocks of migratory birds during the summer months. With the onset of cold weather, most of the tundra fauna either head southward or prepare for hibernation through the winter months.

DESERTS

In the desert regions of northwestern Mexico, the southwestern United States, along the west coast of South America, in north and west central Africa, and in central Asia, the annual rainfall is less than 10 inches. In these areas the sparsely spaced vegetation consists mainly of such species as cactus, sagebrush, and yucca, which are highly adapted to conserve water (Fig. 24–17). Following a sudden, soaking rain, seeds from fast-growing annuals will often germinate, reach maturity, flower, produce seeds, and die within the space of a few weeks. Desert animals also are highly adapted to the demanding conditions of their environment. Pocket mice and kangaroo rats subsist on seeds and other vegetation, obtaining their total water needs from these sources. Snakes obtain their water from the blood of the rodents on which they feed. Desert regions are often characterized by great temperature extremes, the broiling heat of the

Figure 24–17 Cactuses in the desert at Saguaro National Monument near Tucson, Arizona.

daytime alternating with biting cold at night. Most of the activity of desert animals is restricted to the cooler night hours. In some desert regions, such as the Sahara, there is virtually no life at all, except in isolated cases where underground springs foster the growth of lush vegetation.

FRESH-WATER BIOMES

The ponds, lakes, swamps, springs, rivers, and other fresh-water bodies constitute the fresh-water biomes (Fig. 24–18). The shallow water near the shore, called the **littoral zone,** usually contains a different distribution of flora and fauna than is found in the deeper region, the **limnetic zone,** of a lake. Similarly, the life forms of still waters differ from the assortment found in swiftly moving streams. Algae, cattails, pondweeds, water lilies, and numerous other hydrophytes comprise the plant population in fresh-water biomes. Animal species include fish, frogs, water snakes, flatworms, crayfish, and a variety of insects and their larvae. Protists such as *Paramecium* and *Amoeba* and microscopic animals such as hydras and rotifers also abound. Many denizens of fresh-water biomes are

Figure 24–18 A freshwater community: lagoons in Argentina.

adapted to sudden severe changes in environment, such as rapid and extensive temperature changes, turbulent storms resulting in widespread muddying of the watery habitat, freezing in the winter, and even drying up in the summer.

MARINE BIOMES

The total mass of animals and plants (biomass) in the waters of the world far exceeds that found on the land. The marine biomes, like the fresh-water biomes, may be divided into various zones. Important determining factors are the proximity of land masses and major

Figure 24–19 Marine community: a tidal pool group.

ocean currents and the depth of the water, which in turn determine the temperature, amounts of dissolved gases and minerals, and the intensity of sunlight penetrating the water.

As we have seen, the ocean floor at first slopes off gradually from the coasts (the continental shelf), then drops steeply (the continental slope) for perhaps a few thousand feet, until the ocean floor (the abyssal plain) is reached. The depth of the ocean floor varies greatly, as mountains thrust up through the surface as islands and deep canyons extend down for several miles. The region of shallow water extending over the continental shelf is referred to as the **neritic zone.** An interesting portion of this zone is the **intertidal region,** the area between the high-tide and low-tide lines, in which the coastal rocks or sands are alternately laid bare and submerged with the ebb and flow of the tides. The region beyond the continental shelf is known as the **oceanic zone.**

The rays of the sun can penetrate for perhaps 400 feet or so into the depths of the ocean. This region, the **euphotic zone,** is rich in photosynthetic vegetation and in an abundance of animal forms feeding upon the plant life and on each other. In fact, the neritic zone (essentially all of which lies within the boundaries of the euphotic zone) is probably the most heavily populated life zone on earth. Extending below the euphotic zone for about a mile or so is the twilight area of the **bathyal zone.** Below this layer, extending to the bottom of the ocean, is the **abyssal zone,** a world of total darkness and crushing pressures. No photosynthetic plant life can survive in these lower zones, but nourishing bits of matter drift down from the sunlit surface layers, and a variety of animal life thrives in the depths of the sea. Life has even been observed deep in the Marianas Trench in the Pacific Ocean, at depths of more than 35,000 feet from the surface.

Marine organisms that inhabit the bottom of the ocean are called **benthos** (crawling or creeping organisms) or **sessile** (stationary, i.e., attached to some immobile object). The organisms that float passively on the surface **(plankton)** and those that swim freely and actively through the surface waters **(nekton)** are known as **pelagic** organisms. The multitudinous life forms of the sea range from microscopic diatoms to a variety of invertebrates, fish, and even mammals, including the giants of the deep—100-foot long, 150-ton blue whales.

Cyclic Utilization of Matter and Energy

When an animal or plant dies, its body begins to decompose almost immediately. Enzymes within the cells begin to break down their constituents in processes of autolysis, and the breakdown is furthered by scavengers such as pill bugs and beetles and various fungi, bacteria, and other decay microorganisms. Eventually the tissues of the dead organism are broken down into simpler components, which are returned to the soil and can ultimately be utilized by new plant growths. The mechanisms of decay are an integral part of the continuation of life on our planet. They permit precious resources such as carbon, nitrogen, and other minerals to be used over and over again in a cyclic manner. For each element found in living organisms, such a circle of use, release, and reuse can be traced. The most important for living organisms are the carbon, oxygen, water, nitrogen, and phosphorus cycles.

THE CARBON CYCLE

The chemistry of life is a carbon-based chemistry. Thus, all growth, all birth of new life involves an intake of carbon. In a sense, then, the supply of carbon is one of the limiting factors determining the amount of life that our planet can support and the ultimate size of the human population. But we need not fear on this account. The minute fraction (0.04 percent) of CO_2 in the atmosphere could provide enough carbon for more than 10 trillion human beings (i.e., about a three thousand-fold increase in the present human population), and the atmospheric carbon

Figure 24–20 The carbon cycle.

represents only a small proportion of the total amount of carbon in the earth's crust.

The carbon dioxide of the atmosphere, the carbonates and bicarbonates of the earth's crust, and the carbon contained in the bodies of living organisms are interrelated in a cycle of processes (Fig. 24–20) that go on continuously in dynamic equilibrium. The carbon dioxide of the atmosphere is taken in by photosynthesizing plants and protists, which convert a portion of it to more complicated, organic compounds. These plants serve as the carbon source for countless numbers of heterotrophs, some of which in turn serve as carbon sources for other heterotrophs—predators and parasites. Respiration breaks down complex organic chemicals and restores some carbon to the atmosphere in the form of carbon dioxide. Death, followed by putrefaction and fermentation by decay bacteria and fungi, returns carbon dioxide to the atmosphere as well.

This basic carbon cycle has a number of offshoots and side branches. Some carbon is withdrawn from circulation (for a time) by the formation of calcareous shells and coverings by many types of animals. The quantities of this material can be quite substantial. For example, coral reefs laid down by microscopic corals (over periods of millions of years) sometimes reach a thickness of several thousand feet and may extend for hundreds of miles. Carbon is removed from active circulation through the formation of coal and petroleum, and in the carbonates and bicarbonates of the rocks of the earth's crust.

Some of this "incarcerated" carbon is restored to the general circulation by dissolution of carbonates and bicarbonates from deposits on the ocean floor. Fires are a factor of increasing importance in the conversion of organic carbon into carbon dioxide. With the rapid expansion of the industrialized nations of the earth, the amount of fossil fuels consumed has been increasing geometrically. The air in some highly populated industrial areas frequently contains as much as 0.07 percent carbon dioxide, nearly twice the normal level. Some scientists feel that increasing industrialization has led to a slight increase in the overall world level of atmospheric carbon dioxide. If this is true and the trend continues, the ultimate consequences for the earth could be profound. An increase in the carbon dioxide level in the atmosphere would result in a "greenhouse effect" that might lead to a substantial rise in the average temperature of our planet, which in turn could lead to thawing out of the great deposits of snow and ice in the extreme northern and southern regions of the globe and ultimately an inundation of New York, London, and many others of the most heavily populated land areas. The potential imbalance of the carbon cycle is aggravated by another concomitant of expanding urbanization and industrialization: the shrinking of vegetation-covered areas as cities expand and the increasing pollution of the world's waters, killing portions of their photosynthesizing plant and protist populations. These are all factors that ecologists are now weighing seriously in their efforts to determine sane pathways for further development of our planet.

THE OXYGEN CYCLE

Without air, a human being can live for a scant few minutes. The vital ingredient is oxygen, which constitutes only 21 percent of the volume of atmospheric air. This substance is the key ingredient in the energy-yielding processes of respiration, without which all other processes of the body must ultimately come to a halt.

Like carbon, oxygen exists in our world in a cycle of processes (Fig. 24–21) interrelated and balanced in dynamic equilibrium. Oxygen is continuously added to the total atmospheric supply as a by-product of photosynthesis. Simultaneously it is removed from the atmosphere by the respiration of plants and animals. In this process, the oxygen is combined with hydrogen to form water and with carbon to yield carbon dioxide. Thus, the oxygen cycle is intimately interlinked with both the carbon cycle and the water cycle.

Oxygen is the source for the thin but important layer of ozone in the stratosphere. (The

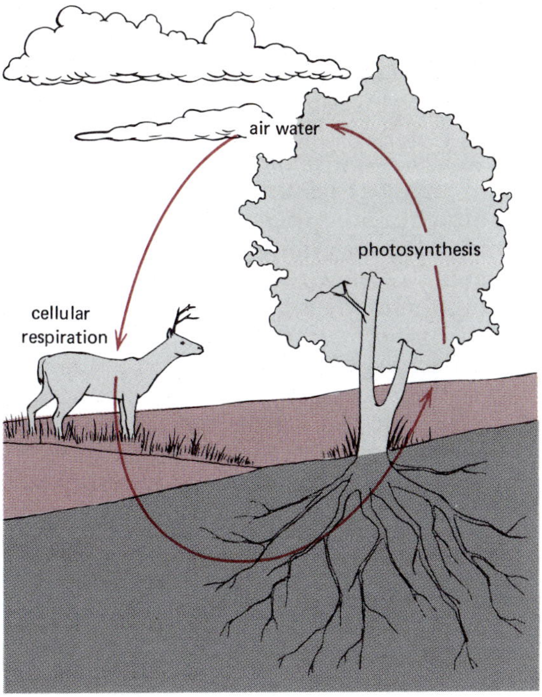

Figure 24–21 The oxygen cycle.

action of ultraviolet rays on oxygen converts it to ozone.) Although ozone is a highly poisonous gas, its presence in the upper atmosphere acts as a protective shield screening out most of the dangerous high-energy radiation coming from the sun.

THE WATER CYCLE

The earth's atmosphere contains more than three thousand cubic miles of water, enough to supply the needs of the entire human population for a generation. Although 10 percent of this water falls daily as rainfall, the atmospheric reserves of water vapor are replenished by evaporation from oceans, rivers, lakes, and other bodies of water.

Another cyclic circulation of water proceeds through an interaction with living organisms. Green plants may utilize water in photosynthesis to build up organic compounds, and these in turn may be consumed by animals. The water may then be returned to the atmosphere as a by-product of respiration, or it may be assimilated into the tissues of the animal. Death and decay liberates water (among other substances) into the soil, and from there the water may find its way to streams or be reabsorbed by plants. In addition to using water for photosynthesis, plants transpire (i.e., evaporate from the leaves) an enormous amount of water into the atmosphere. An acre of corn, for example, will transpire about 350,000 gallons of water during a single growing season.

Figure 24–22 The water cycle. The numbers refer to the volume of water in cubic kilometers.

THE NITROGEN CYCLE

The vast quantity of nitrogen gas present in our atmosphere is essentially useless to most animals and plants. Even though nitrogen comprises about 78 percent of the atmospheric air, it cannot be utilized directly as a food source. For the nitrogen of the atmosphere is in the form of molecular nitrogen, N_2, and most living organisms can utilize nitrogen only in its bound forms—ammonium salts, nitrites, nitrates, and organic nitrogen compounds. The conversion of atmospheric nitrogen into usable forms, a process called **nitrogen fixation,** is accomplished by a variety of bacteria, including representatives of *Azotobacter* and *Clostridium.* These soil-dwelling bacteria convert atmospheric nitrogen to organic nitrogen compounds, such as amino acids and proteins. When the bacteria die, their proteins are decomposed to yield ammonia.

A major share of nitrogen fixation is carried out by bacteria of the genus *Rhizobium,* which live in swellings (nodules) on the roots of leguminous plants such as peas, beans, alfalfa, clover, and soybeans (Fig. 24–23). Together the bacteria and the plants are capable of fixing nitrogen, although neither organism can perform this task in isolation.

A second, but less significant source of nitrogen fixation is through the action of lightning. The static electric discharges of lightning act on the atmospheric components to produce nitrogen oxides, which are washed into the soil and form nitrates. The soil nitrates are utilized by plants for the synthesis of their proteins, which in turn serve as the chief source of nitrogen for the manufacture of animal proteins. In fact, for the most part, plant proteins are merely hydrolyzed (digested) into their component amino acids and then rearranged (assimilated) by the animal cells into the appropriate animal proteins. **Decay bacteria** break down the proteins of dead animals and plants, releasing ammonia. In the soil, this ammonia is converted by **nitrite bacteria** to nitrites, and the latter are converted by **nitrate bacteria** to nitrates. This two-step process of **nitrification** operates similarly on various nitrogenous products of excretion, such as urea and uric acid. Its effects are reversed to some extent by the action of **denitrifying bacteria,** which convert nitrites and nitrates to atmospheric nitrogen. Thus, soil microorganisms play key roles in the intricate nitrogen cycle (Fig. 24–24).

Figure 24–23 Clover roots with nodules containing nitrogen-fixing bacteria.

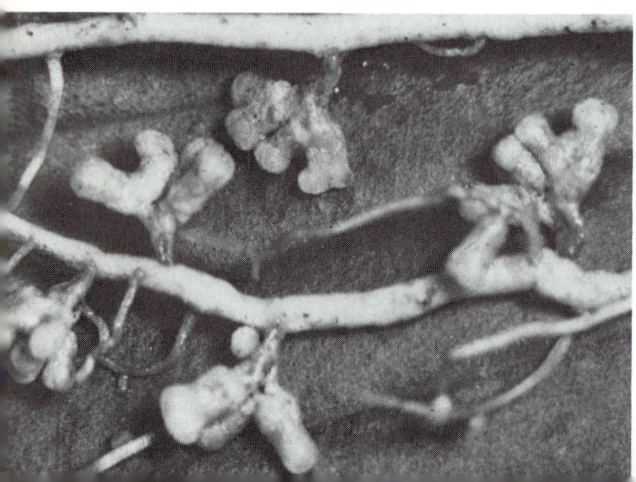

Carolina Biological Supply Co.

About 100 million tons of bound nitrogen is removed from the soil each year with the harvest of the world's crops. Even with the nitrogen-fixing activities of the soil microorganisms and the replacement of 12 million tons in mineral fertilizers, a constant deficit remains. Yet the atmosphere over a single acre of land contains about 35,000 tons of nitrogen. It is not surprising, therefore, that active research efforts have been centered upon developing means of economical nitrogen fixation that do not require extremes of temperature and pressure. The most promising advances thus far have been along the line of complex organometallic catalysts.

THE PHOSPHORUS CYCLE

Carbon, oxygen, hydrogen, and nitrogen are so abundant on our planet that none of these elements can normally be considered to be a limiting factor on the growth of life. But another key component of living matter is found in water and in the solid materials of the earth's crust in far smaller amounts; moreover, it is frequently present in relatively insoluble, poorly accessible forms. This is phosphorus. As a constituent of the nucleic acids and of life's energy currency, ATP, phosphorus is absolutely vital to life. But phos-

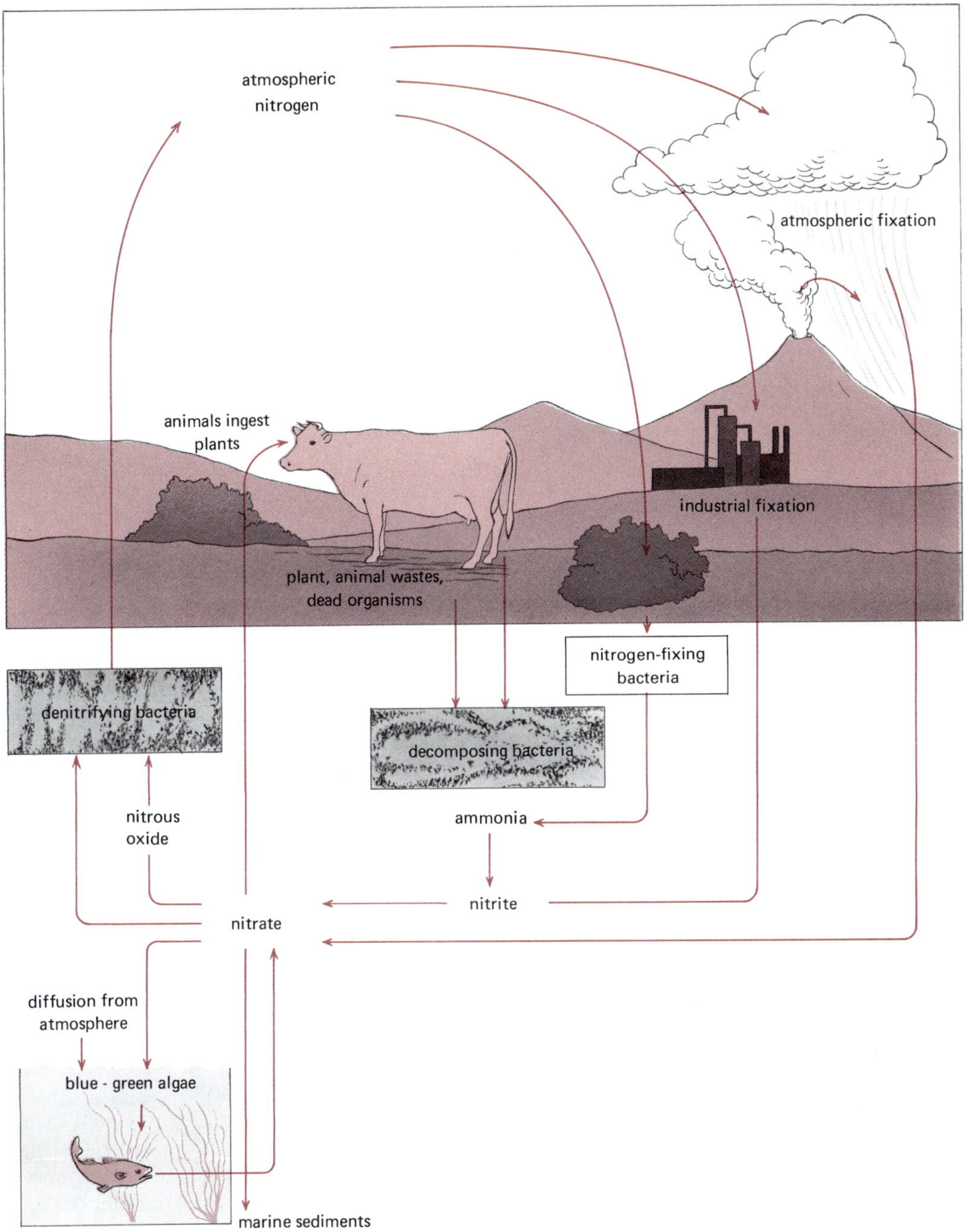

Figure 24–24 The nitrogen cycle.

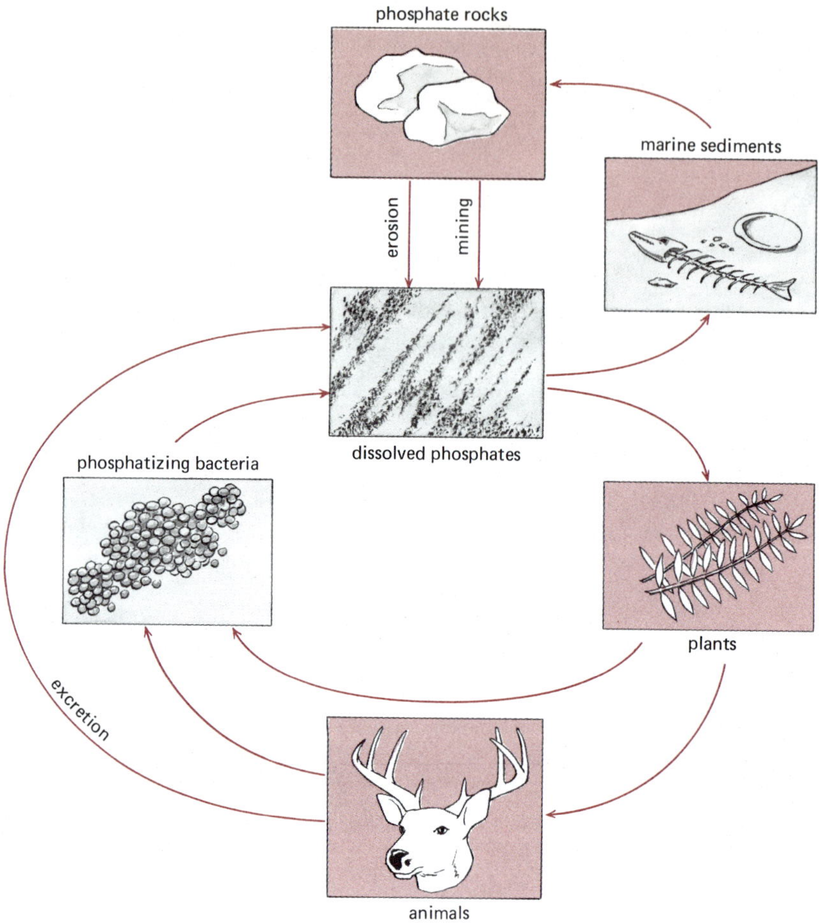

Figure 24–25 The phosphorus cycle.

phates are among the more insoluble of the inorganic salts, and frequently it is the available phosphorus that places the effective limit on the multiplication of life.

In the euphotic zone (sunlit surface waters) of the present-day oceans, for example, there is virtually no inorganic phosphorus; all the phosphorus found in these regions is organic phosphorus, contained in living cells and in their wastes and dead residues. Life has literally multiplied up to the limit of the available phosphorus. In the levels below the euphotic zone, life depends upon an organic rain of dead and living materials from the euphotic zone. Here there is an excess of phosphorus, for in this dark region, barred to photosynthesis, energy has become the bottleneck. Some of the organic phosphorus that drifts down from the euphotic zone is converted to inorganic phosphates, which settle on the bottom and are effectively lost to the general circulation.

On land, local conditions (including the previous history of crop plantings) may result in any of a variety of inorganic ions (trace elements) becoming bottlenecks blocking the further growth of plant and animal life. But again phosphorus is a frequent limiting factor. There is a continuing cycle of phosphorus utilization on land: rains wash down into the soil and dissolve minute quantities of soil phosphates, which are utilized by plants and ultimately by animals. Excretion and decay return this phosphorus to the soil, and much of it is reused by plants and animals. But some of the dissolved minerals in soil are carried off by the seeping soil water to rivers and streams and eventually washed away to sea. There they settle down to form insoluble phosphate deposits on the ocean floor and are lost. It is estimated that the rivers of the world annually carry off about 3.5 million tons of phosphorus in this way, and only a minute fraction is returned in such forms as bird droppings (guano). Man's irrigation systems and his networks of sewage disposal, carrying phosphorus-rich wastes off to rivers and oceans, are accelerating the depletion of this valuable natural resource and increasingly limiting the potential amount of protoplasm that the lands can support with each passing year.

THE ENERGY CYCLE

Essentially every chemical and physical reaction in the universe involves the utilization of energy. The movements of the stars, planets, and satellites, the volcanic eruptions beneath the sea, the violent tornadoes of the Midwest, and even our own speech and thoughts involve the transformation of energy from one form to another. Although energy can be readily converted from one form to another, it can neither be created nor destroyed. (This is a statement of one of the major scientific generalizations, known as the **first law of thermodynamics.**) When energy is converted from one form to another, however, a certain amount of energy is converted irreversibly to a form unavailable for useful work **(second law of thermodynamics).** This unusable energy is known as **entropy.**

The ultimate source of essentially all of the energy utilized on earth is the radiant energy of the sun. The continued operation of the various cycles of nutrient elements is dependent on the energy derived from solar radiation. These cycles may be classified as closed cycles, since the individual elements are utilized over and over again. But the energy that drives these and other reactions is eventually dissipated in the form of heat. Thus energy must be continually resupplied.

The earth receives less than one-billionth of the energy radiated by the sun, and yet this is more than sufficient to provide for all the needs of all the earth's organisms. This is true despite the fact that only a tiny percentage of sunlight is actually absorbed by green plants and utilized in photosynthesis. The organic compounds resulting from the photosynthetic and subsequent reactions represent a form of stored or potential energy. When plants are eaten by animals, much of the stored energy is dissipated in the form of heat. The remainder is either assimilated into various parts of the organisms or is used in the production of highly labile organic compounds (e.g., ATP), which are utilized as readily available sources of energy. As animals are eaten by other animals, again a considerable fraction of their stored energy is lost as heat. Ultimately the organic compounds are reduced by decay bacteria to carbon dioxide, water, and ammonia, with the liberation of heat. This represents a dead end in the energy cycle. A renewed supply of energy must come from the sun. It is comforting to realize that the sun is expected to continue shining for at least several billion years to come.

The Biotic Community

The physical medium in which an organism lives and the interplay of matter and energy do not make up the whole of its environment. No organism lives in complete isolation. Its activities and mode of living are affected by and in turn affect the other living organisms in its immediate vicinity. The organisms that live together in interaction in a particular place or **habitat** collectively comprise a **biotic community.** A meadow interspersed by occasional trees is an example of a biotic community. The deep and extensive root system of the trees aids in water retention, which indirectly aids in the growth of the grass.

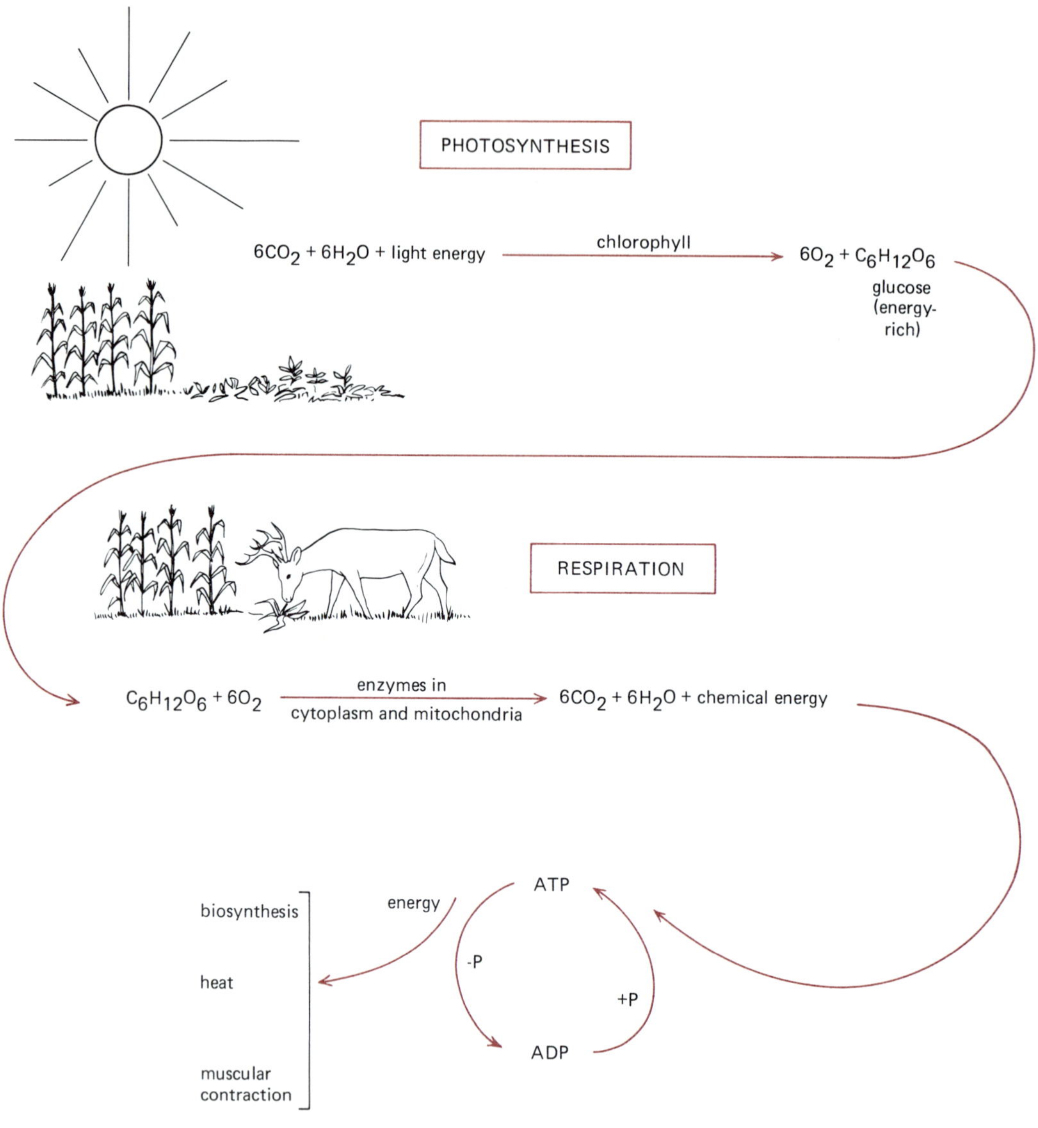

Figure 24–26 The energy cycle.

Field mice feeding on the grass grains may fall prey to a swooping hawk, a darting weasel, or a slithering snake. Cows munching on the grass leave droppings that fertilize the soil and whisk their tails from time to time to swat the bothersome insects. The many insects that inhabit the meadow in turn provide food for the birds nesting in the trees.

The many species comprising a biotic community share the same habitat and affect each others' lives in varied ways. These biological interactions themselves may also be influenced by the physical factors of the environment, such as temperature, light, and the supplies of nutrients. For example, a prolonged drought in the meadow would result in death

Figure 24–27 The living members of a biotic community are interrelated in numerous ways.

of much of the grass. This would not be an isolated consequence, but the start of a long chain reaction. With a reduced supply of grass seeds to eat, the mouse population would drop. The predators that normally feed on mice would be hard pressed to find enough to eat. Some of them, too, would die, and others, driven by hunger, might begin to range farther, perhaps to attack the poultry and sheep of a nearby farmer. The insects would also suffer, and because of the decrease in their numbers the bird population also would drop.

As we shall see in the next chapter, man and his activities represent an important new factor increasingly affecting the ecology of varied biotic communities. Never one to be confined to a narrow habitat, man is now ranging farther and farther, extending his influences into areas previously untouched and—wittingly and unwittingly—affecting the lives of other organisms in numerous ways. A hydroelectric dam built upstream, for example, might turn a peaceful meadow into a wasteland. So might the fumes and noxious chemicals from a nearby factory. Or a highway might cut right through the heart of the meadow, displacing countless organisms from their homes and upsetting the natural economy of the community. But before exploring the pervasive effects of man on the natural world, let us find out more about the normal interactions of organisms with one another, one of the most important facets of ecology.

ECOLOGICAL NICHES

The continent of Australia represents a unique geographical entity, since it became isolated from the rest of the land areas during a time when placental mammals had not yet appeared on the earth. As a result, the marsupials (pouched mammals), which were the dominant form of the time, evolved into numerous and varied forms occupying varied habitats and leading various modes of life: giant herbivorous kangaroos *(Macropus)*, leaping across the great plains and open forests; the pouched "mole" *(Notoryctes)*, burrowing blindly through the soil; the tiny yellow-footed pouched "mouse" *(Phasogale)*, scurrying timidly about the floor of the forests; and larger carnivorous forms such as the "tiger cat" *(Dasyurus)* and the "Tasmanian wolf" *(Thylacinus)*. The common names of these marsupials emphasize the similarities of their general appearance and ways of life to placental mammals of the other continents.

The mode of life of a particular organism and its relationship to the other members of its biotic community or ecological system **(ecosystem)** is referred to as an **ecological niche.** Niches may be extremely broad or narrowly specific. The rat, for example, can live in houses, sewers, and ships, or in barns or out in the fields; it can eat grain, stored processed foods, and it can kill other animals for food (even other rats, if there is no alternative food supply available). The aphid, on the other hand, provides a sharp contrast to the omnivorous rat. Its sole source of nourishment is the juice of plants, which it sucks while perched on the plant stem, leaves, or roots. Most aphid species are so narrowly specialized that they can live on only a few plant species—sometimes only on a single species. If all the plants of this kind should perish, their aphids would die with them.

The ecological niches occupied by marsupials in Australia are not unique to that continent; they have their counterparts in other biotic communities throughout the world. The kangaroos, for example, are herbivores that graze on the grass of the plains, much as the antelopes do in Africa and buffaloes did in the United States. The photosynthesizing grass may be considered as a **food producer** in the ecosystem, while organisms such as the kangaroo, which feed upon it, are **primary consumers.** Other herbivore primary consumers among the marsupials of the Australian ecosystem are the koala or native "bear," which subsists solely on the leaves of the eucalyptus tree, and the phalangers, various opossumlike animals, ranging from tiny "flying squirrels" to the large wombats. One of the **secondary consumers** (eaters of herbivores) in the Australian marsupial ecosystem is the rabbitlike bandicoot, a burrowing animal that feeds mainly on insects, larvae, and earthworms. The pouched "mice" and "rats" are also secondary consumers. The Australian marsupial ecosystem also has its share of **tertiary consumers** (eaters of carnivores). The smaller carnivores and insectivores are preyed upon by such larger forms as the "tiger cat," the "Tasmanian devil" (a badger-sized carnivore with a reputation for an unusually mean disposition), and the "Tasmanian wolf." This last unfortunately developed an appetite not

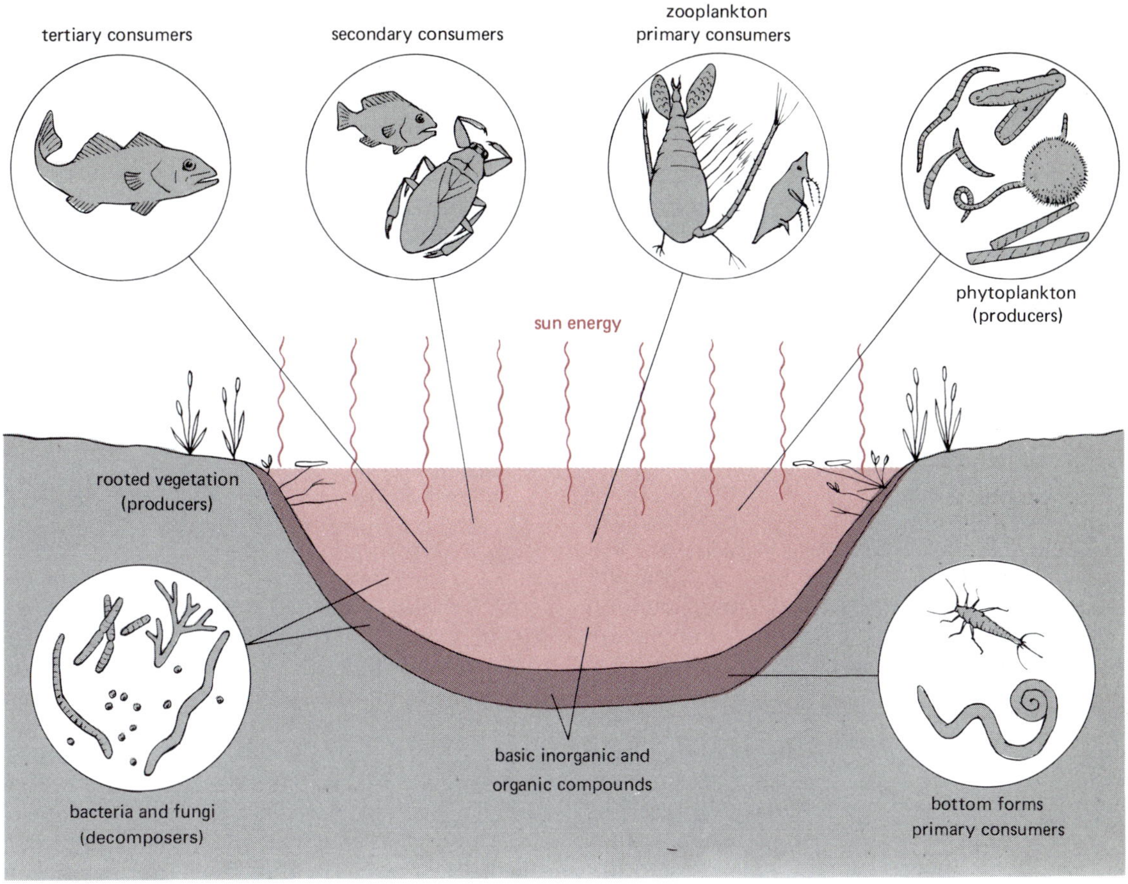

Figure 24–28 A pond community includes a variety of interrelated food producers and consumers.

only for its natural prey, but also for the sheep raised by the Australians, and as a result has now been practically exterminated by bounty hunters.

Like other ecosystems throughout the world, Australia also has numerous species filling other ecological niches: **scavengers,** such as the catfish and various beetles, which live on dead or decaying organic materials;* **parasites,** which gain their nourishment from other living organisms, such as the termites that infest living trees, various plant fungi, and various microbes that cause diseases in man and animals; and **decomposers,** such as numerous soil-dwelling microbes. Like the scavengers, decomposers live on dead and decaying materials.

Thus, we can see that to have a meaningful picture of a species, we must know not only its habitat, but also some information on its ecological niche—its habits and how it fits into the particular ecosystem in which it lives (Fig. 24–28). It has been suggested that the habitat of an organism might be thought of as its "address," while its ecological niche might be regarded as its "profession."

*After imported primary consumers—sheep and cattle—became well established in Australia, it was discovered that the Australian ecosystem was lacking an important group of scavengers, the dung beetles. Blood-sucking flies bred in the mountains of cattle dung that built up. Finally several species of African dung beetles were imported to cope with the problem.

Comparable ecological niches may be occupied by phylogenetically quite varied organisms. For example, the single-celled algae in the seas, the grass in a meadow, and the great pine forests of the taiga are all food producers, and thus are ecological equivalents. Minute crustaceans living on the algae, cows grazing in the meadow, and wood mice scurrying about on the floor of the coniferous forest in search of pine cones with tasty seeds are all primary consumers. The fleas living in the mouse's fur, the enteroparasites living in its intestines, and the fungus growths clinging to the trunks of the trees beneath which it scampers are all parasites.

Ecologists believe that, in general, no two species can occupy exactly the same niche in the same habitat. This principle is most easily visualized in reference to the food supply. If two species sharing the same habitat feed on the same type of food, the situation can be stable as long as their numbers are small and the food supply is ample to support both. But if there were a sudden drop in the food supply, or if the two species multiplied to such an extent that the food supply was no longer sufficient to support both populations, competitive pressures would begin to be felt. As a result, either one or both species would have to modify their habits to take advantage of an alternative food supply (thus assuming a slightly different niche). Or the more successful species would eventually take over the niche, and the other would either move out to another habitat or die out.

Since the coming of settlers from Europe, the continent of Australia has provided a unique "laboratory" for the observation of various ecological pressures. Sheep were able to thrive there because they took over a niche that was not exactly analogous to any already filled in the Australian ecosystem. But when rabbits were introduced, they were so splendidly successful, owing partly to the paucity of natural predators to keep them in check, that they began to threaten the existence both of the sheep and of the native fauna by stripping the grass and foliage on which the latter normally feed.

RELATIONSHIPS AMONG ORGANISMS

A little thought will reveal that there is only a limited number of basic ways in which two organisms, sharing the same habitat, can interact:

1. A and B live together with no apparent effect on one another (neutralism).
2. A and B compete for some basic commodity of life (competition).
3. A and B live together with benefit to one and no apparent effect on the other (commensalism).
4. A and B live together with benefit to one and a negative effect on the other (parasitism).
5. A and B live together or cooperate with mutual benefit (mutualism).
6. A eats B (predation).

All these basic patterns of interaction occur in nature, but the millions of different animal and plant species have produced a virtually infinite number of variations on the basic themes. The first case can be termed **neutralism.** At first glance, a picture of zebras, antelopes, and giraffes peacefully grazing on the grass of an African veldt might seem to be a perfect example of neutralism. But since all three species feed on the same grass, this apparent noninteraction might readily be transformed into a subtle **competition** if a drought caused much of the grass to dry up or if the herds multiplied to such an extent as to outstrip the available food supplies. In such cases other factors, such as the giraffes' ability to utilize supplementary food resources (leaves from overhanging trees) and the relative abilities of the species to propagate themselves and elude capture by predators, would ultimately determine the survival of the competing animals. Cases of subtle competition such as this, which do not involve any definite confrontation between the competing species, are far more frequent in nature than bloody battles in which "winner takes all."

Also widespread in nature are relationships that bring benefit to one or both of the species involved. Such associations, termed

Figure 24–29 Grazing animals compete for food.

symbiosis, range from **commensalism,** in which one species benefits from the association while the other is essentially unaffected, to **mutualism,** in which both partners of the association benefit and often literally cannot live without each other, and **parasitism,** in which one member of the association benefits but the other is harmed. Often it is difficult to draw the line between two types of interaction. One might be tempted to class the bacteria that live within our intestines as commensals: they feed on the undigested food materials in the intestinal contents and generally do not have any significant effect on us. But some of the intestinal flora "earn their keep" by producing valuable B vitamins—a mutualistic relationship; and others may become pathogenic, causing disease—a parasitic relationship.

An organism may participate in a number of different symbiotic relationships of various types. An African bird called the honey guide, for example, has been observed to seek out an animal called the ratel (a member of the weasel family) and lead it to a bees' nest (Fig. 24–30). The ratel, with its long, strong claws, tears open the nest and devours the honey. The honey guide then takes its reward by feeding on the wax of the comb. It is enabled to digest the wax by the action of bacteria living in its intestines. Both these interspecies relationships are clearly mutualistic. But the honey guide also acts as a parasite on another species. Like its counterparts on other continents, the cuckoo and cowbird, the honey guide is a lazy parent. It lays its eggs in the nests of other species and leaves them there for the unsuspecting foster parent to raise. The infant honey guide is equipped with two sharp hooks on its bill, with which it ruthlessly bites the other nestlings to death, thus eliminating possible competition for the mother bird's attentions.

Interestingly, although parasitism is clearly a one-sided relationship, bringing definite harm to one of the "partners," it is the *un*successful parasite that actually kills its host. For then it, too, will die, unless it manages to transfer to another host in time. If a parasitic species were so virulent as to kill each successive host organism invaded, the parasites of that species would soon run out of potential hosts to inhabit, and soon would also become extinct. For a parasite frequently is so specialized that it can live only under the ex-

Figure 24–30 The ratel and the honey guide cooperate in a mutualisitc relationship.

Lynwood M. Chace, Omikron

Figure 24–31 The predator plays an important role in a biotic community.

tremely specific conditions furnished by its accustomed host.

Every biotic community has its share of parasites. Parasitism is so widespread that parasites are represented in every major phylum and most of the minor ones, although parasitic vertebrates are extremely rare. Few organisms are entirely without parasites; indeed, even the parasites often have parasites living on or in them. Frogs, for example, frequently harbor a parasitic protozoan, *Opalina,* in their intestines. This ciliated protozoan in turn is often parasitized by an amoeba—which *itself* can serve as the host for a parasitic bacterium. Presumably there might be a bacteriophage that can parasitize the bacterium.

The last type of interaction among species, **predation,** seems at first glance to offer a clear-cut case of good and evil in the world of nature. To people accustomed to thinking in anthropomorphic terms, the ruthless predators slaughtering their helpless prey seem to be the "bad guys" of nature. The farmer, catching sight of a circling hawk or slinking fox, will quickly reach for his shotgun and, after bringing it down, will have a feeling of satisfaction at having rid the world of another "varmint." Even naturalists occasionally fall into the trap of seeing only one side of the predator's role in the biotic community. Compassion for the suffering of individual animals killed by predators has been the impetus for often misguided attempts to protect the helpless species by eliminating their natural predators. Unfortunately, such attempts have frequently revealed that the predator is not quite the villain he is commonly thought to be.

The role of the predator is an important one in the delicately balanced economy of nature. Predators are much more likely to catch and kill genetically inferior and diseased animals, and thus help to weed them out before they can pass on their inferior genes to offspring or transmit their diseases to others of their species. The processes of evolution, moreover, have generally resulted in an adjustment of breeding levels to allow for a potential excess. Without the pressure of predators, the populations of prey animals can explode to such levels that competition for food and living space become limiting factors. As a result, countless numbers of animals die anyway—but from starvation instead of predation.

Like the successful parasite, the successful predator does not kill all individuals of his potential prey. For if all the prey animals died out, there would soon be no food left for the predator, and his species, too, would become extinct. Of course, the predator (man excluded) does not consciously plan to leave a sufficient number of healthy prey animals to serve as a breeding pool for future generations. The prey animals of the world have evolved a number of adaptations that help them to elude predators and to compensate for the inroads of predation.

One of the most interesting and effective types of adaptation that help prey animals elude predators is **camouflage,** the possession of markings or coloration that help an animal to blend into its surroundings. Adaptations of this type are of great help both to predators and to their potential prey. The stripes of the tiger and the zebra make them almost invisible amid the tall grasses of their habitat. The eye of a casual observer might easily miss a green leafhopper or praying mantis among the bright green leaf blades. The familiar white underside and dark upperside of the flounder is a common pattern among both fish and birds: to another animal passing beneath, the light belly appears to blend into the light background of the sky; to an animal passing above, the dark upperside blends into the dark background of the ground or ocean bottom. Flounders, indeed, like some other species, can even change their colors and markings to some extent to better "match" their surroundings (Fig. 24–32). Field observations of birds catching moths and of other predator-prey interactions have shown that camouflage does have definite survival value —fewer of the individuals that best blend into their surroundings are caught by predators.

Some forms of protective color adaptation seem to contradict the rules of camouflage. For example, it has been observed that birds and other predators rarely attempt to catch certain brightly colored butterflies. It has been found that these butterflies usually are particularly unpleasant-tasting—or *look* very much like other species that are poisonous or bad-tasting. The latter case is an example of **protective mimicry**—the evolution by a potentially defenseless species of an outward similarity to a species that does have potent means of protection. In both cases just mentioned color is used as a **warning signal.** A bird that eats a brightly colored, foul-tasting butterfly will not be eager to sample more of its kind in the future. (Here, as so often in the living world, we see an example of adaptations slanted toward the good of the species, rather than the individual. A foul-tasting, brightly colored bug greedily crunched by an insectivore does not profit as an individual from the fact that the chastened insectivore will never touch another such bug again in its life.)

Color adaptations and other forms of camouflage are often enhanced by behavioral adaptations. Many animals, when startled, "freeze" into immobility, which helps them to better blend into their surroundings. This technique works equally well for many owls, whose mottled feathers blend imperceptibly into the tree trunk on which they perch, and for fawns crouching in a thicket, whose spotted fur resembles a pattern of patches of shadow and sunlight filtering through the overhanging leaves. The walking stick *(Diapheroma femorata)* is an insect shaped exactly like a twig, complete with barklike markings and imitation internodes (Fig. 24–33). If it continually hopped about in rapid motion, the illusion would be destroyed. But the walking stick instead spends much of its time in perfect stillness, in a pose that makes it seem like a natural outgrowth of the branch on which it perches.

Omikron

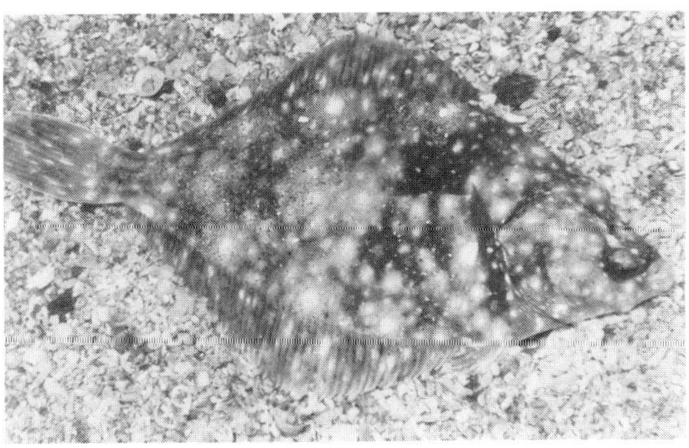

Figure 24–32 The flounder's skin takes on a random pebbled pattern of light and dark spots that blends effectively with the background of the ocean bottom.

Figure 24–33 The walking stick insect resembles a twig.

Other behavioral adaptations in the animal world are centered about the protection of the young. Examples of mothers fiercely defending their young or attempting to decoy predators away from them are well known. Animals that live in groups and herds frequently exhibit a high degree of cooperation that contributes to the welfare of the group.

FOOD CHAINS AND ECOLOGICAL PYRAMIDS

The problems of radioactive fallout have made researchers and the general public more aware of the concept of food chains. Following an atmospheric atomic explosion, the concentrations of various radioactive materials in the atmosphere rise precipitously. Much of the radioactive debris settles to the land below. Radionuclides are taken in (and sometimes selectively concentrated) by the plant life growing on the contaminated soil or in contaminated waters. Such plants may then be directly eaten by man, or the radioactive materials may continue along a more circuitous route, as the plants are eaten by animals and the animals or their milk are consumed by man.

We have already seen that the various organisms of a biotic community are interrelated, especially through the aspect of food production and consumption. Within these relationships, sequences can be discerned in which an organism at one level of the sequence feeds upon those immediately below it and is in turn eaten by those immediately above it. Such a sequence is called a **food chain.** In a given food chain, the energy comes ultimately from the sun; solar energy is captured by plants and then transferred successively to a series of consumers.

In any biotic community there are many food chains, some parallel, some interlinked at various points. In fact, the food chains of a community are usually so interrelated that they may be thought of as a **food web.** In a meadow, for example, the vegetation is eaten by varied primary consumers, such as grasshoppers and other insects, rabbits, and mice. These in turn are preyed upon by secondary consumers. The grasshopper might be gobbled up by a lizard. The mouse might fall victim to a snake. The rabbit might be pounced upon by a hawk. As the food chains interlink, the hawk might also make meals of grasshoppers or mice, or, acting as a tertiary consumer, might find a lizard or a snake a tasty meal. And if not eaten by predators, all of these organisms, from the green plants to the hawk, might die and their stored energy ultimately be transferred to the decomposers that inhabit the soil. Figure 24–34 summarizes these and other relationships in the "web of life" of a typical meadow community.

At each link in the food chain there is a considerable loss of energy. The photosynthesizing plant transforms only a small fraction of the incident solar energy into the stored energy of food materials—frequently less than 1 percent and rarely more than a few percent. As the plants are eaten by animals and these in turn are eaten by other animals, there are further energy losses, mainly in the form of heat.

This progressive loss of energy is reflected in a similar progression of mass at the various levels of producer and consumers in the food chain. In any community, in a relatively stable situation, the total mass (and the total stored energy) of the plant producers is far greater than that of the primary consumers that feed on them; the primary consumers, in turn, support a much smaller mass of secondary consumers (with a much smaller total fund of stored energy), and so on. These progressions are frequently expressed in the form of **ecological pyramids,** such as the one in Figure 24–35. The concept of the ecological pyramid is easy to visualize: because of the constant expenditures of energy of metabolic processes, in addition to the losses due to inefficiency of conversion, a grazing cow, for example, must eat many pounds of grass for every pound of nutrient value that she stores away in her flesh or gives out in her milk. An extreme example is the shrew: feeding mainly on insects, snails, worms, and small mammals, the shrew has such a rapid metabolic rate that it must keep actively feeding almost constantly, and has been estimated to consume its own weight in food *every three hours!*

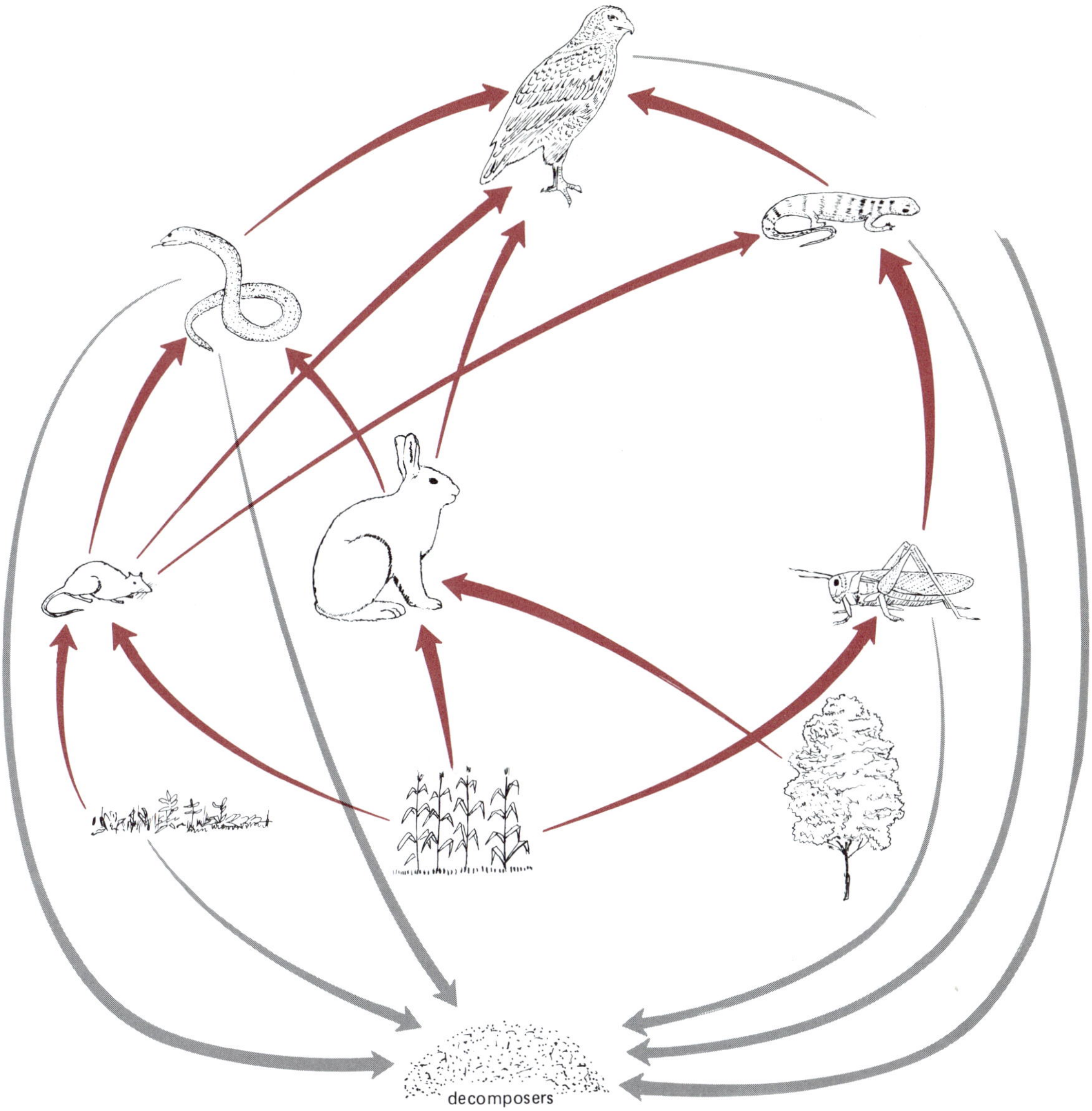

Figure 24–34 Part of the food web in a typical meadow community.

Because a predator is usually larger than its prey, there is often, but not always, a correlation between the biomass at a particular level of the pyramid and the number of organisms: the frenetic shrew, for example, must eat an enormous number of insects and snails to consume the great mass required to keep its metabolic mill turning. Or, taking another example, the ecologist Eugene Odum has estimated that 17,850 pounds of alfalfa (20,000,000 alfalfa plants) is required to support the 2250 pounds of veal (4½ calves) required as a year's food

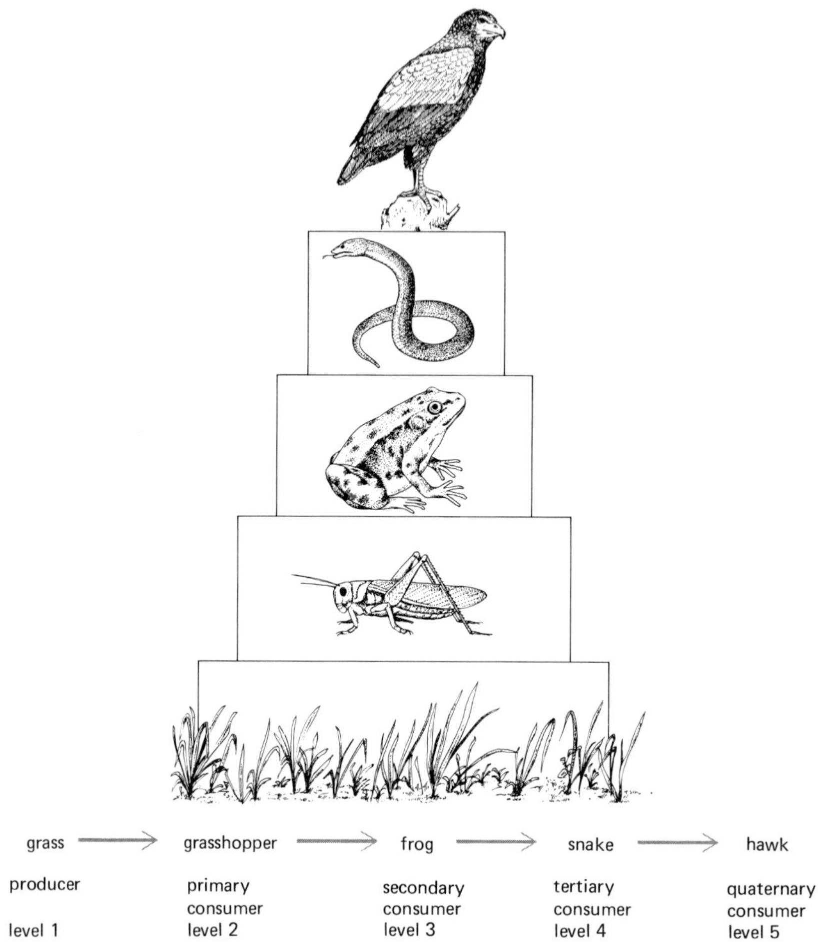

Figure 24–35 The pyramids of life. The members of a biotic community show a basic pyramid relationship, whether compared according to numbers, biomass, or energy content.

supply to support a single 105-pound twelve-year-old boy.

But a little thought will reveal that numbers of organisms at a particular level do not necessarily correspond to the biomass: thousands of tiny insects might be supported by the leaves of a single tree. Or a single lion might in turn provide an ample living for thousands of fleas, ticks, and parasitic worms and millions of intestinal protozoa and bacteria.

Linked together by food webs and by competitive and mutually beneficial relationships, the organisms of a biotic community together contribute to a delicate balance of nature. This precarious balance can be upset by any major change in the environment. And man, since his emergence as a dominant species, has exerted an increasingly pervasive perturbing influence upon his world, a subject we shall take up in the next chapter.

QUESTIONS FOR REVIEW AND THOUGHT

1. What are some of the advantages and disadvantages of life on land as compared to life in the water?
2. What are the normal temperature limits within which organisms can flourish? What are the temperature extremes from which organisms may be revived and resume relatively normal functioning?
3. Name and describe the major terrestrial and marine biomes. What factors delimit their extent?
4. Will man ultimately exhaust his natural resources? Discuss.
5. Describe a particular biotic community and indicate the interrelationships among the various members of the community.
6. Distinguish between the habitat and the ecological niche of an organism. Does the habitat of an organism in any way influence the ecological niche it occupies?
7. Design a balanced aquarium, including appropriate producers, consumers, and scavengers.
8. In what way does the competition in our "free-enterprise system" resemble the competition found among organisms in nature? Why is competition among organisms a necessary part of a stable biotic community?
9. Man lives symbiotically with many different organisms. Explain.
10. Using the concept of food chains, show how radioactive waste products deposited deep in the oceans might one day prove harmful to man.

Omikron

Esaias Baitel, Omikron

CHAPTER 25

Man Changes His World

Introduction

Public awareness of ecology and the interrelatedness of life on our planet is a very recent phenomenon. The history of man's spread through the inhabitable lands of our planet has been characterized in the main by a certain element of ruthlessness. If settlers needed more farmland, they cut down forests, with little awareness of the role of trees in maintaining and enriching the soil and little regard for the animal population that they were displacing. If there was a demand for furs, then seals, leopards, and other fur-bearing animals were killed indiscriminately and often wastefully, with little thought to the question of where the pelts of future years would come from if the breeding population was reduced below maintenance levels. Even major geological reorganizations, such as the diversion of rivers, were undertaken with insufficient knowledge of the full scope of their consequences. Of course there were protests along the way, and occasionally determined efforts by groups of thoughtful citizens were successful in saving a threatened wilderness area or an endangered species. But these protestors represented a small minority, and were often regarded rather condescendingly as "sentimentalists" by the general public. It was the prevailing feeling that the welfare of other species was a consideration secondary to the benefit of mankind.

Now, however, the picture is changing radically. The development of chemical, power, and other technologies has provided man with the means of changing his world to an unprecedented extent. His burgeoning population has spread, encroaching steadily upon the remaining uninhabited areas of the world and threatening ultimately to carpet the lands of our planet with a series of nearly continuous "megalopolises." The effects of man's activities, from the use of pesticides to the effluvia of automobiles and factories, are becoming more and more manifest with each passing year. The balance of nature is being upset in thousands of ways, and man is discovering that effects on the other species of our planet and on our physical environment must ultimately affect man as well.

Figure 25–1 Man changes his world . . .

More and more the feeling is growing that man is not the "lord of creation," but instead a custodian of the welfare of a worldwide biosphere of interrelated life. Larger and larger segments of the general public are becoming aware of the growing ecological problems of our planet and are trying actively to do something about them—through pressure on legislators and through other channels, such as legal suits based on the premise that a healthy environment is a basic public right.

The resultant impetus to ecological studies is leading to a burgeoning of knowledge of the interactions of earth's life forms and their environment. As this knowledge is gained, man is gradually acquiring the means to live in harmony with his environment; it may be hoped that he will also gain the wisdom to put this knowledge to practical use.

Demography: The Study of Populations

Man appears to be on a roller coaster to extinction. He is breeding so prolifically as to endanger not only his food supply, but also his reserve of living space as well. With civilization and the concomitant benefits of technological advance, his birth rate, though declining, has nonetheless far outstripped the rapidly dropping death rate. Figure 25–2 graphically illustrates the ominous future ahead. It is estimated that the world human population was about 250 million in the year A.D. 1. More than sixteen hundred years were required for this population to double to about 500 million in the early seventeeth century. The next doubling was much more rapid: to 1 billion

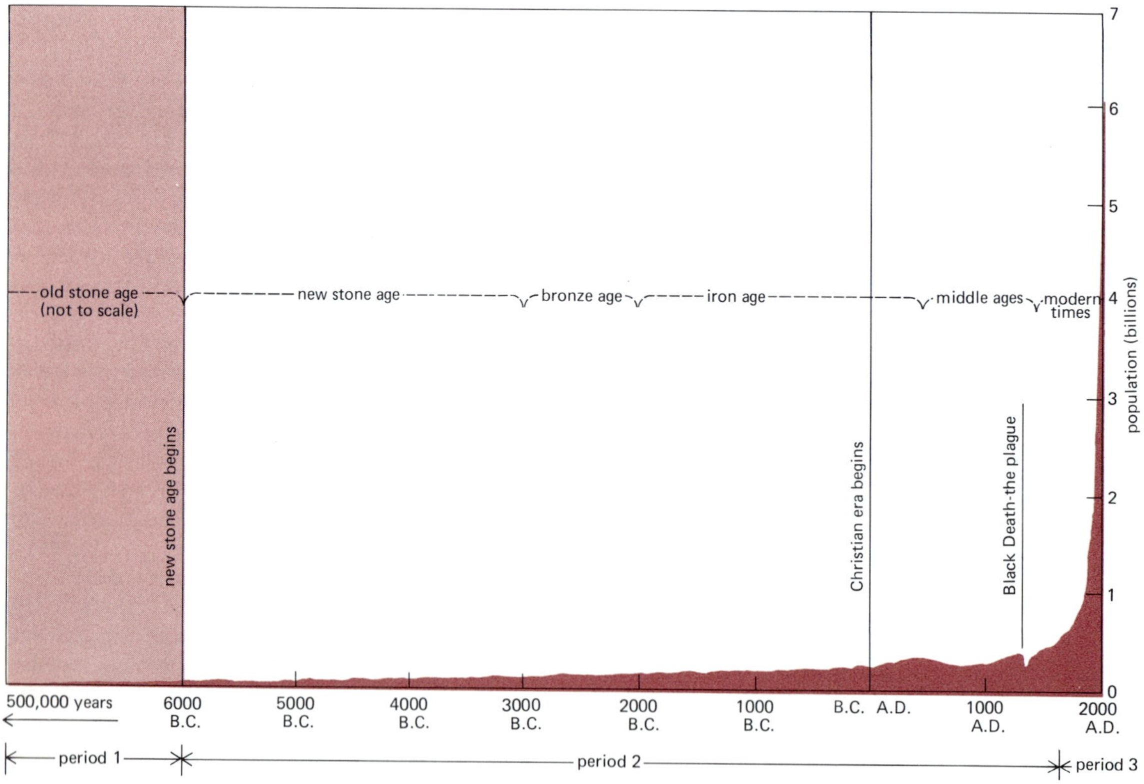

Figure 25–2 Growth of the earth's human population. How and when the current "log phase" will level off is a matter of speculation.

by 1820. The acceleration continued: the next doubling required only a century, and now, a mere half-century later, the world population has essentially doubled again.

At the present rate of increase, the world's population is doubling approximately every forty years. If we extrapolate using this figure (and with the constantly falling world death rate, even this figure may be too conservative), within less than eight hundred years from now there will be more than one person for every square foot of land area over the entire surface of the earth!

Obviously, long before this absurd climax could be reached, living conditions on our planet would become intolerable. For numerous reasons—political, economic, biological, and others—the exponential curve just depicted cannot continue indefinitely (or even for a much longer time). Instead, we can expect the curve to turn downward within the next century, due either to a decrease in the birth rate or to an increase in the death rate, or to some combination of the two.

The principle of the ultimate limitation of the growth of a population was recognized by an English clergyman and economist, Thomas Robert Malthus, at the end of the eighteenth century. In 1798 Malthus published his *Essay on Population*, in which he pointed out that populations, including human populations, have a natural tendency to increase until checked by environmental factors, such as the food supply. He maintained that ultimately the human population would have to be kept in check by famine, disease, and war. His book, at first published anonymously, met with a storm of outraged indignation, and in 1803 Malthus published a revised and enlarged version in which he admitted that "moral restraint" (delayed marriage and sexual abstinence) would act as a further check on human populations.

With the enormous increase in productivity in the century and a half since Malthus's time, the spectacular increase in the human population has been accompanied in many areas, such as the United States, by an overall *rise*—not a decline—in the standard of living. This happy situation has prompted many theorists to hold that Malthusianism is out-of-date, or that it simply does not hold for the "special case" of man.

This optimistic view, however, bolstered by the effects over the last century or two of the opening of new lands and tremendous technological strides, is valid only for the present situation. (And indeed, even now there are many areas of the world that are not so fortunate. In these areas the rapid decline in death rate, brought about by the introduction of new medical techniques, is not being matched by a large enough concomitant increase in environmental potential. Coupled with an essentially undiminished birth rate, it is leading merely to a perpetuation of squalor for larger and larger numbers.) Malthus's basic principle of an ultimate environmental limit on the natural tendency of populations to increase has been observed again and again in nature and must ultimately be faced by man, too. At what level the ultimate restraints will be imposed, and whether they will be imposed from within while there are still sufficient environmental resources for comfortable living conditions for all, or from without, by the impersonal mechanisms of famine, disease, and war—these decisions are in the hands of man himself. It is therefore particularly important that we should be well informed on the principles of populations, how they grow and decline and the factors that act upon them.

In general, a **population** may be defined as a group of organisms of the same species inhabiting a given area. The study of populations is called **demography.** Much has been learned by demographers, not only about human populations, but also about populations of various other organisms.

Demographers study various aspects of populations: their numbers, their distribution and their composition with respect to age and (where pertinent) sex and behavior (social habits, sexual relationships, population movements, etc.). The size of a population is determined by the interaction of the **biotic potential** (the sum of the mechanisms tending to increase the population) and the **environmental resistance** (the sum of all the factors tending to decrease population size). The

Figure 25–3 The biotic potential of a pair of mice . . .

biotic potential of a species, the inherent capacity of a population to increase under optimum environmental conditions, with suitable age and sex ratios, is enormous. In this book we have already cited figures on the reproductive potentials of fruit flies, mice, and humans. It is safe to say that any species on earth, if permitted to multiply unchecked and afforded an unlimited food supply, could literally cover the earth—every square inch of it—in an astonishingly short time. Since we are *not* knee-deep in flies, cabbages, snails, rabbits, elephants, or men, there are obviously factors that counteract the natural tendency of populations to increase. These include the capacity of the environment to supply food and living space, competition among various species for these necessities of life, and the actions of predators and parasites. Various other unfavorable factors for population increase, such as an age or sex disproportion or a low population density, reducing opportunities for mating, also tend to limit population growth. So do factors contributing to mortality, such as disease, accidents, and aging.

The net change in a population—growth or decline—is the resultant of the **birth rate** and the **death rate.** The observed birth rate of a population (the average number of new individuals produced per unit time) is generally considerably smaller than the maximum potential birth rate (the number of new individuals that would be produced under optimum conditions—i.e., if there were no limiting factors). The maximum birth rate may be more or less closely approached in laboratory experiments. The death rate observed in a natural population, analogously, is generally greater than the minimum death rate—the rate that would be observed solely on the basis of death due to old age. In a population living under actual, nonoptimum environmental conditions, deaths occur continually from varied causes other than sheer age, and these may affect organisms of all ages. In some species, such as the oyster, there is an extremely high death rate at very early ages; the few oysters that survive the larval stage usually live on to a ripe old age. In other species, such as many birds, the death rate is relatively constant at all ages. In man, after a relatively high initial death rate during the first few years of life, mortality drops sharply, remaining relatively low until late-middle to old age, when it again rises steeply. The compositions of populations exhibiting such death rate patterns would be quite different. In a population of oysters, in spite of the enormous initial production of offspring, there would be relatively few organisms of reproductive age and postreproductive age at any one time. In a population with a relatively constant death rate, there would be a fairly uniform distribution of the population in the prereproductive, reproductive, and postreproductive age groups. In populations such as the human, there would be high proportions in the reproductive and postreproductive age groups. But with the current trend of medical advances, increasing the numbers of persons reaching older ages, and (at least in the industrially advanced countries) a concomitant slowdown of the rising birth rate, the proportion of the older, postreproductive groups is increasing. Since the birth rate of a population is basically dependent on the number of individuals of reproductive age, it is evident that populations with such different age compositions would exhibit quite different birth rate trends as well.

Food for a Hungry World

If the world's human population continues to grow at the current alarming rate, a further strain will be placed upon the earth's material and energy resources. One of the most frequently voiced fears is: How could we feed an ever-increasing population when already millions are starving to death each year?

It has been pointed out that the main causes of acute food shortages in many areas of the world today are political and economic, rather than biological. Sufficient food for all the earth's inhabitants is being produced, but it is not being distributed effectively.

Scientific advances in recent years have sparked a huge expansion of our food-producing potential. Applied genetics has yielded new, highly productive strains of rice, wheat, barley, and oats that have led to a "green revolution" in agriculture. Often not only productivity but also the food value of crops has been increased. Corn, for example, is normally low in two of the essential amino acids, lysine and tryptophan. In areas where corn makes up a large fraction of the normal diet, protein-deficiency diseases are common. Improved varieties of corn with adequate amounts of these amino acids have been bred, however. Their introduction promises improved health with a minimal change of dietary habits.

Another avenue of providing adequate protein supplies at minimal cost has been the increasing use of soybean additives in a variety of foods. Unlike most plant products, soybeans contain full-valued proteins, with good proportions of all the essential amino acids. Soybean flour is commonly added to wheat flour in bread and macaroni products. Spiraling meat prices have prompted the marketing of soybean-fortified hamburger. The ubiquitous soybean is also to be found in a growing variety of modern processed foods, from soups to cereals, providing a cheap source of additional protein. (Unfortunately, a substantial fraction of the population—some sources have estimated 15 percent—finds that the ingestion of soybean products promotes an explosive multiplication of gas-producing intestinal bacteria. The resulting flatus can be a source of acute social embarrassment, and in certain circumstances even a potential hazard: one of the criteria for future selections of astronauts according to NASA, may be "low flatus production.")

Fish-protein flour, made from "trash fish" that would not ordinarily be consumed by humans, is another type of protein supplement finding increasing use.

Conventional sources of food production alone cannot supply a geometrically increasing population indefinitely. Active research on alternative food sources is now underway. One promising line is the conversion of ordinary weeds, grasses, and other normally untapped plant materials into useful sources of protein. Already tons of such leaf-protein concentrates are being produced commercially. In an hour, a ton of leaves is ground into a pulp, rupturing the cellulose cell walls, and the juice is pressed out. Half to three-quarters of the protein within the leaf is extracted in this way. The protein is coagulated, filtered, and washed, yielding a product that has the texture of cheese and is quite tasteless. At this point flavors are added, and the final product can be processed into a variety of textures and forms.

Significant progress has also been made in growing microorganisms, particularly bacteria and yeasts, on such widely available raw materials as natural gas and crude oil. Tens of thousands of tons of such "single-cell protein" are already being produced from the yeasts fed on crude oil. It has been estimated that by using just 7 percent of the world's annual production of about 2 billion tons of crude oil, sufficient protein could be produced to satisfy the needs of the entire world population in the year 2000.

The revolution in synthetic food production has been paralleled by an equally dramatic development of the food flavoring industry. By combining the tasteless, odorless white protein powder obtained from petroleum-fed yeasts with sophisticated synthetic flavoring agents, a variety of artificial food products have been obtained, ranging from "hamburgers" to cookies, indistinguishable from the natural products.

Further prospects for satisfying the world's food needs are coming from the chemical laboratories. Every year millions of tons of sawdust are burned as a useless by-product of the lumber industry. Sawdust has now been converted into sugar in the laboratory. If this process can be made economical, it will add a huge source of an important food product. In other experiments, sugars have been synthesized from the simplest of raw materials: water and carbon dioxide. As research on photosynthesis continues to provide new insights into the chemistry of food production, such chemical syntheses of foods will doubtless be expanded. The man of the future may ultimately rely primarily on wholly synthetic foods.

Maintaining the Balance of Nature: Conservation

It was noon on a quiet autumn day. A traveler was riding down a country road paralleling the Ohio River. Suddenly he became aware of a growing roar, and he turned to see a great flock of passenger pigeons sweeping in from the north. Soon they were overhead, in a flock so thick that the sun was eclipsed and the air reeked with the odor of the birds. His frightened horse reared and plunged. Still the birds came. As the traveler went on, he passed crowds of men and boys along the riverbank, gleefully shooting at the pigeons. When he reached Louisville at sunset, still no end to the flock could be seen; it took a full three days to pass.

The traveler was the great naturalist John James Audubon, and the year was 1813. Audubon calculated that in the segment of the great flock that had passed while he traveled the river road there were more than a *billion* birds; and this small segment of the total flock would consume *each day* nearly 9 million bushels of food.

Who would have thought that the last passenger pigeon in existence would die only a century later, on September 1, 1914, in a lonely cage at the Cincinnati Zoological Gardens!

Figure 25–4 The passenger pigeon—gone now.

The passenger pigeon is gone. We can discuss its fate and analyze the causes, but we cannot bring it back. Now wildlife specialists fear that literally hundreds of other species are at the brink of the same fate. P. M. Scott, chairman of the World Wildlife Fund, has been quoted as stating that some thousand species of vertebrate animals are threatened with immediate extinction. According to Scott, "We must build a sort of Noah's ark to preserve these animals and birds until humanity recognizes the right of wildlife to live alongside man."

The species now threatened include the Indonesian Komodo dragon, the Tasmanian wolf, the European bison, and the belted sandfish.

Strenuous efforts are now being made to save many wildlife species from extinction. Emergency breeding programs have already brought about some dramatic rescues. The Arabian oryx, a horned ungulate, was disappearing from the desert so rapidly that repeated efforts by British conservationists to locate and capture a few pair to serve as a breeding nucleus were unsuccessful. Finally, in the nick of time, a pair was located in a private zoo belonging to the sheikh of Kuwait. The sheikh was at last persuaded to part with his prizes, and the two oryxes were transported to a new home at the zoo in Phoenix, Arizona.

One of the most widely publicized efforts to save an endangered species that has captured popular interest to an unprecedented extent has been the fight to save the whooping crane. Once common along the Atlantic Gulf coasts, the whooping cranes had been reduced by indiscriminate hunting to a total population of fourteen by 1939. Since then, measures have been instituted to protect the whooping cranes and to foster their recovery. The little flock, now grown to several score (not including a number of young whooping cranes that have been hatched from eggs and raised in captivity), is carefully guarded in its winter home at Aransas National Wildlife Refuge in Texas and in the summer nesting refuge near Great Slave Lake in Canada. During their long semiannual migratory flights—nearly twenty-five hundred miles each way (Fig. 25–5)—hunters are sternly cautioned not to shoot the cranes. Through a combination of applied ecology and creative publicity by conservationists, the plight of the whooping cranes and their hopeful revival are being watched with close interest by scientist and layman alike.

The conservation of endangered species is but one aspect of a greater picture, for man has perturbed the balance of nature in a bewildering variety of ways. The gross effects of such major environmental changes as the building of dams and the stripping of forests, the draining of swamps and the leveling of mountains, are as obvious as they are pervasive. More subtle have been numerous minor side effects of man's activities that have sometimes grown into major problems.

An example of such unintentional meddling with the balance of nature is occurring right now in the United States. A tropical fish dealer in Florida decided to raise his own stock of an imported species, the walking catfish of

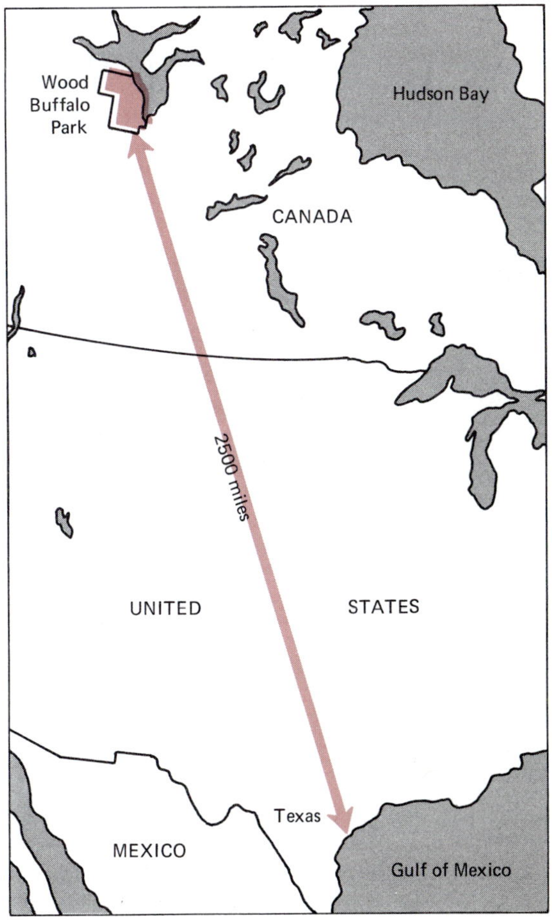

Figure 25–5 Whooping cranes at the Aransas National Wildlife Refuge in Texas (photo) and their migratory route.

Figure 25–6 The walking catfish.

Southeast Asia (Fig. 25–6). The voracious catfish quickly ate all the other fish in his pond and then, utilizing the peculiar ability that earned them their common name, leaped out of the pond and then flopped and waddled their way to other bodies of water. Before anyone realized what was happening, the hardy catfish had proliferated wildly and had spread throughout a thousand-square-mile area of southern Florida. Since the catfish feeds heartily on small fish, shellfish, insects, frogs, and even weeds and roots, and can readily denude a small pond of its animal life and then leave to find greener pastures, it is proving to be an alarming pest. Attempts to poison it have been unsuccessful: while the poison kills other fish in the pond, the walking catfish simply leave. (These adaptive creatures can remain out of water on damp ground for more than twenty-four hours at a time, breathing oxygen through rudimentary lungs.) Now, as though the state of Florida did not have enough troubles with the walking catfish, the process seems to be repeating itself with a new exotic pest, a large species of Hawaiian land snail brought back to Miami by a five-year-old tourist.

A more complicated ecological situation seemed at one point to be imperiling the Pacific islands and large portions of Australia. A large starfish, *Acanthaster planci,* commonly called the crown-of-thorns starfish because of its spiny appearance, underwent a sudden population explosion. This starfish is a voracious eater of corals. By 1970 it was dis-

Figure 25–7 A population explosion of the crown-of-thorns starfish temporarily unbalanced the coral reef communities of the Pacific. Natural predators such as the triton seem to be bringing it under control.

covered that members of the species had already destroyed more than 90 percent of the coral reef that protects the northeastern shore of Guam from the destructive action of waves and had killed the corals over roughly a hundred square miles of the Great Barrier Reef of Australia. The death of the corals in turn led to the destruction of the interrelated community of fish and other water life that inhabit the coral reefs, thus endangering one of the most valuable food resources of such areas. Alarmed scientists hypothesized a complicated variety of causes for the sudden eruption of starfish: tourists had been stripping the area of one of the starfish's few natural enemies, the conchlike triton; and dredging and channel-building projects had disturbed the ecological balance of the area and eliminated many of the micro-feeders that normally would eat the starfish larvae. The situation was believed to be aggravated by residual radiation from atomic tests in the Pacific area and by a gradual accumulation of DDT residues.

In the panic reaction that followed the discovery of the crown-of-thorns menace, frantic searches were made for new potential natural enemies of the starfish, teams of volunteer skindivers injected individual starfish on the reefs with nerve poisons, and the U.S. Congress authorized the appropriation of $5 million for control programs in Hawaii. But by 1972 it was suddenly discovered that the starfish populations were dropping back to former levels—and for no apparent reason. Again ecological relationships had proven far more complicated than anyone had realized.

The growing realization that man's actions can affect the future of the biosphere has led to the emergence of an overriding goal: to seek a workable balance of nature and to restore where possible the delicate balance that man has disturbed or, where this is not feasible, to work out a new balance. The areas of applied ecology include not only efforts to save endangered species, but also measures to utilize more fully the potentials of man's environment—and to utilize them in such a manner as to replace for the use of future generations what man takes from the environment. Soil conservation, forest management, and the management of fish and wildlife resources are important areas of applied ecology in the modern world. As ecologists learn more and more about the interrelationships of organisms, it is being realized that man must learn to anticipate the possible consequences of all his dealings with his environment. Indiscriminate use of pesticides to eliminate crop pests, for example, may also kill off desirable species and start a chain reaction ultimately of far greater harm than the original pests. This problem is aggravated by the widespread phenomenon of biological concentration. (Insects' bodies concentrate some insecticides, and birds eating the insects may receive a much higher dose, and so on up the food chain.) In addition, the tendency of pesticide residues to persist in the environment and to be spread by the waters of rain runoff, streams, rivers, and oceans exacerbates the situation still further. Emphasis is being placed increasingly on biological methods of pest control, utilizing a knowledge of the life cycle of pests and their natural enemies. For example, strips of "decoy crops" can often be planted to attract insect pests away from more economically valuable crops. And sterilization of members of a pest species, with their subsequent release into the area to compete with other members of the breeding population, has proved quite promising with the screwworm fly and other pests.

Studies of the breeding habits of mosquitoes have resulted in an even more ingenious combination of physical, chemical, and biological principles to aid in the control of these pests. At the appropriate time, the surfaces of lakes, ponds, and other breeding areas are treated with a surface-active chemical that lowers the surface tension of the water. The mosquito larvae sink into the water, since its "skin" can no longer support their weight, and they drown before they can develop into mature mosquitoes. Meanwhile, fish and other water animals and plant life are unaffected.

Another promising line of biological pest control is the importation or breeding of

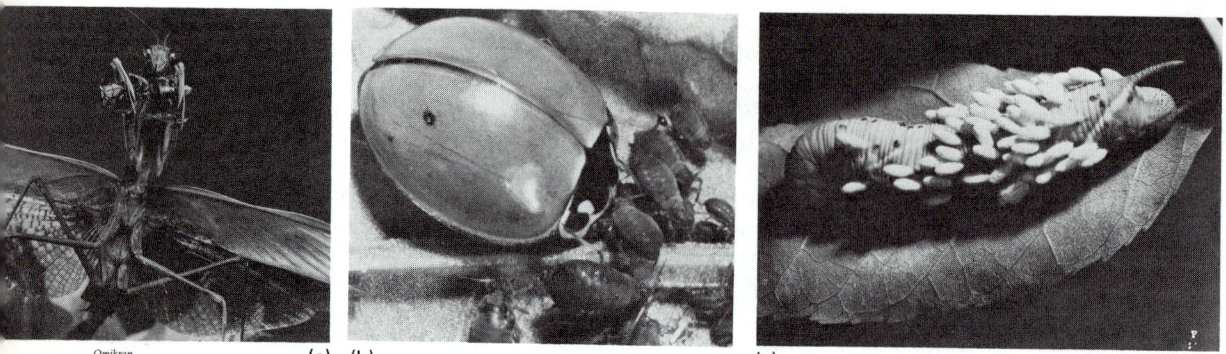

Figure 25–8 Allies in the battle against insect pests: (a) a praying mantis grips a wasp; (b) a ladybug feeds on aphids; (c) cocoons of a parasitic wasp attached to the larva of a sphinx moth.

natural predators or parasites of pest species. Viruses and other pathogens that infect only insect pest species are being developed. The use of ladybugs, praying mantises, and *Trichogramma*, a tiny parasitic wasp that lays its eggs inside the eggs of moths, has become a popular approach for biological control. But here great caution must be exercised. A thorough knowledge of ecological relationships is vital for success, and even then unexpected results may occur. The English sparrow, for example, was originally introduced into the United States in the 1800s to serve as a natural enemy for the gypsy and brown-tailed moths. But hopes of controlling the moths in this way were soon dashed: at that time the horse was a major mode of transportation, and the sparrows found the horse droppings (especially in their large accumulations in the cities) such an attractive food source that they virtually ignored the moths. The sparrow population grew by leaps and bounds, with an adverse effect on the populations of many of the native songbirds. With the advent of the automobile, the sparrow problem was at last brought under control. But gypsy moths still exact an enormous financial toll each year in damaged trees and crops; and in turn the automobile has become a major factor in another of man's greatest present-day ecological problems: the growing pollution of our environment.

Man Tampers with His World: Pollution

It was a bleak December day in 1952. The fog seemed unusually thick to the Londoners as they moved through the streets. With the passing hours, the fog—it was really smog—thickened. People coughed and went their way. For four days the smog was unabated. At last it cleared—but not before it had produced an incredible toll: it is said that four thousand Londoners died as a result of this smog. Ten years later the smog returned once more and claimed the lives of more than a hundred. The following year, in 1963, more than 200 people died in New York City as a result of a similar smog that lingered for ten days.

These and similar events have dramatically focused the attention of the public upon the fact that air pollution *is* a problem. But unfortunately it has only been quite recently that the public has begun to become aware that air pollution is a pervasive, growing, everyday phenomenon, and not just a matter of rare crises. The first U.S. federal government appropriation for a program to alleviate the situation (a meager $186,000) was not passed until 1955. Further strides were taken with the passage of the Clean Air Act in 1963. But rising expenditures for pollution control seem to be swallowed up by more rapidly rising rates of pollution. In 1969, according to federal estimates, air pollution sources were spewing a combined total of more than 140 million tons of contaminants into the atmosphere above the United States each year! Even orbiting astronauts have commented on the ease of recognizing major urban areas by their characteristic smog blankets.

The most successful pollution control programs thus far have been sponsored by the state of California and by Los Angeles County in particular. The city of Los Angeles and its immediate environs are peculiarly subject to a meteorological condition known as **temperature inversion** (Fig. 25–9). This condition results when, for example, a layer of warm air settles over a certain land mass, preferably one enclosed within a valley, as is Los Angeles. The normal "chimney effect," in which air tends to rise, carrying away its load of dust and pollutants from the city below, is effectively stopped by the blanket of warm air above. Thus, the dust and soot and chemicals, trapped in the lower air levels, form what is known as "smog." Throughout a great portion of the year, Los Angeles County suffers frequently from this type of inversion. Thus, it is not surprising that this county and the state of California lead the nation in research and community efforts toward alleviating the problems of air pollution.

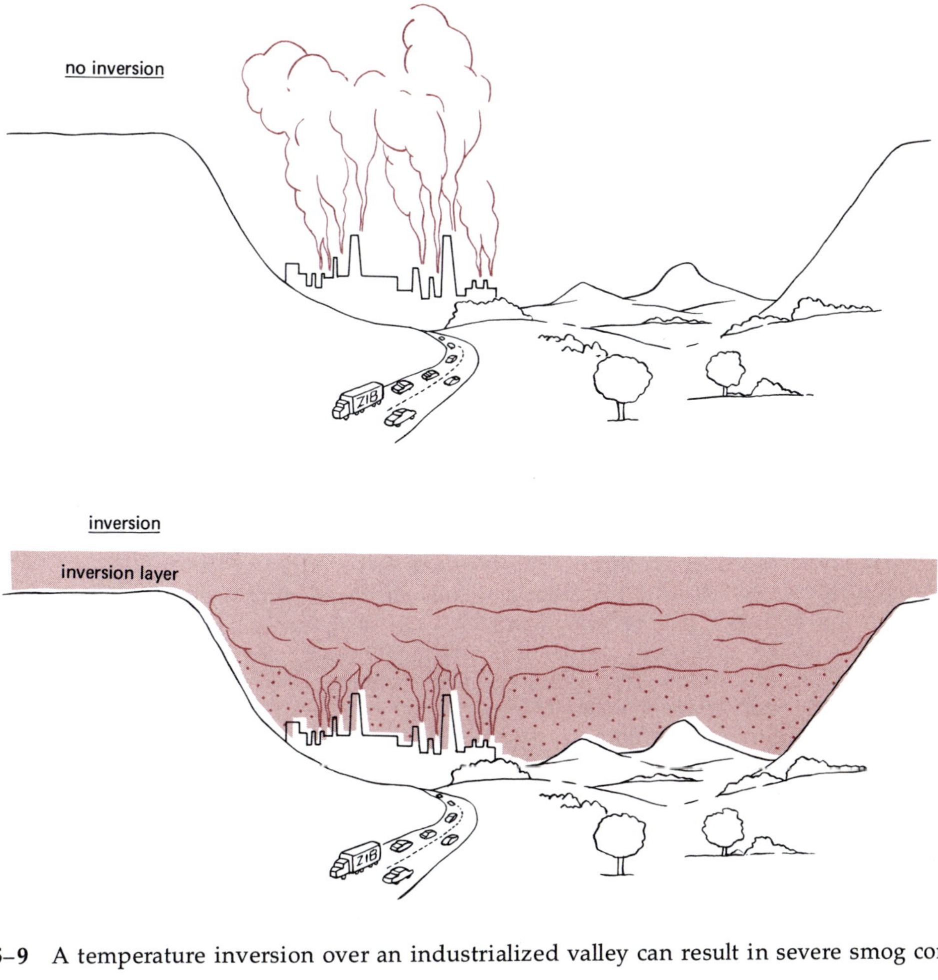

Figure 25–9 A temperature inversion over an industrialized valley can result in severe smog conditions.

The condition of temperature inversion is not at all restricted to the Los Angeles area. It has been estimated by one official of the United States Weather Bureau that temperature inversions in which air is trapped below five hundred feet occur over various parts of the United States more than one-fourth of the time in all seasons. But even without temperature inversions, our atmosphere remains a sewer of noxious and toxic chemicals.

In urban areas, probably the major source of these pollutants is the motor vehicle. The automobile, for example, spouts its poisons in four different ways: from its gas tank, from its carburetor, from its crankcase vent, and from its tail pipe. The unburned exhausts from automobiles contain numerous hydrocarbons, some of which, such as benzpyrene, have been shown to cause cancer in animals. Carbon monoxide, a major by-product of gasoline combustion, is another potential hazard. Small amounts in the atmosphere (or inside the automobile) have been shown to dull the senses and lead to improper judgments; they undoubtedly are the cause of some (and perhaps many) automobile accidents. Lead compounds are another form of pollution discharged into the air by the automobile: it has been estimated that modern urban man has a lead concentration in his blood up to a hundred times the normal level. Devices to reduce the amount of pollutants discharged into the atmosphere, which are required components of all new cars sold in the United States after 1968, can help, although their effect is offset to some extent by the annual increases in the number of motor vehicles on the road. Now new laws are placing increasingly stringent restriction on permissible levels of car emissions in an attempt to bring this source of pollution under control.

Another important source of air pollution is the wastes of industry. Each year more millions of tons of wastes are spewed forth into the atmosphere. The variety of such waste products is enormous. The potential toxicity is evidenced by the miles of blighted vegetation that surround many large industrial complexes. In addition to products of combustion of coal and hydrocarbons, the industrially derived air pollutants include toxic oxides of nitrogen and sulfur, fluorine compounds, various volatile solvents, and a variety of compounds specific to particular industries.

What are the effects upon man's health of all these pollutants produced by motor vehicles, industry, and urban incinerators? The irritating and sometimes deadly smog that accompanies temperature inversions over large cities is only one of the more flamboyant manifestations of air pollution. But every moment of the day and night, our lungs are insidiously subjected to the caustic and debilitating effects of the pollutants that pervade our air. The soot-blackened lungs of the city-dweller are vivid testimony to the dirtiness of our air. As a result, it should not be surprising that the incidence of respiratory diseases of all types is rising alarmingly, particularly in the urban areas. One such lung condition, emphysema, is increasing at a spectacular rate.

Many of the same industries that are polluting our atmosphere are also polluting our waters. Each year, many millions of tons of chemicals are dumped into our waterways. The polluted waters flow downstream where, after some treatment, they are frequently used for drinking water and various industrial purposes. The sewage of this downstream area, in turn, may be dumped further downstream, and may flow on to the next city. Is it any wonder that the fish population along such populated waterways as the Mississippi River has been drastically reduced and that the Great Lakes, too, are dying?

Among the wastes that find their way into the waterways are pesticides, which have been indiscriminately sprayed (often by airplanes) over millions of acres of farmlands. Residues of some of these pesticides, such as dieldrin and DDT, have been found in fish and are known to be toxic to them. The ultimate effects of such pesticide residues on humans who eat such fish (to say nothing of crops on which residues are also detected) are yet to be determined. Birds that feed on insects or fish that have ingested or absorbed pesticides are often affected even more severely through the mechanism of biological concentration. Bald eagles are being debilitated by the effects of pervasive DDT residues, and their eggs are often so thin-shelled that they do not hatch successfully; conservationists thus fear that this species may be doomed. The DDT level in the average American's body has already risen above the content listed as permissible in foodstuffs by the federal Food and Drug Administration (FDA). Public concern ultimately resulted in widespread bans of DDT and its replacement by other insecticides that decompose more rapidly in nature.

In 1971 it was suddenly discovered that many studies of DDT in the environment were actually detecting a different group of chemical, polychlorinated biphenyls (PCB). PCB is an industrial product, manufactured on a large scale and used in chemically treated duplicating papers, in insulating fluids, as a heat transfer agent, and in paints and pesticides. PCB residues have been detected in a variety of wildlife, in turkeys, chickens, and eggs, and even in packaged cereal. The source of the chemical in the last item is ironic: the cardboard packaging of the cereal boxes was made from recycled paper, some of which contained PCB. The ultimate effects of this previously unsuspected pollutant in our food supply are yet to be fully assessed. But large concentrations of PCB have been found to cause serious illness, and PCB residues in waterfowl have caused larger-than-normal percentages of birth defects among their offspring (Fig. 25–10).

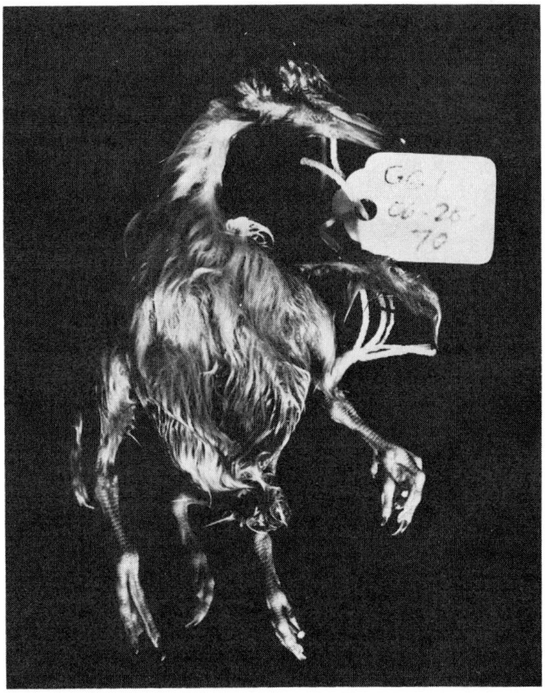

Figure 25–10 The birth defects of this deformed tern chick were traced to the effects of PCB.

Another type of dangerous pollutant now pervading our environment consists of the heavy metals, such as mercury and cadmium. Mercury made the headlines late in 1970, when a high school chemistry teacher found mercury contamination in canned tuna. FDA follow-ups revealed that some lots of tuna and swordfish contained mercury considerably in excess of the permissible limits. Lots of suspect fish were recalled, and panic swept the general public and the fishing industry. Ultimately it was revealed that mercury pollution is not a new phenomenon, although man's industrial activities are obviously contributing to it. Samples of preserved tuna—even prehistoric fish—were found to contain mercury residues. Curiously, only certain types of fish were found to be affected: mainly large fish, such as tuna and swordfish, which are at the top of their respective food chains and thus are subjected to increased amounts of the heavy metal accumulated in the bodies of the smaller fish they eat.

Another environmental pollutant that is becoming an increasing problem in our waters and adjacent land is crude oil. Increased utilization of offshore oil resources, found within the rocks of the continental shelf, has resulted in some spectacular blowouts that have blanketed the surrounding waters and nearby coasts with a layer of oil that spoils beaches, cripples waterbirds (whose feathers become drenched with it), and kills a variety of intertidal life. Man's efforts to remove the contaminating oil, including the use of detergents and hot-water spray, further combine to upset the delicate ecological balance

Figure 25–11 An oil-soaked herring gull.

of the area for years to come. Meanwhile, even after the runaway well has been capped and shut off, oil often continues to seep out through fissures in the disturbed ocean floor. Such has been the case in beleaguered Santa Barbara, California, which one boasted some of the most beautiful white sand beaches in the United States. Offshore drilling there has resulted in spectacular spills and continuing oil leakage. Since this is a region of heavy geological faulting, the constant danger of earthquakes (aggravated by the disturbance of the drilling activity) menaces the entire California coast with further devastating oil spills.

It is not only beaches and coastal areas that are threatened by oil spills. Deliberate dumping of excess oil by ships and leakage due to accidents to huge oil tankers can spread a thin film of oil over vast areas of the ocean surface. Such an oil film can kill phytoplankton, setting off a disastrous chain reaction in the communities of ocean life. Since photosynthesis by marine plant life accounts for the bulk of new oxygen supplied to the atmosphere, some ecologists feel that the oil menace, if unchecked, could ultimately render our planet uninhabitable by man and other air-breathing animals. Some hope for an ultimate solution to the problem of oil spills may be offered by research in another area: the development of microorganisms that grow on petroleum hydrocarbons, breaking them down in the process. Although these microbes are being developed primarily in a search for cheap new sources of food proteins and other nutrients, they might also be used to clean up oil spills before they can do much damage. (Here too, of course, more must be learned about existing and potential ecological interrelationships before new organisms are introduced into the ocean communities on a large scale.)

With the expansion of the peaceful uses of atomic energy (not to mention periodic flare-ups of radioactive contamination of the atmosphere due to testing of nuclear weapons), the problem of radioactive wastes will become an ever-growing source of anxiety. Since a number of these radioactive materials have half-lives measured in terms of years and even decades, they can remain a dangerous source of contamination for generations. Deciding where to dispose of these wastes is already becoming a problem. Wholesale dumping in the depths of the oceans would undoubtedly have serious long-term effects on the life of our planet.

Another pollution problem that has been widely publicized in connection with nuclear power plants is that of thermal pollution: the discharge of the hot waters used to carry off the tremendous amounts of heat generated as a by-product of the generation of electric power. This problem actually is not confined to nuclear power plants, but is a concomitant of any large electrical power plant, and as such will become an increasing problem as our use of power continues to grow and power generation is concentrated in increasingly large networks. Water life is generally adapted to a relatively narrow and relatively constant range of temperature. The sudden influx of hot-water discharges from power plants can kill many fish and other life forms and thus seriously disturb the ecological balance of the area.

Thermal pollution is an area where creative solutions are already being sought and some promising ideas have been suggested. Although influxes of warm water can kill members of the normal aquatic community of an area, other species thrive at the warmer tem-

Figure 25–12 Fishing across from an atomic power plant at Haddam Neck, Connecticut.

peratures. These warm-water species tend to move in and establish a new ecological equilibrium. Thermal pollution can even be utilized to make "fish farming" far more effective and profitable, enabling raising of economically valuable fish, lobsters, and other water species at far greater densities than the waters would normally support. An even more ingenious suggestion has been to pipe the warmed water from power plants under the soil. The warming effect can result in better germination, better growth of plants, and even year-round crops in areas normally too cold for winter farming.

Still another area of pollution has been growing and apparently will continue to grow in intensity and seriousness with the passage of time. The concern about the sonic boom caused by transoceanic supersonic jet planes gives us a glimpse of what the future may hold for us. The term "noise pollution" is being used more and more to describe the ever-growing intensity of noise in our urban life. Physicians have long been aware of the dangers of prolonged exposure to high noise intensities within factories and in other industrial surroundings. They are only now becoming aware of the role that noise plays in the daily life of the average urban citizen. A recent surprising finding is that the noise level in urban communities frequently exceeds the accepted standards of tolerance in industry. Special circumstances, such as those created in discotheques and by rock-and-roll bands, produce an even greater assault upon our eardrums. Deleterious psychological and physiological changes have been found to result from prolonged exposure to high noise levels. In addition to the obvious result of irreversible loss of hearing acuity, more pervasive effects on the major body systems have already been discovered. Exposure to annoyingly high noise levels triggers a general stress reaction involving the adrenal hormones and their effects on other organs and systems of the body. In studies in Europe, men exposed to noise levels greater than 75 decibels (about the level heard close to a high-speed highway) ultimately developed permanent constriction of the blood vessels. Other studies revealed that noise levels in the 90–95-decibel range (about the level heard twenty-five feet from a motorcycle and slightly less than that of a jet plane flying overhead at a thousand feet) produced high blood pressure, which could not be normalized even by drugs. Studies of animals have revealed rises in blood cholesterol levels and sharp increases in birth defects resulting from prolonged exposure to high noise levels.

The pollution picture seems bleak. One might think at first that an obvious step, initiated by an aroused citizenry, would be to eliminate all these extraneous factors that are polluting our environment. But the problem is by no means as simple and one-sided as it might seem at first. In many areas there is a genuine dilemma as to what course is truly the best for man to pursue, and some of man's environmental blunders may ultimately prove to be irreversible, at least within our lifetimes.

For example, the elimination of automobiles before a workable substitute for the internal combustion engine is devised would cause wholesale dislocation in our economy and mode of living. Nor could anyone rationally advocate the elimination of manufacturing industries and electric power. The elimination of pesticides would remove an important source of environmental pollution, but it would have many undesirable results as well. Crops would be stricken with diseases, and insect pests would ravage them unchecked. The severely reduced harvests would rot in the warehouses or become contaminated with insects and rodent droppings.

Many leaders of industry and government are greatly concerned with problems of pollution. In the United States, industry is now spending billions of dollars each year for research on methods of pollution control and the implementation of devices that will reduce pollution.

Progress is being made. The case of polluting detergents is an excellent example of what can be accomplished through a combination of an aroused public and a responsive (and responsible) industry. After the invention of synthetic detergents, the cleansing powers of which showed certain advantages over the soaps that man had used for centuries, the sales of these new synthetic soap-substitutes skyrocketed, as they virtually supplanted their traditional competitors for dishwashing, laundry, and industrial uses. As the use of synthetic detergents increased, an alarming phenomenon began to be observed more and more often. The sewers, streams, and reservoirs were literally frothing. A glass of water from the tap often had a head of foam, like a glass of beer! It was soon discovered that this froth was due to detergent residues. For these synthetic substances were impervious to the actions of the microorganisms that normally degrade soap residues and other refuse flushed away with our sewage. Fortunately, the detergent industry responded quickly to this alarming new problem. Research disclosed ways to modify detergents so that they could be successfully attacked by the scavenger microorganisms of our streams and reservoirs. Now virtually all of the detergents on sale are "biodegradable."

Figure 25–13 That's not snow, it's detergent foam on an upstate New York river.

Omikron

Further progress in this area is urgently needed, however. The high phosphorus content of modern detergents has acted as a potent fertilizer in ponds and lakes, giving rise to the phenomenon of **eutrophication:** overgrowth of a body of water by algae and undesirable forms of plant life, the subsequent decay of which depletes the oxygen reserves of the surface waters and results in death of fish and other water animals. Eutrophication is a natural process in the history of ponds and lakes, but it is drastically accelerated by man's activities. Public concern has resulted in the passage of laws in some areas banning the use of phosphate-based detergents. But it was soon discovered that many of the proposed substitutes that were rushed into the breach were poisonous or caustic substances that posed serious health hazards to users, particularly children. The whole subject of household detergents is still in heated dispute and is an object lesson on the difficulty of devising ecologically sound yet practically feasible solutions to the problems of the modern world.

Though the outlook on pollution may seem discouraging, there are definite hopeful signs. Through the combined efforts of a concerned public and responsible elements in industry, these problems, too, can be solved. And they must be, if man and the biosphere are to survive. In the words of a letter written by a fourteen-year-old boy in oil-drenched Santa Barbara to the president of the United States, there are basic human needs that must be safeguarded: "fresh air, a place to grow, sweet water to drink, and perhaps a consideration for progeny of my own."

Man Tampers with Himself: Food Additives

For twenty-five years, just about all the white bread eaten by the American public was made from flour bleached by a chemical called agene. This chemical produced a fine grade of white flour, and it seemed safe enough. But the suspicions of a British physician were aroused. He fed bread bleached by this chemical to a group of dogs, and the results were appalling: the dogs went wild, running about madly and barking uncontrollably.

More experiments were run on other animals, including cats, rabbits, and mink. After eating the treated bread, the animals went into convulsions. Eventually it was discovered that agene reacts with some of the natural substances in bread during the baking process to produce a poison. It is this poisonous product that caused the running fits in dogs and convulsions in other animals. What were its effects on the generations of humans who had consumed it with their daily bread? Alarmed, the Food and Drug Administration banned the use of agene.

Agene is no longer used in the foods we eat. But each day, Americans consume a fantastic variety of chemicals along with their daily food. Chemicals are added to modern food products as coloring and flavoring agents. You can purchase candies, gelatin desserts, and soft drinks in almost any color of the rainbow. Chemicals are used to change the texture of foods, to make dough rise, and to make aerosol-canned whipped toppings foam. Other chemicals impart flavor or aroma or enhance the natural flavor of foods. Still others keep foods from spoiling, or caking, or turning liquid.

Substances that are added to foods to alter their properties in some way are known as **food additives.** Some additives, such as salt and other spices and preservatives, have been used for millennia. But the use of food additives has burgeoned explosively with the development of modern industrial chemistry and technology. More than three thousand

Figure 25–14 The variety of packaged foods stocked by the modern supermarket would be impossible without food additives.

different food additives are now in regular use. It is difficult to think of a prepared food, from packaged cereals to canned soups, that does not contain some additive. Indeed, most foods today contain a number of food additives.

Considering how chemical additives have permeated the foods we eat, it is rather alarming to realize how ignorant we are of their effects upon our bodies.

To be sure, tests have been run in an attempt to learn how safe these chemicals are. Public concern has led to a considerable tightening of controls in this area. Before 1958, for example, a food manufacturer could legally treat foods with any chemical that caused a desirable change in their properties. (This did not necessarily mean a desirable change in nutritive qualities, or indeed, even in any properties directly benefiting the consumer.) Of course, the prudent manufacturer ran tests to ensure that the additive he proposed to use would not immediately harm anyone. But no standard program of stringent testing was required. And it was the responsibility of the federal Food and Drug Administration (FDA) to establish that a particular substance was actually harmful to human health before it could be banned from further use. In fact, until 1958 a food additive was defined as a substance added to food, the use of which had *not* been proven *safe!*

In practice, this meant that potentially dangerous food additives could be freely used in foods consumed by the public (as long as they were listed on the label or in the FDA standards of identity) until their harmful effects became apparent. The FDA—or any other single organization, for that matter—could not possibly cope with the enormous task of checking out each of the thousands of food additives in use. So that, in fact, the real guinea pigs were often the public themselves.

This frightening situation was changed radically by the 1958 amendment to the Food, Drug, and Cosmetic Act. According to this law, the burden of responsibility was now placed upon the manufacturer. If he wanted to use a new food additive, he would first have to show that it was safe through appropriate tests supervised by the FDA. In addition, the amendment provided for a comprehensive review of all the additives in use at the time.

The problem of determining whether an additive is safe for human consumption is far more complicated than it might at first seem. For example, suppose the safety of a particular food additive is tested on a group of rats. A control group of rats is fed or injected with some innocuous substance, such as a solution of ordinary salt in water, and the experimental group is fed or injected with the food additive being tested. Usually, higher doses are used than those to be used in food that humans will consume.

There are several drawbacks to this procedure, however. First of all, the assumption is made that what is safe for a rat or a mouse is safe for man. This is not necessarily true. (Thus, food additives are finally tested on human volunteers before nationwide distribution is permitted.) Another drawback to the testing procedures used is the fact that only a single additive is tested in any one experiment. But man consumes dozens of different food additives *every day*. It is difficult enough to determine the effects of a single substance upon such a complex organism as man. But what of the effects of the interaction of hundreds or even thousands of chemicals, to say nothing of their interactions with the thousands of natural constituents of foods, the thousands of natural substances in the body, and the hundreds of drugs that are consumed by the public? And what of long-term effects, which might not show up until a substance has been consumed for years or even generations? Only large-scale testing on human beings in a program lasting many years could show whether any particular chemical is truly safe. And the matter is even more complicated.

Present FDA regulations provide for the approval of food additives that have been shown to be safe "at the intended levels of use." (Virtually any chemical can be harmful if it is consumed in large enough amounts. Even common table salt can be poisonous if too much is taken at once. And for persons with special metabolic problems, even a "normal" intake of salt can be harmful.) But modern man in our affluent society takes in a great variety of foods in his normal diet. What happens if he consumes several different foods each of which contains a particular additive at its "safe" level of use? He is now consuming not the "intended level," but several times the level of intake that is assumed to be safe. A similar problem is created with children: since the effect of chemical substances on the body is in general roughly proportional to body weight, a standard portion of a food containing an additive will represent a far greater effective dose of the additive in the body of a child than in the body of an adult who weighs three or four times as much.

Since the passage of the Food Additives Amendment in 1958, a number of food additives which had long been used have been banned as a result of intensive testing and review. Some of these were substances on the "GRAS" ("generally regarded as safe") list. One of the most recent and highly publicized cases was the banning of cyclamates, which had been widely used as "no-calorie" artificial sweeteners. The prohibition, based on laboratory studies indicating that cyclamates cause birth defects and cancer in several species of animals, had an enormous impact on the food industry. Not only had these sweeteners been used in a variety of diet foods, but since cyclamates are cheaper than sugar they had been used in a number of foods for general consumption as well, including soft drinks, puddings, gelatins, salad dressings, jams and jellies, and ice cream. Cyclamates had even found use in such unlikely areas as the curing of bacon and the manufacture of pickles, dog food, and children's vitamins. Studies of cyclamate metabolism in the body emphasized one of the most difficult aspects of the problem of determining the safety of food additives and other substances: the enormous range of normal variability in the population. In many humans, only about 1 percent of the ingested cyclamate is metabolized to a toxic product, cyclohexylamine. But other equally "normal" people convert as much as 40 percent of the cyclamate that they take in to cyclohexylamine. Obviously people of the latter type run a far greater risk from exposure to the same levels of cyclamate. Such individual idiosyncrasies and hypersensitivities are a common phenomenon.

What is the answer to the pervasive use of food additives and the serious health hazards that sometimes result? Should we consider prohibiting food additives altogether? Few rational people would call for such drastic action. In addition to reducing the wide variety of colors, flavors, textures, and other characteristics of our foods, the elimination of food additives would bring many other problems as well. Most foods would literally rot on the shelves, particularly in large urban areas where direct producer-to-consumer supply of foods is no longer feasible. The variety of foods would be sharply curtailed, and many foods would have a bland or even a disagreeable taste. The price of food in general would soar far beyond what it is today.

A dependence upon "natural" food additives is not the answer, either. Two of the major substances banned as a result of FDA testing and review were actually natural substances. Coumarin, found in the tonka bean, had been used for seventy-five years as an artificial vanilla flavoring until it was found to cause liver damage in rats and dogs. Safrole, a substance from sassafras bark that had been used to flavor root beer, was shown to cause cancer in rats.

In addition, the supply of natural food additives could not nearly cover the demand. For example, the entire world supply of natural vanilla cannot even meet the demand for the vanilla ice cream consumed in the United States alone—not to mention the rest of the world—or the vanilla flavoring used in baked goods and beverages.

It is clear that food additives are a necessary part of our society. Since man now will not live without them, he must learn to live with them—rationally.

Drugs in the Modern World

In the early 1960s a modern tragedy shocked the imagination of the world. In several European countries, thousand of babies died; perhaps more tragically, thousands more were born with horrible deformities. Their mothers had taken a new, "harmless" tranquilizer with the generic name thalidomide. The deformed babies were born, for the most part, without arms and/or legs, or with mere stumps. To prevent such a tragedy from recurring, the nations of the world in 1966 formed the International Drug Monitoring System. This organization is functioning as a world drug-screening center to collect, evaluate, and disseminate information on drugs provided by the public health agencies of the world.

Such an undertaking is really incredibly complex. In the United States alone, more than four thousand drugs are currently in use. (These four thousand drugs should not be confused with the more than three thousand food additives used to preserve and modify foods.) Many drugs are potentially hazardous if used in improper doses. And in many instances, deleterious interactions among two or more drugs are already known. Moreover, different people have different sensitivities to various drugs, and sensitivity may vary depending on age, sex, diet, climate, special conditions such as pregnancy and diabetes,

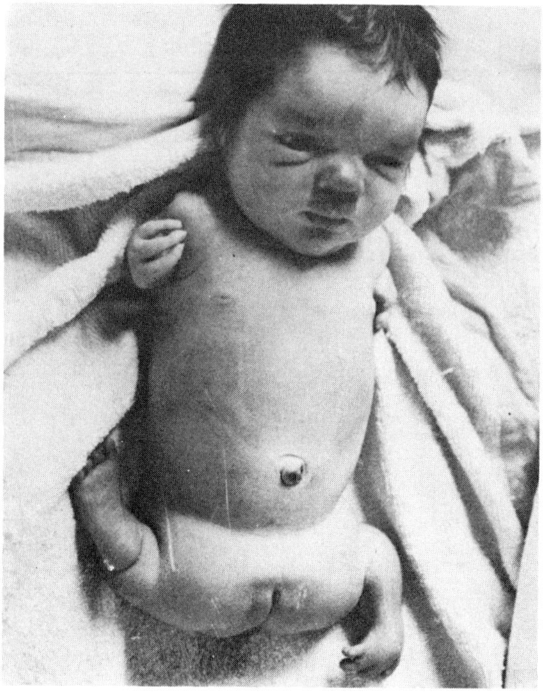

Omikron

Figure 25–15 The tragedy of thalidomide.

and many other factors. As a result, the Food and Drug Administration must keep a watchful eye on and closely regulate the manufacture, sale, and use of drugs.

To complicate matters still further, drugs are also used in agriculture and animal husbandry. For example, antibiotics and other drugs had until recently been routinely added to feed for livestock and poultry to promote more rapid growth and to treat disease and injury. However, careful testing frequently revealed antibiotic residues in edible portions of slaughtered poultry and livestock. Such residues might induce the development of resistant strains of bacteria not only in the animals, but also in humans who consume their meat. Furthermore, in some cases serious antibiotic allergies (especially to penicillin) may develop. It is further feared that the antibiotic residues in foods may interfere with treatments of disease in human patients in which other antibiotics are being used. In the light of these considerations, the FDA has instituted rigorous restrictions of the use of antibiotics in animal husbandry.

It is clear that the effective testing of drugs is an important aspect of the regulation of their use. The normal procedure followed usually involves the testing of a particular drug on a number of animals for both short-term and long-term effects. Then the drug is tested on a small number of human volunteers. The next stage involves controlled testing by physicians throughout the country, involving perhaps hundreds of patients. If a drug successfully negotiates all of these hurdles and it has been shown to be not only safe but also effective, it is given a license by the FDA.

An interesting recent case illustrating some of the difficulties involved in regulating the clinical testing of some drugs was that of DMSO (dimethyl sulfoxide). This drug, first reported by Dr. Stanley Jacob in 1963, at first appeared to be a miracle drug that seemed almost "too good to be true." Early reports noted that this substance kills pain, speeds healing of wounds and burns, reduces inflammation, acts as a tranquilizer, relaxes muscles, purges poisons from the body, relieves arthritis and bursitis, and carries other drugs (such as insulin and various anticancer agents) into the body. After nearly two years of worldwide testing of DMSO, the FDA suddenly withdrew its approval and banned all use (even experimental use) of DMSO. Thousands of physicians were incredulous. The weight of the decision was based on a few reports of changes in the refractive index of the eyes of certain laboratory animals, even though thousands of reports of the use of DMSO in humans seemed to be resounding testimony to its safety and efficacy. Finally in 1966 the FDA changed its ruling to permit resumption of carefully controlled testing of DMSO.

Another area of growing concern not only to the FDA but also to the public at large is the unauthorized use of narcotics and other drugs. Legally obtainable drugs such as alcohol, caffeine, and nicotine; prescription drugs such as tranquilizers, barbiturates, and amphetamines; and illegal drugs such as heroin and LSD are being used—and abused—by growing numbers of people as mood changers in order to produce a feeling of relaxation, or stimulation, or a breakdown of inhibitions, a "trip" of hallucinations, or a euphoric "high." The drugs that are most widely abused today fall into three main classes: **depressants, stimulants,** and **psychedelics** (Table 25–1).

The first of the narcotics, **opium,** has been a problem for centuries. Derived from the opium poppy, this depressant drug can bring about dreamy relief from pain or anxiety, euphoria, and hallucinations. But it is highly **physically addictive:** after the user has taken it for a while, his body comes to depend on a steady supply of the drug. If it is stopped, he will suffer convulsions and other painful **withdrawal** symptoms. Ironically, when **heroin** was first produced, it was touted as a safe, nonaddictive painkiller that could be substituted for opium or for **morphine,** an opium derivative. But it was soon found to be extremely addictive and dangerous (an overdose can kill). The narcotic **methadone** can produce a "high" similar to that of heroin, and it is also an addictive drug. However, it has been found that regular administration of methadone to a heroin addict, in doses too small to produce a "high," blocks the effects of heroin and helps to control the addict's craving for it. It has been objected that addiction-control programs based on methadone maintenance simply substitute one addiction for another. However, such programs have shown some success in enabling addicts to lead relatively normal, productive lives and have probably had a favorable effect in reducing crime rates. (Illegal drugs are generally so expensive that addicts are driven to crime to support their habit.)

With all the current concern about drug addiction, people rarely think of **alcohol** as a drug. Yet it is a depressant drug, one that interacts with the metabolism of the brain and can damage the tissues of the brain and liver. (Recall that the liver plays a major role in detoxifying poisons that have entered the body.) A large proportion of automobile accidents are caused because drivers' coordination, reflexes, and judgment were impaired by a few drinks taken before driving. Alcohol is another drug characterized by the development of both psychological dependence and physical addiction. As with the narcotics and many other drugs, a phenomenon of **tolerance** is observed: the user finds he must take progressively larger doses to achieve the same effect.

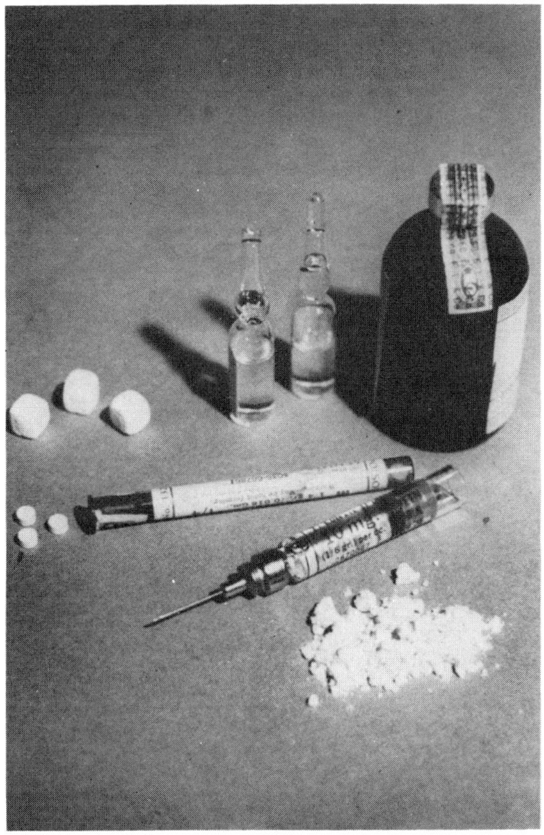

Figure 25–16 Drugs: potential for abuse.

The most commonly used stimulants are **caffeine** (occurring in coffee, tea, cola beverages, and chocolate) and **nicotine** (in cigarettes and other tobacco products). Both have a powerful effect on the central nervous system, but also have harmful side effects (both caffeine and nicotine may be contributing factors in heart disease; excessive use of

TABLE 25-1 Commonly used mood-altering drugs and their effects

TYPE OF DRUG	NAME	ORIGIN	EFFECTS
Depressants			
Alcohol	Beer Liquor Wine	Grain Grain Fruit	Relaxation, breakdown of inhibitions, euphoria, depression, decreased alertness and impairment of coordination; *large doses:* nausea, unconsciousness, hangover, death; *long-term abuse:* malnutrition, psychosis, liver and brain damage, delerium tremens, death
Barbiturates	Chloral hydrate Nembutal Phenobarbital Seconal	Synthetic	Relaxation, euphoria, decreased alertness, drowsiness, impaired coordination, sleep; *large doses:* slurred speech, stupor, hangover, death; *long-term abuse:* sleepiness, confusion, irritability, withdrawal sickness
Inhalants	Aerosols (Freon) Airplane glue Amyl nitrite Nitrous oxide	Synthetic	Relaxation, euphoria, impaired coordination; *large doses:* stupor, death; *long-term abuse:* hallucinations, damage to liver, kidneys, bone marrow, and brain; death
Narcotics	Codeine Demerol Heroin Methadone Morphine Opium	Opium poppy Synthetic Opium poppy Synthetic Opium poppy Opium poppy	Relaxation, relief of pain and anxiety, decreased alertness, euphoria, drowsiness, hallucinations; *large doses:* stupor, death; *long-term abuse:* lethargy, weight loss, temporary impotence, withdrawal sickness
Tranquilizers	Librium Miltown Thorazine	Synthetic	Relief of anxiety and tension, suppression of hallucinations and aggression, sleep; *large doses:* drowsiness, blurred vision, dizziness, stupor; *long-term abuse:* destruction of blood cells, jaundice, coma, death
Stimulants			
Amphetamines	Benzedrine Dexedrine Methedrine	Synthetic	Increased alertness, excitement, euphoria, decreased appetite; *large doses:* restlessness, rapid speech, irritability, insomnia, convulsions; *long-term abuse:* insomnia, excitability, delusions, hallucinations, psychosis

caffeine may produce irritability and insomnia, while nicotine has been implicated in premature birth, lung disease, and cancer). Both caffeine and nicotine are characterized by strong psychological dependence (the question of actual physical addiction is in dispute) and tolerance.

The abuse of **amphetamines** and other "pep pills" is an area of growing concern on the drug front. For a time they were prescribed in weight-control programs to curb appetite, but medical use is now closely restricted. Amphetamines can produce feelings of euphoria and mental alertness. But objective tests indicate that the user's intelligence and creativity have not really been stimulated as much as he believes they have. Amphetamines work by increasing the

Table 25-1 (Cont.)

TYPE OF DRUG	NAME	ORIGIN	EFFECTS
Antidepressants	Elavil Ritalin Tofranil	Synthetic	Relief of anxiety and depression; *large doses:* nausea, hypertension, weight loss, insomnia; *long-term abuse:* stupor, coma, convulsions, heart failure, damage to liver and white blood cells, death
Caffeine	Coffee Cola No-Doz Tea	Coffee bean Kola nut Synthetic Tea leaves	Increased alertness; *large doses:* restlessness, insomnia, upset stomach; *long-term abuse:* may contribute to heart disease and cancer
Cocaine	Cocaine	Coca leaves	Exhilaration, excitation, talkativeness, tremors; *large doses:* irritability, depression, psychosis; *long-term abuse:* damage to blood vessels, convulsions, psychosis
Nicotine	Cigarettes Cigars Pipes Snuff	Tobacco leaves	Relaxation, constriction of blood vessels; *large doses:* headache, loss of appetite, nausea; *long-term abuse:* impairment of breathing, heart and lung diseases, cancer, death
Psychedelics			
Cannabis	Hashish Marijuana THC (tetrahydrocannabinol)	Cannabis plant Cannabis plant Synthetic	Relaxation, breakdown of inhibitions, euphoria, alteration of perception and judgment; *large doses:* panic, stupor; *long-term abuse:* fatigue, psychosis
Hallucinogens	DMT LSD Mescaline Nutmeg Psilocybin Scopolamine STP	Synthetic Synthetic Cactus, peyote Nutmeg tree Psilocybe mushroom Henbane plant or synthetic Synthetic	Perceptual changes, excitation, hallucinations, exhilaration, panic; *large doses:* anxiety, exhaustion, vomiting, psychosis; *long-term abuse:* delusions, panic, psychosis

release of a natural brain stimulant, noradrenaline. Overexcitation may occur to the point where the user cannot sleep, and euphoria alternates with periods of deep despair. Tolerance to the drug builds up quickly, and long use of amphetamines may lead to hallucinations, delusions, and violent behavior.

At this writing, one of the most hotly debated drug topics is whether the use of **marijuana** should be legalized. Some claim that this psychedelic drug is no more dangerous than alcohol (hardly a recommendation) and not addictive (though it does produce some psychological dependence). Opponents of the drug argue that its use promotes a withdrawal from active participation in life. The effects of the drug (from relaxation through euphoria and alteration of perceptions to panic—a "bad

Figure 25–17 A patch of marijuana growing wild in Michigan.

trip"—and even psychosis) seem to vary among users, and even in the same user depending on the external circumstances. Very little about the drug's effects is yet known with certainty, but some studies with marijuana and its active principle, **tetrahydrocannabinol (THC),** indicate that THC and its conversion products persist in the brain for days after it is taken in.

Regardless of the status of marijuana, there is little doubt of the potential dangers of another "consciousness-expanding" drug, **LSD (lysergic acid diethylamide).** LSD is an extremely potent substance, a single ounce of which would provide enough doses for three hundred thousand people. Indeed, it has been estimated that if a few pounds of this drug were well distributed in the water supply of New York, it could affect the mental acuity of the entire population of the city. The LSD molecule is chemically very similar to the natural brain chemical serotonin, and it is believed to interfere in the normal workings of the brain cells. The drug produces hallucinations that may be extremely pleasant fantasies or frightening nightmares. An unfortunate by-product of the use of this and other psychedelic agents is the fact that during the "experience" the user may have his judgment distorted and do harm to himself or others. Serious and sometimes permanent after-effects have been noted: the user may experience recurrences of hallucinations weeks or months after he has stopped taking the drug, susceptible individuals may develop a full-blown psychosis, and there are some indications that the drug can cause chromosome breaks that might result in birth defects in the user's children.

In the coming years, it is likely that problems involving the use and abuse of drugs in our modern world will become increasingly widespread and complex.

QUESTIONS FOR REVIEW AND THOUGHT

1. What should be man's role in the biosphere: custodian of the living species or creative molder of the biosphere to enhance his own ends? Discuss.
2. Scientists argue that the human population explosion is a myth—that in a short time the steep increase in the world population will taper off and a stable equilibrium will be reached. Discuss.

3. Do you believe that the mass destruction of the coral reefs in the Pacific by the crown-of-thorns starfish will ultimately be a major factor in altering the ecological balance of the affected regions? Explain your answer.
4. Is pollution of our air, water, and soil an inevitable concomitant of a highly technological society? Explain.
5. If you possessed dictatorial powers over this nation, how would you go about minimizing the effects of pollution?
6. Is man nearing a point where pollution will produce irreversible damage to the biosphere? Discuss.
7. Describe the consequences of the elimination of all food additives from the marketplace.
8. Could the thalidomide tragedy have been avoided? How?
9. Since drugs must be tested on human volunteers before they are offered to the general public, how can we assure that these volunteers will be reasonably safe from harm?
10. The mood-changing drugs a person takes are his own business, and society has no right to place legal restrictions on them. Do you agree? Discuss.

CHAPTER **26**

Science and Society

Science and the Citizen

Despite the discordant cries to the contrary, we live in a democracy. We, the citizens, have the power to maintain or modify the framework of our government as we see fit. But along with the power we have a responsibility: to use this power wisely. Our society is multistructured, with a degree of complexity that often confounds the average citizen. This has become the age of the specialist. In virtually every area of our lives we are dependent on their services: the physician, the lawyer, the accountant, the plumber, the electrician. The fact that the structure of our society is changing at a headlong pace makes us all the more dependent upon these specialists.

Perhaps the most profound influence responsible for the constant change in our world is the continuing development of the sciences and technologies. Even the scientists themselves in a particular field are only vaguely aware of the progress in areas of science other than their own. Yet the citizen, through his representatives, must often act upon the results of scientific and technological advances and even help to shape their future development.

This seeming paradox has been recognized, and efforts are being made to resolve it. For a number of years the president of the United States had a science advisor, and more recently Congress has sought out scientists of various specialties to serve as advisors for committees. Since Congress is presumably responsive to the needs and wishes of the public, the man in the street, too, should be informed about the meaning of the scientific developments that are reshaping our lives. The long-standing scarcity of interpreters who can make science meaningful to the general public is gradually being remedied by an unprecedented proliferation of science yearbooks and expanded coverage in the newspapers, radio, and television.

Thus, an informed public, one which not only is aware of what is happening in the areas of science and technology but which can also assess the impact of these developments on our society, is an essential requirement for the optimum functioning of our society.

Figure 26–2 Scientists at work.

What better task than to embark upon an all-out war—a war against death. Currently we are spending about $2 billion a year in the various areas of biomedical research. If this sum were raised to $20 billion per year (not as much as we currently spend on tobacco and liquor) for a national crusade against death, we might be able to conquer or at least control death within a generation or two.

An ever-increasing number of scientists believe that this goal can be achieved, given sufficient knowledge. Such a program would be vast and complex, involving the efforts of hundreds of thousands of engineers, physicists, and chemists, as well as biologists and physicians. The central theme of this vast array of brainpower would be a concentration on basic research into the constituents of life and how they interact in the normal as well as the abnormal states. The excitement of such a program would doubtless spread beyond our national borders and infect the world.

The Future

Of all things, we can be certain of one: change. The world of tomorrow will be different from that of today. Each day brings new discoveries and applications. The cumulative effect of these events is having a profound influence upon our society and will ultimately change our lives. The trends that are now shaping our world are so numerous and conflicting that it is becoming increasingly difficult to predict the outcome of their interactions. Indeed, a whole new science has sprung up: the study of the future.

Although it is impossible to predict with certainty what the world will be like even a generation from now, certain trends are clear. If man succeeds in surviving into the twenty-first century, one can envision a society vastly different from that of today. The society of the future will be even more profoundly dependent upon science and technology. Diseases and death will probably be far less common. Even the "poor" of the future will enjoy a standard of living far beyond that attained by the middle class of today.

All this is predicated upon the assumption that man will learn to understand his world, to anticipate future problems, and to solve them wisely.

Figure 26–3 On the moon: the climax of a major national undertaking.

QUESTIONS FOR REVIEW AND THOUGHT

1. In view of the need for specialization in a modern technological society, how can a citizen gain awareness of the background of the varied problems with which he must deal in his role as a citizen?
2. If you were in charge of an information agency, how would you ensure that the public was kept informed not only of important scientific discoveries, but also of the implications that they might have for present and future society?
3. It has been argued that the cyclamates were removed from the market before sufficient scientific evidence was available to establish their status as harmful substances. Is this a valid argument? Explain.
4. Should standards of evidence routinely used in scientific research be applied to practical problems of everyday life? Discuss.
5. Some years ago, while the rate of government sponsorship of research was still increasing, some scientists warned that this trend might lead to governmental control over the direction of research. Have their fears been justified? Explain.
6. In 1969 the federal government actually cut back on funds for research on cancer and heart disease, as part of an effort to stem inflation. In 1973 a substantial fraction of funds appropriated by Congress was held back by the Office of Business Management for similar reasons. Can such actions be justified? Explain.
7. Is it really possible for man to conquer death? Discuss.
8. Outline a comprehensive program that would ensure the maximum use of the available manpower in a "crusade against death."
9. Are the events of the past a valid basis for predicting the future? Discuss.
10. List some potential technological and scientific discoveries that might profoundly alter our world.

Glossary

A

abiogenesis the origin of life from nonliving matter

abscission layer a layer of parenchyma cells growing across the base of a leaf stalk

acid a substance that (*a*) releases hydrogen ions when in a water solution, (*b*) has a pH lower than 7.0

active transport the movement of molecules and ions in a direction opposite to that of the concentration gradient

adaptation any genetically controlled characteristic that helps an organism, or the species to which it belongs, to survive and reproduce in the environment it inhabits

adenosine triphosphate (ATP) the primary energy-carrying molecule in cell metabolism

aerobic in or requiring the presence of oxygen

allele any of a number of alternative gene forms at a given chromosomal locus

anabolism the buildup of complex compounds from simpler substances; anabolic reactions frequently require energy to drive them

analogous structures structures of similar function and often of similar superficial appearance but differing in evolutionary origins

anaerobic without or requiring the absence of oxygen

angiosperm a seed plant which produces flowers for reproduction

annual a plant that lives one year

antheridium a male reproductive organ of a plant; produces sperm cells

antibody a specialized protein which inactivates an antigen

anticodon the unit of genetic coding complementary to the codon

antigen any foreign substance that penetrates the body

archegonium a female reproductive organ of a plant; produces egg cells

artery any thick-walled blood vessel that carries blood away from the heart

asexual reproduction a form of reproduction in which a single individual gives rise to new individuals

atom the smallest unit of an element; not divisible by ordinary chemical means

atomic number the number of protons of an element

autolysis self-digestion

autonomic nervous system a portion of the vertebrate nervous system, consisting of motor neurons under involuntary control that innervate internal organs

autosome any chromosome that is not a sex chromosome

autotroph an organism that can synthesize the organic molecules required for its vital activities from carbon dioxide, minerals, and water, using a source of energy (such as sunlight)

auxin one of a class of plant hormones that promote (in low concentrations) or inhibit (in high concentrations) cell elongation

axenic germfree; reared under sterile conditions

axon a neuron "process" which carries impulses away from the cell body

B

bacteriophage a virus that specifically attacks bacteria

basal metabolic rate (BMR) the minimum amount of energy required to maintain basic cellular and body functions

base a substance that (*a*) releases hydroxyl ions when in a water solution, (*b*) has a pH higher than 7.0

behavior the outwardly directed activities of an animal in response to a stimulus

biennial a plant that lives for two years

bilateral symmetry the property of having two similar sides, with definite upper and lower surfaces and definite anterior and posterior ends

bile a liver secretion which aids in the digestion and absorption of fats

binary fission reproduction by the splitting of a cell into two essentially equal parts by a nonmitotic process

binomial system of classification a system of classification in which all animals and plants are designated by two names: a generic name and a specific name

biome a group of natural communities characterized by distinctive climate and vegetation

biotic community a group of different but interrelated organisms

biotic potential the sum of the mechanisms tending to increase a population

blade the flattened portion of a leaf

blastocyst (blastula) an early embryonic stage composed of a hollow ball of cells

blood a type of connective tissue whose matrix is watery

bolus a ball of food formed in the mouth by the action of chewing and saliva

Brownian movement the erratic jiggling of tiny particles in solution

buffer a substance in solution able to adjust for an influx of either hydrogen or hydroxl ions through an appropriate chemical reaction

C

caecum a blind diverticulum at the anterior end of the large intestine

cambium a type of lateral meristem located along the sides of roots and stems that is responsible for the increase in plant girth

capillary a tiny blood vessel that receives blood from arteries and carries it to veins; all exchanges between blood and tissues occur across the walls of capillaries

carbohydrate any of a class of organic compounds composed of carbon, hydrogen, and oxygen in the proportion $C_yH_{2x}O_x$

carcinogen a cancer-causing chemical

cardiac pertaining to the heart

carnivore a "meat-eating" organism which subsists almost solely on animal flesh, a food source high in fats and proteins

cartilage a type of connective tissue whose matrix consists of a protein-polysaccharide mixture (chondrin)

Casparian strip a lignified, waterproofed thickening in the radial and end walls of endodermal plant cells

catabolism the breakdown or degradation of complex compounds to simpler components, with a release of stored chemical energy

catalyst a substance that accelerates chemical reaction without itself being permanently changed by the reaction

cell membrane (plasma membrane) the outer membrane of a cell

cell sap a fluid composed of sugars, salts, and organic materials such as pigments which is found in plant cell vacuoles

cell wall a tough, rigid boundary around the plasma membrane of a plant cell

cellulose a complex, highly insoluble polysaccharide that is a major constituent of most plant cell walls

central nervous system a portion of the nervous system containing association nerves and exerting control over the entire nervous system

centriole a cytoplasmic organelle, located outside the nucleus, that organizes the spindle during mitosis and meiosis

centromere the constricted region which holds identical twin chromatids together

cephalization a concentration of nerve tissue in the head region

chemosynthesis autotrophic synthesis of organic materials, energy for which is derived from inorganic molecules

chitin a polysaccharide found in the exoskeletons of insects, crabs, lobsters, and other arthropods

chlorophyll the green, light-sensitive pigment of plants necessary for photosynthesis

chromatid a single strand of a duplicated chromosome

chromatin network a diffuse network of thin fibers found within the nucleus of a nondividing cell

material exchange between fetus and mother in placental mammals

plankton mainly microscopic organisms that drift in the upper few hundred feet of the waters of the seas

plasma cell a specialized leukocyte that produces antibodies

plasma membrane the outer membrane of a cell

plasmodium a multinucleated, slimy, unprotected mass in the life cycle of a slime mold

plasmolysis shrinkage of a plant cell away from its wall when the cell is in a hypoosmotic medium

plastid a relatively large organelle in plant cells that functions in photosynthesis and/or storage of nutrients

platelet a specialized formed element in blood—a cell fragment that is a nonnucleated, disclike structure important to the blood-clotting mechanism

pleura the thin membrane surrounding the lungs

poikilothermic an organism incapable of self-regulation of body temperature, being fully dependent on environmental temperature

polar characterized by a separation of electrical charges

pollen the male gametophytes of a seed plant

polymer a large molecule comprised of long chains of smaller units

polyploid possessing more than two of each type of chromosome

polysaccharide many chemically linked units of simple sugars

positive feedback a mechanism in which a reaction is accelerated by its own products; a "vicious cycle"

predation the feeding of free-living organisms on other organisms

progesterone the hormone of pregnancy; a female sex hormone

prosthesis the replacement of a lost or malfunctioning body part with an artificial device

protein a class of organic compounds composed of long polypeptide chains

prothallus the small, heart-shaped, nonvascular gametophyte of a fern

proton a positively charged atomic particle

protoplasm a complex mixture which comprises all living cells

pseudopod a transitory cytoplasmic protrusion of an amoeboid cell

pulmonary circulation the circulation of blood from the right side of the heart to the lungs and back to the left side of the heart

pulse the contractions of artery walls, felt in parts of the body where arteries lie close to the surface

pyloric referring to the junction between the stomach and small intestine

R

race a population the members of which differ significantly from other populations of the same species in the frequency of one or more characteristics or traits

radial symmetry a type of symmetry in which the body parts are arranged regularly around a central axis

radioactivity the emission of radiations by certain isotopes as a result of internal changes in the nuclei of the atoms

recessive the allele that does not express itself in the presence of another allele of the same gene

recon the smallest subunit of the cistron that is interchangeable through genetic recombination

red blood cell see *erythrocyte*

reflex a functional unit of the nervous system, involving the receptor-to-effector pathway

regeneration the replacement of structures of parts that have been lost by accident or mutilation

renal referring to the kidney

reproductive tissue a tissue that produces gametes

respiratory quotient (R.Q.) the ratio of the volume of carbon dioxide eliminated from the body to the volume of molecular oxygen absorbed by the body

reticuloendothelial system (RES) the disease-fighting complex composed of the liver, spleen, thymus, bone marrow, and lymph nodes

rhizoid the rootlike structure of mosses and liverworts that performs a root function

rhizome an underground stem

rhodopsin the light-sensitive pigment in rod cells of the eye

ribonucleic acid (RNA) any of several types of nucleic acids in which the sugar component is the five-carbon ribose

ribosome a granularlike structure associated with the endoplasmic reticulum and involved in protein synthesis

RNA see *ribonucleic acid*

root that portion of the plant below ground that functions in anchorage and absorption

S

salivary glands fluid glands in the oral cavity of some animals

salt an electrically neutral combination of a metal and a nonmetal

saprophytic referring to a form of nutrition in which organic materials are derived from dead and decaying organic matter

sapwood the outer annual rings which function in transport

sclerenchyma a supportive plant tissue in which the cells are smaller and more uniformly thick-walled than collenchyma cells

secretion the process by which products of metabolism are released, usually from specialized glands, to perform useful functions at a site other than the site of origin

seed a structure consisting of the plant embryo, food reserve, and outer covering

semipermeable membrane a membrane permeable to some substances but not to others

senescence the biological process characterized by a gradual deterioration of the structures and functions of an organism

sexual reproduction a form of reproduction in which two individuals are required to produce new individuals

society an independent group of members of the same species

sol the fluid state of a colloidal dispersion

solid matter with a fixed shape and a volume that remains essentially constant even if great pressure is applied

solute the dissolved particles in a solution

solution a homogenous mixture of dissolved particles that will not settle out

solvation the insulation of ions with layers of water or some other polar solvent thus keeping them in solution

solvent the liquid medium in which a solute is dissolved

somatic pertaining to all body cells with the exception of germ cells

speciation the process of new species formation

species the largest unit of population within which effective gene flow occurs or can occur

sperm a male gamete

sphincter a ring of smooth muscle

spongin the horny, fibrous material composing the skeleton of sponges

spontaneous generation a school of thought that states that life arises spontaneously from soil or inanimate objects

stain a substance that imparts color to living and fixed tissues

stamen a male reproductive portion of a flower

starch a glucose polymer; the primary polysaccharide storage substance of vascular plants

stem that portion of the plant above ground, with the exception of the leaves and reproductive structures

stoma an opening, regulated by guard cells, in the epidermis of a leaf

substrate the substance upon which an enzyme acts; also the medium on which an organism lives

superior vena cava the vein which collects the blood from the head, neck, and arm regions

symbiosis the living together of two organisms in an intimate relationship

synapse a region of nonprotoplasmic intercellular material separating two neurons

systemic circulation the circulation of blood from the left side of the heart through the body in general and back to the right side of the heart

systole the phase of contraction of the compartments of the heart

T

taxis the orientation response of animals

taxonomy the science of classification of living things

tendon a type of connective tissue whose matrix is composed of collagen fibers and whose function is to connect one muscle to another muscle or to bones

territory a particular area defended by an individual against any other individual attempting to enter the area

testis a male sex organ which produces sperm

testosterone the male sex hormone

thallus a plant body exhibiting relatively little tissue differentiation and lacking true roots, stems, and leaves

theory hypothesis corroborated by a series of well-planned and -implemented experiments and found to be valid as a basis for further predictions

tissue a group or unit of cells cooperating to perform a uniform function

trace element any inorganic substance necessary only in minute quantities but having a key function in the metabolic reactions of the body

trachea a part of the vertebrate respiratory system running from the pharynx into the thoracic cavity

tracheal tubes a branched respiratory system of air ducts in terrestrial insects

tropism a growth response to a stimulus

turgid swollen with fluid

turgor pressure the increased pressure due to the influx of water which provides support to individual plant cells

U

unit membrane a biological constant common to the surfaces of all kinds of cells; composed of a sandwichlike layered structure of protein-lipid-protein

urea the nitrogen waste products of some vertebrates, formed in the liver by combination of ammonia and carbon dioxide

urethra a duct that leads from the bladder to the outside of the body by way of the penis in the male and via the urethral orifice in the female

V

vacuole a saclike cell structure which functions in temporary storage of materials or in the elimination of excess water and waste products from the cell

vein (blood) any blood vessel that carries blood toward the heart

vein (leaf) a vascular bundle in a leaf

vertebral column a segmented backbone that has replaced or surrounded the primitive notochord

vertebrate an animal with a backbone

villus a highly vascularized fingerlike process from the intestinal lining

virus a submicroscopic, noncellular, obligatorily parasitic particle that exhibits some properties normally associated with living organisms, including the ability to mutate and evolve

vitalism a school of thought which holds that living things are characterized by a "vital force"

vitamin an organic compound, necessary in small quantities, that a given organism cannot synthesize for itself and must consume in prefabricated form in the diet

W

white blood cell see *leukocyte*

X

X chromosome the female sex chromosome

xylem a vascular plant tissue that transports water and dissolved minerals upward through the plant body

Y

Y chromosome the male sex chromosome

yolk stored food material usually found in some types of ova

Z

zoospores a ciliated or flagellated plant spore

zygote a new cell formed by the fusion of an egg and sperm

Suggestions for Further Reading

The Physical World

Baker, J. J. W. and G. E. Allen, *Matter, Energy and Life: An Introduction for Biology Students*, Addison-Wesley, Reading, Mass., 1965.

Fruton, J. S., *Molecules and Life*, Wiley, New York, 1972.

Hardwick, E. R. and C. M. Knobler, *Chemistry, Man and Matter*, Ginn, Waltham, Mass., 1970.

Lenhoff, E. S., *Tools of Biology*, Macmillan, New York, 1966.

Scientific American, *Bio-Organic Chemistry*, Freeman, San Francisco, 1968.

Scientific American, *The Molecular Basis of Life*, Freeman, San Francisco, 1968.

White, E. H., *Chemical Background for the Biological Sciences*, 2nd Edition, Prentice-Hall, Englewood Cliffs, New Jersey, 1970.

The Cell

Coult, D. A., *Molecules and Cells*, Houghton Mifflin, Boston, 1966.

Loewy, A. G. and P. Siekevitz, *Cell Structure and Function*, 2nd Edition, Holt, Rinehart and Winston, New York, 1969.

Scientific American, *The Living Cell*, Freeman, San Francisco, 1965.

Stern, H. and D. L. Nanney, *The Biology of Cells*, Wiley, New York, 1965.

The Kingdoms of Life

Buchsbaum, R., *Animals Without Backbones*, 2nd Edition, University of Chicago Press, Chicago, 1948.

Evans, H. E., *Life on a Little-Known Planet*, Dutton, New York, 1968.

Ketchum, R. M., *The Secret Life of the Forest*, American Heritage Press, New York, 1970.

Milne, L. and M., *The Nature of Life*, Crown, New York, 1970.

Northen, H. and R., *Ingenious Kingdom*, Prentice-Hall, Englewood Cliffs, New Jersey, 1970.

Orr, R. T., *The Animal Kingdom*, Macmillan, New York, 1965.

Rosebury, T., *Life of Man*, Viking, New York, 1969.

Scheffer, V., *The Year of the Whale*, Scribner's, New York, 1969.

Scientific American, *Vertebrate Adaptations*, Freeman, San Francisco, 1969.

Wilson, C., et al., *Botany*, 5th Edition, Holt, Rinehart and Winston, New York, 1971.

Systems of the Body

Calder, N., *The Mind of Man*, Viking, New York, 1970.

Carr, D. E., *The Deadly Feast of Life*, Doubleday, New York, 1971.

Dröscher, V. B., *The Magic of the Senses*, Dutton, New York, 1969.

Guyton, A. C., *Function of the Human Body*, Saunders, Philadelphia, 1969.

Smith, Anthony, *The Body*, Walker, New York, 1968.

Trager, J., *Bellybook*, Grossman, New York, 1972.

Trager, J., *Foodbook*, Grossman, New York, 1970.

Wilentz, J. S., *The Senses of Man*, Crowell, New York, 1968.

Behavior

Alland, A., *The Human Imperative*, Columbia University Press, New York, 1972.

Ardrey, R., *The Social Contract*, Atheneum, New York, 1970.

Ardrey, R., *The Territorial Imperative*, Atheneum, New York, 1966.

Chauvin, R., *Animal Societies*, Hill and Wang, New York, 1968.

Lorenz, K., *King Solomon's Ring*, Crowell, New York, 1952.

Lorenz, K., *On Aggression*, Harcourt Brace Jovanovich, New York, 1966.
Luce, G. G., *Body Time*, Pantheon, New York, 1971.
Luce, G. G. and J. Segal, *Sleep*, Coward-McCann, New York, 1966.
Morris, D., *The Naked Ape*, McGraw-Hill, New York, 1967.
Tinbergen, N., *Curious Naturalists*, Basic Books, New York, 1958.
Scientific American, *Psychobiology*, Freeman, San Francisco, 1967.
Van Lawick-Goodall, J., *In the Shadow of Man*, Houghton Mifflin, Boston, 1971.
Von Frisch, K., *Dancing Bees: An Account of the Life and Senses of the Honey Bee*, Harcourt Brace Jovanovich, New York, 1965.

Genetics

Beadle, G. and M., *The Language of Life*, Doubleday, New York 1967.
Scheinfeld, A., *Your Heredity and Environment*, Lippincott, Philadelphia, 1965.
Scientific American, *Facets of Genetics*, Freeman, San Francisco, 1970.
Srb, A. M., R. D. Owen, and R. S. Edgar, *General Genetics*, Freeman, San Francisco, 1965.
Watson, J. D., *The Double Helix*, Athenum, New York, 1968.
Watson, J. D., *The Molecular Biology of the Gene*, 2nd Edition, W. A. Benjamin, New York, 1969.

Development

Barth, L. J., *Development: Selected Topics*, Addison-Wesley, Reading, Mass., 1964.
de Ropp, Robert, *Man Against Aging*, St. Martin's Press, New York, 1960.
Saunders, J. W. Jr., *Animal Morphogenesis*, Macmillan, New York, 1968.
Saunders, J. W. Jr., *Patterns and Principles of Animal Development*, Macmillan, New York, 1970.
Scientific American, *From Cell to Organism*, Freeman, San Francisco, 1967.
Strehler, B., *Time, Cells, and Aging*, Academic Press, New York, 1962.

Evolution

Ardrey, R., *African Genesis*, Atheneum, New York, 1961.
Calvin, M., *Chemical Evolution*, Oxford University Press, New York, 1969.
Eisley, L., *The Immense Journey*, Random House, New York, 1957.
Irvine, W., *Apes, Angels, and Victorians*, McGraw-Hill, New York, 1972.
Macvey, J. W., *Alone in the Universe?* Macmillan, New York, 1963.
McAlester, A. L., *The History of Life*, Prentice-Hall, Englewood Cliffs, New Jersey, 1968.
Scientific American, *Human Variations and Origins*, Freeman, San Francisco, 1967.
Wallace, B., *Chromosomes, Giant Molecules, and Evolution*, Norton, New York, 1966.

Frontiers of Biological Sciences

Behnke, J. A., *Challenging Biological Problems*, Oxford University Press, New York, 1972.
Bell, D., Editor, *Toward the Year 2000: Work in Progress*, Houghton Mifflin, Boston, 1968.
Chedd, G., *The New Biology*, Basic Books, New York 1972.
de Ropp, R. S., *The New Prometheans*, Delacorte, New York, 1972.
Handler, P., Editor, *Biology and the Future of Man*, Oxford University Press, New York, 1970.
Rosenfeld, A., *The Second Genesis*, Prentice-Hall, Englewood Cliffs, New Jersey, 1969.
Rorvik, D., *Brave New Baby*, Doubleday, New York, 1971.
Scientific American, *The Ocean*, Freeman, San Francisco, 1969.
Taylor, G. R., *The Biological Time Bomb*, World, New York, 1968.
Wallis, E. S., Editor, *Toward Century 21*, Basic Books, New York, 1970.
Warshofsky, F., *The 21st Century: The Control of Life*, Viking, New York, 1969.
Warshofsky, F., *The 21st Century: The New Age of Exploration*, Viking, New York, 1969.

Medicine

Burnet, M. and D. O. White, *Natural History of Infectious Disease*, 4th Edition, Cambridge University Press, Cambridge, 1972.
Cartwright, F. F., *Disease and History*, Crowell, New York, 1972.
De Kruif, P., *Microbe Hunters*, Harcourt Brace Jovanovich, New York, 1932.
Glemser, B., *Man Against Cancer*, Funk and Wagnalls, New York, 1969.
Longmore, D., *Spare-Part Surgery*, Doubleday, New York, 1968.
Moore, F. D., *Transplant*, Simon and Schuster, New York, 1972.
Zinsser, H., *Rats, Lice and History*, Little Brown, Boston, 1935.

Ecology

Benarde, M. A., *Our Precarious Habitat,* Rev. Edition, Norton, New York, 1973.

Benarde, M. A., *The Chemicals We Eat,* American Heritage, New York, 1971.

Brown, L. R., *Seeds of Change,* Praeger, New York, 1970.

Carson, R., *Silent Spring,* Houghton Mifflin, Boston, 1962.

Ehrlich, P. R. and A. H., *Population, Resources, Environment,* Freeman, San Francisco, 1970.

Ehrlich, P. R., *The Population Bomb,* Ballantine Books, New York, 1968.

Hardin, G., Editor, *Population, Evolution, and Birth Control,* 2nd Edition, Freeman, San Francisco, 1969.

Idyll, C. P., *The Sea Against Hunger,* Crowell, New York, 1970.

Maddox, J., *The Doomsday Syndrome,* McGraw-Hill, New York, 1972.

McHale, J., *The Ecological Context,* Braziller, New York, 1970.

Milne, L. and M., *Patterns of Survival,* Prentice-Hall, Englewood Cliffs, New Jersey, 1967.

Odum, E. P., *Ecology,* Holt, Rinehart and Winston, New York, 1963.

Perry, R., *The Unknown Ocean,* Taplinger, New York, 1972.

Scientific American, *The Biosphere,* Freeman, San Francisco, 1970.

Scientific American, *Energy and Power,* Freeman, San Francisco, 1971.

Scientific American, *Man and the Ecosphere,* Freeman, San Francisco, 1971.

Scientific American, *30 Steps to Biology,* Freeman, San Francisco, 1968.

Smith, A., *The Seasons,* Harcourt Brace Jovanovich, New York, 1970.

Taylor, G. R., *The Doomsday Book,* Thames and Hudson, London, 1970.

Young, L. B., Editor, *Population in Perspective,* Oxford University Press, New York, 1968.

Science and the Citizen

Chase, A., *The Biological Imperatives,* Holt, Rinehart and Winston, New York, 1971.

Commoner, B., *Science and Survival,* Viking, New York, 1966.

Dubos, R., *So Human an Animal,* Scribner's, New York, 1968.

Fuller, R. B., *Operating Manual for Spaceship Earth,* Southern Illinois University Press, Carbondale, Ill., 1969.

McHale, J., *The Future of the Future,* Braziller, New York, 1969.

Scientific American, *Science, Conflict and Society,* Freeman, San Francisco, 1969.

Toffler, A., *Future Shock,* Random House, New York, 1970.

Weinberg, A. M., *Reflections on Big Science,* MIT Press, Cambridge, Mass., 1967.

ILLUSTRATION CREDITS
(continued from copyright page)

Chapter 8 8-1 (a) Carl Zeiss; (b) Virus Laboratory, Univ. of California, Berkeley. 8-2 (a) National Institutes of Health; (b), (d) Pfizer Inc.; (c) Merck and Co.; (e), (f), (g) Virus Laboratory, Berkeley. 8-3 (a) Albert V. Crewe; (b) Carl Zeiss. 8-4 R. M. Cole, T. J. Popkin, and J. G. Tully, NIAID, NIH. 8-5 Society of American Bacteriologists. 8-6 (a), (b) Society of American Bacteriologists; (c) Carolina Biological Supply Co. 8-7 Philip Feinberg, Omikron. 8-8 (a) Carolina Biological Supply Co.; (b), (c), (f), (g) Eric V. Gravé; (d), (e) Philip L. Feinberg, Omikron. 8-11 (a), (b) Pfizer Inc.; (c) Leonard Lee Rue III; (d) Jack Dermid. 8-15 (a) Ross E. Hutchins; (b) Lynwood M. Chace, Omikron; (c) Jack Dermid. 8-22 Ross E. Hutchins. 8-23 Jack Dermid. 8-24 (a) Richard F. Trump; (b) Jack Dermid. 8-25 (a) Lynwood M. Chace, Omikron; (b) Richard F. Trump; (c) Jack Dermid. 8-26 (a) Philip L. Feinberg, Omikron; (b) Ripon; (c) Dr. Katherine L. Bick and Dr. George B. Chapman, Georgetown University; (d) USDA. 8-28 Jack Dermid.

Chapter 9 9-2 Douglas P. Wilson. 9-4 Miami Seaquarium. 9-5 (a) Eric V. Grave; (b) Richard F. Trump; (c) Douglas P. Wilson; (d) William M. Stephens, Omikron. 9-8 © Douglas P. Wilson. 9-9 Eric V. Grave. 9-14 National Library of Medicine, Bethesda. 9-16 Richard F. Trump. 9-17 (a), (b) Ross Hutchins; (c) W. M. Stephens, Omikron; (d) M. A. Wilson; (e) Douglas P. Wilson. 9-21 (a) Miami Seaquarium; (b), (d) Douglas P. Wilson; (c) Jack Dermid. 9-23 Carolina Biological Supply Co.; 9-24 (a), (b) Douglas P. Wilson; (c) Jack Dermid; (d) Leonard Lee Rue III. 9-25 (a), (b), (d) Jack Dermid; (c) Ross E. Hutchins. 9-26 Ross E. Hutchins. 9-27 (a) Richard F. Trump; (b) Jack Dermid; (c) Ross E. Hutchins. 9-28 (a) Lynwood M. Chace Omikron; (b), (c) Ross E. Hutchins. 9-29, 9-30, 9-31, 9-32 Ross E. Hutchins. 9-33 (a) Jack Dermid; (b) Lynwood M. Chace, Omikron; (c) Ross E. Hutchins. 9-34 (a), (c) Lynwood M. Chace, Omikron; (b) Leonard Lee Rue III. 9-35 (a) Richard F. Trump; (b) Ross E. Hutchins. 9-36 (a), (b) Ross E. Hutchins; (c) Lynwood M. Chace, Omikron. 9-37 Lynwood M. Chace, Omikron. 9-39 American Museum of Natural History. 9-40 R. Vishniac. 9-41 (a) Michigan Department of Natural Resources; (b), (c), (d), (e), (f), and (g) Miami Seaquarium. 9-43 (a) Leonard Lee Rue III; (b), (c), (d) Jack Dermid. 9-44 (a) Leonard Lee Rue III; (b), (c), (d) Jack Dermid. 9-45 Jack Dermid. 9-46 (a), (b), (c) Jack Dermid; (d), (e) Leonard Lee Rue III. 9-47 (a), (e) Jack Dermid; (b) Leonard Lee Rue III; (c), (d), (f) Lynwood M. Chace, Omikron. 9-48 (a) Australian News and Information Bureau; (b) Jack Dermid.

Chapter 10 10-1 (a) Jack Dermid; (b) Richard F. Trump. 10-6 Dr. L. K. Shumway, Washington State University. 10-7 Richard F. Trump. 10-13 E. Baitel, Omikron. 10-16 Roy Cummings, UNICEF. 10-17 Larry Jordan. 10-18 USDA. 10-20 Eric V. Grave. 10-25 American Museum of Natural History. 10-36 Eric V. Gravé.

Chapter 11 11-14 Philip Feinberg, Omikron. 11-15, 11-18 Eric V. Gravé. 11-19 (a) Dr. William T. Hall, Electro-Nucleonics Laboratories, Inc. 11-22 New York Times. 11-31 American Cancer Society.

Chapter 12 12-6 (a) Ross E. Hutchins; (b) L. M. Beidler; (c) Eric V. Gravé. 12-14 (a) Donald Silberberg, M.D. 12-17 (b) Gabriele Wunderlich. 12-21 Parke, Davis and Co. 12-24 Dr. Stanley Krippner, Maimonides Medical Center. 12-26 Philip Daly. 12-30 (a) Omikron; (b) Dr. John Canary, Georgetown Univ.; (c) Paul Almasy, WHO. 12-31 The Bettmann Archive. 12-34 (a), (c) Dr. Roberta Escamilla, Univ. of California, San Francisco; (b) Dr. Peter H. Forsham, Univ. of California, San Francisco. 12-36 Marineland of Florida. 12-41 Jack Dermid. 12-43 Richard F. Trump. 12-47 USDA. 12-49 S. M. Wittwer, Michigan State University. 12-51 John H. Gerard, National Audubon Society.

Chapter 13 13-1 United Press International and UPI Cablephoto. 13-3 Thomas McAvoy, Time/Life Picture Agency. 13-6 E. R. Willis. 13-9 W. D. Fitzwater, Sport Fisheries and Wildlife. 13-15 Pfizer Inc. 13-16 E. Baitel, Omikron. 13-17, 13-18 Wisconsin Primate Laboratory. 13-19 Adapted from T. C. Schneirla in K. Roeder, Ed., *Insect Physiology*, John Wiley, 1953. 13-22 Richard F. Trump. 13-24 Dr. Stanley Krippner, Maimonides Medical Center. 13-26 Leonard Lee Rue III. 13-28 (a) William M. Stephens, Omikron; (b) William H. Amos; (c) John H. Gerard, National Audubon Society. 13-29 Wisconsin Primate Laboratory. 13-30 Thomas McAvoy, Time/Life Picture Agency. 13-31 Lewis W. Walker, National Audubon Society. 13-34 (a) Lynwood M. Chace, Omikron; (b) USDA. 13-35 Marineland of Florida.

Chapter 14 14-1 Suzanne Szasz. 14-2 Granger Collection. 14-5 (a), 14-7 Carolina Biological Supply Co. 14-8 James Miller, M.D., Univ. of Virginia. 14-9 (a) Willowbrook State School; (b) Omikron; (c) President's Committee on Mental Retardation. 14-10 From *Embryology*, Revised and Enlarged Edition, by Lester George Barth, Copyright, 1949, 1953, by Holt, Rinehart and Winston, Inc. Adapted and reprinted by permission of Holt, Rinehart and Winston. 14-11 Leroy C. Stevens, Jackson Laboratory. 14-12 Carolina Biological Supply Co. 14-20 (a) A. and V. Silverstein. (b), (c), (d) United Nations. 14-25 Modified from Kimball, *Biology*, Addison-Wesley, Reading, Mass. Source for data: Burke's *Peerage*. 14-26 Omikron.

Chapter 15 15-1 Eric V. Gravé. 15-4 The Bettmann Archive. 15-6 USDA. 15-7 Carolina Biological Supply Co. 15-11 (a) Gabriele Wunderlich; (b) Dr. George B. Chapman, Georgetown University. 15-12, 15-14 Society of American Bacteriologists. 15-16, 15-17 Loewy/Siekevitz.

Chapter 16 16-2 (a) Fleischmann Lab; (b) Eric V. Gravé. 16-3 Douglas P. Wilson. 16-4 Grant Haist, Omikron; 16-5 Al Kaplan. 16-7 (a) Carolina Biological Supply Co. (b) Lynwood M. Chace, Omikron. 16-8, 16-9 (a) Lynwood M. Chace, Omikron; (b) Leonard Lee Rue III; (c) Jack Dermid. 16-14 Dr. Landrum B. Shettles, American Museum of Natural History. 16-17 (a) Betsy Cole. 16-20 Dr. Landrum B. Shettles. 16-22 Birth Atlas, Maternity Center Association. 16-23 (a) Roche, Omikron; (b) U.S. Forest Service. 16-25 (a) Ross E. Hutchins; (b) Richard F. Trump. 16-27 (a) Lynwood M. Chace, Omikron; (b) Leonard Lee Rue III; (c) L. M. Beidler. 16-31 Lynwood M. Chace, Omikron.

Chapter 17 17-1 (a) Jack Dermid; (b) Lynwood M. Chace, Omikron. 17-4 (a), (b) Lynwood M. Chace, Omikron; (c)

Jack Dermid. 17-5 (a) Richard F. Trump; (b) L. and M. Milne; (c) Carolina Biological Supply Co.; (d) Lynwood M. Chace, Omikron. 17-6 Y. and M. Vial, Jacana. 17-9 Lynwood M. Chace, Omikron. 17-11 Eric V. Gravé.

Chapter 18 18-1 Lick Observatory. 18-6 (a) Sidney Fox; (b) Abbott Laboratories.

Chapter 19 19-1 Culver Pictures. 19-3 (a) Lynwood M. Chace, Omikron; (b), (c) Leonard Lee Rue III. 19-4 Dunstan, Black Star. 19-6 (a) Richard Frear, U.S. Department of the Interior, National Park Service; (b) Jack Boucher, U.S. Department of the Interior, National Park Service; (c), (d) American Museum of Natural History. 19-8 U.S. Department of the Interior, National Park Service. 19-13 Field Museum of Natural History. 19-14 From E. O. Dodson, *A Textbook of Evolution,* Saunders, 1952, after Haeckel. 19-16 (a), (b) USDA; (c) Bureau of Animal Industry, USDA. 19-17 Australian News and Information Bureau. 19-18 Modified from G. G. Simpson, *Life,* Harcourt Brace, 1969. 19-20 (a), (b) Alvin Staffan, National Audubon Society; (c) W. J. Schoonmaker, National Audubon Society. 19-21 Omikron. 19-23 Keystone Press Agency. 19-27, 19-28, 19-30, 19-31, 19-32 American Museum of Natural History. 19-33 (a) Ross E. Hutchins, (b) Richard F. Trump. 19-34, 19-35 American Museum of Natural History. 19-36 Modified from G. G. Simpson, *Life,* Harcourt Brace, 1969. 19-37 (a) Ross E. Hutchins; (b) Leonard Lee Rue III; (c), (d) American Museum of Natural History. 19-38, 19-39, 19-40 American Museum of Natural History. 19-41 (a), (b) United Nations; (c) Department of Physical Anthropology, Smithsonian Institute; (d) Australian Information Service.

Chapter 20 20-1 UPI. 20-3 Oak Ridge National Laboratory. 20-7 UPI. 20-9 American Cancer Society. 20-10 Argonne National Laboratory. 20-11 (a) F. W. Hoffman, USAEC; (b) Brookhaven National Laboratory. 20-12 USDA. 20-13 Bechtel Corp. 20-14 Lawrence Radiation Laboratory.

Chapter 21 21-1 (a) University of California, San Diego; (b), (d) Bell Labs; (c) Stanford University; (e) American Cancer Society; (f) Pfizer Inc. 21-2 (e) Dr. Habermann, American Cancer Society. 21-3 Frederic Webster, Proceedings of the International Congress of Technology and Blindness. 21-4 (a) Carolina Biological Supply Co.; (b), (c) Ross E. Hutchins; (d) American Museum of Natural History. 21-5 Marineland of Florida. 21-7 (b) Dr. Stanley Krippner, Maimonides Medical Center. 21-8 (b) Lynwood M. Chace, Omikron. 21-10 Bell Telephone Laboratories. 21-11 (a) Elizabeth Wilcox; (b) Carlo Valenti, M.D. 21-12 IBM. 21-13 Fabri-Tek, Inc., Minneapolis, Minn. 21-15 Bell Telephone Laboratories. 21-16, 21-17 IBM.

Chapter 22 22-1 (a), (c) Philip L. Feinberg, Omikron; (b), (d) Eric V. Gravé; (e), (f) Pfizer Inc. 22-2 Lobund Laboratory, Univ. of Notre Dame. 22-3 Thomas D. Luckey. 22-5 NASA. 22-10 (a) Douglas P. Wilson, Marine Biological Library; (b), (d) Miami Seaquarium; (c) W. M. Stephens, Omikron. 22-11 U.S. Navy. 22-12, 22-13, 22-15, 22-16 NASA.

Chapter 23 23-1 Granger Collection. 23-3 Society of American Bacteriologists. 23-4 FDA. 23-6 Merck and Co., Inc. 23-7 USDA. 23-10 Society of American Bacteriologists. 23-11 (a) Eric V. Gravé; (b) Philip L. Feinberg, Omikron. 23-14 American Cancer Society. 23-15 Carlos Odiaga, M.D. 23-17 Elizabeth Wilcox. 23-18 (a) Pace Photo; (b) Pfizer Inc. 23-19 Encino Hospital.

Chapter 24 24-1 (a) Dick Frear, U.S. Department of the Interior, National Park Service; (b), (c) Lynwood M. Chace, Omikron; (d) George A. Grant, U.S. Department of the Interior, National Park Service; (e) Jack Dermid; (f) Leonard Lee Rue III. 24-2 Douglas P. Wilson, Marine Biological Laboratory. 24-3 American Museum of Natural History. 24-5 (a) Jack Dermid; (b) Leonard Lee Rue III. 24-6 Leonard Lee Rue III. 24-7 George Grant, U. S. Department of the Interior, National Park Service. 24-9 Leonard Lee Rue III. 24-10 NASA. 24-12 American Museum of Natural History. 24-13 Dr. W. G. Tomanek, U.S. Department of the Interior, National Park Service. 24-14, 24-15 Richard F. Trump. 24-16, 24-17, 24-18, 24-19 American Museum of Natural History. 24-23 Carolina Biological Supply Co. 24-29 Dade Thornton, National Audubon Society. 24-31 Lynwood M. Chace, Omikron. 24-32 Omikron. 24-33 Ross E. Hutchins.

Chapter 25 25-1 (a) ESSA; (b) Omikron; (c) E. Baitel, Omikron; (d) Gabriele Wunderlich; (e) U. S. Department of the Interior, Bureau of Mines; (f) H. Null, FAO. 25-3 U.S. Department of the Interior, Bureau of Sport Fisheries and Wildlife. 25-4 American Museum of Natural History. 25-5 W. F. Kubichek, National Audubon Society. 25-6 Charles N. Trainor. 25-8 (a) Omikron; (b) Ross E. Hutchins; (c) Richard F. Trump. 25-10 American Museum of Natural History. 25-11 George Komorowski, National Audubon Society. 25-12 Connecticut Yankee Atomic Power Co. 25-13 Omikron. 25-14 USDA. 25-15 Omikron. 25-16 Drug Enforcement Administration. 25-17 William Westrate, USDA-SCS.

Chapter 26 26-1 Official photograph, The White House. 26-2 (a), (b) Abbott Laboratories; (c) Los Alamos Scientific Laboratory; (d) U.S. Atomic Energy Commission. 26-3 NASA.

Index

Boldfaced items refer to illustrations and tables; *italicized* items are definitions or main discussions. (Important terms are also defined in the Glossary.)

A

A, blood type factor 408
aardvark **216**
abiogenesis 497
abscission layer *171*
absorption
  into roots 162
  into bloodstream 255
abstract reasoning 367
abyssal zone *648*
Acanthaster planci 676–677
Acartia clausi (copepod) **194, 197**
accretion *132*, **132**
acetaldehyde 294
acetylcholine 312, **312,** 313
acetylcholinesterase 312–313
  inhibitors 313
achene **470**
acids *28*, 35, 36
acorn, as fruit **470**
acorn worms 192–193, 202, **202**
acromegaly **327**
ACTH 322, **324, 326**
actin **330**
activators, in gene action 434
active immunity 618
active sites, of proteins 41
active transport 23, 83, **84,** 263
  in tubular secretion 264
adaptation, of man to
    environment *135–136*
adaptive radiation *525–526*, **527**
addiction, physical *689*
addition
  mutation 437–438
  reactions *32*
adenine *43*, **44,** 422, **435**
adenosine triphosphate: *see* ATP
adenovirus 142, **143**
adenyl cyclase 323–*324*, **326**
ADH 265
adipose tissue 110, **110,** 232–233
adrenal cortex 322, **325, 326**
adrenal hormones, interactions with sex
    hormones 327
adrenal medulla **325**
adrenaline 273, 321, 323, **325**
adrenals 322, **323**
adrenosterone **325**
adsorption chromatography 59
adventitious roots *164*, **165**
African violet 164
afterbirth *462*
agar *153*
agaves, leaves of 170
agene 685
aging *484–487*
Agnatha 203
air, ecological importance of 638–640
air sacs, of birds 289, **290**
airplane glue **690**
alanine 46, **437**
albinism, inheritance of 413
albumen sac, of chick embryo **461**
alcohol 253, 265, 689, **690**
alcoholic fermentation *294*
aldosterone **325**
alfalfa, roots of 165
algae 149, 400
  animallike 151
  blue-green 88, 151
  brown (kelp) **128,** 152
  brown, plastids of 99
  classification of 79, 127, **128,** 138,
    151–152
  colonial 152
  diatoms 152
  distinguished from mosses and
    liverworts 156
  flame-colored 152
  golden brown 151–152
  green **128,** 151, **152,** 154
  pigments in 226
  red 99, 152
  reproduction of 152, 441
  yellow-green 151
allantois *460*, **461**
alleles *406–407*
  multiple 413
  of fruit flies **408**
allelomorphs *406*
allergies *284*
Alligator mississippiensis **209,** 210
alligators 208–210, **209**
"all-or-none law" 310
Allosaurus 535, **536**
alpha globulins *279*
alpha-helix, of proteins **41**
alpha rays *27*
alternation of generations
  in ferns 158, **159**
  in mosses and liverworts 156, **157**
  in seed plants 159–160, *464*
alternation series, learning of 365
aluminum, as trace element 234
alveoli *290*, 291, **291,** 292, **292,** 460
Alvin, bathyscaph **600**
Amanita muscaria **153**
amantadine hydrochloride 620
amber, fossils in 507, **507**
Ambystoma tigrinum 207
American Q fever 144, **144**
amino acids *40*, **40,** 64, 101, 233–234,
    252–255, *432*, **433,** 436, **436**
  codons for 437
  deamination of 258
  essential 41, *233–234*
  separation of 58
amino group **40**
aminopeptidase **256**
ammonia 219, 234, 261
  excretory product 208, **260**
  in plants 258
ammonites 535, **535**
amnion *460*, **461**
amoeba **86,** 93, **128,** 147, 149
  digestion 239, **239**
  growth of 472
  locomotion of **87,** 333
  nucleus of 88, **88**
  phagocytosis and pinocytosis in 83, **85**
Amoeba proteus **103, 150**
amoebic dysentery 616–617
amoebocytes *240*
amoeboid movement 87, *333*
amphetamines *690–691*, **690**

Amphibia 206
amphibians **129**
  circulatory system of 270
  behavior of **357**
  embryonic development of 478, **513**
  evolution of 206, **207, 525**
  heart of **271**
  kidney of 261
  learning capacities of 357
  respiration of 289
  sex determination in 416
Amphineura 188
amphioxus **201,** 202
ampulla **308**
amygdala **318**
amyl nitrite **690**
amylase **250, 256**
  pancreatic **249,** 253
  salivary **249, 256**
amylopectins *37,* **38**
amyloses *37,* **38**
anabolism *47, 131,* 472
anaerobic pathway *294*
anal pore, of paramecium *239*
anal sphincters, of infants 257
analog computers *579*
analogous structures 509–510, **510, 511**
analytical balance 56
anaphase
  of meiosis 398, **398,** 400
  of mitosis 93, **95**
Anaximander 500
anemia *276*
anesthetic, use of, in surgery 108
Angiospermae *159*
angiosperms **128,** 159, 160, 167, *464,* 466
  life cycle of **163, 467**
  reproduction of 160–161, **163,** 465–467
angler fish **569**
Anguilla 376
Anguina tritici (wheat eelworm) 184
animal kingdom 79, 127
animal pole **458**
animal societies 382
animals 127, **128,** 400
  as heterotrophs 219
  chromosome numbers of **90**
  classification of 138–139
  fresh-water 86
  infection of by viruses 142
  number of species 126, 173
Annelida 184–186
annelids 184, 186
  locomotion of 186
  nephridia of 259, **259**
annual rhythms 369
annuals *167*
Anoplura 198
anoxia, and radiation 559
ant lion **195, 197,** 198
anteaters 216
  spiny 215
antennae, of arthropods 200
anterior pituitary **324, 326**
anther **467, 468**
antheridium 156
anthill **120**
Anthozoa 177
anthrax 199
anthropomorphism *349*
antibiotics 618, 688
antibodies 41, *283,* **284,** 286
anticoagulants 184, *280*
anticodons *436,* **436**
antigen-antibody reaction **283,** 284
antigens *283,* 286
antiserum 618
antitoxins *283*
ants **129, 195, 197,** 199
  as social insects 120, 383, 384
  communication of 355
  development of 200
  harvester **196, 197**
  solving of mazes 357, 365, **365**
  territorial behavior of 373
  use of "light compass" 350
  winged, nuptial flights of 380
Anura 207
anus 242, 243, 244, **251,** *255,* **448, 449**
aorta *272,* 275
apes **129, 216**
  abstract reasoning of 367
  curiosity of 364
  learning potential of 358
aphasia *320*
aphids 198, 200, 221, 442, **442**
apical dominance 342
apical meristem *114–115*
apoenzyme *48*
Apollo, and moon rock assay 592, **592,** 605
appendages 200, 202
appendectomy *255*
appendicitis *255*
appendix, human **251**
appetite, control center in hypothalamus 316
apple 468–469, **470**
Arachnida 197–198
arachnids **194, 197**
archaeology *65*
archegonium 156
Archeopteryx 536, **536**
Archeozoic era 529, **530**
arginine, in genetic code **437**
Argiope aurantica **194, 197**
Aristotle 124, 132, 500
armadillos **216,** 443
arsenic, as trace element 234
arteries **107,** 272–274, **274**
arterioles **263,** *274*
arthritis, rheumatoid 284
Arthropoda 193–200
arthropods **128,** 193–200
  compound eyes of **303**
  exoskeleton 331
  number of species 197
  reproduction of 200
artifacts *66*
"artificial" systems of classification 123
Artiodactyla **216**
Ascaris 183
ascending paper chromatography *60*
Ascomycetes **153,** 154
ascorbic acid 46
asexual reproduction 135, *439–443,* **440**
Asian flu epidemic 615
L-asparaginase 625
asparagine **437**
aspartic acid **437**
aspirin 122
association areas, of cerebral cortex 320
astigmatism *306,* **306**
astronauts 61, 601–603
astronomy *11*
ATA: *see* aurintricarboxylate
atherosclerosis *233,* 626–628
athlete's foot 154
atomic explosion **548**
atomic theory of matter 17
atoms 17, *24,* **24**
atomic numbers *25*
atomic weights *25*
ATP
  and bioluminescence 569
  as energy source 30, 79, 100, **100,** 332
ATP (*cont.*)
  formation of 227, **227, 228,** 294, **295**
  in protein synthesis 436
atria, of heart 269, 270, 272, 273
auditory nerve 307
Audubon, John James 674
aurintricarboxylate 622
Australoid race 545, **545**
Australopithecus 541–542, **542**
Autoanalyzer 632
autolysis, of cells *101*
automation 72
autonomic nervous system *320,* **321**
autoradiography 64
autosomes *414,* 416
autotrophic nutrition *138*
autotrophs *219*
auxin *341,* 342, **342**
Avery, O. T. 421
Aves 210–212
avidin **237**
axenic (germfree) conditions *591*
axolotl 478, **478**
axon *112,* **113,** *310,* **311**
Azoic era *528,* **530**
Azotobacter 652

B

B, blood type factor 408
bacilli 145, **145**
Bacillus pestis 609
backaches 337
backbone 202, 203
back swimmers 198
bacteria 65, 79, 114, 127, **128,** 145–146, 253, 294
  ammonifying 145
  and bioluminescence 568, **569**
  chemosynthetic autotrophic 219
  control of 282, 283
  decay 145, 146
  fermentative 146
  heterotrophic 219
  in feces 255
  in genetics research 428–430, 433–434
  infection by bacteriophage **431**
  locomotion of 101
  nitrogen fixing 36, 145, 652, **653**
  on skin 145, **145, 221**
  parasitic 146
  pathogenic 221, 255
  reproduction of 146, 431
  saprophytic 146, 220
  symbiotic 37, 661
bactericidal agents 619
bacteriology *11*
bacteriophages *142,* **143,** 615, 429, **432**
  in genetics research 428, 432, 437
  life cycle of 431–432, **431**
  T2 **143**
  T4 **143,** 437
bacteriostatic agents 619
balance 307, **308,** 334, 336
balance of nature, man and 137, 669–670
baldness, pattern **418,** 419
bananas 116
banyan trees 161
barbiturates **690**
bark 116
barnacles **194, 197,** 355
Barnard, Christiaan 628
basal metabolic rate (BMR) *229,* **324**
base analogs 428, **428, 429**
base pairing, in DNA 422
bases *28,* 35, 36
basic research 34, 696
Basidiomycetes **153,** 154

bats **129, 216**
  as pollinators 467–468
  flight of 214, 338
  homologous structures of 509
bathyal zone *648*
bathyspheres 599
Beagle, H. M. S., voyage of 501–502, **502,** 515, 516
beans 116, **225,** 233, 469
  germination of 469–470, **471**
bearded lady **328**
bears **216,** 336, 374
Beaumont, William 252, **252**
beavers **216**
Beckwith, Jonathan 433
bedbugs 198
bees 199, 200, **383**
  as pollinators 467–468
  as social insects 120, 383–384
  color discrimination in 357
  communication 355, 387
  eyes of 304
  flight of 338
  learning of 357
  navigation of 369
  sexual practices of 447
  time sense of 369
beetles **128,** 198
  as pollinators 467–468
  potato **196, 197**
  stag **196, 197**
beets 162, 167
behavior *347*–388
  components of, in animals **357**
  stereotyped 350–352
Beijerinck, Martinus W. 613
belted sandfish 674
bends, the *600*
benthos *648*
benzedrine **690**
Benzer, S. 424
beriberi **237**
berries 469, **470**
beta globulins *279*
beta rays *27*
bicarbonate ions 29, 292
bicuspids *247*
biennials *167*, *466*
big bang theory 493, **493**
bilateral symmetry *138–139,* **139**
bile *254*, 255, 258, 277
bile salts **256**
bill, of bird 244
binary code 580, **581**
binary fission *127,* 135, 144, 146, **148,** 149, *441*
binomial classification system *122,* 124
bioelectricity *570–571*
biofeedback conditioning *321*
biogenetic law *513*
biological clocks *367–370*
biological pest control 212, 561, *677–678*
biology *11*
  frontiers of *588–606*
  molecular *422*
bioluminescence 568–570, **569**
biomes 643–648, **644**
bionics *566–570*
biotic community *120, 655–666,* **657**
biotic potential *671–672*
biotin 236, 237
birds **129,** 210–212, **211,** 338
  as pollinators 467–468
  behavior of **357,** 373
  brain **310**
  brooding of eggs 349–350, **349**
  circulatory system of 212
  color vision of 305
birds (*cont.*)
  communication among 380, 386–387
  digestive system of 244, **245**
  embryonic development of 455, **456,** 461, **513**
  evolution of 211
  feeding of 120, 212
  flight of 289–290
  kidney of 261
  learning capacities of 357
  migration of 353, 374, **375**
  nests of 446, **446**
  reproduction of 211, 212, 380, 446
  respiratory systems of 289, **290**
  sex determination in 416
  songs of 353, 373, 389, 386–387
  urine of 261
birth *462–463,* **463**
"birth-control pills" 328, 453, **454**
birth rate *672*
birth, premature 460
bison 374, 674
bit, information unit in computers 580
bitter tastes 246, **246,** 302, 303
bivalves *189*
Black Death 609
bladder **259, 260,** *262,* **263,** 265, **448, 449**
blastocyst *455*
blastula **458**
blood 35, 36, 113, 213, 255, 264, 268, 269
blood clotting 114, 279–280, **280**
blood glucose level: *see* blood sugar level
blood, human 79, 275–280, **276**
  formed elements of *275,* **276**
  in sickle-cell anemia **277**
  pH of 29, 264
  plasma 35, *114,* 275, **276,** 279
  proteins of 61, 284
  types of **278**
blood platelets 114, **114**
blood pressure 274, 316, 321
blood sugar level 231
  in diabetes mellitus 264
  regulation of 232, **232,** 321, **324, 325**
blood supply, and heart disease 626–627
blood transfusions 284–286
blood types 285–286, **285,** 408
blood vessels 274–275, **274,** 299, 315
blood volume, regulation of by kidneys 263
blubber 39
bluegrass, root systems of 165–166
blue whale 460
body rhythms 367
body shape, streamlined 332, **332**
body temperature 212, 214, 316
Bohr atom *24*
Bohr, Niels 24
bologna, nitrite in 428–429
bolus *248,* **249,** 250, 251, 252
bonds
  covalent **30**
  disulfide **42**
  hydrogen **41,** 44
  ionic *29,* **29**
  unsaturated 33
bone marrow 280–281
  red *281*
  yellow *281*
bone tissue, human **107**
bones 36, 39, 47, 63, 101, 110, **110,** 234
bony fishes **204,** 205, 532, **532**
book lungs *289*
boron 234
Boston ivy 167
bower bird 380
Bowman's capsule **259,** *263,* **264**
Boysen-Jensen, P. 339, **340**
brachiopods *532,* **532**
Bragg spectrometer 70, **71**
brain 39, 47, 112, 265, 307, **314,** 485
  and reflex actions 314
  chordate 202
  electrostimulation of (ESB) 318
  evolution of **310**
  human 313, **314,** *315,* **316**
  maps of **319**
  memory molecules in 359
  neurons in 360
  of vertebrates 309
  olfactory bulb of 302
  primitive 309
  reflex centers in 315
  respiratory center in 293
  role in learning 356
  sensitivity to blood sugar level 231
  visual center of **304,** 305
  vomiting center of 253
brain stem 315, **316,** 318
brain waves 320, **320,** 488–489
"brainwashing" 372
branches, of trees 167
bread molds **128,** 154
  Neurospora crassa, in genetic studies 429
breast-feeding *462–463*
breathing 293, **294,** 315
brittle stars 191, 192, **192**
5-bromouracil 428, **428, 429**
bronchi *290,* **291,** 321
bronchioles *290,* **291,** 321
Bronze Age 16
Brontosaurus *535,* **536**
brooding of eggs 349–350, **349**
Brown, Robert 76
Brownian movement *21,* **22**
bubonic plague 199, 284, 610
budding 175, *441,* **441**
buffer system *29,* **29**
Buffon, G. L. 500
bulbs, of plants 166
bullfrog tadpole **447**
bumblebee **196, 197, 468**
"burping" of babies 251
Butler box 364–365, **364**
butter 146, 232
buttercup 159, **160**
butterflies **129,** 198, 478
  as pollinators **468**
  chemoreceptors of 303
  flight of 338
  grayling (*Eumenis semele*), response to stimulus 350
  head of, mouthparts **245**
  monarch **196, 197, 468**
  painted lady **479**
  sex determination in 416
  silver-spotted **196, 197**
  swallow-tailed, number of wingbeats 338

C

cactus 166, **166,** 170, **647**
  saguaro 640
cadmium (pollutant) 681
caecum *255*
caffeine 428, *689,* **691**
calcitonin **324**
calcium 262
  in blood clotting 280
  in bones 234
  in muscle contraction 36
  in shells 36
  in teeth 234
calcium apatite 247
calcium carbonate 110, 147

calcium phosphate 100, 247
Callitroga hominivorax 561, **561**
calorie *229*
Calorie *229*
Calvin, Melvin 227, 497
cambium *115*, 116
Cambrian period **530**
camels 216, 516–517, **517**
camouflage *662–663*
canary 357, **357**
cancer 284, 292, **293**, 477, *622–626*, **623**, **624**
canine teeth *247*, **247**, 248
Canis familiaris 122, **123**, **503**
Cannabis **691**, *692*
capillaries *272*, *255*, *263*, **263**, 269, *272*, 274, **274**, 275, 291
  lymphatic 281
permeability of **325**
capillary action 162, **164**
capsule *104*
carbaminohemoglobin *292*
carbohydrates 34, *37*, 99, 114, 162, 227, 231–232, 233, 279, 316–317
  as energy sources 39, 230
  in human diet 229, 231
  R. Q. of 229
carbon 25
  covalent bonds of **31**
  in organic compounds 30
carbon-14 63, 64, 65, 559, 560, **560**
carbon cycle 648–650, **649**
carbon dioxide (CO_2) 114, 227, 258
  combustion product 222, 650
  excretion of 261
  in atmosphere 291, 648, 650
  in blood 29
  in photosynthesis 63, 168, 221, 223
  in respiration 257–258, 292, 293, **295**, 296
carbon monoxide 292, 680
carbonic acid 29
Carboniferous period 158, **530**
carboxyl group **40**
carboxypeptidase 254, **256**
Carchesium **67**
carcinogens *429*, *624*
cardiac cycle *273*
cardiac muscle tissue 111–112, **112**
cardiac sphincter **249**, *251*, 252–253
cardinal, territorial display of **373**
carnation leaf **115**
cartilage 110, **110**
Carnivora **216**
carnivores 216, *220*, **220**
  R. Q. of 230
  teeth of 243, **244**, 248
carotenes 223, 226
carotenoids 58
carrageenan 153
Carrel, Alexis 473–474
carrier gas, in chromatography 61
carrots 162, 165, 167, 466
cartilage 110, **110**
cartilaginous fishes **204**, *205*
cast, and fossil formation 507, **508**
castings, of earthworm *184*, 243
cats **129**, 135, **216**, 220, 317, 353, 354, **390**, 452
  chromosome number of 395
  curiosity of 364
  learning capacities of 358, 365
  nematode parasites of 183
  Thorndike's learning experiments on 365–366
catabolism *47*, *131*, 258, 472
catalase 49
catalysts *32*, *40*
caterpillars 220, 353, 355, 478
cattle **90**, 183, **216**
Caucasoid race 544, **545**
caudal fins, of tadpoles *207*
cedars 159
cell body of neurons 112, **113**, 310
cell membranes *77*, 323
  and active transport **84**
  lipid content of 39
  of animal cells 79
  of freshwater organisms 259
  proteins in 39
cell metabolism 35
cell plate 96
cell sap *103*
cell theory *75–76*
cell walls 104
  of bacteria 146
  of plants 79, 127
  of rickettsias 144
cells *74–105*
  animal **78**, **96**
  dividing 77, **94**, **95**, 96
  growth of in culture 474
  liver, of newborn rat **78**
  of human cheek **77**
  of human trachea **77**
  plant **80**, **96**
cellular level of organization in Porifera 138
cellular respiration 100, 101, 294–296, **295**
cellulose *37*, 96, 99, 116, 255
  chemical structure **38**
  in cell wall 79, 104
cement, of teeth 247
Cenozoic era **530**, *537–539*
center of gravity 334, 336, 337
centipedes **128**, **194**, **197**, 198, 200
centrifugation 61
centrifuge 62, **62**
centrioles 93
centromere 91, 93
Centruroides sculpturatus **194**, **197**
cephalization *184*, 199, 203
Cephalochordata 202
Cephalopoda 189–191
cephalopods 189–191, **511**
cereal grasses, grains of 469
cerebellum 315, **316**, 318
cerebral cortex 316, **316**
  electrical stimulation of 358
  mapping of 319, **319**
  part of limbic system **318**
cerebrospinal fluid 313
cerebrum 315, **316**, 318, 319, 320
Cerosium, in prosthetic devices 630
cervix **449**, 451, 462, **463**
Cestoda, representatives of 180
Cetacea **216**
chaffinch, song of 353
chalk 147
chalones **325**
chambered nautilus 189
cheese 146
cheetah 334
Chelonia 208
Chelydra 209
chemical analysis 57
chemical nomenclature 122
chemical reactions *32–33*
chemical warfare agents 313
chemistry *11*
chemoreceptors *135*, *299*, 301, 302
chemotherapy *619*
chemotropism *339*
cherry 469
chewing 243, 248
chick
chick (*cont.*)
  embryo heart, growth of in Carrel's experiment 473–474
  embryonic development of 455, **461**, **513**
  experiments on dietary tryptophan restriction 485
chickadee, black-capped **211**
chicken pox 142
chickens 202, 212, **516**
chihuahua 90
Chilopoda 198
chimaeras 205
chimpanzee
  and evolution **540**
  behavior of 350, 386
  chromosome number of 395
  problem solving 358, 364, 366, **366**
chinch bugs 198
chipmunks 373
Chiroptera **216**
chitin *37*, 199, 476
chiton **187**, 188, **188**
Chlamydomonas *102*, **152**
chloral hydrate **690**
Chlorella 227, 603
chlorine 27, 29, **30**, 234
chlorophyll *79*, *98*, 116, 127, 151, 171, **224**
  absorption spectrum **226**
  in photosynthesis 168, 223, 226
  structure of 47, 225, **225**
chlorophyll a 58, 225, **225**, **226**, **227**
chlorophyll b 58, **227**
Chlorophyta 151, **151**
chloroplasts *98*, **98**, 99, *116*, *223*, **224**
  accessory pigments in 226
  in *Euglena* 348
  in photosynthesis 168, 223–224
  structure of **224**
cholecystokinin **256**
cholera 221
cholesterol **46**, 232, 627
choline 236
Chondrichthyes 205
chondrin 110
Chordata **129**, 200–216
chordates 200–216
  evolution of 193
  nervous system of 202
  notochord of **201**
  respiratory system of 202
chorio-allantoic membrane, of chick embryo **461**
chorion *460*, **461**
chorionic gonadotropin **325**
choroid, of human eye 305
chromatids *91*, 93
chromatin network *89*, 93, 96
chromatograms 60, **60**, 64
chromatography 47, 53, *58–61*
chromatophores, of cephalopods 191
chromoplasts *98–99*, **98**, 116
chromosome maps 92
chromosome puffs, of insect larvae 480–481
chromosome set, human 91, **401**
chromosome theory, of W. S. Sutton 406
chromosomes 66, *77*, *89–91*, 113, 439
  DNA in 420, **420**
  giant, of *Drosophila* 395
  homologous 91, 400
  human **91**, **401**, 414, **414**
  in meiosis **396**, **397**
  in mitosis **396**, **397**
  in nucleus *77–79*
  mutations of **401**, **427**
  number of **90**, 395, 426
Chrysophyta 151, **151**, 152

chyme **249,** *251,* 253, 254, 255, 257
chymotrypsin **249,** 254, **256**
cicadas 198, **476**
cigarettes **691**
cilia *101,* **102,** 109, 138, 147, 149, 332, **332,** 352
Ciliata 149, **150**
ciliates *101*
cinnamon fern **158**
circadian rhythms 329, *368–370*
circulation
  in unicellular organisms 268
  of blood, seen by Leeuwenhoek 65
circulatory systems *108, 118,* 268–281
  accessory organs of 280–281
  amphibian 270
  blood vessels of 274–275, **274**
  closed 269, **270**
  earthworm 268, **268**
  fish 269, **270**
  frog **270**
  grasshopper **269**
  human **268, 271,** 271–275, **274**
  open 269, **269**
  pulmonary *270,* **271**
  systemic *270,* **271**
  vertebrate 202, 269
cistron *424,* **424**
clams 36, **129,** 186, 187, **188,** 269, 289, 333, **334,** 368, 478
  giant bear's paw 189
  Lima **187**
classification of life, criteria for 137
classification systems, categories of 123, 124, 126
clavicle, bone marrow in 281
cleavage *455,* **456**
clinical death 488–489
clitoris *450–451,* **449**
cloaca **260,** *261*
clone *431*
cloning *596–597,* **597**
Clostridium 652
Clostridium botulinum 612, **612**
clots *114, 279,* **280,** 284
clover **90,** 171
  nodules in roots of 652, **652**
clumping, of blood *284,* **285**
Cnidaria 176–178
cnidoblasts *177, 241,* **241**
CO_2: *see* carbon dioxide
coal 158
cobalt, as trace element 234
cobalt-60 64, 561
cocaine **691**
cocci 145, **145**
cochlea *307,* **307, 308**
cockleburs 469
cockroaches **195, 197,** 198, 200, 220, 297, 350, 351, 355, 357
coconuts 469
cocoon **201**
codeine **690**
codons *436,* **436,** 437
coelacanths 205, 509
coelenterates *138,* 176–178, 240–241, 259, 268, 309
coelenteron *241*
cofactors *48*
coffee **691**
cola **691**
cold, perception of 301, **301**
colds, common 221, 614, 622
Coleoptera **196, 197,** 198
coleoptile, oat
  plant growth experiments 339–341
  test for auxin 341
collagen 110
collar cells *240*
collateral blood supply *626*
collecting tubule **259, 263**
collenchyma 116
colloidal dispersion *19,* 19–20, 87
colloidal particles 88, **88**
colon **249,** *255,* 257, **257,** 262
colonies
  of bacteria 430–431, **430**
  of phages 432
color-blindness 305, 416–417
color discrimination, in bees 357
color vision 212, 305
colostrum *463*
column chromatography 58, **59**
columnar epithelium **77,** 109
combination reactions *32*
commensalism 660, *661*
communication in animals *384–387*
community *108,* **119**
companion cells 117
comparative anatomy, and evolution *509–510*
comparative embryology, and evolution *512–514*
competition, among organisms 660, **661**
compounds *17, 30*
computers 71, 72, 318, *579–586*
  and education 584, **584**
  biomedical applications of 581–583
  in medicine 632–633
  uses in society 583–586
concentration gradient *23*
conditional reflex 314–315, 362, 363
conditioned stimulus *362*
conditioning 366
condom 453
conductive tissues, in plants 117, 138
cones, in retina *305*
cones, of gymnosperms 181, 161, 464, **464, 465**
conifers 158, 159, 171, 464
conjugation *443*
  in bacteria 146, *431*
  in *Paramecium* **148,** 149, **444**
connective tissue 107, 108, *109–111*
  dense 111
  loose 111
connector neurons *112,* **113,** *310,* 314
conservation *674–678*
constipation *257*
consumers
  primary *658,* **659**
  secondary *658,* **659**
  tertiary *658,* **659**
contraception *452*
contraceptives, types of 453
contractile proteins in locomotion 333
contractile vacuoles 101, 102, 147, 259
contractility **330**
control group *7*
control rods, in nuclear reactor 550
convergent evolution *526,* **528**
convolutions, of brain *319*
copepods *153,* **194, 197**
copper, as trace element 234
copperheads 210
copulation *451–453*
copy error, in DNA replication **428**
coral reef 178, 677
coral snakes 210
corals 177, **177,** 178, **638**
cork **75,** 76, 115, 116, 167
corn (*Zea mays*) 126, 159, **160,** 171, 231, **475**
  chromosome number of **90**
  flower 466, **466**
  fruit **470**
corn (*cont.*)
  protein, amino acid content in 234, 673
  roots 164–166
  seed 470, **471**
cornea 305, 629
corona radiata *452,* **451**
coronary circulation 626–628, **627**
corpus allatum *480,* **481**
corpus callosum **316**
corpus cardiacum **481**
corpus cavernosum **448**
corpus luteum *450,* **450, 454**
Correns, Karl 391
cortical sex hormones **325**
cortisol **325, 326**
cortisone, secreted by adrenal cortex **325**
cotton boll weevil **196, 197**
cotyledons 469, 470, **471**
courtship 212, 373, *379–381*
Cousteau, Jacques 601
coughing 291, 315
coumarin 687
covalent bonds *30,* **30, 31**
cowbird, song of 353
cowpox 283
Cowper's glands *448*
cows 37, 120, 220, 248, 383, 463
crabs 37, 197, 482
Crab nebula **492**
cranium, of infant during birth **463**
crayfish **194,** 197, **197,** 199
creativity, center of in brain 320
crepuscular organisms *368*
Cretaceous period **531**
cretinism **327**
Crick, Francis H. C. 43, 422, **422, 423,** 437
crickets 198, 373
cristae *100*
critical mass, for nuclear reaction *549–550*
crocodiles **129,** 208, 210
Crocodilia 208
Crocodilus americanus 210
Cro-Magnon man 543–544, **543, 544**
crop
  of bird 244, **245**
  of earthworm *243*
  of grasshopper 244, **245**
cross-pollination *467*
crosses 409–412
  dihybrid **411**
  predicting results of **410**
  test **412**
crossing-over **404,** *406,* **407**
croton, taproot of **165**
crowding, effects of 348–349
crown, of teeth *247,* **248**
crown-of-thorns starfish, depredations of *676–677,* **676**
Crustacea 197
crustaceans **194,** 197, 199, 200
cryogenics *576–577*
Cryptocercus punctulatus **150**
cryptoorchidism *447*
ctenosaur, false iguana **209**
cuboidal epithelium 109
cuckoo 353
cucumber 469
curiosity 364
Curtis, H. J. 486
cutaneous receptors 299–303
cutin *115–116*
cuttlebone, canary food supplement 188
cuttlefish 187, 188, 189
cyanide 40, 100, 101
Cyanophyta **151**
cycads **531, 535**
Cylert 360

cyclamates 686–687
cycles of matter and energy 648–655
cyclic AMP *323,* 324, **326**
cyclosis 87, **87,** *268*
cyclostomes 203
cyst 149
cysteine **437**
cytochrome oxidase 40
cytochrome system **295,** 296
cytokinesis *92*
cytologists *92*
cytoplasm *76,* 433, *77,* 82, 87, 104, 141
cytoplasmic inclusions *103*
cytopons 623, **623**
cytoscope **53**
cytosine *43,* **44,** 46, 422, **435**

D

2,4-D 341, **341**
Dacron, in prosthetic devices 630
damsel flies 198
dandelions 165, 469
Dart, Raymond 542
Darwin, Charles 184, 339, **340,** 500, *501–506,* **501,** 515
Darwin, Erasmus 500
Darwin, Francis 339, **340**
Dasyurus 658
data *6*
daughter chromosomes 93
dayflower leaf **115**
daylight, as triggering factor in migration 375
"day-neutral" plants 343, **344**
DDT, as a pollutant 681
death *487–489,* 696–698
death rate 672
DeBakey, Michael 631
decay bacteria *652*
decomposers *659*
decomposition reactions *32*
decompression, and deep-sea diving 600
deep-sea diving 599–601, **600, 601**
deer **129, 216,** 373, 374, 382
defecation 257, 258
degenerative diseases 626
dehydroepiandrosterone **325**
dehydrogenase 48–49
deletion, type of chromosome mutation **427,** 437–438
Demerol **690**
Democritus 17
demography *670–672*
dendrites *112,* **113,** *310,* **311**
denitrifying bacteria *652*
density gradient, technique of centrifugation 62
dental caries *248*
Dentalium 188
dentin, of teeth *247,* **248**
deoxycorticosterone **325**
deoxyribonucleic acid *43*
  see also DNA
deoxyribose *43,* **44,** 422, **422,** 435
depressants (drugs) *688,* **689–690**
Dermaptera 198
dermis *111*
Dermochelys 209
desalting mechanisms, of bony fishes 86
descending paper chromatography *60*
desert life, adaptation to 137
deserts *646–647,* **647**
Desmacidon fruticosum **174**
detergents, as pollutants 684, **684**
deuterium 27, 64
Deuteromycetes 154
Devonian period **530**
Dewar flask 574
dexedrine **690**
diabetes 327, 453, 593, 594
  mellitus 41; urine test for 264
dialysis *23, 83*
diapedesis *277,* **278**
Diapheroma femorata 663, **663**
diaphragm *213,* 251, **251,** 253, 293, **294**
diarrhea *257*
diastole *273*
diatoms 99, **128,** 153
dicots **128,** 171
  and monocots, distinctions between **161**
  herbaceous stem **167**
  seeds *469,* **471**
  see also dicotyledons
dicotyledons *159,* **160,** 167
dictyosomes: *see* Golgi complex
Didelphis virginiana 215
diet, human 231–238, 627
differential reproduction *525*
differentially permeable membranes of cells 82
differentiation 173, 174, 456
diffraction grating **70**
diffraction pattern 70, **70**
diffusion **22,** *23,* 82, 162, **164,** 259
digestion 103, *238–257*
  extracellular *239*
  intracellular *239*
digestive juices **256,** 321
digestive system 118
  human 243, *244–257,* **251**
  of amoeba 239
  of bird 244, **245**
  of earthworm 242–243, **242**
  of grasshopper 244, **245**
  of hydra 241
  of insects 243–244
  of mouse **242**
  of paramecium **240**
  of planaria 241–242
  of sponges 240
  of starfish 243
digestive tract, length of 244, 246
digital computers *579*
dihybrid cross *394,* **394, 411**
dimethyl sulfoxide (DSMO) 577, 688
dinoflagellates **128,** 149, 151, 153
dinosaurs 208
  age of 212–213
diphtheria 284
Diplococcus pneumoniae **588**
Diplodocus 535, **536**
diploid number 400, *455*
  in meiosis 396
  in mitosis 396, 397
  in somatic cells 420
Diplopoda 198
Diptera **196, 197,** 198
disaccharides, hydrolysis of **250,** 253
discrimination, in conditioned responses 362
diseases
  bacterial *609–612*
  body defenses against *281–284*
  degenerative 626
  fungus 615–616
  infectious 618–622
  parasitic *616–618*
  viral *612–615*
distal convoluted tubule **264**
Dittmer, H. J. 166
diurnal animals *368*
divergent evolution *526*
dizziness, symptom of motion sickness 308
DMSO: *see* dimethyl sulfoxide
DNA 58, 62, 91, 92, 254, 323, 432–435
  genetic function of 46, 420
DNA (*cont.*)
  histones associated with 434
  in bacteria 145
  in bacteriophages 431, **431**
  in chromosomes 89, 395
  in rickettsias 144
  in viruses 141, 142
  information-storing capacity 424–425
  molecular weight of 43
  radiation effects on 428
  replication of 93, 423, *434–435,* **435**
  structure 44, **45,** 422, **422–423**
  test-tube synthesis of 594, **595**
  transcription of **433**
  Watson-Crick model of 422, **422–423,** 435, **435**
DNA bases, possible combinations and permutations 424
DNA nucleotide, structure of **422**
DNase (deoxyribonuclease) **256**
dogs 122, **123,** 127, **129, 216,** 220, 306, 354, 358, 452
  chromosome number of **90,** 395
  domestic 502, **503**
  English setter **213**
  learning capacities of 358
  life span of 484
  nematode parasites of 183
  Pavlov's experiments on 315, 362
dolphins 214, **216,** 333, **386,** 509
  bottle-nosed (*Tursiops truncatus*) 386
  communication of 386
  intelligence of 358
Domagk, Gerhard 619
dominance, among animals 382–383
dominance, in genetics *393*
  incomplete 408, **409**
donkey, interspecies mating **126**
dormancy, of angiosperms 467
dorsal surface *180*
double displacement reactions *32,* 33
double helix 422, **422–423, 435**
Double Helix, The, by James D. Watson 43
Dover, white cliffs of 147
Down's syndrome 401–402, **401**
dragonflies **195, 197,** 198, 220, 480
dreaming 372
drives 363, 364
Drosophila melanogaster (fruit fly)
  chromosomes of **396,** 416
  giant chromosomes of 395
  in genetic studies 395, 403, **403,** 405–406, 429
  linkage groups of 406
  mutations of **396,** 426, **426**
  salivary gland preparation **396**
  sex determination in 416
  sex-linked genes in 417
  white-eye gene of 416, **417**
drugs 599, *687–692*
drupe **470**
duckbilled platypus 215, **215**
ducks 202, 212, 261
  and imprinting 378
  instinctive behavior of 354, **355,** 380
  mallard **211**
"duct glands": *see* exocrine glands
"ductless glands": *see* endocrine glands
Dugesia tigrina 179, **180**
dugongs **216**
duodenal mucosa **324**
duodenum **249,** *253,* 254, **256, 323,** 328
Dutrochet, René 75
dysentery amoeba 147

E

ear 306–308, **307**
  and equilibrium 299, 307–308, **308**
  and sound reception 306–307

eardrum *307*, **307**
earthworms 124, **128**, 139, **185**, 186, 242, **242**, 444
circulatory system of 268, **268**
digestive system of 243
effects on soil 120, 243
excretory system of **259**
learning in 356, **356**
locomotion of **185**, 334
muscles of **185**
regeneration of 186
reproduction of 186, **186**
respiration of 289
earwigs 198
Easter lily **521**
eating 99
Eaton, M. D. 142
echysone *480*, **481**
echidna **517**
Echinodermata 191–193
echinoderms **129**, *191*, **192**, 193
larvae of 192–193, 478
reproduction of 193
skeleton of 193
water vascular system 192, **193**
ecological isolation *520*
ecological niches *658–662*
ecological pyramids 664–666, **666**
ecology *637–666*
ecosystem *658–662*
ectoderm, of embryo *456*
ectoplasm *87*
Edelman, Gerald **284**
Edentata **216**
EEG (electroencephalogram) **320**, 371, **371**, **572**
eels, migration of 376–377, **376**
effector *314*
egestion **240**, *258*
egg-laying, of hens 351
eggs *76*, *439*, 444, 466
bird 211, 212, 445, 455, **456**
brooding of, by herring gull **349**
fertilization by sperm 414, **452**
formation of 400, **402**, 403, 449
haploid chromosome number of 455
human **451**
in diet 232, 233
in sexual reproduction 135, 443
insect **479**
mammals, 214, 455, **456**
reptile 208
sea urchin 455
Ehrlich, Paul 619
Einstein, Albert 5
ejaculation *451*
EKG (electrocardiogram) 572
elasmobranchs 205
elastin 111
Elavil **691**
electric catfish 571
electric eel 570–571, **570**
electric fish 571
electrocardiogram **572**
electrocardiography *571*
electrochemical reactions 311
electroencephalogram **320**, **371**, **572**
electroencephalography *370*, 571
electrolyte *61*
electromagnetic radiation 27
electronics 566–586, 631–633
electrons *24*
electrophoresis 53, 58, 61, **61**
elements *17*
elephantiasis 183, **183**
elephants **129**, 202, **216**, 220, 248, 334
African **213**
gestation period of 456
herd organization of 382
elms 166
Elodea 103
elution *59*
elvers *376*, **376**
embryo
human 457, **460**, **461**, 474
implantation of 451, 457
of blue whale 460
plant 138, **467**, 468, 469
embryology 193
embryonic development *453–462*, **458**, **459**
Embryophyta *138*
emotional deprivation 377
emotional development 378
emphysema *292*, **292**
emulsion *20*
enamel, of teeth *247*, 248, **248**
encephalitis 199
Enders, J. F. 614
endocardium **108**
endocrine glands *322*, **323**, 448–449
endocrine system 118, *322–329*
endoderm, of embryo *456*
endogenous rhythms 367–368, 370
endoplasm *87*
endoplasmic reticulum (ER) *79*, 82, 96, **97**, 99
rough 82, 97
smooth 82
endoskeleton *193*, 202, *331*
endosperm 466, **467**, 468, 469, **471**
endotoxins *612*
energy *17*, 17–19, 30, 99, 211
in cellular respiration 294
in foods 131, 230
energy cycle 655, **655**
energy metabolism 294
engram *360*, **361**
enterokinase **256**
entropy 655
environment
adaptation of life forms to **136**
man and 669–670
environmental resistance *671*
enzymes *40*, *48–51*
and metal ions 36
digestive 250, **250**, 252, **256**
temperature sensitivity of 32
Eocene epoch **531**, 538
Eohippus **538**
Ephemeroptera 198
epicotyl *469*, **471**
epidermis of plants 115, **115**
epididymis *448*, **448**, **449**
epiglottis *250*, 291
epilepsy 319, 320, 358
epithelial tissue 107, *108*, 109, **254**
equatorial plane of cell 93
Equus **538**
erythroblastosis fetalis *286*
erythrocytes *276*, **276**, **278**, 284
effect of cigarette smoke on 292
production of 281
ESB (electrical stimulation of the brain) *318*
Escherichia coli 48, *142*, **143**, 220, 429
in genetics research 429, **429**, 514–515
esophagus 250–252, *291*
human **249**, **251**
of earthworm 243
of grasshopper 244
Essay on the Principle of Population, by Thomas Malthus 501, 671
estrogens **325**, *450*, **450**, **454**
estrous cycles 329, 369, 452
ethology *347*
ethyl alcohol 294
ethylurethane 428
etiolation *224*
Ettinger, Robert 577
eucaryotes *434*
Euglena 79, 101, 127, **128**, 149, 151, 303, 332, 348, **348**
Euglenophyta **151**
Eumycophyta 153–154
euphotic zone *648*
Euplotes **150**
Eutheria 214, 215
eutrophication *684*
Evans, Ronald E., astronaut **602**
evaporation **164**
evolution 124, 173, *500–546*
convergent *526*, **528**
divergent *526*
excretion *257–265*
in man 261–265
in plants 103, 258–259
excretory systems 118
human **261**
vertebrate **260**
exobiology *603–606*
exocrine glands 322
exoskeletons 199, *331*, 334, 476, **476**
exotoxins *612*
experimental group *7*
exteroceptors *299*
extinction, of conditioned reflex *362*
extinction, of species *674–675*
extracts *57*
extraembryonic membranes 460, **461**
eye color, human, inheritance of 413
eyes 39–40, 82, 314
compound, of arthropods 200
human 304–306, **304**
of birds 212
of cephalopod **511**
of fish **511**
of insect **511**

F

facilitation *361–362*
Fahrenheit scale *16*, **16**
see also temperature scale
fainting, as protective mechanism 275
fallen arches 337
fallopian tubes *450*, 451, 453
family, in classification of organisms 124
family life, patterns of 380
farsightedness: *see* hypermetropia
Fasciola hepatica (sheep liver fluke) 180
fat depots **232**
fatigue
of muscles 112
of olfactory receptors 302
fats *37–39*, 111, 316–317, 460
digestion of 238–239, **238**
energy content of 39, 230
in diet 229, 232
R. Q. of 229
saturated 37–39, **39**
unsaturated 37–39
fatty acids 37–39, 101, 232, 253, 255
FDA *685–687*
fear 317, 364
feathers, of birds 211
feces **249**, *255*, 256, 258
feedback mechanisms 48, 296
in hormone action 325, **326**
negative *48*, **50**
positive *48*, **50**
feeding currents 149, 175, 240, 352
feet, soles of, and equilibrium 299
Felicinae 158
femur, bone marrow in 281
ferns 158
climbing **128**

ferns (*cont.*)
  common **158**
  lady **128**
  life cycle of 158, **159**
  sporulation in 442
ferroheme **225**
ferrous salts 219
fertile material, and nuclear fission *549*
fertility, enhancement of 452–453
fertilization **396, 397,** *445–447*, **452**
  and meiosis 400
  external 445
  human *451–453*
  in angiosperms 466–467
  internal *445*
fetus 110, 457, **460, 461,** 462, **463**
Feulgen's reagent 58, *420*, **420**
fibrin *280*
fibrinogen *279*, 280, **280**
fibroblast cells **420**
fibrous roots 164, **165**
fiddler crabs 368, 370, **370**, 373
filament, of flower **468**
filaments, of algal cells *152*
filarial worms 183
filariasis 199
finches, and Darwin 505–506, **505**
fins 203, 205
fireflies 198, 568, **569**
firs 159, **160**
fish farming 683
fish-protein flour 673
fishes 127, **129**, 137, 243, 305, **332**, 416, 446, 484
  behavior of **357**, 382
  brain of **310**
  chemoreceptors of 303
  circulatory system of 269, **270**, **271**
  embryonic development of **513**
  excretory system of **260**
  eye of **511**
  eyeless **512**
  flying 205, 338
  homeostasis in **86**
  learning capacities of 357
  luminous 205
  orientation of 350–351, **351**
  respiratory system of 288–289, **288**
  teeth of 248
fission bomb *549*, **551**
fissionable material *549*
fistula *252*, **252**
flagella *79*, 101, **102**, 138, 146, 147, 152, 332, **332**
Flagellata 149, **150**
flame cells *259*
flatworms **128**, 138, 179–182, **180**, 268, 287, 303, 445
  digestive systems of 251–242
  experiments on memory and learning 356, 358–359
  flame cells of 259
  nervous system of 309, **309**
  regeneration of 174, 181–182, **181**, 442, 482
flax, retting of 146
fleas 144, 198, 199, **221**, 337, **337**
Fleming, Alexander 4–5
flight 210, 337–338, **337**, **338**
flies **129, 196, 197**, 198–200
flora, bacterial, in human body 145
florigen *343*
flounder 205, 663, **663**
flowering, photoperiodism 466
flowering hormone 343
flowers 116, 118, **118**, 161, **163**, *467–468*, **468**
  adaptation to pollinators 468
flowers (*cont.*)
  imperfect *466*, **466**
  perfect *466*, **466**
flukes *179*
fluorine 234
folic acid 236, **237**
follicle, ovarian *450*, **450**
food 28, *218*, 229, 598
  and population dynamics 673–674
  production of, in plants 116, 168
  use of in living organisms 130–131
food additives 685–687, **685**
food chains 152–153, 197, *664–666*
Food, Drug, and Cosmetic Act of 1958 686
food producer *658*, **659**
food storage, in plants 162, 166
food vacuoles 102, *239*
food web *664–666*
foraminiferans, outer coverings of 147
forgetting 362, 363
formed elements, of blood *114*
fossil fuels 562
fossils 506–509, **507**, **508**
four-o'clock, incomplete dominance in **409**
Fox, Sidney 498, **499**
foxes **90**, 380
fragmentation 152, 175, 441
frankfurters, nitrite in 428–429
fraternal twins *443*
fresh-water biome 647, **647**
Frisch, Karl von 347, 386–387
frogs **129**, 132, 202, 207–208, 220, 289, 337, 443, 483
  brain of **310**
  bullfrog **207**; tadpole **477**
  embryonic development of 455, **456 458–459**
  green **520**
  green tree **207**
  leopard **520**
  life cycle of **206**
  mating of 445, **445**
  metamorphosis of 327, 477–478, **477**
  wood **477**, **520**
fronds, of ferns *158*
frontal lobes of brain 320
fruit fly (*Drosophila melanogaster*) 113, **404**
  chromosome number of **90**
  in genetic studies 395, 403–406, **403**, **408**
  eye of **303**
  Y chromosome of 416
  see also Drosophila melanogaster
fruits 118, *161*, *467–470*, **470**
FSH **324**, **450**, **454**
FSH-RF **454**
Fucus 152
fungi 219, 220, 221
  algal 154
  bracket, 154
  classification of 79, 127,**128**, 138
  cup 154
  gill 154
  reproduction of 154
  sac 154
  sporulation in 442
Fungi Imperfecti 154
fur seals 373, 374, 380
future, prediction of 698

G

Gabor, Dennis 574, 575
β-galactosidase **434**
Galápagos Islands, and Darwin 505, 516
Galápagos tortoise 127, **473**
gall *184*
gall bladder, human **251**, 254, **256**
gametes *113*, 135, 152, 156, 392, *439*
  human **113**
  in meiosis **399**
gametogenesis **402**
gametophytes *464*
  of angiosperms **467**
  of ferns 158, **159**
  of gymnosperms **465**
  of mosses and liverworts *156*, **157**
gamma globulins *279*
gamma rays *27*, 428
ganglion *309*
gas chromatography 58, 61
gas exchange, respiratory
  in fish **288**, 289
  in humans 273, 291–292
  in insects 289
gases 35, 36, 114
gastric juice 28, **249**, 252, 253, **324**
gastrin **256**, 322, **324**
Gastropods 188
gastrula, stage of development *455*
geese 212, 354, **355**
Geiger counter 64
gel, in colloidal dispersion 87
gelation *87*
gemmule **176**
gene migration *524–525*
generic name 124
genes *91*, 92, *395*, 406–407, 424, **424**
  activation by hormones 481
  recessive 416
  regulation of 89, 433–434
  Rex, in mice 411
  sex-linked, in *Drosophila* 417
  test-tube synthesis of 438, 595–596, **595**
  white-eye, of *Drosophila* 416, **417**
genetic code *435–438*, **437**
genetic determination of life span, theory of 485
genetic drift *522*, **523**
genetic engineering 438, *593–597*
genetic isolation *520*
genetic recombination 431, 432
genetics *11*, *91*, 139, *390*
  and laws of probability 412
  microbial *429–432*
  molecular *420–438*
genus 124
genotype *408*
geographic isolation *518*
geology *11*
geòtaxis *351*
geotropism *339*, **342**
gerbils 214
German measles 457
germ theory of disease, 610–611
germ-free chamber 590–592, **590**, **591**, **592**
germ-free research *589–592*
germination *160*, *469–470*, 471
gestation period *456*
ghost crab **194**, **197**
gibberellins 342, **343**
gibbon 540, **540**
Gila monster 210
gill capillaries 288, **288**
gill pouches 202, 457
gill slits *288*
gills 205, 269, 288, **288**, 289
gingkos **535**
giraffes **216**
gizzard
  of birds 244, **245**
  of earthworm *243*
  of grasshopper 244, **245**
  of insects 244

glands 320, *322*, **323**
Glaser, Donald 565
gliding *338*
glomerulus **259**, *263*, **263**
glucagon **232**, **324**
glucocorticoids **325**
gluconeogenesis, and cancer 625
glucose *37*
  absorption into bloodstream 253
  as energy source 39, 231, 296
  digestion product **250**
  in glycolysis 294
  in photosynthesis 221, 227, **228**
  metabolism of **324**, **325**, 327
glue 153
glutamic acid **437**, 438
glutamine **437**
glyceraldehyde phosphate **228**
glycerol 37, 232, 253, 255
glycine **437**
glycogen 37, 231, **324**, **325**
glycolysis *294*, **295**
gnotobiology *589–593*
goats 282, 283
Gödel, Kurt 9
goiter **327**
goldfish 202, 360
Golgi, Camillo 99
Golgi bodies 87, 96, 112
Golgi complex 99
gonads 322, **323**, 375, *443*
goose, grayleg 378
gophers **216**
gorilla 395, 540
grains 233, **470**
grana *223*, **224**
Grand Canyon, fossil record in 508, **508**
grapes 425, 469
GRAS 686
grasses 120, 159
grasshoppers **128**, **195**, **197**, 198, 200, 220
  digestive system of 244
  eyes of 304
  head of, mouthparts **245**
  incomplete metamorphosis of **479**, 480
  locomotion of 337
  respiratory system of **289**
grasslands 644–645, **645**
Gray, Asa 506
Great Barrier Reef 677
Great Dane, chromosome number of 90
Great Lakes, lampreys in 203
green revolution 673
greenhouse effect 650
Griffith, F. 421
ground substance *87*
growth 97, **132**, 472–484
  differential 474–475, **475**
  in nonliving world 132
  indeterminate 127
  mitosis in 97
  of plants, 341, 342
growth hormone 322, **324**
guanine *43*, 44, **44**, 422, **435**
guano, in phosphorus cycle 655
guard cells 116, 168, **169**
guinea pigs **216**, 354, 369
  behavior of 363–364
  genetics of **411**, 413
  gestation period of 456
  inability to synthesize vitamin C 235
Guinea worm 183
gullet, of paramecium *239*
Gulliver, life detection device 605–606, **605**
gums, of mammals 246
gunpowder 17
gustatory receptors 303
gut *239*, 241
Gymnarchus 571, **571**
Gymnospermae 159
gymnosperms **128**, 159, 167, *464*, *466*
  comparison with angiosperms 161
  life cycle of **162**
  reproduction of 160–161, **163**
gypsy moths 355, 678

H

habit *363*, **363**
habitats 655
habituation *360*
Haeckle, Ernst 506, 512–513
hagfishes 203
hair
  of mammals 213
  sensitivity to touch stimuli 300
hair cells of ear 307, 308
half-life *27*
hamsters 359, **359**
hand, temperature receptors on **300**
handedness, determination of 320
haploid number *396*, *439*, *455*
  of chromosomes in gametes 420
haploid state 400
  of daughter cells, in meiosis **396**, **397**
hares **216**
Harlow, H. F. 377, 378
hashish **691**
haw **466**
hawk **129**, 354
hay bacillus **588**
HCl 249
  See also hydrochloric acid
healing of wounds 97, *482*, **483**
hearing 319, 320
heart **108**
  age effects on 485
  evolution of 208, **271**
  human 271–275
  of amphibians 270, **271**
  of birds 212, 270
  of earthworm 268
  of fish 270, **271**
  of mammals 213, 270, **271**
  of reptiles 270, **271**
  of vertebrates 202
  sounds of 273
  three-chambered **270**
  transplants of 628
  two-chambered **270**
  valves of **273**
heart attacks 233, 280, 292, 327
heart disease 232–233, 626–628
heart-lung machines 488, **628**
heartbeat 273, 315, 316, 321
heartburn 253
heartwood *168*
heat
  ecological effects of 641–643
  perception of 301, **301**
heat capacity *35*
Heisenberg uncertainty principle *9*, 24
Heisenberg, Werner 9
Helianthus and hybridization 524, 525
Helmont, Jan Baptista van 222, **222**
helium, in deep-sea diving 600
helix, of proteins 41
heme 225, **225**
Hemichordata 202
hemispheres, of brain 319, 320
Hemiptera 198
hemocyanins 40, *269*
hemoglobin 202, 254, 292
  functions of 276, 292
  in blood of earthworm 268
hemoglobin (*cont.*)
  in erythrocytes *276*
  in sickle cell anemia 438
  iron in 36, 63
  molecular weight 40
  structure of 42, 47, 277
hemophilia *280*, 416–417, **418**
hemp, fibers of 117
hens
  egg-laying, stimulated by light 351
  eggs, size of 445
  pecking order of 382–383, **382**
Henle, loop of, **259**, **263**, **264**
heparin (anticoagulant) 280
herbivores *219–220*, **220**
  herd organization of 382
  R. Q. of 230
  teeth of 243, **244**, 248
heredity *390–419*
herring **204**
herring gull **349**
hermaphrodites *444*
herpes simplex 620
heroin *689*, **690**
herons, courtship of 379, **379**
heterotrophic nutrition, of protists *138*
heterotrophs *219*
  holozoic **220**
heterozygote *408*
hexapodal locomotion **335**
hibernation 39, *642*, **643**
hibiscus, pollen grains of **468**
hindbrain **316**
Hipparion **538**
hippocampus **318**
Hippocrates 616
hippopotamuses **216**
hirudin 184
Hirudo medicinalis 184, **184**
histamine **325**
histidine **437**
histochemistry *58*
histones 89, *434*
hogs, parasites of 182
holdfasts, of algae 152
Holley, Robert W. 46
holographic theory of memory 576
holography *574–576*, **575**
holozoic nutrition *219*
homeostasis *262*, 298, 364
homing behavior 374–377
homing pigeons 374
Homo erectus 542
homoiotherms *212*, 214, *642*, **642**
homology *395*
Homoptera 198
Homo sapiens 124, 396, 543–546
Homo sapiens neanderthalensis: see Neanderthal man
homogenates 57, 62
homologous structures 509–510, **509**
homozygote *408*
honey guide 661, **661**
honeybees
  communication among 386, **387**
  head of, mouthpart **245**
  use of "light compass" 350
  honeycomb **383**
Hooke, Robert 65, 75
hookworms 183
hormones 34, *46*, 75, 79, 322, 323, **324**
  in activation of DNA 434
  in sterotyped response of insects 384
  insect 480
  juvenile 480, **481**
  mechanism of action **326**
hornet 303
horsefly 338

horses **129**, **216**, 219, 306, 509
chromosome number of 395
interspecies mating with donkey **126**; with zebra **524**
learning capacities of 358
locomotion of 334, **335**
nematode parasites of 183
teeth of 248
host range 432
host preferences of parasites 221
hot springs 640, **641**
houseflies **245**, 338
howling monkeys 380, 382, 386
Hoyle, Fred 492
Huggins, Charles B. 622, 624
human body
differential growth of 475, **475**
water content of 35
human embryonic development **461**
human fetal development **460**
human societies, compared to insect societies 384
humerus, bone marrow in 281
hunger 251, 364
Huxley, Thomas 506
hyacinth
bulbs of 166
mitosis of **92**, **94**, **95**
water 155
hyaluronidase 452
hybrid *393*
hybridization 524–525
hydra **128**, 177, **177**, **178**
digestive system of 241, **241**
locomotion of 178
nerve net of 309, **309**
reproduction of 441, **441**
hydraulic systems 333
Hydrazoa 177
hydrocarbons, as a pollutant 680
hydrochloric acid 252, 253, **256**, 322
hydrocortisone 325
hydroelectric power 562
hydrogen 25, 30, 64, 294
hydrogen bonds 41, 44
hydrogen ions *28*, 29
hydrogen sulfide 219
hydrolysis *33*
hydrolytic enzymes, in lysosomes 101
hydrotropism *339*
hydroxyl ion 28
Hymenoptera **196**, **197** 199
hypermetropia *305–306*, **306**
hypertensin **325**
hypertonic medium *85*, **85**
hyphae, of fungi *154*
hypocotyl *469*, **471**
Hypohippus **538**
hypothalamus 318, **316**, **326**
and body temperature 299
as endocrine gland 323, **324**
as part of limbic system 318, **318**
control centers in 316–317, **317**, 318
in female reproductive cycle **450**, **454**
in regulation of blood sugar level **232**
hypothesis *6–7*
hypotonic medium *85*, **85**, 259
Hyracoidea **216**
hyraxes **216**
hysterectomy 455

I

ichthyology *11*
ichthyosaurs **534**, *535*
identical twins 443, **443**
idiosomes: *see* Golgi complex
ileocolic sphincter *254*
ileum *253*, 255
imaginal discs *480*
imagination, center of in brain 320
imago *480*
immunity 283, 618
immunologic mechanism 284
implantation, of embryo 451, 457
imprinting 377–379, **378**
incisors *247*, **247**
of cows 248
of horses 248
of rodents 248
incompleteness theorem, Goedel's *9*
incubation period *612*
incus (anvil) 307, **307**
indole-3-acetic acid (auxin) 341
inducer, in *lac* operon **434**
infertility, treatment of 453
influenza 142, 284, 620
virus 142, **143**
information explosion 2–3, **3**
ingestion, of paramecium **240**
inheritance, of human skin color **413**
inorganic compounds 30
inositol, as water-soluble vitamin 236
insect societies 199, 383–384, **383**
Insecta 198
insecticides 313
Insectivora **216**
insects 120, 193–200, 289, 305, 338, 443, 446
behavior of **357**
blood of 269
brains of 309
compound eyes of 304
digestive systems of 243–244
embryonic development of **456**
exoskeleton of 37, 331, 334
eyes of **511**
heads of, mouthparts of **245**
learning in 357
locomotion of **335**
metamorphosis of **479**, 480, 488
provision for young 446
respiratory systems of 289, **289**
swarming of 382
insight 320, *366–367*
instinct *352–355*, **357**
insulin 40, 322, 327
in diabetes mellitus 264
in regulation of blood sugar level **232**, **325**
structure of 41, **42**
integument *37*, *202*
integumentary system 118
interbreeding 126
interdependence, of species 118–120
interferon *282*, 621–622, **621**
International Drug Monitoring System 687
interneurons **113**
interoceptors *299*
interphase
in meiosis 400
in mitosis 93, **94**, **96**
interspecies mating **126**
intertidal region *648*
intestinal glands, source of digestive secretions **256**
intestinal juice *254*
intestines
large **251**, 255–257, **261**
of earthworm 243
of grasshopper 244
of planaria 242
small **249**, 251, **251**, 253–255, **254**, **256**
intracellular digestion 101
intrauterine device (IUD) 453, **455**
intussusception *132*, **132**
inversion, type of chromosome mutation **427**
invertebrates
behavior of **357**
hemocyanins in 40
learning capacities of 357
iodine 63, 64, 234, 478
5-iodo-2-deoxyuridine 620
ion-exchange chromatography 58, 59
ionic bonds *29*, **29**
ions *20*, 20–21, 28
iris diaphragm, of human eye *305*
Irish moss 153
iron 234, 262
deficiency of 276
in hemoglobin 47, 63
Iron Age 17
iron bacteria 219
irradiation *554*, 555, **555**
irritability *135*
isolation
ecological *520*
effects of 298–299
genetic *520*
geographic *518*
isolator, germ free *590–591*, **590**
isoleucine **437**
isomerization reactions **32**, 33
isomers *33*
Isoptera 198
isotonic equilibrium 86
isotonic medium **85**, 86, 259
isotopes *25*, 27
itching 301

J

jackdaw 378–379
Jacob, François 433–434
Jacob, Stanley 688
Janssen, Zaccharias 65
jawbone 247
jaws 203, 205, 246
jejenum *253*
jellies 153
jellyfish **128**, 177, **177**, 178, 309
Jenner, Edward 283, 614
jet exhaustion *367*
jet propulsion 333, **333**
JOHNNIAC 582
jungle life, adaptation to 137
Jurassic period **531**

K

Kalanchoe pinnata 114, **115**
kangaroo **129**, 215, 337, **337**, **517**, 658
kangaroo rat 261
karate 316
karyokinesis *92*
Kaufman, H. E. 620
kelp, giant 151
Kendrew, John C. 41
Khorana, Har Gobind 438, 595–596, **595**
kidney tissue, of cat **107**
kidney tubules 109, **259**, **263**
kidneys **325**, 485
adjustment of blood pH 264
as endocrine glands **323**
as excretory organ **261**, 262
human 262, **263**
of amphibians 261
of birds 261
of fish 203, **260**, 261
of reptiles 261
of vertebrates 259–261
transplants of 628
kilocalorie (kcal) *229*
kinesis *351*
kinetic energy *17*, 32

kinetic theory *21*
kinetin 343
kinins 343
Kjeldahl nitrogen determination 57–58, **57**
Kleitman, Nathaniel 370–371
knee-jerk reflex 314, **315,** 351
Knoll and Ruska 67
knowledge, acquisition of 2–4, **3**
Koch, Robert 611
Koch's postulates 611
Kohler, W. 366, **366**
Komodo dragon 674
Kornberg, Arthur 435, 594–595, **594,** 597
Krebs citric acid cycle **295,** 296, 625
Kuhn, R. 58
kwashiokor *233,* **233**

L

labium majorum **449**
labium minorum **449**
labor *462*
labor, division of, in insect societies 120, **383,** 384
lac operon 433, **434**
lacewings 198
lactase **256**
lactation **324,** *462–463*
lacteal *255*
lactic acid 294
lactose 533
ladybugs 198, 678, **678**
"lag" phase in plant growth *475*
Lagomorpha **216**
Lamarck, Jean Baptiste de 75, 500
lamellae *223*
lampreys 203, **203,** 205, 351
land animals, water conservation in 261
Landau, Lev 488
Landsteiner, Karl 284–285
Lansing, A. I. 484–485
larva
of insect **479,** 480, **481**
of sponge 175–176
of tunicate **477**
larynx *290,* 291, **291**
laser **564,** *573*
lateral buds 342
lateral meristem *115*
Laue diffraction pattern 70
Laue spectrometer **71**
law of independent assortment *394*
law of segregation *392–393*
lead as a pollutant 680
leadership among animals 382–383
leaf blades 158, 171
leaf insects, classification of 198
leaf spot 154
leaf hoppers 198
Learm 582, **582**
learning 355–360
center in brain 319, 320
comparison of different groups of animals **357**
in bees 357
in insects 357
molecular basis of 358–359
transfer of 359
learning capacities
of amphibians 357
of cats 358
of dogs 358
of fish 357
of horses 358
of invertebrates 357
of mammals 357
of mice 358
of rats 358
learning capacities (*cont.*)
of reptiles 357
learning machines 581–583, **582**
leather, tanning of 146
leaves 115, 116, 118, **118,** *168–171*
colors of, in autumn 171
compound **170,** 171
fall of 171
of ferns 158
of seed plants 161
simple **170,** 171
structure of **169**
Lederberg, Joshua 597
Lederer, E. 58
leeches 184, **184**
Leeuwenhoek, Anton van 65, **65,** 75, 146–147
legumes 469, **470**
Leith, Emmett 574
lemmings 135
lemurs **216**
length 54, **54**
lens
of eye, proteins in 39–40
of human eye **304,** 305
Lepas fascicularis **194,** 197, **197**
Lepidoptera **196, 197,** 198
leptocephalus larva 376, **376**
leucine **437**
Leuconostoc mesenteroides **430**
leucoplasts *98,* **98,** 116
Leucosolenia complicata **174**
leukemia *281*
leukocytes **276,** *277*
antibody production by 277
differential leukocyte count 279
phagocytosis of bacteria **278**
production of 281
LH **324, 450, 454**
LH-RF **454**
Librium **690**
lice 144, 198
life
characteristics of 130
kingdoms of **128**
on other worlds: *see* exobiology
origin of 2, 135, 495–499, **496**
synthesis of 34, 593
life expectancy 485, 487–488
life sciences *11*
life span
of dog 484
of man 484
of rat 484
of rotifer 484
potential 487–488
ligaments 110
light
absorption by chlorophyll 225, **226**
and bird migration 640–641
and photoperiodism 640
and photosynthesis 223–227, 640
receptors in animals and protists 303
visible *303*
light saturation point *226*
lignin 104
lily 159, **160,** 171
chromosome number of **90**
meiosis of **398–399**
limbic system of brain *318,* **318**
limnetic zone 647
Lindbergh, Charles 473–474
linen, fibers of 117
linkage groups **404,** 406
Linnaeus, Carolus 124, 127, 509
linoleic acid 232
linolenic acid 232
lionfish **204**
lions 243, 355, 374
lipase **249,** 253, **256**
lipids *37–30,* **46**
hydrolysis of 253
in bacteria 146
in blood plasma 114
in cell membrane 83
in chloroplasts 223
storage of, in Golgi complex 99
synthesis of 82
transport of 279
lipochondria: *see* Golgi complex
lipofucsin, senility pigment 486
lips of mammals 246
Lissmann, H. W. 571
liter *29*
litmus paper *28*
littoral zone *647*
liver
and deamination of excess amino acids 234
as accessory organ of circulation 277, 280, 281, 282
as excretory organ 234, 261–262, **261**
glycogen storage in 37, 232
human 117, **251,** 253, 255
in regulating blood sugar level 231
source of digestive secretions 254, **256**
liverworts **128,** 156, **156**
lizards **129,** 202, 208–210
regeneration in 210, 482, **482**
lobe-fin fish 532, **532**
lobsters 37, **128,** 197, 199, 269
locomotion *330–338,* 353
bipedal **335**
hexapodal 334, **335**
modes of 331
of insect **335**
of horse **335**
of human **335**
quadrupedal 334, **335**
locus of genes *406*
locusts 198
"log" phase in plant growth *475*
lock-and-key theory of enzyme action *49,* **50**
"long-day" plants 343, **344**
loop of Henle **259, 263, 264**
Lorenz, Konrad 347, 378–379, **378**
louse **195, 197**
LSD **691,** *692*
LTH **324**
luciferase 569–570
luciferin 569–570
Lucky Dragon 553
Luidia ciliaris **442**
Lumbricus terrestris 124, 184
See also earthworm
lumen of blood vessels 626
lungs 29, 291–293, 460
as excretory organ 261, **261**
human **291, 292, 293**
of birds 212
of fish 203
of mammals 213
of reptiles 208
of vertebrates *289,* **290**
lungfishes 205, **525,** 532, **532**
Lycoperdon sp. **153**
Lyell, Charles 501
lymph 35, *281*
lymph nodes *281,* 282
lymphatic system, human 255, 281, **282**
lymphocytes **324**
lysergic acid diethylamide (LSD) **691,** *692*
lysine 234, **437,** 673
lysosomes *79,* 87, 101, 239
lysozyme 71

M

M, human blood type factor 408
macrophages *280*
Macropus 658
maggots 132
magnesium 36, 47, 234
maidenhair fern **158**
malaria 149, 199, 616
males, death rate of 414, 416
malleus (hammer) 307, **307**
maltase **250, 256**
Malthus, Thomas 501, 671
maltose **250**, 253
Mammalia 212, 214, **215**
mammals **129,** 202, 212–215, **213**
behavior of 214, **357,** 373, 380
characteristics of 213
circulatory system of 213–214
communication among 386
embryonic development of 455, **456, 513**
evolution of 214–215
heart of **271**
learning capacities of 357
locomotion of 214
nervous system of 214
reproduction of 214
respiratory systems of 289
urine of 261
mammary glands *213*
man 229, 238
abstract reasoning of 367
adaptability to habitats 214
and balance of nature 137, 669–670
behavior of **357**
bipedal locomotion of 336, **337**
chromosome number of **90,** 395
classification of 126, **129, 216**
embryonic development of **513**
evolution of brain 310
gestation period of 456
learning potential of 358, 365
life span of 484
nematode parasites of 183
races of *544–546,* **545**
manatees **216**
manganese, as trace element 234
mantids 198
mantle 187, 189
maple sugar 168
maples, 159, **160,** 171, 469
Marchantia 156, **156**
margarine 232
marijuana *691–692,* **691, 692**
marine biomes *647–648,* **647**
Mars, manned mission to 601
marsupium *215*
marsupials 214–215
evolution of 215, 516
reproduction of 214
Martin, A. J. P. 58
mass, measurement of 55
mass spectrometer **53,** 64
maternal age, effect on life expectancy of progeny 485
maternal instinct **324**
mating
and external fertilization 445
behavior of sticklebacks 381, **381**
role of pheromones in 355
matrix 104, 110, 113
matter *15*
maturation, and stereotyped behavior 353
mayfly **195, 197,** 198
mazes 356, **356,** 365, **365**
solving by ants 357, **365**
solving by cockroaches 357
solving by rats **365**
McCarty, M. 421
McCay, Clive 476, 485
McElroy, W. C. 569
McLeod, C. M. 421
meadow community 120
measles 221, 284
measles virus **143**
measurement, units of **54**
meats 232, 233
mechanism, theory of 130
medical electronics 631–633
medulla oblongata 253, 293, *315–316,* **316**
medusa 178, **179**
Megascolides australis 186
meiosis *396–409,* **397, 398, 399,** 400, **407,** 414
comparison with mitosis **396, 397**
during human fertilization **452**
errors of 426
in angiosperms 466, **467**
in lily **398, 399**
in vertebrates 400
phases of **397, 398, 399,** 400
melanin **413**
melanocyte-stimulating hormone **324**
melanoma *624,* **624**
melatonin **324,** 328, 329
memory
center of, in brain 319, 320
long-term 358
molecular basis of 358–360
short-term 358
memory storage, theory of 360
memory trace *360,* **361**
memory transfer **359**
Mendel, Gregor *391–395,* **391,** 504
Mendel's first law: *see* law of segregation
Mendel's second law: *see* law of independent assortment
Mendelian ratios, deviations from 411–413
meninges 313
menstrual cycle 328, 368–369, **450**
menstruation *450,* **454**
Menten, M. L. 49
Merck, Sharp, and Dohme 42
mercury, as a pollutant 681
Merrifield, Robert B. 42, **43**
meristematic tissue *114,* **115**
Merychippus **538**
mesentery 253
mesoderm of embryo 456
Mesozoic era *534–537*
messenger RNA (mRNA) *46, 435,* 432, **433,** 436
metabolic poisons, theory of aging 485–486
metabolism *47–48, 131*
calcium, stimulation of **324**
oxidative, stimulation of **324**
sodium-potassium, regulation of **325**
See also cell metabolism
metabolites 35
metamorphosis *200, 477–481,* **477, 479**
complete, of insects **479,** *480*
hormonal control in silkmoth **481**
in frogs 101, 327, 477
incomplete, of insects **479,** 480
types of, in insects **479,** 489
metaphase
in meiosis **398,** 400
in mitosis 93, **95**
metastasis *623*
Metatheria 214
metazoa 173
meteorology *11*
Methisazone 620
methionine **437**
methyl mercaptan 302
metric system 54
mice 109, **129,** 202, **216,** 220, 223, **223,** 235, 474, 485, **672**
adaptability to habitats 214
chromosome number of **90**
excretion in **260**
fertilized ovum of **403**
gestation period of 456
learning capacities of 358, 359
pecking order in 383
reproductive potential of 113
Rex gene in 411, **412**
role of pheromones in mating activities 355
Michaelis, Leonor 49
microbes and spontaneous generation 133
microflora *589*
micrographs
high-resolution electron **68**
scanning electron **69**
microorganisms in genetics research 429–430
microprojector **53**
microscope
comparison of light and electron microscopes **69**
compound 65–66, **66**
electron 67–68, **67, 69, 564**
scanning electron 69
simple 65
stereoscopic 67
use of in measurements of size 54
microscopy
dark-field 66–67
phase-contrast 67
ultraviolet 67
microsomal fraction 97
microtome **53,** 66
microvilli **254,** *255*
midbrain 315, 316, **316,** 318
midge 338
midribs of leaves 171
migration 353, 374–377
of birds 369, 374–376, **375,** 640-641
mildews, powdery 154
milk
human, production of 463; compared to cows' milk 463
in diet 232, 233
of monotremes 215
milk glands, of pigeons 381
milkweed 469, **470**
Milky Way Galaxy **604**
Miller, Stanley 497
Miller-Urey experiment 497, **498**
millipedes **128, 194, 197,** 198, 200
Miltown **690**
mimicry 663
mineral ions, reabsorption of 263
mineral salts in diet 234
mineralocorticoids **325**
minerals 221, 229, 598
mint 223
Miocene epoch **531, 538**
Miohippus 538
Mirbel 75
mitochondria 66, *79,* 87, 100, 141
mites 144, 197, 198
mitral valve 272
mitosis *77,* 92–97
comparison with meiosis **396, 397**
in algae 441
in hyacinth **94, 95**
in whitefish embryo **94, 95**
of animal cell **94, 95**
of plant cell **94, 95**
phases of 93, **94, 95,** 96
significance of 96–97

mitosis (*cont.*)
time duration of 96
mixtures *19*, 19–20, **20**
mobile phase, in chromatography 59
moderator, in nuclear reactors *550*
molars *247*, **247**
of cows 248
of horses 248
molds
mutations of, induced by nitrous acid 428
See also fungi; *Neurospora crassa*
moles **129**, 214, 216
star-nosed **213**
molecular genetics *391*
molecular sieves 58
molecular weights *27*
molecules *17*
biological, three-dimensional structures of 70
Mogula manhattensis (tunicate) **203**
Mollusca 186–191
mollusks **129**, 186–191
body structure of **188**
calcium in shells of 36
larvae of 478
learning ability of 356
locomotion of 188
reproduction of 191
shells of 36, **187**, 188
molting *199*, *476*, **476**
molting hormone *480*
mongolism: *see* Down's syndrome
Mongoloid race 545, **545**
monkeys **129**, 214, **216**, 318, 540
abstract reasoning of 367
cebus **90**
communication among 386
curiosity of 364
in Butler box 364–365, **364**
learning potential of 358, 365
rhesus **90**, **213**, 286, 377, 378
squirrel **540**
velvet **540**
monocots **128**
comparison with dicots **161**
herbaceous stem **167**
seeds 469, **471**
venation of leaves 171
See also monocotyledons
monocotyledons *159*, **160**, 167
See also monocots
Monod, Jacques 433–434
monogamy among animals 380
Monoplacophora 188
monosaccharides *37*, 253
mons veneris **449**, *451*
Monsanto Chemical Company 485
monotremes 214, 215, 516
monthly cycles of marine animals 368
moray eel **204**
Morgan, C. L. 349
Morgan, T. H. 395, 403, **404**, 416, 426
Morgan's canon 349, 350, 365
morphine *689*, **690**
morning glories 167
"morning sickness" 457
Morse code, comparison to DNA 424
mortality
of fruit fly 487, **487**
of man 487, **487**
of oyster 487, **487**
mosquitoes **196**, **197**, 198–199, 245
anopheles, and malaria 149
as disease carriers 149, 199
control of 677
male, mating behavior of 353, **353**
mosses **128**, 156, **156**, 442
life cycle of **156**
moths 135, 198, 200, 416
as pollinators 467–468
Cecropia, caterpillar of **201**
life cycle of **201**
luna **196**, **197**
tomato sphinx **196**, **197**
"mother surrogate" 377–378
motion sickness 253, 308
motivation 364, **364**
motor area of brain **319**, 320
motor neurons *112*, **113**, *310*, **311**, 314
mouth
digestive secretions of **256**
human 246, **251**
in digestive process **249**
of earthworm 242
of grasshopper 244
of mammals 246
of planaria 241
movements 330
in plants 135
mucin **256**
mucous membranes 248, 252
mule **126**
Muller, H. J. 426–428
multicellular organisms 75
multivator, life-detection device **605**
muscarine 273
muscle cells *330*
muscle tissue *108*, 111–112, **112**
muscle tone 318
muscles *330*, **330**
cardiac, innervation of 320
contraction of 36, 87, 100
human, compared with insect **331**
lipid content of 39
skeletal 330
smooth, innervation of 320
muscular system 118, 135
mushrooms **128**, 153, **153**, 154
mustard gas 428
mutagenic agents 515
mutagens 428–429, 432
Mutation Theory, The, by Hugo de Vries 425–426
mutations *135*, 424–429, 437
and evolution 520–522
in chromosomes **401**
in *Drosophila melanogaster* **396**, **426**
in genes *406*
in *lac* genes 433
temperature and 448
muton *424*, **424**
mutualism 660, *661*, **661**
mycelium *154*
Mycobacterium tuberculosis 610, 611
mycoplasmas *76*, **128**, 142, 144, **144**
mycoses *615*
myelin sheath 39
myocardium **108**
myoglobin 41, **42**
myopia *305*, **306**
myosin **330**
Myrmecia gulosa **120**
Myxomycophyta 154–155

N

N, human blood type factor 408
nacre *189*
natural law *8*
natural sciences *10*
NADPH 227, **227**, **228**
naiads *480*
β-naphthaleneacetic acid 341
nasal epithelium **261**, 290, 301, 302, **302**
nasopharynx *250*
natural selection *504*
"natural" systems of classification 123
nausea 308
navigation
during migration 374–375
of bees 369
of migratory birds 369
Neanderthal man 542–544, **543**
nearsightedness: *see* myopia
nebular hypothesis 495
neck, of teeth 247, **248**
Necturus 207
Negroid race 545, **545**
Neisseria gonorrhoeae 278
nekton *648*
Nematoda 182–184
nematodes 182–183, **183**
Nembutal **690**
neon goby **204**
nephridia 259, **259**
nephridiopore **259**
nephron *259*, **259**, *262*, 263, **263**, 264, 265
nephrostome **259**
neritic zone *648*
nerve cells 76, 107, **113**
See also neurons
nerve gases *313*
nerve impulse transmission 100, 311, **312**
nerve net *309*
nerve tissue **107**, *109*, *112*
nerves
human 83
stimulation of regeneration 483–484
nervous system 118, 135, 298, 308–322
autonomic 320–321, **321**
central **309**, 313
coordination with endocrine system 322
evolution of 309
human **313**
of annelid worms 356
of chordates 202
of flatworms 356
parasympathetic 321, **321**
peripheral 313, 320
sympathetic 321, **321**
nest building 355, **355**
nests, birds' 446, **446**
neural tube, in embryonic development of frog **459**
neurohumors *312*
neurons *112*, **113**, *310*, **311**, 313, 321, **321**
in brains of rats 360
in memory formation 360
of animals 298
of mouse spinal cord ganglion **77**
Neuroptera 198
Neurospora crassa 48, 429, **429**
neutrons *24*, 428
Newton's third law of motion *333*
newts 207, 478–479, **478**
niacin 236, **237**
nickel, as trace element 234
nicotine 292, *689*, **690**
nictitating membrane *212*
night blindness 237
nightingale, song of 380
ninhydrin 60
Nirenberg, Marshall W. 436–437
Nissl bodies 112
nitrates 219, 652, 653
nitrate bacteria (*Nitrobacter*) 219, *652*
nitrification *652*
nitrite bacteria (*Nitrosomonas*) 219, *652*
nitrites 219, 429
nitrogen 36, 58, 64, 435
nitrogen cycle 652, **652**
nitrogen fixation *652*
nitrogen mustard 428
nitrogenous wastes, of animals 258
nitrosamines *429*
nitrous acid 428, 432

nitrous oxide **690**
nocturnal animals *368*
No-Doz **691**
noise pollution 683
Nomeus 176, **176**
nondisjunction *401*, 415
"nonsense" codons *437*, **437**
noradrenaline **325**
northern coniferous forest 645–646, **646**
nostrils 290
notochord *200*, 201, **201**, 202, **203**
Notoryctes 658
nuclear fission *27*, *549–550*, **549**
nuclear fusion *550–552*
nuclear membrane 79, 82, 93, 96
nuclear power 561–563
nuclear reactor 550, **550**, **562**
nuclear regions, in bacteria and blue-green algae 88
nuclear weapons 27
nucleases 254, **256**
nucleic acids 34, *43–46*, 64
  See also DNA and RNA
nucleoid bodies 88
nucleolar organization *92*
nucleolus 92, 93, 96
nucleoproteins *46*, 89, 395, 434
nucleosides 255
nucleotides *43*, 58, 434
nucleus (of atoms) *24*
nucleus (of cells) 66, 76, *77*, *88–89*, **89**, 97
  absence of, in viruses 141
  DNA content 46
  experimental removal of, in amoeba 88
  in protozoa 147
  nucleolus in 92
nudibranchs *188*
nurse (Sertoli) cell **449**
nuts 232, 233, 469, **470**
nutmeg **691**
nutrients
  carried by blood 79
  in roast beef sandwich 230–231, **231**
  organic, transport of in plants 117
nutrition *227–238*
  effects on developing embryo 457
  in protists 138
  modes of 219–221
nymphs **479**, *480*

O

O, human blood type factor 408
Oak Ridge 550, **560**
oaks 149, **160**, 171
oat coleoptiles 339–341
Obelia 177, 178, **179**
Occam's razor **6**, 349
oceanic zone *648*
oceanography *597–601*
oceans 597–601, **598**
octopuses **129**, 186, **187**, 189, **190**
  care of young 191
  courtship of 191
  experiments on learning 356–357
  learning capacity of 309
  locomotion of 191
  reproduction of 191
  tentacles, use of 191
oddity principle 357, 367
Odonata 198
odors
  as releasers of innate behavior patterns 354–355
  perception of 302
Odum, Eugene 665
Oedogonium **152**
oil, as a pollutant 681–682, **682**
oils, in diet 232
Oken, Lorenz 75
olfactory receptors *302*, **302**
Oligocene epoch **531**, **538**
oligosaccharides *37*
ommatidia *303*, **303**
omnivores *220*, **220**
  R.Q. of 230
  teeth of 244
On the Origin of Origin of Species (Darwin) 502, 539
"one gene—one protein" hypothesis *424*
onions **90**, 166, 168
oöcyte **402**
oögenesis *402*, **402**
oögonia **402**
Opalina 662
Oparin, A. I. 497
operator genes 433, **434**
operculum 205
operon 433
Ophioglossum 395
opium 689, **690**
opossum 202, 215
opsonins 283
optic nerve **304**, 305, 316
optical microscope 53
oral cavity
  of grasshopper *244*
  of mammals *246*
oral groove, of paramecium *239*
oranges, navel 425
orangutan 395, 540
orbitals, electron *24*, **25**, **31**
orchids 161, 219
order, in classification of organisms 124
Ordovician period **530**
organ-system level of development, of flatworms 180
organ systems *108*, *118*, 199
organ transplants 628–630
organelles *77*, 87, 147
organic compounds 30, 36–47
organic matter, content in natural waters 239
organisms *108*, *118*, 130, **131**
  parasitic 154
  pathogenic 221
  saprophytic 154
organs *108*, *117*
  electric, of bony fishes 205
  formation of during embryonic development 456
  of plant vegetative system 118
orgasm 451
origin of life 495–499, **496**
ornithosis virus 142
Orthoptera 195, **197**, **198**
oryx, Arabian 674–675
oscillating universe **493**, 494
Oscillatoria **128**
osculum *240*
osmosis *23*, **23**, *83*, **83**, 85, 86, 103, 263
Osteichthyes 205
Ostracoderm *532*, **532**
ostrich **129**, **211**, 212, 445
otoliths *307*, **308**
outbreeding *464*
oval window *307*, **307**
ovarian follicle **454**
ovaries *113*, **325**, *443*, 449
  alternate production of ova **449**, 450
  as endocrine glands 322, **323**
  in angiosperms 465–467
  in human menstrual cycle **450**
  of flowers **163**, **468**
  of plants 160, 469
  of rats 329
overpopulation, of man 670–671
oviducts 109, **449**, *450*
ovulation *450*, **450**, 451
ovules
  in female cones of pine trees 464, 465
  in life cycle of angiosperm **467**
ovum 113, **113**, *439*, 443
  fertilization of **452**
  human, formation of 402, 449, 450
  size of 444–445
owl, screech **211**
oxalic acid 103, 258–259
oxidation *33*
  in respiration 287
oxidation-reduction reactions *33*
oxygen 25, 36, 64
  amount of in atmosphere 291
  and respiration 287, 292
  and photosynthesis 223, 258
  and R.Q. 229
  transport of 79, 114, 288
oxygen cycle 650–651, **651**
oxyhemoglobin 292
oxytocin **324**, 462, 463
oyster drill 189
oysters 186, 189, 191, 368, 444–445, 478
ozone layer of atmosphere 650

P

PABA *see:* para-aminobenzoic acid
pacemaker, cardiac 273, **630**, 631
paddling 333
pain 316, 300–301, **301**
pain receptors 300
palate 246
Paleocene epoch **531**
paleontologists *248*
paleontology *65*, *506*
Paleozoic era 529, **530**
pancreas **107**, **251**, 253, 322, **323**
  alpha cells of 324
  and blood sugar level **232**
  and insulin secretion 41, 231, 327
  beta cells of 325
  enzymes secreted by 253–254, **256**
  in diabetes mellitus 264
pancreatic juice **324**
pancreozymin **256**
pangolins **216**
pantothenic acid 236, **237**
papain 239
paper chromatography 58, 60
para-aminobenzoic acid **236**, 620, **620**
paralytics, possibility of rehabilitation through ESB 318
Paramecium **102**, **128**, **239**
  cilia of 149
  conjugation in 149, 443, **444**
  digestion in 239, **240**
  egestion in **240**
  ingestion in **240**
  locomotion of *101*, 332, **332**
  pellicle of 147
  reproduction of **148**, 443
  response to stimuli 297–298, **298**
Paramecium multimicronucleatum **150**
parasites *221*, **221**, 616–618, *659*
parasitism 660, *661*, 662
parathormone **324**
parathyroid glands 322, **324**
parenchyma, in plants 116, 117, 171
parotid gland **249**
parrots, communication in 386
parsimony, principle of 349
parthenogenesis 200, **442**
particle size analyzer **53**
partition chromatography 58–59
parturition *462*
passenger pigeon 674, **674**

passive immunity *618*
Pasteur, Louis 133, **134,** 590, **609,** 614
Pasteurella pestis 609, 610
pathogens 145
Pauling, Linus **564,** 565
Pavlov, Ivan 315, 362
PCB, as a pollutant 681
peas, garden, genetics of **90,** 391–394, **393, 394,** 395, 409–410, **410**
peas 233, 469
peach 469, **470**
peacock 380
pearls 189
pears 117, 469
peat mosses 157
pecking order 382–383, **382**
pectin *104*
pelagic zone *648*
Pelecypoda 189
pelicans **129, 211**
pellagra **237**
pellicle 79, 147
pelvis
  bone marrow in 281
  during labor 462, **463**
penguins 212, 381
penicillin 5, 70, 265, 620
Penicillium 153, 620
Penicillium chrysogenum **153**
penis *214,* 262, *445,* 448, **448,** 451
pentose phosphate shunt **295,** 296
pentoses *43,* **44**
pepsin **249,** 252, 253, **256**
pepsinogen *252*
peptide bonds *40,* **40,** 254
peptide chains **433**
peptides 252, 253, 254, **436**
Peranema trichophorum **150**
perennials *167*
pericardium **108**
pericarp *469*
periodic table **26**
Peripatus **194, 197**
peripheral nerves **313**
Perissodactyla **216**
peristalsis *251,* 253, 254, 257, **257,** 265, 321
permafrost *646*
Permian period **530**
permease **434**
pernicious anemia **237**
Perutz, M. F. 42
pest control 355, *677–678*
pesticide residues 677
petals 161, **468**
petrified forest 507
petroleum, formation of 158
Petromyzon marinus 203
pH *29,* **53,** 252, 253
Phaeophyta 151, **151,** 152
phages 142; *see also* bacteriophages
phagocytes 282, 283
phagocytosis *83*
  by amoeba **84, 85,** 239, **239**
  by white blood cells 277, **278**
pharynx 250, 251, *290,* 291
  of earthworm 243
  of planaria *241*
Phasogale 658
phenobarbital **690**
phenol 428
phenotype *408*
phenylalanine 437, **437**
phenylketonuria 409, 438
pheromones **325,** 355, 374, 452
Philodina citrina 484, 485
phloem 117, 168
Pholidota **216**
phonoreceptors *135, 299,* 306–308
phosphate **422,** 684
phosphodiesterase, inhibition of 323–324
phosphoglyceric acid **228**
phospholipids, of unit membrane 81
phosphorus
  in bones 63, 234
  in detergent pollution, and eutrophication 684
  in teeth 234
  radioactive, as tracer 64
phosphorus cycle 652, 654–655, **654**
photoperiodism *343,* 344, *640*
photoreceptors 135, 299, 303–306
photosynthesis 36, 127, 168, 219, *221–222,* 303
  action spectrum of **226**
  and light intensity 226
  and Melvin Calvin 227
  as part of carbon cycle **649,** 650
  carbohydrates produced by 162
  in cactus stems 166, **166**
  in *Euglena,* 127, 348
  in food production 116, 219
  light and dark reactions of 227, **227, 228**
phototaxis *350–351*
phototropism *339,* **339,** 340, **340,** 348
Phycomycetes **153,** 154
phylum *124*
physical sciences *11*
physics *11*
physiological saline 86
physiology *11*
Phytophthora infestans 154
piezoelectric crystals 577
pigeons 362, 381
pigments
  in algae 226
  in cell sap 103
pigs 214, **216,** 220
pikas **216**
pill bug 197, 350
Piltdown man 542
pine **128,** 159, **160,** 464
  Austrian, pistillate (female) cones of **464**
  cones **464**
  cone scale **163**
  life cycle of **465**
  reproduction of 464–465
  seed 465, **465**
  slash, staminate (male) cones of **464**
  wood, structure of **104**
pineal gland **323,** 324, *328,* **329,** 375
  pineapple 469, **470**
pinocytosis *83,* **85**
Pioneer 10 **606**
Pisces 203
pistil *161, 466,* **468**
pistillate cones (female), of Austrian pine **464**
pit vipers 572, **572**
Pithecanthropus **543**
pituitary gland
  and hypothalamus **316, 324**
  anterior **324, 450, 454**
  as endocrine gland 322, 323
  hormones of **324,** 325–327
  posterior **324**
placenta 214, 215, **325,** *456–462,* **461**
  as endocrine gland **323,** 328
placental mammals 214–216, **216**
placoderms 532
planarians
  digestive system of 241–242
  learning in 179, 356, 358–359
  regeneration in 483
  reproduction of 180–181
p[...]
plan[...]ont.)
plant g[...]180, **181**
plant king[...]ms [...]603–604
plants 127, **12**[...]7, **197,** 648
  as photosynthe[...] (auxin) **341**
  chromosome num[...]55
  classification of 138 [...]s 219
  excretion in 258–259, **258**
  flowering 158, 159
  flowering of 343–344
  fungus diseases of 615–616
  growth of 115, 222–223, **222,** 475–476
  hormones of 298
  intracellular digestion in 239
  nematode parasites of 183–184
  number of species 126
  organ systems of 118
  pigments of 58
  reproduction of 441, *463–471*
  responsiveness of 338–345
  seed **160**
  sleep movements of 369
  structure of **118**
  tissues of 114–117
  virus infections of 142
plaque, on teeth 247–248
plaques (phage colonies) *432,* **432**
plasma, blood, proteins in 284
plasma cells, in production of antibodies 283
plasma membrane *77,* 81, 127, 147
  See also cell membrane
plasmodia *154,* **155**
Plasmodium 149
plasmolysis 85, 86
plastids 66, 79, 98
platelets **278,** *279*
platform balance 55, **55**
Platyhelminthes 179–182
platypus **129**
Pleistocene epoch **531, 538**
plesiosaurs 534, 535
pleura *293*
pleuropneumonialike organisms (PPLO) *142*
Pliocene epoch **531, 538**
Pliohippus **538**
ploidy 400
Plowshare Program 561
Pneumococcus, Griffith experiment with 421, **421**
pneumonia 292
poikilotherms *642,* **642**
poinsettia **344**
poisons, detoxification by liver 255
poison ivy 164, **165**
polar podies **402,** 403, **403,** 452, **452**
polar life, adaptation to 137
polar nuclei 466, **467**
polarity
  conservation of, in regeneration 483
  of molecules 21
polio 284, 614
pollen *160*
  cones, of pine trees 464, **465**
  grains, of hibiscus 468; of pine **69,** 464, **465**
  in life cycle of angiosperm **467**
pollen tubes
  in angiosperm 466, **467**
  in pine 464–465, **465**
pollination 198, 199, 464, **465, 467**
pollution 1, 257, *678–684*
  air 678–680

pollution (*cont.*)
control of 678–679, 682, 684
noise 5–6, 683
thermal 682–683
water 680–682
Poly I:C 620
polyembryony 443
polymers **38,** *40*, ..21
polyp 178, **179**
polypeptide...
polyploid..es 37, 146, 284
polyrib..s *97*, **97,** *436*, **436**
Poly...*chium* **156**
polyuridylic acid 437
pompano **204**
pons 315, *316*, **316**
poppy 159, **160,** 689, **690**
population 670–672, **670**
porcupines **216**
porcupine grass **165**
Porifera 138, 174–176
porphyrins 47, *225*
porpoises **216,** 333
Portuguese man-of-war 176–177, **176,** 178, 333
posterior pituitary gland 265, **324**
posture, maintenance of 318
potassium 36, 234, 311
potato blight 154, **588**
potatoes **69,** 116, 166, **166,** 168, 231
potential energy *17*
powder method, in X-ray diffraction study 70
praying mantis **195, 197,** 678, **678**
preadaptation, and evolution 522
Precambrian eras 528–529, **530**
predation 660, *662*, **662**
predator 665
pressure filtration *263*
pregnancy *452*, *456–462*
and curvature of spine 337
test for, detection of chorionic gonadotropin **325**
pregnenolone **326**
premolars *247*, **247**
pressure points *274*, **274**
Price, Charles C. 34, 593
Priestley, Joseph 222–223, **223**
primary consumers *658*, **659**
primary root *164*
Primates **216**
primates **357,** 358, 540, **540**
primrose, evening 425
probability, laws of, and genetics 412
problem solving 363–366, **366**
Proboscidea 216
proboscis, of moth **196, 197**
procaryotes 433–434
progeny, selection of, in milk cows 419
progesterone 322, **325, 326,** *450*, **450, 454,** 462
progestin-estrogen "pill" **454**
proline **437**
promine *477*
Prontosil 619
prophase
of meiosis **393,** 400
of mitosis *93*, 96
proplastids *224*
proprioceptors *308*
proportional counter 64
Prospect of Immortality, The, by Robert Ettinger 577
prostaglandins **325**
prostate *448*, **448**
prosthetic devices 630–631, **630**

...or bacteriophages 431, **431**
...39–42, 233–234, **249, 433**
...amino acids in 64, 435
bacterial 146, 284
contractile 332
digestion of 238, **238,** 254
energy content of 230
hormones 322
hydrolysis of 253
in blood plasma 114
in chloroplasts 223
in diet 229, 233, **233**
in human body 74
in memory formation 360
in neurons 360
in unit membrane 81
primary structure *40*, 41
R.Q. of 229
secondary structure of *41*
storage of, in Golgi complex 99
structure determination 70
synthesis of 42, 82, 92, 100, 432, 435–438, **436**
tertiary structure *41*
three-dimensional structure 41
proteinases 254, **256**
proteinoid microspheres 498, **499**
proteolytic enzymes 239, 457
Proterozoic era 529, **530**
Proteus vulgaris **588**
prothallus *158*
Protheria 214
prothoracic gland **481**
prothrombin *280*
protist kingdom 79
Protista *127*
protists 47, **128,** 101, 138, 331–332, 400
protons *24*
protoplasm *34*, 76
protoplast *76*
Protozoa 146–150
protozoa 76, 79, 127, **128,** 219
and cellulose digestion in termites 37
and Leeuwenhoek 65, 147
behavior of **357**
digestion in 239
marine, excretion in 259
pathogenic 221
reproduction of 149
proximal convoluted tubule 264
pseudopods *87*, 138, 141, 147, 149, 239
psilocybin **691**
psychedelics *688*, **689–690**
psychosomatic illness *321–322*
PTC tasting 303
puberty 448, 449–450
puffballs **153,** 154
puffer fish **599**
pulmonary trunk 272–273
pulp, of teeth *247*, 248
Punnett, R.C. 410
Punnett square 410, **410, 411**
pupil, of human eye 305
pupa **479,** 480, **481**
purines *43*, 44, 46, 422, **422**
Purkinje, Johannes 76
pus *277*
pyloric sphincter **249,** *251*, 253, 254
pyrimidines *43*, **44,** 46, 422, **422**
Pyrrophyta 151, **151,** 152
pyruvic acid 101, *294*, 296

Q

quack grass, rhizomes of 166
quadrupedal locomotion **335**
quadruplets 453
quasars 494
Quaternary period **531**
Queen Victoria, royal hemophilia pedigree 417, **418**

R

rabbits **129, 216,** 220
chromosome number of **90**
copulation of 452
in Australia **504**
parthenogenesis 443
scent marking of territories 355
rabies 614
races of man 518, **519,** *544–546*, **545**
raccoons 365
radial symmetry *138*, **139**
radiation background, natural 554–556
radiation counter 64
radiations 27, 554
biological effects of 65, 428, 553–554, *556–557*
electromagnetic 299, 303
gamma rays **558**
in biological pest control 65
ultraviolet 428
X rays 559
radioactive fallout 27, **551,** 552–553, **553**
radioactive tracers 47, 63, 64, 226–227
radioactive wastes, as a pollutant 682
radioactivity 27
radioisotopes *27*, 63, 65
radiolaria 90, **90,** 147, **150**
radionuclides 552
radiosensitivity, of tissues 555, **555**
radiotoxins *558*
radishes 116, **116,** 162, 165, **344**
radula *188*, 191
rage 317
ragweed 344
Rana clamitans 520, **520**
Rana pipiens 520, **520**
Rana sylvatica 520, **520**
raspberry 469
rats **216,** 253, 353–354, 364
and control of autonomic nervous systems 321
and vitamins 235, 236, 238
brains of, neurons in 360
chromosome number of **90**
effects of continuous exposure to light 328–329
gestation period of 456
hypothalamus of 317, **318**
learning in 358, **359,** 362–363, **362, 365** 367
life span of 476, 484, 485
nest building 355
Norway, with litter **355**
ratel and honey guide 661, **661**
rattlesnakes, classification of 210
ray-finned fish 532, **532**
rays (fish) 205
electric 205
stingrays 205
RBE: *see* relative biological effectiveness
reasoning
center in brain 319, 320
comparison of different groups of animals **357**
deductive *367*
inductive *367*
Recent epoch **531**
recessiveness, in genetics *393–394*
recon *424*
reconstitution of tobacco mosaic virus 432
rectum 244, **248, 249, 251,** *255*, 257, **260**
red, as releaser of instinctive behavior 354, **354**

red blood cells 65, 76, 107
"ghosts" **81**
in formed elements of blood 114, **114**
in hypotonic medium 85
in isotonic medium 86
life span of 63, 64, 488
mature mammalian 89
role of hemoglobin in 36
see also erythrocytes
red tides 151
Redi, Francesco 133, **133**
reduction *33*
reduction division *396–400*
redwood trees 132, 166
Reed, Walter 613–614
"referred pain" *300*
reflexes *351*, **357**
reflex actions *313*
reflex arc *314*
regeneration *481–484*
and mitosis 97
as reproductive process 442
in flatworms 174, 181–182, **181,** 359, 442, 482, 483
in lower animals 441
in sponges 175
of tail, in lizard **482**
regulator genes *433*, **434**
relative biological effectiveness (RBE) *554*
relativity, theory of 5
"releasers" *354*
REM sleep *372*, **372**
remission, of cancer 625
renal cortex *262*
renal medulla *262*
renal pelvis *262*, 265
rennin **256**
repair of damaged body parts *481–484*
repressors 434, **434**
reproduction *132–135, 439–455, 463–471*
and evolution 503
asexual 439–443
in animals 439–455
in plants 463–471
seasonal 452
sexual 400, 443–455
types of 135
reproductive cells 113
reproductive potential 112–113
reproductive system 118
reproductive tissue 109
reptiles **129,** 208–210, **209,** 305
behavior of **357**
brain of **310**
care for young 446
continuous growth of 484
embryonic development of 455, **513**
excretion in 208, **260,** 261
learning capacities of 357
locomotion of 334
reproduction of 208
respiratory system of 287
water retention in 208
Reptilia 208–210
research 4, 34, 696
resolving power *68*
respiration *287–296*
aerobic *296*
anaerobic *294*
cellular *294–296*, **295**
respiratory quotient (R.Q.) *229*
of carnivores **231**
of herbivores **231**
of omnivores **231**
of tissues 230
respiratory system 118, 267
human 290–296, **291**
of amphibians 289
respiratory system (*cont.*)
of birds 289, **290**
of chordates 202
of fish **288**
of grasshopper **289**
of insects 289, **289**
of mammals 289
of reptiles 289
of spiders 289
resting stage of cells 77, 93
reticular activating system (RAS) *316*
reticular formation *316*, **318**
reticuloendothelial system (RES) *281–282*
retina, of human eye **304,** 305
retine *477*
Rex gene, in mouse **412**
Reyniers, J.A. 590
rheumatic fever 610, 626
Rh conflict 286, **287**
Rh factor *286*
rhinoceroses **216**
Rhizobium 652
rhizoids *156*
rhizomes *158*, 166
Rhizopus nigricans **153,** 154
Rhodophyta 151, **151,** 152
rhodopsin *305*
"rhythm method" 453
ribonuclease (RNase) 42, **256,** 359
ribonucleic acid *43*
See also RNA
ribose *37*, *43*, 44, 435
ribosomal RNA *46*, *436*
ribosomes *46*, 79, *82*, 87, 97
protein synthesis at 432, **433,** 436, **436**
ribs 281, 293
ribulose diphosphate **228**
rice 116, 159, **160,** 231
rickets **237**
rickettsias **128,** *144*, **144**
Rime of the Ancient Mariner 568
ringworm 154, 615
rings, annual, of tree **167,** 168
Ritalin **691**
RNA 37, 58, 254, 323
deactivation of, in learning experiments on planarians 359
in chromosomes 89
in memory formation 360
in neurons 112, 360
in nucleolus 92
in protein synthesis 435
in rickettsias 144
in viruses 141, 142
messenger (mRNA) 46, 97, 432, **433,** *435*, 436; in *lac* operon **434;** template function of 436
molecular weight of 43
of tobacco mosaic virus 432
ribosomal *46*, 92, 433, *436*
transfer (tRNA) 46, 432, **433**; in protein synthesis *436*, **436**
types of 432, **433**
RNA polymerase, in protein synthesis 435
RNase: *see* ribonuclease
roaches 480
Robbins, F. C. 614
Robertson, J. D. 81
robin **129,** 354, 373
rock mosses 157
Rockefeller University 42
Rocky Mountain spotted fever 144, **144,** 198
Rodentia **216**
rodents 248, 452, 456
rods, in retina *305*
roentgen 557
e[illegible]16, **116,** 162, 224–225
fibro[illegible] **118,** 162
of ferns 164, **165**
of seed pla[illegible] 115
primary 164, [illegible]
prop 164
secondary 164
taproot 164, **165**
types of 164–165, **165**
root (of teeth) 247, **248**
rot, of plants, cause of 154
rotifiers 442, 484, **484,** 485
rough ER 82
round dance, of bees 386, **387**
round window, of ear *307*
roundworm **90, 128,** 182–184, **183**
Rous, Peyton 622, 623
royal fern **158**
R.Q.: *see* respiratory quotient
rubber 168
rubidium-87 65
rumen, of cows 37
running 334, **335,** 336
runners 166, **166**
Ruska 67
rusts, of plants 154
rye, winter 166

S

Sabin, Albert 614
saccule 307, **308**
"safe period" 453
safrole 687
Saint Augustine 124
salamanders **129,** 207, 297, 354, **478**
embryonic development of **513**
regeneration in 482, 483
spotted **207**
salami, nitrite in 428–429
saliva 244, 248, 250
salivary glands 322
as excretory organs **261**
human 248, **249**
of grasshopper 244
preparations of, from *Drosophila melanogaster* **396**
source of digestive secretions **256**
salivation 135, 315, 362
Salix 466
Salk, Jonas 614
salmon **129,** 428
Salmonella 610
Salmonella typhimurium **611**
Salmonella typhosa 610, 612
salts 27
in blood plasma 114
in living organisms 35, 36
in protoplasm 34
in sea water 36
inorganic, conduction of 117
salty tastes 246, **246,** 302
salvarsan 619
sand dollars 191, **192**
Sanger, Frederick 41
saprophytes *220*
sapwood *168*
Sarcodina 149, **150**
Sargassum 152
saturated fats, and heart disease 627
sauerkraut 146
scabs (healing of wounds) 482, **483**
in plants 154
scale insects 198
scales of fish 205

INDEX

scallops 189, 191–192, 333
scalp tissue, human **107**
scanning electron microscope [illegible]
Scaphopoda 188
scarlet fever 610
scars 110, 482, **483**
scavengers *658*
scent markers, of [illegible]17
territories *145–146*
schistosome[illegible] *145–146*
schistoso[illegible]be 382–383
Schizo[illegible], Mathias 75
Schi[illegible]
S[illegible]dt, J. 376
[illegible]chooling, of fish 382
Schwann, Theodor 75
science *4*
and society *694–698*
and technology *4*
limitations of *8–10*
scientific attitudes 695–696
scientific method *5–8*, **6**
control group *7*
experimental group *7*
hypothesis *6–7*
theory *8*
scientific revolution 1–4
scintillation counter 64
sclera, of human eye *305*
sclerenchyma 117
Scopoloamine **691**
scorpions **194,** 197, **197,** 198
scotophobin, memory chemical *360*
Scott, P.M. 674
screwworm fly 561, **561,** 677
scrotal sac *447*, 448
scrotum **448**
scurvy **237**
Scyphozoa 177
sea anemones 177, **177,** 178
sea cucumbers 191, 192
sea fans 177
sea horses **204,** 380–381
sea lilies 191, 192
sea lions **216**
sea pens 177
sea slug **599**
sea squirt **129,** 202
sea stars 191
sea urchins **129,** 191, 192, **192**
embryonic development 455
sea water 19, 86
seagulls **129,** 354
seals 214, **216,** 243, 244
harbor **213**
seaweed bladderwrack **599**
sebaceous glands, of mammals *214*
Seconal **690**
secondary consumers 658, **659**
secondary roots *164*
secretin **256,** 322, **324**
secretion *258*
sedimentation, in centrifugation 62
seed coat 160, **467,** 469, **471**
seed cones, of pine trees 464, **465**
seed leaves *469*
seed plants 464
seeds 118, *468–471*
annual rhythms of germinating capacity 370
bean, germination of **165**
dispersal of 468–469
of angiosperms **163,** 467, **467**
of gymnosperms **163**
of plants *160*
wind dispersal of 469
segmentation 139

[illegible]eabilit[illegible] *82*
[illegible]eabsorption *263*
[illegible]um, as trace element 234
[illegible]lye, Hans 485, 486, **486**
semen *448*
semicircular canals of ear *307*, 307, 308
semilunar valve 273
seminal vesicles 325, *448*, **448**
seminiferous tubules *448*, **448, 449**
semipermeable membranes *23*, *77*, 82, 83, **83**
Seneca 65
senescence *484–487*
senility pigments 485–486
sense organs 118, 298–299
senses, relative reliance on by
bats 299
birds 299
dogs 299
man 299
monkeys 299
rodents 299
sensitivity of analytical balance *56*
sensory area of brain **319,** 320
sensory neurons *112*, **113,** *310*, 314
sepals 161, **468**
septal area, of brain **318**
septuplets 453
sequoias 155, 159, 167, 472–473, **473,** 484
serine **473**
serological techniques, and evolution 514, **515**
serum albumins *279*
serum globulins *279*
setae **185,** *186*
sewage, and pollution 680
sex attractants (pheromones) **325,** 452
sex characteristics
female **325**
secondary 325, *327*
sex chromosomes *91*
human *414*, **414**
nonsexual information in 416
of *Drosophila* 416
sex determination
in amphibia 416
in birds 416
in butterflies 416
in fish 416
in humans *414–416*, **415**
in moths 416
sex drive, as source of motivation 364
sex hormones 323, 327, 328
female 450
male 448
sex-influenced heredity 417–419
sex-limited heredity *417–419*, **418**
sex linkage *416–419*, **417**
sex reversal 328
sexual intercourse 451
sexual reproduction 135, *439*, *443–447*
compared with asexual reproduction 440, **440**
of bacteria 146
of plants 463
S-shaped growth curves **475**
shapes, as releasers of instinctive behavior 354, **355**
sharks **129,** 202, 205, 248
tiger **204**
sheep 382, 425, **425**
shells of mollusks 187, **187,** 188
shipworms 186, 189
shoes, high-heeled, and curvature of spine 337
"short-day" plants 343, **344**
shrews 214, **216, 310,** 664

shrimps 197, 199
shrubs 115, *167*
sickle cell anemia 276–277, **277,** 438, 521
sieve tube 117
sight 299, 319, 320
sign language, used by chimpanzees 386
sign stimuli *354*, 384
silicon, as trace element 234
silkmoth **481**
silkworm **90,** 198
Silurian period **530**
silverfish **195, 197,** 198, **479,** 480
single-celled organisms 74
single displacement reactions *32*, 33
single-pan analytical balance 56, **57**
Sinsheimer, Robert 593, 597
sinuses, of invertebrates 269
Siphonaptera 198
Sirenia **216**
skates 205
roundel **204**
skeletal-muscle activity, control of 318
skeletal muscle tissue 111, **112**
skeletal system 118
of vertebrates 330–331
skeletons
human, compared with monkey **336**
modifications for bipedal locomotion 336–337
of birds 211
skin 39, 76, 109
as excretory organ **261,** 262
color, inheritance of 413
in disease fighting 281
skin diver **36**
Skinner, B.F. 362, 364
Skinner box 362–363, **362**
skull (cranial bones) *202*, 281, 313, **314**
sleep 370–372
control of 315, 316
deep 371, **372**
deprivation, consequences of 372
dreaming *372*, **372**
EEG in **371**
electrical induction of 372
laboratory **371**
patterns of 371–372, **372**
REM *372*, **372**
sleeping sickness 616
slime molds 79, 127, **128,** 154, **155**
Sloan-Kettering Institute for Cancer Research 629
sloths 151, **216**
slugs 186, 188, 189, **189**
smallpox 283
smell receptors 301–302
smog 678, **679**
smoking 292, **293**
smooth ER 82
smooth-muscle activity 316
smooth muscle tissue 111, **112**
smuts, of plants 154
snails **129,** 188, 189, **189**
calcium in shells 36
garden 186, **187, 188**
hermaphroditic reproduction of 191
learning ability of 356
shells of 187, **187**
snakes **129,** 202, 208, 210, 220
corn **210**
pit vipers 210
scarlet king **209**
skin of, in molting **476**
swallowing of prey **210**
venom of 210
sneezing 291, 315
soap 21

social behavior 381, 382
social hierarchy 382
social insects 383–384
socialization 378
social sciences 10
society *120*
sodium 27, 29, **30,** 36, 234
sodium bicarbonate **256**
sodium carbonate 253
sodium chloride 27–28
sodium ions, transport of 311
sodium nitrite 428
sodium-potassium pump 83
soil 120, 184
sol 87
sol-gel transformations in amoeboid motion 333
solar energy, use in photosynthesis 221
solar system, origin of 494–495
solation *87*
solute *20*
solutions *20*
solvation *21*
solvent *20*
somatic cells 135
somatic mutation theory of aging 486–487
sonar, and bats 566–568, **567**
songs, of birds 212
sonogram **578**
sour tastes 246, **246,** 302
soybeans 470, 673
space, exploration of 601–603
space flight, life support systems in 602–603
Spallanzani, Lazzaro 133, **134**
sparrow, English 678
Spartina and polyploidy 522
spawning 207
specialization, of cell function 173, 174
speciation *127*
species 36, 124, 126
specific name 124
speech, function of tongue in 246
sperm 76, 113, **113,** 135, **402,** *439,* 445, **449,** 451
  fertilization of egg 414, 447, 452, **452**
  formation of **402,** 443, 448
  functions of 444
  haploid chromosome number of 455
  human **445**
  hyaluronidase in 452
  in angiosperm 466
  locomotion of 101, 445
spermatids **402, 449**
spermatocytes **402, 449**
spermatogenesis *402,* **402**
spermatogonia **402, 449**
spermicidal foams 453
spermicidal jellies 453
spermicidal liquids 453
Sphagnum 157
sphygmomanometer *274*
spiders 37, **128,** 197, 200, 289
  black widow 198
  mating of 380
  orange garden **194, 197**
  silk of 198
  web spinning of 353
spinach 103, 258–259, 344
spinal cord *309,* 313, **314, 316**
  and reflex actions 314
  in human nervous system **313**
  of vertebrates 202
spindle *93, 96*
spiny anteater **517**
spiracles, of insects *289,* **289**
spirillum 145, **145**
spleen
  as accessory organ of circulation 280, 282
  in disease fighting 277, 281
sponges **128,** 174–176, **174, 176**
  digestive system of 239, **240**
  elimination of waste products 259
  feeding currents 175, 240
  larvae of 478
  regeneration of 175
  reproduction of 175–176
  respiration of 268, 287
spongin 175
spongocoel *240*
spontaneous generation of life 132–134
sporangia **158**
spores 135, 442
  of algae 152
  of bacteria *146*
  of ferns 158
  of slime molds 154
sporophyte 161, 464, 469
  of angiosperm **467**
  of ferns 158, **159**
  of gymnosperm **465**
  of mosses and liverworts *156,* **157**
Sporozoa 149
sporulation *442*
spring balance 55, **55**
spruces 159, **160,** 464
sputum *291*
Squamata 208
squamous epithelium **77,** 109
squids 186–189, **188, 190,** 289
  giant 200
  locomotion of 191, 333, **333**
squirrels **216,** 373
  flying **213,** 338
squirrel monkey **540**
stage micrometer 55
staining, living specimens 66
stamens *161, 466,* **468**
staminate cones (male) of slash pine 464
stapes (stirrup) 307, **307**
Staphylococcus aureus **432, 499**
starch *37*
  digestion of 238, **238, 250**
  granules, scanning EM of **69**
  hydrolysis of 253
starfish **129,** 191, 192, **192**
  crown-of-thorns 676—677, **676**
  digestive system of **193,** 243
  locomotion of 193, **193,** 333
  Luidia ciliaris 442
  regeneration of 174, 191, 441–442, **442,** 482
starlings 386
stationary phase
  in chromatography 59
  in plant growth *475*
steady-state theory 492, **493**
stegocephalians 533, **533**
stem cells 278
stems 115, 116, 118, **118,** 166–168
  herbaceous *167*
  of ferns 158
  of seed plants 161
  uses of 168
  woody, *167,* **167**
Stentor 149, *352*
  behavior of 352, **352,** 356
stereotyped behavior *309, 350–355*
sternum, bone marrow in 281
steroids 34, 46, 82, 114, 322
stethoscope 273
stickleback
  care of young 381, 446
  mating behavior of 381, **381**
stickleback (*cont.*)
  territorial behavior of 354, **354,** 373
"sticky-end" technique, of DNA synthesis 595–596, **595**
stigma **467, 468**
stimulants (drugs) *688,* **689–691**
stimuli 298
  effects of absence of 299
  plant response to 339
  response to 135, **136,** 310–311
stink bugs 198
St. Martin, Alexis 252, **252**
stomach 28, *251*
  acid in 36, 281
  as endocrine gland **323,** 328
  gastric juice in 253, **256**
  human **249,** *251,* **251**
  mucosa of **324**
  of bird 244
  of grasshopper 244
stomata **115,** *116, 168,* **169,** *225*
stones, of fruits *469*
stony corals 178
stop signals, in genetic code **437**
STP **691**
strawberries 166, **166, 470**
Strehler, Bernard 485
strength, decrease with age 485
Streptococcus pyogenes 610, **611**
stress theory of aging 485
striated muscle tissue: *see* skeletal muscle tissue
strokes 233, 280, 292, 318
stroma *223*
strontium-90 552, 553, **553, 558**
structural genes *433,* **434**
struggle for survival 503
style, of flower **467, 468**
suberin 104
sublingual salivary gland **249**
submaxillary salivary gland **249**
substrate *48,* 49
sucrase 48, **256**
sucrose *37,* **256**
Suctoria 149
sugar cane 166, 168
sugars 37, 231, **295**
  and dental caries 248
sulfanilamide 620, **620**
sulfide bacteria 219
sulfur, radioactive, as tracer 64
Summerlin, William 629
Sumner, James 49
sunflower **470**
sunlight, spectrum of 225
supernatant, in centrifugation 62
support
  function of plant stems 116, 166
  in walking and running 334
survival of the fittest 504
suspension *19,* **20**
Sutherland, E. W. 323
Sutton, W. S. 406
swallowing 250–251, 315
swan **446**
swarming, of insects 382
sweat 262
sweat glands 214, 262
sweating 35, 321
sweat pores 35
sweet tastes 246, **246,** 302
sweeteners, as food additives 686–687
swim bladder, of bony fishes 205
swimming 332, **332,** 333
swordfish 205
symbiosis *179,* 660, *661*
symmetry *138,* **139**
  bilateral *138,* **139**

symmetry (*cont.*)
  in animal classification **139**
  of arthropods 199
  radial *138*, **139**
synapse *112*, *310*, 311–312
synapsis **427**
Synge, R. L. M. 58
synthesis, of organic materials, in plants 131
synthetic function, of Golgi complex 99
syphilis 610
systole *273*
Szent-Gyorgyi, Albert 477

T

tadpoles **206**, 207
  metamorphosis of 478, 488
  of tunicates 477, **477**
Taenia 180, **180**
TAF: *see* tumor angiogenesis factor
taiga *645*
tail
  detachment as protective mechanism of lizard 481–482, **482**
  of human embryo 457
  of vertebrates 202
  regeneration by lizard 481–482, **482**
tapeworms 180, **180**
taproots 164–165, **165**
target tissue *322*
tarsiers **216**
Tasmanian wolf 674
taste buds *246*, **246**, 485
taste receptors, types of 302
tastes
  as releasers of innate behavior patterns 354–355
  relationship to smell 301
  types of 246
taxis *350*, 351, **357**
taxonomists *79*
taxonomy *123*, 124, 126, **509**, **514**
tea **691**
teaching machines 584
technology *4*
teeth
  calcium in 234
  composition of human set 246–247, **247**
  decay of 248
  deciduous ("baby") 247, **247**
  function in digestion 243
  human **244**, **247**
  milk *247*
  of carnivores 243, **244**, 248
  of elephants 248
  of fish 248
  of herbivores 243, **244**, 248
  of mammals 214, 246
  of omnivores **244**
  of sharks 247
  of walruses 248
  permanent **247**
  phosphorus in 234
  structure of **248**
  water content of 35
  wisdom *247*
Teflon in prosthetic devices 630
Tektite II **601**
telophase 96
  in meiosis **399**, 400
  in mitosis **95**, **96**
temperate deciduous forest 645, **645**
temperature
  as mutagenic factor 428
  body, control of 299, 315
  inversion 678–680, **679**
  units of **54**
temperature scale
  Centigrade *16*, **16**
  Farenheit *16*, **16**
tendons 110
tendrils *167*
Teredo (shipworm) 189
termites 37, 120, **195**, **197**, 383
terminal branches of neurons 112, 310
terminal bud of plant 342
Terrapene 209
territorial behavior 373–374
  display by cardinal **373**
  of herons 379
  pheromones and **325**
territory *373*
Tertiary period **531**
"test tube babies" *455*
testes *113*, **325**, *443*, 447, **448**
  as endocrine glands 322, **323**, 448
testosterone **325**, **326**, *448*
Testudo 209
tetanus 284, 618, **618**
tetrahydrocannabinol **691**, 692
thalamus 315, 316, **316**, 318, **318**
thalidomide 457, 687, **688**
thallus, of liverworts *156*
THC: *see* tetrahydrocannabinol
Theophrastus 124
therapsid 533, **534**
thermal neutrons *549*
thermodynamics
  first law of 655
  second law of 655
thermograms **564**
thermonuclear fusion 562–563
thermonuclear reactions 551–552
thermoreceptors *299*, 301, **301**
thigmotropism *339*
thin-layer chromatography 58, 61
third eyelid 510, **512**
thirst 364
thorazine **690**
Thorndike, E. L. 365, 366
Thorndike puzzle box 365, **365**
threonine **437**
thrombin *280*
thrombophlebitis *453*
thromboplastin *280*
thromboplastinogen *280*
thromboplastinogenase 279
Thylacinus 658
thymidylic acid 435
thymine *43*, **44**, 46, 422, 428, **428**, 435, **435**
thymosin, secreted by thymus **324**
thymus, in disease fighting 281, **324**
thyroid gland 63, 322, **323**, **324**, 327
thyroid hormone
  effect on metamorphosis 478, 480
  See also thyroxin
thyrotropic hormone **324**
thyroxin 322, **324**
  See also thyroid hormone
Thysanura 198
ticks **144**, 197, 198, **221**
  dog **194**, **197**
tickling, sensation of 301
tidal rhythms of marine animals 368
tigers 202, 243
time sense 369
time sharing, in computers 585
time zones, effects of rapid changes of 367, **368**
Tinbergen, Niko 347, 354
tissues *107*, 107–117
  formation of, in embryonic development 456
  in animals 107–114
  in plants 114–117
tissues (*cont.*)
  regeneration of 174
  staining of 58
  types of 108–109, **108**
tissue level of development, in coelenterates 138
toads 202, 207–208
  Fowler's **207**
toadstools 154
tobacco mosaic virus (TMV) 142, **142**, *432*
Tofranil **691**
tolerance, and drug addiction *689*
tomato **90**, 98, **98**, 116, 159, **344**, 469, **470**
tongue 246, **246**
tooth shell 188
tortoise 209, **473**, **513**
touch *299*, 300, 319
toxins 277, 284, *612*
trace elements *234*
tracers, radioactive 63
trachea **251**, *290*, **291**
tracheal tubes *289*
tracheophyta 157–171
tracheophytes *158*, 533
transacetylase **434**
transfer RNA (tRNA) *46*, 432, **433**, *436*, **436**
transformation of bacteria **421**
transforming principle *421*
translocation (type of chromosome mutation) **427**
transpiration *166*, *168*, 258
transpiration pull *162*, **164**, *166*
transplants of organs 629–630
trees 115, 117, 120, 167, **167**
Trematoda 179–180
trench fever 144
Treponema pallidum 610, **620**
trial-and-error learning 365–367
Triassic period **531**
Trichogramma 678
Trichinella spiralis 182, **182**
trichinosis 182
tricuspid valve 272
tricuspids 247
Tridacna 189
triiodothyronine **324**
trilobites *529*, 532, **532**
tristearin **39**
triton 677
tritium 27, 64
tropical rain forest 644, **644**
tropisms *339*, 342, *350*
trout 203
true fungi, characteristics of 153–154
trunk, of tree 167, **167**
Trypanosoma gambiense **620**
trypanosomes 147, 149, 616, 619, 620
trypsin **249**, 254, **256**
trypsinogen 254
tryptophan 234, 341, **437**, 485, 673
Tschermak, Erich von 391
tsetse flies 616
Tswett, M. S. 58
tube feet 192, **193**, 333
tube-in-tube body plan *242*, **242**, 243
tuberculosis 221, 610
tubers 166
tubular secretion *264*
Tubulidentata **216**
tularemia 199
tulip tree, stem of **117**
tumor angiogenesis factor (TAF) 623
tuna, and mercury pollution 681
tundra *646*, **646**
Tunicata 202
tunicates 83, 202, **203**, 477, **477**, 478
Turbellaria 179

turgor, of cell *85*
turgor pressure 85, 103
turkeys 212, 443
turnips 116, 164, 165
turpentine, source of 168
turtles **129,** 202, 208–209
box **209**
distinction between turtle and tortoise 209
growth of 473
locomotion of 333
tusks 248
twins
and fertility drugs 453
fraternal 443
identical 401, 443, **443**
twiners *167*
typhoid fever 199, 284, 610
typhus 144, **144**
Tyrannosaurus rex 536, **537**
tyrosine **437**

U

UCLM II 582
UFOs 604
Ulothrix **152**
ultracentrifuge *76*
ultramicrobalances 57
ultramicroscope 66
ultrasonics *557–579*
umbilical arteries 457, 460
umbilical cord *457,* 460
umbilical vein 457, 460
unconditioned stimulus *362*
ungulates, even-toed **216**
ungulates, odd-toed **216**
unit membrane *81,* 82, 142
univalves *188*
unsaturated bonds 33
Upatnieks, Juris 574
uracil *43,* **44,** 435
urea 208, 234, **260,** 261
ureters **260,** *262,* **263,** 265, **448, 449**
urethra **260,** *262,* **263,** 265, *448,* **448, 449,** *450*
Urey, Harold C. 497, **564,** 565
uric acid 208, 212, **260,** 261
uridylic acid 435
urinary system **263**
urination *265*
uriniferous ducts, of frog **260**
urine **260,** 262–264, **264**
Urodela 207
Ussher, Archbishop 396
utricle *307,* **308**
uterus 109, 328, **449,** *450,* 451, 460–462
contractions of **325,** 462
embryo in 457, 460, **461**
lining of, in human menstrual cycle **450**
uvula *250,* 251

V

vaccina virus 142
vaccines 283, 618
vacuoles *79,* 99
in bacterial cells 145
in plant cells 103
in protozoa 147
toxic substances in 258
vagina 214, 262, **449,** *450,* 451
valence electrons **26, 30**
valine **437,** 438
valves, of heart 272, 273
vanadium 83
variation and evolution 502–503
vas deferens *448,* **448, 449**
vasa efferentia **448, 449**
vascular bundles *168*
vascular system, of plant **118**
vascular tissue (blood tissue) *108*
vasectomy *453*
vasoconstriction **325**
vasopressin **324**
vectors, disease 144
vegetal pole **458**
vegetative system, in plants 118
veins 274, **275**
lymphatic 281
of leaves 116, 171
pulmonary *272,* 273
valves in 275, **275**
velvet monkey **540**
vena cava *272*
venation
netted (palmate) **170,** 171
netted (pinnate) **170,** 171
parallel **170,** 171
ventral surface, of flatworm *180*
ventricles, of heart
of birds and mammals 270
of fish 269
of humans 272, 273
venules **263,** *274*
Venus flytrap 135, 168, 170, **170,** 349
vermiform appendix *255,* 510, **512**
vernalization *643*
vertebrae 281, **314**
vertebral column *202,* 313
Vertebrata 202
vertebrates 201, 202, 400
behavior of **357**
brain of 309
circulatory system of 269–275
excretory system of 259, **259, 260**
eyes of 304
heart of, development of **271**
regeneration in 482
respiratory system of 289
support and balance in 334, 336
vertigo, cause of 308
vesicles 99
vestigial structures *510,* **512**
Vibrio comma 610, **611**
villi **249, 254,** 255
Vinci, Leonardo da 210
vinegar 146
Virchow, Rudolph 76, 134
Virchow's postulate 76
vireo, red-eyed, nest of **446**
viruses 127, **128,** 141–142, **143,** 144
control of 282
culture of 615–616, **615**
filterable 613
mutations of 428
pathogenic 221
protection from 114
tobacco mosaic 141
visual purple *305*
vitalism, theory of 130
Vitallium 630
vitamin A (retinol) 37, 205, 235, 236, **237**
vitamin B complex 236, **237**
vitamin B_1 (thiamine) 236, **237**
vitamin B_2 (riboflavin) 236, **237**
vitamin B_6 (pyridoxine) 236, **237**
vitamin B_{12} (cobalamin) 236, **237,** 238
vitamin C (ascorbic acid) 46, 235, 236, **237**
vitamin D (calciferol) 37, 235, **237**
vitamin E (alpha-tocopherol) 37, 235, **237**
vitamin K 37, 235, **237,** 238
vitamin supplements 234–235, **235**
vitamins 34, 46, 145, *235*
deficiency diseases 235
fat-soluble 232, *235–236,* **237**
food sources of 229, 235, **237**
vitamins (*cont.*)
production by intestinal bacteria 237–238, 255
water-soluble 236, **237**
vitelline membrane, of chick embryo **461**
vocal cords 291
voiding reflex 265
volume, units of **54**
Volvox 147
vomiting *253,* 315
Vorticella 149, **150**
Vries, Hugo de 391, 425–426

W

waggle dance, of bees 386–387, **387**
wakefulness, control of 315
walking 334, **335,** 336
walking catfish *675–676,* **676**
walking sticks 198, 663, **663**
walking worm **194,** 197, **197**
Wallace, Alfred Russel 502
walnut 117, 171
walrus 248
warblers 375
wasps 199, 381, 383–384
mud **196, 197**
waste vacuoles 103
water 35
as a solvent 21, 35
boiling point of 16
buoyant effects of 638
conduction of, in plants 117
ecological importance of 637–638
excretion of, in man 261
freezing point of 16
heat capacity of 35
in cell respiration **295,** *296*
in diet 229, 234
in photosynthesis 168, 221, 223
molecular weight 27, 40
polarity of **21**
reabsorption in large intestine 255
retention of 214, 261, 639
reutilization by plants 258
water balance 316
water boatmen 198
water fleas 197
water bugs, giant **195, 197,** 198, 220
water cycle 651, **651**
water lily 640
water moccasins 210
water molds 154
water vascular system, of starfish 333
watermelons 469
Watson, James D. 43, 422, **422, 423,** 424
Watson-Crick model of DNA 422, **422–423,** 435, **435**
weasels **216**
weevils 198
weight, units of **54**
weights and measures, "English" system of 54
Weller, T. H. 614
Went, F. W. 340–341, **341**
whales **129,** 214, 216, **332,** 333
blue 200, 214
wheat 159, **160,** 231
durum, chromosome number of **90**
wheat eelworm 184
white blood cells 76, 114, **114**
and phagocytosis 83, 101
See also leukocytes
white Leghorn chicken **516**
white-crested black Polish chicken **516**
whitefish embryo, mitosis of **94, 95**
whooping crane 374, *675,* **675**
wild boars 374
wild type (*Drosophila*) 406

Wilkins, Maurice H. C. 422, 424
willows (*Salix*) 159, 160, 222, **222,** 466
wings
  of arthropods 200
  of bat **510**
  of bird **510**
  of insect **510**
witch hazel 469
withdrawal from drugs 689
Wöhler, Friedrich 9, 130
Wolf trap, life detection device 605, **605**
wolves, packs of 382
wombat **517**
wood, cell walls of 104
woodchucks, in hibernation 39
wood frog **445, 477**
wood sorrel (*Oxalis grandis*) 369
wood thrush **446**
Worm Runners' Digest 179
worms 132, 139, 174
  behavior of **357**
  in intestines **221**
wounds, healing of *482*, **483**

X

X chromosomes *414*, **414**
xanthophylls 226
xerophthalmia **237**
X-ray crystallography *70*, **71**
X-ray diffraction 40–41, 422
X rays 65, 426–428, 559, 624
xylem *117*, **164,** 168, 225

Y

Y chromosome *414*, **414,** 415, 416
yeast **128,** 153, 154, 294, 441, **441, 475**
yellow fever 199, 284, **613**
yogis, and control of autonomic nervous systems 321, **322**
yolk 113, *444*
  of bird eggs 211, 455
  of insect and fish eggs 446
  of mammal eggs 214
  of reptile eggs 208, 455
yolk sac
  of birds *460*, **461**
  of human embryo *460*, **461**
  of reptiles *460*
young, care of 380, 446, **446**

Z

Zea mays 466, **466**
zebras **216, 524**
zedonk **524**
Zernike, Fritz 67
zinc 36, 234
zoos, behavior of animals in 348
zoospores *152*
zygotes *90*, *113*, 135, **402,** *439–440*, 455
  of algae 152
  of angiosperms 466–467
  of pine tree 465
  of plants 160
  of slime molds 154
  of sponge 175–176